Molecular, Genetic, and Nutritional Aspects of Major and Trace Minerals

Molecular, Genetic, and Nutritional Aspects of Major and Trace Minerals

Edited by

James F. Collins

AMSTERDAM • BOSTON • HEIDELBERG • LONDON • NEW YORK • OXFORD • PARIS
SAN DIEGO • SAN FRANCISCO • SINGAPORE • SYDNEY • TOKYO

Academic Press is an imprint of Elsevier

Academic Press is an imprint of Elsevier
125 London Wall, London EC2Y 5AS, United Kingdom
525 B Street, Suite 1800, San Diego, CA 92101-4495, United States
50 Hampshire Street, 5th Floor, Cambridge, MA 02139, United States
The Boulevard, Langford Lane, Kidlington, Oxford OX5 1GB, United Kingdom

Notices
Knowledge and best practice in this field are constantly changing. As new research and experience broaden our understanding, changes in research methods, professional practices, or medical treatment may become necessary.

Practitioners and researchers must always rely on their own experience and knowledge in evaluating and using any information, methods, compounds, or experiments described herein. In using such information or methods they should be mindful of their own safety and the safety of others, including parties for whom they have a professional responsibility.

To the fullest extent of the law, neither the Publisher nor the authors, contributors, or editors, assume any liability for any injury and/or damage to persons or property as a matter of products liability, negligence or otherwise, or from any use or operation of any methods, products, instructions, or ideas contained in the material herein.

Library of Congress Cataloging-in-Publication Data
A catalog record for this book is available from the Library of Congress

British Library Cataloguing-in-Publication Data
A catalogue record for this book is available from the British Library

ISBN: 978-0-12-802168-2

For information on all Academic Press publications
visit our website at https://www.elsevier.com/

Working together
to grow libraries in
developing countries

www.elsevier.com • www.bookaid.org

Publisher: Nikki Levy
Acquisition Editor: Megan Ball
Editorial Project Manager: Jaclyn Truesdell
Production Project Manager: Lisa Jones
Designer: Victoria Pearson

Typeset by TNQ Books and Journals

Contents

Part X
Selenium

37. Selenium: Basic Nutritional Aspects

Wen-Hsing Cheng and Xin Gen Lei

38. Selenium and Cancer

*Dolph Lee Hatfield, Bradley Allen Carlson,
Petra Akiko Tsuji, Ryuta Tobe
and Vadim N. Gladyshev*

39. Could Selenium Be a Double-Edged Sword?

Nishi Karunasinghe and Lynnette Robyn Ferguson

Part XI
Electrolytes

40. Sodium: Basic Nutritional Aspects

*Daniel Laubitz, Fayez K. Ghishan and
Pawel R. Kiela*

41. Potassium Channel Mutations and Human Disease: Focus on Adrenal Hypertension

Michael Stowasser

Part XII
Nonessentials

42. Chromium: Basic Nutritional and Toxicological Aspects

John Bertram Vincent

List of Contributors

Abedalrazaq Alkukhun Yale University School of Medicine, New Haven, CT, United States

Gregory Jon Anderson QIMR Berghofer Medical Research Institute, Australia

Tayze T. Antunes University of Ottawa, Ottawa, ON, Canada

Michael Aschner Albert Einstein College of Medicine, New York, NY, United States

Terry J. Aspray Newcastle University, Newcastle upon Tyne, Tyne and Wear, United Kingdom; Freeman Hospital, Newcastle upon Tyne, Tyne and Wear, United Kingdom

Thomas Bartnikas Brown University, Providence, Rhode Island, United States

Abdel A. Belaidi The University of Melbourne, Parkville, VIC, Australia

Roberto Bravo-Sagua Universidad de Chile, Santiago, Chile

Gregory A. Brent David Geffen School of Medicine at UCLA, Los Angeles, CA, United States

George J. Brewer University of Michigan, Ann Arbor, MI, United States

Mona S. Calvo U.S. Food and Drug Administration, Laurel, MD, United States

Bradley Allen Carlson National Institutes of Health, Bethesda, MD, United States

Wen-Hsing Cheng Mississippi State University, Mississippi, MS, United States

Sylvia Christakos Rutgers, The State University of New Jersey, New Jersey Medical School, Newark, NJ, United States

Mariana Cifuentes Universidad de Chile, Santiago, Chile

James F. Collins University of Florida, Gainesville, FL, United States

Puneet Dhawan Rutgers, The State University of New Jersey, New Jersey Medical School, Newark, NJ, United States

Ralph Marsland Duckworth Teesside University, Middlesbrough, United Kingdom; Newcastle University, Newcastle-upon-Tyne, United Kingdom

Lynnette Robyn Ferguson University of Auckland, Auckland, New Zealand

David Michael Frazer QIMR Berghofer Medical Research Institute, Australia

Toshiyuki Fukada Tokushima Bunri University, Tokushima, Japan; Showa University, Tokyo, Japan; RIKEN Center for Integrative Medical Sciences, Yokohama, Japan

Kazuhisa Fukue Kyoto University, Kyoto, Japan

Priyanka V. Gangodkar GenePath Dx (Causeway Healthcare Private Limited), Pune, India

Eduardo Garcia-Fuentes Institute of Biomedical Research of Malaga (IBIMA), Regional University Hospital, Malaga, Spain; CIBEROBN, Institute of Health Carlos III, Malaga, Spain

Michael D. Garrick University at Buffalo, Buffalo, NY, United States

John P. Geibel Yale University School of Medicine, New Haven, CT, United States

Fayez K. Ghishan University of Arizona, Tucson, AZ, United States

Vadim N. Gladyshev Harvard Medical School, Boston, MA, United States

A. Grubman The University of Melbourne, Parkville, VIC, Australia

Thomas E. Gunter University of Rochester, Rochester, NY, United States

Hajo Haase Berlin Institute of Technology, Berlin, Germany

Dolph Lee Hatfield National Institutes of Health, Bethesda, MD, United States

Ka He Indiana University, Bloomington, IN, United States

Carolina Herrera Brown University, Providence, Rhode Island, United States

Kayo Ikuta Tokushima University, Tokushima, Japan

Francisco J. Rios University of Glasgow, Glasgow, Scotland

Sami Judeeba Yale University School of Medicine, New Haven, CT, United States

Lillian J. Juttukonda Vanderbilt University Medical Center, Nashville, TN, United States

Taiho Kambe Kyoto University, Kyoto, Japan

Ichiro Kaneko Tokushima University, Tokushima, Japan

Yujian James Kang Sichuan University, Chengdu, Sichuan, China; University of Louisville, School of Medicine, Louisville, KY, United States

Nishi Karunasinghe University of Auckland, Auckland, New Zealand

Anuradha V. Khadilkar Jehangir Medical Research Institute Jehangir Hospital, Pune, India

Pawel R. Kiela University of Arizona, Tucson, AZ, United States

Katerine S. Knust Universidade Federal do Estado do Rio de Janeiro (UNIRIO), Rio de Janeiro, Brazil

Mitchell D. Knutson University of Florida, Gainesville, FL, United States

Yuko Komiya Rutgers-Robert Wood Johnson Medical School, Piscataway, NJ, United States

Daniel Laubitz University of Arizona, Tucson, AZ, United States

Sergio Lavandero Universidad de Chile, Santiago, Chile; University of Texas Southwestern Medical Center, Dallas, TX, United States

Xin Gen Lei Cornell University, Ithaca, NY, United States

Angela M. Leung UCLA David Geffen School of Medicine, Los Angeles, CA, United States; VA Greater Los Angeles Healthcare System, Los Angeles, CA, United States

Anna Milanesi David Geffen School of Medicine at UCLA, Los Angeles, CA, United States

Ken-ichi Miyamoto Tokushima University, Tokushima, Japan

Augusto C. Montezano University of Glasgow, Glasgow, Scotland

Stefano Mora IRCCS San Raffaele Scientific Institute, Milan, Italy

Armando Salim Munoz-Abraham Yale University School of Medicine, New Haven, CT, United States

Forrest Harold Nielsen USDA, ARS, Grand Forks Human Nutrition Research Center, Grand Forks, ND, United States

Thirayost Nimmanon Cardiff University, Cardiff, United Kingdom; Phramongkutklao College of Medicine, Bangkok, Thailand

Yukina Nishito Kyoto University, Kyoto, Japan

Tanara Vieira Peres Albert Einstein College of Medicine, New York, NY, United States

Anne-Laure Perraud National Jewish Health, Denver, CO, United States; University of Colorado Denver, Denver, CO, United States

Michael Pettiglio Brown University, Providence, Rhode Island, United States

Nikhil D. Phadke GenePath Dx (Causeway Healthcare Private Limited), Pune, India

Ananda S. Prasad Wayne State University School of Medicine, Detroit, MI, United States

Vijayababu M. Radhakrishnan University of Arizona, Tucson, AZ, United States

Marcela Reyes Universidad de Chile, Santiago, Chile

Loren Warren Runnels Rutgers-Robert Wood Johnson Medical School, Piscataway, NJ, United States

Carsten Schmitz University of Colorado Denver, Denver, CO, United States; National Jewish Health, Denver, CO, United States

Guenter Schwarz University of Cologne, Cologne, Germany

Hiroko Segawa Tokushima University, Tokushima, Japan

Yatrik Madhukar Shah University of Michigan, Ann Arbor, MI, United States

Eric P. Skaar Vanderbilt University Medical Center, Nashville, TN, United States

Laura Soldati Università degli Studi of Milan, Milan, Italy

Michael Stowasser The University of Queensland, School of Medicine, Brisbane, QLD, Australia

Sawako Tatsumi Tokushima University, Tokushima, Japan

Kathryn M. Taylor Cardiff University, Cardiff, United Kingdom

Ryuta Tobe National Institutes of Health, Bethesda, MD, United States

Rhian M. Touyz University of Glasgow, Glasgow, Scotland

Cari Lewis Tsinovoi Indiana University, Bloomington, IN, United States

Petra Akiko Tsuji Towson University, Towson, MD, United States

Jaime Uribarri The Icahn School of Medicine at Mount Sinai, New York, NY, United States

Inés Velasco Hospital Riotinto, Huelva, Spain

Vaishali Veldurthy Rutgers, The State University of New Jersey, New Jersey Medical School, Newark, NJ, United States

Giuseppe Vezzoli IRCCS San Raffaele Scientific Institute, Milan, Italy

John Bertram Vincent The University of Alabama, Tuscaloosa, AL, United States

Tao Wang Sichuan University, Chengdu, Sichuan, China

Ran Wei Rutgers, The State University of New Jersey, New Jersey Medical School, Newark, NJ, United States

Marianne Wessling-Resnick Harvard T.H. Chan School of Public Health, Boston, MA, United States

A.R. White The University of Melbourne, Parkville, VIC, Australia

Ying Xiao Sichuan University, Chengdu, Sichuan, China

Xiang Xue University of Michigan, Ann Arbor, MI, United States

Hironori Yamamoto Jin-ai University, Fukui, Japan

Wen Yin Sichuan University, Chengdu, Sichuan, China

Wenjing Zhang Sichuan University, Chengdu, Sichuan, China

Fatemeh Vida Zohoori Teesside University, Middlesbrough, United Kingdom

Series Preface

In this series on Molecular Nutrition, the editors of each book aim to disseminate important material pertaining to molecular nutrition in its broadest sense. The coverage ranges from molecular aspects to whole organs, and the impact of nutrition or malnutrition on individuals and whole communities. It includes concepts, policy, preclinical studies, and clinical investigations relating to molecular nutrition. The subject areas include molecular mechanisms, polymorphisms, SNPs, genomic wide analysis, genotypes, gene expression, genetic modifications, and many other aspects. Information given in the Molecular Nutrition series relates to national, international, and global issues.

A major feature of the series that sets it apart from other texts is the initiative to bridge the transintellectual divide so that it is suitable for novices and experts alike. It embraces traditional and nontraditional formats of nutritional sciences in different ways. Each book in the series has both overviews and detailed and focused chapters.

Molecular Nutrition is designed for nutritionists, dieticians, educationalists, health experts, epidemiologists, and health-related professionals such as chemists. It is also suitable for students, graduates, postgraduates, researchers, lecturers, teachers, and professors. Contributors are national or international experts, many of whom are from world-renowned institutions or universities. It is intended to be an authoritative text covering nutrition at the molecular level.

Victor R. Preedy
Series Editor

Part I

Calcium

Chapter 1

Calcium-Sensing Receptor Polymorphisms and Human Disease

Giuseppe Vezzoli[1], Laura Soldati[2], Stefano Mora[1]
[1]IRCCS San Raffaele Scientific Institute, Milan, Italy; [2]Università degli Studi of Milan, Milan, Italy

INTRODUCTION

Circulating calcium ions can directly modulate cell activity in humans by means of a plasma membrane receptor that is sensitive to extracellular calcium, the calcium-sensing receptor (CaSR). CaSR was firstly cloned from bovine parathyroid cells in 1993 (Brown et al., 1992) and then in human parathyroid cells and renal tubular cells (Aida et al., 1995; Garrett et al., 1995). CaSR is a 1078-amino–acid protein that belongs to the third class of G-protein-coupled receptor (GPCR) family. It is expressed as a disulfide-linked homodimer in caveolin-rich areas of the plasma membrane, although it may also form heterodimers with other members of the GPCR family (Kifor et al., 1998). As an environmental sensor, CaSR elicits the paracrine or autocrine adaptive responses of human cells to changes in local or serum calcium concentrations. This adaptive response is fundamental for the physiological effect of parathyroid and kidney cells in human calcium homeostasis. The parathyroid glands and renal distal tubules are the tissues with the highest expression of CaSR, and its presence enables them to regulate calcium excretion and parathyroid hormone (PTH) secretion in response to serum calcium changes (Fig. 1.1). CaSR stimulation by the increase of serum calcium is followed by the inhibition of calcium reabsorption in the renal tubules and PTH secretion to restore normal serum calcium levels (Riccardi and Brown, 2010; Riccardi and Kemp, 2012). CaSR was also shown to be essential for osteoblast-mediated bone remodeling (Dvorak et al., 2004). Therefore CaSR is a key factor in calcium homeostasis (Riccardi and Kemp, 2012).

The CaSR molecule includes a large bilobed Venus-flytrap–like extracellular domain of 612 amino acids, a seven-membrane–spanning domain of 250 amino acids, and a C-terminal intracellular domain of 216 amino acids (Riccardi and Kemp, 2012). Calcium binding to the negatively charged residues in the pocket of the CaSR extracellular domain induces a conformational change of the CaSR molecule that causes the transmembrane and intracellular domains to activate intracellular signaling. Calcium ions are the main CaSR agonists, but CaSR also responds to other divalent (Ba, Cd, Co, Mg) and trivalent (Gd, La) cations and to polycationic compounds such as polyamines, aminoglycosides (neomycin, gentamycin), and polypeptides (poly-L-arginine, β-amyloid) (Riccardi and Kemp, 2012). The signaling cascade induced by CaSR activation is tissue specific and mediated by G-proteins (Fig. 1.2) (Magno et al., 2011). However, CaSR has also been identified in many organs not directly involved in calcium homeostasis and is now considered as ubiquitously expressed in mammalian cells. It has been implicated in insulin secretion, adipocyte metabolism, smooth muscle cell activity, and gastric function (Table 1.1), although its effects in these tissues is not as crucial as that in calcium-regulating organs (Riccardi and Kemp, 2012).

The human *CaSR* gene (3q13.3–21) spans 103 kb and comprises eight exons with two promoters, P1 and P2, having unknown functional differences (Fig. 1.3) (Canaff and Hendy, 2002). Loss-of-function mutations cause familial hypocalciuric hypercalcemia (FHH; OMIM #145980) in heterozygous patients and severe neonatal hyperparathyroidism (SNH; OMIM #239200) in homozygous patients (Hofer and Brown, 2003; Pearce et al., 1995). In these patients, CaSR cannot inhibit PTH production and renal tubular calcium reabsorption appropriately and patient phenotype is characterized by hypercalcemia and low calcium excretion. Serum PTH and calcium are slightly or moderately high in FHH, but severely high in SNH. SNH patients also develop bone demineralization and failure to thrive in the first 6 months of life. Mutations of two other genes, *GNA11* (19p13) and *AP2S1* (19q13), may also cause FHH. Gain-of-function mutations of *CaSR* cause autosomal dominant hypercalcemia (ADH; OMIM #601198), a disorder characterized by high urinary calcium excretion and inappropriately low serum PTH and hypocalcemia. ADH in patients with highly activating mutations is associated with Bartter syndrome type 5 because of a urinary sodium and potassium leak resulting in renal hypokalemia (Vezzoli et al., 2006).

Molecular, Genetic, and Nutritional Aspects of Major and Trace Minerals. http://dx.doi.org/10.1016/B978-0-12-802168-2.00001-4

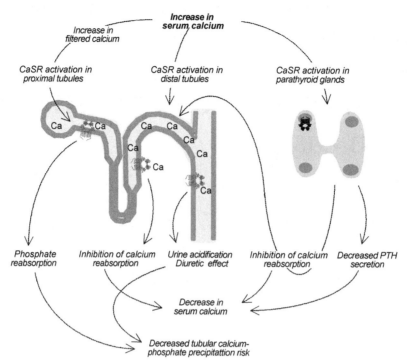

FIGURE 1.1 Renal and parathyroid response to increases in serum calcium concentrations. Calcium-sensing receptor (CaSR) is sensitive to tubular fluid calcium in the proximal tubule and in the collecting duct whereas it responds to serum calcium in the ascending limb and the cortical convoluted tubule.

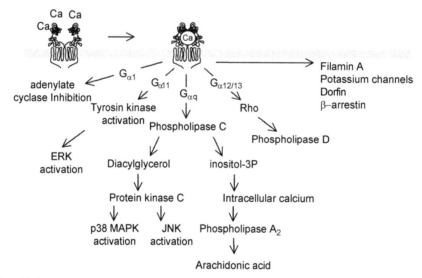

FIGURE 1.2 Calcium ions bind to the negatively charged pocket of the calcium-sensing receptor (CaSR) extracellular domain and activate different signaling cascades mediated by G-proteins. Various protein-binding partners of the CaSR's C-terminal region have been recognized, including filamin, potassium channels, dorfin, and β-arrestin.

CaSR activity may be modified by genetic polymorphisms (Table 1.2). Two nonsynonymous polymorphisms of the intracellular domain, rs1801725 and rs1402636, have a significant frequency in the general population and are associated with specific benign phenotypes. In addition, polymorphisms of noncoding regions of *CaSR* have been associated with specific phenotypes; among them, the rs6776158 polymorphism of promoter 1 may have particular relevance in human disorders.

TABLE 1.1 Functions of Different Cells Influenced by Extracellular Calcium Ions and Calcium-Sensing Receptor Activities (Riccardi and Kemp, 2012).

Organ	Effect	Related Human Disorders or Phenotype
Parathyroid glands	Inhibition of parathyroid hormone secretion	Autosomal-dominant hypocalcemia, familial hypocalciuric hypercalcemia
Kidney		
Glomeruli	Prevention of proteinuria	Glomerulonephritis
Macula densa	Inhibition of renin secretion	Arterial hypertension
Ascending limb	Inhibition of calcium reabsorption	Kidney stones
	Inhibition of sodium reabsorption	Bartter syndrome type 5
Convoluted distal tubule	Inhibition of calcium reabsorption	Autosomal-dominant hypercalcemia, familial hypocalciuric hypercalcemia
Proximal tubule	Increased phosphate reabsorption	Kidney stones
Collecting duct	Proton excretion	Kidney stones
	Diuresis	Kidney stones
Bone	Osteoblast activation	Fracture risk
	Chondrocyte activation	Delayed growth plate development
Ovary	Oocyte maturation	
Testis	Spermatozoa motility	
Colon	Epithelial cell differentiation	Colon cancer
	Fluid secretion	
Duodenum	Cholecystokinin secretion	Increased appetite
Intestine	Secretion of antidiabetic peptides	Diabetes mellitus
	Fluid reabsorption	Diarrhea
Stomach	Gastrin secretion	
	Acid secretion	
Breast	Milk production	
Pancreas	Insulin secretion	
Central nervous system		Alzheimer disease
Adipose tissue	Lipolysis inhibition	Fat accumulation
Tongue	Kokumi taste	
Arteries		
Smooth muscle cells	Prevention of vascular calcification	Coronary disease
Endothelial cells	Nitric oxide production	Artery wall calcification
		Arterial hypertension
Prostate		Prostate cancer
Heart	Cardiomyocyte hypertrophy	Cardiac hypertrophy
Blood	B lymphocyte activation	
Skin	Epidermal differentiation	
	Cell adhesion	
Thyroid gland	Calcitonin production	

See the text for specification on renal effects.

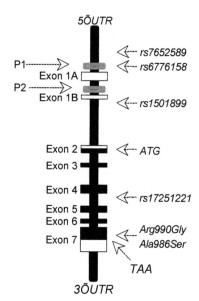

FIGURE 1.3 Promoters and the most significant polymorphisms in the calcium-sensing receptor gene.

TABLE 1.2 Calcium-Sensing Receptor Gene Polymorphisms Involved in Human Disorders

Single-Nucleotide Polymorphism	Position	Domain	Alleles	Minor Allele Frequency	Amino Acid Change	Disorder Associated With the Minor Allele
rs1801725	123,486,447	Exon 7	G>T	20%	986 A/S	Higher serum calcium
						Lower serum phosphate
rs1042636	122,284,922	Exon 7	A>G	5%	990 R/G	Hypercalciuria, Higher calcium excretion, Calcium kidney stones
				50%[a]		
rs1801726	122,284,985	Exon 7	C>G	2%	1011 q/E	Higher serum calcium
rs17251221	123,475,937	Intron 4	A>G	19%	None	Higher serum calcium
						Higher serum magnesium
rs6776158	122,183,002	Promoter 1	A>G	26%	None	Calcium kidney stones
						Lower serum phosphate
rs1501899	123,390,018	Intron 1	G>a	26%	None	Calcium kidney stones
rs7652589	123,371,778	5'UTR	G>a	27%	None	Calcium kidney stones

[a]*Frequency in Asian people.*

Calcium-Sensing Receptor in the Parathyroid Glands

PTH has a pivotal role in calcium homeostasis because of its effects on renal tubular calcium reabsorption and bone metabolism. CaSR is the key of the negative feedback regulating PTH secretion according to the circulating calcium concentration. CaSR stimulation leads to adenylate cyclase inhibition, protein kinase C activation, and calcium mobilization from intracellular stores, which decrease PTH gene transcription and PTH secretion (Kifor et al., 1997). The opposite occurs during hypocalcemia, which may stimulate parathyroid cell proliferation to maintain high levels of PTH synthesis and secretion (Corbetta et al., 2002). In the case of chronic hypocalcemia, this mechanism may give rise to nodular proliferation of parathyroid cells in patients with chronic kidney disease or hypophosphatemic rickets (Yano et al., 2000a).

Parathyroid adenomas isolated from patients with primary hyperparathyroidism are characterized by a reduced CaSR expression; therefore they are insensitive to the normal negative feedback regulation of PTH secretion that contributes to the shift to the right of the PTH–calcium curve and to the autonomous PTH secretion of the gland. The defect in CaSR expression was associated with decreased transcriptional activity of promoter 1 (Chikatsu et al., 2000). In addition, hyperplastic glands from patients with chronic kidney disease may have decreased expression of CaSR (Yano et al., 2000a).

Calcium-Sensing Receptor in the Kidney

CaSR has been detected in glomerular podocytes, mesangial cells, and in cells of all renal tubular segments from humans and laboratory animals (Riccardi and Brown, 2010; Riccardi and Kemp, 2012). The highest expression was observed in the thick ascending limb of the loop of Henle, where approximately 25% of calcium is filtered. Here, CaSR is located in the basolateral cell membrane and is activated by the increase in circulating calcium concentrations. Microperfusion studies in renal tubules from laboratory animals showed that CaSR inhibits paracellular calcium reabsorption in the ascending limb (Riccardi and Brown, 2010). This effect is sustained by CaSR-induced production of two microRNAs, miR-9 and miR-374, that upregulate claudin-14 expression and downregulate claudin-16 expression (Fig. 1.4). Claudin-16 and claudin-19 form tight junction channels, which drive paracellular reabsorption of calcium and magnesium, whereas claudin-14 inhibits permeability of this channel (Gong and Hou, 2014; Toka et al., 2012). To enhance calcium transport in the ascending limb, CaSR reduces the transepithelial electric gradient driving paracellular calcium reabsorption (Fig. 1.4) by inhibiting sodium-potassium-chloride cotransport (NKCC2) activity and potassium recycling channels (which generate the transepithelial electric gradient; Gamba and Friedman, 2009). In addition to passive reabsorption in the ascending limb, experiments in cultured canine cells showed that CaSR inhibits active calcium reabsorption in cortical convoluted tubules where CaSR is located in the basolateral cell membrane (Gamba and Friedman, 2009; Blankenship et al., 2001).

In the collecting duct, CaSR is expressed on the luminal membrane of tubular cells and is sensitive to the tubular fluid calcium concentration that results from CaSR-modulated calcium reabsorption in the ascending limb (Topala et al., 2009).

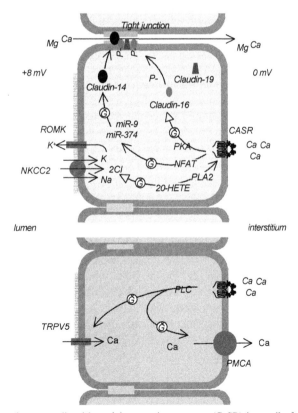

FIGURE 1.4 Activities and signaling pathways mediated by calcium-sensing receptor (CaSR) in a cell of the ascending limb of the Henle's loop (*upper panel*) and the cortical convoluted tubule (*lower panel*). 20-*HETE*, 20-hydroxyeicosatetraenoic acid; *NFAT*, nuclear factor of activated T cells; *NKCC2*, sodium–potassium–chloride cotransporter; *PKA*, protein kinase A; *PLA2*, the phospholipase A2; *PMCA*, plasma membrane calcium-transporting ATPase; *ROMK*, renal outer-medullary potassium channel; *TRPV5*, transient receptor potential cation channel subfamily V member 5. miR-9 and miR-374 are two microRNA molecules that target claudin-14 gene transcription.

Its activation may decrease cAMP production and aquaporin-2 trafficking, which downregulates the tubular response to antidiuretic hormone and water reabsorption. Furthermore, CaSR may stimulate urine acidification in this tubular segment by enhancing proton pump function (Bustamante et al., 2008; Casare et al., 2014). In the proximal tubule, CaSR is expressed on the luminal cell membrane and is sensitive to calcium present in the glomerular filtrate. Cell culture experiments showed that CaSR may antagonize the effect of PTH and promote phosphate reabsorption via the sodium-phosphate cotransporter (NPT2) by inhibiting cAMP production (Ba et al., 2004). Here, it also promotes proton secretion through the sodium-hydrogen exchanger (NHE3; Capasso et al., 2013).

These findings demonstrate the integrated effects of CaSR in the kidney. CaSR activation by serum calcium results in a calciuretic effect in the ascending limb and cortical distal tract that increases the risk of calcium-phosphate precipitation in the tubular lumen. This risk is counterbalanced by the fact that tubular fluid calcium may activate CaSR in the collecting duct leading to water and proton excretion and increased calcium-phosphate solubility. Furthermore, activation of CaSR in the proximal tubule decreases phosphate load to the distal tubule (Renkema et al., 2009).

Calcium-Sensing Receptor in Bones

Extracellular calcium is important for chondrocyte differentiation and normal development of the growth plate. Calcium deficiency may cause rickets as a result of delayed chondrocyte differentiation and reduced matrix synthesis and mineralization (Thacher, 2003). Several studies showed that the extracellular calcium sensing of chondrocytes is mediated by CaSR (Chang et al., 1999, 2008). CaSR has been localized to maturing chondrocytes in the growth plate and its expression was found to increase as the cells hypertrophy, suggesting a role of the receptor in mediating terminal differentiation (Chang et al., 1999). Hypertrophic chondrocytes from homozygous knockout mice showed a markedly reduced expression of insulin growth-like factor (IGF)-1 and IGF1 receptor, suggesting that CaSR promotes chondrocyte differentiation by enhancing IGF1 production or signaling (Chang et al., 2008).

Studies performed on cells of the osteoblastic lineage demonstrated that changes in extracellular calcium affect cell proliferation, differentiation, and mineralization via the CaSR (Dvorak et al., 2004; Chattopadhyay et al., 2004). Despite the evidence that CaSR mediates extracellular calcium sensing in osteoblasts, its role in bone development in vivo has been controversial. The generation of a conditional CaSR knockout model, in which the receptor was broadly deleted across the osteoblast lineage (including preosteoblasts, proliferating, differentiating, and mature osteoblasts and osteocytes) showed an early lethality and skeletal abnormalities (retarded skeletal development and growth; Chang et al., 2008; Dvorak-Ewell et al., 2011). From the second week of life, conditional knockout mice suffered long bone fractures along with deficient skeletal mineralization. Microcomputed tomography studies showed markedly reduced mineral content in trabecular and cortical bone, decreased bone volume, and altered trabecular architecture (Dvorak-Ewell et al., 2011). Fluorescently labeled mineralizing surfaces exhibited a markedly delayed and disorganized mineralization of trabecular and cortical bone. Conversely, osteoclast numbers and activity were increased in the knockout mice, and receptor activator of nuclear factor kappa-B ligand (RANK-L) mRNA expression was increased in femoral cortices. Cultured osteoblasts isolated from the knockouts showed an elevated osteoclastogenesis potential, stimulating the development of osteoclasts that were up to fivefold more positive to tartrate-resistant acid phosphatase compared with cells from control mice (Dvorak-Ewell et al., 2011). During synthesis, CaSR seems to result in bone anabolism in the presence of calcium by promoting commitment, survival, and differentiation of early osteoblasts and by suppressing local RANK-L–dependent osteoclastogenesis.

During bone resorption, the local calcium concentration reaches 8–40 mM, and the surrounding cells are exposed to these high calcium levels. This stimulus is sensed by CaSR present in cells populating the bone marrow. If stimulated with calcium, mesenchymal stromal cells derived from rat bone marrow migrate and proliferate in a concentration-dependent manner and overexpress osteogenic markers (Gonzales-Vazquez et al., 2014). During fetal life in mice, the unusual calcium level in bone interstitial fluid also triggers hematopoietic stem cells to home to the endosteal niche within the bone marrow after migration from liver or spleen. Hematopoietic stem cells that express CaSR were able to home to bone marrow whereas those from CaSR knockout mice could not (Adams et al., 2006). This mechanism leads to the preferential localization of adult mammalian hematopoiesis in bones.

CALCIUM-SENSING RECEPTOR GENE POLYMORPHISMS AND CALCIUM–PHOSPHATE HOMEOSTASIS

Approximately 200 nonsynonymous, single-nucleotide polymorphisms of the human CaSR gene have been detected. Two apparently benign nonsynonymous polymorphisms—rs1801725 (G>T; Ala986Ser) and rs1042636 (A>G; Arg-990Gly)—have been most extensively studied in humans because of their significant minor allele frequency in the European population (~20% for rs1801725% and 5% for rs1402636; Cole et al., 1998). Another polymorphism, rs1801726

(C>G, Gln1011Glu), has a minor allele frequency of approximately 2% (Scillitani et al., 2004). These polymorphisms are located in exon 7, a region that codes for the CaSR intracellular domain. Other polymorphisms not changing amino acid sequence have been studied in humans. rs17251221 (A>G) is located in intron 4 and is in linkage with rs1801725 (Kapur et al., 2010); rs6776158 (A>G) is a polymorphism of the *CaSR* gene promoter 1 with a minor allele frequency of approximately 27%; it is in linkage with rs1501899 (G>A), located in intron 1, and rs7652589 (G>A) located upstream promoter 1 in the 5′-untranslated region (Fig. 1.3; Vezzoli et al., 2013). These polymorphisms were in the first haplotype block of the CaSR gene including the 3′ untranslated region and exon 1; another haplotype block includes the coding part of the gene (Yun et al., 2007).

In 1999 a Canadian study documented the association of serum calcium with the genotype at rs1801725; the minor allele T was associated with higher serum calcium values that rose with the number of copies of the allele (Cole et al., 1998). These findings were generally confirmed in subsequent studies (Scillitani et al., 2004), including two genome-wide association studies (GWASs) that identified rs1801725 and rs17251221 as the *CaSR* polymorphisms having the most significant association with serum calcium. Their minor allele was associated with higher serum calcium and explained 0.54% of serum calcium variance. One of these GWASs found that rs1402636 also contributed to serum calcium levels because the minor allele G was associated with lower serum calcium (Kapur et al., 2010). Although the phenotype suggested a functional effect, no alterations of CaSR activity were observed when the minor allele at rs1801725 was tested in embryonic renal cells transfected with the CaSR gene (Vezzoli et al., 2007).

An effect on calcium excretion could be exerted by rs1402636 because the minor allele G was associated with idiopathic hypercalciuria and greater calcium excretion in patients with primary hyperparathyroidism (Vezzoli et al., 2007; Corbetta et al., 2006). Functional tests in embryonic renal cells transfected with the *CaSR* showed an activating effect of the G minor allele at rs1402636; it may increase CaSR sensitivity to calcimimetic, cell calcium oscillations and activity of the endoplasmic reticulum (ER) calcium pump that leads to calcium accumulation in the ER (Terranegra et al., 2010; Ranieri et al., 2013). This allele could encode a more efficient CaSR that depresses tubular calcium reabsorption in the ascending limb and decreases serum calcium in individuals with this genotype (Kapur et al., 2010; Scillitani et al., 2004).

CALCIUM-SENSING RECEPTOR GENE POLYMORPHISMS AND DISORDERS OF THE PARATHYROID GLANDS

CaSR gene polymorphisms were shown to possibly influence the phenotype of patients with primary hyperparathyroidism in different case series. Concentrations of serum ionized calcium and PTH were higher in patients carrying the minor alleles at polymorphisms of the regulatory region (rs7652589 and rs1501899), which are in linkage disequilibrium with rs6776158 of promoter 1, causing decreased CaSR expression (Vezzoli et al., 2011). In addition, kidney stone risk was higher in patients carrying the minor allele of these polymorphisms (Vezzoli et al., 2011). Calcium excretion and stone risk were instead promoted by rs1402636, which increases CaSR activity (Corbetta et al., 2006; Scillitani et al., 2007). The contemporary presence of the minor allele at polymorphisms rs1501899 and rs1402636 led to an increase of stone risk despite their apparently different effects on CaSR function (Vezzoli et al., 2014). Moreover, rs1402636 may influence the phenotype in uremic patients with secondary hyperparathyroidism. The minor allele of this polymorphism may be associated with lower serum PTH levels (Yano et al., 2000b) and suppression of PTH levels in response to the increase of postdialysis serum calcium (Yokoama et al., 2002), suggesting greater control of PTH secretion in patients carrying the minor allele.

CALCIUM-SENSING RECEPTOR GENE POLYMORPHISMS AND DISORDERS OF THE KIDNEY

CaSR has been implicated in the control of glomerular function because calcimimetics can stabilize the cytoskeleton of podocytes and decrease glomerular sclerosis, proteinuria, blood pressure, and the progression of renal failure in rat kidney (Ogata et al., 2003). However, no data are available relating CaSR to renal filtration in humans. On the contrary, various CaSR genotypes, including rs1402638, rs6776158, rs1501899, and rs7652589, were hypothesized to play a role in calcium nephrolithiasis. Minor allele G at rs1402638 was shown to have an activating effect and was associated with hypercalciuria in calcium stone formers and osteoporotic women without kidney stones (Vezzoli et al., 2002). These findings may be explained by an increased inhibitory effect of the *CaSR* polymorphic variant on calcium reabsorption in the ascending limb (Vezzoli et al., 2007). Hypercalciuria may predispose individuals to urinary stones, and the polymorphism rs1402636 was associated with stones in some populations but not in others (O'Seaghdha et al., 2010). Minor alleles at polymorphisms of the regulatory region of the *CaSR*, rs6776158, rs1501899, and rs7652589, were associated with idiopathic calcium stones (Vezzoli et al., 2010, 2013) and with stone formers with primary hyperparathyroidism (Vezzoli et al., 2011). Furthermore, the minor allele at rs6776158 was shown to reduce promoter 1 transcriptional activity in two renal cell models.

Accordingly, CaSR expression in medulla samples was decreased in patients carrying the minor allele at rs6776158 (Vezzoli et al., 2013). This polymorphism was not associated with calcium excretion and could promote stones by decreasing the protective effects of CaSR on calcium–phosphate precipitation within the tubular lumen or the interstitium. Taken together, these findings suggest a complex effect of CaSR on calcium stone risk; rs1402636 may predispose hypercalciuric subjects to stones by amplifying their tendency for high calcium excretion whereas rs6776158 may predispose subjects to stones by decreasing protection against intratubular calcium precipitation.

CALCIUM-SENSING RECEPTOR GENE POLYMORPHISMS AND BONE MINERAL DENSITY

Several studies investigated the association between *CaSR* polymorphisms and bone mineral density (BMD). Many studies included the genotyping of the rs1801725 (Arg986ser) polymorphism in different cohorts. A study performed on 97 Swedish girls and young women found a negative association with BMD measured at the lumbar spine (Lorentzon et al., 2001). Subjects carrying the minor allele T had lower BMD measurements, but the relationship was biased by the simultaneous finding of a lower degree of physical activity within the same group of subjects. Conversely, other studies did not find an association between the *CaSR* rs1801725 polymorphism and BMD measured at the lumbar spine or the femur in large groups of postmenopausal Caucasian women (Takacs et al., 2002), or measured at the distal forearm in healthy Finnish adults (Laaksonen et al., 2009). Another study found no association of three polymorphisms located in exon 7 of *CaSR* (rs18017125, rs1402636, and rs18017126) with BMD measured at the lumbar spine, left forearm, and left total hip in 110 dizygotic female twins (Harding et al., 2006). Moreover, a recent GWAS did not show any association between BMD and the CaSR gene locus in 2211 women and 1633 men. These authors measured BMD at the lumbar spine, femoral neck, trochanter, and whole body and recorded calcaneal quantitative ultrasound (Gupta et al., 2011). Conversely, a recent study in 266 patients with primary hyperparathyroidism showed that the minor allele T at the rs1801725 polymorphism was more frequent in patients who suffered from vertebral fractures regardless of gender, BMD measurements, or serum calcium concentration. Furthermore, carriers of the minor allele at rs1801725 plus two of the following three factors—age, serum calcium concentration, and spinal BMD—showed a risk of fracture that was 4.7-fold higher than homozygotes for the more common allele (Eller-Vainicher et al., 2014).

CONCLUSIONS

Findings reported here indicate that environmental calcium may influence development and function of different organs, especially those involved in calcium homeostasis. CaSR mediates the effect of environmental calcium, and its polymorphic variants may contribute to individual variability. *CaSR* polymorphisms may be a determinant of serum calcium and calcium excretion and may predispose to calcium nephrolithiasis. Two mechanisms may explain *CaSR* polymorphism contribution to stone risk: (1) variants decreasing CaSR expression in the kidney may reduce its protective effect against calcium-phosphate precipitation and (2) variants associated with hypercalciuria may trigger calcium–phosphate precipitation in the tubular lumen. It has been hypothesized that the protective effect of CaSR in relation to kidney stones could reflect its protective effect against soft tissue and vascular calcification (Vezzoli et al., 2013). Therefore *CaSR* polymorphisms may influence individual phenotype in terms of serum calcium and calcium excretion. These changes may increase risk for bone fracture and kidney stones as observed in patients with primary and secondary hyperparathyroidism, although the effect of *CaSR* polymorphisms on fracture risk and BMD remains controversial. According to the current data, clinicians could use *CaSR* polymorphisms as markers to explain hypercalciuria, hypercalcemia, and predisposition to kidney stones, or to predict the response to calcimimetic drugs that are used to cure secondary hyperparathyroidism in patients undergoing dialysis and to correct hypercalcemia in patients with primary hyperparathyroidism.

SUMMARY POINTS

- CaSR is a 1078-amino–acid protein that belongs to the third class of the GPCR family, expressed as a disulfide-linked homodimer in caveolin-rich areas of the plasma membrane.
- CaSR is an environmental sensor that elicits paracrine or autocrine adaptive responses of human cells to local or serum variations of extracellular calcium concentration.
- CaSR is mainly expressed in the parathyroid glands and renal distal tubules.
- CaSR enables the parathyroid glands and renal tubules to regulate PTH secretion and calcium excretion in response to serum calcium changes.

- Two apparently benign nonsynonymous *CaSR* polymorphisms, rs1801725 and rs1042636, have been most extensively studied in humans.
- The genotype at rs1801725 was associated with serum calcium that was higher in carriers of the minor allele.
- The minor allele at rs1402636 upregulates in vitro CaSR activity and was associated with idiopathic hypercalciuria.
- The genotype at rs1402636 may modify the response to calcimimetic drugs.
- Polymorphisms associated with decreased CaSR expression—rs6776158, rs1501899, and rs7652589—were associated with idiopathic calcium stones.
- The phenotype of patients with primary hyperparathyroidism may be modified by *CaSR* polymorphisms.
- The minor allele at rs1801725 could have a negative effect on BMD values at the lumbar spine and fracture risk.
- Clinicians could use *CaSR* polymorphisms as markers to explain hypercalciuria, hypercalcemia, or kidney stone production, or to predict the response to calcimimetic drugs.

KEY FACTS

- CaSR is the extracellular calcium sensor that enables parathyroid glands and renal distal tubules to regulate calcium excretion and PTH secretion in response to serum calcium changes.
- *CaSR* gene polymorphisms were found to be associated with serum calcium and urine calcium excretion.
- CaSR activity in the renal tubules protects against calcium–phosphate precipitation in the tubular fluids.
- Polymorphisms decreasing CaSR expression in renal tubular cells may predispose individuals to calcium kidney stones.
- *CaSR* polymorphisms may modify the phenotype of healthy subjects and patients with primary and secondary hyperparathyroidism.

MINI DICTIONARY OF TERMS

Bone mineral density (BMD) Variable used to express the amount of mineral matter per square centimeter at different bone segments, usually forearm, lumbar spine, and femur.

Calcimimetics Drugs that are allosteric activators of CaSR, thereby decreasing PTH secretion and its hypercalcemic effect.

Caveolin A membrane protein associated with microdomains in the plasma membrane that are rich in signal-transducing proteins.

Hyperparathyroidism Condition of excessive PTH secretion from the parathyroid glands. Patients with primary hyperparathyroidism are characterized by hypercalcemia whereas those with secondary hyperparathyroidism display hypocalcemia or normocalcemia.

Osteoclasts Bone cells that express RANK-ligand and are responsible for bone resorption.

Osteoblasts Bone cells responsible for bone production and mineralization. They activate bone remodeling by producing RANK.

Polymorphism Difference in DNA sequence among individuals that may contribute to individual variability and predisposition for diseases.

RANK Receptor activator of nuclear factor kappa-B; produced by osteoblasts, it plays a critical role in triggering osteoclast activity.

Tight junction A junctional element connecting adjacent epithelial cells. It is a barrier to maintain cell polarity.

REFERENCES

Adams, G.B., Chabner, K.T., Alley, I.R., Olson, D.P., Szczepiorkowski, Z.M., Poznansky, M.C., Kos, C.H., Pollak, M.R., Brown, E.M., Scadden, D.T., 2006. Stem cell engraftment at the endosteal niche is specified by the calcium-sensing receptor. Nature 439, 599–603.

Aida, K., Koishi, S., Tawata, M., Onaya, T., 1995. Molecular cloning of a putative Ca-sensing receptor cDNA from human kidney. Biochem. Biophys. Res. Commun. 214, 524–529.

Ba, J., Brown, D., Friedman, P.A., 2004. Calcium-sensing receptor regulation of PTH-inhibitable proximal tubule phosphate transport. Am. J. Physiol. Ren. Physiol. 285, F1233–F1243.

Blankenship, K.A., Williams, J.J., Lawrence, M.S., McLeish, K.R., Dean, W.L., Arthur, J.M., 2001. The calcium-sensing receptor regulates calcium absorption in MDCK cells by inhibition of PMCA. Am. J. Physiol. Ren. Physiol. 280, F815–F822.

Brown, E.M., Gamba, G., Riccardi, D., Lombardi, M., Butters, R., Kifor, O., Sun, A., Hediger, M.A., Lytton, J., Hebert, S., 1992. Cloning and characterization of an extracellular Ca-sensing receptor from bovine parathyroid. Nature 366, 575–580.

Bustamante, M., Hasler, U., Leroy, V., deSeigneux, M., Dimitrov, M., Mordasini, D., Rousselot, M., Martin, P.Y., Feraille, E., 2008. Calcium-sensing receptor attenuates AVP-induced aquaporin-2 expression via a calmodulin-dependent mechanism. J. Am. Soc. Nephrol. 19, 109–116.

Canaff, L., Hendy, G.N., 2002. Human calcium-sensing receptor gene. J. Biol. Chem. 277, 30337–30350.

Capasso, G., Geibel, P.J., Damiano, S., Jaeger, P., Richards, W.G., Geibel, J.P., 2013. The calcium sensing receptor modulates fluid reabsorption and acid secretion in the proximal tubule. Kidney Int. 84, 277–284.

Casare, F., Milan, D., Fernandez, R., 2014. Stimulation of calcium-sensing receptor increases biochemical H-ATPase activity in mouse cortex and outer medullary regions. Can. J. Physiol. Pharmacol. 92, 181–188.

Chang, W., Tu, C., Chen, T.H., Komuves, L., Oda, Y., Pratt, S.A., Miller, S., Shoback, D., 1999. Expression and signal transduction of calcium-sensing receptors in cartilage and bone. Endocrinology 140, 5883–5893.

Chang, W., Tu, C., Chen, T.H., Bikle, D., Shoback, D., 2008. The extracellular calcium-sensing receptor (CaSR) is a critical modulator of skeletal development. Sci. Signal. 1, ra1.

Chattopadhyay, N., Yano, S., Tfelt-Hansen, J., Rooney, P., Kanuparthi, D., Bandyopadhyay, S., Ren, X., Terwillinger, E., Brown, E.M., 2004. Mitogenic action of calcium-sensing receptor on rat calvarial osteoblasts. Endocrinology 145, 3451–3462.

Chikatsu, N., Fukumoto, S., Takeuchi, Y., Suzawa, M., Obara, T., Matsumoto, T., Fujita, T., 2000. Cloning and characterization of two promoters for the human calcium-sensing receptor (CaSR) and changes of CaSR expression in parathyroid adenomas. J. Biol. Chem. 275, 7553–7557.

Cole, D.E.C., Peltekova, V.D., Rubin, L.A., Hawker, G.A., Vieth, R., Liew, C.C., Hwang, D.M., Evrovski, J., Hendy, G.N., 1998. A986S polymorphism of the calcium-sensing receptor and circulating calcium concentrations. Lancet 353, 112–115.

Corbetta, S., Lania, A., Filopanti, M., Vicentini, L., Ballarè, E., Spada, A., 2002. Mitogen-activated protein kinase cascade in human normal and tumoral parathyroid cells. J. Clin. Endocrinol. Metab. 87, 2201–2205.

Corbetta, S., Eller-Vainicher, C., Filopanti, M., Saeli, P., Vezzoli, G., Arcidiacono, T., Loli, P., Syren, M.L., Soldati, L., Beck-Peccoz, P., Spada, A., 2006. R990G polymorphism of the calcium-sensing receptor and renal calcium excretion in patients with primary hyperparathyroidism. Eur. J. Endocrinol. 155, 687–692.

Dvorak, M.M., Siddiqua, A., Ward, D.T., Carter, D.H., Dallas, S.L., Nemeth, E.F., Riccardi, D., 2004. Physiological changes in extracellular calcium concentration directly control osteoblast function in the absence of calciotropic hormones. Proc. Natl. Acad. Sci. U.S.A. 101, 5140–5145.

Dvorak-Ewell, M.M., Chen, T.H., Liang, N., Garvey, C., Liu, B., Tu, C., Chang, W., Bikle, D.D., Shoback, D.M., 2011. Osteoblast extracellular Ca2+sensing receptor regulates bone development, mineralization, and turnover. J. Bone Min. Res. 26, 2935–2947.

Eller-Vainicher, C., Battista, C., Guarnieri, V., Muscarella, S., Palmieri, S., Salcuni, A.S., Guglielmi, G., Corbetta, S., Minisola, S., Spada, A., Hendy, G.N., Cole, D.E., Chiodini, I., Scillitani, A., 2014. Factors associated with vertebral fracture risk in patients with primary hyperparathyroidism. Eur. J. Endocrinol. 171, 399–406.

Gamba, G., Friedman, P.A.F., 2009. Thick ascending limb: the Na+:K+:2Cl− co-transporter, NKCC2, and the calcium-sensing receptor, CaSR. Eur. J. Physiol. 458, 61–76.

Garrett, J.E., Capuano, I.V., Hammerland, L.G., Hung, B.C., Brown, E.M., Hebert, S.C., Nemeth, E.F., Fuller, F., 1995. Molecular cloning and functional expression of human parathyroid calcium receptor cDNAs. J. Biol. Chem. 270, 12919–12925.

Gong, Y., Hou, J., 2014. Claudin-14 underlies Ca-sensing receptor–mediated Ca metabolism via NFAT-microRNA–based mechanisms. J. Am. Soc. Nephrol. 25, 745–760.

Gonzales-Vazquez, A., Planell, J.A., Engel, E., 2014. Extracellular calcium and CaSR drive osteoinduction in mesenchymal stromal cells. Acta Biomater. 10, 2824–2833.

Gupta, M., Cheung, C.L., Hsu, Y.H., Demissie, S., Cupples, L.A., Kiel, D.P., Karasik, D., 2011. Identification of homogeneous genetic architecture of multiple genetically correlated traits by block clustering of genome-wide associations. J. Bone Min. Res. 26, 1261–1271.

Harding, B., Curley, A.J., Hannan, F.M., Christie, P.T., Bowl, M.R., Turner, J.J.O., Barbert, M., Gillham-Nasenyat, I., Hamson, G., Spector, T.D., Thakker, R.V., 2006. Functional characterization of calcium sensing receptor polymorphisms and absence of association with indices of calcium homeostasis and bone mineral density. Clin. Endocrinol. 65, 598–605.

Hofer, A.M., Brown, E.M., 2003. Extracellular calcium sensing and signaling. Nature 4, 530–538.

Kapur, K., Johnson, T., Beckmann, N.D., Sehmi, J., Tanaka, T., Kutalik, Z., Styrkarsdottir, U., Zhang, W., Marek, D., Gudbjartsson, D.F., Milaneschi, Y., Holm, H., DiIorio, A., Waterworth, D., Li, Y., Singleton, A.B., Bjornsdottir, U.S., Sigurdsson, G., Hernandez, D.G., DeSilva, R., Elliott, P., Eyjolfsson, G.I., Guralnik, J.M., Scott, J., Thorsteinsdottir, U., Bandinelli, S., Chambers, J., Stefansson, K., Waeber, G., Ferrucci, L., Kooner, J.S., Mooser, V., Vollenweider, P., Beckmann, J.S., Bochud, M., Bergmann, S., 2010. Genome-wide meta-analysis for serum calcium identifies significantly associated SNPs near the calcium-sensing receptor (CASR) gene. PLoS Genet. 6, e1001035.

Kifor, O., Diaz, R., Butters, R., Brown, E.M., 1997. The Ca-sensing receptor (CaR) activates phospholipases C, A$_2$, and D in bovine parathyroid and CaR-transfected, human embryonic kidney (HEK293) cells. J. Bone Min. Res. 12, 715–722.

Kifor, O., Diaz, R., Butters, R., Kifor, I., Brown, E.M., 1998. The calcium-sensing receptor is localized in caveolin-rich plasma membrane domains of bovine parathyroid cells. J. Biol. Chem. 273, 21708–21713.

Laaksonen, M.M.L., Outila, T.A., Kärkkäinen, M.U.M., Kemi, V.E., Rita, H.J., Perola, M., Valsta, L.M., Lamberg-Allardt, C.J.E., 2009. Associations of vitamin D receptor, calcium-sensing receptor and papathyroid hormone gene polymorphisms with calcium homeostasis and peripheral bone density in adult Finns. J. Nutr. Nutr. 2, 55–63.

Lorentzon, M., Lorentzon, R., Lerner, U.H., Nordstrom, P., 2001. Calcium sensing recepto polymorphysm, circulating calcium concentrations and bone mineral density in healthy adolescent girls. Eur. J. Endocrinol. 144, 257–261.

Magno, A.L., Ward, B.K., Ratajczak, T., 2011. The calcium-sensing receptor: a molecular perspective. Endocr. Rev. 32, 3–30.

O'Seaghdha, C.M., Yang, Q., Glazer, N.L., Leak, T.S., Dehghan, A., Smith, A.V., Kao, W.H.L., Lohman, K., Hwang, S.J., Johnson, A.D., Hofman, A., Uitterlinden, A.G., Chen, Y.D.I., The GEFOS Consortium, Brown, E.M., Siscovick, D.S., Harris, T.B., Psaty, B.M., Coresh, J., Gudnason, V., Witteman, J.C., Liu, Y.M., Kestenbaum, B.R., Fox, C.S., Kottgen, A., 2010. Common variants in the calcium-sensing receptor gene are associated with total serum calcium levels. Hum. Mol. Genet. 19, 4296–4303.

Ogata, H., Ritz, E., Odoni, G., Amman, K., Orth, S.R., 2003. Beneficial effects of calcimimetics on progression of renal failure and cardiovascular risk factors. J. Am. Soc. Nephrol. 14, 959–967.

Pearce, S.H.S., Trump, D., Wooding, C., Besser, G.M., Chew, S.L., Grant, D.B., Heath, D.A., Hughes, I.A., Paterson, C.R., Whyte, M.P., Thakker, R.V., 1995. Calcium-sensing receptor mutations in familial benign hypercalcemia and neonatal hyperparathyroidism. J. Clin. Invest. 96, 2683–2692.

Ranieri, M., Tamma, G., Di Mise, A.M., Vezzoli, G., Soldati, L., Svelto, M., Valenti, G., 2013. Excessive signal transduction of gain-of-function variants of the calcium-sensing receptor (CaSR) are associated with increased ER to cytosol calcium gradient. PLoS One 8, e79113.

Renkema, K.Y., Velic, A., Dijkman, H.B., Verkaart, S., van der Kemp, A.W., Nowk, M., Timmermans, K., Doucet, A., Wagner, C.A., Bindels, R.J., Hoenderop, J.G., 2009. The calcium-sensing receptor promotes urinary acidification to prevent nephrolithiasis. J. Am. Soc. Nephrol. 20, 1705–1713.

Riccardi, D., Brown, E.M., 2010. Physiology and pathophysiology of the calcium-sensing receptor in the kidney. Am. J. Physiol. Ren. Physiol. 298, F485–F499.

Riccardi, D., Kemp, P.J., 2012. The calcium-sensing receptor beyond extracellular calcium homeostasis: conception, development, adult physiology, and disease. Annu. Rev. Physiol. 74, 271–297.

Scillitani, A., Guarnieri, V., De, G.S., Muscarella, L.A., Battista, C., D'Agruma, L., Bertoldo, F., Florio, C., Minisola, S., Hendy, G.N., Cole, D.E.C., 2004. Blood ionized calcium is associated with clustered polymorphisms in the carboxyl-terminal tail of the calcium-sensing receptor. J. Clin. Endocrinol. Metab. 89, 5634–5638.

Scillitani, A., Guarnieri, V., De, G.S., Battista, C., DeGironimo, S., Muscarella, L.A., Chiodini, J., Cignarelli, M., Minisola, S., Bertoldo, F., Francucci, C.M., Malavolta, N., Piovesan, A., Mascia, M.L., Muscarella, S., Hendy, G.N., D'Agruma, L., Cole, D.E.C., 2007. Primary hyperparathyroidism and the presence of kidney stones are associated with different haplotypes of the calcium-sensing receptor. J. Clin. Endocrinol. Metab. 92, 277–283.

Takacs, I., Speer, G., Banjnok, E., Tabak, A., Nagy, Z., Horvath, C., Kovacs, K., Lakatos, P., 2002. Lack of association between calcium-sensing receptor gene "A986S" polymorphism and bone mineral density in Hungarian postmenopausal women. Bone 30, 849–852.

Terranegra, A., Ferraretto, A., Dogliotti, E., Scarpellini, M., Corbetta, S., Barbieri, A.M., Spada, A., Arcidiacono, T., Rainone, F., Aloia, A., Cusi, D., Vezzoli, G., Soldati, L., 2010. Calcimimetic R-568 effects on activity of R990G polymorphism of calcium-sensing receptor. J. Mol. Endocrinol. 25, 245–256.

Thacher, T.D., 2003. Calcium-deficiency rickets. Endocr. Dev. 6, 105–125.

Toka, H.R., Al-Romaih, K., Koshy, J.M., DiBrartolo, S., Kos, C.H., Quinn, S.J., Curhan, G.C., Mount, D.B., Brown, E.M., Pollak, M.R., 2012. Deficiency of the calcium-sensing receptor in the kidney causes parathyroid hormone–independent hypocalciuria. J. Am. Soc. Nephrol. 23, 1879–1890.

Topala, C.N., Schoeber, J.P.H., Searchfield, L.E., Riccardi, D., Hoenderop, J.G.J., Bindels, R.J.M., 2009. Activation of the Ca-sensing receptor stimulates the activity of the epithelial Ca channel TRPV5. Cell Calcium 45, 331–339.

Vezzoli, G., Tanini, A., Ferrucci, L., Soldati, L., Bianchin, C., Franceschelli, F., Malentacchi, C., Porfirio, B., Adamo, D., Terranegra, A., Falchetti, A., Cusi, D., Bianchi, G., Brandi, M.L., 2002. Influence of calcium-sensing receptor gene on urinary calcium excretion in stone-forming patients. J. Am. Soc. Nephrol. 13, 2517–2523.

Vezzoli, G., Arcidiacono, T., Paloschi, V., Terranegra, A., Biasion, R., Weber, G., Mora, S., Syren, M.L., Coviello, D., Cusi, D., Bianchi, G., Soldati, L., 2006. Autosomal dominant hypocalcemia with mild type 5 Bartter syndrome. J. Nephrol. 19, 525–528.

Vezzoli, G., Terranegra, A., Arcidiacono, T., Biasion, R., Coviello, D., Syren, M.L., Paloschi, V., Giannini, S., Mignogna, G., Rubinacci, A., Ferraretto, A., Cusi, D., Bianchi, G., Soldati, L., 2007. R990G polymorphism of calcium-sensing receptor does produce a gain-of-function and predispose to primary hypercalciuria. Kidney Int. 71, 1155–1162.

Vezzoli, G., Terranegra, A., Arcidiacono, T., Gambaro, G., Milanesi, L., Mosca, E., Soldati, L., 2010. Calcium kidney stones are associated with a haplotype of the calcium-sensing receptor gene regulatory region. Nephrol. Dial. Transpl. 25, 2245–2252.

Vezzoli, G., Scillitani, A., Corbetta, S., Terranegra, A., Dogliotti, E., Guarnieri, V., Arcidiacono, T., Paloschi, V., Rainone, F., Eller-Vainicher, C., Borghi, L., Nouvenne, A., Guerra, A., Meschi, T., Allegri, F., Cusi, D., Spada, A., Cole, D.E.C., Hendy, G.N., Spotti, D., Soldati, L., 2011. Polymorphisms at the regulatory regions of the calcium-sensing receptor gene influence stone risk in primary hyperparathyroidism. Eur. J. Endocrinol. 164, 421–427.

Vezzoli, G., Terranegra, A., Aloia, A., Arcidiacono, T., Milanesi, L., Mosca, E., Mingione, A., Spotti, D., Cusi, D., Hou, J., Hendy, G.N., Soldati, L., 2013. Decreased transcriptional activity of calcium-sensing receptor gene promoter 1 is associated with calcium nephrolithiasis. J. Clin. Endocrinol. Metab. 98, 3839–3847.

Vezzoli, G., Scillitani, A., Corbetta, S., Terranegra, A., Dogliotti, E., Guarnieri, V., Arcidiacono, T., Macrina, L., Mingione, A., Brasacchio, C., Eller-Vainicher, C., Cusi, D., Spada, A., Cole, D.E., Hendy, G.N., Spotti, D., Soldati, L., 2014. Risk of nephrolithiasis in primary hyperparathyroidism is associated with two polymorphisms of the calcium-sensing receptor gene. J. Nephrol. 40, 499–506.

Yano, S., Sugimoto, T., Tsukamoto, T., Chihrara, K., Kobayashi, A., Kitazawa, S., Maeda, S., Kitazawa, R., 2000a. Association of decreased calcium-sensing receptor expression with proliferation of parathyroid cells in secondary hyperparathyroidism. Kidney Int. 58, 1980–1986.

Yano, S., Sugimoto, T., Kanzawa, M., Tsukamoto, T., Hattori, T., Hattori, S., Chihara, K., 2000b. Association of polymorphic alleles of the calcium-sensing receptor gene with parathyroid hormone secretion in hemodialysis patients. Nephron 85, 317–323.

Yokoama, K., Shigematsu, T., Tsukada, T., Hara, S., Yamada, A., Kowagushi, Y., Hosoya, T., 2002. Calcium-sensing receptor gene polymorphism affects the parathyroid response to moderate hypercalcemic suppression in patients with ebd-stage renal disease. Clin. Nephrol. 57, 131–135.

Yun, F.H.J., Wong, B.Y.L., Chase, M., Shuen, A.Y., Canaff, L., Thongthai, K., Siminovitch, K., Hendy, G.N., Cole, D.E.C., 2007. Genetic variation at the calcium-sensing receptor (CASR) locus: implications for clinical molecular diagnostics. Clin. Biochem. 40, 551–561.

Chapter 2

Molecular Aspects of the Calcium-Sensing Receptor and Calcium Homeostasis

Abedalrazaq Alkukhun, Armando Salim Munoz-Abraham, Sami Judeeba, John P. Geibel
Yale University School of Medicine, New Haven, CT, United States

INTRODUCTION

Since the first description of the calcium-sensing receptor (CaSR) was published more than 20 years ago (Alfadda et al., 2014), significant progress in understanding its diverse physiologic roles has been achieved. Many homeostatic pathways have been illustrated based on CaSR function, and now it is recognized as a key modulator of the physiology and pathophysiology of multiple organ systems (Alfadda et al., 2014). Ongoing research on the CaSR at the cellular and molecular level has revealed new functions of the receptor as new disease models are analyzed. Moreover, new therapies are being developed to treat a wide variety of diseases based on modulation of the CaSR. In this chapter we describe the role of the CaSR in various organ systems and outline how it modulates various physiological and pathophysiological functions.

STRUCTURAL COMPONENTS OF THE CALCIUM-SENSING RECEPTOR

The CaSR was originally cloned from bovine parathyroid glands (Brown et al., 1993). It is now well known that the receptor is expressed in the kidney, parathyroid gland, thyroid gland, gastrointestinal (GI) tract, bone, and skin, among other tissues (Brown, 2007b; see Fig. 2.1).

The CaSR is a Class C, G-protein–coupled receptor that comprises three main structures: an extracellular domain (ECD), a seven-transmembrane–spanning domain, and an intracellular domain (Alfadda et al., 2014). The receptor has a large ECD that includes a "Venus flytrap" (VFT) sequence that contains dimerization and orthosteric sites for binding of endogenous agonists (Alfadda et al., 2014). The ECD contains 612 amino acid residues, and the majority of the disease-causing mutations are located in this domain (Alfadda et al., 2014). The ECD is linked to a transmembrane domain (TMD) rich in cysteine residues (Alfadda et al., 2014) that composes the majority of the protein. These cysteine residues form disulfide bridges within the TMD and with the VFT, which are required for the proper physiological function of the receptor (Alfadda et al., 2014). The CaSR ECD is formed by two lobes (LB1 and LB2) that are separated by a cavity delineating the ligand-binding site representing the VFT. Calcium (Ca^{2+}) molecules bind in the cleft between the two lobes of each VFT, causing the lobes to close in on one another and the VFT to rotate, leading to receptor activation. The CaSR is a homodimer that is capable of coupling to multiple G-proteins, such as $G_{q/11}$. The CaSR regulates different pathways in different tissues where it is expressed. It can stimulate phospholipases, mitogen-activated protein kinases (MAPKs), ion transport, fluid secretion, and bicarbonate and acid transport (Alfadda et al., 2014; Brown, 2007a; Magno et al., 2011).

CALCIUM-SENSING RECEPTOR FUNCTION IN THE PARATHYROID GLAND

The role of the CaSR in the parathyroid gland was one of the first physiological functions discovered. Since the successful cloning of the CaSR from parathyroid cells, it has been shown to regulate parathyroid hormone (PTH) secretion and thus influence calcium homeostasis (Brown, 2007a). Subsequently, CaSR has been linked to many inherited diseases with mutant forms of the receptor perturbing the balance between PTH secretion and extracellular Ca^{2+} levels (Brown et al., 1993).

The CaSR may be responsible for maintaining a constant level of ionized calcium (Ca^{2+}) within the extracellular fluids surrounding chief cells of the parathyroid gland. In the parathyroid gland, CaSR regulates the production and secretion of PTH depending on whether Ca^{2+} is low or high (Brown, 2007a). When extracellular Ca^{2+} concentrations are low, PTH

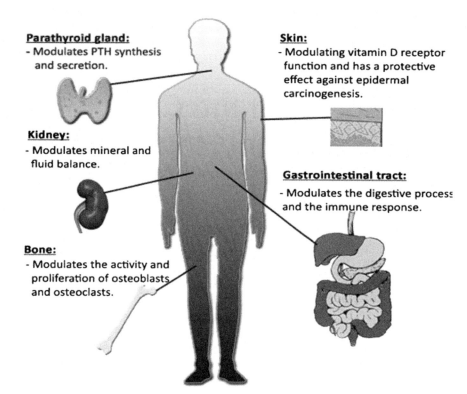

Parathyroid gland:
- Modulates PTH synthesis and secretion.

Skin:
- Modulating vitamin D receptor function and has a protective effect against epidermal carcinogenesis.

Kidney:
- Modulates mineral and fluid balance.

Gastrointestinal tract:
- Modulates the digestive process and the immune response.

Bone:
- Modulates the activity and proliferation of osteoblasts and osteoclasts.

FIGURE 2.1 Schematic representation of known calcium-sensing receptor modulated organs. *PTH*, parathyroid hormone.

secretion increases due to downregulation of the CaSR. PTH produces two actions: (1) enhanced bone resorption to increase the availability of Ca^{2+} in serum and (2) increased renal production of 1,25-dihydroxyvitamin D_3, which stimulates Ca^{2+} absorption in the GI tract. Concurrently, the CaSR located in the thyroid gland prevents calcitonin secretion. Calcitonin is a hormone responsible for stimulating renal Ca^{2+} excretion (Magno et al., 2011).

When extracellular Ca^{2+} increases, CaSR expression increases, which decreases PTH released into the circulation by acting at multiple levels (Brown, 2007a). The CaSR in the parathyroid gland has three main regulatory actions: (1) PTH synthesis, (2) PTH secretion, and (3) cell proliferation (Brown, 2007a). These actions of the CaSR are mainly produced by a negative feedback regulation that extends to the gene level. The CaSR expressed in the parathyroid regulates multiple signaling pathways, including MAPK, phospholipase C (PLC), phospholipase A_2, and phospholipase D (Huang et al., 2007; Kifor et al., 2001). Kifor et al. (2001) described a high extracellular Ca^{2+} environment that produces CaSR activation (in a bovine parathyroid model) via protein kinase C (PKC), phospholipase A_2, and phospholipase $DG_{q/11}$-mediated activation of phosphatidylinositol-specific PLC that leads to intracellular Ca^{2+} mobilization and PKC activation. This chain of events stimulates a MAPK cascade that activates extracellular signal-regulated kinases (ERK1/2) and calcium-dependent phospholipase A_2, causing the release of arachidonic acid (Kifor et al., 2001). Arachidonic acid is then metabolized into biologically active lipid mediators such as hydroperoxyeicosatetraenoic acid and hydroxyeicosatetranoic acid, which suppress PTH secretion (Kifor et al., 2001; Marie, 2010). This reduced PTH production is thought to occur via downregulation of PTH mRNA expression and reduced proliferation of parathyroid cells by the increased intracellular Ca^{2+} concentration (Magno et al., 2011; see Fig. 2.2).

The role of CaSR in Ca^{2+} homeostasis can be further explained by examining the consequences of receptor inactivation or stimulation in the parathyroid gland. Receptor inactivation in CaSR knockout mice, in which exaggerated PTH secretion occurs even with hypercalcemia, is usually fatal. Unrestrained PTH secretion may be driven by a higher set point at which Ca^{2+} triggers PTH release. Marked parathyroid hyperplasia is also observed in the knockout mice, exemplifying another pathway by which CaSR regulates parathyroid gland function and PTH secretion (Brown, 2007b). In humans, inactivating CaSR mutations occur in familial hypocalciuric hypercalcemia (FHH), an autosomal-dominant disease (heterozygous). In FHH, PTH release increases because the parathyroid gland cannot properly sense extracellular Ca^{2+} levels (Brown, 2007a; Magno et al., 2011). Also in this disease, renal Ca^{2+} reabsorption is excessive, causing hypocalciuria and thus exacerbating

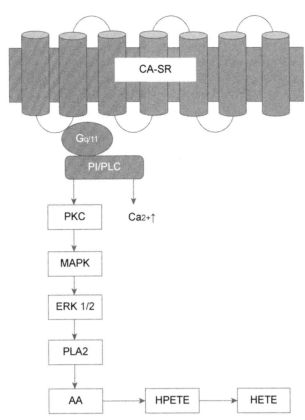

FIGURE 2.2 Calcium-sensing receptor (CaSR) activation mechanism via phosphatidylinositol (PI)-specific phospholipase C (PLC) pathway. *AA*, ara-chidonic acid; *ERK1/2*, extracellular signal-regulated kinases; *HETE*, hydroxyeicosatetranoic acid; *HPETE*, hydroperoxyeicosatetraenoic acid; *MAPK*, mitogen-activated protein kinase; *PKC*, protein kinase C; *PLA$_2$*, phospholipase A$_2$, *Originally proposed by Kifor et al.*

the hypercalcemia. Other findings include hypermagnesemia. FHH is relatively benign and asymptomatic in comparison to neonatal severe hyperparathyroidism (NSHPT), which is the homozygous form of FHH. In NSHPT, there is complete loss of CaSR function. It presents with severe hypercalcemia (>15 mg/dL), hyperparathyroidism, and skeletal demineralization. NSHPT can be fatal if it is not properly detected and treated (Brown, 2007b; Magno et al., 2011). Furthermore, autosomal-dominant hypocalcemia (ADH) occurs from activating CaSR mutations. ADH is characterized by an increased sensitivity of parathyroid cells to extracellular Ca^{2+} levels. This disorder leads to a spectrum of disease phenotypes, ranging from less (e.g., muscle cramps and paresthesia) to more severe (e.g., seizures; Huang et al., 2007).

The CaSR can also be stimulated by other compounds in addition to Ca^{2+} (Magno et al., 2011). Compounds that have agonist effects on the receptor by allosteric activation are known as calcimimetics. Examples of calcimimetics are synthetic, small-molecule compounds such as R-568 and R-467, which are able to stimulate the CaSR in the parathyroid (Magno et al., 2011). CaSR stimulation in the parathyroid gland increases the intracellular concentration of Ca^{2+} and inhibits PTH secretion in the experimental setting. The use of these calcimimetics in multiple laboratory models has led to the development of drugs for clinically targeting PTH regulation. Currently, calcimimetic agents such as cinacalcet are used for the treatment of secondary hyperparathyroidism in patients with end-stage renal disease and those with parathyroid cancer (Block et al., 2004). Conversely, calcilytics are compounds that act as CaSR antagonists. Examples include NPS-2143 and Calhex 231 (Huang et al., 2007; Magno et al., 2011). NPS-2143 increases PTH secretion and reverses the effect of calci-mimetics such as R-467 in animal models. NPS-2143 has also been studied in osteoporosis models and for the treatment of ADH (Nemeth and Shoback, 2013).

Further understanding of CaSR function in PTH-dependent pathways, which regulate Ca^{2+} metabolism, will facilitate the discovery of new drugs (or small molecules) for the treatment of PTH- and calcium-related diseases in humans. Pharmacologic agents must be carefully examined to ensure that disruption of other physiologic systems regulated by the CaSR does not occur. Because high receptor homology is maintained across various organs and tissues, this is a major concern when considering the use of systemically administered agents that alter CaSR function.

CALCIUM-SENSING RECEPTOR FUNCTION IN BONE

Bone remodeling is closely related to body Ca^{2+} homeostasis. High extracellular Ca^{2+} inhibits bone resorption and favors bone formation; conversely, decreased extracellular Ca^{2+} promotes bone resorption to increase Ca^{2+} availability in the circulation to support metabolic processes throughout the body (Magno et al., 2011). As previously mentioned, CaSR exerts its effects on bone metabolism by hormonal regulation through PTH. Furthermore, evidence has shown that CaSR (along with other calcium-regulating receptors) is present in the bone and that it might exert local control on bone mineralization (Marie, 2010). Thus the CaSR, as a key player in Ca^{2+} homeostasis, plays a pivotal role in bone maintenance and remodeling (Marie, 2010).

The main role of CaSR in bone appears to be in the recruitment, differentiation, function, and survival of osteoblasts and osteoclasts (Brown, 2007b; Marie, 2010). In osteoclasts, elevated extracellular calcium during bone resorption activates CaSR, leading to downregulation of osteoclastic activity (Marie, 2010). CaSR is also expressed in osteoblasts, and exposure to high Ca^{2+} levels stimulates the receptor, which promotes cell migration, proliferation, and DNA synthesis (Marie, 2010). However, it is possible that other receptors can sense and respond to extracellular calcium in the bone, suggesting that the CaSR might not be the only regulator of bone remodeling (Marie, 2010). For example, some studies have shown that CaSR knockout mice still respond to changes in extracellular calcium in the local bone environment (Marie, 2010). In addition, hormone-independent cation-sensing receptors in bone may be involved in remodeling and metabolism (Marie, 2010). G-protein–coupled receptor family C group 6 (GPRC6A; a CaSR-like receptor) is an example of a calcimimetic cation receptor that has the ability to sense extracellular Ca^{2+}, amino acids, R-568, and osteocalcin in osteoblasts (Marie, 2010). GPRC6A has a reduced sensitivity to Ca^{2+}, allowing it to appropriately sense the local bone Ca^{2+} concentration, which can be higher than normal Ca^{2+} blood levels (Brown, 2007a). More research is needed to identify and further understand the specific cation-responsive receptors in bone and whether they have independent functions and/or synergistic roles with other receptors.

Although the bone receptors remain to be further analyzed, multiple pathways that are activated by Ca^{2+} in the bone environment have been described, mostly in in vitro experiments. In osteoblasts, Ca^{2+} acting via the CaSR increases osteoblast replication, differentiation, and survival through the ERK1/2 and AKT (protein kinase B) pathways by increasing the production of prostaglandin E2 and insulin-like growth factor (IGF)-I and -II (Marie, 2010). In osteoclasts, the CaSR mainly promotes osteoclast differentiation or apoptosis via the nuclear factor-κB, PLC, and inositol triphosphate (IP_3) pathways (Marie, 2010).

Various bone diseases related to alterations of CaSR expression or activity have been studied in animal models. In mice, tissue-specific deletion of the CaSR in osteoblasts led to decreased bone mass, decreased osteoblast differentiation, and increased apoptosis (Chang et al., 2010). In the clinical setting, the CaSR in bone can be targeted to prevent and treat osteoporosis. For example, strontium ranelate, a potent CaSR agonist, increases bone formation and decreases resorption by reducing osteoclast differentiation and activity and stimulating preosteoblast differentiation (Meunier et al., 2004), which eventually increases osteoblast numbers and increases bone formation. In the last decade, clinical trials with strontium ranelate showed successful treatment of osteoporosis by reducing the risk of vertebral fracture in postmenopausal women (Meunier et al., 2004).

More research on the physiologic role of CaSR in bone metabolism may lead to the development of promising new therapies for diseases such as osteoporosis. With the increase in the fractional elderly population worldwide, targeted therapies are needed to prevent and treat osteoporosis.

CALCIUM-SENSING RECEPTOR FUNCTION IN THE KIDNEY

In this section we will review the function of the CaSR in the kidneys based on the individual nephron segments starting at the proximal tubule and ending in the collecting ducts (CDs). We will then discuss the pathophysiological role of CaSR and kidney diseases as well as the current clinically relevant target therapy that modulates CaSR activity. We will also provide evidence that the receptor can modulate magnesium transport along the nephron and there can be an important regulatory site for magnesium.

In the proximal convoluted tubules (PCTs) of the kidney, the CaSR is present on the apical surface of epithelial cells (Riccardi and Brown, 2010; Riccardi et al., 1998), where it facilitates PTH-dependent phosphorus metabolism. Activation of CaSR in the S3 segment of the PCT by luminal gadolinium, a trivalent cationic agonist of CaSR, regulates PTH-dependent phosphorus absorption, thereby increasing urinary phosphorus secretion (Riccardi and Brown, 2010).

Maiti and Beckman conducted studies on renal proximal-tubule–derived cells and demonstrated that an increase in extracellular fluid calcium concentration increases renal proximal tubule vitamin D receptor expression (Riccardi and Brown, 2010). This response is due to activation of CaSR and signal mediation by the p38α MAPK pathway (Alfadda et al., 2014).

Moreover, Egbuna et al. studied mice lacking full-length CaSR and showed that the mice have increased sensitivity to the calcemic action of 1α-25-dihydroxy-vitamin D_3 (1α-$25(OH)_2D_3$) independent of PTH levels (Riccardi and Brown, 2010). Furthermore, it has also been shown that 1α-$25(OH)_2D_3$ upregulates renal CaSR expression in the PCT (Riccardi and Brown, 2010).

Interestingly, acute PTH injection and a chronic phosphate-rich diet in rats can also reduce the expression of the CaSR in the brush-border membrane of renal epithelial cells (Alfadda et al., 2014). Recent studies have also shown that an increase in luminal calcium or the use of calcimimetic agents such as R-568 activate CaSR in the PCT, leading to enhanced fluid reabsorption and H^+ secretion (Capasso et al., 2013).

The thick ascending limb (TAL) and loop of Henle are important for the homeostasis of several minerals. The TAL actively reabsorbs sodium and chloride (Rocha and Kokko, 1973). Hebert et al. have shown that the sodium potassium 2 chloride cotransporter (NKCC2) and the renal outer medullary potassium K^+ (ROMK) channel contribute to the lumen positive charge (Hebert et al., 1997; Riccardi and Brown, 2010). As a consequence of Na^+ and Cl^- reabsorption, the net lumen positive charge acts as a driving force for passive absorption of Ca^{2+} and Mg^{2+} (Alfadda et al., 2014). It has been shown that almost 25% of calcium filtered by the glomeruli is reabsorbed in the cortical portion of the TAL (Hebert et al., 1997). The TAL is also the major site for Mg^{2+} reabsorption (Alfadda et al., 2014).

Throughout the TAL, CaSR is abundantly expressed (Alfadda et al., 2014; Riccardi et al., 1998). CaSR controls the metabolism of Ca^{2+} and other minerals in this renal tubule segment. Increased systemic Ca^{2+} and/or Mg^{2+} levels activate basolateral CaSR receptors in the TAL, leading to lower Ca^{2+} and Mg^{2+} reabsorption via the paracellular route (Alfadda et al., 2014; Hebert et al., 1997). Moreover, the activation of basolateral CaSR induced by high Ca^{2+} levels or neomycin (a CaSR agonist) in the TAL, via cytochrome p450 metabolites and/or prostaglandins, inhibits the ROMK channels (Alfadda et al., 2014). This leads to a decrease in K^+ levels in the TAL (Riccardi and Brown, 2010). As a consequence, activity of the NKCC2 cotransporter decreases because it requires K^+ for normal function. This response (decreased NKCC2 activity) will diminish the net positive lumen voltage and thus inhibit Ca^{2+} and Mg^{2+} absorption. Some authors also noted that extracellular Ca^{2+} and neomycin decreased chloride reabsorption in the cortical TAL by inhibiting the cyclic AMP (cAMP) pathway (Alfadda et al., 2014). Furthermore, Loupy et al. showed that CaSR inhibitors increased Ca^{2+} reabsorption and paracellular pathway permeability without having an effect on NaCl reabsorption (Alfadda et al., 2014).

Claudin channels 14 and 16 expressed in the tight junction region between renal epithelial cells also play important roles in Ca^{2+} homeostasis in the TAL. In CaSR knockout mice, downregulation of paracellular claudin-14 channels was noted (which negatively regulate paracellular cation permeability in the TAL) as well as upregulation of the paracellular claudin-16 channels (which positively regulate paracellular cation permeability). Thus the loss of CaSR function in the TAL decreases Ca^{2+} excretion independent of PTH action (Alfadda et al., 2014).

Activation of CaSR by Mg^{2+}, Ca^{2+}, gadolinium, and neomycin has also been shown to reduce Mg^{2+} reabsorption in the TAL (Riccardi and Brown, 2010). In addition, PTH has been shown to regulate Ca^{2+} transport in the cortical TAL (Alfadda et al., 2014; Riccardi and Brown, 2010), again highlighting the regulatory feedback occurring between the CaSR and PTH. Gadolinium and NPS R-467 (an organic CaSR agonist) activate the CaSR, resulting in reduction of PTH-induced Ca^{2+} absorption across the apical membrane of renal epithelial cells (Alfadda et al., 2014; Riccardi and Brown, 2010). CaSR activity can also be affected by changes in pH as noted by Quinn et al. (1997). The CaSR becomes more sensitive to changes in Ca^{2+} levels as the pH rises (alkaline environment) whereas sensitivity decreases as the pH drops (acidic environment; Riccardi and Brown, 2010). These studies made the seminal observations that the CaSR was modulated not only by its primary ligand calcium but that other environmental factors can also allosterically modify the receptor and the binding pocket for Ca^{2+}, resulting in changes in receptor activation kinetics.

The distal convoluted tubules (DCTs) and CDs are nephron segments that have an important effect on the mineral metabolism in general and specifically on Ca^{2+} and Mg^{2+}. The electrophysiological ion gradient inhibits passive calcium reabsorption via the paracellular route, thus requiring active transport via a transcellular route (Riccardi and Brown, 2010). Ca^{2+} enters the DCT cells via Ca^{2+} channel transient receptor potential vanilloid member-5 (TRPV5) whereas Mg^{2+} is transported via the apical channel transient receptor channel potential melstatin member-6 (TRPM6). Vitamin D_3 and PTH control the activity of both of these channels (Riccardi and Brown, 2010). After Ca^{2+} enters cells of the DCT, it is transported to the basolateral side by calbindin D28K. Ca^{2+} subsequently exits cells through the sodium–calcium exchanger and the plasma membrane Ca^{2+}-ATPase (Riccardi and Brown, 2010).

The CaSR is present in the DCT on the basolateral side (Alfadda et al., 2014; Riccardi et al., 1998). Topala et al. (2009) showed that the activation of the CaSR in the DCT stimulates the activity of the Ca^{2+} TRPV5, leading to Ca^{2+} influx. However, TRPV6, the closest homologue to TRPV5, was not activated by CaSR stimulation (Alfadda et al., 2014).

There is evidence that the CaSR is expressed in the CD, as reported by multiple authors (Riccardi and Brown, 2010; Riccardi et al., 1998). Hypercalcemia leads to hypercalciuria and increases the risk of nephrolithiasis. The kidney should

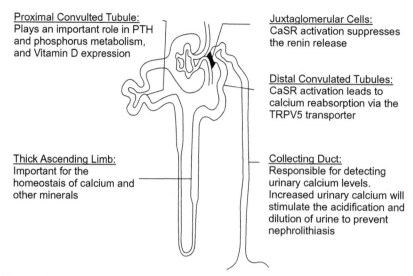

Proximal Convulted Tubule:
Plays an important role in PTH
and phosphorus metabolism,
and Vitamin D expression

Juxtaglomerular Cells:
CaSR activation suppresses
the renin release

Distal Convulated Tubules:
CaSR activation leads to
calcium reabsorption via the
TRPV5 transporter

Thick Ascending Limb:
Important for the
homeostais of calcium and
other minerals

Collecting Duct:
Responsible for detecting
urinary calcium levels.
Increased urinary calcium will
stimulate the acidification and
dilution of urine to prevent
nephrolithiasis

FIGURE 2.3   Known calcium-sensing receptor (CaSR) actions along the renal nephron. *PTH*, parathyroid hormone; *TRPV5*, transient receptor potential vanilloid member-5.

normally be able to adapt to the increase in Ca^{2+} levels by acidifying the urine. The mechanism involves activation of the CaSR in the CD via increased luminal Ca^{2+} levels leading to urinary acidification by increasing H^+-ATPase activity. Polyuria also occurs by CaSR-mediated inactivation of aquaporin-2 channels (Riccardi and Brown, 2010). In the medullary portion of the CD, the CaSR is expressed on the apical side and is responsible for sensing urinary Ca^{2+} levels. The CaSR colocalizes with aquaporin-2 channels, providing further evidence that receptor modulation will affect the flow of water in addition to modulating Ca^{2+} secretion and absorption. The increase in Ca^{2+} levels in the medullary CDs will result in the activation of the apical CaSR, thus reducing the activity of aquaporin-2 water channels (Riccardi and Brown, 2010).

Expanding our discussion to include other aspects of renal physiology, in the 1980s Fray, Park, and Valentine showed an inverse relation between systolic Ca^{2+} levels and renin secretion (Riccardi and Brown, 2010). Later in 2007 Ortiz-Capisano et al. were able to express CaSR on juxtaglomerular cells. They have shown that the activation of CaSR resulted in the inhibition of adenylyl cyclase-V activity, thereby decreasing cAMP formation and resulting in the suppression of renin release from the kidneys (Riccardi and Brown, 2010). In addition, stimulation of the Ca^{2+}-dependent phosphodiesterase 1C expressed in juxtaglomerular cells was shown to contribute to the inhibition of renin release (Riccardi and Brown, 2010; Fig. 2.3).

PATHOPHYSIOLOGICAL DISORDERS OF THE KIDNEY AND CALCIUM-SENSING RECEPTOR MODULATION

Chronic kidney disease (CKD) remains a global public health burden (Jha et al., 2013). One of the major consequences of CKDs is secondary hyperparathyroidism (Alfadda et al., 2014). As CKD progresses, disturbances in Ca^{2+} and vitamin D metabolism become more prominent, thus decreasing CaSR activity (Riccardi and Brown, 2010). The decline in CaSR activity results in increased PTH production and secretion and parathyroid gland hyperplasia (Riccardi and Brown, 2010). Studies showed that reduced CaSR activity can result in mineral abnormalities such as hypocalciuria (Riccardi and Brown, 2010). Other authors suggested the pathophysiology behind the increase of PTH secretion by the parathyroid gland is a result of decreased CaSR and vitamin D receptors within the parathyroid gland (Alfadda et al., 2014). Moreover, CaSR can also be used as a target to treat CKDs (Block et al., 2004). Cinacalcet, a calcimemetic agent, activates the CaSR in CKDs and results in lower PTH levels and maintenance of residual renal function (Block et al., 2004).

Increased urinary Ca^{2+} levels is a major cause of the formation of kidney stones (nephrolithiasis). Idiopathic hypercalciuria (IH) is found in almost 40% of patients presenting with Ca^{2+} nephrolithiasis (Pak et al., 1980). In a study conducted on seven families with autosomal-dominant inheritance of IH, there was no mutation found in the CaSR gene (Alfadda et al., 2014). However, other studies reported that CaSR Arg990Gly polymorphism was associated with increased risk of developing Ca^{2+} nephrolithiasis and IH by increasing Ca^{2+} excretion (Vezzoli et al., 2012). Conversely, Harding et al. showed that there was no association between this polymorphism and an increased risk of nephrolithiasis (Alfadda et al., 2014).

Thus the association between mutations in CaSR and the risk of developing Ca^{2+} nephrolithiasis remains a controversial topic with more research expected in this area.

FHH is an autosomal-dominant condition defined by symptomless, nonprogressive hypercalcemia with relatively low urinary Ca^{2+} excretion (Alfadda et al., 2014; Riccardi and Brown, 2010). PTH levels are within the normal range in FHH whereas Ca^{2+} levels are typically higher than normal (~11 mg/dL; Riccardi and Brown, 2010). Even with elevated serum Ca^{2+} levels in FHH, the kidney excretes less Ca^{2+} compared to normal (Alfadda et al., 2014). Mutations in CaSR, most often in the ECD and the signal transduction pathway, can result in FHH (Hendy et al., 2009; Ward et al., 2012).

NSHPT is characterized by severe hypercalcemia. It typically manifests in neonates presenting with polyuria, hypotonia, dehydration, respiratory distress, bone fractures, and failure to thrive (Riccardi and Brown, 2010; Ward et al., 2012). There is limited evidence on the relationship between serum and urinary Ca^{2+} levels and the role of CaSR in the kidneys of patients with NSHPT.

As in the parathyroid gland and in bone, there are relevant pharmacological interactions that modulate CaSR in the kidney. Calcimimetic agents of relevance include aminoglycosides, peptides, and amino acids, which are physiological activators of the CaSR (Alfadda et al., 2014). Cinacalcet (mentioned previously) is mainly used to treat patients with primary hyperparathyroidism in whom surgery is contraindicated, secondary hyperparathyroidism in patients with CKD, and parathyroid tumors (Alfadda et al., 2014; Block et al., 2004; Vestergaard and Thomsen, 2011). More recently, cinacalcet has also been used to treat patients with NSHPT and it has shown promising results (Gannon et al., 2014). In addition, cinacalcet showed promising results in treating FHH with CaSR mutations (Timmers et al., 2006).

In conclusion, CaSR in the kidney plays an important role in the homeostasis of several minerals. Thus interference with the physiological pathways of the receptor is expected to result in various pathophysiological conditions. Further investigations to understand the roles of the CaSR in the kidney are imperative.

CALCIUM-SENSING RECEPTOR FUNCTION IN THE GASTROINTESTINAL SYSTEM

As previously mentioned, the CaSR has been identified in various organs within the body. A wide variety of organs are influenced by CaSR modulation and play an important role in Ca^{2+} metabolism. These organs necessarily participate in the regulation of calcium homeostasis, including nutrient-sensing organs of the GI tract (Geibel and Hebert, 2009). The activity of the CaSR in the GI tract is allosterically modulated by Ca^{2+}, amino acids, and pH. Strong experimental evidence suggested that the role of the CaSR in the GI tract extends beyond Ca^{2+} metabolism to modulate complex GI physiology (Geibel and Hebert, 2009). CaSR in the GI tract influences gut hormone secretion, neuronal response, epithelial transport, and satiety (Geibel and Hebert, 2009). In this section, we will focus on the role of the CaSR in maintaining physiological functions of the alimentary tract and its crucial role in digestion.

Previous and ongoing efforts to determine the expression patterns and understand physiologic roles of CaSR in the GI tract clearly illustrate that the CaSR is vital for digestion. Starting with the tongue, the taste buds express CaSR; furthermore, these receptors may sense not only Ca^{2+} and amino acids but also play a role in the identification of taste (Bystrova et al., 2010). In addition, the tongue senses salinity, which also modulates CaSR activity (Geibel and Hebert, 2009). CaSR expression in the human esophagus exists on epithelial cell lines (Mulder et al., 2009). The receptor has been shown to be particularly active in the esophageal epithelial cell line, HET-1A (Mulder et al., 2009). Esophageal CaSR has been linked to the pathogenesis of esophagitis (Mulder et al., 2009). Esophagitis is characterized by production of mediators such as eosinophil-released major basic protein (Mulder et al., 2009), which activates the CaSR in the esophagus. The activation of the CaSR leads to elevation of fibroblast growth factor-9 (FGF9) in esophageal epithelial cells. Elevated FGF9 may lead to basal cell proliferation and hyperplasia, which is a clinical feature of esophagitis (Mulder et al., 2009).

In the stomach, CaSR plays a critical role in food digestion. The investigation of gastric CaSR has shown that it is expressed on the apical surface of human gastric mucous epithelial cells, gastrin-secreting cells (called G cells), and parietal cells (Feng et al., 2010). Parietal cells are responsible for secreting hydrochloric acid into the lumen of the gastric gland and then into the stomach (Feng et al., 2010). In parietal cells, activation of CaSR causes intracellular Ca^{2+} mobilization (Feng et al., 2010). Intracellular mobilization of Ca^{2+} in the parietal cells occurs after hormonal or neuronal stimulation of acid secretion (Feng et al., 2010). Gastrin, a regulator of gastric acid secretion, is released by gastric G cells as one of the first steps in classical hormone-regulated acid secretion (Feng et al., 2010). The CaSR located on the gastric G cell has been shown to regulate gastrin secretion leading to acid secretion in vivo (Feng et al., 2010). In addition, activation of the CaSR on the gastric G cells activates the proton pump (H^+, K^+ ATPase) in parietal cells (Geibel and Hebert, 2009).

Intriguingly, amino acid activation of CaSR appears to be stereoselective. The most potent agonists are the aromatic amino acids, L-phenylalanine and L-tryptophan, followed by aliphatic and polar amino acids. The other acidic, basic, and branched-chain amino acids are weak stimulators of CaSR or are inactive (Conigrave and Brown, 2006). In the presence of physiological concentrations of extracellular Ca^{2+}, activation of the CaSR via L-type amino acids leads to gastric acid secretion. The activation of CaSR here is independent of hormonal stimulation and potent secretagogues such as histamine (Geibel et al., 2001). This is exemplified nicely in the CaSR knockout mice. Gastrin release was negligible after administration of Ca^{2+} by oral gavage in the CaSR knockout mice. However, wild-type and heterozygous mice had a significantly higher gastrin response after oral administration of Ca^{2+} (Feng et al., 2010). The insensitivity of the G cell, in the absence of CaSR, to pH and luminal contents further supports the role of CaSR in gastric acid secretion (Geibel and Hebert, 2009). Reduction in the extracellular Ca^{2+} concentration leads to inactivation of CaSR, resulting in downregulation of acid secretion even with continuous stimulation via potent secretagogues such as histamine (Geibel et al., 2001).

CaSR plays an important role in the modulation of gastric secretion in response to luminal chyme. Moreover, CaSR agonists promote gastric mucosal defense mechanisms, including increasing mucus production, stimulating blood flow, and promoting bicarbonate secretion (Akiba et al., 2009). In addition, CaSR in gastric mucous epithelial cells has been identified as a growth factor. In cultured goblet cells, CaSR activation increased intracellular Ca^{2+} levels and enhanced proliferation (Geibel and Hebert, 2009). This observation supported the hypothesis that the CaSR is important for mucosal repair. It is also possible that the CaSR coordinates gastric acid secretion with body calcium requirements (Geibel and Hebert, 2009). When the body calcium demand is increased, CaSR is activated, resulting in stimulation of gastric acid production, thereby promoting calcium absorption. Therefore continued CaSR activation in the stomach by dietary Ca^{2+} may lead to downregulation of acid secretion. This might promote gastric mucosal repair by elevating internal gastric pH (Geibel and Hebert, 2009).

Zollinger–Ellison syndrome (ZES) is caused by a tumor in the non-β-cells of pancreatic islets and is characterized by increased gastrin production. High gastrin levels stimulate maximal acid production by parietal cells, eventually leading to severe mucosal ulcerations of the upper GI tract. When ZES occurs in a patient with multiple endocrine neoplasia type 1, the result is hyperparathyroidism (Epelboym and Mazeh, 2014). Gastrin blood levels tend to correlate with the severity of the associated hyperparathyroidism (Geibel and Hebert, 2009). The expression of CaSR is significantly increased (up to 80-fold) in all gastrinomas, yet expression is variable (Geibel and Hebert, 2009).

The CaSR in the liver is expressed in hepatocytes but not in nonparenchymal stellate, sinusoidal endothelial, or Kupffer cells (Conigrave and Brown, 2006). Activation of CaSR in an in vitro model in hepatocytes results in stimulation of bile flow via activation of the IP_3-sensitive store (Canaff, 2001). In the small intestine lumen, phenylalanine activates the CaSR enteroendocrine cell line STC-1. This stimulates cholecystokinin production and secretion. Furthermore, CaSR was stimulated by phenylalanine to a greater degree when more extracellular Ca^{2+} was added (Hira et al., 2008). This response suggests that Ca^{2+} stimulates bile flow via the CaSR (Conigrave and Brown, 2006).

In the pancreas, acinar cells express the CaSR on the basolateral membranes whereas the pancreatic duct cells express the CaSR on apical membranes (Conigrave and Brown, 2006). CaSR expression was also detected in human islets of Langerhans and in the intrapancreatic nerves and blood vessels (Racz et al., 2002). The receptor may be involved in the control of pancreatic fluid and enzyme secretion (Conigrave and Brown, 2006), but the mechanism is still not completely understood.

The CaSR in the small intestine is expressed on the basal membranes of villous and crypt epithelial cells, enterochromaffin cells, and in epithelial cells of Brunner's submucosal glands (Geibel and Hebert, 2009). In Brunner's submucosal glands, it is involved in the secretion of an alkali fluid into the lumen of the duodenum that is rich in bicarbonate in response to the calcium stimulus (Geibel and Hebert, 2009). CaSR is also expressed on neurons of the submucosal and myenteric plexuses of the GI (Geibel and Hebert, 2009). This particular expression pattern suggests that the CaSR may be involved in control of GI motility and electrophysiological functions (Conigrave and Brown, 2006). With regards to enterochromaffin cells, Muramatsu et al. described that the CaSR might be involved in postprandial glycemic control. This is possibly associated with the fact that CaSR is expressed and functions in 5-hydroxytryptamine (5-HT)-producing enterochromaffin cells (Liu et al., 2003). CaSR in these cells is activated by extracellular Ca^{2+}, which induces 5-HT secretion in 5-HT-producing parafollicular cells, which might delay gastric emptying. CaSR agonists could potentially reduce the rapid elevation of plasma glucose after an oral glucose load (Muramatsu et al., 2014).

Another interesting aspect of intestinal CaSR is that it interacts with luminal microorganisms and nutrients of the small intestine. This interaction is important to modulate intestinal barrier function and immune responses; thus the CaSR maintains intestinal physiological processes (Cheng et al., 2014). Deficiency in epithelial CaSR of the small intestine is associated with proinflammatory immune responses, which diminish intestinal barrier function and alter microbiota composition.

In addition, CaSR deficiency in the small intestine is associated with autoinflammatory disorders, such as inflammatory bowel disease (Cheng et al., 2014). Moreover, activation of the intestinal CaSR by dietary agonists (e.g., Ca^{2+}) protects against intestinal inflammation. Conversely, inhibition of the CaSR by reduction of dietary calcium promotes further inflammation in animal models with induced colitis (Pele et al., 2007). The effects of CaSR activation by Ca^{2+}, L-amino acids, and other nutrient activators in the small intestine is currently uncertain.

CaSR in colon is expressed on the apical and basolateral membranes of crypt epithelial cells (Geibel and Hebert, 2009). The CaSR in the colon is activated by Ca^{2+}, other divalent ions (e.g., Gd^{3+}, the trivalent Gadolinium ion), and the antibiotic neomycin. Activation of the receptor results in an increase in intracellular Ca^{2+} in surface and crypt cells of the colonic mucosa (Geibel and Hebert, 2009). This response could relate to colonic water and electrolyte balance. Net fluid absorption in colonic crypts occurs under physiological conditions (Geibel and Hebert, 2009). This physiological function can be abolished by cAMP-mediated induction of fluid secretion. When the colonic crypts are exposed to secretagogues such as forskolin, cholera toxin (causing cAMP generation), or *Escherichia coli* heat-stable enterotoxin (causing cyclic GMP generation), there is an induction of fluid and electrolyte secretion (Geibel and Hebert, 2009). Reversal of forskolin-induced fluid shift happens when CaSR is stimulated by Ca^{2+} or spermine (a polyamine; Geibel and Hebert, 2009). CaSR activation in intestinal and colonic epithelial cells by calcimimetic drugs could potentially decrease electrolyte and fluid secretion after exposure to cholera toxin or heat-stable *E. coli* enterotoxin (Singh et al., 2014). Thus CaSR was targeted to prevent the electrophysiological disturbance and intestinal barrier dysfunction after infectious diarrheal disease by host–pathogen interactions. Moreover, Cheng et al. found that activation of neural CaSR inhibits basal and cAMP-stimulated fluid secretion after exposure to cholera toxin or forskolin in proximal and distal colon via inhibition of the enteric nervous system (ENS; Cheng, 2012). Therefore CaSR agonists could potentially be used as a unique therapy for secretory diarrhea, acting as antidiarrheal agents at the level of the colonic epithelium as well as the ENS.

Recent observations suggested a potential antitumorigenic role for CaSR in the colon. CaSR activation decreases proliferation and promotes differentiation of several different colonic cell types (Fetahu et al., 2014). CaSR expression is lost in colorectal carcinoma (Geibel and Hebert, 2009). Recent prospective cohort studies have shown a beneficial effect of calcium supplements on colorectal carcinoma (Heine-Broring et al., 2014). In addition, randomized double-blind trials clearly showed a significant reduction in colorectal adenoma recurrence using calcium supplements (Geibel and Hebert, 2009). CaSR in response to increasing levels of extracellular Ca^{2+} could be a potential mechanism underlying the antitumorigenic action of Ca^{2+} in colon cancer. A potential mechanism involves CaSR-mediated activation of E-cadherin expression and suppression of β-catenin (Geibel and Hebert, 2009).

In conclusion, CaSR in the GI tract is important for nutrient digestion and absorption. Future work will focus on regulation of the CaSR in the GI tract to understand how it relates to various pathological conditions, which result in GI malfunction (see Fig. 2.4).

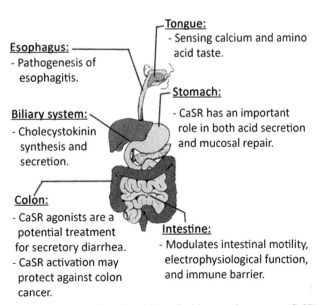

FIGURE 2.4   Summarization of the known pathophysiological activities of calcium-sensing receptor (CaSR) along the gastrointestinal system.

CALCIUM-SENSING RECEPTOR FUNCTION IN THE SKIN

Skin cells also express CaSR. In epidermal keratinocytes CaSR is vital for Ca^{2+} homeostasis and the immune function of the skin (Tu et al., 2012). CaSR in the skin is also critically important for the normal function of the vitamin D receptor. In animal models keratinocytes that are deficient in CaSR have low vitamin D receptor levels (Tu et al., 2012). Remarkably, activation of the vitamin D receptor and CaSR have protective effects against epidermal carcinogenesis, but additional studies are needed to elucidate the underlying mechanisms (Bikle, 2014).

CONCLUSION

In this chapter we reviewed the function of CaSR in various organs and tissues, and we considered its role in some pathophysiological states. When considered together, it is apparent that the CaSR is important from a physiological and pathophysiological perspective. Any disruption of CaSR activity significantly affects the normal homeostasis of divalent ions within the body. As discussed, the CaSR is responsible for sensing changes in extracellular Ca^{2+}; thus it is important for regulating the function of the parathyroid gland, kidney, bone, GI, and skin. For more than 2 decades, basic and clinical research has allowed us to achieve a better understanding of the functional roles of the CaSR. We are now able to treat a wide variety of diseases and disorders by directed therapy toward this receptor. Further research is still needed to discover other specific functions of the CaSR in each organ, and with the technological advances that have occurred over these last 2 decades we look forward to further clarification of receptor function and the development of new targeted therapies that will improve the quality of life for a great variety of patients.

REFERENCES

Akiba, Y., Watanabe, C., Mizumori, M., Kaunitz, J.D., 2009. Luminal L-glutamate enhances duodenal mucosal defense mechanisms via multiple glutamate receptors in rats. Am. J. Physiol. Gastrointest. Liver Physiol. 297 (4), G781–G791. http://dx.doi.org/10.1152/ajpgi.90605.2008.

Alfadda, T.I., Saleh, A.M., Houillier, P., Geibel, J.P., 2014. Calcium-sensing receptor 20 years later. Am. J. Physiol. Cell Physiol. 307 (3), C221–C231. http://dx.doi.org/10.1152/ajpcell.00139.2014.

Bikle, D.D., 2014. Vitamin d receptor, a tumor suppressor in skin. Can. J. Physiol. Pharmacol. 1–6. http://dx.doi.org/10.1139/cjpp-2014-0367.

Block, G.A., Martin, K.J., de Francisco, A.L., Turner, S.A., Avram, M.M., Suranyi, M.G., et al., 2004. Cinacalcet for secondary hyperparathyroidism in patients receiving hemodialysis. N. Engl. J. Med. 350 (15), 1516–1525. http://dx.doi.org/10.1056/NEJMoa031633.

Brown, E.M., Gamba, G., Riccardi, D., Lombardi, M., Butters, R., Kifor, O., et al., 1993. Cloning and characterization of an extracellular Ca(2+)-sensing receptor from bovine parathyroid. Nature 366 (6455), 575–580. http://dx.doi.org/10.1038/366575a0.

Brown, E.M., 2007a. The calcium-sensing receptor: physiology, pathophysiology and car-based therapeutics. Subcell. Biochem. 45, 139–167. Retrieved from: http://www.ncbi.nlm.nih.gov/pubmed/18193637.

Brown, E.M., 2007b. Clinical lessons from the calcium-sensing receptor. Nat. Clin. Pract. Endocrinol. Metab. 3 (2), 122–133. http://dx.doi.org/10.1038/ncpendmet0388.

Bystrova, M.F., Romanov, R.A., Rogachevskaja, O.A., Churbanov, G.D., Kolesnikov, S.S., 2010. Functional expression of the extracellular-Ca^{2+}-sensing receptor in mouse taste cells. J. Cell Sci. 123 (Pt 6), 972–982. http://dx.doi.org/10.1242/jcs.061879.

Canaff, L., Petit, J.L., Kisiel, M., Watson, P.H., Gascon-Barré, M., Hendy, G.N., February 9, 2001. Extracellular calcium-sensing receptor is expressed in rat hepatocytes. Coupling to intracellular calcium mobilization and stimulation of bile flow. J. Biol. Chem 276 (6), 4070–4079.

Capasso, G., Geibel, P.J., Damiano, S., Jaeger, P., Richards, W.G., Geibel, J.P., 2013. The calcium sensing receptor modulates fluid reabsorption and acid secretion in the proximal tubule. Kidney Int. 84 (2), 277–284. http://dx.doi.org/10.1038/ki.2013.137.

Chang, W., Dvorak, M., Shoback, D., 2010. Assessing constitutive activity of extracellular calcium-sensing receptors in vitro and in bone. Methods Enzymol. 484, 253–266. http://dx.doi.org/10.1016/B978-0-12-381298-8.00013-7.

Cheng, S.X., Lightfoot, Y.L., Yang, T., Zadeh, M., Tang, L., Sahay, B., et al., 2014. Epithelial CaSR deficiency alters intestinal integrity and promotes proinflammatory immune responses. FEBS Lett. 588 (22), 4158–4166. http://dx.doi.org/10.1016/j.febslet.2014.05.007.

Cheng, S.X., 2012. Calcium-sensing receptor inhibits secretagogue-induced electrolyte secretion by intestine via the enteric nervous system. Am. J. Physiol. Gastrointest. Liver Physiol. 303 (1), G60–G70. http://dx.doi.org/10.1152/ajpgi.00425.2011.

Conigrave, A.D., Brown, E.M., 2006. Taste receptors in the gastrointestinal tract. II. L-Amino acid sensing by calcium-sensing receptors: implications for GI physiology. Am. J. Physiol. Gastrointest. Liver Physiol. 291 (5), G753–G761. http://dx.doi.org/10.1152/ajpgi.00189.2006.

Epelboym, I., Mazeh, H., 2014. Zollinger–Ellison syndrome: classical considerations and current controversies. Oncologist 19 (1), 44–50. http://dx.doi.org/10.1634/theoncologist.2013-0369.

Feng, J., Petersen, C.D., Coy, D.H., Jiang, J.K., Thomas, C.J., Pollak, M.R., Wank, S.A., 2010. Calcium-sensing receptor is a physiologic multimodal chemosensor regulating gastric g-cell growth and gastrin secretion. Proc. Natl. Acad. Sci. USA 107 (41), 17791–17796. http://dx.doi.org/10.1073/pnas.1009078107.

Fetahu, I.S., Hobaus, J., Aggarwal, A., Hummel, D.M., Tennakoon, S., Mesteri, I., et al., 2014. Calcium-sensing receptor silencing in colorectal cancer is associated with promoter hypermethylation and loss of acetylation on histone 3. Int. J. Cancer 135 (9), 2014–2023. http://dx.doi.org/10.1002/ijc.28856.

Gannon, A.W., Monk, H.M., Levine, M.A., 2014. Cinacalcet monotherapy in neonatal severe hyperparathyroidism: a case study and review. J. Clin. Endocrinol. Metab. 99 (1), 7–11. http://dx.doi.org/10.1210/jc.2013-2834.

Geibel, J.P., Hebert, S.C., 2009. The functions and roles of the extracellular Ca^{2+}-sensing receptor along the gastrointestinal tract. Annu. Rev. Physiol. 71, 205–217. http://dx.doi.org/10.1146/annurev.physiol.010908.163128.

Geibel, J.P., Wagner, C.A., Caroppo, R., Qureshi, I., Gloeckner, J., Manuelidis, L., et al., 2001. The stomach divalent ion-sensing receptor scar is a modulator of gastric acid secretion. J. Biol. Chem. 276 (43), 39549–39552. http://dx.doi.org/10.1074/jbc.M107315200.

Hebert, S.C., Brown, E.M., Harris, H.W., 1997. Role of the Ca(2+)-sensing receptor in divalent mineral ion homeostasis. J. Exp. Biol. 200 (Pt 2), 295–302. Retrieved from: http://www.ncbi.nlm.nih.gov/pubmed/9050237.

Heine-Broring, R.C., Winkels, R.M., Renkema, J.M., Kragt, L., van Orten-Luiten, A.C., Tigchelaar, E.F., et al., 2014. Dietary supplement use and colorectal cancer risk: a systematic review and meta-analyses of prospective cohort studies. Int. J. Cancer. http://dx.doi.org/10.1002/ijc.29277.

Hendy, G.N., Guarnieri, V., Canaff, L., 2009. Calcium-sensing receptor and associated diseases. Prog. Mol. Biol. Transl. Sci. 89, 31–95. http://dx.doi.org/10.1016/s1877-1173(09)89003-0.

Hira, T., Nakajima, S., Eto, Y., Hara, H., 2008. Calcium-sensing receptor mediates phenylalanine-induced cholecystokinin secretion in enteroendocrine STC-1 cells. FEBS J. 275 (18), 4620–4626. http://dx.doi.org/10.1111/j.1742-4658.2008.06604.x.

Huang, Y., Zhou, Y., Yang, W., Butters, R., Lee, H.W., Li, S., et al., 2007. Identification and dissection of Ca(2+)-binding sites in the extracellular domain of Ca(2+)-sensing receptor. J. Biol. Chem. 282 (26), 19000–19010. http://dx.doi.org/10.1074/jbc.M701096200.

Jha, V., Garcia-Garcia, G., Iseki, K., Li, Z., Naicker, S., Plattner, B., et al., 2013. Chronic kidney disease: global dimension and perspectives. Lancet 382 (9888), 260–272. http://dx.doi.org/10.1016/s0140-6736(13)60687-x.

Kifor, O., MacLeod, R.J., Diaz, R., Bai, M., Yamaguchi, T., Yao, T., et al., 2001. Regulation of MAP kinase by calcium-sensing receptor in bovine parathyroid and CaR-transfected HEK293 cells. Am. J. Physiol. Renal Physiol. 280 (2), F291–F302. Retrieved from: http://www.ncbi.nlm.nih.gov/pubmed/11208605.

Liu, K.P., Russo, A.F., Hsiung, S.C., Adlersberg, M., Franke, T.F., Gershon, M.D., Tamir, H., 2003. Calcium receptor-induced serotonin secretion by parafollicular cells: role of phosphatidylinositol 3-kinase-dependent signal transduction pathways. J. Neurosci. 23 (6), 2049–2057. Retrieved from: http://www.ncbi.nlm.nih.gov/pubmed/12657663.

Magno, A.L., Ward, B.K., Ratajczak, T., 2011. The calcium-sensing receptor: a molecular perspective. Endocr. Rev. 32 (1), 3–30. http://dx.doi.org/10.1210/er.2009-0043.

Marie, P.J., 2010. The calcium-sensing receptor in bone cells: a potential therapeutic target in osteoporosis. Bone 46 (3), 571–576. http://dx.doi.org/10.1016/j.bone.2009.07.082.

Meunier, P.J., Roux, C., Seeman, E., Ortolani, S., Badurski, J.E., Spector, T.D., et al., 2004. The effects of strontium ranelate on the risk of vertebral fracture in women with postmenopausal osteoporosis. N. Engl. J. Med. 350 (5), 459–468. http://dx.doi.org/10.1056/NEJMoa022436.

Mulder, D.J., Pacheco, I., Hurlbut, D.J., Mak, N., Furuta, G.T., MacLeod, R.J., Justinich, C.J., 2009. FGF9-induced proliferative response to eosinophilic inflammation in oesophagitis. Gut 58 (2), 166–173. http://dx.doi.org/10.1136/gut.2008.157628.

Muramatsu, M., Hira, T., Mitsunaga, A., Sato, E., Nakajima, S., Kitahara, Y., et al., 2014. Activation of the gut calcium-sensing receptor by peptide agonists reduces rapid elevation of plasma glucose in response to oral glucose load in rats. Am. J. Physiol. Gastrointest. Liver Physiol. 306 (12), G1099–G1107. http://dx.doi.org/10.1152/ajpgi.00155.2014.

Nemeth, E.F., Shoback, D., 2013. Calcimimetic and calcilytic drugs for treating bone and mineral-related disorders. Best Pract. Res. Clin. Endocrinol. Metab. 27 (3), 373–384. http://dx.doi.org/10.1016/j.beem.2013.02.008.

Pak, C.Y., Britton, F., Peterson, R., Ward, D., Northcutt, C., Breslau, N.A., et al., 1980. Ambulatory evaluation of nephrolithiasis. Classification, clinical presentation and diagnostic criteria. Am. J. Med. 69 (1), 19–30.

Pele, L.C., Thoree, V., Mustafa, F., He, S., Tsaprouni, L., Punchard, N.A., et al., 2007. Low dietary calcium levels modulate mucosal caspase expression and increase disease activity in mice with dextran sulfate sodium induced colitis. J. Nutr. 137 (11), 2475–2480. Retrieved from: http://www.ncbi.nlm.nih.gov/pubmed/17951488.

Quinn, S.J., Ye, C.P., Diaz, R., Kifor, O., Bai, M., Vassilev, P., Brown, E., October 1997. The Ca2+-sensing receptor: a target for polyamines. Am. J. Physiol 273 (4 Pt 1), C1315–C1323.

Racz, G.Z., Kittel, A., Riccardi, D., Case, R.M., Elliott, A.C., Varga, G., 2002. Extracellular calcium sensing receptor in human pancreatic cells. Gut 51 (5), 705–711. Retrieved from: http://www.ncbi.nlm.nih.gov/pubmed/12377811.

Riccardi, D., Brown, E.M., 2010. Physiology and pathophysiology of the calcium-sensing receptor in the kidney. Am. J. Physiol. Renal Physiol. 298 (3), F485–F499. http://dx.doi.org/10.1152/ajprenal.00608.2009.

Riccardi, D., Hall, A.E., Chattopadhyay, N., Xu, J.Z., Brown, E.M., Hebert, S.C., 1998. Localization of the extracellular Ca^{2+}/polyvalent cation-sensing protein in rat kidney. Am. J. Physiol. 274 (3 Pt 2), F611–F622. Retrieved from: http://www.ncbi.nlm.nih.gov/pubmed/9530279.

Rocha, A.S., Kokko, J.P., 1973. Sodium chloride and water transport in the medullary thick ascending limb of Henle. Evidence for active chloride transport. J. Clin. Invest. 52 (3), 612–623. http://dx.doi.org/10.1172/jci107223.

Singh, V., Yang, J., Chen, T.E., Zachos, N.C., Kovbasnjuk, O., Verkman, A.S., Donowitz, M., 2014. Translating molecular physiology of intestinal transport into pharmacologic treatment of diarrhea: stimulation of Na+ absorption. Clin. Gastroenterol. Hepatol. 12 (1), 27–31. http://dx.doi.org/10.1016/j.cgh.2013.10.020.

Topala, C.N., Schoeber, J.P., Searchfield, L.E., Riccardi, D., Hoenderop, J.G., Bindels, R.J., April 2009. Activation of the Ca2+-sensing receptor stimulates the activity of the epithelial Ca2+ channel TRPV5. Cell Calcium 45 (4), 331–339.

Timmers, H.J., Karperien, M., Hamdy, N.A., de Boer, H., Hermus, A.R., 2006. Normalization of serum calcium by cinacalcet in a patient with hyper-calcaemia due to a de novo inactivating mutation of the calcium-sensing receptor. J. Intern. Med. 260 (2), 177–182. http://dx.doi.org/10.1111/j.1365-2796.2006.01684.x.

Tu, C.L., Crumrine, D.A., Man, M.Q., Chang, W., Elalieh, H., You, M., et al., 2012. Ablation of the calcium-sensing receptor in keratinocytes impairs epidermal differentiation and barrier function. J. Invest. Dermatol. 132 (10), 2350–2359. http://dx.doi.org/10.1038/jid.2012.159.

Vestergaard, P., Thomsen, S., 2011. Medical treatment of primary, secondary, and tertiary hyperparathyroidism. Curr. Drug Saf. 6 (2), 108–113.

Vezzoli, G., Terranegra, A., Soldati, L., 2012. Calcium-sensing receptor gene polymorphisms in patients with calcium nephrolithiasis. Curr. Opin. Nephrol. Hypertens. 21 (4), 355–361. http://dx.doi.org/10.1097/MNH.0b013e3283542290.

Ward, B.K., Magno, A.L., Walsh, J.P., Ratajczak, T., 2012. The role of the calcium-sensing receptor in human disease. Clin. Biochem. 45 (12), 943–953. http://dx.doi.org/10.1016/j.clinbiochem.2012.03.034.

Chapter 3

New Developments in Our Understanding of the Regulation of Calcium Homeostasis by Vitamin D

Puneet Dhawan, Ran Wei, Vaishali Veldurthy, Sylvia Christakos
Rutgers, The State University of New Jersey, New Jersey Medical School, Newark, NJ, United States

INTRODUCTION

Calcium is the fifth most abundant element in the human body. Calcium is essential for bone formation and for many physiological processes, including blood coagulation, muscle contraction, nerve-pulse transmission, and hormone secretion. A principal factor involved in the maintenance of normal calcium homeostasis is vitamin D. In recent years, through mouse models as well as in vitro studies, new insight has been obtained related to vitamin D metabolism and the mechanisms by which vitamin D acts in intestine, kidney, and bone to maintain calcium homeostasis. In addition, genome-wide studies have provided a new perspective on the complex mechanisms involved in the regulation of calcium homeostasis by vitamin D. This chapter will provide a review of vitamin D action and will focus on recent new developments that have changed our understanding of the regulation of calcium homeostasis by vitamin D.

BIOACTIVATION OF VITAMIN D

Vitamin D is produced in the skin upon ultraviolet irradiation from 7-dehydrocholesterol. The synthesis of vitamin D in the skin is the most important source of vitamin D and varies with season and latitude. Few foods, which include fortified dairy products and fish oils, contain appreciable amounts of vitamin D. For vitamin D to regulate calcium homeostasis it must first be metabolized to its active form (Fig. 3.1). Vitamin D is transported in the blood by vitamin D binding protein (DBP; which binds vitamin D and its metabolites) to the liver, where it is hydroxylated at C-25 to produce 25-hydroxyvitamin D_3 ($25(OH)D_3$; Bikle et al., 2013). Although several cytochrome P450 enzymes had been considered as candidate enzymes mediating this C-25 hydroxylation, CYP2R1, a microsomal vitamin D 25-hydroxylase, is now considered the key enzyme responsible for the conversion of vitamin D to $25(OH)D_3$ (Cheng et al., 2004). Patients with a mutation in CYP2R1 have $25(OH)D_3$ deficiency as well as symptoms of vitamin-D–dependent rickets, further indicating that CYP2R1 is the biologically relevant vitamin D 25-hydroxylase (Thacher et al., 2015). Recent studies in CYP2R1 knockout mice confirmed the role of CYP2R1 in the hydroxylation of vitamin D at C-25 (Zhu et al., 2013). However, studies in these mice showed that although synthesis of $25(OH)D_3$ is significantly reduced, it is not eliminated, suggesting that other 25-hydroxylases, yet to be identified, are involved in the catalysis of vitamin D to $25(OH)D_3$. 25-Hydroxylation of vitamin D has not been reported to be highly regulated. $25(OH)D_3$ is transported by DBP from the liver to the kidney. In the renal proximal tubule, megalin and cubulin, cell surface receptors for DBP, internalize the complex of DBP and $25(OH)D_3$ (Willnow and Nykjaer, 2010). $25(OH)D_3$ is hydroxylated in the kidney at the position of C-1 by CYP27B1 ($25(OH)D_3$ 1-α hydroxylase), resulting in the formation of 1,25-dihydroxyvitamin D_3 ($1,25(OH)_2D_3$), the hormonally active from of vitamin D (Fig. 3.1). In humans CYP27B1 mutations, resulting in nonfunctional or deleted CYP27B1, cause vitamin D dependency rickets type-1 (VDDR1; or pseudovitamin D deficiency rickets), which is characterized by hypocalcemia, hypophosphatemia, hyperparathyroidism, decreased bone mineralization, and low circulating levels of $1,25(OH)_2D_3$ (Kitanaka et al., 1998). Replacement therapy with $1,25(OH)_2D_3$ results in the correction of this abnormal phenotype. In addition to the kidney, CYP27B1 has been reported during pregnancy in placenta and in monocytes and macrophages as well as other immune cells after disease-specific activation (Adams and Hewison, 2012). $25(OH)D_3$ can also be converted in the kidney to $24,25(OH)_2D_3$ by hydroxylation at C-24 by the mitochondrial P450 enzyme CYP24A1 ($25(OH)D_3$ 24-hydroxylase; Jones et al., 2014).

Molecular, Genetic, and Nutritional Aspects of Major and Trace Minerals. http://dx.doi.org/10.1016/B978-0-12-802168-2.00003-8

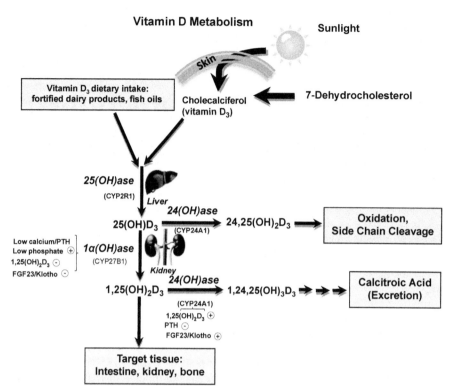

FIGURE 3.1   The pathway of vitamin D metabolism. Vitamin D derived from the diet or synthesized from 7-dehydrocholesterol by ultraviolet irradiation of the skin is converted to its active form, 1,25-dihydroxyvitamin D_3 (1,25$(OH)_2D_3$), by a series of hydroxylations at C-25 by CYP2R1 in the liver followed by hydroxylation at C-1 by CYP27B1 in the kidney. Hydroxylation of either 25-hydroxyvitamin D_3 (25$(OH)D_3$) or 1,25$(OH)_2D_3$ by CYP24A1 initiates the catabolic pathway. Human CYP24A1 is capable of C-23 and C-24 hydroxylation. C-23 hydroxylation can also initiate inactivation of 1,25$(OH)_2D_3$ and 25$(OH)D_3$ (not shown). 1,25$(OH)_2D_3$, parathyroid hormone (PTH), fibroblast growth factor-23 (FGF23)/α-klotho, and serum calcium and phosphate act together to maintain optimal 1,25$(OH)_2D_3$ levels. *1,24,25$(OH)_3D_3$,* 1,24,25-trihydroxyvitamin D_3; *1α(OH)ase,* 25-hydroxyvitamin $D_3$1α-hydroxylase; *24(OH)ase,* 25-hydroxyvitamin D_3 24-hydroxylase; *24,25$(OH)_2D_3$,* 24,25-dihydroxyvitamin D_3; *25(OH)ase,* vitamin D 25-hydroxylase.

CYP24A1 can also hydroxylate 1,25$(OH)_2D_3$, resulting in the formation of 1,24,25$(OH)_3D_3$ (Fig. 3.1). This reaction constitutes the first step in the catabolism of 1,25$(OH)_2D_3$ to calcitroic acid, which is excreted in the bile. Thus CYP24A1 limits 1,25$(OH)_2D_3$ levels by catabolizing 1,25$(OH)_2D_3$, resulting in products targeted for excretion and by producing 24,25$(OH)_2D_3$, thus decreasing the pool of 25$(OH)D_3$ available for conversion to 1,25$(OH)_2D_3$. 24,25$(OH)_2D_3$ can be further converted to different metabolites and finally to calcitroic acid. It has been noted that human CYP24A1 is capable of C-23 as well as C-24 hydroxylation and that C-23 hydroxylation can also initiate inactivation of the vitamin D molecule (Jones et al., 2014). CYP24A1 is present in all cells containing the VDR. CYP24A1 is induced by 1,25$(OH)_2D_3$. Thus 1,25$(OH)_2D_3$ induces its own deactivation, resulting in the control of the biological activity of 1,25$(OH)_2D_3$ in cells, thus preventing hypercalcemia resulting from high circulating 1,25$(OH)_2D_3$ levels.

REGULATION OF VITAMIN D METABOLISM

Parathyroid hormone (PTH), released in response to hypocalcemia, is the major stimulator of 1,25$(OH)_2D_3$ production in the kidney. PTH has been shown to induce CYP27B1 transcription. Nuclear receptor-related-1 protein, also known as nuclear orphan receptor-4A2, has been reported to be one factor mediating the PTH induction of CYP27B1 transcription (Zierold et al., 2007). In turn, 1,25$(OH)_2D_3$ indirectly suppresses PTH production in the parathyroid gland by increasing serum calcium, which acts via the calcium-sensing receptor in the parathyroid gland to suppress PTH release and directly by inhibiting the transcription of PTH (Canaff and Hendy, 2002; Demay et al., 1992). 1,25$(OH)_2D_3$ suppresses its own production by inhibiting CYP27B1 (Brenza and DeLuca, 2000). The mechanisms involved in 1,25$(OH)_2D_3$ inhibition of CYP27B1 remain to be determined. In contrast to CYP27B1, CYP24A1 is reciprocally regulated; it is induced by 1,25$(OH)_2D_3$ and inhibited by low calcium (Christakos et al., 2012; Plum and DeLuca, 2010). In addition to 1,25$(OH)_2D_3$, PTH, and low calcium, fibroblast growth factor-23 (FGF23), which promotes phosphate excretion and requires klotho, a transmembrane protein, as a cofactor, is also an important regulator of vitamin D metabolism. Together, FGF23 and α-klotho inhibit $1,25(OH)_2D_3$ production and increase expression of CYP24A1 (Hu et al., 2013). Thus, it is important to note that $1,25(OH)_2D_3$, PTH, FGF23/α-klotho, and serum calcium and phosphate act together as regulators of calcium homeostasis (Fig. 3.1).

NEW INSIGHTS INTO MECHANISMS INVOLVED IN THE EFFECTS OF $1,25(OH)_2D_3$ IN INTESTINE, KIDNEY, AND BONE

Intestine

The principal action of vitamin D in the maintenance of calcium homeostasis is enhancing intestinal calcium absorption. If, in response to low dietary calcium, pregnancy, or lactation (when the demand for calcium increases), serum calcium is unable to be maintained by enhanced calcium absorption, then PTH and $1,25(OH)_2D_3$ act to increase calcium reabsorption from the distal tubules of the kidney and to mobilize calcium from bone. Two major mechanisms have been proposed for intestinal calcium absorption: facilitated diffusion and paracellular movement through tight junctions (Fig. 3.2). Facilitated diffusion is a saturable process involving three $1,25(OH)_2D_3$ regulated steps: calcium entry into the enterocytes through an apical membrane calcium channel, transcellular movement of calcium by binding to the calcium-binding protein calbindin-D_{9k}, and extrusion of calcium from the cell by the plasma membrane CaATPase (PMCA1b). In the facilitated diffusion model, uptake into the enterocyte is mediated by the vitamin-D–inducible epithelial calcium channel transient receptor potential vanilloid type-6 (TRPV6). In addition, it has been suggested that calbindin-D_{9k} (the $1,25(OH)_2D_3$ regulated calcium-binding protein in mammalian intestine) participates in calcium absorption by regulating intracellular calcium transport. The association of calbindin-D_{9k} and TRPV6 with transcellular intestinal calcium absorption is supported by their colocalization in the intestine, the coinduction of both proteins before the peak of intestinal calcium absorption in response to $1,25(OH)_2D_3$, and their reduced expression in VDR knockout (null) mice (Christakos, 2012). However, studies in calbindin-D_{9k} and TRPV6 null mice have challenged the traditional facilitated diffusion model. For example, there are no phenotypical differences between calbindin-D_{9k} or TRPV6 null and wild-type (WT) mice when dietary calcium is normal (Kutuzova et al., 2006, 2008). These findings indicate that under normal calcium conditions, calbindin-D_{9k} and TRPV6 are redundant for intestinal calcium absorption, suggesting compensation by other channels or proteins yet to be identified. However, under conditions of dietary calcium restriction, an increase in bone turnover as well as a decrease in the efficiency of intestinal calcium absorption have been reported in TRPV6 null mice (as compared with WT mice; Benn et al., 2008; Lieben et al., 2010). In addition, using TRPV6/calbindin-D_{9k} double null mice, it has been shown that under low dietary calcium conditions intestinal calcium absorption is least efficient in the absence of both proteins (Lieben et al.,

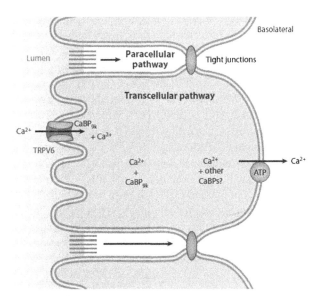

FIGURE 3.2 1,25-Dihydroxyvitamin D_3 ($1,25(OH)_2D_3$)-mediated calcium transport in the intestine. $1,25(OH)_2D_3$ enhances transcellular intestinal calcium transport by increasing the expression of the calcium-binding protein calbindin-D_{9k} (CaBP_{9k}) and the apical membrane calcium channel transient receptor potential vanilloid type-6 (TRPV6). Studies in CaBP_{9k} and TRPV6 null mice suggest that in the absence of these proteins there is compensation by other channels or proteins yet to be identified. Under conditions of high dietary calcium, the paracellular calcium transport pathway predominates. Some reports indicate that paracellular calcium transport may also be regulated by $1,25(OH)_2D_3$. *CaBP*, calcium-binding protein.

2010). Taken together, these findings suggest that TRPV6 is required for proper bone mineralization under conditions of low dietary calcium, that calbindin-D_{9k} and TRPV6 can act together in certain aspects of the absorptive process, and that the process of $1,25(OH)_2D_3$-mediated intestinal calcium absorption is more complex than the three-step facilitated diffusion model. Although it remains to be determined, calbindin may associate with TRPV6 and modulate the activity of the calcium channel. Calbindin may also act as a calcium buffer preventing toxic levels from accumulating in the intestinal cell. Calcium may diffuse through the cell bound to calbindin or other calcium-binding proteins. Furthermore, transgenic mice overexpressing TRPV6 throughout the intestine develop hypercalcemia and soft tissue calcification, further indicating a role for TRPV6 in the calcium absorptive process (Cui et al., 2012). It is possible that in future studies calcium channels involved in calcium entry into the enterocyte in addition to TRPV6 will be identified.

In addition to the active transcellular pathway, it has been suggested that $1,25(OH)_2D_3$ can also affect paracellular calcium transport, a passive process that occurs between epithelial cells at intercellular tight junctions, predominately in more distal regions of the intestine (Fig. 3.2). Tight junctions form a barrier to the movement of macromolecules by maintaining size and charge selectivity. $1,25(OH)_2D_3$ has been reported to regulate the expression of intestinal claudin-2 and claudin-12 (transmembrane components of tight junctions; Fujita et al., 2008). In addition, $1,25(OH)_2D_3$ suppresses cadherin-17 (a cell adhesion protein) and aquaporin-8 (a tight junction channel) in the intestine (Kutuzova and Deluca, 2004). It has been suggested that $1,25(OH)_2D_3$, by regulating the expression of certain tight junction components in the intestine, alters barrier structure, thus increasing permeability to calcium. However, it should be noted that further in vivo studies are needed to determine the physiological significance with regard to intestinal calcium transport of the regulation of tight junction components by $1,25(OH)_2D_3$.

Kidney

In addition to the intestine, the kidney also plays a central role in maintaining calcium homeostasis. Sixty-five percent of the filtered calcium is reabsorbed in the proximal tubule, whereas the thick ascending loop of Henle absorbs another 20%. In these segments, calcium is reabsorbed by passive, paracellular pathways that are $1,25(OH)_2D_3$ and PTH independent. The remaining 10–15% of filtered calcium is reabsorbed in the distal convoluted and connecting tubules (Boros et al., 2009). In addition, in this part of the nephron, calcium reabsorption is hormonally regulated, and it occurs via a transcellular pathway involving active transport (Boros et al., 2009). Similar to studies in the intestine, in the distal tubule apical calcium entry is facilitated by a highly calcium-selective epithelial calcium channel transient receptor potential vanilloid member-5 (TRPV5; which shares 75% sequence homology with TRPV6; Hoenderop and Bindels, 2008). Studies in TRPV5 null mice, which show hypercalciuria and impaired calcium reabsorption in the distal tubule, provide evidence for a critical role for TRPV5 in renal calcium reabsorption (Hoenderop et al., 2003). After calcium entry, it has been suggested that active transcellular calcium transport in the distal tubule (similar to active intestinal calcium transport) involves binding of calcium to calbindins [both calbindin-D_{9k} (9000 M_r) and calbindin-D_{28k} (28,000 M_r) are present in mouse kidney; only calbinidn-D_{28k} is present in rat and human kidney]. Subsequently, calcium diffuses through the cytosol and is extruded via PMCA1b and the Na^+/Ca^{2+} exchanger (NCX1; Ajibade et al., 2010). Both $1,25(OH)_2D_3$ and PTH stimulate active calcium reabsorption from the distal tubule. PTH has been reported to activate TRPV5 via protein kinase A-dependent phosphorylation (de Groot et al., 2009). With regard to $1,25(OH)_2D_3$, in vivo studies using *Cyp27b1* null mice (that are defective in the synthesis of $1,25(OH)_2D_3$) have shown that the expression of renal TRPV5, calbindin-D_{9k}, calbindin-D_{28k}, and NCX1 mRNA is reduced in these mice and that this reduced expression is normalized by $1,25(OH)_2D_3$ (Hoenderop et al., 2002). These findings suggest that $1,25(OH)_2D_3$ affects renal calcium reabsorption by regulating the expression of these calcium transport proteins. However, it should be noted that compared with intestinal TRPV6, renal TRPV5 is less sensitive to regulation by $1,25(OH)_2D_3$ (Song et al., 2003). This tissue-specific regulation may indicate that multiple factors, in addition to $1,25(OH)_2D_3$, play a significant role in the regulation of renal TRPV5. Although little information is available with regard to the exact role of the vitamin-D–inducible calbindins in the process of renal calcium transport, it has been suggested that calbindin-D_{28k} acts to ferry calcium in the process of transcellular calcium diffusion or to buffer calcium, preventing calcium-mediated cell death (Ajibade et al., 2010). It has also been reported that calbindin-D_{28k} binds to TRPV5 and modulates calcium influx through the apical membrane (Lambers et al., 2006). The suggested role for calbindin-D_{9k} is as an enhancer of ATP-dependent calcium transport at the basolateral membrane (Bouhtiauy et al., 1994). Recent studies have reported that in addition to $1,25(OH)_2D_3$ and PTH, FGF23 can also promote renal calcium reabsorption by increasing TRPV5 protein abundance in the plasma membrane through different intracellular signaling cascades (Andrukhova et al., 2014).

Bone

The body contains approximately 1000 g of calcium. Most body calcium (99%) is localized in the mineral phase of bone. Bone can serve as a source of calcium for calcium-dependent physiological processes. In addition, in response to low serum calcium, calcium can be removed from bone to maintain calcium homeostasis (Maes and Kronenberg, 2015). Decreased serum calcium is detected by the calcium-sensing receptor in the parathyroid gland, which results in increased PTH synthesis (Chen and Goodman, 2004). Physiologic effects of PTH include enhanced renal calcium reabsorption, bone resorption, and increased production of $1,25(OH)_2D_3$. Osteoclastogenesis is stimulated by PTH and $1,25(OH)_2D_3$ (Maes and Kronenberg, 2015). The effect of both hormones on osteoclastogenesis is indirect and has been reported in in vitro studies to involve upregulation of receptor activator of nuclear factor κB ligand (RANKL) in osteoblastic cells, and it requires cell-to-cell contact between osteoblasts and osteoclast precursors (Maes and Kronenberg, 2015; Yasuda et al., 1998). However, studies have suggested that the osteocyte is the major source of RANKL in vivo (Nakashima et al., 2011; Xiong et al., 2014). Further investigations are needed to define the regulation of RANKL by $1,25(OH)_2D_3$ and PTH in vivo.

Although in vitro studies have supported a direct effect of $1,25(OH)_2D_3$ on bone, in vivo studies indicate an indirect role resulting from stimulation of intestinal calcium absorption by $1,25(OH)_2D_3$. VDR null mice develop hyperparathyroidism, hypocalcemia, and rickets after weaning. When VDR null mice are fed a rescue diet containing lactose and high calcium, serum calcium, PTH levels, and bone mineral are normalized (Amling et al., 1999). In addition, transgenic expression of VDR only in the intestine of VDR null mice also results in normalization of serum calcium and bone mineralization (Xue and Fleet, 2009). These findings suggest that a major effect of $1,25(OH)_2D_3$/VDR is to provide calcium to bone by stimulating intestinal absorption. In summary, data from in vivo and in vitro studies suggest that the effects of $1,25(OH)_2D_3$ on bone are diverse and that $1,25(OH)_2D_3$ can affect bone formation and resorption.

GENOMIC MECHANISM OF 1,25-DIHYDROXYVITAMIN D₃ ACTION

The molecular mechanism of $1,25(OH)_2D_3$ action is similar to that of other steroid hormones. The biological actions of $1,25(OH)_2D_3$ are mediated by the VDR, which belongs to the steroid receptor family. VDR-mediated transcription involves the formation of a VDR/RXR heterodimeric complex upon ligand activation, which interacts with specific sequences in DNA [vitamin D response elements (VDREs)] in and around target genes. Although significant variability in the sequence of VDREs has been reported, in general these VDREs consist of two direct imperfect repeats of the hexanucleotide sequence GGGTGA separated by a spacer of three nucleotides. The initiation of transcription is a multifactorial process that involves other proteins in a large complex. Primary coactivators, such as the p160 coactivators and steroid receptor coactivators 1, 2, and 3 (SRC-1, SRC-2, and SRC-3), which possess histone acetylase activity, bind to VDR, and recruit secondary coactivators such as CBP/p300 (which also has histone acetylase activity; Fig. 3.3). The result of histone acetylation is the opening up of the tightly packed nucleosome, allowing the transcription complex to come in contact with the DNA (Christakos, 2008; Pike and Meyer, 2014). Recent studies have noted that certain methyltransferases, which methylate core histone tails, also play a role as secondary coactivators in VDR-mediated transcription (Christakos et al., 2007). In addition, $1,25(OH)_2D_3$ action involves the interaction of VDR with basal transcription factors (TFIIB and TAT binding protein-associated factors) as well as with Mediator, a multiprotein complex that functions through recruitment of RNA pol II (Christakos, 2008; Pike and Meyer, 2014; Fig. 3.3). Several specific transcription factors including C/EBP β and YY1 have been reported to modulate VDR-mediated transcription (Dhawan et al., 2005; Guo et al., 1997; Raval-Pandya et al., 2001). The SWI/SNF complex, which functions to remodel chromatin using the energy of ATP hydrolysis, has also been shown to be important for the regulation of specific $1,25(OH)_2D_3$ target genes (Christakos et al., 2006; Fig. 3.3). Because VDR coregulatory proteins are master regulators of $1,25(OH)_2D_3$ action, these proteins may represent a new class of drug targets for therapeutic intervention. Recent genome-wide studies have noted that the actions of $1,25(OH)_2D_3$, similar to other steroids, are complex, involving the regulation of gene activity at a range of locations many kilobases upstream as well as downstream of the transcription start site (Pike and Meyer, 2010). These genome-wide studies have provided new insights related to the mechanisms involved in the tissue-specific actions of $1,25(OH)_2D_3$. In addition to genomic actions, nongenomic effects of $1,25(OH)_2D_3$ have also been suggested (reviewed in Ref. Bikle , 2014).

SUMMARY POINTS AND KEY FACTS

- Vitamin D is a principal factor involved in the maintenance of calcium homeostasis.
- For vitamin D to regulate calcium homeostasis, it must first be metabolized to its active form, $1,25(OH)_2D_3$, by two successive hydroxylations; in the liver at C-25 and in the kidney at C-1.
- CYP2R1, a microsomal 25-hydroxylase, is now considered the key enzyme responsible for the conversion of vitamin D to $25(OH)D_3$.
- $1,25(OH)_2D_3$, PTH, FGF23/α-klotho, and serum calcium act together as important regulators of calcium homeostasis.

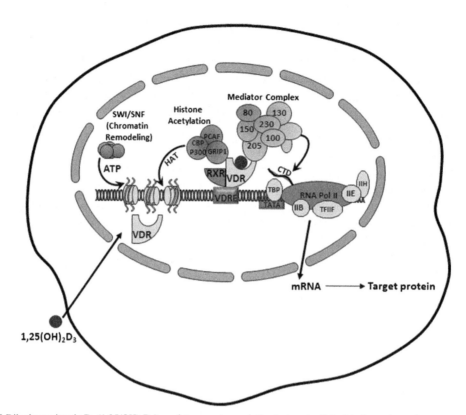

FIGURE 3.3 1,25-Dihydroxyvitamin D_3 (1,25(OH)$_2D_3$) regulates gene transcription in target cells by binding to its nuclear receptor, vitamin D receptor (VDR). This activated VDR heterodimerizes with retinoid X receptor (RXR) and binds to vitamin D response elements (VDREs) in and around target genes. In addition to basal transcription factors, liganded VDR interacts with steroid receptor coactivator (SRC)/p160 coactivators [shown here is GRIP1 (or SRC2)], which has histone acetylase activity (HAT) as the primary coactivators. Members of the p160 family recruit proteins as secondary coactivators such as calcium-binding protein (CBP)/P300 (which also have HAT activity). VDR also interacts with mediator complex, which facilitates the activation of the RNA polymerase II (RNA Pol II) holoenzyme through its C-terminal domain (CTD), thus promoting formation of the preinitiation complex. The SWI/SNF complex, which remodels chromatin using the energy of ATP hydrolysis, also contributes to activation by VDR. *GRIP1*, glucocorticoid receptor interacting protein 1; *PCAF*, p300/CBP-associated factor; *SWI/SNF*, SWItch/sucrose non-fermentable; *TBP*, TATA-binding protein.

- In intestine, the vitamin D target proteins TRPV6, an epithelial calcium channel, and the calcium-binding protein calbindin-D_{9k} may act together in the calcium-absorptive process.
- Active intestinal calcium absorption independent of TRPV6 or calbindin-D_{9k} can occur.
- The epithelial calcium channel TRPV5 in the distal renal tubules has been reported to have a critical role in renal calcium reabsorption and is regulated by 1,25(OH)$_2D_3$, PTH, and FGF23.
- Data from in vivo and in vitro studies suggest that the effects of 1,25(OH)$_2D_3$ in bone are diverse and that 1,25(OH)$_2D_3$ can affect bone formation and resorption.
- Genome-wide studies have noted that actions of 1,25(OH)$_2D_3$ mediated by VDR are complex and involve coactivators, gene-specific transcription factors, and regulation of gene transcription at sites many kilobases from the transcription start site.

MINI DICTIONARY OF TERMS

1,25-Dihydroxyvitamin D_3 (1,25(OH)$_2D_3$), the hormonally active form of vitamin D.

Calbindin-D_{9k} A 9000-molecular weight calcium-binding protein inducible by 1,25(OH)$_2D_3$ in mammalian intestine and mouse kidney.

Calbindin-D_{28k} A 28,000-molecular weight calcium-binding protein inducible by 1,25(OH)$_2D_3$ in avian intestine and in human, rat, mouse, and avian kidney.

CYP2R1 A key enzyme involved in the conversion of vitamin D to 25(OH)D_3 in the liver.

CYP27B1 Also called 25-hydroxyvitamin D_3 1α hydroxylase, this enzyme is involved in the conversion of 25(OH)D_3 to 1,25(OH)$_2D_3$.

CYP24A1 Also called 25-hydroxyvitamin D_3 24 hydroxylase, this enzyme is involved in the catabolism of 1,25(OH)$_2D_3$ and 25(OH)D_3. It is induced by 1,25(OH)$_2D_3$ as an autoregulatory mechanism to prevent hypercalcemia.

RANKL Receptor activator of nuclear factor kappa-β ligand is synthesized in osteoblasts and serves to activate osteoclasts.

TRPV6 $1,25(OH)_2D_3$-inducible epithelial calcium channel involved in intestinal calcium entry.

TRPV5 $1,25(OH)_2D_3$-inducible epithelial calcium channel in the renal distal tubule that plays a role in calcium reabsorption.

VDR The vitamin D receptor, which heterodimerizes with the retinoid X receptor (RXR) and binds to hormone response elements on DNA, resulting in the expression of target gene products. The biological actions of $1,25(OH)_2D_3$ are mediated by the VDR.

REFERENCES

Adams, J.S., Hewison, M., 2012. Extrarenal expression of the 25-hydroxyvitamin D-1-hydroxylase. Arch. Biochem. Biophys. 523, 95–102.

Ajibade, D., Benn, B.S., Christakos, S., 2010. Mechanism of action of 1,25-dihydroxyvitamin D3 in intestinal calcium absorption and renal calcium transport. In: Holick, M.F. (Ed.), Nutrition and Health: Vitamin D. Springer Science, New York, NY, pp. 175–187.

Amling, M., Priemel, M., Holzmann, T., Chapin, K., Rueger, J.M., Baron, R., Demay, M.B., 1999. Rescue of the skeletal phenotype of vitamin D receptor-ablated mice in the setting of normal mineral ion homeostasis: formal histomorphometric and biomechanical analyses. Endocrinology 140, 4982–4987.

Andrukhova, O., Smorodchenko, A., Egerbacher, M., Streicher, C., Zeitz, U., Goetz, R., Shalhoub, V., Mohammadi, M., Pohl, E.E., Lanske, B., Erben, R.G., 2014. FGF23 promotes renal calcium reabsorption through the TRPV5 channel. EMBO J. 33, 229–246.

Benn, B.S., Ajibade, D., Porta, A., Dhawan, P., Hediger, M., Peng, J.B., Jiang, Y., Oh, G.T., Jeung, E.B., Lieben, L., Bouillon, R., Carmeliet, G., Christakos, S., 2008. Active intestinal calcium transport in the absence of transient receptor potential vanilloid type 6 and calbindin-D9k. Endocrinology 149, 3196–3205.

Bikle, D., Adams, J., Christakos, S., 2013. Vitamin D: production, metabolism and clinical requirements. In: Rosen, C. (Ed.), Primer on Metabolic Bone Diseases, eighth ed. John Wiley and Sons, Inc., Hoboken, NJ, pp. 235–245.

Bikle, D., 2014. Vitamin D metabolism. Mech. action Clin. Appl. Chem. Biol. 21, 319–329.

Boros, S., Bindels, R.J., Hoenderop, J.G., 2009. Active Ca(2+) reabsorption in the connecting tubule. Pflugers Arch. 458, 99–109.

Bouhtiauy, I., Lajeunesse, D., Christakos, S., Brunette, M.G., 1994. Two vitamin D3-dependent calcium binding proteins increase calcium reabsorption by different mechanisms. II. Effect of CaBP 9K. Kidney Int. 45, 469–474.

Brenza, H.L., DeLuca, H.F., 2000. Regulation of 25-hydroxyvitamin D3 1alpha-hydroxylase gene expression by parathyroid hormone and 1,25-dihydroxyvitamin D3. Arch. Biochem. Biophys. 381, 143–152.

Canaff, L., Hendy, G.N., 2002. Human calcium-sensing receptor gene. Vitamin D response elements in promoters P1 and P2 confer transcriptional responsiveness to 1,25-dihydroxyvitamin D. J. Biol. Chem. 277, 30337–30350.

Chen, R.A., Goodman, W.G., 2004. Role of the calcium-sensing receptor in parathyroid gland physiology. Am. J. Physiol. Ren. Physiol. 286, F1005–F1011.

Cheng, J.B., Levine, M.A., Bell, N.H., Mangelsdorf, D.J., Russell, D.W., 2004. Genetic evidence that the human CYP2R1 enzyme is a key vitamin D 25-hydroxylase. Proc. Natl. Acad. Sci. U.S.A. 101, 7711–7715.

Christakos, S., Dhawan, P., Shen, Q., Peng, X., Benn, B., Zhong, Y., 2006. New insights into the mechanisms involved in the pleiotropic actions of 1,25dihydroxyvitamin D3. Ann. N. Y. Acad. Sci. 1068, 194–203.

Christakos, S., Dhawan, P., Benn, B., Porta, A., Hediger, M., Oh, G.T., Jeung, E.B., Zhong, Y., Ajibade, D., Dhawan, K., Joshi, S., 2007. Vitamin D: molecular mechanism of action. Ann. N. Y. Acad. Sci. 1116, 340–348.

Christakos, S., Ajibade, D.V., Dhawan, P., Fechner, A.J., Mady, L.J., 2012. Vitamin D: metabolism. Rheum. Dis. Clin. North Am. 38, 1–11 vii.

Christakos, S., 2008. Vitamin D gene regulation. In: Bilezikian, J., Raisz, L.G., Martin, T.J. (Eds.), Principles of Bone Biology. Elsevier-Academic Press, New York, NY, pp. 779–794.

Christakos, S., 2012. Recent advances in our understanding of 1,25-dihydroxyvitamin D(3) regulation of intestinal calcium absorption. Arch. Biochem. Biophys. 523, 73–76.

Cui, M., Li, Q., Johnson, R., Fleet, J.C., 2012. Villin promoter-mediated transgenic expression of transient receptor potential cation channel, subfamily V, member 6 (TRPV6) increases intestinal calcium absorption in wild-type and vitamin D receptor knockout mice. J. Bone Min. Res. 27, 2097–2107.

de Groot, T., Lee, K., Langeslag, M., Xi, Q., Jalink, K., Bindels, R.J., Hoenderop, J.G., 2009. Parathyroid hormone activates TRPV5 via PKA-dependent phosphorylation. J. Am. Soc. Nephrol. 20, 1693–1704.

Demay, M.B., Kiernan, M.S., DeLuca, H.F., Kronenberg, H.M., 1992. Sequences in the human parathyroid hormone gene that bind the 1,25-dihydroxyvitamin D3 receptor and mediate transcriptional repression in response to 1,25-dihydroxyvitamin D3. Proc. Natl. Acad. Sci. U.S.A. 89, 8097–8101.

Dhawan, P., Peng, X., Sutton, A.L., MacDonald, P.N., Croniger, C.M., Trautwein, C., Centrella, M., McCarthy, T.L., Christakos, S., 2005. Functional cooperation between CCAAT/enhancer-binding proteins and the vitamin D receptor in regulation of 25-hydroxyvitamin D3 24-hydroxylase. Mol. Cell Biol. 25, 472–487.

Fujita, H., Sugimoto, K., Inatomi, S., Maeda, T., Osanai, M., Uchiyama, Y., Yamamoto, Y., Wada, T., Kojima, T., Yokozaki, H., Yamashita, T., Kato, S., Sawada, N., Chiba, H., 2008. Tight junction proteins claudin-2 and -12 are critical for vitamin D-dependent Ca2+ absorption between enterocytes. Mol. Biol. Cell 19, 1912–1921.

Guo, B., Aslam, F., van Wijnen, A.J., Roberts, S.G., Frenkel, B., Green, M.R., DeLuca, H., Lian, J.B., Stein, G.S., Stein, J.L., 1997. YY1 regulates vitamin D receptor/retinoid X receptor mediated transactivation of the vitamin D responsive osteocalcin gene. Proc. Natl. Acad. Sci. U.S.A. 94, 121–126.

Hoenderop, J.G., Bindels, R.J., 2008. Calciotropic and magnesiotropic TRP channels. Physiology (Bethesda) 23, 32–40.

Hoenderop, J.G., Dardenne, O., Van Abel, M., Van der Kemp, A.W., Van Os, C.H., St-Arnaud, R., Bindels, R.J., 2002. Modulation of renal Ca2+ transport protein genes by dietary Ca2+ and 1,25-dihydroxyvitamin D3 in 25-hydroxyvitamin D3-1alpha-hydroxylase knockout mice. FASEB J. 16, 1398–1406.

Hoenderop, J.G., van Leeuwen, J.P., van der Eerden, B.C., Kersten, F.F., van der Kemp, A.W., Merillat, A.M., Waarsing, J.H., Rossier, B.C., Vallon, V., Hummler, E., Bindels, R.J., 2003. Renal Ca2+ wasting, hyperabsorption, and reduced bone thickness in mice lacking TRPV5. J. Clin. Invest. 112, 1906–1914.

Hu, M.C., Shiizaki, K., Kuro-o, M., Moe, O.W., 2013. Fibroblast growth factor 23 and Klotho: physiology and pathophysiology of an endocrine network of mineral metabolism. Annu. Rev. Physiol. 75, 503–533.

Jones, G., Prosser, D.E., Kaufmann, M., 2014. Cytochrome P450-mediated metabolism of vitamin D. J. Lipid Res. 55, 13–31.

Kitanaka, S., Takeyama, K., Murayama, A., Sato, T., Okumura, K., Nogami, M., Hasegawa, Y., Niimi, H., Yanagisawa, J., Tanaka, T., Kato, S., 1998. Inactivating mutations in the 25-hydroxyvitamin D3 1alpha-hydroxylase gene in patients with pseudovitamin D-deficiency rickets. N. Engl. J. Med. 338, 653–661.

Kutuzova, G.D., Deluca, H.F., 2004. Gene expression profiles in rat intestine identify pathways for 1,25-dihydroxyvitamin D(3) stimulated calcium absorption and clarify its immunomodulatory properties. Arch. Biochem. Biophys. 432, 152–166.

Kutuzova, G.D., Akhter, S., Christakos, S., Vanhooke, J., Kimmel-Jehan, C., Deluca, H.F., 2006. Calbindin D(9k) knockout mice are indistinguishable from wild-type mice in phenotype and serum calcium level. Proc. Natl. Acad. Sci. U.S.A. 103, 12377–12381.

Kutuzova, G.D., Sundersingh, F., Vaughan, J., Tadi, B.P., Ansay, S.E., Christakos, S., Deluca, H.F., 2008. TRPV6 is not required for 1alpha,25-dihydroxyvitamin D3-induced intestinal calcium absorption in vivo. Proc. Natl. Acad. Sci. U.S.A. 105, 19655–19659.

Lambers, T.T., Mahieu, F., Oancea, E., Hoofd, L., de Lange, F., Mensenkamp, A.R., Voets, T., Nilius, B., Clapham, D.E., Hoenderop, J.G., Bindels, R.J., 2006. Calbindin-D28K dynamically controls TRPV5-mediated Ca2+ transport. EMBO J. 25, 2978–2988.

Lieben, L., Benn, B.S., Ajibade, D., Stockmans, I., Moermans, K., Hediger, M.A., Peng, J.B., Christakos, S., Bouillon, R., Carmeliet, G., 2010. Trpv6 mediates intestinal calcium absorption during calcium restriction and contributes to bone homeostasis. Bone 47, 301–308.

Maes, C., Kronenberg, H.M., 2015. Bone development and remodeling. In: Jameson, J.L., De Groot, L. (Eds.), Endocrinology: Adult and Pediatric. Elsevier, New York, NY, pp. 1038–1062.

Nakashima, T., Hayashi, M., Fukunaga, T., Kurata, K., Oh-Hora, M., Feng, J.Q., Bonewald, L.F., Kodama, T., Wutz, A., Wagner, E.F., Penninger, J.M., Takayanagi, H., 2011. Evidence for osteocyte regulation of bone homeostasis through RANKL expression. Nat. Med. 17, 1231–1234.

Pike, J.W., Meyer, M.B., 2010. The vitamin D receptor: new paradigms for the regulation of gene expression by 1,25-dihydroxyvitamin D(3). Endocrinol. Metab. Clin. North Am. 39, 255–269.

Pike, J.W., Meyer, M.B., 2014. Fundamentals of vitamin D hormone-regulated gene expression. J. Steroid Biochem. Mol. Biol. 144 (Pt A), 5–11.

Plum, L.A., DeLuca, H.F., 2010. Vitamin D, disease and therapeutic opportunities. Nat. Rev. Drug Discov. 9, 941–955.

Raval-Pandya, M., Dhawan, P., Barletta, F., Christakos, S., 2001. YY1 represses vitamin D receptor-mediated 25-hydroxyvitamin D(3)24-hydroxylase transcription: relief of repression by CREB-binding protein. Mol. Endocrinol. 15, 1035–1046.

Song, Y., Peng, X., Porta, A., Takanaga, H., Peng, J.B., Hediger, M.A., Fleet, J.C., Christakos, S., 2003. Calcium transporter 1 and epithelial calcium channel messenger ribonucleic acid are differentially regulated by 1,25 dihydroxyvitamin D3 in the intestine and kidney of mice. Endocrinology 144, 3885–3894.

Thacher, T.D., Fischer, P.R., Singh, R.J., Roizen, J., Levine, M.A., 2015. CYP2R1 mutations impair generation of 25-hydroxyvitamin D and cause an atypical form of vitamin D deficiency. J. Clin. Endocrinol. Metab. 100, E1005–E1013.

Willnow, T.E., Nykjaer, A., 2010. Cellular uptake of steroid carrier proteins–mechanisms and implications. Mol. Cell Endocrinol. 316, 93–102.

Xiong, J., Piemontese, M., Thostenson, J.D., Weinstein, R.S., Manolagas, S.C., O'Brien, C.A., 2014. Osteocyte-derived RANKL is a critical mediator of the increased bone resorption caused by dietary calcium deficiency. Bone 66, 146–154.

Xue, Y., Fleet, J.C., 2009. Intestinal vitamin D receptor is required for normal calcium and bone metabolism in mice. Gastroenterology 136, 1317–1327 e1311–1312.

Yasuda, H., Shima, N., Nakagawa, N., Yamaguchi, K., Kinosaki, M., Mochizuki, S., Tomoyasu, A., Yano, K., Goto, M., Murakami, A., Tsuda, E., Morinaga, T., Higashio, K., Udagawa, N., Takahashi, N., Suda, T., 1998. Osteoclast differentiation factor is a ligand for osteoprotegerin/osteoclastogenesis-inhibitory factor and is identical to TRANCE/RANKL. Proc. Natl. Acad. Sci. U.S.A. 95, 3597–3602.

Zhu, J.G., Ochalek, J.T., Kaufmann, M., Jones, G., Deluca, H.F., 2013. CYP2R1 is a major, but not exclusive, contributor to 25-hydroxyvitamin D production in vivo. Proc. Natl. Acad. Sci. U.S.A. 110, 15650–15655.

Zierold, C., Nehring, J.A., DeLuca, H.F., 2007. Nuclear receptor 4A2 and C/EBPbeta regulate the parathyroid hormone-mediated transcriptional regulation of the 25-hydroxyvitamin D3-1alpha-hydroxylase. Arch. Biochem. Biophys. 460, 233–239.

Chapter 4

Calcium in Obesity and Related Diseases: The Calcium-Sensing Receptor as a Novel Mediator

Roberto Bravo-Sagua[1], Marcela Reyes[1], Sergio Lavandero[1,2], Mariana Cifuentes[1]

[1]Universidad de Chile, Santiago, Chile; [2]University of Texas Southwestern Medical Center, Dallas, TX, United States

INTRODUCTION

Calcium nutrition has been traditionally implicated in bone health, and recommendations for dietary calcium intake have aimed mainly at preventing long-term bone mineralization deficit. In 1984 (McCarron et al., 1984) and later in the 2000s (Zemel and Miller, 2004), a pool of epidemiological and clinical data initially suggested that poor calcium nutrition is associated with obesity or fat mass gain. This evidence was later supported by mechanistic studies addressing the pathophysiology of the association (Zemel and Miller, 2004). However, conflicting results were also found, generating a lack of consensus. The heterogeneity of study designs and different views on what constitutes a relevant effect may explain the contradictions (Villarroel et al., 2014). There needs to be careful attention to methodological aspects and definitions of physiological relevance before drawing any conclusions.

It is currently accepted that the development of obesity-related diseases does not depend on the level of obesity itself but rather on the degree of adipose tissue dysfunction. Knowledge of the cellular pathways that preclude adipose tissue from properly managing energy surplus and consequently generate whole-body cardiometabolic alterations is fundamental to rationally approach the health costs of the obesity epidemic. Among numerous mechanisms involved, the role of calcium, as a nutrient in the diet and as an intracellular signaling component, is emerging as an important mediator of adipose function. Intracellular Ca^{2+} regulates key processes in healthy adipose tissue, such as differentiation (Shi et al., 2000; Szabo et al., 2008), lipid accumulation (Xue et al., 1998, 2001), and insulin signaling (Tebar et al., 1996; Goko and Matsuoka, 1999). Alternatively, the generation of pathological Ca^{2+} signals activates multiple events that result in acute stress responses (Kennedy et al., 2010) and adipose tissue inflammation (Sun and Zemel, 2007). These are pathological hallmarks of damage in obesity, which contribute to compromised cell function and the development of chronic pathologies (Guerrero-Hernandez and Verkhratsky, 2014). In the following sections, we address the problem of obesity and how it is influenced by Ca^{2+} as a dietary component. We also discuss the possible role of Ca^{2+} as a mediator of the intracellular mechanisms of adipose tissue dysfunction and obesity-related diseases.

OBESITY: A SERIOUS PUBLIC HEALTH PROBLEM WORLDWIDE

Definition and Epidemiology

Obesity is one of the main health problems worldwide, affecting more than 600 million adults (~2 billion, including overweight), almost doubling the prevalence as compared with 35 years ago (WHO, 2015[1]). In several developed and developing countries, most of the adult population is either obese or overweight (WHO, 2015). Overweight and obesity are also present in childhood, affecting more than 40 million children younger than 5 years of age in 2013, including those in developing countries, where this disorder coexists with poor nutrition (WHO, 2015).

Obesity is considered an illness, defined as "an abnormal or excessive fat accumulation that may impair health" (WHO, 2015). It is also an important risk factor for several ailments such as cardiometabolic alterations and cancers [defined under the concept of noncommunicable diseases (NCDs)]. NCDs are currently affecting younger populations, even at the

---

1. WHO. Fact sheet N°311, updated January 2015. http://www.who.int/mediacentre/factsheets/fs311/en/.

Molecular, Genetic, and Nutritional Aspects of Major and Trace Minerals. http://dx.doi.org/10.1016/B978-0-12-802168-2.00004-X

preschool age (Corvalan et al., 2009). Overall, the burden associated with obesity and related NCDs is important, especially considering that it is a preventable condition.

Healthy Versus Unhealthy Obesity

Overnutrition is associated with cardiometabolic alterations; however, increasing evidence shows the existence of an obese phenotype that is dissociated from these consequences (Fabbrini et al., 2015). In fact, approximately 20–35% of obese individuals have little or no associated alterations (the percentage depends on the number and nature of the alterations studied), whereas about the same proportion of normal-weight persons have cardiometabolic alterations (Wildman et al., 2008; Bluher, 2010; Oliveros et al., 2014). It is currently accepted that it is not the amount of adipose tissue, but the quality of the tissue that leads to healthy or unhealthy phenotypes (Bays, 2011), in which dysfunctional or "sick" adipose tissue mediates the development of obesity-related diseases (Fig. 4.1). The location of adipose tissue depots is also an important determinant of adipose tissue-related pathogenesis (Karpe and Pinnick, 2015). A functional adipose tissue is located mainly in the subcutaneous depot (and particularly, the gluteofemoral region). It is efficient in storing lipids and expands predominantly through the increase in adipose cell number and not cell size (Tchoukalova et al., 2010). This enables it to handle the nutritional surplus without changing its secretory profile or developing the inflammatory changes that would impair function in other tissues. On the other hand, visceral adipose tissue depots have been reported to be more metabolically active and to have important differences in function (triglyceride handling, secretion, and receptor expression) as compared with subcutaneous depots (Bays, 2011). It is widely accepted that the visceral depot is the one associated with adipose tissue pathogenesis whereas the subcutaneous depot may even be protective (Kloting et al., 2010; Karpe and Pinnick, 2015). Compared with obese subjects with altered cardiometabolic status, insulin-sensitive obese persons have larger subcutaneous adipose tissue mass and smaller visceral depots. Moreover, the latter do not show an involvement of inflammatory pathways nor alterations at the protein expression level (Karpe and Pinnick, 2015), supporting the idea that such alterations are in the causal line of NCDs.

Calcium Nutrition and Obesity Hypothesis

The association between calcium nutrition and obesity emerged from observations made more than 30 years ago, in which a large cross-sectional study (>20,000 US adults) observed lower calcium intake in the upper percentiles of body mass index (BMI; kg/m^2) and a negative association between BMI and calcium intake (McCarron et al., 1984). Additional studies in adults (Pereira et al., 2002) and pediatric populations (Carruth and Skinner, 2001; Abreu et al., 2012) have supported the idea; however, other studies have found no association (Rajpathak et al., 2006) or the opposite association (Berkey et al., 2005).

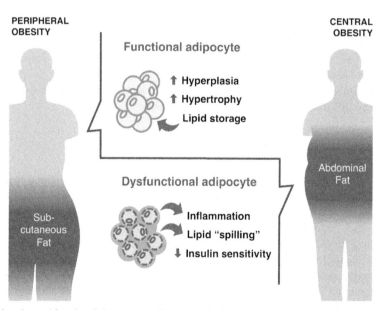

FIGURE 4.1 Adipose tissue location and function. Subcutaneous adipose tissue, mainly located in the lower body, is associated with "healthy" fat accumulation, undergoing expansion without compromising its function. Visceral adipose tissue, located in the upper body, is associated with the development of metabolic diseases because it promotes an inflammatory state that leads to impaired function and insulin resistance.

Almost two decades after the initial report, the work by Zemel et al. in humans, rodents, and cell models proposed an antiobesity effect of dietary calcium (Zemel and Miller, 2004). According to this hypothesis, dietary calcium acts through the modulation of circulating calcitriol (vitamin D), which regulates cytosolic free Ca^{2+} in the adipocyte (Fig. 4.2A). Low dietary calcium elevates blood calcitriol, which promotes the elevation of cytosolic free Ca^{2+} in the adipocyte. This in turn stimulates lipogenesis (triglyceride synthesis and/or storage) and inhibits lipolysis (triglyceride breakdown; Shi et al., 2001). Calcitriol was also reported to inhibit adipocyte uncoupling protein-2 expression, which may decrease fat oxidation and thermogenesis (Shi et al., 2002). Together, these processes may contribute to overall elevated fat accumulation. Other fat-conserving mechanisms of poor calcium nutrition may operate as well, such as modifications in fecal fat excretion and appetite control, as discussed previously (Villarroel et al., 2014). As reviewed recently, this hypothesis has stimulated an important debate; however, it is likely that the controversy is more related to the expectations set on an antiobesity nutrient than to questioning whether the data indeed support the mechanism (Villarroel et al., 2014).

Calcium Nutrition and Obesity-Related Diseases: The Role of the Calcium-Sensing Receptor

A study that evaluated calcium intake in healthy women determined that low habitual dietary calcium could be associated with higher cardiovascular risk factors, such as excessive abdominal adiposity, insulin resistance, and elevated blood pressure as well as lower serum levels of adiponectin and high-density lipoprotein cholesterol (da Silva Ferreira et al., 2013). There are inconsistent data among the few calcium-only supplementation studies in humans, but some have observed trends toward a decrease in serum triglycerides (Chai et al., 2013) and a decrease in blood pressure, particularly in subjects with low initial calcium intake (Reid et al., 2010). As mentioned before, the wide variability in study designs and subject characteristics make it difficult to draw any unifying conclusions (Challoumas et al., 2013). The data are controversial to the point that although adequate dietary calcium or calcium supplementation have been associated with cardiovascular protection, some reports have suggested that supplemental calcium is associated with increased risk of cardiovascular events (Mao et al., 2013). The need for further and well-controlled studies is evident to clarify these observations.

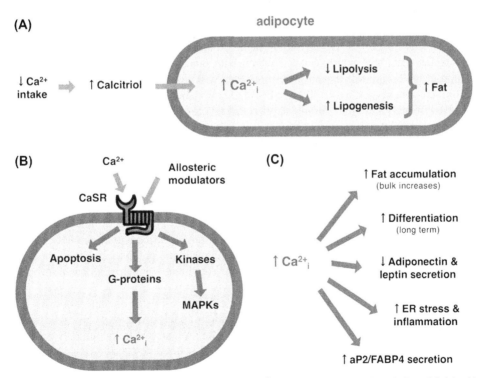

FIGURE 4.2 Calcium at the center of the obesity problem. (A) Low levels of Ca^{2+} intake lead to increased circulating calcitriol, which induces increases in intracellular Ca^{2+} in adipocytes. In the cytosol, Ca^{2+} has an inhibitory effect on lipolysis and promotes lipogenesis, thus promoting lipid accumulation. (B) The calcium-sensing receptor (CaSR) is a complex regulator of signaling cascades. Its activation depends on extracellular Ca^{2+} and can be modulated by multiple allosteric factors. Its downstream signaling has been shown to positively regulate apoptosis, to increase intracellular Ca^{2+} via heterotrimeric G-proteins, and to activate stress-related kinases such as mitogen-activated protein kinases (MAPKs). (C) In adipocytes, increased intracellular Ca^{2+} alters lipid metabolism: It augments fat accumulation, promotes adipogenesis, activates proinflammatory cascades, and decreases adipokine secretion while increasing the secretion of adipocyte protein-2 (aP2)/fatty acid binding protein-4 (FABP4). *ER*, endoplasmic reticulum.

In the search for the cellular mechanisms involved in the calcium–obesity hypothesis, our laboratory focuses on the role of the CaSR in adipocyte physiology. The CaSR belongs to the C family of G-protein–coupled receptors, and its main ligand is extracellular Ca^{2+}. Its most relevant role is the regulation of circulating Ca^{2+} levels; accordingly, its presence and function were originally described in parathyroid cells (Brown et al., 1993). However, the receptor was later described in many other tissues with numerous regulatory roles aside from Ca^{2+} homeostasis, including antilipolytic (Cifuentes and Rojas, 2008; He et al., 2011a,b), proadipogenic (He et al., 2012; Villarroel et al., 2013), and proinflammatory (Cifuentes et al., 2012) actions in human adipose tissue and adipose cells. Upon activation by several possible agonists (Chakravarti et al., 2012), the CaSR stimulates a wide variety of intracellular signaling cascades (Fig. 4.2B) involving various G-proteins, generation of second messengers (e.g., Ca^{2+} itself), and activation of kinases [e.g., the mitogen-activated protein kinase (MAPK) cascade; Chakravarti et al., 2012], among others. It is not surprising then that the calcium-sensing receptor (CaSR) is involved in regulating many cellular processes. Evidence gathered thus far suggests that the activation of the CaSR in adipose tissue may be associated with a dysfunctional phenotype, favoring obesity-related diseases, and may be a novel component consistent with Zemel's hypothesis (Villarroel et al., 2014).

Regarding the role of the CaSR in obesity-related diseases, the evidence is promising, albeit very scarce. Moreover, the local concentrations of extracellular Ca^{2+} or other CaSR agonists in obesity-relevant tissues may differ from circulating levels, thus increasing the complexity of the possible participation of the CaSR. Of note, an interaction between CaSR and calcitriol has been reported (Canaff and Hendy, 2002; He et al., 2011b), providing a direct link with the mechanism previously discussed. In addition, our laboratory has reported that exposure to obesity-related proinflammatory cytokines upregulates CaSR expression in human adipocytes, and that secretory products of human adipose tissue explants elicit this same effect, notably more strongly because the adipose tissue donor has a larger BMI (Cifuentes et al., 2010). We have also shown that treatment with a CaSR-specific allosteric activator elevates inflammatory cytokine expression in preadipocytes, differentiated adipocytes, and human adipose tissue explants (Cifuentes et al., 2012). This upregulation may result in vivo in an enhanced proinflammatory environment in obesity, which would sustain the dysfunctional state and the complications of the disease. More recently, a greater expression of CaSR was observed in peripheral blood monocytes of individuals with diabetes and peripheral artery disease as compared with individuals with atherosclerosis alone (Malecki et al., 2013). Furthermore, CaSR expression on the monocyte surface directly correlated with fasting glucose levels (Malecki et al., 2013), possibly suggesting an association with insulin resistance. In a murine model of diabetes-induced cardiac injury, CaSR was upregulated and its activation led to cardiomyocyte apoptosis, in part through increasing cytosolic Ca^{2+} (Qi et al., 2013). CaSR activation attenuated the expression of antiapoptotic proteins and induced proapoptotic proteins, suggesting that CaSR induces cell death through the mitochondrial pathway.

Since the first report in 1997 proposing the involvement of the CaSR in insulin secretion (Kato et al., 1997), the accumulated, although still limited, data suggest a role of the CaSR in glucose homeostasis (Wauson et al., 2013). CaSR activation elevates insulin secretion in islets and different models of pancreatic β-cells, in part by facilitating cell-to-cell communication within the pancreatic islet and likely through an intracellular pathway that differs from that used by glucose (Hills et al., 2012; Wauson et al., 2013). At this point the clinical implications of these observations are far from being understood.

The CaSR has also been directly linked to blood pressure modulation, and hence with hypertension. As reviewed by Peterlik et al. (2013), the CaSR is functionally expressed in all cells of the vascular wall, and several lines of evidence support a role in regulating vascular tone and blood pressure. CaSR activation has been related to vascular repair, integrity, relaxation, and lower calcification (Peterlik et al., 2013). The receptor is functionally expressed in renal juxtaglomerular cells, and its activation inhibits the release of renin (Ortiz-Capisano et al., 2013), the proteolytic enzyme that produces angiotensin I, thereby favoring a decrease in angiotensin II and aldosterone, with an expected decrease in blood pressure. Furthermore, treatment with CaSR activators has been shown to reduce blood pressure in murine models of hypertension (Smajilovic et al., 2011). Moreover, genetic studies in humans revealed that common CaSR variants are associated with blood pressure in certain populations (Smajilovic et al., 2011).

Taken together, from the few studies available, it is clear that CaSR is a promising mediator for many of the mechanisms that link calcium to obesity and its related diseases. The observations suggest roles that are tissue and context specific, with CaSR activation linked to beneficial and detrimental effects. More research will undoubtedly contribute to gain more insight and take advantage of this knowledge to design specific therapies based on CaSR modulation.

CELLULAR MECHANISMS FOR FAT DYSFUNCTION: ROLE OF CALCIUM

Cellular Aspects

Excess adipose tissue results from a positive energy balance, sustained over time. The energy surplus is stored in the form of triglycerides inside of the predominant cells in adipose tissue, the adipocytes. Upon the need for extra storage, these cells increase in size (hypertrophy) and number (hyperplasia), and changes occur in the architecture of the adipose tissue (including extracellular matrix and blood vessels). The current paradigm proposes that if this remodeling is occurring properly in the subcutaneous adipose tissue, then the energy excess can be handled without deleterious consequences at the cellular, tissue, and systemic level. However, when subcutaneous adipose tissue plasticity is impaired or exceeded, fatty acids are directed to other adipose tissue locations (i.e., visceral) or other organs and tissues that become overloaded, thus altering their functions (from lipotoxicity; Bays, 2011).

The increase in adipocyte size is accompanied by stress signaling and changes in the expression of genes involved in inflammation and lipid metabolism. These alterations lead to a metabolic remodeling that ultimately increases lipolysis and therefore fatty acid release to the bloodstream (Michaud et al., 2014). Thus at some point the energy surplus does not uniquely affect adipose cells, but virtually every cell type throughout the body. At the intracellular level, in nonadipose cells such as pancreatic β-cells, the lipid excess affects the structure and function of the endoplasmic reticulum (ER; Kharroubi et al., 2004), which in turn activates a signaling cascade responsible for a cellular inflammatory response (Hotamisligil, 2010). The activation of such pathways impairs insulin signaling, yielding an insulin-resistant cell. As mentioned, because every organ is exposed to the excess circulating free fatty acids, multiple manifestations of metabolic alteration, impaired insulin signaling, inflammation, and dysfunction can be observed throughout the body.

Aside from inducing proinflammatory cytokine production, adipocyte stress signaling also reduces adiponectin production/secretion (Lim et al., 2014). This adipokine is a peptide hormone normally produced by healthy adipose tissue and known for its insulin-sensitizing action. It stimulates glucose uptake and use by skeletal muscle, reduces inflammation in the cardiovascular system, and decreases glucose production and fat accumulation in the liver (Lim et al., 2014). Circulating levels of adiponectin are usually inversely correlated with adiposity, and an impairment in its secretion greatly contributes to the pathogenesis of different cardiometabolic alterations, such as type II diabetes mellitus (Lihn et al., 2005).

Many intracellular mechanisms have been proposed as responsible for the deleterious effects of obesity as a consequence of the impaired function of adipose tissue. In this context, Ca^{2+} may be an insufficiently acknowledged common mediator that will be discussed in the following sections (Fig. 4.2C).

Calcium-Directed Adipocyte Pathogenesis

Being a highly dynamic intracellular messenger, Ca^{2+} regulates several processes within cells, including cell homeostasis, proliferation, differentiation, death, and communication with other cells and the environment. These processes directly determine adipose tissue physiology, thus greatly impacting the development of obesity and its comorbidities. After briefly reviewing basic concepts of calcium regulation at the intracellular level (Fig. 4.3), the role of calcium in several aspects of adipocyte function and pathogenesis will be discussed.

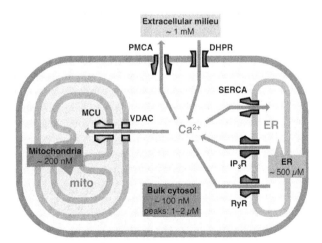

FIGURE 4.3 Cellular machinery mediating calcium homeostasis. Extracellular Ca^{2+} entry to the cytoplasm is regulated by plasma membrane channels such as the dihydropyridine receptor (DHPR) (voltage-gated); it can then extrude back out of cells via ATPase pumps such as plasma membrane Ca^{2+}-ATPase (PMCA). The endoplasmic reticulum (ER) acts as an intracellular Ca^{2+} reservoir by accumulation via pumps [sarco/endoplasmic reticulum Ca^{2+}-ATPase (SERCA)]. Upon stimulation, the ER can release Ca^{2+} back into the cytoplasm through channels such as inositol triphosphate receptor (IP_3R) and ryanodine receptor (RyR). In addition, mitochondria act as intracellular Ca^{2+} buffers. To enter to the mitochondrial matrix, Ca^{2+} crosses both mitochondrial membranes through specialized channels [voltage-dependent anion channel (VDAC), mitochondrial Ca^{2+} uniporter (MCU)].

Intracellular Ca^{2+} Homeostasis

Ca^{2+} concentration is 10,000 times higher in the extracellular milieu (~1 mM) compared with the bulk cytosol (~100 nM). This difference is actively maintained by a series of pumps that require energy (ATP) to transport Ca^{2+} from the cytoplasm to the extracellular environment by the plasma membrane Ca^{2+}-ATPase (PMCA), or to the ER by the sarco/endoplasmic reticulum Ca^{2+}-ATPase (SERCA). The ER is the main intracellular Ca^{2+} storage organelle, reaching levels close to the extracellular medium (~0.5 mM; Berridge et al., 2003; Rizzuto and Pozzan, 2006).

The large difference in Ca^{2+} concentration of the cytosol with respect to the interstitial fluids and the ER provides a driving force for rapid Ca^{2+} entry. Selective channels at the plasma and ER membranes allow for Ca^{2+} translocation to the cytosol without ATP consumption. Consequently, the gating of these channels needs to be highly regulated. For instance, at the plasma membrane the dihydropyridine receptor (DHPR) opens upon changes in transmembrane voltage, induced by depolarizing cues, such as electrical stimulation or excitatory neurotransmitters. On the other hand, at the ER surface the inositol triphosphate receptors (IP$_3$Rs) open upon elevations of inositol triphosphate (IP$_3$), which is a second messenger generated in response to extracellular signals (including those acting via CaSR activation). This gating mechanism allows for Ca^{2+} release from ER stores in response to extracellular cues. Also at the ER surface are the ryanodine receptors (RyRs), the gating of which is regulated by cytosolic Ca^{2+} itself, creating waves of Ca^{2+} throughout the cell in response to Ca^{2+} elevations. These mechanisms of Ca^{2+} storage and release allow for rapid peaks in cytosolic Ca^{2+}, with 10-fold concentration increases (~1 μM; Berridge et al., 2003; Rizzuto and Pozzan, 2006). In the case of CaSR, upon activation, it stimulates IP$_3$ production via G-protein mediators located at the plasma membrane. Hence, it elicits intracellular Ca^{2+} elevations through IP$_3$R (Chakravarti et al., 2012).

Mitochondria are also key players in Ca^{2+} homeostasis. The mitochondrial matrix has a negative transmembrane potential with respect to the cytosol, which provides electrostatic attraction for positively charged Ca^{2+} ions. Therefore Ca^{2+} readily enters the mitochondrial matrix, crossing both mitochondrial membranes through dedicated channels, namely the voltage-dependent anion channel (VDAC) and the mitochondrial Ca^{2+} uniporter (MCU), without requiring ATP consumption. This renders mitochondria as relevant intracellular Ca^{2+} signaling regulators that limit or buffer cytosolic Ca^{2+} elevations (Rizzuto and Pozzan, 2006).

Adipocyte Differentiation

The generation of new adipocytes, termed adipogenesis, occurs throughout the human lifespan and requires preadipocyte differentiation, a process that is modulated by Ca^{2+}. In human preadipocytes, intracellular Ca^{2+} elevations during the early stages of adipogenesis (0–1 h) inhibit cell differentiation. In contrast, at later stages (48–72 h), intracellular Ca^{2+} potentiates adipogenesis (Shi et al., 2000). This dual role suggests that intracellular Ca^{2+} regulation is key for promoting adipogenesis and thus adipose tissue function. The relevance of the timing within the differentiation process at which Ca^{2+} influences adipogenesis has been documented by other authors (Szabo et al., 2008), highlighting the complex regulation of the process. These data also correlate with results from our laboratory that show that acute CaSR activation decreases whereas long-term activation increases the expression of adipogenic markers (Villarroel et al., 2013). In addition, a study in neonatal piglets (Li and Stahl, 2014) showed that excessive and deficient calcium intake increase the adipogenic potential of mesenchymal stem cells, highlighting the importance of adequate calcium nutrition for cell differentiation.

Adipocyte Fat Metabolism

In adipocytes, cytosolic Ca^{2+} acts as a promoter of fat accumulation via two complementary mechanisms: it inhibits lipolysis by antagonizing the hormone-sensitive lipolytic pathway (Xue et al., 1998, 2001); conversely, it increases de novo lipogenesis by enhancing fatty acid synthase expression and activity (Jones et al., 1996). However, the role of Ca^{2+} as an inhibitor of adipocyte lipolysis appears to be more complex. Contrasting evidence shows that a pathway involving Ca^{2+} and the Ca^{2+}-dependent protein calmodulin contributes to hormone-stimulated and exercise-induced lipolysis (Kawai, 1985; Izawa and Komabayashi, 1994). These opposite roles in the regulation of fat metabolism imply a very tight regulation of Ca^{2+} signals. Of note, the evidence for the antilipolytic role emanates from bulk increases in cytosolic Ca^{2+} generated by cell depolarization (Xue et al., 2001). In contrast, evidence for the prolipolytic role stems from G-protein–derived signaling; thus it may be more localized (Kawai, 1985; Izawa and Komabayashi, 1994; van Harmelen et al., 2008). Thus these findings highlight the importance of the source and scale of Ca^{2+} elevations in the regulation of fat accumulation and the specific context that is being studied.

Mitochondria participate in lipid metabolism through fatty acid β-oxidation, although this contribution to adipocyte energy partitioning is rather small (Wang et al., 2003). Ca^{2+} entry into the mitochondrial matrix is known to increase oxidative metabolism, thus favoring this process (Mick et al., 1991). In a more prominent mechanism, the previously mentioned role of mitochondria as intracellular Ca^{2+} buffers limits excessive cytosolic Ca^{2+}; thus it negatively regulates lipid anabolism. Accordingly, dissipation of mitochondrial transmembrane potential (which abolishes the buffer capacity) leads to Ca^{2+}-dependent increases in lipogenesis (Sun and Zemel, 2003) and triglyceride accumulation (Vankoningsloo et al., 2005). Therefore mitochondrial Ca^{2+} uptake in adipocytes might have an important role in regulating fat accumulation. Interestingly, it has been observed that adipocytes from obese humans have reduced mitochondrial function as compared with nonobese controls (Yin et al., 2014). In addition, mice with experimentally induced adipose-specific mitochondrial dysfunction show insulin resistance, hypertension, and cardiac dysfunction (Vernochet et al., 2014), strongly supporting the relevant role of adipose tissue mitochondria in whole-body cardiometabolic health.

Endoplasmic Reticulum Stress and Inflammation

The ER is a key organelle for intracellular Ca^{2+} handling, lipid metabolism, and protein synthesis and folding. Disruption of ER homeostasis is termed ER stress, and it leads to impaired protein folding capacity, thus leading to aggregation of misfolded proteins and general cell damage (Bravo et al., 2013). In obese individuals adipocytes are subject to ER stress-inducing conditions, such as high levels of free fatty acids (Guo et al., 2007; Jiao et al., 2011), hypoxia (Hosogai et al., 2007), and an inflammatory environment (Zhou et al., 2010). In turn, ER stress heavily contributes to adipocyte pathogenesis because it increases inflammation (Kawasaki et al., 2012) and antagonizes insulin signaling (Xu et al., 2010). Excessive Ca^{2+} release from ER stores has been pinpointed as a mediator of the detrimental effects generated by lipotoxicity-induced ER stress in human adipocytes (Kennedy et al., 2010). Via this mechanism, cytosolic Ca^{2+} becomes a dual regulator—not only modulating the normal function of the adipose tissue but also mediating the stress response that leads to inflammation. This inflammatory state in adipose tissue decreases adiponectin secretion and contributes to the development of various manifestations of the metabolic syndrome (Lim et al., 2014). In adipocyte protein-2 (aP2)-agouti transgenic mice (a model of human obesity), high calcium intake decreases body adipose tissue as well as oxidative stress and proinflammatory cytokine production induced by obesogenic diet (Sun and Zemel, 2007). In the same study the blocker of extracellular Ca^{2+} entry, nifedipine, decreased inflammation in cultured adipocytes in response to calcitriol, suggesting that calcitriol-induced elevated cytosolic Ca^{2+} may be associated with the effects of poor calcium nutrition at the intracellular level. Accordingly, CaSR is elevated during inflammation (Cifuentes et al., 2010), and its activation contributes to the adipocyte inflammatory state in vitro (Cifuentes et al., 2012), which may occur through the same mechanisms (i.e., elevations in cytosolic Ca^{2+}).

Novel Adipose-Derived Mediators

In recent years novel mediators derived from adipocytes have been identified, further confirming the role of adipose tissue function in obesity-related diseases. Adipocyte protein 2 (aP2; also known as fatty acid-binding protein-4) was first considered a cytosolic lipid chaperone. However, adipocytes have been shown to secrete it in response to fasting and lipolysis-inducing stimuli (Cao et al., 2013). Once released, it acts as an endocrine protein that increases hepatic glucose production (gluconeogenesis). Consistent with this, aP2 is elevated in obese patients and correlates with other markers of the metabolic syndrome, thus becoming a circulating biomarker for obesity-related diseases. Recent findings on aP2 show that its secretory mechanism does not proceed through Golgi-mediated vesicular trafficking but rather consists of a Ca^{2+}-dependent noncanonical pathway (Schlottmann et al., 2014). Although this topic still requires further research, it reveals new scenarios in which Ca^{2+} may contribute to adipose tissue dysfunction.

Insulin Signaling

In the fed state, insulin signaling in adipocytes inhibits lipolysis while promoting lipogenesis, triglyceride uptake, and protein synthesis (Dimitriadis et al., 2011). Obesity is frequently associated with insulin resistance, which is, in part, a consequence of the inflammatory state in adipose tissue and cytokine imbalance (Ouchi et al., 2011). In turn, decreased insulin sensitivity contributes to adipose tissue dysfunction by lowering its capacity to accumulate fat, thus leading to whole-body lipotoxicity.

It has been reported that the antilipolytic effect of insulin on rat adipocytes requires Ca^{2+} and its associated signaling pathways (Tebar et al., 1996; Goko and Matsuoka, 1999), thereby highlighting the importance of this ion in mediating insulin actions. Nevertheless, the role of Ca^{2+} on other aspects of insulin signaling in adipocytes is less clear. For instance, extracellular Ca^{2+} has been shown to be necessary for insulin-stimulated secretion of the adipokine leptin by regulating glucose uptake (Cammisotto and Bukowiecki, 2004; Wang et al., 2014). Protein anabolism and glucose incorporation into triglycerides are also known to require extracellular Ca^{2+} (Emami and Perry, 1986). Conversely, intracellular Ca^{2+} reportedly inhibits leptin secretion (Cammisotto and Bukowiecki, 2004) and has been implicated in the negative regulation of insulin signaling, therefore acting in cell desensitization (Worrall and Olefsky, 2002). These results suggest that Ca^{2+} is a complex mediator of insulin actions; therefore its regulation must be finely tuned. Indeed, in adipocytes, insulin reportedly influences multiple pathways that shape Ca^{2+} signaling. Insulin elicits an increase in cytosolic Ca^{2+} through voltage-dependent Ca^{2+} channels (Draznin et al., 1987), and the RyRs negatively influence the expression of the insulin-sensitizing adipokine adiponectin (Tsai et al., 2013). Knockdown of RyR3 is proposed to elevate adiponectin gene expression and protein (and thus insulin sensitivity), in part mediated by decreased cytosolic Ca^{2+} and the gain of mitochondrial mass and function (Tsai et al., 2013). In addition, diacylglycerol and IP_3 derived from phospholipase-C–dependent cell signaling have been linked to activation of intracellular Ca^{2+} accumulation, inflammatory signaling, loss of lipid-storing capacity, and insulin resistance in human primary adipocytes (Kennedy et al., 2010; Shen et al., 2013).

CONCLUSION

The studies reviewed in this chapter highlight the relevance of calcium in obesity and related diseases. From perspectives that range from the nutritional to the cell signaling aspects of the pathophysiology of obesity and its related diseases, it is clear that calcium plays a key role. There is strong scientific evidence for a relevant role of deficient calcium nutrition in the development of obesity and its complications, suggesting that this ion has a protective role against fat accumulation. Moreover, intracellular Ca^{2+} signaling is at the very core of adipose tissue function and dysfunction, leading to completely different outcomes depending on temporality, spatiality, and cellular context. These observations imply that Ca^{2+} signals are finely modulated and decoded, relying on several transductional units to determine adipocyte fate.

Emerging evidence positions CaSR as a promising mediator that links Ca^{2+} with different aspects in the pathophysiology of obesity and its related diseases, thereby constituting a potential therapeutic target for a disorder that is largely lacking in pharmacological tools. CaSR interacts with multiple signaling partners, including agonists, allosteric modulators, second messengers, and downstream targets that allow it to act as a beneficial or detrimental agent in cell physiology. More research is undoubtedly necessary to uncover the exact mechanisms by which CaSR can induce or prevent adipocyte dysfunction to design specific therapies based on its modulation.

REFERENCES

Abreu, S., et al., 2012. Milk intake is inversely related to body mass index and body fat in girls. Eur. J. Pediatr. 171, 1467–1474.

Bays, H.E., 2011. Adiposopathy is "sick fat" a cardiovascular disease? J. Am. Coll. Cardiol. 57, 2461–2473.

Berkey, C.S., et al., 2005. Milk, dairy fat, dietary calcium, and weight gain: a longitudinal study of adolescents. Arch. Pediatr. Adolesc. Med. 159, 543–550.

Berridge, M.J., et al., 2003. Calcium signalling: dynamics, homeostasis and remodelling. Nat. Rev. Mol. Cell Biol. 4, 517–529.

Bluher, M., 2010. The distinction of metabolically 'healthy' from 'unhealthy' obese individuals. Curr. Opin. Lipidol. 21, 38–43.

Bravo, R., et al., 2013. Endoplasmic reticulum and the unfolded protein response: dynamics and metabolic integration. Int. Rev. Cell Mol. Biol. 301, 215–290.

Brown, E.M., et al., 1993. Cloning and characterization of an extracellular Ca(2+)-sensing receptor from bovine parathyroid. Nature 366, 575–580.

Cammisotto, P.G., Bukowiecki, L.J., 2004. Role of calcium in the secretion of leptin from white adipocytes. Am. J. Physiol. Regul. Integr. Comp. Physiol. 287, R1380–R1386.

Canaff, L., Hendy, G.N., 2002. Human calcium-sensing receptor gene. Vitamin D response elements in promoters P1 and P2 confer transcriptional responsiveness to 1,25-dihydroxyvitamin D. J. Biol. Chem. 277, 30337–30350.

Cao, H., et al., 2013. Adipocyte lipid chaperone AP2 is a secreted adipokine regulating hepatic glucose production. Cell Metab. 17, 768–778.

Carruth, B.R., Skinner, J.D., 2001. The role of dietary calcium and other nutrients in moderating body fat in preschool children. Int. J. Obes. Relat. Metab. Disord. 25, 559–566.

Chai, W., et al., 2013. Effects of calcium and vitamin D supplementation on blood pressure and serum lipids and carotenoids: a randomized, double-blind, placebo-controlled, clinical trial. Ann. Epidemiol. 23, 564–570.

Chakravarti, B., et al., 2012. Signaling through the extracellular calcium-sensing receptor (CaSR). Adv. Exp. Med. Biol. 740, 103–142.

Challoumas, D., et al., 2013. Effects of calcium intake on the cardiovascular system in postmenopausal women. Atherosclerosis 231, 1–7.

Cifuentes, M., Rojas, C.V., 2008. Antilipolytic effect of calcium-sensing receptor in human adipocytes. Mol. Cell Biochem. 319, 17–21.

Cifuentes, M., et al., 2010. Obesity-associated proinflammatory cytokines increase calcium sensing receptor (CaSR) protein expression in primary human adipocytes and LS14 human adipose cell line. Arch. Biochem. Biophys. 500, 151–156.

Cifuentes, M., et al., 2012. Calcium sensing receptor activation elevates proinflammatory factor expression in human adipose cells and adipose tissue. Mol. Cell Endocrinol. 361, 24–30.

Corvalan, C., et al., 2009. Effect of growth on cardiometabolic status at 4y of age. Am. J. Clin. Nutr. 90, 547–555.

da Silva Ferreira, T., et al., 2013. Dietary calcium intake is associated with adiposity, metabolic profile, inflammatory state and blood pressure, but not with erythrocyte intracellular calcium and endothelial function in healthy pre-menopausal women. Br. J. Nutr. 110, 1079–1088.

Dimitriadis, G., et al., 2011. Insulin effects in muscle and adipose tissue. Diabetes Res. Clin. Pract. 93 (Suppl. 1), S52–S59.

Draznin, B., et al., 1987. Insulin and glyburide increase cytosolic free-Ca^{2+} concentration in isolated rat adipocytes. Diabetes 36, 174–178.

Emami, S., Perry, M.C., 1986. Requirement for bivalent cations in the actions of insulin and sodium nitroprusside on metabolism in rat adipocytes. FEBS Lett. 200, 51–57.

Fabbrini, E., et al., 2015. Metabolically normal obese people are protected from adverse effects following weight gain. J. Clin. Invest. 125, 787–795.

Goko, H., Matsuoka, A., 1999. W-5 and quin 2-AM reverse the inhibitory effect of insulin on lipolysis due to dibutyryl cAMP. Diabetes Res. Clin. Pract. 44, 101–106.

Guerrero-Hernandez, A., Verkhratsky, A., 2014. Calcium signalling in diabetes. Cell Calcium 56, 297–301.

Guo, W., et al., 2007. Palmitate modulates intracellular signaling, induces endoplasmic reticulum stress, and causes apoptosis in mouse 3T3-L1 and rat primary preadipocytes. Am. J. Physiol. Endocrinol. Metab. 293, E576–E586.

He, Y., et al., 2011a. Involvement of calcium-sensing receptor in inhibition of lipolysis through intracellular cAMP and calcium pathways in human adipocytes. Biochem. Biophys. Res. Commun. 404, 393–399.

He, Y.H., et al., 2011b. The calcium-sensing receptor affects fat accumulation via effects on antilipolytic pathways in adipose tissue of rats fed low-calcium diets. J. Nutr. 141, 1938–1946.

He, Y.H., et al., 2012. The calcium-sensing receptor promotes adipocyte differentiation and adipogenesis through PPARgamma pathway. Mol. Cell Biochem. 361, 321–328.

Hills, C.E., et al., 2012. Calcium-sensing receptor activation increases cell-cell adhesion and beta-cell function. Cell Physiol. Biochem. 30, 575–586.

Hosogai, N., et al., 2007. Adipose tissue hypoxia in obesity and its impact on adipocytokine dysregulation. Diabetes 56, 901–911.

Hotamisligil, G.S., 2010. Endoplasmic reticulum stress and the inflammatory basis of metabolic disease. Cell 140, 900–917.

Izawa, T., Komabayashi, T., 1994. Ca^{2+} and lipolysis in adipocytes from exercise-trained rats. J. Appl. Physiol. 77 (1985), 2618–2624.

Jiao, P., et al., 2011. FFA-induced adipocyte inflammation and insulin resistance: involvement of ER stress and IKKbeta pathways. Obes. (Silver Spring) 19, 483–491.

Jones, B.H., et al., 1996. Upregulation of adipocyte metabolism by agouti protein: possible paracrine actions in yellow mouse obesity. Am. J. Physiol. 270, E192–E196.

Karpe, F., Pinnick, K.E., 2015. Biology of upper-body and lower-body adipose tissue–link to whole-body phenotypes. Nat. Rev. Endocrinol. 11, 90–100.

Kato, M., et al., 1997. Calcium-evoked insulin release from insulinoma cells is mediated via calcium-sensing receptor. Surgery 122, 1203–1211.

Kawai, A., 1985. The role of calmodulin in hormone-stimulated lipolysis. Metabolism 34, 303–308.

Kawasaki, N., et al., 2012. Obesity-induced endoplasmic reticulum stress causes chronic inflammation in adipose tissue. Sci. Rep. 2, 799.

Kennedy, A., et al., 2010. Inflammation and insulin resistance induced by trans-10, cis-12 conjugated linoleic acid depend on intracellular calcium levels in primary cultures of human adipocytes. J. Lipid Res. 51, 1906–1917.

Kharroubi, I., et al., 2004. Free fatty acids and cytokines induce pancreatic beta-cell apoptosis by different mechanisms: role of nuclear factor-kappaB and endoplasmic reticulum stress. Endocrinology 145, 5087–5096.

Kloting, N., et al., 2010. Insulin-sensitive obesity. Am. J. Physiol. Endocrinol. Metab. 299, E506–E515.

Li, Y., Stahl, C.H., 2014. Dietary calcium deficiency and excess both impact bone development and mesenchymal stem cell lineage priming in neonatal piglets. J. Nutr. 144, 1935–1942.

Lihn, A.S., et al., 2005. Adiponectin: action, regulation and association to insulin sensitivity. Obes. Rev. 6, 13–21.

Lim, S., et al., 2014. Modulation of adiponectin as a potential therapeutic strategy. Atherosclerosis 233, 721–728.

Malecki, R., et al., 2013. Altered monocyte calcium-sensing receptor expression in patients with type 2 diabetes mellitus and atherosclerosis. J. Physiol. Pharmacol. 64, 521–527.

Mao, P.J., et al., 2013. Effect of calcium or vitamin D supplementation on vascular outcomes: a meta-analysis of randomized controlled trials. Int. J. Cardiol. 169, 106–111.

McCarron, D.A., et al., 1984. Blood pressure and nutrient intake in the United States. Science 224, 1392–1398.

Michaud, A., et al., 2014. Abdominal subcutaneous and omental adipocyte morphology and its relation to gene expression, lipolysis and adipocytokine levels in women. Metabolism 63, 372–381.

Mick, G.J., et al., 1991. Selective stimulation of in situ intermediary metabolism by free calcium in permeabilized rat adipocytes. Biochim. Biophys. Acta 1133, 73–80.

Oliveros, E., et al., 2014. The concept of normal weight obesity. Prog. Cardiovasc. Dis. 56, 426–433.

Ortiz-Capisano, M.C., et al., 2013. Juxtaglomerular cell CaSR stimulation decreases renin release via activation of the PLC/IP(3) pathway and the ryanodine receptor. Am. J. Physiol. Ren. Physiol. 304, F248–F256.

Ouchi, N., et al., 2011. Adipokines in inflammation and metabolic disease. Nat. Rev. Immunol. 11, 85–97.

Pereira, M.A., et al., 2002. Dairy consumption, obesity, and the insulin resistance syndrome in young adults: the CARDIA Study. JAMA 287, 2081–2089.

Peterlik, M., et al., 2013. Calcium nutrition and extracellular calcium sensing: relevance for the pathogenesis of osteoporosis, cancer and cardiovascular diseases. Nutrients 5, 302–327.

Qi, H., et al., 2013. Crucial role of calcium-sensing receptor activation in cardiac injury of diabetic rats. PLoS One 8, e65147.

Rajpathak, S.N., et al., 2006. Calcium and dairy intakes in relation to long-term weight gain in US men. Am. J. Clin. Nutr. 83, 559–566.

Reid, I.R., et al., 2010. Effects of calcium supplementation on lipids, blood pressure, and body composition in healthy older men: a randomized controlled trial. Am. J. Clin. Nutr. 91, 131–139.

Rizzuto, R., Pozzan, T., 2006. Microdomains of intracellular Ca^{2+}: molecular determinants and functional consequences. Physiol. Rev. 86, 369–408.

Schlottmann, I., et al., 2014. Calcium-dependent release of adipocyte fatty acid binding protein from human adipocytes. Int. J. Obes. (Lond.) 38, 1221–1227.

Shen, W., et al., 2013. The phospholipase C inhibitor U73122 attenuates trans-10, cis-12 conjugated linoleic acid-mediated inflammatory signaling and insulin resistance in human adipocytes. J. Nutr. 143, 584–590.

Shi, H., et al., 2000. Role of intracellular calcium in human adipocyte differentiation. Physiol. Genomics 3, 75–82.

Shi, H., et al., 2001. 1alpha,25-Dihydroxyvitamin D3 modulates human adipocyte metabolism via nongenomic action. FASEB J. 15, 2751–2753.

Shi, H., et al., 2002. 1alpha,25-dihydroxyvitamin D3 inhibits uncoupling protein 2 expression in human adipocytes. FASEB J. 16, 1808–1810.

Smajilovic, S., et al., 2011. The calcium-sensing receptor and calcimimetics in blood pressure modulation. Br. J. Pharmacol. 164, 884–893.

Sun, X., Zemel, M.B., 2003. Effects of mitochondrial uncoupling on adipocyte intracellular Ca(2+) and lipid metabolism. J. Nutr. Biochem. 14, 219–226.

Sun, X., Zemel, M.B., 2007. Calcium and 1,25-dihydroxyvitamin D3 regulation of adipokine expression. Obesity (Silver Spring) 15, 340–348.

Szabo, E., et al., 2008. Calreticulin inhibits commitment to adipocyte differentiation. J. Cell Biol. 182, 103–116.

Tchoukalova, Y.D., et al., 2010. Regional differences in cellular mechanisms of adipose tissue gain with overfeeding. Proc. Natl. Acad. Sci. U.S.A. 107, 18226–18231.

Tebar, F., et al., 1996. The antilipolytic effects of insulin and epidermal growth factor in rat adipocytes are mediated by different mechanisms. Endocrinology 137, 4181–4188.

Tsai, S.H., et al., 2013. Knockdown of RyR3 enhances adiponectin expression through an atf3-dependent pathway. Endocrinology 154, 1117–1129.

van Harmelen, V., et al., 2008. Vascular peptide endothelin-1 links fat accumulation with alterations of visceral adipocyte lipolysis. Diabetes 57, 378–386.

Vankoningsloo, S., et al., 2005. Mitochondrial dysfunction induces triglyceride accumulation in 3T3-L1 cells: role of fatty acid beta-oxidation and glucose. J. Lipid Res. 46, 1133–1149.

Vernochet, C., et al., 2014. Adipose tissue mitochondrial dysfunction triggers a lipodystrophic syndrome with insulin resistance, hepatosteatosis, and cardiovascular complications. FASEB J. 28, 4408–4419.

Villarroel, P., et al., 2013. Adipogenic effect of calcium sensing receptor activation. Mol. Cell Biochem. 384, 139–145.

Villarroel, P., et al., 2014. Calcium, obesity, and the role of the calcium-sensing receptor. Nutr. Rev. 72, 627–637.

Wang, T., et al., 2003. Metabolic partitioning of endogenous fatty acid in adipocytes. Obes. Res. 11, 880–887.

Wang, Y., et al., 2014. Insulin-stimulated leptin secretion requires calcium and PI3K/Akt activation. Biochem. J. 458, 491–498.

Wauson, E.M., et al., 2013. Minireview: nutrient sensing by G protein-coupled receptors. Mol. Endocrinol. 27, 1188–1197.

Wildman, R.P., et al., 2008. The obese without cardiometabolic risk factor clustering and the normal weight with cardiometabolic risk factor clustering: prevalence and correlates of 2 phenotypes among the US population (NHANES 1999–2004). Arch. Intern Med. 168, 1617–1624.

Worrall, D.S., Olefsky, J.M., 2002. The effects of intracellular calcium depletion on insulin signaling in 3T3-L1 adipocytes. Mol. Endocrinol. 16, 378–389.

Xu, L., et al., 2010. ER stress in adipocytes inhibits insulin signaling, represses lipolysis, and alters the secretion of adipokines without inhibiting glucose transport. Horm. Metab. Res. 42, 643–651.

Xue, B., et al., 1998. The agouti gene product inhibits lipolysis in human adipocytes via a Ca^{2+}-dependent mechanism. FASEB J. 12, 1391–1396.

Xue, B., et al., 2001. Mechanism of intracellular calcium ([Ca2+]i) inhibition of lipolysis in human adipocytes. FASEB J. 15, 2527–2529.

Yin, X., et al., 2014. Adipocyte mitochondrial function is reduced in human obesity independent of fat cell size. J. Clin. Endocrinol. Metab. 99, E209–E216.

Zemel, M.B., Miller, S.L., 2004. Dietary calcium and dairy modulation of adiposity and obesity risk. Nutr. Rev. 62, 125–131.

Zhou, Q.G., et al., 2010. Advanced oxidation protein products induce inflammatory response and insulin resistance in cultured adipocytes via induction of endoplasmic reticulum stress. Cell Physiol. Biochem. 26, 775–786.

Chapter 5

Calcium: Basic Nutritional Aspects

Terry J. Aspray

Newcastle University, Newcastle upon Tyne, Tyne and Wear, United Kingdom; Freeman Hospital, Newcastle upon Tyne, Tyne and Wear, United Kingdom

INTRODUCTION

In addressing calcium as a nutrient, we often focus on bone health and osteoporosis. However, its role in physiology, as a cofactor for blood clotting and in nerve and muscle function, means that calcium intake and excretion need to be tightly regulated to maintain a consistent extracellular and intracellular concentration. In this chapter we will discuss the implications of calcium's varied functions and the relationship to several specific organs, including the skeleton, kidneys, endocrine system (specifically the parathyroid glands), and cardiovascular system. Although reference is made at the end of the chapter to nutritional intakes, which are usually derived from calcium balance studies, there is further reflection on potential benefits and negative effects of dietary calcium with specific reference to calcium as a foodstuff or as a dietary supplement. The optimal dietary source of dietary calcium, the effects of nonfood mineral sources, and the separate effects of calcium intakes to achieve an adequate intake from potential pharmacological or pathological effects of supplementation are active and unresolved questions. However, the balance of benefit and risk in calcium nutrition and its place along with vitamin D (or not) is currently a hot topic in clinical nutrition.

Functions of Calcium

Calcium is the fifth most abundant element in the human body, with the majority found in skeletal tissue. Calcium has important roles (with phosphate) in promoting the deposition of hydroxyapatite in bone and a mechanical role in strengthening bones and teeth. Furthermore, calcium is essential in animal physiology because it supports the function of excitable tissues, including nerves and heart muscle, as well as blood clotting (see Table 5.1), and the regulation of aspects of cell and organelle function, including autoregulation of calcium levels, excitability, exocytosis, motility, apoptosis, and transcription. Whereas extracellular fluid, such as plasma, maintains a concentration of calcium of approximately 2.45 mmol/L, intracellular levels are much lower, on the order of 10^{-7} mmol/L. This large concentration gradient across cell membranes is maintained by calcium/magnesium ATPase activity. It is important that levels of calcium in the plasma are tightly regulated, particularly because changes in extracellular levels can influence the physiological functions previously mentioned. As mentioned, total plasma calcium levels are normally approximately 2.45 mmol/L, with the largest proportion (48%) present in its ionized form, whereas 46% of the cation is protein bound (80% albumin and 20% globulin) and a further 7% is complexed with anions such as bicarbonate and inorganic phosphate (Blaine et al., 2014). The total plasma calcium level may change in disease states, especially with changes in serum protein levels, but ionized calcium remains relatively stable.

Calcium Homeostasis, Vitamin D, and Parathyroid Hormone

Humans are exposed to a wide range of dietary calcium intakes with potential for seasonal fluctuation as well as evidence of diurnal variation in calcium status because homeostasis is maintained through sleep–wake cycles and pulsed intakes, related to meals (Redmond et al., 2014). Serum calcium levels are controlled by several factors responsible for optimizing absorption and maintaining eucalcemia. Classically we think in terms of

- calcium absorption from the diet, particularly with reference to vitamin D—25-hydxoxy-vitamin D (25(OH)D) supply and 1,25-dihydroxy-vitamin D (1,25(OH)$_2$D) action;
- excretion or reabsorption of calcium by the kidney, influenced by parathyroid hormone (PTH); and
- buffering of calcium in the skeleton (Redmond et al., 2014).

Molecular, Genetic, and Nutritional Aspects of Major and Trace Minerals. http://dx.doi.org/10.1016/B978-0-12-802168-2.00005-1

TABLE 5.1 Some of the Essential Functions of Calcium in the Body

Function	Comment
Neuromuscular activity	Nerves and muscle fibers use calcium ions as their main regulatory and signaling molecule.
Membrane function	Calcium-dependent channels are found throughout the body, from phospholipases to mitochondria.
Hormone secretion	Calcium ions move across membranes in a range of secretory organs, including the parathyroid gland, thyroid, and β cell.
Enzyme activity	Several enzymes are sensitive to ambient calcium levels in the intracellular or extracellular space.
Blood coagulation	Calcium levels are important at several stages in the intrinsic pathway of coagulation.
Skeletal mineralization and strength	The mineralization of bone matrix and formation of calcium hydroxyapatite is an important contributor to skeletal strength.
Mineral reservoir	Calcium within bone mineral acts as a useful buffer to be released when serum calcium levels are low, and the resorption of calcium at the kidney is insufficient to meet the needs.

When dietary calcium intake is high, it crosses the intestinal epithelium by a paracellular pathway that involves the passive movement of calcium through tight junctions between epithelial cells down a concentration gradient; this process is only affected indirectly by $1,25(OH)_2D$. Conversely, with moderate- or low-intake levels, active, transcellular transport is promoted by $1,25(OH)_2D$, which increases the synthesis of calbindin D_{9k} to control movement of calcium across enterocytes (Blaine et al., 2014).

In the kidney, 100–200 mg of calcium is excreted every day, which is less than 2% of the total filtered load of calcium, the vast majority of which is reabsorbed by the renal tubules (60–70% occurring in the proximal convoluted tubule, mostly by passive transport). In the loop of Henle, 20% of calcium reabsorption occurs by a combination of passive and active transport, and then approximately 10% occurs in the distal convoluted tubule, entirely by active transport because it is against an electrochemical gradient. Although the distal part of the nephron reabsorbs less than 10% of the filtered calcium load, it is the focus for regulating calcium excretion, under the predominant influence of PTH (Blaine et al., 2014). In renal disease, the control of calcium homeostasis (together with phosphate and acid-base balance) becomes important because the kidney is the site of calcium excretion and the synthesis of $1,25(OH)_2D$.

More than 99% of calcium is retained within the skeleton, although only 1% of this is freely exchangeable with extracellular fluid. Hence the physiological buffering capacity of bone is dependent on a relatively small proportion of skeletal mass. There is evidence of diurnal variation in calcium homeostasis, with low serum calcium being noted at night and higher PTH in the afternoon and at night. There is also evidence of diurnal variation in bone turnover, with apparent rises in markers of bone resorption after fasting, possibly under the influence of PTH (Redmond et al., 2014). The implications are that gut absorption and renal retention of calcium are important in the maintenance of blood calcium levels, with some involvement of the skeleton, in response to diurnal change. The evidence suggests that bone can be accessed to maintain extracellular calcium levels, and this is certainly important in disease states. For example, hyperparathyroidism, whether due to a primary parathyroid adenoma or an increase in PTH secondary to renal disease or vitamin D deficiency, may result in significant bone resorption (Fig. 5.1).

Vitamin D status is important for calcium homeostasis, but a detailed discussion is beyond the immediate scope of this chapter. Although $1,25(OH)_2D$ is required to optimize calcium absorption, its precursor, $25(OH)D$, is generally agreed to be the best biomarker for assessing vitamin D status. However, the threshold of $25(OH)D$ defining vitamin D deficiency is less generally agreed upon, ranging from 25 to 75 nmol/L (Francis et al., 2013). Reduction in plasma $25(OH)D$ is associated with the mineralization defect of bone seen in children as rickets and in adults as osteomalacia; less severe deficiency may still lead to secondary hyperparathyroidism, increased bone resorption, bone loss, impaired muscle function, and an increased risk of falls and fragility fractures (Bischoff-Ferrari, 2012; Bischoff-Ferrari et al., 2009a; Rejnmark et al., 2012). Particularly for children, deficiency effects of vitamin D and calcium can be closely linked, with nutritional rickets appearing to be a response to varying levels of vitamin D deficiency and low dietary calcium intake, depending on the population affected (Pettifor, 2013).

Dietary Calcium Requirements

Dietary calcium intake recommendations should be defined as the level of intake below or above which there is a risk to health. The Institute of Medicine (IOM) report on calcium and vitamin D is a comprehensive assessment of the evidence and discusses at length the methods used in its risk assessment that uses a framework to assess a range of factors: from

(A)

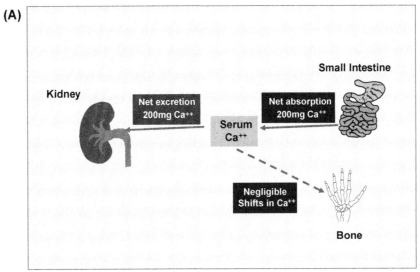

(B)

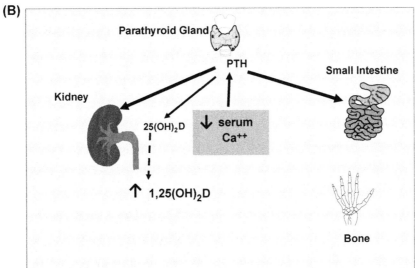

(C)

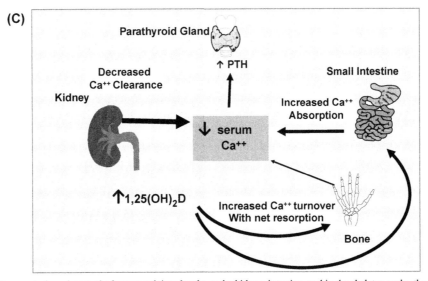

FIGURE 5.1 Calcium homeostasis and control of serum calcium levels, at the kidney, intestine, and in the skeleton under the influence of parathyroid hormone (PTH), 25-hydroxy-vitamin D (25(OH)D), and 1,25-dihydroxy-vitamin D (1,25(OH)$_2$D). (A) Net changes in calcium in a balanced state, with equal absorption and renal reabsorption. (B) Lower serum calcium levels result in an increase in PTH, which also promotes conversion of 25(OH)D to 1,25(OH)$_2$D. (C) An increase in PTH stimulates renal reabsorption of calcium and an increase in 1,25(OH)$_2$D, which promotes active absorption of calcium via calmodulin and more chronically active resorption of skeleton to release calcium (as does an increase in PTH—not presented in the figure). See text for more details.

Hazard Identification to Hazard Characterization, Intake Assessment, and Risk Identification. The choice of biomarkers presents a potential issue, with a dominance in this area of vitamin D status markers. However, the IOM report suggests that a range of health outcome indicators exist that might be considered in evaluating calcium (and vitamin D) nutrition, including cancer risk, cardiovascular disease (CVD), diabetes, pregnancy outcomes, and skeletal health (Ross and Institute of Medicine (U.S.) Committee to Review Dietary Reference Intakes for Vitamin D and Calcium, 2011).

Dietary calcium requirements might be physiologically defined in terms of an intake threshold at which neutral calcium balance is achieved (i.e., where there is no difference between the total calcium intake and the sum of urinary and endogenous fecal calcium loss). Studies to evaluate calcium balance are difficult to perform, but results from such studies in adults aged 50 years or younger did form the basis for the IOM recommendations on dietary calcium (Ross and Institute of Medicine (U.S.) Committee to Review Dietary Reference Intakes for Vitamin D and Calcium, 2011). Throughout the course of life, we are in varying stages of calcium balance, through skeletal growth in infancy and childhood and on to menopausal bone loss for women and continued bone loss for both sexes in older adult life. Interactions between calcium balance and other factors, not the least being vitamin D, but also phosphate, protein, and endogenous sex hormones, will make individual requirements vary. However, estimates are remarkably robust, giving balance for adults at levels between 700 and 750 mg calcium per day.

In this chapter we shall not focus solely on physiological calcium balance but on aspects of relevance to calcium status and health, including skeletal health and other issues such as cardiovascular and renal stone disease. When considering the health implications of calcium intake, in addition to choice of biomarkers, there is a need to consider developmental stage of the life so that recommendations are likely to differ between the sexes and at different ages. However, further consideration of the identification of dietary calcium requirements will be left until later in the chapter.

Skeletal Health

Bone Density

Focusing on skeletal health, it is difficult to tease out the independent effects of calcium and vitamin D status in many trials, considering their combined effects on PTH secretion, intestinal calcium absorption, calcium balance, bone mineral content, fracture risk, and even risk for developing rickets/osteomalacia (see IOM Report Section 4-4; Ross and Institute of Medicine (U.S.) Committee to Review Dietary Reference Intakes for Vitamin D and Calcium, 2011). There is conflicting evidence about the impact of calcium on bone mineral density (BMD). Epidemiological data from North America suggest a link between both calcium and vitamin D and BMD, but the relationship between calcium and higher BMD is only evident at lower levels of vitamin D (Bischoff-Ferrari, Kiel, et al., 2009). In meta-analyses and systematic reviews, calcium supplementation in children and postmenopausal women has shown a small increase in BMD, but this is not evident in premenopausal women or men (Uusi-Rasi et al., 2013; Winzenberg et al., 2006). However, there is one study from New Zealand that shows higher BMD in men supplemented with 1200 mg/day of calcium but not with 600 mg/day (Reid et al., 2008). There are also interesting data from communities with low calcium intakes such as the Gambia, West Africa, where habitual dietary calcium intakes are less than 250 mg per day and BMD is lower than in Europe and North America (Aspray et al., 1996). However, despite such low calcium intakes, bone mineral loss in women during prolonged lactation recovered within 12 months of discontinuing breast feeding (Sawo et al., 2013). A further study of calcium supplementation in pregnancy resulted in a paradoxically lower maternal bone mineral content after breast feeding (Jarjou et al., 2013), and in prepubertal Gambian boys with low dietary calcium intakes supplementation resulted in an earlier adolescent growth spurt but no lasting effect on bone mineral or bone size (Ward et al., 2014).

The international variation in calcium intakes and unpredicted outcomes of supplementation must raise concerns about the potential for variation in effects of calcium (particularly as supplements) with unexpected consequences of increasing dietary calcium in communities that may have adapted to low intakes. Similar concerns have been expressed about the presumed benefits of calcium in Asia (Lee and Jiang, 2008), and these results highlight the importance of evaluating the effects of supplementation in different populations, which may have adapted to habitually low nutritional intakes (Fig. 5.2).

Fractures

BMD is a proxy measure for the clinically significant risk of fragility fracture. However, it is again interesting that low BMD does not appear to predict increased risk of fractures in some populations, such as rural Gambians (Aspray et al., 1996). Epidemiological data from across the world have shown variable links of dietary calcium intake to fracture risk. For example, in 5022 Swedish postmenopausal women followed up over 19 years, hip fracture risk was highest at the lowest calcium intake and rose again in the highest intakes, although it is possible that the highest intakes may have been due to

FIGURE 5.2 Osteoporotic fractures are not seen in rural West Africa, with dietary calcium intakes throughout life less than 250 mg/day, and women who have multiple pregnancies with prolonged lactation periods and low bone mineral density.

calcium supplementation and indicate patients with underlying risk factors for fracture (Warensjo et al., 2011). In another analysis of 170,991 women and 68,606 men participating in cohort studies and clinical trials, no relationship was seen between calcium supplementation and hip fracture risk (Bischoff-Ferrari et al., 2007). The most recent Cochrane review concluded that calcium supplementation had a small positive effect on bone density with a trend toward reduction in vertebral fractures, but no clear evidence on the prevention of nonvertebral fractures (Shea et al., 2004). However, calcium given with vitamin D supplementation has been associated with a decrease in fracture risk. This benefit is primarily seen in the old and frail and particularly those living in nursing homes: meta-analyses showed that the overall reduction in hip fracture was on the order of 12%, comprising a reduction by 21% for care home residents, but a nonsignificant reduction of 8% for older people living in their own home (Ross and Institute of Medicine (U.S.) Committee to Review Dietary Reference Intakes for Vitamin D and Calcium, 2011) and a similar trend is seen for all-fracture risk (Chung et al., 2011). In another study of ambulatory patients aged more than 70 years who had sustained a fragility fracture, there was no difference between calcium or vitamin D or both or neither on the risk of subsequent fracture (Grant et al., 2005).

Falls

Falls risk is commonly linked to osteoporosis and fractures, with one study showing a combination calcium plus vitamin D supplement reducing the odds of falling by 46–65% in older women (Bischoff-Ferrari et al., 2006). The dissection of calcium effects independent of vitamin D is difficult as we have already seen with prediction of BMD and fracture risk. Although vitamin D deficiency is associated with a myopathy, and can thus be associated with falling, there appears to be no evidence that calcium is an independent risk factor for this. The separate effects of calcium and vitamin D supplementation are difficult to discern, with meta-analyses differing in their conclusions. One study suggests an independent and dose response of vitamin D on falls risk (Bischoff-Ferrari et al., 2009a). However, others have concluded that vitamin D monotherapy or vitamin D with calcium show no beneficial effect when compared with placebo, whereas there is a modest effect of vitamin D with calcium compared with calcium alone (relative risk 0.84, 95% confidence interval 0.76–0.92; Bolland et al., 2014). The results remain inconsistent because of heterogeneity of study design, but there appears to be no suggestion that calcium (dietary or as a supplement) has an independent effect on falls risk, although it may have a neutral or positive impact on the effect of vitamin D.

Cardiovascular Disease

Calcium supplementation has been associated with short-term improvements in cholesterol and blood pressure, which, over the longer term, might be expected to be protective with regard to cardiovascular health (Reid et al., 2002, 2005). This has been confirmed with lower blood pressure associated with higher calcium intakes in a large cohort from the 1999–2010 National Health and Nutrition Examination Survey (Chen et al., 2015). In an Australian study of more than 41,000 middle-aged and older men and women followed up for 12 years, calcium intakes of up to 1348 mg per day from food were associated with decreased risk of nonfatal CVD and stroke as well as fracture and all-cause mortality (Khan et al., 2015). An American epidemiological study of almost 400,000 subjects found no association between dietary calcium intake and CVD. However, supplemental calcium intake was associated in men (but not women) with a 20% increase in cardiovascular death but not significantly with cerebrovascular disease death (Xiao et al., 2013). Because diabetes is defined in terms of its associated microvascular and macrovascular risks, we can also consider it here. There appears to be a mixed picture suggesting that increased dietary calcium intakes are associated with increased insulin sensitivity (which would be protective against diabetes mellitus; Ma et al., 2006) but also studies showing no relationship to diabetes incidence (de Boer et al., 2008) or an increased risk of diabetes associated with higher serum calcium levels (Lorenzo et al., 2014).

Reid and colleagues have reviewed the effects of calcium on cardiovascular outcomes and arrived at the conclusion that higher calcium intakes (particular when taken as supplements) may be harmful. It is observed, in vitro, that vascular smooth muscle exhibits calcification with increasing calcium concentration and there are further suggestions in vivo of higher serum calcium levels being associated with carotid plaque and abdominal aortic calcification (Reid, 2013). Pathological calcium homeostasis, as seen in hyperparathyroidism, is a risk factor for CVD in endocrinology and renal medicine. However, evidence of the independent effects of higher calcium intake on vascular (or any other) risk are difficult to extract from many clinical trials because calcium has commonly been used as an adjunct to vitamin D, particularly in osteoporosis studies. One piece of supporting evidence, without the confounding effects of vitamin D, is the observation that calcium-based phosphate binders, used in treating kidney disease, were associated with a 22% greater mortality when compared with noncalcium-containing phosphate binders (Jamal et al., 2013). In other studies, a secondary analysis of a randomized controlled trial of calcium supplementation on BMD and fracture incidence showed that myocardial infarction was more commonly seen in the supplemented group (Bolland et al., 2008), and in a meta-analysis of calcium supplementation trials (without vitamin D), 5 studies with patient-level data (8151 participants, median follow-up 3.6 years) and 11 studies with trial-level data (11,921 participants, mean duration 4.0 years), myocardial infarction rates were 27–31% higher (Bolland et al., 2010). A further review of more than 36,000 postmenopausal women participating in the Women's Health Initiative (WHI) Calcium and Vitamin D study found that calcium supplementation was associated with increased cardiovascular risk only in those who had not previously been taking supplements. The associated hazard ratio ranged from 1.13 to 1.22 (significance $p = 0.04–0.05$; Bolland et al., 2011). However, an editorial expressed concern about the cardiovascular adjudication or study design of underlying trials, suggesting that, although it is not possible to provide reassurance that calcium supplements do not cause adverse cardiovascular events or to link them with certainty to increased cardiovascular risk (Abrahamsen and Sahota, 2011).

Preeclampsia and Hypertension in Pregnancy

Hypertension disorders in pregnancy (HDP) include gestational diabetes and preeclampsia, a condition that threatens the life of mother and infant. Relative calcium deficiency has been implicated in the etiology of preeclampsia. A meta-analysis, including 16 cohort studies, found that dietary calcium intakes were associated with a 24% lower risk of HDP. However, low dietary energy and magnesium intakes were also associated with increased risk of HDP, making clear conclusions about the independent effects of dietary calcium more difficult (Schoenaker et al., 2014). Although some recommendations for dietary calcium supplementation are 1.5–2.0 g/day, as little as 1 g/day has been shown to decrease rates of preeclampsia by more than 60% (Hofmeyr et al., 2014).

This question of whether calcium increases cardiovascular risk highlights the clear difficulties in identifying harm from proper nutrition, based on clinical trials of supplements, which are designed to show beneficial effects rather than rarer adverse events. For pregnancy-associated hypertension, there are clear primary clinical outcomes (blood pressure control, symptoms of preeclampsia, successful delivery, and mortality—both maternal and infant), and most of the evidence supports the importance of calcium in the pathogenesis of this condition, its prevention, and therapy. However, in other contexts evaluating the role of calcium is more difficult because it has often been used as a co-supplement with vitamin D. What evidence we have for the potential to harm from calcium is where it is given as a supplement and not as habitual dietary intakes. Therefore caution is warranted in the promotion of very high intakes using supplements in light of this evidence, although the evidence (even from meta-analyses) is not consistent (Bolland et al., 2010; Lewis et al., 2015).

Cancer

In the Melbourne study cohort looking at dietary calcium intakes and a range of outcomes, rates of cancer have been reported to be lower in those with a higher dietary calcium intake (Khan et al., 2015). In cancer-specific analyses, higher calcium intakes have been associated with lower rates of colonic tumors (Heine-Broring et al., 2015; Keum et al., 2015). However, no association was found between dietary calcium intakes and breast cancer incidence or mortality (Abbas et al., 2013). For prostate cancer, total calcium and dairy calcium intakes, but not supplemental calcium intakes, were associated with increased risk (Aune et al., 2015). Some of this evidence has been translated into practice, with calcium supplementation recommended for at least colonic cancer reduction in those at risk. However, a meta-analysis using individual patient data from ten calcium supplement trials (without vitamin D co-supplementation) has shown no significant effect on cancer rates (Bristow et al., 2013), although it should be recognized that the power of these studies to show a significant effect is limited. Overall, it is reassuring that there is no suggestion of cancer risk from dietary calcium intakes and there may be a benefit, although results are inconclusive at present.

Calcium and the Kidney

Hypercalcemia and hypercalciuria are potential adverse outcomes from higher calcium intakes, and elevated serum calcium levels are certainly seen in disorders such as primary hyperparathyroidism and granulomatous conditions such as sarcoidosis, with peripheral conversion of 25(OH)D to its active metabolite $1,25(OH)_2D$. Elevated urinary calcium excretion may occur in some of these cases, sometimes resulting in the development of renal stone disease. The relationship of calcium intake to stone formation is complex. Assuming adequate vitamin D status and normal renal function, dietary calcium intake (whether from food or supplements) should suppress PTH levels in the blood, thus promoting normocalcemia. In primary hyperparathyroidism, there may be a considerable burden on the kidney to excrete excess calcium, and one feature associated with this can be renal stone formation. In contrast, impaired renal function is associated with poorer resorption of calcium by the kidney, resulting in hypocalcemia, and secondary hyperparathyroidism. This perturbation of the parathyroid–renal axis and secondary hyperparathyroidism is exacerbated by raised blood phosphate concentration, which is a marker for renal functional decline and increased cardiovascular mortality, whereas plasma calcium and PTH are not (Palmer et al., 2011).

Renal Stones

The impact of nutrition on renal stone formation depends on the type of stone and the underlying pathological process. Stones can develop in the renal tract for several reasons, including inborn errors of metabolism or infection. Other dietary and metabolic influences include calcium, oxalate, phosphate, and urate homeostasis. Calcium oxalate stones are the most common forms seen in Europe and North America, and risk factors for them include low fluid intake and high dietary oxalate and sodium. The formation of calcium oxalate crystals in the urinary tract is caused by the urinary concentration of oxalate, more so than by calcium, which is dependent on dietary oxalate amounts and efficiency or intestinal absorption. Higher dietary calcium intakes (particularly from food) may cause calcium to bind to oxalate in the gastrointestinal tract, which decreases oxalate absorption. This results in less oxalate being presented to the kidney and prevents renal stone formation. There is contradictory evidence from studies of calcium intakes and stone formation. The WHI study of postmenopausal women found a 17% greater rate of stone formation in women taking calcium and vitamin D supplements. However, the stone types were not identified, and approximately half of the women were taking the calcium supplement without food, which would diminish any beneficial effects of binding oxalate from the diet. The more detailed observational WHI study, which included an assessment of dietary calcium intake, found that higher calcium consumption was associated with a decrease in kidney stone formation by 5–28% ($p=0.01$), with evidence of a beneficial effect of calcium supplementation (Sorensen et al., 2012). In another analysis of the Health Professionals Follow-up Study of 30,762 men and the Nurses' Health Studies I and II of 94,164 and 101,701 women, respectively, the highest quintile for calcium intake was associated with a 17–24% lower rate of kidney stone formation (Taylor and Curhan, 2013). Promoting a higher calcium intake in calcium oxalate stone-formers seems rational, and promoting supplements to be taken with food should also help to bind oxalate before it is absorbed in the gut.

Mortality

Unifying many of the issues already discussed in this chapter, the effect of dietary calcium intakes on longevity or mortality (irrespective of the cause) is an important consideration. There are few analyses of epidemiological data that address this factor (independent of the effect of vitamin D status). Michaelsson et al. (2013) looked at the outcomes for 61,433 postmenopausal Swedish women followed up for 19 years, of who 38,984 had a dietary assessment. They found that, compared with a reference calcium intake of 600–1000 mg/day, lower intakes were associated with a 38% greater mortality, and higher calcium intakes were associated with a 40% greater mortality. They also found that calcium supplement use was associated with no significant effect in those with low dietary calcium intakes (<600 mg/day) but mortality was increased by 157% in those who already had a higher dietary calcium intake (>1400 mg/day). While focusing on cardiovascular mortality, the National Institute of Health American Association of Retired Persons Diet and Health Study followed up 388,229 men and women, aged 50–71 years, with 7904 men and 3874 women dying from CVD over 12 years of follow-up. They found that calcium supplement use was associated with a 20% greater mortality in men but not in women, concluding that "supplemental but not dietary calcium intake was associated with an increased CVD mortality in men but not in women" (Xiao et al., 2013). If we compare the Swedish study with the US study, we might infer that a cumulative total calcium intake threshold of 1200–1500 mg is associated with significantly increased mortality, which is most likely to be due to CVD. However, another analysis of the WHI study found no significant difference in CVD or cancer, and another study by the Vitamin D Individual Patient Analysis of Randomized Trials group found that calcium supplementation (with vitamin D) was associated with a significantly lower mortality, although this benefit was confined to the oldest old (DIPART Group, 2010).

Source of Calcium: Does It Matter?

The evidence on the effects of calcium shows a clear difference between calcium-rich foods, for which there is no epidemiological evidence of harm, and the potential of calcium supplementation to have negative (potentially pathological) effects. Bauer has highlighted that nutritional sources of calcium are widely available and associated with few adverse effects (other than possible lactose intolerance), and more ingested calcium is absorbed from dietary sources than from supplements (Bauer, 2013). The evidence for oxalate kidney stones clearly suggests that calcium-rich food or supplements should be taken with meals (Taylor and Curhan, 2013). However, clinical outcomes are generally lacking (particularly relating to fracture reduction) and metabolically it is difficult to prove any difference between dietary and supplement intakes.

Recommendations for Dietary Calcium Intakes

The dietary reference intake (DRI) framework for recommending nutrient intakes, as adopted by the IOM in North America and Canada, aims to provide quantitative reference values for recommended intakes and safe upper intakes of nutrients (Ross and Institute of Medicine (U.S.) Committee to Review Dietary Reference Intakes for Vitamin D and Calcium, 2011). The resultant estimated average requirement is used to calculate a recommended dietary allowance, which ensures that 97.5% of a population are adequately nourished, and the tolerable upper intake level, which defines the upper intake limit to avoid possible toxicity. It is important to consider whether the use of biomarkers of adequacy or clinical outcomes of the presence (or absence) of nutritional disease states are the best indicators defining optimal nutrition. The IOM report notes that it is particularly difficult to assess calcium and vitamin D requirements separately because "they act in concert and are often administered together in experimental studies" (see Section 4-1; Ross and Institute of Medicine (U.S.) Committee to Review Dietary Reference Intakes for Vitamin D and Calcium, 2011). Inadequate vitamin D status is associated with poorer dietary calcium absorption, such that calcium balance studies are included among potential biomarkers of calcium and vitamin D status (Ross and Institute of Medicine (U.S.) Committee to Review Dietary Reference Intakes for Vitamin D and Calcium, 2011). Therefore, in determining DRIs, the IOM review looked at several factors relevant to calcium and vitamin D nutrition, many of which we have already discussed, including the following:

- cancer risk,
- CVD and falls,
- type 2 diabetes and obesity,
- immune response,
- neuropsychological function,
- physical performance,
- preeclampsia and hypertension in pregnancy, and
- skeletal health.

The IOM guidance eventually uses evidence from calcium balance studies to set the DRI, rationalizing that negative calcium balance, in the long term, is detrimental to skeletal health. The details are available in the report (Ross and Institute of Medicine (U.S.) Committee to Review Dietary Reference Intakes for Vitamin D and Calcium, 2011). In Europe, more recently, the European Food Safety Authority (EFSA) has reviewed dietary calcium guidelines intending to recommend an average requirement, population reference intake (PRI), and an adequate intake of dietary calcium, using a slightly different methodology from that of the IOM. However, EFSA also depends heavily on evidence from calcium balance studies. In addition, the report notes that several factors including vitamin D, fruits, vegetables, vitamin K, moderate alcohol, protein, and inulin can increase calcium absorption whereas sodium, phytate, and high alcohol can decrease it (EFSA NDA Panel (EFSA Panel on Dietetic Products), 2015 (draft)). In developing national and international dietary recommendations for calcium intakes (as shown in Table 5.2), some common themes include the following:

- dominance of dairy produce as a source of dietary calcium,
- variation in dietary calcium requirements across the life course, and
- differences in recommended intakes between the genders.

Table 5.2 shows a summary of recommended intakes (referred to as PRIs) from several international guidelines and taken from the draft EFSA report (EFSA NDA Panel (EFSA Panel on Dietetic Products), 2015 (draft)). Intakes are generally lower in infants, ranging from 260 to 540 mg/day. Up to the age of 3 years, there is a range of values from 350 to 700 mg daily and up to 10 years a similar range of 450–1000 mg. The highest recommended intake is for teenagers to young adulthood, in which intakes in excess of 1000 mg are commonly recommended. Guidelines differ in their recommended intakes for young adults, although there is no gender difference. However, higher intakes are recommended for older women (Afssa (Agence française de sécurité sanitaire des aliments), 2001; D-A-CH, 2013; DH (Department of Health), 1991; NNR (Nordic Nutrition Recommendations), 2004; Ross and Institute of Medicine (U.S.) Committee to Review Dietary Reference Intakes for Vitamin D and Calcium, 2011; SCF (Scientific Committee for Food), 1993; WHO/FAO (World Health Organization/Food and Agriculture Organization of the United Nations), 2004).

Calcium Intakes

Nutritional surveys estimate the habitual intakes of populations, and thus the adequacy of calcium nutrition, using the recommended nutrient intake (RNI), where one exists. Surveys indicate that dietary calcium intakes are generally comparable to recommended intakes; for example, only 6.6% of postmenopausal Swedish women were consuming less than 600 mg calcium a day (Michaelsson et al., 2013). In the United States, although concerns have been expressed about poorer calcium intakes among the older population, throughout adult life they appear to have average intakes greater than 900 mg/day in men and 650–700 mg/day in women whereas older women have the lowest intakes of less than 600 mg/day but the greatest likelihood of taking calcium supplements (Mangano et al., 2011). One might conclude that calcium undernutrition is not a major problem in these communities; however, we need to balance the dietary calcium recommendations based on balance studies with the growing evidence base of risk and benefit for individuals, particularly if strategies for optimizing calcium intakes are likely to use nonfood supplements. We also know that worldwide there are communities with much lower habitual dietary calcium intakes, especially in Africa and Asia, where typical calcium intakes are 200–400 mg/day and predominantly from plant origin. Although typical intakes in excess of 700 mg/day may be expected in Northern Europe and North America, there may also be significant differences between ethnic groups in the same geographical area. For example, only 8% of White British women have a calcium intake below the lower reference nutritional intake, compared with 42% of women of Black African or Caribbean origin. A similar pattern is seen in North America, comparing African Americans with White Americans (Redmond et al., 2014).

How Are Other Minerals Affected by Calcium?

As already discussed, the most important mineral interacting with calcium homeostasis is probably inorganic phosphate, because its handling is intimately linked to calcium through PTH and $1,25(OH)_2D$. Calcium salts complex with phosphate in the gut and are used clinically to decrease blood phosphate levels in chronic kidney disease to prevent hyperphosphatemia. There is also increasing evidence of the actions of phosphotonins (e.g., fibroblast growth factor-23), which increase renal phosphate excretion and suppress the actions of PTH and $1,25(OH)_2D$ with consequential impairment of calcium homeostasis. Magnesium is the second most common intracellular cation and shares some physiological effects with calcium, being important as a cofactor in intracellular signaling, neuromuscular excitability, and bone health. It is regulated by PTH and calcitonin in a similar fashion to calcium; therefore derangement of these hormonal axes will have similar effects on magnesium and calcium absorption and excretion. Moreover, diseases affecting small bowel epithelial function will affect both. Magnesium deficiency has a variable effect on PTH, with acute lowering of magnesium levels promoting a transient increase in PTH levels but chronically lower serum magnesium levels suppressing PTH. Low serum magnesium has been implicated in adverse effects seen with proton pump inhibitors, with deleterious effects on calcium homeostasis and causing hypocalcaemia.

TABLE 5.2 Overview of Dietary Reference Values for Adults and Children Based on European Food Safety Authority Panel on Dietetic Products

	Nordic Council of Ministers (2014)	D-A-CH (2013)	Institute of Medicine (2011)	World Health Organization/Food and Agriculture Organization of the United Nations (2004)	Afssa (2001)	Health Council of the Netherlands (2000)ᵃ	SCF (1993)	DH (1991)
Adults								
Age (years)	18–20	≥19	19–50	19–65 (m) 19–Menopause (f)	20–65 (m) 20–55 (f)	19–50	≥18	≥19
PRI								
Men (mg/day)	900	1000	1000	1000	900	1000	700	700
Women (mg/day)	900	1000	1000	1000	900	1000	700	700
Age (years)	≥21		51–70	≥65 (m) Postmenopausal (f)	≥66 (m) ≥56 (f)	51–70		
Men (mg/day)	800		1000	1300	1200	1100		
Women (mg/day)	800		1200	1300	1200	1100		
Age (years)			≥70			≥70		
PRI								
Men (mg/day)			1200			1200		
Women (mg/day)			1200			1200		
Children								
Age (months)	6–11	4–<12	6–12	7–12		6–11	6–11	0–12
PRI (mg/day)	540	330	260ᵃ	400		450	400	525
Age (years)	1–5	1–<4	1–3	1–3	1–3	1–3	1–3	1–3
PRI (mg/day)	600	600	700	500	500	500	400	350
Age (years)	6–9	4–<7	4–8	4–6	4–6	4–8	4–6	4–6
PRI (mg/day)	700	750	1000	600	700	700	450	450
Age (years)	10–17	7–<10	9–18	7–9	7–9	9–18	7–10	7–10
PRI (mg/day)	900	900	1300	700	900	1200 (m) 1100 (f)	550	550
Age (years)		10–<13		10–18	10–19		11–17	11–18
PRI (mg/day)		1100		1300	1200		1000 (m)	1000 (m)
Age (years)		13–<19						
PRI (mg/day)		1200					800 (f)	800 (f)

m, males; *f*, females; *PRI*, population reference intake.
ᵃAdequate intake.
From the European Food Safety Authority Panel on Dietetic Products draft report 2015.

There is also some evidence that lower dietary magnesium intakes (e.g., calcium) may be associated with lower BMD, although there is no evidence that there is an associated increased risk of fragility fracture (Orchard et al., 2014).

In bone, divalent cations such as strontium or magnesium may compete with calcium in forming hydroxyapatite crystals, with varying effects on bone strength (Ilich and Kerstetter, 2000), although more recently strontium (as its ranelate salt) has been used as a therapeutic agent in osteoporosis with proven efficacy in fracture prevention. Calcium can also interfere with the absorption of cations in the gut, particularly iron, whereas the evidence of an effect on zinc absorption may be confined to animals (Ilich and Kerstetter, 2000). Sodium has a negative effect on calcium absorption at the intestine as well as a tendency to promote renal stone formation as discussed earlier in the chapter, and potassium may be beneficial to skeletal integrity. One review advises to "decrease sodium intake, increase potassium intake, and consume more fresh fruits and vegetables," which appears to be a reasonable nutritional guide to complement bone health (Prentice, 2004).

SUMMARY POINTS

- Calcium is important for bone health, and the current RNI is adequate for most people.
- Much of the evidence about calcium nutrition is confused by the tendency to focus on
  - supplementation and
  - studies with vitamin D cotherapy.
- Optimizing dietary calcium intake is generally preferably to supplementation.
- Higher calcium intakes are associated with increased bone mass.
- Calcium does not reduce fracture risk (other than when given with vitamin D to frail older people).
- Higher and lower calcium intakes may be associated with increased cardiovascular risk, with the greatest evidence of risk associated with the use of supplements, resulting in high intakes.
- There may be a beneficial effect of high calcium intakes on risk of developing renal stones, at least for calcium oxalate stones.
- The relationship of calcium intake to cancer risk is unclear, but it appears generally to be beneficial.
- Targeted supplementation in people who are most likely to benefit is probably the best strategy.

KEY FACTS

- Calcium homeostasis is regulated by vitamin D and PTH.
- Dietary calcium interacts with vitamin D, obscuring the evidence.
- Calcium deficiency can cause rickets and osteomalacia.
- Higher dietary calcium intakes can improve
  - longevity;
  - BMD; and
  - risk of
    - fractures,
    - falls.
    - renal stones, and
    - hypertension in pregnancy.
- High and low calcium intakes may increase cardiovascular risk.

REFERENCES

Abbas, S., Linseisen, J., Rohrmann, S., Chang-Claude, J., Peeters, P.H., Engel, P., Brustad, M., Lund, E., Skeie, G., Olsen, A., Tjonneland, A., Overvad, K., Boutron-Ruault, M.C., Clavel-Chapelon, F., Fagherazzi, G., Kaaks, R., Boeing, H., Buijsse, B., Adarakis, G., Ouranos, V., Trichopoulou, A., Masala, G., Krogh, V., Mattiello, A., Tumino, R., Sacerdote, C., Buckland, G., Suarez, M.V., Sanchez, M.J., Chirlaque, M.D., Barricarte, A., Amiano, P., Manjer, J., Wirfalt, E., Lenner, P., Sund, M., Bueno-de-Mesquita, H.B., van Duijnhoven, F.J., Khaw, K.T., Wareham, N., Key, T.J., Fedirko, V., Romieu, I., Gallo, V., Norat, T., Wark, P.A., Riboli, E., 2013. Dietary intake of vitamin D and calcium and breast cancer risk in the European prospective investigation into cancer and nutrition. Nutr. Cancer 65, 178–187.

Abrahamsen, B., Sahota, O., 2011. Do calcium plus vitamin D supplements increase cardiovascular risk? BMJ 342, d2080.

Afssa (Agence française de sécurité sanitaire des aliments), 2001. Apports nutritionnels conseillés 1618 pour la population française. Afssa, Paris.

Aspray, T.J., Prentice, A., Cole, T.J., Sawo, Y., Reeve, J., Francis, R.M., 1996. Low bone mineral content is common but osteoporotic fractures are rare in elderly rural Gambian women. J. Bone Min. Res. 11, 1019–1025.

Aune, D., Navarro Rosenblatt, D.A., Chan, D.S., Vieira, A.R., Vieira, R., Greenwood, D.C., Vatten, L.J., Norat, T., 2015. Dairy products, calcium, and prostate cancer risk: a systematic review and meta-analysis of cohort studies. Am. J. Clin. Nutr. 101, 87–117.

Bauer, D.C., 2013. Clinical practice. Calcium supplements and fracture prevention. N. Engl. J. Med. 369, 1537–1543.

Bischoff-Ferrari, H.A., 2012. Vitamin D and fracture prevention. Rheum. Dis. Clin. North Am. 38, 107–113.

Bischoff-Ferrari, H.A., Dawson-Hughes, B., Baron, J.A., Burckhardt, P., Li, R., Spiegelman, D., Specker, B., Orav, J.E., Wong, J.B., Staehelin, H.B., O'Reilly, E., Kiel, D.P., Willett, W.C., 2007. Calcium intake and hip fracture risk in men and women: a meta-analysis of prospective cohort studies and randomized controlled trials. Am. J. Clin. Nutr. 86, 1780–1790.

Bischoff-Ferrari, H.A., Dawson-Hughes, B., Staehelin, H.B., Orav, J.E., Stuck, A.E., Theiler, R., Wong, J.B., Egli, A., Kiel, D.P., Henschkowski, J., 2009a. Fall prevention with supplemental and active forms of vitamin D: a meta-analysis of randomised controlled trials. BMJ 339, b3692.

Bischoff-Ferrari, H.A., Kiel, D.P., Dawson-Hughes, B., Orav, J.E., Li, R., Spiegelman, D., Dietrich, T., Willett, W.C., 2009b. Dietary calcium and serum 25-hydroxyvitamin D status in relation to BMD among U.S. adults. J. Bone Min. Res. 24, 935–942.

Bischoff-Ferrari, H.A., Orav, E.J., Dawson-Hughes, B., 2006. Effect of cholecalciferol plus calcium on falling in ambulatory older men and women: a 3-year randomized controlled trial. Arch. Intern Med. 166, 424–430.

Blaine, J., Chonchol, M., Levi, M., 2014. Renal control of calcium, phosphate, and magnesium homeostasis. Clin. J. Am. Soc. Nephrol. 10 (7).

Bolland, M.J., Avenell, A., Baron, J.A., Grey, A., MacLennan, G.S., Gamble, G.D., Reid, I.R., 2010. Effect of calcium supplements on risk of myocardial infarction and cardiovascular events: meta-analysis. BMJ 341, c3691.

Bolland, M.J., Barber, P.A., Doughty, R.N., Mason, B., Horne, A., Ames, R., Gamble, G.D., Grey, A., Reid, I.R., 2008. Vascular events in healthy older women receiving calcium supplementation: randomised controlled trial. BMJ 336, 262–266.

Bolland, M.J., Grey, A., Avenell, A., Gamble, G.D., Reid, I.R., 2011. Calcium supplements with or without vitamin D and risk of cardiovascular events: reanalysis of the Women's Health Initiative limited access dataset and meta-analysis. BMJ 342, d2040.

Bolland, M.J., Grey, A., Reid, I.R., 2014. Differences in overlapping meta-analyses of vitamin D supplements and falls. J. Clin. Endocrinol. Metab. 99, 4265–4272.

Bristow, S.M., Bolland, M.J., MacLennan, G.S., Avenell, A., Grey, A., Gamble, G.D., Reid, I.R., 2013. Calcium supplements and cancer risk: a meta-analysis of randomised controlled trials. Br. J. Nutr. 110, 1384–1393.

Chen, Y., Strasser, S., Cao, Y., Wang, K.S., Zheng, S., 2015. Calcium intake and hypertension among obese adults in United States: associations and implications explored. J. Hum. Hypertens. 29 (9).

Chung, M., Lee, J., Terasawa, T., Lau, J., Trikalinos, T.A., 2011. Vitamin D with or without calcium supplementation for prevention of cancer and fractures: an updated meta-analysis for the U.S. Preventive Services Task Force. Ann. Intern Med. 155, 827–838.

D-A-CH, 2013. Referenzwerte für die Nährstoffzufuhr. Deutsche Gesellschaft für Ernährung - Österreichische Gesellschaft für Ernährung – Schweizerische Gesellschaft für Ernährungsforschung – Schweizerische Vereinigung für Ernährung. Neuer Umschau Buchverlag, Frankfurt/Main, Germany. 292 pp.

de Boer, I.H., Tinker, L.F., Connelly, S., Curb, J.D., Howard, B.V., Kestenbaum, B., Larson, J.C., Manson, J.E., Margolis, K.L., Siscovick, D.S., Weiss, N.S., Women's Health Initiative Investigators, 2008. Calcium plus vitamin D supplementation and the risk of incident diabetes in the Women's Health Initiative. Diabetes Care 31, 701–707.

DH (Department of Health), 1991. Dietary Reference Values for Food Energy and Nutrients for the United Kingdom. Report of the Panel on Dietary Reference Values of the Committee on Medical Aspects of Food Policy. HMSO, London, UK. 212 pp.

EFSA NDA Panel (EFSA Panel on Dietetic Products), N.a.A., (draft), 2015. Dietary Reference Values for Calcium. EFSA.

Francis, R.M., Aspray, T.J., Fraser, W., Gittoes, N., Javaid, K., Macdonald, H., Patel, S., Selby, P., Tanna, N., Bowring, C., 2013. Vitamin D and Bone Health: A Practical Clinical Guideline for Patient Management. The National Osteoporosis Society.

Grant, A.M., Avenell, A., Campbell, M.K., McDonald, A.M., MacLennan, G.S., McPherson, G.C., Anderson, F.H., Cooper, C., Francis, R.M., Donaldson, C., Gillespie, W.J., Robinson, C.M., Torgerson, D.J., Wallace, W.A., Group, R.T., 2005. Oral vitamin D3 and calcium for secondary prevention of low-trauma fractures in elderly people (Randomised Evaluation of Calcium or vitamin D, RECORD): a randomised placebo-controlled trial. Lancet 365, 1621–1628.

Group, D., 2010. Patient level pooled analysis of 68 500 patients from seven major vitamin D fracture trials in US and Europe. BMJ 340, b5463.

Health Council of the Netherlands, 2000. Dietary reference intakes: calcium, vitamin D, thiamin, riboflavin, niacin, pantothenic acid, and biotin. publication no. 2000/12. ISBN 90-5549-323-6, 180 pp.

Heine-Bröring, R.C., Winkels, R.M., Renkema, J.M., Kragt, L., van Orten-Luiten, A.C., Tigchelaar, E.F., Chan, D.S., Norat, T., Kampman, E., 2015. Dietary supplement use and colorectal cancer risk: a systematic review and meta-analyses of prospective cohort studies. Int. J. Cancer 136, 2388–2401.

Hofmeyr, G.J., Belizan, J.M., von Dadelszen, P., Calcium and Pre-eclampsia (CAP) Study Group, 2014. Low-dose calcium supplementation for preventing pre-eclampsia: a systematic review and commentary. BJOG 121, 951–957.

Ilich, J.Z., Kerstetter, J.E., 2000. Nutrition in bone health revisited: a story beyond calcium. J. Am. Coll. Nutr. 19, 715–737.

Jamal, S.A., Vandermeer, B., Raggi, P., Mendelssohn, D.C., Chatterley, T., Dorgan, M., Lok, C.E., Fitchett, D., Tsuyuki, R.T., 2013. Effect of calcium-based versus non-calcium-based phosphate binders on mortality in patients with chronic kidney disease: an updated systematic review and meta-analysis. Lancet 382, 1268–1277.

Jarjou, L.M., Sawo, Y., Goldberg, G.R., Laskey, M.A., Cole, T.J., Prentice, A., 2013. Unexpected long-term effects of calcium supplementation in pregnancy on maternal bone outcomes in women with a low calcium intake: a follow-up study. Am. J. Clin. Nutr. 98, 723–730.

Keum, N., Lee, D.H., Greenwood, D.C., Zhang, X., Giovannucci, E.L., 2015. Calcium intake and colorectal adenoma risk: dose-response meta-analysis of prospective observational studies. Int. J. Cancer 136, 1680–1687.

Khan, B., Nowson, C.A., Daly, R.M., English, D.R., Hodge, A.M., Giles, G.G., Ebeling, P.R., 2015. Higher dietary calcium intakes are associated with reduced risks of fractures, cardiovascular events, and mortality: a prospective cohort study of older men and women. J. Bone Min. Res. 30 (10).

Lee, W.T., Jiang, J., 2008. Calcium requirements for Asian children and adolescents. Asia Pac. J. Clin. Nutr. 17 (Suppl. 1), 33–36.

Lewis, J.R., Radavelli-Bagatini, S., Rejnmark, L., Chen, J.S., Simpson, J.M., Lappe, J.M., Mosekilde, L., Prentice, R.L., Prince, R.L., 2015. The effects of calcium supplementation on verified coronary heart disease hospitalization and death in postmenopausal women: a collaborative meta-analysis of randomized controlled trials. J. Bone Min. Res. 30, 165–175.

Lorenzo, C., Hanley, A.J., Rewers, M.J., Haffner, S.M., 2014. Calcium and phosphate concentrations and future development of type 2 diabetes: the Insulin Resistance Atherosclerosis Study. Diabetologia 57, 1366–1374.

Ma, B., Lawson, A.B., Liese, A.D., Bell, R.A., Mayer-Davis, E.J., 2006. Dairy, magnesium, and calcium intake in relation to insulin sensitivity: approaches to modeling a dose-dependent association. Am. J. Epidemiol. 164, 449–458.

Mangano, K.M., Walsh, S.J., Insogna, K.L., Kenny, A.M., Kerstetter, J.E., 2011. Calcium intake in the United States from dietary and supplemental sources across adult age groups: new estimates from the National Health and Nutrition Examination Survey 2003-2006. J. Am. Diet. Assoc. 111, 687–695.

Michaelsson, K., Melhus, H., Warensjo Lemming, E., Wolk, A., Byberg, L., 2013. Long term calcium intake and rates of all cause and cardiovascular mortality: community based prospective longitudinal cohort study. BMJ 346, f228.

NNR (Nordic Nutrition RecoNNR (Nordic Nutrition Recommendations), 2004. Integrating Nutrition and Physical Activity. Council of Ministers, Copenhagen, Denmark. 435 pp.

Nordic Council of Ministers, 2014. Nordic Nutrition Recommendations 2012. Integrating nutrition and physical activity, fifth ed. 627 pp.

Orchard, T.S., Larson, J.C., Alghothani, N., Bout-Tabaku, S., Cauley, J.A., Chen, Z., LaCroix, A.Z., Wactawski-Wende, J., Jackson, R.D., 2014. Magnesium intake, bone mineral density, and fractures: results from the Women's Health Initiative Observational Study. Am. J. Clin. Nutr. 99, 926–933.

Palmer, S.C., Hayen, A., Macaskill, P., Pellegrini, F., Craig, J.C., Elder, G.J., Strippoli, G.F., 2011. Serum levels of phosphorus, parathyroid hormone, and calcium and risks of death and cardiovascular disease in individuals with chronic kidney disease: a systematic review and meta-analysis. JAMA 305, 1119–1127.

Pettifor, J.M., 2013. Nutritional rickets: pathogenesis and prevention. Pediatr. Endocrinol. Rev. 10 (Suppl. 2), 347–353.

Prentice, A., 2004. Diet, nutrition and the prevention of osteoporosis. Public Health Nutr. 7, 227–243.

Redmond, J., Jarjou, L.M., Zhou, B., Prentice, A., Schoenmakers, I., 2014. Ethnic differences in calcium, phosphate and bone metabolism. Proc. Nutr. Soc. 73, 340–351.

Reid, I.R., 2013. Cardiovascular effects of calcium supplements. Nutrients 5, 2522–2529.

Reid, I.R., Ames, R., Mason, B., Reid, H.E., Bacon, C.J., Bolland, M.J., Gamble, G.D., Grey, A., Horne, A., 2008. Randomized controlled trial of calcium supplementation in healthy, nonosteoporotic, older men. Arch. Intern. Med. 168, 2276–2282.

Reid, I.R., Horne, A., Mason, B., Ames, R., Bava, U., Gamble, G.D., 2005. Effects of calcium supplementation on body weight and blood pressure in normal older women: a randomized controlled trial. J. Clin. Endocrinol. Metab. 90, 3824–3829.

Reid, I.R., Mason, B., Horne, A., Ames, R., Clearwater, J., Bava, U., Orr-Walker, B., Wu, F., Evans, M.C., Gamble, G.D., 2002. Effects of calcium supplementation on serum lipid concentrations in normal older women: a randomized controlled trial. Am. J. Med. 112, 343–347.

Rejnmark, L., Amstrup, A.K., Mollerup, C.L., Heickendorff, L., Mosekilde, L., 2012. Further insights into the pathogenesis of primary hyperparathyroidism: a Nested Case-control study. J. Clin. Endocrinol. Metab. 98 (1).

Ross, A.C., Institute of Medicine (U.S.) Committee to Review Dietary Reference Intakes for Vitamin D and Calcium, 2011. DRI, Dietary Reference Intakes: Calcium, Vitamin D. National Academies Press, Washington, DC.

Sawo, Y., Jarjou, L.M., Goldberg, G.R., Laskey, M.A., Prentice, A., 2013. Bone mineral changes after lactation in Gambian women accustomed to a low calcium intake. Eur. J. Clin. Nutr. 67, 1142–1146.

SCF (Scientific Committee for Food), 1993. Nutrient and Energy Intakes for the European 2157 Community, Reports of the Scientific Committee for Food, 31st Series. Food - Science and Technique. European Commission, Luxembourg. 248 pp.

Schoenaker, D.A., Soedamah-Muthu, S.S., Mishra, G.D., 2014. The association between dietary factors and gestational hypertension and pre-eclampsia: a systematic review and meta-analysis of observational studies. BMC Med. 12, 157.

Shea, B., Wells, G., Cranney, A., Zytaruk, N., Robinson, V., Griffith, L., Hamel, C., Ortiz, Z., Peterson, J., Adachi, J., Tugwell, P., Guyatt, G., Osteoporosis Methodology, G., Osteoporosis Research Advisory, G, 2004. Calcium supplementation on bone loss in postmenopausal women. Cochrane Database Syst. Rev. (1) CD004526.

Sorensen, M.D., Kahn, A.J., Reiner, A.P., Tseng, T.Y., Shikany, J.M., Wallace, R.B., Chi, T., Wactawski-Wende, J., Jackson, R.D., O'Sullivan, M.J., Sadetsky, N., Stoller, M.L., Group, W.H.I.W., 2012. Impact of nutritional factors on incident kidney stone formation: a report from the WHI OS. J. Urol. 187, 1645–1649.

Taylor, E.N., Curhan, G.C., 2013. Dietary calcium from dairy and nondairy sources, and risk of symptomatic kidney stones. J. Urol. 190, 1255–1259.

Uusi-Rasi, K., Karkkainen, M.U., Lamberg-Allardt, C.J., 2013. Calcium intake in health maintenance - a systematic review. Food Nutr. Res. 57.

Ward, K.A., Cole, T.J., Laskey, M.A., Ceesay, M., Mendy, M.B., Sawo, Y., Prentice, A., 2014. The effect of prepubertal calcium carbonate supplementation on skeletal development in Gambian boys-a 12-year follow-up study. J. Clin. Endocrinol. Metab. 99, 3169–3176.

Warensjo, E., Byberg, L., Melhus, H., Gedeborg, R., Mallmin, H., Wolk, A., Michaelsson, K., 2011. Dietary calcium intake and risk of fracture and osteoporosis: prospective longitudinal cohort study. BMJ 342, d1473.

WHO/FAO (World Health Organization/Food and Agriculture Organization of the United Nations), 2004. Vitamin and Mineral Requirements in Human Nutrition: Report of a Joint FAO/WHO Expert Consultation. WHO/FAO, Bangkok, Thailand. 341 pp.

Winzenberg, T., Shaw, K., Fryer, J., Jones, G., 2006. Effects of calcium supplementation on bone density in healthy children: meta-analysis of randomised controlled trials. BMJ 333, 775.

Xiao, Q., Murphy, R.A., Houston, D.K., Harris, T.B., Chow, W.H., Park, Y., 2013. Dietary and supplemental calcium intake and cardiovascular disease mortality: the National Institutes of Health-AARP diet and health study. JAMA Intern. Med. 173, 639–646.

Chapter 6

Molecular Aspects of Calcium and Bone Mineralization

Anuradha V. Khadilkar[1], Nikhil D. Phadke[2], Priyanka V. Gangodkar[2]

[1]Jehangir Medical Research Institute Jehangir Hospital, Pune, India; [2]GenePath Dx (Causeway Healthcare Private Limited), Pune, India

INTRODUCTION

Calcium is the most abundant mineral in the human body, and 99% of the body's calcium is found in bones and teeth whereas the remaining 1% is distributed in cells, blood, and body fluids (Bonjour, 2011). Dietary sources of calcium include milk and other dairy products such as yogurt and cheese. Smaller amounts are also found in certain fish and seafood, vegetables such as broccoli and kale, dry fruits, and enriched foods (e.g., fruit juices, milk, and bread; Gropper et al., 2008). The recommended daily allowances for calcium intake range from 500 to 800 mg/day up to 8 years of age and 1000–1300 mg/day in adolescents and adults (Institute of Medicine, 2011). Mineralization of bone is one of the most important functions of calcium in the body; however, calcium is also essential for several other physiological processes, including nerve impulse transmission, regulation of blood clotting, membrane permeability, and muscle contraction (Rosol et al., 1995).

Calcium in the skeleton is in the form of hydroxyapatite and other calcium–phosphate complexes (e.g., calcium phospholipid phosphate complexes). Although appearing static, bone is a metabolically active tissue and constantly undergoes remodeling through the action of osteoclasts and osteoblasts. Whereas the osteoblasts replace new tissue at sites of damage or where remodeling has occurred, osteoclasts are responsible for degrading bone tissue (Seibel, 2005). Apart from providing skeletal strength, bone serves as a calcium reservoir to maintain tightly controlled circulating concentrations vital to cellular signaling, muscle contraction, and blood coagulation (Peacock, 2010). When the diet is low in calcium or when serum calcium concentrations are low due to other causes (e.g., vitamin D deficiency or disorders of the parathyroid gland), calcium is released from the bones into the bloodstream to meet the body's needs of calcium.

METABOLISM AND REGULATION

Absorption of dietary calcium is crucial for the maintenance of bone health and serum calcium concentrations. The transcellular pathway of calcium absorption from the intestine is important for the maintenance of calcium homeostasis, particularly so when the intake of dietary calcium is restricted (Favus, 1985). The amount of calcium consumed and the efficiency of the calcium absorption process determines the amount of calcium absorbed from the diet.

Serum calcium concentrations are kept stable, with variations that do not exceed 2–3% that are mainly due to a circadian rhythm (Rejnmark et al., 2002). To maintain concentrations within the reference range, renal calcium excretion is reduced, the efficiency of dietary calcium absorption is altered in the intestine, or calcium is withdrawn from bones (Fleet, 2006). Thus calcium homeostasis is regulated by the three major hormones acting on their corresponding receptors in the intestine, kidneys, and bone: parathyroid hormone (PTH), the active metabolite of vitamin D [1,25-dihydroxyvitamin D3 ($1,25(OH)_2D_3$)], and calcitonin.

PTH is an important endocrine regulator of calcium concentrations in the extracellular fluid. When serum calcium concentrations drop, PTH mobilizes calcium from bone by stimulating osteoclasts to resorb bone mineral, thus liberating calcium into blood (Åkerström et al., 2005). PTH also facilitates calcium absorption from the small intestine (by stimulating production of the active form of vitamin D in the kidney) to elevate blood levels of calcium. PTH reduces calcium excretion from the kidneys by increasing tubular reabsorption of calcium. PTH also stimulates calcium release from bone; low PTH concentrations cause serum calcium levels to fall and serum phosphate levels to increase, whereas increased levels of PTH can result in hypercalcemia. PTH mediates conversion of 25-hydroxyvitamin D_3 to $1,25(OH)_2D_3$, thus enabling renal tubular calcium reabsorption. Increased concentrations of serum calcium inhibit PTH secretion and bone resorption whereas low concentrations stimulate PTH secretion. Actions of PTH are performed through activation of the PTH/PTH-related peptide type 1 receptor (Juppner et al., 1991).

Molecular, Genetic, and Nutritional Aspects of Major and Trace Minerals. http://dx.doi.org/10.1016/B978-0-12-802168-2.00006-3

Vitamin D acts through its receptor [vitamin D receptor (VDR)] in the intestine, bone cells, parathyroid glands, and kidneys to regulate concentrations of serum calcium (St-Arnaud, 2008; Christakos et al., 2003). The active form of vitamin D $(1,25(OH)_2D_3)$ affects the process of transcellular calcium transport. It has been proposed that the process of transcellular calcium transport involves entry of calcium via the apical calcium channel, transient receptor potential vanilloid type-6 (TRPV6), followed by translocation of calcium through the enterocyte and then basolateral extrusion by the plasma membrane calcium pump-1b (Wasserman, 2005). TRPV6 and calbindin-D_{9k} (encoding an intracellular calcium binding protein) mRNA levels are regulated by $1,25(OH)_2D_3$ in duodenum, thus allowing increases in calcium absorption (Song et al., 2003a,b).

In individuals with habitually low intake of dietary calcium, the efficiency of calcium absorption is increased via an adaptive increase in renal production of $1,25(OH)_2 D_3$. This process is mediated by transcriptional activation through the VDR gene (Favus et al., 1974; Song et al., 2003a,b). Studies in mice lacking TRPV6 (which mediates dietary calcium import into enterocytes) suggest that TRPV6 is not required for intestinal calcium absorption when dietary calcium content is normal or high. In these situations, passive diffusion of dietary calcium contributes to the maintenance of normal serum calcium concentrations (Lieben et al., 2010).

Excessive calcium levels can also be detrimental, resulting in metabolic disturbances and metastatic calcification. Calcitonin, a hormone that is secreted by the C cells of the thyroid gland in response to an increase in serum calcium concentrations, is one of the factors that is believed to protect against the development of hypercalcemia. Calcitonin stimulates production of the active vitamin D metabolite in the renal proximal tubules. During pregnancy and lactation, secretion of calcitonin and $1,25(OH)_2D_3$ are increased. This increase in calcitonin and vitamin D production may play a role in the transfer of maternal calcium to the fetus and in the prevention of maternal bone loss (Kovacs and Kronenberg, 1997). Calcitonin also decreases osteoclastic activity and has been used for treatment of hypercalcemia (Felsenfeld and Levine, 2015).

CALCIUM, GENETICS, AND GENETIC DISORDERS

Parathyroid Hormone Axis

Genetic defects in parathyroid secretion have an effect on the calcium homeostasis. For example, in DiGeorge syndrome there is abnormal development of the parathyroid glands, hypoparathyroidism, and resultant hypocalcemia. There are also other anomalies such as T cell immune deficiency and abnormal facial appearance. The pathognomonic genetic defect in patients with DiGeorge syndrome is a deletion in the 22q11.2 genomic region (Kitsiou-Tzeli et al., 2004).

The calcium-sensing receptor (CaSR) protein, encoded by *CaSR* (3q13), is found in abundance in cells of the parathyroid glands as well as in the kidneys. When large amounts of calcium bind to CaSR in the parathyroid glands, the production of PTH is blocked, and this prevents the release of calcium into the blood. Likewise, increased calcium binding to CaSR in the kidneys reduces the reabsorption of calcium from the body fluids. Mutations in *CaSR* (i.e., autosomal-dominant hypocalcemia type 1), result in low levels of calcium in the blood (hypocalcemia) and hypoparathyroidism. Hypocalcemia can result in muscle cramps or seizures.

Mutations in *CaSR* can also result in familial isolated hyperparathyroidism, a condition characterized by hypercalcemia. This condition presents with nausea, vomiting, kidney stones, osteoporosis, and hypertension due to increased concentrations of serum calcium. Familial isolated hyperparathyroidism has also been associated with mutations in the multiple endocrine neoplasia type 1 and cell division cycle-73 genes.

Defects in PTH action may result in a condition described as pseudohypoparathyroidism with hypocalcemia, and patients may develop cataracts (Levine, 2012). In this disorder, mutations in the genes encoding PTH receptor-1 and G protein alpha S (encoded by *GNAS*) lead to peripheral resistance to PTH action, also leading to severe bone anomalies (Duchatelet et al., 2005). Additional genetic syndromes, such as autoimmune polyglandular syndrome type 1 (patients present with hypoparathyroidism, adrenal insufficiency, and chronic mucocutaneous candidiasis); hypoparathyroidism, deafness, and renal dysplasia syndrome; and hypoparathyroidism, retardation, and dysmorphism syndrome, also result in hypocalcemia and bone abnormalities as a part of the clinical spectrum. Moreover, mutations in GATA binding protein-3 (which is a transcription factor involved in the development of the kidney) lead to hypoparathyroidism, deafness, and renal dysplasia. In this autosomal-dominant disorder, hypocalcemia and very low levels of PTH are associated with renal dysplasia and sensorineural deafness (Ali et al., 2007).

Vitamin D Axis

There has been great interest in *VDR* and its association with calcium metabolism and bone health. Deletion of *VDR* causes reduction in calcium absorption by more than 70%. *VDR* deletion also results in poor growth, low serum Ca, high serum PTH levels, and severe osteomalacia (Van Cromphaut et al., 2001). However, a recent study on *VDR* knockout mice revealed that there is VDR-independent regulation of intestinal calcium absorption, and that a high-calcium diet promotes a VDR-independent anabolic effect on calcium absorption and bone formation (Song et al., 2003a,b).

$1,25(OH)_2D_3$ binds to the VDR receptor to mediate its actions. In the absence of a functional VDR receptor, calcium absorption is impaired; thus bones are inadequately mineralized, leading to rickets in children and osteomalacia in adults (Malloy and Feldman, 2010). *CYP27B1* encodes 25-hydroxyvitamin D-1α-hydroxylase [which converts calcidiol (i.e., 25-hydroxyvitamin D) to calcitriol (i.e., $1,25(OH)_2D_3$, the active form of vitamin D)]; mutations of this gene result in vitamin-D–dependent rickets (Demir et al., 2015). Several other reports have also described resistant rickets (resistant to treatment with vitamin D) as due to a defect in *VDR* in children (Zhou et al., 2009; Malloy et al., 2011). The association of *VDR* polymorphisms and osteoporosis or osteomalacia in adults is less clearly defined (Ray et al., 2009; Ak et al., 2005). Even so, there are several studies reporting *VDR* polymorphisms and their associations with bone mineralization. In a study on different restriction enzyme polymorphisms in *VDR* (*Bsm1*, *Apa1*, *Taq 1*, and *Fok1* sites) and genes with estrogen and calcitonin receptors, it was observed that there was no association of these polymorphisms with calcium absorption, bone mass, or mineral accretion (Hansen et al., 1997). In another study in which elderly Caucasians were supplemented with 800 mg of calcium per day for 18 months, only the Bb genotype of the *BsmI* single-nucleotide polymorphism (SNP) was found to be associated with differences in bone mineral density (BMD); however, for other genotypes (*Apa1*, *Taq1*, and *Fok1*), calcium intake did not play a significant role (Ferrari et al., 1995). Moreover, among postmenopausal women, *Fok 1* polymorphisms were associated with lower femur BMD (Zajíčková et al., 2002). Similar observations were also reported in a Japanese study on pubertal girls (Katsumata et al., 2002). Another study in postmenarchal Indian girls with habitually low calcium intake suggested that *VDR* polymorphism (defined by the *Bsm1* genotype) influenced total body and lumbar spine bone mass indices (Sanwalka et al., 2013). However, the *Fok1* genotype had no positive influence on bone mass accrual in response to calcium supplementation (Sanwalka et al., 2015).

The discrepancy in the results of the various studies are possibly due to the differing levels of calcium intake in the populations studied as well as the age of participants. Thus reports suggest that if calcium intake is not considered, then the association between *VDR* genotype and BMD may not be appropriately interpreted. Furthermore, studies also suggest that the association between *VDR* alleles and bone mineralization may become less evident in older subjects (Kiel et al., 1997).

Apart from *VDR*, various other genes influencing vitamin D metabolism have been studied for their association with calcium homeostasis and bone mineralization. Osteoporosis (i.e., brittle bone disease with reduced bone mass) has also been linked to the vitamin D binding protein (DBP). DBP binds to and transports vitamin D to target tissues and helps in maintaining calcium homeostasis. A recent study focused on two SNPs in the DBP gene (SNPs rs7041 = Glu416Asp and rs4588 = Thr420Lys) in more than 6000 elderly Caucasians. The intent was to assess interactions of *DBP* genotype with *VDR* genotype and dietary calcium intake in relation to fracture risk. Haplotypes of the *DBP* SNPs corresponding to protein variations were identified (referred to as haplotypes 1, 2, and 3). The *DBP* genotype was found to be associated with increased and decreased serum 25-dihydroxyvitamin D_3 and $1,25-(OH)_2D_3$ for haplotype 1 and 2, respectively. Interactions were observed between *DBP* and *VDR* haplotypes in determining fracture risk. Moreover, in a subgroup with dietary calcium intake less than 1.09 g/day, fracture risk of *DBP* haplotype 1 homozygotes was higher. Thus these authors concluded that the *DBP* genotype was associated with fracture risk in combination with other genetic factors and dietary calcium intake levels (Fang et al., 2009). Furthermore, the *CYP24A1* gene regulates the hydroxylation of $1,25(OH)_2D_3$ and targets it for further catabolism. *CYP24A1* mutations can cause idiopathic infantile hypercalcemia, bone metabolism disturbances, and nephrocalcinosis (Cools et al., 2015).

OTHER GENETIC FACTORS AFFECTING CALCIUM HOMEOSTASIS

In addition to vitamin-D–related genes, other genes have also been found to be associated with calcium intake and bone mineralization. Some years ago the low-density lipoprotein receptor-related protein-5 (LRP5) gene was reported to be involved in rare bone disorders, and the role of the Wnt/β-catenin pathway was described in the regulation of bone homeostasis. Reports suggest that variants in these genes (i.e., *LRP5* and *Wnt/β-catenin*) also contribute to variations in bone traits (Lara-Castillo and Johnson, 2015). In a study on 9- to 18-year-old Iranian children, the effect of *LRP5* polymorphisms on bone density and biochemical markers were studied. These authors found that one *LRP5* polymorphism (rs556442) did not have a significant influence on serum calcium; however, lean body mass was greater in the *GG* genotype and total body bone area, spine bone density, and femoral bone mineral content (BMC) were lower in the *AG* heterozygous genotype. Thus it was concluded that this *LRP5* polymorphism was positively associated with bone density in this group of children (Ashouri et al., 2014).

Four common polymorphisms of *LRP5* (rs1784235, rs491347, rs4988321, and rs4988330) were also assessed for their association with BMD and calcium intake in Greek postmenopausal women. Although none of the polymorphisms were associated with the presence of osteoporosis, all polymorphisms were associated with spine BMD. These authors also found an interaction of the rs4988321 SNP with dietary intake of calcium. The carriers of the *A* allele had significantly lower spine BMD in comparison with the *GG* homozygotes in the lowest calcium intake group (<680 mg/day); however, in the highest calcium intake group no differences were found in the genotypes. It was concluded that a higher calcium intake attenuated the negative effect of these polymorphisms on bone density (Stathopoulou et al., 2010).

Associations have also been described between interleukin (IL)-6 promoter polymorphisms and bone mass. In a cross-sectional study in premenarchal Chinese girls, the authors examined the association among IL-6 promoter alleles (*-174G/C* and *-634C/G*), bone mass, and calcium intake. However, these authors did not detect polymorphisms in the *IL-6 -174G/C* locus; all were *GG* homozygotes. The *IL-6 -634C/G* polymorphism was significantly associated with bone mass. Girls with the *CC* genotype had higher levels of bone mass than those with *G* allele carriers. Calcium had a greater impact on total body BMC in *G* allele carriers than in *CC* carriers, and the G-allele–associated lower total body BMC was found only in subjects with low dietary calcium intake. Thus it was concluded that the *IL-6 -634C/G* polymorphism was significantly associated with BMD and that the association was modified by calcium intake in these young Chinese girls (Li et al., 2008).

Associations between dietary factors and *IL-6* polymorphisms with bone mass have also been investigated in the Framingham Offspring Cohort. From the study cohort, 1574 unrelated men and women were genotyped for *IL-6 -572* and *-174* alleles, and interaction analyses with lifestyle factors and BMD measurements at the hip were performed. In models in which the main effects of *IL-6* polymorphisms and association with BMD were studied, no associations were found in either gender. However, interactions between *IL-6 -174* genotypes and years since menopause, estrogen status, dietary calcium, and vitamin D intake were observed in women, and BMD was significantly lower with the *-174 GG* genotype compared with *CC*, and intermediate with *GC*, in women with calcium intake less than 940 mg/day. In estrogen-deficient women with poor calcium intake, there were significant differences in the BMD between genotypes *CC* and *GG*. Thus these authors concluded that *IL-6* genetic variations were prominently associated with hip BMD in late postmenopausal women, especially in those with inadequate calcium intake. However, *IL-6* polymorphisms were not significant determinants of bone mass in women or men in another study. However, these associations were not seen in men (Ferrari et al., 2004).

Thus several studies have examined the association between *IL-6* polymorphisms and BMD. A meta-analysis suggested that there are modest effects of the *-634C/G* and *-174G/C* polymorphisms on bone density. Even so, there were insufficient data on the interactions with diet; hence these authors suggested that large-scale and well-designed studies are required to further investigate gene–gene and gene–environment interactions related to *IL-6* polymorphisms and bone density in various populations (Zhao et al., 2013; Wang et al., 2013).

Genetic disorders of lactose intolerance have also been associated with reduced calcium intake, bone density, and fractures in postmenopausal women. Lactase deficiency, an autosomal-recessive condition due to *A -13910 T/C* dimorphism (*LCT*) near the lactase phlorizin hydrolase gene, is reported to have an effect on calcium supply and osteoporotic fractures in the elderly. Reports suggested that the *LCT (T/C-13910)* polymorphism is associated with milk intolerance and thus reduced milk calcium intake and reduced BMD (Obermayer-Pietsch et al., 2004). In a further report by the same group, the authors evaluated whether lactose also decreases intestinal calcium absorption in subjects with lactase deficiency and whether lactose-reduced diet and lactose-free calcium supplementation could maintain BMD. They concluded that decreased calcium intake as well as lactose-associated impaired calcium absorption may predispose subjects with reduced lactase to osteoporosis and that calcium supplementation may help to maintain BMD (Obermayer-Pietsch et al., 2007).

Given the number of genes that are associated with regulation of blood calcium concentrations and bone mineralization, genome-wide association studies (GWAS) have explored loci influencing calcium concentrations and hence bone health. Studying the relationship between calcium intake and genetic makeup may lead to a better understanding of the scientific basis of the interaction of calcium-related genes and the biological response triggered by calcium intake. Thus in a recent study genetic polymorphisms influencing serum calcium levels in east Asians were investigated. These authors found the strongest relationship between an SNP in *CaSR* and calcium levels. Furthermore, the cystatin A (*CSTA*), diacylglycerol kinase (*DGKD*), and glucokinase regulatory protein (*GCKR*) loci were also found to be significantly associated with calcium levels in Europeans and Indians (Vinayagamoorthy et al., 2015). Little is known about the relationship of *DGKD* with calcium-associated disorders; members of this large family of enzymes bind to and regulate proteins activated by either diacylglycerol or phosphatidic acid (Luo et al., 2004). DGKD has been shown to have calcium-binding activity and it is more active in the presence of calcium (Yamada et al., 1997). In one of the first reports on the association of *DGKD* with bone, the DGKD axis was described to be a critical regulator of bone homeostasis via its actions on osteoclast differentiation and transcription factor c-Fos expression (Zamani et al., 2015). *GCKR* has also been reported to be associated with calcium levels in African-American and European-American children (Chang et al., 2015). In another GWAS study, the authors identified and replicated one known and six new loci influencing serum calcium near genes linked to bone metabolism.

Of these, *DGKD*, *GCKR*, *CaSR*, and *CYP24A1* polymorphisms were associated with BMD in the general population. In their supporting mouse studies, these authors demonstrated a relationship between these loci and gene expression in response to dietary calcium intake (O'Seaghdha et al., 2013).

CONCLUSIONS

Calcium is the most abundant mineral in the human body, and 99% of the body's calcium is found in bones and teeth. Serum calcium concentrations are kept stable, with variations that do not exceed 2–3%, by PTH, vitamin D, and calcitonin. Various genetic mutations/polymorphisms affecting the action of PTH, vitamin D, or their receptors may affect the absorption of calcium and mineralization of bones. The varying associations of genes with calcium and bone metabolism are possibly due to the differing levels of calcium intake in the populations studied as well as the age of participants and the genetic makeup. Studying the relationship between calcium intake and genetic makeup may lead to a better understanding of the scientific basis of the interaction between calcium-related genes and the biological response triggered by calcium intake. GWAS studies have identified six to seven loci, including *DGKD*, *GCKR*, *CaSR*, and *CYP24A1*, which are associated with bone density in the general population. Associations of genes with calcium homeostasis and bone mineralization may be modified by calcium intake; thus gene-related health conditions may be modified to some degree by modifying dietary calcium intake. Therefore knowledge of genes related to calcium intake may help in modifying lifestyle for improving bone health.

REFERENCES

Ak, D.G., Kahraman, H., Dursun, E., Duman, B.S., Erensoy, N., Alagöl, F., Tanakol, R., Yılmazer, S., 2005. Polymorphisms at the ligand binding site of the vitamin D receptor gene and osteomalacia. Dis. Markers 21, 191–197.

Åkerström, G., Hellman, P., Hessman, O., Segersten, U., Westin, G., 2005. Parathyroid glands in calcium regulation and human disease. Ann. NY Acad. Sci. 1040, 53–58.

Ali, A., Christie, P.T., Grigorieva, I.V., Harding, B., Van Esch, H., Ahmed, S.F., Thakker, R.V., 2007. Functional characterization of GATA3 mutations causing the hypoparathyroidism-deafness-renal (HDR) dysplasia syndrome: insight into mechanisms of DNA binding by the GATA3 transcription factor. Hum. Mol. Genet. 16, 265–275.

Ashouri, E., Meimandi, E.M., Saki, F., Dabbaghmanesh, M.H., Omrani, G.R., Bakhshayeshkaram, M., 2014. The impact of LRP5 polymorphism (rs556442) on calcium homeostasis, bone mineral density, and body composition in Iranian children. J. Bone Miner. Metab. 1–7.

Bonjour, J., 2011. Calcium and phosphate: a duet of ions playing for bone health. J. Am. Coll. Nutr. 30, 438–448.

Chang, X., Li, J., Guo, Y., Wei, Z., Mentch, F.D., Hou, C., Zhao, Y., Qiu, H., Kim, C., Sleiman, P.M., Hakonarson, H., 2015. Genome-wide association study of serum minerals levels in children of different ethnic background. PLoS One 10, e0123499.

Christakos, S., Dhawan, P., Liu, Y., Peng, X., Porta, A., 2003. New insights into the mechanisms of vitamin D action. J. Cell. Biochem. 88, 695–705.

Cools, M., Goemaere, S., Baetens, D., Raes, A., Desloovere, A., Kaufman, J.M., De Schepper, J., Jans, I., Vanderschueren, D., Billen, J., De Baere, E., Fiers, T., Bouillon, R., 2015. Calcium and bone homeostasis in heterozygous carriers of CYP24A1 mutations: a cross-sectional study. Bone (15), 00254–00259 pii:S8756-3282.

Demir, K., Kattan, W.E., Zou, M., Durmaz, E., BinEssa, H., Nalbantoğlu, Ö., Al-Rijjal, R.A., Meyer, B., Özkan, B., Shi, Y., 2015. Novel CYP27B1 gene mutations in patients with vitamin D-dependent rickets type 1A. PLoS One 10, e0131376.

Duchatelet, S., Ostergaard, E., Cortes, D., Lemainque, A., Julier, C., 2005. Recessive mutations in PTHR1 cause contrasting skeletal dysplasias in Eiken and Blomstrand syndromes. Hum. Mol. Genet. 14, 1–5.

Fang, Y., van Meurs, J.B., Arp, P., van Leeuwen, J.P., Hofman, A., Pols, H.A., Uitterlinden, A.G., 2009. Vitamin D binding protein genotype and osteoporosis. Calcif. Tissue Int. 85, 85–93.

Favus, M.J., Walling, M.W., Kimberg, D.V., 1974. Effects of dietary calcium restriction and chronic thyroparathyroidectomy on the metabolism of [3H] 25-hydroxyvitamin D3 and the active transport of calcium by rat intestine. J. Clin. Invest. 53, 1139–1148.

Favus, M., 1985. Factors that influence absorption and secretion of calcium in the small intestine and colon. Am. J. Physiol. 248, 147–157.

Felsenfeld, A.J., Levine, B.S., 2015. Calcitonin, the forgotten hormone: does it deserve to be forgotten? Clin. Kidney J. 8, 180–187.

Ferrari, S., Rizzoli, R., Chevalley, T., Eisman, J.A., Bonjour, J.P., Slosman, D., 1995. Vitamin-D-receptor-gene polymorphisms and change in lumbar-spine bone mineral density. Lancet 345, 423–424.

Ferrari, S.L., Karasik, D., Liu, J., Karamohamed, S., Herbert, A.G., Cupples, L.A., Kiel, D.P., 2004. Interactions of Interleukin-6 promoter polymorphisms with dietary and lifestyle factors and their association with bone mass in men and women from the Framingham osteoporosis study. J. Bone Miner. Res. 19, 552–559.

Fleet, J., 2006. Molecular regulation of calcium metabolism. In: Weaver, C.M., Heaney, R.P. (Eds.), Calcium in Human Health. Humana Press, Totowa, NJ, pp. 163–190.

Gropper, S., Smith, J., Groff, J., 2008. Advanced Nutrition and Human Metabolism, fifth ed. Wadsworth, Belmont, California, USA.

Hansen, T.S., Abrahamsen, B., Henriksen, F.L., Hermann, A.P., Jensen, L.B., Hørder, M., Gram, J., 1997. Vitamin D receptor alleles do not predict bone mineral density or bone loss in Danish perimenopausal women. Bone 22, 571–575.

Institute of Medicine, 2011. Dietary Reference Intakes for Calcium and Vitamin D. The National Academies Press, Washington, DC.

Juppner, H., Abou-Samra, A.B., Freeman, M., Kong, X.F., Schipani, E., Richards, J., Kronenberg, H.M., 1991. AG protein-linked receptor for parathyroid hormone and parathyroid hormone-related peptide. Science 254, 1024–1026.

Katsumata, K., Nishizawa, K., Unno, A., Fujita, Y., Tokita, A., 2002. Association of gene polymorphisms and bone density in Japanese girls. J. Bone Miner. Metab. 20, 164–169.

Kiel, D.P., Myers, R.H., Cupples, L.A., Kong, X.F., Zhu, X.H., Ordovas, J., Schaefer, E.J., Felson, D.T., Rush, D., Wilson, P.W., Eisman, J.A., Holick, M.F., 1997. The BsmI vitamin D receptor restriction fragment length polymorphism (bb) influences the effect of calcium intake on bone mineral density. J. Bone Miner. Res. 12, 1049–1057.

Kitsiou-Tzeli, S., Kolialexi, A., Fryssira, H., Galla-Voumvouraki, A., Salavoura, K., Kanariou, M., Th Tsangaris, G., Kanavakis, E., Mavrou, A., 2004. Detection of 22q11.2 deletion among 139 patients with Di George/Velocardiofacial syndrome features. In Vivo 18, 603–608.

Kovacs, C.S., Kronenberg, H.M., 1997. Maternal-fetal calcium and bone metabolism during pregnancy, puerperium, and lactation 1. Endocr. Rev. 18, 832–872.

Lara-Castillo, N., Johnson, M.L., 2015. LRP receptor family member associated bone disease. Rev. Endocr. Metab. Disord. 1–8.

Levine, M.A., 2012. An update on the clinical and molecular characteristics of pseudohypoparathyroidism. Curr. Opin. Endocrinol. Diabetes Obes. 19, 443–451.

Li, X., He, G.P., Zhang, B., Chen, Y.M., Su, Y.X., 2008. Interactions of interleukin-6 gene polymorphisms with calcium intake and physical activity on bone mass in pre-menarche Chinese girls. Osteoporos. Int. 19, 1629–1637.

Lieben, L., Benn, B.S., Ajibade, D., Stockmans, I., Moermans, K., Hediger, M.A., Peng, J.B., Christakos, S., Bouillon, R., Carmeliet, G., 2010. Trpv6 mediates intestinal calcium absorption during calcium restriction and contributes to bone homeostasis. Bone 47, 301–308.

Luo, B., Regier, D.S., Prescott, S.M., Topham, M.K., 2004. Diacylglycerol kinases. Cell Signal 16, 983–989.

Malloy, P.J., Feldman, D., 2010. Genetic disorders and defects in vitamin D action. Endocrinol. Metab. Clin. North Am. 39, 333–346.

Malloy, P.J., Zhou, Y., Wang, J., Hiort, O., Feldman, D., 2011. Hereditary vitamin D–resistant rickets (HVDRR) owing to a heterozygous mutation in the vitamin D receptor. J. Bone Miner. Res. 26, 2710–2718.

Obermayer-Pietsch, B.M., Bonelli, C.M., Walter, D.E., Kuhn, R.J., Fahrleitner-Pammer, A., Berghold, A., Goessler, W., Stepan, V., Dobnig, H., Leb, G., Renner, W., 2004. Genetic predisposition for adult lactose intolerance and relation to diet, bone density, and bone fractures. J. Bone Miner. Res. 19, 42–47.

Obermayer-Pietsch, B.M., Gugatschka, M., Reitter, S., Plank, W., Strele, A., Walter, D., Bonelli, C., Goessler, W., Dobnig, H., Högenauer, C., Renner, W., Fahrleitner-Pammer, A., 2007. Adult-type hypolactasia and calcium availability: decreased calcium intake or impaired calcium absorption? Osteoporos. Int. 18, 445–451.

O'Seaghdha, C.M., Wu, H., Yang, Q., Kapur, K., Guessous, I., Zuber, A.M., Köttgen, A., Stoudmann, C., Teumer, A., Kutalik, Z., Mangino, M., Dehghan, A., Zhang, W., Eiriksdottir, G., Li, G., Tanaka, T., Portas, L., Lopez, L.M., Hayward, C., Lohman, K., Matsuda, K., Padmanabhan, S., Firsov, D., Sorice, R., Ulivi, S., Brockhaus, A.C., Kleber, M.E., Mahajan, A., Ernst, F.D., Gudnason, V., Launer, L.J., Mace, A., Boerwinckle, E., Arking, D.E., Tanikawa, C., Nakamura, Y., Brown, M.J., Gaspoz, J.M., Theler, J.M., Siscovick, D.S., Psaty, B.M., Bergmann, S., Vollenweider, P., Vitart, V., Wright, A.F., Zemunik, T., Boban, M., Kolcic, I., Navarro, P., Brown, E.M., Estrada, K., Ding, J., Harris, T.B., Bandinelli, S., Hernandez, D., Singleton, A.B., Girotto, G., Ruggiero, D., d'Adamo, A.P., Robino, A., Meitinger, T., Meisinger, C., Davies, G., Starr, J.M., Chambers, J.C., Boehm, B.O., Winkelmann, B.R., Huang, J., Murgia, F., Wild, S.H., Campbell, H., Morris, A.P., Franco, O.H., Hofman, A., Uitterlinden, A.G., Rivadeneira, F., Völker, U., Hannemann, A., Biffar, R., Hoffmann, W., Shin, S.Y., Lescuyer, P., Henry, H., Schurmann, C., SUNLIGHT Consortium, GEFOS Consortium, Munroe, P.B., Gasparini, P., Pirastu, N., Ciullo, M., Gieger, C., März, W., Lind, L., Spector, T.D., Smith, A.V., Rudan, I., Wilson, J.F., Polasek, O., Deary, I.J., Pirastu, M., Ferrucci, L., Liu, Y., Kestenbaum, B., Kooner, J.S., Witteman, J.C., Nauck, M., Kao, W.H., Wallaschofski, H., Bonny, O., Fox, C.S., Bochud, M., 2013. Meta-analysis of genome-wide association studies identifies six new Loci for serum calcium concentrations. PLoS Genet. 9, e1003796.

Peacock, M., 2010. Calcium metabolism in health and disease. Clin. J. Am. Soc. Nephrol. 5, 23–30.

Ray, D., Goswami, R., Gupta, N., Tomar, N., Singh, N., Sreenivas, V., 2009. Predisposition to vitamin D deficiency osteomalacia and rickets in females is linked to their 25 (OH) D and calcium intake rather than vitamin D receptor gene polymorphism. Clin. Endocrinol. 71, 334–340.

Rejnmark, L., Lauridsen, A., Vestergaard, P., Heickendorff, L., Andreasen, F., Mosekilde, L., 2002. Diurnal rhythm of plasma 1, 25-dihydroxyvitamin D and vitamin D-binding protein in postmenopausal women: relationship to plasma parathyroid hormone and calcium and phosphate metabolism. Eur. J. Endocrinol. 146, 635–642.

Rosol, T., Chew, D., Nagode, L., Capen, C., 1995. Pathophysiology of calcium metabolism. Vet. Clin. Pathol. 24, 49–63.

Sanwalka, N., Khadilkar, A., Chiplonkar, S., Khatod, K., Phadke, N., Khadilkar, V., 2013. Vitamin D receptor gene polymorphisms and bone mass indices in post-menarchal Indian adolescent girls. J. Bone Miner. Metab. 31, 108–115.

Sanwalka, N., Khadilkar, A., Chiplonkar, S., Khatod, K., Phadke, N., Khadilkar, V., 2015. Influence of vitamin D receptor gene Fok1 polymorphism on bone mass accrual post calcium and vitamin D supplementation. Indian J. Pediatr. 1–6.

Seibel, M., 2005. Biochemical markers of bone turnover part I: biochemistry and variability. Clin. Biochem. Rev. 26, 97–122.

Song, Y., Kato, S., Fleet, J.C., 2003a. Vitamin D receptor (VDR) knockout mice reveal VDR-independent regulation of intestinal calcium absorption and ECaC2 and calbindin-D_{9k} mRNA. J. Nutr. 133, 374–380.

Song, Y., Peng, X., Porta, A., Takanaga, H., Peng, J.B., Hediger, M.A., Fleet, J.C., Christakos, S., 2003b. Calcium transporter 1 and epithelial calcium channel messenger ribonucleic acid are differentially regulated by 1,25-dihydroxyvitamin D_3 in the intestine and kidney of mice. Endocrinology 144, 3885–3894.

St-Arnaud, R., 2008. The direct role of vitamin D on bone homeostasis. Arch. Biochem. Biophys. 473, 225–230.

Stathopoulou, M.G., Dedoussis, G.V., Trovas, G., Katsalira, A., Hammond, N., Deloukas, P., Lyritis, G.P., 2010. Low-density lipoprotein receptor-related protein 5 polymorphisms are associated with bone mineral density in Greek postmenopausal women: an interaction with calcium intake. J. Am. Diet. Assoc. 110, 1078–1083.

Van Cromphaut, S.J., Dewerchin, M., Hoenderop, J.G., Stockmans, I., Van Herck, E., Kato, S., Carmeliet, G., 2001. Duodenal calcium absorption in vitamin D receptor–knockout mice: functional and molecular aspects. Proc. Natl. Acad. Sci. 98, 13324–13329.

Vinayagamoorthy, N., Yim, S.H., Jung, S.H., Park, S.W., Kim, Y.J., Kim, B.J., Chung, Y.J., 2015. Association of common variants in the calcium-sensing receptor gene with serum calcium levels in East Asians. J. Hum. Genet. 60 (8), 407–412.

Wang, Z., Yang, Y., He, M., Wang, R., Ma, J., Zhang, Y., Zhao, L., Yu, K., 2013. Association between interleukin-6 gene polymorphisms and bone mineral density: a meta-analysis. Genet. Test. Mol. Biomarkers 17, 898–909.

Wasserman, R., 2005. Vitamin D and the intestinal absorption of calcium: a view and overview. In: Feldman, D., Pike, J.W., Glorieux, F. (Eds.), Vitamin D. Elsevier Academic Press, San Diego, CA, pp. 411–428.

Yamada, K., Sakane, F., Matsushima, N., Kanoh, H., 1997. EF-hand motifs of α, β and γ isoforms of diacylglycerol kinase bind calcium with different affinities and conformational changes. Biochem. J. 321, 59–64.

Zajíčková, K., Ofková, I., Bahbouh, R., Křepelová, A., 2002. Vitamin D receptor gene polymorphisms, bone mineral density and bone turnover: FokI genotype is related to postmenopausal bone mass. Physiol. Res. 51, 501–509.

Zamani, A., Decker, C., Cremasco, V., Hughes, L., Novack, D.V., Faccio, R., 2015. Diacylglycerol kinase ζ (DGKζ) is a critical regulator of bone homeostasis via modulation of c-Fos levels in osteoclasts. J. Bone Miner. Res. 30 (10), 1852–1863.

Zhao, Y., Shen, L., Ji, H.F., January 2013. Osteoporosis risk and bone mineral density levels in patients with Parkinson's disease: a meta-analysis. Bone 52 (1), 498–505. http://dx.doi.org/10.1016/j.bone.2012.09.013.

Zhou, Y., Wang, J., Malloy, P.J., Dolezel, Z., Feldman, D., 2009. Compound heterozygous mutations in the vitamin D receptor in a patient with hereditary 1,25-dihydroxyvitamin D-resistant rickets with alopecia. J. Bone Miner. Res. 24, 643–651.

Part II

Copper

Chapter 7

Copper: Basic Physiological and Nutritional Aspects

James F. Collins
University of Florida, Gainesville, FL, United States

COPPER: HISTORICAL FACTS

Copper has been used therapeutically in humans for over one thousand years, with various uses peaking in the 19th century (Mason, 1979). Since that time, the use of copper for medicinal purposes has become less frequent as rigorous proof of its effectiveness was not established.

That copper is a constituent of plants and marine invertebrates was first recognized almost two centuries ago; however, it was not until the early 1920s that it was firmly established that copper was also present in human tissues when it was shown that copper was present in the brain (Bodansky, 1921). It was also noted around this time that copper and iron salts were required to cure experimental anemia in rats, which was the first identification of a specific physiological role for copper (Cohn et al., 1927). Subsequently, the requirement for copper and iron in preventing anemia was confirmed in several mammalian species.

Irrefutable evidence of the involvement of copper in human disease was first documented around 1900, when Wilson's disease (WD) was described (Wilson, 1912). The fact that WD was due to an inborn error of metabolism was not, however, recognized until several decades later (Mason, 1979). A link between copper depletion and anemia in humans was hypothesized in the 1930s, but definitive proof of such a relationship would not be established until much later. Moreover, altered serum or tissue copper concentrations were noted in numerous human pathologies in the 1950s, but it was not clear whether perturbed copper homeostasis was primarily or secondarily related to the underlying pathology. Human copper deficiency was initially documented in 1962 when Menkes disease (MD) was first described (Menkes et al., 1962), but it was more than 10 years thereafter before the underlying genetic defect was discovered (Danks et al., 1972).

As of today, it is firmly established that copper is an essential dietary nutrient for humans and other mammalian species. Consistent with its definition as a trace mineral, human tissues and body fluids contain copper at concentrations in the microgram per gram (i.e., parts per million) to the nanogram per gram (i.e., parts per billion) range. It is also now clear that inadequate or excess copper intake leads to severe pathological consequences. Not surprisingly, several well-established symptoms of copper deficiency relate to alterations in the activity of copper-dependent enzymes. Moreover, as is typical with heavy metal intoxication, excess copper may interfere with several enzyme systems. Thus, mammals have developed regulated systems to control all aspects of copper metabolism.

COPPER BIOCHEMISTRY

Copper is a member of the first transition series of elements, with an atomic mass of approximately 63.5 Da. Copper has two predominant oxidation states in biological systems, Cu^{2+} (cupric) and Cu^{+} (cuprous); this redox property is critical for the function of numerous cuproenzymes. Cuprous copper (Cu^{+}) is highly insoluble in aqueous solutions and is thus strongly complexed with various biomolecules (Dyer and Leddicotte, 1961). Cu^{2+} occurs most frequently in live animals. Most copper in biological systems is bound to proteins via specific interactions with amino acid side chains. These interactions reveal significant stereorestraints that limit the number of potential copper-binding sites on proteins. Observable magnetic parameters and spectrophotometric properties allow copper-binding sites to be classified into three distinct types, types I–III (Dyer and Leddicotte, 1961). Cu^{2+} binds to proteins in these three distinct forms, giving rise to unique physiochemical properties. Type I cuproenzymes have a deep blue color characteristic of many copper-containing oxidases; type II cuproenzymes, contained within several multicopper oxidases, lack the blue color but are detectable by electron paramagnetic resonance (EPR); and type III proteins, which are neither blue nor detectable by EPR, contain a binuclear copper center that exists as a Cu^{+}–Cu^{+} couple. A single protein may have more than one type of copper-binding site (Turnlund, 2006).

BIOCHEMICAL AND PHYSIOLOGIC FUNCTIONS OF COPPER

Copper serves a predominant role in mammalian biology as an enzymatic cofactor for a host of cuproenzymes, but no structural roles for copper have been described. These enzymes, most of which are oxidases, are collectively involved in single electron transfer reactions between a substrate and molecular oxygen using either oxidized (cupric, Cu^{2+}) or reduced copper (cuprous, Cu^+) (Harris, 1997; Linder, 2003) atoms. Furthermore, there are well-recognized nonenzymatic functions of copper in diverse physiological processes such as angiogenesis, gas transport, neurohormone homeostasis, and regulation of gene expression (Dyer and Leddicotte, 1961). Several copper-dependent enzymes and copper-binding proteins have been identified in mammals, as described in detail in the following sections and listed in Table 7.1.

CATALYTIC FUNCTIONS OF COPPER (COPPER-DEPENDENT PROTEINS)

Ferroxidases: Ceruloplasmin and Hephaestin

Ceruloplasmin (CP) is a liver-derived, circulating multicopper ferroxidase (FOX). CP mediates iron release from many tissues by oxidizing ferrous iron (Fe^{2+}) to allow the binding of ferric (Fe^{3+}) iron to transferrin in the interstitial fluids. CP has a molecular weight of approximately 130 kD, and it contains six to seven copper atoms/molecule. A recently discovered cell membrane-associated, glycophosphatidylinositol (GPI)-anchored isoform of CP (GPI-CP) is expressed in hepatocytes, brain, and macrophages, and possibly other tissues. GPI-CP may be required for iron efflux from macrophages via interaction with the predominant iron export protein, ferroportin 1 (De Domenico et al., 2007). Most copper in the blood is associated with CP (65 to >95% depending upon the species) CP is not, however, required for copper transport in the blood. Accordingly, patients with aceruloplasminemia, an autosomal-recessive disorder in which no CP is produced (Xu et al., 2004), have no noted perturbations in copper homeostasis, but rather exhibit pathological iron accumulation in pancreas, retina, brain, and other tissues.

Hephaestin (HEPH) is a membrane-anchored intestinal FOX which may be required for optimal intestinal iron absorption in mice. HEPH may be particularly important under physiologic circumstances in which intestinal iron absorption is enhanced (e.g., the rapid postnatal growth period and pregnancy). HEPH shares approximately 50% homology with CP. It was discovered as the mutant gene in the sex-linked anemia (*sla*) mouse (Vulpe et al., 1999). HEPH is also expressed in the antrum of the stomach, the enteric nervous system, and pancreatic β-cells (Hudson et al., 2010). HEPH expression responds to body copper levels via modulation of its activity with concomitant changes in intestinal iron absorption (Chen et al., 2006). Moreover, another multicopper FOX, called zyklopen, was recently discovered in the mammalian placenta (Chen et al., 2010).

Superoxide Dismutases

The three known superoxide dismutase (SOD) proteins function to scavenge superoxide free radicals to protect against oxidative damage. Two SOD proteins, intracellular Cn/Zn-superoxide dismutase (SOD1) and extracellular SOD (SOD3), require copper (and zinc) for function; copper is involved in catalysis, while zinc plays a structural role. SOD2 requires manganese and is located in the mitochondrial matrix. SOD3, the predominant extracellular dismutase, is located in lymph, synovial fluid, and plasma (Fattman et al., 2003). SOD3 dysfunction may relate to the development of chronic obstructive pulmonary disease (COPD) in humans (Oberley-Deegan et al., 2009). Moreover, SOD3 inactivation in mice suggests an interaction with nitric oxide, which modulates vascular tone (Jung et al., 2003).

Cytochrome C Oxidase

Cytochrome C oxidase (CCO) is composed of 13 protein subunits. It contains three copper ions, which are necessary for its function. It also contains zinc, magnesium, and two heme groups. CCO functions as the terminal constituent of the electron transport chain in the inner mitochondrial membrane, where it reduces molecular oxygen to form water, thus allowing ATP production. Copper chaperones are required for assembly of the CCO complex, and mutations in the genes encoding these chaperones lead to various pathologies (Robinson and Winge, 2010). Lack of CCO function is likely lethal.

Dopamine β-Monooxygenase

This copper-dependent enzyme allows dopamine to be converted to norepinephrine. Dopamine β-monooxygenase (DBM) is expressed in noradrenergic and adrenergic brain neurons, the medulla of the adrenal gland, and in sympathetic neurons of the peripheral nervous system (Prohaska, 2006). It requires copper in each of its four subunits. DBM knockout in mice causes embryonic lethality, exemplifying its essentiality in the nervous system (Thomas et al., 1995).

TABLE 7.1 Copper-Dependent and Copper-Binding Proteins in Mammals

Copper-Dependent Enzymes	Physiologic Function	Functional Alterations in Copper Deficiency
Amine oxidases	Deamination of mono- and diamines	Unclear
Ceruloplasmin	Iron release from storage sites; ferroxidase	Decreased activity; Iron accumulation in liver, brain, pancreas
Cu/Zn Superoxide dismutase-1 (intracellular)	Antioxidant defense	Increased susceptibility to oxygen free radicals
Cytochrome C oxidase (CCO)	Electron transport chain; ATP production	Impaired oxidative phosphorylation; hypomyelination
Dopamine β-monooxygenase	Catecholamine metabolism	Neuropathologies; hypomyelination of nerve fibers
Hephaestin	Intestinal iron transport; ferroxidase	Decreased activity; impaired iron efflux from duodenal enterocytes
Lysyl oxidase	Production of elastin and collagen	Cardiovascular system abnormalities; bone and cartilage instability
Monoamine oxidases A and B	Degradation of amine neurotransmitters	Unclear
Peptidylglycine α-amidating monooxygenase (PAM)	Activation of biologically-active peptides	Altered central nervous system hormone production; cardiovascular system dysfunction
Superoxide dismutase-3 (extracellular)	Antioxidant defense	Increased susceptibility to oxygen free radicals
Tyrosinase	Pigmentation; melanin biosynthesis	Hypopigmentation; abnormal keratinization of hair
Zyklopten	Placental iron efflux	Unclear

Copper-Binding Proteins	Copper-Dependent Function
Albumin	Transport of copper in portal circulation from intestine to liver
Amyloid precursor protein (APP)	Copper transport to brain
Antioxidant protein 1 (ATOX1)	Copper chaperone for ATP7A and ATP7B; copper-dependent transcription factor
Blood Clotting Factors V and VIII	Possible functional role of copper unknown
Copper chaperone for superoxide dismutase (CCS)	Copper Chaperone for SOD1 in cytosol
Copper metabolism domain containing 1 (COMMD1)	Biliary copper excretion; interaction with ATP7b; mutated gene in Bedlington terrier copper toxicosis
Copper transporter 1 (Ctr1)	Copper import protein; mediates copper uptake in intestine, liver, heart and other tissues
Copper transporter 2 (Ctr2)	Plasma membrane uptake and vesicular copper transport
Copper-transporting ATPase 1 (ATP7A)	Transports copper into trans-Golgi network and out of cells; Menkes disease gene
Copper-transporting ATPase 2 (ATP7B)	Transports copper into trans-Golgi network; biliary copper excretion Wilson's disease gene
COX11 cytochrome C oxidase copper chaperone (COX11)	Mitochondrial chaperone for CCO
COX17 cytochrome C oxidase copper chaperone (COX17)	Copper delivery from cytosol to mitochondrial intermembrane space
Metallothionein I and II	Intracellular copper storage proteins (also binds zinc and cadmium)
Prion Protein	Unknown; several possible functions proposed
SCO1 cytochrome C oxidase assembly protein (SCO1)	Mitochondrial copper chaperone
SCO2 cytochrome C oxidase assembly protein (SCO2)	Mitochondrial copper chaperone
X-linked inhibitor of apoptosis (XIAP)	Ubiquitination of COMMD1 and CCS
α2-Macroglobulin	Transport of copper in portal circulation from intestine to liver

Tyrosinase

Tyrosinase (TYR) functions in melanin biosynthesis, thus being required for pigmentation. Loss of activity causes albinism. TYR catalyzes the conversion of tyrosine to dopamine and subsequent oxidation of dopamine to dopaquinone in the melanin synthesis pathway. Copper deprivation of domestic and laboratory animals leads to achromotrichia (i.e., absence of pigment in hair), exemplifying the copper dependency of melanin synthesis (Prohaska, 2006).

Amine Oxidases

Amine oxidases (AOs), some of which are copper-dependent proteins, function in the oxidative deamination of biogenic primary amines. AOs exist in plasma and tissues as dimers and require an organic cofactor (2,4,5-trihydroxyphenylalanine quinone) (Brazeau et al., 2004). In the plasma, AOs catabolize physiologically-active amines, including histidine, tyramine, and polyamines. AOs may also be involved in intracellular signaling via production of H_2O_2 (McDonald et al., 2007). An example of a copper-dependent AO is vascular adhesion protein-1 (VAP-1), which may be involved in leukocyte trafficking (Dunkel et al., 2008). VAP-1 mediates leukocyte extravasation and ameliorates inflammatory responses, as demonstrated by the phenotype of VAP-1 KO mice (Stolen et al., 2005). AO activity is enhanced in liver fibrosis, congestive heart failure, and hyperthyroidism, all conditions in which connective tissue activation and deposition occurs (Linder, 2003).

Monoamine Oxidase

Monoamine oxidases (MAOs) are copper-containing proteins involved in the catabolism of catecholamines. These enzymes react with serotonin, norephinephrine, tyramine, and dopamine. MAOA and MAOB are two identified isoforms with unique tissue localization. MAO dysfunction is associated with attention deficit disorder, depression, irregular sexual maturation, and substance abuse. Moreover, MAO activity is attenuated by some antidepressant drugs (Turnlund, 2006).

Diamine Oxidases

Multiple diamine oxidases (DOAs) exist and are expressed in many cells and tissues. One DAO catalyzes the catabolism of histamine, which attenuates allergic reactions and inhibits gastric acid production. DAOs also catabolize polyamines, thus limiting excessive cell growth and having potential relevance to cancer and apoptosis (Tininello et al., 2006). DAO activity is highest in the small bowel, the kidney, and in the human placenta, where it inactivates biologically-active amines of fetal origin (Turnlund, 2006).

Lysyl Oxidase

Lysyl oxidase (LOX) is another example of a copper-dependent AO; it functions to crosslink and stabilize collagen and elastin fibers, thus being involved in connective tissue formation. LOX uses lysyl tyrosyl quinone as a cofactor, which is required for its catalytic activity in the extracellular matrix (Kagan and Li, 2003). LOX has been linked to various pathological processes, including fibrosis, neurodegenerative and cardiovascular diseases, and tumor progression and metastasis, and it is, in fact, considered a potential therapeutic target in these diseases (Rodriguez et al., 2008). At least four LOX-encoding genes have now been identified. All proteins encoded by these genes have similar catalytic domains and putative copper and cofactor binding sites. LOXs may also be involved in the transition to malignant phenotypes, chemotactic responses, and cell proliferation (Lucero and Kagan, 2006).

Peptidylglycine α-Amidating Monooxygenase

Peptidylglycine α-amidating monooxygenase (PAM) is a copper-dependent enzyme that activates several bioactive peptides including neuropeptide Y, substance P, vasopressin, cholecystokinin, gastrin and others (Prohaska, 2006). PAM catalyzes two sequential steps in the C-terminal α-amidation of peptides, which are mediated by two active sites with distinct functional properties: peptidylglycine α-hydroxylating monooxygenase and peptidyl-α-hydroxyglycine α-amidating lyase. These enzymatic events modify neuroendocrine peptides and convert them into active α-amidated peptides. Binding of copper to PAM affects the 3-D structure of the protein and allows endocytic trafficking. Moreover, studies in laboratory rodents showed that decreased PAM expression and mild copper deficiency similarly alter the function of several neuropeptides, thus leading to perturbations in thermal regulation, seizure sensitivity, and anxiety-like behavior (Bousquet-Moore et al.,

2010). PAM is an essential, nonredundant enzyme in mice because PAM knockout leads to embryonic lethality (Czyzyk et al., 2003). Interestingly, heterozygosity for PAM ablation increases incidence of seizures and impairs thermal regulation (Bousquet-Moore et al., 2010).

Other Copper-Dependent Proteins

The activity of several other proteins responds to changes in dietary copper intake, although this does not necessarily mean that they are cuproenzymes (Prohaska, 1988). Other possible copper-dependent enzymes include S-adenosylhomocysteine hydrolase, prion protein, and blood clotting factors V and VIII (Prohaska, 2006).

Copper-Binding Proteins

Several copper-binding proteins have been recently identified (Table 7.1). Many are copper chaperones, which is logical given that unbound copper is highly reactive. Other copper-binding proteins transiently interact with copper, and may, for example, mediate the flux of copper across biological membranes. Moreover, various proteins bind copper in the blood (e.g., albumin, α_2-macroglobulin) or serve as intracellular copper storage proteins [e.g., metallothionein (MT)]. New functions of these proteins are frequently described in the scientific literature. For example, copper metabolism MURR1 domain-containing protein-1 (COMMD1) and copper chaperone for SOD1 (CCS) alter the activity of hypoxia-inducible factor-1α, a hypoxia-responsive transcription factor (Feng et al., 2009; Van De Sluis et al., 2007). Furthermore, COMMD1 can regulate SOD1 activity (Vonk et al., 2010), and X-linked inhibitor of apoptosis may regulate CCS and COMMD1 expression levels (Brady et al., 2010).

PHYSIOLOGICAL FUNCTIONS OF COPPER

Connective Tissue Formation

LOX, a copper-dependent enzyme, mediates the formation of connective and bone tissue, specifically maintaining the structure of connective tissue in the cardiovascular system. Copper deprivation logically then causes connective tissue dysfunction, osteoporosis, and skeletal defects. Copper-deficient neonates develop bone abnormalities reminiscent of scurvy (Uauy et al., 1998). This is also noted in copper-deprived, weanling rats in which LOX activity is diminished (Werman et al., 1995). Moreover, copper supplementation may prevent bone loss in adult humans. For example, a placebo-controlled study noted that administration of copper glycinate (2 mg/day) to young adult women improved collagen crosslinking (Disilvestro et al., 2010). Conversely, copper supplementation (3–6 mg Cu/day) of healthy males and females for 6 weeks did not affect markers of bone health (Baker et al., 1999). Nonetheless, copper supplementation may be effective in reducing bone loss, particularly in the elderly with marginal copper status.

Iron Metabolism

Copper–iron interactions have been recognized for more than a century (Collins et al., 2010). One noted link between these two essential trace minerals is the multicopper FOXs, CP, and HEPH. Dietary iron and copper status affects expression/activity of both proteins. Because CP requires copper for its catalytic activity, activity is extremely low during copper deprivation (Broderius et al., 2010). Low CP results in pathological iron overload given that the predominant function of CP is in iron efflux from sites of iron storage (Thackeray et al., 2010). HEPH activity also decreases during copper deprivation (Chen et al., 2006, 2009), possibly impairing intestinal iron absorption given that HEPH functions in iron export from enterocytes. Moreover, a placental multicopper FOX was discovered (zyklopten; Chen et al., 2010), but how copper affects its activity/expression is currently unknown. Furthermore, copper deficiency, similar to iron deficiency, causes microcytic, hypochromic anemia. Although the reason for this has not been experimentally determined, it could be that iron utilization for mitochondrial heme synthesis requires adequate copper levels.

Central Nervous System

Copper is required for brain development, and it further supports the functions of the central nervous system (CNS). Thus, copper deprivation of rodents during gestational and perinatal development causes behavioral abnormalities, which may persist into adulthood even if the early copper deficiency is corrected (Prohaska, 2006). This may be explained by the fact that copper deficiency alters the activity CCO, DBM, SOD1, and PAM in the CNS (Prohaska, 2006). In humans, a similar phenomenon is observed in patients with the genetic copper deficiency disorder MD; infants suffering from this disease display a neuropathological phenotype (Kaler et al., 2008). Furthermore, copper-deficiency–related tremors, ataxia, hypomyelination or demyelination of nerve fibers, and decreases in neurotransmitter levels are likely the result of impaired sphingolipid production (as mediated by CCO) and decreased activity of DBM and MAO. Adequate dietary copper is thus required for normal development and function of the CNS.

Cardiac Function and Cholesterol Metabolism

Several pathological anomalies associated with neonatal copper deficiency in rodents are noted in the cardiovascular system. For example, copper-deficient chicks and pigs succumb to aortic rupture due to defects in vascular connective tissue function (Harris, 1997). In mice, it was also shown that dietary copper deprivation causes cardiac hypertrophy, an abnormality that was quickly prevented by copper supplementation (Zhou et al., 2008). Furthermore, when the Ctr1 copper transporter was specifically ablated in the cardiac tissue of mice, severe cardiomyopathy was observed (Kim et al., 2010), supporting a role for copper in cardiovascular health.

Some human clinical trials have shown no cardiovascular effects associated with copper deprivation whereas others have demonstrated that cardiac arrhythmias develop (Prohaska, 2006). An investigation of elderly subjects with heart failure demonstrated that micronutrient supplementation, including copper, improved cardiovascular function (Witte et al., 2005). It was suggested that copper was the antioxidant nutrient in this treatment that led to the positive physiologic changes (Klevay, 2006). Interestingly, numerous anatomical, chemical, and physiological similarities between copper-deficient animals and humans with heart disease have been identified (Klevay, 2002). Moreover, an investigation in healthy young women revealed that copper supplementation (3–6 mg/day) improved at least one cardiovascular disease (CVD) risk factor (Bugel et al., 2005). Additional observational human studies have suggested a relationship between high serum copper and decreased incidence of CVD (Easter et al., 2010). Furthermore, copper deficiency also alters blood lipid profiles and blood pressure and causes anemia. Perturbed lipid metabolism includes hypercholesterolemia and hypertriglyceridemia, with noted increases in low-density lipoprotein and decreases in high-density lipoprotein cholesterol levels as well as glucose intolerance (which was observed in copper-deficient rats). Therefore, because copper deficiency alters lipid metabolism, thus increasing risk for CVD, copper probably plays a vital role in atherogenesis (Aliabadi, 2008).

Immune Function

Copper may be important for the function of the immune system. Copper deficiency is frequently associated with an increased risk of infection, and perturbations in copper homeostasis alter immune system function in rodents (Prohaska and Failla, 1993). Thus, cellular and humoral immune system factors may be altered by copper deficiency. For example, the composition of circulating and bone and lymphoid immune cells is altered, and functions of macrophages and lymphocytes are suppressed. Moreover, copper may also be required for bacterial killing by macrophages (White et al., 2009). Copper deficiency has also been associated with neutropenia in humans. Macrophage and lymphocyte function are adversely affected under these conditions. A recent investigation in adult males demonstrated that in vitro stimulation of T lymphocytes was diminished after consumption of a low-copper diet for 6 weeks (Kelley et al., 1995). Moreover, in vitro studies in human T cell and monocyte cell lines showed that secretion of proinflammatory cytokines was suppressed after exposure to a copper chelator (Hopkins and Failla, 1999). These observations suggest that copper influences the ability of immune cells to respond to infectious stimuli.

Other Physiologic Functions of Copper

Copper may be important for additional physiological functions, such as thermal regulation and glucose metabolism. Copper may also influence blood clotting factors V and VIII, but a clear connection with copper deficiency has not been definitively established. Furthermore, copper plays a key role in the antioxidant SOD enzymes, but free copper promotes formation of prooxidants (via Fenton chemistry). SOD enzyme activity is impaired when copper concentrations are low, thus possibly impairing antioxidant defense (Broderius and Prohaska, 2009).

COPPER BIOAVAILABILITY AND NUTRIENT INTERACTIONS

Several dietary factors may have adverse effects on the bioavailability of ingested copper, including carbohydrates, iron, zinc, certain amino acids and proteins, molybdenum, and vitamin C (Lonnerdal, 1998). High intake of zinc from the excessive use of zinc-containing denture creams induced symptoms of systemic copper deficiency in several patients (Nations et al., 2008). Moreover, moderate zinc intake levels may alter copper homeostasis in postmenopausal women (Milne et al., 2001). The influence of dietary constituents on copper absorption may be more significant during early postnatal life because function of the digestive system and regulation of biliary copper excretion are immature.

Copper and iron interact in numerous documented ways (Collins et al., 2010). Iron–copper interactions in the mammalian intestine include the regulation of HEPH by dietary copper levels and the regulation of copper-transporting ATPase 1 (ATP7A) expression by iron levels (Collins et al., 2005; Ravia et al., 2005). Liver copper concentrations also vary inversely according to iron status. Moreover, CP expression correlates with serum and liver copper levels, with increased CP activity observed during iron deprivation (Ranganathan et al., 2011). Furthermore, the FOX activity of GPI-CP is necessary for iron release from brain and reticuloendothelial macrophages (which recycle iron from senescent red blood cells). Interestingly, bone marrow iron utilization requires copper because copper deficiency impairs hemoglobin production despite normal serum iron levels (Collins et al., 2010).

High dietary zinc ingestion impairs intestinal copper absorption. This may be explained by induction of MT, a copper-binding protein, in intestinal epithelial cells. Moreover, copper depletion has been observed in humans consuming 50 mg of zinc daily for extended periods. This is the basis for the established upper tolerable intake levels (ULs) for zinc of 40 mg/day for adults (Institute of Medicine, 2002).

Studies in ruminants have shown interactions between copper and molybdenum (Davis and Mertz, 1987). High Mo intake caused secondary copper deficiency, which was corrected by copper supplementation. Similar observations have been reported in rats, but whether such an interaction occurs in humans is unknown.

Vitamin C supplementation may induce copper deficiency in experimental animals, but whether this also occurs in humans is unclear. Plasma ascorbate levels correlate with serum antioxidant activity. In premature infants, plasma AA levels were negatively associated with serum CP and antioxidant activity (Powers et al., 1995). Additional studies in humans also suggested that AA supplementation perturbs serum FOX activity (presumably mediated by CP).

Copper depletion in rats is influenced by the type of dietary carbohydrates. However, studies in pigs showed opposite results (O'Dell, 1990). Furthermore, erythrocyte SOD levels were diminished in humans consuming a high-fructose diet as compared with a high-corn–starch diet, but unexpectedly, copper retention increased (Reiser et al., 1985).

Copper: Food Sources

The average diet of an adult in the United States provides slightly more copper than the recommended dietary allowance (RDA) of 0.9 mg/day. The best dietary sources of copper are whole-grain products, seeds, nuts, organ meats, shellfish, wheat bran cereal, and chocolate-containing foods. Vegan diets supply adequate copper, but absorption seems to be lower from plant foods (Hunt et al., 1998). Additional dietary sources of copper include vitamin and mineral supplements, which often contain copper as cupric oxide, which has lower bioavailability (Baker, 1999).

COPPER METABOLISM

Genetic Regulation

Genetic variants in copper-homeostasis–related genes likely alter copper requirements in individual humans. A single copper-responsive transcription factor has been identified in mammals. The copper chaperone Atox1 has been shown to translocate to the nucleus and control expression of cell cycle control-related genes (Itoh et al., 2008). Moreover, transcripts encoding many proteins involved in copper homeostasis do not respond to changes in dietary copper intake, demonstrating a lack of transcriptional control (Prohaska and Gybina, 2004). Regulation is predominantly at the level of protein trafficking (Van Den Berghe and Klomp, 2010). However, Atp7a gene transcription was induced during iron deprivation in intestinal epithelial cells (Collins et al., 2009).

Overall Body Copper Homeostasis

The average intake of copper is approximately 1.3 mg/day (Fig. 7.1). Of that, approximately 0.8 mg is absorbed daily, which is delivered first to the liver. Excretion (~0.4 mg/day) occurs predominantly into the bile via the copper exporter ATP7B, with total losses in the feces of approximately 1 mg/day. Copper is incorporated into CP in the liver and secreted into the blood along with atomic copper, which binds to serum proteins. Total body copper content is approximately 100 mg, with the distribution being as follows: bone and muscle (~63%), liver (~9%), brain (~8%),

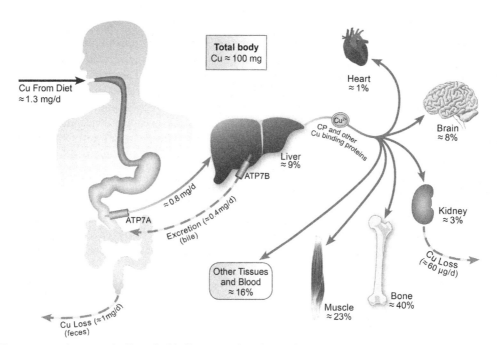

FIGURE 7.1   Human copper homeostasis. Shown in this diagram are the major regulatory mechanisms that control body copper levels, from absorption in the diet, to distribution to various body tissues, to excretory mechanisms. Copper-transporting ATPases play critical roles in intestinal absorption (ATP7A) and biliary excretion (ATP7B). Numbers under various organs indicate the approximate percentage of body copper that is present in that organ/tissue. Copper is predominantly in the cupric form (Cu^{2+}) in the diet and within the serum, but it has to be reduced for absorption into cells lining the small intestine and other cells of the body. Once copper exits cells, the oxidizing environment of the interstitial fluids causes it to reoxidize to Cu^{2+}. Although most copper in the serum is bound to ceruloplasmin (CP), other copper-binding proteins exist because the absence of CP (in aceruloplasminemia) does not cause copper deficits in peripheral tissues.

kidney (~3), heart (~1%), and approximately 16% in blood and other tissues (Fig. 7.1). Homeostatic control of body copper homeostasis includes modulation of intestinal copper absorption and hepatic copper excretion, as described in the following sections.

Copper Transport Across the Intestinal Mucosa

Dietary copper and endogenous copper from saliva, pancreatic juice, and bile contribute to the intestinal copper pool; however, biliary copper may be complexed with bile salts and thus unavailable for absorption. Dietary copper is reduced from the Cu^{2+} to Cu^{+} for transport across the brush-border membrane (BBM) of enterocytes in the proximal small bowel (Fig. 7.2). This is presumably mediated by one of three cupric reductases that have been identified: cytochrome b [558] ferric/cupric reductase, STEAP2 metalloreductase, and cytochrome B reductase 1 [Cybrd1; also called duodenal cytochrome b (Dcytb)]. The exact contribution of each has not been definitively established (Collins et al., 2010). Once reduced, cuprous copper (Cu^{+}) is transported into enterocytes by copper transporter-1 (CTR1). CTR1 is likely endocytosed into enterocytes, where it then mediates copper export from the endosome into the cytosol (Nose et al., 2006). Ctr1 knockout mice develop severe systemic copper deficiency; therefore Ctr1 is required for optimal copper absorption. Ctr1 has been definitively localized to the enterocyte BBM in three mammalian species, and expression was shown to be induced by dietary copper deprivation (Nose et al., 2010). Divalent metal-ion transporter-1 (DMT1), a well-established intestinal iron and manganese importer, may also be involved in absorption of dietary copper (Arredondo et al., 2003), but this possibility remains unproven to date. Copper transport by DMT1 seems plausible, particularly during iron deficiency, when DMT1 expression is strongly upregulated in the setting of low dietary iron (Ravia et al., 2005). Copper may also enter enterocytes via an ATP-dependent mechanism that was described in DMT1-mutant Belgrade rats (Knopfel et al., 2005).

After absorption into enterocytes, copper is bound by one of several chaperone proteins: COX17 cytochrome C oxidase copper chaperone (COX17), a chaperone for CCO, delivers copper to the mitochondria; antioxidant 1 copper chaperone (ATOX1), a chaperone for ATP7A, delivers copper to the trans-golgi network (TGN); and CCS, a copper chaperone for SOD1, delivers copper to the cytosol to support the biosynthesis of Cu, Zn-SOD. An alternatively spliced form of the *ATP7A* transcript

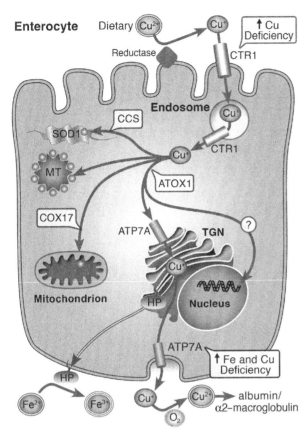

Enterocyte

FIGURE 7.2 Copper homeostasis in enterocytes. Shown in this diagram is a single enterocyte with the processes that are involved in copper absorption from the diet. Dietary copper is first reduced and then transported across the brush-border membrane by copper transporter-1 (CTR1). Once copper enters the cytosol, it is rapidly bound to chaperones for distribution to various cellular compartments. Copper chaperone for superoxide dismutase (CCS) delivers copper to the cytosolic Cu/Zn superoxide dismutase (SOD1); ATOX1 delivers copper to the Menkes copper-transporting ATPase (ATP7A) in the trans-golgi network (TGN) and may also deliver copper to the nucleus where ATOX1 can also act as a transcription factor to regulate genes related to cell cycle control; COX17 delivers copper to the mitochondria. Excess copper can be bound to metallothionein (MT) under certain conditions, equating to a mucosal block to copper absorption. In the TGN, copper is incorporated into copper-containing proteins bound for the secretory pathway, possibly including hephaestin (HEPH), a multicopper ferroxidase that is important to oxidize transported iron on the basolateral surface for binding to transferrin. Under conditions of copper excess, ATP7A traffics to the basolateral membrane and functions in copper export. Once cuprous copper exits cells, it spontaneously oxidizes and is then bound by albumin and α2-macroglobulin for transport through the portal blood to the liver. Expression of the predominant players in both the import and export processes may be upregulated by copper deficiency (CTR1 and ATP7A), and interestingly, the copper export process may also be increased during iron deficiency because ATP7A expression is strongly induced.

has also been hypothesized to possibly be a nuclear chaperone (Reddy et al., 2000). ATOX1 may also be a copper-activated transcription factor (Itoh et al., 2008). Excess intracellular copper can be stored in an inert form in MT. Finally, copper is transported out of enterocytes by ATP7A, which traffics from the TGN to the basolateral membrane when copper is in excess. Once copper is effluxed from enterocytes, the oxidizing environment of the interstitial fluids converts Cu^+ to cupric copper (Cu^{2+}), which binds to albumin or α_2-macroglobulin, which then enters the portal circulation.

Copper Transport and Transfer

Studies in HepG2 cells suggest that albumin and α_2-macroglobulin may differentially deliver copper to proteins on the surface of hepatocytes (Collins et al., 2010). Cu^{2+} derived from the portal blood has to first be reduced because Ctr1, the copper importer on hepatocytes, interacts with Cu^+ (Fig. 7.3). Ablation of Ctr1 in mouse hepatocytes demonstrated the importance of Ctr1 in hepatic copper uptake (Kim et al., 2009). Inside hepatocytes, copper is bound to chaperones and distributed to various intracellular compartments. ATP7B transports copper into the TGN, where it is incorporated into CP and other cuproproteins. ATP7B translocates from the TGN to the canalicular membrane of the hepatocyte when copper is in excess, facilitating excretion of copper into the bile. Biliary copper secretion requires the interaction between ATP7B and COMMD1. Moreover, COMMD1 levels are regulated by the ubiquitin ligase X-linked inhibitor of apoptosis (Collins et al., 2010).

Copper Excretion

Endogenous copper is excreted via hepatocytes into bile, a process that involves ATP7B. Excreted copper is bound to bile salts and is thus unavailable for reabsorption. Copper excretion is not fully functional during the fetal and neonatal periods, which may explain why hepatic copper levels are increased during these developmental stages. Cholestasis in the elderly may also increase hepatic copper concentrations. Copper loss in urine is insignificant (30–50 µg/day), but renal dysfunction may increase copper losses (Danks, 1988).

FIGURE 7.3   **Copper homeostasis in hepatocytes.** Shown is a single hepatocyte with the major proteins involved in copper homeostasis in the liver. Copper from the blood has to be reduced before being transported into the cell by copper transporter-1 (CTR1). A similar host of copper chaperones bind copper (as detailed in the legend to Fig 7.2) and facilitate movement throughout the cell; excess copper can also be stored bound to metallothionein (MT). During early life, both copper-exporting ATPases (ATP7A, ATP7B) are expressed in hepatocytes and may be necessary for the production of cuproteins in the trans-golgi network (TGN). After the neonatal period, ATP7A expression decreases dramatically. A large percentage of copper is incorporated into ceruloplasmin (CP), which is secreted into the blood stream. A GPI-anchored version of CP also exists in hepatocytes. Both CP proteins are ferroxidases that play critical roles in iron release from certain tissues and cell types. Under conditions of excess body copper, ATP7B translocates to the canalicular membrane and facilitates copper excretion in the bile. ATP7B protein levels are controlled by copper metabolism MURR1 domain-containing protein-1 (COMMD1) and X-linked inhibitor of apoptosis (XIAP) via the proteasome pathway. Additional copper export pathways may exist in hepatocytes because a percentage of serum copper is not bound to CP and may therefore not proceed through the secretory pathway; thus a potential unknown copper transported is depicted on the lower surface of the cell.

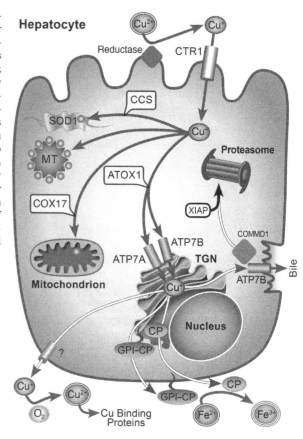

Copper: Homeostatic Mechanisms

Stable isotope studies in humans show that efficient adaptive mechanisms have evolved to protect against copper deficiency and toxicity regardless of dietary intake levels. These mechanisms include regulation of intestinal copper absorption and biliary copper excretion. Dietary copper absorption is regulated according to dietary intake levels and body copper status. Increased absorption during copper deprivation results from increases in Ctr1 expression on the BBM of enterocytes (Nose et al., 2010) and Atp7a on the basolateral surface (Kim et al., 2010). When copper intake is high, copper may be bound by MT in enterocytes, equating to a mucosal block to absorption. Moreover, biliary excretion may increase with high intake levels. Normally, approximately 10% of dietary copper is absorbed (Harvey et al., 2005). Copper retention in brain and heart increases significantly in copper-deprived rats (Levenson and Janghorbani, 1994). These adaptive mechanisms maintain tissue copper concentrations to ensure normal homeostasis, but they become ineffective when copper intake approaches less than 0.7 mg/day (which is less likely in the U.S.).

Genetic Defects in Copper Metabolism

Copper-related pathologies most frequently arise due to defects in two copper-transporting ATPases, ATP7a and ATP7B, resulting in MD and WD. In MD, the copper export function of ATP7A is impaired, thus leading to impaired copper absorption. In WD, a mutation in the predominant copper transporter involved in biliary excretion (ATP7B) is mutated, resulting in hepatic copper overload. Each of these genetic disorders will be discussed briefly in the following sections. Detailed reviews have been published (Tumer and Moller, 2010; Mak and Lam, 2008).

Menkes Disease

MD is an X-linked, recessive disorder of copper metabolism affecting multiple organs. Most patients are males (Tumer et al., 2001). Typical symptoms include neurodegeneration, connective tissue disturbances, and atypical kinky hair. MD is usually fatal by 3 years of age; no cure exists, although subcutaneous injection of copper histidine early in life has shown promise (Kaler et al., 2008).

The prevalence of MD varies with a lower occurrence noted in Japan and European countries (1:300,000–1:360,000), but a much higher incidence observed in Australia (1:50,000–100,000) (Tumer and Moller, 2010). The underlying genetic defect is in the *ATP7A* gene, which encodes a copper-transporting ATPase. Variable forms of MD exist, with occipital horn syndrome being the least severe and classic MD the most severe. To date, more than 170 different mutations in the *ATP7A* gene have been documented, presumably leading to the noted heterogeneity in the disease phenotype.

Defective copper export from cells is the basic physiological disturbance in MD, with most tissues accumulating excessive copper. However, copper does not accumulate to toxic levels, presumably because intestinal copper absorption is impaired. Brain copper remains low because Atp7a delivers copper to the CNS; in MD, copper accumulates in the blood–brain barrier, resulting in copper deprivation of neuronal and glial cells (Nishihara et al., 1998). Low copper in some peripheral tissues, despite accumulation in others, leads to the characteristic systemic copper deficiency in MD, including low serum copper and CP activity, and impaired synthesis of SOD1 and CCS. LOX activity is also impaired, leading to defective artery formation in the CNS and osteoporosis. Progressive neurodegeneration is noted in the brain, causing the classic neurological symptoms of MD.

Wilson's Disease

WD is an autosomal-recessive disease leading to abnormal copper storage. The underlying defect is in *ATP7B*, which encodes a copper-transporting ATPase. The worldwide prevalence of WD is approximately 1:30,000 (Scheinberg and Sternlieb, 1984). In WD patients, copper accumulation in liver, brain, and cornea (Kayser–Fleisher rings) results in multi-organ damage, presumably as a result of oxidative damage caused by unbound, highly reactive copper. Neurologic damage and cirrhosis, acute hepatitis, hemolytic crisis, and hepatic failure may occur if left untreated.

Permanent organ damage can be avoided in WD with proper medical treatment. Typical treatment utilizes decoppering chelation therapy with penicillamine and/or trientine or high zinc dosing (which impairs the absorption of dietary copper). The most widely accepted therapeutic approaches currently rely mostly on zinc and/or trientine (Lewitt, 1999). The effectiveness of treatment can be monitored by 24-h urinary free copper excretion. Importantly, side effects should be carefully followed because copper deficiency may result from excessive copper chelation (Macintyre et al., 2004). Liver transplantation may be necessary for those with advanced disease.

COPPER DEFICIENCY IN HUMANS

Severe copper deficiency in mammals results in abnormalities in the immune, skeletal, and cardiovascular systems; hypochromic anemia (which does not respond to iron supplementation); hypopigmentation; thrombocytopenia; and neutropenia (Uauy et al., 1998). Copper deficiency during the gestational and perinatal periods has particularly severe consequences, relating to neurological abnormalities. Detailed reviews of copper deficiency in humans have been published (Danks, 1988; Uriu-Adams et al., 2010; Tapiero et al., 2003). Systemic copper deficiency in humans can result from ineffective absorption of dietary copper or excessive copper loss via bile. Copper deficiency may also be the result of MD. In addition to MD patients, copper deficiency in humans is rare, but it occurs most frequently in: (1) individuals chronically receiving total parenteral nutrition without proper copper supplementation; (2) premature infants consuming milk-based formulas lacking adequate copper; (3) neonates experiencing chronic diarrhea or malnutrition; (4) hospitalized patients undergoing long-term peritoneal dialysis; (5) severe burn patients; (6) renal dialysis patients; and (7) persons consuming large doses of supplemental zinc, antacids, or copper chelators (Prohaska, 2006). Moreover, recent studies suggest an association between surgical bowel resection in the management of morbid obesity and acquired copper deficiency (Griffith et al., 2009). Individuals undergoing this surgical intervention may require supplemental copper, but even so, manifestations of copper deficiency may only be partly corrected.

Copper deficiency in humans is accompanied by low serum copper levels and reduced serum FOX activity, presumably as mediated by CP. Usual pathophysiological features include anemia, leukopenia, and neutropenia. During periods of rapid growth, osteoporosis is a common feature, often accompanied by flaring of and fractures at the margins of the metaphyses. Furthermore, moderate copper deficiency may result in additional manifestations, including arthritis, arterial disease, depigmentation, myocardial disease, neurological abnormalities (Danks, 1988), cardiac arrhythmias, increased serum cholesterol, and glucose intolerance (Davis and Mertz, 1987). It is also important to note that these observations were not duplicated in other studies, indicating that future work in humans is necessary.

COPPER: DIETARY CONSIDERATIONS AND REQUIREMENTS

Dietary reference intakes for copper were established in 2001 (Trumbo et al., 2001). Adequate intake levels for copper have been established as 200 µg/day for infants 0–6 months of age and 220 µg/day for those between 7 and 12 months. RDAs are lower for individuals 18 years old or younger (varying from 340 µg/day in 1–3 year olds to 890 µg/day for 14–18 year olds). For adults, the RDA is 900 µg/day. During pregnancy and lactation, the RDAs increase to 1 and 1.3 mg/day, respectively. On the basis of rodent studies, it has been proposed that the human RDA for copper is low (Prohaska and Brokate, 2002). ULs have also been established for copper, ranging from 1 mg/day in children 1–3 years of age to 10 mg/day in adults. Interestingly, analyses have suggested that an intake of 2.6 mg/day is the most appropriate intake level for adults to decrease the risk of deficiency while simultaneously avoiding symptoms of copper overload (Chambers & Krewski, 2010).

EVALUATION OF COPPER STATUS

Analytical Methods

A common method to quantify the copper content of biologic samples or food is spectroscopy (Turnlund, 2006). Most commonly frequently used to quantify copper concentrations are inductively coupled plasma (ICP) emission spectroscopy and atomic absorption spectroscopy (AAS). For AAS, samples are atomized with a graphite furnace [graphite furnace AAS (GFAAS)] or with an air-acetylene flame (flame AAS) for electrothermal ionization. ICP is often utilized when more than one mineral is being quantified. The most sensitive of these methods is GFAAS, with a detection limit of 0.014 µg/mL for copper. The next most sensitive is ICP, with a detection limit of 0.04 µg/mL, followed by AAS with a limit of 1.5 µg/mL.

Studies to determine the distribution of copper in living animals typically use stable isotopes of copper (Patterson and Veillon, 2001). Studies using this methodology can track the absorption, utilization, excretion, and turnover of copper in biological systems. The most common method to measure the copper isotope ratios is mass spectrometry (MS), with ICP-MS and thermal ionization MS being the most commonly used techniques.

Assessment of Copper Status

Identifying biomarkers of copper status that are sensitive to even marginal deficiency, and are noninvasive and consistently reliable, have been the foci of experimental work (Harvey et al., 2009). The effects of copper deprivation on tissue metal concentrations and the activity of various cuproenzymes vary according to dietary Cu levels, gender, strain, species, and the specific organ being analyzed. The most commonly utilized method in humans has been to quantify copper levels and the activity of various cuproenzymes in blood (Prohaska, 2006). Reductions in plasma copper and CP activity are noted in severely copper-deficient humans. However, observed reductions in serum Cu and CP activity are complicated by the fact that several physiological alterations can increase copper content and CP activity of blood, including the acute-phase response to infection and inflammation, pregnancy and other hormonal perturbations, and some carcinogenic phenotypes (Prohaska, 2006). Other utilized markers, including SOD1 activity in erythrocytes, CCO activity in platelets and mononuclear cells, and copper content of various circulating blood cells, have shown limited usefulness.

Recent studies have evaluated other potential biomarkers of copper status. Studies in rats revealing alterations in serum and tissue PAM activity (Prohaska and Gybina, 2005) have correlated with similar observations in copper-deficient humans (Prohaska, 1997). Other recent studies have suggested that plasma DAO activity could be a useful biomarker (Kehoe et al., 2000), but a problematic additional finding is that the activity of this enzyme may be altered by other physiological perturbations (Failla, 1999). In addition, the ratio of SOD1 to CCS protein levels in erythrocytes was suggested to be a useful biomarker in copper-deficient rodents (Prohaska and Gybina, 2004), but its applicability in humans is unknown.

COPPER TOXICITY AND UPPER LIMITS

Given the tight homeostatic control of body copper levels and the ability of the liver to excrete excess copper in the bile, it is predictable that copper toxicity is generally low. Mammals have evolved adequate homeostatic control of copper. Ingestion of high copper levels may, however, override the innate checkpoints designed to regulate overall body copper levels. Consumption of foods or fluids contaminated with high copper levels results in a metallic taste and gastrointestinal discomfort. In medicine, copper was historically used as an emetic. Because of the possible adverse consequences of high copper ingestion, a UL of 10 mg/day has been established. A study determined that no adverse effects were noted when drinking water contained up to 5 mg copper per liter (Araya et al., 2003). This is in contrast to recommendations from the World Health Organization, which recommends that copper in water not exceed 2 mg/L. These differing recommendations highlight the fact that there is controversy regarding what constitutes a safe level of copper in drinking water (Brewer et al., 2010).

Copper is often included in micronutrient supplements without consequence. A study supplemented adults with 10 mg cupric gluconate daily for 12 weeks without evidence of liver damage or gastrointestinal distress (Pratt et al., 1985).

However, copper toxicity risks are higher for neonates and infants given an immature biliary excretion system. Past reports of copper accumulation in the liver of children with idiopathic copper toxicosis, Austrian tyrol, and Indian childhood cirrhosis were linked to the preparation of infant formulas with copper-contaminated drinking water and consumption of other foods stored in brass containers (Araya et al., 2003). Copper loading is still observed clinically today in the setting of WD and other disorders in which biliary copper excretion is impaired, such as biliary cirrhosis and biliary atresia. These individuals should obviously carefully monitor their copper intakes. Copper chelation therapy is the treatment of choice in cases in which hepatic copper accumulation has already occurred.

ACKNOWLEDGMENT

The writing of this chapter was supported by NIH grant DK074867.

REFERENCES

Aliabadi, H., 2008. Med. Hypoth 6, 1163–1166.

Araya, M., Koletzko, B., Uauy, R., 2003. J. Pediatr. Gastroenterol. Nutr. 37, 422–429.

Arredondo, M., Cambiazo, V., Tapia, L., et al., 2003. Am. J. Physiol. Cell Physiol. 284, C1525–C1530.

Baker, A., Harvey, L., Kajask-Newman, G., et al., 1999. J. Nutr. 53, 408–412.

Baker, H., 1999. J. Nutr. 129, 2278–2279.

Bodansky, M.J., 1921. Biol. Chem. 48, 361.

Bousquet-Moore, D., Mains, R.E., Eipper, B.A., 2010. J. Neurosci. Res. 88, 2535–2545.

Brady, G.F., Galban, S., Liu, X., et al., 2010. Mol. Cell Biol. 30, 1923–1963.

Brazeau, B.J., Johnson, B.J., Wilmot, C.M., et al., 2004. Arch. Biochem. Biophys. 428, 22–31.

Brewer, G.J., Danzeisen, R., Stern, B.R., et al., 2010. Letter to the editor and reply: toxicity of copper in drinking water. J. Toxicol. Environ. Health B Crit. Rev. 13, 449–459.

Broderius, M.A., Prohaska, J.R., 2009. Nutr. Res. 29, 494–502.

Broderius, M., Mostad, E., Wendroth, K., et al., 2010. Comp. Biochem. Physiol. C Toxicol. Pharmacol. 151, 473–479.

Bugel, S., Harper, A., Rock, E., et al., 2005. Br. J. Nutr. 94, 231–236.

Chambers, A., Krewski, D., Birkett, N., Plunkett, L., Hertzberg, R., Danzeisen, R., Aggett, P.J., Starr, T.B., Baker, S., Dourson, M., Jones, P., Keen, C.L., Meek, B., Schoeny, R., Slob, W., October 2010. An exposure-response curve for copper excess and deficiency. J. Toxicol. Environ. Health B Crit. Rev 13(7–8), 546–578. http://dx.doi.org/10.1080/10937404.2010.538657.

Chen, H., Huang, G., Su, T., et al., 2006. J. Nutr. 136, 1236–1241.

Chen, H., Attieh, Z.K., Dang, T., et al., 2009. J. Cell Biochem. 107, 803–808.

Chen, H., Attieh, Z.K., Syed, B.A., 2010. J. Nutr. 140, 1728–1735.

Cohn, E.J., Minot, G.R., Fulton, J.F., et al., 1927. J. Biol. Chem. 74: 1xix.

Collins, J.F., Franck, C.A., Kowdley, K.V., et al., 2005. Am. J. Physiol. Gastrointest. Liver Physiol. 288, G964–G971.

Collins, J.F., Hua, P., Lu, Y., et al., 2009. Am. J. Physiol. Gastroint. Liver Physiol. 297, G695–G707.

Collins, J.F., Prohaska, J.R., Knutson, M.D., 2010. Nutr. Rev. 68, 133–147.

Czyzyk, T.A., Morgan, D.J., Peng, B., et al., 2003. J. Neurosci. Res. 74, 446–455.

Danks, D.M., Campbell, P.E., Stevens, B.J., et al., 1972. Pediatrics 50, 188–201.

Danks, D.M., 1988. Annu. Rev. Nutr. 8, 235–257.

Davis, G.K., Mertz, W., 1987. Copper. In: Mertz, W. (Ed.), Trace Elements in Human and Animal Nutrition, vol. 1. fifth ed. Academic Press, San Diego, pp. 301–364.

De Domenico, I., Ward, D.M., Di Patti, M.C., et al., 2007. EMBO J. 26, 2823–2831.

Disilvestro, R.A., Selsby, J., Siefker, K., 2010. J. Trace Elem. Med. Biol. 23, 165–168.

Dunkel, P., Gelain, A., Barlocco, D., et al., 2008. Curr. Med. Chem. 15, 1827–1839.

Dyer, F.F., Leddicotte, G.W., 1961. The Radiochemistry of Copper. National Academy of Sciences, National Research Council, Washington, DC.

Easter, R.N., Chan, Q., Lai, B., et al., 2010. Vas. Med. 15, 61–69.

Failla, M.L., 1999. Proc. Nutr. Soc. 58, 497–505.

Fattman, C.L., Schaefer, L.M., Oury, T.D., 2003. Free Radic. Biol. Med. 35, 236–256.

Feng, W., Ye, F., Xue, W., et al., 2009. Mol. Pharmacol. 75, 174–182.

Griffith, D.P., Liff, D.A., Ziegler, T.R., et al., 2009. Obesity 17, 827–831.

Harris, E.D., 1997. Copper. In: O'Dell, B.L., Sunde, R.A. (Eds.), Clinical Nutrition in Health and Disease: Handbook of Nutritionally Essential Mineral Elements, vol. 2. Marcel Dekker, New York, pp. 231–273.

Harvey, L.J., Dainty, J.R., Hollands, W.J., et al., 2005. Am. J. Clin. Nutr. 81, 807–813.

Harvey, L.J., Ashton, K., Hooper, L., et al., 2009. Am. J. Clin. Nutr. 89, 2009S–2024S.

Hopkins, R.G., Failla, M.L., 1999. J. Nutr. 129, 596–601.

Hudson, D.M., Curtis, S.B., Smith, V.C., et al., 2010. Am. J. Physiol. Gastrointest. Liver Physiol. 298, G425–G432.

Hunt, J.R., Matthys, L.A., Johnson, L.K., 1998. Am. J. Clin. Nutr. 67, 421–430.

Institute of Medicine, 2002. Zinc. In: Food and Nutrition Board (Ed.), Dietary Reference Intakes for Vitamin A, Vitamin K, Arsenic, Boron, Chromium, Copper, Iodine, Manganese, Molybdenum, Nickel, Silicon, Vanadium and Zinc. National Academy Press, Washington, DC, pp. 442–501.

Itoh, S., Kim, H.W., Nakagawa, O., et al., 2008. J. Biol. Chem. 283, 9157–9167.

Jung, O., Marklund, S.L., Geiger, H., et al., 2003. Circ. Res. 93, 622–629.

Kagan, H.M., Li, W., 2003. J. Cell Biochem. 88, 660–672.

Kaler, S.G., Holmes, C.S., Goldstein, D.S., et al., 2008. N. Engl. J. Med. 358, 605–614.

Kehoe, C.A., Turley, E., Bonham, M.P., et al., 2000. Br. J. Nutr. 84, 151–156.

Kelley, D.S., Dauda, P.A., Taylor, P.C., et al., 1995. Am. J. Clin. Nutr. 62, 412–416.

Kim, H., Son, H.Y., Bailey, S.M., et al., 2009. Am. J. Physiol. Gastrointest. Liver Physiol. G356–G364.

Kim, B.E., Turski, M.L., Nose, Y., et al., 2010. Cell Metab. 11, 353–363.

Klevay, L.M., 2002. Advances in cardiovascular-copper research. In: Schrauzer, G.N. (Ed.), Trace Elements in Nutrition, Health and Disease. Institute Rosell, Montreal, Canada, pp. 64–71.

Klevay, L.M., 2006. Eur. Heart J. 27, 117.

Knopfel, M., Smith, C., Solioz, M., 2005. Biochem. Biophys. Res. Commun. 330, 345–352.

Levenson, C.W., Janghorbani, M., 1994. Anal. Biochem. 221, 243–249.

Lewitt, P.A., 1999. Mov. Disord. 14, 555–556.

Linder, M.C., 2003. Biochemistry and molecular biology of copper in mammals. In: Massaro, E.J. (Ed.), Handbook of Copper Pharmacology and Toxicology. Humana Press, Totowa, NJ, pp. 3–32.

Lonnerdal, B., 1998. Am. J. Clin. Nutr. 67, 1046S–1053S.

Lucero, H.A., Kagan, H.M., 2006. Cell Mol. Life Sci. 63, 3204–3216.

Macintyre, G., Gutfreund, K.S., Martin, W.R., et al., 2004. J. Lab. Clin. Med. 144, 294–301.

Mak, C.M., Lam, C.W., 2008. Crit. Rev. Clin. Lab. Sci. 45, 263–290.

Mason, K.E., 1979. J. Nutr. 109, 1979–2066.

McDonald, A., Tipton, K., O'Sullivan, J., et al., 2007. J. Neural Transm. 114, 783–786.

Menkes, J.H., Alter, M., Steigleder, G.K., et al., 1962. Pediatrics 29, 764–779.

Milne, D.B., Davis, C.D., Nielsen, F.H., 2001. Nutr 17, 701–708.

Nations, S.P., Boyer, P.J., Love, L.A., et al., 2008. Neurol 71, 639–643.

Nishihara, E., Furuyama, T., Yamashita, S., et al., 1998. Neuroreport 9, 3259–3263.

Nose, Y., Kim, B.E., Thiele, D.J., 2006. Cell Metab. 4, 235–244.

Nose, Y., Wood, L.K., Kim, B.E., et al., 2010. J. Biol. Chem. 285, 32385–32392.

O'Dell, B.L., 1990. Nutr. Rev. 48, 425–434.

Oberley-Deegan, R.E., Regan, E.A., Kinnula, V.L., et al., 2009. COPD 6, 307–312.

Patterson, K.Y., Veillon, C., 2001. Exp. Biol. Med. 226, 271–282.

Powers, H.J., Loban, A., Silvers, K., et al., 1995. Free Radic. Res. 22, 57–65.

Pratt, W.B., Omdahl, J.L., Sorenson, J.R., 1985. Am. J. Clin. Nutr. 42, 681–682.

Prohaska, J.R., Brokate, B., 2002. J. Nutr. 132, 3142–3145.

Prohaska, J.R., Failla, M.L., 1993. Copper and immunity. In: Klurfeld, D.M. (Ed.), Human Nutrition: A Comprehensive Treatise. Plenum Press, New York, pp. 309–332.

Prohaska, J.R., Gybina, A.A., 2004. J. Nutr. 134, 1003–1006.

Prohaska, J.R., Gybina, A.A., 2005. J. Neurochem. 93, 698–705.

Prohaska, J.R., 1988. Biochemical roles of copper in animals. In: Prasad, A.S. (Ed.), Essential and Toxic Trace Elements in Human Health and Disease. Alan R. Liss, Inc., New York, NY.

Prohaska, J.R., 1997. Neurochemical roles of copper as antioxidant or prooxidant. In: Conner, J.R. (Ed.), Metals and Oxidative Damage in Neurological Disorders. Plenum Press, New York, NY.

Prohaska, J.R., 2006. Copper. In: Bowman, B.A., Russell, R.M. (Eds.), Present Knowledge in Nutrition, ninth ed. ILSI Press, Washington, DC, pp. 458–470.

Ranganathan, P.N., Lu, Y., Jiang, L., Kim, C., Collins, J.F., September 15, 2011. Serum ceruloplasmin protein expression and activity increases in iron-deficient rats and is further enhanced by higher dietary copper intake. Blood 118 (11), 3146–3153. http://dx.doi.org/10.1182/blood-2011-05-352112. Epub 2011 Jul 18.

Ravia, J.J., Stephen, R.M., Fk, G., et al., 2005. J. Biol. Chem. 280, 63221–63227.

Reddy, M.C., Majumdar, S., Harris, E.D., 2000. Biochem. J. 350, 855–863.

Reiser, S., Smith, J.C., Mertz, W., et al., 1985. Am. J. Clin. Nutr. 42, 242–251.

Robinson, N.J., Winge, D.R., 2010. Annu. Rev. Biochem. 79, 537–562.

Rodriguez, C., Rodriguez-Sinovas, A., Martinez-Gonzales, J., 2008. Drug News Perspect. 21, 218–224.

Scheinberg, I., Sternlieb, I., 1984. Major Prob. Intern. Med. 23, 1–24.

Stolen, C.M., Martilla-Ichihara, F., Koskinen, K., et al., 2005. Immunity 22, 105–115.

Tapiero, H., Townsend, D.M., Tew, K.D., 2003. Biomed. Pharmacother. 57, 386–398.

Thackeray, E.W., Sanderson, S.O., Fox, J.C., et al., May 24, 2010. J. Clin. Gastrenterol.

Thomas, S.A., Matsumoto, A.M., Palmiter, R.D., 1995. Nature 374, 643–646.

Tininello, A., Pietrangeli, P., De Marchi, U., et al., 2006. Biochim. Biophys. Acta 1765, 1–13.

Trumbo, P., Yates, A.A., Schlicker, S., et al., 2001. J. Am. Diet. Assoc. 101, 294–301.

Tumer, Z., Moller, L.B., 2010. Eur. J. Hum. Genet. 18, 511–518.

Tumer, Z., Moller, L.B., Horn, N., 2001. Am. J. Med. Genet. 99, 217–222.

Turnlund, J.R., 2006. Copper. In: Shills, M.E., Shike, M., Ross, A.C., et al. (Eds.), Modern Nutrition in Health and Disease, tenth ed. Lippincott, Williams & Wilkins, Baltimore, MD, pp. 286–299.

Uauy, R., Olivares, M., Gonzalez, M., 1998. Am. J. Clin. Nutr. 67, 952S–959S.

Uriu-Adams, J.Y., Scherr, R.E., Lanoue, L., et al., 2010. Biofactors 36, 136–152.

Van De Sluis, B., Muller, P., Duran, K., et al., 2007. Mol. Cell Biol. 27, 4142–4156.

Van Den Berghe, P.V., Klomp, L.W., 2010. J. Biol. Inorg. Chem. 15, 37–46.

Vonk, W.I., Wijmenga, C., Berger, R., et al., 2010. J. Biol. Chem. 285, 28991–29000.

Vulpe, C.D., Kuo, Y.M., Murphy, T.L., et al., 1999. Nat. Genet. 21, 195–199.

Werman, M.J., Barat, E., Bhathena, S.J., 1995. J. Nutr. 15, 857–863.

White, C., Lee, J., Kambe, T., et al., 2009. J. Biol. Chem. 284, 33949–33956.

Wilson, S.A.K., 1912. Brain 34, 295–509.

Witte, K.K.A., Nikitin, N.P., Parker, A.C., et al., 2005. Eur. Heart J. 26, 2238–2244.

Xu, X., Pin, S., Gathinji, M., et al., 2004. Ann. N. Y. Acad. Sci. 1012, 299–305.

Zhou, Z., Johnson, W.T., Kang, Y.J., 2008. J. Nutr. Biochem. 20, 621–628.

Chapter 8

Copper and Molecular Aspects of Cell Signaling

A. Grubman, A.R. White

The University of Melbourne, Parkville, VIC, Australia

INTRODUCTION

Copper is an enigmatic ion that has a far greater role in living systems than would be suggested from its categorization as a trace element. Although its levels in tissues and cells is normally maintained in the nanomolar to low micromolar range, the reactive transition between oxidative states (primarily CuII and CuI) has been harnessed through evolutionary biology as a key driving force in many complex and critical cell processes, including energy production and protection against oxidative stress. The last point is somewhat ironic because copper can also be one of the strongest catalysts of oxidative damage because of the same reactive changes that underlie its utility in biological processes. Cells have accordingly developed a range of key molecules, transporters, chaperones, storage sites, and high-affinity copper binding sites to ensure that copper is carefully managed. When tissue or cell copper levels become too low or highly elevated, or if copper is mislocalized, the outcome can lead to cell toxicity and in some cases pathological changes to the organism. A rapidly growing body of research now demonstrates an important role for abnormal copper handling in a range of disorders including cancer and Alzheimer's disease (AD). However, this has also opened up major potential new targets for therapeutic intervention. This research is advancing rapidly because of highly sensitive new approaches to mapping copper to subcellular organelles and even proteins using X-ray microscopic and metalloproteomic techniques. New insights into copper functions are also being generated by the development of probes that can distinguish between tightly bound and more loosely associated (labile) copper in cells, with the latter likely having a key role in cell signaling processes. Although other ions have been well accepted as key modulators of cell signaling processes, including Ca, Mg, and more recently Zn, until recently copper was not readily accepted into this group. A large body of data now demonstrates that copper controls cell signaling processes at many sites and thus contributes to most aspects of cell function in this manner. As detailed in this chapter, copper interacts with cell receptors, kinases, and phosphatases; induces epigenetic control of nuclear gene expression; and has broader effects in cell communication. The studies demonstrate that copper serves both as a signal itself and can modulate signaling processes. This review will cover the available evidence for copper control of cell signaling processes, including how copper is handled within the metalloproteome, copper interaction with multiple sites along intracellular signaling pathways, how copper modulates cell-to cell signaling, how copper signaling affects other minerals, and the cell fates controlled by copper-mediated signaling. Although copper signaling has been identified in most organisms, for the purposes of this chapter, only studies performed in mammals or mammalian cells have been included.

THE COPPER METALLOPROTEOME

Copper binds with high affinity to proteins, second only to zinc, as indicated by the Irving Williams series, which describes the relative stabilities of transition metal complexes, yet less than 1% of the mammalian proteome is predicted to bind copper in vivo (Andreini et al., 2008). Thus the problem of allocation of copper to the correct cellular protein destination and exclusion from incorrect sites is solved in eukaryotic cells by chaperone-mediated delivery to terminal protein targets (Foster et al., 2014). Sequential transfer of copper down increasing gradients of binding affinity thereby provides specificity.

Copper is highly redox active; thus, to prevent potentially damaging mismetallation or oxidative Fenton or Haber–Weiss chemistry, cellular free copper levels are maintained at less than one copper ion per cell. Copper sequestration is achieved through binding to cysteine residues in the antioxidant peptides metallothionein (MT) or glutathione (GSH). It is this readily exchangeable copper pool that is likely to be involved in cellular signaling processes.

Copper is a structural component of many enzymes, dominated by oxidoreductases, which are able to stabilize cupric and cuprous copper through S-donor ligands (i.e., methionine, cysteine) and N-donor ligands (histidine), respectively (Andreini et al.,

Molecular, Genetic, and Nutritional Aspects of Major and Trace Minerals. http://dx.doi.org/10.1016/B978-0-12-802168-2.00008-7

2008). Protein crystallography has yielded structures of biologically important copper transporters, chaperones, and enzymes such as superoxide dismutase-1 (SOD1), antioxidant-1 (ATOX1), copper chaperone for SOD, ATP7A/B, SCO1 (synthesis of cytochrome c oxidase 1) and COX17 (cytochrome c oxidase 17) (Banci et al., 2010). In addition, various human proteins have been cocrystallized with cupric or cuprous cofactors, including protein deglycase (DJ-1), vascular adhesion protein 1 (VAP-1), coagulation factor VIII, myoglobin, synaptotagmin, kallikrein-7, tumor susceptibility gene-101, S100A12-copper complex, and human DNA polymerase-β (based on copper-containing protein structures that were deposited to the RCSB protein data bank). However, it is unclear whether all of these proteins are bona fide copper-binding proteins in vivo because copper may not gain access to these proteins because of inadequate subcellular availability or spatial separation. Immobilized metal affinity chromatography (IMAC) has been commonly utilized to study copper-binding proteins in plants and algae, but large-scale metalloproteomic approaches in mammalian systems are limited. Copper-IMAC has been used to identify RNA- and DNA-binding nuclear cuproproteins in mouse brain (Watanabe and Tezuka, 2006) and for identification of high-affinity copper proteins in hepatoma cell lines (including albumin, enolase, calcium binding proteins, DNA-/RNA-binding proteins, heat-shock proteins, and copper-dependent redox enzymes; She et al., 2003). The latter study also identified the metal-binding patterns of abundant mammalian cuproproteins, $C(X)_m C$ ($m = 2$–4) and $H(X)n H$ ($n = 0$–5), where C represents cysteine and H histidine.

A combination of experimental approaches and bioinformatics has vastly improved our understanding of copper proteomes, primarily led by the Florence group (Professors Ivano Bertini et al.). The bioinformatic approach interrogates libraries, such as Pfam for metal-binding domains, that can be examined genome wide to identify potential metalloproteins (Andreini et al., 2009). This powerful approach is not without constraints because metal-binding domains may have evolutionarily lost critical copper-binding residues. In addition, as yet uncharacterized copper-binding domains may not be captured in Pfam; however, hidden Markov models of metal-binding patterns have vastly improved prediction of copper proteomes (Andreini et al., 2009). Nonetheless, identification of transient or low-affinity, biologically important interactions, such as those involved in signaling mediated by the GSH exchangeable pool, remains a challenge.

COPPER MODULATION OF CELL SIGNALING

Intracellular signaling pathways control cell responses to external stimuli; therefore they are normally mediated by outside-in transduction of messages from external membrane receptors via specific kinases leading to modulation of nuclear gene transcription. Recent research has led to a growing understanding that copper is an important modulator of cell signal transduction pathways, with points of control at all steps in the process, as we recently reviewed (Grubman and White, 2014). Copper can control receptor–ligand interactions at the cell membrane, modulate kinase and associated phosphatase activities, and ultimately regulate nuclear expression of many important genes.

External Cell Receptors and Associated Pathways

Growth Factors, Receptors, and Associated Tyrosine Kinase Pathways

Outside-in signaling is initiated by peptide or polypeptide binding to an external site on a membrane receptor. Several studies have described a key role for copper in controlling ligand interaction with cell membrane receptors. Brain-derived neurotrophic factor (BDNF) has an N-terminal copper-binding domain, which controls its conformation and subsequent binding to tropomyosin receptor kinase B (TrkB), the cognate receptor for BDNF on neurons (Travaglia et al., 2012). BDNF with bound copper inhibits BDNF-induced proliferation in cultures of SH-SY5Y neuroblastoma cells (Travaglia et al., 2012). Conversely, copper has been shown to have stimulatory effects on nerve growth factor (NGF)-mediated neuronal proliferation (Travaglia et al., 2011). In this case copper binds to His4 and His8 on one monomer and His84 and Asp105 on the second monomer of the NGF active dimer, and this enhances cell proliferation in SH-SY5Y cultures exposed to 10 μM copper (Travaglia et al., 2011). However, this is in contrast with a report from Wang et al. (1999). who described copper-mediated inhibition of NGF neuroprotection against oxidative stress in PC12 pheochromocytoma cells. The reason for this discrepancy may reflect use of different cell lines, higher copper concentration (100 compared with 10 μM), or contrasting effects of copper on NGF-mediated growth compared with antioxidant action. Further complexity in the action of copper on NGF was demonstrated by Birkaya and Aletta (2005), who observed that NGF promotes cellular accumulation of copper in PC12 cells with subsequent stimulatory action on copper-dependent protein methylation and neurite outgrowth.

Similarly, copper has a key modulatory action on additional growth factors, including fibroblast growth factor (FGF). Copper induces release of FGF-1 from melanoma cells in culture via activation of phosphoinositol-3-kinase (PI3K)-Akt signaling (Di Serio et al., 2008). Copper also promotes FGF-1 release via direct interaction with sphingosine kinase-1 (SK-1). Copper binding induces release of SK-1, which binds to FGF-1 and mediates its release. This mechanism may have a role in copper-dependent promotion of angiogenesis (Soldi et al., 2007).

Copper can regulate membrane receptor activity through structural modification or phosphorylation. Wu et al. (1999) demonstrated a key role for copper and other metals in modulation of cell signaling in human bronchial epithelial cells. Activation of mitogen-activated protein kinase (MAPK) pathways by copper requires phosphorylation of Tyr845 of the epidermal growth factor receptor (EGFR), leading to downstream upregulation of MAPK and extracellular signal-regulated kinase (ERK) kinase (MEK) activity. Studies by Price et al. confirmed that copper activates EGFR in a human astroglial cell line (U87MG) that overexpresses EGFR (Price et al., 2009). Copper-dependent phosphorylation of multiple EGFR tyrosine residues (Tyr1068 and Tyr845) occurred through a mechanism independent of native EGFR ligands [including epidermal growth factor (EGF), leukemia inhibitory factor (LIF), transforming growth factor (TGF), etc.). Instead, this process required inhibition of protein tyrosine phosphatase (PTP)-1b and mediated prolonged EGFR phosphorylation and induction of growth factor secretion in culture medium (Price et al., 2009).

Integrins

Outside-in signal transduction by cells also occurs via membrane integrin receptors. These transmembrane receptors read the extracellular environment and respond to changes activated by common matrix proteins, including collagen, fibronectin, and vitronectin. Receptor activation initiates signal transduction via kinases such as integrin-linked kinase (ILK) and focal adhesion kinase (FAK), resulting in downstream modulation of MAPK and other signaling pathways (Hehlgans et al., 2007). These pathways control cell migration, tissue invasion, replication, or differentiation and are associated with development, tissue repair, and angiogenesis (Hehlgans et al., 2007). Copper-mediated regulation of integrins occurs through several key processes, but overall it appears to have a net potentiating effect on integrin signaling. Studies have shown that copper (as inorganic copper complexes) can stimulate apoptosis in H322 cancer cells, keratinocytes, and fibroblasts (involved in wound healing). Copper modulates apoptosis via upregulation of integrin receptor subunits. Copper complexes of the salicylaldehydepyrazole hydrazone derivative (E)-N'-(2-hydroxybenzylidene)-1-(4-tert-butylbenzyl)-3-phenyl-1H-pyrazole-5-carbo hydrazide (Cu-16; 10 μM) induced integrin-β4 expression in cancer cells, increasing integrin-mediated apoptosis (Fan et al., 2010). (E)-N'-(2-hydroxybenzylidene)−1−benzyl−3−phenyl-1H−pyrazole5-carbohydrazide (Cu-15) also activated integrin-β1 in human umbilical cord endothelial cells, demonstrating a potential role for copper–integrin interactions in angiogenesis (Fan et al., 2009). Copper promotes keratinocyte migratory processes required in wound healing by stimulating integrins-α2, -α3, -αV, -α6, and -β1 (~15–60 μM copper; Tenaud et al., 2000). Additional copper complexes such as Cu-Glycyl-L-histidyl-L-lysyl induce upregulation of integrin-α6 and -β1 in fibroblasts, leading to increased collagen synthesis and tissue repair (Kang et al., 2009). A role for copper in integrin signaling has also been reported in human lens. The peptide secreted protein acidic and rich in cysteine (SPARC) induces activation of ILK, leading to enhanced survival of lens cells, and it is dependent on direct interactions between SPARC and integrin-β1 during cell stress as well as SPARC and copper (Weaver et al., 2008).

A downstream target of integrin signaling is growth-factor–mediated responses. Copper alters expression of vascular endothelial growth factor (VEGF) receptor-1, leading to modulation of cyclic-GMP–dependent protein kinase-1 signaling and regression of cardiomyocyte hypertrophy (Zhou et al., 2009). Other reports show that copper chelation by tetrathiomolybdate in rats reduced FAK activity and VEGF expression in the lungs (Mizuno et al., 2012). Work from our group has demonstrated that copper complexes such as clioquinol-copper (10 μM) upregulate integrin-α2 and -α5 and induce src-dependent phosphorylation of EGFR in U87MG cells (Price et al., 2008). These studies demonstrate that copper can act at multiple external sites to modulate signaling pathways.

Downstream Kinase Pathways

PI3K-Akt

There is strong evidence that copper is a key regulator of intracellular signaling kinase cascades, particularly of MAPK pathways. Ostrakhovitch and colleagues demonstrated that copper-induced activation of a key cell survival pathway via PI3K and its downstream partner Akt (Ostrakhovitch et al., 2002). The study reported copper-dependent phosphorylation of Akt in skin fibroblasts (using 10–100 μM $CuSO_4$) or HeLa epithelial cancer cells (3 μM $CuSO_4$), leading to increased inhibitory phosphorylation of its downstream target, glycogen synthase kinase-3 (GSK3) (Ostrakhovitch et al., 2002). Interestingly, that study and others found that the ability of copper to activate the kinase pathways was independent of the generation of reactive oxygen species (ROS; Eckers et al., 2009), providing further important evidence that copper in itself is an essential modulator of cell signaling, and cellular responses to copper are not just in the context of oxidative stress. Activation of PI3K also increased phosphorylation and nuclear exclusion of FoxO transcription factor (TF) in HepG2 cells, independent of insulin receptor activation (Hamann et al., 2014). Copper-mediated Akt activation also involved FGF release in melanoma cells (Di Serio et al., 2008).

Extracellular Signal-Regulated Kinase

Recent studies have shown that copper may have an essential role in ERK activity. Modulation of copper transporter-1 (Ctr1) expression in cells revealed that when cells were deficient in this transporter (thus reducing copper uptake), ERK activity was substantially inhibited after exposure to receptor tyrosine kinase (RTK) ligands, FGF, platelet-derived growth factor (PDGF), and EGF (Tsai et al., 2012). Moreover, PDGF and EGF increased activation of PI3K in normal cells but not in Ctr1$^{-/-}$ cells. These findings were further supported by copper chelation (reducing ERK activity) and supplementation (enhancing activity), and they strongly indicate that copper modulated RTK-mediated cell signaling upstream of Ras. Tsai et al. also showed that lack of Ctr1 induced a dramatic decrease in SOD1 levels leading to reduced generation of H_2O_2 (Tsai et al., 2012). Thus copper-dependent H_2O_2 normally promotes RTK signaling through inhibition of phosphatases that restrict sustained RTK activation. Controversy regarding whether copper-dependent activation of signaling requires ROS generation may be attributed to an indirect role of H_2O_2 in activation of signal transduction by copper. Similar findings in *Drosophila* and mice (Turski et al., 2012) suggest that reduced MAPK kinase, MEK1 activity, and ERK phosphorylation can be induced by copper chelation or knockout or mutation of Ctr1. A recent study elegantly demonstrated for the first time that direct Cu binding to MEK1 was required for B-Raf (BRAF) kinase signaling (Brady et al., 2014). BRAF pathways control tumorigenesis in melanoma and other cancers; thus this study demonstrates the therapeutic potential of copper chelation to prevent BRAF signaling in tumors with constitutive-on BRAF mutations (Brady et al., 2014). The results indicate that copper is important in inducing MAPK signaling at several key checkpoints, including SOD1-mediated H_2O_2 production to sustain normal RTK activation and MEK1-mediated phosphorylation of ERK.

c-Jun N-Terminal Kinase

Studies on activation of the stress response kinase, c-Jun N-terminal kinase (JNK), have provided additional support for a key modulatory role by copper in cell signaling. As with the studies on PI3K signaling, there have been contrasting reports on whether copper control of JNK is ROS dependent or independent. ROS generation by copper was found to be associated with JNK activation in HL-60 cells by copper delivered using the cell permeable copper ionophore pyrrolidine dithiocarbamate (PDTC; EC$_{50}$ 100 μM together with 0.1 μM basal copper in the medium; Chen et al., 2008). Copper increased JNK activity and increased DNA binding of the TFs, nuclear factor-κB (NF-κB), and activator protein-1 (AP-1), although it was not shown in that paper whether this effect was JNK dependent. Treatment of Hep62 cells (hepatic cell line) with copper nanoparticles also induced JNK stimulation and AP-1 activation (Piret et al., 2012). In lung epithelial cells treated with PDTC-Cu (0.5 μM PDTC and 1 μM CuCl$_2$), JNK-dependent caspase-7 and -9 activation led to mitochondrial-mediated apoptosis (Chen et al., 2010). Disulfiram-mediated copper delivery to doxorubicin-resistant HL-60 cells (1 μM disulfiram and 1 μM CuCl$_2$) induced ROS, JNK activation, and cell death (Liu et al., 2012). In addition, oxidative stress and cell death induced by copper (10 μM CuCl$_2$ and 0.1 μM of the chelator neurocuproine) in astrocytes was dependent on JNK (Chen et al., 2008). Alternatively, copper treatment of COS-7 cells (400 μM CuSO$_4$) induced ROS production and lipid peroxidation but independently of copper-mediated JNK activation and downstream protooncogene targets (Mattie et al., 2008). Analogous findings have been reported for activation of p38 by copper (Moon et al., 2004). Unlike ERK signaling, in which copper directly binds to the upstream kinase, MEK1 (Brady et al., 2014), the precise molecular mechanism by which copper controls JNK (or p38) activation, whether involving ROS or not, has not yet been elucidated. In addition, the upstream components of copper-dependent JNK activation are currently unknown.

Intracellular Mechanisms of Copper-Mediated Cell Signaling

Phosphatase–Kinase Dynamics

There are several potential mechanisms by which copper may control kinase and related cell signaling processes. In some cases this may be due to direct binding to kinases and/or receptors, resulting in structural conformation and hence activity changes (Turski et al., 2012). In addition, copper can potently inhibit the activity of cellular phosphatases that are essential for modulation of kinase (and receptor) activation and deactivation. Copper inhibits PTP1b activity at submicromolar concentrations (Wang et al., 2010). We have shown that copper delivered into cells with a bis(thiosemicarbazonato) complex (Cu^{II}(gtsm)) inhibits PTP1b, resulting in prolonged EGFR-dependent growth factor and cytokine secretion (Price et al., 2009). This is consistent with a report by Urquiza et al., who demonstrated that copper inhibits alkaline phosphatase activity (ED$_{50}$ 42 μM), and additional studies reporting copper-mediated reductions in brain cytosolic phosphatase activity (Hanahisa and Yamaguchi, 1998; Urquiza et al., 2010).

Nuclear Signaling

Copper also controls cell signaling through modulation of nuclear TF activity and localization. Copper sensors containing copper-binding motifs are common in lower organisms, but copper-responsive TFs in mammals, such as MTF1 and Nrf2, lack clear copper binding sites. Copper was shown to inhibit Sp1 by binding in place of zinc to the zinc binding domain of this TF (Song et al., 2008). Studies have also shown that copper can modulate the key inflammatory TF, NF-κB (Chen et al., 2008; Song et al., 2009), either through upstream signal transduction pathways or via direct interactions of labile copper[(II)] with the NF-κB–inhibiting protein copper metabolism (Murr1) domain containing-1 (Sarkar and Roberts, 2011). Additional TFs activated by copper include AP-1 (Mattie et al., 2008) and the important cancer TF p53 (Hoti et al., 2004; Ostrakhovitch and Cherian, 2004). Copper also influences the nuclear receptors FXR/NR1H4 and GR/NR3C1 in hepatocytes, which may lead to copper control of gene expression linked to these receptors, including lipid and carbohydrate metabolism (Wilmarth et al., 2012).

COPPER AS A REGULATOR OF EPIGENETICS

Epigenetics is the study of heritable gene regulatory traits acquired during an individual's lifetime in response to environmental stimuli without alterations to the DNA sequence. The epigenetic code controls gene expression or silencing through (1) a series of reversible chemical modifications to the histones, proteins that package DNA into chromatin; (2) methylation of CpG dinucleotides in DNA; and (3) gene silencing mediated by noncoding RNAs (Goldberg et al., 2007).

The contribution of copper to epigenetic regulatory mechanisms has primarily been examined in the context of cancer using high micromolar concentrations of copper or copper-containing drugs as a means to selectively kill cancerous cells. Therefore our understanding of the physiological role of endogenous copper pools in epigenetic control of gene expression is still in its infancy. However, there is evidence that deregulation of copper homeostasis may be linked to dysfunction on an epigenetic level (Alsadany et al., 2013).

Histone Modification

Acetylation

A plethora of posttranslational modifications to histone proteins, including acetylation, phosphorylation, ubiquitination, and sumoylation, forms a code to direct highly coordinated chromatin expansion or condensation, to regulate gene activation or silencing, respectively, of the DNA bound to modified histones. Copper has been repeatedly associated with inhibition of histone acetylation (Cheng et al., 2012). Acetylation of positively charged histone proteins impairs their DNA-binding ability, thereby facilitating chromatin expansion, allowing TF binding to promoters and ultimately gene expression. Therefore global histone hypoacetylation is likely to repress gene transcription, as was described in a microarray study in human hepatoma HepG2 cells in response to high copper doses (Song et al., 2009). Levels of histone acetylation are controlled by the relative activities of histone acetyl transferase (HAT) and histone deacetylase (HDAC) proteins, and disruption of this balance had been linked to cancer. Emerging evidence indicates that under different experimental paradigms copper can directly or indirectly affect expression or activity of either class of enzymes.

Histone Deacetylase Activity

Copper toxicity in HepG2 cells involved upregulation of *HDAC4* and *HDAC10* expression (Song et al., 2009). A study in a cohort of 25 AD patients and 25 age-matched controls reported that copper and HDAC enzymes are both increased and significantly correlated in AD patient serum; thus they may represent a new diagnostic marker for AD (Alsadany et al., 2013). Interestingly, meta-analyses have determined that copper levels are reduced in AD brains (Schrag et al., 2011), and global histone acetylation was reported to be increased in pathologically affected regions in human postmortem AD brains (Narayan et al., 2015), which is consistent with a role for copper in inhibition of histone acetylation.

Histone Acetyl Transferase Activity

Several studies from Lanzhou University have demonstrated that inhibition of global histone acetylation at subcytotoxic copper doses occurs through modulation of HAT activity. Kang and colleagues found that 100–200 μM copper induced global hypoacetylation of histones and hypoacetylation of histones H3 and H4 in Hep3B hepatoma cells in a dose- and time-dependent manner that was mediated by bathocuproine-sulfonate–chelatable copper pools. Enzyme activity assays in cell and histone extracts in the presence of copper demonstrated that HAT and not HDAC activity was specifically affected (Kang et al., 2004). In a follow-up study, Lin et al. (2005) confirmed that copper (from 50 μM) also reduced histone acetylation in HL-60 human leukemia cells. The mechanism was dependent on generation of oxidative hydrogen peroxide and of superoxide radicals, potentially because of the redox properties of copper (Lin et al., 2005). However, both studies used $CuCl_2$ for cell treatments (which does not efficiently enter cells) and did not measure increases in cellular copper load by inductively coupled plasma–mass spectroscopy or alternative methods. Thus subtle increases in intracellular copper concentrations may also be expected to affect histones. Indeed, Chen et al. (2008) showed that in the presence of the copper ionophore PTDC, as little as 1 μM copper was sufficient to rapidly inhibit HAT-dependent histone acetylation in HL-60 cells and hepatoma Hep3B cells. Thus PTDC is likely to efficiently transport copper into cells, bypassing CTR1-dependent entry. Taken together, these studies suggest that an increase in intracellular labile copper pools may inhibit histone acetylation via HAT.

Histone Interactions

Alternatively, Cu-dependent epigenetic regulation may occur through modulation of histone gene expression, as reported in genetic copper overload $Atp7b^{-/-}$ mice modeling Wilson disease (Huster et al., 2007), or through direct copper binding to histone proteins. Inhibition of acetylation may involve binding of copper to histidine residues in the N-terminal tail of histone H4 and both N- and C-terminal tails of histones H2A and H2B, but these interactions have only been examined in cell-free in vitro conditions (i.e., Karavelas et al., 2005). Oxidation and hydroxyl radical production as a result of copper binding to histones may also play a role in copper-mediated DNA damage (Mylonas et al., 2001). The biological consequences of copper–histone interactions have not been directly investigated. It is also unclear how physiological copper pools gain access to histones as part of homeostatic gene regulation. Endogenous copper has been directly visualized in healthy nuclei by X-ray fluorescence microscopy techniques (Ralle et al., 2010), and a readily exchangeable copper pool in the nucleus may exist through MT or Atox1-dependent delivery (Cherian and Apostolova, 2000; Itoh et al., 2008). Nuclear-targeted fluorescent copper sensors may yield unprecedented advances in our understanding of copper in dynamic epigenetic processes.

Copper may also play a role in maintenance of histone acetylation. Several studies have indicated that Cu is required for HDAC inhibition by an endoplasmic-reticulum–localized protein with copper-dependent nuclease activity, reticulon, which is being developed as a potential anticancer target. Copper–reticulon complexes induced degradation of recombinant HDAC8 and reduced the activity of HDAC enzymes, as demonstrated by high-performance liquid chromatography enzyme activity assays on HeLa cell nuclear extracts and purified HDAC (Nepravishta et al., 2010). Interestingly, incubation of HDAC8 with 100 μM $CuCl_2$ did not result in the same fragmentation pattern as that induced by reticulon–copper complexes.

Histone Phosphorylation

Aside from its involvement in modulation of histone acetylation, copper was implicated in histone phosphorylation by human vaccinia-related kinase-1 (VRK1), which regulates cell cycle, DNA damage, and MAPK responses to cellular stress. Copper (100 μM) was found to inhibit VRK1-dependent H3 phosphorylation (Barcia-Sanjurjo et al., 2013). The biological consequences of H3 phosphorylation are complex and associated with condensed and extended chromatin configurations; therefore the effect of copper-dependent inhibition of H3 phosphorylation may depend on the cellular context. Nonetheless, modulation of histone acetylation and phosphorylation represent additional mechanisms of copper-mediated gene regulation.

DNA Methylation

Methylation of cytosine residues on CpG dinucleotides in DNA results in stable gene silencing of methylated regions by either interfering with TF binding or forming complexes with proteins involved in condensation of chromatin structures (reviewed in Goldberg et al., 2007). DNA methylation plays many roles in mammalian biology and development, including silencing repetitive sequences and X-chromosome inactivation in females, performed by the methyl transferase family of genes. Only one study has thus far evaluated the role of copper on DNA methylation in mammals. Medici and colleagues reported repression of the de novo DNA methyl transferase, *Dnmt3b*, which correlated with global DNA hypomethylation in a mouse model of Wilson disease (tx-j mice; Medici et al., 2013). Chelation of excess copper via penicillamine treatment restored DNA methylation levels, but it paradoxically resulted in further reductions in *Dnmt3b* expression. Thus the relationship between *Dnmt* expression and DNA methylation in response to copper warrants further exploration.

Noncoding RNAs

Noncoding RNAs can induce stable gene expression changes through translational repression by microRNAs (miRNAs) or mRNA degradation by short interfering RNAs. Although there is a wealth of evidence that copper modulates expression of multiple miRNAs in many plant species, studies in mammalian systems are entirely lacking. One study determined that copper exposure resulted in 28 differentially expressed miRNAs in the zebrafish olfactory system with functions relating to neurogenesis, morphology, and cell fate (Wang et al., 2013).

Given the importance of copper in control of multiple epigenetic mechanisms, and the role of copper in disease, there is a need to study the stable effects of copper on epigenetic regulation by examining global or targeted gene expression changes after a copper stimulus is removed. In addition, because it is becoming clear that a plethora of human diseases involve copper deficiency in various subcellular organelles, the effects of chelation of loosely bound endogenous copper pools on epigenetics warrant examination.

CELL–CELL SIGNALING REGULATED BY COPPER

Neuronal Signaling

Neuronal transmission occurs at synapses and is orchestrated through membrane-bound voltage-gated ion transporters. These transporters generate electrochemical action potentials across neuronal membranes through concentration gradients of ions including calcium, sodium, potassium, and chloride. Copper plays an integral but complex role in synaptic signaling and has been shown to directly bind multiple postsynaptic receptors, including excitatory N-methyl-D-aspartate (NMDA) and α-amino-3-hydroxy-5-methyl-4-isoxazolepropionic acid (AMPA) glutamate receptors as well as and inhibitory γ-aminobutyric acid (GABA) ionotropic type A or metabotropic $GABA_B$ receptors. Two decades of work demonstrate the potent effects of exogenously applied copper on neuronal signaling through modulatory effects on NMDA, AMPA, and GABA receptors and direct interactions with several neurotransmitters, as reviewed in Gaier et al. (2013) and Opazo et al. (2014) (Figs. 8.1–8.3). The copper content of human cerebrospinal fluid (CSF) is generally reported to be below 1 μM, although it is higher in hippocampi, consistent with the described role of copper in physiological memory functions (Gaier et al., 2013). High copper concentrations (~100 μM) have been detected in synaptic vesicles and synaptosomes and can be mobilized into the synapse upon depolarization with GABA receptor agonists (Kardos et al., 1989). Extracellular calcium-dependent copper release occurs through ATP7A trafficking to synapses in hippocampal neuron cultures in response to NMDA receptor stimulation (Schlief et al., 2005). The importance of vesicular copper on behavior was recently demonstrated in mice heterozygous for the peptide amidating monooxygenase (Pam) gene. These mice demonstrate compromised fear responses and long-term potentiation induction due to amygdala-specific loss of the copper proteins ATOX-1 and ATP7A (Gaier et al., 2014). This results in decreased vesicular copper, an effect rescued by copper supplementation, which has important implications for the role of dietary copper on neural signaling and behavior including memory and fear responses.

A seminal paper by Dodani et al. (2014) demonstrated for the first time that physiological labile copper modulates neuronal signaling, thus refuting the dogma that cellular copper stores are static and do not control neuronal signaling processes. Their study used a fluorescent copper-specific sensor, CF3, to show that endogenous, chelatable copper reversibly depresses calcium-dependent network excitability in hippocampal neurons and intact retinas. The same group had previously used a combination of X-ray fluorescence microscopy and fluorescent copper sensors to visualize calcium-dependent trafficking of copper toward neurites, elicited by depolarization (Dodani et al., 2011). Consistent with this, Castro et al. (2014) reported that the rapid and transient neocuprine-dependent reduction of excitability, observed by calcium transients in hippocampal neurons, required copper delivery.

The effects of copper application on neural activity appear to be biphasic. Rapid responses to copper have a net inhibitory effect on GABAergic and glutamatergic currents, an effect that has been replicated in neurons isolated from multiple central nervous system (CNS) sites, including hippocampus, cortex, olfactory bulb, and cerebellum (Opazo et al., 2014). However, Peters and colleagues noted that copper application for 3 h enhances neuronal transmission (Peters et al., 2011). Copper may additionally affect neuronal signaling through upregulation of the GluA1 subunit of AMPA receptors and increased membrane clustering (Peters et al., 2011). Nonetheless, there is evidence that the copper-dependent effects on synaptic transmission are temporary and are homeostatically regulated after 24 h (Peters et al., 2011). The effects of copper on NMDA receptors are concentration dependent in neonatal cerebellar granule cells: high copper is inhibitory whereas lower copper concentrations (<30 μM) facilitate calcium-dependent signaling (Marchetti et al., 2014). The effects of copper on GABA receptors have been attributed to direct copper binding to a specific motif in the second transmembrane domain of the α and β receptor subunits (Gaier et al., 2013). Oxidative mechanisms have been implicated in the effects of copper

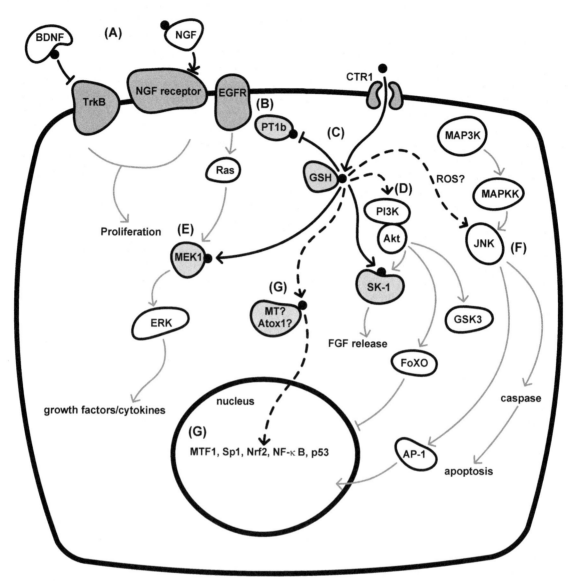

FIGURE 8.1 Copper-dependent modulation of cellular signal transduction pathways. (A) Extracellular copper can modulate cell signaling by directly binding to growth factors such as brain-derived neurotrophic factor (BDNF) or nerve growth factor (NGF). Copper binding to BDNF blocks tropomyosin receptor kinase B (TrkB)-dependent cell proliferation whereas NGF binding has been shown to have stimulatory or inhibitory effects on NGF-dependent proliferation. (B) Copper enters cells through the copper transporter-1 (Ctr1). (C) Loosely bound copper is stored in glutathione (GTH) within the cytoplasm. This exchangeable pool may modulate signaling cascades including phosphoinositol-3-kinase (PI3K) and mitogen-activated protein kinase (MAPK) pathways. (D) Copper has been shown to stimulate the PI3K/Akt pathway, leading to FoxO nuclear exclusion and glycogen synthase kinase-3 (GSK3) phosphorylation. Akt signaling may also occur through direct interactions of copper with the downstream sphingosine kinase-1 (SK-1) to promote cellular fibroblast growth factor (FGF) release. (E) Copper promotes epidermal growth factor receptor (EGFR) signaling through at least two mechanisms: direct activation of MAPK extracellular signal-regulated kinase kinase (MEK)1 and direct inhibition of PT1b, resulting in increased ERK-dependent growth factor secretion. (F) Copper induces c-Jun N-terminal kinase (JNK) activation by an unknown mechanism that may involve reactive oxygen species and results in activator protein-1 (AP-1) activation and caspase-dependent apoptosis. (G) Copper is able to modulate activity of numerous transcription factors, including metal-responsive zinc finger transcription factor (MTF1), Sp1, Nrf2, nuclear factor-κB (NF-κB), and p53, by as yet unknown mechanisms. Nuclear copper delivery may occur through antioxidant-1 (Atox1) or metallothionein (MT). ***Black circle***, copper; ***solid arrow***, stimulates, involving direct copper binding; ***solid T***, inhibits, involving direct copper binding; ***dashed arrow***, indirectly stimulates; ***lines*** culminating with both an ***arrowhead*** and a T, has been reported to either stimulate or inhibit, depending on the experimental paradigm. *FoXO*, forkhead box protein O1; *MAP3K*, mitogen activated protein kinase kinase kinase; *Nrf2*, nuclear factor erythroid 2 [NF-E2]-related factor 2; *PT1b*, protein tyrosine phosphatase 1b; *Sp1*, specificity protein 1.

on AMPA receptors whereas those on NMDA receptors are oxidation independent and occur because of nitrosylation of the receptor (Gaier et al., 2013). In addition to its effects on glutamate and GABA receptors, copper also inhibits P2X ATP-gated purinergic receptors (Huidobro-Toro et al., 2008), voltage-gated calcium channels (VGCCs; Gaier et al., 2013), and P2Y receptors by a mechanism that does not involve copper entry into postsynaptic neurons (Dolovcak et al., 2010).

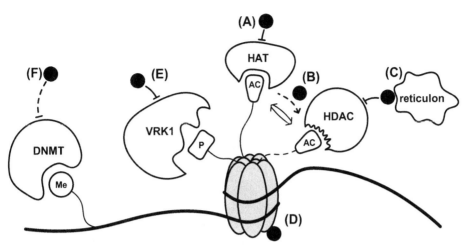

FIGURE 8.2 Copper-dependent modulation of epigenetics. (A) Copper induces hypoacetylation of histones H3 and H4. Copper can directly bind to and inhibit histone acetyl transferase (HAT) proteins. (B) Copper can induce upregulation of histone deacetylase (HDAC) proteins, but (C) copper reticulon complexes inhibit HDAC proteins, and (D) copper can directly bind to histone proteins H2, H3, and H4, although it is currently unclear how this affects epigenetic regulation. (E) Copper can inhibit histone H3 phosphorylation by vaccinia-related kinase-1 (VRK1). (F) Excess copper reduces DNA methylation through repression of DNA methyl transferase transcription. *Black circle*, copper; *solid T*, inhibits, involving direct copper binding; *dashed arrow*, indirectly stimulates; *dashed T*, indirectly inhibits. *AC*, acetyl co-enzyme A; *DNMT*, DNA methyltransferase.

Together, these studies demonstrate the key role of copper in neuronal signaling and clarify why diseases of copper overload or deficiency exert such profound effects on brain function.

Amyloid-β

Amyloid-β (Aβ), the pathological hallmark of AD, is a copper-binding neuronal peptide that is processed from amyloid precursor protein (APP) by sequential secretase-mediated cleavage. Copper binding to Aβ produces ROS and induces neurotoxic oligomerization, as reviewed in (Hung et al., 2010). APP itself also binds copper, which enhances APP trafficking to the plasma membrane and neurites.

Prion Protein

The cell membrane-localized cellular prion protein (PrP^c) binds copper via octapeptide repeat regions (Hornshaw et al., 1995). PrP^c-deficient mice display reduced membrane-bound copper and cannot elicit appropriate synaptic responses in cerebellar Purkinje cells in response to copper exposure (Brown et al., 1997). This may be due to altered copper import by loss of membrane-bound PrP^c. This is supported by the fact that cleavage of PrP^c from the cell surface reduced copper content in cerebellar cell membranes in wild-type cells, indicating that membrane-bound copper may be largely associated with PrP^c (Brown et al., 1997a). It has been recently shown that PrP^c is also cleaved to produce N-terminal fragments that have protective action against oxidative stress. This protective process appears to be regulated by copper binding to the octarepeat domain, resulting in modulation of MEK1 (Haigh et al., 2015). An additional role for copper and PrP^c in desensitization of NMDA signaling upon prolonged glutamate exposure has been reported as a physiological mechanism to prevent excitotoxicity (You et al., 2012), a process that can be blocked by Aβ.

HOW OTHER MINERALS ARE AFFECTED OR BEHAVE

Calcium

As described in *Neuronal Signaling*, copper profoundly influences calcium dynamics within neurons during synaptic transmission. A complex bidirectional relationship exists, whereby copper is released from cells via ATP7A trafficking to neurites upon calcium-dependent depolarization (Dodani et al., 2011) and copper release at the synapse in turn affects calcium flux into postsynaptic neurons (Gaier et al., 2013). This is achieved through modulation of multiple calcium-permeant receptors, including L-type and high VGCCs, NMDA, P2Y, and P2X receptors, in a context-specific manner (Gaier et al., 2013). Thus copper not only intricately affects calcium concentrations but also long-term behavioral changes triggered by subcellular calcium-dependent responses such as induction of long-term potentiation (Gaier et al., 2013). In addition to its

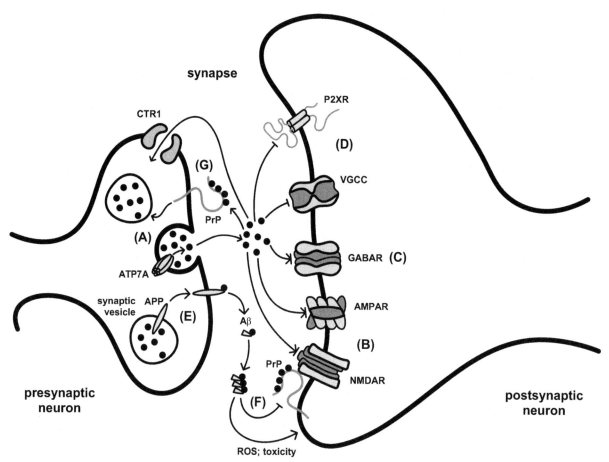

FIGURE 8.3 Copper-dependent modulation of neurotransmissions. (A) Copper is trafficked to vesicles via ATP7A, where it may interact with neurotransmitters. Extracellular release of copper is required for neurotransmission. (B) High copper concentrations or acute application of copper inhibits neurotransmission through α-amino-3-hydroxy-5-methyl-4-isoxazolepropionic acid (AMPA) and *N*-methyl-D-aspartate (NMDA) glutamate receptors (AMPAR and NMDAR, respectively) via oxidative or nitrosative mechanisms, respectively. (C) Copper directly interacts with and modulates signaling at γ-aminobutyric acid (GABA) receptors (GABAR). The biphasic activity of copper at glutamate and GABA receptors is represented by *lines* culminating with both an *arrowhead* and a *T*. (D) Copper can also inhibit P2X and voltage-gated calcium channel (VGCC) receptors. (E) Amyloid precursor protein (APP) is trafficked to the cell membrane in a copper-dependent manner, where it can be cleaved to amyloid-β (Aβ). Aβ interactions with copper induce oligomerization and reactive oxygen species (ROS)-dependent neurotoxicity in Alzheimer's disease. (F) Aβ can also interfere with the copper and prion protein (PrP)-dependent inhibition of NMDA-dependent excitotoxicity. (G) PrP may be implicated in copper import. Alternatively, copper reuptake may occur through copper transporter-1 (CTR1). *P2XR*, P2X purinergic receptor. *Adapted from Gaier, E.D., Eipper, B.A., Mains, R.E., 2013. Copper signaling in the mammalian nervous system: synaptic effects. J. Neurosci. Res. 91, 2–19.*

effect on calcium channels, copper was shown to robustly affect two two-pore–domain potassium channels that regulate neuronal excitability. Specifically, copper inhibited the TASK-3 channel and activated the TREK1 channel in HEK293 cells stably transfected with these receptors (Gruss et al., 2004).

Zinc

Subcellular zinc metabolism is closely intertwined with that of copper. Both metals play complementary roles because both are key second messengers and modulators of kinase function and neuronal signaling, an effect that is likely attributable to their propensity for binding biological molecules (Irving Williams series). Not only are the steady-state concentrations of copper and zinc correlated in the brains of human subjects without neurological disease (Krebs et al., 2014), but zinc and copper can be coregulated, cotransported, and functionally required by the same proteins. Indeed, several zinc transporters of the Zip family (Zip1, Zip2, and Zip4) can transport copper (Dempski, 2012), both metals are stored in exchangeable MT-bound pools in the cytoplasm, and both are required for antioxidant function as cofactors of SOD1. Moreover, either metal can regulate transcriptional programs that influence their own and each other's concentration, including expression of MT through the metal-responsive zinc finger transcription factor MTF1 (Mattie and Freedman, 2004). Thus even transient perturbation of one metal may be accompanied by alterations to the concentration and/or localization of the other, as is evident in neurological diseases in which the metabolism of both metals is commonly disrupted (Barnham and Bush, 2014).

Iron

The multifaceted intersection of copper and iron metabolism has been extensively reviewed elsewhere (Gulec and Collins, 2014). Importantly, copper can control master iron regulators at the systemic and cellular levels. Two of the mechanisms responsible for this are direct interactions of copper with the iron-regulatory hormone hepcidin (Tselepis et al., 2010) and stabilization of the TF hypoxia-inducible factor during normoxic but iron-deficient conditions (Martin et al., 2005). In an analogous manner to a subset of zinc transporters, iron transporter divalent metal-ion transporter-1 is reported to be copper permeable, although whether this is physiological is as yet unclear (reviewed in Gulec and Collins, 2014). Another direct example of copper involvement in iron biology is that of ceruloplasmin (Cp; Holmberg and Laurell, 1947), a ferroxidase requiring six copper ions for function, which plays a key role in iron efflux from tissues such as the CNS. An aberrant signaling role for Cp in Parkinson's disease was recently observed, involving oxidative-stress–induced deamidation of arginine residues in the asparagine-glycine-arginine motifs of Cp in CSF (Barbariga et al., 2015). Deamidation of Cp promoted loss of ferroxidase activity and gained integrin-$\alpha\nu\beta6$–binding properties, thereby switching on FAK1/ERK/Akt/MAPK signaling pathways to control transcriptional programs involved in cell proliferation and cytoskeletal reorganization (Barbariga et al., 2014).

PERSPECTIVES

This chapter described the role of copper in intracellular signal transduction, resulting in regulation of a wide range of processes through direct interactions with membrane receptor signaling components, modulation of kinase/phosphatase activity as a second messenger, and regulation of gene expression through transcriptional and epigenetic mechanisms. Copper is also critical for a range of intercellular signaling processes, as epitomized by its emerging role in synaptic signaling, through coordinated vesicular release and direct modulation of receptors at the synapse. Effects on long-term potentiation and behavior carry important implications for neurological disorders associated with copper deregulation.

Continued improvement of specific and sensitive fluorescent copper sensors, such as subcellular, organelle-specific targeting; ratiometric properties; and a range of affinities to probe numerous intracellular pools, will continue to enrich our understanding of endogenous copper dynamics and copper-dependent control of cell signaling pathways. It will be of utmost importance to include the relevant controls such as sensors without copper-binding domains to eliminate the potential effects of dye leakage or photobleaching (Dodani et al., 2014). In addition, studies are needed to more accurately/completely define a physiological and abnormal copper proteome. This could be achieved through a combination of copper-IMAC and bioinformatic approaches in mammalian systems on different genetic backgrounds affecting copper homeostasis (Wilson's and Menkes' diseases and CTR1 knockout) as well as biochemical copper-deficient (i.e., bathocuproine sulfonate chelation) and copper-replete conditions.

SUMMARY POINTS

- Copper is an important dietary ion with essential functions as a structural part of enzymes.
- Free copper is toxic because of its ability to cycle between two oxidation states; thus most cellular copper is tightly bound.
- Copper is directly transferred from membrane transporter proteins to carrier proteins, which deliver copper to copper enzymes.
- A loosely bound (readily exchangeable) copper pool exists in cells, and it is this copper pool that most likely participates in cell signaling processes.
- Copper can affect multiple outside-in signal transduction pathways by direct or indirect interactions with proteins at each step of the cascade.
- Copper has been shown to directly bind signaling receptors, kinases, phosphatases, and nuclear TF proteins.
- Copper can control epigenetics, or stable changes to gene expression in the absence of changes to an organism's genome sequence, by controlling how tightly DNA is coiled.
- Copper control of epigenetics can occur through modification of the DNA packaging histone proteins or through changes to DNA methylation.
- Copper is normally released at neuronal synapses, and this is required for neurotransmission, memory formation, and fear responses.

- Copper can modulate neurotransmission by direct interactions with glutamate and GABA receptors on postsynaptic neurons.
- Copper metabolism profoundly affects the metabolism of other metals.
- Continued development of specific and sensitive fluorescent copper sensors to probe subcellular copper pools involved in signaling is essential for advancement of this field.

MINI DICTIONARY OF TERMS

AMPA receptor α-Amino-3-hydroxy-5-methyl-4-isoxazolepropionic acid; an excitatory receptor expressed on postsynaptic neurons responding to the neurotransmitter glutamate.

Copper binding site A specific sequence of amino acids that directly binds to and stabilizes copper in proteins, usually containing a histidine, cysteine, or methionine residues.

Epigenetics Stable changes to gene expression in the absence of changes to DNA sequence (e.g., tissue-specific gene expression). Controlled by DNA methylation, histone modification, and noncoding RNAs.

GABA receptor γ-Aminobutyric acid type A; an inhibitory receptor expressed on postsynaptic neurons responding to the neurotransmitter GABA.

Glutathione Small cellular antioxidant peptide that loosely binds copper.

HAT Histone acetyl transferase, a protein that adds an acetyl group to particular lysine amino acids on the N-terminal tails of histone proteins.

HDAC Histone deacetylase, a protein that removes acetyl groups from particular lysine amino acids on the N-terminal tails of histone proteins; performs the opposite reaction to a histone acetyl transferase.

Histone Small positively charged proteins that tightly wind around DNA and package DNA into chromatin. Specific combinations of chemical modifications on histone proteins direct whether the chromatin is in an open or closed state and whether the DNA is actively transcribed.

IMAC Immobilized metal affinity chromatography, a technique used to selectively capture metal-bound proteins from an organism's proteome.

Kinase A protein that adds a phosphate group to other proteins; kinases are integral players in signal transduction pathways.

NMDA receptor N-methyl-D-aspartate; an excitatory receptor expressed on postsynaptic neurons responding to the neurotransmitter, glutamate.

Phosphatase A protein that removes phosphate groups from other proteins; performs the opposite reaction to kinases.

Prion protein A neuronal protein expressed on the cell surface that strongly interacts with copper. Improperly folded prion protein is the infectious agent in several diseases termed transmissible spongiform encephalopathies, including scrapie and mad cow disease.

Signal transduction An organism's response to extracellular stimuli, such as growth factors, involving binding of a chemical signal to a membrane receptor, which generally induces a conformational change that causes intracellular adaptor proteins to bind to the receptor. This then initiates a cascade of protein–protein interactions that result in a change in gene expression, phenotype, and/or secretion of effectors in response to the initiating signal.

REFERENCES

Alsadany, M.A., Shehata, H.H., Mohamad, M.I., Mahfouz, R.G., 2013. Histone deacetylases enzyme, copper, and IL-8 levels in patients with Alzheimer's disease. Am. J. Alzheimers Dis. Other Demen. 28, 54–61.

Andreini, C., Banci, L., Bertini, I., Rosato, A., 2008. Occurrence of copper proteins through the three domains of life: a bioinformatic approach. J. Proteome Res. 7, 209–216.

Andreini, C., Bertini, I., Rosato, A., 2009. Metalloproteomes: a bioinformatic approach. Acc. Chem. Res. 42, 1471–1479.

Banci, L., Bertini, I., Cantini, F., Ciofi-Baffoni, S., 2010. Cellular copper distribution: a mechanistic systems biology approach. Cell Mol. Life Sci. 67, 2563–2589.

Barbariga, M., Curnis, F., Andolfo, A., Zanardi, A., Lazzaro, M., Conti, A., Magnani, G., Volonte, M.A., Ferrari, L., Comi, G., Corti, A., Alessio, M., 2015. Ceruloplasmin functional changes in Parkinson's disease-cerebrospinal fluid. Mol. Neurodegener. 10, 59.

Barbariga, M., Curnis, F., Spitaleri, A., Andolfo, A., Zucchelli, C., Lazzaro, M., Magnani, G., Musco, G., Corti, A., Alessio, M., 2014. Oxidation-induced structural changes of ceruloplasmin foster NGR motif deamidation that promotes integrin binding and signaling. J. Biol. Chem. 289, 3736–3748.

Barcia-Sanjurjo, I., Vazquez-Cedeira, M., Barcia, R., Lazo, P.A., 2013. Sensitivity of the kinase activity of human vaccinia-related kinase proteins to toxic metals. J. Biol. Inorg. Chem. 18, 473–482.

Barnham, K.J., Bush, A.I., 2014. Biological metals and metal-targeting compounds in major neurodegenerative diseases. Chem. Soc. Rev. 43, 6727–6749.

Birkaya, B., Aletta, J., 2005. NGF promotes copper accumulation required for optimum neurite outgrowth and protein methylation. J. Neurobiol. 63, 49–61.

Brady, D.C., Crowe, M.S., Turski, M.L., Hobbs, G.A., Yao, X., Chaikuad, A., Knapp, S., Xiao, K., Campbell, S.L., Thiele, D.J., Counter, C.M., 2014. Copper is required for oncogenic BRAF signalling and tumorigenesis. Nature 509, 492–496.

Brown, D.R., Qin, K., Herms, J.W., Madlung, A., Manson, J., Strome, R., Fraser, P.E., Kruck, T., von Bohlen, A., Schulz-Schaeffer, W., Giese, A., Westaway, D., Kretzschmar, H., 1997. The cellular prion protein binds copper in vivo. Nature 390, 684–687.

Castro, P.A., Ramirez, A., Sepulveda, F.J., Peters, C., Fierro, H., Waldron, J., Luza, S., Fuentealba, J., Munoz, F.J., De Ferrari, G.V., Bush, A.I., Aguayo, L.G., Opazo, C.M., 2014. Copper-uptake is critical for the down regulation of synapsin and dynamin induced by neocuproine: modulation of synaptic activity in hippocampal neurons. Front. Aging Neurosci. 6, 319.

Chen, J., Du, C., Kang, J., Wang, J., 2008. Cu2+ is required for pyrrolidine dithiocarbamate to inhibit histone acetylation and induce human leukemia cell apoptosis. Chem. Biol. Interact. 171, 26–36.

Chen, S.H., Lin, J.K., Liang, Y.C., Pan, M.H., Liu, S.H., Lin-Shiau, S.Y., 2008. Involvement of activating transcription factors JNK, NF-kappaB, and AP-1 in apoptosis induced by pyrrolidine dithiocarbamate/Cu complex. Eur. J. Pharmacol. 594, 9–17.

Chen, S.H., Lin, J.K., Liu, S.H., Liang, Y.C., Lin-Shiau, S.Y., 2008. Apoptosis of cultured astrocytes induced by the copper and neocuproine complex through oxidative stress and JNK activation. Toxicol. Sci. 102, 138–149.

Chen, Y., Chen, K., Chen, C., Wu, H., Su, C., Wu, C., Way, T., Hung, D., Yen, C., Yang, Y., Lu, T., 2010. Pyrrolidine dithiocarbamate (PDTC)/Cu complex induces lung epithelial cell apoptosis through mitochondria and ER-stress pathways. Toxicol. Lett. 199, 333–340.

Cheng, T.F., Choudhuri, S., Muldoon-Jacobs, K., 2012. Epigenetic targets of some toxicologically relevant metals: a review of the literature. J. Appl. Toxicol. 32, 643–653.

Cherian, M.G., Apostolova, M.D., 2000. Nuclear localization of metallothionein during cell proliferation and differentiation. Cell Mol. Biol. (Noisy-le-grand) 46, 347–356.

Dempski, R.E., 2012. The cation selectivity of the ZIP transporters. Curr. Top. Membr. 69, 221–245.

Di Serio, C., Doria, L., Pellerito, S., Prudovsky, I., Micucci, I., Massi, D., Landriscina, M., Marchionni, N., Masotti, G., Tarantini, F., 2008. The release of fibroblast growth factor-1 from melanoma cells requires copper ions and is mediated by phosphatidylinositol 3-kinase/Akt intracellular signaling pathway. Cancer Lett. 267, 67–74.

Dodani, S.C., Domaille, D.W., Nam, C.I., Miller, E.W., Finney, L.A., Vogt, S., Chang, C.J., 2011. Calcium-dependent copper redistributions in neuronal cells revealed by a fluorescent copper sensor and X-ray fluorescence microscopy. Proc. Natl. Acad. Sci. U.S.A. 108, 5980–5985.

Dodani, S.C., Firl, A., Chan, J., Nam, C.I., Aron, A.T., Onak, C.S., Ramos-Torres, K.M., Paek, J., Webster, C.M., Feller, M.B., Chang, C.J., 2014. Copper is an endogenous modulator of neural circuit spontaneous activity. Proc. Natl. Acad. Sci. U.S.A. 111, 16280–16285.

Dolovcak, S., Waldrop, S.L., Fitz, J.G., Kilic, G., 2010. Copper inhibits P2Y(2)-dependent Ca(2+) signaling through the effects on thapsigargin-sensitive Ca(2+) stores in HTC hepatoma cells. Biochem. Biophys. Res. Commun. 397, 493–498.

Eckers, A., Reimann, K., Klotz, L., 2009. Nickel and copper ion-induced stress signaling in human hepatoma cells: analysis of phosphoinositide 3′-kinase/Akt signaling. Biometals 22, 307–316.

Fan, C., Su, H., Zhao, J., Zhao, B., Zhang, S., Miao, J., 2010. A novel copper complex of salicylaldehyde pyrazole hydrazone induces apoptosis through up-regulating integrin beta4 in H322 lung carcinoma cells. Eur. J. Med. Chem. 45, 1438–1446.

Fan, C., Zhao, J., Zhao, B., Zhang, S., Miao, J., 2009. Novel complex of copper and a salicylaldehyde pyrazole hydrazone derivative induces apoptosis through up-regulating integrin beta 4 in vascular endothelial cells. Chem. Res. Toxicol. 22, 1517–1525.

Foster, A.W., Osman, D., Robinson, N.J., 2014. Metal preferences and metallation. J. Biol. Chem. 289, 28095–28103.

Gaier, E.D., Eipper, B.A., Mains, R.E., 2013. Copper signaling in the mammalian nervous system: synaptic effects. J. Neurosci. Res. 91, 2–19.

Gaier, E.D., Rodriguiz, R.M., Zhou, J., Ralle, M., Wetsel, W.C., Eipper, B.A., Mains, R.E., 2014. In vivo and in vitro analyses of amygdalar function reveal a role for copper. J. Neurophysiol. 111, 1927–1939.

Goldberg, A.D., Allis, C.D., Bernstein, E., 2007. Epigenetics: a landscape takes shape. Cell 128, 635–638.

Grubman, A., White, A.R., 2014. Copper as a key regulator of cell signalling pathways. Expert Rev. Mol. Med. 16, e11.

Gruss, M., Mathie, A., Lieb, W.R., Franks, N.P., 2004. The two-pore-domain K(+) channels TREK-1 and TASK-3 are differentially modulated by copper and zinc. Mol. Pharmacol. 66, 530–537.

Gulec, S., Collins, J.F., 2014. Molecular mediators governing iron-copper interactions. Annu. Rev. Nutr. 34, 95–116.

Haigh, C.L., McGlade, A.R., Collins, S.J., 2015. MEK1 transduces the prion protein N2 fragment antioxidant effects. Cell Mol. Life Sci. 72, 1613–1629.

Hamann, I., Petroll, K., Grimm, L., Hartwig, A., Klotz, L.O., 2014. Insulin-like modulation of Akt/FoxO signaling by copper ions is independent of insulin receptor. Arch. Biochem. Biophys. 558, 42–50.

Hanahisa, Y., Yamaguchi, M., 1998. Inhibitory effect of zinc and copper on phosphatase activity in the brain cytosol of rats: involvement of SH groups. Biol. Pharm. Bull. 21, 1222–1225.

Hehlgans, S., Haase, M., Cordes, N., 2007. Signalling via integrins: implications for cell survival and anticancer strategies. Biochim. Biophys. Acta 1775, 163–180.

Holmberg, C.G., Laurell, C.B., 1947. Investigations in serum copper; nature of serum copper and its relation to the iron-binding protein in human serum. Acta Chem. Scand. 1, 944–950.

Hornshaw, M.P., McDermott, J.R., Candy, J.M., Lakey, J.H., 1995. Copper binding to the N-terminal tandem repeat region of mammalian and avian prion protein: structural studies using synthetic peptides. Biochem. Biophys. Res. Commun. 214, 993–999.

Hoti, N., Zhu, D., Song, Z., Wu, Z., Tabassum, S., Wu, M., 2004. p53-dependent apoptotic mechanism of a new designer bimetallic compound tri-phenyl tin benzimidazolethiol copper chloride (TPT-CuCl2): in vivo studies in Wistar rats as well as in vitro studies in human cervical cancer cells. J. Pharmacol. Exp. Ther. 311, 22–33.

Huidobro-Toro, J.P., Lorca, R.A., Coddou, C., 2008. Trace metals in the brain: allosteric modulators of ligand-gated receptor channels, the case of ATP-gated P2X receptors. Eur. Biophys. J. 37, 301–314.

Hung, Y.H., Bush, A.I., Cherny, R.A., 2010. Copper in the brain and Alzheimer's disease. J. Biol. Inorg. Chem. 15, 61–76.

Huster, D., Purnat, T.D., Burkhead, J.L., Ralle, M., Fiehn, O., Stuckert, F., Olson, N.E., Teupser, D., Lutsenko, S., 2007. High copper selectively alters lipid metabolism and cell cycle machinery in the mouse model of Wilson disease. J. Biol. Chem. 282, 8343–8355.

Itoh, S., Kim, H.W., Nakagawa, O., Ozumi, K., Lessner, S.M., Aoki, H., Akram, K., McKinney, R.D., Ushio-Fukai, M., Fukai, T., 2008. Novel role of antioxidant-1 (Atox1) as a copper-dependent transcription factor involved in cell proliferation. J. Biol. Chem. 283, 9157–9167.

Kang, J., Lin, C., Chen, J., Liu, Q., 2004. Copper induces histone hypoacetylation through directly inhibiting histone acetyltransferase activity. Chem. Biol. Interact. 148, 115–123.

Kang, Y., Choi, H., Na, J., Huh, C., Kim, M., Youn, S., Kim, K., Park, K., 2009. Copper-GHK increases integrin expression and p63 positivity by keratinocytes. Arch. Dermatol. Res. 301, 301–306.

Karavelas, T., Mylonas, M., Malandrinos, G., Plakatouras, J.C., Hadjiliadis, N., Mlynarz, P., Kozlowski, H., 2005. Coordination properties of Cu(II) and Ni(II) ions towards the C-terminal peptide fragment -ELAKHA- of histone H2B. J. Inorg. Biochem. 99, 606–615.

Kardos, J., Kovacs, I., Hajos, F., Kalman, M., Simonyi, M., 1989. Nerve endings from rat brain tissue release copper upon depolarization. A possible role in regulating neuronal excitability. Neurosci. Lett. 103, 139–144.

Krebs, N., Langkammer, C., Goessler, W., Ropele, S., Fazekas, F., Yen, K., Scheurer, E., 2014. Assessment of trace elements in human brain using inductively coupled plasma mass spectrometry. J. Trace Elem. Med. Biol. 28, 1–7.

Lin, C., Kang, J., Zheng, R., 2005. Oxidative stress is involved in inhibition of copper on histone acetylation in cells. Chem. Biol. Interact. 151, 167–176.

Liu, P., Brown, S., Goktug, T., Channathodiyil, P., Kannappan, V., Hugnot, J., Guichet, P., Bian, X., Armesilla, A., Darling, J., Wang, W., 2012. Cytotoxic effect of disulfiram/copper on human glioblastoma cell lines and ALDH-positive cancer-stem-like cells. Br. J. Cancer 107, 1488–1497.

Marchetti, C., Baranowska-Bosiacka, I., Gavazzo, P., 2014. Multiple effects of copper on NMDA receptor currents. Brain Res. 1542, 20–31.

Martin, F., Linden, T., Katschinski, D.M., Oehme, F., Flamme, I., Mukhopadhyay, C.K., Eckhardt, K., Troger, J., Barth, S., Camenisch, G., Wenger, R.H., 2005. Copper-dependent activation of hypoxia-inducible factor (HIF)-1: implications for ceruloplasmin regulation. Blood 105, 4613–4619.

Mattie, M., McElwee, M., Freedman, J., 2008. Mechanism of copper-activated transcription: activation of AP-1, and the JNK/SAPK and p38 signal transduction pathways. J. Mol. Biol. 383, 1008–1018.

Mattie, M.D., Freedman, J.H., 2004. Copper-inducible transcription: regulation by metal- and oxidative stress-responsive pathways. Am. J. Physiol. Cell Physiol. 286, C293–C301.

Medici, V., Shibata, N.M., Kharbanda, K.K., LaSalle, J.M., Woods, R., Liu, S., Engelberg, J.A., Devaraj, S., Torok, N.J., Jiang, J.X., Havel, P.J., Lonnerdal, B., Kim, K., Halsted, C.H., 2013. Wilson's disease: changes in methionine metabolism and inflammation affect global DNA methylation in early liver disease. Hepatology 57, 555–565.

Mizuno, S., Yasuo, M., Bogaard, H., Kraskauskas, D., Alhussaini, A., Gomez-Arroyo, J., Farkas, D., Farkas, L., Voelkel, N., 2012. Copper deficiency induced emphysema is associated with focal adhesion kinase inactivation. PLoS One 7, e30678.

Moon, S., Jung, S., Choi, Y., Lee, Y., Patterson, C., Kim, C., 2004. PDTC, metal chelating compound, induces G1 phase cell cycle arrest in vascular smooth muscle cells through inducing p21Cip1 expression: involvement of p38 mitogen activated protein kinase. J. Cell Physiol. 198, 310–323.

Mylonas, M., Malandrinos, G., Plakatouras, J., Hadjiliadis, N., Kasprzak, K.S., Krezel, A., Bal, W., 2001. Stray Cu(II) may cause oxidative damage when coordinated to the -TESHHK- sequence derived from the C-terminal tail of histone H2A. Chem. Res. Toxicol. 14, 1177–1183.

Narayan, P.J., Lill, C., Faull, R., Curtis, M.A., Dragunow, M., 2015. Increased acetyl and total histone levels in post-mortem Alzheimer's disease brain. Neurobiol. Dis. 74, 281–294.

Nepravishta, R., Bellomaria, A., Polizio, F., Paci, M., Melino, S., 2010. Reticulon RTN1-C(CT) peptide: a potential nuclease and inhibitor of histone deacetylase enzymes. Biochemistry 49, 252–258.

Opazo, C.M., Greenough, M.A., Bush, A.I., 2014. Copper: from neurotransmission to neuroproteostasis. Front. Aging Neurosci. 6, 143.

Ostrakhovitch, E., Cherian, M., 2004. Differential regulation of signal transduction pathways in wild type and mutated p53 breast cancer epithelial cells by copper and zinc. Arch. Biochem. Biophys. 423, 351–361.

Ostrakhovitch, E., Lordnejad, M., Schliess, F., Sies, H., Klotz, L., 2002. Copper ions strongly activate the phosphoinositide-3-kinase/Akt pathway independent of the generation of reactive oxygen species. Arch. Biochem. Biophys. 397, 232–239.

Peters, C., Munoz, B., Sepulveda, F.J., Urrutia, J., Quiroz, M., Luza, S., De Ferrari, G.V., Aguayo, L.G., Opazo, C., 2011. Biphasic effects of copper on neurotransmission in rat hippocampal neurons. J. Neurochem. 119, 78–88.

Piret, J., Jacques, D., Audinot, J., Mejia, J., Boilan, E., Noel, F., Fransolet, M., Demazy, C., Lucas, S., Saout, C., Toussaint, O., 2012. Copper(II) oxide nanoparticles penetrate into HepG2 cells, exert cytotoxicity via oxidative stress and induce pro-inflammatory response. Nanoscale 4, 7168–7184.

Price, K., Filiz, G., Caragounis, A., Du, T., Laughton, K., Masters, C., Sharples, R., Hill, A., Li, Q., Donnelly, P., Barnham, K., Crouch, P., White, A., 2008. Activation of epidermal growth factor receptor by metal-ligand complexes decreases levels of extracellular amyloid beta peptide. Int. J. Biochem. Cell Biol. 40, 1901–1917.

Price, K.A., Caragounis, A., Paterson, B.M., Filiz, G., Volitakis, I., Masters, C.L., Barnham, K.J., Donnelly, P.S., Crouch, P.J., White, A.R., 2009. Sustained activation of glial cell epidermal growth factor receptor by bis(thiosemicarbazonato) metal complexes is associated with inhibition of protein tyrosine phosphatase activity. J. Med. Chem. 52, 6606–6620.

Ralle, M., Huster, D., Vogt, S., Schirrmeister, W., Burkhead, J.L., Capps, T.R., Gray, L., Lai, B., Maryon, E., Lutsenko, S., 2010. Wilson disease at a single cell level: intracellular copper trafficking activates compartment-specific responses in hepatocytes. J. Biol. Chem. 285, 30875–30883.

Sarkar, B., Roberts, E.A., 2011. The puzzle posed by COMMD1, a newly discovered protein binding Cu(II). Metallomics 3, 20–27.

Schlief, M.L., Craig, A.M., Gitlin, J.D., 2005. NMDA receptor activation mediates copper homeostasis in hippocampal neurons. J. Neurosci. 25, 239–246.

Schrag, M., Mueller, C., Oyoyo, U., Smith, M.A., Kirsch, W.M., 2011. Iron, zinc and copper in the Alzheimer's disease brain: a quantitative meta-analysis. Some insight on the influence of citation bias on scientific opinion. Prog. Neurobiol. 94, 296–306.

She, Y.M., Narindrasorasak, S., Yang, S., Spitale, N., Roberts, E.A., Sarkar, B., 2003. Identification of metal-binding proteins in human hepatoma lines by immobilized metal affinity chromatography and mass spectrometry. Mol. Cell Proteomics 2, 1306–1318.

Soldi, R., Mandinova, A., Venkataraman, K., Hla, T., Vadas, M., Pitson, S., Duarte, M., Graziani, I., Kolev, V., Kacer, D., Kirov, A., Maciag, T., Prudovsky, I., 2007. Sphingosine kinase 1 is a critical component of the copper-dependent FGF1 export pathway. Exp. Cell Res. 313, 3308–3318.

Song, I.S., Chen, H.H., Aiba, I., Hossain, A., Liang, Z.D., Klomp, L.W., Kuo, M.T., 2008. Transcription factor Sp1 plays an important role in the regulation of copper homeostasis in mammalian cells. Mol. Pharmacol. 74, 705–713.

Song, M.O., Li, J., Freedman, J.H., 2009. Physiological and toxicological transcriptome changes in HepG2 cells exposed to copper. Physiol. Genomics 38, 386–401.

Tenaud, I., Leroy, S., Chebassier, N., Dreno, B., 2000. Zinc, copper and manganese enhanced keratinocyte migration through a functional modulation of keratinocyte integrins. Exp. Dermatol. 9, 407–416.

Travaglia, A., Arena, G., Fattorusso, R., Isernia, C., La Mendola, D., Malgieri, G., Nicoletti, V., Rizzarelli, E., 2011. The inorganic perspective of nerve growth factor: interactions of Cu^{2+} and Zn^{2+} with the N-terminus fragment of nerve growth factor encompassing the recognition domain of the TrkA receptor. Chemistry 17, 3726–3738.

Travaglia, A., La Mendola, D., Magri, A., Nicoletti, V., Pietropaolo, A., Rizzarelli, E., 2012. Copper, BDNF and its N-terminal domain: inorganic features and biological perspectives. Chemistry 18, 15618–15631.

Tsai, C., Finley, J., Ali, S., Patel, H., Howell, S., 2012. Copper influx transporter 1 is required for FGF, PDGF and EGF-induced MAPK signaling. Biochem. Pharmacol. 84, 1007–1013.

Tselepis, C., Ford, S.J., McKie, A.T., Vogel, W., Zoller, H., Simpson, R.J., Diaz Castro, J., Iqbal, T.H., Ward, D.G., 2010. Characterization of the transition-metal-binding properties of hepcidin. Biochem. J. 427, 289–296.

Turski, M., Brady, D., Kim, H., Kim, B., Nose, Y., Counter, C., Winge, D., Thiele, D., 2012. A novel role for copper in Ras/mitogen-activated protein kinase signaling. Mol. Cell Biol. 32, 1284–1295.

Urquiza, N., Manca, S., Moyano, M., Dellmans, R., Lezama, L., Rojo, T., Naso, L., Williams, P., Ferrer, E., 2010. Copper(II) complexes of methimazole, an anti Grave's disease drug. Synthesis, characterization and its potential biological behavior as alkaline phosphatase inhibitor. Biometals 23, 255–264.

Wang, L., Bammler, T.K., Beyer, R.P., Gallagher, E.P., 2013. Copper-induced deregulation of microRNA expression in the zebrafish olfactory system. Environ. Sci. Technol. 47, 7466–7474.

Wang, Q., Lu, L., Yuan, C., Pei, K., Liu, Z., Guo, M., Zhu, M., 2010. Potent inhibition of protein tyrosine phosphatase 1B by copper complexes: implications for copper toxicity in biological systems. Chem. Commun. (Camb.) 46, 3547–3549.

Wang, W., Post, J., Dow, K., Shin, S., Riopelle, R., Ross, G., 1999. Zinc and copper inhibit nerve growth factor-mediated protection from oxidative stress-induced apoptosis. Neurosci. Lett. 259, 115–118.

Watanabe, M., Tezuka, M., 2006. Copper binding of nuclear RNA- and DNA-binding proteins p54nrb/NonO and PSF. Biomed. Res. Trace Elem. 17, 328–331.

Weaver, M., Workman, G., Sage, E., 2008. The copper binding domain of SPARC mediates cell survival in vitro via interaction with integrin beta1 and activation of integrin-linked kinase. J. Biol. Chem. 283, 22826–22837.

Wilmarth, P.A., Short, K.K., Fiehn, O., Lutsenko, S., David, L.L., Burkhead, J.L., 2012. A systems approach implicates nuclear receptor targeting in the Atp7b(−/−) mouse model of Wilson's disease. Metallomics 4, 660–668.

Wu, W., Graves, L., Jaspers, I., Devlin, R., Reed, W., Samet, J., 1999. Activation of the EGF receptor signaling pathway in human airway epithelial cells exposed to metals. Am. J. Physiol. 277, L924–L931.

You, H., Tsutsui, S., Hameed, S., Kannanayakal, T.J., Chen, L., Xia, P., Engbers, J.D., Lipton, S.A., Stys, P.K., Zamponi, G.W., 2012. Abeta neurotoxicity depends on interactions between copper ions, prion protein, and N-methyl-D-aspartate receptors. Proc. Natl. Acad. Sci. U.S.A. 109, 1737–1742.

Zhou, Y., Bourcy, K., Kang, Y., 2009. Copper-induced regression of cardiomyocyte hypertrophy is associated with enhanced vascular endothelial growth factor receptor-1 signalling pathway. Cardiovasc. Res. 84, 54–63.

Chapter 9

Copper and Hypoxia-Inducible Transcription Factor Regulation of Gene Expression

Wenjing Zhang[1], Ying Xiao[1], Wen Yin[1], Tao Wang[1], Yujian James Kang[1,2]

[1]Sichuan University, Chengdu, Sichuan, China; [2]University of Louisville, School of Medicine, Louisville, KY, United States

INTRODUCTION TO HYPOXIA-INDUCIBLE FACTORS

Molecular Features of Hypoxia-Inducible Factor-1

Hypoxia-inducible factors (HIFs) are master regulators of oxygen homeostasis, which control angiogenesis, erythropoiesis, glycolysis, and cell survival/proliferation under normal and pathologic conditions. HIFs are DNA-binding heterodimers composed of an α subunit (HIF-1α or HIF-2α) and a 91- to 94-kDa HIF-1β subunit (also known as aryl hydrocarbon receptor nuclear translocater; Wang and Semenza, 1993a,b; 1995). Both subunits are basic helix-loop-helix proteins containing a PAS domain (first identified in Per, Arnt, and Sim proteins; Wang et al., 1995). HIF-1α and HIF-1β subunits dimerize via helix-loop-helix and PAS domains and bind to DNA via basic domains located in the amino-terminal half of each protein (Jiang et al., 1996).

The carboxy-terminal half of HIF-1α contains a transactivation domain (Jiang et al., 1997) that interacts with the coactivator CBP/p300 (Arany et al., 1996; Kallio et al., 1998) and an oxygen-dependent degradation domain (O'Rourke et al., 1999). As a result of the oxygen-dependent domain, the level of HIF-1α is a rate-limiting factor in HIF-1 activation (Huang et al., 1998; Wang and Semenza, 1993b). To activate target gene transcription, HIF-1 interacts with CBP/p300 and binds to hypoxia-response elements (HREs) that contain the core HIF-1 binding site sequence 5'-RCGTG-3' (where R stands for G or A; Semenza et al., 1996).

Under normoxia, the HIF-1α protein is continuously synthesized and simultaneously degraded by the ubiquitin-proteasome system, resulting in a stable level of HIF-1α protein, which in most cases is undetectable (Huang et al., 1998; Salceda and Caro, 1997). The process of HIF-1α degradation is oxygen dependent and involves iron-dependent prolyl hydroxylase domain proteins (PHDs). There are three PHDs (PHD1, PHD2, and PHD3) known in mammals. PHDs require oxygen for hydroxylation as well as the tricarboxylic acid cycle intermediate, 2-oxoglutarate (α-ketoglutarate), iron (Fe^{2+}), and ascorbate as cofactors (Schofield and Zhang, 1999). In the presence of oxygen, PHDs hydroxylate two proline residues (P564 and P402) located within the oxygen-dependent degradation domain of HIF-1α (Huang et al., 1998; Ivan et al., 2001; Jaakkola et al., 2001). These hydroxylation motifs act as the docking site for the von Hippel-Lindau tumor suppressor protein (VHL). VHL forms part of a complex with elongins B and C, cullin-2, and Rbx1 to constitute a functional E3 ubiquitin–protein ligase (Iwai et al., 1999). Subsequently, the hydroxylated prolines become ubiquitinated and the entire HIF-1α protein undergoes proteasomal degradation.

In addition to PHDs, another defense strategy involves the asparaginyl hydroxylase factor inhibiting HIF-1 (FIH-1), which is located in the nucleus to prevent the remaining HIF-1α from activating transcription of target genes during normoxia (Mahon et al., 2001). FIH-1 hydroxylates the N803 asparagine residue located within the C-terminal transactivation domain of HIF-1α, which prevents recruitment of the coactivator CBP/p300 and the subsequent transactivation of target genes (Lando et al., 2002a,b; Linke et al., 2004).

Under hypoxic conditions, the PHDs and FIH-1 are inactivated by low O_2 levels, which enable the accumulation of the HIF-1α protein. Under these conditions, HIF-1α assembles with the partner protein HIF-1β in the nucleus. It then recognizes HREs located in the enhancer regions of various target genes; interacts with coactivators CBP/p300, steroid receptor coactivator-1, nuclear redox regulator Ref-1, and molecular chaperone heat-shock protein-90 (HSP90; Arany et al., 1996; Carrero et al., 2000; Ema et al., 1999; Minet et al., 1999); and initiates transcription. Interestingly, FIH-1 remains active at a relatively lower oxygen concentration as compared with PHDs; thus FIH-1 is capable of inhibiting the activity of HIF-1α subunits that escape proteasomal degradation under moderate hypoxia (Dayan et al., 2006).

Molecular, Genetic, and Nutritional Aspects of Major and Trace Minerals. http://dx.doi.org/10.1016/B978-0-12-802168-2.00009-9

Targets of Hypoxia-Inducible Factor-1

The known target genes of the HIF-1 complex are involved in a wide spectrum of cellular events (Tables 9.1 and 9.2; Manalo et al., 2005), including angiogenesis-related genes (Berra et al., 2000; Conway et al., 2001; Eckhart et al., 1997; Josko et al., 2000); red blood cell metabolism-related genes (Lok and Ponka, 1999; Tacchini et al., 1999; Wang and Semenza, 1996; Wanner et al., 2000); glucose metabolism-related genes (Chen et al., 2001; Gatenby and Gillies, 2004; Ullah et al., 2006; Wenger, 2000; Wykoff et al., 2000); apoptosis-related genes (Carmeliet et al., 1998; Krick et al., 2005); and cell growth, proliferation, and differentiation-associated genes (Feldser et al., 1999; Krishnamachary et al., 2003). Vascular endothelial cell growth factor (VEGF) is one of the major angiogenesis-related target genes. Specifically, VEGF promotes proliferation of endothelial cells and increases the number and density of blood vessels (Berra et al., 2000; Conway et al., 2001;

TABLE 9.1 Selected Genes Induced by Hypoxia-Inducible Factor-1

Gene Name	Symbol	Gene Name	Symbol
Oxidoreductases			
Prostaglandin I2 (prostacyclin) synthase	PTGIS	Inositol-1,4,5-triphosphate receptor, type 2	ITPR2
Prolyl hydroxylase domain-containing protein-3	EGLN3	AXL receptor tyrosine kinase	AXL
N-myc downstream regulated gene-1	NDRG1	Leptin receptor	LEPR
Prostaglandin-endoperoxide synthase-1	PTGS1	G-protein–coupled receptor kinase-5	GRK5
Prolyl hydroxylase domain-containing protein-2	EGLN1	**Other signal transduction**	
Peptidylglycine α-amidating monooxygenase	PAM	Cyclin-dependent kinase inhibitor-1C (p57, Kip2)	CDKN1C
ERO1-like	ERO1L	Phospholipase C, γ2	PLCG2
Thioredoxin interacting protein	TXNIP	AMP-activated protein kinase family member 5	ARK5
Lysyl oxidase-like-2	LOXL2	PDZ domain containing-3	PDZK3
Collagens/modifying enzymes		Ectonucleotide pyrophosphatase/phosphodiesterase-1	ENPP1
Lysyl oxidase	LOX	SPRY domain-containing SOCS box protein SSB-1	SSB1
Procollagen proline 4-hydroxylase, α2	P4HA2	Dual specificity phosphatase-6	DUSP6
Collagen, type 1, α2	COL1A2	Pellino homolog-2	PELI2
Procollagen lysine hydroxylase-2	PLOD2	Ribosomal protein S6 kinase, 90 kDa, polypeptide-2	RPS6KA2
Collagen, type V, α1	COL5A1	Adenylate kinase-3	AK3
Procollagen lysine hydroxylase-2	PLOD2	Rho-related BTB domain containing-1	RHOBTB1
Collagen, type XVIII, α1	COL18A1	Calcium/calmodulin-dependent serine protein kinase	CASK
Collagen, type IX, α1	COL9A1	CK2 interacting protein-1	CKIP-1
Collagen, type IV, α1	COL4A1	Regulator of G-protein signalling-3	RGS3
Collagen, type IV, α2	COL4A2	LIM and senescent cell antigen-like domains-1	LIMS1
Procollagen lysine hydroxylase	PLOD	Ras association (RalGDS/AF-6) domain family-2	RASSF2
Protein tyrosine phosphatase, receptor type, F	PTPRF		
Growth factors/cytokines			
Angiopoietin-like-4	ANGPTL4	WAS protein family, member-2	WASF2
Stanniocalcin-1	STC1	Growth arrest and DNA-damage-inducible, β	GADD45B

TABLE 9.1 Selected Genes Induced by Hypoxia-Inducible Factor-1—cont'd

Gene Name	Symbol	Gene Name	Symbol
Inhibin, β A (activin A, activin AB α)	INHBA	Cyclin G2	CCNG2
Vascular endothelial growth factor	VEGF	Cyclin-dependent kinase (CDC2-like)-11	CDK11
Inhibin, βE	INHBE	Triple functional domain (PTPRF interacting)	TRIO
Stanniocalcin-2	STC2	Related RAS viral (r-ras) oncogene homolog	RRAS
Endothelin-1	EDN1	Myocytic induction/differentiation originator	MIDOR1
Growth differentiation factor-10	GDF10	FK506 binding protein-9, 63 kDa	FKBP9
Insulin-like growth factor binding protein-3	IGFBP3	**Transcription factors**	
Chemokine (C-X3-C motif) ligand-1	CX3CL1	V-maf oncogene homolog F	MAFF
Vascular endothelial growth factor C	VEGFC	Peroxisome proliferative activated receptor, γ	PPARG
Relaxin-1	RLN1	Transcription factor-8	TCF8
Platelet-derived growth factor B polypeptide	PDGFB	Jumonji domain containing-1	JMJD1
Placental growth factor	PGF	Zinc finger and BTB domain containing-1	ZBTB1
Adrenomedullin	ADM	Hypoxia-inducible factor-3, α subunit	HIF3A
Receptors			
cAMP responsive element binding protein-3-like 2	CREB3L2	SRY (sex-determining region Y)-box-4	SOX4
γ-Aminobutyric acid (GABA) A receptor, π	GABRP	Basic helix-loop-helix domain containing, class B, 2	BHLHB2
Chemokine orphan receptor-1	CMKOR1	RelA-associated inhibitor	RAI
Very-low–density lipoprotein receptor	VLDLR	Neuronal PAS domain protein-2	NPAS2
Protein tyrosine phosphatase, receptor type, R	PTPRR	Notch homolog-4	NOTCH4
Cubilin (intrinsic factor-cobalamin receptor)	CUBN	MAX interacting protein-1	MXI1
Chemokine (C-X-C motif) receptor-4	CXCR4	Myocyte enhancer factor-2A	MEF2A
Adenosine A2a receptor	ADORA2A	Transducin-like enhancer of split 1 homolog	TLE1
Natriuretic peptide receptor A/guanylate cyclase A	NPR1	HIV-1 enhancer binding protein-2	HIVEP2
Insulin receptor	INSR	Transcription factor-7-like-1	TCF7L1
Protein tyrosine phosphatase, receptor type, B	PTPRB	Zinc finger protein-292	ZNF292
Activin A receptor, type IB	ACVR1B	Short stature homeobox-2	SHOX2
Protein tyrosine phosphatase, receptor type, R	PTPRR	Thioredoxin interacting protein	TXNIP
Integral membrane protein-2A	ITM2A	Hairy and enhancer of split-1	HES1
Erythropoietin receptor	EPOR	Zinc fingers and homeoboxes-2	ZHX2
Cadherin-2, type 1, N-cadherin (neuronal)	CDH2	B-cell CLL/lymphoma-6 (zinc finger protein-51)	BCL6
TNF receptor superfamily, member 10b	TNFRSF10B	Sirtuin (sir2 homolog)-3	SIRT3
TNF receptor superfamily, member 14	TNFRSF14	MAD, mothers against decapentaplegic homolog-6	MADH6
Opsin 3 (encephalopsin, panopsin)	OPN3	Nuclear factor of activated T cells-4	NFATC4
Leptin receptor	LEPR		

TABLE 9.2 Selected Genes Repressed by Hypoxia-Inducible Factor-1

Gene Name	Symbol	Gene Name	Symbol
CDC/cyclins		**RNA binding/metabolism**	
Cyclin E2	CCNE2	DEAD/H box polypeptide-11	DDX11
Cyclin E1	CCNE1	Small nuclear ribonucleoprotein polypeptide A'	SNRPA1
Cyclin F	CCNF	Heterogeneous nuclear ribonucleoprotein U	HNRPU
Cell division cycle-2	CDC2	RNA, U3 small nucleolar interacting protein-2	RNU3IP2
Cell division cycle-6	CDC6	DEAD box polypeptide-18	DDX18
Cell division cycle-7	CDC7	Small nuclear ribonucleoprotein polypeptide A'	SNRPA1
MCM proteins		POP7 (processing of precursor) homolog	RPP20
Minichromosome maintenance deficient-4	MCM4	Small nuclear RNA activating complex-5	SNAPC5
Minichromosome maintenance deficient-2	MCM2	Heterogeneous nuclear ribonucleoprotein D-like	HNRPDL
Minichromosome maintenance deficient-5	MCM5	DEAD box polypeptide-23	DDX23
Minichromosome maintenance deficient-6	MCM6	DEAH (Asp-Glu-Ala-his) box polypeptide-35	DHX35
Minichromosome maintenance deficient-10	MCM10	Splicing factor, arginine/serine-rich-2	SFRS2
Minichromosome maintenance deficient-3	MCM3	Heterogeneous nuclear ribonucleoprotein D	HNRPD
DNA/RNA polymerases		DEAD box polypeptide-21	DDX21
Polymerase (RNA) III polypeptide K	POLR3K	Small nuclear ribonucleoprotein D1 polypeptide	SNRPD1
Polymerase (RNA) II polypeptide L	RNAPOL2	Small nuclear ribonucleoprotein D3 polypeptide	SNRPD3
Polymerase (DNA directed), ε2	POLE2	Ribonuclease P1	RNASEP1
Polymerase (RNA) III (32kD)	RPC32	Ribonuclease P (30 kDa)	RPP30
Polymerase (DNA directed), δ2	POLD2	**Ubiquitin/proteasome**	
Polymerase (DNA directed), γ	POLG	Proteasome activator subunit-3	PSME3
Replication factors		Ubiquitin-conjugating enzyme E2S	UBE2S
Chromatin assembly factor 1, subunit A	CHAF1A	Ubiquitin carboxyl-terminal hydrolase L5	UCHL5
DNA replication complex GINS protein PSF2	PFS2	Proteasome 26S subunit, non-ATPase, 11	PSMD11
DNA replication helicase 2-like	DNA2L	Proteasome activator subunit-4	PSME4
Replication factor C5	RFC5	Ubiquitin specific protease-39	USP39
Replication factor C (activator-1) 3	RFC3	Ubiquitin specific protease-1	USP1
Origin recognition complex, subunit 6-like	ORC6L	**Transcription factors**	
Replication factor C (activator-1) 2	RFC2	Myeloid/lymphoid leukemia; translocated to, 10	MLLT10
Other cell cycle		BRCA1 associated RING domain-1	BARD1
S-phase kinase-associated protein-2	SKP2	Breast cancer-1, early onset	BRCA1
MAD2 mitotic arrest deficient-like-1	MAD2L1	ELK1, member of ETS oncogene family	ELK1
Block of proliferation-1	BOP1	Transcription factor A, mitochondrial	TFAM
ZW10 interactor	ZWINT	MYC-associated zinc finger protein	MAZ
M-phase phosphoprotein-9	MPHOSPH9	Peroxisome proliferative activated receptor, δ	PPARD
Nucleolar and coiled-body phosphoprotein-1	NOLC1	Retinoblastoma-1	RB1
Proliferation-associated-2G4	PA2G4	Thyroid hormone receptor interactor-13	TRIP13
Activator of S-phase kinase	ASK	Transcription factor 6-like 1	TF6M

TABLE 9.2 Selected Genes Repressed by Hypoxia-Inducible Factor-1—cont'd

Gene Name	Symbol	Gene Name	Symbol
Protein regulator of cytokinesis-1	PRC1	Inhibitor of DNA binding-1	ID1
NIMA (never in mitosis a)-related kinase-4	NEK4	**Ribosomal/mitochondrial biogenesis**	
Ribonucleotide metabolism		Mitochondrial ribosomal protein S12	MRPS12
Ribonucleotide reductase M2 polypeptide	RRM2	Ts translation elongation factor, mitochondrial	TSFM
Dihydrofolate reductase	DHFR	Ribosomal RNA processing-4	RRP4
CTP synthase	CTPS	Ribosome biogenesis regulator homolog	RRS1
Phosphoribosyl pyrophosphate synthetase-1	PRPS1	Ribosomal protein L39	RPL39
Uridine monophosphate synthetase	UMPS	Interleukin enhancer binding protein-3	ILF3
Dihydrofolate reductase	DHFR	Ribosomal RNA processing-4	RRP4
5-Aminoimidazole-4-carboxamide ribonucle-otide transformylase	ATIC	Zinc finger protein-500	ZNF500
Poly(ADP-ribose) polymerase	ADPRT	Mitochondrial ribosomal protein L40	MRPL40
GTP-binding protein	GTPBP4	Nucleolar protein (KKED repeat)	NOL5A
ADP-ribosyltransferase (NAD+; poly(ADP-ribose) polymerase)-like-2	ADPRTL2	Mitofusin-2	MFN2
Ribonucleotide reductase M1 polypeptide	RRM1	Mitochondrial ribosomal protein L46	MRPL46

Josko et al., 2000). In addition, local blood flow under pathophysiological conditions is controlled by modulation of vascular tone through production of NO (inducible nitric oxide synthase), CO (heme oxygenase-1), endothelin-1, adrenomedulin, or activation of the α1β-adrenergic receptor. All of these events involve HIF-1 target genes (Eckhart et al., 1997; Hu et al., 1998; Lee et al., 1997; Melillo et al., 1995; Nguyen and Claycomb, 1999; Palmer et al., 1998; Wenger, 2002). The red blood cell metabolism-related genes include erythropoietin (Wang and Semenza, 1996; Wanner et al., 2000), ceruloplasmin (Lok and Ponka, 1999; Mukhopadhyay et al., 2000), and the transferrin receptor (Tacchini et al., 1999). Erythropoietin promotes differentiation of stem cells to red blood cells and their proliferation and maturation. Ceruloplasmin is required to oxidize ferrous iron to ferric iron. The transferrin receptor enables cellular uptake of diferric transferrin. Many genes involved in glucose uptake and glycolysis were also identified as HIF-1 target genes (Wenger, 2000), including glucose transporters GLUT1 and GLUT3, which mediate cellular glucose uptake (Chen et al., 2001); transmembrane carbonic anhydrases (Wykoff et al., 2000); lactate dehydrogenase A (Gatenby and Gillies, 2004); and the plasma membrane lactate transporter MCT4 (Ullah et al., 2006). HIF-responsive apoptosis-related genes include caspase-3 and cytochrome c (Carmeliet et al., 1998; Krick et al., 2005; Magnon et al., 2007). Cell growth, proliferation, and differentiation-associated genes targeted by the HIF-1 complex include fibroblast growth factor (FGF), transforming growth factor-α, insulin-like growth factor (IGF)-2, and epidermal growth factor receptor (Feldser et al., 1999; Krishnamachary et al., 2003; Semenza, 2003).

RELATIONSHIP BETWEEN COPPER AND HYPOXIA-INDUCIBLE FACTOR-1

The requirement for Cu in angiogenesis has been known since the first demonstration of Cu induction of intraocular vascularization in rats (McAuslan and Gole, 1980) more than 30 years ago. Cu was later found to stimulate microvessel formation in the avascular cornea of rabbits, thus dismissing species specificity or uniqueness as a factor in the response (Ziche et al., 1982). Further studies proved that Cu is a pleiotropic agent capable of affecting numerous components of the angiogenic response system, including endothelial cell proliferation (Hu, 1998; Li et al., 2012; Narayanan et al., 2013), migration (Alessandri et al., 1984; McAuslan et al., 1983; Narayanan et al., 2013), tube formation (Li et al., 2014; Narayanan et al., 2013), and vessel maturation (Soncin et al., 1997).

Cu stimulation of angiogenesis involves its influence on the bioactivity or production of several factors involved in the process of angiogenesis (Harris, 2004; Zheng et al., 2015). Among these factors, the most studied is VEGF. Direct evidence of Cu regulation of VEGF expression was first demonstrated in keratinocytes in cultures and in vivo (Sen et al., 2002) and further confirmed by a sustained VEGF expression in hypertrophic myocardium, which was associated with enhanced myocardial angiogenesis (Jiang et al., 2007). Furthermore, an anti-VEGF antibody effectively blunted Cu stimulation of angiogenesis, suggesting that Cu-induced angiogenesis works through regulation of VEGF expression (Jiang et al., 2007).

The wide ranging effects of Cu on angiogenesis suggest that Cu may be involved in the regulation of transcription factors that control expression of many of the angiogenic genes, such as HIF-1 (Xie and Kang, 2009; Zheng et al., 2015). A direct relationship between HIF-1 and Cu-induced angiogenesis was demonstrated by an in vitro study using small interfering RNA targeting HIF-1α (Jiang et al., 2007). In the absence of HIF-1α, VEGF expression and angiogenesis stimulated by Cu was effectively suppressed (Jiang et al., 2007). Therefore the effect of Cu on angiogenesis is mainly dependent on the regulation of HIF-1 activity. Further studies revealed a role for Cu in the regulation of HIF-1 transcriptional activity. Physiologically relevant Cu levels do not influence the stability of HIF-1α in cytosol but rather are required for formation of the HIF-1 transcriptional complex and for the binding of HIF-1 to HREs (Feng et al., 2009). Moreover, Cu is related to the selectivity of HIF-1–mediated regulation of gene expression (Zhang et al., 2014).

MECHANISMS BY WHICH CU REGULATES HYPOXIA-INDUCIBLE FACTOR-1 ACTIVITY

Hypoxia-Inducible Factor-1α Stabilization by Copper

As previously noted, activation of HIF-1α is censored by two systems, PHDs in cytosol and FIH-1 in the nucleus. Some transition metals such as cobalt and nickel enhance HIF-1 transcriptional activity by inhibiting these two systems, leading to stabilization of the HIF-1α protein (Hirsila et al., 2005; Ke et al., 2005; Maxwell and Salnikow, 2004). High Cu levels are also capable of stabilizing HIF-1α by inhibition of PHDs (Martin et al., 2005; van Heerden et al., 2004). Under normoxic conditions, treatment with 100 μM $CuCl_2$ was shown to stabilize nuclear HIF-1α in different cell lines, resulting in HRE-dependent reporter gene expression and increased mRNA levels of the HIF-1 target genes for ceruloplasmin, VEGF, and GLUT1. In vitro hydroxylation assays showed that $CuCl_2$ inhibited prolyl-4-hydroxylation caused by PHDs (Martin et al., 2005). However, unlike other transition metals that are not essential for HIF-1 activation, Cu at physiologically relevant levels is required for HIF-1 regulation of gene expression.

The Requirement of Copper for Hypoxia-Inducible Factor-1 Transcriptional Activity

The Cu requirement for HIF-1 activation was initially defined by the effect of a Cu chelator, tetraethylenepentamine (TEPA), on HIF-1 activation in cultured human cardiomyocytes induced by IGF-1, a well-known growth factor that stimulates VEGF production by activating HIF-1 (Jiang et al., 2007). HIF-1 activity was measured by an enzyme-linked immunoabsorbent assay. IGF-1 and Cu activated HIF-1 along with a stimulation of VEGF expression to the same extent, but there was no additive or synergistic effect of IGF-1 and Cu in combination. Importantly, TEPA equally blocked the stimulatory effect of IGF-1 or Cu, suggesting that Cu was required for HIF-1 activation. This inhibitory effect of TEPA could be relieved by the addition of excess Cu to the culture medium (Jiang et al., 2007).

The Cu requirement for HIF-1 activation was also demonstrated in cells other than cardiomyocytes. Treatment of HepG2 cells with TEPA suppressed hypoxia-induced activation of HIF-1 (Feng et al., 2009). The addition of excess Cu to the cell cultures relieved the suppression by TEPA, further indicating the requirement of Cu for the activation of HIF-1. Cu deprivation affected neither the production nor the stability of HIF-1α, but it reduced HIF-1 binding to the HRE of the target genes and inhibited the recruitment of p300 and thus assembly of the transcriptional complex. Cu likely inhibits FIH-1 to ensure the integrity of the HIF-1 transcriptional complex. Thus these observations demonstrated that Cu is required for HIF-1 activation through regulation of HIF-1 binding to the HRE and the formation of the HIF-1 transcriptional complex.

In addition to hypoxia-induced activation of HIF-1, it has been noted that Cu is also required for cobalt activation of HIF-1 transcriptional activity (Qiu et al., 2012). Human umbilical vein endothelial cells (HUVECs) were treated with cobalt, leading to accumulation of HIF-1α with subsequently increased VEGF expression. TEPA did not significantly affect the accumulation of HIF-1α, but it blocked an increase in the VEGF mRNA and protein levels. The effect of TEPA was reversed by an addition of Cu. This study demonstrated that Cu was required for cobalt-activated transcriptional activity of HIF-1, although Cu did not affect cobalt-induced accumulation of HIF-1α in the cells.

Selectivity of Copper Regulation of Hypoxia-Inducible Factor-1

A recent study (Zhang et al., 2014) revealed that Cu is not required for the expression of all HIF-1-regulated genes. In this study HUVECs were treated with cobalt to induce HIF-1 activation. TEPA treatment suppressed the expression of

BCL2/adenovirus E1B 19-kDa protein-interacting protein 3 (*BNIP3*) but did not affect another HIF-1 target gene, *IGF-2*. Further screening identified two groups of genes: those whose expression required Cu and those that did not. This study demonstrated there exists Cu-dependent and -independent HIF-1 regulation of gene expression, indicating that Cu manipulation may serve as a mechanism by which the selectivity of HIF-1 regulation of gene expression can be achieved (Zhang et al., 2014).

Proteins That May Be Involved in Cu Regulation of Hypoxia-Inducible Factor-1

The Cu ion is highly reactive; thus it is tightly regulated in the cell. There is virtually no detectable free Cu ion in cells. Cu chaperones within the cell coordinate Cu to prevent severe oxidative damage by free Cu and to direct Cu localization. There is also evidence that Cu chaperones, such as the Cu chaperone for superoxide dismutase-1 (CCS), may mediate the action of Cu on HIF-1 (Wang et al., 2016; Feng et al., 2009; Jiang et al., 2007; Qiu et al., 2012). In a study of cultured human cardiomyocytes, CCS gene silencing blocked the binding of IGF-1–induced HIF-1 to the HRE and VEGF expression. These results mimicked the effect of Cu chelation. Furthermore, CCS directly interacts with HIF-1, as revealed by an immunoprecipitation assay (Jiang et al., 2007). Other studies using different cell lines yielded similar results. CCS gene silencing in HepG2 cells suppressed hypoxia-induced activation of HIF-1, which could not be relieved by addition of excess Cu to the cultures (Feng et al., 2009). In addition, gene silencing of CCS also blocked VEGF expression induced by cobalt activation of HIF-1 transcriptional activity in HUVECs with little effect on cobalt-induced HIF-1α accumulation (Qiu et al., 2012), indicating that Cu regulation of HIF-1 is CCS dependent.

In addition to CCS, Cu metabolism MURR1 domain containing-1 protein (COMMD1) has been reported as a negative regulator of HIF-1 signaling. Mouse embryos deficient for COMMD1 have increased expression of HIF–1-regulated genes [i.e., *VEGF*, *PGK* (Phosphoglycerate kinase 1), and *BNIP3*], corresponding to increased HIF-1α protein stability (van de Sluis et al., 2007). Conversely, overexpression of COMMD1 resulted in downregulation of HIF-1 activity (van de Sluis et al., 2007). COMMD1 competes with HSP90β for binding to HIF-1α to facilitate HIF-1α degradation, which is VHL/O_2 and ubiquitin independent (van de Sluis et al., 2009).

Because there are multiple sites of Cu regulation of HIF-1, there should be multiple proteins involved in the regulation process (Fig. 9.1). Most of these proteins have not been identified and further work, especially identification of possible Cu-binding proteins, is required to elucidate how Cu regulates HIF-1 activity and its target gene expression.

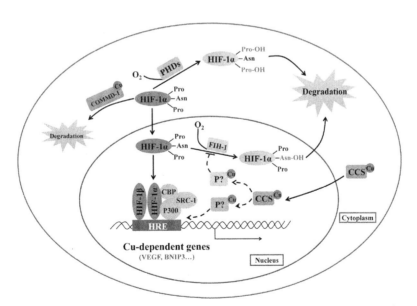

FIGURE 9.1 Cu regulation of hypoxia-inducible factor (HIF)-1 activity. HIF-1 is composed of HIF-1α and HIF-1β. Under normoxia, HIF-1α is hydroxylated by prolyl hydroxylase domain proteins (PHDs) in the cytosol and factor inhibiting HIF-1 (FIH-1) in the nucleus, and then it undergoes proteasomal degradation. In addition, Cu metabolism MURR1 domain containing-1 protein (COMMD-1) mediates a von Hippel-Lindau tumor suppressor protein (VHL)/O_2 and ubiquitin-independent degradation of HIF-1α. Under hypoxia conditions, HIF-1α escapes degradation as a result of the inhibition of PHDs and FIH-1 and dimerizes with HIF-1β in the nucleus. The heterodimer then recruits cofactors such as p300, CREB-binding protein (CBP) and steroid receptor co-activator 1 (SRC1) to form the transcriptional complex. Cu chaperone for superoxide dismutase-1 (CCS) brings Cu into the nucleus. Cu is required for the interaction of HIF-1 and hypoxia-responsive elements (HREs) to mediate Cu-dependent expression of genes such as vascular endothelial cell growth factor (VEGF) and BCL2/adenovirus E1B 19-kDa protein-interacting protein 3 (BNIP3), and Cu also acts as an inhibitor of FIH-1. These effects would be mediated by unidentified putative Cu-binding proteins (P?).

CLINICAL IMPLICATIONS

Given the relationship between Cu and HIF-1, it is possible to modulate HIF-1 activity by manipulating Cu levels in diseases associated with inappropriate HIF-1 activity, such as ischemia/hypoxia-related diseases and cancer.

Cardiovascular Diseases

HIF-1 has been an experimental therapeutic target for ischemic heart disease because it plays an important role in repair of heart tissue by inducing angiogenesis during ischemic/hypoxic injury. The beneficial effect of HIF-1 activation on acute myocardial ischemia has been demonstrated through the inhibition of PHD function or transgenic expression of HIF-1α in animal models (Eckle et al., 2008; Huang et al., 2008; Kido et al., 2005; Philipp et al., 2006). However, studies also linked the long-term activation of HIF to the pathogenesis of chronic ischemic cardiomyopathy (Bekeredjian et al., 2010; Lei et al., 2008; Moslehi et al., 2010). Cardiac-specific VHL deletion in mice caused cardiomyopathy, which was prevented by concomitant deletion of HIF-1α (Lei et al., 2008). Long-term cardiac-specific inactivation of PHD2 and PHD3, and subsequent HIF activation, resulted in ischemic cardiomyopathy (Moslehi et al., 2010). In studies with ischemic myocardial infarction, insufficient angiogenesis was found to be accompanied by accumulation of HIF-1α (Zhang et al., 2016; He and Kang, 2013). These findings suggest that simply increasing the amount of HIF-1 is insufficient for angiogenesis and tissue repair. Thus proper modulation of HIF-1 activity is necessary for long-term treatment. Importantly, Cu efflux from ischemic heart tissue has been demonstrated in several studies (Berenshtein et al., 1997; Chevion et al., 1993; Shokrzadeh et al., 2009). Therefore, under the condition of HIF-1α accumulation but Cu depression, HIF-1–controlled gene expression would become unregulated. Expression of Cu-dependent HIF-1 target genes, such as *VEGF* and other angiogenesis-related genes, would be suppressed, whereas that of Cu-independent HIF-1 target genes would remain active, leading to depressed angiogenesis and increased fibrosis in the ischemic myocardium. Fragmentary information can be retrieved from the literature regarding Cu treatment and ischemic heart disease (Zhang et al., 2016; He and Kang, 2013; Jiang et al., 2007; Zheng et al., 2015). Cu regulation of HIF-1 activity plays an important role in the recovery from cardiovascular disease. Dietary supplementation of Cu reversed depression of VEGF expression via HIF-1 activation, leading to regression of cardiac hypertrophy (Jiang et al., 2007). Moreover, many other HIF-1–regulated genes such as VEGFR-1 were also involved in this process (Jiang et al., 2007; Zheng et al., 2015). Zhang et al. found that the treatment of diabetic cardiomyopathy with trientine caused an increase in Cu concentrations in the heart (Zhang et al., 2013), suggesting that trientine cannot only serve as a Cu chelator but also function to deliver or redistribute Cu. Regardless of the underlying mechanisms in target tissues, Cu supplementation may constitute an important approach for activation of HIF-1 regulation of gene expression to promote angiogenesis during ischemic heart disease.

Cancer

In 1971 Folkman and his associates first developed the concept that the growth and metastasis capacity of tumors is dependent upon angiogenesis (Folkman, 1971). In 1980 McAuslan and Reilly noted the role of Cu in angiogenesis and found that Cu salts were the simplest angiogenic components of tumor extracts (McAuslan and Reilly, 1980). Several reports have shown that serum and tumor Cu levels were significantly elevated in cancer patients (Goyal et al., 2006) whereas HIF-1α protein was also shown to be overexpressed in various human cancers (Talks et al., 2000; Zhong et al., 1999). These findings suggest that Cu chelators might be useful for cancer chemotherapy.

Cu reduction therapy is currently under investigation as an anticancer strategy. Cu chelators such as D-penicillamine (Brem et al., 1990), tetrathiomolybdate (Pan et al., 2002), clioquinol (Ding et al., 2005), and trientine (Hayashi et al., 2007) have been shown to inhibit angiogenesis in tumors in vitro and in vivo (Table 9.3). However, these drugs need to be combined with other antiangiogenic therapies for improved efficacy. Moreover, studies have shown that several tumor types expressed elevated ceruloplasmin (Kunapuli et al., 1987), the main Cu transport protein in the plasma. The degradation of ceruloplasmin as a consequence of Cu chelation may also have an important role in reducing angiogenesis. Indeed, clinical trials with Cu chelators combined with other agents are ongoing.

OPEN QUESTIONS

Although Cu regulation of HIF-1 activity has been extensively studied, many questions remain unanswered regarding the selectivity of Cu-dependent expression of genes under the control of HIF-1. How does Cu or Cu chaperones enter the nucleus? How many Cu-binding proteins are involved in the Cu regulation of HIF-1 or Cu regulation of gene expression, and what are their identities? For example, in addition to HIF-1, a study published in 2008 documented a novel function

TABLE 9.3 Clinical Trials With Copper Chelators

Tumor Type	Therapy	Results	References
Gliomas implanted intracerebrally	Low copper diet and penicillamine	Reduced tumor growth and vascularization but no difference in survival.	Brem et al. (1990)
Gliosarcoma xenograft	Low copper diet and penicillamine	Reduction in tumor weight and vascular density.	Yoshida et al. (1995)
Hepatocellular carcinoma	Penicillamine	Decreased tumor growth and interleukin-8 production.	Moriguchi et al. (2002)
Her-2 transgenic mice	Tetrathiomolybdate	82% control mice developed mammary cancer vs. 0% treated with tetrathiomolybdate.	Pan et al. (2002)
Inflammatory breast cancer	Tetrathiomolybdate	70% reduction in tumor growth. Decreased CD31 vascular staining.	Pan et al. (2002)
Murine lung cancer	Tetrathiomolybdate	Reduction in tumor growth. Effect enhanced by radiotherapy.	Khan et al. (2002)
Inflammatory breast cancer (murine SUM149 xenografts)	Tetrathiomolybdate + doxorubicin	Synergistic reduction in tumor growth.	Pan et al. (2003)
Human leukemia cancer cells	8-Hydroxyquinoline (8-OHQ)	Cancer-specific proteasome inhibition. Apoptotic tumor cells.	Daniel et al. (2004)
Head and neck tumors	Tetrathiomolybdate	Reduction in tumor growth.	Teknos et al. (2005)
Colorectal carcinoma	Trientine and methotrexate	Synergistic effects on tumor growth. Decreased vascular endothelial growth factor and interleukin-8.	Yoshiji et al. (2005)
Breast cancer cells	Clioquinol	Cancer-specific proteasome inhibition. Apoptotic tumor cells.	Daniel et al. (2005)
Breast cancer cells	Pyrrolidinedithiocarbamate	Cancer-specific proteasome inhibition. Apoptotic tumor cells.	Daniel et al. (2005)
Mice bearing MDA-MB-231 tumor xenografts	Disulfiram	74% reduction in tumor growth. Cancer-specific proteasome inhibition. Apoptotic tumor cells.	Chen et al. (2006)
Human breast and prostate cancer cells	Diethyldithiocarbamate	Cancer-specific proteasome inhibition. Apoptotic tumor cells.	Pang et al. (2007)
Human SK-N-MC neuroepithelioma cells	Dp44mT	Lysosomal damage in cancer cell.	Lovejoy et al. (2011)

of antioxidant-1, a Cu chaperone, as a Cu-dependent transcription factor involved in cell proliferation (Itoh et al., 2008). Together with what we have discussed in this chapter, more evidence emerges to indicate the role of Cu in the regulation of gene expression. Further identification of Cu-binding proteins in the nucleus and elucidation of their physiologic roles will help us understand better the role of Cu as a regulator of gene expression.

ACKNOWLEDGMENT

The work cited in this manuscript was partially supported by the National Science Foundation of China (grant number: 81230004 to Y. J. Kang).

REFERENCES

Alessandri, G., Raju, K., Gullino, P.M., 1984. Angiogenesis in vivo and selective mobilization of capillary endothelium in vitro by heparin-copper complex. Microcirc. Endothel. Lymphat. 1, 329–346.

Arany, Z., Huang, L.E., Eckner, R., Bhattacharya, S., Jiang, C., Goldberg, M.A., Bunn, H.F., Livingston, D.M., 1996. An essential role for p300/CBP in the cellular response to hypoxia. Proc. Natl. Acad. Sci. U.S.A. 93, 12969–12973.

Bekeredjian, R., Walton, C.B., MacCannell, K.A., Ecker, J., Kruse, F., Outten, J.T., Sutcliffe, D., Gerard, R.D., Bruick, R.K., Shohet, R.V., 2010. Conditional HIF-1α expression produces a reversible cardiomyopathy. PLoS One 5, e11693.

Berenshtein, E., Mayer, B., Goldberg, C., Kitrossky, N., Chevion, M., 1997. Patterns of mobilization of copper and iron following myocardial ischemia: possible predictive criteria for tissue injury. J. Mol. Cell Cardiol. 29, 3025–3034.

Berra, E., Pages, G., Pouyssegur, J., 2000. MAP kinases and hypoxia in the control of VEGF expression. Cancer Metastasis Rev. 19, 139–145.

Brem, S.S., Zagzag, D., Tsanaclis, A.M., Gately, S., Elkouby, M.P., Brien, S.E., 1990. Inhibition of angiogenesis and tumor growth in the brain. Suppression of endothelial cell turnover by penicillamine and the depletion of copper, an angiogenic cofactor. Am. J. Pathol. 137, 1121–1142.

Carmeliet, P., Dor, Y., Herbert, J.M., Fukumura, D., Brusselmans, K., Dewerchin, M., Neeman, M., Bono, F., Abramovitch, R., Maxwell, P., Koch, C.J., Ratcliffe, P., Moons, L., Jain, R.K., Collen, D., Keshert, E., 1998. Role of HIF-1α in hypoxia-mediated apoptosis, cell proliferation and tumour angiogenesis. Nature 394, 485–490.

Carrero, P., Okamoto, K., Coumailleau, P., O'Brien, S., Tanaka, H., Poellinger, L., 2000. Redox-regulated recruitment of the transcriptional coactivators CREB-binding protein and SRC-1 to hypoxia-inducible factor 1α. Mol. Cell. Biol. 20, 402–415.

Chen, C., Pore, N., Behrooz, A., Ismail-Beigi, F., Maity, A., 2001. Regulation of glut1 mRNA by hypoxia-inducible factor-1. Interaction between H-ras and hypoxia. J. Biol. Chem. 276, 9519–9525.

Chen, D., Cui, Q.C., Yang, H., Dou, Q.P., 2006. Disulfiram, a clinically used anti-alcoholism drug and copper-binding agent, induces apoptotic cell death in breast cancer cultures and xenografts via inhibition of the proteasome activity. Cancer Res. 66, 10425–10433.

Chevion, M., Jiang, Y., Har-El, R., Berenshtein, E., Uretzky, G., Kitrossky, N., 1993. Copper and iron are mobilized following myocardial ischemia: possible predictive criteria for tissue injury. Proc. Natl. Acad. Sci. U.S.A. 90, 1102–1106.

Conway, E.M., Collen, D., Carmeliet, P., 2001. Molecular mechanisms of blood vessel growth. Cardiovasc. Res. 49, 507–521.

Daniel, K.G., Gupta, P., Harbach, R.H., Guida, W.C., Dou, Q.P., 2004. Organic copper complexes as a new class of proteasome inhibitors and apoptosis inducers in human cancer cells. Biochem. Pharmacol. 67, 1139–1151.

Daniel, K.G., Chen, D., Orlu, S., Cui, Q.C., Miller, F.R., Dou, Q.P., 2005. Clioquinol and pyrrolidine dithiocarbamate complex with copper to form proteasome inhibitors and apoptosis inducers in human breast cancer cells. Breast Cancer Res. 7, R897–R908.

Dayan, F., Roux, D., Brahimi-Horn, M.C., Pouyssegur, J., Mazure, N.M., 2006. The oxygen sensor factor-inhibiting hypoxia-inducible factor-1 controls expression of distinct genes through the bifunctional transcriptional character of hypoxia-inducible factor-1α. Cancer Res. 66, 3688–3698.

Ding, W.Q., Liu, B., Vaught, J.L., Yamauchi, H., Lind, S.E., 2005. Anticancer activity of the antibiotic clioquinol. Cancer Res. 65, 3389–3395.

Eckhart, A.D., Yang, N., Xin, X., Faber, J.E., 1997. Characterization of the α1B-adrenergic receptor gene promoter region and hypoxia regulatory elements in vascular smooth muscle. Proc. Natl. Acad. Sci. U.S.A. 94, 9487–9492.

Eckle, T., Kohler, D., Lehmann, R., El Kasmi, K., Eltzschig, H.K., 2008. Hypoxia-inducible factor-1 is central to cardioprotection: a new paradigm for ischemic preconditioning. Circulation 118, 166–175.

Ema, M., Hirota, K., Mimura, J., Abe, H., Yodoi, J., Sogawa, K., Poellinger, L., Fujii-Kuriyama, Y., 1999. Molecular mechanisms of transcription activation by HLF and HIF1α in response to hypoxia: their stabilization and redox signal-induced interaction with CBP/p300. EMBO J. 18, 1905–1914.

Feldser, D., Agani, F., Iyer, N.V., Pak, B., Ferreira, G., Semenza, G.L., 1999. Reciprocal positive regulation of hypoxia-inducible factor 1α and insulin-like growth factor 2. Cancer Res. 59, 3915–3918.

Feng, W., Ye, F., Xue, W., Zhou, Z., Kang, Y.J., 2009. Copper regulation of hypoxia-inducible factor-1 activity. Mol. Pharmacol. 75, 174–182.

Folkman, J., 1971. Tumor angiogenesis: therapeutic implications. N. Engl. J. Med. 285, 1182–1186.

Gatenby, R.A., Gillies, R.J., 2004. Why do cancers have high aerobic glycolysis? Nat. Rev. Cancer 4, 891–899.

Goyal, M.M., Kalwar, A.K., Vyas, R.K., Bhati, A., 2006. A study of serum zinc, selenium and copper levels in carcinoma of esophagus patients. Indian J. Clin. Biochem. 21, 208–210.

Harris, E.D., 2004. A requirement for copper in angiogenesis. Nutr. Rev. 62, 60–64.

Hayashi, M., Nishiya, H., Chiba, T., Endoh, D., Kon, Y., Okui, T., 2007. Trientine, a copper-chelating agent, induced apoptosis in murine fibrosarcoma cells in vivo and in vitro. J. Vet. Med. Sci. 69, 137–142.

He, W., Kang, Y.J., 2013. Ischemia-induced copper loss and suppression of angiogenesis in the pathogenesis of myocardial infarction. Cardiovasc. Toxicol. 13, 1–8.

Hirsila, M., Koivunen, P., Xu, L., Seeley, T., Kivirikko, K.I., Myllyharju, J., 2005. Effect of desferrioxamine and metals on the hydroxylases in the oxygen sensing pathway. FASEB J. 19, 1308–1310.

Hu, J., Discher, D.J., Bishopric, N.H., Webster, K.A., 1998. Hypoxia regulates expression of the endothelin-1 gene through a proximal hypoxia-inducible factor-1 binding site on the antisense strand. Biochem. Biophys. Res. Commun. 245, 894–899.

Hu, G.F., 1998. Copper stimulates proliferation of human endothelial cells under culture. J. Cell. Biochem. 69, 326–335.

Huang, L.E., Gu, J., Schau, M., Bunn, H.F., 1998. Regulation of hypoxia-inducible factor 1α is mediated by an O_2-dependent degradation domain via the ubiquitin-proteasome pathway. Proc. Natl. Acad. Sci. U.S.A. 95, 7987–7992.

Huang, M., Chan, D.A., Jia, F., Xie, X., Li, Z., Hoyt, G., Robbins, R.C., Chen, X., Giaccia, A.J., Wu, J.C., 2008. Short hairpin RNA interference therapy for ischemic heart disease. Circulation 118, S226–S233.

Itoh, S., Kim, H.W., Nakagawa, O., Ozumi, K., Lessner, S.M., Aoki, H., Akram, K., McKinney, R.D., Ushio-Fukai, M., Fukai, T., 2008. Novel role of antioxidant-1 (Atox1) as a copper-dependent transcription factor involved in cell proliferation. J. Biol. Chem. 283, 9157–9167.

Ivan, M., Kondo, K., Yang, H., Kim, W., Valiando, J., Ohh, M., Salic, A., Asara, J.M., Lane, W.S., Kaelin Jr., W.G., 2001. HIFα targeted for VHL-mediated destruction by proline hydroxylation: implications for O_2 sensing. Science 292, 464–468.

Iwai, K., Yamanaka, K., Kamura, T., Minato, N., Conaway, R.C., Conaway, J.W., Klausner, R.D., Pause, A., 1999. Identification of the von Hippel-lindau tumor-suppressor protein as part of an active E3 ubiquitin ligase complex. Proc. Natl. Acad. Sci. U.S.A. 96, 12436–12441.

Jaakkola, P., Mole, D.R., Tian, Y.M., Wilson, M.I., Gielbert, J., Gaskell, S.J., von Kriegsheim, A., Hebestreit, H.F., Mukherji, M., Schofield, C.J., Maxwell, P.H., Pugh, C.W., Ratcliffe, P.J., 2001. Targeting of HIF-α to the von Hippel-Lindau ubiquitylation complex by O$_2$-regulated prolyl hydroxylation. Science 292, 468–472.

Jiang, B.H., Rue, E., Wang, G.L., Roe, R., Semenza, G.L., 1996. Dimerization, DNA binding, and transactivation properties of hypoxia-inducible factor 1. J. Biol. Chem. 271, 17771–17778.

Jiang, B.H., Zheng, J.Z., Leung, S.W., Roe, R., Semenza, G.L., 1997. Transactivation and inhibitory domains of hypoxia-inducible factor 1α. Modulation of transcriptional activity by oxygen tension. J. Biol. Chem. 272, 19253–19260.

Jiang, Y., Reynolds, C., Xiao, C., Feng, W., Zhou, Z., Rodriguez, W., Tyagi, S.C., Eaton, J.W., Saari, J.T., Kang, Y.J., 2007. Dietary copper supplementation reverses hypertrophic cardiomyopathy induced by chronic pressure overload in mice. J. Exp. Med. 204, 657–666.

Josko, J., Gwozdz, B., Jedrzejowska-Szypulka, H., Hendryk, S., 2000. Vascular endothelial growth factor (VEGF) and its effect on angiogenesis. Med. Sci. Monit. 6, 1047–1052.

Kallio, P.J., Okamoto, K., O'Brien, S., Carrero, P., Makino, Y., Tanaka, H., Poellinger, L., 1998. Signal transduction in hypoxic cells: inducible nuclear translocation and recruitment of the CBP/p300 coactivator by the hypoxia-inducible factor-1α. EMBO J. 17, 6573–6586.

Ke, Q., Kluz, T., Costa, M., 2005. Down-regulation of the expression of the FIH-1 and ARD-1 genes at the transcriptional level by nickel and cobalt in the human lung adenocarcinoma A549 cell line. Int. J. Environ. Res. Public Health 2, 10–13.

Khan, M.K., Miller, M.W., Taylor, J., Gill, N.K., Dick, R.D., Van Goled, K., Brewert, G.J., Merajver, S.D., 2002. Radiotherapy and antiangiogenic TM in lung cancer. Neoplasia 4, 164–170.

Kido, M., Du, L., Sullivan, C.C., Li, X., Deutsch, R., Jamieson, S.W., Thistlethwaite, P.A., 2005. Hypoxia-inducible factor 1-α reduces infarction and attenuates progression of cardiac dysfunction after myocardial infarction in the mouse. J. Am. Coll. Cardiol. 46, 2116–2124.

Krick, S., Eul, B.G., Hanze, J., Savai, R., Grimminger, F., Seeger, W., Rose, F., 2005. Role of hypoxia-inducible factor-1α in hypoxia-induced apoptosis of primary alveolar epithelial type II cells. Am. J. Respir. Cell Mol. Biol. 32, 395–403.

Krishnamachary, B., Berg-Dixon, S., Kelly, B., Agani, F., Feldser, D., Ferreira, G., Iyer, N., LaRusch, J., Pak, B., Taghavi, P., Semenza, G.L., 2003. Regulation of colon carcinoma cell invasion by hypoxia-inducible factor 1. Cancer Res. 63, 1138–1143.

Kunapuli, S.P., Singh, H., Singh, P., Kumar, A., 1987. Ceruloplasmin gene expression in human cancer cells. Life Sci. 40, 2225–2228.

Lando, D., Peet, D.J., Gorman, J.J., Whelan, D.A., Whitelaw, M.L., Bruick, R.K., 2002a. FIH-1 is an asparaginyl hydroxylase enzyme that regulates the transcriptional activity of hypoxia-inducible factor. Genes Dev. 16, 1466–1471.

Lando, D., Peet, D.J., Whelan, D.A., Gorman, J.J., Whitelaw, M.L., 2002b. Asparagine hydroxylation of the HIF transactivation domain a hypoxic switch. Science 295, 858–861.

Lee, P.J., Jiang, B.H., Chin, B.Y., Iyer, N.V., Alam, J., Semenza, G.L., Choi, A.M., 1997. Hypoxia-inducible factor-1 mediates transcriptional activation of the heme oxygenase-1 gene in response to hypoxia. J. Biol. Chem. 272, 5375–5381.

Lei, L., Mason, S., Liu, D., Huang, Y., Marks, C., Hickey, R., Jovin, I.S., Pypaert, M., Johnson, R.S., Giordano, F.J., 2008. Hypoxia-inducible factor-dependent degeneration, failure, and malignant transformation of the heart in the absence of the von Hippel-Lindau protein. Mol. Cell. Biol. 28, 3790–3803.

Li, S., Xie, H., Kang, Y.J., 2012. Copper stimulates growth of human umbilical vein endothelial cells in a vascular endothelial growth factor-independent pathway. Exp. Biol. Med. (Maywood) 237, 77–82.

Li, Q.F., Ding, X.Q., Kang, Y.J., 2014. Copper promotion of angiogenesis in isolated rat aortic ring: role of vascular endothelial growth factor. J. Nutr. Biochem. 25, 44–49.

Linke, S., Stojkoski, C., Kewley, R.J., Booker, G.W., Whitelaw, M.L., Peet, D.J., 2004. Substrate requirements of the oxygen-sensing asparaginyl hydroxylase factor-inhibiting hypoxia-inducible factor. J. Biol. Chem. 279, 14391–14397.

Lok, C.N., Ponka, P., 1999. Identification of a hypoxia response element in the transferrin receptor gene. J. Biol. Chem. 274, 24147–24152.

Lovejoy, D.B., Jansson, P.J., Brunk, U.T., Wong, J., Ponka, P., Richardson, D.R., 2011. Antitumor activity of metal-chelating compound Dp44mT is mediated by formation of a redox-active copper complex that accumulates in lysosomes. Cancer Res. 71, 5871–5880.

Magnon, C., Opolon, P., Ricard, M., Connault, E., Ardouin, P., Galaup, A., Metivier, D., Bidart, J.M., Germain, S., Perricaudet, M., Schlumberger, M., 2007. Radiation and inhibition of angiogenesis by canstatin synergize to induce HIF-1α-mediated tumor apoptotic switch. J. Clin. Invest 117, 1844–1855.

Mahon, P.C., Hirota, K., Semenza, G.L., 2001. FIH-1: a novel protein that interacts with HIF-1α and VHL to mediate repression of HIF-1 transcriptional activity. Genes Dev. 15, 2675–2686.

Manalo, D.J., Rowan, A., Lavoie, T., Natarajan, L., Kelly, B.D., Ye, S.Q., Garcia, J.G., Semenza, G.L., 2005. Transcriptional regulation of vascular endothelial cell responses to hypoxia by HIF-1. Blood 105, 659–669.

Martin, F., Linden, T., Katschinski, D.M., Oehme, F., Flamme, I., Mukhopadhyay, C.K., Eckhardt, K., Troger, J., Barth, S., Camenisch, G., Wenger, R.H., 2005. Copper-dependent activation of hypoxia-inducible factor (HIF)-1: implications for ceruloplasmin regulation. Blood 105, 4613–4619.

Maxwell, P., Salnikow, K., 2004. HIF-1: an oxygen and metal responsive transcription factor. Cancer Biol. Ther. 3, 29–35.

McAuslan, B.R., Gole, G.A., 1980. Cellular and molecular mechanisms in angiogenesis. Trans. Ophthalmol. Soc. U.K. 100, 354–358.

McAuslan, B.R., Reilly, W., 1980. Endothelial cell phagokinesis in response to specific metal ions. Exp. Cell. Res. 130, 147–157.

McAuslan, B.R., Reilly, W.G., Hannan, G.N., Gole, G.A., 1983. Angiogenic factors and their assay: activity of formyl methionyl leucyl phenylalanine, adenosine diphosphate, heparin, copper, and bovine endothelium stimulating factor. Microvasc. Res. 26, 323–338.

Melillo, G., Musso, T., Sica, A., Taylor, L.S., Cox, G.W., Varesio, L., 1995. A hypoxia-responsive element mediates a novel pathway of activation of the inducible nitric oxide synthase promoter. J. Exp. Med. 182, 1683–1693.

Minet, E., Mottet, D., Michel, G., Roland, I., Raes, M., Remacle, J., Michiels, C., 1999. Hypoxia-induced activation of HIF-1: role of HIF-1α-Hsp90 interaction. FEBS Lett. 460, 251–256.

Moriguchi, M., Nakajima, T., Kimura, H., Watanabe, T., Takashima, H., Mitsumoto, Y., Katagishi, T., Okanoue, T., Kagawa, K., 2002. The copper chelator trientine has an antiangiogenic effect against hepatocellular carcinoma, possibly through inhibition of interleukin-8 production. Int. J. Cancer 102, 445–452.

Moslehi, J., Minamishima, Y.A., Shi, J., Neuberg, D., Charytan, D.M., Padera, R.F., Signoretti, S., Liao, R., Kaelin Jr., W.G., 2010. Loss of hypoxia-inducible factor prolyl hydroxylase activity in cardiomyocytes phenocopies ischemic cardiomyopathy. Circulation 122, 1004–1016.

Mukhopadhyay, C.K., Mazumder, B., Fox, P.L., 2000. Role of hypoxia-inducible factor-1 in transcriptional activation of ceruloplasmin by iron deficiency. J. Biol. Chem. 275, 21048–21054.

Narayanan, G., Bharathidevi, S.R., Vuyyuru, H., Muthuvel, B., Konerirajapuram Natrajan, S., 2013. CTR1 silencing inhibits angiogenesis by limiting copper entry into endothelial cells. PLoS One 8, e71982.

Nguyen, S.V., Claycomb, W.C., 1999. Hypoxia regulates the expression of the adrenomedullin and HIF-1 genes in cultured HL-1 cardiomyocytes. Biochem. Biophys. Res. Commun. 265, 382–386.

O'Rourke, J.F., Tian, Y.M., Ratcliffe, P.J., Pugh, C.W., 1999. Oxygen-regulated and transactivating domains in endothelial PAS protein 1: comparison with hypoxia-inducible factor-1α. J. Biol. Chem. 274, 2060–2071.

Palmer, L.A., Semenza, G.L., Stoler, M.H., Johns, R.A., 1998. Hypoxia induces type II NOS gene expression in pulmonary artery endothelial cells via HIF-1. Am. J. Physiol. 274, L212–L219.

Pan, Q., Kleer, C.G., van Golen, K.L., Irani, J., Bottema, K.M., Bias, C., De Carvalho, M., Mesri, E.A., Robins, D.M., Dick, R.D., Brewer, G.J., Merajver, S.D., 2002. Copper deficiency induced by tetrathiomolybdate suppresses tumor growth and angiogenesis. Cancer Res. 62, 4854–4859.

Pan, Q., Bao, L.W., Merajver, S.D., 2003. Tetrathiomolybdate inhibits angiogenesis and metastasis through suppression of the NFκB signaling cascade1 1 NIH grants R01CA77612 (SDM), P30CA46592, and M01-RR00042, head and neck SPORE P50CA97248, Susan G. Komen Breast Cancer Foundation, NIH Cancer Biology Postdoctoral Fellowship T32 CA09676 (QP), Department of defense breast Cancer Research Program Postdoctoral Fellowship (QP), and Tempting Tables Organization, Muskegon, MI. Mol. Cancer Res. 1, 701–706.

Pang, H., Chen, D., Cui, Q.C., Dou, Q.P., 2007. Sodium diethyldithiocarbamate, an AIDS progression inhibitor and a copper-binding compound, has proteasome-inhibitory and apoptosis-inducing activities in cancer cells. Int. J. Mol. Res. 19, 809–816.

Philipp, S., Jurgensen, J.S., Fielitz, J., Bernhardt, W.M., Weidemann, A., Schiche, A., Pilz, B., Dietz, R., Regitz-Zagrosek, V., Eckardt, K.U., Willenbrock, R., 2006. Stabilization of hypoxia inducible factor rather than modulation of collagen metabolism improves cardiac function after acute myocardial infarction in rats. Eur. J. Heart Fail 8, 347–354.

Qiu, L., Ding, X., Zhang, Z., Kang, Y.J., 2012. Copper is required for cobalt-induced transcriptional activity of hypoxia-inducible factor-1. J. Pharmacol. Exp. Ther. 342, 561–567.

Salceda, S., Caro, J., 1997. Hypoxia-inducible factor 1α (HIF-1α) protein is rapidly degraded by the ubiquitin-proteasome system under normoxic conditions. Its stabilization by hypoxia depends on redox-induced changes. J. Biol. Chem. 272, 22642–22647.

Schofield, C.J., Zhang, Z., 1999. Structural and mechanistic studies on 2-oxoglutarate-dependent oxygenases and related enzymes. Curr. Opin. Struct. Biol. 9, 722–731.

Semenza, G.L., Jiang, B.H., Leung, S.W., Passantino, R., Concordet, J.P., Maire, P., Giallongo, A., 1996. Hypoxia response elements in the aldolase A, enolase 1, and lactate dehydrogenase A gene promoters contain essential binding sites for hypoxia-inducible factor 1. J. Biol. Chem. 271, 32529–32537.

Semenza, G.L., 2003. Targeting HIF-1 for cancer therapy. Nat. Rev. Cancer 3, 721–732.

Sen, C.K., Khanna, S., Venojarvi, M., Trikha, P., Ellison, E.C., Hunt, T.K., Roy, S., 2002. Copper-induced vascular endothelial growth factor expression and wound healing. Am. J. Physiol. Heart Circ. Physiol. 282, H1821–H1827.

Shokrzadeh, M., Ghaemian, A., Salehifar, E., Aliakbari, S., Saravi, S.S., Ebrahimi, P., 2009. Serum zinc and copper levels in ischemic cardiomyopathy. Biol. Trace Elem. Res. 127, 116–123.

Soncin, F., Guitton, J.D., Cartwright, T., Badet, J., 1997. Interaction of human angiogenin with copper modulates angiogenin binding to endothelial cells. Biochem. Biophys. Res. Commun. 236, 604–610.

Tacchini, L., Bianchi, L., Bernelli-Zazzera, A., Cairo, G., 1999. Transferrin receptor induction by hypoxia. HIF-1-mediated transcriptional activation and cell-specific post-transcriptional regulation. J. Biol. Chem. 274, 24142–24146.

Talks, K.L., Turley, H., Gatter, K.C., Maxwell, P.H., Pugh, C.W., Ratcliffe, P.J., Harris, A.L., 2000. The expression and distribution of the hypoxia-inducible factors HIF-1α and HIF-2α in normal human tissues, cancers, and tumor-associated macrophages. Am. J. Pathol. 157, 411–421.

Teknos, T.N., Islam, M., Arenberg, D.A., Pan, Q., Carskadon, S.L., Abarbanell, A.M., Marcus, B., Paul, S., Vandenberg, C.D., Carron, M., 2005. The effect of tetrathiomolybdate on cytokine expression, angiogenesis, and tumor growth in squamous cell carcinoma of the head and neck. Arch. Otolaryngol. 131, 204–211.

Ullah, M.S., Davies, A.J., Halestrap, A.P., 2006. The plasma membrane lactate transporter MCT4, but not MCT1, is up-regulated by hypoxia through a HIF-1α-dependent mechanism. J. Biol. Chem. 281, 9030–9037.

van de Sluis, B., Muller, P., Duran, K., Chen, A., Groot, A.J., Klomp, L.W., Liu, P.P., Wijmenga, C., 2007. Increased activity of hypoxia-inducible factor 1 is associated with early embryonic lethality in Commd1 null mice. Mol. Cell. Biol. 27, 4142–4156.

van de Sluis, B., Groot, A.J., Vermeulen, J., van der Wall, E., van Diest, P.J., Wijmenga, C., Klomp, L.W., Vooijs, M., 2009. COMMD1 promotes pVHL and O$_2$-independent proteolysis of HIF-1α via HSP90/70. PLoS One 4, e7332.

van Heerden, D., Vosloo, A., Nikinmaa, M., 2004. Effects of short-term copper exposure on gill structure, metallothionein and hypoxia-inducible factor-1α (HIF-1α) levels in rainbow trout (*Oncorhynchus mykiss*). Aquat. Toxicol. 69, 271–280.

Wang, G.L., Semenza, G.L., 1993a. Characterization of hypoxia-inducible factor 1 and regulation of DNA binding activity by hypoxia. J. Biol. Chem. 268, 21513–21518.

Wang, G.L., Semenza, G.L., 1993b. General involvement of hypoxia-inducible factor 1 in transcriptional response to hypoxia. Proc. Natl. Acad. Sci. U.S.A. 90, 4304–4308.

Wang, G.L., Semenza, G.L., 1995. Purification and characterization of hypoxia-inducible factor 1. J. Biol. Chem. 270, 1230–1237.

Wang, G.L., Semenza, G.L., 1996. Molecular basis of hypoxia-induced erythropoietin expression. Curr. Opin. Hematol. 3, 156–162.

Wang, G.L., Jiang, B.H., Rue, E.A., Semenza, G.L., 1995. Hypoxia-inducible factor 1 is a basic-helix-loop-helix-PAS heterodimer regulated by cellular O_2 tension. Proc. Natl. Acad. Sci. U.S.A. 92, 5510–5514.

Wang, L., Ge, Y., Kang, Y.J., 2016. Featured Article: Effect of copper on nuclear translocation of copper chaperone for superoxide dismutase-1. Exp. Biol. Med. (Maywood) 241 (14), 1483–1488.

Wanner, R.M., Spielmann, P., Stroka, D.M., Camenisch, G., Camenisch, I., Scheid, A., Houck, D.R., Bauer, C., Gassmann, M., Wenger, R.H., 2000. Epolones induce erythropoietin expression via hypoxia-inducible factor-1α activation. Blood 96, 1558–1565.

Wenger, R.H., 2000. Mammalian oxygen sensing, signalling and gene regulation. J. Exp. Biol. 203, 1253–1263.

Wenger, R.H., 2002. Cellular adaptation to hypoxia: O_2-sensing protein hydroxylases, hypoxia-inducible transcription factors, and O_2-regulated gene expression. FASEB J. 16, 1151–1162.

Wykoff, C.C., Beasley, N.J., Watson, P.H., Turner, K.J., Pastorek, J., Sibtain, A., Wilson, G.D., Turley, H., Talks, K.L., Maxwell, P.H., Pugh, C.W., Ratcliffe, P.J., Harris, A.L., 2000. Hypoxia-inducible expression of tumor-associated carbonic anhydrases. Cancer Res. 60, 7075–7083.

Xie, H., Kang, Y.J., 2009. Role of copper in angiogenesis and its medicinal implications. Curr. Med. Chem. 16, 1304–1314.

Yoshida, D., Ikeda, Y., Nakazawa, S., 1995. Suppression of tumor growth in experimental 9L gliosarcoma model by copper depletion. Neurol.Med. Chir. 35, 133–135.

Yoshiji, H., Yoshii, J., Kuriyama, S., Ikenaka, Y., Noguchi, R., Yanase, K., Namisaki, T., Kitade, M., Yamazaki, M., Fukui, H., 2005. Combination of copper-chelating agent, trientine, and methotrexate attenuates colorectal carcinoma development and angiogenesis in mice. Oncol. Rep. 14, 213–218.

Zhang, L., Ward, M.L., Phillips, A.R., Zhang, S., Kennedy, J., Barry, B., Cannell, M.B., Cooper, G.J., 2013. Protection of the heart by treatment with a divalent-copper-selective chelator reveals a novel mechanism underlying cardiomyopathy in diabetic rats. Cardiovasc. Diabetol. 12, 123.

Zhang, Z., Qiu, L., Lin, C., Yang, H., Fu, H., Li, R., Kang, Y.J., 2014. Copper-dependent and -independent hypoxia-inducible factor-1 regulation of gene expression. Metallomics.

Zhang, W., Zhao, X., Xiao, Y., Chen, J., Han, P., Zhang, J., Fu, H., Kang, Y.J., 2016. The association of depressed angiogenic factors with reduced capillary density in the Rhesus monkey model of myocardial ischemia. Metallomics 8 (7), 654–662.

Zheng, L., Han, P., Liu, J., Li, R., Yin, W., Wang, T., Zhang, W., Kang, Y.J., 2015. Role of copper in regression of cardiac hypertrophy. Pharmacol. Ther. 148, 66–84.

Zhong, H., De Marzo, A.M., Laughner, E., Lim, M., Hilton, D.A., Zagzag, D., Buechler, P., Isaacs, W.B., Semenza, G.L., Simons, J.W., 1999. Overexpression of hypoxia-inducible factor 1α in common human cancers and their metastases. Cancer Res. 59, 5830–5835.

Ziche, M., Jones, J., Gullino, P.M., 1982. Role of prostaglandin E1 and copper in angiogenesis. J. Natl. Cancer. Inst. 69, 475–482.

Chapter 10

Copper in Wilson's and Alzheimer's Diseases, Copper-Lowering Therapy in Cancer and Other Diseases, and Copper Deficiency

George J. Brewer

University of Michigan, Ann Arbor, MI, United States

INTRODUCTION

In this chapter, several topics indicated in the chapter title involving copper are covered. The treatment of Wilson's disease (WD) will be a little more detailed than the other topics. It is designed to be a standalone review of WD, so that a physician, or other interested reader, can learn about clinical presentations, diagnosis, and treatment in some detail. Alzheimer's disease (AD) is a much more common disease, extensively written about, and there is no need to extensively consider the disease itself. Rather, a new chapter in the story of this disease, copper toxicity, will be dealt with. The copper-lowering therapy section will focus on the promise of this type of therapy in various diseases, such as cancer. Finally, there will be a section on copper deficiency, which has become more common as a result of excessive swallowing of denture adhesive containing large quantities of zinc.

WILSON'S DISEASE

Introduction

WD is an autosomal-recessive inherited disease of copper accumulation and copper toxicity (Brewer, 2000, 2001, 2008; Scheinberg and Sternlieb, 1984; Brewer and Yuzbaysian-Gurkan, 1992). The normal diet contains approximately 1.0 mg of copper, which is approximately 0.25 mg in excess of requirements. The liver controls copper balance by excretion of the excess copper in the bile for loss in the stool. In WD, both copies of the gene that produces the protein ATP7B, which is critical in the biliary copper excretion pathway, are mutated (Yamaguchi et al., 1993).

The WD patient accumulates a small amount of copper every day, and for many years the excess is stored in the liver, which eventually damages the liver beginning early in childhood. In approximately half of the patients the liver damage presents clinically during the late teenage years or early 20s in Western countries, earlier in India and the Orient, but occasionally much later in life.

In approximately half of the patients, the liver disease does not clinically present. Excess copper spills out into the bloodstream and begins to damage those parts of the brain which coordinate movement, and the patient presents with a neurologic movement disorder, often accompanied by behavioral abnormalities.

In addition to the liver and brain damage, other parts of the body may be affected. In women, menstruation often stops and abortion is frequent. Arthritis in the knee, sunflower cataracts and cardiac arrhythmias may occur. Patients may have thrombocytopenia, leucopenia, and bleeding from esophageal or gastric varices from hypersplenism due to cirrhosis.

Clinical Presentations

In the liver presentation the patient may have acute hepatitis, with elevated transaminase enzymes, and if severe enough, jaundice, often misdiagnosed as viral hepatitis. The diagnosis of WD is made by elevated urine copper, elevated liver copper

Molecular, Genetic, and Nutritional Aspects of Major and Trace Minerals. http://dx.doi.org/10.1016/B978-0-12-802168-2.00010-5

if a biopsy is done, and usually a very low blood ceruloplasmin (Cp). If the diagnosis is not made, then often times the hepatitis will remit spontaneously, only to recur again later. At this phase the disease is often misdiagnosed as relapsing hepatitis. The patient will ultimately develop cirrhosis and may present clinically with one or more of its complications.

An alternative occasional hepatic presentation is explosive acute liver failure, often in a teenager. This is often accompanied by hemolytic anemia because the liver's release of large quantities of copper causes destruction of erythrocytes. The combination of liver disease and hemolytic anemia almost always indicates a diagnosis of WD.

The neurologic presentation, occurring in approximately half of patients, may begin with a behavioral disturbance a year or so before any neurologic symptom. This can take many forms, such as depression, manicky behavior, sexual exhibitionism, etc., and even psychotic behavior. The movement disorder (Fink et al., 1999) may start with one symptom alone for several months to a year or two, such as tremor. The tremor is not distinctive for WD. It is often misdiagnosed as essential tremor. Additional movement disorder symptoms are related to dystonia and incoordination. Handwriting may become sloppy, leading to very small writing, micrographia, in compensation. Speech is often affected, sometimes severely. Swallowing may be affected. If there is danger of aspiration, then it is important to have a gastric feeding tube placed to avoid pneumonia and lung damage. Patients often drool. The dystonia gradually pulls more and more parts of the body out of physiologic positions so that the patient assumes increasingly grotesque positions and abnormal facial expressions. Patients with tremor and dystonia may be misdiagnosed with Parkinson's disease. However, Parkinson's generally occurs in older people whereas WD is a young person's disease. A very useful diagnostic feature of patients with neurological WD is the presence of Kayser-Fleischer (KF) rings—copper deposits in a brownish ring around the outer portion of the cornea. A slit lamp examination by an ophthalmologist is required to definitively establish whether they are present. They are present in 99% of neurologic WD patients.

Some patients, usually siblings of a patient, are diagnosed while presymptomatic. When a new diagnosis is made, it is important to evaluate the siblings of the affected. On average one in four siblings of an affected individual will also have the disease and should be prophylactically treated.

Diagnosis

Regarding the various tests used to diagnose WD, the blood Cp is low in approximately 90% of WD patients because ATP7B has a second role of attaching copper to the Cp molecule. Without a normal copper complement, Cp is rapidly cleared from the blood. However, approximately 10% of WD patients have a normal Cp and approximately 10% of heterozygotes have a low Cp. These factors make the Cp level only moderately useful for diagnosis.

Serum copper can be measured and it can provide a little additional information. Blood Cp accounts for approximately 90% of total serum copper in normal people; however, in WD, the non-Cp serum copper, called free copper, is elevated. Each milligram of Cp contains 3 µg of copper. Therefore the free copper can be calculated by allotting 3 µg of copper per milligram/deciliter of Cp and subtracting from the total serum copper. For example if serum copper is 100 µg/dL and serum Cp is 30 mg/dL, then $3 \times 30 = 90$ µg of copper in Cp and $100 - 90 = 10$ µg/dL of serum-free copper.

The 24-h urine copper is a very useful diagnostic tool. It is almost always elevated over 100 µg/24 h in a symptomatic patient. It is relatively specific because only in severe obstructive liver disease will the urine copper be elevated over 100 µg/24 h. In approximately half of presymptomatic patients the urine copper is over 100 µg/24 h. In the other half the level will be elevated between 50 and 100 µg/24 h. Because heterozygous carriers can be elevated into this same range, further workup is needed to differentiate affected individuals from carrier individuals. For reasons of possible contamination or errors, it is best not to use a single test of urine copper as criteria for making or ruling out the diagnosis.

The quantitative liver copper value from a liver biopsy is generally the gold standard for definitive diagnosis of WD. It has been partially replaced by DNA mutation analysis, but in many patients DNA analysis is not adequate for diagnosis (see paragraph below), and in those patients quantitative liver copper is still the ultimate definitive test. It is always elevated over 200 µg of copper per gram of dry weight of tissue. The only false positive is in severe obstructive liver disease. Heterozygous carriers can have mildly elevated liver copper, but they do not have values of 200 or greater. Again, because of possible errors in any procedure, it is always best to have more than one positive (or negative) test to rule the diagnosis in or out. Another caution is that copper staining of the liver is unreliable as a diagnostic method.

Many WD-causing mutations in the ATP7B gene can be identified, and when such mutations are found in both copies of *ATP7B* this is a definitive, gold standard test. The problem is that several hundred different WD-causing mutations have now been found, and most patients are compound heterozygotes, meaning they have two different mutations on their two ATP7B alleles. Most DNA laboratories offering this test can only detect a portion of the many different mutations; therefore they can only make a definitive diagnosis of WD in a portion of patients, ranging from 50% to 95%, depending on the population and the variety of causative mutations in that population.

However, once a definitive diagnosis of WD is made by any method, DNA analysis can be used to definitively genotype siblings of the affected individual. This can be done by direct sequencing if the point mutations have been identified in the affected patient or by haplotype analysis, which can be done as follows. DNA markers on the chromosome on each side of the ATP7B gene can be typed on both chromosomes so that the region of the two chromosomes can be identified if present in siblings. Any sibling who shares both sets of markers is diagnosed as presymptomatic affected, whereas siblings with one are heterozygous carriers and siblings who have neither are clear.

The detection of KF rings is a valuable adjunct to diagnosis when present. They are nearly specific for WD, occurring rarely because of copper accumulation in severe obstructive liver disease and even more rarely from no known cause. They are present 99% of the time in neurologically presenting WD, in approximately half of patients with the liver presentations, and in approximately one-third of presymptomatic patients. Again, because of possible errors it is good to have other data to corroborate the presence or absence of the diagnosis.

The key to the early diagnosis of WD is for the doctor to think of the possibility of the diagnosis. Early diagnosis is critical because the disease is very treatable, and permanent liver and brain damage can be minimized by early treatment.

Treatment

Regarding the various anticopper treatments for WD, penicillamine was the first oral drug introduced, and a whole generation of patients owe their lives to this drug (Walshe, 1956). It is a copper chelator and acts by mobilizing copper from various tissues, binding it, and causing its excretion in the urine. It is usually given in doses of 1 g/day, in two divided doses, and separated from food by 2 h. It is fully efficacious. Pyridoxine, 25 mg/day, should be given when penicillamine is used because penicillamine interferes with pyridoxine metabolism and can cause pyridoxine deficiency.

The major problem with penicillamine is its many side effects (Brewer and Yuzbaysian-Gurkan, 1992). These include an initial hypersensitivity reaction in 25–35% of patients as well as bone marrow depression, proteinuria, skin damage, and autoimmune diseases, sometimes fatal. If given as initial treatment to neurologic patients, then it will make 25% of them permanently worse on the basis of a retrospective survey (Brewer et al., 1987). This high rate of neurological worsening when penicillamine is used to treat neurologically presenting patients has been recently confirmed by a prospective study from India, which found a 30% rate of neurologic worsening (Kalita et al., 2014). It probably does this because it mobilizes copper into the blood stream for excretion in the urine, and the elevation of copper in the blood results in elevation of copper in the brain and neurologic worsening. Because of all of its side effects, penicillamine has no place in the modern treatment of WD, although it is still widely used because a whole generation of physicians have used it and are unfamiliar with newer and better treatments.

Trientine is another copper chelator with a similar mechanism of action, causing excretion of copper in the urine, but it is much safer (Walshe, 1982). It also is used in doses of 1.0 g/day, in two divided doses, and separated from food by 2 h. It is fully efficacious.

Side effects from trientine include proteinuria and some of the same side effects as penicillamine, except much less frequently. It does not produce the side effects on skin seen with penicillamine. It also produces permanent neurologic worsening in approximately 25% of patients if used as initial treatment in the neurologic presentation (Brewer et al., 2006).

Trientine is recommended in combination with zinc for the initial treatment of the hepatic presentation if hepatic failure is present (Askari et al., 2003); as a second choice to zinc for maintenance; and for treatment of presymptomatic, pregnant, and pediatric patients.

Zinc was approved by the US Food and Drug Administration (FDA) for WD therapy in 1997 after a long series of studies to show its efficacy and lack of toxicity (references in Brewer, 2000). Zinc is not a copper chelator, unlike penicillamine and trientine. It acts by blocking the intestinal absorption of copper by induction of intestinal cell metallothionein, which binds copper from food and endogenous secretions, and holds it in the intestinal cell until the cell sloughs into the stool with a few days of turnover time. Thus it increases stool and not urine copper. Because it blocks the absorption of copper from endogenous secretions, which are normally reabsorbed, it puts the patient into a negative copper balance. In adults it is given in doses of 50 mg of elemental zinc 3 times/day, each dose separated from food and beverages other than water by at least 1, and even better, 2 h. It is fully efficacious in 100% of patients, having shown in 40 of 40 straight patients that it produced negative copper balances and full blockade of oral copper-64 absorption (Brewer et al., 1990).

The only side effect of zinc is gastric intolerance. Because it is taken on an empty stomach, the capsule containing the zinc salt dissolves rapidly and deposits a high concentration of zinc salt in one location in the stomach, causing gastric irritation and symptoms in perhaps 25% of patients. In perhaps 10%, symptoms are severe enough that they have difficulty taking zinc reliably and they need to be switched to trientine. However, because of the gastric intolerance, additional patients besides the 10% may comply poorly with the zinc regimen. This has led to claims of "zinc failures" in the literature (for examples see Weiss et al., 2011; Gunther, 2013). However, the authors publishing these papers did not evaluate compliance. Therefore these papers are not reporting zinc failures, but rather a higher rate of noncompliance than seen with other treatments. There are no zinc failures because it was shown efficacious in 40 of 40 patients (Brewer et al., 1990).

Zinc is recommended in combination with trientine for the initial treatment of the hepatic presentation when hepatic failure is present; as the first choice for maintenance therapy, pregnancy (Brewer et al., 2000a), and pediatric patients; and for the treatment of the presymptomatic patient from the beginning (Brewer et al., 1994). Zinc is also recommended for the therapy of the neurologic presentation from the beginning until tetrathiomolybdate (TM) becomes available.

TM was developed because of the lack of appropriate therapy for the neurologic presentation of WD. As previously indicated, the chelators penicillamine and trientine make a high proportion (~25%) of neurologic patients permanently neurologically worse if used as initial therapy. Zinc is a little slow acting for these acute patients, taking up to 6 months to eliminate copper toxicity. During this period, the disease may progress because of its own natural history. TM was developed to fill this therapeutic hiatus.

TM is a marvelous drug for the initial treatment of the neurologically presenting patient. A 55-patient open-label study was done using a neurological function evaluating instrument to assess neurologic deterioration, and only 3.6% of patients deteriorated (Brewer et al., 2003a). A double-blind study was performed evaluating TM as initial therapy versus trientine; 26% of trientine patients deteriorated versus only 4% of TM patients deteriorating, a statistically significant ($p = 0.05$) difference (Brewer et al., 2006). This low rate of deterioration with TM treatment is dramatically better than the high rate with penicillamine and trientine.

TM can be used to produce two mechanisms of action. If given away from food, then it is well absorbed and forms a tripartite complex with albumin and free copper. The copper is bound tightly in this complex, and the complex is slowly metabolized by the liver. In this way the free, toxic copper in the blood can be brought down to very low levels in approximately 2 weeks. If TM is given with food, then it forms the tripartite complex with food copper and food proteins and prevents copper absorption. In this way it can cause an immediate negative copper balance. No other drug can neutralize copper toxicity as rapidly as TM. In the studies previously reported, TM was used in 20-mg doses three times daily away from food and 20-mg doses three times daily with meals.

The problem with TM is that it is not commercially available. An initial attempt at FDA approval failed. A Swedish company is now working on efforts to get it approved for WD. Until TM is available, zinc is recommended for initial treatment of the neurologically presenting patient. Although zinc is slow, it does not have the drug-catalyzed worsening seen in a high proportion of patients when penicillamine or trientine are used.

Summarizing appropriate treatment choices, it is important for the physician to remember that there are now treatment choices in addition to penicillamine. Far too many patients are still being started on, and treated with, penicillamine. For the initial treatment of the neurologically presenting patient, TM is by far the best choice. Until TM becomes available, zinc is the best choice. Because a high percentage (~25%) of patients become disastrously and permanently neurologically worse, penicillamine and trientine should be avoided. Unfortunately, guidelines for therapy, written by hepatologists for liver disease for liver disease societies, still recommend a chelator for initial therapy of all newly presenting patients (European Association for the Study of the Liver, 2012; Roberts and Schilsky, 2008). This is a serious mistake for neurologically presenting patients. Zinc should be indefinitely given. Recovery that takes place may take 1–2 years. It is important to maintain as much physical and speech activity as possible. Many of these patients recover remarkably well.

For the hepatically presenting patient with severe liver failure, the first thing to consider is whether liver transplantation is necessary to save the patient's life. The prognostic index of Nazer et al. (1986), later updated by Dhawan et al. (2005), can be consulted. If these indices indicate the patient is highly unlikely to survive with medical treatment, then hepatic transplantation should be performed.

If medical therapy is undertaken in a patient with liver failure, then the best choice is a combination of zinc and trientine. The chelator will help put the patient into a quick and significant negative copper balance, and zinc will help to reduce copper toxicity in the liver by inducing hepatocyte metallothionein, which will bind some of the toxic copper. After 4–6 months, when liver function tests have significantly normalized, the trientine can be stopped and zinc maintenance therapy continued. In patients with a hepatic presentation without liver failure, zinc therapy can be used from the beginning and carried on into maintenance therapy.

Zinc is the first choice for maintenance therapy because it has fewer side effects than trientine, but trientine is the second choice. If gastric intolerance is significant enough with zinc to threaten compliance, then trientine should be immediately substituted. Zinc therapy should be monitored to corroborate compliance. This is easily done by collecting 24-h urines and measuring copper and zinc. Urine copper, after 6–12 months of zinc therapy, should come down to 125 μg or less/24 h. Urine zinc should be at least 2 mg or more/24 h if the patient is properly taking their zinc. This should be done every 3 months for

1 year, then every 6 months for 2 years, and then each year in patients proven to be well complying. Maintenance therapy should be continued for the rest of the patient's life.

Zinc can be used as therapy from the beginning in presymptomatic patients and is very good in pregnant and pediatric (Brewer et al., 2001) patients. The adult dose of zinc is 50 mg for three times per day, with each dose separated from food and beverages by at least 1 and better 2 h. The pediatric dose is 25 mg twice a day until age 5 years, 25 mg for three times per day until age 15 years or 125 pounds body weight, and then the adult dose.

Trientine is second choice for maintenance, pregnancy, and pediatric therapy. It is more difficult to monitor it for compliance because the drug acts to increase copper in the urine; therefore urine copper cannot be used to evaluate copper status. Free copper levels in the blood can be followed, but they are more difficult and prone to error. It is important to regularly check blood counts, blood biochemistries, and urinalysis during the first 1 or 2 years of trientine therapy to evaluate possible trientine toxicities.

If diagnosis and initiation of treatment are accomplished early enough, and compliance is good, then the prognosis is very good in these patients. Many, perhaps most, can live a normal lifespan.

ALZHEIMER'S DISEASE

AD is a second disease, WD being the first, in which copper toxicity plays a major role. The important role of copper in AD has only come to light over the last few years. A puzzling aspect has been that AD involves brain damage, and WD involves brain damage in half of the patients, but the neurologic diseases are completely medically different. AD involves cognition loss, and there is no hint of a neurologic movement disorder. When WD involves the brain it causes a neurologic movement disorder, but it does not cause cognition loss. Furthermore, the parts of the brain that are involved in the two diseases are different. WD involves the basal ganglia and those parts of the brain that coordinate movement. AD involves the parts of the brain involved in memory and cognition, such as the hippocampus and the cerebrum. Finally, the pathology is completely different. AD brains have extracellular amyloid plaques and intracellular neurofibrillary tangles. WD brains have neuronal damage and loss, but no extracellular or intracellular bodies. If the brain damage in both diseases is caused by copper toxicity, then why are the clinical pictures, the areas of brain involvement, and the pathology so completely different?

The apparent answer to these puzzles has only come to light in 2014, and it makes a fascinating story, involving the two valence states of copper, which are Cu^{1+} or copper-1 and Cu^{2+} or copper-2. It has been shown by Ceko et al. (2014) that food copper is almost completely copper-1. This is a surprise and unexpected because copper in live tissues is in copper-1 and copper-2 forms, forming a redox couplet, and allowing many important, redox-dependent, life reactions. However, apparently after death, or at harvest, in the absence of oxygen, most copper-2 assumes the reduced, or copper-1 state. Of course inorganic copper, such as copper in drinking water and in supplement pills, is copper-2. It turns out that copper-2 ingestion is responsible for the huge epidemic of AD we are experiencing; thus the AD copper toxicity appears to be from copper-2. In contrast, WD is a copper overload disease, primarily from copper-1, with no special role for copper-2—thus the differences between the two diseases. Part of the fascinating story alluded to earlier is that the identification of copper-2 ingestion as a major risk factor for AD is that copper-2 ingestion can be avoided, and most of the major epidemic of AD now going on can be aborted.

The first realization that AD is a copper toxicity disease must be credited to Rosanna Squitti and her group in Italy. As pointed out in the section on WD, the blood copper can be divided into two pools, with the majority covalently bound to Cp, and the remainder (10–25% depending on how Cp is measured), called free copper, loosely bound to albumin and small molecules. It is the free copper that is toxic, when the free copper pool expands, as it does in WD.

The Squitti group has focused on the free copper pool in AD. First, they have shown that the free copper pool is significantly larger in AD than age-matched controls (Squitti et al., 2005). Second, they have shown that the size of the AD pool correlates with a measure of cognition (i.e., the larger the pool, the poorer the cognition; Squitti et al., 2006). Third, they have shown that the larger the pool, the greater the rate of cognition loss over time (Squitti et al., 2009). Finally, they have shown that the larger the pool, the greater the risk that mild cognitively impaired patients, a precursor state to AD, will convert to full AD (Squitti et al., 2014). These sets of studies show that the size of the free copper pool is intimately involved with the pathogenesis of AD, and establish AD as, at least in part, a copper toxicity disease.

It is important to examine the epidemiology of AD in the world today to obtain a better understanding of what has been happening to AD prevalence and how copper-2 fits in. A major epidemic of AD is underway in the United States and Western countries. For example, the prevalence of AD in the United States is 10% in those older than 60 years of age, 20% in those older than 70 years of age, and 30% in those older than 80 years of age (Alzheimer's Association, 2010). In contrast, the prevalence is only approximately 1% in the elderly in underdeveloped countries. For example, in rural India, in those age 65 years and older it is 1.07% (Chandra et al., 1998), and in Nigeria, Africa, in those ages 65–74 years it is 0.52%

(Hendrie et al., 1995). Another interesting fact, disputed by some, is that even in Western countries a disease such as AD was rare before 1900. It appears that Waldman and Lamb first pointed this out in their book *Dying for a Hamburger* (Waldman and Lamb, 2005). They state that very competent clinicians in the late 1800s—Osler an internist (Osler, 1910), Gowers a neurologist (Gowers, 1888), and Freud a psychiatrist (Strachey et al., 1966), all of who published extensively—did not mention an AD-like disease. Furthermore, Boyd, a pathologist (Boyd, 1938), who wrote a textbook of pathology during the period, updated over several decades, did not describe the characteristic AD pathology (i.e., amyloid plaques and neurofibrillary tangles) in brains at autopsy. Some dispute this claim of AD rarity before 1900, saying that because AD is a disease of aging, there were not enough old people during that time or that the disease just was not noticed. However, Waldman and Lamb showed that half of the population of France were age 60 years and older in 1911, and the US census for 1900 showed 3.6 million people age 60 years and older, enough to generate 36,000 AD cases at today's rate. The objection that the disease was not noticed would not seem to explain the lack of plaques and tangles at autopsy. This controversy might be settled by reexamining brains of people age 60 years and older who died during the period 1850–1925, if such brain tissue is still available, and checking for plaques and tangles using modern techniques.

Settling this controversy is of more than academic importance. If positive (plaques and tangles at a 10% rate), then it means that the apparent new epidemic is an artifact of the disease not being noticed in the past. If negative (plaques and tangles at a low 1% or so rate), then it means that the epidemic is new. If it is new, then it requires an adjustment in thinking about risk factors, namely that a new and important environmental risk factor has emerged in developed but not undeveloped countries. On the basis of the totality of the evidence, it appears highly likely that the epidemic is new, and that will be assumed here. Likewise, on the basis of the weight of evidence, it seems highly likely that copper-2 ingestion is the new environmental risk factor, the evidence for which is summarized in the following.

First, there is the paper of Sparks and Schreurs (2003), in which they showed that tiny amounts of copper (0.12 ppm) added to drinking water in a rabbit model of AD greatly enhanced AD-type pathology in the brain and memory loss in the rabbits. This work was replicated in other AD models, including the mouse model, by Sparks et al. (2006), and in another laboratory using the mouse model (Singh et al., 2013). This tiny amount of copper, if added in food, would have no effect. Animal chow generally contains 3–6 ppm copper. Changing the chow from 3 to 6 ppm, a 25-fold greater change than the 0.12 ppm added to the water, would cause no toxicity in rabbits, mice, etc. What this demonstrates is that copper in drinking water, which is known to be copper-2, is exquisitely more toxic than copper in food. For reference, the US Environmental Protection Agency allows 1.3 ppm copper in US drinking water (National Research Council, 2000), more than 10 times the amount toxic in animal models.

This Sparks and Schreurs (2003) paper was followed by a paper by Morris and colleagues (2006), in which they reported a study in a large Chicago population. They looked at nutrient intake and measured cognition over time. They found that if those in the highest quintile of copper intake (and these people were in the highest quintile of copper intake because they were taking a supplement pill containing copper, which is of course copper-2) also ate a high-fat diet, then they lost cognition at six times the rate of other groups.

A clue as to why the inorganic copper (i.e., the copper-2) of drinking water and supplement pills is so much more toxic than the organic copper of food is given by older studies in which a copper-64 label, given orally as an inorganic copper salt (copper-2), was used to evaluate whether zinc therapy was blocking copper absorption in WD (Hill et al., 1986). In the absence of zinc therapy, a portion of the label (up to ~25% of the administered dose) appeared within 1–2 h in the blood. If the copper in food was labeled, then it would be 1–2 days before the label appeared in the blood, and it would appear covalently (safely) linked to a protein secreted into the blood by the liver. What this shows is that the organic copper of food is taken up by the liver and put into safe channels, whereas some inorganic copper (or copper-2) bypasses the liver and appears immediately in the blood as a part of the free copper pool. This copper-2 appears to be very toxic to the parts of the brain involved in cognition.

With these factors in mind the epidemic of AD can be examined in relation to the spread of copper plumbing in developed countries. Copper plumbing is not used in undeveloped countries because of the expense. Copper plumbing began to be used in the early 1900s, but the use was curtailed by World War I and then by World War II. After 1950 the use of copper plumbing exploded so that now 80% or more US homes have copper plumbing (Foley, 1985). This history of use of copper plumbing parallels rather closely the AD epidemic, which increased rather slowly during the first half of the 20th century and then has exploded since 1950. Japan is especially interesting in regard to the copper-2 in drinking water hypothesis. It is a developed country with a low prevalence of AD (Ueda et al., 1992), but it has shunned copper plumbing for fear of toxicity. However, when Japanese migrate to Hawaii, where copper plumbing is used, they develop AD at the rate seen in other developed countries (White et al., 1996).

Does copper leach from copper plumbing in large enough amounts to cause toxicity? The answer is yes. In a study of drinking-water copper from 280 households all across North America, it was found that approximately one-third of the samples had copper levels of 0.1 ppm or above, the levels found toxic in AD animal models; approximately one-third had values 0.01 ppm or lower, values low enough to be deemed safe; and approximately one-third had copper levels between these values, levels of unknown safety (Brewer, 2012). Therefore one-third of the drinking-water samples were unsafe, and another one-third was of unknown safety if the animal model studies are an appropriate guide.

The studies by Ceko et al. (2014) alluded to earlier offer an explanation for the toxicity of copper-2. They find that the copper in food, here called organic copper, is predominantly copper-1. Copper-1 can be absorbed by a specific copper transporter called Ctr1 (Ohnik and Thiele, 2014). After absorption it is bound to a copper chaperone called ATOX, then handed off to a protein called ATP7A, which loads it into vesicles for exit from the intestinal cell. From there it ends up in the liver, which puts the copper into safe channels. Copper-2 cannot be taken up by Ctr1 unless it is reduced to copper-1. If copper-2 is not immediately reduced, then it can be absorbed by diffusion by a nonenergy-dependent carrier (Ceko et al., 2014) and possibly by the divalent metal ion transporter. When copper-2 goes by way of these routes, some of it ends up bypassing the liver and appears immediately in the blood. This copper-2 appears to be toxic to those parts of the brain involved in cognition. Humans may have evolved to safely handle copper-1, the predominant copper in food, but not copper-2.

Additional data have developed to support the AD copper toxicity hypothesis. Shen et al. (2014) have studied the soil copper concentrations in soil in relation to AD prevalence across all of the provinces of China. There is good correlation of high soil copper concentration with high AD prevalence. For example, if soil copper concentration is 60–80 ppm, then there is 2.6 times as much risk of AD as in provinces where soil copper concentration is 20–40 ppm. It seems very likely that drinking-water copper concentration will be highly correlated with soil copper concentration. For another piece of evidence in support, James et al. (2012) have shown that in autopsied AD brains there is a high level of toxicity from "mobile" copper as compared with control brains. In addition, Mursa et al. (2011) have shown in the Iowa Women's Health Study that all-cause mortality is significantly increased in women taking copper supplement pills. This study does not sort out mortality related to AD, but it shows the general toxicity of copper supplements.

At this point it should be made clear that it is not being suggested that copper-2 is the cause of AD. It is being asserted that copper-2 ingestion is a major risk factor of AD and the cause of the huge increase in prevalence leading to the current epidemic of AD. The general consensus is that the underlying cause of AD is β-amyloid accumulation and aggregation into amyloid plaques, which cause toxicity to neurons. This is called the amyloid cascade hypothesis (Hardy and Selkoe, 2002). What causes the accumulation and aggregation is unknown. Age is a risk factor, as are certain genetic alleles. Metals such as copper can cause β-amyloid aggregation; therefore copper-2 could fit in here (Sarell et al., 2010). The plaques can emit damaging oxidant radicals, particularly if they bind iron or copper; therefore copper-2 could fit in here also (Sayre et al., 2000). What is being suggested by the copper-2 hypothesis, in relation to the amyloid cascade hypothesis, is that copper-2 increases the risk of aggregation and/or the toxicity of the plaques.

Although the amyloid cascade hypothesis is the majority opinion, not all agree with it. An alternative hypothesis is called the oxidant damage hypothesis (Lee et al., 2005; Nunomura et al., 2012). In this scenario, oxidant damage occurs first, and β-amyloid acts as an antioxidant. As β-amyloid accumulates as a protective measure, it aggregates into plaques. In this hypothesis the plaques are a result of the disease, not the cause (Lee et al., 2005). In this hypothesis, copper-2 could be a primary causal agent, perhaps one of several, causing initial oxidant damage.

Earlier it was suggested that if the copper-2 hypothesis is valid, then much of the current AD epidemic can be aborted. This can be done by simply avoiding copper-2 ingestion by avoiding ingestion of drinking-water copper and supplement-pill copper. It is easy to avoid supplement-pill copper simply by not taking the pills. It is completely unnecessary for the general population to take supplemental copper because idiopathic copper deficiency is extremely rare. Copper deficiency from specific causes, such as gastrointestinal surgery or excessive zinc ingestion, can be easily identified and treated.

The drinking water for the home should be tested for copper levels. If the level is 0.01 ppm or lower, then it is safe. If it is higher, then copper plumbing need not be removed. A device such as a reverse osmosis device can be installed to remove copper on the tap providing drinking and cooking water. Even if copper plumbing is not present the drinking-water copper level should be tested because sometimes source water can have a high copper level.

Admittedly, copper-2 as a major risk factor for AD is not a finally proven hypothesis yet. However, the evidence is strong enough such that prudence dictates avoiding copper-2 ingestion.

THE PROMISE OF COPPER-LOWERING THERAPY IN CANCER AND OTHER DISEASES

Folkman (1972) is credited with coming up with the concept of antiangiogenic therapy for cancer. The idea is that after a cluster of cancer cells reaches approximately 2 mm in diameter, they must develop a blood supply, called angiogenesis, to grow further. Because adult humans do not require much angiogenesis, therapy aimed at inhibiting angiogenesis might be effective against cancer. We now know that there are many angiogenic-promoting agents [e.g., vascular endothelial growth factor (VEGF); fibroblast growth factor (FGF); and secreted protein, acidic and rich in cysteine (SPARC)], which cancers can recruit to stimulate angiogenesis and allow growth. Today there are therapies for cancer with some efficacy such as antibodies against specific angiogenic promoters, such as VEGF.

Copper is required for the activation of many angiogenic promoters (e.g., FGF and SPARC), suggesting that anticopper drugs might be effective antiangiogenic therapies for cancer. Penicillamine was shown to have efficacy in inhibiting a cancer in rabbits (Brem et al., 1990). Later, TM, discussed earlier in the section on WD, was shown to have excellent efficacy in a series of mouse cancer models (Pan et al., 2002; Cox et al., 2001, 2003; Khan et al., 2002; Pan et al., 2003; van Golen et al., 2002). One particularly impressive example was the HER2/neu mammary cancer mouse model, in which newborn mice are genetically programmed to develop mammary cancer during the first year of life (Pan et al., 2002). TM therapy completely prevented the tumors whereas control mice developed obvious mammary tumors, often multiple. On pathological examination, the TM mice had small clusters of cancer cells in their mammary glands, which did not grow into visible tumors because of a lack of angiogenesis.

After that there were multiple TM trials in human cancer at the University of Michigan, all of these in bulky advanced cancer and all showing very limited efficacy (Brewer et al., 2000b; Redman et al., 2003; Henry et al., 2006; Gartner et al., 2009; Pass et al., 2008). The way TM was used in the mouse and human studies was to lower copper levels to an intermediate status, avoiding clinical copper deficiency, but lowering copper availability, which appears to inhibit angiogenic promoters activated by copper. The proper copper status is obtained by monitoring blood Cp levels. The amount of Cp in the blood is determined by copper availability in the liver. The normal Cp level in the human is 20–35 mg/dL. If the Cp is reduced to 5 mg/dL or lower, then clinical copper deficiency appears, beginning with anemia. To avoid this the target Cp is between approximately 8 and 15 mg/dL.

The University of Michigan licensed the cancer use of TM to a company, which elected to use the choline salt of TM because it had more stability. All of the WD work and the cancer work at the University of Michigan had been done with the ammonium salt, which shows enough air instability that capsules were only used for 60 days after preparation, whereas bulk supplies were protected from air in containers with argon replacing the air. The company essentially repeated the University of Michigan human cancer work with two trials of choline TM against advanced bulky cancer and found very limited efficacy (Lin et al., 2013; Lowndes et al., 2008).

What was forgotten in all of the human work was that the mouse trials that had shown great efficacy were all against what might be called microscopic cancer. The HER2/neu study, which showed such dramatic results, involved small clusters of cells. Likewise, the other mouse models involved injections of cancer cells into the mice, again involving microscopic cancer. It seems likely that small clusters of cancer cells can use or recruit only one, or a very limited number, of angiogenic promoters, and that the promoter or promoters available are all dependent on copper. In contrast, advanced cancers cause inflammation and attract all kinds of inflammatory cells, many of which synthesize angiogenic promoters that can be used by the cancer to promote angiogenesis, and many of these are not dependent on copper, limiting TM efficacy against advanced cancers.

The key to TM efficacy against human cancer may be to remember the mouse examples and to use it against micrometastatic disease. An oncologist on the US West Coast has used TM successfully in this manner. He gets the tumor to no evidence of disease (NED) status by conventional therapy. However, the various tumors have close to 100% probability of recurrence because of micrometastatic disease. He treats with TM, keeping the Cp below 15 for 3 years. When the TM is stopped, the disease does not recur, and the cancer is cured. It appears that the micrometastatic clusters of cells die or are killed over that 3-year period. He has successfully cured patients with a variety of tumors. Unfortunately, all of this work is anecdotal because none of it is published. The author has reviewed some of the patient records and is convinced of the authenticity of the findings, but without publication the work is necessarily viewed as anecdotal.

There is one study, with breast cancer, in which this kind of TM work has been published. Breast cancer offers a good chance to test TM against micrometastatic disease. After removal of the primary (if there are multiple) positive lymph nodes, micrometastatic disease is present, and these cancers have a very high probability of recurrence. Jain et al. (2013) from Vahdat's group have recently published very encouraging preliminary data on TM treatment of breast cancer. They have found a type of circulating cell, called endothelial progenitor cells (EPCs), which are important in initiating angiogenesis and are a key to predict relapse in breast cancer. Forty-one breast cancer patients were enrolled: 29 stage 2/3 and 12 stage 4 NED. TM (100 mg) was given orally to maintain Cp at less than 17 mg/dL for 2 years or until relapse. Seventy-five percent of patients achieved the copper depletion target within 1 month. In copper-depleted patients, but not in those patients who did not achieve copper depletion, there was a significant reduction in EPCs. Of six patients who relapsed, only one had EPC levels below baseline. Eighty-five percent of the patients were still relapse free at 10 months, a remarkable time in this group of patients with severe disease, and the study is ongoing. They concluded that TM is safe, can maintain EPC levels below baseline if copper depletion is achieved, promotes tumor dormancy, and ultimately may prevent relapse.

A study to further evaluate the efficacy of TM against micrometastatic cancer has been initiated by the author and his colleagues and involves a double-blind study of osteosarcoma in dogs. After removal of the primary, most of these dogs have micrometastatic cancer in the lungs. Endpoints are time to macroscopic lung metastases and survival.

In summary, it appears there is much promise for TM therapy for the cure of many cases of human cancer, in which the remaining cancer is micrometastatic, or the disease can be brought to NED status, with only micrometastatic disease remaining. It would appear these patients can often be cured by TM. Unfortunately, the early human trials were all against bulky, advanced cancer, and the minimal efficacy caused pharmaceutical companies to turn away. That is extremely unfortunate because the promise of TM to cure many cancers is very real.

There is also great promise that many diseases in addition to cancer can be treated by copper-lowering therapy. These include diseases of fibrosis, inflammation, and autoimmunity. Discussing fibrosis first, there are many serious, often fatal, diseases of fibrosis such as liver cirrhosis, idiopathic pulmonary fibrosis, fibrotic disease of the kidneys, and others. Fibrosis results from pathological activation of the cytokines of the fibrotic pathway, which operate physiologically for wound healing, as an example. This pathway involves activation of transforming growth factor-β (TGF-β), which activates connective tissue growth factor (CTGF), which turns on many genes involved in fibrosis, such as collagen genes. Many steps in this pathway including activators of TGF-β, and perhaps CTGF itself, are dependent on copper.

The bleomycin mouse model of pulmonary fibrosis is a reasonably good model of human idiopathic pulmonary fibrosis. Bleomycin is injected via the trachea of mice. In a study using TM versus placebo, TM strongly prevented the fibrosis (Brewer et al., 2003b) and prevented the large increase in TGF-β and an inflammatory cytokine, tumor necrosis factor-α (TNF-α; Brewer et al., 2004). TM also was very effective in preventing cirrhosis in mice from carbon tetrachloride injection (Askari et al., 2004). In addition, TM prevented cirrhosis from bile duct ligation in another mouse model of cirrhosis (Song et al., 2008). TM also speeded up recovery from carbon-tetrachloride–induced cirrhosis in mice offering hope for treatment of human cirrhosis (Hou et al., 2009). The concept is that the fibrotic process and the repair process are going on simultaneously, and TM inhibits the fibrotic process, allowing recovery to go on much faster.

In summary, from the mouse work it appears that TM therapy offers promise of substantial benefit in human fibrotic conditions. There are no known treatments for fibrotic diseases. In particular, cirrhosis is very common—from alcoholic liver disease, hepatitis C, primary biliary cirrhosis, and others—and there are no treatments that offer recovery from any of these.

Excess inflammation is part of many disease processes, including fibrotic and autoimmune diseases. The inflammatory process begins with some type of tissue injury, the accumulation of inflammatory cells at the site, and the release of excess damaging inflammatory cytokines, such as TNF-α, interleukin (IL)-1β, IL-2, and nuclear factor kappa B, a transcription factor that causes release of inflammatory cytokines, which can further damage tissue cells. A mouse model of immune-modulated tissue damage is injection of concanavilin A (Con A) into mice. The Con A binds to liver cells, which causes an immune attack on the liver cells. The tissue damage can be measured by the release of the transaminase enzymes, alanine transaminase (ALT) and aspartate transaminase (AST), from the damaged liver. TM strongly prevents the release of AST and ALT after Con A injection in mice and inhibits the usual elevation of TNF-α in the blood (Askari et al., 2004).

Another example of tissue damage from excess inflammation is cardiac damage from the chemotherapeutic drug, doxorubicin, widely used as a treatment of cancer, but with a side effect of heart damage. A toxic dose of doxorubicin in mice causes release of the enzymes lactic dehydrogenase (LDH), creatinine kinase (CK), and troponin I into the blood from the damaged heart. TM therapy was able to prevent heart damage as measured by preventing the release of LDH, CK, and troponin I into the blood and preventing the marked increases in the blood of TNF-α, IL-1β, and IL-2, an indicator of cytotoxic T cell activation (Hou et al., 2005).

Another example of tissue damage from excess inflammation is the liver damage produced by excess doses of acetaminophen (ACAP), also known as Tylenol. Overdoses of ACAP, accidental or with suicide intent, are the most common cause of acute liver failure seen in the emergency room. TM therapy was able to almost completely prevent liver damage from a toxic dose of ACAP in the mouse with a very strong inhibition of AST, ALT, and IL-1β appearance in the blood (Ma et al., 2004). Furthermore, it was shown that TM could be given by injection some time after ACAP injection and still prevent much of the liver damage, indicating that TM might be effective in the emergency room for ACAP poisoning.

There are multiple examples of TM therapy in mouse models greatly mitigating immune-modulated disease, modeling human autoimmune diseases. The Con A immune-modulated model for hepatitis was discussed earlier (Askari et al., 2004). There is also an immune-modulated arthritis mouse model involving the injection of bovine collagen II. TM therapy greatly inhibited swelling and redness of joints, and it markedly reduced histologic abnormalities (McCubbin et al., 2006), suggesting that TM should be evaluated in the very common human autoimmune disease, rheumatoid arthritis. TM also markedly reduced the great increase in TNF-α, IL-1β, and IL-2 seen in controls. TM also markedly protected against the great increase in urinary isoprostanes (a marker of oxidant damage) seen in controls.

Multiple sclerosis (MS) is a human autoimmune disease involving an immune attack on myelin of the central nervous system. A mouse model of MS involves injection of a protein, myelin proteolipid protein, which causes clinical symptoms and lesions in the mouse brain and spinal cord that can be counted and graded. TM therapy strongly prevented clinical symptoms; reduced the number of central nervous system lesions; inhibited elevated IL-2, TNF-α, and interferon-γ; and prevented the marked increase in urine isoprostane levels seen in controls (Hou et al., 2008).

Summarizing, it appears that TM therapy has great promise in the treatment of human diseases of excess inflammation and autoimmunity. Some of these diseases are very common, and treatments currently available are far from optimal. Symptomatic treatments include aspirin and nonsteroidal antiinflammatory drugs, which can help alleviate pain but do not inhibit inflammation and the resulting tissue damage, at least not very much. Corticosteroids are the best treatment, but they usually are only partially effective and have a long list of serious side effects, particularly if used long term. In general, it appears that anything steroids can do, TM can do better and without all of the side effects.

Given how promising for human medicine TM would appear to be based on mouse models of fibrosis, inflammation, and autoimmunity, it is surprising that only one human clinical trial has been conducted. This was a double-blind clinical trial in primary biliary cirrhosis (Askari et al., 2010). This disease involves all three causation elements under discussion. It is an autoimmune attack on the bile ducts, which leads to inflammation and fibrosis (cirrhosis). The study was designed to be a 2-year double-blind trial of TM versus placebo. In TM patients the approach was similar to that used in the cancer trials to keep the Cp between 8 and 15. Endpoints were improved blood levels of AST, ALT, and TNF-α in TM-treated patients versus controls. The study ended prematurely because of a cutoff in funding. Nevertheless, 13 TM patients and 15 controls were followed for an average of 13 months. Endpoints were reached in that AST, ALT, and TNF-α levels were significantly improved in TM-treated patients versus controls (Askari et al., 2010). There was a trend in a favorable direction for bilirubin levels that did not reach statistical significance. Serum-free copper was also significantly reduced by TM.

In summary of this section, there is great promise in TM use for curing cancer if the remaining disease is micrometastatic, or it can be reduced to micrometestatic, NED, status. There are a plethora of mouse studies that indicate TM should be evaluated in human fibrotic, inflammatory, and autoimmune diseases. There is even one positive human trial in such a disease, primary biliary cirrhosis. As the title of this section indicates, there is much promise of copper-lowering therapy with TM in cancer and other diseases.

CLINICAL COPPER DEFICIENCY

Introduction

First, what is meant by the title of this section, *Clinical Copper Deficiency*, especially because the previous section discussed deliberate copper lowering as a therapeutic tool? The stages of copper deficiency can be thought of as follows. When Cp is lowered into the range 8–15 mg/dL, it means that liver copper has become less available. From the previous section, it should be clear that this copper status significantly reduces the activity of some angiogenic promoters as well as fibrotic and inflammatory pathway cytokines, and this effect can be used for therapeutic purposes. This copper status can be thought of as chemical copper deficiency, but it is not clinical copper deficiency. There are no clinical (medical), negative effects from this copper status.

The next stage of copper deficiency begins what might be called mild clinical copper deficiency, because anemia, and sometimes leukopenia, appear. This is a type of bone marrow depression. It occurs because copper is required at a certain stage of hemoglobin synthesis, causing the anemia, which is hypochromic and microcytic, resembling iron deficiency. Copper is also required for cellular production, causing the leucopenia. The blood Cp level is low, often approximately 2–5, in mild copper deficiency.

The next stage of copper deficiency is severe clinical copper deficiency. The blood Cp is 0–2, anemia is always present, and often leucopenia and neurologic symptoms appear. The neurologic syndrome is called myelopolyneuropathy. It is characterized by numbness and parasthesias affecting the lower extremities first, and then rapidly progressing to the upper extremities, and loss of balance. Neurologic examination reveals deficits involving motor-sensory polyneuropathy and myelopathy (corticospinal tract and dorsal columns).

There are certain rules to remember in diagnosing severe clinical copper deficiency. First, anemia is always present. Thus if a clinician is faced with an undiagnosed neurologic syndrome without anemia, then the diagnosis and the underlying cause are not copper deficiency. The second rule is that the Cp is always very low. There are scientists who worry that because Cp is an acute-phase reactant, that in the face of an acute-phase stimulant such as infection, inflammatory disease, cancer, etc., which cause Cp levels to be very high, that Cp is not a reliable indicator of copper deficiency. However, this is a false concern. As the literally hundreds of patients with cancer treated with copper-lowering therapy by TM show (discussed in detail in the last section), Cp is an excellent measure of copper status despite an acute-phase reaction, which all of these patients have. They all start with a very high Cp, but as TM therapy ensues, Cp levels come down, and when it reaches 8–15, it is just as reliable of an indicator of copper status as in anyone without an acute-phase reaction. Therefore the bottom line is that Cp is an excellent indicator of copper status when copper deficiency is present. As with the absence of anemia, it is an ironclad rule that if the Cp is not low, the patient does not have copper deficiency.

Zinc-Induced Copper Deficiency

As was made clear in the first section of this chapter on WD, zinc in an adequate dose will block copper absorption and put the WD patient into a negative copper balance. Of course these patients have large stores of copper in their liver, and it takes many years of zinc therapy to deplete this copper. However, the dose of zinc effective in WD would cause clinical copper deficiency in a normal person within a few months. However, zinc is being taken perhaps by upward of half of the population. In general this is safe because it is a question of dose and concomitant food ingestion. Therefore the safety of zinc supplementation dose in normal (non-WD) people is important to understand.

The first thing to make clear in discussing zinc dose is that all of the doses being discussed are where zinc is ingested away from food and beverages other than water by at least 2h. Many people take zinc with food, which renders the zinc ineffective. The zinc binds to indigestible substances in food and is never seen by the intestinal cell; therefore high zinc doses can be taken, but they are meaningless because they are ineffective.

The studies on zinc dose in WD show that up to 75mg of zinc as a single dose can be taken without producing a dangerous negative copper balance. Seventy-five milligrams daily is probably near the upper limit; therefore it is probably not a good idea to take more than approximately 50mg as a single daily dose. In addition, the WD work made it clear that divided doses were more effective at producing negative copper balance than single doses. For example, 37.5mg twice a day, or 25mg three times a day produced much larger and consistent negative copper balances than 75mg once a day.

Although some recommend monitoring copper status if zinc supplements are used, that is probably not necessary if no more than 50mg once a day is the dose. If at annual checkup anemia has appeared, then a Cp can be done to make sure it is not due to copper deficiency, but a copper deficiency cause at a zinc dose of 50mg daily or less would be very rare.

Most people take a zinc dose of 25mg or less daily to prevent zinc deficiency, and this is completely safe. It has been suggested that a daily zinc dose of 45mg/day in elderly will prevent approximately two-thirds of infections (Prasad et al., 2007), probably by bolstering the immune system, and this dose (or a 50-mg dose) is safe.

The anemia that develops in mild clinical copper deficiency is a great warning signal. If properly diagnosed and treated, usually simply by stopping the zinc dose, then it rapidly reverses. If this warning signal is seen, recognized, and treated, then it prevents the patient from progressing into severe clinical copper deficiency and a neurological syndrome that is often only partially reversible.

Dental-Adhesive–Induced and Idiopathic Copper Deficiency

Over the last 25years there has been a marked increase in myelopolyneuropathy, the neurological syndrome from severe copper deficiency. The author was involved in one of these cases (Hedera et al., 2003). The patient had myelopolyneuropathy, was severely anemic, had no measurable Cp, and had a very high serum zinc. The source of the zinc was a mystery. The anemia was cured by copper therapy, but the neurologic symptoms did not improve; they were irreversible. Then Nations et al. (2008) published a paper that brought sudden clarity to the cause. They published on several patients with the syndrome, and every one had ill-fitting dentures and were using and swallowing large quantities of denture adhesive that had high concentrations of zinc. The patient previously referred to had also been doing this. Our group then identified 10 additional patients with myelopolyneuropathy who had been swallowing large amounts of dental adhesive and published this series of cases (Hedera et al., 2009). The estimated amount of zinc being swallowed by these patients was usually 200 or 300mg/day or more—more than enough zinc to cause severe clinical copper deficiency. In all cases, when the patient quit swallowing large amounts of dental adhesive the zinc levels returned to normal. In all cases anemia disappeared with copper treatment, but in general the neurological symptoms did not improve.

Before the spate of cases from dental adhesive there had been occasional case reports of myelopolyneuropathy—some of them zinc induced and some of them due to malabsorption of various types, a topic to be considered in the next section; however, a few of them were idiopathic (i.e., from no known cause). It is possible that some of the idiopathic cases were due to dental adhesive because this cause has only come to light in the last decade. Whether truly idiopathic cases of severe clinical copper deficiency causing myelopolyneuropathy exist seems unlikely. It seems likely that they all result from something causing poor absorption of copper or something (such as zinc) blocking the absorption of copper. Of course a diet very deficient in copper content would do it, but that does not appear to happen, at least in Western societies.

Copper Deficiency From Poor Absorption

Malabsorption of nutrients can arise from various causes. In malabsorption syndromes many nutrients, including many micronutrients, which include many minerals and vitamins, can be poorly absorbed. Therefore copper is just one of many nutrients that is poorly absorbed in these syndromes.

Malabsorption can occur in diseases of the small intestine, such as regional enteritis (Crohn's disease), or with surgical removal of part of the small intestine. Celiac disease, or sprue, can cause malabsorption. It appears that gastric bypass surgery, possibly in combination with the diet that follows, can cause micronutrient, including copper, inadequacy. Many micronutrients, such as the minerals copper, zinc, and iron, are more poorly absorbed from vegetable foods than meats. Therefore people on vegetarian diets are at slightly higher risk of deficiency, but in the case of copper, probably only if some other causative factor also intervenes.

Copper Deficiency in the Healthy General Population

Almost all multivitamin/multimineral and multimineral supplement pills contain approximately 1 mg of copper. The rationale for this is the view that some diets are inadequate in copper and that there is some level of prevalence of copper deficiency in the general population. Probably approximately half of the adult population in the United States takes one of these pills each day.

Of course it is clear that there are known causes of copper deficiency, such as those discussed in the previous sections. However, these are easily identified and appropriately treated. However, some scientists believe (Klevay, 2011, is an example) there is a reasonably high prevalence of copper deficiency in the general, healthy, population and that these people benefit from the copper supplementation in multimineral supplement pills. These people further argue that the level of prevalence of copper deficiency in the general population is unknown because there is no good test for copper deficiency. They argue that Cp levels and serum copper levels are unreliable because Cp is an acute-phase reactant, causing high levels of Cp and serum copper irrespective of copper status.

These views are completely wrong. As discussed in the section on copper-lowering therapy, the Cp is always elevated in cancer patients because of an acute-phase reaction, but as TM is used to lower copper availability, the Cp comes down and becomes a completely reliable indicator of copper status when under 15–20 mg/dL. This has been studied in hundreds of patients with a known acute-phase reaction, proving Cp is completely reliable as an indicator of copper deficiency. Because unexplained low Cp levels are just not found in the general healthy population, it follows that unexplained, or undetected, copper deficiency is extremely rare in the general, healthy, population.

It follows from this that there is no rationale for including copper in multimineral supplement pills intended for the general, healthy, population. Furthermore, going back to the section on AD and the paper of Morris et al. (2006), it seems clear that the copper-2 in these pills is causing cognition loss and is involved in AD causation. Therefore the copper in these pills is not only not useful, it appears that it is at great risk for causing harm. It is important, perhaps urgent, that the copper be removed from supplement pills intended for the general populations.

A final point should be made that people who are copper deficient need to take a copper supplement. However, copper supplement pills are all copper-2, known now to be toxic. It would be very useful for a supplement maker to make a supplement pill containing copper-1.

REFERENCES

Alzheimer's Association, 2010. Alzheimer's Disease Facts and Figures. pp. 1–74.
Askari, F.K., Greenson, J., Dick, R.D., Johnson, V.D., Brewer, G.J., 2003. Treatment of Wilson's disease with zinc. XVIII. Initial treatment of the hepatic decompensation presentation with trientine and zinc. J. Lab. Clin. Med. 142, 385–390.
Askari, F.K., Dick, R., Mao, M., Brewer, G.J., 2004. Tetrathiomolybdate therapy protects against concanavalin a and carbon tetrachloride hepatic damage in mice. Exp. Biol. Med. (Maywood) 229, 857–863.
Askari, F., Innis, D., Dick, R.B., Hou, G., Marrero, J., Greenson, J., Brewer, G.J., 2010. Treatment of primary biliary cirrhosis with tetrathiomolybdate: results of a double-blind trial. Trans. Res. 155, 123–130.
Brewer, G.J., 2000. Recognition, diagnosis, and management of Wilson's disease. Exp. Biol. Med. 223, 39–46.
Brewer, G.J., 2001. Wilson's Disease: A Clinician's Guide to Recognition, Diagnosis, and Management. Kluwer Academic Publishers, Boston.

Brewer, G.J., 2008. Wilson disease. In: Fauci, A.S., Braunwald, E., Kasper, D.L., Hauser, S.L., Longo, D.L., Jameson, J.L., Loscalzo, J. (Eds.), Harrison's Principles of Internal Medicine, seventeenth ed. McGraw-Hill Companies, Inc., New York, pp. 2449–2452.

Brewer, G.J., 2012. Copper excess, zinc deficiency, and cognition loss in Alzheimer's disease. Biofactors 38, 107–113.

Brewer, G.J., Yuzbasiyan-Gurkan, V., 1992. Wilson disease. Medicine 71, 139–164.

Brewer, G.J., Terry, C.A., Aisen, A.M., Hill, G.M., 1987. Worsening of neurologic syndrome in patients with Wilson's disease with initial penicillamine therapy. Arch. Neurol. 44, 490–493.

Brewer, G.J., Yuzbasiyan-Gurkan, V., Dick, R., 1990. Zinc therapy of Wilson's disease: VIII. Dose response studies. J. Trace Elem. Exp. Med. 3, 227–234.

Brewer, G.J., Dick, R.D., Yuzbasiyan-Gurkan, V., Johnson, V., Wang, Y., 1994. Treatment of Wilson's disease with zinc. XIII: therapy with zinc in presymptomatic patients from the time of diagnosis. J. Lab. Clin. Med. 123, 849–858.

Brewer, G.J., Johnson, V.D., Dick, R.D., Fink, K.J., Kluin, K.J., Hedera, P., 2000a. Treatment of Wilson's disease with zinc. XVII: treatment during pregnancy. Hepatology 31, 364–370.

Brewer, G.J., Dick, R.D., Grover, D.K., LeClaire, V., Tseng, M., Wicha, M., Pienta, K., Redman, B.G., Thierry, J., Sondak, V.K., Strawderman, M., LeCarpentier, G., Merajver, S.D., 2000b. Treatment of metastic cancer with tetrathiomolybdate, an anticopper, antiangiogenic agent: phase I study. Clin. Cancer Res. 6, 1–10.

Brewer, G.J., Dick, R.D., Johnson, V.D., Fink, J.K., Kluin, K.J., Daniels, S., 2001. Treatment of Wilson's disease with zinc. XVI: treatment during the pediatric years. J. Lab. Clin. Med. 137, 191–198.

Brewer, G.J., Hedera, P., Kluin, K.J., Carlson, M.D., Askari, F., Dick, R.B., Sitterly, J.A., Fink, J.K., 2003a. Treatment of Wilson's disease with tetrathiomolybdate III. Initial therapy in a total of 55 neurologically affected patients and follow-up with zinc therapy. Arch. Neurol. 60, 378–385.

Brewer, G.J., Ullenbruch, M.R., Dick, R., Olivarez, I., Phan, S.H., 2003b. Tetrathiomolybdate therapy protects against bleomycin-induced pulmonary fibrosis in mice. J. Lab. Clin. Med. 141, 210–216.

Brewer, G.J., Dick, R., Ullenbruch, M.R., Jin, H., Phan, S.H., 2004. Inhibition of key cytokines by tetrathiomolybdate in the bleomycin model of pulmonary fibrosis. J. Inorg. Biochem. 98, 2160–2167.

Brewer, G.J., Askari, F., Lorincz, M.T., Carlson, M., Schilsky, M., Kluin, K.J., Hedera, P., Moretti, P., Fink, J.K., Tankanow, R., Dick, R.B., Sitterly, J., 2006. Treatment of Wilson disease with ammonium tetrathiomolybdate: IV. Comparison of tetrathiomolybdate and trientine in a double-blind study of treatment of the neurologic presentation of Wilson disease. Arch. Neurol. 63, 521–527.

Boyd, W.A., 1938. Textbook of Pathology: An Introduction to Medicine. Lea and Febiger, Philadelphia.

Brem, S., Zagzag, D., Tsanaelis, A.M.C., Gatley, S., Elkouby, M.P., Brian, S.E., 1990. Inhibition of angiogenesis and tumor growth in the brain. Suppression of endothelial cell turnover by penicillamine and the depletion of copper, an angiogenic cofactor. Am. J. Pathol. 137, 1121–1142.

Ceko, M.J., Aitken, J.B., Harris, H.H., 2014. Speciation of copper in a range of rood types by X-ray absorption spectroscopy. Food Chem. 164, 50–54.

Chandra, V., Ganguli, M., Panda, V., Johnston, J., Belle, S., Dekosky, S.T., 1998. Prevalence of Alzheimer's disease & other dementias in rural India. Neurology 51, 1000–1008.

Cox, C., Teknos, T.N., Barrios, M., Brewer, C.J., Dick, R.D., Merajver, S.D., 2001. The role of copper suppression as an antiangiogenic strategy in head and neck squamous cell carcinoma. Laryngoscope 111, 696–701.

Cox, C., Merajver, S.D., Yoo, S., Dick, R.D., Brewer, G.J., Lee, J.S., Teknos, T.N., 2003. Inhibition of the growth of squamous cell carcinoma by tetrathiomolybdate-induced copper suppression in a murine model. Arch. Otolaryngol. Head. Neck Surg. 129, 781–785.

Dhawan, A., Taylor, R.M., Cheeseman, P., 2005. Wilson's disease in children: 37-year experience and revised King's score for liver transplantation. Liver Transplant. 11, 441–448.

European Association for the Study of the Liver, 2012. EASL clinical practice guidelines: Wilson's disease. J. Hepatol. 56, 671–685.

Fink, J.K., Hedera, P., Brewer, G.J., 1999. Hepatolenticular degeneration (Wilson's disease). Neurologist 5, 171–185.

Foley, P.T., 1985. International copper Demand Patterns – The Case of Plumbing Tube, CRU Consultants Inc., New York, NY, Economics of Internationally Traded Minerals, Economics of Copper, Section 5.2. pp. 183–186.

Folkman, J., 1972. Anti-angiogenesis: new concept for therapy of solid tumors. Ann. Surg. 175, 409–416.

Gunther, P., Kuhn, H.J., Herman, W., 2013. Monitoring therapy in Wilson's disease by the oral radiocopper test. Aktuelle Neurol. 40, 195–199.

Gowers, W.R., 1888. A Manual of Diseases of the Nervous System. P Blakiston, Son, and Co., Philadelphia.

van Golen, K.L., Bao, L., Brewer, G.J., Pienta, K.J., Kamradt, J.M., Livant, D., Merajver, S., 2002. Suppression of tumor recurrence and metastasis by a combination of the PHSCN sequence and the antiangiogenic compound tetrathiomolybdate in prostate carcinoma. Neoplasia 4, 373–379.

Gartner, E.M., Griffith, K.A., Pan, Q., Brewer, G.J., Henja, G.F., Merajver, S.D., Zalupski, M.M., 2009. A pilot trial of the anti-angiogenic copper lowering agent tetrathiomolybdate in combination with irinotecan, 5-flourouracil, and leucovorin for metastic colorectal cancer. Invest. New Drugs 27, 159–165.

Hardy, J., Selkoe, D.J., 2002. The amyloid hypothesis of Alzheimer's disease: progress and problems on the road to therapeutics. Science 297, 353–356.

Hedera, P., Fink, J.K., Bockenstadt, P.L., Brewer, G.J., 2003. Myelopolyneuropathy and pancytopenia due to copper deficiency and high zinc levels of unknown source: further support for existence of a new zinc overload syndrome. Arch. Neurol. 60, 1303–1306.

Hedera, P., Peltier, A., Fink, J.F., Wilcock, S., London, Z., Brewer, G.J., 2009. Myelopolyneuropathy and pancytopenia due to copper deficiency and high zinc levels of unknown origin. II the denture cream is a primary source of unknown zinc. Neurotox 30, 996–999.

Hendrie, H., Osuntokun, O., Hall, K.S., Ogunniyi, A.O., Hui, S.L., Unverzagt, F.W., Gureje, O., Rodenberg, C.A., Baiyewu, O., Musick, B.S., 1995. Prevalence of Alzheimer's disease and dementia in two communities: Nigerian Africans and African Americans. Am. J. Psychiatry 152, 1485–1492.

Henry, N.I., Dunn, R., Merjaver, S., Pan, Q., Pienta, K.J., Brewer, G.J., Smith, D.C., 2006. Phase II trial of copper depletion with tetrathiomolybdate as an antiangiogenesis strategy in patients with hormone-refractory prostate cancer. Oncology 71, 168–175.

Hill, G.M., Brewer, G.J., Juni, J.E., Prasad, A.S., Dick, R.D., 1986. Treatment of Wilson's disease with zinc. II. Validation of oral 64 copper with copper balance. Am. J. Med. Sci. 292, 344–349.

Hou, G., Dick, R., Abrams, G.D., Brewer, G.J., 2005. Tetrathiomolybdate protects against cardiac damage by doxorubicin in mice. J. Lab. Clin. Med. 146, 299–303.

Hou, G., Abrams, G.D., Dick, R., Brewer, G.J., 2008. Efficacy of tetrathiomolybdate in a mouse model of multiple sclerosis. Transl. Res. 152, 239–244.

Hou, G., Dick, R., Brewer, G.J., 2009. Improvement in dissolution of liver fibrosis in an animal model by tetrathiomolybdate. Exp. Biol. Med. 234, 662–665.

Jain, S., Cohen, J., Ward, M.M., Kornhauser, N., Chuang, E., Cigler, T., Moore, A., Donovan, D., Lam, C., Cobham, M.V., Schneider, S., Hurtado Rua, S.M., Benkert, S., Mathijsen Greenwood, C., Zelkowitz, R., Warren, J.D., Lane, M.E., Mitall, V., Rafii, S., Vahdat, L.T., 2013. Tetrathiomolybdate-associated copper depletion decreases circulating endothelial progenitor cells in women with breast cancer at high risk of relapse. Ann. Oncol. 24, 1491–1498.

James, S.A., Voritakis, I., Adlard, P.A., Duce, J.A., Masters, C.L., Cherny, R.A., Bush, A.I., 2012. Elevated labile Cu is associated with oxidative pathology in Alzheimer disease. Free Radic. Biol. Med. 52, 298–302.

Kalita, J., Kumar, V., Chandra, S., Kumar, B., Misra, U.K., 2014. Worsening of Wilson's disease following penicillamine therapy. Eur. Neurol. 71, 126–131.

Khan, M.K., Miller, M.W., Taylor, J., Navkiranjit, K.G., Dick, R.D., Van Golen, K., Brewer, G.J., Merajver, S.D., 2002. Radiotherapy and antiangiogenic TM in lung cancer. Neoplasia 4, 1–7.

Klevay, L.M., 2011. Is the Western diet adequate in copper? J. Trace Elem. Med. Biol. 4, 204–212.

Lee, H.-G., Castellani, R.J., Zhu, X., Perry, G., Smith, M.A., 2005. Amyloid-B in Alzheimer's disease: the horse or the cart? Pathogenic or protective? Int. J. Exp. Pathol. 86, 133–138.

Lin, J., Zahurak, M., Beer, T.M., Ryan, C.J., Wilding, G., Mathew, P., Morris, M., Callahan, J.A., Gordon, G., Reich, S.D., Carducci, M.A., Antonavakis, E.S., 2013. A non-comparative randomized phase II study of 2 doses of ATN-224, a copper/zinc superoxide dismutase inhibitor, in patients with a biochemically recurrent hormone-naïve prostate cancer. Urol. Oncol. 31, 581–588.

Lowndes, S.A., Adams, A., Timma, A., Fisher, N., Smythe, J., Watt, S.M., Joel, S., Donate, F., Hayward, C., Reich, S., Middleton, M., Mazar, A., Harris, R.L., 2008. Phase I study of copper-binding agent ATN-224 in patients with advanced solid tumors. Clin. Cancer Res. 14, 7526–7534.

Ma, S., Hou, G., Dick, R., Brewer, G.J., 2004. Tetrathiomolybdate protects against liver injury from acetaminophen in mice. J. Appl. Res. Clin. Ext. Ther. 4, 419–426.

McCubbin, M.D., Hou, G., Abrams, G.D., Dick, R., Zhang, Z., Brewer, G.J., 2006. Tetrathiomolybdate is effective in a mouse model of arthritis. J. Rheumatol. 33, 2501–2506.

Morris, M.C., Evans, D.A., Tangney, C.C., Bienias, J.L., Schneider, J.A., Wilson, R.S., Scherr, P.A., 2006. Dietary copper and high saturated and trans fat intakes associated with cognitive decline. Arch. Neurol. 63, 1085–1088.

Mursa, J., Robien, K., Hamack, L.J., Park, K., Jacobs, D.R., 2011. Dietary supplements and mortality rate in older women: the Iowa Women's Health Study. Arch. Int. Med. 445, 1625–1633.

National Research Council, 2000. Copper in Drinking Water.

Nations, S.P., Boyer, P.J., Love, L.A., Burritt, M.F., Butz, J.A., Wolfe, G.I., Hynan, L.S., Reisch, J., Trevedi, J.R., 2008. Denture cream: an unusual source of excess zinc, leading to hypocupremia and neurologic disease. Neurology 71, 639–643.

Nazer, H., Ede, R.J., Mowat, A.P., Williams, R., 1986. Wilson's disease: clinical presentation and use of prognostic index. Gut 27, 1377–1381.

Nunomura, A., Tamaoki, T., Motohashi, N., Nakumura, M., McKeel Jr., D.W., Tabaton, M., Lee, H.-G., Smith, M.A., Perry, G., Zhu, X., 2012. The earliest stage of cognitive impairment in transition from normal aging to Alzheimer's disease is marked by prominent RNA oxidation in vulnerable neurons. J. Neuropathol. Exp. Neurol. 71, 233–241.

Ohnik, H., Thiele, D.J., 2014. How copper transverses cellular membranes through the copper transporter 1, Ctrl. Ann. N.Y. Acad. Sci. 1314, 32–41.

Osler, W., 1910. Modern Medicine in Theory and Practice. Lea and Febiger, Philadelphia and New York.

Pan, Q., Kleer, C.G., van Golen, K.I., Irani, J., Bottema, K.M., Bias, C., DeCarvalho, M., Mesri, E., Robins, D., Dick, R., Brewer, G., Merajver, S., 2002. Copper deficiency induced by tetrathiomolybdate suppresses tumor growth and angiogenesis. Cancer Res. 62, 4854–4859.

Pan, Q., Bao, L.W., Kleer, C.G., Brewer, G.J., Merajver, S.D., 2003. Antiangiogenic tetrathiomolybdate enhances the efficacy of doxorubicin against breast carcinoma. Mol. Cancer Ther. 2, 617–622.

Pass, H.I., Brewer, G.J., Dick, R., Carbone, M., Merajver, S., 2008. A phase II trial of tetrathiomolybdate after surgery for malignant mesothelioma: final results. Ann. Thorac. Surg. 86, 383–389 (Discussion 90).

Prasad, A.S., Beck, F.W., Bao, B., Fitzgerald, S.T., Snell, D.C., Steinberg, J.D., Cardozo, L.J., 2007. Zinc supplementation decreases incidence of infections in the elderly: effect of zinc on generation of cytokines and oxidative stress. Am. J. Clin. Nutr. 85, 837–844.

Redman, B.G., Esper, P., Pan, Q., Dunn, R.I., Hussain, H.G., Chenevert, T., Brewer, G.J., Merajver, S.D., 2003. Phase II trial of tetrathiomolybdate in patients with advanced kidney cancer. Clin. Cancer Res. 9, 1666–1672.

Roberts, E.A., Schilsky, M.L., 2008. Aasld practice guidelines, diagnosis and treatment of Wilson's disease: an update. Hepatology 47, 2089–2111.

Sarell, C.J., Wilkinson, S.R., Viles, J.H., 2010. Sub-stochiometric levels of copper ions accelerate the kinetics of fibre formation and promote cell toxicity of amyloid beta from Alzheimer's disease. J. Biol. Chem. 285, 41533–41540.

Sayre, L.M., Perry, G., Harris, P.L., Liu, Y., Schubert, K.A., Smith, M.A., 2000. In situ oxidative catalysis by neurofibrillary tangles and senile plaques in Alzheimer's disease: a central role for bound transition metals. J. Neurochem. 74, 270–279.

Scheinberg, I.H., Sternlieb, I., 1984. Wilson's disease. In: Smith, L.H.J. (Ed.), Major Problems in Internal Medicine. W.B. Saunders Company, Philadelphia.

Shen, X.L., Yu, J.H., Zhang, D.F., Xie, J.X., Jiang, H., 2014. Positive relationship between mortality from Alzheimer's disease and soil metal concentrations in mainland China. J. Alzheimers Dis. 42, 893–900.

Singh, I., Sagare, A.P., Coma, M., Perlmutter, D., Gelein, R., Bell, R.D., Deane, R.J., Zhong, E., Parisi, M., Ciszewski, J., Kasper, R.T., Deane, R., 2013. Low levels of copper disrupt brain amyloid-beta homeostasis by altering its production and clearance. Proc. Natl. Acad. Sci. U.S.A. 110, 14471–14476.

Song, M., Song, Z., Barve, S., Zhang, J., Chen, T., Liu, M., Arteel, G.E., Brewer, G.J., McClain, C.J., 2008. Tetrathiomolybdate protects against bile duct ligation-induced cholestatic liver injury and fibrosis. J. Pharmacol. Exp. Ther. 325, 409–416.

Sparks, D.L., Schreurs, B.G., 2003. Trace amounts of copper in water induce beta-amyloid plaques and learning deficits in a rabbit model of Alzheimer's disease. Proc. Natl. Acad. Sci. U.S.A. 100, 11065–11069.

Sparks, D.L., Friedland, R., Petanceska, S., Schreurs, B.G., Shi, J., Perry, G., Smith, M.A., Sharma, A., Derosa, S., Ziolkowski, C., Stankovic, G., 2006. Trace copper levels in the drinking water, but not zinc or aluminum, influence CNS Alzheimer-like pathology. J. Nutr. Health Aging 10, 247–254.

Squitti, R., Pasqualetti, P., Dal Forno, G., Moffa, F., Cassetta, E., Lupoi, D., Vernieri, F., Rossi, L., Baldassini, M., Rossini, P.M., 2005. Excess of serum copper not related to ceruloplasmin in Alzheimer disease. Neurology 64, 1040–1046.

Squitti, R., Barbati, G., Rossi, L., Ventriglia, M., Dal Forno, G., Cesaretti, S., Moffa, F., Caridi, I., Casseta, P., Pasqualetti, P., Calabrese, L., Lupoi, D., Rossini, P.M., 2006. Excess of nonceruloplasmin serum copper in AD correlates with MMSE, CSF [beta]-amyloid, and h-tau. Neurology 67, 76–82.

Squitti, R., Bressi, F., Pasqualetti, P., Bonomini, C., Ghidoni, R., Binetti, G., Cassetta, E., Moffa, F., Ventriglia, M., Vernieri, F., Rossini, P.M., 2009. Longitudinal prognostic value of serum "free" copper in patients with Alzheimer disease. Neurology 72, 50–55.

Squitti, R., Ghidoni, R., Siotto, M., Ventriglia, M., Benussi, L., Paterlini, A., Magri, M., Binetti, G., Cassetta, E., Caprara, D., Vernieri, F., Rossini, P.M., Pasqualetti, P., 2014. Value of serum nonceruloplasmin copper for prediction of mild cognitive impairment conversion to Alzheimer disease. Ann. Neurol. 75, 574–580.

Strachey, J., Freud, A., Strachey, A., Tyson, A., 1966. 24 Volumes Entitled, the Standard Edition of the Complete Psychological Works of Sigmund Freud, Written between 1895 and 1939. The Hogarth Press and the Institute of Psycho-Analysis, London.

Ueda, K., Kawano, H., Hasuo, Y., Fujishima, M., 1992. Prevalence and etiology of dementia in a Japanese community. Stroke 23, 798–803.

Waldman, M., Lamb, M., 2005. Dying for a Hamburger: Modern Meat Processing and the Epidemic of Alzheimer's Disease, first US ed. Thomas Dune Books/St Martin's Press, New York.

Walshe, J.M., 1956. Penicillamine, a new oral therapy for Wilson's disease. Am. J. Med. 21, 487–495.

Walshe, J.M., 1982. Treatment of Wilson's disease with trientiine (triethylene tetramine) dihydrochloride. Lancet 1, 643–647.

Weiss, K.H., Gotthard, D., Klemn, D., Merle, U., Ferenci-Foerster, D., Schaefer, M., Ferenci, P., Stremmel, W., 2011. Zinc Monotherapy is not an effective as chelating agents in treatments of Wilson's disease. Gastroenterology 140, 1189–1198.

White, L., Petrovitch, H., Ross, G.W., Masaki, K.H., Abbott, R.D., Teng, E.L., Rodriguez, B.L., Blanchette, P.L., Havlik, R.J., Wergowske, G., Chiu, D., Foley, D.J., Murdaugh, C., Curb, J.D., 1996. Prevalence of dementia in older Japanese-American men in Hawaii: the Honolulu-Asia aging study. J. Am. Med. Assoc. 276, 955–960.

Yamaguchi, Y., Helny, M.E., Gitlin, J.D., 1993. Isolation and characterization of a human liver cDNA as a candidate gene for Wilson disease. Biochem. Biophys. Res. Commun. 197, 271–277.

Part III

Iodine

Chapter 11

Iodine: Basic Nutritional Aspects

Katerine S. Knust[1], Angela M. Leung[2],[3]

[1]Universidade Federal do Estado do Rio de Janeiro (UNIRIO), Rio de Janeiro, Brazil; [2]UCLA David Geffen School of Medicine, Los Angeles, CA, United States; [3]VA Greater Los Angeles Healthcare System, Los Angeles, CA, United States

ROLE OF IODINE

The main function of iodine is the synthesis of the thyroid hormones, thyroxine (T4) and triiodothyronine (T3). Four iodine atoms are required for each T4 molecule and three atoms for each T3 molecule produced. Thyroid-stimulating hormone (TSH, also called thyrotropin), released from the pituitary gland, helps to regulate thyroid hormone production and secretion (World Health Organization, 2007; Zimmermann, 2009).

The thyroid hormones have multiple effects on metabolism and are associated with direct or indirect effects on many systems and organs. Thyroid hormones function in lipolysis in adipose tissue, muscle contraction, bone growth, heart rate control, digestion and absorption of nutrients in the gastrointestinal system, oxygen consumption, body heat production, nervous system development, and linear growth.

IODINE METABOLISM

Iodine is ingested in various chemical forms. Iodide is rapidly and almost completely absorbed (>90%) in the stomach and duodenum (Marwaha and Gopalakrishnan, 2011). Iodate, which is used in many countries in the iodization of salt, is reduced in the intestine and absorbed in the form of iodide. Organically bound iodine is typically digested and the released iodide absorbed (Gropper and Smith, 2015; World Health Organization, 2007; Zimmermann, 2008). Some iodine-containing compounds (e.g., thyroid hormones and amiodarone) are absorbed intact (Institute of Medicine, 2001; Zimmermann et al., 2008).

Once absorbed, iodide is removed from the circulation primarily by the kidney and thyroid gland. The sodium-iodide symporter (NIS) in the thyroidal basolateral membrane is responsible for iodine concentration. It transfers iodide from the circulation into the thyroid gland at a concentration gradient of approximately 20–50 times that of the plasma to ensure that the thyroid gland obtains adequate amounts of iodine for hormone synthesis (Institute of Medicine, 2001). The thyroid selectively concentrates iodide in quantities for proper synthesis of thyroid hormones, and most of the remaining iodine is excreted in the urine. Clearance of circulating iodine varies with the intake of iodine. In situations in which there is adequate nutritional iodine, 10% or less of the iodine absorbed is taken up by the thyroid, but this percentage can exceed 80% among individuals with chronic iodine deficiency (Yarrington and Pearce, 2011). Various tissues can concentrate iodide, including the breast, salivary glands, choroid plexus, and gastric mucosa, but the role of iodine in most of these extrathyroidal tissues is unknown. However, one assumption is that during breastfeeding, the concentration of iodine into the mammary gland provides a dietary source of iodine for the nursing infant (Institute of Medicine, 2001; World Health Organization, 2001; Zimmermann et al., 2008).

The body of a healthy adult contains 15–20 mg of iodine, of which 70–80% is in the thyroid. In chronic iodine deficiency, the iodine content of the thyroid might fall to less than 20 µg. In iodine-sufficient areas, the adult thyroid traps approximately 60 µg of iodine per day to balance losses and maintain synthesis of thyroid hormone (Dillon and Milliez, 2000; Zimmermann et al., 2008). In normal situations, plasma iodine has a half-life of approximately 10 h, but this time is reduced in iodine deficiency (Zimmermann et al., 2008). The kidney has no mechanism to conserve iodide; thus it represents the major route (~80% to >90%) of iodide excretion. Usual urinary iodine (UI) excretion exceeds 100 µg/L if intake is adequate. Fecal excretion of iodide (up to 20% of total excreted) is relatively low and ranges from 6.7 to 42.1 µg/day (Vought et al., 1963). Some iodide is also lost in sweat—a loss that can be of consequence in hot, tropical regions where iodide intake may be only marginally adequate (Gropper and Smith, 2015; Institute of Medicine, 2001).

Iodine makes up 65% and 59% of the weights of T4 and T3, respectively. Thyroid hormone turnover is slow; the half-life of T4 is approximately 5 days and 1.5–3 days for T3.

Molecular, Genetic, and Nutritional Aspects of Major and Trace Minerals. http://dx.doi.org/10.1016/B978-0-12-802168-2.00011-7

SOURCES OF IODINE NUTRITION

The concentration of iodine in foods is highly variable. The iodine concentration in most foods and beverages is low, and most consumed foods provide 3–75 µg per serving. A detailed survey of the iodine content in foods of the US food supply has been reported (Pennington et al., 1995). In this study, the authors included fruit-flavored cereal, dairy products or mixed foods (chocolate milk, chocolate milk shake, low-fat milk, cheese pizza, low-fat plain yogurt, buttermilk, macaroni, and cheese), cod/haddock, frozen chicken pot pie, fried chicken frozen dinner, corn grits, homemade lasagna, white rice, pancakes, homemade chicken noodle casserole, canned spaghetti in tomato sauce, apple pie, fish sticks, chocolate pudding, and mashed potatoes as the foods highest in iodine (0.054–0.451 mg per serving).

Most of the iodine on Earth is found in the oceans (50 g/L). Iodide ions in seawater are oxidized to form elemental iodine, which is volatile, evaporates into the atmosphere, and returns to the soil in the form of rain, completing the cycle. However, the iodine cycle is slow and incomplete in some areas, and soil and groundwater become deficient in iodine. Soils deficient in iodine are common in inland regions and mountainous regions, but they can also occur in coastal areas. The iodine concentrations of some dietary sources are related to the iodine content of soil in various regions. For example, in plant cultivation, seeds germinated in soil poor in iodine generate vegetables, grains, and fruits deficient in iodine. Furthermore, the concentration of this element in foods can vary according to the amount and type of fertilizer that is used during cultivation. The amount of iodine contained in the meat depends on the soil and plants in animal feeds. The concentration of iodine in the water reflects the concentration of iodine in soil and rocks of the region and is associated with the incidence of iodine deficiency among the inhabitants residing in the local area (Gropper and Smith, 2015; Pennington et al., 1995; Zimmermann et al., 2008).

Iodine is found in seafood, but there is also wide variation of its concentrations among freshwater and saltwater fish. Foods of marine origin have higher concentrations of iodine because marine animals concentrate iodine from seawater (Institute of Medicine, 2001). Seaweed (such as kelp, nori, kombu, and wakame) is one of the best food sources of iodine, but it is highly variable in its content (16–8165 µg/g; Teas et al., 2004). Many factors, such as geographic variation, season, preparation, and storage conditions, affect the iodine content of seaweed (Teas et al., 2004; Zimmermann, 2009).

Iodine-containing disinfectants used for cleaning milk cans and teats can additionally increase the native iodine content of dairy products (Pearce et al., 2004). For example, milk provides approximately 40–60 µg of iodide/200 mL, low-fat plain yogurt provides 32 µg/100 g, and cheddar cheese provides 36 µg/100 g. Other foods high in protein also provide iodine. An egg provides approximately 20 µg of iodide, and meats generally provide approximately 15–40 µg/100 g. Fish sticks provide 52 µg/100 g. Additional sources of iodine are breads and grain products made from bread dough. Dough oxidizers or conditioners contain iodates (IO_3^-), food additives that improve cross-linking of the gluten. Conditioners are added to store-bought bread to maintain freshness and prolong shelf life. A 100-g serving of white enriched bread contains approximately 73 µg of iodide (Gropper and Smith, 2015; Pearce et al., 2004).

Iodine can be added to salt in the form of potassium iodide or iodate, and iodized salt (1/4 teaspoon or 1.5 g) supplies approximately 70 µg of iodide. Iodized salt was introduced in the United States in 1924 (Carpenter, 2005) and was a primary factor in significantly reducing the incidence of iodine deficiency. However, in the United States, processed foods are typically manufactured with noniodized salt, and iodine intake in the United States overall has varied over the years because of changes in the iodine content of foodstuffs. Possible reductions in the concentration of iodine in dairy products because of the discontinued use of iodate conditioners in store-bought breads, new recommendations for reducing salt intake for blood pressure control, and the increase in noniodinated salts in processed and prepackaged foods have been described as possible causes of the decrease in iodine intake in the United States in recent years (Pearce et al., 2004).

In parts of the world where iodine deficiency is endemic, iodized oil supplements and water iodination represent other means of iodine supplementation. However, many countries still have insufficient iodine supplementation programs (World Health Organization, 2007).

RECOMMENDED DIETARY INTAKE OF IODINE

Recommendations by the US Institute of Medicine from 2001 are summarized in Table 11.1. The minimum amount of iodide to prevent goiter is estimated at 50–75 µg/day, or approximately 1 µg/kg of body weight (Hublin and Richards, 2009). In infants younger than 12 months of age, the adequate intake (AI) for iodine is used instead of the recommended dietary allowance (RDA). An AI is that used when insufficient scientific evidence is available to calculate the estimated average requirement for a nutrient (from which the RDA is based). The AI is a value based on experimentally derived intake levels or approximations of observed mean nutrient intakes by a group (or groups) of healthy people. The AI for children and adults is expected to meet or exceed the amount needed to maintain a defined nutritional state or criterion of adequacy in essentially all members of a specific healthy population (Institute of Medicine, 2005).

TABLE 11.1 Recommended Dietary Allowances for Iodine (Andersson et al., 2007; Institute of Medicine, 2001)

Age	Nonpregnancy, Nonlactating (µg)	Pregnancy (µg)	Lactation (µg)
Birth to 6 months	110[a]		
7–12 months	130[a]		
1–3 years	90		
4–8 years	90		
9–13 years	120		
14–18 years	150	220	290
≥19 years	150	220	290

[a]Adequate intake.

Although the recommendations apply equally to both sexes, iodine needs are higher during pregnancy and lactation (Andersson et al., 2007). There is an increased iodine requirement during pregnancy because of higher amounts of maternal T4 production required for the fetus; iodine transfer to the fetus, especially in late gestation; and an increase in maternal renal iodine clearance (Dillon and Milliez, 2000; Glinoer, 2006; Zimmermann et al., 2008). During lactation, the physiology of thyroid hormone production and UI excretion returns to normal, but to ensure that the infant gets enough iodine from breast milk to build reserves in the thyroid gland, the iodine recommendations during lactation are still increased (Andersson et al., 2007).

The World Health Organization (WHO), the United Nations Children's Emergency Fund (UNICEF), and the International Council for the Control of Iodine Deficiency Disorders (ICCIDD; now the Iodine Global Network) recommend a daily iodine intake of 250 µg for pregnant and 150 µg for lactating women, or a single annual dose of 400 mg of iodized oil for all women pregnant, lactating, or of reproductive age in countries where less than 20% of households have access to iodized salt (World Health Organization, 2014). In North America, the American Thyroid Association recommends that all women supplement with a 150-µg iodine-containing multivitamin during preconception, pregnancy, and lactation (Stagnaro-Green et al., 2011).

FACTORS INFLUENCING IODINE BIOAVAILABILITY

Iodine occurs in foods mainly as inorganic iodide, which is readily and almost completely absorbed (90%) by the stomach and duodenum (Yarrington and Pearce, 2011). Food components do not appear to influence iodine absorption, but they can reduce its utilization for the production of thyroid hormones (Hurrell, 1997). Goitrogens are substances that interfere with iodine metabolism and the production of thyroid hormones. They are so termed because these substances contain agents to potentially inhibit iodine uptake into the thyroid to decrease thyroid hormone production. Table 11.2 summarizes the major types of goitrogenic foods and their mechanisms for decreasing iodine availability to the thyroid and/or thyroid hormone production. The potential adverse effects of goitrogenic substances are ameliorated in the setting of adequate iodine nutrition (Gropper and Smith, 2015; Hurrell, 1997; Institute of Medicine, 2001).

Glucosinolates, one example of a goitrogen, are found as a natural component of some foods, such as cruciferous vegetables that include cabbage, kale, cauliflower, broccoli, rutabaga, turnips, Brussel sprouts, and mustard greens. Cassava, which is a tuberous root within this genus, is native to South America and is rich in starch. Cassava contains the cyanogenic glucoside, linamar, which, once hydrolyzed within the gastrointestinal tract, releases cyanide, which is then metabolized to thiocyanate to compete with iodide for uptake into the thyroid gland. Cyanogenic glucosides are also found in lima beans, flaxseed, linseed, sorghum, and sweet potatoes. Other goitrogenic substances that interfere with iodide metabolism include polycyclic hydrocarbons and phenol compounds derived from coal; halide ions such as bromide (Br^-) and astatide (At^-); perchlorate (ClO_4^-); and lithium (Li^-), a drug used as a mood stabilizer (Gropper and Smith, 2015; Institute of Medicine, 2001; World Health Organization, 2007).

TABLE 11.2 Dietary Goitrogens, Environmental Pollutants, and Nutritional Deficiencies With Potential to Decrease Thyroid Hormone Production

	Mechanism
Foods	
Cassava, lima beans, linseed, sorghum, sweet potato	Contain cyanogenic glucosides, which are metabolized to thiocyanates that compete with iodine for uptake into the thyroid
Cruciferous vegetables such as cabbage, kale, cauliflower, broccoli, turnips, rapeseed	Contain glucosinolates, metabolites that compete with iodine into the thyroid
Soy isoflavones, millet	Contain flavonoids which impair thyroid peroxidase activity
Environmental Pollutants	
Perchlorate	Competitively inhibits the transport of iodide into the thyroid by the sodium-iodide symporter
Thiocyanate in cigarette smoke	Competes with iodine for active transport into the thyroid and secretory epithelium of the lactating breast
Disulfides from coal processes	Reduces thyroidal iodine uptake
Nutrients	
Selenium deficiency	Decreases glutathione peroxidase and the accumulation of H_2O_2, which induces thyroid cell destruction and thyroid fibrosis, resulting in thyroid failure
Iron deficiency	Reduces heme-dependent thyroperoxidase activity in the thyroid
Vitamin A deficiency	Increases thyroid-stimulating hormone stimulation and goiter through decreased vitamin-A-mediated suppression of the pituitary thyroid-stimulating hormone-β gene

Adapted from Zimmermann, M.B., 2008. Iodine requirements and the risks and benefits of correcting iodine deficiency in populations. J. Trace Elem. Med. Biol. 22, 81–92.

Coal is a source of a large variety of antithyroid and goitrogenic compounds such as disulfides and thiocyanates (Lindsay et al., 1992). Perchlorate is a chemical that is produced by humans (used in rocket fuel, explosives, fireworks, and other products) or it can be naturally produced through atmospheric processes and then settle on the surface water or land (Leung et al., 2014). Cigarette smoke contains cyanide that is metabolized to thiocyanate. Thiocyanate is also naturally found in *Brassica* genus vegetables, such as cauliflower, broccoli, kale, and Brussels sprouts (Leung et al., 2012). Along with thiocyanates and nitrates, perchlorates can potentially affect thyroid function by competitive inhibition of the NIS that is responsible for the uptake of iodide in thyrocytes (Leung et al., 2010).

Deficiencies of vitamin A, selenium, or iron can each exacerbate the effects of iodine deficiency. The metabolism of the thyroid hormones is largely dependent upon a group of three selenium-dependent iodothyronine 5′-deiodinases. Iron deficiency impairs thyroid hormone synthesis and function; iron is a component of the enzyme thyroperoxidase, which attaches iodine to tyrosine residues on thyroglobulin for the production of the thyroid hormones. Iron also may be involved in the binding of T3 to nuclear receptors. Vitamin A deficiency reduces iodine uptake by the thyroid gland and decreases the synthesis of thyroglobulin and the coupling of iodotyrosine residues (the joining of thyroid hormone precursors) to form T4 (Gropper and Smith, 2015; Hess, 2010).

ASSESSMENT OF IODINE NUTRITION

Iodine nutritional status cannot be assessed in individuals because of the day-to-day variation of iodine intake, but various methods can be used to assess iodine nutrition at the population level. Nutritional evaluation of iodine status should be used in populations in areas at risk for iodine deficiency. Four methods are generally recommended for evaluating a population's nutritional status of iodine: median UI concentrations, the goiter rate, serum TSH concentrations, and serum thyroglobulin concentrations (Zimmermann, 2004, 2008). These methods are complementary and can be used in different situations. UI is a sensitive indicator of recent iodine intake (days), serum TSH and thyroglobulin concentrations reveal intermediate

TABLE 11.3 Epidemiological Criteria for Assessing Iodine Nutrition Based on Median Urinary Iodine Concentrations in School-Age Children

Median Urinary Iodine (µg/L)	Iodine Intake	Iodine Nutrition
<20	Insufficient	Severe iodine deficiency
20–49	Insufficient	Moderate iodine deficiency
50–99	Insufficient	Mild iodine deficiency
100–199	Adequate	Optimal
200–299	More than adequate	Risk of iodine-induced hyperthyroidism within 5–10 years after introduction of iodized salt in susceptible groups
≥300	Excessive	Risk of adverse health consequences (iodine-induced hyperthyroidism, autoimmune thyroid diseases)

responses (weeks to months), and the goiter rate reflects long-term iodine nutrition (Zimmermann, 2008). Table 11.3 summarizes the median UI levels used to assess the adequacy of iodine nutrition in a population.

Daily iodine intake for population estimates can be extrapolated from UI concentrations using estimates of mean 24-h urine volume and on the assumptions of an average iodine bioavailability of 92% according to the following formula: *daily iodine intake = urinary iodine* (µg/L) × 0.0235 × *body weight* (kg) (simplified from estimating a median 24-h urine volume of 0.9 mL/h/kg and average iodine bioavailability of 92%; Institute of Medicine, 2001). As an example, UI excretion of 100 µg/L in a 57-kg girl would indicate a daily intake of 134 µg (Institute of Medicine, 2001). Median UI concentrations exceeding 100 µg/L indicate that population iodine intake is adequate, whereas median UI concentrations of less than 100 µg/L suggest inadequate iodine intake, including mild iodine deficiency, in a population. Median UI concentrations less than 50 µg/L are usually associated with insufficient thyroid hormone secretion and are indicative of moderate iodine deficiency, whereas concentrations less than 20 µg/L categorize a population as severely iodine deficient (Gropper and Smith, 2015; Zimmermann et al., 2008). When iodine intake increases, the urinary excretion of iodine also increases. Median UI concentrations equal to or in excess of 500 µg/L have been associated with increasing thyroid volume, which in turn increases the risk of thyroid dysfunction (World Health Organization, 2007).

Nutritional iodine status can also be evaluated by the prevalence of goiter in a population. Two methods are available to determine the presence of goiter: inspection/palpitation of the neck and thyroid ultrasonography. When inspection/palpitation is used, goiter is present when each thyroid lobe has a volume greater that the terminal phalanx of the thumbs of the patient being examined (Zimmermann, 2008; Zimmermann et al., 2008). WHO categorizes goiter as grade 0 if the thyroid is not palpable or visible, grade 1 goiter when the thyroid is palpable but not visible when the neck is in a normal position, and grade 2 goiter when the thyroid is clearly visible when the neck is in a normal position (World Health Organization, 2001; Zimmermann, 2008; Zimmermann et al., 2008). During treatment the goiter rate is difficult to interpret because it can indicate the history of the nutritional status of iodine as well as the current nutritional status. Throughout the treatment of goiter by iodine supplementation, the thyroid can take months to return to its normal size; thus the goiter rate reflects long-term nutrition. Moreover, inspection/palpitation of goiter in regions of mild iodine deficiency has poor sensitivity and specificity; in such areas, measurement of thyroid volume by ultrasound is preferable for the classification of goiter (Zimmermann et al., 2000, 2008; Zimmermann, 2008).

Serum concentrations of thyroid hormones are generally poor indicators of iodine nutritional status, especially in older children and adults. In iodine-deficient populations, serum concentrations of T3 and TSH increase or remain unchanged, whereas T4 generally decreases. This increase in T3 concentration is an adaptive response of the thyroid to iodine deficiency; however, these changes are often within the normal range, and the overlap with iodine-sufficient populations is large enough to render thyroid hormone concentration an insensitive measure of iodine nutrition (Institute of Medicine, 2001; World Health Organization, 2001; Zimmermann et al., 2008). Serum TSH concentration is a sensitive indicator of intermediate response (weeks to months) of the nutritional status of iodine in newborns in at-risk populations (Gropper and Smith, 2015; Sullivan et al., 1997; Zimmermann, 2008; Zimmermann et al., 2008).

Compared with adults, the newborn's thyroid contains less iodine but is associated with higher rates of iodine turnover. In newborns deficient in iodine, serum TSH concentrations increase in the first weeks of life. To maintain the uptake of iodine into the thyroid when the iodine supply is inadequate, there is an increased stimulation of TSH. Therefore the concentration of serum TSH obtained in a newborn 3–4 days after birth can be considered a sensitive indicator to evaluate the nutritional status of iodine (Sullivan et al., 1997; Zimmermann et al., 2008). A frequency of more than 3% of newborn serum TSH concentrations greater than 5 mIU/L in the population is suggestive of iodine deficiency.

Finally, serum thyroglobulin can be used for evaluating the nutritional status of iodine because thyroglobulin is a protein synthesized by the thyroid gland. The primary clinical use of the serum thyroglobulin concentration is surveillance of postoperative, differentiated thyroid cancer, but it is typically elevated in thyroidal hyperplasia from any cause, including the endemic goiter of iodine deficiency. Serum thyroglobulin concentrations greater than 10 μg/L suggest inadequate iodine intake (Institute of Medicine, 2001; World Health Organization, 2007; Zimmermann, 2008).

IODINE DEFICIENCY

Thyrotropin-releasing hormone (TRH), secreted by the hypothalamus, acts on the pituitary gland to stimulate TSH, which increases the activity of the thyroid gland to generate T4. TSH output is regulated by T4 through negative feedback to the pituitary. When the intake of iodine is insufficient and iodine stores in the thyroid gland are depleted, there is a decrease in the production of the thyroid hormones. The decline in blood T4 levels triggers the secretion of TSH in an effort to increase uptake of available iodine, thus resulting in hyperplasia of the thyroid. Elevated T4 levels then inhibit the release of TSH and TRH. This condition of physiological adaptation to iodine deficiency is called goiter, in which thyroid hyperplasia and hypertrophy occur in response. Goiter can occur at any age, including in newborns.

Iodine deficiency has many consequences, including impairment of growth and development in animals and humans. The effects of iodine deficiency are collectively called iodine-deficiency disorders (IDDs). The IDDs result from inadequate production of thyroid hormones due to iodine deficiency, causing mental retardation, hypothyroidism, goiter, cretinism, and varying degrees of other growth and development abnormalities (Institute of Medicine, 2001; Zimmermann, 2008; Zimmermann et al., 2008).

During fetal development, iodine deficiency gives rise to serious adverse effects. Although goiter is the most visible effect of iodine deficiency, the most serious adverse effect is damage to the fetus. The thyroid hormones play an important role in the processes of early growth and development of most of organs, especially of the brain (Delange, 2001). Severe dietary maternal iodine deficiency in pregnancy has the potential to cause maternal and fetal hypothyroidism. Severe iodine deficiency is associated with poor obstetric outcomes, including spontaneous abortion, prematurity, stillbirth, and congenital abnormalities (Yarrington and Pearce, 2011).

Mental retardation and endemic cretinism result from an insufficient supply of thyroid hormones to the developing brain. Maternal T4 crosses the placenta before onset of fetal thyroid function at 10–12 weeks and represents up to 20–40% of T4 measured in cord blood at birth (Sack, 2003; Zimmermann, 2008; Zimmermann et al., 2008). Normal thyroid hormone amounts are required for neural cell migration and myelination of the central nervous system of the fetal brain, which mostly occurs in the perinatal period and during fetal and early postnatal development (Dillon and Milliez, 2000; Institute of Medicine, 2001; Sack, 2003; Zimmermann, 2008; Zimmermann et al., 2008). Iodine treatment of pregnant women in areas of severe deficiency reduces fetal and perinatal mortality and improves motor and cognitive performance of the offspring (Connolly et al., 1979; Zimmermann, 2008).

TREATMENT AND PREVENTION OF IODINE DEFICIENCY

In areas where iodine intake is marginal, iodine has been orally and intramuscularly administered as iodized oil; introduced into the water supply; used in crop irrigation; incorporated into animal fodder; and introduced into food through salt iodization, bread iodophors, and other products (Leung and Braverman, 2014).

In nearly all regions affected by iodine deficiency, the most effective way to control iodine deficiency is through salt iodization (Sullivan et al., 1997; Zimmermann, 2008; Zimmermann et al., 2008). Universal salt iodization is used to describe the iodization of all salt for human (food industry and household) and livestock consumption. Although the ideal method is universal salt iodization, even the countries with successful salt iodization programs can have inadequate iodine nutrition because food industries are often reluctant to mandate iodized salt, and many countries do not iodize salt for livestock.

The addition of iodine to table salt is simple and inexpensive; moreover, iodine does not affect the color or taste of salt. Common salt is easily found and consumed in virtually every country in the world; therefore it becomes an easy method for population-level access to iodine (Zimmermann, 2008; Zimmermann et al., 2008).

WHO, UNICEF, and ICCIDD recommend that iodine be added at a concentration of 20–40 mg of iodine (or 34–66 mg potassium iodate) per kilogram of salt, depending upon local salt intake. When all salt used in processed food is iodized, the lower limit (20 mg) is recommended (World Health Organization, 1996). Iodine can be added to salt in the form of potassium iodide (KI) or potassium iodate (KIO_3).

Salt iodization remains the most cost-effective way to deliver iodine and to improve cognition in iodine-deficient populations. Worldwide, the cost of salt iodization per year is estimated as $0.02–0.05 per child covered, and the cost per childhood deaths prevented is $1000 and per disability-adjusted life year gained is $34–36 (Zimmermann, 2008; Zimmermann et al., 2008). Alternatively, before widespread salt iodization, the yearly potential losses attributable to iodine deficiency in the developing world have been estimated as $35.7 billion, compared with an estimated $0.5 billion yearly cost for salt iodization (i.e., a 70:1 benefit to cost ratio; Zimmermann, 2008; Zimmermann et al., 2008).

However, in some regions, iodization of salt might not be practical for control of iodine deficiency, at least in the short term. This difficulty might arise in remote areas where communication is poor or where there are many small-scale salt producers. In these areas, iodized oil supplements can be used (World Health Organization, 2001). Iodized oil is prepared by esterification of the unsaturated fatty acids in seed or vegetable oils and addition of iodine to the double bonds (Untoro et al., 2006; Zimmermann, 2008). It can be given orally or by intramuscular injection. Oral administration is simpler and more common that the intramuscular route, but the injection has a longer duration of action. Usual oral doses are 200–499 mg iodine per year and are often targeted at women of childbearing age, pregnant women, and children (Connolly et al., 1979). In one study, iodized oil given in the first and second trimesters of pregnancy reduced the frequency of neurological abnormalities and improved developmental test scores after 7 years compared with supplementation later in pregnancy or treatment after birth (O'Donnell et al., 2002). However, the disadvantages are an uneven concentration of iodine in the body over time and the need for direct contact with individuals, resulting in increased cost (Zimmermann, 2008).

Bread can be effective vehicle for iodine intake by including baker's salt enriched with iodine. Although iodizing drinking water or irrigation water can also be effective, the high cost and complexity of monitoring are disadvantages. Iodine-containing milk is a major dietary source in countries such as Switzerland and the United States (Pearce et al., 2004) because of the use of iodophors in the dairy industry rather than from the deliberate addition of iodine. In Finland, iodine-fortified animal fodder has increased the iodine content of foods derived from animal sources. In countries affected by IDDs, whenever possible, iodine should be routinely added to complementary foods (i.e., foods other than breastmilk or infant formula among infants) to provide 90 μg of iodine per day.

Iodine can also be given as potassium iodine or potassium iodate as drops or tablets. Single oral doses of potassium iodide monthly (30 mg) or every 2 weeks (8 mg) can provide adequate iodine for school-aged children (Todd and Dunn, 1998; Zimmermann, 2008). In iodine-deficient countries or regions that have a low distribution of iodized salt, supplements should be given to pregnant women, lactating women, and infants (Institute of Medicine, 2001; World Health Organization, 2014).

TOXIC EFFECTS OF EXCESS IODINE INTAKE

The tolerable upper intake level is the highest level of daily nutrient intake that is likely to pose no risk of adverse health effects in almost all individuals. A tolerable upper intake level for iodine has been set at 1100 μg (1.1 mg)/day (Gropper and Smith, 2015; Institute of Medicine 2001). European and US expert committees have recommended tolerable upper intake levels for iodine, but they caution that individuals with chronic iodine deficiency may respond adversely to intakes lower than these (Institute of Medicine, 2001; World Health Organization, 2007). The American Thyroid Association cautions against the ingestion of any supplements containing more than 500 mcg of iodine per day (Leung et al., 2015).

In 1948, Wolff and Chaikoff observed a reduction in the synthesis of hormones in rats exposed to high amounts of iodide intraperitoneally administered. The acute Wolff–Chaikoff effect is not completely understood, but it is thought to be at least partially explained by the generation of several inhibitory substances (such as intrathyroidal iodolactones, iodoaldehydes, and/or iodolipids) on thyroid activity (Leung and Braverman, 2014; Wolff and Chaikoff, 1948; Zimmermann, 2008). In most individuals this decreased production of the thyroid hormones is only transient (lasting ~24 h), and production resumes after adaptation to the acute iodine administration. This adaptation is associated with decreased expression of the NIS present on the basolateral membrane of thyroid follicular cells that mediates the transport of iodine from the circulation into the thyroid. The decreased expression of NIS results in reduced intrathyroidal iodine concentrations. In turn, reduced iodine levels lead to a decrease in levels of the iodinated substances that inhibit synthesis of thyroid hormones, which results in resumption of normal production of thyroid hormones. In areas of iodine sufficiency, healthy individuals are remarkably tolerant to iodine intakes of up to 1 mg/day because the thyroid is able to adjust to a wide range of intakes to regulate the synthesis and release of thyroid hormones (Chow et al., 1991; Zimmermann, 2008). However, excess iodine exposure or even doses of iodine in the milligram range may cause thyroid dysfunction, which might be transient or permanent, in those with damaged thyroid glands or dysregulation of thyroid follicular cells because normal downregulation of iodine transport into the gland does not occur (Leung and Braverman, 2014; Zimmermann, 2008).

Poor monitoring and oversupplementation of iodine in several countries with supplementation programs can result in excessive iodine intake. In addition, in some countries, excessive intake results from overconsumption of foods naturally high in iodine, such as seaweed. Some signs of acute iodide toxicity include burning of the mouth, throat, and stomach; nausea; vomiting; diarrhea; and fever (Institute of Medicine, 2001; Zimmermann, 2008; Zimmermann et al., 2008).

Finally, iodine is also used in some medical applications and can be a source of potentially supraphysiologic iodine exposure. Iodine tinctures (dilute mixtures of alcohol and iodine) and Betadine are used as iodine-containing antiseptics. Iodinated contrast in radiographic studies is a source of iodine exposure that is often several thousand-fold higher than recommended daily nutritional intake amounts. In some instances of severe hyperthyroidism (increased thyroid hormone production), saturated solutions of potassium iodine may be preoperatively used in conjunction with other medical treatments for thyroid surgery. There is also a role for prophylactic potassium iodide consumption in the event of a nuclear accident if indicated.

CONCLUSION

Iodine is an important micronutrient that is required for the synthesis of the thyroid hormones, T4 and T3. The diet is the primary method of achieving adequate iodine nutrition, with dairy products, some breads, seaweed and other seafood, and iodized salt as the most common iodine-containing foods. Adolescents and adults require iodine in amounts of 150 μg/day, with higher amounts required during pregnancy and lactation (250 and 290 μg, respectively). Iodine deficiency is the main cause of endemic goiter, but environmental toxicants and dietary goitrogens can aggravate the effect. Iodine deficiency symptoms resulting from inadequate production of thyroid hormones due to iodine deficiency include mental retardation, hypothyroidism, goiter, cretinism, and varying degrees of other growth and developmental abnormalities. In nearly all regions affected by iodine deficiency, the most effective way to supplement iodine is through salt iodization, but iodine can also be orally and intramuscularly administered as iodized oil, introduced into the water supply, used in crop irrigation, incorporated into animal fodder, and introduced into other foodstuffs. A tolerable upper intake level for iodine is recommended at 1100 μg/day to minimize the risks of iodine-induced thyroid dysfunction.

ACKNOWLEDGEMENT

This work was supported by NIH 7K23HD068552 (AML).

REFERENCES

Andersson, M., de Benoist, B., Delange, F., Zupan, J., 2007. Prevention and control of iodine deficiency in pregnant and lactating women and in children less than 2-years-old: conclusions and recommendations of the Technical Consultation. Public Health Nutr. 10, 1606–1611.

Carpenter, K.J., 2005. David Marine and the problem of goiter. J. Nutr. 135, 675–680.

Chow, C.C., Phillips, D.I., Lazarus, J.H., Parkes, A.B., 1991. Effect of low dose iodide supplementation on thyroid function in potentially susceptible subjects: are dietary iodide levels in Britain acceptable? Clin. Endocrinol. (Oxf) 34, 413–416.

Connolly, K.J., Pharoah, P.O., Hetzel, B.S., 1979. Fetal iodine deficiency and motor performance during childhood. Lancet 2, 1149–1151.

Delange, F., 2001. Iodine deficiency as a cause of brain damage. Postgrad. Med. J. 77, 217–220.

Dillon, J.C., Milliez, J., 2000. Reproductive failure in women living in iodine deficient areas of West Africa. BJOG 107, 631–636.

Glinoer, D., 2006. Iodine nutrition requirements during pregnancy. Thyroid 16, 947–948.

Gropper, S.S., Smith, J.L., 2015. Nutrition – Advanced Nutrition and Human Metabolism, sixth ed. Cengage Learning.

Hess, S.Y., 2010. The impact of common micronutrient deficiencies on iodine and thyroid metabolism: the evidence from human studies. Best Pract. Res. Clin. Endocrinol. Metab. 24, 117–132.

Hublin, J.J., Richards, M.P., 2009. The Evolution of Hominin Diets. Integrating Approaches to the Study of Palaeolithic Subsistence. Springer.

Hurrell, R.F., 1997. Bioavailability of iodine. Eur. J. Clin. Nutr. 51 (Suppl. 1), S9–S12.

Institute of Medicine, 2001. Dietary Reference Intakes: Vitamin A, Vitamin K, Arsenic, Boron, Chromium, Copper, Iodine, Iron, Manganese, Molybdenum, Nickel, Silicon, Vanadium, and Zinc, vol. 101, pp. 294–301.

Institute of Medicine, 2005. Dietary Reference Intakes for Energy, Carbohydrate, Fiber, Fat, Fatty Acids, Cholesterol, Protein, and Amino Acids (Macronutrients). The National Academies Press.

Leung, A.M., Braverman, L.E., 2014. Consequences of excess iodine. Nat. Rev. Endocrinol. 10, 136–142.

Leung, A.M., Braverman, L.E., He, X., Schuller, K.E., Roussilhes, A., Jahreis, K.A., Pearce, E.N., 2012. Environmental perchlorate and thiocyanate exposures and infant serum thyroid function. Thyroid 938–943.

Leung, A.M., Pearce, E.N., Braverman, L.E., 2010. Perchlorate, iodine and the thyroid. Best Pract. Res. Clin. Endocrinol. Metab. 24, 133–141.

Leung, A.M., Pearce, E.N., Braverman, L.E., 2014. Environmental perchlorate exposure: potential adverse thyroid effects. Curr. Opin. Endocrinol. Diabetes Obes. 21, 372–376.

Leung, A.M., et al., 2015. Potential risks of excess iodine ingestion and exposure: statement by the American Thyroid Association Public Health Committee. Thyroid 25(2), 145–146.

Lindsay, R.H., Hill, J.B., Gaitan, E., Cooksey, R.C., Jolley, R.L., 1992. Antithyroid effects of coal-derived pollutants. J. Toxicol. Environ. Health 37, 467–481.

Marwaha, R.K., Gopalakrishnan, S., 2011. Facts of iodine supplementation. J. Assoc. Physicians India 59 (Suppl.), 7–10.

O'Donnell, K.J., Rakeman, M.A., Zhi-Hong, D., Xue-Yi, C., Mei, Z.Y., DeLong, N., Brenner, G., Tai, M., Dong, W., DeLong, G.R., 2002. Effects of iodine supplementation during pregnancy on child growth and development at school age. Dev. Med. Child Neurol. 44, 76–81.

Pearce, E.N., Pino, S., He, X., Bazrafshan, H.R., Lee, S.L., Braverman, L.E., 2004. Sources of dietary iodine: bread, cows' milk, and infant formula in the Boston area. J. Clin. Endocrinol. Metab. 89, 3421–3424.

Pennington, J.A.T., Sshoen, S.A., Salmon, G.D., Young, B., Johnson, R.D., Marts, R.W., 1995. Composition of core foods of the U.S. food supply, 1982–1991: III. Copper, manganese, selenium, and iodine. J. Food Compos. Anal. 8, 171–217.

Sack, J., 2003. Thyroid function in pregnancy – maternal-fetal relationship in health and disease. Pediatr. Endocrinol. Rev. 1 (Suppl. 2), 170–176 Discussion 176.

Stagnaro-Green, A., Abalovich, M., Alexander, E., Azizi, F., Mestman, J., Negro, R., Nixon, A., Pearce, E.N., Soldin, O.P., Sullivan, S., Wiersinga, W., 2011. Guidelines of the American Thyroid Association for the diagnosis and management of thyroid disease during pregnancy and postpartum. Thyroid 21, 1081–1125.

Sullivan, K.M., May, W., Nordenberg, D., Houston, R., Maberly, G.F., 1997. Use of thyroid stimulating hormone testing in newborns to identify iodine deficiency. J. Nutr. 127, 55–58.

Teas, J., Pino, S., Critchley, A., Braverman, L.E., 2004. Variability of iodine content in common commercially available edible seaweeds. Thyroid 14, 836–841.

Todd, C.H., Dunn, J.T., 1998. Intermittent oral administration of potassium iodide solution for the correction of iodine deficiency. Am. J. Clin. Nutr. 67, 1279–1283.

Untoro, J., Schultink, W., West, C.E., Gross, R., Hautvast, J.G., 2006. Efficacy of oral iodized peanut oil is greater than that of iodized poppy seed oil among Indonesian schoolchildren. Am. J. Clin. Nutr. 84, 1208–1214.

Vought, R.L., London, W.T., Lutwak, L., Dublin, T.D., 1963. Reliability of estimates of serum inorganic iodine and daily fecal and urinary iodine excretion from single casual specimens. J. Clin. Endocrinol. Metab. 23, 1218–1228.

Wolff, J., Chaikoff, I.L., 1948. Plasma inorganic iodide as a homeostatic regulator of thyroid function. J. Biol. Chem. 174, 555–564.

World Health Organization, 2014. Iodine Supplementation in Pregnant and Lactating Women. Available at: http://www.who.int/elena/titles/guidance_summaries/salt_iodization/en/.

World Health Organization, ICCIDD, UNICEF, 1996. Recommended Iodine Levels in Salt and Guidelines for Monitoring Their Adequacy and Effectiveness. WHO/NUT/96.13.

World Health Organization, UNICEF, 2007. In: Anderson, M., de Benoist, B., Darton-Hill, I., Delange, F. (Eds.), Iodine Deficiency in Europe.

World Health Organization, UNICEF, ICCIDD, 2001. Assessment of Iodine Deficiency Disorders and Monitoring Their Elimination. A Guide for Programme Managers, third ed. Available at: http://whqlibdoc.who.int/publications/2007/9789241595827_eng.pdf.

Yarrington, C., Pearce, E.N., 2011. Iodine and pregnancy. J. Thyroid Res. 2011.

Zimmermann, M., Saad, A., Hess, S., Torresani, T., Chaouki, N., 2000. Thyroid ultrasound compared with World Health Organization 1960 and 1994 palpation criteria for determination of goiter prevalence in regions of mild and severe iodine deficiency. Eur. J. Endocrinol. 143, 727–731.

Zimmermann, M.B., 2004. Assessing iodine status and monitoring progress of iodized salt programs. J. Nutr. 134, 1673–1677.

Zimmermann, M.B., 2008. Iodine requirements and the risks and benefits of correcting iodine deficiency in populations. J. Trace Elem. Med. Biol. 22, 81–92.

Zimmermann, M.B., 2009. Iodine deficiency. Endocr. Rev. 30, 376–408.

Zimmermann, M.B., Jooste, P.L., Pandav, C.S., 2008. Iodine-deficiency disorders. Lancet 372, 1251–1262.

Chapter 12

Iodine and Thyroid Hormone Synthesis, Metabolism, and Action

Anna Milanesi, Gregory A. Brent

David Geffen School of Medicine at UCLA, Los Angeles, CA, United States

IODIDE TRANSPORT

Iodine Absorption From the Gut

The primary sources of iodine are foods naturally abundant in iodine, such as seafood and dairy products, or foods fortified with iodine, such as salt and bread (discussed in Chapter 11). Iodide is rapidly, and almost completely (>90%), absorbed in the small intestine and stomach. Iodate is used for salt iodization and is reduced to iodide in the gut before being absorbed. Most organically bound iodine is digested and released as iodide before absorption, except for some forms, such as thyroxine, that can be absorbed intact (Rohner et al., 2014).

Iodide is transported by the sodium/iodide symporter (NIS), a plasma glycoprotein that mediates transport of iodide into all iodide-concentrating tissues including the thyroid gland, intestine, salivary gland, lactating breast, and stomach (Dai et al., 1996) (Fig. 12.1). The transport of iodide by NIS across the membrane requires an electrochemical sodium gradient from outside to inside of cells, which is maintained by the Na^+/K^+ ATPase. NIS is the only transport protein for iodide and is expressed in all tissues that transport iodide, although there is some variation in the extent of glycosylation of the NIS protein. The intestinal NIS is expressed in enterocytes and has been shown to have a K_m for Na^+-dependent, ClO_4- sensitive iodide transport of $13.4 \pm 2.0 \mu M$, similar to the one expressed in thyroid follicular cells (Nicola et al., 2009). NIS in the small intestine is located apically in the brush-border membrane, in contrast to the other NIS-expressing tissues, such as thyroid, lactating breast, and salivary gland, where it is expressed on the basolateral membrane, adjacent to blood vessels. NIS expression in the intestine, as has also been described in the thyroid, is downregulated by excess iodide, through an autoregulatory mechanism that functions to avoid toxic levels of intracellular iodide (Nicola et al., 2009). High intracellular iodide in enterocytes downregulates NIS expression, increases NIS degradation, and reduces NIS-mediated uptake of iodide (Nicola et al., 2015). There is then an "escape" from this effect where normal iodide transport is restored, generally after 7–10 days. NIS plays the major role in iodide absorption in the small intestine, but there may be contribution of other iodide permeable transporters or channels, such as chloride channels and pendrin, a member of the multifunctional anion exchanger family (de Carvalho and Quick, 2011).

Iodine Absorption Through Skin

Iodide can also be absorbed through the skin or other mucous membranes. It has been reported that the use of topical-iodine antiseptics in premature infants correlates with increased urine iodine levels and is associated with thyroid dysfunction (Linder et al., 1997; Aitken and Williams, 2014). Moreover, iodide-induced thyrotoxicosis has been described in a paraplegic woman who was using a topical iodine prior to self-catheterization (Pramyothin et al., 2011).

Iodide Transport in the Thyroid

The ability of the thyroid to concentrate iodide, which is required for thyroid hormone synthesis, was first reported by Baumann in 1896 (Baumann, 1896). A robust iodine concentrating ability and capacity for iodine storage in the thyroid was essential for individuals with low or variable iodine intake, when diet was restricted to food from a local region, such as mountainous areas that are naturally low in iodine (Dohan et al., 2003). The thyroid gland is located anterolaterally to the trachea. The thyroid follicles are the functional units of the thyroid, consisting of spherical structures of a monolayer

Molecular, Genetic, and Nutritional Aspects of Major and Trace Minerals. http://dx.doi.org/10.1016/B978-0-12-802168-2.00012-9

143

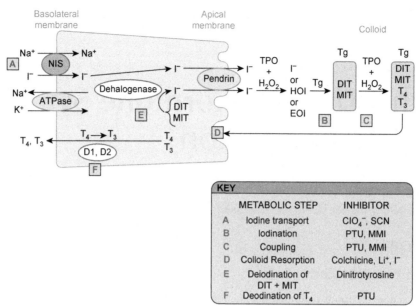

FIGURE 12.1 **Major pathways of thyroid hormone biosynthesis and release.** TSH is the primary stimulus for thyroid hormone synthesis. NIS transports iodine from the blood stream into the basolateral membrane of the thyroid follicular cell, requiring a sodium gradient, high extracellular, and low intracellular, maintained by ATPase. Iodide is transported from the apical membrane into the colloid by the action of pendrin. Thyroglobulin, the substrate for thyroid hormone synthesis and contained in the colloid, is iodinated by the action of TPO. Coupling reactions join MIT and DIT to form triiodothyronine (T_3) or DIT and DIT to form thyroxine (T_4). The D1 and D2 enzymes convert T_4 to T_3. The dehalogenase enzymes remove iodide from MIT and DIT, likely to conserve and recycle iodide. *ATPase*, sodium potassium ATPase; *D1 and D2*, 5′-deiodinases types 1 and 2; *DIT*, diiodotyrosine; *EOI*, enzyme-linked species; *HOI*, hypoiodous acid; *MIT*, monoiodotyrosine; *MMI*, methimazole; *NIS*, sodium iodide symporter; *PTU*, propylthiouracil; *Tg*, thyroglobulin; *TPO*, thyroid peroxidase; *TSH*, thyroid stimulating hormone. *Used with permission from Brent, G.A., Koenig, R.J., 2011. Thyroid and antithyroid drugs. In: Brunton, L. (Ed.), Goodman and Gilman's the Pharmacological Basis of Therapeutics, twelfth ed. McGraw-Hill, New York, pp. 1129–1161.*

of polarized cells with the basolateral side facing the bloodstream and the apical side toward the follicular lumen (Smanik et al., 1996; Fig. 12.1). NIS concentrates iodide 30- to 60-fold, relative to the concentration in the blood, into the thyroid follicular cells (Kaminsky et al., 1994). Thyroid stimulating hormone (TSH) regulates iodide transport into the thyroid (Kogai et al., 2006), stimulating NIS transcription (Saito et al., 1997; Kogai et al., 2000) and promoting correct insertion of NIS into the plasma membrane (Riedel et al., 2001). TSH, secreted from the anterior pituitary, is primarily regulated by T_4/T_3 feedback to the pituitary and hypothalamus. With prolonged low-iodine intake, TSH stimulates NIS expression/activity and also thyroid growth and enlargement, referred to as a goiter. NIS-mediated iodide transport is directly inhibited by thyiocyanate and perchlorate, found in water and foods. Iodide regulates its own accumulation in the thyroid, as was discussed earlier in enterocytes. This effect was described in 1948, when it was reported that high doses of iodide block iodide organification in vivo, referred to as the Wolff–Chaikoff effect (Wolff and Chaikoff, 1948; Pramyothin et al., 2011). This effect is self-limited, because it is followed by an "escape" mechanism secondary to the downregulation of iodide uptake. Mutation of SLC45A, the gene encoding NIS, leads to congenital iodide transport defect which is characterized by hypothyroidism and goiter secondary to an inability to concentrate iodine in the thyroid (Wolff, 1983).

Iodide in the follicular cell then moves into the follicular lumen. Traditionally, it was thought that iodide simply crosses the apical membrane by an electrochemical gradient (Wolff, 2005); however, studies showed that TSH stimulates transport of iodide through the apical membrane (Nilsson et al., 1990), consistent with the presence of a specific transporter, pendrin (Yoshida et al., 2002; Gillam et al., 2004). Pendrin belongs to the SLC26A family, and is encoded by the *SLC26A4* gene, first cloned in 1997 (Everett et al., 1997). Pendrin is present in the apical membrane of the follicular cells (Scott et al., 1999; Royaux et al., 2000). Pendred syndrome, characterized by sensorineural deafness and occasionally goiter, is associated with mutations in the *SLC26A4* gene (Everett et al., 1997). Interestingly, thyroid function studies in Pendred syndrome patients are normal. Normal thyroid function tests in individuals with Pendred syndrome, as well as studies of pendrin knockout mice, suggest that pendrin is not the only apical transporter mediating iodide efflux (Fong, 2011). Another possible candidate for apical iodide transport is chloride channel 5 (ClCn5), but this is still under investigation (van den Hove et al., 2006).

Iodide, after translocation into the follicular lumen, is oxidized and incorporated into thyroglobulin (Tg) by thyroid peroxidase (TPO), with hydrogen peroxide (H_2O_2) produced through the action of dual oxidase-2 (DUOX2). This process of incorporation of iodide into protein is called "organification" and results in the formation of mono- and diiodotyrosines (MIT and DIT). Iodine is covalent bound at the 3 and 4 positions of tyrosyl residues on Tg and the iodinized thyroid residues are used for thyroid hormone synthesis. Iodinated Tg is stored in the follicular lumen and it is internalized via endocytosis or micropinocytosis, followed by proteolytic cleavage in the lysosomes with subsequent release of T_3 and T_4 into the bloodstream (Bizhanova and Kopp, 2009). In this process, partially iodized thyroid hormone intermediates reenter the cycle of organification. Mutation of the *Duox2* gene, or of its maturation factor, DuoxA2, results in congenital hypothyroidism secondary to an iodide organification defect (Maruo et al., 2008; Donko et al., 2014).

Iodide Transport in Tissues Outside the Thyroid

Iodide uptake has been shown in several tissues, in addition to the thyroid and intestine, including the lactating mammary gland, salivary gland, gastric mucosa, lacrimal gland, choroid plexus, and ciliary body of the eye (Wolff, 1964; Wright, 1974; De La Vieja et al., 2000; Tazebay et al., 2000; Dohan and Carrasco, 2003; Riesco-Eizaguirre and Santisteban, 2006). The only tissues where iodide is organified into protein, however, are the thyroid and lactating breast. TSH regulation of NIS expression and insertion has been demonstrated only in the thyroid. The iodide transport in extrathyroid tissues is mediated by NIS, as it is in the thyroid (Aceves et al., 2013).

NIS expression in the breast increases with pregnancy and lactation and expression is also stimulated in some breast cancers (Tazebay et al., 2000). Breast milk contains about 150–180 mcg/L of I⁻ (Semba and Delange, 2001) and for infants solely receiving breast milk, it is the only source of iodide for thyroid hormone production. In the lactating breast, NIS is expressed on the basolateral membrane of mammary alveolar cells (Cho et al., 2000) and it actively concentrates iodide into milk (Mountford et al., 1986). A lactoperoxidase, similar to thyroid peroxidase, is expressed in the lactating breast, which mediates organification of iodine. The expression of NIS in the mammary gland during gestation and lactation suggests that lactogenic hormones regulate NIS expression in this tissue. Several studies in fact showed that estrogen, prolactin, oxytocin, and insulin stimulate the expression of functional NIS in the mammary gland (Cho et al., 2000; Rillema et al., 2002). Estrogen may have a direct effect on NIS transcription, because the NIS promoter contains estrogen responsive elements (Ohno et al., 1999). Moreover, estrogen potentiates NIS induction by oxytocin (Zingg et al., 1998; Tazebay et al., 2000). Both prolactin and oxytocin are released with suckling and enhance NIS expression in the mammary gland after delivery (Cho et al., 2000). NIS mRNA and protein are also expressed in the majority of breast cancers, but not sufficient levels to concentrate a significant amount of iodide (Tazebay et al., 2000; Rudnicka et al., 2003). NIS upregulation in breast cancer offers a possible diagnostic and therapeutic option for breast cancer. If NIS activity could be upregulated, oral radioiodine could then be used to image and treat breast cancer, as is done for thyroid cancer (Kogai et al., 2006).

In the salivary gland and gastric mucosa, iodide is transported though the basolateral NIS in epithelial cells, and secreted in the saliva and gastric juice through chloride channels, best demonstrated for the Slc26a4 transporter (Dohan et al., 2003; Bruno et al., 2004; Josefsson et al., 2006). Iodide secreted in the saliva and gastric juice is likely reabsorbed in the small intestine and this may be an important mechanism for conservation and reutilization of iodide, especially during low-iodine intake (Venturi and Venturi, 2009). Physiologic iodide uptake by the salivary gland explains the damage that can be induced by radioactive iodine treatment for thyroid cancer, such as sialoadenitis, xerostomia, and dental and periodontal diseases (Mandel and Mandel, 2003; Van Nostrand, 2011).

The supply of iodide for fetal thyroid hormones biosynthesis is dependent on maternal iodine intake and placenta transfer of iodide to the fetal compartment (Glinoer, 1997; Delange, 1998). Iodide transport through the placenta, however, is not fully understood. The placental enzyme, 5-deiodinase type 3 (D3), removes an inner ring iodine and converts T_4 to the inactive reverse T_3. D2 is also expressed in the placenta and both enzymes may have a role in iodine supply to the fetus (Chan et al., 2003; Fig. 12.2). Low expression of NIS has been reported in the human placenta (Mitchell et al., 2001; Di Cosmo et al., 2006). The choriocarcinoma cell line, BeWo, expresses NIS in the apical (maternal) membrane, accumulates iodide, and possibly releases iodide in the fetal circulation through pendrin (Manley et al., 2005). Rats consuming a low-iodine diet showed upregulation of NIS in the placenta (Chan et al., 2003), suggesting a compensatory mechanism to support fetal development. Human chorionic gonadotropin (hCG), which is at high circulating levels during pregnancy, stimulates iodide accumulation and NIS expression in two choriocarcinoma cell lines (Arturi et al., 2002; Li et al., 2012). Moreover, hCG, prolactin, and oxytocin significantly increase iodine uptake in placental primary culture, and this can be attributed to increased NIS expression (Burns et al., 2013).

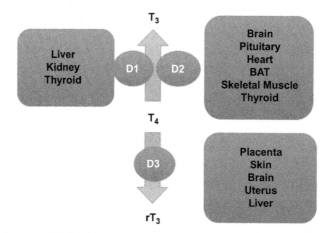

FIGURE 12.2 Pathways of iodothyronine deiodination and tissue distribution. Thyroxine (T_4) can be converted to the active form of thyroid hormone, T_3, by removal of an outer ring iodine. This reaction is catalyzed by the deiodinase isozymes, D1, type I 5′ iodothyronine deiodinase, and D2, type II 5′ iodothyronine deiodinase. The tissue distribution of D1 and D2 enzymes is shown. When serum T_4 levels are low, D1 is downregulated and D2 upregulated. D2 is the major contributor to circulating T_3. T_4 is inactivated by removal of an inner ring iodine to reverse T_3 (r T_3), which is catalyzed by the enzyme D3, type III 5 iodothyronine deiodinase. The tissue distribution of the D3 enzyme is shown. *BAT*, brown adipose tissue.

IODINE CLEARANCE

Absorbed iodine is cleared by the circulation, mostly by the thyroid and kidney. Thyroid clearance varies with iodine intake: in an iodine sufficient condition about 10% of the iodide is taken up by the thyroid, but in chronic iodine deficiency, the iodide uptake by the thyroid can exceed 80% (Rohner et al., 2014). Plasma iodine has a physiologic half-life of about 10 h, but in case of iodine deficiency or hyperthyroidism, conditions with increased thyroid uptake of iodine, plasma half-life is shortened.

In humans, iodide is eliminated primarily through the kidney (Cavalieri, 1997) and average urinary iodine concentration in a population reflects overall iodine intake. Given the day-to-day variation in dietary iodine intake, however, a spot urinary iodine level may not accurately reflect overall iodine intake in an individual. The mechanism of iodide excretion, however, is not fully understood. Iodine clearance depends mostly on glomerular filtration without tubular reabsorption (Bricker and Hlad, 1955). NIS mRNA has been detected in the human kidney and basolateral staining for NIS was most notable in proximal tubular cells (Spitzweg et al., 2001). However, these findings were not confirmed by other groups (Vayre et al., 1999; Lacroix et al., 2001). Moreover, there are no studies showing iodine concentrating capability via NIS in the kidney, with the exception of a human kidney tumor cell line (Spitzweg et al., 2001). Iodine excreted into the gastrointestinal tract, by the salivary gland and gastric mucosa, is reabsorbed through the small intestine, likely representing a mechanism for conservation and reutilization of iodide (Venturi and Venturi, 2009).

DIRECT ACTIONS OF IODINE

In addition to thyroid hormone biosynthesis, iodide has other biological actions that include innate antimicrobial defense, antioxidant action, and effects on cell proliferation.

Several studies have suggested that iodide may have an antimicrobial action. The lacrimal, salivary glands, stomach, and lactating breast tissue express components of the lactoperoxidase system that has bactericidal activities, through the generation of H_2O_2 and hypoiodite (IO^-) (Bosch et al., 2000). Lactoperoxidase in these tissues oxidizes iodide to IO^-, and this could explain the antibacterial action of iodine. In this context, there are studies showing low rates of dental caries in populations living in iodine rich areas (Venturi and Venturi, 2009). Moreover, iodine has been shown to have an antiviral affect when applied to human airway epithelial cells in vitro (Fischer et al., 2011).

Very small amounts of iodine decrease oxygen-free radical production and also have some antiinflammatory effects in humans (Winkler et al., 2000; Soriguer et al., 2011). Iodine, in addition to the well described antibacterial effect, has an antiinflammatory action important for wound healing (Beukelman et al., 2008). These antioxidant and antibacterial activities of iodide may have a role in the biological action of iodide in gastric mucosa and salivary gland (Geiszt et al., 2003).

Iodine plays a key role in the regulation of thyroid function and thyrocyte proliferation, and this phenomenon is called "autoregulation." Excess iodine causes reduced hormone secretion, thyroglobulin proteolysis, and thyroid blood flow, and it also inhibits thyroid follicular cell growth in vivo and thyrocyte proliferation in vitro. Most of the apoptotic and antiproliferative effects on thyrocytes is thyroid peroxidase (TPO) activity-dependent, because they can be reversed by inhibiting the TPO enzyme. The apoptotic action of iodine has been shown also in extra-thyroidal tissues. Several groups have reported a proapoptotic action of iodide in cancer cell lines mediated by different mechanisms, including a direct toxic effect of iodine, activation, of mitochondrial-mediated apoptosis or indirect activation of the bel-2-like protein 4 (BAX)-caspase apoptotic pathway (Zhang et al., 2003; Singh et al., 2011).

THYROID HORMONE ACTION

Nuclear Receptor

Thyroid hormone is a peptide-derived hormone, but its action in the nucleus is most similar to steroid-derived hormones. The majority of thyroid hormone action is mediated by a nuclear receptor, thyroid hormone nuclear receptors (TR) α and β (Brent, 2012). These receptor isoforms have both developmental and tissue-specific patterns of expression. TRα is expressed predominantly in the brain, heart, intestine, and bone, while TRβ is expressed in the pituitary and liver. Even within a specific tissue, such as the heart, bone, or brain, there is differential expression of receptor types in various regions. TR binds to specific DNA sequences, referred to as thyroid hormone response elements. TR generally binds to the response element as a heterodimer, with the retinoid X receptor. In the absence of ligand, TR binds to a nuclear corepressor protein, which suppresses activation of genes positively regulated by thyroid hormone. Addition of a ligand, T_3, disrupts binding of TR to the corepressor protein and attracts coactivator proteins to bind to TR, which then activates gene transcription.

Thyroid Hormone Metabolism

Activation of the prohormone T_4, to the active hormone T_3, is an important element of the regulation of thyroid hormone action (Brent, 2012). The enzymes that convert T_4 to T_3 are the 5′-deiodinases D1 and D2 (Fig. 12.2). There are significant differences in the enzymes with respect to developmental expression, tissues-specific expression, and regulation (Greben et al., 2008). Both enzymes have a requirement for the trace element selenium for normal function, as the amino acid sele-nocyteine is required for normal enzyme function. D2 is expressed early in development in tissues where thyroid hormone action is critical, including the brain, pituitary, heart, brown adipose tissue, skeletal muscle, and thyroid. D1 is expressed later developmentally and found in the liver, kidney, and thyroid. Most significantly, D1 is directly regulated by T_3; D1 is elevated when T_3 is elevated and low when T_3 levels are low. D2 is reduced when T_4 levels are high, and activated, through a ubiquitination–deubiquitination system, when T_4 levels are low. This is especially important for maintaining adequate tissue levels of T_3 in critical tissues, when T_4 levels may fall, as occurs in times of prolonged insufficient iodine intake. Since selenium is required for D2 activity, combined iodine and selenium deficiency impairs D2 activity and the ability to compensate for low iodine through this pathway (Zimmermann, 2009).

Thyroid Hormone Action in Development

Thyroid hormone acts as a pleiotropic factor in various tissues during development by regulating genes involved in growth and differentiation (Mishra et al., 2010; Chatonnet et al., 2011; Qiu et al., 2012). Also, during postnatal life, it regulates metabolism, and has significant impacts on other organs including the brain, heart, liver, skeletal muscle, and bone (Brent, 2012). A sufficient T_3 supply during fetal development is critical for axon elongation, dendrite branching, cortical development, and neuronal differentiation and migration (Cheng et al., 2010).

Compensation for Iodine Insufficiency

Iodine insufficiency affects a significant amount of people worldwide. According to WHO, over 30% of the population has inadequate iodine intake (UI < 100 µg/L), and are at risk to develop iodine deficiency disorders (de Benoist et al., 2008) (discussed in Chapter 11). The impact of iodine deficiency is most significant during fetal life and early childhood development, with manifestation of goiter, abortion, stillbirth, increased infant mortality, delayed physical development, impaired mental function, and reduced school performance (Zimmermann, 2009). Adults can manifest impaired mental function, reduced work performance, hypothyroidism, and goiter. Severe dietary iodine insufficiency can result in hypothyroidism and goiter at all ages.

A healthy adult has 15–20 mg of stored iodine, mostly in the thyroid (Zimmermann, 2009). The mechanisms for compensation of reduced iodine include an increase in TSH and enhanced T_4 to T_3 conversion by D2. Generally, iodine intake below 100 µg/day results in an increase in serum TSH, which increases NIS expression and membrane insertion. Associated with increased iodide clearance by the thyroid is a progressive reduction in excretion of iodide by the kidney. The thyroid also breaks down thyroglobulin to secrete more T_3. These mechanisms can compensate for low-iodine intake until the daily intake falls below $50\,\mu g$/day. At this level, the thyroid gland enlarges, serum thyroglobulin increases, and there is increased T_4 to T_3 conversion due to activation of D2. The final change, with prolonged profound iodine deficiency, is a fall in T_3. Thyroid enlargement, as determined by examination or thyroid ultrasound, and elevated serum thyroglobulin, are markers used to assess the iodine status of a population.

Iodine in Pregnancy

Despite the strategy of universal salt iodization, mild iodine deficiency is still a problem in pregnant and lactating woman. Maternal low iodine can translate in a wide range of growth and mental impairments in newborns. The importance of maternal iodine was first recognized in the Swiss Alps and Himalayas, where severe endemic goiter and cretinism was reported (Morreale de Escobar et al., 2000; de Escobar et al., 2004). Low iodine is the most frequent cause of maternal hypothyroxinemia (Glinoer and Delange, 2000), and can be associated with modest changes in maternal thyroxine and TSH (Morreale de Escobar et al., 2000; Stagnaro-Green et al., 2011). Epidemiological studies showed that offspring of mothers with hypothyroxinemia in pregnancy had reduced mental performance and psychomotor deficits (Henrichs et al., 2013). These findings from studies around the world emphasize the importance of iodine supplementation in pregnancy.

ACKNOWLEDGMENTS

This work was supported by NIH grant 1K08DK097295 (to AM), NIH RO1DK98576 (to GAB), and Veterans Affairs Merit Review Grant (to GAB).

REFERENCES

Aceves, C., Anguiano, B., Delgado, G., 2013. The extrathyronine actions of iodine as antioxidant, apoptotic, and differentiation factor in various tissues. Thyroid 23, 938–946.

Aitken, J., Williams, F.L., 2014. A systematic review of thyroid dysfunction in preterm neonates exposed to topical iodine. Arch. Dis. Child. Fetal Neonatal Ed. 99, F21–F28.

Arturi, F., Lacroix, L., Presta, I., Scarpelli, D., Caillou, B., Schlumberger, M., Russo, D., Bidart, J.M., Filetti, S., 2002. Regulation by human chorionic gonadotropin of sodium/iodide symporter gene expression in the JAr human choriocarcinoma cell line. Endocrinology 143, 2216–2220.

Baumann, E., 1896. Uber das Thyrojodin. Munch. Med. Wochenschr. 43, 309–312.

Beukelman, C.J., van den Berg, A.J., Hoekstra, M.J., Uhl, R., Reimer, K., Mueller, S., 2008. Anti-inflammatory properties of a liposomal hydrogel with povidone-iodine (Repithel) for wound healing in vitro. Burns 34, 845–855.

Bizhanova, A., Kopp, P., 2009. Minireview: the sodium-iodide symporter NIS and pendrin in iodide homeostasis of the thyroid. Endocrinology 150, 1084–1090.

Bosch, E.H., van Doorne, H., de Vries, S., 2000. The lactoperoxidase system: the influence of iodide and the chemical and antimicrobial stability over the period of about 18 months. J. Appl. Microbiol. 89, 215–224.

Brent, G.A., 2012. Mechanisms of thyroid hormone action. J. Clin. Invest. 122, 3035–3043.

Bricker, N.S., Hlad Jr., C.J., 1955. Observations on the mechanism of the renal clearance of I131. J. Clin. Invest. 34, 1057–1072.

Bruno, R., Giannasio, P., Ronga, G., Baudin, E., Travagli, J.P., Russo, D., Filetti, S., Schlumberger, M., 2004. Sodium iodide symporter expression and radioiodine distribution in extrathyroidal tissues. J. Endocrinol. Invest. 27, 1010–1014.

Burns, R., O'Herlihy, C., Smyth, P.P., 2013. Regulation of iodide uptake in placental primary cultures. Eur. Thyroid. J. 2, 243–251.

Cavalieri, R.R., 1997. Iodine metabolism and thyroid physiology: current concepts. Thyroid 7, 177–181.

Chan, S., Kachilele, S., Hobbs, E., Bulmer, J.N., Boelaert, K., McCabe, C.J., Driver, P.M., Bradwell, A.R., Kester, M., Visser, T.J., Franklyn, J.A., Kilby, M.D., 2003. Placental iodothyronine deiodinase expression in normal and growth-restricted human pregnancies. J. Clin. Endocrinol. Metab. 88, 4488–4495.

Chatonnet, F., Picou, F., Fauquier, T., Flamant, F., 2011. Thyroid hormone action in cerebellum and cerebral cortex development. J. Thyroid. Res. 2011, 145762.

Cheng, S.Y., Leonard, J.L., Davis, P.J., 2010. Molecular aspects of thyroid hormone actions. Endocr. Rev. 31, 139–170.

Cho, J.Y., Leveille, R., Kao, R., Rousset, B., Parlow, A.F., Burak Jr., W.E., Mazzaferri, E.L., Jhiang, S.M., 2000. Hormonal regulation of radioiodide uptake activity and Na+/I− symporter expression in mammary glands. J. Clin. Endocrinol. Metab. 85, 2936–2943.

Dai, G., Levy, O., Carrasco, N., 1996. Cloning and characterization of the thyroid iodide transporter. Nature 379, 458–460.

de Benoist, B., McLean, E., Andersson, M., Rogers, L., 2008. Iodine deficiency in 2007: global progress since 2003. Food Nutr. Bull. 29, 195–202.

de Carvalho, F.D., Quick, M., 2011. Surprising substrate versatility in SLC5A6: Na+-coupled I− transport by the human Na+/multivitamin transporter (hSMVT). J. Biol. Chem. 286, 131–137.

de Escobar, G.M., Obregon, M.J., del Rey, F.E., 2004. Maternal thyroid hormones early in pregnancy and fetal brain development. Best. Pract. Res. Clin. Endocrinol. Metab. 18, 225–248.

De La Vieja, A., Dohan, O., Levy, O., Carrasco, N., 2000. Molecular analysis of the sodium/iodide symporter: impact on thyroid and extrathyroid pathophysiology. Physiol. Rev. 80, 1083–1105.

Delange, F., 1998. Screening for congenital hypothyroidism used as an indicator of the degree of iodine deficiency and of its control. Thyroid 8, 1185–1192.

Di Cosmo, C., Fanelli, G., Tonacchera, M., Ferrarini, E., Dimida, A., Agretti, P., De Marco, G., Vitti, P., Pinchera, A., Bevilacqua, G., Naccarato, A.G., Viacava, P., 2006. The sodium-iodide symporter expression in placental tissue at different gestational age: an immunohistochemical study. Clin. Endocrinol. (Oxf.) 65, 544–548.

Dohan, O., Carrasco, N., 2003. Advances in Na(+)/I(−) symporter (NIS) research in the thyroid and beyond. Mol. Cell. Endocrinol. 213, 59–70.

Dohan, O., De la Vieja, A., Paroder, V., Riedel, C., Artani, M., Reed, M., Ginter, C.S., Carrasco, N., 2003. The sodium/iodide Symporter (NIS): characterization, regulation, and medical significance. Endocr. Rev. 24, 48–77.

Donko, A., Morand, S., Korzeniowska, A., Boudreau, H.E., Zana, M., Hunyady, L., Geiszt, M., Leto, T.L., 2014. Hypothyroidism-associated missense mutation impairs NADPH oxidase activity and intracellular trafficking of Duox2. Free Radic. Biol. Med. 73, 190–200.

Everett, L.A., Glaser, B., Beck, J.C., Idol, J.R., Buchs, A., Heyman, M., Adawi, F., Hazani, E., Nassir, E., Baxevanis, A.D., Sheffield, V.C., Green, E.D., 1997. Pendred syndrome is caused by mutations in a putative sulphate transporter gene (PDS). Nat. Genet. 17, 411–422.

Fischer, A.J., Lennemann, N.J., Krishnamurthy, S., Pocza, P., Durairaj, L., Launspach, J.L., Rhein, B.A., Wohlford-Lenane, C., Lorentzen, D., Banfi, B., McCray Jr., P.B., 2011. Enhancement of respiratory mucosal antiviral defenses by the oxidation of iodide. Am. J. Respir. Cell Mol. Biol. 45, 874–881.

Fong, P., 2011. Thyroid iodide efflux: a team effort? J. Physiol. 589, 5929–5939.

Geiszt, M., Witta, J., Baffi, J., Lekstrom, K., Leto, T.L., 2003. Dual oxidases represent novel hydrogen peroxide sources supporting mucosal surface host defense. FASEB J. 17, 1502–1504.

Gillam, M.P., Sidhaye, A.R., Lee, E.J., Rutishauser, J., Stephan, C.W., Kopp, P., 2004. Functional characterization of pendrin in a polarized cell system. Evidence for pendrin-mediated apical iodide efflux. J. Biol. Chem. 279, 13004–13010.

Glinoer, D., Delange, F., 2000. The potential repercussions of maternal, fetal, and neonatal hypothyroxinemia on the progeny. Thyroid 10, 871–887.

Glinoer, D., 1997. The regulation of thyroid function in pregnancy: pathways of endocrine adaptation from physiology to pathology. Endocr. Rev. 18, 404–433.

Greben, B., Zavacki, A.M., Ribich, S., Kim, B.W., Huang, S.A., Simonides, W.S., Zeold, A., Bianco, A.C., 2008. Cellular and molecular basis of deiodinase-regulated thyroid hormone signaling. Endocr. Rev. 29, 898–938.

Henrichs, J., Ghassabian, A., Peeters, R.P., Tiemeier, H., 2013. Maternal hypothyroxinemia and effects on cognitive functioning in childhood: how and why? Clin. Endocrinol. (Oxf.) 79, 152–162.

Josefsson, M., Evilevitch, L., Westrom, B., Grunditz, T., Ekblad, E., 2006. Sodium-iodide symporter mediates iodide secretion in rat gastric mucosa in vitro. Exp. Biol. Med. (Maywood) 231, 277–281.

Kaminsky, S.M., Levy, O., Salvador, C., Dai, G., Carrasco, N., 1994. Na(+)-I− symport activity is present in membrane vesicles from thyrotropin-deprived non-I(−)-transporting cultured thyroid cells. Proc. Natl. Acad. Sci. U.S.A. 91, 3789–3793.

Kogai, T., Curcio, F., Hyman, S., Cornford, E.M., Brent, G.A., Hershman, J.M., 2000. Induction of follicle formation in long-term cultured normal human thyroid cells treated with thyrotropin stimulates iodide uptake but not sodium/iodide symporter messenger RNA and protein expression. J. Endocrinol. 167, 125–135.

Kogai, T., Taki, K., Brent, G.A., 2006. Enhancement of sodium/iodide symporter expression in thyroid and breast cancer. Endocr. Relat. Cancer 13, 797–826.

Lacroix, L., Mian, C., Caillou, B., Talbot, M., Filetti, S., Schlumberger, M., Bidart, J.M., 2001. Na(+)/I(−) symporter and Pendred syndrome gene and protein expressions in human extra-thyroidal tissues. Eur. J. Endocrinol. 144, 297–302.

Li, H., Patel, J., Mortimer, R.H., Richard, K., 2012. Ontogenic changes in human placental sodium iodide symporter expression. Placenta 33, 946–948.

Linder, N., Davidovitch, N., Reichman, B., Kuint, J., Lubin, D., Meyerovitch, J., Sela, B.A., Dolfin, Z., Sack, J., 1997. Topical iodine-containing antiseptics and subclinical hypothyroidism in preterm infants. J. Pediatr. 131, 434–439.

Mandel, S.J., Mandel, L., 2003. Radioactive iodine and the salivary glands. Thyroid 13, 265–271.

Manley, S.W., Li, H., Mortimer, R.H., 2005. The BeWo choriocarcinoma cell line as a model of iodide transport by placenta. Placenta 26, 380–386.

Maruo, Y., Takahashi, H., Soeda, I., Nishikura, N., Matsui, K., Ota, Y., Mimura, Y., Mori, A., Sato, H., Takeuchi, Y., 2008. Transient congenital hypothyroidism caused by biallelic mutations of the dual oxidase 2 gene in Japanese patients detected by a neonatal screening program. J. Clin. Endocrinol. Metab. 93, 4261–4267.

Mishra, A., Zhu, X.G., Ge, K., Cheng, S.Y., 2010. Adipogenesis is differentially impaired by thyroid hormone receptor mutant isoforms. J. Mol. Endocrinol. 44, 247–255.

Mitchell, A.M., Manley, S.W., Morris, J.C., Powell, K.A., Bergert, E.R., Mortimer, R.H., 2001. Sodium iodide symporter (NIS) gene expression in human placenta. Placenta 22, 256–258.

Morreale de Escobar, G., Obregon, M.J., Escobar del Rey, F., 2000. Is neuropsychological development related to maternal hypothyroidism or to maternal hypothyroxinemia? J. Clin. Endocrinol. Metab. 85, 3975–3987.

Mountford, P.J., Coakley, A.J., Fleet, I.R., Hamon, M., Heap, R.B., 1986. Transfer of radioiodide to milk and its inhibition. Nature 322, 600.

Nicola, J.P., Basquin, C., Portulano, C., Reyna-Neyra, A., Paroder, M., Carrasco, N., 2009. He Na⁺/I⁻ symporter mediates active iodide uptake in the intestine. Am. J. Physiol. Cell Physiol. 296, C654–C662.

Nicola, J.P., Carrasco, N., Masini-Repiso, A.M., 2015. Dietary I(−) absorption: expression and regulation of the Na(+)/I(−) symporter in the intestine. Vitam. Horm. 98, 1–31.

Nilsson, M., Bjorkman, U., Ekholm, R., Ericson, L.E., 1990. Iodide transport in primary cultured thyroid follicle cells: evidence of a TSH-regulated channel mediating iodide efflux selectively across the apical domain of the plasma membrane. Eur. J. Cell Biol. 52, 270–281.

Ohno, M., Zannini, M., Levy, O., Carrasco, N., di Lauro, R., 1999. The paired-domain transcription factor Pax8 binds to the upstream enhancer of the rat sodium/iodide symporter gene and participates in both thyroid-specific and cyclic-AMP-dependent transcription. Mol. Cell. Biol. 19, 2051–2060.

Pramyothin, P., Leung, A.M., Pearce, E.N., Malabanan, A.O., Braverman, L.E., 2011. Clinical problem-solving. A hidden solution. N. Engl. J. Med. 365, 2123–2127.

Qiu, J., Ma, X.L., Wang, X., Chen, H., Huang, B.R., 2012. Insulin-like growth factor binding protein-6 interacts with the thyroid hormone receptor alpha1 and modulates the thyroid hormone-response in osteoblastic differentiation. Mol. Cell. Biochem. 361, 197–208.

Riedel, C., Levy, O., Carrasco, N., 2001. Post-transcriptional regulation of the sodium/iodide symporter by thyrotropin. J. Biol. Chem. 276, 21458–21463.

Riesco-Eizaguirre, G., Santisteban, P., 2006. A perspective view of sodium iodide symporter research and its clinical implications. Eur. J. Endocrinol. 155, 495–512.

Rillema, J.A., Williams, C.H., Moulden, J., Golden, K.L., 2002. Effect of insulin on iodide uptake in mouse mammary gland explants. Exp. Biol. Med. (Maywood) 227, 32–35.

Rohner, F., Zimmermann, M., Jooste, P., Pandav, C., Caldwell, K., Raghavan, R., Raiten, D.J., 2014. Biomarkers of nutrition for development – iodine review. J. Nutr. 144, 1322S–1342S.

Royaux, I.E., Suzuki, K., Mori, A., Katoh, R., Everett, L.A., Kohn, L.D., Green, E.D., 2000. Pendrin, the protein encoded by the Pendred syndrome gene (PDS), is an apical porter of iodide in the thyroid and is regulated by thyroglobulin in FRTL-5 cells. Endocrinology 141, 839–845.

Rudnicka, L., Sinczak, A., Szybinski, P., Huszno, B., Stachura, J., 2003. Expression of the Na(+)/I(−) symporter in invasive ductal breast cancer. Folia Histochem Cytobiol. 41, 37–40.

Saito, T., Endo, T., Kawaguchi, A., Ikeda, M., Nakazato, M., Kogai, T., Onaya, T., 1997. Increased expression of the Na⁺/I⁻ symporter in cultured human thyroid cells exposed to thyrotropin and in Graves' thyroid tissue. J. Clin. Endocrinol. Metab. 82, 3331–3336.

Scott, D.A., Wang, R., Kreman, T.M., Sheffield, V.C., Karniski, L.P., 1999. The Pendred syndrome gene encodes a chloride–iodide transport protein. Nat. Genet. 21, 440–443.

Semba, R.D., Delange, F., 2001. Iodine in human milk: perspectives for infant health. Nutr. Rev. 59, 269–278.

Singh, P., Godbole, M., Rao, G., Annarao, S., Mitra, K., Roy, R., Ingle, A., Agarwal, G., Tiwari, S., 2011. Inhibition of autophagy stimulate molecular iodine-induced apoptosis in hormone independent breast tumors. Biochem. Biophys. Res. Commun. 415, 181–186.

Smanik, P.A., Liu, Q., Furminger, T.L., Ryu, K., Xing, S., Mazzaferri, E.L., Jhiang, S.M., 1996. Cloning of the human sodium iodide symporter. Biochem. Biophys. Res. Commun. 226, 339–345.

Soriguer, F., Gutierrez-Repiso, C., Rubio-Martin, E., Linares, F., Cardona, I., Lopez-Ojeda, J., Pacheco, M., Gonzalez-Romero, S., Garriga, M.J., Velasco, I., Santiago, P., Garcia-Fuentes, E., 2011. Iodine intakes of 100–300 mug/d do not modify thyroid function and have modest anti-inflammatory effects. Br. J. Nutr. 105, 1783–1790.

Spitzweg, C., Dutton, C.M., Castro, M.R., Bergert, E.R., Goellner, J.R., Heufelder, A.E., Morris, J.C., 2001. Expression of the sodium iodide symporter in human kidney. Kidney Int. 59, 1013–1023.

Stagnaro-Green, A., Abalovich, M., Alexander, E., Azizi, F., Mestman, J., Negro, R., Nixon, A., Pearce, E.N., Soldin, O.P., Sullivan, S., Wiersinga, W., The American Thyroid Association Taskforce on Thyroid Disease During Pregnancy and Postpartum, 2011. Guidelines of the American Thyroid Association for the diagnosis and management of thyroid disease during pregnancy and postpartum. Thyroid 21, 1081–1125.

Tazebay, U.H., Wapnir, I.L., Levy, O., Dohan, O., Zuckier, L.S., Zhao, Q.H., Deng, H.F., Amenta, P.S., Fineberg, S., Pestell, R.G., Carrasco, N., 2000. The mammary gland iodide transporter is expressed during lactation and in breast cancer. Nat. Med. 6, 871–878.

van den Hove, M.F., Croizet-Berger, K., Jouret, F., Guggino, S.E., Guggino, W.B., Devuyst, O., Courtoy, P.J., 2006. The loss of the chloride channel, ClC-5, delays apical iodide efflux and induces a euthyroid goiter in the mouse thyroid gland. Endocrinology 147, 1287–1296.

Van Nostrand, D., 2011. Sialoadenitis secondary to (1)(3)(1)I therapy for well-differentiated thyroid cancer. Oral Dis. 17, 154–161.

Vayre, L., Sabourin, J.C., Caillou, B., Ducreux, M., Schlumberger, M., Bidart, J.M., 1999. Immunohistochemical analysis of Na⁺/I⁻ symporter distribution in human extra-thyroidal tissues. Eur. J. Endocrinol. 141, 382–386.

Venturi, S., Venturi, M., 2009. Iodine in evolution of salivary glands and in oral health. Nutr. Health 20, 119–134.

Winkler, R., Griebenow, S., Wonisch, W., 2000. Effect of iodide on total antioxidant status of human serum. Cell Biochem. Funct. 18, 143–146.

Wolff, J., Chaikoff, I.L., 1948. Plasma inorganic iodide, a chemical regulator of normal thyroid function. Endocrinology 42, 468–471.

Wolff, J., 1964. Transport of iodide and other anions in the thyroid gland. Physiol. Rev. 44, 45–90.

Wolff, J., 1983. Congenital goiter with defective iodide transport. Endocr. Rev. 4, 240–254.

Wolff, J., 2005. What is the role of pendrin? Thyroid 15, 346–348.

Wright, E.M., 1974. Active transport of iodide and other anions across the choroid plexus. J. Physiol. 240, 535–566.

Yoshida, A., Taniguchi, S., Hisatome, I., Royaux, I.E., Green, E.D., Kohn, L.D., Suzuki, K., 2002. Pendrin is an iodide-specific apical porter responsible for iodide efflux from thyroid cells. J. Clin. Endocrinol. Metab. 87, 3356–3361.

Zhang, L., Sharma, S., Zhu, L.X., Kogai, T., Hershman, J.M., Brent, G.A., Dubinett, S.M., Huang, M., 2003. Nonradioactive iodide effectively induces apoptosis in genetically modified lung cancer cells. Cancer Res. 63, 5065–5072.

Zimmermann, M.B., 2009. Iodine deficiency. Endocr. Rev. 30, 376–408.

Zingg, H.H., Grazzini, E., Breton, C., Larcher, A., Rozen, F., Russo, C., Guillon, G., Mouillac, B., 1998. Genomic and non-genomic mechanisms of oxytocin receptor regulation. Adv. Exp. Med. Biol. 449, 287–295.

Chapter 13

Iodine and Adipocytokines: Cellular Aspects

Inés Velasco[1], Eduardo Garcia-Fuentes[2,3]

[1]Hospital Riotinto, Huelva, Spain; [2]Institute of Biomedical Research of Malaga (IBIMA), Regional University Hospital, Malaga, Spain; [3]CIBEROBN, Institute of Health Carlos III, Malaga, Spain

INTRODUCTION

Iodine is required for the synthesis of iodine-containing thyroid hormones in the thyroid gland. These hormones are important for correct neurodevelopment. An adequate intake of iodine is necessary for the correct function of the thyroid and normal levels of thyroid hormones. The urinary iodine (UI) concentration is the most widely used biochemical marker of iodine intake.

However, recent studies have demonstrated that iodine can influence metabolism in the mammary gland, adipose tissue, liver, and other tissues in ways unrelated to thyroid hormone (Poncin et al., 2008; Smyth, 2003; Venturi, 2001; Gutiérrez-Repiso et al., 2014). Adipose tissue is an organ involved in different metabolic processes, such as adipogenesis, angiogenesis, regulation of lipid and carbohydrate metabolism, and immune response. The correct function of this tissue is important for adequate regulation of body homeostasis. Adipocytes secrete adipocytokines, such as leptin and adiponectin, which have distinct endocrine functions.

It is known that iodide inhibits cyclic AMP (cAMP) signaling (by an unknown mechanism; Van Sande et al., 1975), which is involved in the regulation of leptin synthesis and secretion (Fried et al., 2000). Conversely, iodine seems to be related to insulin resistance and adiponectin. In some studies UI has been negatively associated with insulin resistance (Soriguer et al., 2011a). However, the relation with adiponectin is more complex, with positive, negative, or no association with UI being reported.

THE ROLE OF IODINE IN THE HUMAN BODY

Iodine is an essential mineral for the normal development of the body. Its biological importance is indicated by the fact that its main function is to participate in the synthesis of iodine-containing thyroid hormones in the thyroid gland. It is an essential component of the thyroid hormones, which play important roles in development and metabolism, especially of the brain. Iodine deficiency is the world's single greatest cause of preventable brain damage.

The effects of iodine deficiency are designated by the term *iodine deficiency disorders*, which refers to all of the effects of iodine deficiency in a population that can be prevented with an adequate iodine intake. Iodine deficiency in humans can cause several problems, such as spontaneous abortion, increased infant mortality, cretinism, goiter, neurologic dysfunction, and growth retardation. The criteria for iodine deficiency have been established by the World Health Organization (WHO), stating that the median UI concentration in a population should be greater than 100 μg/L, with less than 20% of the population excreting 50 μg/L (Tables 13.1 and 13.2). The estimation of worldwide iodine deficiency is based on UI data. UI concentration is the most widely used biochemical marker of iodine intake because most ingested iodine is excreted in the urine (>90%). Moreover, an equilibrium between dietary iodine intake and UI excretion exists. WHO, the United Nations Children's Emergency Fund (UNICEF), and the International Council for the Control of Iodine Deficiency Disorders (ICCIDD; WHO, 2007) recommend the following iodine intakes:

- *0–5 years of age:* 90 μg/day
- *6–12 years of age:* 120 μg/day
- *12 years of age or older:* 150 μg/day
- *pregnant and lactating women:* 250 μg/day

Molecular, Genetic, and Nutritional Aspects of Major and Trace Minerals. http://dx.doi.org/10.1016/B978-0-12-802168-2.00013-0

The UI concentration establishes whether a population is at risk of iodine deficiency and requires correction of its iodine levels. It is well known that UI excretion is highly influenced by changes in dietary iodine intake and changes in thyroid function, leading to wide interindividual variability. But changes in UI excretion over time were not influenced just by dietary iodine intake. UI excretion in a population is stable over time, as has been recently demonstrated (Gutiérrez-Repiso et al., 2015), suggesting that endogenous factors affect iodine metabolism. Such factors may include genes/proteins related to iodide metabolism (sodium-iodide symporter (NIS) and sodium multivitamin transporter; Dohan et al., 2003; Delmondes de Carvalho and Quick, 2011), thyroidal uptake (NIS; potassium voltage-gated channel subfamily Q member 1 (KCNQ1) and potassium voltage-gated channel subfamily E regulatory subunit 2 (KCNE2), subunits of a potassium channel; Dohan et al., 2003; Purtell et al., 2012), metabolism (dehalogenase-1, the iodotyrosine deiodinase or "dehalogenase"; Pendrin, the apical membrane chloride-iodide transporter; Moreno et al., 2008; Kim et al., 2009), and renal excretion (Pendrin; Kim et al., 2009).

WHO has recommended a different iodine intake depending on the age and status of the population, with an important emphasis in pregnant women and infants. Dietary iodine is found in the form of iodide or iodate. This iodine is ingested with the diet mainly through food, iodized salt, drinking water, and milk. The main sources of iodine intake seem to be associated with the increase in iodized salt intake and probably with the increased iodine concentration in dairy products (Soriguer et al., 2011a,b). Milk and dairy products are the most important source of iodine in some countries. Milk has high iodine content because of iodine given to dairy cattle in mineral supplements. Furthermore, iodine intake also depends on the level of fortification of iodized products. Many countries iodize household salt and/or salt used by the food industry.

When iodine intake is low, the production of thyroid hormones is reduced, with consequent alteration of thyroid gland function. However, iodide, the main substrate in the synthesis of thyroid hormones, can also exert direct physiological and signaling effects on different organs. Different studies demonstrate that iodine can exert some activity unrelated to thyroid hormone. Iodine is found in higher concentrations not only in the thyroid, but also in other tissues, such as salivary glands, lacrimal glands, stomach mucous membrane, choroid plexus, lactating mammary gland, pancreatic islets of Langerhans, and in the ciliary muscle of the eye. Its distribution is related to the action of NIS, an iodine transporter. An adequate iodine concentration in the mammary gland is important for the development of the newborn. Iodine concentration in breast milk is in a similar range to UI (50–200 µg/L).

TABLE 13.1 Criteria for Assessing Iodine Nutrition Based on Median Urinary Iodine Concentration of School-Age Children (≥6 Years) and Adults (WHO, 2007)[a]

Median Urinary Iodine (µg/L)	Type of Iodine Intake	Iodine Nutrition
<20	Insufficient	Severe iodine deficiency
20–49	Insufficient	Moderate iodine deficiency
50–99	Insufficient	Mild iodine deficiency
100–199	Adequate	Optimal
200–299	More than adequate	Slight risk of iodine-induced hyperthyroidism
>300	Excessive	Risk of adverse health consequences (iodine-induced hyperthyroidism, autoimmune thyroid diseases)

[a]Applies to adults but not to pregnant and lactating women.

TABLE 13.2 Criteria for Assessing Iodine Nutrition Based on Median Urinary Iodine Concentration of Pregnant Women (WHO, 2007)

Median Urinary Iodine (µg/L)	Type of Iodine Intake
<150	Insufficient
150–249	Adequate
250–499	Above requirements
≥500	Excessive

POSSIBLE MECHANISM OF IODINE ACTION

In addition to contributing to thyroid hormone synthesis, inorganic iodides may act as antioxidants or have different direct actions in different tissues (Venturi, 2001). These actions would require that iodide penetrates cells. It has been hypothesized that an inhibitory effect of iodide is mediated by an iodinated molecule (Van Sande et al., 1975; Fig. 13.1). There are different candidates for this iodinated molecule, such as I_2 (Nunez and Pommier, 1982), iodine and oxygen radical iodide (Denef et al., 1996), iodinated peptides (Lissitzky et al., 1961), or iodinated lipids (Pereira et al., 1990). Some iodinated lipids, such as aldehyde 2-iodohexadecanal, reproduce two known effects of iodide: inhibition of adenylyl cyclase (Panneels et al., 1994) and inhibition of H_2O_2 generation (Ohayon et al., 1994). It is known that iodide inhibits cAMP cascades, but the mechanism is unknown. Inhibition of cholera toxin and forskolin-activated cAMP signaling influences the G_s–adenylyl cyclase couple and cAMP generation (Cochaux et al., 1987; Heldin et al., 1985). The inhibition of cAMP formation was demonstrated with 10^{-5} M NaI (Van Sande et al., 1975). Because cAMP has been shown to play important roles in almost all aspects of biological functions, this inhibition of cAMP synthesis may produce different effects downstream of cAMP, at least in rats (Yamamoto et al., 1972) and mice (Bagchi et al., 1985).

Conversely, antioxidant properties of iodide have been demonstrated in mammary gland and in other tissues, such as liver, kidney, muscle, and adipose tissue (Poncin et al., 2008; Smyth, 2003; Venturi, 2001; Li et al., 2013). Iodide was shown to effectively scavenge reactive oxygen species in human blood cells (Küpper et al., 2008), decreasing damage by free oxygen radicals. Antioxidant effects of NaI at concentrations as low as 15 μM have been shown to be equivalent to the effects of the established antioxidant ascorbic acid at much higher concentrations (50 μM; Smyth, 2003). In addition, iodine intake also seems to be related to the activity of the enzymes glutathione peroxidase, superoxide dismutase, and catalase (Smyth, 2003; Soriguer et al., 2011c; Giray and Hincal, 2002; Maier et al., 2007). In addition, it is known that I_2 participates in peroxidase reactions. The antioxidative reaction of I^- can be directed against peroxides, especially lipid peroxides. Hence, the synthesized iodine can be reduced to iodide by the relatively alkaline environment of the body or it can be incorporated into organic compounds. The rate of synthesis of I_2 or hypoiodous acid (HOI) from iodide can be increased or decreased when iodide is increased (Gottardi, 1999). The oxidation of iodide is a prerequisite to the formation of bioactive molecules that contain iodine. Iodide oxidation is catalyzed by thyroid peroxidase and lactoperoxidase and depends upon iodide concentration. I_2 is formed and released by these enzymes at elevated iodide concentrations (Magnusson et al., 1984), which could influence some aspects of iodide regulation. However, I_2 and HOI cannot be synthesized independently in an aqueous environment. The synthesis of I_2 necessarily leads to the formation of HOI under physiological conditions, and this has potential consequences for iodination of biomolecules.

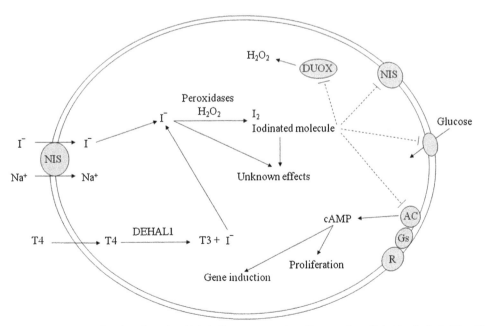

FIGURE 13.1 Possible mechanisms of action through which iodine can exert its effects at the cellular level. I^-, iodide; *NIS*, sodium/iodide symporter; *T4*, thyroxine; *T3*, triiodothyronine; *DEHAL1*, dehalogenase-1; *DUOX*, dual oxidase; *R*, receptor; G_s, stimulatory G protein of adenylyl cyclase; *AC*, adenylyl cyclase; *cAMP*, 3′,5′-cyclic AMP; – – – –⊣, inhibition; →, stimulation.

IODINE AND ADIPOCYTOKINES

Adipose tissue is involved in different metabolic processes, such as adipogenesis, angiogenesis, regulation of lipid and carbohydrate metabolism, and immune response. The correct function of this tissue is important for an adequate regulation of body homeostasis. Adipocytes secrete bioactive molecules, called adipocytokines, which include hormones, cytokines, and other proteins with functions not well studied. These molecules, such as leptin and adiponectin, have different endocrine paracrine and autocrine functions. The regulation of the secretion of these hormones is important because they are involved in the regulation of many pathophysiological processes, such as obesity and insulin resistance/type 2 diabetes mellitus (T2DM). Previous studies suggest that iodine could influence the biosynthesis of these adipocytokines, but there is little information on this possible relationship (Gutiérrez-Repiso et al., 2014; Herter-Aeberli et al., 2015).

Iodine and Leptin

Leptin is a 16-kDa protein that is synthesized mainly in adipose tissue. It is related to body fat mass. Leptin acts in the central nervous system, particularly in the hypothalamus, suppressing food intake and increasing energy expenditure. It is known that leptin levels are increased in obesity. Leptin synthesis is regulated by different mechanisms. For example, inhibition of leptin synthesis occurs in response to fasting, cAMP, and agonists of the β3-adrenergic receptor. Moreover, cAMP-mediated signaling pathways are important for maintaining metabolic homeostasis and have been implicated in the regulation of leptin production and secretion (Fried et al., 2000). In mammals, many cAMP functions are mediated by cAMP-dependent protein kinase A and exchange proteins directly activated by cAMP. In this sense iodine could be influence these processes because it inhibits cAMP signaling cascades. Leptin also stimulates thyroid-stimulating hormone (TSH) production by the hypothalamic–pituitary axis in rats (Ortiga-Carvalho et al., 2002). Moreover, TSH may stimulate leptin production by adipocytes (Menendez et al., 2003), suggesting crosstalk between these two hormones (Santini et al., 2014). A trend toward a reduction in leptin as TSH decreased with iodine treatment was also reported (Herter-Aeberli et al., 2015).

Iodine and Adiponectin

T2DM and thyroid disease are conditions that are globally widespread. Both disorders have been shown to influence each other, and associations between both diseases have been observed (Herter-Aeberli et al., 2015; Asvold et al., 2008). Indeed, the prevalence of thyroid dysfunction in patients with T2DM is higher than in the general population (Monzani et al., 2004). Thyroid hormones are important determinants of glucose homeostasis (Zimmermann et al., 2009), and they affect glucose metabolism via several mechanisms. Moreover, hyperthyroidism has been recognized to promote hyperglycemia (Klop et al., 2013). On the other hand, subclinical hypothyroidism can affect adipocytokine levels, probably because of increased weight gain and fat deposition, although evidence suggests that changes in thyroid hormones could be the consequence, rather than the cause, of weight gain (Soriguer et al., 2011a).

Although there are established relationships between thyroid hormones and T2DM, the role of iodine in T2DM has not been established. In a recent study, UI was noted to markedly decrease in T2DM patients (as compared with control subjects), and it correlated not only with cardiovascular risk factors, such as glucose and insulin levels, and insulin resistance (HOMA-IR), but also with biomarkers of obesity and inflammation (Al-Attas et al., 2012). Given these observations, a critical role for iodine in the development of insulin resistance in T2DM individuals has been suggested. This could be mediated by the inhibition of the pentose phosphate pathway by iodide (Corvilain et al., 1991). Other effects of iodide include the inhibition of glucose transport, presumably by reduction of the number of available glucose carriers, as documented in pig cells (Filetti et al., 1986). However, in other cross-sectional studies, no significant relations were found between UI content and glucose, insulin, or insulin resistance (Gutiérrez-Repiso et al., 2014; Skrypnyk, 2010). Thus further studies are needed to elucidate the functional and mechanistic roles of iodine in mediating insulin resistance. On the other hand, NIS expression and iodide uptake can be decreased by metformin treatment (a medication used in T2DM patients), at least in thyrocytes. This compound exerts its effect by modulation of AMP-activated protein kinase (AMPK), which in turn regulates the activation of the cAMP-response element (CRE) in the NIS promoter. This suggests that the use of AMPK-modulating compounds may be useful for the enhancement of iodide uptake by thyrocytes, with potential benefit in the treatment of thyroid cancer patients (Abdulrahman et al., 2014).

A common issue found in insulin resistance and T2DM is a low level of adiponectin. This is an adipocytokine recognized as a key regulator of insulin sensitivity and tissue inflammation. It is produced by adipose tissue and has direct actions in liver, skeletal muscle, and the vasculature, with functions related to enhancing hepatic insulin sensitivity, increasing fuel oxidation, and decreasing vascular inflammation. Although the serum adiponectin level was negatively associated with TSH levels in some studies, UI was not associated with serum adiponectin levels (Skrypnyk, 2010). However, the insulin resistance level

TABLE 13.3 Studies on the Possible Association Between Adiponectin and Urinary Iodine Levels

Study	Type of Association	Type of Study	Subjects
Al-Attas et al. (2012)	No association in serum	Cross-sectional	266 adults
Skrypnyk, 2010	No association in serum	Cross-sectional	68 adults
Gutierrez-Repiso et al. (2014)	No association in serum	Cross-sectional	508 adults
	Positive in serum	Intervention	30 adults
	Negative in breast milk	Cross-sectional	88 women
Lewinski et al. (2013)	Positive in serum	Intervention	23 adults

(HOMA-IR index) and UI were negatively associated (Skrypnyk, 2010). In this study the authors suggest that there is a connection between insulin resistance and iodine deficiency in the biosphere. However, the relationship between iodine and adiponectin is more complex (Table 13.3). When iodine is present at levels similar to serum iodine levels (normal iodine in plasma in euthyroid humans is 10^{-7} M with optimal daily iodine intakes of 100–200 µg; Fradkin and Wolff, 1983), adiponectin is not significantly associated with UI levels (Gutiérrez-Repiso et al., 2014; Skrypnyk, 2010). Similar results were found at baseline in a study performed in adult humans with different doses of iodide (100–300 µg/day, in the form of KI; Gutiérrez-Repiso et al., 2014). However, after 2 months of treatment, serum adiponectin and UI levels were positively associated after adjusting for glucose level. Moreover, low iodine doses can have a slight and chronic antioxidant effect, as was previously demonstrated (Soriguer et al., 2011c). This can indirectly produce other beneficial effects, as exemplified by the therapeutic effects of seaweed baths or iodine-rich solutions known as thalassotherapy, used historically to treat ocular diseases, thyroid disease, diabetes, cardiac and respiratory disease, and arteriosclerosis (Smyth, 2003; Moser et al., 1991).

In contrast, at higher iodine concentration, and similar to those found in breast milk (50–200 µg/L), iodine levels are negatively associated with adiponectin levels (Gutiérrez-Repiso et al., 2014). Moreover, when adipocytes are treated with iodine at these levels in vitro, a decrease of adiponectin synthesis is noted (Gutiérrez-Repiso et al., 2014). In addition, at this concentration iodine seems to have an antioxidant effect, producing an inhibition of the expression of different oxidative stress markers, such as glutathione peroxidase and superoxide dismutase 1 and 2 (Gutiérrez-Repiso et al., 2014). Previous in vitro studies further suggest that antioxidants can regulate adiponectin expression through a reduction in oxidative stress. This possible reduction of oxidative stress, or a direct effect of iodine, could be responsible for the lower adiponectin synthesis. However, data are inconclusive. Moreover, when iodine is administered at a very high dose, such as after radioiodine administration, a significant and sustained increase in serum adiponectin was noted (Lewinski et al., 2013).

Interference Between Iodine and Other Minerals

It is known that other minerals can compete with iodine in different aspects of iodine metabolism. Fluorine is much more chemically active than iodine and an excessive intake and accumulation of fluorine affects the adult brain and thyroid structure and their development, especially in the case of iodine deficiency (Wang et al., 2009). Moreover, a very high bromide intake produces a decrease of iodide accumulation in the mammary gland and an increase of urinary iodine. This could also lead to hypothyroxinemia (Pavelka, 2009). Iodine and selenium intake may affect the activity of deiodinases, which act as catalysts in the deiodination of thyroid hormones. Selenium is part of more than 20 selenoproteins, such as glutathione peroxidases, iodothyronine deiodinases, thioredoxin reductases, and selenophosphate synthetases (Moreno-Reyes, 2009). However, to date there is no information on the possible interference of these minerals on the effect of iodine on the synthesis and secretion of adipocytokines.

SUMMARY POINTS

- Iodine plays an important role in development and metabolism, especially of the brain.
- Iodine deficiency is the world's greatest single cause of preventable brain damage.
- The main sources of iodine intake are iodized salt and dairy products.
- Iodine has antioxidant/prooxidant effects in different tissues, depending on its concentration.
- Leptin, a protein involved in the regulation of food intake and energy expenditure, could be regulated by iodine through cAMP levels.
- Iodine at a dose similar to those found in serum or breast milk seems to have an antioxidant effect.
- Iodine produces a decrease of adiponectin synthesis at doses similar to those found in breast milk.

REFERENCES

Abdulrahman, R.M., Boon, M.R., Sips, H.C., Guigas, B., Rensen, P.C., Smit, J.W., Hovens, G.C., 2014. Impact of metformin and compound C on NIS expression and iodine uptake in vitro and in vivo: a role for CRE in AMPK modulation of thyroid function. Thyroid 24, 78–87.

Al-Attas, O.S., Al-Daghri, N.M., Alkharfy, K.M., Alokail, M.S., Al-Johani, N.J., Abd-Alrahman, S.H., Yakout, S.M., Draz, H.M., Sabico, S., 2012. Urinary iodine is associated with insulin resistance in subjects with diabetes mellitus type 2. Exp. Clin. Endocrinol. Diabetes 120, 618–622.

Asvold, B.O., Bjoro, T., Nilsen, T.I.L., Gunnell, D., Vatten, L.J., 2008. Thyrotropin levels and risk of fatal coronary heart disease. Arch. Intern. Med. 168, 855–860.

Bagchi, N., Shivers, B., Brown, T.R., 1985. Studies on the mechanism of acute inhibition of thyroglobulin hydrolysis by iodine. Acta Endocrinol. Copenh. 108, 511–517.

Cochaux, P., Van Sande, J., Swillens, S., Dumont, J.E., 1987. Iodide-induced inhibition of adenylate cyclase activity in horse and dog thyroid. Eur. J. Biochem. 170, 435–442.

Corvilain, B., van Sande, J., Laurent, E., Dumont, J.E., 1991. The H$_2$O$_2$-generating system modulates protein iodination and the activity of the pentose phosphate pathway in dog thyroid. Endocrinology 128, 779–785.

Delmondes de Carvalho, F., Quick, M., 2011. Surprising substrate versatility in SLC5A6. Na+-coupled transport by the human Na*/multivitamin transporter (hSMVT). J. Biol. Chem. 286, 131–137.

Denef, J.F., Many, M.C., van den Hove, M.F., 1996. Iodine-induced thyroid inhibition and cell necrosis: two consequences of the same free-radical mediated mechanism? Mol. Cell Endocrinol. 121, 101–103.

Dohan, O., De la Vieja, A., Paroder, V., Riedel, P., Artani, M., Reed, M., Ginter, C.S., Carrasco, N., 2003. The sodium/iodide Symporter (NIS): characterization, regulation and medical significance. Endocr. Rev. 24, 48–77.

Filetti, S., Vetri, M., Damante, G., Belfiore, A., 1986. Thyroid autoregulation: effect of iodine on glucose transport in cultured thyroid cells. Endocrinology 118, 1395–1400.

Fradkin, J.E., Wolff, J., 1983. Iodide-induced thyrotoxicosis. Medicine 62, 1–20.

Fried, S.K., Ricci, M.R., Russel, C.D., Laferrere, B., 2000. Regulation of leptin production in humans. J. Nutr. 130, 3127S–3131S.

Giray, B., Hincal, F., 2002. Oxidative DNA base damage, antioxidant enzyme activities and selenium status in highly iodine-deficient goitrous children. Free Radic. Res. 36, 55–62.

Gottardi, W., 1999. Iodine and disinfection: theoretical study on mode of action, efficiency, stability, and analytical aspects in the aqueous system. Arch. Pharm. Weinh. 332, 151–157.

Gutiérrez-Repiso, C., Velasco, I., Garcia-Escobar, E., Garcia-Serrano, S., Rodríguez-Pacheco, F., Linares, F., Ruiz de Adana, M.S., Rubio-Martin, E., Garrido-Sanchez, L., Cobos-Bravo, J.F., Priego-Puga, T., Rojo-Martinez, G., Soriguer, F., García-Fuentes, E., 2014. Does dietary iodine regulate oxidative stress and adiponectin levels in human breast milk? Antioxid. Redox Signal 20, 847–853.

Gutiérrez-Repiso, C., Colomo, N., Rojo-Martinez, G., Valdés, S., Tapia, M.J., Esteva, I., Ruiz de Adana, M.S., Rubio-Martin, E., Lago-Sampedro, A., Santiago, P., Velasco, I., Garcia-Fuentes, E., Moreno, J.C., Soriguer, F., 2015. Evolution of urinary iodine excretion over eleven years in an adult population. Clin. Nutr. 34, 712–718.

Heldin, N.E., Karlsson, F.A., Westermark, B., 1985. Inhibition of cyclic AMP formation by iodide in suspension cultures of porcine thyroid follicle cells. Mol. Cell Endocrinol. 41, 61–67.

Herter-Aeberli, I., Cherkaoui, M., El Ansari, N., Rohner, R., Stinca, S., Chabaa, L., von Eckardstein, A., Aboussad, A., Zimmermann, M.B., 2015. Iodine supplementation decreases hypercholesterolemia in iodine-deficient, overweight women: a randomized controlled trial. J. Nutr. 145, 2067–2075.

Kim, Y.H., Pham, T.D., Zheng, W., Hog, S., Baylis, C., Pech, V., Beierwaltes, W.H., Farley, D.B., Braverman, L.E., Verlander, J.W., Wall, S.M., 2009. Role of pendrin in iodide balance: going with the flow. Am. J. Physiol. Ren. Physiol. 297, 1069–1079.

Klop, B., Elte, J.W.F., Cabezas, M.C., 2013. Dyslipidemia in obesity: mechanisms and potential targets. Nutrients 5, 1218–1240.

Küpper, F.C., Carpenter, L.J., McFiggans, G.B., Palmer, C.J., Waite, T.J., Boneberg, E.M., Woitsch, S., Weiller, M., Abela, R., Grolimund, D., Potin, P., Butler, A., Luther 3rd, G.W., Kroneck, P.M., Meyer-Klaucke, W., Feiters, M.C., 2008. Iodide accumulation provides kelp with an inorganic antioxidant impacting atmospheric chemistry. Proc. Natl. Acad. Sci. U. S. A. 105, 6954–6958.

Lewinski, A., Brona, A., Lewandowski, K.C., Jedrzejuk, D., Bohdanowicz-Pawlak, A., Skowronska-Jozwiak, E., Bienkiewicz, M., Milewicz, A., 2013. Effects of radioiodine administration on serum concentrations of matrix metalloproteinases, adiponectin and thrombospondin-1. Thyroid. Res. 6, 9.

Li, Q., Mair, C., Schedle, K., Hellmayr, I., Windisch, W., 2013. Effects of varying dietary iodine supplementation levels as iodide or iodate on thyroid status as well as mRNA expression and enzyme activity of antioxidative enzymes in tissues of grower/finisher pigs. Eur. J. Nutr. 52, 161–168.

Lissitzky, S., Gregoire, J., Gregoire, J., Limozin, N., 1961. The presence and in vitro activity of free iodinated peptides in the thyroid gland of mammals and man. Gen. Comp. Endocrinol. 1, 519–533.

Magnusson, R.P., Taurog, A., Dorris, M.L., 1984. Mechanisms of thyroid peroxidase- and lactoperoxidase-catalyzed reactions involving iodide. J. Biol. Chem. 259, 13783–13790.

Maier, J., van Steeg, H., van Oostrom, C., Paschke, R., Weiss, R.E., Krohn, K., 2007. Iodine deficiency activates antioxidant genes and causes DNA damage in the thyroid gland of rats and mice. Biochim. Biophys. Acta 1773, 990–999.

Menendez, C., Baldelli, R., Camina, J.P., Escudero, B., Peino, R., Dieguez, C., Casanueva, F.F., 2003. TSH stimulates leptin secretion by a direct effect on adipocytes. J. Endocrinol. 176, 7–12.

Monzani, F., Caraccio, N., Kozakowa, M., Dardano, A., Vittone, F., Virdis, A., Taddei, S., Palombo, C., Ferrannini, E., 2004. Effect of levothyroxine replacement on lipid profile and intima-media thickness in subclinical hypothyroidism: a double-blind, placebo-controlled study. J. Clin. Endocrinol. Metab. 89, 2099–2106.

Moreno, J.C., Klootwijk, W., van Toor, H., Pinto, G., D'Alessandro, M., Lèger, A., Goudie, D., Polak, M., Grüters, A., Visser, T.J., 2008. Mutations in the iodotyrosine deiodinase gene and hypothyroidism. N. Engl. J. Med. 358, 1811–1818.

Moreno-Reyes, R., 2009. Iodine, selenium deficiency and Kashin–Beck disease. In: Preedy, V.R., Burrow, G.N., Watson, R. (Eds.), Comprehensive Handbook of Iodine. Nutritional, Biochemical, Pathological and Therapeutic Aspects. Academic Press, Burlington, pp. 687–700.

Moser, M., Buchberger, W., Mayer, H., Winkler, R., 1991. Influence of an iodine-drinking cure on the antioxidative status of diabetic patients. Wien. Klin. Wochenschr. 103, 183–186.

Nunez, J., Pommier, J., 1982. Formation of thyroid hormones. Vitam. Horm. 39, 175–229.

Ohayon, R., Boeynaems, J.M., Braekman, J.C., Van den Bergen, H., Gorin, Y., Virion, A., 1994. Inhibition of thyroid NADPH-oxidase by 2-iodohexadecanal in a cell-free system. Mol. Cell Endocrinol. 99, 133–141.

Ortiga-Carvalho, T.M., Oliveira, K.J., Soares, B.A., Pazos-Moura, C.C., 2002. The role of leptin in the regulation of TSH secretion in the fed state: in vivo and in vitro studies. J. Endocrinol. 174, 121–125.

Panneels, V., Van Sande, J., Van den Bergen, H., Jacoby, C., Braekman, J.C., Dumont, J.E., Boeynaems, J.M., 1994. Inhibition of human thyroid adenylyl cyclase by 2-iodoaldehydes. Mol. Cell Endocrinol. 106, 41–50.

Pavelka, S., 2009. Bromide interference with iodine metabolism: goitrogenic and whole-body effects of excessive inorganic bromide in the rat. In: Preedy, V.R., Burrow, G.N., Watson, R. (Eds.), Comprehensive Handbook of Iodine. Nutritional, Biochemical, Pathological and Therapeutic Aspects. Academic Press, Burlington, pp. 587–595.

Pereira, A., Braekman, J.C., Dumont, J.E., Boeynaems, J.M., 1990. Identification of a major iodolipid from the horse thyroid gland as 2-iodohexadecanal. J. Biol. Chem. 265, 17018–17025.

Poncin, S., Gérard, A.C., Boucquey, M., Senou, M., Calderon, P.B., Knoops, B., Lengelé, B., Many, M.C., Colin, I.M., 2008. Oxidative stress in the thyroid gland: from harmlessness to hazard depending on the iodine content. Endocrinology 149, 424–433.

Purtell, K., Paroder-Belenitsky, M., Reyna-Neyra, A., Nicola, J.P., Koba, W., Fine, E., Carrasco, N., Abbot, G.W., 2012. The KCNQ1-KCNE2 K+ channel is required for adequate thyroid I-uptake. FASEB J. 26, 3252–3259.

Santini, F., Marzullo, P., Rotondi, M., Ceccarini, G., Pagano, L., Ippolito, S., Chiovato, L., Biondi, B., 2014. Mechanisms in endocrinology: the crosstalk between thyroid gland and adipose tissue: signal integration in health and disease. Eur. J. Endocrinol. 171, R137–R152.

Skrypnyk, N.V., 2010. Correlation between adipocytokines with iodine deficiency criteria in patients with metabolic syndrome and type 2 diabetes mellitus in Carpathians region. Lik. Sprava 1–2, 70–77.

Smyth, P.P., 2003. Role of iodine in antioxidant defence in thyroid and breast disease. Biofactors 19, 121–130.

Soriguer, F., Valdes, S., Morcillo, S., Esteva, I., Almaraz, M.C., de Adana, M.S., Tapia, M.J., Dominguez, M., Gutierrez-Repiso, C., Rubio-Martin, E., Garrido-Sanchez, L., Perez, V., Garriga, M.J., Rojo-Martinez, G., Garcia-Fuentes, E., 2011a. Thyroid hormone levels predict the change in body weight: a prospective study. Eur. J. Clin. Invest. 41, 1202–1209.

Soriguer, F., Gutierrez-Repiso, C., Gonzalez-Romero, S., Olveira, G., Garriga, M.J., Velasco, I., Santiago, P., de Escobar, G.M., Garcia-Fuentes, E., Iodine Deficiency Disorders Group of Spanish Society of Endocrinology and Nutrition, 2011b. Iodine concentration in cow's milk and its relation with urinary iodine concentrations in the population. Clin. Nutr. 30, 44–48.

Soriguer, F., Gutiérrez-Repiso, C., Rubio-Martin, E., Linares, F., Cardona, I., López-Ojeda, J., Pacheco, M., González-Romero, S., Garriga, M.J., Velasco, I., Santiago, P., García-Fuentes, E., 2011c. Iodine intakes of 100–300 µg/d do not modify thyroid function and have modest anti-inflammatory effects. Br. J. Nutr. 25, 1–8.

Van Sande, J., Grenier, C., Willems, C., Dumont, J.E., 1975. Inhibition by iodide of the thyroid cyclic AMP-system. Endocrinology 96, 781–786.

Venturi, S., 2001. Is there a role for iodine in breast diseases? Breast 10, 379–382.

Wang, J., Ge, Y., Ning, H., Niu, R., 2009. Dna damage in brain and thyroid gland cells due to high fluoride and low iodine. In: Preedy, V.R., Burrow, G.N., Watson, R. (Eds.), Comprehensive Handbook of Iodine. Nutritional, Biochemical, Pathological and Therapeutic Aspects. Academic Press, Burlington, pp. 643–649.

WHO, 2007. Assessment of iodine deficiency disorders and monitoring their elimination. In: A Guide for Programme Managers, third ed. WHO Press, Geneva.

Yamamoto, K., Onaya, T., Yamada, T., Kotani, M., 1972. Inhibitory effect of excess iodide on thyroid hormone release as measured by intracellular colloid droplets. Endocrinology 90, 986–991.

Zimmermann, M.B., Aeberli, I., Melse-Boonstra, A., Grimci, L., Bridson, J., Chaouki, N., Mbhenyane, X., Jooste, P.L., 2009. Iodine treatment in children with subclinical hypothyroidism due to chronic iodine deficiency decreases thyrotropin and C-peptide concentrations and improves the lipid profile. Thyroid 19, 1099–1104.

Part IV

Iron

Chapter 14

Iron: Basic Nutritional Aspects

Marianne Wessling-Resnick

Harvard T.H. Chan School of Public Health, Boston, MA, United States

INTRODUCTION

Iron is one of the major elements comprising our planet. This transition metal exists in several different oxidation states with Fe^{2+} (ferrous) and Fe^{3+} (ferric) being the most common. The ability of iron to move between oxidation states underlies its biological importance in different catalytic functions. With few exceptions, iron is necessary for all living organisms and it plays a particularly critical role in human health. It is an essential nutrient, supporting oxygen binding and transport (e.g., hemoglobin), electron transport (e.g., heme-containing enzymes like cytochromes), and oxidative metabolism (e.g., iron-sulfur containing enzymes like NADH dehydrogenase). It is necessary for DNA synthesis and cellular proliferation. Due to these key roles, deficiency of iron is manifested as anemia, leading to reduced work capacity, impaired mental function and lowered immunity. Fortunately, iron supplements can help to resolve iron-deficiency anemia in patients, while food fortification programs can prevent health problems at the population level. However, iron also plays an essential physiological role as a micronutrient; inherited and acquired human disorders that perturb iron homeostasis can cause conditions of iron overload. Iron loading due to primary or secondary hemochromatosis can produce liver damage, leading to fibrosis, cirrhosis, and increased risk of hepatic cancer. Iron overload also may result in endocrinopathies and cardiac dysfunction. Thus, the balance between iron's role as an essential mineral and a powerful toxicant must be achieved by nutritional homeostasis. Whether one considers this element to be friend or foe, iron homeostasis represents a unique challenge to human metabolism. This chapter covers the basic nutritional aspects of iron and gives a broad overview of its regulation.

PHYSIOLOGICAL FUNCTIONS OF IRON

Iron and Oxygen

The intimate relationship between iron and oxygen forms the basis of human life. Iron-containing heme serves as a cofactor for the proteins hemoglobin and myoglobin. Both provide essential oxygen-carrying roles in our body. Red blood cell hemoglobin harbors about two-thirds of the total body iron content, helping to transport oxygen from the lungs to the rest of the body. Myoglobin provides short-term storage of oxygen in muscle to provide a supply at times of high demand. Common tests for iron deficiency involve measuring hemoglobin as well as hematocrit—the percentage of red blood cells circulating in blood by volume. The physical symptoms of iron deficiency include fatigue and weakness, directly reflecting the essential nature of its function in respiration. The body responds to iron deficiency by increasing oxygen delivery from hemoglobin, redistributing blood flow to protect the brain and heart, and enhancing cardiac output (Camaschella, 2015). Iron absorption from the diet also becomes enhanced. Signs of iron deficiency are summarized in Table 14.1.

Interestingly, iron is also important to sense hypoxia, a condition arising due to inadequate oxygen supply (Hentze et al., 2010). Hypoxia can be induced by high altitudes or by lung disease, and our body responds by increasing the production of red blood cells (erythropoiesis), blood vessel growth (angiogenesis), and anaerobic metabolism (glycolysis). Hypoxia-inducible factors (HIFs) bind to a specific element in genes involved in these responses. An iron-dependent enzyme called prolyl hydroxylase regulates HIF1alpha (HIF1α). Under normoxic conditions, oxygen together with iron, ascorbate, and 2-oxoglutarate support the modification of HIF1α with hydroxyl groups by prolyl hydroxylases. Hydroxylated HIF1α is then degraded and cannot induce gene expression. Under hypoxic conditions, when oxygen levels are low, hydroxylation is blocked and HIF1α is stabilized to activate genes involved in compensating responses (Fig. 14.1).

Reciprocal effects of oxygen are known to regulate iron levels. Iron-regulatory proteins (IRPs) bind to iron-responsive elements (IREs) in a number of mRNAs encoding proteins involved in maintaining iron balance. Table 14.2 lists several of these factors, and whether the IRE is placed in the 5′ or 3′ untranslated regions of their transcripts. When the IRE is present in the 5′ region, translation of the message is blocked by IRP binding to reduce protein levels. If the IRE is at the 3′ end, degradation of

TABLE 14.1 Signs and Indicators of Iron Deficiency Conditions

Iron Deficiency Symptoms	
Fatigue, weakness (tired)	
Decreased work capacity (poor endurance)	
Delays in cognitive and social development (children)	
Impaired motor development (children)	
Inability to maintain body temperature (cold)	
Poor mental concentration (decreased performance)	
Diminished immune function (infections)	
Iron Parameters	
Iron depletion	Reduced serum iron
	Reduced transferrin iron saturation
	Low hepcidin
Iron deficiency	Low serum iron
	Low transferrin iron saturation
	Low hepcidin
Iron-deficiency anemia	Low serum iron
	Low transferrin saturation
	Reduced ferritin
	Reduced hemoglobin
	Reduced mean corpuscular volume
	Hypochromic red cells
	Low hepcidin
Anemia of chronic disease (inflammation)	Low serum iron
	Low-to-normal transferrin saturation
	High ferritin
	Low hemoglobin
	High hepcidin

the mRNA is prevented by IRP binding so that message stability (and therefore protein level) is enhanced. Since many of the IRE-containing factors play direct roles in iron balance, this feedback mechanism helps to maintain homeostasis in a manner such that "iron controls iron" (Kuhn, 2015). The two IRPs that interact with these mRNAs are known to have slightly different activities. IRP1 contains an iron-sulfur cluster (4Fe-4S), which is best assembled under low-oxygen conditions. However, when iron is low, cluster assembly is impaired. Under these conditions, IRP1 acts as an RNA-binding protein and binds to IREs. IRP2, on the other hand, lacks the iron-sulfur cluster but is degraded under high-iron conditions in an oxygen-dependent manner. Dioxygen binds to a factor called FBXL5, stabilizing this protein allowing it to bind to IRP2, promoting its degradation. Thus, when oxygen is low, IRP2 may be more active than IRP1 even under low-iron conditions, since the latter's iron-sulfur cluster may assemble and block RNA-binding activity. Conversely, under high-oxygen conditions, cluster assembly may be impaired to activate IRP1 more readily while IRP2 is degraded. This process is depicted in Fig. 14.2.

These interwoven relationships are perhaps best revealed by the presence of an unusual IRE in the 5′ untranslated region of HIF2α mRNA. This element confers regulation of HIF2α translation through IRP1. In the kidney, HIF2α regulates gene expression of erythropoietin (EPO), a secreted factor that helps to promote erythropoiesis (production of red blood cells) in response to hypoxia. Under hypoxia, repression of HIF2alpha translation by IRP1 is reversed by enhanced iron-cluster assembly, which reduces the regulator's RNA-binding activity to relieve the block on HIF2α translation. Increased levels

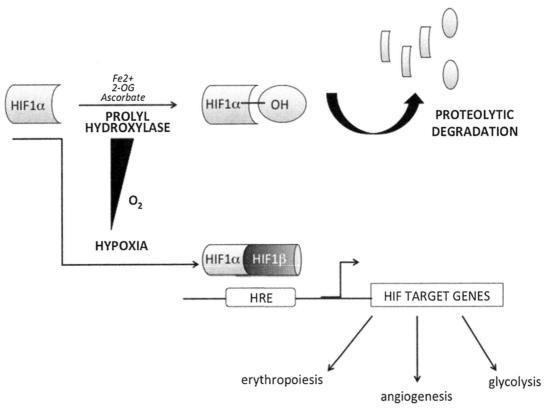

FIGURE 14.1 Iron and hypoxia. Hypoxia-inducible factors (HIFs) are transcription factors regulated by iron. In the presence of oxygen, with iron, 2-oxoglutarate, and ascorbate, prolyl hydroxylase modifies HIF1α and it becomes targeted for degradation. Under low-iron/hypoxic conditions, HIF1α is stable and binds to HIF1β to interact with HIF-response elements (HRE) to induce gene expression of factors involved in the response to hypoxia.

TABLE 14.2 Metabolic Factors That Contain Iron-Responsive Elements

Iron Metabolism Factors	Iron-Responsive Elements	Response to Iron Deficiency
Human ferritin (H-chain)	5'UTR	Reduce iron storage
Human ferritin (L-chain)	5'UTR	Reduce iron storage
Human erythrocyte aminolevulinic acid synthase	5'UTR	Increase heme synthesis
Human aconitase (mitochondrial)	5'UTR	Increase energy production (citric acid cycle)
Drosophila succinate dehydrogenase (subunit B)	5'UTR	Increase energy production (citric acid cycle)
Human HIF2α	5' UTR	Increase expression of hypoxia-inducible genes
Human ferroportin (splice variant)	5' UTR	Increase iron export
Human transferrin receptor (5 IREs)	3'UTR	Increase iron delivery
Human divalent metal transporter-1 (splice variant)	3'UTR	Increase iron uptake

of HIF2α then can help to upregulate EPO synthesis. However, under low-iron conditions, iron-cluster synthesis would be impaired to attenuate this response. Thus, oxygen-sensing is coordinated with iron-sensing to limit induction of red blood cell production by EPO when iron supplies are insufficient (Kuhn, 2015; Hentze et al., 2010).

Iron and Energy Metabolism

Many iron proteins are essential for oxygen consumption and metabolism. Among these key enzymes are heme proteins including the cytochromes involved in the mitochondrial electron transport chain. These factors are necessary for the generation of ATP, serving as electron carriers along this pathway. Nonheme iron-containing proteins like NADH dehydrogenase

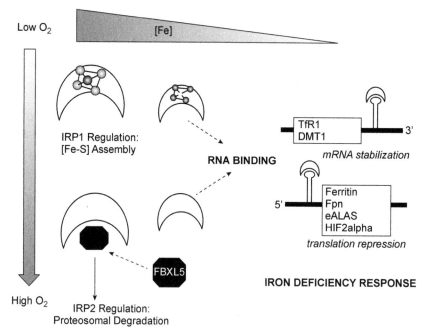

FIGURE 14.2 Iron-responsive proteins. Two iron-responsive proteins (IRPs) exert posttranscriptional control over iron metabolic factors listed in Table 14.2. IRP1 is an iron-sulfur cluster protein. Under high-iron and low-oxygen conditions, the cluster assembles to block RNA binding. With iron deficiency and subsequent cluster disassembly, IRP1 becomes more active as an RNA-binding protein and will bind either the 5' or 3' untranslated regions of mRNAs containing iron-response elements (IREs) to block translation or to stabilize the message, respectively. IRP2 is regulated by FBXL5, which associates with the factor under high O$_2$ conditions to induce proteosomal degradation. Under normoxia, IRP2 binds RNA similar to IRP1. Thus, both iron and oxygen contribute to the iron deficiency response by differentially regulating RNA binding activities of IRP1 and IRP2.

and succinate dehydrogenase also function in energy metabolism. Such iron-sulfur cluster-containing proteins are involved in oxidation/reduction reactions. Both mitochondrial complex I and II, which support oxidative phosphorylation, contain Fe-S clusters. Inadequate iron supply clearly can impair the pathways involved in energy metabolism by disabling the enzymes involved (Fig. 14.3). In fact, iron is required for thyroid peroxidase, a key synthetic enzyme necessary to produce thyroid hormones that regulate energy metabolism, growth, development, and body temperature (Table 14.1). Poor control of body temperature under cold conditions often accompanies iron deficiency due to its primary role in energy metabolism and its regulation (Camaschella, 2015). As iron-deficiency anemia evolves, work performance becomes impaired with diminished oxidative capacity of muscle, reduced oxidative energy production, and in more severe cases, acidosis with increased lactic acid production. Iron deficiency has been causally linked to reduced aerobic capacity and physical work (Haas and Brownlie, 2001).

Generally, the amount of iron in the diet is more or less related to energy intake, since iron is ubiquitously present in foods (Zimmermann and Hurrell, 2007). However, it is important to understand that iron deficiency can arise when requirements for this micronutrient simply outpace energy demands. Iron-deficiency anemia is a particularly significant problem in young children and women. Infancy and early childhood represent periods where rapid growth places a particularly high demand on the supply of iron to fulfill requirements for development. In adolescents, continued rapid growth during puberty with the expansion of muscle mass and additional blood loss due to menstruation at puberty continues the high-iron demand. Finally, during pregnancy the increased blood mass and iron necessary to support fetal growth increases needs for the micronutrient. Lack of sufficient iron is associated with pregnancy complications and increased incidences of pre-term delivery and low birth weights, along with the risk of maternal and child mortality (Zimmermann and Hurrell, 2007; Camaschella, 2015 #6).

Iron and the Brain

Among all tissues, the brain is perhaps the most critical for the full experience of human life. Iron is necessary to support the function of the human brain but the precise mechanism(s) underlying the requirements for this micronutrient in intellectual development are rather poorly understood. Roles for iron are clearly complex and multifactorial. Motor, cognitive, and behavioral development are affected by low-iron status (Prado and Dewey, 2014). Auditory and optic nerve impulses

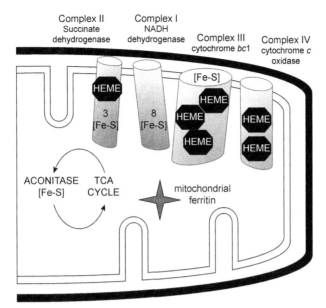

FIGURE 14.3 **Mitochondrial iron.** The electron transport chain relies on many iron-sulfur (Fe-S) cluster proteins and hemoproteins (cytochromes) to function. Complex I (NADH dehydrogenase) contains eight Fe-S clusters while complex II (succinate dehydrogenase) contains three Fe-S clusters and a heme moiety. Complex II also participates in the tricarboxylic acid cycle (TCA) which requires the activity of mitochondrial aconitase, another Fe-S cluster protein. Cytochrome $bc1$ (complex II) has several heme groups plus an Fe-S cluster used for its function and cytochrome c oxidase (complex IV) also has two heme moieties. Mitochondrial ferritin can store iron in the matrix, but it is unique from cytosolic or serum ferritin and does not appear to regulated by IRPs.

are slower in iron-deficient infants, pointing to poor myelination, a process which is known to require iron. These impairments may reduce the stimulation a child receives from their environment, a necessary component for proper development. In fact, anemic children are known to explore their environment to a lesser extent, and this lack of activity may lead to developmental delays.

Animal studies have shown that early life iron deficiency does not alter the overall brain size but does significantly reduce the size of the hippocampus, a region of the brain involved in learning and memory (Beard, 2008). Moreover, it has been shown that brain iron is necessary for dendritic branching in the hippocampus. These effects are persistent even after iron repletion. Animal studies also show that neurotransmitters are disrupted by iron deficiency, and in particular, dopamine metabolism is significantly impaired. In fact, tyrosine hydroxylase, which is required for synthesis of the neurotransmitter dopamine, is another iron-requiring enzyme. Some longitudinal studies suggest that cognitive effects resulting from iron deficiency in early childhood may not be reversed by later iron repletion (Lozoff et al., 2006). For these reasons, and those discussed earlier, it is important to identify and correct iron-deficiency anemia in children and to encourage the use of micronutrient supplements during pregnancy.

Much evidence also points to the role of iron in emotional and mental health during adult life (Kim and Wessling-Resnick, 2014). Serum iron levels are reduced in schizophrenics and individuals suffering from depression have lower hematocrit and serum transferrin values. The relationship between iron and emotional behavior is determined by many factors discussed earlier, including effects on neurotransmitters, brain iron levels in different regions, nutritional status at different periods of time, etc. The sum of data suggests that people with mood symptoms may require greater supplementation than healthy individuals (Lomagno et al., 2014).

Iron and Immunity

Iron plays important roles during inflammation and the immune response to infection. Since invading pathogens also require iron for survival, the human body undergoes "iron withdrawal" as part of the innate immune response. Iron absorption is reduced and the recycling of iron after erythrophagocytosis of red blood cells is circumvented as a means to reduce circulating levels of the metal. Thus, during the acute-phase response, serum iron levels rapidly decline while levels of the iron storage protein ferritin increase (Wessling-Resnick, 2010).

Many significant immune functions are carried out by heme-containing proteins, including myeloperoxidase, a neutrophil enzyme that synthesizes reactive oxygen species (ROS) to kill invading pathogens. Other heme-containing enzymes, such as catalase and peroxidase, protect us against damaging ROS. Importantly, iron is required for ribonucleotide reductase, a rate-limiting enzyme in the synthesis of nucleic acids. Thus, iron is necessary for the proliferation of T-lymphocytes that defend our body against viruses and other pathogens. For all of these reasons, it is critical to appropriately balance iron nutriture with the response to infection. It has been shown that high iron exerts a strong negative influence on infectious diseases like malaria (Sazawal et al., 2006). On the other hand, iron deficiency is associated with poor outcomes in other infections. A consensus view is that low-iron status can be protective while high iron exacerbates infection. Whether or not iron supplements should be used during inflammation and infection, when the body undergoes iron restriction, is debatable (Drakesmith and Prentice, 2012).

Other Functions of Iron

Iron deficiency not only affects the brain during fetal development but also development of the kidney, liver, and placenta (Rao and Georgieff, 2007). Due to the fetal origins of adult diseases, lack of sufficient iron can lead to complications later in life. For example, intrauterine growth restriction impairs nephrogenesis, leading to adult hypertension (Mackenzie and Brenner, 1995). Restless legs syndrome is associated with low iron in certain brain regions, and patients often benefit from iron supplementation (Allen and Earley, 2007). Iron plays a primary role in maintaining health across the lifespan.

ADVERSE CONSEQUENCES OF TOO MUCH IRON

Inherited and Acquired Iron-loading Diseases

As harmful as iron deficiency might be, too much iron has adverse impacts on human health as well. While iron loading can be acquired, it is more typically associated with inherited disease. Hereditary hemochromatosis (iron overload) arises due to defects in the *HFE*, *TfR2*, *HJV*, *HAMP*, and *FPN* genes. In addition, mutations in the DMT-1 and ferritin heavy chain genes also result in iron overload. All of these factors play important roles in the maintenance of iron homeostasis, and loss of their activities can have major consequences on the amount of iron taken up by the human body. Unlike most other micronutrient metals, levels of iron appear to be regulated at the point of absorption since strong evidence for regulatory mechanisms to excrete iron is lacking. The large store of body iron present in red blood cells will be reutilized after it is scavenged by the reticuloendothelial system, which is responsible for the erythrophagocytosis and release of iron from senescent red cells. The largest demand for iron is driven by erythropoiesis, due to the need for iron in the production of heme. Given that very little iron is lost, surfeit iron is stored in the storage protein ferritin, largely in the liver. While the ferritin storage pool can expand, excess loading does occur and creates health problems in these inherited disorders. Greater than 80% of hereditary hemochromatosis cases in people of northern European descent are due to homozygosity in the C282Y allele of *HFE*, with 5–14% of the population affected (Distante et al., 2004). However, despite the prevalence of this allele in the northern European population, many carriers do not display clinical symptoms, suggesting that other factors can influence the degree of iron loading and its associated problems, including the diet.

Acquired or secondary hemochromatosis also occurs. Patients that are dependent on blood transfusions, including those with hemolytic, sideroblastic, and thalassemic anemia, often suffer from iron loading. Patients with chronic liver disease experience iron overload, regardless of whether the disorder arises due to alcoholism, viruses, or metabolic dysfunction. As the primary storage site for excess iron, the liver has a great capacity to hold and then release iron at times of high demand. However, when this capacity is exceeded, hepatoxicity can occur due to the ability of iron to promote the formation of ROS. Iron's ability to undergo redox cycling creates so-called Fenton chemistry that produces the most damaging free hydroxyl radical. A number of iron-requiring enzymes contribute to the generation and control of oxygen free radicals (Fig. 14.4).

ROS can induce lipid peroxidation, DNA damage, and cell death. Ultimately, iron-catalyzed ROS results in fibrosis, cirrhosis, and increased risk of hepatocellular carcinoma. Iron's toxicity due to oxidative stress also causes problems with the cardiac and endocrine systems, which have more mitochondria and may take up free iron in the serum by tissue-specific iron transporters. Heart failure and endocrinopathies like diabetes are more common in patients with primary and secondary hemochromatosis. To improve outcomes associated with iron-loading disorders, these patients are commonly subject to iron depletion by phlebotomy or treated with iron chelators to reduce the overload (Pietrangelo, 2004).

Iron Toxicity and Children

Certain inherited diseases affect iron accumulation in children. Many of these disorders impair mitochrondrial function (Mena et al., 2015). Friedreich's Ataxia, a genetic disorder disrupting the function of the mitochondrial frataxin protein,

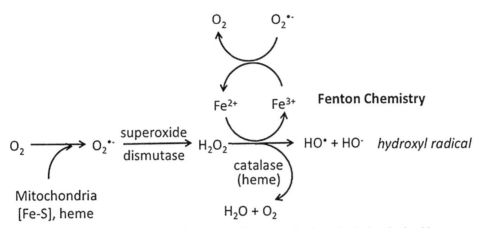

FIGURE 14.4 Reactive oxygen species and iron. Iron-sulfur cluster and heme-associated proteins in the mitochondria generate superoxide radicals during respiration. Superoxide dismutase produces hydrogen peroxide, which is reduced to water by catalase, another heme-containing enzyme. Fenton chemistry resulting from the presence of ferrous iron can produce the most damaging free radical, the hydroxyl radical. Oxidized ferric iron may then interact with oxygen radicals to cycle back to the ferrous form.

results in neuronal impairment and eventual cardiomyopathy and heart failure at an early age. Mitochondrial ferritin is implicated in sideroblastic anemia, which develops when iron is present but unable to become incorporated into heme. Several inherited forms of sideroblastic anemia have been described, including an X-linked disorder due to mutation in the transporter ABCB7, which is involved in Fe-S cluster assembly (Fleming, 2011). In general, due to the key role of mitochondria in both heme and Fe-S cluster biosynthesis, mitochondrial dysfunction can impair iron homeostasis. These problems usually appear in early stages of life.

An important note about iron toxicity and children is that the most common cause of fatal poisoning under the age of six is accidental overdose of iron. Acute poisoning symptoms may subside within a day but serious organ damage may arise later on in the tissues discussed earlier (liver, heart, and the central nervous system). The tolerable upper intake level for iron intake in children is 40 mg/day and 45 mg/day in adults.

Iron and Aging

While older adults are susceptible to anemia (Goodnough and Schrier, 2014), increased brain iron over time contributes to neurodegenerative diseases. Individuals with *HFE*-associated hemochromatosis, for example, are at increased risk for Alzheimer's disease (Ali-Rahmani et al., 2014). Iron chelation is one avenue that has been explored to treat both Alzheimer's and Parkinson's disease (Ward et al., 2015). Parkinson's disease is associated with high-iron accumulation in the substantia nigra, where dopamine metabolism is affected. Neurodegeneration with brain iron accumulation is also characterized by iron accumulation in the basal ganglia (Xu et al., 2013). Thus, specific brain regions appear to be affected. Other diseases implicated to involve metal accumulation in aging are cerebral hemorrhage, amotrophic lateral sclerosis, Huntington's disease, multiple sclerosis, pantothenate kinase type 2 deficiency, aceruloplasminemia, HIV-1 encephalitis, and neuroferritinopathy.

The anemia of aging is also a concern in the elderly (Guralnik et al., 2004). Aging is associated with progressive decline in kidney filtration, which plays an important role in clearance of the iron-regulatory hormone hepcidin. Kidney dysfunction can reduce EPO synthesis. With age, there is also a progressive loss in the intestines' ability to absorb nutrients including iron. However, caution must be employed with the use of iron supplements to treat anemia in older people. It is important to determine the cause of anemia, which may arise due to inadequate absorption of other micronutrients like folate or vitamin B12. Nondietary environmental factors also can influence the balance of iron, including heat stress and "inflammaging" stressors (Bloomer et al., 2008). In general, formulations of over-the-counter multivitamin mineral supplements for adults over the age of 50 do not contain iron.

Other Toxicities Associated With Iron

Although many animal studies support the idea that oxidative stress caused by iron is associated with atherosclerosis and heart disease, human epidemiological studies have been equivocal. In general, dietary heme iron rather than total iron intake is associated with increased risk of coronary heart disease and heart attack (van der et al., 2005). Since absorption of nonheme iron is suppressed by high-iron status, heme iron from animal sources may play a greater role in cardiovascular disease (Danesh and Appleby, 1999). Likewise, dietary heme iron is associated with increased risk of colorectal cancer (Kato et al., 1999), but this effect may be associated with carcinogens that arise when meat is cooked. It may be that the increased presence of iron in the colon itself may produce ROS to cause disease. Finally, type 2 diabetes and metabolic syndrome are associated with increased iron levels (Ford and Cogswell, 1999). A very simple recommendation for people who may be concerned about toxicities arising from excess iron is to donate blood—in fact, frequent blood donors represent a population with reduced risk of myocardial infarction (Salonen et al., 1998).

FINDING THE RIGHT BALANCE

Dietary Iron Absorption

Since iron homeostasis is regulated to limit body iron levels, the mechanisms of intestinal absorption must control iron balance. Heme iron is more effectively absorbed than nonheme iron, and these distinct sources of iron appear to be taken up by independent pathways. While heme iron sources represent only 10% of total iron in the typical diet, it provides the majority of absorbed iron since its bioavailability is 4–8 times higher than nonheme iron. Unfortunately, heme iron assimilation by the intestine is poorly characterized. One proposed membrane transporter, heme carrier protein 1 (HCP1), was subsequently found to be a folate carrier although it might have minimal low-affinity heme transport capacity (Qiu et al., 2006). Moreover, only one heme exporter has been identified but it is not clear whether or not this factor is expressed in the intestine (Quigley et al., 2004). This has led to speculations that intact heme may be transcytosed directly across the enterocyte in a targeted fashion via vesicular trafficking, much like vitamin B12, or that perhaps intact heme is exported into lymph after absorption from the diet (Korolnek and Hamza, 2014).

Uptake of dietary nonheme iron is much better understood. Nonheme iron is absorbed in the upper duodenum where it crosses the luminal surface of enterocytes with the help of the transmembrane protein divalent metal transporter-1 (DMT-1) (Gulec et al., 2014). This is a proton-coupled transporter than mediates the import of several different metals; as a consequence, iron uptake can be affected by the competing presence of other metals. DMT-1 specifically recognizes the ferrous form of iron, thus its activity is promoted by exogenous factors that help to reduce dietary nonheme iron like vitamin C. Additional metal transporters, for example members of the Zip family (Jenkitkasemwong et al., 2012), may also provide mechanisms of iron uptake across the intestine.

Because ferric iron is the predominant form of nonheme iron in our diet, reduction to ferrous iron is an important first step in assimilation. This can be facilitated by an endogenous ferrireductase called DcytB (duodenal cytochrome B). This heme-containing enzyme is highly expressed at the apical surface of duodenal enterocytes and levels are coordinately regulated by iron, hypoxia, erythropoietic activity, and increased demand for iron (Lane et al., 2015). While knockout mice lacking DcytB do not show signs of anemia or iron deficiency (Gunshin et al., 2005), this may be due to the compensating presence of ascorbate and other ferrireductants. Its importance in human iron metabolism is supported by a known polymorphism in the *DCYTB* gene promoter that is a genetic modifier for *HFE*-associated hemochromatosis (Constantine et al., 2009).

After iron is imported across the apical surface, it can be either used metabolically (e.g., mitochondrial heme synthesis) or stored (e.g., deposited in intestinal ferritin). To enter circulation, iron must be exported across the basolateral surface and reoxidized to the ferric form to bind transferrin for delivery to peripheral tissues. These functions are carried out by the membrane transporter ferroportin (Fpn) and the ferroxidase hephaestin, respectively (Gulec et al., 2014). To date, Fpn is the only known iron export protein and it is highly expressed in the intestine, liver, and macrophages. The intestine also expresses hephaestin, a copper-containing oxidase homologous to ceruloplasmin. Both of these ferroxidases can oxidize Fe^{2+} to Fe^{3+}. Ferric iron (Fe^{3+}) is then available to bind to transferrin for delivery to peripheral tissues. Two major destinations are: (1) the bone marrow, to supply iron for heme production during erythropoiesis, and (2) the liver, where any excess iron can be stored in ferritin for future needs (Fig. 14.5).

Iron-Regulatory Hormones

Communication between the intestine, bone marrow, and liver must provide homeostatic control to balance the amount of dietary iron absorption. With no known mechanism of excretion, excessive assimilation would lead to liver iron-loading as observed in patients with inherited and acquired hemochromatosis. On the other hand, bone marrow must signal the demand for iron necessary to keep up with red blood cell production. Under hypoxic conditions due to hemorrhage or blood loss, for example, the red blood cell mass must expand to increase oxygen delivery.

The major regulatory hormone that orchestrates the absorption of iron in response to the body's needs and existing storage levels is called hepcidin (Ganz and Nemeth, 2012). Hepcidin is primarily synthesized by the liver. The amount of

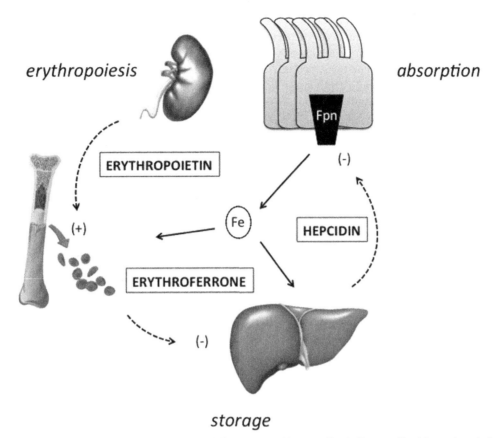

FIGURE 14.5   Hormonal control of iron metabolism. Absorbed dietary iron is either stored in the liver or utilized for erythropoiesis. Iron absorption across the intestinal mucosa is negatively regulated by hepcidin, a liver-derived peptide hormone, which downregulates Fpn to reduce iron intake. In response to hypoxia, the kidney-derived hormone erythropoietin upregulates erythropoiesis. Erythroblasts in turn produce erythroferrone, a newly discovered factor that negatively regulates hepcidin synthesis by the liver. These coordinated control mechanisms ensure the iron response to physiological needs while preventing excess iron accumulation.

hepcidin produced directly reflects the degree of body iron loading. A second factor called erythroferrone has been discovered (Kautz et al., 2014). This regulatory factor is produced by erythroblasts in response to EPO's stimulation of erythropoiesis, and it suppresses hepcidin (Fig. 14.5).

The mechanism of hepcidin function is quite simple. This peptide hormone is a ligand for its "receptor" Fpn. Hepcidin binds to Fpn and induces its downregulation by stimulating internalization and degradation (Nemeth et al., 2004). Thus, there is a direct impact on the amount of iron "exsorbed" by the intestine and released into circulation. Since enterocytes turn over very rapidly and slough off the microvillous surface of the intestine, hepcidin's block on dietary iron assimilation diverts any "absorbed" iron present inside enterocytes away from the body (Donovan et al., 2005).

Levels of hepcidin are set by the liver (Steinbicker and Muckenthaler, 2013). Bone morphogenic proteins (BMPs) activate signal transduction through specific hepatic receptors. In particular, BMP6 expression is regulated by iron levels. BMP6 acts with a specific coreceptor called hemojuvelin (HJV) to activate specific cell surface receptors that function in signaling through SMAD transcription factors to activate hepcidin gene expression (Meynard et al., 2014). A negative regulator of these interactions, TMPRSS6 (matripase-2), blocks BMP/SMAD signaling and mutations in its gene give rise to iron-refractory, iron-deficiency anemia (Finberg et al., 2008). Conversely, defects in the genes encoding HJV and hepcidin (HAMP) give rise to juvenile hemochromatosis (Papanikolaou et al., 2004). *HFE* and *TFR2* mutations also give rise to inherited forms of hemochromatosis, but mostly in adults in the fourth or fifth decade of life. The latter factors are thought to "fine-tune" iron-sensing by detecting the iron saturation of circulating transferrin, which directly reflects the availability of iron for the body's immediate needs. Transferrin receptor-1 and transferrin receptor-2 (TfR2) both interact with *HFE* to modify hepcidin expression but the signaling mechanism has yet to be fully elucidated (Worthen and Enns, 2014). Note once again that genetic defects in humans in Hfe, HJV, TfR2, Fpn, hepcidin, and DMT-1 all contribute to iron overload disease.

Also as discussed earlier, iron withdrawal from circulation is part of the acute-phase response to infection and inflammation. Inflammatory control of hepcidin gene expression results from activation of the STAT signaling pathway by interleukin-6 and other cytokines (Ganz and Nemeth, 2012). This inflammatory response promotes the so-called anemia of chronic disease or anemia of inflammation. Not only is dietary iron absorption reduced, since Fpn is present in hepatocytes and macrophages, the inflammatory production of hepcidin also blocks the release of iron from stores. Recently, hepcidin-independent regulation of Fpn has been reported in response to some inflammatory stimuli (Guida et al., 2015). This effect appears to involve downregulation of Fpn transcription. These self-imposed states of hypoferremia (iron deficiency) help us to defend against infectious diseases by depriving pathogens of iron.

The need for iron necessarily reflects the demands for oxygen. Rapid blood loss due to hemorrhage or parasitic diseases necessarily fuels a demand for increased iron to match increased erythropoiesis. One milliliter of blood contains approximately 0.5 mg of iron, so even a small amount of bleeding can lead to rapid and significant loss of iron. "Erythroid regulators" appear to negatively regulate hepcidin production to help the body recover from blood loss independent of its iron status. Two bone morphogenetic BMP family members have been implicated as suppressors of hepcidin production: growth differentiation factor-15 (GDF15) and twisted-gastrulation 1 (TWSG1). Such activities might modulate the SMAD signaling pathway described earlier to control hepcidin levels. In addition, erythroferrone is made by erythroblasts in response to EPO (Kautz et al., 2014). Since EPO controls the rate of erythropoiesis and reflects oxygen levels in the body (due to HIF2α regulation), erythroferrone directly links red blood cell production and its demand for hemoglobinization with the amount of iron absorbed. It is a negative regulator of hepcidin production (Fig. 14.5).

Transcriptional and Posttranscriptional Control of Iron Absorption

HIF2α not only regulates EPO production in the kidney, but this transcription factor also regulates gene expression of DMT-1, DcytB, and Fpn in the intestine. Under low-iron or low-oxygen conditions, HIF2α is stabilized and dimerizes with HIF1β to induce transcription of these iron transport factors. DMT-1 (with the help of DcytB) promotes apical iron uptake and Fpn then mediates its release across the basolateral surface of duodenal enterocytes to promote iron absorption. Another layer of regulation occurs posttranscriptionally since the DMT-1 and Fpn transcripts have IREs. Thus, IRP-binding activity, which is regulated by iron, can stabilize or repress translation of the two transporters (Table 14.1; Fig. 14.2). These reciprocal effects can balance the entry and egress of iron within the enterocyte itself.

Cellular levels of iron within the enterocyte are controlled by spatial organization of the intestinal mucosa (Gulec et al., 2014). The duodenal luminal surface at the villous tip confronts dietary iron while the serosal surface of duodenal enterocytes in the crypt has input about the body's iron status from circulating transferrin. Transferrin receptors are abundant in crypt enterocytes, which differentiate along the axis of intestinal villi to become absorptive enterocytes at the tip. Here, transferrin receptors are not highly expressed. "Crypt programming" is thought to coordinate differentiation of duodenal enterocytes so that expression of iron-regulatory factors is influenced early by circulating iron status reported by transferrin saturation. Under low-iron conditions, less iron would be taken up or sensed by crypt cells, therefore upregulation of DMT-1 levels via transcriptional and posttranscriptional responses would occur. Similarly, upon high-iron status, circulating hepcidin would reduce egress of iron due to reduced ferroportin levels. Under these high intracellular iron conditions, DMT-1 levels would be lower. The responses exerted through IRP activity are also directly reflected in HIF2α transcriptional regulation.

An interesting effect occurs with a high dietary iron challenge. The so-called "mucosal block" has been described in response to increased luminal iron content. This effect limits the amount of iron absorbed immediately after a bolus of iron is delivered. Unlike crypt programming, the mucosal block responds to dietary iron and not body iron status. The mucosal block is thought to arise from local control of ferritin, since once iron is taken up by absorptive enterocytes it can be sequestered in the iron storage protein. In fact, because its IRE appears in the 5' region of the ferritin transcript, translation would be blocked by IRP activity until released upon high intracellular iron. This mechanism short circuits the transcriptional control set-point exerted in the crypt and enables a direct response to the dietary environment. IRP control helps to adjust the enterocyte's iron absorptive capacity to reduce excessive iron assimilation across epithelial cells at the villous tip. The role of IRP control over the mucosal block imposed by posttranscriptionally regulated ferritin synthesis has only recently been characterized (Galy et al., 2013). Since duodenal enterocytes are exfoliated, the excess iron stored in ferritin will ultimately be removed from the body. This pathway might be one mechanism that could be considered to "regulate" iron excretion (Donovan et al., 2005).

IRON AND OUR DIET

Recommended Dietary Allowance or Dietary Reference Intake

The necessary amount of iron required on a daily basis varies with age, gender, and other health factors (Camaschella, 2015). For adult men, 8 mg/day is recommended while for menstruating women this value is higher, 18 mg/day. Pregnant women are recommended to intake 27 mg/day. As discussed earlier, young children and women who are pregnant or at childbearing age are at the most risk for iron deficiency. However, other factors increase the risk of iron deficiency. Hemolytic anemia and parasitic diseases, *Helicobacter pylori* infection, bariatric surgery, chronic kidney disease, and other disorders cause a higher need for iron. Not only do sick people require more iron, but healthy individuals who adhere to a strict vegetarian diet or who engage in intensive athletic training may also require more iron.

Sources of Iron

As described earlier, heme iron is more readily absorbed than nonheme iron; thus iron from meat sources including poultry and fish can fulfill up to 1/3 the total iron absorbed from the diet. For these reasons, the amount of iron acquired from a plant-based diet may be less than from a plant- and animal-based diet. One recommendation from the US Food and Nutrition Board is that individuals with a completely vegetarian diet should adjust their dietary intake to 14 mg/day for adult men and postmenopausal women, 33 mg/day for premenopausal women, and 26 mg/day for adolescent women.

Since refined grain products are fortified with vitamins and minerals in the United States and other countries (Allen, 2002), cereal and other products have a high iron content. Beef, chicken, and fish are high in heme iron, and other factors in meat are thought to promote nonheme iron uptake as well. Many readers of this chapter will be pleased to learn that dark chocolate is also listed as a food source high in iron in the USDA nutrient database.

Enhancers and Inhibitors of Iron Absorption

As already described above, vitamin C (ascorbic acid) is a strong enhancer of dietary iron absorption since it is a ferrireductant. Other organic acids (citric, malic, tartaric, and lactic) can have enhancing effects; generally, acidic gastric secretions are thought to help promote iron absorption, particularly since DMT-1 is a proton-coupled transporter. Other food factors can block nonheme iron absorption, for example, phytic acid. Even small amounts (5–10 mg) inhibit iron absorption by 50% (Food and Nutrition Board, 2001). Polyphenols, which are found in fruit and vegetables, coffee and tea, as well as in wine, are also potent inhibitors of iron uptake. Certain drugs may reduce iron absorption, including antacids and proton pump inhibitors that decrease acid in the stomach. Since metals often share transport pathways, competition between cations can occur, especially between zinc and iron. On the other hand, because copper is required for ferroxidases that promote iron uptake (hephaestin, ceruloplasmin), copper-deficiency anemia can arise (Gulec et al., 2014).

In surveys of the US population, it has been reported that men intake an average of 16–18 mg/day and that women take in about 15 mg/day (Food and Nutrition Board, 2001). Of course, strong dietary influences such as those described above factor into the final amount of iron absorbed, but generally men consume more than the recommended amount of iron while women take in less. Thus, iron-deficiency anemia during pregnancy is of particular concern. Iron supplements are available to prevent and treat iron deficiency but their use should be discussed with a health care provider. An over-the-counter multivitamin/multimineral supplement typically contains about 18 mg of iron, which is above the recommended amount. Iron fortification programs have been implemented to prevent iron deficiency in many countries, but are controversial due to the extent that high iron can cause oxidative stress and place people who have inherited or acquired overload diseases at risk despite benefit to the population as a whole (Allen, 2002). In the age of personalized care and precision medicine, given the significant health impact of iron it is important for one to know and keep track of their individual iron status since iron deficiency can be reversed through changes in diet and supplementation, while iron overload can be limited through these measures as well.

REFERENCES

Ali-Rahmani, F., Schengrund, C.L., Connor, J.R., 2014. HFE gene variants, iron, and lipids: a novel connection in Alzheimer's disease. Front. Pharmacol. 5, 165.

Allen, L.H., 2002. Iron supplements: scientific issues concerning efficacy and implications for research and programs. J. Nutr. 132, 813S–819S.

Allen, R.P., Earley, C.J., 2007. The role of iron in restless legs syndrome. Mov. Disord. 22 (Suppl. 18), S440–S448.

Beard, J.L., 2008. Why iron deficiency is important in infant development. J. Nutr. 138, 2534–2536.

Bloomer, S.A., Brown, K.E., Buettner, G.R., Kregel, K.C., 2008. Dysregulation of hepatic iron with aging: implications for heat stress-induced oxidative liver injury. Am. J. Physiol. Regul. Integr. Comp. Physiol. 294, R1165–R1174.

Camaschella, C., 2015. Iron-deficiency anemia. N. Engl. J. Med. 372, 1832–1843.

Constantine, C.C., Anderson, G.J., Vulpe, C.D., McLaren, C.E., Bahlo, M., Yeap, H.L., Gertig, D.M., Osborne, N.J., Bertalli, N.A., Beckman, K.B., Chen, V., Matak, P., McKie, A.T., Delatycki, M.B., Olynyk, J.K., English, D.R., Southey, M.C., Giles, G.G., Hopper, J.L., Allen, K.J., Gurrin, L.C., 2009. A novel association between a SNP in CYBRD1 and serum ferritin levels in a cohort study of HFE hereditary haemochromatosis. Br. J. Haematol. 147, 140–149.

Danesh, J., Appleby, P., 1999. Coronary heart disease and iron status: meta-analyses of prospective studies. Circulation 99, 852–854.

Distante, S., Robson, K.J., Graham-Campbell, J., Arnaiz-Villena, A., Brissot, P., Worwood, M., 2004. The origin and spread of the HFE-C282Y haemo-chromatosis mutation. Hum. Genet. 115, 269–279.

Donovan, A., Lima, C.A., Pinkus, J.L., Pinkus, G.S., Zon, L.I., Robine, S., Andrews, N.C., 2005. The iron exporter ferroportin/Slc40a1 is essential for iron homeostasis. Cell Metab. 1, 191–200.

Drakesmith, H., Prentice, A.M., 2012. Hepcidin and the iron-infection axis. Science 338, 768–772.

Finberg, K.E., Heeney, M.M., Campagna, D.R., Aydinok, Y., Pearson, H.A., Hartman, K.R., Mayo, M.M., Samuel, S.M., Strouse, J.J., Markianos, K., Andrews, N.C., Fleming, M.D., 2008. Mutations in TMPRSS6 cause iron-refractory iron deficiency anemia (IRIDA). Nat. Genet. 40, 569–571.

Fleming, M.D., 2011. Congenital sideroblastic anemias: iron and heme lost in mitochondrial translation. Hematology 2011, 525–531.

Food and Nutrition Board, I.O.M., 2001. Dietary Reference Intakes for Vitamin A, Vitamin K, Boron, Chromium, Opper, Iodine, Iron, Manganese, Molyb-denum, Nickel, Silicon, Vanadium, and Zinc. National Academy Press, Washington, DC, pp. 290–393.

Ford, E.S., Cogswell, M.E., 1999. Diabetes and serum ferritin concentration among U.S. adults. Diabetes Care 22, 1978–1983.

Galy, B., Ferring-Appel, D., Becker, C., Gretz, N., Grone, H.J., Schumann, K., Hentze, M.W., 2013. Iron regulatory proteins control a mucosal block to intestinal iron absorption. Cell Rep. 3, 844–857.

Ganz, T., Nemeth, E., 2012. Hepcidin and iron homeostasis. Biochim. Biophys. Acta 1823, 1434–1443.

Goodnough, L.T., Schrier, S.L., 2014. Evaluation and management of anemia in the elderly. Am. J. Hematol. 89, 88–96.

Guida, C., Altamura, S., Klein, F.A., Galy, B., Boutros, M., Ulmer, A.J., Hentze, M.W., Muckenthaler, M.U., 2015. A novel inflammatory pathway mediat-ing rapid hepcidin-independent hypoferremia. Blood 125, 2265–2275.

Gulec, S., Anderson, G.J., Collins, J.F., 2014. Mechanistic and regulatory aspects of intestinal iron absorption. Am. J. Physiol. Gastrointest. Liver Physiol. 307, G397–G409.

Gunshin, H., Starr, C.N., Direnzo, C., Fleming, M.D., Jin, J., Greer, E.L., Sellers, V.M., Galica, S.M., Andrews, N.C., 2005. Cybrd1 (duodenal cytochrome b) is not necessary for dietary iron absorption in mice. Blood 106, 2879–2883.

Guralnik, J.M., Eisenstaedt, R.S., Ferrucci, L., Klein, H.G., Woodman, R.C., 2004. Prevalence of anemia in persons 65 years and older in the United States: evidence for a high rate of unexplained anemia. Blood 104, 2263–2268.

Haas, J.D., Brownlie 4th, T., 2001. Iron deficiency and reduced work capacity: a critical review of the research to determine a causal relationship. J. Nutr. 131, 676S–688S, discussion 688S–690S.

Hentze, M.W., Muckenthaler, M.U., Galy, B., Camaschella, C., 2010. Two to tango: regulation of mammalian iron metabolism. Cell 142, 24–38.

Jenkitkasemwong, S., Wang, C.Y., Mackenzie, B., Knutson, M.D., 2012. Physiologic implications of metal-ion transport by ZIP14 and ZIP8. Biometals 25, 643–655.

Kato, I., Dnistrian, A.M., Schwartz, M., Toniolo, P., Koenig, K., Shore, R.E., Zeleniuch-Jacquotte, A., Akhmedkhanov, A., Riboli, E., 1999. Iron intake, body iron stores and colorectal cancer risk in women: a nested case-control study. Int. J. Cancer 80, 693–698.

Kautz, L., Jung, G., Valore, E.V., Rivella, S., Nemeth, E., Ganz, T., 2014. Identification of erythroferrone as an erythroid regulator of iron metabolism. Nat. Genet. 46, 678–684.

Kim, J., Wessling-Resnick, M., 2014. Iron and mechanisms of emotional behavior. J. Nutr. Biochem. 25, 1101–1107.

Korolnek, T., Hamza, I., 2014. Like iron in the blood of the people: the requirement for heme trafficking in iron metabolism. Front. Pharmacol. 5, 126.

Kuhn, L.C., 2015. Iron regulatory proteins and their role in controlling iron metabolism. Metallomics 7, 232–243.

Lane, D.J., Bae, D.H., Merlot, A.M., Sahni, S., Richardson, D.R., 2015. Duodenal cytochrome b (DCYTB) in iron metabolism: an update on function and regulation. Nutrients 7, 2274–2296.

Lomagno, K.A., Hu, F., Riddell, L.J., Booth, A.O., Szymlek-Gay, E.A., Nowson, C.A., Byrne, L.K., 2014. Increasing iron and zinc in pre-menopausal women and its effects on mood and cognition: a systematic review. Nutrients 6, 5117–5141.

Lozoff, B., Beard, J., Connor, J., Barbara, F., Georgieff, M., Schallert, T., 2006. Long-lasting neural and behavioral effects of iron deficiency in infancy. Nutr. Rev. 64, S34–S43, discussion S72–S91.

Mackenzie, H.S., Brenner, B.M., 1995. Fewer nephrons at birth: a missing link in the etiology of essential hypertension? Am. J. Kidney Dis. 26, 91–98.

Mena, N.P., Urrutia, P.J., Lourido, F., Carrasco, C.M., Nunez, M.T., 2015. Mitochondrial iron homeostasis and its dysfunctions in neurodegenerative disorders. Mitochondrion 21, 92–105.

Meynard, D., Babitt, J.L., Lin, H.Y., 2014. The liver: conductor of systemic iron balance. Blood 123, 168–176.

Nemeth, E., Tuttle, M.S., Powelson, J., Vsghn, M.B., Donovan, A., Ward, D.M., Ganz, T., Kaplan, J., 2004. Hepcidin regulates cellular iron efflux by binding to ferroportin and inducing its internalization. Science 306, 2090–2093.

Papanikolaou, G., Samuels, M.E., Ludwig, E.H., MacDonald, M.L., Franchini, P.L., Dube, M.P., Andres, L., MacFarlane, J., Sakellaropoulos, N., Politou, M., Nemeth, E., Thompson, J., Risler, J.K., Zaborowska, C., Babakaiff, R., Radomski, C.C., Pape, T.D., Davidas, O., Christakis, J., Brissot, P., Lock-itch, G., Ganz, T., Hayden, M.R., Goldberg, Y.P., 2004. Mutations in HFE2 cause iron overload in chromosome 1q-linked juvenile hemochromatosis. Nat. Genet. 36, 77–82.

Pietrangelo, A., 2004. Hereditary hemochromatosis–a new look at an old disease. N. Engl. J. Med. 350, 2383–2397.

Prado, E.L., Dewey, K.G., 2014. Nutrition and brain development in early life. Nutr. Rev. 72, 267–284.

Qiu, A., Jansen, M., Sakaris, A., Min, S.H., Chattopadhyay, S., Tsai, E., Sandoval, C., Zhao, R., Akabas, M.H., Goldman, I.D., 2006. Identification of an intestinal folate transporter and the molecular basis for hereditary folate malabsorption. Cell 127, 917–928.

Quigley, J.G., Yang, Z., Worthington, M.T., Phillips, J.D., Sabo, K.M., Sabath, D.E., Berg, C.L., Sassa, S., Wood, B.L., Abkowitz, J.L., 2004. Identification of a human heme exporter that is essential for erythropoiesis. Cell 118, 757–766.

Rao, R., Georgieff, M.K., 2007. Iron in fetal and neonatal nutrition. Semin. Fetal Neonat. Med. 12, 54–63.

Salonen, J.T., Tuomainen, T.P., Salonen, R., Lakka, T.A., Nyyssonen, K., 1998. Donation of blood is associated with reduced risk of myocardial infarction. The Kuopio Ischaemic heart disease risk factor study. Am. J. Epidemiol. 148, 445–451.

Sazawal, S., Black, R.E., Ramsan, M., Chwaya, H.M., Stoltzfus, R.J., Dutta, A., Dhingra, U., Kabole, I., Deb, S., Othman, M.K., Kabole, F.M., 2006. Effects of routine prophylactic supplementation with iron and folic acid on admission to hospital and mortality in preschool children in a high malaria transmission setting: community-based, randomised, placebo-controlled trial. Lancet 367, 133–143.

Steinbicker, A.U., Muckenthaler, M.U., 2013. Out of balance–systemic iron homeostasis in iron-related disorders. Nutrients 5, 3034–3061.

van der, A.D., Peeters, P.H., Grobbee, D.E., Marx, J.J., van der Schouw, Y.T., 2005. Dietary haem iron and coronary heart disease in women. Eur. Heart J. 26, 257–262.

Ward, R.J., Dexter, D.T., Crichton, R.R., 2015. Neurodegenerative diseases and therapeutic strategies using iron chelators. J. Trace Elem. Med. Biol. 31, 267–273.

Wessling-Resnick, M., 2010. Iron homeostasis and the inflammatory response. Annu. Rev. Nutr. 30, 105–122.

Worthen, C.A., Enns, C.A., 2014. The role of hepatic transferrin receptor 2 in the regulation of iron homeostasis in the body. Front. Pharmacol. 5, 34.

Xu, W., Barrientos, T., Andrews, N.C., 2013. Iron and copper in mitochondrial diseases. Cell Metab. 17, 319–328.

Zimmermann, M.B., Hurrell, R.F., 2007. Nutritional iron deficiency. Lancet 370, 511–520.

Chapter 15

Hepcidin and the Hormonal Control of Iron Homeostasis

David Michael Frazer, Gregory Jon Anderson
QIMR Berghofer Medical Research Institute, Australia

INTRODUCTION

Hepcidin is the master regulator of body iron homeostasis. It was first identified as a cysteine-rich defensin-like peptide that was isolated from human blood, and it showed antimicrobial activity against a range of bacteria and yeast species (Krause et al., 2000; Park et al., 2001). A link with iron was first suggested when iron loading was found to increase hepatic hepcidin expression (Pigeon et al., 2001). However, it was not until the accidental disruption of the *Hamp* gene, which encodes the hepcidin peptide, caused severe iron loading in the *USF2* knockout mouse that the central role of hepcidin in iron homeostasis was recognized (Nicolas et al., 2001). Subsequent manipulation of hepcidin expression by targeted disruption and overexpression in mouse models, and identification of *HAMP* mutations in human patients with the severe iron-loading disorder hemochromatosis type 2 (juvenile hemochromatosis), have confirmed its important role in modulating body iron levels (Darshan et al., 2010; Lesbordes-Brion et al., 2006). In fact, inappropriate hepcidin production is now known to cause the disordered iron homeostasis in conditions such as hereditary hemochromatosis, β-thalassemia, and the anemia of inflammation (Darshan et al., 2010). This chapter discusses how hepcidin regulates iron homeostasis as well as the mechanisms by which hepcidin production is regulated.

BIOLOGY OF HEPCIDIN

The mature hepcidin peptide is 25 amino acids in length and is encoded by the *HAMP* gene situated on chromosome 19 in humans (Krause et al., 2000). Two *Hamp* genes, *Hamp1* and *Hamp2*, encode two hepcidin peptides in mice, although only the *Hamp1* gene product plays an active role in iron homeostasis, with the role of the *Hamp2* gene product currently unknown (Lou et al., 2004; Nicolas et al., 2002). The human *HAMP* gene encodes an 84-amino–acid precursor protein consisting of a 24-amino–acid signal sequence targeting the peptide to the endoplasmic reticulum and a 60-amino–acid prohepcidin region (Krause et al., 2000; Park et al., 2001). Prohepcidin is cleaved to the 25-amino–acid mature peptide by the prohormone convertase furin before being secreted into the circulation (Valore and Ganz, 2008). Additional N-terminal cleavage produces two smaller isoforms of 22 or 20 amino acids (Krause et al., 2000; Park et al., 2001), although the function of these peptides remains unknown. However, they are unlikely to be involved in iron homeostasis because all five N-terminal amino acids have been shown to be important for hepcidin function (Nemeth et al., 2006). Hepcidin is expressed predominantly by hepatocytes, but low-level expression has also been detected in other cells such as macrophages and adipocytes, where it possibly plays a role in regulating local iron levels (Ganz and Nemeth, 2012).

The mature hepcidin peptide contains eight cysteine residues, all of which are involved in disulfide bond formation (Park et al., 2001). The 25-amino–acid peptide consists of a β-sheet element and a β-hairpin–loop motif stabilized by four disulfide bonds formed between cysteines at position 7 and 23, 10 and 13, 14 and 22, and 11 and 19 (Jordan et al., 2009; Fig. 15.1A). This structure appears to be highly conserved during evolution, with all eight cysteine residues in the same position in species as diverse as humans and zebrafish (Park et al., 2001; Shike et al., 2004). The removal of individual disulfide bonds reduces the activity of hepcidin, but, surprisingly, it does not entirely abolish it (Preza et al., 2011).

Interaction With Ferroportin

Hepcidin regulates iron homeostasis by binding to the iron transporter ferroportin on the surface of target cells (Nemeth et al., 2004). Ferroportin is the only known mammalian iron export protein (Frazer and Anderson, 2014) and is essential

(A)

Human DTHFPICIFCCGCCHRSKCGMCCKT
Mouse DTNFPICIFCCKCCNNSQCGICCKT
Rat DTNFPICLFCCKCCKNSSCGLCCIT
Zebrafish QSHLSLCRFCCKCCRNKGCGYCCKF

(B)

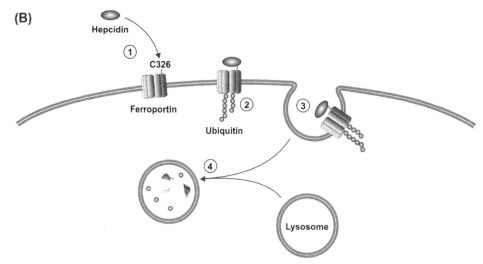

Hepcidin

C326

Ferroportin

Ubiquitin

Lysosome

FIGURE 15.1 Hepcidin and the regulation of ferroportin-dependent iron export. (A) Alignment of the mature hepcidin peptides of human, mouse (hepcidin1), rat, and zebrafish. The eight conserved cysteine residues are shaded. The disulfide bonds identified in the human peptide are indicated. (B) The proposed regulation of ferroportin protein by hepcidin. Circulating hepcidin binds to ferroportin by forming a disulfide bond with the free cysteine at position 326 of ferroportin (1) leading to the polyubiquitination of the ferroportin protein (2). This leads to the internalization of the ferroportin/hepcidin complex (3) and its degradation in lysosomes (4).

for dietary iron absorption, the recycling of iron from senescent red blood cells, and the mobilization of storage iron. The binding of hepcidin to cell-surface ferroportin causes the iron transporter to be internalized and degraded, thereby reducing iron export (Nemeth et al., 2004; Fig. 15.1B). A cysteine residue at position 326 (C326) of ferroportin is located in an extracellular loop, is important for hepcidin binding (Fernandes et al., 2009), and it likely participates in a disulfide bond exchange with one of the four disulfide bridges in the hepcidin peptide (Preza et al., 2013). Mutations in C326 cause a severe autosomal-dominant form of iron loading because of a failure to appropriately downregulate ferroportin (Sham et al., 2005). Mice carrying a C326S mutation in ferroportin die of severe iron overload between the ages of 7 and 14 months (Altamura et al., 2014). Other residues in the same extracellular loop are thought to directly interact with hepcidin, including F324 and Y333 (Preza et al., 2011). The binding of hepcidin to ferroportin causes the rapid polyubiquitination of lysines between residues 229 and 269 in the third intracellular loop of ferroportin (Qiao et al., 2012). Ubiquitination of ferroportin is essential for internalization (Qiao et al., 2012) and targets the ferroportin/hepcidin complex for degradation in lysosomes (Preza et al., 2013).

THE REGULATION OF HEPCIDIN EXPRESSION

Because hepcidin binding to ferroportin leads to its degradation, hepcidin acts as an inhibitor of cellular iron export (Frazer and Anderson, 2014). Therefore a decrease in hepcidin production leads to increases in dietary iron absorption, red blood cell iron recycling, and storage iron release. Hepcidin production is reduced when the body has a perceived need for more iron, most obviously when iron stores are low or after blood loss. In contrast, elevated hepcidin reduces iron absorption, promotes the intracellular storage of iron, and occurs when the body has adequate or excessive iron stores. Hepcidin is also increased as part of the immune response to inflammation (see Section: Inflammation).

The central role of hepcidin in body iron homeostasis and its inappropriate expression in conditions such as hereditary hemochromatosis, β-thalassemia, and the anemia of inflammation has resulted in much interest in the molecular basis of hepcidin production by hepatocytes. There are several signaling pathways regulating hepcidin expression, with complex arrays of molecules involved, not all of which are well characterized. The central and probably best characterized pathway is the bone morphogenetic protein (BMP)/SMAD signaling cascade.

The Bone Morphogenetic Protein/SMAD Pathway

The first indication that the SMAD pathway was involved in hepcidin regulation came with the reporting that the liver-specific SMAD4 knockout mouse developed severe iron loading because of a lack of hepcidin production (Wang et al., 2005). Further studies showed that a range of BMPs, including BMP-2, BMP-4, and BMP-9, were the activators of SMAD signaling to induce hepcidin expression (Babitt et al., 2006; Truksa et al., 2006). BMPs form part of the transforming growth factor (TGF)-β superfamily of cytokines and bind to specific BMP receptor complexes on the surface of target cells leading to the phosphorylation of SMAD1, SMAD5, and SMAD8 in the cytosol (Babitt et al., 2006). Phosphorylated SMAD then binds to SMAD4, and the complex is translocated to the nucleus where it binds to specific BMP-responsive elements in the promoters of target genes. Analysis of the *HAMP* promoter has revealed that at least two highly conserved BMP-response elements are required for the gene to fully respond to BMP stimulation (Parrow and Fleming, 2014).

In addition to stimulating hepcidin expression, BMPs also stimulate the production of several other molecules in hepatocytes, including SMAD6, SMAD7, inhibitor of DNA binding 1 (ID1), and atonal bHLH transcription factor 8 (ATOH8; Kautz et al., 2008; Vujic Spasic et al., 2013). SMAD6 and SMAD7 are inhibitory SMADs induced by SMAD signaling, with SMAD7 being a general antagonist of the TGFβ superfamily and SMAD6 a specific inhibitor of BMP signaling (Vujic Spasic et al., 2013). Both inhibitory SMADs have been shown to inhibit the expression of hepcidin in cultured cells, and their expression is increased in response to iron loading in mice (Mleczko-Sanecka et al., 2010; Vujic Spasic et al., 2013). However, in vivo evidence that the inhibitory SMADs play a direct role in hepcidin regulation is yet to be presented. Likewise, ATOH8 and ID1 have also been shown to regulate hepcidin expression in cultured cells (Meynard et al., 2011; Patel et al., 2014), although their role in vivo is yet to be defined.

The Role of Bone Morphogenetic Protein-6

Although several recombinant BMPs are able to stimulate hepcidin expression in vitro and in vivo, BMP6 is the only endogenously produced BMP that has been shown to be important for hepcidin regulation. Hepatic BMP6 expression is reduced when iron stores are low and increases with dietary iron loading (Kautz et al., 2008). However, the most conclusive evidence comes from *Bmp6* knockout mice, which develop severe iron loading in the liver, heart, and pancreas due to very low hepcidin production (Andriopoulos et al., 2009; Meynard et al., 2009). Furthermore, hepcidin expression is reduced after treatment with neutralizing anti-BMP6 antibodies and injections of recombinant BMP6 increase hepcidin production (Andriopoulos et al., 2009).

Hepcidin is synthesized predominantly in hepatocytes, and it is these cells that are responsive to BMP6 stimulation. However, it is the nonparenchymal cells of the liver that produce BMP6 in response to increased iron stores, implying that BMP6 acts in a paracrine manner to regulated hepcidin expression (Enns et al., 2013). Kupffer cells, sinusoidal endothelial cells, and hepatic stellate cells have all been shown to express Bmp6 in response to body iron loading in mice, with sinusoidal endothelial cells appearing to show the highest expression levels (Enns et al., 2013). Precisely how these cells respond to changes in hepatocyte iron loading is not known. It has been suggested that hepatocytes produce a yet to be identified cytokine that relays iron levels to nonparenchymal cells, which respond by altering BMP6 production (Enns et al., 2013). Indeed, a recent study has showed that iron-loaded ferritin can stimulate Bmp6 production in 10-day-old mice (Feng et al., 2012). The injected ferritin localized to the sinusoidal endothelial cells of the liver, and it was in these cells that Bmp6 induction was greatest, leading to the suggestion that ferritin produced by hepatocytes and Kupffer cells may be the cytokine responsible for BMP6 regulation. Others have shown that nonparenchymal cells in the liver can directly detect iron-loaded transferrin in the circulation (Enns et al., 2013) and suggested that this, rather than iron stores, determines the level of BMP6 produced. Further research is clearly required to determine how the nonparenchymal cells of the liver regulate BMP6 production.

Bone Morphogenetic Protein Receptor Complex

Regardless of its source, BMP6 stimulates hepcidin production by binding to the BMP receptor complex on the surface of hepatocytes. This complex consists of two type I receptors and two type II receptors. Of the four type I receptors [activin-like kinase (Alk)1, Alk2, Alk3, and Alk6], Alk2 and Alk3 are essential for hepcidin regulation because mice lacking either of these receptors develop iron loading (Steinbicker et al., 2011). There are also three type II receptors [bone morphogenetic protein receptor (BMPR)2, activin type IIA receptor (ActRIIa), and ActRIIb], with ActRIIa and BMPR2 playing redundant roles in hepcidin regulation because mice require both receptors to be knocked out for iron loading to develop (Mayeur et al., 2014a). The binding of BMP to the receptor complex induces the phosphorylation of the type I receptor by the type II receptor. This in turn allows the receptor complex to phosphorylate SMAD1, SMAD5, and SMAD8 in the cytosol, activating the signaling pathway (Darshan et al., 2010).

Several other molecules also associate with the receptor complex and are essential for BMP signaling and hepcidin production. The GPI-linked hemojuvelin is a member of the repulsive guidance molecule family of proteins that enhance BMP signaling (Babitt et al., 2006). A specific role for hemojuvelin in hepcidin regulation was revealed when it was reported to be mutated in patients with juvenile hemochromatosis, a severe inherited iron-loading disorder characterized by a reduction in hepcidin production (Darshan et al., 2010). Hemojuvelin knockout mice have lower levels of phosphorylated Smad1, Smad5, and Smad8 and very low *Hamp1* mRNA levels in the liver compared with wild-type mice despite significant iron loading and an increase in Bmp6 expression (Babitt et al., 2006; Gkouvatsos et al., 2014; Huang et al., 2005). It is thought that hemojuvelin acts as a coreceptor for the BMP receptor complex and binds BMPs (Babitt et al., 2006), including BMP6 (Andriopoulos et al., 2009), enhancing their ability to induce SMAD phosphorylation.

The membrane-bound form of hemojuvelin can be cleaved by furin to produce a soluble protein that retains the ability to bind BMP6 (Silvestri et al., 2008a). Soluble hemojuvelin competes with membrane-bound hemojuvelin for BMP6 binding; therefore it is an inhibitor of hepcidin expression (Lin et al., 2005). The production of soluble hemojuvelin is regulated by iron, with some studies suggesting that increasing intracellular iron levels reduce soluble hemojuvelin production by inhibiting furin expression (Lin et al., 2005; Silvestri et al., 2008a) whereas others have suggested that diferric transferrin inhibits soluble hemojuvelin production independently of intracellular iron (Zhang et al., 2007). The reason for these conflicting results is unknown; however, all studies were performed in cultured cells and so the differences may be due to the cell lines and conditions used for each study. This has yet to be investigated in detail in vivo.

The ability of hemojuvelin to enhance BMP signaling is modulated by neogenin. Neogenin binds to hemojuvelin (Zhang et al., 2005), and knockdown studies in cultured cells show that neogenin enhances BMP-induced hepcidin expression (Zhang et al., 2009). Confirmation of a role for neogenin in vivo came with the analysis of the neogenin knockout mouse, which, although short lived because of developmental issues unrelated to iron, shows significant iron loading (Lee et al., 2010). Surprisingly, hemojuvelin overexpression in cultured cells lacking neogenin inhibited hepcidin expression further than in cells lacking neogenin alone, implying that hemojuvelin in the absence of neogenin has an inhibitory effect on hepcidin expression (Zhang et al., 2009). A recent study has provided a possible explanation for this by showing that hemojuvelin competes with the type I BMP receptor for BMP binding (Healey et al., 2015). This study also suggested that the binding between hemojuvelin and BMP was pH dependent whereas BMP binding to the BMP receptor was not. A model for BMP signaling was presented in which BMPs bind to the hemojuvelin/neogenin complex at the cell surface, inducing clustering of the ternary complex and subsequent internalization along with the BMP receptor complex (Fig. 15.2). Acidification of the endosome releases BMPs from the hemojuvelin/neogenin complex, allowing it to bind to BMP receptors, inducing SMAD phosphorylation and hepcidin expression. In the absence of neogenin, hemojuvelin remains at the cell surface, thereby reducing BMP availability.

Another modulator of hemojuvelin function is matriptase2. Encoded by the *transmembrane protease serine-6 (TMPRSS6)* gene, this transmembrane serine protease is expressed predominantly in the liver and, when mutated, causes iron-deficiency anemia in mice (Du et al., 2008) and humans (Finberg et al., 2008) because of elevated hepcidin production. Mice lacking matriptase2 and hemojuvelin exhibit reduced hepcidin expression and develop iron loading similar to that seen in mice lacking hemojuvelin alone, suggesting that hemojuvelin is required for matriptase2 function (Finberg et al., 2010). This is supported by cell culture experiments showing that matriptase2 degrades hemojuvelin to produce fragments that, unlike the cleavage mediated by furin, are unable to bind BMPs (Maxson et al., 2010; Silvestri et al., 2008b). In addition, membrane-bound hemojuvelin protein levels are reduced in cells overexpressing matriptase2 (Silvestri et al., 2008b); however, in vivo data have questioned this model. Surprisingly, hepatic hemojuvelin levels are decreased in mice lacking functional matriptase2 (Krijt et al., 2011). In addition, hemojuvelin heterozygous knockout mice show no iron phenotype (Huang et al., 2005), indicating that hemojuvelin is not rate limiting for BMP signaling. Further studies showing an interaction between matriptase2 and neogenin suggest that the protease is internalized with the hemojuvelin/neogenin complex after BMP binding and that matriptase2-mediated degradation of hemojuvelin occurs in an internal compartment (Enns et al., 2012). Indeed, evidence suggests that the production of soluble hemojuvelin by furin also originates from an internal pool of hemojuvelin formed by the retrograde trafficking of hemojuvelin from the cell surface (Enns et al., 2012; Silvestri et al., 2008a).

The expression of matriptase2 is regulated by iron in rodents, although conflicting results exist. Meynard et al. (2011) showed that *TMPRSS6* mRNA and matriptase2 protein levels are stimulated by iron loading or exogenous BMP6 injection in cultured cells and in mice. This increase was due to the BMP6-dependent production of ID1 because small interfering RNA-mediated knockdown of ID1 in cultured cells greatly diminished BMP6 stimulation of *TMPRSS6* mRNA. Thus it was

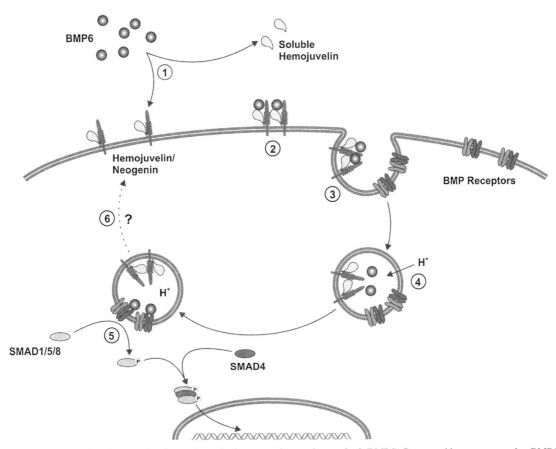

FIGURE 15.2   Activation of the SMAD signaling pathway by bone morphogenetic protein-6 (BMP6). Recent evidence suggests that BMP6 binds to the hemojuvelin/neogenin complex on the surface of hepatocytes (1), causing clustering of the ternary complex (2). The complex, along with the BMP receptor complex, is then internalized (3) and the endosome acidified (4), leading to the release of BMP6 from hemojuvelin. BMP6 is then free to bind to its receptor complex, thereby inducing SMAD signaling (5). Whether the proteins within the endosome are recycled back to the cell membrane is not known (6).

proposed that BMP6-mediated matriptase2 induction forms a negative feedback loop to tightly control hepcidin expression in a similar manner to SMAD6 and SMAD7. In contrast, Zhang et al. (2011) showed that acute iron deficiency in rats increases hepatic matriptase2 protein levels without altering *Tmprss6* mRNA. Further studies by this group showed that the increase in matriptase2 results from a decrease in lysosomal degradation of the protein (Zhao et al., 2015). They saw no evidence of *TMPRSS6* mRNA regulation by BMP6 or ID1 in cultured cells nor by iron loading in rat liver. There is clearly a need for further research into the regulation of matriptase2 function.

HFE and Transferrin Receptor-2

A role for the major histocompatibility complex class I molecule HFE in iron homeostasis became evident when it was reported as the gene mutated in the most common form of the hereditary iron-loading disorder hemochromatosis (Frazer and Anderson, 2003). It was subsequently shown in hemochromosis patients and *Hfe* knockout mice that disruption of HFE causes a reduction in hepcidin production (Bridle et al., 2003). Indeed, iron loading is reduced in *Hfe* knockout mice if they are crossed with mice constitutively overexpressing hepcidin (Nicolas et al., 2003). Although BMP6 responds appropriately to iron loading when HFE is defective, studies in humans and mice suggest that SMAD phosphorylation is reduced, indicating that HFE modulates other components of the BMP/SMAD signaling pathway (Kautz et al., 2009; Ryan et al., 2010).

The function of HFE remains poorly understood. Shortly after its discovery, HFE was found to interact with transferrin receptor (TFR)-1 (Parkkila et al., 1997), a cell membrane protein involved in the uptake of transferrin-bound iron from the circulation. The HFE binding site on TFR1 overlaps with that of transferrin (West et al., 2001), and HFE and transferrin have been shown to compete for binding to TFR1 at the cell surface (Giannetti and Bjorkman, 2004). Rodent studies have confirmed the importance of this competition because mice engineered to prevent Hfe/Tfr1 interaction develop iron deficiency due to elevated hepcidin production, and those in which binding is favored develop iron loading due to hepcidin suppression (Schmidt et al., 2008). It is currently thought that the HFE/TFR1 complex detects diferric transferrin in the circulation as an indicator of body iron usage (Frazer and Anderson, 2003). Higher iron requirements decrease diferric transferrin levels, favoring HFE/TFR1 interaction and hepcidin suppression. In contrast, elevated diferric transferrin out-competes HFE for TFR1 binding, thus signaling an increase in hepcidin expression. Supporting this hypothesis are several studies showing a correlation between circulating diferric transferrin levels and hepcidin expression (Frazer et al., 2002; Wilkins et al., 2006).

Also implicated in the detection of circulating diferric transferrin is TFR2. Mutations in this TFR1 homolog cause a rare form of hereditary hemochromatosis (Darshan et al., 2010). This is due to reduced SMAD signaling and lower hepcidin production, despite iron loading and increased BMP6 expression (Corradini et al., 2011). TFR2 binds diferric transferrin with a 25-fold lower affinity than TFR1 (West et al., 2000) and, unlike TFR1, is not regulated by intracellular iron levels (Fleming et al., 2000). In contrast, the TFR2 protein is stabilized by diferric transferrin binding (Johnson and Enns, 2004; Robb and Wessling-Resnick, 2004), suggesting that increased TFR2 protein levels somehow stimulate hepcidin production. However, overexpressing Tfr2 in mice has no effect on hepcidin expression, implying that TFR2 protein levels are not a limiting factor in hepcidin regulation (Gao et al., 2010).

It has been suggested that increasing diferric transferrin levels causes the HFE/TFR1 complex to dissociate, allowing HFE to interact with TFR2 (Goswami and Andrews, 2006; Fig. 15.3). Indeed, interactions between HFE and TFR2 have been detected in cultured cells (Goswami and Andrews, 2006), and mapping of the binding sites suggests that HFE and diferric transferrin occupy different sites on TFR2 (Chen et al., 2007), allowing HFE to bind when diferric transferrin is elevated. Further cell culture studies suggest that TFR2, HFE, and hemojuvelin form a complex, with hemojuvelin and TFR2 competing with TFR1 for HFE binding (D'Alessio et al., 2012). However, complex formation is controversial because other studies have failed to find an association between HFE and TFR2 (Rishi et al., 2013; West et al., 2000). Furthermore, the loss of function of HFE and TFR2 in humans and mice produces a more pronounced iron loading and reduction in hepcidin expression than the loss of either molecule alone (Pietrangelo et al., 2005; Wallace et al., 2009), suggesting that HFE and TFR2 belong to separate hepcidin regulatory pathways. A recent study suggests that HFE, when not bound to TFR1, binds and stabilizes the BMP receptor ALK3, providing an alternative explanation for hepcidin regulation by HFE (Wu et al., 2014).

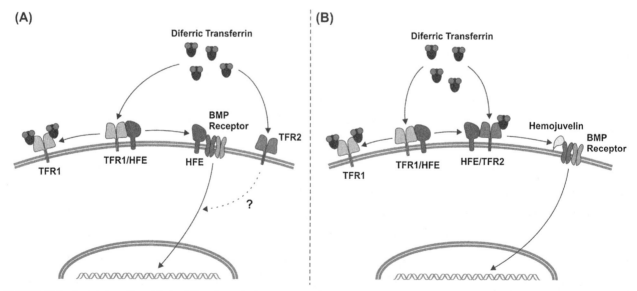

FIGURE 15.3 Regulation of hepcidin by HFE and transferrin receptor (TFR)-1. Two main hypotheses describing the regulation of hepcidin by HFE and TFR2 have been proposed. (A) The hypothesis in which HFE and TFR2 belong to separate pathways. Diferric transferrin disrupts the HFE/TFR1 complex by binding to TFR1. Free HFE then affects SMAD signaling, possibly by stabilizing the type I bone morphogenetic protein (BMP) receptor activin like kinase 3 (ALK3). The binding of diferric transferrin to TFR2 affects SMAD signaling by an unknown mechanism. The opposing hypothesis suggests that HFE and TFR2 are part of the same pathway as depicted in panel B. TFR2 is stabilized by diferric transferrin. In addition, diferric transferrin disrupts the HFE/TFR1 complex, allowing HFE to also bind to TFR2. The HFE/TFR2 complex affects SMAD signaling, possibly by interacting with hemojuvelin and the BMP receptor complex.

The Erythroid Regulator

Developing erythrocytes consume more than 80% of the iron that enters the circulation (Finch, 1994); therefore it is not surprising that erythropoiesis has a strong influence on hepcidin expression. However, it is not the rate of erythropoiesis per se that is important but rather the adequacy of the iron supply to developing blood cells (Finch, 1994). The effect of this so-called "erythroid regulator" is seen in conditions such as β-thalassemia, in which continued premature destruction of erythrocytes stimulates erythropoiesis, greatly increasing the iron demands of the erythroid marrow. When these demands cannot be met by circulating iron, the erythroid regulator decreases hepatic hepcidin production, causing increased dietary iron absorption and subsequent body iron loading (Frazer et al., 2012).

The molecular signals by which the iron requirements of the erythroid marrow signal changes in hepcidin expression are poorly understood. Bmp6 expression is elevated appropriately in response to iron loading in β-thalassemic mice; however, Smad phosphorylation is not increased (Frazer et al., 2012), indicating that the erythroid regulator inhibits the ability of BMP6 to stimulate hepcidin expression. Functional matriptase2 is also important for erythroid regulator signaling because decreasing *Tmprss6* expression in β-thalassemic mice increases hepcidin expression and reduces iron loading (Nai et al., 2012). HFE, TFR2, and hemojuvelin all appear to be dispensable for erythroid regulator function because hepcidin can still be suppressed by an erythropoietic stimulus when these molecules are not functional (Kautz et al., 2014; Ramos et al., 2011), although, because hepcidin is already decreased in these knockout mice, it is difficult to determine if the effect of the erythroid regulator is blunted in the absence of these molecules.

The identity of the molecule(s) signaling erythroid marrow iron requirements to the liver has been the focus of much research. The first real contender for the erythroid regulator was growth differentiation factor-15 (GDF15). Similar to BMP6, GDF15 is a member of the TGFβ superfamily and is expressed highly in developing erythrocytes (Tanno et al., 2007). Serum levels are increased in patients with β-thalassemia, and GDF15 decreases hepcidin production in primary human hepatocytes, making it a likely erythroid regulator. However, other studies have not found any correlation between GDF15 and hepcidin production (Ashby et al., 2010; Li et al., 2015), and a study using knockout mice showed that Gdf15 was not required for hepcidin reduction after phlebotomy (Casanovas et al., 2013). GDF15 is now thought to be necessary for normal erythrocyte development (Ramirez et al., 2009). Another molecule expressed by developing erythrocytes is twisted gastrulation-1 (TWSG1). TWSG1 is a BMP agonist/antagonist highly expressed in the bone marrow of β-thalassemic mice and has been shown to inhibit BMP-induced hepcidin expression in primary cultures of hepatocytes (Tanno et al., 2009). However, TWSG1 has not yet been detected in the circulation of humans or mice, and its role in hepcidin regulation in vivo has not been fully defined.

The most promising candidate for the erythroid regulator is the recently discovered erythroferrone. Similar to GDF15 and TWSG1, erythroferrone is highly expressed in developing erythrocytes (Kautz et al., 2014). Knockout mouse models show that erythroferrone is required for the reduction in hepcidin that occurs after stimulated erythropoiesis. In addition, exogenous erythroferrone reduces hepcidin expression in mice and cultured cells, although how erythroferrone signals changes in hepcidin expression are not known. β-Thalassemic mice lacking erythroferrone have higher hepcidin levels and reduced iron loading than β-thalassemic mice with functional erythroferrone, although β-thalassemic mice lacking erythroferrone still show significantly greater iron loading than wild-type mice. This suggests that other erythroid regulators may exist, with each playing a role in regulating hepcidin expression during stimulated erythropoiesis.

Closely related to stimulated erythropoiesis is the effect of hypoxia, which also reduces hepcidin production (Frazer and Anderson, 2014). However, because hypoxia increases erythropoietin production and stimulates erythropoiesis, it is unclear whether hypoxia can directly affect hepcidin expression or whether it acts by stimulating erythropoiesis and inducing the erythroid regulator. Hypoxia-inducible factors (HIFs) are transcription factors implicated in most aspects of hypoxia-induced gene expression (Mastrogiannaki et al., 2012). Studies in cultured cells and iron-deficient mice demonstrated that HIFs can bind to hypoxia-response elements in the hepcidin promoter and inhibit transcription (Peyssonnaux et al., 2007). Other potential mechanisms of hypoxia-induced hepcidin repression include a decrease in SMAD4 expression (Chaston et al., 2011), increase in the expression of furin (Silvestri et al., 2008a) and TMPRSS6 (Lakhal et al., 2011), and the induction of a novel platelet-derived growth factor-BB pathway (Sonnweber et al., 2014). However, studies using knockout mice have suggested that HIF-1α is the only HIF to affect hepcidin production and that its effect in vivo is minor (Mastrogiannaki et al., 2012; Peyssonnaux et al., 2007). Further studies in mice suggest that the inhibition of hepcidin expression by hypoxia is dependent on increased erythropoietic activity and does not occur if erythropoiesis is suppressed (Liu et al., 2012; Mastrogiannaki et al., 2012), consistent with an indirect effect of hypoxia on hepcidin production.

Inflammation

Hepcidin expression is increased by inflammatory stimuli (Pigeon et al., 2001). This produces a hypoferremic state that is thought to play an important role in host defense by withholding iron from invading pathogens (Ganz and Nemeth, 2015).

Prolonged inflammation and hypoferremia can also withhold iron from developing erythrocytes and contribute to the anemia of inflammation that occurs in conditions such as chronic kidney disease and infections (Kim et al., 2014). Indeed, in some mouse models of anemia of inflammation, inhibiting hepcidin production can restore hematological parameters to normal levels (Theurl et al., 2011), although others suggest that whereas hepcidin contributes to the development of the anemia, other factors such as reduced erythrocyte lifespan and depressed erythropoiesis also play a role (Kim et al., 2014).

Hepcidin expression is upregulated by several inflammatory cytokines, including interleukin (IL)-6, IL-1, IL-22, and interferon-α (Schmidt, 2015; Wallace and Subramaniam, 2015), although IL-6 appears to be the main cytokine responsible for hepcidin induction during inflammation. The IL-6 signaling pathway has been well characterized, with the ligand binding to a receptor complex consisting of IL-6R and gp130 (Ataie-Kachoie et al., 2014). Upon binding and activation of the receptor complex, janus kinases are phosphorylated, which in turn phosphorylate gp130 and subsequently cytoplasmic signal transducer and activator of transcription-3 (STAT3). STAT3 then dimerizes, translocates to the nucleus, and binds to specific response elements in the promoters of target genes. gp130 and STAT3 are essential for IL-6 stimulation of hepcidin expression (Pietrangelo et al., 2007; Wrighting and Andrews, 2006), and potential STAT3 response elements have been identified in the hepcidin promoter (Verga Falzacappa et al., 2008; Wrighting and Andrews, 2006).

Interestingly, IL-6 fails to stimulate hepcidin production in liver-specific Smad4 knockout mice (Wang et al., 2005), implying that a functional BMP6/SMAD pathway is required for IL-6 signaling. In fact, the deletion of the BMP response element in close proximity to the STAT3 response element in the proximal hepcidin promoter abrogates IL-6-induced hepcidin expression in cultured cells (Verga Falzacappa et al., 2008). In addition, the BMP type I receptor ALK3 was recently shown to be essential for IL-6 induction of hepcidin (Mayeur et al., 2014b). Modulation of the BMP6/SMAD signaling pathway by inflammation has also been observed. Lipopolysaccharide (LPS) treatment of mice stimulates the production of activin B, which binds to the BMP receptor complex, inducing SMAD phosphorylation and hepcidin expression (Besson-Fournier et al., 2012). *Tmprss6* expression has also been shown to be downregulated by LPS in mice, potentially increasing BMP6/SMAD signaling (Meynard et al., 2013).

CONCLUSION

Since the discovery of hepcidin in 2000, our understanding of how the body maintains iron homeostasis has enormously grown. The importance of hepcidin in this process is reflected in the complexity of the pathways regulating its expression. Central to this is the BMP6/SMAD signaling pathway, with most regulators of hepcidin expression either modulating this pathway or requiring its activity to function. Despite our increased knowledge, there are still several molecules that are known to affect hepcidin expression but for which the pathways are poorly characterized, and other pathways in which the molecules involved remain unclear. Indeed, there are likely still molecules to be discovered, suggesting that the network of pathways regulating hepcidin expression will continue to become more complex.

REFERENCES

Altamura, S., Kessler, R., Grone, H.J., Gretz, N., Hentze, M.W., Galy, B., Muckenthaler, M.U., 2014. Resistance of ferroportin to hepcidin binding causes exocrine pancreatic failure and fatal iron overload. Cell Metab. 20, 359–367.

Andriopoulos Jr., B., Corradini, E., Xia, Y., Faasse, S.A., Chen, S., Grgurevic, L., Knutson, M.D., Pietrangelo, A., Vukicevic, S., Lin, H.Y., Babitt, J.L., 2009. BMP6 is a key endogenous regulator of hepcidin expression and iron metabolism. Nat. Genet. 41, 482–487.

Ashby, D.R., Gale, D.P., Busbridge, M., Murphy, K.G., Duncan, N.D., Cairns, T.D., Taube, D.H., Bloom, S.R., Tam, F.W., Chapman, R., Maxwell, P.H., Choi, P., 2010. Erythropoietin administration in humans causes a marked and prolonged reduction in circulating hepcidin. Haematologica 95, 505–508.

Ataie-Kachoie, P., Pourgholami, M.H., Richardson, D.R., Morris, D.L., 2014. Gene of the month: interleukin 6 (IL-6). J. Clin. Pathol. 67, 932–937.

Babitt, J.L., Huang, F.W., Wrighting, D.M., Xia, Y., Sidis, Y., Samad, T.A., Campagna, J.A., Chung, R.T., Schneyer, A.L., Woolf, C.J., Andrews, N.C., Lin, H.Y., 2006. Bone morphogenetic protein signaling by hemojuvelin regulates hepcidin expression. Nat. Genet. 38, 531–539.

Besson-Fournier, C., Latour, C., Kautz, L., Bertrand, J., Ganz, T., Roth, M.P., Coppin, H., 2012. Induction of activin B by inflammatory stimuli up-regulates expression of the iron-regulatory peptide hepcidin through Smad1/5/8 signaling. Blood 120, 431–439.

Bridle, K.R., Frazer, D.M., Wilkins, S.J., Dixon, J.L., Purdie, D.M., Crawford, D.H., Subramaniam, V.N., Powell, L.W., Anderson, G.J., Ramm, G.A., 2003. Disrupted hepcidin regulation in HFE-associated haemochromatosis and the liver as a regulator of body iron homoeostasis. Lancet 361, 669–673.

Casanovas, G., Vujic Spasic, M., Casu, C., Rivella, S., Strelau, J., Unsicker, K., Muckenthaler, M.U., 2013. The murine growth differentiation factor 15 is not essential for systemic iron homeostasis in phlebotomized mice. Haematologica 98, 444–447.

Chaston, T.B., Matak, P., Pourvali, K., Srai, S.K., McKie, A.T., Sharp, P.A., 2011. Hypoxia inhibits hepcidin expression in HuH7 hepatoma cells via decreased SMAD4 signaling. Am. J. Physiol. Cell Physiol. 300, C888–C895.

Chen, J., Chloupkova, M., Gao, J., Chapman-Arvedson, T.L., Enns, C.A., 2007. HFE modulates transferrin receptor 2 levels in hepatoma cells via interactions that differ from transferrin receptor 1-HFE interactions. J. Biol. Chem. 282, 36862–36870.

Corradini, E., Rozier, M., Meynard, D., Odhiambo, A., Lin, H.Y., Feng, Q., Migas, M.C., Britton, R.S., Babitt, J.L., Fleming, R.E., 2011. Iron regulation of hepcidin despite attenuated Smad1,5,8 signaling in mice without transferrin receptor 2 or Hfe. Gastroenterology 141, 1907–1914.

D'Alessio, F., Hentze, M.W., Muckenthaler, M.U., 2012. The hemochromatosis proteins HFE, TfR2, and HJV form a membrane-associated protein complex for hepcidin regulation. J. Hepatol. 57, 1052–1060.

Darshan, D., Frazer, D.M., Anderson, G.J., 2010. Molecular basis of iron-loading disorders. Expert Rev. Mol. Med. 12, e36.

Du, X., She, E., Gelbart, T., Truksa, J., Lee, P., Xia, Y., Khovananth, K., Mudd, S., Mann, N., Moresco, E.M., Beutler, E., Beutler, B., 2008. The serine protease TMPRSS6 is required to sense iron deficiency. Science 320, 1088–1092.

Enns, C.A., Ahmed, R., Wang, J., Ueno, A., Worthen, C., Tsukamoto, H., Zhang, A.S., 2013. Increased iron loading induces Bmp6 expression in the non-parenchymal cells of the liver independent of the BMP-signaling pathway. PLoS One 8, e60534.

Enns, C.A., Ahmed, R., Zhang, A.S., 2012. Neogenin interacts with matriptase-2 to facilitate hemojuvelin cleavage. J. Biol. Chem. 287, 35104–35117.

Feng, Q., Migas, M.C., Waheed, A., Britton, R.S., Fleming, R.E., 2012. Ferritin upregulates hepatic expression of bone morphogenetic protein 6 and hepcidin in mice. Am. J. Physiol. Gastrointest. Liver Physiol. 302, G1397–G1404.

Fernandes, A., Preza, G.C., Phung, Y., De Domenico, I., Kaplan, J., Ganz, T., Nemeth, E., 2009. The molecular basis of hepcidin-resistant hereditary hemochromatosis. Blood 114, 437–443.

Finberg, K.E., Heeney, M.M., Campagna, D.R., Aydinok, Y., Pearson, H.A., Hartman, K.R., Mayo, M.M., Samuel, S.M., Strouse, J.J., Markianos, K., Andrews, N.C., Fleming, M.D., 2008. Mutations in TMPRSS6 cause iron-refractory iron deficiency anemia (IRIDA). Nat. Genet. 40, 569–571.

Finberg, K.E., Whittlesey, R.L., Fleming, M.D., Andrews, N.C., 2010. Down-regulation of Bmp/Smad signaling by Tmprss6 is required for maintenance of systemic iron homeostasis. Blood 115, 3817–3826.

Finch, C., 1994. Regulators of iron balance in humans. Blood 84, 1697–1702.

Fleming, R.E., Migas, M.C., Holden, C.C., Waheed, A., Britton, R.S., Tomatsu, S., Bacon, B.R., Sly, W.S., 2000. Transferrin receptor 2: continued expression in mouse liver in the face of iron overload and in hereditary hemochromatosis. Proc. Natl. Acad. Sci. U.S.A. 97, 2214–2219.

Frazer, D.M., Anderson, G.J., 2003. The orchestration of body iron intake: how and where do enterocytes receive their cues? Blood Cells Mol. Dis. 30, 288–297.

Frazer, D.M., Anderson, G.J., 2014. The regulation of iron transport. Biofactors 40, 206–214.

Frazer, D.M., Wilkins, S.J., Becker, E.M., Vulpe, C.D., McKie, A.T., Trinder, D., Anderson, G.J., 2002. Hepcidin expression inversely correlates with the expression of duodenal iron transporters and iron absorption in rats. Gastroenterology 123, 835–844.

Frazer, D.M., Wilkins, S.J., Darshan, D., Badrick, A.C., McLaren, G.D., Anderson, G.J., 2012. Stimulated erythropoiesis with secondary iron loading leads to a decrease in hepcidin despite an increase in bone morphogenetic protein 6 expression. Br. J. Haematol. 157, 615–626.

Ganz, T., Nemeth, E., 2012. Hepcidin and iron homeostasis. Biochim. Biophys. Acta 1823, 1434–1443.

Ganz, T., Nemeth, E., 2015. Iron homeostasis in host defence and inflammation. Nat. Rev. Immunol. 15 (8), 500–510.

Gao, J., Chen, J., De Domenico, I., Koeller, D.M., Harding, C.O., Fleming, R.E., Koeberl, D.D., Enns, C.A., 2010. Hepatocyte-targeted HFE and TFR2 control hepcidin expression in mice. Blood 115, 3374–3381.

Giannetti, A.M., Bjorkman, P.J., 2004. HFE and transferrin directly compete for transferrin receptor in solution and at the cell surface. J. Biol. Chem. 279, 25866–25875.

Gkouvatsos, K., Fillebeen, C., Daba, A., Wagner, J., Sebastiani, G., Pantopoulos, K., 2014. Iron-dependent regulation of hepcidin in Hjv-/- mice: evidence that hemojuvelin is dispensable for sensing body iron levels. PLoS One 9, e85530.

Goswami, T., Andrews, N.C., 2006. Hereditary hemochromatosis protein, HFE, interaction with transferrin receptor 2 suggests a molecular mechanism for mammalian iron sensing. J. Biol. Chem. 281, 28494–28498.

Healey, E.G., Bishop, B., Elegheert, J., Bell, C.H., Padilla-Parra, S., Siebold, C., 2015. Repulsive guidance molecule is a structural bridge between neogenin and bone morphogenetic protein. Nat. Struct. Mol. Biol. 22, 458–465.

Huang, F.W., Pinkus, J.L., Pinkus, G.S., Fleming, M.D., Andrews, N.C., 2005. A mouse model of juvenile hemochromatosis. J. Clin. Invest. 115, 2187–2191.

Johnson, M.B., Enns, C.A., 2004. Diferric transferrin regulates transferrin receptor 2 protein stability. Blood 104, 4287–4293.

Jordan, J.B., Poppe, L., Haniu, M., Arvedson, T., Syed, R., Li, V., Kohno, H., Kim, H., Schnier, P.D., Harvey, T.S., Miranda, L.P., Cheetham, J., Sasu, B.J., 2009. Hepcidin revisited, disulfide connectivity, dynamics, and structure. J. Biol. Chem. 284, 24155–24167.

Kautz, L., Jung, G., Valore, E.V., Rivella, S., Nemeth, E., Ganz, T., 2014. Identification of erythroferrone as an erythroid regulator of iron metabolism. Nat. Genet. 46, 678–684.

Kautz, L., Meynard, D., Besson-Fournier, C., Darnaud, V., Al Saati, T., Coppin, H., Roth, M.P., 2009. BMP/Smad signaling is not enhanced in Hfe-deficient mice despite increased Bmp6 expression. Blood 114, 2515–2520.

Kautz, L., Meynard, D., Monnier, A., Darnaud, V., Bouvet, R., Wang, R.H., Deng, C., Vaulont, S., Mosser, J., Coppin, H., Roth, M.P., 2008. Iron regulates phosphorylation of Smad1/5/8 and gene expression of Bmp6, Smad7, Id1, and Atoh8 in the mouse liver. Blood 112, 1503–1509.

Kim, A., Fung, E., Parikh, S.G., Valore, E.V., Gabayan, V., Nemeth, E., Ganz, T., 2014. A mouse model of anemia of inflammation: complex pathogenesis with partial dependence on hepcidin. Blood 123, 1129–1136.

Krause, A., Neitz, S., Magert, H.J., Schulz, A., Forssmann, W.G., Schulz-Knappe, P., Adermann, K., 2000. LEAP-1, a novel highly disulfide-bonded human peptide, exhibits antimicrobial activity. FEBS Lett. 480, 147–150.

Krijt, J., Fujikura, Y., Ramsay, A.J., Velasco, G., Necas, E., 2011. Liver hemojuvelin protein levels in mice deficient in matriptase-2 (Tmprss6). Blood Cells Mol. Dis. 47, 133–137.

Lakhal, S., Schodel, J., Townsend, A.R., Pugh, C.W., Ratcliffe, P.J., Mole, D.R., 2011. Regulation of type II transmembrane serine proteinase TMPRSS6 by hypoxia-inducible factors: new link between hypoxia signaling and iron homeostasis. J. Biol. Chem. 286, 4090–4097.

Lee, D.H., Zhou, L.J., Zhou, Z., Xie, J.X., Jung, J.U., Liu, Y., Xi, C.X., Mei, L., Xiong, W.C., 2010. Neogenin inhibits HJV secretion and regulates BMP-induced hepcidin expression and iron homeostasis. Blood 115, 3136–3145.

Lesbordes-Brion, J.C., Viatte, L., Bennoun, M., Lou, D.Q., Ramey, G., Houbron, C., Hamard, G., Kahn, A., Vaulont, S., 2006. Targeted disruption of the hepcidin 1 gene results in severe hemochromatosis. Blood 108, 1402–1405.

Li, X.Y., Ying, J., Li, J.H., Zhu, S.L., Li, J., Pai, P., 2015. Growth differentiation factor GDF-15 does not influence iron metabolism in stable chronic haemodialysis patients. Ann. Clin. Biochem. 52, 399–403.

Lin, L., Goldberg, Y.P., Ganz, T., 2005. Competitive regulation of hepcidin mRNA by soluble and cell-associated hemojuvelin. Blood 106, 2884–2889.

Liu, Q., Davidoff, O., Niss, K., Haase, V.H., 2012. Hypoxia-inducible factor regulates hepcidin via erythropoietin-induced erythropoiesis. J. Clin. Invest. 122, 4635–4644.

Lou, D.Q., Nicolas, G., Lesbordes, J.C., Viatte, L., Grimber, G., Szajnert, M.F., Kahn, A., Vaulont, S., 2004. Functional differences between hepcidin 1 and 2 in transgenic mice. Blood 103, 2816–2821.

Mastrogiannaki, M., Matak, P., Mathieu, J.R., Delga, S., Mayeux, P., Vaulont, S., Peyssonnaux, C., 2012. Hepatic hypoxia-inducible factor-2 down-regulates hepcidin expression in mice through an erythropoietin-mediated increase in erythropoiesis. Haematologica 97, 827–834.

Maxson, J.E., Chen, J., Enns, C.A., Zhang, A.S., 2010. Matriptase-2- and proprotein convertase-cleaved forms of hemojuvelin have different roles in the down-regulation of hepcidin expression. J. Biol. Chem. 285, 39021–39028.

Mayeur, C., Leyton, P.A., Kolodziej, S.A., Yu, B., Bloch, K.D., 2014a. BMP type II receptors have redundant roles in the regulation of hepatic hepcidin gene expression and iron metabolism. Blood 124, 2116–2123.

Mayeur, C., Lohmeyer, L.K., Leyton, P., Kao, S.M., Pappas, A.E., Kolodziej, S.A., Spagnolli, E., Yu, B., Galdos, R.L., Yu, P.B., Peterson, R.T., Bloch, D.B., Bloch, K.D., Steinbicker, A.U., 2014b. The type I BMP receptor Alk3 is required for the induction of hepatic hepcidin gene expression by interleukin-6. Blood 123, 2261–2268.

Meynard, D., Kautz, L., Darnaud, V., Canonne-Hergaux, F., Coppin, H., Roth, M.P., 2009. Lack of the bone morphogenetic protein BMP6 induces massive iron overload. Nat. Genet. 41, 478–481.

Meynard, D., Sun, C.C., Wu, Q., Chen, W., Chen, S., Nelson, C.N., Waters, M.J., Babitt, J.L., Lin, H.Y., 2013. Inflammation regulates TMPRSS6 expression via STAT5. PLoS One 8, e82127.

Meynard, D., Vaja, V., Sun, C.C., Corradini, E., Chen, S., Lopez-Otin, C., Grgurevic, L., Hong, C.C., Stirnberg, M., Gutschow, M., Vukicevic, S., Babitt, J.L., Lin, H.Y., 2011. Regulation of TMPRSS6 by BMP6 and iron in human cells and mice. Blood 118, 747–756.

Mleczko-Sanecka, K., Casanovas, G., Ragab, A., Breitkopf, K., Muller, A., Boutros, M., Dooley, S., Hentze, M.W., Muckenthaler, M.U., 2010. SMAD7 controls iron metabolism as a potent inhibitor of hepcidin expression. Blood 115, 2657–2665.

Nai, A., Pagani, A., Mandelli, G., Lidonnici, M.R., Silvestri, L., Ferrari, G., Camaschella, C., 2012. Deletion of TMPRSS6 attenuates the phenotype in a mouse model of beta-thalassemia. Blood 119, 5021–5029.

Nemeth, E., Preza, G.C., Jung, C.L., Kaplan, J., Waring, A.J., Ganz, T., 2006. The N-terminus of hepcidin is essential for its interaction with ferroportin: structure-function study. Blood 107, 328–333.

Nemeth, E., Tuttle, M.S., Powelson, J., Vaughn, M.B., Donovan, A., Ward, D.M., Ganz, T., Kaplan, J., 2004. Hepcidin regulates cellular iron efflux by binding to ferroportin and inducing its internalization. Science 306, 2090–2093.

Nicolas, G., Bennoun, M., Devaux, I., Beaumont, C., Grandchamp, B., Kahn, A., Vaulont, S., 2001. Lack of hepcidin gene expression and severe tissue iron overload in upstream stimulatory factor 2 (USF2) knockout mice. Proc. Natl. Acad. Sci. U.S.A. 98, 8780–8785.

Nicolas, G., Bennoun, M., Porteu, A., Mativet, S., Beaumont, C., Grandchamp, B., Sirito, M., Sawadogo, M., Kahn, A., Vaulont, S., 2002. Severe iron deficiency anemia in transgenic mice expressing liver hepcidin. Proc. Natl. Acad. Sci. U.S.A. 99, 4596–4601.

Nicolas, G., Viatte, L., Lou, D.Q., Bennoun, M., Beaumont, C., Kahn, A., Andrews, N.C., Vaulont, S., 2003. Constitutive hepcidin expression prevents iron overload in a mouse model of hemochromatosis. Nat. Genet. 34, 97–101.

Park, C.H., Valore, E.V., Waring, A.J., Ganz, T., 2001. Hepcidin, a urinary antimicrobial peptide synthesized in the liver. J. Biol. Chem. 276, 7806–7810.

Parkkila, S., Waheed, A., Britton, R.S., Bacon, B.R., Zhou, X.Y., Tomatsu, S., Fleming, R.E., Sly, W.S., 1997. Association of the transferrin receptor in human placenta with HFE, the protein defective in hereditary hemochromatosis. Proc. Natl. Acad. Sci. U.S.A. 94, 13198–13202.

Parrow, N.L., Fleming, R.E., 2014. Bone morphogenetic proteins as regulators of iron metabolism. Annu. Rev. Nutr. 34, 77–94.

Patel, N., Varghese, J., Masaratana, P., Latunde-Dada, G.O., Jacob, M., Simpson, R.J., McKie, A.T., 2014. The transcription factor ATOH8 is regulated by erythropoietic activity and regulates HAMP transcription and cellular pSMAD1,5,8 levels. Br. J. Haematol. 164, 586–596.

Peyssonnaux, C., Zinkernagel, A.S., Schuepbach, R.A., Rankin, E., Vaulont, S., Haase, V.H., Nizet, V., Johnson, R.S., 2007. Regulation of iron homeostasis by the hypoxia-inducible transcription factors (HIFs). J. Clin. Invest. 117, 1926–1932.

Pietrangelo, A., Caleffi, A., Henrion, J., Ferrara, F., Corradini, E., Kulaksiz, H., Stremmel, W., Andreone, P., Garuti, C., 2005. Juvenile hemochromatosis associated with pathogenic mutations of adult hemochromatosis genes. Gastroenterology 128, 470–479.

Pietrangelo, A., Dierssen, U., Valli, L., Garuti, C., Rump, A., Corradini, E., Ernst, M., Klein, C., Trautwein, C., 2007. STAT3 is required for IL-6-gp130-dependent activation of hepcidin in vivo. Gastroenterology 132, 294–300.

Pigeon, C., Ilyin, G., Courselaud, B., Leroyer, P., Turlin, B., Brissot, P., Loreal, O., 2001. A new mouse liver-specific gene, encoding a protein homologous to human antimicrobial peptide hepcidin, is overexpressed during iron overload. J. Biol. Chem. 276, 7811–7819.

Preza, G.C., Pinon, R., Ganz, T., Nemeth, E., 2013. Cellular catabolism of the iron-regulatory peptide hormone hepcidin. PLoS One 8, e58934.

Preza, G.C., Ruchala, P., Pinon, R., Ramos, E., Qiao, B., Peralta, M.A., Sharma, S., Waring, A., Ganz, T., Nemeth, E., 2011. Minihepcidins are rationally designed small peptides that mimic hepcidin activity in mice and may be useful for the treatment of iron overload. J. Clin. Invest. 121, 4880–4888.

Qiao, B., Sugianto, P., Fung, E., Del-Castillo-Rueda, A., Moran-Jimenez, M.J., Ganz, T., Nemeth, E., 2012. Hepcidin-induced endocytosis of ferroportin is dependent on ferroportin ubiquitination. Cell Metab. 15, 918–924.

Ramirez, J.M., Schaad, O., Durual, S., Cossali, D., Docquier, M., Beris, P., Descombes, P., Matthes, T., 2009. Growth differentiation factor 15 production is necessary for normal erythroid differentiation and is increased in refractory anaemia with ring-sideroblasts. Br. J. Haematol. 144, 251–262.

Ramos, P., Guy, E., Chen, N., Proenca, C.C., Gardenghi, S., Casu, C., Follenzi, A., Van Rooijen, N., Grady, R.W., de Sousa, M., Rivella, S., 2011. Enhanced erythropoiesis in Hfe-KO mice indicates a role for Hfe in the modulation of erythroid iron homeostasis. Blood 117, 1379–1389.

Rishi, G., Crampton, E.M., Wallace, D.F., Subramaniam, V.N., 2013. In situ proximity ligation assays indicate that hemochromatosis proteins Hfe and transferrin receptor 2 (Tfr2) do not interact. PLoS One 8, e77267.

Robb, A., Wessling-Resnick, M., 2004. Regulation of transferrin receptor 2 protein levels by transferrin. Blood 104, 4294–4299.

Ryan, J.D., Ryan, E., Fabre, A., Lawless, M.W., Crowe, J., 2010. Defective bone morphogenic protein signaling underlies hepcidin deficiency in HFE hereditary hemochromatosis. Hepatology 52, 1266–1273.

Schmidt, P.J., 2015. Regulation of iron metabolism by hepcidin under conditions of inflammation. J. Biol. Chem. 290, 18975–18983.

Schmidt, P.J., Toran, P.T., Giannetti, A.M., Bjorkman, P.J., Andrews, N.C., 2008. The transferrin receptor modulates Hfe-dependent regulation of hepcidin expression. Cell Metab. 7, 205–214.

Sham, R.L., Phatak, P.D., West, C., Lee, P., Andrews, C., Beutler, E., 2005. Autosomal dominant hereditary hemochromatosis associated with a novel ferroportin mutation and unique clinical features. Blood Cells Mol. Dis. 34, 157–161.

Shike, H., Shimizu, C., Lauth, X., Burns, J.C., 2004. Organization and expression analysis of the zebrafish hepcidin gene, an antimicrobial peptide gene conserved among vertebrates. Dev. Comp. Immunol. 28, 747–754.

Silvestri, L., Pagani, A., Camaschella, C., 2008a. Furin-mediated release of soluble hemojuvelin: a new link between hypoxia and iron homeostasis. Blood 111, 924–931.

Silvestri, L., Pagani, A., Nai, A., De Domenico, I., Kaplan, J., Camaschella, C., 2008b. The serine protease matriptase-2 (TMPRSS6) inhibits hepcidin activation by cleaving membrane hemojuvelin. Cell Metab. 8, 502–511.

Sonnweber, T., Nachbaur, D., Schroll, A., Nairz, M., Seifert, M., Demetz, E., Haschka, D., Mitterstiller, A.M., Kleinsasser, A., Burtscher, M., Trubsbach, S., Murphy, A.T., Wroblewski, V., Witcher, D.R., Mleczko-Sanecka, K., Vecchi, C., Muckenthaler, M.U., Pietrangelo, A., Theurl, I., Weiss, G., 2014. Hypoxia induced downregulation of hepcidin is mediated by platelet derived growth factor BB. Gut 63, 1951–1959.

Steinbicker, A.U., Bartnikas, T.B., Lohmeyer, L.K., Leyton, P., Mayeur, C., Kao, S.M., Pappas, A.E., Peterson, R.T., Bloch, D.B., Yu, P.B., Fleming, M.D., Bloch, K.D., 2011. Perturbation of hepcidin expression by BMP type I receptor deletion induces iron overload in mice. Blood 118, 4224–4230.

Tanno, T., Bhanu, N.V., Oneal, P.A., Goh, S.H., Staker, P., Lee, Y.T., Moroney, J.W., Reed, C.H., Luban, N.L., Wang, R.H., Eling, T.E., Childs, R., Ganz, T., Leitman, S.F., Fucharoen, S., Miller, J.L., 2007. High levels of GDF15 in thalassemia suppress expression of the iron regulatory protein hepcidin. Nat. Med. 13, 1096–1101.

Tanno, T., Porayette, P., Sripichai, O., Noh, S.J., Byrnes, C., Bhupatiraju, A., Lee, Y.T., Goodnough, J.B., Harandi, O., Ganz, T., Paulson, R.F., Miller, J.L., 2009. Identification of TWSG1 as a second novel erythroid regulator of hepcidin expression in murine and human cells. Blood 114, 181–186.

Theurl, I., Schroll, A., Sonnweber, T., Nairz, M., Theurl, M., Willenbacher, W., Eller, K., Wolf, D., Seifert, M., Sun, C.C., Babitt, J.L., Hong, C.C., Menhall, T., Gearing, P., Lin, H.Y., Weiss, G., 2011. Pharmacologic inhibition of hepcidin expression reverses anemia of chronic inflammation in rats. Blood 118, 4977–4984.

Truksa, J., Peng, H., Lee, P., Beutler, E., 2006. Bone morphogenetic proteins 2, 4, and 9 stimulate murine hepcidin 1 expression independently of Hfe, transferrin receptor 2 (Tfr2), and IL-6. Proc. Natl. Acad. Sci. U.S.A. 103, 10289–10293.

Valore, E.V., Ganz, T., 2008. Posttranslational processing of hepcidin in human hepatocytes is mediated by the prohormone convertase furin. Blood Cells Mol. Dis. 40, 132–138.

Verga Falzacappa, M.V., Casanovas, G., Hentze, M.W., Muckenthaler, M.U., 2008. A bone morphogenetic protein (BMP)-responsive element in the hepcidin promoter controls HFE2-mediated hepatic hepcidin expression and its response to IL-6 in cultured cells. J. Mol. Med. Berl. 86, 531–540.

Vujic Spasic, M., Sparla, R., Mleczko-Sanecka, K., Migas, M.C., Breitkopf-Heinlein, K., Dooley, S., Vaulont, S., Fleming, R.E., Muckenthaler, M.U., 2013. Smad6 and Smad7 are co-regulated with hepcidin in mouse models of iron overload. Biochim. Biophys. Acta 1832, 76–84.

Wallace, D.F., Subramaniam, V.N., 2015. Analysis of IL-22 contribution to hepcidin induction and hypoferremia during the response to LPS in vivo. Int. Immunol. 27 (6), 281–287.

Wallace, D.F., Summerville, L., Crampton, E.M., Frazer, D.M., Anderson, G.J., Subramaniam, V.N., 2009. Combined deletion of Hfe and transferrin receptor 2 in mice leads to marked dysregulation of hepcidin and iron overload. Hepatology 50, 1992–2000.

Wang, R.H., Li, C., Xu, X., Zheng, Y., Xiao, C., Zerfas, P., Cooperman, S., Eckhaus, M., Rouault, T., Mishra, L., Deng, C.X., 2005. A role of SMAD4 in iron metabolism through the positive regulation of hepcidin expression. Cell Metab. 2, 399–409.

West Jr., A.P., Bennett, M.J., Sellers, V.M., Andrews, N.C., Enns, C.A., Bjorkman, P.J., 2000. Comparison of the interactions of transferrin receptor and transferrin receptor 2 with transferrin and the hereditary hemochromatosis protein HFE. J. Biol. Chem. 275, 38135–38138.

West Jr., A.P., Giannetti, A.M., Herr, A.B., Bennett, M.J., Nangiana, J.S., Pierce, J.R., Weiner, L.P., Snow, P.M., Bjorkman, P.J., 2001. Mutational analysis of the transferrin receptor reveals overlapping HFE and transferrin binding sites. J. Mol. Biol. 313, 385–397.

Wilkins, S.J., Frazer, D.M., Millard, K.N., McLaren, G.D., Anderson, G.J., 2006. Iron metabolism in the hemoglobin-deficit mouse: correlation of diferric transferrin with hepcidin expression. Blood 107, 1659–1664.

Wrighting, D.M., Andrews, N.C., 2006. Interleukin-6 induces hepcidin expression through STAT3. Blood 108, 3204–3209.

Wu, X.G., Wang, Y., Wu, Q., Cheng, W.H., Liu, W., Zhao, Y., Mayeur, C., Schmidt, P.J., Yu, P.B., Wang, F., Xia, Y., 2014. HFE interacts with the BMP type I receptor ALK3 to regulate hepcidin expression. Blood 124, 1335–1343.

Zhang, A.S., Anderson, S.A., Meyers, K.R., Hernandez, C., Eisenstein, R.S., Enns, C.A., 2007. Evidence that inhibition of hemojuvelin shedding in response to iron is mediated through neogenin. J. Biol. Chem. 282, 12547–12556.

Zhang, A.S., Anderson, S.A., Wang, J., Yang, F., DeMaster, K., Ahmed, R., Nizzi, C.P., Eisenstein, R.S., Tsukamoto, H., Enns, C.A., 2011. Suppression of hepatic hepcidin expression in response to acute iron deprivation is associated with an increase of matriptase-2 protein. Blood 117, 1687–1699.

Zhang, A.S., West Jr., A.P., Wyman, A.E., Bjorkman, P.J., Enns, C.A., 2005. Interaction of hemojuvelin with neogenin results in iron accumulation in human embryonic kidney 293 cells. J. Biol. Chem. 280, 33885–33894.

Zhang, A.S., Yang, F., Wang, J., Tsukamoto, H., Enns, C.A., 2009. Hemojuvelin-neogenin interaction is required for bone morphogenic protein-4-induced hepcidin expression. J. Biol. Chem. 284, 22580–22589.

Zhao, N., Nizzi, C.P., Anderson, S.A., Wang, J., Ueno, A., Tsukamoto, H., Eisenstein, R.S., Enns, C.A., Zhang, A.S., 2015. Low intracellular iron increases the stability of matriptase-2. J. Biol. Chem. 290, 4432–4446.

Chapter 16

Genetic Rodent Models of Systemic Iron Homeostasis

Thomas Bartnikas, Carolina Herrera, Michael Pettiglio

Brown University, Providence, Rhode Island, United States

INTRODUCTION

Iron is essential for metabolic processes such as gas transport, oxidative phosphorylation, and nucleotide metabolism. It is found in heme groups and iron-sulfur clusters or bound directly to enzymes such as ribonucleotide reductase, nitric oxide synthase, and transport proteins such as transferrin. However, iron is toxic when present in excess. Toxicity is presumed to reflect iron-catalyzed formation of reactive oxygen species and other radicals deleterious to cellular function. To ensure that each cell, tissue, and organ is provided with sufficient yet not excessive amounts of iron, multiple pathways govern iron absorption, distribution, recycling, and excretion (Fig. 16.1).

Iron homeostasis begins with maternal–fetal iron transfer. After delivery, a neonate relies on milk intake for nutrient acquisition but after weaning must acquire iron from other foods. Dietary iron absorption is a major point of regulation of iron levels. Once iron is absorbed, it is trafficked through the body for storage or use in key metabolic processes. During protein, organelle, and cell turnover, iron is reclaimed, redistributed for reuse, stored, or excreted from the organism. In contrast to iron absorption, iron excretion is not viewed as a regulated pathway.

Genetic rodent models have been invaluable to the study of iron biology. Gene inactivation allows one to infer the role of a gene by observing the resulting phenotype, an advantage not conferred by pure dietary studies. This is not to say that dietary studies are not useful. A combination of genetic and dietary manipulation is often the best approach to dissect the function and mechanism of action of a particular gene in iron homeostasis.

This chapter summarizes most of the currently available models (Table 16.1). Models are arranged largely by the processes impacted by their genetic defects. Their origins are described as is relevant. If mice are referred to as deficient in

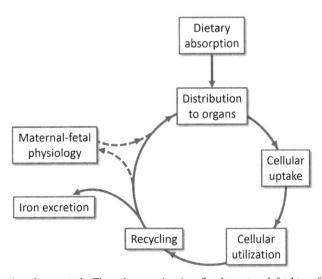

FIGURE 16.1 Schematic of rodent iron homeostasis. The rodent acquires iron first by maternal–fetal transfer. Once weaned, the rodent acquires iron by dietary iron absorption, a highly regulated process. Iron is distributed throughout the body to dependent cells, tissues, and organs, where specific pathways mediate cellular iron uptake, utilization, and recycling. Iron is then either excreted from the body or redistributed for reuse.

Molecular, Genetic, and Nutritional Aspects of Major and Trace Minerals. http://dx.doi.org/10.1016/B978-0-12-802168-2.00016-6

TABLE 16.1 Genes Mutated in Rodent Models of Aberrant Iron Homeostasis

Gene	Full Name and/or Established/Putative Role
Abcb7	• ATP-binding cassette, subfamily B (MDR/TAP), member 7. • Exports unidentified iron-sulfur cluster precursor out of mitochondria.
Abcb8	• ATP-binding cassette, subfamily B (MDR/TAP), member 8. • Essential for mitochondrial iron export.
Abcb10	• ATP-binding cassette, subfamily B (MDR/TAP), member 10. • Component of mitochondrial inner membrane protein complex. • Essential for heme synthesis and erythroid maturation.
Acvr1	• Activin A receptor, type 1. • Required for hepcidin expression.
Acvr2a	• Activin receptor IIA. • Required for hepcidin expression.
Ahsg	• Alpha-2-HS-glycoprotein. • May stimulate hepcidin expression.
Atp4a	• ATPase, H$^+$/K$^+$ exchanging, gastric, alpha polypeptide. • Catalytic α-subunit of gastric hydrogen-potassium ATPase. • Required for gastric acidification.
B2m	• Beta-2 microglobulin. • Required for hepcidin expression.
Bmpr1a	• Bone morphogenetic protein receptor, type 1A. • Required for hepcidin expression.
Bmpr2	• Bone morphogenetic protein, type II (serine/threonine kinase). • Required for hepcidin expression.
Bmp6	• Bone morphogenetic protein 6. • Essential for hepcidin expression.
Cp	• Ceruloplasmin, multicopper oxidase. • Essential for cellular iron efflux.
Cybrd1	• Cytochrome b reductase 1. • Present on luminal surface of duodenal enterocytes.
Dmt1 (Slc11a2)	• Divalent metal transporter 1. • Officially known as Slc11a2 (solute carrier family 11 (protein-coupled divalent metal ion transporters), membrane 2). • Essential for dietary iron absorption and erythroid iron acquisition.
Exoc6	• Exocyst complex component 6. • Required for erythroid vesicular trafficking and iron assimilation.
Fam132b	• Family with sequence similarity 132, member B. • Mediates suppression of hepcidin expression in β-thalassemia.
Fbxl5	• F-box and leucine-rich repeat protein 5. • Component of E3 ubiquitin ligase complex that degrades Irp2.
Ferroportin (Slc40a1)	• Officially known as Slc40a1 (solute carrier family 40 (iron-regulated transporter), member 1). • Essential for cellular iron export from duodenal enterocytes and macrophages.
Fth1, Ftl1 (ferritin)	• Ferritin heavy chain 1 (Fth1) and light chain 1 (Ftl1). • Components of ferritin, large multi-subunit iron storage complex.
Fech	• Ferrochelatase. • Catalyzes incorporation of iron into protoporphyrin IX to form heme.
Flvcr	• Feline leukemia virus subgroup C receptor. • Officially known as Mfsd7b, major facilitator superfamily domain containing 7B. • Essential for heme export in erythroid precursors.

TABLE 16.1 Genes Mutated in Rodent Models of Aberrant Iron Homeostasis — cont'd

Gene	Full Name and/or Established/Putative Role
Fxn	• Frataxin. • Mitochondrial protein essential for iron-sulfur cluster synthesis.
Gast	• Gastrin, hormone that stimulates gastric acid secretion. • May be required for response to iron deficiency.
Grk6	• G protein-coupled receptor kinase 6. • Regulates engulfment of red cells by macrophages.
Hp	• Haptoglobin, plasma protein with high affinity for hemoglobin. • Required only in conditions of hemolysis or other stresses.
Hamp	• Hepcidin antimicrobial peptide. • Posttranslationally down-regulates ferroportin expression.
Hmox1	• Heme oyxgenase 1. • Catalyzes conversion of heme to biliverdin, carbon monoxide, and iron. • Essential for red cell breakdown.
Hfe2	• Hemochromatosis type 2 (juvenile). • Also known as hemojuvelin. • Bmp co-receptor protein essential for hepcidin expression.
Hpx	• Hemopexin, plasma protein with high affinity for heme. • Required only in conditions of hemolysis or other stresses. • Not to be confused with hpx mutation in transferrin gene!
Heph	• Hephaestin, multicopper oxidase. • Essential for duodenal enterocyte iron export.
Hfe	• Hemochromatosis gene. • Nonclassical major histocompatibility complex class I molecule. • Essential for hepcidin expression. • Modulates transferrin import into erythroid cells.
Hif1a Hif2a	• Alpha and beta subunit of hypoxia inducible factors (Hif). • Hif2a, officially known as Epas1 (endothelial PAS domain protein 1). • Hif: Transcription factor that mediates response to hypoxia. • Hif2 regulates dietary iron absorption.
Hri	• Heme-regulated eIF2α kinase. • Officially known as Eif2ak1 (eukaryotic transcription initiation factor 2 alpha kinase 1). • Inhibits globin gene transcription under conditions of heme deficiency. • Required for red pulp macrophage development.
Iop1	• Iron-only hydrogenase-like protein 1. • Officially known as Narfl (nuclear prelaminin A recognition factor-like). • Component of cytoplasmic iron-sulfur cluster assembly complex.
Irp1 Irp2	• Iron regulatory protein 1 and 2. • Irp1 officially known as Aco1 (aconitase 1). • Irp2 officially known as Ireb2 (iron responsive element binding protein 2). • Posttranscriptionally regulates expression of genes relevant to iron homeostasis.
Iscu	• IscU iron-sulfur cluster scaffold homolog (E. coli). • Mitochondrial protein that serves as scaffold during iron-sulfur cluster synthesis.
Kcne2	• Potassium voltage-gated channel, Isk-related subfamily, gene 2. • Required for gastric acidification.
Mbd5	• Methyl-CpG binding domain protein 5. • Regulates ferritin heavy chain expression.
Mitoferrin 1	• Officially known as Slc25a37 (solute carrier family 25, member 37). • Essential for mitochondrial iron import.

Continued

TABLE 16.1 Genes Mutated in Rodent Models of Aberrant Iron Homeostasis—cont'd

Gene	Full Name and/or Established/Putative Role
Mms19	• Nucleotide excision repair protein homolog. • Component of cytoplasmic iron-sulfur cluster assembly complex.
Ncoa4	• Nuclear receptor coactivator 4. • Targets ferritin for degradation by autolysosomes. • Required for erythrophagocytosis.
Ndfip1	• Nedd4 family interacting protein 1. • Mediates interaction between Dmt1 and ubiquitin ligase targeting Dmt1 for degradation.
Neo1	• Neogenin. • Membrane protein that interacts with hemojuvelin. • Protects hemojuvelin from cleavage.
Nme1 and *2*	• NME/NM23 nucleoside diphosphate kinase 1 and 2. • Required for erythroid transferrin receptor expression.
Nramp1	• Officially known as Slc11a1 (solute carrier family 11 (proton-coupled divalent metal ion transporters), member 1). • Natural resistance-associated macrophage protein 1. • Required for iron transport across phagolysosomal membranes.
Picalm	• Phosphatidylinositol-binding clathrin assembly protein. • Required for transferrin internalization.
Smad4	• SMAD family member 4. • Mediator of bone morphogenetic protein signaling. • Essential for hepcidin expression.
Spic	• Spi-C transcription factor (Spi-1/PU.1 related). • Essential for red pulp macrophage development.
Stat5	• Officially known as Stat5a (signal transducer and activation of transcription 5A). • Regulates erythroid transferrin receptor expression.
Steap3	• STEAP family member 3. • Steap: Six-transmembrane epithelial antigen of the prostate. • Ferrireductase essential for internalization of transferrin-bound iron.
Tmprss6	• Transmembrane serine protease 6. • Inhibits hepcidin expression by cleaving hemojuvelin.
Trf	• Transferrin. • Serum iron-binding protein. • Essential for iron delivery for erythropoiesis and regulation of dietary iron absorption.
Tfrc	• Transferrin receptor. • Required for transferrin internalization.
Tfr2	• Transferrin receptor 2. • Minor role in transferrin internalization. • Essential for hepcidin expression. • Regulates red cell production.
Zip14	• Officially known as Slc39a14 (solute carrier family 39 (zinc transporter), member 14). • Imports nontransferrin-bound iron in liver and pancreas.

a particular factor, these mice were generated transgenically. Otherwise, models are denoted as generated by irradiation or chemical mutagenesis or identified during routine animal maintenance. Official gene symbols and full names are used unless noted. Most of the topics covered in this chapter can be found in greater detail elsewhere (Bartnikas et al., 2012; Beilschmidt and Puccio, 2014; Evstatiev and Gasche, 2012; Ganz, 2013; Korolnek and Hamza, 2015; Mastrogiannaki et al., 2013; Perdomini et al., 2013; Smith and Thévenod, 2009; Veuthey and Wessling-Resnick, 2014; Wilkinson and Pantopoulos, 2014; Zhang et al., 2014). This chapter does not cover studies using in vitro or cell-based models, studies on relationships between iron homeostasis and inflammation, infection, neoplasia and ineffective erythropoiesis, or studies on iron homeostasis in organs without established roles in systemic iron homeostasis.

MATERNAL–FETAL AND MATERNAL–NEONATAL IRON PHYSIOLOGY

Maternal–fetal iron physiology requires mobilization of maternal iron stores, their transfer to the fetus, and distribution within and utilization by the fetus. (The words fetus and embryo are used interchangeably here.) Most genetic rodent-based studies do not focus on this topic. Its investigation is not straightforward. Both maternal and fetal genotypes must be considered. Studies can be expensive and time-consuming given that breeding rodents must often be sacrificed for analysis. If a gene is essential in utero, studies must be performed on embryos prior to their demise.

Studies that address fetal iron physiology are more common, albeit largely unintentionally. Any experiment in which gene mutation leads to embryonic lethality indicates that the gene is essential in utero. Examples of such genes are included throughout this chapter. In most studies demonstrating embryonic lethality, the role of the gene in maternal iron physiology is often not studied. Such studies require development of homozygous adult female mice using inducible or temporal gene ablation. In those studies where embryonic lethality is not noted, the gene of interest may still play a role in maternal–fetal iron physiology. In cases where genetic ablation does not lead to embryonic lethality, a role for that gene in utero is usually not explored. Such studies could be of great value and can utilize preexisting models.

In contrast to maternal–fetal iron physiology, maternal–neonatal iron physiology involves mobilization of maternal iron stores to milk and absorption of milk iron by the developing neonate. This process is not well understood.

OVERVIEW OF DIETARY IRON ABSORPTION

As total body iron levels reflect a balance between dietary iron absorption and iron excretion, regulation of iron absorption is paramount to maintenance of physiologic body iron levels. Dietary iron absorption is incompletely understood. Dietary iron exists in heme and nonheme forms but rodent models of aberrant heme iron absorption do not exist. The following sections address key steps in dietary absorption of nonheme iron (Fig. 16.2).

GASTRIC ACIDIFICATION

Nonheme iron is liberated from ferritin, other proteins, and nonprotein carriers by gastric acidity, which also assures that iron remains soluble and available for absorption. Gastric acidification requires activity of several factors. First, Kcne2 and Kcnq1 form a channel that supplies potassium for the gastric hydrogen-potassium ATPase that acidifies stomach contents. Kcne2-deficient mice develop iron deficiency although only male mutant mice develop anemia, suggesting an effect of sex hormones on erythropoiesis and iron homeostasis (Salsbury et al., 2014). Second, the gastric ATPase consists of catalytic α and structural β subunits encoded respectively by *Atp4a* and *Atp4b*. Sublytic mice, generated in a chemical mutagenesis screen for hematologic abnormalities, are homozygous for a mutation in the Atp4a ATPase domain. These mice develop anemia secondary to impaired gastrointestinal iron absorption. Third, gastrin (Gast) is a peptide hormone that stimulates acid secretion. Gastrin-deficient mice are phenotypically normal on an iron-replete diet but develop anemia on an iron-deficient diet more severe than the anemia observed in wild-type mice on an iron-deficient diet.

IRON IMPORT INTO DUODENUM

After liberation from food, nonheme iron is reduced from the ferric (Fe^{3+}) to ferrous (Fe^{2+}) state prior to absorption. A candidate for this step is cytochrome b reductase 1 (Cybrd1), a ferric reductase present on the luminal surface of duodenal enterocytes. However, Cybrd1-deficient mice do not develop altered body iron levels even when placed on an iron-deficient diet and only develop a modest phenotype under hypoxic conditions.

After reduction, nonheme iron is transported across the luminal surface of duodenal enterocytes. The metal transporter Slc11a2, more commonly known and referred to here as divalent metal-ion transporter 1 (Dmt1), mediates this process. The Belgrade rat and the microcytic anemia (mk) mouse both develop anemia secondary to impaired dietary iron absorption and erythroid iron uptake. Generated by irradiation, the Belgrade rat is homozygous for a loss-of-function G185R substitution in Dmt1. Originating during routine mouse breeding, the mk mouse is also homozygous for a G185R substitution in Dmt1. Compared to Belgrade rats and mk mice, Dmt1-deficient mice display a more pronounced phenotype: severe iron-deficiency anemia and death by one week. Mice with intestine-specific Dmt1 deficiency do not exhibit early postnatal lethality but develop severe iron deficiency and anemia secondary to impaired intestinal iron absorption. The difference between transgenic and nontransgenic

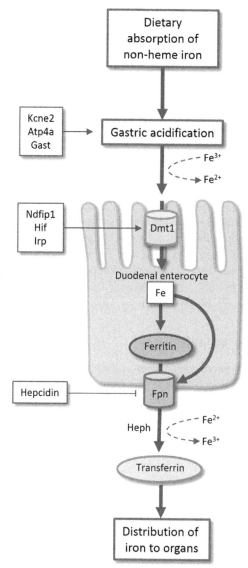

FIGURE 16.2 **Rodent models of aberrant dietary nonheme iron absorption.** Dietary absorption of nonheme iron occurs via a multistep process. Rodent models bearing discrete genetic defects currently exist for some but not all steps; this figure depicts those steps. (This figure does not depict those steps lacking firmly established genetic rodent models.) Iron is liberated from food and maintained in a soluble form by gastric acid. Gastric acidification is achieved by potassium voltage-gated channel Kcne2, H^+/K^+ exchanging gastric ATPase Atp4a, and gastrin hormone Gast. Iron is reduced from ferric (Fe^{3+}) to ferrous (Fe^{2+}) at the luminal surface of duodenal enterocytes before transport into the enterocyte by divalent metal transporter 1 (Dmt1). Dmt1 levels are regulated by Ndfip1 (targets Dmt1 for degradation), hypoxia inducible factors (Hifs), and iron regulatory proteins (Irps). Once in the enterocyte, iron can be stored in ferritin, utilized within the enterocyte or trafficked across the cell body to the basolateral face of the cell for export by ferroportin (Fpn). Iron export is coupled to oxidation from Fe^{2+} to Fe^{3+}, catalyzed by multicopper oxidase hephaestin (Heph). Fe^{3+} then binds to transferrin for distribution throughout the body.

phenotypes is attributed to partial retention of function by G185R Dmt1. Dmt1 is not only essential for iron absorption under physiologic conditions—inducible Dmt1 ablation decreases iron accumulation in mouse models of β-thalassemia, an inherited disease of anemia and iron overload due to insufficient globin chain production (Anderson et al., 2013).

Dmt1 expression is regulated by multiple factors:

- Ndfip1, an adaptor protein that mediates interaction between Dmt1 and an ubiquitin ligase that targets Dmt1 for degradation. Ndfip1-deficient mice have increased hepatic Dmt1 protein levels and activity and hepatic iron loading and display impaired Dmt1 regulation on an iron-deficient diet.
- Hypoxia inducible factors (Hif), heterodimeric transcription factors essential for the response to hypoxia. In oxygen-replete conditions, Hif α subunits are targeted for proteasomal degradation by iron-dependent enzymes known as prolyl hydroxylases. In conditions of hypoxia or iron deficiency, Hif α subunits are spared from degradation, heterodimerize with β subunits, and stimulate gene transcription. Hif2α plays a pivotal role in the regulation of iron absorption. Mice

with intestine-specific Hif2α deficiency exhibit decreased duodenal Dmt1 mRNA levels and plasma and hepatic iron levels, while mice with intestine-specific Hif1α deficiency do not. Hif2 also mediates increased iron absorption and tissue iron loading observed in a mouse model of β-thalassemia (Anderson et al., 2013) and in iron-loaded mice with inherited deficiency of the iron regulatory hormone hepcidin.

● Iron regulatory proteins (Irp), posttranscriptional regulators that modulate gene expression by binding to iron responsive elements (IREs) in mRNAs. Binding of Irps to IREs in 5′ or 3′ untranslated regions inhibits mRNA translation or degradation, respectively. Irp1 acts as a cytosolic aconitase in iron-replete conditions but is converted into an RNA-binding protein by loss of an iron-sulfur cluster in iron-deficient conditions. In contrast, Irp2 is targeted for proteasomal degradation in iron-replete conditions but accumulates in iron-deficient conditions. Proteasomal degradation of Irp2 is dependent upon Fbxl5, a component of an E3 ubiquitin ligase complex. Fbxl5-deficient mice exhibit embryonic lethality with excessive iron accumulation, while mice with deficiency of both Fbxl5 and Irp2 are viable. Studies using multiple mouse models with engineered inactivation of Irp1 and/or Irp2 indicate that Irp function is complex. Global deletion of Irp1 and Irp2 is lethal, although mice with deletion of Irp1 or Irp2 are viable. Intriguingly, independently generated Irp2-deficient mouse models differ in their level of neurologic dysfunction. Studies in mouse models suggest that Irp1 indirectly suppresses duodenal Dmt1 expression: Irp1 deficiency leads to excessive Hif2 activity which, in turn, leads to excessive Dmt1 levels. Multiple forms of Dmt1 mRNAs exist, not all of which possess IREs.

Once transported into the duodenal enterocyte, iron is utilized by the cell, exported out of the cell for distribution to other organs, or stored in ferritin. Ferritin consists of multiple subunits of ferritin heavy (Fth1) and light (Ftl1) chains arranged in a shell encasing several thousand iron atoms. Fth1 possesses ferroxidase activity required for iron uptake into ferritin; Ftl1 facilitates formation of the iron core within ferritin. Fth1 and Ftl1 expression are inhibited by the Irp system under iron-deficient conditions. Fth1-deficient mice exhibit early embryonic lethality, suggesting that Ftl1 cannot compensate for Fth1 deficiency. Inducible global *Fth1* ablation in adult animals results in a minimal phenotype, but leads to severe liver damage when mice are exposed to a high-iron diet (Darshan et al., 2009). Intestine-specific Fth1 deficiency results in increased iron absorption and tissue iron levels, suggesting that intestinal Fth1 is essential for modulating dietary iron absorption. Mice with intestine-specific deficiency of Mbd5, a member of the methyl-CpG-binding domain family of proteins, develop aberrant Fth1 expression and iron levels (Tao et al., 2014). The phenotype of Ftl1-deficient mice is less severe than Fth1-deficient mice: Ftl1-deficient mice are viable but do develop increased serum iron levels and decreased hepatic iron levels (Li et al., 2015).

IRON EXPORT FROM DUODENUM TO BLOOD

Iron export from enterocytes into blood is a key regulatory point in iron biology. Often cited as the only known vertebrate cellular iron exporter, ferroportin, officially known as Slc40a1, is a cellular iron transporter essential for enterocyte iron export. Ferroportin-deficient mice die in utero. Mice with postnatal intestinal ferroportin deficiency develop iron-deficiency anemia and enterocyte iron loading. Ferroportin expression is regulated by several factors. Both Hif2 and Irp2 are essential for stimulation of ferroportin expression. The Irp system is also a prominent regulator of ferroportin expression. One particularly intriguing mouse model is the polycythemia (pcm) mouse, harboring a radiation-induced deletion in the ferroportin promoter which leads to aberrant transcription and loss of an IRE in the 5′ untranslated region. Duodenal ferroportin expression varies in pcm heterozygous mice depending on their age, suggesting complex developmental Irp-dependent ferroportin regulation. As with Dmt1, not all ferroportin mRNAs possess IREs.

Ferroportin is also posttranslationally negatively regulated by hepcidin (Hamp), a peptide hormone secreted largely by the liver (Fig. 16.3). While mice carry two hepcidin genes, *Hamp1* and *Hamp2*, only mice transgenically overexpressing Hamp1 display defects in iron metabolism. Hepcidin regulates ferroportin by binding to it and stimulating its internalization and degradation. Hamp1-deficient mice demonstrate increased duodenal ferroportin expression and severe tissue iron loading with relative sparing of splenic macrophages. The splenic iron deficiency is attributed to inappropriately high macrophage ferroportin levels. Mice with liver-specific *Hamp1* inactivation recapitulate the phenotype of mice with global Hamp1 deficiency, indicating that the liver is the main source of systemically active hepcidin (Zumerle et al., 2014). Hamp1 deficiency recapitulates hereditary hemochromatosis, a disease of iron overload caused by mutations in hepcidin or factors essential for its expression. The potent role of hepcidin in iron homeostasis is further underscored by the systemic iron deficiency and anemia of mice transgenically overexpressing hepcidin. This overexpression phenotype is similar to the characteristics of anemia of inflammation, a condition associated with excessive hepcidin levels. The central role for hepcidin in ferroportin regulation is also highlighted by the enterocyte iron deficiency observed in mice engineered to express a hepcidin-resistant ferroportin mutant (Altamura et al., 2014).

Just as iron is reduced prior to absorption, iron is oxidized during export from enterocytes into blood. This oxidation is mediated by the multicopper ferroxidase hephaestin (Heph). Sex-linked anemia (sla) mice, harboring an irradiation-induced hephaestin deletion, exhibit impaired enterocyte iron export. Mice with global or intestine-specific hephaestin

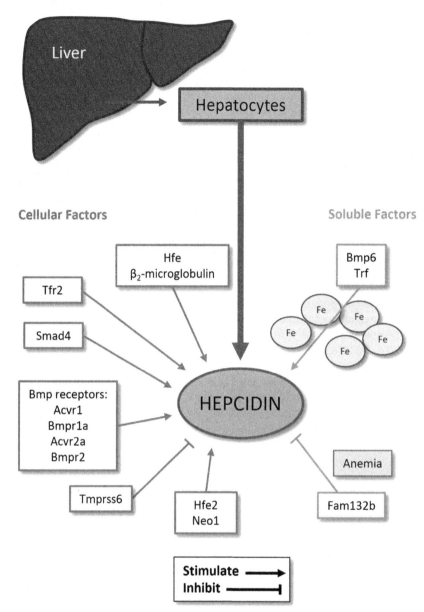

FIGURE 16.3   Rodent models of aberrant hepcidin expression. Hepcidin expression is regulated by multiple factors. Rodent models bearing discrete genetic defects exist for many of those factors, as depicted in this figure. (This figure does not depict those steps or factors that lack firmly established genetic rodent models.) Hepcidin is synthesized and secreted largely by the liver. In conditions of iron excess, hepcidin expression is stimulated by bone morphogenetic protein 6 (Bmp6) and diferric transferrin (Trf). In conditions of anemia, hepcidin expression is inhibited by Fam132b, a tumor necrosis factor protein family member secreted by erythroid precursors, also known as erythroferrone. Multiple membrane-bound proteins contribute to hepcidin regulation. Hemojuvelin (Hfe2) is a Bmp co-receptor protein; it is cleaved from the membrane by transmembrane serine protease 6 (Tmprss6) and protected from cleavage by neogenin (Neo1). Hepcidin expression also requires Bmp type I receptors Acvr1 and Bmpr1a and type II receptors Acvr2a and Bmpr2. Smad4 is an intracellular mediator of Bmp signaling. Like Hfe2, Hfe (hemochromatosis protein), and Tfr2 (transferrin receptor 2) are also required for stimulation of hepcidin expression in conditions of iron excess. β_2-microglobulin is required for Hfe expression.

deficiency develop a similar phenotype as sla mice—anemia, impaired iron absorption, and iron retention in duodenal enterocytes (Fuqua et al., 2014)—indicating that the intestine is a key site of hephaestin function.

REGULATION OF HEPCIDIN EXPRESSION

Hepcidin expression is stimulated in conditions of iron excess and inflammation, and inhibited in conditions of anemia and hypoxia. Most rodent models of aberrant hepcidin regulation serve as models of hereditary hemochromatosis. Studies using these models have identified key regulatory pathways that modulate hepcidin expression although the functional and physical relationship between pathway components is not always clear. In lieu of summarizing the known relationships between these components, many of which are established from in vitro and cell culture studies, the components of these pathways are summarized by category (Fig. 16.3).

- *Soluble stimulators of hepcidin expression*: Bone morphogenetic protein 6 (Bmp6) and transferrin (Trf) stimulate hepcidin expression in conditions of iron excess. The tissue origin of Bmp6 is contested. Transferrin is synthesized mainly by the liver. Bmp6-deficient mice and hpx mice (hypotransferrinemic mice, a model of transferrin deficiency) are hepcidin-deficient and severely iron overloaded. Bmp6 and transferrin may alter hepcidin expression in response to chronic or acute iron loading, respectively, although the means by which these two factors regulate hepcidin expression appear to overlap.

- *Soluble inhibitors of hepcidin expression*: In conditions of anemia, hepcidin expression is inhibited by Fam132b, also known as erythroferrone, a member of the tumor necrosis factor-related protein family secreted by erythroid precursors. Fam132b-deficient mice fail to suppress hepcidin expression after phlebotomy (Kautz et al., 2014). *Fam132b* inactivation in a mouse model of β-thalassemia intermedia, a disease of anemia and iron overload due to inappropriately low hepcidin expression, increases hepcidin levels, and decreases tissue iron levels. The mechanism of action of Fam132b is not known.

- *Cellular factors required for hepcidin expression*:
  - *Hemojuvelin (HJV), neogenin*: HJV (Hfe2) is a Bmp co-receptor. Hemojuvelin-deficient mice are hepcidin-deficient and severely iron-overloaded. Tissue-specific deletions indicate that the liver is the key site of HJV expression. HJV modulates both Bmp6-and transferrin-dependent hepcidin expression. The multifunctional transmembrane receptor protein neogenin (Neo1) is proposed to protect HJV against cleavage by proteases. Neogenin-deficient mice develop hepcidin deficiency and iron overload.
  - *Bmp type I and II receptors*: Liver-specific deletion models demonstrate that Bmp type I receptors Acvr1 and Bmpr1a are required for hepcidin expression but do not play identical roles, while Bmp type II receptors Acvr2a and Bmpr2 play redundant roles in hepcidin expression (Mayeur et al., 2014).
  - *Hfe, β_2-microglobulin*: Hfe is a nonclassical major histocompatibility complex molecule. Mice with genetic ablation or knock-in mutations of *Hfe* develop hepcidin deficiency and iron overload, although these phenotypes are mild compared to those observed in other models such as hemojuvelin-deficient mice. Hfe modulates Bmp-, transferrin-, and inflammation-dependent hepcidin expression, but its precise function is unclear. Mice deficient in β_2-microglobulin, a protein required for Hfe assembly and localization, also develop iron overload and hepcidin deficiency.
  - *Transferrin receptor 2*: While similar in sequence to transferrin receptor, transferrin receptor 2 (Tfr2) plays a minor role in transferrin internalization. Hepcidin deficiency and tissue iron loading are observed in mice with global or hepatocyte-specific Tfr2 inactivation, mice homozygous for a hereditary hemochromatosis Tfr2 mutation, and outbred rats homozygous for a Tfr2 polymorphism (Bartnikas et al., 2013b).
  - *Others*: Mice with hepatocyte-specific disruption of Smad4, an intracellular mediator of Bmp signaling, develop hepcidin deficiency and tissue iron excess.

- *Cellular factors required for inhibition of hepcidin expression*:
  - *Tmprss6*: Generated in a chemical mutagenesis screen, the mask mouse harbors a recessive splicing mutation in *Tmprss6* (transmembrane serine protease 6). Mask mice, along with two independently established Tmprss6-deficient mice lines and another mutagenesis-derived mouse line carrying a hypomorphic Tmprss6 allele (Bartnikas et al., 2013a), exhibit iron-deficiency anemia secondary to increased hepcidin expression. Mouse models of Tmprss6 deficiency recapitulate the clinical characteristics of iron refractory iron-deficiency anemia, a condition due to Tmprss6 mutations and hepcidin excess. Tmprss6 negatively regulates hepcidin expression by inhibiting the Bmp6-hemojuvelin-hepcidin signaling pathway. Tmprss6 is believed to cleave hemojuvelin from the hepatocyte membrane. Tmprss6 is also proposed to inhibit activity of Ahsg (alpha-2-HS-glycoprotein), a plasma protein with multiple functions. Ahsg-deficient mice develop hepcidin deficiency (Stirnberg et al., 2015).

DISTRIBUTION OF IRON TO ORGANS

Once exported from duodenum into blood, iron is distributed to organs throughout the body. Transferrin is the most well-established serum/plasma iron transporter. Its essential role in iron biology is underscored by the severe phenotype of hpx mice. Hpx mice carry an autosomal recessive splice site mutation in the transferrin gene. (This mutant transferrin allele, referred to as the *hpx* allele, is not to be confused with the *hpx* gene which encodes the serum heme-binding protein hemopexin.) The *hpx* mutation arose during routine colony maintenance and causes severe transferrin deficiency, anemia, and multitissue iron excess. The anemia highlights that transferrin is essential for iron delivery to bone marrow for

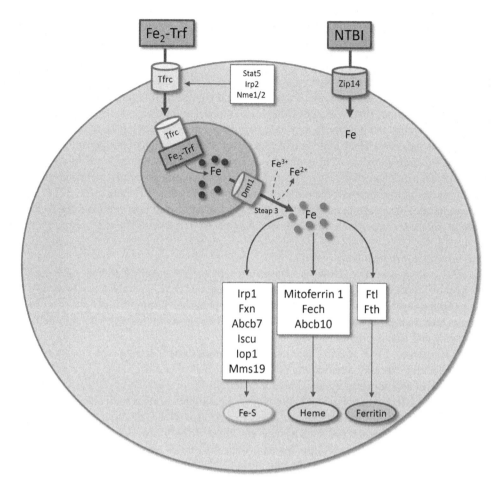

FIGURE 16.4 Rodent models of aberrant cellular iron acquisition and utilization. Circulating iron can exist as transferrin-bound iron (diferric transferrin, Fe_2-Trf) or nontransferrin-bound iron (NTBI). NTBI is imported into some but not all cell types by membrane-bound metal transport protein Zip14. Diferric transferrin is imported into cells by receptor-mediated endocytosis. (This figure depicts a generic cell. Note that only erythroid precursors rely solely on the transferrin cycle for iron acquisition.) The diferric transferrin–transferrin receptor (Tfrc) complex is internalized. Endosomal acidification leads to liberation of iron from transferrin. Iron is exported into the cytoplasm by Dmt1 (divalent metal transporter 1). Export is coupled to reduction from Fe^{3+} to Fe^{2+} catalyzed by ferrireductase Steap3. Iron is then utilized for synthesis of iron-sulfur clusters (Fe-S) and heme groups or other processes or stored in ferritin. Factors required for iron-sulfur cluster, heme, and ferritin synthesis/assembly are indicated. (This figure does not depict those steps/processes lacking firmly established genetic rodent models.)

erythropoiesis. Tissue iron excess is attributed to secondary deficiency in hepcidin, as anemia suppresses hepcidin expression and transferrin is essential for stimulation of hepcidin expression by iron. Tissue iron excess also reflects the fact that most tissues can acquire iron through uptake of nontransferrin-bound iron (NTBI). The metal transport protein Zip14 (Slc39a14) is essential for NTBI uptake into liver and pancreas (Jenkitkasemwong et al., 2015) (Fig. 16.4).

Although most tissues can import NTBI, NTBI is presumed to be present largely under pathophysiologic conditions. Transferrin-bound iron uptake is believed to be the dominant means of iron acquisition under conditions of health. The role of transferrin is best established in the context of erythroid iron uptake. Diferric transferrin is internalized by the transferrin cycle which is initiated by binding of diferric transferrin to transferrin receptor (Tfrc). Global Tfrc deficiency leads to embryonic lethality; mice with tissue-specific deficiency have yet to be reported. Tfrc expression is regulated by the transcription factor Stat5, Irp2, and Nme1/2. Stat5 stimulates Irp2 transcription; Nme1/2 are multifunctional enzymes with roles in nucleotide metabolism and other processes although their mechanistic link to Tfrc expression is unclear.

Once the transferrin-Tfrc complex is internalized, endosomal acidification leads to liberation of iron from transferrin. Ferric iron is then reduced to ferrous iron by the ferrireductase Steap3. Deficiency of Steap3 results in anemia, as observed in Steap3-deficient mice (Blanc et al., 2015, p. 3) and nm1054 mice harboring a spontaneously arisen multigene deletion. Fragile-red mice, bearing a chemical mutagen-induced Steap3 Y228H substitution, also develop anemia. The Y228H mutation affects a motif required for appropriate intracellular Steap3 targeting (Lambe et al., 2009). Ferrous iron is exported into the cytoplasm by Dmt1. The transferrin cycle is also adversely affected by the hemoglobin deficit (hbd) mutation that arose spontaneously in a mouse line decades ago. The hbd mutation is an autosomal recessive deletion in Sec15l1 (Exoc6) that leads to impaired erythroid vesicular trafficking and iron assimilation (Lim et al., 2005). Fit mice, carrying chemical mutagen-induced mutations in the phosphatidylinositol-binding clathrin assembly protein (Picalm), and Picalm-deficient mice exhibit severe anemia attributed to impaired erythroid maturation and transferrin internalization (Ishikawa et al., 2014; Klebig et al., 2003; Suzuki et al., 2012). Finally, studies in Hfe-deficient mice indicate that Hfe, a factor essential for hepcidin expression, modulates diferric transferrin import into erythroid cells (Ramos et al., 2011).

CELLULAR IRON UTILIZATION

Iron that is not stored intracellularly in ferritin can be incorporated into iron-sulfur clusters, heme groups, or enzymes. Iron-sulfur clusters are essential cofactors in pathways such as the electron transport chain and in the activity of proteins such as Irp1. Iron-sulfur cluster assembly occurs via two pathways. For mitochondrial iron-sulfur cluster-binding proteins, clusters are assembled in mitochondria. For cytoplasmic proteins, cluster assembly begins in mitochondria then continues in the cytoplasm where cluster incorporation occurs. Iron-sulfur cluster biosynthesis is better understood in microbial systems but several mouse models have been established that are deficient in essential factors (Fig. 16.4). Global deficiency in these factors often leads to embryonic lethality. Frataxin, a protein mutated in the degenerative disease Friedreich ataxia, is a mitochondrial protein essential for iron-sulfur cluster biosynthesis although its precise role has yet to be established (Perdomini et al., 2013). Several mouse models deficient in factors essential for cytosolic iron-sulfur cluster assembly have also been reported. Abcb7 is a mitochondrial protein that mediates export of an unidentified iron-sulfur cluster precursor from mitochondria to cytoplasm (Pondarre et al., 2007). Iscu is a mitochondrial protein that acts as a scaffold during cluster synthesis (Nordin et al., 2011). Mms19 and Iop1 are components of a cytoplasmic iron-sulfur cluster assembly complex (Gari et al., 2012; Song and Lee, 2011). Finally, Abcb8, a mitochondrial inner membrane protein, is essential for mitochondrial iron export. Abcb8-deficient mice develop mitochondrial iron accumulation, cardiomyopathy, and deficiency in the activity of cytosolic but not mitochondrial iron-sulfur cluster-containing enzymes (Ichikawa et al., 2012).

Iron can also be incorporated into heme, a cofactor in globins, cytochromes, and other proteins. Heme synthesis is an orchestrated, multistep process occurring in mitochondria and cytoplasm. Glycine and succinyl-CoA are converted to protoporphyrin IX which then undergoes iron insertion to form heme. Mice with genetic deficiencies in factors essential for heme synthesis display severe phenotypes (Fig. 16.4). Models deficient in mitoferrin 1 (Slc25a37), ferrochelatase, and Abcb10 have been reported. Mitoferrin 1 is essential for mitochondrial iron import. Mice with global mitoferrin 1 deficiency are not viable; mice with hematopoietic mitoferrin 1 deficiency develop anemia secondary to impaired erythroid maturation. Ferrochelatase catalyzes the incorporation of iron into protoporphyrin IX. Mice homozygous for a chemically induced recessive loss-of-function ferrochelatase mutation develop anemia and protoporphyrin accumulation. Abcb10 forms part of a mitochondrial inner membrane protein complex with mitoferrin 1 and ferrochelatase. Hematopoietic Abcb10 inactivation leads to protoporphyrin accumulation and impaired erythroid maturation (Yamamoto et al., 2014).

ERYTHROPOIESIS

Most of the iron in the body is dedicated to erythropoiesis and found within bone marrow, red blood cells, or red cell-scavenging macrophages. Readers are referred to other reviews for detailed information on this central process in iron biology. Many of the relevant mouse models have been already described. In addition, Irp1 is essential for regulation of Hif2, which stimulates expression of erythropoietin, a largely kidney-derived hormone that inhibits apoptosis of erythroid precursors. Irp2 is required for expression of transferrin receptors and factors essential for heme biosynthesis. Heme export by the heme transport protein Flvcr is essential for viability of erythroid precursors. Globin chain expression is coordinated with heme availability by the heme-regulated eIF2α kinase Hri which inhibits globin gene transcription under conditions of heme deficiency. Hri-deficient mice develop heme aggregates in red cell precursors. Factors essential for regulation of hepcidin expression have also been implicated in erythropoiesis. In addition to the role of Hfe in regulation of transferrin internalization mentioned earlier, studies in mice with bone marrow–specific transferrin receptor 2 inactivation indicate that Tfr2 regulates red cell production in response to iron availability (Nai et al., 2015).

RECYCLING OF IRON FROM RED CELLS

Two pathways reclaim iron from damaged or senescent red blood cells and provide the bone marrow with most of its iron needs. The first consists of reclamation of heme and hemoglobin released into blood by intravascular lysis of red blood cells or tissue damage. This is mediated by hemopexin and haptoglobin, plasma proteins with high affinity for heme and hemoglobin respectively. Heme-hemopexin and hemoglobin-haptoglobin complexes are cleared from circulation after binding to specific receptors expressed on macrophages and other cell types in the liver and other organs. Mice with deficiencies in these factors have minimal phenotypes unless stressed. The second, more dominant pathway is erythrophagocytosis, the phagocytosis of red blood cells by reticuloendothelial macrophages resident in the spleen, liver, and other organs. Several mouse models harbor defects in macrophage development, erythrophagocytosis or macrophage iron processing and export:

- *Macrophage development*: Mice with homozygous ablation of the *Spic* transcription factor gene exhibit increased splenic iron concentrations, attributed to impaired Spi-C-dependent macrophage development in the spleen (Kohyama et al., 2009). Hri, a kinase essential for regulation of heme synthesis, is also required for macrophage maturation. Hri-deficient mice exhibit impaired erythrophagocytosis.

- *Red cell engulfment*: Mice with deficiency in G protein-coupled receptor kinase 6 (Grk6) develop increased splenic iron levels in red pulp macrophages. Grk6 is proposed to regulate engulfment of red cells by macrophages (Nakaya et al., 2013).

- *Heme degradation*: The degradation of heme is a critical step in red cell breakdown. Heme oxygenase 1 catabolizes heme to biliverdin, carbon monoxide, and iron. Heme oxygenase 1–deficient mice are anemic, iron-deficient in serum, yet iron-loaded in liver and kidney. They are largely devoid of splenic and hepatic macrophages, most likely due to the toxicity of accumulated heme.

- *Ferritin degradation*: Autophagy of ferritin relies on Ncoa4, a factor that targets ferritin for degradation by autolysosomes, with Ncoa4-deficient mice developing splenic iron accumulation (Dowdle et al., 2014). The splenic iron accumulation is believed to reflect an essential role for Ncoa4 in erythrophagocytosis.

- *Iron transport*: Nramp1 (Slc11a1), a membrane-bound iron transporter with functional and sequence similarity to Dmt1, is required for iron transport across phagolysosomal membranes. After transport by Nramp1, cytoplasmic iron is available for export out of macrophages by ferroportin. Mice with macrophage-specific ferroportin deficiency develop iron accumulation in macrophages in spleen and other tissues. Flatiron (ffe) mice, generated in a chemical mutagenesis screen and harboring a dominant missense mutation that impairs membrane localization of ferroportin, also develop iron loading in liver-resident macrophages (Zohn et al., 2007).

- *Iron oxidation*: The multicopper oxidase ceruloplasmin oxidizes iron upon export from reticuloendothelial macrophages before iron binds to transferrin for redistribution to bone marrow and other organs. Ceruloplasmin-deficient mice accumulate iron progressively in reticuloendothelial macrophages, hepatocytes, and in the central nervous system.

IRON EXPORT

A role for ferroportin in cellular iron export from cells other than enterocytes and macrophages has not been thoroughly explored. Even if ferroportin activity in these other cell types is not essential for systemic iron homeostasis, ferroportin may play a role in regional iron homeostasis. For example, ferroportin expression in cardiomyocytes is essential for cardiac iron homeostasis but dispensable for systemic iron homeostasis (Lakhal-Littleton et al., 2015). Alternatively, an unidentified transporter may be responsible for cellular iron export in other cell types and tissues throughout the body. It is not improbable that cellular iron export is mediated by other factors in addition to ferroportin.

Iron excretion at the organismal level is very poorly understood. Not often studied, this process is not believed to be regulated in response to iron levels. Fecal iron excretion may represent hepatobiliary iron excretion and sloughing of iron-laden cells lining the gastrointestinal tract. Urinary iron excretion is likely a complicated process, given that iron transport in the kidney must accommodate renal iron homeostasis and urinary iron excretion. While analysis of Hamp1-deficient mice and hemojuvelin-deficient mice indicates that hepcidin regulates renal iron homeostasis (Moulouel et al., 2013), a potential role for iron transporters in urinary iron excretion has not been thoroughly investigated.

IRON TOXICITY

Rodents are largely resistant to toxicity of iron or other metals for reasons unknown. Furthermore, the rodent strain upon which a mutation is maintained can influence the severity of the resulting phenotype. Nevertheless, end organ toxicity can be induced through genetic, dietary, and pharmacologic interventions. Mice with deficiency in both Hfe and transferrin receptor 2 more closely recapitulate liver disease observed in hereditary hemochromatosis than do mice with single gene deficiencies (Delima et al., 2012). Hamp1-deficient mice fed an iron-rich diet develop liver injury and fibrosis (Lunova et al., 2014). Hemojuvelin-deficient mice are more sensitive to CCl_4, an inducer of liver fibrosis (Sebastiani et al., 2011).

AREAS FOR ADDITIONAL MODELS AND FUTURE RESEARCH

Listed here are topics to consider for future study; some have already been mentioned. Rodents with defects in these processes may already exist; iron-dependent phenotypes are not always immediately obvious unless directly assessed.

- microbiomes and iron homeostasis;
- dietary heme iron absorption;
- maternal–fetal iron homeostasis; and
- cellular and organismal iron export.

OTHER MINERALS

As molecular pathways of metal homeostasis are not always metal-specific, genetic rodent models of aberrant iron homeostasis can display defects in the homeostasis of other metals. Two better known overlaps are iron-manganese and iron-copper. For example, Belgrade rats have defective duodenal manganese absorption. Flatiron mice, a model of inherited ferroportin deficiency, exhibit reduced intestinal manganese absorption (Seo and Wessling-Resnick, 2015). Hfe-deficient mice develop aberrant iron and manganese absorption, possibly due to aberrant regulation of Dmt1, ferroportin, and other transporters (Kim et al., 2013). Iron–copper interactions center largely around the copper-dependent oxidases ceruloplasmin and hephaestin, both required for cellular iron efflux (Gulec and Collins, 2014).

SUMMARY POINTS

- Iron is an essential dietary nutrient that is toxic in excess.
- Several pathways regulate levels of iron in cells, tissues, organs, and the body.
- Dietary iron absorption is highly regulated.
- Most iron is dedicated to erythropoiesis.
- Iron is recycled from red blood cells for reuse.
- Iron excretion is not believed to be regulated in response to iron levels.

KEY FACTS

- Rodent models with genetic defects are critical to the study of iron homeostasis.
- Rodent models do not always recapitulate human disease.

MINI DICTIONARY OF TERMS

Dmt1 (divalent metal-ion transporter 1) Transport protein; essential for iron import into duodenal enterocytes and iron acquisition by erythroid precursors.
Ferritin Large intracellular, multi-subunit, iron storage protein.
Ferroportin Transport protein; essential for iron export from duodenal enterocytes and macrophages.
Hemochromatosis Inherited human disease of iron overload; caused by mutations in genes required for hepcidin expression such as Hfe, transferrin receptor 2, and hemojuvelin.
Hepcidin (Hamp) Hormone; secreted mainly by the liver; binds to ferroportin and induces its internalization and degradation.
Hephaestin, ceruloplasmin Copper-dependent enzymes; oxidize iron upon cellular export.
Hypoxia-inducible factor (Hif) Transcription factors; stimulate gene expression in conditions of hypoxia.
Iron regulatory protein (Irp) RNA-binding protein; modulates translation or degradation of RNA species containing iron responsive elements.
Transferrin Soluble serum iron-binding protein; essential for iron delivery to erythroid precursors; internalized into cell by binding to transferrin receptor.

REFERENCES

Altamura, S., Kessler, R., Gröne, H.-J., Gretz, N., Hentze, M.W., Galy, B., Muckenthaler, M.U., 2014. Resistance of ferroportin to hepcidin binding causes exocrine pancreatic failure and fatal iron overload. Cell Metab. 20, 359–367. http://dx.doi.org/10.1016/j.cmet.2014.07.007.
Anderson, E.R., Taylor, M., Xue, X., Ramakrishnan, S.K., Martin, A., Xie, L., Bredell, B.X., Gardenghi, S., Rivella, S., Shah, Y.M., 2013. Intestinal HIF2α promotes tissue-iron accumulation in disorders of iron overload with anemia. Proc. Natl. Acad. Sci. U.S.A. 110, E4922–E4930. http://dx.doi.org/10.1073/pnas.1314197110.

Bartnikas, T.B., Fleming, M.D., Schmidt, P.J., 2012. Murine mutants in the study of systemic iron metabolism and its disorders: an update on recent advances. Biochim. Biophys. Acta 1823, 1444–1450. http://dx.doi.org/10.1016/j.bbamcr.2012.01.011.

Bartnikas, T.B., Steinbicker, A.U., Campagna, D.R., Blevins, S., Woodward, L.S., Herrera, C., Bloch, K.D., Justice, M.J., Fleming, M.D., 2013a. Identification and characterization of a novel murine allele of Tmprss6. Haematologica 98, 854–861. http://dx.doi.org/10.3324/haematol.2012.074617.

Bartnikas, T.B., Wildt, S.J., Wineinger, A.E., Schmitz-Abe, K., Markianos, K., Cooper, D.M., Fleming, M.D., 2013b. A novel rat model of hereditary hemochromatosis due to a mutation in transferrin receptor 2. Comp. Med. 63, 143–155.

Beilschmidt, L.K., Puccio, H.M., 2014. Mammalian Fe-S cluster biogenesis and its implication in disease. Biochimie 100, 48–60. http://dx.doi.org/10.1016/j.biochi.2014.01.009.

Blanc, L., Papoin, J., Debnath, G., Vidal, M., Amson, R., Telerman, A., An, X., Mohandas, N., 2015. Abnormal erythroid maturation leads to microcytic anemia in the TSAP6/Steap3 null mouse model. Am. J. Hematol. 90 (3), 235–241.

Darshan, D., Vanoaica, L., Richman, L., Beermann, F., Kühn, L.C., 2009. Conditional deletion of ferritin H in mice induces loss of iron storage and liver damage. Hepatology 50, 852–860. http://dx.doi.org/10.1002/hep.23058.

Delima, R.D., Chua, A.C.G., Tirnitz-Parker, J.E.E., Gan, E.K., Croft, K.D., Graham, R.M., Olynyk, J.K., Trinder, D., 2012. Disruption of hemochromatosis protein and transferrin receptor 2 causes iron-induced liver injury in mice. Hepatology 56, 585–593. http://dx.doi.org/10.1002/hep.25689.

Dowdle, W.E., Nyfeler, B., Nagel, J., Elling, R.A., Liu, S., Triantafellow, E., Menon, S., Wang, Z., Honda, A., Pardee, G., Cantwell, J., Luu, C., Cornella-Taracido, I., Harrington, E., Fekkes, P., Lei, H., Fang, Q., Digan, M.E., Burdick, D., Powers, A.F., Helliwell, S.B., D'Aquin, S., Bastien, J., Wang, H., Wiederschain, D., Kuerth, J., Bergman, P., Schwalb, D., Thomas, J., Ugwonali, S., Harbinski, F., Tallarico, J., Wilson, C.J., Myer, V.E., Porter, J.A., Bussiere, D.E., Finan, P.M., Labow, M.A., Mao, X., Hamann, L.G., Manning, B.D., Valdez, R.A., Nicholson, T., Schirle, M., Knapp, M.S., Keaney, E.P., Murphy, L.O., 2014. Selective VPS34 inhibitor blocks autophagy and uncovers a role for NCOA4 in ferritin degradation and iron homeostasis in vivo. Nat. Cell Biol. 16 (11), 1069–1079. http://dx.doi.org/10.1038/ncb3053.

Evstatiev, R., Gasche, C., 2012. Iron sensing and signalling. Gut 61, 933–952. http://dx.doi.org/10.1136/gut.2010.214312.

Fuqua, B.K., Lu, Y., Darshan, D., Frazer, D.M., Wilkins, S.J., Wolkow, N., Bell, A.G., Hsu, J., Yu, C.C., Chen, H., Dunaief, J.L., Anderson, G.J., Vulpe, C.D., 2014. The multicopper ferroxidase hephaestin enhances intestinal iron absorption in mice. PLoS One 9, e98792. http://dx.doi.org/10.1371/journal.pone.0098792.

Ganz, T., 2013. Systemic iron homeostasis. Physiol. Rev. 93, 1721–1741. http://dx.doi.org/10.1152/physrev.00008.2013.

Gari, K., León Ortiz, A.M., Borel, V., Flynn, H., Skehel, J.M., Boulton, S.J., 2012. MMS19 links cytoplasmic iron-sulfur cluster assembly to DNA metabolism. Science 337, 243–245. http://dx.doi.org/10.1126/science.1219664.

Gulec, S., Collins, J.F., 2014. Molecular mediators governing iron-copper interactions. Annu. Rev. Nutr. 34, 95–116. http://dx.doi.org/10.1146/annurev-nutr-071812-161215.

Ichikawa, Y., Bayeva, M., Ghanefar, M., Potini, V., Sun, L., Mutharasan, R.K., Wu, R., Khechaduri, A., Jairaj Naik, T., Ardehali, H., 2012. Disruption of ATP-binding cassette B8 in mice leads to cardiomyopathy through a decrease in mitochondrial iron export. Proc. Natl. Acad. Sci. U.S.A. 109, 4152–4157. http://dx.doi.org/10.1073/pnas.1119338109.

Ishikawa, Y., Maeda, M., Pasham, M., Aguet, F., Tacheva-Grigorova, S.K., Masuda, T., Yi, H., Lee, S.-U., Xu, J., Teruya-Feldstein, J., Ericsson, M., Mullally, A., Heuser, J., Kirchhausen, T., Maeda, T., 2014. Role of the clathrin adaptor PICALM in normal hematopoiesis and polycythemia vera pathophysiology. Haematologica 100 (4), 439–451. http://dx.doi.org/10.3324/haematol.2014.119537.

Jenkitkasemwong, S., Wang, C.-Y., Coffey, R., Zhang, W., Chan, A., Biel, T., Kim, J.-S., Hojyo, S., Fukada, T., Knutson, M.D., 2015. SLC39A14 is required for the development of hepatocellular iron overload in murine models of hereditary hemochromatosis. Cell Metab. 22 (1), 138–150. http://dx.doi.org/10.1016/j.cmet.2015.05.002.

Kautz, L., Jung, G., Valore, E.V., Rivella, S., Nemeth, E., Ganz, T., 2014. Identification of erythroferrone as an erythroid regulator of iron metabolism. Nat. Genet. 46, 678–684. http://dx.doi.org/10.1038/ng.2996.

Kim, J., Buckett, P.D., Wessling-Resnick, M., 2013. Absorption of manganese and iron in a mouse model of hemochromatosis. PLoS One 8, e64944. http://dx.doi.org/10.1371/journal.pone.0064944.

Klebig, M.L., Wall, M.D., Potter, M.D., Rowe, E.L., Carpenter, D.A., Rinchik, E.M., 2003. Mutations in the clathrin-assembly gene Picalm are responsible for the hematopoietic and iron metabolism abnormalities in fit1 mice. Proc. Natl. Acad. Sci. U.S.A. 100, 8360–8365. http://dx.doi.org/10.1073/pnas.1432634100.

Kohyama, M., Ise, W., Edelson, B.T., Wilker, P.R., Hildner, K., Mejia, C., Frazier, W.A., Murphy, T.L., Murphy, K.M., 2009. Role for Spi-C in the development of red pulp macrophages and splenic iron homeostasis. Nature 457, 318–321. http://dx.doi.org/10.1038/nature07472.

Korolnek, T., Hamza, I., 2015. Macrophages and iron trafficking at the birth and death of red cells. Blood 125, 2893–2897. http://dx.doi.org/10.1182/blood-2014-12-567776.

Lakhal-Littleton, S., Wolna, M., Carr, C.A., Miller, J.J.J., Christian, H.C., Ball, V., Santos, A., Diaz, R., Biggs, D., Stillion, R., Holdship, P., Larner, F., Tyler, D.J., Clarke, K., Davies, B., Robbins, P.A., 2015. Cardiac ferroportin regulates cellular iron homeostasis and is important for cardiac function. Proc. Natl. Acad. Sci. U.S.A. 112, 3164–3169. http://dx.doi.org/10.1073/pnas.1422373112.

Lambe, T., Simpson, R.J., Dawson, S., Bouriez-Jones, T., Crockford, T.L., Lepherd, M., Latunde-Dada, G.O., Robinson, H., Raja, K.B., Campagna, D.R., Villarreal Jr., G., Ellory, J.C., Goodnow, C.C., Fleming, M.D., McKie, A.T., Cornall, R.J., 2009. Identification of a Steap3 endosomal targeting motif essential for normal iron metabolism. Blood 113, 1805–1808. http://dx.doi.org/10.1182/blood-2007-11-120402.

Lim, J.E., Jin, O., Bennett, C., Morgan, K., Wang, F., Trenor, C.C., Fleming, M.D., Andrews, N.C., 2005. A mutation in Sec15l1 causes anemia in hemoglobin deficit (hbd) mice. Nat. Genet. 37, 1270–1273. http://dx.doi.org/10.1038/ng1659.

Li, W., Garringer, H.J., Goodwin, C.B., Richine, B., Acton, A., VanDuyn, N., Muhoberac, B.B., Irimia-Dominguez, J., Chan, R.J., Peacock, M., Nass, R., Ghetti, B., Vidal, R., 2015. Systemic and cerebral iron homeostasis in ferritin knock-out mice. PLoS One 10, e0117435. http://dx.doi.org/10.1371/journal.pone.0117435.

Lunova, M., Goehring, C., Kuscuoglu, D., Mueller, K., Chen, Y., Walther, P., Deschemin, J.-C., Vaulont, S., Haybaeck, J., Lackner, C., Trautwein, C., Strnad, P., 2014. Hepcidin knockout mice fed with iron-rich diet develop chronic liver injury and liver fibrosis due to lysosomal iron overload. J. Hepatol. 61, 633–641. http://dx.doi.org/10.1016/j.jhep.2014.04.034.

Mastrogiannaki, M., Matak, P., Peyssonnaux, C., 2013. The gut in iron homeostasis: role of HIF-2 under normal and pathological conditions. Blood 122, 885–892. http://dx.doi.org/10.1182/blood-2012-11-427765.

Mayeur, C., Leyton, P.A., Kolodziej, S.A., Yu, B., Bloch, K.D., 2014. BMP type II receptors have redundant roles in the regulation of hepatic hepcidin gene expression and iron metabolism. Blood 124, 2116–2123. http://dx.doi.org/10.1182/blood-2014-04-572644.

Moulouel, B., Houamel, D., Delaby, C., Tchernitchko, D., Vaulont, S., Letteron, P., Thibaudeau, O., Puy, H., Gouya, L., Beaumont, C., Karim, Z., 2013. Hepcidin regulates intrarenal iron handling at the distal nephron. Kidney Int. 84, 756–766. http://dx.doi.org/10.1038/ki.2013.142.

Nai, A., Lidonnici, M.R., Rausa, M., Mandelli, G., Pagani, A., Silvestri, L., Ferrari, G., Camaschella, C., 2015. The second transferrin receptor regulates red blood cell production in mice. Blood 125, 1170–1179. http://dx.doi.org/10.1182/blood-2014-08-596254.

Nakaya, M., Tajima, M., Kosako, H., Nakaya, T., Hashimoto, A., Watari, K., Nishihara, H., Ohba, M., Komiya, S., Tani, N., Nishida, M., Taniguchi, H., Sato, Y., Matsumoto, M., Tsuda, M., Kuroda, M., Inoue, K., Kurose, H., 2013. GRK6 deficiency in mice causes autoimmune disease due to impaired apoptotic cell clearance. Nat. Commun. 4, 1532. http://dx.doi.org/10.1038/ncomms2540.

Nordin, A., Larsson, E., Thornell, L.-E., Holmberg, M., 2011. Tissue-specific splicing of ISCU results in a skeletal muscle phenotype in myopathy with lactic acidosis, while complete loss of ISCU results in early embryonic death in mice. Hum. Genet. 129, 371–378. http://dx.doi.org/10.1007/s00439-010-0931-3.

Perdomini, M., Hick, A., Puccio, H., Pook, M.A., 2013. Animal and cellular models of Friedreich ataxia. J. Neurochem. 126 (Suppl. 1), 65–79. http://dx.doi.org/10.1111/jnc.12219.

Pondarre, C., Campagna, D.R., Antiochos, B., Sikorski, L., Mulhern, H., Fleming, M.D., 2007. Abcb7, the gene responsible for X-linked sideroblastic anemia with ataxia, is essential for hematopoiesis. Blood 109, 3567–3569. http://dx.doi.org/10.1182/blood-2006-04-015768.

Ramos, P., Guy, E., Chen, N., Proenca, C.C., Gardenghi, S., Casu, C., Follenzi, A., Van Rooijen, N., Grady, R.W., de Sousa, M., Rivella, S., 2011. Enhanced erythropoiesis in Hfe-KO mice indicates a role for Hfe in the modulation of erythroid iron homeostasis. Blood 117, 1379–1389. http://dx.doi.org/10.1182/blood-2010-09-307462.

Salsbury, G., Cambridge, E.L., McIntyre, Z., Arends, M.J., Karp, N.A., Isherwood, C., Shannon, C., Hooks, Y., The Sanger Mouse Genetics Project, Ramirez-Solis, R., Adams, D.J., White, J.K., Speak, A.O., 2014. Disruption of the potassium channel regulatory subunit Kcne2 causes iron-deficient anemia. Exp. Hematol. 42 (12), 1053–1058. http://dx.doi.org/10.1016/j.exphem.2014.07.269.

Sebastiani, G., Gkouvatsos, K., Maffettone, C., Busatto, G., Guido, M., Pantopoulos, K., 2011. Accelerated CCl4-induced liver fibrosis in Hjv-/- mice, associated with an oxidative burst and precocious profibrogenic gene expression. PLoS One 6, e25138. http://dx.doi.org/10.1371/journal.pone.0025138.

Seo, Y.A., Wessling-Resnick, M., 2015. Ferroportin deficiency impairs manganese metabolism in flatiron mice. FASEB J. 29 (7), 2726–2733. http://dx.doi.org/10.1096/fj.14-262592.

Smith, C.P., Thévenod, F., 2009. Iron transport and the kidney. Biochim. Biophys. Acta 1790, 724–730. http://dx.doi.org/10.1016/j.bbagen.2008.10.010.

Song, D., Lee, F.S., 2011. Mouse knock-out of IOP1 protein reveals its essential role in mammalian cytosolic iron-sulfur protein biogenesis. J. Biol. Chem. 286, 15797–15805. http://dx.doi.org/10.1074/jbc.M110.201731.

Stirnberg, M., Maurer, E., Arenz, K., Babler, A., Jahnen-Dechent, W., Gütschow, M., 2015. Cell surface serine protease matriptase-2 suppresses fetuin-A/AHSG-mediated induction of hepcidin. Biol. Chem. 396 (1), 81–93.

Suzuki, M., Tanaka, H., Tanimura, A., Tanabe, K., Oe, N., Rai, S., Kon, S., Fukumoto, M., Takei, K., Abe, T., Matsumura, I., Kanakura, Y., Watanabe, T., 2012. The clathrin assembly protein PICALM is required for erythroid maturation and transferrin internalization in mice. PLoS One 7, e31854. http://dx.doi.org/10.1371/journal.pone.0031854.

Tao, Y., Wu, Q., Guo, X., Zhang, Z., Shen, Y., Wang, F., 2014. MBD5 regulates iron metabolism via methylation-independent genomic targeting of Fth1 through KAT2A in mice. Br. J. Haematol. 166, 279–291. http://dx.doi.org/10.1111/bjh.12863.

Veuthey, T., Wessling-Resnick, M., 2014. Pathophysiology of the Belgrade rat. Front. Pharmacol. 5, 82. http://dx.doi.org/10.3389/fphar.2014.00082.

Wilkinson, N., Pantopoulos, K., 2014. The IRP/IRE system in vivo: insights from mouse models. Front. Pharmacol. 5, 176. http://dx.doi.org/10.3389/fphar.2014.00176.

Yamamoto, M., Arimura, H., Fukushige, T., Minami, K., Nishizawa, Y., Tanimoto, A., Kanekura, T., Nakagawa, M., Akiyama, S.-I., Furukawa, T., 2014. Abcb10 role in heme biosynthesis in vivo: Abcb10 knockout in mice causes anemia with protoporphyrin IX and iron accumulation. Mol. Cell. Biol. 34, 1077–1084. http://dx.doi.org/10.1128/MCB.00865-13.

Zhang, D.-L., Ghosh, M.C., Rouault, T.A., 2014. The physiological functions of iron regulatory proteins in iron homeostasis – an update. Front. Pharmacol. 5, 124. http://dx.doi.org/10.3389/fphar.2014.00124.

Zohn, I.E., De Domenico, I., Pollock, A., Ward, D.M., Goodman, J.F., Liang, X., Sanchez, A.J., Niswander, L., Kaplan, J., 2007. The flatiron mutation in mouse ferroportin acts as a dominant negative to cause ferroportin disease. Blood 109, 4174–4180. http://dx.doi.org/10.1182/blood-2007-01-066068.

Zumerle, S., Mathieu, J.R.R., Delga, S., Heinis, M., Viatte, L., Vaulont, S., Peyssonnaux, C., 2014. Targeted disruption of hepcidin in the liver recapitulates the hemochromatotic phenotype. Blood 123, 3646–3650. http://dx.doi.org/10.1182/blood-2014-01-550467.

Chapter 17

Iron, Cancer, and Hypoxia-Inducible Factor Signaling

Xiang Xue, Yatrik Madhukar Shah

University of Michigan, Ann Arbor, MI, United States

INTRODUCTION

Iron is an essential micronutrient required for numerous biological and cellular processes: oxygen transport (hemoglobin and myoglobin), oxygen sensing (hypoxia-inducible factor prolyl hydroxylases), iron storage and transport (transferrin and ferritin), host defense and inflammation (NADPH oxidase, indoleamine 2,3-dioxygenase, nitric oxide synthases, and lipoxygenases), energy production (cytochrome c, cytochrome c oxidase, and NADH dehydrogenase), xenobiotic detoxification (cytochrome P450), and DNA synthesis (ribonucleotide reductase; Ganz and Nemeth, 2015). Iron also catalyzes free-radical formation through the Fenton and Haber–Weiss reactions (Jomova and Valko, 2011). Therefore, cellular iron levels are tightly regulated and iron dysregulation can facilitate tumor initiation, growth, and tumor metastasis. Altered intratumoral iron metabolism results in enhanced oxidative stress, lipid peroxidation, protein modification, and DNA damage, and consequently promotes oncogene activation and rapid tumor cell proliferation. This chapter will focus on the integration of hypoxia and iron signaling critical for tumor progression.

OXYGEN-SENSING TRANSCRIPTION FACTORS

Tissues maintain a narrow range of O_2 for respiration and homeostasis. Cells respond to reduced O_2 tension (hypoxia) by altering expression of genes encoding proteins critical in erythropoiesis, angiogenesis, iron transport, glycolysis, cell proliferation, survival and apoptosis, tumorigenesis, and metastasis (Wang and Semenza, 1993b). These changes are initiated by a nuclear basic-helix-loop-helix-per-ARNT-sim heterodimeric transcription factor HIF. Hypoxia-inducible factor (HIF) comprises an oxygen-dependent α-subunit (HIF-α) and a constitutively expressed β-subunit, aryl hydrocarbon nuclear translocator (ARNT or HIF-1β; Wang et al., 1995). In cells, adequate levels of oxygen and iron cause rapid degradation of HIF-α subunits initiated by a conserved family of prolyl hydroxylase domain-containing enzymes (PHD; Fong and Takeda, 2008). Four PHDs family members have been identified, PHD1/EGLN2/HPH3, PHD2/EGLN1/HPH2, PHD3/EGLN3/HPH1, and P4HTM (Bruick and McKnight, 2001; Epstein et al., 2001; Koivunen et al., 2007). PHDs are an important link between oxygen levels and iron since the 2-oxoglutarate-dependent dioxygenase activity requires iron and oxygen as substrates (Myllyharju and Koivunen, 2013). At normal levels of oxygen, PHDs hydroxylate two conserved prolines in the oxygen-dependent degradation domain of the HIF-α subunit. Oxygen-dependent prolyl-hydroxylation of HIF-α subunits is required for binding by the von Hippel–Lindau tumor suppressor protein (VHL), a recognition component of an E3 ubiquitin ligase complex, resulting in rapid proteasomal degradation (Ivan et al., 2001; Jaakkola et al., 2001; Maxwell et al., 1999). Inactivation of VHL in normoxic cells results in HIF activation demonstrating that VHL is essential for HIF degradation (Pugh and Ratcliffe, 2003).

A decrease in cellular oxygen or iron availability inhibits PHD-dependent proline hydroxylation of HIF-α subunits. VHL can thus no longer bind to HIF-α subunits, resulting in their stabilization. Following stabilization, HIF-α subunits interact with ARNT and other coactivators, such as p300/CREB binding protein, to activate transcription of genes encoding proteins that regulate the hypoxic response (Arany et al., 1996). There are three isoforms of HIF-α, HIF-1α (Semenza and Wang, 1992; Wang and Semenza, 1993a), HIF-2α (Tian et al., 1997), and HIF3α (Makino et al., 2001). HIF-1α and HIF-2α share significant amino acid sequence homology and both have two prolyl-hydroxylation sites and two transactivation domains (TADs; N-terminal and C-terminal), whereas HIF-3α has only one prolyl-hydroxylation site and the N-terminal TAD (Heikkila et al., 2011). HIFs regulate distinct and overlapping target genes. Moreover, HIF-α activation by low iron

Molecular, Genetic, and Nutritional Aspects of Major and Trace Minerals. http://dx.doi.org/10.1016/B978-0-12-802168-2.00017-8

or a reduction in O_2 leads to an increase in the expression of different subsets of HIF target genes. The mechanisms of these differential gene expression patterns are not clear. In addition to its role in transcriptional regulation, HIF-2α forms a complex with RNA-binding proteins and increases polysomal translation of mRNAs code for proteins such as epidermal growth factor receptor, platelet-derived growth factor receptor α, and insulin-like growth factor 1 receptor (Uniacke et al., 2012). Genetic disruption of *Hif1a*, *Hif2a*, *Arnt*, or *Vhl* leads to embryonic lethality in mice, demonstrating the importance of an appropriate hypoxic response in vivo (Gnarra et al., 1997; Iyer et al., 1998; Maltepe et al., 1997; Tian et al., 1998).

OXYGEN AND IRON SIGNALING

Hepcidin functions as the major regulator of tissue iron export. Hepcidin binds to the iron exporter ferroportin (FPN) on highly conserved residues leading to internalization and degradation (Nemeth et al., 2004). Under normal serum iron levels, the relatively high expression of hepatic hepcidin leads to consistent degradation of FPN and represses iron mobilization from the tissue stores to the blood circulation. Under low systemic iron levels, hepcidin expression is robustly decreased leading to enhanced intestinal iron absorption and mobilization from tissue stores, which increase serum iron levels (Frazer and Anderson, 2003). In addition, systemic hypoxia leads to a rapid and robust inhibition of hepcidin expression, which can be recapitulated in mice with HIF overexpression in the kidney or liver (Liu et al., 2012; Anderson et al., 2012). Hypoxic repression of hepcidin is through an indirect mechanism via enhancing the synthesis of erythropoietin (EPO) in liver and kidney, which increases erythropoietic activity and iron demand in bone marrow (Liu et al., 2012). The transcription factor CCAAT-enhancer-binding protein alpha (C/EBPα) is required for EPO-independent hepcidin repression by HIF and basal hepcidin expression (Shah and Xie, 2014). These findings are consistent with a study of indigenous populations who live at high altitudes, in which EPO-dependent and independent pathways for hepcidin repression were identified (Lundgrin et al., 2013).

In addition to the iron and oxygen link via hepcidin, hypoxia is also a major regulator of apical, basolateral, and intracellular iron handling by duodenal enterocytes. Divalent metal transporter 1 (DMT-1, also known as SLC11A2, NRAMP2, and DCT1) is highly expressed in the proximal duodenum and is a major mediator of intestinal iron absorption (Gunshin et al., 1997). Hypoxia increases DMT-1-mediated ferrous iron uptake (Qian et al., 2011). HIF-2α (but not HIF-1α) is the major transcriptional regulator of DMT-1 (Mastrogiannaki et al., 2009). HIF-2α also directly regulates the expression of ferric reductase duodenal cytochrome b (DcytB) and FPN during iron deficiency (Shah et al., 2009; Taylor et al., 2011). Moreover, HIF-1α is a critical regulator of transferrin receptor 1 (TfR1) (Bianchi et al., 1999; Lok and Ponka, 1999; Tacchini et al., 1999) and heme oxygenase-1 (HO-1) (Lee et al., 1997) expression during hypoxia as well as during iron deficiency.

HYPOXIA AND CANCER

Due to the rapid proliferation of tumor epithelial cells, the relatively slow-growing blood vessels do not provide sufficient nutrients and oxygen to the tumor. This leads to hypoxia in solid tumors. Hypoxia is a negative prognostic and predictive factor in several cancers due to its role in chemoresistance, radioresistance, angiogenesis, invasiveness, metastasis, and altered metabolism (Wilson and Hay, 2011). The major mechanism by which hypoxia potentiates tumorigenesis is through activation of the HIF signaling pathway (Mabjeesh and Amir, 2007; Ruan et al., 2009). HIF promotes neovascularization via induction of vascular endothelial growth factor (VEGF) (Garcia-Heredia et al., 2015) and alters tumor metabolism by promoting aerobic glycolysis (i.e., the Warburg effect) (Semenza, 2002). Overexpression of the HIF-1α or HIF-2α is positively associated with poor treatment responses and outcomes in an extensive range of human tumors (Semenza, 2007). Based on studies with human cancer cell lines as well as animal models, overexpression of HIF-1α or HIF-2α is associated with increased tumor growth, angiogenesis, and metastasis, whereas disruption of HIF-1α or HIF-2α leads to decreased tumorigenesis and cell growth (Semenza, 2007). However, the role of hypoxic regulation of iron handling in tumorigenesis has not been clearly defined to date. Recent work demonstrates that HIF-2α-induced colon tumorigenesis is partially mediated via an increase in intracellular iron (Xue et al., 2012).

IRON AND CANCER RISK

Iron overload, due to hereditary mutations which cause excessive intestinal iron absorption, is a risk factor for multiple tumors including colorectal cancer (CRC) (Chua et al., 2010; Xue and Shah, 2013), breast cancer (Moore et al., 2009), lung cancer (Xiong et al., 2014), liver cancer (Ba et al., 2011), prostate cancer (Torti and Torti, 2013), and glioblastoma (Schonberg et al., 2015). Red meat, which is a major dietary component in Western countries, is enriched in heme iron (Theil, 2011). Red and processed meat intake is associated with increased risk of several cancers including colorectal

(Aune et al., 2013), lung (Xue et al., 2014), pancreatic (Larsson and Wolk, 2012), and gastric cancer (Larsson et al., 2006). These observations are consistent with a systemic review and meta-analysis results demonstrating that heme iron uptake is positively correlated with increased cancer risk (Fonseca-Nunes et al., 2014). The strongest association with red meat intake and cancer is observed in CRC; a large meta-analysis demonstrated that red meat intake correlates with an increase in colorectal adenomas, suggesting that iron may play a key role in the early stages of the colorectal carcinogenesis. In large population-based studies, the risks for CRC and gastric cancer are positively associated with red and processed meat consumption (Cross et al., 2010; Gonzalez et al., 2006; Norat et al., 2005). Moreover, increased risk of CRC in red meat consumers is entirely dependent on heme iron abundance, and independent of fat intake (Sesink et al., 1999, 2000). Feeding rats with heme damages the colonic mucosal surface, causing compensatory hyperplasia (Sesink et al., 1999). Trapping heme with calcium phosphate or green vegetable derived chlorophyll inhibits the cytolytic and hyperproliferative effects of dietary heme and reduces CRC risk in meat eaters (Balder et al., 2006; de Vogel et al., 2005; Pierre et al., 2008). In addition, ferrous iron is widely supplemented in the American diet (Nelson, 2001). Supplemental iron is positively associated with distal CRC, which may cause enhanced free-radical production in the colon (Lee et al., 2004; Lund et al., 1999). Tumoral iron accumulation mediated by HIF-2α is involved in colon tumorigenesis in mice (Xue et al., 2012). In addition, dietary iron uptake also increases the risk of lung cancer (Zhou et al., 2005) and breast cancer (Moore et al., 2009). Iron reduction by phlebotomy reduces cancer risk and mortality (Zacharski et al., 2008). Similarly, cancer incidence is significantly reduced in frequent blood donors (Merk et al., 1990) and the mortality rate is lower in donor cancer patients compared with nondonor cancer patients (Vahidnia et al., 2013).

MECHANISMS REGULATING INTRATUMORAL IRON

Hepcidin: Hepcidin is mainly synthesized in the liver, but epithelial cells from other tissues such as prostate and breast can also synthesize hepcidin (Pinnix et al., 2010; Tesfay et al., 2015). In tumor tissues including prostate cancer, breast cancer, and CRC, the expression of hepcidin is robustly increased (Tesfay et al., 2015; Ward et al., 2008; Zhang et al., 2014). This tumoral hepcidin leads to decreased cell-surface FPN levels and increased intratumoral iron retention, which supports cancer cell survival and rapid growth. Cell signaling pathways including BMP4/7, IL6, Wnt, and the dual BMP and Wnt antagonist, SOSTDC1, are reported to be responsible for inducing hepcidin overexpression in prostate cancer (Tesfay et al., 2015). Interestingly, prostate cancer patients with bone metastases have higher levels of serum hepcidin compared with those without metastasis (Tanno et al., 2011). Urinary hepcidin level is positively associated with increasing T-stage of CRC (Ward et al., 2008). High FPN and low hepcidin gene expression in breast cancer cells identifies an extremely favorable cohort of breast cancer patients (Pinnix et al., 2010), indicating potentially important roles for the hepcidin-FPN regulatory axis in oncogenic signaling. Plasma hepcidin is a good predictive marker of malignant breast lesions (Orlandi et al., 2014).

TfR: TfR1 is an HIF-1α target gene that regulates cellular iron uptake in most cells. TfR1 expression is correlated with the rate of cell proliferation (Okazaki et al., 2010). Cancer cells have higher TfR1 levels compared to normal cells. A decrease in TfR1 expression or function reduces cell proliferation and inhibits tumor growth (Daniels et al., 2011; Jiang et al., 2010), whereas overexpression confers a growth advantage to cells (O'Donnell et al., 2006). A large cohort study found that a high transferrin saturation level is associated with excess risk for CRC and lung cancers (Knekt et al., 1994). In colon tissue, TfR1 is expressed in normal colonocytes, and further increased in CRC of Dukes A or B grade (Prutki et al., 2006). Eighty-eight percent of nonsmall-cell lung cancer patients are positive for TfR1 expression (Kukulj et al., 2010). Increased TfR1 cell-surface expression in tumors is correlated to a higher degree of malignancy and thus may provide prognostic information regarding malignant tumors (Habeshaw et al., 1983; Wrba et al., 1986). A second transferrin receptor, TfR2 was also shown to mediate cellular iron uptake (Kawabata et al., 1999). TfR2 is frequently expressed in various human cancer cell lines such as ovarian, colon, hepatoma, and glioblastoma (Calzolari et al., 2007; Herbison et al., 2009). In primary colon tumors, TfR2 is expressed in about 26% of CRC patients (Calzolari et al., 2009). The expression of TfR2, together with increased TfR1, may represent a mechanism by which cancer cells accumulate iron derived from circulating diferric transferrin.

DMT-1: DMT-1 is expressed in many different tissues including normal and tumor astroglial cells (Lis et al., 2004). DMT-1 expression is increased in esophageal adenocarcinoma, as compared to Barrett's metaplasia (Boult et al., 2008). Overexpression of DMT-1 was further associated with metastatic adenocarcinoma. DMT-1 expression is increased in human breast cancer MCF-7 cells compared to normal human mammary epithelial MCF-12A cells (Jiang et al., 2010). DMT-1 is normally expressed at low levels in the colon but is highly induced in colon adenomas and carcinomas isolated from patients (Sabates-Bellver et al., 2007; Skrzypczak et al., 2010; Zou et al., 2002). Two independent groups have demonstrated that the expression of DMT-1 protein is increased in mouse intestinal tumors, and this increase is dependent on the HIF-2α (Xue et al., 2012) and Wnt (Radulescu et al., 2012) signaling pathways.

MECHANISMS OF IRON ACTION

There are eight distinguishing hallmarks of cancer: (1) sustaining proliferative signaling; (2) evading growth suppressors; (3) activating invasion and metastasis; (4) enabling replicative immortality; (5) inducing angiogenesis; (6) resisting cell death; (7) deregulating cellular energetics; and (8) avoiding immune destruction. In addition, two enabling factors are critical in the initiation and maintenance of these cancer-related phenomenon: (1) genome instability and (2) inflammation (Hanahan and Weinberg, 2011). Iron is involved in each hallmark and enabling characteristic.

Oxidative Stress

The presence of free iron generates highly reactive hydroxyl radicals from hydrogen peroxide (Fenton reaction), leading to DNA oxidation, mitochondrial damage, and peroxidation of membrane lipids. Increased oxidative stress is genotoxic and can lead to DNA damage and subsequent mutations important for the initiation of tumorigenesis (Glei et al., 2002; Knobel et al., 2007; Ribeiro et al., 2008). Moreover, lipid peroxidation propagates free-radical reactions and facilitates the formation of DNA-adducts that may contribute to tumorigenesis (Marnett, 1999). Heme iron from red meat also promotes colon carcinogenesis through its catalytic effects on the formation of carcinogenic N-nitroso compounds or cytotoxic and genotoxic secondary lipid oxidation products, such as 4-hydroxy-2(E)-nonenal (HNE; Baradat et al., 2011; Bastide et al., 2011; Iyer et al., 1998; Joosen et al., 2009; Knobel et al., 2006). Lastly, oxidative stress induces a pleiotropic effect in cancers and can regulate proliferation, apoptosis, mitochondrial metabolism, and inflammation. The involvement of iron-induced oxidative damage in cancer is further demonstrated by the fact that the antioxidant N-acetylcysteine inhibits dietary iron-induced CRC progression and nitrotyrosine-positive cell numbers in the long-term administration of dextran sulfate sodium-induced CRC model (Seril et al., 2002a,b).

p53 Signaling

p53 is a DNA-binding protein essential in preserving the integrity of the genome in cells. p53 is the most widely mutated gene in cancer, underscoring its importance in tumorigenesis. Iron downregulates, whereas iron chelation with deferoxamine (DFO) increases, expression of mouse double minute gene 2 (MDM2), which is an ubiquitin ligase involved in the degradation of p53 (Dongiovanni et al., 2010). Iron-mediated reduction in MDM2 levels leads to p53 upregulation and induction of antioxidant enzymes, thereby providing a mechanism to counteract iron-related oxidative stress. At toxic doses, iron can also increase nuclear protein level and DNA-binding activity of p53, which causes cell cycle arrest (Fukuchi et al., 1995; Liang and Richardson, 2003). However, other studies have shown a negative association of iron and p53 expression. Hypoxia and DFO induce accumulation of wild-type p53 through HIF-1α-dependent binding and stabilization of the p53 protein (An et al., 1998). In addition, iron chelation induces the phosphorylation of p53 at serines 15 and 37 leading to decreased interaction with MDM2 and reduced degradation (Le and Richardson, 2002). Iron depletion also inhibits ribonucleotide reductase activity that results in reduced deoxyribonucleotide levels, which increases p53 transcriptional activity. One report shows that heme directly binds to the p53 protein, modulates p53-DNA interactions, and facilitates nuclear export and cytosolic degradation of p53 (Shen et al., 2014).

Activation of Oncogenic Signaling

Wnt/β-catenin Signaling: Wnt is a protein ligand that can activate three major pathways involved in tumor progression: planar cell polarity pathway, calcium pathway, and the canonical Wnt pathway (Pez et al., 2013). Wnt functions by binding to a cell-surface receptor leading to an increase in intracellular signaling. The best-characterized protumorigenic pathway downstream of Wnt is the β-catenin pathway (Niehrs, 2012). Under low Wnt stimulation, β-catenin is rapidly degraded, whereas Wnt stimulation triggers β-catenin stabilization and regulation of genes essential for tumor growth. Similar to p53, the Wnt/β-catenin pathway is frequently mutated in several cancers, leading to constitutive β-catenin signaling. Iron can further increase β-catenin signaling in tumors which have aberrantly high levels of β-catenin activation, but not in normal cells which do not have constitutive β-catenin signaling (Brookes et al., 2008). Consistent with these data, iron chelators can block Wnt signaling and decrease the expression of cyclin D1 and c-Myc to suppress cell growth (Coombs et al., 2012; Song et al., 2011).

Cell Cycle Proteins: Currently there is no direct evidence that increases in intratumoral iron can regulate the cell cycle in tumor cells. However, iron chelation decreases cyclins and cyclin-dependent kinases (CDK), and increases CDK inhibitors independent of p53 activation (Fu and Richardson, 2007). A well-characterized pathway in several cancer cell lines of iron chelation-mediated cell cycle arrest is through an increase expression of the CDK inhibitor p21 (Le and Richardson, 2002).

c-Myc and STAT3: c-Myc and STAT3 are essential transcription factors in regulating growth of tumors (Chai et al., 2015; Li and Simon, 2013). c-Myc is critical for regulating tumor growth and metabolism in several different cancers and is highly amplified in many tumors (Dang et al., 2009). STAT3 signaling is highly activated following inflammatory activation and it functions as a critical transcription factor in initiating and maintaining the growth of tumors (Hirano et al., 2000). Iron chelation decreases c-Myc expression through a STAT3-dependent mechanism in leukemic, pancreatic, and prostate-derived cell lines (Lui et al., 2015b). Moreover, one study demonstrates that altering intracellular iron mobilization through ferritin depletion significantly abrogates STAT3 signaling in glioblastoma stem-like cells (Schonberg et al., 2015).

Oncogenic Kinases: Iron has a profound role in intracellular kinase signaling (Lui et al., 2015a). Mitogen-activated protein kinases are a family of kinases including extracellular signal-regulated kinase (ERK), p38, and c-Jun N-terminal kinase (JNK) (Rose et al., 2010). ERK is an essential kinase for the proliferative effects downstream of growth factor stimulation, whereas p38 and JNK are stress-activated kinases. Although the data can vary depending on cell type, iron can regulate ERK, p38, and JNK to modulate tumor growth (Markel et al., 2007; Yu and Richardson, 2011).

Metastasis

Metastasis is a coordinated process in which advanced tumor cells gain the ability to disseminate via the circulatory system and invade and propagate at distant sites. Epithelial cells can transition to mesenchymal-like cells to gain motile and invasive properties. This is termed epithelial-to-mesenchymal transition, which is an essential mechanism driving metastasis (Kalluri and Weinberg, 2009). The HIF and Wnt pathways are essential in this transition and as stated earlier, these pathways are regulated by iron (Richardson et al., 2013; Zhang et al., 2013). Moreover, matrix metalloproteinases (MMPs) are essential for metastatic cells to degrade the extracellular matrix and invade through the basement membrane. Several MMPs are regulated directly by iron, and iron chelators block their activity (Brenneisen et al., 1998).

POTENTIAL IN CANCER THERAPY

Targeting hypoxia signaling and iron metabolism holds great promise in cancer therapy. Bioreductive prodrugs targeting hypoxic cells are in active clinic trials (Chawla et al., 2014). Hypoxia initiates the activation of bioreductive prodrugs by preventing redox cycling of the prodrug radical anions generated by one-electron reductases. The resulting cytotoxic drugs typically induce DNA replication fork damage, exacerbated by suppression of homologous recombination in hypoxic cells, leading to cell death. Disruption of HIF signaling has long been known to hold high potential for cancer therapy based on preclinical studies (Kung et al., 2000; Semenza, 2003), and molecules that inhibit various components of this pathway have been reported (Kung et al., 2004; Lee et al., 2009; Miranda et al., 2013; Nordgren and Tavassoli, 2011). Iron chelation therapy is efficacious in animal studies of carcinogenesis. Two separate studies using genetic mouse models of CRC demonstrated that mice on low-iron diets had decreased colon tumors compared to mice on an iron-replete diet (Radulescu et al., 2012; Xue et al., 2012). Moreover, iron chelators, such as DFO and deferasirox, have antineoplastic activities, as mentioned earlier. DFO and deferasirox are attractive compounds since they are currently used clinically for iron overload, are orally bioavailable and have limited deleterious side effects. However, there is need to assess the efficacy of these compounds in cancer-specific clinical trials. Moreover, there are several other promising compounds that can limit intratumoral iron. Hepcidin antagonists that block the biological activity of hepcidin would be expected to allow continuous iron export from cells (Montalbetti et al., 2013). Anti-TfR1 antibodies and DMT-1 inhibitors may also be useful in limiting intratumoral iron (Cadieux et al., 2012; Montalbetti et al., 2015).

OTHER MINERALS IN CANCER

The role of intratumoral iron in tumor progression is beginning to be explored, however, details of the molecular mechanism are still not understood. Even less is known about the role of other micronutrients. Currently, large cancer genome initiatives such as *The Cancer Genome Atlas* are starting to provide evidences that several additional cellular metal-ion transporters are dysregulated in cancer. Future goals will be to understand the micronutrient requirements for tumors and to assess whether proteins handling these micronutrients are useful targets for cancer treatment.

SUMMARY POINTS AND KEY FACTS

- Iron is a key micronutrient for cellular growth.
- Hypoxia is a major regulator of cellular iron levels.

- Iron accumulation is observed in several cancers.
- Iron triggers heightened oxidative damage leading to tumor progression.
- Iron is essential for several oncogenic signaling pathways.
- Iron chelation therapy may be an efficacious approach to limit tumor growth.

MINI DICTIONARY OF TERMS

Angiogenesis Formation of new blood vessels.
Apoptosis Programmed cell death mechanism in multicellular organisms.
Chemoresistance The resistance of cancer cells to chemicals used for cancer treatments.
Erythropoiesis Synthesis of red blood cells.
Genotoxic Agents that can lead to damage of DNA.
Hypoxia Decreased oxygen in the body, organ, or cell.
Meta-analysis Analytical approach to combine results from multiple studies.
Metastasis The spread of cancer to a distal site.
Oxidative stress Cellular imbalance between the production of reactive oxygen species and antioxidant defenses leading to cellular and organ injury.
Radioresistance The resistance of cancer cells to ionizing radiation.
Tumorigenesis Transformation of normal cells into cancer cells.

REFERENCES

An, W.G., Kanekal, M., Simon, M.C., Maltepe, E., Blagosklonny, M.V., Neckers, L.M., 1998. Stabilization of wild-type p53 by hypoxia-inducible factor 1alpha. Nature 392, 405–408.

Anderson, E.R., Taylor, M., Xue, X., Martin, A., Moons, D.S., Omary, M.B., Shah, Y.M., 2012. The hypoxia-inducible factor-C/EBPalpha axis controls ethanol-mediated hepcidin repression. Mol. Cell Biol. 32, 4068–4077.

Arany, Z., Huang, L.E., Eckner, R., Bhattacharya, S., Jiang, C., Goldberg, M.A., Bunn, H.F., Livingston, D.M., 1996. An essential role for p300/CBP in the cellular response to hypoxia. Proc. Natl. Acad. Sci. U. S. A. 93, 12969–12973.

Aune, D., Chan, D.S., Vieira, A.R., Navarro Rosenblatt, D.A., Vieira, R., Greenwood, D.C., Kampman, E., Norat, T., 2013. Red and processed meat intake and risk of colorectal adenomas: a systematic review and meta-analysis of epidemiological studies. Cancer Causes Control 24, 611–627.

Ba, Q., Hao, M., Huang, H., Hou, J., Ge, S., Zhang, Z., Yin, J., Chu, R., Jiang, H., Wang, F., Chen, K., Liu, H., Wang, H., 2011. Iron deprivation suppresses hepatocellular carcinoma growth in experimental studies. Clin. Cancer Res. 17, 7625–7633.

Balder, H.F., Vogel, J., Jansen, M.C., Weijenberg, M.P., van den Brandt, P.A., Westenbrink, S., van der Meer, R., Goldbohm, R.A., 2006. Heme and chlorophyll intake and risk of colorectal cancer in the Netherlands cohort study. Cancer Epidemiol. Biomarkers Prev. 15, 717–725.

Baradat, M., Jouanin, I., Dalleau, S., Tache, S., Gieules, M., Debrauwer, L., Canlet, C., Huc, L., Dupuy, J., Pierre, F.H., Gueraud, F., 2011. 4-Hydroxy-2(E)-nonenal metabolism differs in Apc(+/+) cells and in Apc(Min/+) cells: it may explain colon cancer promotion by heme iron. Chem. Res. Toxicol. 24, 1984–1993.

Bastide, N.M., Pierre, F.H., Corpet, D.E., 2011. Heme iron from meat and risk of colorectal cancer: a meta-analysis and a review of the mechanisms involved. Cancer Prev. Res. (Phila) 4, 177–184.

Bianchi, L., Tacchini, L., Cairo, G., 1999. HIF-1-mediated activation of transferrin receptor gene transcription by iron chelation. Nucleic Acids Res. 27, 4223–4227.

Boult, J., Roberts, K., Brookes, M.J., Hughes, S., Bury, J.P., Cross, S.S., Anderson, G.J., Spychal, R., Iqbal, T., Tselepis, C., 2008. Overexpression of cellular iron import proteins is associated with malignant progression of esophageal adenocarcinoma. Clin. Cancer Res. 14, 379–387.

Brenneisen, P., Wenk, J., Klotz, L.O., Wlaschek, M., Briviba, K., Krieg, T., Sies, H., Scharffetter-Kochanek, K., 1998. Central role of Ferrous/Ferric iron in the ultraviolet B irradiation-mediated signaling pathway leading to increased interstitial collagenase (matrix-degrading metalloprotease (MMP)-1) and stromelysin-1 (MMP-3) mRNA levels in cultured human dermal fibroblasts. J. Biol. Chem. 273, 5279–5287.

Brookes, M.J., Boult, J., Roberts, K., Cooper, B.T., Hotchin, N.A., Matthews, G., Iqbal, T., Tselepis, C., 2008. A role for iron in Wnt signalling. Oncogene 27, 966–975.

Bruick, R.K., McKnight, S.L., 2001. A conserved family of prolyl-4-hydroxylases that modify HIF. Science 294, 1337–1340.

Cadieux, J.A., Zhang, Z., Mattice, M., Brownlie-Cutts, A., Fu, J., Ratkay, L.G., Kwan, R., Thompson, J., Sanghara, J., Zhong, J., Goldberg, Y.P., 2012. Synthesis and biological evaluation of substituted pyrazoles as blockers of divalent metal transporter 1 (DMT1). Bioorg. Med. Chem. Lett. 22, 90–95.

Calzolari, A., Oliviero, I., Deaglio, S., Mariani, G., Biffoni, M., Sposi, N.M., Malavasi, F., Peschle, C., Testa, U., 2007. Transferrin receptor 2 is frequently expressed in human cancer cell lines. Blood Cells Mol. Dis. 39, 82–91.

Calzolari, A., Deaglio, S., Maldi, E., Cassoni, P., Malavasi, F., Testa, U., 2009. TfR2 expression in human colon carcinomas. Blood Cells Mol. Dis. 43, 243–249.

Chai, E.Z., Shanmugam, M.K., Arfuso, F., Dharmarajan, A., Wang, C., Kumar, A.P., Samy, R.P., Lim, L.H., Wang, L., Goh, B.C., Ahn, K.S., Hui, K.M., Sethi, G., 2015. Targeting transcription factor STAT3 for cancer prevention and therapy. Pharmacol. Ther. 162, 86–97.

Chawla, S.P., Cranmer, L.D., Van Tine, B.A., Reed, D.R., Okuno, S.H., Butrynski, J.E., Adkins, D.R., Hendifar, A.E., Kroll, S., Ganjoo, K.N., 2014. Phase II study of the safety and antitumor activity of the hypoxia-activated prodrug TH-302 in combination with doxorubicin in patients with advanced soft tissue sarcoma. J. Clin. Oncol. 32, 3299–3306.

Chua, A.C., Klopcic, B., Lawrance, I.C., Olynyk, J.K., Trinder, D., 2010. Iron: an emerging factor in colorectal carcinogenesis. World J. Gastroenterol. 16, 663–672.

Coombs, G.S., Schmitt, A.A., Canning, C.A., Alok, A., Low, I.C., Banerjee, N., Kaur, S., Utomo, V., Jones, C.M., Pervaiz, S., Toone, E.J., Virshup, D.M., 2012. Modulation of Wnt/beta-catenin signaling and proliferation by a ferrous iron chelator with therapeutic efficacy in genetically engineered mouse models of cancer. Oncogene 31, 213–225.

Cross, A.J., Ferrucci, L.M., Risch, A., Graubard, B.I., Ward, M.H., Park, Y., Hollenbeck, A.R., Schatzkin, A., Sinha, R., 2010. A large prospective study of meat consumption and colorectal cancer risk: an investigation of potential mechanisms underlying this association. Cancer Res. 70, 2406–2414.

Dang, C.V., Le, A., Gao, P., 2009. MYC-induced cancer cell energy metabolism and therapeutic opportunities. Clin. Cancer Res. 15, 6479–6483.

Daniels, T.R., Ortiz-Sanchez, E., Luria-Perez, R., Quintero, R., Helguera, G., Bonavida, B., Martinez-Maza, O., Penichet, M.L., 2011. An antibody-based multifaceted approach targeting the human transferrin receptor for the treatment of B-cell malignancies. J. Immunother. 34, 500–508.

Dongiovanni, P., Fracanzani, A.L., Cairo, G., Megazzini, C.P., Gatti, S., Rametta, R., Fargion, S., Valenti, L., 2010. Iron-dependent regulation of MDM2 influences p53 activity and hepatic carcinogenesis. Am. J. Pathol. 176, 1006–1017.

Epstein, A.C., Gleadle, J.M., McNeill, L.A., Hewitson, K.S., O'Rourke, J., Mole, D.R., Mukherji, M., Metzen, E., Wilson, M.I., Dhanda, A., Tian, Y.M., Masson, N., Hamilton, D.L., Jaakkola, P., Barstead, R., Hodgkin, J., Maxwell, P.H., Pugh, C.W., Schofield, C.J., Ratcliffe, P.J., 2001. C. elegans EGL-9 and mammalian homologs define a family of dioxygenases that regulate HIF by prolyl hydroxylation. Cell 107, 43–54.

Fong, G.H., Takeda, K., 2008. Role and regulation of prolyl hydroxylase domain proteins. Cell Death Differ. 15, 635–641.

Fonseca-Nunes, A., Jakszyn, P., Agudo, A., 2014. Iron and cancer risk–a systematic review and meta-analysis of the epidemiological evidence. Cancer Epidemiol. Biomarkers Prev. 23, 12–31.

Frazer, D.M., Anderson, G.J., 2003. The orchestration of body iron intake: how and where do enterocytes receive their cues? Blood Cells Mol. Dis. 30, 288–297.

Fu, D., Richardson, D.R., 2007. Iron chelation and regulation of the cell cycle: 2 mechanisms of posttranscriptional regulation of the universal cyclin-dependent kinase inhibitor p21CIP1/WAF1 by iron depletion. Blood 110, 752–761.

Fukuchi, K., Tomoyasu, S., Watanabe, H., Kaetsu, S., Tsuruoka, N., Gomi, K., 1995. Iron deprivation results in an increase in p53 expression. Biol. Chem. Hoppe Seyler 376, 627–630.

Ganz, T., Nemeth, E., 2015. Iron homeostasis in host defence and inflammation. Nat. Rev. Immunol. 15, 500–510.

Garcia-Heredia, J.M., Felipe-Abrio, B., Cano, D.A., Carnero, A., 2015. Genetic modification of hypoxia signaling in animal models and its effect on cancer. Clin. Transl. Oncol. 17, 90–102.

Glei, M., Latunde-Dada, G.O., Klinder, A., Becker, T.W., Hermann, U., Voigt, K., Pool-Zobel, B.L., 2002. Iron-overload induces oxidative DNA damage in the human colon carcinoma cell line HT29 clone 19A. Mutat. Res. 519, 151–161.

Gnarra, J.R., Ward, J.M., Porter, F.D., Wagner, J.R., Devor, D.E., Grinberg, A., Emmert-Buck, M.R., Westphal, H., Klausner, R.D., Linehan, W.M., 1997. Defective placental vasculogenesis causes embryonic lethality in VHL-deficient mice. Proc. Natl. Acad. Sci. U. S. A. 94, 9102–9107.

Gonzalez, C.A., Jakszyn, P., Pera, G., Agudo, A., Bingham, S., Palli, D., Ferrari, P., Boeing, H., del Giudice, G., Plebani, M., Carneiro, F., Nesi, G., Berrino, F., Sacerdote, C., Tumino, R., Panico, S., Berglund, G., Siman, H., Nyren, O., Hallmans, G., Martinez, C., Dorronsoro, M., Barricarte, A., Navarro, C., Quiros, J.R., Allen, N., Key, T.J., Day, N.E., Linseisen, J., Nagel, G., Bergmann, M.M., Overvad, K., Jensen, M.K., Tjonneland, A., Olsen, A., Bueno-de-Mesquita, H.B., Ocke, M., Peeters, P.H., Numans, M.E., Clavel-Chapelon, F., Boutron-Ruault, M.C., Trichopoulou, A., Psaltopoulou, T., Roukos, D., Lund, E., Hemon, B., Kaaks, R., Norat, T., Riboli, E., 2006. Meat intake and risk of stomach and esophageal adenocarcinoma within the European Prospective Investigation into Cancer and Nutrition (EPIC). J. Natl. Cancer Inst. 98, 345–354.

Gunshin, H., Mackenzie, B., Berger, U.V., Gunshin, Y., Romero, M.F., Boron, W.F., Nussberger, S., Gollan, J.L., Hediger, M.A., 1997. Cloning and characterization of a mammalian proton-coupled metal-ion transporter. Nature 388, 482–488.

Habeshaw, J.A., Lister, T.A., Stansfeld, A.G., Greaves, M.F., 1983. Correlation of transferrin receptor expression with histological class and outcome in non-Hodgkin lymphoma. Lancet 1, 498–501.

Hanahan, D., Weinberg, R.A., 2011. Hallmarks of cancer: the next generation. Cell 144, 646–674.

Heikkila, M., Pasanen, A., Kivirikko, K.I., Myllyharju, J., 2011. Roles of the human hypoxia-inducible factor (HIF)-3alpha variants in the hypoxia response. Cell Mol. Life Sci. 68, 3885–3901.

Herbison, C.E., Thorstensen, K., Chua, A.C., Graham, R.M., Leedman, P., Olynyk, J.K., Trinder, D., 2009. The role of transferrin receptor 1 and 2 in transferrin-bound iron uptake in human hepatoma cells. Am. J. Physiol. Cell Physiol. 297, C1567–C1575.

Hirano, T., Ishihara, K., Hibi, M., 2000. Roles of STAT3 in mediating the cell growth, differentiation and survival signals relayed through the IL-6 family of cytokine receptors. Oncogene 19, 2548–2556.

Ivan, M., Kondo, K., Yang, H., Kim, W., Valiando, J., Ohh, M., Salic, A., Asara, J.M., Lane, W.S., Kaelin Jr., W.G., 2001. HIFalpha targeted for VHL-mediated destruction by proline hydroxylation: implications for O_2 sensing. Science 292, 464–468.

Iyer, N.V., Kotch, L.E., Agani, F., Leung, S.W., Laughner, E., Wenger, R.H., Gassmann, M., Gearhart, J.D., Lawler, A.M., Yu, A.Y., Semenza, G.L., 1998. Cellular and developmental control of O_2 homeostasis by hypoxia-inducible factor 1 alpha. Genes Dev. 12, 149–162.

Jaakkola, P., Mole, D.R., Tian, Y.M., Wilson, M.I., Gielbert, J., Gaskell, S.J., von Kriegsheim, A., Hebestreit, H.F., Mukherji, M., Schofield, C.J., Maxwell, P.H., Pugh, C.W., Ratcliffe, P.J., 2001. Targeting of HIF-alpha to the von Hippel-Lindau ubiquitylation complex by O_2-regulated prolyl hydroxylation. Science 292, 468–472.

Jiang, X.P., Elliott, R.L., Head, J.F., 2010. Manipulation of iron transporter genes results in the suppression of human and mouse mammary adenocarcinomas. Anticancer Res. 30, 759–765.

Jomova, K., Valko, M., 2011. Importance of iron chelation in free radical-induced oxidative stress and human disease. Curr. Pharm. Des. 17, 3460–3473.

Joosen, A.M., Kuhnle, G.G., Aspinall, S.M., Barrow, T.M., Lecommandeur, E., Azqueta, A., Collins, A.R., Bingham, S.A., 2009. Effect of processed and red meat on endogenous nitrosation and DNA damage. Carcinogenesis 30, 1402–1407.

Kalluri, R., Weinberg, R.A., 2009. The basics of epithelial-mesenchymal transition. J. Clin. Invest. 119, 1420–1428.

Kawabata, H., Yang, R., Hirama, T., Vuong, P.T., Kawano, S., Gombart, A.F., Koeffler, H.P., 1999. Molecular cloning of transferrin receptor 2. A new member of the transferrin receptor-like family. J. Biol. Chem. 274, 20826–20832.

Knekt, P., Reunanen, A., Takkunen, H., Aromaa, A., Heliovaara, M., Hakulinen, T., 1994. Body iron stores and risk of cancer. Int. J. Cancer 56, 379–382.

Knobel, Y., Glei, M., Osswald, K., Pool-Zobel, B.L., 2006. Ferric iron increases ROS formation, modulates cell growth and enhances genotoxic damage by 4-hydroxynonenal in human colon tumor cells. Toxicol. In Vitro 20, 793–800.

Knobel, Y., Weise, A., Glei, M., Sendt, W., Claussen, U., Pool-Zobel, B.L., 2007. Ferric iron is genotoxic in non-transformed and preneoplastic human colon cells. Food Chem. Toxicol. 45, 804–811.

Koivunen, P., Tiainen, P., Hyvarinen, J., Williams, K.E., Sormunen, R., Klaus, S.J., Kivirikko, K.I., Myllyharju, J., 2007. An endoplasmic reticulum transmembrane prolyl 4-hydroxylase is induced by hypoxia and acts on hypoxia-inducible factor alpha. J. Biol. Chem. 282, 30544–30552.

Kukulj, S., Jaganjac, M., Boranic, M., Krizanac, S., Santic, Z., Poljak-Blazi, M., 2010. Altered iron metabolism, inflammation, transferrin receptors, and ferritin expression in non-small-cell lung cancer. Med. Oncol. 27, 268–277.

Kung, A.L., Wang, S., Klco, J.M., Kaelin, W.G., Livingston, D.M., 2000. Suppression of tumor growth through disruption of hypoxia-inducible transcription. Nat. Med. 6, 1335–1340.

Kung, A.L., Zabludoff, S.D., France, D.S., Freedman, S.J., Tanner, E.A., Vieira, A., Cornell-Kennon, S., Lee, J., Wang, B., Wang, J., Memmert, K., Naegeli, H.U., Petersen, F., Eck, M.J., Bair, K.W., Wood, A.W., Livingston, D.M., 2004. Small molecule blockade of transcriptional coactivation of the hypoxia-inducible factor pathway. Cancer Cell 6, 33–43.

Larsson, S.C., Wolk, A., 2012. Red and processed meat consumption and risk of pancreatic cancer: meta-analysis of prospective studies. Br. J. Cancer 106, 603–607.

Larsson, S.C., Orsini, N., Wolk, A., 2006. Processed meat consumption and stomach cancer risk: a meta-analysis. J. Natl. Cancer Inst. 98, 1078–1087.

Le, N.T., Richardson, D.R., 2002. The role of iron in cell cycle progression and the proliferation of neoplastic cells. Biochim. Biophys. Acta 1603, 31–46.

Lee, P.J., Jiang, B.H., Chin, B.Y., Iyer, N.V., Alam, J., Semenza, G.L., Choi, A.M., 1997. Hypoxia-inducible factor-1 mediates transcriptional activation of the heme oxygenase-1 gene in response to hypoxia. J. Biol. Chem. 272, 5375–5381.

Lee, D.H., Jacobs Jr., D.R., Folsom, A.R., 2004. A hypothesis: interaction between supplemental iron intake and fermentation affecting the risk of colon cancer. The Iowa Women's Health Study. Nutr. Cancer 48, 1–5.

Lee, K., Zhang, H., Qian, D.Z., Rey, S., Liu, J.O., Semenza, G.L., 2009. Acriflavine inhibits HIF-1 dimerization, tumor growth, and vascularization. Proc. Natl. Acad. Sci. U. S. A. 106, 17910–17915.

Li, B., Simon, M.C., 2013. Molecular Pathways: targeting MYC-induced metabolic reprogramming and oncogenic stress in cancer. Clin. Cancer Res. 19, 5835–5841.

Liang, S.X., Richardson, D.R., 2003. The effect of potent iron chelators on the regulation of p53: examination of the expression, localization and DNA-binding activity of p53 and the transactivation of WAF1. Carcinogenesis 24, 1601–1614.

Lis, A., Barone, T.A., Paradkar, P.N., Plunkett, R.J., Roth, J.A., 2004. Expression and localization of different forms of DMT1 in normal and tumor astroglial cells. Brain Res. Mol. Brain Res. 122, 62–70.

Liu, Q., Davidoff, O., Niss, K., Haase, V.H., 2012. Hypoxia-inducible factor regulates hepcidin via erythropoietin-induced erythropoiesis. J. Clin. Invest. 122, 4635–4644.

Lok, C.N., Ponka, P., 1999. Identification of a hypoxia response element in the transferrin receptor gene. J. Biol. Chem. 274, 24147–24152.

Lui, G.Y., Kovacevic, Z., Richardson, V., Merlot, A.M., Kalinowski, D.S., Richardson, D.R., 2015a. Targeting cancer by binding iron: dissecting cellular signaling pathways. Oncotarget 6, 18748–18779.

Lui, G.Y., Kovacevic, Z., Menezes, S.V., Kalinowski, D.S., Merlot, A.M., Sahni, S., Richardson, D.R., 2015b. Novel thiosemicarbazones regulate the signal transducer and activator of transcription 3 (STAT3) pathway: inhibition of constitutive and interleukin 6-induced activation by iron depletion. Mol. Pharmacol. 87, 543–560.

Lund, E.K., Wharf, S.G., Fairweather-Tait, S.J., Johnson, I.T., 1999. Oral ferrous sulfate supplements increase the free radical-generating capacity of feces from healthy volunteers. Am. J. Clin. Nutr. 69, 250–255.

Lundgrin, E.L., Janocha, A.J., Koch, C.D., Gebremedhin, A., Di Rienzo, A., Alkorta-Aranburu, G., Brittenham, G.M., Erzurum, S.C., Beall, C.M., 2013. Plasma hepcidin of Ethiopian highlanders with steady-state hypoxia. Blood 122, 1989–1991.

Mabjeesh, N.J., Amir, S., 2007. Hypoxia-inducible factor (HIF) in human tumorigenesis. Histol. Histopathol. 22, 559–572.

Makino, Y., Cao, R., Svensson, K., Bertilsson, G., Asman, M., Tanaka, H., Cao, Y., Berkenstam, A., Poellinger, L., 2001. Inhibitory PAS domain protein is a negative regulator of hypoxia-inducible gene expression. Nature 414, 550–554.

Maltepe, E., Schmidt, J.V., Baunoch, D., Bradfield, C.A., Simon, M.C., 1997. Abnormal angiogenesis and responses to glucose and oxygen deprivation in mice lacking the protein ARNT. Nature 386, 403–407.

Markel, T.A., Crisostomo, P.R., Wang, M., Herring, C.M., Lahm, T., Meldrum, K.K., Lillemoe, K.D., Rescorla, F.J., Meldrum, D.R., 2007. Iron chelation acutely stimulates fetal human intestinal cell production of IL-6 and VEGF while decreasing HGF: the roles of p38, ERK, and JNK MAPK signaling. Am. J. Physiol. Gastrointest. Liver Physiol. 292, G958–G963.

Marnett, L.J., 1999. Lipid peroxidation-DNA damage by malondialdehyde. Mutat. Res. 424, 83–95.

Mastrogiannaki, M., Matak, P., Keith, B., Simon, M.C., Vaulont, S., Peyssonnaux, C., 2009. HIF-2alpha, but not HIF-1alpha, promotes iron absorption in mice. J. Clin. Invest. 119, 1159–1166.

Maxwell, P.H., Wiesener, M.S., Chang, G.W., Clifford, S.C., Vaux, E.C., Cockman, M.E., Wykoff, C.C., Pugh, C.W., Maher, E.R., Ratcliffe, P.J., 1999. The tumour suppressor protein VHL targets hypoxia-inducible factors for oxygen-dependent proteolysis. Nature 399, 271–275.

Merk, K., Mattsson, B., Mattsson, A., Holm, G., Gullbring, B., Bjorkholm, M., 1990. The incidence of cancer among blood donors. Int. J. Epidemiol. 19, 505–509.

Miranda, E., Nordgren, I.K., Male, A.L., Lawrence, C.E., Hoakwie, F., Cuda, F., Court, W., Fox, K.R., Townsend, P.A., Packham, G.K., Eccles, S.A., Tavassoli, A., 2013. A cyclic peptide inhibitor of HIF-1 heterodimerization that inhibits hypoxia signaling in cancer cells. J. Am. Chem. Soc. 135, 10418–10425.

Montalbetti, N., Simonin, A., Kovacs, G., Hediger, M.A., 2013. Mammalian iron transporters: families SLC11 and SLC40. Mol. Asp. Med. 34, 270–287.

Montalbetti, N., Simonin, A., Simonin, C., Awale, M., Reymond, J.L., Hediger, M.A., 2015. Discovery and characterization of a novel non-competitive inhibitor of the divalent metal transporter DMT1/SLC11A2. Biochem. Pharmacol. 96, 216–224.

Moore, A.B., Shannon, J., Chen, C., Lampe, J.W., Ray, R.M., Lewis, S.K., Lin, M., Stalsberg, H., Thomas, D.B., 2009. Dietary and stored iron as predictors of breast cancer risk: a nested case-control study in Shanghai. Int. J. Cancer 125, 1110–1117.

Myllyharju, J., Koivunen, P., 2013. Hypoxia-inducible factor prolyl 4-hydroxylases: common and specific roles. Biol. Chem. 394, 435–448.

Nelson, R.L., 2001. Iron and colorectal cancer risk: human studies. Nutr. Rev. 59, 140–148.

Nemeth, E., Tuttle, M.S., Powelson, J., Vaughn, M.B., Donovan, A., Ward, D.M., Ganz, T., Kaplan, J., 2004. Hepcidin regulates cellular iron efflux by binding to ferroportin and inducing its internalization. Science 306, 2090–2093.

Niehrs, C., 2012. The complex world of WNT receptor signalling. Nat. Rev. Mol. Cell Biol. 13, 767–779.

Norat, T., Bingham, S., Ferrari, P., Slimani, N., Jenab, M., Mazuir, M., Overvad, K., Olsen, A., Tjonneland, A., Clavel, F., Boutron-Ruault, M.C., Kesse, E., Boeing, H., Bergmann, M.M., Nieters, A., Linseisen, J., Trichopoulou, A., Trichopoulos, D., Tountas, Y., Berrino, F., Palli, D., Panico, S., Tumino, R., Vineis, P., Bueno-de-Mesquita, H.B., Peeters, P.H., Engeset, D., Lund, E., Skeie, G., Ardanaz, E., Gonzalez, C., Navarro, C., Quiros, J.R., Sanchez, M.J., Berglund, G., Mattisson, I., Hallmans, G., Palmqvist, R., Day, N.E., Khaw, K.T., Key, T.J., San Joaquin, M., Hemon, B., Saracci, R., Kaaks, R., Riboli, E., 2005. Meat, fish, and colorectal cancer risk: the European Prospective Investigation into cancer and nutrition. J. Natl. Cancer Inst. 97, 906–916.

Nordgren, I.K., Tavassoli, A., 2011. Targeting tumour angiogenesis with small molecule inhibitors of hypoxia inducible factor. Chem. Soc. Rev. 40, 4307–4317.

O'Donnell, K.A., Yu, D., Zeller, K.I., Kim, J.W., Racke, F., Thomas-Tikhonenko, A., Dang, C.V., 2006. Activation of transferrin receptor 1 by c-Myc enhances cellular proliferation and tumorigenesis. Mol. Cell Biol. 26, 2373–2386.

Okazaki, F., Matsunaga, N., Okazaki, H., Utoguchi, N., Suzuki, R., Maruyama, K., Koyanagi, S., Ohdo, S., 2010. Circadian rhythm of transferrin receptor 1 gene expression controlled by c-Myc in colon cancer-bearing mice. Cancer Res. 70, 6238–6246.

Orlandi, R., De Bortoli, M., Ciniselli, C.M., Vaghi, E., Caccia, D., Garrisi, V., Pizzamiglio, S., Veneroni, S., Bonini, C., Agresti, R., Daidone, M.G., Morelli, D., Camaschella, C., Verderio, P., Bongarzone, I., 2014. Hepcidin and ferritin blood level as noninvasive tools for predicting breast cancer. Ann. Oncol. 25, 352–357.

Pez, F., Lopez, A., Kim, M., Wands, J.R., Caron de Fromentel, C., Merle, P., 2013. Wnt signaling and hepatocarcinogenesis: molecular targets for the development of innovative anticancer drugs. J. Hepatol. 59, 1107–1117.

Pierre, F., Santarelli, R., Tache, S., Gueraud, F., Corpet, D.E., 2008. Beef meat promotion of dimethylhydrazine-induced colorectal carcinogenesis biomarkers is suppressed by dietary calcium. Br. J. Nutr. 99, 1000–1006.

Pinnix, Z.K., Miller, L.D., Wang, W., D'Agostino Jr., R., Kute, T., Willingham, M.C., Hatcher, H., Tesfay, L., Sui, G., Di, X., Torti, S.V., Torti, F.M., 2010. Ferroportin and iron regulation in breast cancer progression and prognosis. Sci. Transl. Med. 2, 43ra56.

Prutki, M., Poljak-Blazi, M., Jakopovic, M., Tomas, D., Stipancic, I., Zarkovic, N., 2006. Altered iron metabolism, transferrin receptor 1 and ferritin in patients with colon cancer. Cancer Lett. 238, 188–196.

Pugh, C.W., Ratcliffe, P.J., 2003. The von Hippel-Lindau tumor suppressor, hypoxia-inducible factor-1 (HIF-1) degradation, and cancer pathogenesis. Semin. Cancer Biol. 13, 83–89.

Qian, Z.M., Wu, X.M., Fan, M., Yang, L., Du, F., Yung, W.H., Ke, Y., 2011. Divalent metal transporter 1 is a hypoxia-inducible gene. J. Cell Physiol. 226, 1596–1603.

Radulescu, S., Brookes, M.J., Salgueiro, P., Ridgway, R.A., McGhee, E., Anderson, K., Ford, S.J., Stones, D.H., Iqbal, T.H., Tselepis, C., Sansom, O.J., 2012. Luminal iron levels govern intestinal tumorigenesis after Apc loss in vivo. Cell Rep. 2, 270–282.

Ribeiro, M.L., Priolli, D.G., Miranda, D.D., Arcari, D.P., Pedrazzoli Jr., J., Martinez, C.A., 2008. Analysis of oxidative DNA damage in patients with colorectal cancer. Clin. Colorectal Cancer 7, 267–272.

Richardson, A., Kovacevic, Z., Richardson, D.R., 2013. Iron chelation: inhibition of key signaling pathways in the induction of the epithelial mesenchymal transition in pancreatic cancer and other tumors. Crit. Rev. Oncog. 18, 409–434.

Rose, B.A., Force, T., Wang, Y., 2010. Mitogen-activated protein kinase signaling in the heart: angels versus demons in a heart-breaking tale. Physiol. Rev. 90, 1507–1546.

Ruan, K., Song, G., Ouyang, G., 2009. Role of hypoxia in the hallmarks of human cancer. J. Cell Biochem. 107, 1053–1062.

Sabates-Bellver, J., Van der Flier, L.G., de Palo, M., Cattaneo, E., Maake, C., Rehrauer, H., Laczko, E., Kurowski, M.A., Bujnicki, J.M., Menigatti, M., Luz, J., Ranalli, T.V., Gomes, V., Pastorelli, A., Faggiani, R., Anti, M., Jiricny, J., Clevers, H., Marra, G., 2007. Transcriptome profile of human colorectal adenomas. Mol. Cancer Res. 5, 1263–1275.

Schonberg, D.L., Miller, T.E., Wu, Q., Flavahan, W.A., Das, N.K., Hale, J.S., Hubert, C.G., Mack, S.C., Jarrar, A.M., Karl, R.T., Rosager, A.M., Nixon, A.M., Tesar, P.J., Hamerlik, P., Kristensen, B.W., Horbinski, C., Connor, J.R., Fox, P.L., Lathia, J.D., Rich, J.N., 2015. Preferential iron trafficking characterizes glioblastoma stem-like cells. Cancer Cell 28, 441–455.

Semenza, G.L., Wang, G.L., 1992. A nuclear factor induced by hypoxia via de novo protein synthesis binds to the human erythropoietin gene enhancer at a site required for transcriptional activation. Mol. Cell Biol. 12, 5447–5454.

Semenza, G.L., 2002. HIF-1 and tumor progression: pathophysiology and therapeutics. Trends Mol. Med. 8, S62–S67.

Semenza, G.L., 2003. Targeting HIF-1 for cancer therapy. Nat. Rev. Cancer 3, 721–732.

Semenza, G.L., 2007. Evaluation of HIF-1 inhibitors as anticancer agents. Drug Discov. Today 12, 853–859.

Seril, D.N., Liao, J., Ho, K.L., Warsi, A., Yang, C.S., Yang, G.Y., 2002a. Dietary iron supplementation enhances DSS-induced colitis and associated colorectal carcinoma development in mice. Dig. Dis. Sci. 47, 1266–1278.

Seril, D.N., Liao, J., Ho, K.L., Yang, C.S., Yang, G.Y., 2002b. Inhibition of chronic ulcerative colitis-associated colorectal adenocarcinoma development in a murine model by N-acetylcysteine. Carcinogenesis 23, 993–1001.

Sesink, A.L., Termont, D.S., Kleibeuker, J.H., Van der Meer, R., 1999. Red meat and colon cancer: the cytotoxic and hyperproliferative effects of dietary heme. Cancer Res. 59, 5704–5709.

Sesink, A.L., Termont, D.S., Kleibeuker, J.H., Van Der Meer, R., 2000. Red meat and colon cancer: dietary haem, but not fat, has cytotoxic and hyperproliferative effects on rat colonic epithelium. Carcinogenesis 21, 1909–1915.

Shah, Y.M., Xie, L., 2014. Hypoxia-inducible factors link iron homeostasis and erythropoiesis. Gastroenterology 146, 630–642.

Shah, Y.M., Matsubara, T., Ito, S., Yim, S.H., Gonzalez, F.J., 2009. Intestinal hypoxia-inducible transcription factors are essential for iron absorption following iron deficiency. Cell Metab. 9, 152–164.

Shen, J., Sheng, X., Chang, Z., Wu, Q., Wang, S., Xuan, Z., Li, D., Wu, Y., Shang, Y., Kong, X., Yu, L., Li, L., Ruan, K., Hu, H., Huang, Y., Hui, L., Xie, D., Wang, F., Hu, R., 2014. Iron metabolism regulates p53 signaling through direct heme-p53 interaction and modulation of p53 localization, stability, and function. Cell Rep. 7, 180–193.

Skrzypczak, M., Goryca, K., Rubel, T., Paziewska, A., Mikula, M., Jarosz, D., Pachlewski, J., Oledzki, J., Ostrowski, J., 2010. Modeling oncogenic signaling in colon tumors by multidirectional analyses of microarray data directed for maximization of analytical reliability. PLoS One 5.

Song, S., Christova, T., Perusini, S., Alizadeh, S., Bao, R.Y., Miller, B.W., Hurren, R., Jitkova, Y., Gronda, M., Isaac, M., Joseph, B., Subramaniam, R., Aman, A., Chau, A., Hogge, D.E., Weir, S.J., Kasper, J., Schimmer, A.D., Al-awar, R., Wrana, J.L., Attisano, L., 2011. Wnt inhibitor screen reveals iron dependence of beta-catenin signaling in cancers. Cancer Res. 71, 7628–7639.

Tacchini, L., Bianchi, L., Bernelli-Zazzera, A., Cairo, G., 1999. Transferrin receptor induction by hypoxia. HIF-1-mediated transcriptional activation and cell-specific post-transcriptional regulation. J. Biol. Chem. 274, 24142–24146.

Tanno, T., Rabel, A., Alleyne, M., Lee, Y.T., Dahut, W.L., Gulley, J.L., Miller, J.L., 2011. Hepcidin, anaemia, and prostate cancer. BJU Int. 107, 678–679.

Taylor, M., Qu, A., Anderson, E.R., Matsubara, T., Martin, A., Gonzalez, F.J., Shah, Y.M., 2011. Hypoxia-inducible factor-2alpha mediates the adaptive increase of intestinal ferroportin during iron deficiency in mice. Gastroenterology 140, 2044–2055.

Tesfay, L., Clausen, K.A., Kim, J.W., Hegde, P., Wang, X., Miller, L.D., Deng, Z., Blanchette, N., Arvedson, T., Miranti, C.K., Babitt, J.L., Lin, H.Y., Peehl, D.M., Torti, F.M., Torti, S.V., 2015. Hepcidin regulation in prostate and its disruption in prostate cancer. Cancer Res. 75, 2254–2263.

Theil, E.C., 2011. Iron homeostasis and nutritional iron deficiency. J. Nutr. 141, 724S–728S.

Tian, H., McKnight, S.L., Russell, D.W., 1997. Endothelial PAS domain protein 1 (EPAS1), a transcription factor selectively expressed in endothelial cells. Genes Dev. 11, 72–82.

Tian, H., Hammer, R.E., Matsumoto, A.M., Russell, D.W., McKnight, S.L., 1998. The hypoxia-responsive transcription factor EPAS1 is essential for catecholamine homeostasis and protection against heart failure during embryonic development. Genes Dev. 12, 3320–3324.

Torti, S.V., Torti, F.M., 2013. Iron and cancer: more ore to be mined. Nat. Rev. Cancer 13, 342–355.

Uniacke, J., Holterman, C.E., Lachance, G., Franovic, A., Jacob, M.D., Fabian, M.R., Payette, J., Holcik, M., Pause, A., Lee, S., 2012. An oxygen-regulated switch in the protein synthesis machinery. Nature 486, 126–129.

Vahidnia, F., Hirschler, N.V., Agapova, M., Chinn, A., Busch, M.P., Custer, B., 2013. Cancer incidence and mortality in a cohort of US blood donors: a 20-Year study. J. Cancer Epidemiol. 2013, 814842.

de Vogel, J., Jonker-Termont, D.S., van Lieshout, E.M., Katan, M.B., van der Meer, R., 2005. Green vegetables, red meat and colon cancer: chlorophyll prevents the cytotoxic and hyperproliferative effects of haem in rat colon. Carcinogenesis 26, 387–393.

Wang, G.L., Semenza, G.L., 1993a. Characterization of hypoxia-inducible factor 1 and regulation of DNA binding activity by hypoxia. J. Biol. Chem. 268, 21513–21518.

Wang, G.L., Semenza, G.L., 1993b. General involvement of hypoxia-inducible factor 1 in transcriptional response to hypoxia. Proc. Natl. Acad. Sci. U. S. A. 90, 4304–4308.

Wang, G.L., Jiang, B.H., Rue, E.A., Semenza, G.L., 1995. Hypoxia-inducible factor 1 is a basic-helix-loop-helix-PAS heterodimer regulated by cellular O_2 tension. Proc. Natl. Acad. Sci. U. S. A. 92, 5510–5514.

Ward, D.G., Roberts, K., Brookes, M.J., Joy, H., Martin, A., Ismail, T., Spychal, R., Iqbal, T., Tselepis, C., 2008. Increased hepcidin expression in colorectal carcinogenesis. World J. Gastroenterol. 14, 1339–1345.

Wilson, W.R., Hay, M.P., 2011. Targeting hypoxia in cancer therapy. Nat. Rev. Cancer 11, 393–410.

Wrba, F., Ritzinger, E., Reiner, A., Holzner, J.H., 1986. Transferrin receptor (TrfR) expression in breast carcinoma and its possible relationship to prognosis. An immunohistochemical study. Virchows Arch. A Pathol. Anat. Histopathol. 410, 69–73.

Xiong, W., Wang, L., Yu, F., 2014. Regulation of cellular iron metabolism and its implications in lung cancer progression. Med. Oncol. 31, 28.

Xue, X., Shah, Y.M., 2013. Intestinal iron homeostasis and colon tumorigenesis. Nutrients 5, 2333–2351.

Xue, X., Taylor, M., Anderson, E., Hao, C., Qu, A., Greenson, J.K., Zimmermann, E.M., Gonzalez, F.J., Shah, Y.M., 2012. Hypoxia-inducible factor-2alpha activation promotes colorectal cancer progression by dysregulating iron homeostasis. Cancer Res. 72, 2285–2293.

Xue, X.J., Gao, Q., Qiao, J.H., Zhang, J., Xu, C.P., Liu, J., 2014. Red and processed meat consumption and the risk of lung cancer: a dose-response meta-analysis of 33 published studies. Int. J. Clin. Exp. Med. 7, 1542–1553.

Yu, Y., Richardson, D.R., 2011. Cellular iron depletion stimulates the JNK and p38 MAPK signaling transduction pathways, dissociation of ASK1-thioredoxin, and activation of ASK1. J. Biol. Chem. 286, 15413–15427.

Zacharski, L.R., Chow, B.K., Howes, P.S., Shamayeva, G., Baron, J.A., Dalman, R.L., Malenka, D.J., Ozaki, C.K., Lavori, P.W., 2008. Decreased cancer risk after iron reduction in patients with peripheral arterial disease: results from a randomized trial. J. Natl. Cancer Inst. 100, 996–1002.

Zhang, Q., Bai, X., Chen, W., Ma, T., Hu, Q., Liang, C., Xie, S., Chen, C., Hu, L., Xu, S., Liang, T., 2013. Wnt/beta-catenin signaling enhances hypoxia-induced epithelial-mesenchymal transition in hepatocellular carcinoma via crosstalk with HIF-1alpha signaling. Carcinogenesis 34, 962–973.

Zhang, S., Chen, Y., Guo, W., Yuan, L., Zhang, D., Xu, Y., Nemeth, E., Ganz, T., Liu, S., 2014. Disordered hepcidin-ferroportin signaling promotes breast cancer growth. Cell Signal. 26, 2539–2550.

Zhou, W., Park, S., Liu, G., Miller, D.P., Wang, L.I., Pothier, L., Wain, J.C., Lynch, T.J., Giovannucci, E., Christiani, D.C., 2005. Dietary iron, zinc, and calcium and the risk of lung cancer. Epidemiology 16, 772–779.

Zou, T.T., Selaru, F.M., Xu, Y., Shustova, V., Yin, J., Mori, Y., Shibata, D., Sato, F., Wang, S., Olaru, A., Deacu, E., Liu, T.C., Abraham, J.M., Meltzer, S.J., 2002. Application of cDNA microarrays to generate a molecular taxonomy capable of distinguishing between colon cancer and normal colon. Oncogene 21, 4855–4862.

Chapter 18

Iron Transporters and Iron Homeostasis

Mitchell D. Knutson

University of Florida, Gainesville, FL, United States

OVERVIEW OF WHOLE-BODY IRON HOMEOSTASIS AND SCOPE OF REVIEW

In contrast to all other minerals, iron cannot be actively excreted from the body. As a consequence, whole-body iron balance is maintained primarily by regulating how much dietary iron is absorbed by enterocytes of the small intestine. Once absorbed, iron in blood plasma is taken up mostly by erythroid precursor cells of the bone marrow and incorporated into erythrocyte hemoglobin, which accounts for nearly two-thirds of total body iron. Erythrocytes live for approximately 120 days and then are cleared from the circulation by reticuloendothelial macrophages of the liver, spleen, and bone marrow that catabolize erythrocyte hemoglobin, releasing iron from heme. Most of the liberated iron is returned to the plasma and delivered to the bone marrow, where it is reincorporated into new erythroid precursor cells. Iron is stored in reticuloendothelial macrophages and hepatocytes of the liver. Hepatocytes play another important role in iron metabolism by synthesizing the iron-regulatory hormone hepcidin. Thus overall whole-body iron homeostasis can generally be viewed as the integration of iron metabolism of four main cell types: the enterocyte, the erythroid precursor cell, the reticuloendothelial macrophage, and the hepatocyte (Fig. 18.1). The aim of this review is to summarize current knowledge of iron metabolism in these four cell types. Emphasis is placed on iron transporters and molecular mechanisms that direct the movement of iron (and heme) into, within, and out of cells. The role of hepcidin in controlling the flow of iron among various cell types, along with the main drivers of hepcidin expression, will also be discussed.

DIETARY IRON ABSORPTION BY THE ENTEROCYTE

Dietary iron is classified as heme or nonheme. Heme iron is from meat and derives mainly from hemoglobin and myoglobin. Nonheme iron, essentially inorganic iron, derives from cereals, vegetables, and fruits. Although meat is a good source of heme iron, roughly half of its iron is nonheme derived from the iron-storage protein ferritin, iron-sulfur cluster proteins, and other iron-containing enzymes. Accordingly, most dietary iron comprises nonheme iron, even in individuals who consume a lot of meat.

Nonheme iron in the proximal small intestine is taken up into the enterocyte via divalent metal-ion transporter-1 (DMT1/SLC11A1), a proton-coupled transporter located on the apical membrane (Fleming et al., 1997; Gunshin et al., 1997) (Fig. 18.2). Mice with intestine-specific loss of DMT1 are normal at birth, but they develop severe iron-deficiency anemia by 8 weeks of age because of markedly impaired intestinal iron absorption (Gunshin et al., 2005a; Shawki et al., 2015).

Four different DMT1 mRNAs encoding distinct DMT1 proteins have been identified: DMT1A, DMT1A-IRE, DMT1B, and DMT1B-IRE (Hubert and Hentze, 2002). The four DMT1 isoforms are nearly the same except for small differences in the N- and C-termini that yield proteins ranging from 557 to 590 amino acids (human DMT1; Shawki et al., 2012). Transport studies in *Xenopus* oocytes concluded that all four DMT1 isoforms transport iron with equal efficiency after adjusting for expression level at the plasma membrane (Mackenzie et al., 2007). The isoforms differ in their tissue expression, subcellular distribution, and responsiveness to cellular iron status. In the intestine, the iron-responsive element (IRE)-containing DMT1 isoforms are most abundant (Hubert and Hentze, 2002), and their protein levels increase markedly in response to iron deficiency (Canonne-Hergaux et al., 1999). The variants containing an IRE, a stem–loop structure in the mRNA, are regulated at the posttranscriptional level. Indeed, the regulation of DMT1 and other iron-related proteins through IREs is central to the control of cellular iron metabolism, but they are beyond the scope of the present review; the reader is referred elsewhere (Anderson et al., 2012).

Whereas DMT1 transports only divalent reduced iron (Fe^{2+}; Gunshin et al., 1997), most dietary iron in the lumen of the small intestine exists as low-molecular–weight chelates of oxidized Fe^{3+}. Therefore before intestinal uptake iron must be either enzymatically or nonenzymatically reduced. Enzymatic reduction is believed to be catalyzed by the ferrireductase

Molecular, Genetic, and Nutritional Aspects of Major and Trace Minerals. http://dx.doi.org/10.1016/B978-0-12-802168-2.00018-X

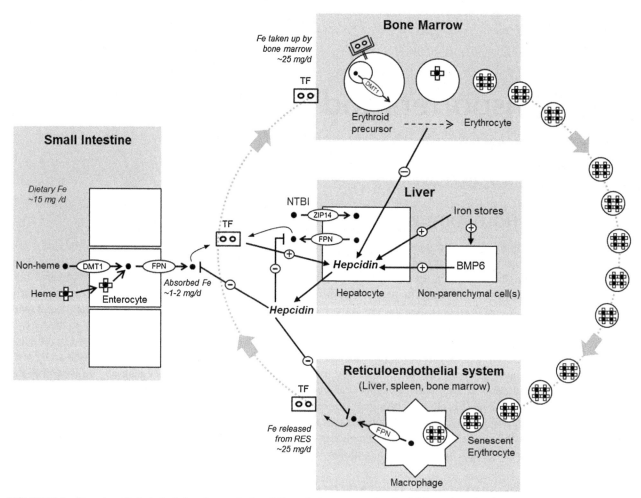

FIGURE 18.1   Overview of whole-body iron homeostasis and the main transporters involved. Dietary iron absorbed by the small intestine is bound by the plasma protein transferrin (TF), which delivers most of its iron to erythroid precursor cells of the bone marrow. The iron incorporated into erythrocyte hemoglobin is eventually reclaimed by macrophages of the reticuloendothelial system (RES) and released to the plasma, where it binds to TF and returns to the bone marrow. The uptake of iron by the enterocyte and assimilation of iron by the erythroid precursor cell requires the iron-import protein divalent metal-ion transporter-1 (DMT1). The release of iron into the plasma is mediated by the iron-export protein ferroportin (FPN) in the enterocyte, reticuloendothelial macrophage, and hepatocyte. Hepcidin negatively regulates FPN; therefore it controls how much iron is released into the plasma. Hepcidin expression is positively regulated by iron via TF saturation and bone morphogenetic protein-6 (BMP6) and negatively regulated by erythropoiesis and related factors. Details of the roles of specific transporters and the regulation of hepcidin are provided in the text. *ZIP14*, ZRT- and IRT-like protein-14.

duodenal cytochrome B (DCYTB, encoded by the *CYBRD1* gene) expressed on the enterocyte apical membrane (McKie et al., 2000). The observation that *Cybrd1*-null mice have normal hemoglobin levels and iron stores has led to the conclusion that DCYTB is not essential for dietary iron absorption in mice (Gunshin et al., 2005b). However, such a conclusion seems premature because iron absorption per se was not measured in these animals, especially in the context of iron deficiency when DCYTB is normally markedly upregulated (Frazer et al., 2005; McKie et al., 2001). It is possible that luminal iron reduction is catalyzed by ascorbate secreted into gastric juice (Rathbone et al., 1989). Consistent with this possibility, one study reported that duodenal ascorbate levels are increased in iron-deficient/hypoxic mice (Atanasova et al., 2004).

Iron in the form of heme is readily taken up by the intestine, but the molecular mechanisms involved have been elusive. In 2005 a low-affinity, pH-independent heme transporter in the duodenum was identified and named heme carrier protein-1 (HCP1; Shayeghi et al., 2005). In 2006 the same protein was independently identified as the protein responsible for high-affinity uptake of folate and named protein-coupled folate transporter-1 (PCFT1; Qiu et al., 2006). Loss-of-function mutations in *PCFT1* were found to be associated with hereditary folate malabsorption, thereby confirming its physiologic role as an intestinal folate transporter. Accordingly, *Pcft1*-null mice show systemic folate deficiency, but otherwise normal iron status (Salojin et al., 2011). Although these mice could conceivably be used to assess PCFT1/HCP1-mediated heme absorption, mice are not a good model to study heme absorption because their capacity to take up heme/hemoglobin into the enterocyte is extremely low (Fillebeen et al., 2015).

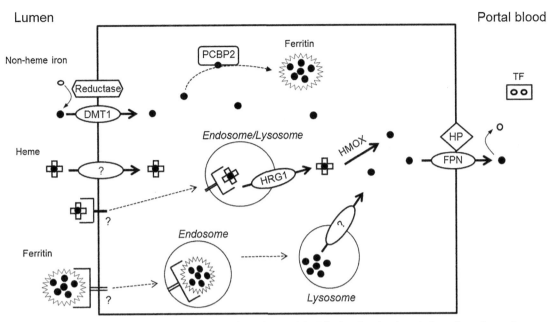

FIGURE 18.2 Iron metabolism in the enterocyte. Dietary nonheme iron, mainly Fe^{3+}, is reduced at the apical membrane by a reductase, most likely duodenal cytochrome b (DCYTB). The resulting Fe^{2+} is transported into the enterocyte via divalent metal-ion transporter-1 (DMT1). Iron in the cytosol can be stored in ferritin, possibly facilitated by the iron chaperone poly C binding protein 2 (PCBP2), or exported into portal blood via ferroportin (FPN) at the basolateral membrane. Efficient basolateral transfer of iron is believed to involve the ferroxidase hephaestin (HP), which converts Fe^{2+} to Fe^{3+} so that it can bind to plasma transferrin (TF). Dietary heme is taken up into the enterocyte either directly at the cell surface or via receptor-mediated endocytosis followed by heme-responsive gene-1 (HRG1)-mediated transport into the cytosol and catabolism by heme oxygenase (HMOX). Dietary ferritin is taken up via endocytosis and traffics to the lysosome. Iron released from heme or ferritin into the cytosol is handled in a similar fashion as is dietary nonheme iron.

It is possible that heme uptake by the intestine is mediated by heme-responsive gene-1 (HRG-1) protein, a bona fide high-affinity heme transporter expressed in multiple tissues, including human small intestine (Rajagopal et al., 2008). HRG-1 could function at the apical membrane or in intracellular vesicles, analogous to its function in macrophage phagolysosomes (see "Iron Metabolism in the Macrophage" section). The later possibility seems most likely as heme is taken up into endosomes/lysosomes (Parmley et al., 1981; Wyllie and Kaufman, 1982). Biochemical studies in the 1970s and 1980s suggested the presence of high-affinity heme receptors in the intestine (Grasbeck et al., 1979; Tenhunen et al., 1980), but little progress in this area has been made since then.

Dietary iron can be obtained additionally from ferritin in meat and foods of plant origin (Layrisse et al., 1975; Lonnerdal et al., 2006). Studies in rats have shown that purified horse spleen ferritin or soybean meal ferritin as the sole source of dietary iron can correct iron-deficiency anemia nearly as well as ferrous sulfate (Beard et al., 1996). Some dietary ferritin is likely degraded by normal digestive processes in the gastrointestinal tract, but ferritin is relatively resistant to acid hydrolysis and proteolytic digestion, suggesting that the protein is present intact in the duodenum. Biochemical and kinetic analyses of ferritin binding to intestinal cells suggest that ferritin binds to specific receptors at the apical surface (Kalgaonkar and Lonnerdal, 2009; San Martin et al., 2008). Once bound, ferritin is taken up into the cell via clathrin-mediated endocytosis and sorted through a degradation pathway involving early endosomes, autophagosomes, and lysosomes (Antileo et al., 2013). Within endosomes/lysosomes, the ferritin protein cage is eventually digested and the iron is dissolved and released into the cytosol, likely via a transmembrane transporter.

Dietary iron taken up from the apical membrane as nonheme iron, heme, or ferritin ultimately becomes part of a cytosolic pool of ferrous iron that is either utilized by the cell, stored in ferritin, or delivered to the basolateral membrane for transport via ferroportin into portal blood (Abboud and Haile, 2000; Donovan et al., 2000; McKie et al., 2000). Mice with intestine-specific inactivation of the *ferroportin* gene become severely anemic and accumulate iron in enterocytes, indicating that ferroportin is essential for intestinal iron export (Donovan et al., 2005). How iron is directed across the enterocyte is unknown, but it may involve one or more members of the poly C binding protein (PCBP) family of cytosolic metallo-chaperone proteins. PCBP1 and PCBP2 have been shown to serve as metallochaperones, delivering cytosolic iron to ferritin (Leidgens et al., 2013; Shi et al., 2008). A recent study reported that PCBP2 interacts with DMT1 and ferroportin, suggesting that PCBP2 may function as a gatekeeper for iron import and export (Yanatori et al., 2014).

Efficient basolateral transfer of iron is believed to require ferroxidase activity to convert ferrous to ferric iron, because transferrin binds only the ferric form. Part of this activity is provided by hephaestin, a multicopper oxidase that is mutated in the sex-linked anemia (*sla*) mouse (Vulpe et al., 1999). Similar to the *sla* mouse, the intestine-specific hephaestin knockout mouse develops anemia by 7 weeks of age and accumulates iron in enterocytes (Fuqua et al., 2014). However, the anemia in both mutant mouse models resolves over time, indicating that hephaestin is not essential for iron absorption. Indeed, intestine-specific hephaestin knockout mice show only modest impairments in iron absorption and are able to upregulate iron absorption in response to iron deficiency. Therefore it appears that enterocyte hephaestin is required only in early life when body iron needs are the highest because of rapid growth. It is possible that ceruloplasmin, a circulating homologue of hephaestin, might be able to compensate for loss of hephaestin (Cherukuri et al., 2005).

IRON METABOLISM IN THE HEPATOCYTE

Iron absorbed by the intestine enters the portal blood stream, which connects the intestine to the liver. It is generally assumed—although unproven—that nonheme iron in portal blood is bound to the plasma glycoprotein transferrin. Transferrin-bound iron (TBI) can be taken up by the liver via transferrin receptor-1 (TFR1) expressed at the hepatocyte sinusoidal membrane (Morgan et al., 1986; De Vos et al., 1988). After TBI binds to TFR1, the complex is internalized into endosomes (Fig. 18.3). Endosomal acidification causes transferrin to release its ferric iron, which is acted upon by a reductase to yield ferrous iron that can be transported via DMT1 into the cytosol. Although hepatocytes have all of the components of the TFR1-mediated iron uptake pathway, studies of perfused rat liver have found that less than 5% of TBI is taken up by the liver (Zimelman et al., 1977), indicating that the contribution of the transferrin cycle to the overall iron economy of the liver is minimal. In conditions of iron overload, the contribution would be even less because hepatic TFR1 levels are

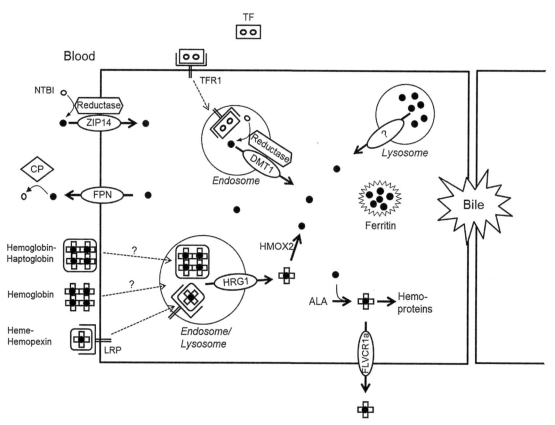

FIGURE 18.3 Iron metabolism in the hepatocyte. Hepatocytes can take up iron from transferrin (TF), hemoglobin-haptoglobin, free hemoglobin, heme-hemopexin, and nontransferrin-bound iron (NTBI). Holo-TF is taken up via transferrin receptor-1 (TFR1) and receptor-mediated endocytosis. In the acidic endosome, Fe^{3+} is released and reduced to Fe^{2+} before transport into the cytosol via divalent metal-ion transporter-1 (DMT1). Hemoglobin-haptoglobin and free hemoglobin can be taken up via endocytosis or perhaps pinocytosis and into lysosomes. Heme-hemopexin binds to low-density lipoprotein receptor-related protein (LRP) receptor and is taken up by endocytosis. Internalized heme may be transported via heme-responsive gene-1 (HRG1) into the cytosol, where inducible heme oxygenase-2 (HMOX2) liberates iron from heme. Hepatocyte iron is stored in cytosolic ferritin or in hemosiderin in lysosomes. In iron overload, ZRT- and IRT-like protein-14 (ZIP14) mediates the uptake of NTBI as Fe^{2+}, likely after reduction by a cell-surface reductase. Iron is released from hepatocytes via ferroportin (FPN), with the ferroxidase ceruloplasmin (CP) being required for efficient release. Excess heme newly synthesized from δ-aminolevulinic acid (ALA) can be exported from the hepatocyte via feline leukemia virus subgroup C cellular receptor-1a (FLVCR1a).

downregulated by iron loading (Nam and Knutson, 2012). Moreover, DMT1 is dispensable for hepatic iron homeostasis because mice with hepatocyte-specific deletion of DMT1 have normal hepatic nonheme iron concentrations and display no impairments in hepatic iron loading (Wang and Knutson, 2013).

During iron overload the uptake of nontransferrin-bound iron (NTBI) assumes a primary role in hepatic iron loading. The term *NTBI* refers to a heterogeneous mixture of redox-active and chelatable forms of iron that appears in the plasma when transferrin saturations exceed 75% (Cabantchik, 2014; Le Lan et al., 2005). NTBI is taken up via ZRT- and IRT-like protein-14 (ZIP14; encoded by the *SLC39A14* gene), a metal-ion transporter located on the sinusoidal membrane of hepatocytes (Nam et al., 2013). *Slc39a14*-knockout mice display markedly impaired hepatic uptake of plasma NTBI and fail to develop hepatocellular iron overload when crossed with mouse models of hemochromatosis, indicating that ZIP14-mediated NTBI uptake is a major contributor to hepatic iron overload (Jenkitkasemwong et al., 2015). Although NTBI is most often associated with iron overload, a recent study concluded that NTBI is a normal constituent of plasma and appears to play a dominant role in importing iron into liver, especially during the first few weeks of life (Chakrabarti et al., 2015).

The transport of iron by ZIP14 and DMT1 requires a reduction step from Fe^{3+} to Fe^{2+}, presumably catalyzed by a reductase, but the molecular identity of such a protein is unknown. A logical candidate is six transmembrane epithelial antigen of the prostate-3 (Steap3), a ferrireductase essential for iron uptake by developing erythroid cells (Ohgami et al., 2005; Fig. 18.4). Among mouse tissues, Steap3 is most abundantly expressed in liver, suggesting an important function in this organ. However, hepatic iron concentrations in Steap3-null mice are higher than those in wild-type controls (Zhang et al.,

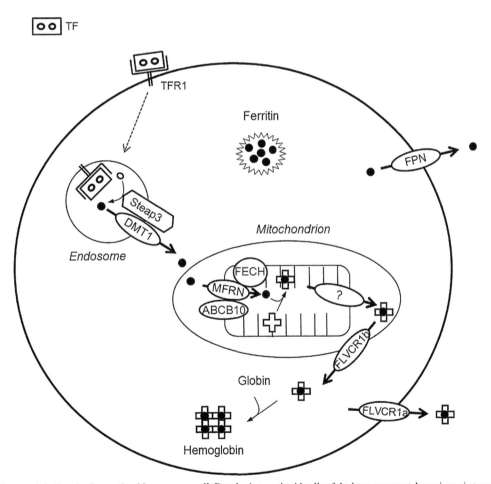

FIGURE 18.4   Iron metabolism in the erythroid precursor cell. Developing erythroid cells of the bone marrow take up iron via transferrin receptor-1 (TFR1)-mediated endocytosis of transferrin (TF). In the acidic endosome, Fe^{3+} dissociates from TF and is reduced to Fe^{2+} by six transmembrane epithelial antigen of the prostate-3 (Steap3) before transport into the cytosol via divalent metal-ion transporter-1 (DMT1). Most iron taken up into erythroid cells is destined for the mitochondrion, the site of heme synthesis. Iron crosses the inner mitochondrial membrane via mitoferrin-1 (MFRN1), which forms a complex with ATP-binding cassette, subfamily B, member 10 (ABCB10) and ferrochelatase (FECH). Heme is exported from the mitochondria via feline leukemia virus receptor-1b (FLVCR1b) and then incorporated into hemoglobin. Iron and heme can be exported via cell-surface ferroportin (FPN) and feline leukemia virus subgroup C cellular receptor-1a (FLVCR1a), respectively.

2012a), suggesting that Steap3 is not required for hepatic iron uptake. A more likely candidate for the hepatic reductase is prion protein (PrP^c). PrP-null mice display diminished hepatic iron concentrations and reduced hepatic uptake of intraperitoneally administered ^{59}Fe-citrate (Singh et al., 2009; Tripathi et al., 2015).

The liver additionally acquires iron via the routine scavenging and clearance of hemoglobin and heme in the plasma. Hemoglobin enters the plasma during enucleation of erythroblasts or from intravascular or intramedullary hemolysis. Studies in rats have demonstrated that most intravenously administered labeled hemoglobin is taken up by hepatocytes (Kino et al., 1980). Although it is widely believed that free hemoglobin is rapidly bound to the plasma protein haptoglobin and taken up into the liver as the hemoglobin–haptoglobin complex, haptoglobin knockout mice display normal plasma clearance of intravenously administered ^{125}I-labeled hemoglobin (Lim et al., 1998). This finding is consistent with the observation that isolated rat hepatocytes can readily take up free hemoglobin independently of haptoglobin (Weinstein and Segal, 1984). Heme appears in the plasma when the capacity of haptoglobin to bind hemoglobin becomes exceeded, such as after extensive hemolysis. Unbound hemoglobin in the plasma is oxidized to ferrihemoglobin, which dissociates into ferriheme and globin. Ferriheme is then bound by hemopexin and transported to the liver, where it is taken up via the low-density lipoprotein receptor-related protein (LRP) receptor (also known as CD91), located on hepatocytes and Kupffer cells (Hvidberg et al., 2005). Studies in hemopexin-knockout mice provide evidence that hemopexin is required for heme uptake by hepatocytes but not Kupffer cells (Vinchi et al., 2008). At the cellular level, hemoglobin and heme–hemopexin bound to LRP are taken up into the hepatocyte through endocytosis. In the endosome, heme is dissociated and transported into the cytosol likely via the heme transporter HRG-1, which is expressed in the liver (Rajagopal et al., 2008). In the cytosol, heme is degraded through the action of heme oxygenase-2, and the liberated iron is incorporated into ferritin.

The contribution of hemoglobin and heme iron to the overall iron economy of the liver is unknown. Mice deficient in haptoglobin or hemoglobin show no alterations in hepatic nonheme iron concentrations, suggesting that these proteins and related heme-uptake pathways are dispensable for the maintenance of hepatic iron levels under normal conditions (Morello et al., 2009; Tolosano et al., 2005). Interestingly, mice deficient in haptoglobin and the hemochromatosis protein Hfe show less hepatic iron accumulation than do *Hfe* knockout mice, suggesting that haptoglobin-mediated heme-iron recovery contributes to hepatocyte iron loading in *Hfe* knockout mice (Tolosano et al., 2005). Hemoglobin and heme clearance by the liver is probably more significant in humans than in mice because in humans, 10–20% of red cells normally undergo intravascular hydrolysis (Garby and Noyes, 1959).

Most iron not utilized by the hepatocyte is stored in cytosolic ferritin, although some is also found in lysosomes, particularly in iron-loaded cells, where partially degraded ferritin aggregates into what is called hemosiderin. As the main repository of body iron stores and producer of the iron-regulatory hormone hepcidin, the liver plays a central role in maintaining whole-body iron balance. It is well known that liver storage iron can be mobilized to where it is needed (e.g., to erythroid precursor cells in the bone marrow), although the steps in this process are not well delineated. When needed, iron in ferritin/hemosiderin can be dissolved and chelated by endogenous molecules such as glutathione and ascorbate. If this occurs in the lysosome, then the iron must be transported across the lysosomal membrane and into the cytosol. Possible transporters that may fulfill this function include DMT1, ZIP14, ZIP8, and TRPML1 (transient receptor potential mucolipin 1), which all have been detected in lysosomes (Begum et al., 2002; Dong et al., 2008; Tabuchi et al., 2000; Zhao et al., 2010). However, of these, only DMT1 and TRPML1 are known to transport iron below pH 6.0; therefore they represent the best candidates for mediating iron transport out of the acidic lysosome (Dong et al., 2008; Gunshin et al., 1997).

Iron mobilization from the liver requires ferroportin. Mice with selective loss of *ferroportin* in either hepatocytes or macrophages show defective hepatic iron mobilization and anemia when subjected to an iron-deficient diet or phlebotomy (Zhang et al., 2012b). Mouse hepatocyte ferroportin localizes to the sinusoidal membrane and is most predominantly expressed in periportal regions (Ramey et al., 2010). Efficient mobilization of iron from hepatocytes/liver additionally requires the circulating ferroxidase ceruloplasmin (Harris et al., 1999; Osaki et al., 1971; Richardson et al., 1999), which likely serves to oxidize ferrous iron to ferric iron for loading onto transferrin.

Iron can also exit the liver as a part of heme via the heme exporter feline leukemia virus receptor (FLVCR1a). Studies of mice selectively lacking FLVCR1a in hepatocytes suggest that FLVCR1a exports newly synthesized excess heme and not excess heme acquired from hemolysis (i.e., via hemoglobin-haptoglobin or heme-hemopexin; Vinchi et al., 2015). The liver has a high rate of heme synthesis, with more than 50% of heme committed for cytochrome P450s. It also appears to synthesize more heme than is actually required, and it is this excess heme that is transported via FLVCR1a.

IRON METABOLISM IN THE DEVELOPING ERYTHROID CELL

Requiring approximately 25 mg of iron per day, developing erythroid cells of the bone marrow are the major iron-consuming cells in the body. Erythroid precursor cells take up iron via endocytosis of transferrin via TFR1. In the adult human, 80%

of total body cellular TFR1 is found in the erythroid marrow (Cook, 1999). After internalization of the transferrin–TFR1 complex into the acidic endosome, ferric iron dissociates from transferrin and is reduced by the reductase Steap3 before transport into the cytosol via DMT1 (Fig. 18.4). The observation that Steap3 and DMT1-null animals display microcytic hypochromic anemia demonstrates that these proteins are required for efficient erythroid iron utilization (Gunshin et al., 2005a; Ohgami et al., 2005). However, the requirement is not absolute: 8-week-old Steap3 and DMT1-null mice have hemoglobin levels of approximately 8–9 g/dL, suggesting that alternative means of endosomal reduction and transport exist in erythroid cells.

After transport into the cytosol, it is generally assumed that iron becomes bound to low-molecular–weight ligands and then passes freely across the permeable outer mitochondrial membrane via porins. Alternatively, some data indicate that iron does not pass through the cytosol but rather is delivered directly from the endosome to the mitochondria through close physical contact (Sheftel et al., 2007). Iron crosses the inner mitochondrial membrane via mitoferrin (MFRN)-1 (Shaw et al., 2006). In nonerythroid cells, MFRN1 and its paralogue MFRN2 contribute to mitochondrial iron uptake (Paradkar et al., 2009). MFRN1 has been shown to form a complex with the ATP-binding cassette transporter ATP-binding cassette, subfamily B, member 10 (ABCB10), a protein required for hemoglobinization of erythroid cells (Tang et al., 2012), and ferrochelatase, which catalyzes the insertion of iron into protoporphyrin IX to produce heme (Chen et al., 2010). How newly synthesized heme crosses the inner mitochondrial membrane into the intermembrane space is not known. Several reports have suggested that ABCB10 (also known as ABC-me) functions as a mitochondrial heme exporter (Chen et al., 2010; Mena et al., 2015; Shirihai et al., 2000), but a recent report provides strong evidence that it is not (Bayeva et al., 2013). In any case, ABCB10 expression in erythroid cells has been shown to stabilize MRFN1 protein and thereby promotes iron import into mitochondria (Chen et al., 2009).

Heme likely exits the mitochondria via FLVCR1b, a mitochondrial isoform of FLVCR1. The *FLVCR1* gene codes for two proteins: FLVCR1a, a plasma membrane protein, and FLVCR1b, a mitochondrial protein (Chiabrando et al., 2012). The amino acid sequence of FLVCR1b corresponds to residues 277–555 of FLVCR1a and contains an N-terminal mitochondrial targeting sequence. Although mitochondrial heme export via FLVCR1b has not been measured directly, overexpression of FLVCR1b results in intracellular heme accumulation whereas suppression of FLVCR1b results in mitochondrial heme accumulation (Chiabrando et al., 2012). As in hepatocytes, which have the second highest rate of heme synthesis after erythroid cells, the FLVCR1a isoform in erythroid cells functions in cellular export of excess heme (Mercurio et al., 2015).

Unexpectedly, despite being the most avid consumers of iron in the body, erythroid cells have been found to express ferroportin (Zhang et al., 2009). Moreover, the protein is expressed at the plasma membrane and regulated by hepcidin (Zhang et al., 2011b). It has been proposed that the presence of ferroportin at the plasma membrane during iron deficiency serves to downregulate erythropoiesis and provide iron to other cells that are more sensitive to iron deprivation (Zhang et al., 2011b).

Iron Metabolism in the Macrophage

Macrophages of the liver, spleen, and bone marrow—collectively known as the reticuloendothelial system—play a central role in iron metabolism by recycling iron (~25 mg/day) from senescent erythrocytes (Knutson and Wessling-Resnick, 2003). Most recycled iron is returned to the bone marrow for reincorporation into hemoglobin in erythroid precursor cells. Erythrocyte iron recycling begins when macrophages recognize and bind effete or damaged red cells in the circulation. After binding to the macrophage, the erythrocyte is internalized into the phagosome, which fuses with lysosomes to form a phagolysosome (Fig. 18.5). Hydrolytic enzymes in the phagolysosome degrade the protein moiety of hemoglobin, and the remaining heme is exported into the cytosol via HRG1 (White et al., 2013). It is possible that some heme in the phagolysosome is degraded to release ionic iron, which may be transported into the cytosol via natural resistance-associated macrophage protein-1 (Nramp1). Under normal circumstances the contribution of macrophage Nramp1 to erythrocyte–iron recycling appears to be small because $Nramp1^{-/-}$ mice have normal hematocrits and only minor alterations in splenic iron concentrations (i.e., 15–20% higher than those in wild-type mice; Soe-Lin et al., 2009). However, Nramp1 does seem to be important for efficient erythrocyte iron recycling after induced hemolytic stress or conditions of enhanced erythrophagocytosis (Soe-Lin et al., 2009).

Macrophages can also acquire heme by endocytosis of hemoglobin-haptoglobin and heme-hemopexin via CD163 and LRP, respectively (Hvidberg et al., 2005; Kristiansen et al., 2001). Interestingly, although CD163 is abundantly expressed in macrophages, it appears to be largely dispensable for hemoglobin–haptoglobin clearance as indicated by studies of CD163-deficient mice (Etzerodt et al., 2013). Studies of hemopexin-null mice have revealed that macrophages also have a hemopexin-independent heme uptake pathway (Vinchi et al., 2008). Heme in the cytosol is degraded by heme oxygenase-1 (HMOX1), the inducible form of heme oxygenase. The critical importance of HMOX1 for reticuloendothelial macrophages

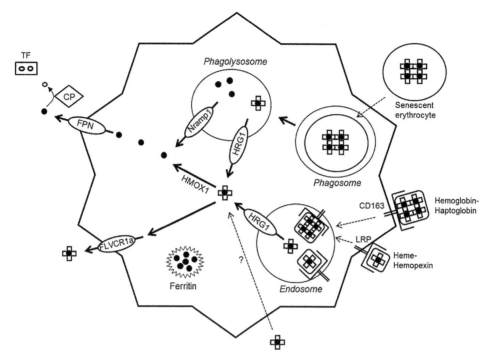

FIGURE 18.5 Iron metabolism in the macrophage. Reticuloendothelial macrophages acquire most of their iron via erythrophagocytosis. Internalized erythrocytes are catabolized in the phagolysosome to release heme, which is exported into the cytosol via heme-responsive gene-1 (HRG1). Heme is also acquired from hemoglobin-haptoglobin via the CD163 receptor and heme-hemopexin via the low-density lipoprotein receptor-related protein (LRP) receptor. The macrophage can additionally take up heme through an unknown hemopexin-independent pathway. Heme in the cytosol is catabolized by heme oxygenase-1 (HMOX1) to release iron, which can be stored in ferritin or exported from the cell via ferroportin (FPN). Cellular iron export and iron loading onto transferrin (TF) are facilitated by ceruloplasmin (CP). During conditions of increased erythrophagocytosis, natural resistance-associated macrophage protein-1 (Nramp1) promotes efficient iron recycling, presumably by transporting iron from the phagolysosome into the cytosol. Heme export is mediated by feline leukemia virus subgroup C cellular receptor-1a (FLVCR1a).

is illustrated by the fact that HMOX1-deficient macrophages cannot survive after erythrophagocytosis, most likely because of heme overload/toxicity (Kovtunovych et al., 2010). An alternative fate of cytosolic heme is export from the cell via FLVCR1a. Such a heme export pathway not only provides a safety valve for heme excess but is also essential for normal heme–iron recycling and systemic iron balance (Keel et al., 2008).

Iron liberated from heme is either stored in ferritin in the cytosol or exported from the macrophage via ferroportin (Knutson et al., 2005). Accordingly, mice with selective inactivation of ferroportin in macrophages display iron accumulation in hepatic and splenic macrophages (Zhang et al., 2011c). Surprisingly, loss of macrophage ferroportin resulted in only mild anemia, indicating that other iron-supplying pathways (e.g., intestinal iron absorption, heme export/recycling) can compensate to support erythropoiesis.

REGULATION OF IRON HOMEOSTASIS BY HEPCIDIN

The circulating peptide hormone hepcidin controls body iron balance by regulating intestinal iron absorption and iron release from macrophages of the reticuloendothelial system (Fig. 18.1). At the cellular level, hepcidin binds to ferroportin, causing its internalization and degradation, thereby inhibiting cellular iron export (Nemeth et al., 2004). Hepcidin synthesis by hepatocytes is positively regulated by plasma iron concentrations (transferrin saturation) and hepatic iron stores (Detivaud et al., 2005; Lin et al., 2007) and negatively regulated by erythropoiesis (Nicolas et al., 2002). Hepcidin regulation by iron stores involves bone morphogenetic protein-6 (Andriopoulos et al., 2009) produced in hepatic nonparenchymal cells (Fig. 18.1), most likely sinusoidal endothelial cells (Zhang et al., 2011a). The negative regulation of hepcidin by erythropoiesis involves erythroferrone, a hormone produced by erythroid precursor cells of the bone marrow (Kautz et al., 2014). The complex molecular pathways that govern hepcidin expression are comprehensively covered elsewhere (Ganz, 2013).

REFERENCES

Abboud, S., Haile, D.J., 2000. A novel mammalian iron-regulated protein involved in intracellular iron metabolism. J. Biol. Chem. 275, 19906–19912.

Anderson, C.P., Shen, M., Eisenstein, R.S., Leibold, E.A., 2012. Mammalian iron metabolism and its control by iron regulatory proteins. Biochim. Biophys. Acta 1823, 1468–1483.

Andriopoulos Jr., B., Corradini, E., Xia, Y., Faasse, S.A., Chen, S., Grgurevic, L., Knutson, M.D., Pietrangelo, A., Vukicevic, S., Lin, H.Y., Babitt, J.L., 2009. BMP6 is a key endogenous regulator of hepcidin expression and iron metabolism. Nat. Genet. 41, 482–487.

Antileo, E., Garri, C., Tapia, V., Munoz, J.P., Chiong, M., Nualart, F., Lavandero, S., Fernandez, J., Nunez, M.T., 2013. Endocytic pathway of exogenous iron-loaded ferritin in intestinal epithelial (Caco-2) cells. Am. J. Physiol. Gastrointest. Liver Physiol. 304, G655–G661.

Atanasova, B., Mudway, I.S., Laftah, A.H., Latunde-Dada, G.O., McKie, A.T., Peters, T.J., Tzatchev, K.N., Simpson, R.J., 2004. Duodenal ascorbate levels are changed in mice with altered iron metabolism. J. Nutr. 134, 501–505.

Bayeva, M., Khechaduri, A., Wu, R., Burke, M.A., Wasserstrom, J.A., Singh, N., Liesa, M., Shirihai, O.S., Langer, N.B., Paw, B.H., Ardehali, H., 2013. ATP-binding cassette B10 regulates early steps of heme synthesis. Circulation Res. 113, 279–287.

Beard, J.L., Burton, J.W., Theil, E.C., 1996. Purified ferritin and soybean meal can be sources of iron for treating iron deficiency in rats. J. Nutr. 126, 154–160.

Begum, N.A., Kobayashi, M., Moriwaki, Y., Matsumoto, M., Toyoshima, K., Seya, T., 2002. Mycobacterium bovis BCG cell wall and lipopolysaccharide induce a novel gene, BIGM103, encoding a 7-TM protein: identification of a new protein family having Zn-transporter and Zn-metalloprotease signatures. Genomics 80, 630–645.

Cabantchik, Z.I., 2014. Labile iron in cells and body fluids: physiology, pathology, and pharmacology. Front. Pharmacol. 5, 45.

Canonne-Hergaux, F., Gruenheid, S., Ponka, P., Gros, P., 1999. Cellular and subcellular localization of the Nramp2 iron transporter in the intestinal brush border and regulation by dietary iron. Blood 93, 4406–4417.

Chakrabarti, M., Barlas, M.N., McCormick, S.P., Lindahl, L.S., Lindahl, P.A., 2015. Kinetics of iron import into developing mouse organs determined by a pup-swapping method. J. Biol. Chem. 290, 520–528.

Chen, W., Dailey, H.A., Paw, B.H., 2010. Ferrochelatase forms an oligomeric complex with mitoferrin-1 and Abcb10 for erythroid heme biosynthesis. Blood 116, 628–630.

Chen, W., Paradkar, P.N., Li, L., Pierce, E.L., Langer, N.B., Takahashi-Makise, N., Hyde, B.B., Shirihai, O.S., Ward, D.M., Kaplan, J., Paw, B.H., 2009. Abcb10 physically interacts with mitoferrin-1 (Slc25a37) to enhance its stability and function in the erythroid mitochondria. Proc. Natl. Acad. Sci. USA 106, 16263–16268.

Cherukuri, S., Potla, R., Sarkar, J., Nurko, S., Harris, Z.L., Fox, P.L., 2005. Unexpected role of ceruloplasmin in intestinal iron absorption. Cell Metab. 2, 309–319.

Chiabrando, D., Marro, S., Mercurio, S., Giorgi, C., Petrillo, S., Vinchi, F., Fiorito, V., Fagoonee, S., Camporeale, A., Turco, E., Merlo, G.R., Silengo, L., Altruda, F., Pinton, P., Tolosano, E., 2012. The mitochondrial heme exporter FLVCR1b mediates erythroid differentiation. J. Clin. Invest. 122, 4569–4579.

Cook, J.D., 1999. The measurement of serum transferrin receptor. Am. J. Med. Sci. 318, 269–276.

De Vos, R., Sciot, R., van Eyken, P., Desmet, V.J., 1988. Immunoelectron microscopic localization of hepatic transferrin receptors in human liver with and without iron overload. Virchows Arch. B Cell Pathol. Incl. Mol. Pathol. 55, 11–17.

Detivaud, L., Nemeth, E., Boudjema, K., Turlin, B., Troadec, M.B., Leroyer, P., Ropert, M., Jacquelinet, S., Courselaud, B., Ganz, T., Brissot, P., Loreal, O., 2005. Hepcidin levels in humans are correlated with hepatic iron stores, hemoglobin levels, and hepatic function. Blood 106, 746–748.

Dong, X.P., Cheng, X., Mills, E., Delling, M., Wang, F., Kurz, T., Xu, H., 2008. The type IV mucolipidosis-associated protein TRPML1 is an endolysosomal iron release channel. Nature 455, 992–996.

Donovan, A., Brownlie, A., Zhou, Y., Shepard, J., Pratt, S.J., Moynihan, J., Paw, B.H., Drejer, A., Barut, B., Zapata, A., Law, T.C., Brugnara, C., Lux, S.E., Pinkus, G.S., Pinkus, J.L., Kingsley, P.D., Palis, J., Fleming, M.D., Andrews, N.C., Zon, L.I., 2000. Positional cloning of zebrafish ferroportin1 identifies a conserved vertebrate iron exporter. Nature 403, 776–781.

Donovan, A., Lima, C.A., Pinkus, J.L., Pinkus, G.S., Zon, L.I., Robine, S., Andrews, N.C., 2005. The iron exporter ferroportin/Slc40a1 is essential for iron homeostasis. Cell Metab. 1, 191–200.

Etzerodt, A., Kjolby, M., Nielsen, M.J., Maniecki, M., Svendsen, P., Moestrup, S.K., 2013. Plasma clearance of hemoglobin and haptoglobin in mice and effect of CD163 gene targeting disruption. Antioxid. Redox Signal. 18, 2254–2263.

Fillebeen, C., Gkouvatsos, K., Fragoso, G., Calve, A., Garcia-Santos, D., Buffler, M., Becker, C., Schumann, K., Ponka, P., Santos, M.M., Pantopoulos, K., 2015. Mice are poor heme absorbers and do not require intestinal Hmox1 for dietary heme iron assimilation. Haematologica 100.

Fleming, M.D., Trenor 3rd, C.C., Su, M.A., Foernzler, D., Beier, D.R., Dietrich, W.F., Andrews, N.C., 1997. Microcytic anaemia mice have a mutation in Nramp2, a candidate iron transporter gene. Nat. Genet. 16, 383–386.

Frazer, D.M., Wilkins, S.J., Vulpe, C.D., Anderson, G.J., 2005. The role of duodenal cytochrome b in intestinal iron absorption remains unclear. Blood 106, 4413 author reply 4414.

Fuqua, B.K., Lu, Y., Darshan, D., Frazer, D.M., Wilkins, S.J., Wolkow, N., Bell, A.G., Hsu, J., Yu, C.C., Chen, H., Dunaief, J.L., Anderson, G.J., Vulpe, C.D., 2014. The multicopper ferroxidase hephaestin enhances intestinal iron absorption in mice. PLoS One 9, e98792.

Ganz, T., 2013. Systemic iron homeostasis. Physiol. Rev. 93, 1721–1741.

Garby, L., Noyes, W.D., 1959. Studies on hemoglobin metabolism. II. Pathways of hemoglobin iron metabolism in normal man. J. Clin. Invest. 38, 1484–1486.

Grasbeck, R., Kouvonen, I., Lundberg, M., Tenhunen, R., 1979. An intestinal receptor for heme. Scand. J. Haematol. 23, 5–9.

Gunshin, H., Fujiwara, Y., Custodio, A.O., Direnzo, C., Robine, S., Andrews, N.C., 2005a. Slc11a2 is required for intestinal iron absorption and erythropoiesis but dispensable in placenta and liver. J. Clin. Invest. 115, 1258–1266.

Gunshin, H., Mackenzie, B., Berger, U.V., Gunshin, Y., Romero, M.F., Boron, W.F., Nussberger, S., Gollan, J.L., Hediger, M.A., 1997. Cloning and characterization of a mammalian proton-coupled metal-ion transporter. Nature 388, 482–488.

Gunshin, H., Starr, C.N., Direnzo, C., Fleming, M.D., Jin, J., Greer, E.L., Sellers, V.M., Galica, S.M., Andrews, N.C., 2005b. Cybrd1 (duodenal cytochrome b) is not necessary for dietary iron absorption in mice. Blood 106, 2879–2883.

Harris, Z.L., Durley, A.P., Man, T.K., Gitlin, J.D., 1999. Targeted gene disruption reveals an essential role for ceruloplasmin in cellular iron efflux. Proc. Natl. Acad. Sci. USA 96, 10812–10817.

Hubert, N., Hentze, M.W., 2002. Previously uncharacterized isoforms of divalent metal transporter (DMT)-1: implications for regulation and cellular function. Proc. Natl. Acad. Sci. USA 99, 12345–12350.

Hvidberg, V., Maniecki, M.B., Jacobsen, C., Hojrup, P., Moller, H.J., Moestrup, S.K., 2005. Identification of the receptor scavenging hemopexin-heme complexes. Blood 106, 2572–2579.

Jenkitkasemwong, S., Wang, C.Y., Coffey, R., Zhang, W., Chan, A., Biel, T., Kim, J.S., Hojyo, S., Fukada, T., Knutson, M.D., 2015. SLC39A14 is required for the development of hepatocellular iron overload in murine models of hereditary hemochromatosis. Cell Metab. 22, 138–150.

Kalgaonkar, S., Lonnerdal, B., 2009. Receptor-mediated uptake of ferritin-bound iron by human intestinal Caco-2 cells. J. Nutr. Biochem. 20, 304–311.

Kautz, L., Jung, G., Valore, E.V., Rivella, S., Nemeth, E., Ganz, T., 2014. Identification of erythroferrone as an erythroid regulator of iron metabolism. Nat. Genet. 46, 678–684.

Keel, S.B., Doty, R.T., Yang, Z., Quigley, J.G., Chen, J., Knoblaugh, S., Kingsley, P.D., De Domenico, I., Vaughn, M.B., Kaplan, J., Palis, J., Abkowitz, J.L., 2008. A heme export protein is required for red blood cell differentiation and iron homeostasis. Science 319, 825–828.

Kino, K., Tsunoo, H., Higa, Y., Takami, M., Hamaguchi, H., Nakajima, H., 1980. Hemoglobin-haptoglobin receptor in rat liver plasma membrane. J. Biol. Chem. 255, 9616–9620.

Knutson, M., Wessling-Resnick, M., 2003. Iron metabolism in the reticuloendothelial system. Crit. Rev. Biochem. Mol. Biol. 38, 61–88.

Knutson, M.D., Oukka, M., Koss, L.M., Aydemir, F., Wessling-Resnick, M., 2005. Iron release from macrophages after erythrophagocytosis is up-regulated by ferroportin 1 overexpression and down-regulated by hepcidin. Proc. Natl. Acad. Sci. USA 102, 1324–1328.

Kovtunovych, G., Eckhaus, M.A., Ghosh, M.C., Ollivierre-Wilson, H., Rouault, T.A., 2010. Dysfunction of the heme recycling system in heme oxygenase 1-deficient mice: effects on macrophage viability and tissue iron distribution. Blood 116, 6054–6062.

Kristiansen, M., Graversen, J.H., Jacobsen, C., Sonne, O., Hoffman, H.J., Law, S.K., Moestrup, S.K., 2001. Identification of the haemoglobin scavenger receptor. Nature 409, 198–201.

Layrisse, M., Martinez-Torres, C., Renzy, M., Leets, I., 1975. Ferritin iron absorption in man. Blood 45, 689–698.

Le Lan, C., Loreal, O., Cohen, T., Ropert, M., Glickstein, H., Laine, F., Pouchard, M., Deugnier, Y., Le Treut, A., Breuer, W., Cabantchik, Z.I., Brissot, P., 2005. Redox active plasma iron in C282Y/C282Y hemochromatosis. Blood 105, 4527–4531.

Leidgens, S., Bullough, K.Z., Shi, H., Li, F., Shakoury-Elizeh, M., Yabe, T., Subramanian, P., Hsu, E., Natarajan, N., Nandal, A., Stemmler, T.L., Philpott, C.C., 2013. Each member of the poly-r(C)-binding protein 1 (PCBP) family exhibits iron chaperone activity toward ferritin. J. Biol. Chem. 288, 17791–17802.

Lim, S.K., Kim, H., bin Ali, A., Lim, Y.K., Wang, Y., Chong, S.M., Costantini, F., Baumman, H., 1998. Increased susceptibility in Hp knockout mice during acute hemolysis. Blood 92, 1870–1877.

Lin, L., Valore, E.V., Nemeth, E., Goodnough, J.B., Gabayan, V., Ganz, T., 2007. Iron transferrin regulates hepcidin synthesis in primary hepatocyte culture through hemojuvelin and BMP2/4. Blood 110, 2182–2189.

Lonnerdal, B., Bryant, A., Liu, X., Theil, E.C., 2006. Iron absorption from soybean ferritin in nonanemic women. Am. J. Clin. Nutr. 83, 103–107.

Mackenzie, B., Takanaga, H., Hubert, N., Rolfs, A., Hediger, M.A., 2007. Functional properties of multiple isoforms of human divalent metal-ion transporter 1 (DMT1). Biochem. J. 403, 59–69.

McKie, A.T., Barrow, D., Latunde-Dada, G.O., Rolfs, A., Sager, G., Mudaly, E., Mudaly, M., Richardson, C., Barlow, D., Bomford, A., Peters, T.J., Raja, K.B., Shirali, S., Hediger, M.A., Farzaneh, F., Simpson, R.J., 2001. An iron-regulated ferric reductase associated with the absorption of dietary iron. Science 291, 1755–1759.

McKie, A.T., Marciani, P., Rolfs, A., Brennan, K., Wehr, K., Barrow, D., Miret, S., Bomford, A., Peters, T.J., Farzaneh, F., Hediger, M.A., Hentze, M.W., Simpson, R.J., 2000. A novel duodenal iron-regulated transporter, IREG1, implicated in the basolateral transfer of iron to the circulation. Mol. Cell 5, 299–309.

Mena, N.P., Urrutia, P.J., Lourido, F., Carrasco, C.M., Nunez, M.T., 2015. Mitochondrial iron homeostasis and its dysfunctions in neurodegenerative disorders. Mitochondrion 21, 92–105.

Mercurio, S., Aspesi, A., Silengo, L., Altruda, F., Dianzani, I., Chiabrando, D., 2015. Alteration of heme metabolism in a cellular model of Diamond-Blackfan Anemia. Eur. J. Haematol. 96, 367–374.

Morello, N., Tonoli, E., Logrand, F., Fiorito, V., Fagoonee, S., Turco, E., Silengo, L., Vercelli, A., Altruda, F., Tolosano, E., 2009. Haemopexin affects iron distribution and ferritin expression in mouse brain. J. Cell Mol. Med. 13, 4192–4204.

Morgan, E.H., Smith, G.D., Peters, T.J., 1986. Uptake and subcellular processing of 59Fe-125I-labelled transferrin by rat liver. Biochem. J. 237, 163–173.

Nam, H., Knutson, M.D., 2012. Effect of dietary iron deficiency and overload on the expression of ZIP metal-ion transporters in rat liver. Biometals 25, 115–124.

Nam, H., Wang, C.Y., Zhang, L., Zhang, W., Hojyo, S., Fukada, T., Knutson, M.D., 2013. ZIP14 and DMT1 in the liver, pancreas, and heart are differentially regulated by iron deficiency and overload: implications for tissue iron uptake in iron-related disorders. Haematologica 98, 1049–1057.

Nemeth, E., Tuttle, M.S., Powelson, J., Vaughn, M.B., Donovan, A., Ward, D.M., Ganz, T., Kaplan, J., 2004. Hepcidin regulates cellular iron efflux by binding to ferroportin and inducing its internalization. Science 306, 2090–2093.

Nicolas, G., Chauvet, C., Viatte, L., Danan, J.L., Bigard, X., Devaux, I., Beaumont, C., Kahn, A., Vaulont, S., 2002. The gene encoding the iron regulatory peptide hepcidin is regulated by anemia, hypoxia, and inflammation. J. Clin. Invest. 110, 1037–1044.

Ohgami, R.S., Campagna, D.R., Greer, E.L., Antiochos, B., McDonald, A., Chen, J., Sharp, J.J., Fujiwara, Y., Barker, J.E., Fleming, M.D., 2005. Identification of a ferrireductase required for efficient transferrin-dependent iron uptake in erythroid cells. Nat. Genet. 37, 1264–1269.

Osaki, S., Johnson, D.A., Frieden, E., 1971. The mobilization of iron from the perfused mammalian liver by a serum copper enzyme, ferroxidase I. J. Biol. Chem. 246, 3018–3023.

Paradkar, P.N., Zumbrennen, K.B., Paw, B.H., Ward, D.M., Kaplan, J., 2009. Regulation of mitochondrial iron import through differential turnover of mitoferrin 1 and mitoferrin 2. Mol. Cell. Biol. 29, 1007–1016.

Parmley, R.T., Barton, J.C., Conrad, M.E., Austin, R.L., Holland, R.M., 1981. Ultrastructural cytochemistry and radioautography of hemoglobin–iron absorption. Exp. Mol. Pathol. 34, 131–144.

Qiu, A., Jansen, M., Sakaris, A., Min, S.H., Chattopadhyay, S., Tsai, E., Sandoval, C., Zhao, R., Akabas, M.H., Goldman, I.D., 2006. Identification of an intestinal folate transporter and the molecular basis for hereditary folate malabsorption. Cell 127, 917–928.

Rajagopal, A., Rao, A.U., Amigo, J., Tian, M., Upadhyay, S.K., Hall, C., Uhm, S., Mathew, M.K., Fleming, M.D., Paw, B.H., Krause, M., Hamza, I., 2008. Haem homeostasis is regulated by the conserved and concerted functions of HRG-1 proteins. Nature 453, 1127–1131.

Ramey, G., Deschemin, J.C., Durel, B., Canonne-Hergaux, F., Nicolas, G., Vaulont, S., 2010. Hepcidin targets ferroportin for degradation in hepatocytes. Haematologica 95, 501–504.

Rathbone, B.J., Johnson, A.W., Wyatt, J.I., Kelleher, J., Heatley, R.V., Losowsky, M.S., 1989. Ascorbic acid: a factor concentrated in human gastric juice. Clin. Sci. 76, 237–241.

Richardson, D.R., Chua, A.C., Baker, E., 1999. Activation of an iron uptake mechanism from transferrin in hepatocytes by small-molecular-weight iron complexes: implications for the pathogenesis of iron-overload disease. J. Lab. Clin. Med. 133, 144–151.

Salojin, K.V., Cabrera, R.M., Sun, W., Chang, W.C., Lin, C., Duncan, L., Platt, K.A., Read, R., Vogel, P., Liu, Q., Finnell, R.H., Oravecz, T., 2011. A mouse model of hereditary folate malabsorption: deletion of the PCFT gene leads to systemic folate deficiency. Blood 117, 4895–4904.

San Martin, C.D., Garri, C., Pizarro, F., Walter, T., Theil, E.C., Nunez, M.T., 2008. Caco-2 intestinal epithelial cells absorb soybean ferritin by mu2 (AP2)-dependent endocytosis. J. Nutr. 138, 659–666.

Shaw, G.C., Cope, J.J., Li, L., Corson, K., Hersey, C., Ackermann, G.E., Gwynn, B., Lambert, A.J., Wingert, R.A., Traver, D., Trede, N.S., Barut, B.A., Zhou, Y., Minet, E., Donovan, A., Brownlie, A., Balzan, R., Weiss, M.J., Peters, L.L., Kaplan, J., Zon, L.I., Paw, B.H., 2006. Mitoferrin is essential for erythroid iron assimilation. Nature 440, 96–100.

Shawki, A., Anthony, S.R., Nose, Y., Engevik, M.A., Niespodzany, E.J., Barrientos, T., Ohrvik, H., Worrell, R.T., Thiele, D.J., Mackenzie, B., 2015. Intestinal DMT1 is critical for iron absorption in the mouse but is not required for the absorption of copper or manganese. Am. J. Physiol. Gastrointest. Liver Physiol 309, 367–374.

Shawki, A., Knight, P.B., Maliken, B.D., Niespodzany, E.J., Mackenzie, B., 2012. H(+)-coupled divalent metal-ion transporter-1: functional properties, physiological roles and therapeutics. Curr. Top. Membr. 70, 169–214.

Shayeghi, M., Latunde-Dada, G.O., Oakhill, J.S., Laftah, A.H., Takeuchi, K., Halliday, N., Khan, Y., Warley, A., McCann, F.E., Hider, R.C., Frazer, D.M., Anderson, G.J., Vulpe, C.D., Simpson, R.J., McKie, A.T., 2005. Identification of an intestinal heme transporter. Cell 122, 789–801.

Sheftel, A.D., Zhang, A.S., Brown, C., Shirihai, O.S., Ponka, P., 2007. Direct interorganellar transfer of iron from endosome to mitochondrion. Blood 110, 125–132.

Shi, H., Bencze, K.Z., Stemmler, T.L., Philpott, C.C., 2008. A cytosolic iron chaperone that delivers iron to ferritin. Science 320, 1207–1210.

Shirihai, O.S., Gregory, T., Yu, C., Orkin, S.H., Weiss, M.J., 2000. ABC-me: a novel mitochondrial transporter induced by GATA-1 during erythroid differentiation. EMBO J. 19, 2492–2502.

Singh, A., Kong, Q., Luo, X., Petersen, R.B., Meyerson, H., Singh, N., 2009. Prion protein (PrP) knock-out mice show altered iron metabolism: a functional role for PrP in iron uptake and transport. PLoS One 4, e6115.

Soe-Lin, S., Apte, S.S., Andriopoulos Jr., B., Andrews, M.C., Schranzhofer, M., Kahawita, T., Garcia-Santos, D., Ponka, P., 2009. Nramp1 promotes efficient macrophage recycling of iron following erythrophagocytosis in vivo. Proc. Natl. Acad. Sci. USA 106, 5960–5965.

Tabuchi, M., Yoshimori, T., Yamaguchi, K., Yoshida, T., Kishi, F., 2000. Human NRAMP2/DMT1, which mediates iron transport across endosomal membranes, is localized to late endosomes and lysosomes in HEp-2 cells. J. Biol. Chem. 275, 22220–22228.

Tang, L., Bergevoet, S.M., Bakker-Verweij, G., Harteveld, C.L., Giordano, P.C., Nijtmans, L., de Witte, T., Jansen, J.H., Raymakers, R.A., van der Reijden, B.A., 2012. Human mitochondrial ATP-binding cassette transporter ABCB10 is required for efficient red blood cell development. Br. J. Haematol. 157, 151–154.

Tenhunen, R., Grasbeck, R., Kouvonen, I., Lundberg, M., 1980. An intestinal receptor for heme: its parital characterization. Int. J. Biochem. 12, 713–716.

Tolosano, E., Fagoonee, S., Garuti, C., Valli, L., Andrews, N.C., Altruda, F., Pietrangelo, A., 2005. Haptoglobin modifies the hemochromatosis phenotype in mice. Blood 105, 3353–3355.

Tripathi, A.K., Haldar, S., Qian, J., Beserra, A., Suda, S., Singh, A., Hopfer, U., Chen, S.G., Garrick, M.D., Turner, J.R., Knutson, M.D., Singh, N., 2015. Prion protein functions as a ferrireductase partner for ZIP14 and DMT1. Free Radic. Biol. Med. 84, 322–330.

Vinchi, F., Gastaldi, S., Silengo, L., Altruda, F., Tolosano, E., 2008. Hemopexin prevents endothelial damage and liver congestion in a mouse model of heme overload. Am. J. Pathol. 173, 289–299.

Vinchi, F., Ingoglia, G., Chiabrando, D., Mercurio, S., Turco, E., Silengo, L., Altruda, F., Tolosano, E., 2015. Heme exporter FLVCR1a regulates heme synthesis and degradation and controls activity of cytochromes P450. Gastroenterology 146, 1325–1338.

Vulpe, C.D., Kuo, Y.M., Murphy, T.L., Cowley, L., Askwith, C., Libina, N., Gitschier, J., Anderson, G.J., 1999. Hephaestin, a ceruloplasmin homologue implicated in intestinal iron transport, is defective in the sla mouse. Nat. Genet. 21, 195–199.

Wang, C.Y., Knutson, M.D., 2013. Hepatocyte divalent metal-ion transporter-1 is dispensable for hepatic iron accumulation and non-transferrin-bound iron uptake in mice. Hepatology 58.

Weinstein, M.B., Segal, H.L., 1984. Uptake of free hemoglobin by rat liver parenchymal cells. Biochem. Biophys. Res. Commun. 123, 489–496.

White, C., Yuan, X., Schmidt, P.J., Bresciani, E., Samuel, T.K., Campagna, D., Hall, C., Bishop, K., Calicchio, M.L., Lapierre, A., Ward, D.M., Liu, P., Fleming, M.D., Hamza, I., 2013. HRG1 is essential for heme transport from the phagolysosome of macrophages during erythrophagocytosis. Cell Metab. 17, 261–270.

Wyllie, J.C., Kaufman, N., 1982. An electron microscopic study of heme uptake by rat duodenum. Lab. Invest. 47, 471–476.

Yanatori, I., Yasui, Y., Tabuchi, M., Kishi, F., 2014. Chaperone protein involved in transmembrane transport of iron. Biochem. J. 462, 25–37.

Zhang, A.S., Anderson, S.A., Wang, J., Yang, F., DeMaster, K., Ahmed, R., Nizzi, C.P., Eisenstein, R.S., Tsukamoto, H., Enns, C.A., 2011a. Suppression of hepatic hepcidin expression in response to acute iron deprivation is associated with an increase of matriptase-2 protein. Blood 117, 1687–1699.

Zhang, D.L., Hughes, R.M., Ollivierre-Wilson, H., Ghosh, M.C., Rouault, T.A., 2009. A ferroportin transcript that lacks an iron-responsive element enables duodenal and erythroid precursor cells to evade translational repression. Cell Metab. 9, 461–473.

Zhang, D.L., Senecal, T., Ghosh, M.C., Ollivierre-Wilson, H., Tu, T., Rouault, T.A., 2011b. Hepcidin regulates ferroportin expression and intracellular iron homeostasis of erythroblasts. Blood 118, 2868–2877.

Zhang, F., Tao, Y., Zhang, Z., Guo, X., An, P., Shen, Y., Wu, Q., Yu, Y., Wang, F., 2012a. Metalloreductase Steap3 coordinates the regulation of iron homeostasis and inflammatory responses. Haematologica 97, 1826–1835.

Zhang, Z., Zhang, F., An, P., Guo, X., Shen, Y., Tao, Y., Wu, Q., Zhang, Y., Yu, Y., Ning, B., Nie, G., Knutson, M.D., Anderson, G.J., Wang, F., 2011c. Ferroportin1 deficiency in mouse macrophages impairs iron homeostasis and inflammatory responses. Blood 118, 1912–1922.

Zhang, Z., Zhang, F., Guo, X., An, P., Tao, Y., Wang, F., 2012b. Ferroportin1 in hepatocytes and macrophages is required for the efficient mobilization of body iron stores in mice. Hepatology 56, 961–971.

Zhao, N., Gao, J., Enns, C.A., Knutson, M.D., 2010. ZRT/IRT-like protein 14 (ZIP14) promotes the cellular assimilation of iron from transferrin. J. Biol. Chem. 285, 32141–32150.

Zimelman, A.P., Zimmerman, H.J., McLean, R., Weintraub, L.R., 1977. Effect of iron saturation of transferrin on hepatic iron uptake: an in vitro study. Gastroenterology 72, 129–131.

Chapter 19

Regulation of Divalent Metal-Ion Transporter-1 Expression and Function

Michael D. Garrick

University at Buffalo, Buffalo, NY, United States

INTRODUCTION—HISTORICAL PERSPECTIVE

The author views this chapter as an opportunity not only to tell the reader about what is known on the regulation of divalent metal-ion transporter-1 (DMT1) function and expression but also about what needs to be known next. In covering the less well-known and yet-to-be-known aspects, the author will endeavor to label clearly what is speculative or less certain.

DMT1 is the product of the solute carrier family 11 member 2 (*SLC11A2*) gene on human chromosome 12. This gene is the second member of its family with the solute carrier family 11 member 1 gene having been identified first as having alleles that provide resistance or susceptibility to particular pathogens in mice (and ultimately in other species also). This related gene was initially named natural resistance-associated macrophage protein *(NRAMP)-1*. Therefore DMT1's first name was NRAMP2 because it represented a cDNA/mRNA sequence initially detected as similar (paralogous) to *NRAMP1*. Because its expression was not limited to macrophages, calling it NRAMP2 was clearly not justified, but its function initially remained unidentified; therefore Nramp2 was applied to the predicted protein sequence until functions could be assigned. Ultimately, one function assigned to DMT1 relates to natural resistance (see below); therefore we now have come full circle.

The first function assigned to DMT1 involved iron transport as a result of three papers that appeared in rapid succession (Fleming et al., 1997, 1998; Gunshin et al., 1997). The earlier two papers implied that the protein was a cellular importer of iron likely for duodenal acquisition of the nutrient whereas the later one indicated that it was also part of endosomal release of iron during the transferrin (Tf) cycle (also see Chapters 14–18). One group (Gunshin et al., 1997) also showed that the transporter involved cotransport of protons, a property very appropriate for the duodenal expression where the local environment of the enterocytes would be mildly acidic; this property was also consistent with endosomal release of iron from Tf given that a mildly acidic environment aids the Tf/transferrin receptor (TfR) complex in releasing its iron to DMT1. They also showed that a current (later attributed to the proton flux) entered the *Xenopus* oocytes in which the Slc11a2 was expressed not only on exposure of the cells to Fe^{2+} but also on exposure to multiple other metal ions. This property led them to suggest that each of the metal ions was transported and assign the name divalent cation transporter-1 to the transporter. The name ultimately evolved to DMT1, and the suggestion that other cations were transported led to debate about which ones (reviewed below). In any case the ability of DMT1 to transport iron and other metals certainly guarantees its relevance to mineral homeostasis and places DMT1 as deserving a chapter in this text. The author has previously reviewed related topics (Garrick, 2011; Garrick and Garrick, 2009; Mackenzie and Garrick, 2005; Roth and Garrick, 2003) as have others (Gulec et al., 2014; Shawki et al., 2012).

ONE DIVALENT METAL-ION TRANSPORTER-1 FUNCTION—NUTRITIONAL IRON ENTRY INTO DUODENAL ENTEROCYTES

Fig. 19.1 illustrates how DMT1 acts as the major intestinal iron importer. Convincing evidence supports this role because the microcytic mouse (Fleming et al., 1997) and the Belgrade rat (Fleming et al., 1998), in a remarkable coincidence, have an identical mutation in DMT1 (G185R) leading to greatly diminished gastrointestinal iron uptake. Moreover, mice in which intestinal DMT1 is selectively inactivated recapitulate this difficulty, perhaps a little more severely than in the mutants with the amino acid substitution (Gunshin, Fujiwara, et al., 2005; Shawki et al., 2015). Entry requires reduction of Fe^{3+} in the intestinal lumen to Fe^{2+} and is facilitated by the local pH being mildly acidic, a cell-surface proton gradient locally attributable to an apical sodium/hydrogen exchanger (Shawki et al., 2016) and the basolateral Na^+-K-ATPase and probably other

Molecular, Genetic, and Nutritional Aspects of Major and Trace Minerals. http://dx.doi.org/10.1016/B978-0-12-802168-2.00019-1

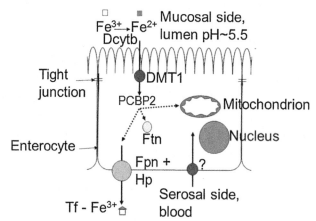

FIGURE 19.1 A model for divalent metal-ion transporter-1 (DMT1)'s participation in duodenal iron assimilation. Enterocytes have tight junctions such that most nutrients cannot pass between cells. Most iron entering the lumen of the gut is ferric (Fe^{3+}, also represented by a white square); hence it is reduced to ferrous (Fe^{2+} also represented by a red or dark square) for entry via DMT1. Duodenal cytochrome B (Dcytb; also known as Cybrd 1 for cytochrome b reductase-1) clearly acts as a reductase for this function, although there is controversy on its necessity (see text). The pH of the proximal duodenal lumen reflects incomplete neutralization of the acid gastric contents, providing a proton motive force for Fe^{2+} entry via DMT1. After that entry, DMT1 might pass Fe^{2+} on to poly(rC)-binding protein-2 (PCBP2), from which it can follow three potential paths (see text). Ordinarily most flux will go to the basolateral part of the enterocyte where ferroportin (Fpn) serves as the iron exporter facilitated by reoxidation to Fe^{3+} dependent on a ferroxidase such as hephaestin (Hp) and loading onto transferrin (Tf; also represented by the *triangle*). Although this path is the main one for nutritional iron uptake, flux can also go to the mitochondria where there is a high demand for iron in enterocytes, or when entry exceeds demand, flux can also go to storage in ferritin (Ftn). *Modified from Fig. 2 of a prior review (Garrick, 2011) and reproduced with permission from Springer-Verlag.*

factors such as the duodenum being immediately distal to more extreme gastric acidity. Iron is a major mineral nutrient, with much of its value to mammals residing in its ability to undergo redox reactions. Although reduction before entry via DMT1 can be catalyzed by Dcytb, Dcytb's function is dispensable as shown by engineering mice with intestinal Dcytb specifically ablated (Gunshin, Starr, et al., 2005). Nevertheless, Dcytb does stimulate iron uptake by intestinal epithelial cells (Latunde-Dada et al., 2008); however, other ferrireductases make it at least partially redundant (see "Reduction, Divalent Metal-Ion Transporter-1 Function, and Iron Metabolism" section below).

DMT1 has 12 transmembrane domains. Its N- and C-termini are intracellular; therefore there are many regions of this multipass protein that could interact with other proteins within or on either side of the plasma membrane. Recently, poly(rC)-binding protein (PCBP)-2, a multifunctional protein that binds to poly C stretches in mRNA but also apparently acts as an iron chaperone, has been found to interact with DMT1 to facilitate entrance of Fe^{2+} into HEp-2 cells (Yanatori et al., 2014). Although this function needs confirmation, as well as to ask whether it occurs in enterocytes, this finding is very exciting because the authors also demonstrated that PCBP2 aids in getting Fe^{2+} delivered to ferroportin (Fpn) for export from cells. In the enterocyte, export includes reoxidation by hephaestin and binding to Tf. It remains to be determined if PCBP2 also participates in getting Fe^{2+} delivered to mitochondria, but a substantial portion of iron flux in enterocytes is into that organelle to meet a high energy demand. There is also evidence that PCBP2 aids in getting excess Fe^{2+} delivered to ferritin for storage and eventual loss when the enterocyte sloughs off into the intestinal lumen.

Enterocytes also need their own supply of iron. Apical DMT1 probably functions to some degree for such acquisition, but there are not much data to address whether this pathway occurs and how, nor are there data on whether basolateral DMT1 could also function in a minor way for the cell's acquisition or if iron uptake on this side of the cell occurs via the Tf cycle.

A SECOND DMT1 FUNCTION—ENDOSOMAL EXPORT AFTER IRON ACQUISITION IN MOST CELLS VIA THE TF CYCLE

Fig. 19.2 illustrates the role of DMT1 in the Tf cycle, a process that delivers most iron to cells in mammals. Normally, iron in circulation binds to Tf but does not exceed the capacity of Tf for carrying iron, and most cells have the TfR-1 present on the plasma membrane. The pH dependence of the Fe–Tf–TfR1 interaction is such that mono- or diferric Tf (Fe_1Tf or Fe_2Tf) both remain tightly associated with TfR1 at the pH of plasma (7.4). Fe_1Tf and Fe_2Tf cooperatively release Fe under mildly acidic conditions when Tf binds to TfR1, and that release is probably facilitated when a reductase such as six transmembrane epithelial antigen of the prostate (Steap)-3 (Ohgami et al., 2005, 2006) provides stimulation. Association

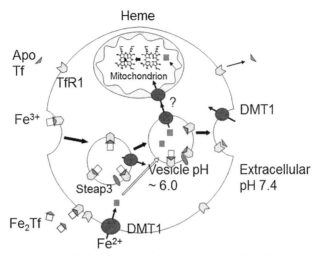

FIGURE 19.2 Divalent metal-ion transporter-1 (DMT1) and the transferrin (Tf) cycle. Apo Tf (on the *left*) has very little affinity for transferrin recep-tor-1 (TfR1; also represented by a *hexagon*) on the cell surface; however, Fe_2Tf binds tightly to TfR1. Invagination of the plasma membrane endocytoses the Fe_2Tf-TfR1 complex. In the endocytic vesicle/endosome, this complex is in close proximity to DMT1 and six-transmembrane epithelial antigen of the prostate-3 (Steap3; a ferrireductase). The pH of the endosome becomes more acidic than the cellular environment because of a proton pump on the endosomal membrane often called a vacuolar ATPase. Under these conditions, release of the Fe^{3+} from the Tf-TfR1 complex occurs, likely due to the pH and the reduction to Fe^{2+}, which is suited for exit from the endosome via DMT1. Although the Fe^{2+} could move on to poly(rC)-binding protein-2 (PCBP2), another possibility is direct contact with an entry transporter on the outer mitochondrial membrane (see text). In the mitochondrion, iron will participate in heme synthesis (illustrated) and Fe-S cluster formation (not shown). Meanwhile the apo Tf–TfR1 (apo Tf refers to Tf without any Fe bound to it) complex exocytoses to the plasma membrane where the higher pH leads to dissociation so that apo Tf is free to collect more Fe^{3+}. *Modified from Fig. 3 of a prior review (Garrick, 2011) and reproduced with permission from Springer-Verlag.*

of apo Tf-TfR1 (apo Tf refers to Tf without any Fe bound to it) is much weaker at pH 7.4; therefore apo Tf essentially does not bind to TfR1 before endocytosis and is readily released from TfR1 after exocytosis. Hence TfR1 acts as a receptor for receptor-mediated endocytosis of Fe-Tf with the physiological acidification of the endosome thus formed leading to release of ferrous iron so that DMT1 can serve as an exit transporter for it. Exocytosis releases the apo Tf to collect more iron. Although PCBP2 could be involved in chaperoning Fe^{2+} on its way to the mitochondrion, another hypothesis (Sheftel et al., 2007) is that direct interorganellar transfer occurs between the endosome and the mitochondrion—called the "kiss and run" hypothesis. A recent finding that favors this hypothesis is that DMT1 is present on the outer mitochondrial membrane (Wolff, Garrick, et al., 2014; Wolff, Ghio, et al., 2014), making it possible that DMT1 also serves as the entrance to mito-chondria and might be "the lips of the kiss." A question mark is applied to this possibility to remind the reader that much more data need to be developed before one can consider the path shown to be well supported. It is attractive here because it would mean that DMT1 has a third function—a gateway for metals to traffic to mitochondria.

Fig. 19.2 also illustrates a role that emerges for cellular iron uptake when circulating Tf nears or exceeds saturation with Fe^{3+}. The excess Fe is referred to as nontransferrin-bound iron (NTBI). NTBI is a challenge for cells, particularly because it readily participates in reactions such as Fenton chemistry (Eqs. (19.1) and (19.2)) to generate reactive oxygen species (ROS). To minimize the danger from ROS, NTBI can enter cells via DMT1 or another iron importer and probably merge into pathways utilized by Tf-bound iron.

$$Fe^{2+} + H_2O_2 \rightarrow Fe^{3+} + HO\bullet + OH^- \qquad (19.1)$$

$$Fe^{3+} + H_2O_2 \rightarrow Fe^{2+} + HOO\bullet + H^+ \qquad (19.2)$$

METAL-ION TRANSPORT BY DIVALENT METAL-ION TRANSPORTER-1

The previous two sections establish that DMT1 is a major iron transporter. What criteria permit such a statement? One answer is to ask about the phenotype when a mutation greatly diminishes the activity of the gene product. The relative inability of the microcytic mouse (Fleming et al., 1997) or the Belgrade rat (Fleming et al., 1998), each with the same muta-tion (G185R) in DMT1, to acquire iron gastrointestinally supports this statement. In addition, as previously noted, mice in which intestinal DMT1 is selectively ablated also have this difficulty (Gunshin, Fujiwara, et al., 2005). This phenotypic criterion also applies to the role of DMT1 as endosomal iron exporter in the Tf cycle (Fleming et al., 1998) in the Belgrade

TABLE 19.1 Metal Ion Transport by Divalent Metal-Ion Transporter-1

Metal Ion	Capability	Physiological or Pathological Significance
Fe^{2+}	Yes shown multiple ways	Major cellular import; major endosomal export; possible mitochondrial import
Mn^{2+}	Yes shown multiple ways	Cellular import, endosomal export, mitochondrial import. Relevance relative to other Mn transporters uncertain
Cd^{2+}	Yes shown multiple ways	Pathological, may contribute particularly to nephrotoxicity
Co^{2+}	Yes shown multiple ways	Probably physiological and pathological but limited data to date
Ni^{2+}	Yes shown multiple ways	May contribute to Ni carcinogenicity
Zn^{2+}	Yes shown multiple ways	Unlikely to contribute to normal Zn homeostasis because other Zn transporters appear to be more important
VO^{2+}	Yes	Unclear if divalent metal-ion transporter-1 is part of vanadium metabolism
Pb^{2+}	Yes, but some data are difficult to interpret	Divalent metal-ion transporter-1 may not contribute to its toxicity, yet iron deficiency may exacerbate Pb pathology
Cu^{1+}	Yes shown multiple ways	Divalent metal-ion transporter-1 may not contribute to normal Cu homeostasis but could contribute when divalent metal-ion transporter-1 levels are high as in iron deficiency

rat, particularly in view of earlier experiments identifying a defect in iron delivery with most other steps of the cycle operating normally. A second answer comes from overexpressing DMT1 in a cell and detecting an increase in iron or other metal flux into that cell as a consequence. This approach really asks whether a transporter can transport a particular metal ion whereas looking at the mutant phenotype asks more about the extent to which it does transport the substrate physiologically (or pathologically if the substrate is toxic). The second approach revealed that DMT1 could transport iron when cRNA was injected into *Xenopus* oocytes (Gunshin et al., 1997; Mackenzie et al., 2007) or the cDNA, transiently transfected into a cell line (Fleming et al., 1998). This approach extends to permanent transfection of a cell line and is particularly informative when one can regulate the overexpression (Garrick et al., 2006).

Using primarily these criteria, Table 19.1 summarizes the current status of knowledge for the metal ions transported by DMT1. DMT1 clearly has the capability to transport Mn^{2+} (Garrick et al., 2006; Mackenzie et al., 2007; also see Chapters 30–32 for manganese). Defects in Mn uptake from Mn-Tf and from "free" Mn in serum by Belgrade rat reticulocytes (Chua and Morgan, 1997) support the argument that there is a detectable phenotype making DMT1 a physiologically relevant transporter of Mn. DMT1 is present in the microvilli of the olfactory epithelium, and Belgrade rats take up much less Mn after nasal installation than appropriate control rats in which nasal Mn uptake tracks DMT1 levels (Thompson et al., 2007), associating DMT1 with Mn physiology in another context. Given that it is difficult to make rats Mn deficient (Roth and Garrick, 2003), it is not surprising that researchers remain uncertain if Belgrade rats are Mn deficient, but it is clear that DMT1 transports Mn. What remains unclear is the extent to which Mn transport depends on DMT1, as opposed to other transporters, but intestinal loss of DMT1 in mice does not impair Mn absorption (Shawki et al., 2015), suggesting that adequate Mn uptake occurs via other transporters or other routes. A mutation in the fruit fly called Malvolio turns out to be in the single gene that corresponds to DMT1 and Nramp1 and presents as abnormal taste behavior corrected by supplementation with excess Mn or Fe (Orgad et al., 1998). This remarkable observation makes one wonder about DMT1's role in the olfactory epithelium and its relation to taste and odor detection.

As for iron, one must consider toxic consequences when too much Mn accumulates. Excess Mn can lead to manganism, a neurological disorder characterized by gait modification similar but not identical to that occurring in Parkinson's disease (Roth and Garrick, 2003; see also Chapters 30–32). Predictably iron deficiency increases olfactory DMT1 levels to increase Mn uptake by that route (Veuthey and Wessling-Resnick, 2014). The microcytic mouse and Belgrade rat have decreased susceptibility to toxic compounds that induce Parkinsonian symptoms (Salazar et al., 2008), supporting an etiological role for DMT1 in at least some forms of that disease.

Given that Cd^{2+} is toxic and has no physiological role, it is surprising that DMT1 exhibits a higher affinity for this ion than any other (Illing et al., 2012). Cd^{2+} enters through the intestines and probably other tissues. It is a source of nephrotoxicity, a topic recently reviewed (Thevenod and Wolff, 2015) and very likely related to the high levels of DMT1 present in this organ, being expressed particularly in the proximal tubule. DMT1 likely plays a role in recovering filtered Fe with DMT1's levels increased during iron deficiency. This regulation could increase the sensitivity to Cd. The Belgrade rat or microcytic mouse provides mutants in which one could test whether DMT1 plays a critical part in this pathology, and intestinal- or kidney-specific ablation of DMT1 could aid in seeing where DMT1 contributes but the situation is more complex than one might expect because the Belgrade rat did not exhibit a clear defect in recovering filtered Fe (Ferguson et al., 2003). Other mechanisms possibly account for this discrepancy (Thevenod and Wolff, 2015), and Cd toxicity is clearly a different concern than recovering iron from filtrate.

Co^{2+} also exhibits a high affinity for DMT1 (Illing et al., 2012). This observation likely accounts for the competitive nature of intestinal Fe and Co transport (Thomson et al., 1971). DMT1 may account for the entry of cobalt to meet the modest needs of humans for this mineral. The transporter is unlikely to play a role in meeting cobalamin needs because vitamin B12 is a separate nutritional requirement. There is an interesting indirect demonstration of the role of DMT1 in Co toxicity (Howitt et al., 2009). A ubiquitin E3-ligase system targets DMT1 for proteolysis (discussed in that context below under "Divalent Metal-Ion Transporter-1 Regulation—Posttranslationally at the Protein Turnover Level") and appears to limit Co neurotoxicity by doing so. Co metabolism is yet another situation in which rodent mutants could inform us more on DMT1's role.

Four of the remaining cations have a lesser affinity for DMT1. Ni^{2+} competes less effectively than Fe^{2+} for DMT1, but it is clearly transported by that protein (Chen et al., 2005; Davidson et al., 2005; Illing et al., 2012). This metal ion is a known carcinogen, affecting oxygen sensing and iron homeostasis at multiple levels (also see Chapter 17). Zn^{2+} is a low-affinity ligand for DMT1 (Illing et al., 2012). Although Zn^{2+} metabolism is important (also see Chapters 20–24), mutations in other transporters of the metal do have deficiency phenotypes; for example, Zrt-/Irt-like protein (ZIP) transporter-4 presenting as acrodermatitis enteropathica (Wang et al., 2002) and ZIP14, in which ablation in mice leads to a defect in G-protein-coupled receptor-mediated signaling required for systemic growth. Interestingly, a hypomorphic mouse mutant of ZIP8, another member of the ZIP family, exhibits defects in hematopoiesis attributed to Zn^{2+} metabolism but equally consistent with a disturbance in Fe homeostasis. DMT1 also binds VO^{2+} (Illing et al., 2012), with the physiological significance being uncertain in humans, where V acquisition may rely on vanadate VO_4^-. Moreover, an ascidian (a sea squirt; an invertebrate filter feeder) that does accumulate VO^{2+} relies on a DMT1-related transporter to do so (Ueki et al., 2011). Pb^{2+} in relation to DMT1 poses a conundrum. Transport may occur, but at least one assay was considered difficult to interpret (Illing et al., 2012), and knocking down DMT1 mRNA affected Fe^{2+} transport but not that for Pb^{2+} (Bannon et al., 2003). There is a well-known association of lead poisoning with iron deficiency in the inner cities of some countries, but it is unclear if that association relates only to pica (eating dirt, paint, and other nonfood substances), which occurs during iron deficiency, or if the induction of duodenal DMT1 by iron deficiency also contributes to the problem.

DMT1 transport of Cu is particularly controversial, thus deserving separate coverage (also see Chapters 7–10 for Cu). Ablating intestinal DMT1 in mice did not cause Cu deficiency (Shawki et al., 2015), and the same group could not find evidence for DMT1 taking up Cu in *Xenopus* oocytes (Illing et al., 2012). They also cite many of the papers supporting or arguing against DMT1 participation in Cu metabolism. Nevertheless, there is strong evidence that DMT1 can transport Cu^{1+} (Arredondo et al., 2014; Jiang et al., 2013). DMT1 apparently comes into play when highly expressed, a state that is particularly relevant physiologically during iron deficiency. Because Cu and Fe homeostasis are strongly intertwined, such occasions may reflect the additional demand for Cu during iron deficiency. Indeed, copper deficiency in swine was classically noted to resemble iron-deficiency anemia (Lee et al., 1968). An interesting question is whether the high levels of DMT1 are the key to its transporting Cu^{1+} or if iron deficiency induces a change in state for DMT1 that permits Cu import. Another curiosity is that it appears that DMT1's ligand is Cu^{1+}, not Cu^{2+} (Arredondo et al., 2003, 2014; Jiang et al., 2013), making the "divalent" in DMT1's name perhaps slightly inappropriate.

REDUCTION, DIVALENT METAL-ION TRANSPORTER-1 FUNCTION, AND IRON METABOLISM

During iron metabolism, Fe^{3+} is frequently converted to Fe^{2+} and vice versa. These frequent reduction and oxidation reactions not only take advantage of how freely iron undergoes the loss or gain of an electron but also set up concentration gradients to help drive the metal from one intracellular locale to another and create favorable conditions for iron transport. In the intestinal tract, DMT1's role in entry requires reduction facilitated by Dcytb; whereas in the Tf cycle Steap3 is clearly critical for generating Fe^{2+}. Given that Dcytb is at least partially redundant, it appears that other ferrireductases remain to be identified. What are some of the possibilities?

The Steap family has four members; of these, only Steap1 lacks the structural capability for Fe reduction (Ohgami et al., 2006); therefore Steap2 and Steap4 can also serve as ferrireductases. Stromal cell-derived receptor-2 (a relative of Dcytb) also has this capability (Vargas et al., 2003) and appears to carry out such a function (Ji and Kosman, 2015). Clearly we need to learn more about the distribution and functional role of these ferrireductases.

Some remarkable candidates for DMT1-associated ferrireductases have recently emerged. Knockout of prion protein leads to markedly diminished hepatic iron stores (Tripathi et al., 2015); this property and coexpression with DMT1 (or ZIP14) led the authors to postulate that prion protein is a ferrireductase or that it indirectly promotes Fe^{3+} reduction. Remarkably, prion protein is a relative of the large ZIP family of transporters, some of which clearly transport Fe as well as Zn (Singh et al., 2015), leading to the argument that ferrireductase capability is closely related to this origin. The authors note that an octapeptide repeat domain of the prion protein is critical for the postulated function (Singh et al., 2013), but most of the evidence for ferrireductase activity is inferential. Other surprising, speculative candidates are anion exchange protein-2 (Ghio et al., 2003) and voltage-dependent anion channel-1 (VDAC1), a mitochondrial protein for which the major physiological function relates to its regulated release of ATP across the outer mitochondrial membrane (Baker, Lane, et al., 2004; Baker, Ly, et al., 2004). The evidence is only suggestive in each case; for example, VDAC1 acts as a ferricyanide reductase, but it is unlikely that ferricyanide is the physiological substrate and there is an ample supply of reductant capacity in the mitochondrion to fuel ferrireduction. The key take-home message is that the issue of how cells generate Fe^{2+} to supply this metal ion for DMT1 and other transporters deserves a lot more analysis than has occurred thus far.

DIVALENT METAL-ION TRANSPORTER-1 ISOFORMS, INTERACTIONS, AND MODIFICATIONS

Fig. 19.3 illustrates the four major protein isoforms of DMT1. These occur because of alternative promoters so that transcription starts in exon 1A or exon 1B (Hubert and Hentze, 2002) and alternative polyadenylation sites in exons 16 or 17 (Fleming et al., 1998). The choice of exon 16 results in the presence of an iron-responsive element (IRE) in the 3′ untranslated portion of the mRNA, whereas exon 17 leads to its absence. Although the IRE is in the mRNA, the custom has been to apply the terms +IRE and −IRE to the mRNA and protein isoforms. In any case, distinct amino acid sequences precede the termination codon for −IRE and +IRE DMT1. There is an initiation codon in exon 1A; therefore a distinct amino acid sequence precedes the common polypeptide portion in the 1A isoform whereas the 1B isoform does not initiate until well into exon 2 in the common polypeptide portion. Hence the four isoforms are 1A/−IRE, 1B/−IRE, 1A/+IRE, and 1B/+IRE. The 1A/+IRE form predominates in the duodenum; therefore it is the main nutritional iron importer. The 1B/−IRE form is the main form in most other cells, although there is a substantial contribution of 1A forms in the kidney. There is also evidence that the different C-termini target the +IRE and −IRE DMT1, respectively, to early endosomes and late endosomes/lysosomes (Tabuchi et al., 2002). Although these variant forms of DMT1 mRNA and protein are sometimes misattributed to alternate splicing, they are clearly due to alternative transcription initiation sites and polyadenylation sites. There are also variant DMT1 transcripts that are due to alternate splicing events, last reviewed in Mims and Prchal (2005).

Although it is now clear that PCBP2 acts as an iron chaperone (Fig. 19.1), how iron transits the enterocyte to end up on Tf still has several observations that remain to be well integrated. There is evidence that DMT1 endocytoses from the

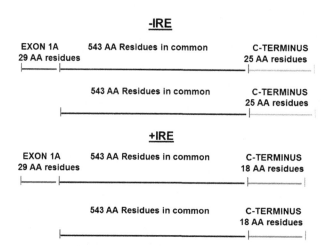

FIGURE 19.3 Divalent metal-ion transporter-1 (DMT1) protein isoforms. The four major isoforms differ at the N- and C-termini while sharing a larger central region. The 1A forms have an N-terminal extension of 29 amino acid residues in humans whereas the 1B forms do not; the iron-response element (−IRE) forms have C-terminal extensions of 25 residues that differ from the +IRE forms' C-terminal extensions of 18 residues. *Modified from Fig. 4b of a prior review (Garrick, 2011) and reproduced with permission from Springer-Verlag.*

apical surface whereas apo Tf endocytoses from the basolateral surface to come together in a vesicle (Ma et al., 2006). The authors argued that DMT1 was actively involved in the transcytosis of iron. It was also argued that endocytosis of DMT1 after intestinal exposure to iron might be part of the so-called mucosal block, in which a dose of iron suppresses iron uptake for a period afterward. The same group also identified as potentially involved a DMT1-associated protein that turned out to be peripheral benzodiazepine receptor-associated protein-7 (PAP7; Okazaki et al., 2012). In their study, PAP7 associated with +IRE DMT1. The benzodiazepine receptor is named for its ability to bind a tranquilizer. Although it clearly does that, it is unlikely that it evolved for that purpose. Interestingly, DMT1 also interacts with PAP7 and a series of small signaling GTPases in neurons resulting in altered iron homeostasis and affecting neurotoxicity (Chen et al., 2013; Choi et al., 2013). It is likely that a clear picture will emerge in the future for the physiological significance of these interactions in very different cell types, but it is not yet here.

The standard model for maintenance of iron homeostasis (see Chapter 18) has hepcidin, a peptide hormone produced primarily in the liver and the master regulator of iron metabolism, responding to systemic iron levels (and inflammation) then binding to Fpn as the hepcidin receptor and causing internalization of Fpn and ultimately Fpn turnover. Therefore hepcidin levels are usually inversely related to systemic iron levels. This model is extremely well supported, but it leaves out any role for DMT1. Nevertheless, hepcidin also inhibits apical iron entry via DMT1 in enterocytes (Mena et al., 2008) even more rapidly than it affects Fpn, and this process involves DMT1 internalization and turnover (Brasse–Lagnel et al., 2011). These data suggest that the standard model should also involve a mechanism for these DMT1 responses to the master regulator despite the emphasis thus far on Fpn/hepcidin interaction.

Glycosylation on the fourth extracellular domain of DMT1 may contribute to DMT1's localizing to the apical plasma membrane (Tabuchi et al., 2002), but this observation did not take into account the knowledge that there were N-terminal isoforms; therefore it needs to be reevaluated in terms of how those relate to plasma membrane versus vesicle localization. DMT1 also has sites where it could be phosphorylated or subsequently acted upon by phosphatases, but these too have not been carefully evaluated for whether regulation, localization, or functional distinctions depend on their status.

DIVALENT METAL-ION TRANSPORTER-1 REGULATION—AN OVERVIEW

Gene expression in general is regulated at multiple stages. DNA→RNA→protein involves transcription, RNA itself, translation, and the protein itself. Our knowledge for DMT1 is sparse at some of these levels but more extensive than many acknowledge at others.

DMT1 REGULATION—TRANSCRIPTION

The two promoters of DMT1, located 5′ proximal on the DNA to exons 1A and 1B, exhibit very different properties. The 1B promoter responds to nuclear factor kappa light-chain enhancer of activated B cells (NF-κB; Paradkar and Roth, 2006). Although the paper cited places this property in the context of a nitric oxide response and downregulation, its data show that 1B DMT1 transcription is exquisitely sensitive to upregulation during the inflammatory NF-κB response. Thus this process accounts for upregulation of DMT1 during many conditions in which iron must be sequestered; some of these processes are covered in publications (Ghio et al., 2006; Wang et al., 2005). Nevertheless, the timing of the publications has obstructed a full linkage. One can conclude from these observations that 1B DMT1 likely participates in iron sequestration during infection and inflammation and that its role is similar to the hepcidin/Fpn relationship covered in Chapter 18 and discussed briefly herein. The same authors also showed that nuclear factor Y (NF-Y) regulates 1B DMT1 via a CCAAT box in the promoter (Paradkar and Roth, 2007). The physiological significance is less clear; however, with cell cycle regulation and apoptosis among the functions controlled by NF-Y, it is possible that this effect on 1B DMT1 overlaps with those mediated by NF-κB.

The 1A DMT1 promoter clearly responds to hypoxia (Lis et al., 2005). Anemia associated with iron deficiency makes this effect equally a response to demand for nutritional iron (Shah et al., 2009). Shah's group also demonstrated that Dcytb expression involves the same hypoxia-inducible factor-2α signaling system, and Atp7A, a Cu-ATPase that increases expression during iron deficiency, also responds to this regulator (Xie and Collins, 2011; also see Chapter 17). Although an early characterization of the 1B promoter proposed similar regulation (Lee et al., 1998), convincing evidence to support such a role is lacking. Remarkably then, the 1A promoter responds to iron deficiency, primarily in the duodenum, to allow increased entry of iron nutritionally to aid in managing that problem, and the 1B promoter addresses tissue exposure to iron (often associated with inflammation or a risk of infection) to allow increased entry of iron to sequester the metal to limit risk of tissue damage due to ROS or an infectious agent.

DIVALENT METAL-ION TRANSPORTER-1 REGULATION—POSTTRANSCRIPTIONALLY AT THE MRNA LEVEL

The IRE/iron-regulatory protein (IRP) system is a major contributor to iron homeostasis. IREs are stem–loop structures with consensus sequences that occur in mRNA to which one of two IRPs may bind. If the IRE is in the 5′ portion of mRNA, then IRP binding will prevent initiation of translation. If the IRE is in the 3′ portion of the mRNA, then IRP binding will stabilize the mRNA. IRPs bind effectively when cells are iron deficient; therefore a 5′ IRE blocks translation whereas a 3′ IRE stabilizes mRNA during iron deficiency. DMT1's IRE is in the 3′ portion of the mRNA; therefore it should stabilize +IRE DMT1 mRNA under iron-deficient conditions. Although this rationale seemed likely, time passed before experimental evidence supported it, at first from intestinal-specific ablation of IRPs (Galy et al., 2008) and more recently from observing increased +IRE DMT1 mRNA after overexpression and chelation of iron (Jiang et al., 2013).

 –IRE DMT1 cannot respond in this fashion to iron status, but it does have a regulatory response to a microRNA (miRNA) designated Let-7d at least in erythroid cells (Andolfo et al., 2010). There is an inverse correlation of Let-7d levels and –IRE DMT1 mRNA; overexpression of Let-7d impairs erythroid differentiation of K562 cells with iron accumulating in endosomes. These observations suggest that Let-7d destabilizes the mRNA, leading to a deficit in the Tf cycle's release of iron to the mitochondria.

DIVALENT METAL-ION TRANSPORTER-1 REGULATION—AT THE TRANSLATIONAL LEVEL

Let-7d miRNA (Andolfo et al., 2010) modulation of –IRE DMT1 mRNA could also directly affect its translation. The initial discovery of miRNAs showed that they frequently block translation. This possibility still exists at least hypothetically even if the miRNA binds to the 3′ end of the mRNA; therefore one would like to see experiments supporting or ruling out such a mechanism.

DIVALENT METAL-ION TRANSPORTER-1 REGULATION—POSTTRANSLATIONALLY AT THE PROTEIN TURNOVER LEVEL

Degradation of most proteins occurs in an organelle called the proteasome or in the lysosome. A protein will frequently be marked for such targeting by ubiquitination. There are three types of enzymes involved: E1, E2, and E3. The E3-ubiquitin ligases provide most of the target specificity. One should use care in assuming that ubiquitination is a tag for turnover because it can also affect intracellular locale, but all of the examples that follow do appear to relate to turnover. Remarkably, 1B DMT1 ubiquitination is the responsibility of Parkin (Roth et al., 2010). The authors clearly demonstrated that 1A DMT1 remains unaffected. The observations leave us with a little puzzle because 1A DMT1 differs from 1B by having an N-terminal extension, suggesting that the 29 amino acid residues (in humans, other species have 28 or 30) block a site recognized by Parkin. This targeting provides a separate line of evidence to support a role for DMT1 in the etiology of Parkinson's disease as noted earlier (Salazar et al., 2008). Because Parkin interacts with most of its targets in the mitochondrion, the data were also the first hint that DMT1 might be present in that organelle.

 There are also observations that a second E3-ligase system targets DMT1 (Foot et al., 2008; Howitt et al., 2009). It involves a family of E3-ligases that interact with accessory proteins called Ndfips (Nedd4 family interacting proteins). Mice in which Ndfip1 has been ablated exhibit unusually high levels of duodenal DMT1 (Foot et al., 2011). Incomplete data (Garrick et al., 2012) point toward this family and accessories targeting +IRE DMT1, a result consistent with the effect of knocking out Ndfip1. Thus we potentially know how all major DMT1 isoforms except 1A/–IRE get targeted for turnover. Remarkably, there are also hints that Ndfips could link to Parkinson's disease through DMT1 (Howitt et al., 2014; Jia et al., 2015). Remarkably, the Nedd4-2 ubiquitin E3-ligase system also interacts with DMT1 and budding adaptor proteins designated Arrdc1 and Arrdc4 to release DMT1 from the cell in extracellular vesicles rather than target DMT1 to the proteasome or lysosome (Mackenzie et al., 2016). This mechanism is obviously a novel one for maintaining iron homeostasis and may have been detected years earlier (Sharp et al., 2002; Simovich et al., 2003), although the earlier observations occurred before there was sufficient knowledge to understand the detection of extracellular DMT1.

SUMMARY POINTS

- DMT1 is the major duodenal iron importer.
- DMT1 is the major endosomal iron exporter.

- DMT1 can transport multiple metals other than iron.
- As its name implies, most metal ions must be divalent to be transported by DMT1.
- DMT1 has four major isoforms, differing in their N- and C-termini.
- A hypoxia-responsive element renders transcription of the 1A isoform of DMT1 responsive to iron deficiency and hypoxia.
- The 1B promoter for DMT1 senses inflammation and related stimuli via an NF-κB–responsive element.
- Iron deficiency stabilizes DMT1 +IRE mRNA via IRP binding to the IRE.
- Turnover of 1B DMT1 depends on Parkin, an E3-ubiquitin ligase.

MINI DICTIONARY OF TERMS

Cybrd-1 Cytochrome b reductase-1; see Dcytb.

DCT1 Divalent cation transporter-1, a synonym for DMT1.

Dcytb Duodenal cytochrome B, also known as Cybrd-1 for cytochrome b reductase-1; a ferrireductase that makes iron ready for transport by DMT1 or other ferrous transporters.

DMT1 Divalent metal-ion transporter-1, this chapter's main topic; it transports at least nine cations and is most physiologically relevant for Fe^{2+}. Iron import into cells and iron export from endosomes after entry there via Tf are two roles. Although nominally proton coupled, it can actually aid in ion movement even at neutral pH, albeit less efficiently.

Fpn Ferroportin; the only known cellular exporter of iron, making it a critical protein for regulation of iron homeostasis.

Hepcidin A peptide hormone synthesized mainly in the liver, encoded by the *HAMP* gene. In humans, a larger preprohormone (84 residues) is cleaved to a prohormone (60) and that to the actual hormone (25) that binds to Fpn leading to internalization of the iron exporter and subsequently to its degradation.

HIF-2α Hypoxia-inducible factor-2α; a part of a transcription factor system that responds to hypoxia and equivalently to iron deficiency. A response element for it exists in the promoter for 1A DMT1, ensuring upregulation of duodenal DMT1 transcription during iron deficiency.

IRE Iron-response element; a stem–loop structure in some mRNAs that renders the mRNA responsive to intracellular labile iron status. When the IRE is in the 5′ part of the mRNA, an IRP binding to the IRE can block translation, whereas when the IRE is in the 3′ part of the mRNA, an IRP binding to the IRE can stabilize the mRNA against degradation.

IRP Iron-regulatory protein. There are two IRPs, and both are available to bind IREs when there is iron deficiency but not when the cell is iron replete.

Let-7d The second miRNA to be discovered, –IRE DMT1 lists among the many mRNAs the function of which it regulates.

Manganism A neurological disorder characterized by gait modification similar but not identical to that occurring in Parkinson's disease.

NF-κB Nuclear factor kappa-light-chain-enhancer of activated B cells; a protein complex that controls transcription of genes related to cytokine production and cell survival in response to stimuli such as stress, cytokines, free radicals, inflammation, and infection. A response element for it exists in the promoter for 1B DMT1, supporting DMT1's role in sequestering iron under such challenges.

NRAMP1 Natural resistance-associated macrophage protein-1; identified initially as a gene expressed in macrophages and responsible for innate resistance to several pathogens. The protein turned out to be a metal-ion transporter closely related to DMT1.

Nramp2 Natural resistance associated macrophage protein-2; the initial name assigned to DMT1 because of its apparent similarity to NRAMP1. The name was inappropriate largely because DMT1 was not restricted in expression to macrophages.

NTBI Nontransferrin-bound iron; when the level of iron exceeds the carrying capacity of Tf, this term refers to the state of that excess iron.

PAP7 Peripheral benzodiazepine receptor-associated protein-7; a protein that interacts with +IRE DMT1, perhaps involved in transcytosis of iron in the enterocyte but also interacting with DMT1 and small GTPases in neuronal tissues.

PCBP2 Poly(rC)-binding protein-2, a protein initially identified by its ability to bind stretches of C's in mRNA that turns out to function as an iron chaperone interacting with DMT1 to help deliver Fe^{2+} to intracellular locales.

Prion Protein infection; a term coined by Stanley Prusiner for a proteinaceous infectious particle after finding that certain infections involved transmission by agents that had no DNA nor RNA. The relation to DMT1 derives from the ability of prion protein to stimulate iron uptake by the transporter, leading to the postulate that prion protein is a ferrireductase.

ROS Reactive oxygen species; highly reactive forms of oxygen frequently generated in the presence of iron such as hydroxyl radicals, singlet oxygen, and superoxides.

SLC11A2 Solute carrier family 11 member 2; a synonym for DMT1 with its designation based on a classification scheme for solute carriers for channels and transporters. When italicized and capitalized, the term refers to the gene that encodes it. Number 1 in this family is NRAMP1.

Steap3 Six-transmembrane epithelial antigen of the prostate-3; a ferrireductase that functions primarily in the endosome.

Tf Transferrin, the main carrier protein for Fe^{3+} in serum; having binding sites for Fe^{3+} on its N- and C-terminal lobes, transferrin can carry 0 (apo), 1 (monoferric), or 2 (diferric) Fe^{3+}.

TfR1 Transferrin receptor-1; a dimeric, single membrane pass protein that binds mono- or diferric Tf on the cell surface and undergoes endocytosis, vesicle acidification, reduction, and release of the iron in what is referred to as the Tf cycle. DMT1 participates as the main way iron can exit the vesicle, making DMT1 also an iron exporter (but not at the level of the whole cell).

VDAC1 Voltage-dependent anion channel-1; a mitochondrial protein for which the major physiological function relates to its regulated release of ATP across the outer mitochondrial membrane. There is evidence that VDAC1 is a ferricyanide reductase, consistent with speculation that it could be a mitochondrial ferrireductase.

ZIP Zrt-/Irt-like transporter; an acronym that requires definition of Zrt (zinc-related transporter) and Irt (iron-related transporter) with both terms applied to a plant metal transporter family that turned out also to have related transporters in mammals. There are many members (see Chapter 22).

ACKNOWLEDGMENTS

Dr. Laura M. Garrick critically reviewed this chapter, and her comments are greatly appreciated.

REFERENCES

Andolfo, I., De Falco, L., Asci, R., Russo, R., Colucci, S., Gorrese, M., Zollo, M., Iolascon, A., 2010. Regulation of divalent metal transporter 1 (DMT1) non-IRE isoform by the microRNA Let-7d in erythroid cells. Haematologica 95, 1244–1252.

Arredondo, M., Mendiburo, M., Flores, S., Singleton, S., Garrick, M., 2014. Mouse divalent metal transporter 1 is a copper transporter in HEK293 cells. BioMetals 27, 115–123.

Arredondo, M., Munoz, P., Mura, C.V., Nuñez, M.T., 2003. DMT1, a physiologically relevant apical Cu^{1+} transporter of intestinal cells. Am. J. Physiol. Cell Physiol. 284, C1525–C1530.

Baker, M.A., Lane, D.J.R., Ly, J.D., De Pinto, V., Lawen, A., 2004. VDAC1 is a Transplasma membrane NADH-ferricyanide reductase. J. Biol. Chem. 279, 4811–4819.

Baker, M.A., Ly, J.D., Lawen, A., 2004. Characterization of VDAC1 as a plasma membrane NADH-oxidoreductase. BioFactors 21, 215–221.

Bannon, D.I., Abounader, R., Lees, P., Bressler, J.P., 2003. Effect of DMT1 knockdown on iron, cadmium, and lead uptake in Caco-2 cells. Am. J. Physiol. Cell Physiol. 284, C44–C50.

Brasse-Lagnel, C., Karim, Z., Letteron, P., Bekri, S., Bado, A., Beaumont, C., 2011. Intestinal DMT1 cotransporter is down-regulated by hepcidin via proteasome internalization and degradation. Gastroenterology 140, 1261–1271. e1261.

Chen, H., Davidson, T., Singleton, S., Garrick, M.D., Costa, M., 2005. Nickel decreases cellular iron level and converts cytosolic aconitase to iron-regulatory protein 1 in A549 cells. Toxicol. Appl. Pharmacol. 206, 275–287.

Chen, Y., Khan, R.S., Cwanger, A., Song, Y., Steenstra, C., Bang, S., Cheah, J.H., Dunaief, J., Shindler, K.S., Snyder, S.H., Kim, S.F., 2013. Dexras1, a small GTPase, is required for glutamate-NMDA neurotoxicity. J. Neurosci. 33, 3582–3587.

Choi, B.R., Bang, S., Chen, Y., Cheah, J.H., Kim, S.F., 2013. PKA modulates iron trafficking in the striatum via small GTPase, Rhes. Neuroscience 253, 214–220.

Chua, A., Morgan, E.H., 1997. Manganese metabolism is impaired in the Belgrade laboratory rat. J. Comp. Physiol. B 167, 361–369.

Davidson, T., Chen, H., Garrick, M., D'Angelo, G., Costa, M., 2005. Soluble nickel interferes with cellular iron homeostasis. Mol. Cell. Biochem. 279, 157–162.

Ferguson, C.J., Wareing, M., Delannoy, M., Fenton, R., Mclarnon, S.J., Ashton, N., Cox, A.G., McMahon, R., Garrick, L.M., Green, R., Smith, C.P., Riccardi, D., 2003. Iron handling and gene expression of the divalent metal transporter, DMT1, in the kidney of the anemic Belgrade (b) rat. Kidney Int. 64, 1755–1764.

Fleming, M.D., Romano, M.A., Su, M.A., Garrick, L.M., Garrick, M.D., Andrews, N.C., 1998. Nramp2 is mutated in the anemic Belgrade (b) rat: evidence of a role for Nramp2 in endosomal iron transport. Proc. Natl. Acad. Sci. U. S. A. 95, 1148–1153.

Fleming, M.D., Trenor, C.I., Su, M.A., Foernzler, D., Beier, D.R., Dietrich, W.F., Andrews, N.C., 1997. Microcytic anaemia mice have a mutation in Nramp2, a candidate iron transporter gene. Nat. Genet. 16, 383–386.

Foot, N.J., Dalton, H.E., Shearwin-Whyatt, L.M., Dorstyn, L., Tan, S.-S., Yang, B., Kumar, S., 2008. Regulation of the divalent metal ion transporter DMT1 and iron homeostasis by a ubiquitin-dependent mechanism involving Ndfips and WWP2. Blood 112, 4268–4275.

Foot, N.J., Leong, Y.A., Dorstyn, L.E., Dalton, H.E., Ho, K., Zhao, L., Garrick, M.D., Yang, B., Hiwase, D., Kumar, S., 2011. Ndfip1-deficient mice have impaired DMT1 regulation and iron homeostasis. Blood 117, 638–646.

Galy, B., Ferring-Appel, D., Kaden, S., Gröne, H., Hentze, M., 2008. Iron regulatory proteins are essential for intestinal function and control key iron absorption. Cell Metab. 7, 79–85.

Garrick, M., 2011. Human iron transporters. Genes Nutr. 6, 45–54.

Garrick, M.D., Garrick, L.M., 2009. Cellular iron transport. Biochim. Biophys. Acta (BBA) General Subj. 1790, 309–325.

Garrick, M.D., Kuo, H.-C., Vargas, F., Singleton, S., Zhao, L., Smith, J.J., Paradkar, P., Roth, J.A., Garrick, L.M., 2006. Comparison of mammalian cell lines expressing distinct isoforms of divalent metal transporter 1 in a tetracycline-regulated fashion. Biochem. J. 398, 539–546.

Garrick, M.D., Zhao, L., Roth, J.A., Jiang, H., Feng, J., Foot, N.J., Dalton, H., Kumar, S., Garrick, L.M., 2012. Isoform specific regulation of divalent metal (ion) transporter (DMT1) by proteasomal degradation. Biometals 25, 787–793.

Ghio, A.J., Nozik-Grayck, E., Turi, J., Jaspers, I., Mercatante, D.R., Kole, R., Piantadosi, C.A., 2003. Superoxide-dependent iron uptake: a new role for anion exchange protein 2. Am. J. Respir. Cell Mol. Biol. 29, 653–660.

Ghio, A.J., Turi, J., Yang, F., Garrick, L.M., Garrick, M.D., 2006. Iron homeostasis in the lung. Biol. Res. 39, 67–77.

Gulec, S., Anderson, G.J., Collins, J.F., 2014. Mechanistic and regulatory aspects of intestinal iron absorption. Am. J. Physiol. Gastrointest. Liver Physiol. 307, G397–G409.

Gunshin, H., Fujiwara, Y., Custodio, A.O., DiRenzo, C., Robine, S., Andrews, N.C., 2005. Slc11a2 is required for intestinal iron absorption and erythropoiesis but dispensable in placenta and liver. J. Clin. Invest. 115, 1258–1266.

Gunshin, H., Mackenzie, B., Berger, U.V., Gunshin, Y., Romero, M.F., Boron, W.F., Nussberger, S., Gollan, J.L., Hediger, M.A., 1997. Cloning and characterization of a mammalian proton-coupled metal-ion transporter. Nature 388, 482–488.

Gunshin, H., Starr, C.N., DiRenzo, C., Fleming, M.D., Jin, J., Greer, E.L., Sellers, V.M., Galica, S.M., Andrews, N.C., 2005. Cybrd1 (duodenal cytochrome b) is not necessary for dietary iron absorption in mice. Blood 106, 2879–2883.

Howitt, J., Gysbers, A.M., Ayton, S., Carew-Jones, F., Putz, U., Finkelstein, D.I., Halliday, G.M., Tan, S.-S., 2014. Increased Ndfip1 in the substantia nigra of parkinsonian brains is associated with elevated iron levels. PLoS One 9, e87119.

Howitt, J., Putz, U., Lackovic, J., Doan, A., Dorstyn, L., Cheng, H., Yang, B., Chan-Ling, T., Silke, J., Kumar, S., Tan, S.S., 2009. Divalent metal transporter 1 (DMT1) regulation by Ndfip1 prevents metal toxicity in human neurons. Proc. Natl. Acad. Sci. U. S. A. 106, 15489–15494.

Hubert, N., Hentze, M.W., 2002. Previously uncharacterized isoforms of divalent metal transporter (DMT)-1: implications for regulation and cellular function. Proc. Natl. Acad. Sci. U. S. A. 99, 12345–12350.

Illing, A.C., Shawki, A., Cunningham, C.L., Mackenzie, B., 2012. Substrate profile and metal-ion selectivity of human divalent metal-ion Transporter-1. J. Biol. Chem. 287, 30485–30496.

Ji, C., Kosman, D.J., 2015. Molecular mechanisms of non-transferrin-bound and transferrin-bound iron uptake in primary hippocampal neurons. J. Neurochem. 133, 668–683.

Jia, W., Xu, H., Du, X., Jiang, H., Xie, J., 2015. Ndfip1 attenuated 6-OHDA-induced iron accumulation via regulating the degradation of DMT1. Neurobiol. Aging. 36, 1183–1193.

Jiang, L., Garrick, M.D., Garrick, L.M., Zhao, L., Collins, J.F., 2013. Divalent metal transporter 1 (Dmt1) mediates copper transport in the duodenum of iron-deficient rats and when overexpressed in iron-deprived HEK-293 cells. J. Nutr. 143, 1927–1933.

Latunde-Dada, G.O., Simpson, R.J., McKie, A.T., 2008. Duodenal cytochrome B expression stimulates iron uptake by human intestinal epithelial cells. J. Nutr. 138, 991–995.

Lee, G.R., Nacht, S., Lukens, J.N., Cartwright, G.E., 1968. Iron metabolism in copper-deficient swine. J. Clin. Invest. 47, 2058–2069.

Lee, P.L., Gelbart, T., West, C., Halloran, C., Beutler, E., 1998. The human Nramp2 gene: characterization of the gene structure, alternative splicing, promoter region and polymorphisms. Blood Cells Mol. Dis. 24, 199–215.

Lis, A., Paradkar, P.N., Singleton, S., Kuo, H.C., Garrick, M.D., Roth, J.A., 2005. Hypoxia induces changes in expression of isoforms of the divalent metal transporter (DMT1) in rat pheochromocytoma (PC12) cells. Biochem. Pharmacol. 69, 1647–1655.

Ma, Y., Yeh, M., Yeh, K.Y., Glass, J., 2006. Iron Imports. V. Transport of iron through the intestinal epithelium. Am. J. Physiol. Gastrointest. Liver Physiol. 290, G417–G422.

Mackenzie, B., Garrick, M.D., 2005. Iron Imports. II. Iron uptake at the apical membrane in the intestine. Am. J. Physiol. Gastrointest. Liver Physiol. 289, G981–G986.

Mackenzie, B., Takanaga, H., Hubert, N., Rolfs, A., Hediger, M.A., 2007. Functional properties of multiple isoforms of human divalent metal-ion transporter 1 (DMT1). Biochem. J. 403, 59–69.

Mackenzie, K.D., Foot, N.J., Anand, S., Dalton, H.E., Chaudhary, N., Collins, B.M., Mathivanan, S., Kumar, S., 2016. Regulation of the divalent metal ion transporter via membrane budding. Cell Discov. 2, 16011.

Mena, N.P., Esparza, A., Tapia, V., Valdes, P., Nunez, M.T., 2008. Hepcidin inhibits apical iron uptake in intestinal cells. Am. J. Physiol. Gastrointest. Liver Physiol. 294, G192–G198.

Mims, M.P., Prchal, J.T., 2005. Divalent metal transporter 1. Hematology 10, 339–345.

Ohgami, R.S., Campagna, D.R., Greer, E.L., Antiochos, B., McDonald, A., Chen, J., Sharp, J.J., Fujiwara, Y., Barker, J.E., Fleming, M.D., 2005. Identification of a ferrireductase required for efficient transferrin-dependent iron uptake in erythroid cells. Nat. Genet. 37, 1264–1269.

Ohgami, R.S., Campagna, D.R., McDonald, A., Fleming, M.D., 2006. The Steap proteins are metalloreductases. Blood 108, 1388–1394.

Okazaki, Y., Ma, Y., Yeh, M., Yin, H., Li, Z., Yeh, K.Y., Glass, J., 2012. DMT1 (IRE) expression in intestinal and erythroid cells is regulated by peripheral benzodiazepine receptor-associated protein 7. Am. J. Physiol. Gastrointest. Liver Physiol. 302, G1180–G1190.

Orgad, S., Nelson, H., Segal, D., Nelson, N., 1998. Metal ions suppress the abnormal taste behavior of the Drosophila mutant malvolio. J. Exp. Biol. 201, 115–120.

Paradkar, P.N., Roth, J.A., 2006. Nitric oxide transcriptionally down-regulates specific isoforms of divalent metal transporter (DMT1) via NF-kappaB. J. Neurochem. 96, 1768–1777.

Paradkar, P.N., Roth, J.A., 2007. Expression of the 1B isoforms of divalent metal transporter (DMT1) is regulated by interaction of NF-Y with a CCAAT-box element near the transcription start site. J. Cell Physiol. 211, 183–188.

Roth, J.A., Garrick, M.D., 2003. Iron interactions and other biological reactions mediating the physiological and toxic actions of manganese. Biochem. Pharmacol. 66, 1–13.

Roth, J.A., Singleton, S., Feng, J., Garrick, M., Paradkar, P.N., 2010. Parkin regulates metal transport via proteasomal degradation of the 1B isoforms of divalent metal transporter 1. J. Neurochem. 113, 454–464.

Salazar, J., Mena, N., Hunot, S., Prigent, A., Alvarez-Fischer, D., Arredondo, M., Duyckaerts, C., Sazdovitch, V., Zhao, L., Garrick, L.M., Nuñez, M.T., Garrick, M.D., Raisman-Vozari, R., Hirsch, E.C., 2008. Divalent metal transporter 1 (DMT1) contributes to neurodegeneration in animal models of Parkinson's disease. Proc. Natl. Acad. Sci. U.S.A 105, 18578–18583.

Shah, Y.M., Matsubara, T., Ito, S., Yim, S.H., Gonzalez, F.J., 2009. Intestinal hypoxia-inducible transcription factors are essential for iron absorption following iron deficiency. Cell Metab. 9, 152–164.

Sharp, P., Tandy, S., Yamaji, S., Tennant, J., Williams, M., Srai, S., 2002. Rapid regulation of divalent metal transporter (DMT1) protein but not mRNA expression by non-haem iron in human intestinal Caco-2 cells. FEBS Lett. 510, 71–76.

Shawki, A., Anthony, S.R., Nose, Y., Engevik, M.A., Niespodzany, E.J., Barrientos, T., Ohrvik, H., Worrell, R.T., Thiele, D.J., Mackenzie, B., 2015. Intestinal DMT1 is critical for iron absorption in the mouse but is not required for the absorption of copper or manganese. Am. J. Physiol. Gastrointest. Liver Physiol. 309, G635–G647.

Shawki, A., Engevik, M.A., Kim, R.S., Knight, P.B., Baik, R.A., Anthony,, S.R., Worrell, R.T., Shull, B., 2016. Intestinal brush-border Na$^+$/H$^+$ exchanger-3 drives H$^+$-coupled iron absorption in the mouse. Am. J. Physiol. Gastrointest. Liver Physiol.

Shawki, A., Knight, P.B., Maliken, B.D., Niespodzany, E.J., Mackenzie, B., 2012. H(+)-coupled divalent metal-ion transporter-1: functional properties, physiological roles and therapeutics. Curr. Top. Membr. 70, 169–214.

Sheftel, A.D., Zhang, A.S., Brown, C., Shirihai, O.S., Ponka, P., 2007. Direct interorganellar transfer of iron from endosome to mitochondrion. Blood 110, 125–132.

Simovich, M., Hainsworth, L.N., Fields, P.A., Umbreit, J.N., Conrad, M.E., 2003. Localization of the iron transport proteins mobilferrin and DMT-1 in the duodenum: the surprising role of mucin. Am. J. Hematol. 74, 32–45.

Singh, A., Haldar, S., Horback, K., Tom, C., Zhou, L., Meyerson, H., Singh, N., 2013. Prion protein regulates iron transport by functioning as a ferrireductase. J. Alzheimers Dis. 35, 541–552.

Singh, N., Asthana, A., Baksi, S., Desai, V., Haldar, S., Hari, S., Tripathi, A.K., 2015. The prion-ZIP connection: from cousins to partners in iron uptake. Prion 9, 420–428.

Tabuchi, M., Tanaka, N., Nishida-Kitayama, J.O., Kishi, F., 2002. Alternative splicing regulates the subcellular localization of divalent metal transporter 1 isoforms. Mol. Biol. Cell 13, 4371–4387.

Thevenod, F., Wolff, N.A., 2015. Iron transport in the kidney: implications for physiology and cadmium nephrotoxicity. Metallomics 8, 17–42.

Thompson, K., Molina, R., Donaghey, T., Schwob, J., Brain, J., Wessling-Resnick, M., 2007. Olfactory uptake of manganese requires DMT1 and is enhanced by anemia. FASEB J. 21, 223–230.

Thomson, A.B., Valberg, L.S., Sinclair, D.G., 1971. Competitive nature of the intestinal transport mechanism for cobalt and iron in the rat. J. Clin. Invest. 50, 2384–2394.

Tripathi, A.K., Haldar, S., Qian, J., Beserra, A., Suda, S., Singh, A., Hopfer, U., Chen, S.G., Garrick, M.D., Turner, J.R., Knutson, M.D., Singh, N., 2015. Prion protein functions as a ferrireductase partner for ZIP14 and DMT1. Free Radic. Biol. Med. 84, 322–330.

Ueki, T., Furuno, N., Michibata, H., 2011. A novel vanadium transporter of the Nramp family expressed at the vacuole of vanadium-accumulating cells of the ascidian *Ascidia sydneiensis samea*. Biochim. Biophys. Acta 1810, 457–464.

Vargas, J.D., Herpers, B., McKie, A.T., Gledhill, S., McDonnell, J., van den Heuvel, M., Davies, K.E., Ponting, C.P., 2003. Stromal cell-derived receptor 2 and cytochrome b561 are functional ferric reductases. Biochim. Biophys. Acta 1651, 116–123.

Veuthey, T., Wessling-Resnick, M., 2014. Pathophysiology of the Belgrade rat. Front. Pharmacol. 5, 82.

Wang, K., Zhou, B., Kuo, Y.M., Zemansky, J., Gitschier, J., 2002. A novel member of a zinc transporter family is defective in acrodermatitis enteropathica. Am. J. Hum. Genet. 71, 66–73.

Wang, X., Garrick, M.D., Yang, F., Dailey, L.A., Piantadosi, C.A., Ghio, A.J., 2005. TNF, IFN-gamma, and endotoxin increase expression of DMT1 in bronchial epithelial cells. Am. J. Physiol. Lung Cell Mol. Physiol. 289, L24–L33.

Wolff, N.A., Garrick, L.M., Zhao, L., Garrick, M.D., Thévenod, F., 2014. Mitochondria represent another locale for the divalent metal transporter 1 (DMT1). Channels 8, 458–466.

Wolff, N.A., Ghio, A.J., Garrick, L.M., Garrick, M.D., Zhao, L., Fenton, R.A., Thévenod, F., 2014. Evidence for mitochondrial localization of divalent metal transporter 1 (DMT1). FASEB J. 28, 2134–2145.

Xie, L., Collins, J.F., 2011. Transcriptional regulation of the Menkes copper ATPase (Atp7a) gene by hypoxia-inducible factor (HIF2{alpha}) in intestinal epithelial cells. Am. J. Physiol. Cell Physiol. 300, C1298–C1305.

Yanatori, I., Yasui, Y., Tabuchi, M., Kishi, F., 2014. Chaperone protein involved in transmembrane transport of iron. Biochem. J. 462, 25–37.

Part V

Zinc

Chapter 20

Discovery of Zinc for Human Health and Biomarkers of Zinc Deficiency

Ananda S. Prasad

Wayne State University School of Medicine, Detroit, MI, United States

INTRODUCTION

In 1869, Raulin reported for the first time that zinc was essential for the growth of microorganisms. Many years later in 1926, zinc was recognized as a growth factor for plants. In 1934, zinc was shown to be essential for the growth of rats (Todd et al., 1933). In 1958, zinc was shown to be essential for the growth of poultry. However, until 1961, zinc was not considered to be essential for human health and most scientists considered it improbable that zinc deficiency in humans would lead to a significant clinical problem. This dogma has now changed. Our studies in 1963 established, for the first time, that zinc was essential for human health and that its deficiency occurred in the Middle East (Prasad et al., 1963). The current estimate of the World Health Organization (WHO) is that nearly 2 billion subjects in the developing world may be affected by zinc deficiency. Populations living in villages of the developing world subsist on a cereal protein diet high in phytate, an organic phosphate compound which renders zinc unavailable for absorption. Acquired deficiency of zinc has now been reported in many diseases, such as cirrhosis of the liver, chronic renal disease, malabsorption syndrome, chronic alcoholism, sickle cell disease, and other chronic diseases including malignancies (Prasad, 1993).

The major clinical effects of zinc deficiency in humans include severe growth retardation, hypogonadism, cell-mediated immune dysfunction, increased oxidative stress, upregulation of inflammatory cytokine production, problems with healing, and impaired cognitive function. In the early 1960s, we knew of only three enzymes, carbonic anhydrase, alcohol dehydrogenase, and carboxypeptidase, which required zinc for their activities (Prasad, 1993, 2013). Today, we know of over 300 enzymes that are zinc dependent. At present, we know that over 2000 transcription factors required for gene expression of various proteins need zinc for maintenance of their structures and for their binding to DNA. Moreover, zinc is now known to be a second messenger for immune cells. Zinc is essential for cell-mediated immunity, and zinc is both an antioxidant and antiinflammatory agent (Prasad, 2013). The intracellular zinc level is tightly controlled and for its homeostasis, we now know that there are 14 ZIP and 10 ZNT transporters (Prasad, 2013). In this chapter, I will present a brief history of how zinc was discovered as an essential element for humans, the impact of this discovery on human health, and the biomarkers of zinc deficiency.

DISCOVERY OF ZINC AS AN ESSENTIAL ELEMENT FOR HUMAN HEALTH

I was trained as a clinical-scientist at the University of Minnesota, Department of Medicine under Dr. C.J. Watson. The clinical-scientist program was started in a few medical schools after World War II. The purpose was to train physicians not only in clinical medicine but also in basic sciences so that the clinical-scientist could investigate clinical problems in research laboratories to understand the basic mechanisms involved in clinical disorders. This type of training program truly advanced our knowledge rapidly, and the United States trained many leaders in medicine around the world.

Following my training under Dr. Watson, I was contacted by Prof. H.A. Reimann, Chief of Medicine at the Jefferson Medical School in Philadelphia. Prof. Reimann, who was a personal friend and physician of the Shah of Iran, had accepted a position as Chief of Medicine at the University of Shiraz Medical School in Shiraz, Iran. Professor Reimann wanted me to join him in Shiraz and help him set up a medical curriculum at the Shiraz Medical School patterned after an American medical school. Initially I was reluctant to make a move from Minneapolis to Shiraz, but Prof. Reimann was very persuasive and I accepted his offer.

The story of zinc began when an Iranian physician presented to me at the medical center grand rounds, a 21-year-old male, who looked like a 10-year-old boy and was severely anemic. His genitalia were infantile. He had rough and dry skin, mental lethargy, hepatosplenomegaly, and geophagia. He ate only bread (made of whole wheat flour) and he had no intake of animal protein. He consumed 0.5 kg of clay daily. He was severely iron deficient but had no blood loss. Later, I discovered that this syndrome was common in the villages of Shiraz, Iran (Prasad et al., 1961). Iron deficiency alone could not account for all the features we observed in this case, inasmuch as growth retardation and testicular atrophy are not seen in iron-deficient, experimental animals. An examination of the periodic table suggested to me that deficiency of another transitional element, perhaps zinc, may have been also present, which could account for growth retardation and hypogonadism. We considered the possibility that the high phosphate content of the diet and geophagia may have decreased the bioavailability of both iron and zinc, which resulted in deficiency of both elements (Prasad et al., 1961). Our later studies in Egypt documented conclusively that zinc deficiency occurred in humans and that zinc supplementation resulted in 5–6 inches of growth in 1 year and that genitalia became normal within 3–6 months of zinc supplementation (Prasad et al., 1963; Sandstead et al., 1967). For nearly one decade, the possibility that zinc deficiency occurred in humans remained very controversial. Several reports, however, supported our idea and in 1974, the National Research Council of the National Academy of Sciences declared zinc as an essential element for humans and established a recommended dietary allowance (RDA) for zinc (Prasad, 1993, 2013). In 1978, the US Food and Drug Administration (FDA) made it mandatory to include zinc in total parenteral nutrition fluids (Prasad, 1993, 2013). The details of circumstances leading to the discovery of human zinc deficiency in the Middle East have been published in Sandstead (2012).

Severe Zinc Deficiency

Acrodermatitis Enteropathica

In 1973, Barnes and Moynahan reported a 2-year-old girl with severe acrodermatitis enteropathica (AE) who was being treated with diiodohydroxy quinolone and a lactose-deficient, synthetic diet, but she was not showing any response to this therapy. The serum zinc concentration was significantly decreased. They administered oral zinc sulfate to correct this deficiency. Surprisingly, the skin lesions and gastrointestinal symptoms cleared up after this therapy. When zinc was inadvertently omitted from the child's regimen, the child suffered a relapse; however, she again completely responded to oral zinc therapy. The authors then realized that zinc might have been fundamental to the pathogenesis of this rare inherited disorder and that the clinical improvement reflected correction of zinc status in the patient. This original observation was quickly confirmed in other patients with AE throughout the world. The underlying pathogenesis of the zinc deficiency in these patients is due to malabsorption of zinc caused by a mutation in ZIP4, an intestinal zinc transporter (Wang et al., 2002).

AE is a lethal, autosomal, recessive trait which usually occurs in infants of Italian, American, or Iranian lineage (Prasad, 1993). The disease develops in the early months of life soon after weaning from breast feeding. The dermatologic manifestations of severe zinc deficiency in patients with AE include bullous pustular dermatitis of the extremities and the oral, anal, and genital areas around the orifices, paronychia, and alopecia. Ophthalmic signs include blepharitis, conjunctivitis, photophobia, and corneal opacities. Neuropsychiatric signs include irritability, emotional instability, tremors, and occasional cerebellar ataxia. Weight loss, growth retardation, and male hypogonadism are also prominent clinical features. Congenital malformation of fetuses and infants born to pregnant women with AE has been observed commonly (Prasad, 1993, 2013).

AE patients have an increased susceptibility to infections. Thymic hypoplasia, absence of germinal centers in lymph nodes and plasmacytosis in the spleen are seen consistently. All T-cell-mediated functional abnormalities are completely corrected with zinc supplementation. Clinical course is downhill with failure to thrive and complicated by intercurrent bacterial, fungal, viral, and other opportunistic infections. Gastrointestinal disturbances are severe including diarrhea, malabsorption, steatorrhea, and lactose intolerance. The disease, if unrecognized and untreated, is fatal. Zinc supplementation results in complete recovery.

The AE gene has been localized to a ~3.5 cm region on 8q24 chromosome. The gene encodes a histidine-rich protein, which is now referred to as ZIP-4, which is a member of a large family of transmembrane proteins, known as zinc transporters. In patients with AE, mutations in this gene have been demonstrated (Wang et al., 2002).

Total Parenteral Nutrition

Kay and Tasman-Jones in 1975 reported the occurrence of severe zinc deficiency in subjects receiving total parenteral nutrition (TPN) for prolonged periods without zinc. Okada et al. (1976) also reported similar results without zinc. These observations were documented by several investigators and indeed, in the United States, zinc is now being routinely included in TPN fluids for subjects who are likely to receive such therapy for extended periods.

Penicillamine Therapy

A severe deficiency of zinc has also been observed in patients with Wilson's disease who received penicillamine therapy as a decoppering agent. This treatment may induce excessive zinc loss and result in severe deficiency of zinc (Klingberg et al., 1976).

In summary, the manifestations of severe zinc deficiency in humans include bullous pustular dermatitis, alopecia, diarrhea, emotional disorder, weight loss, intercurrent infections due to cell-mediated immune dysfunctions, hypogonadism in males, neurosensory disorders, and problems with healing of ulcers. Severe deficiency of zinc, if untreated, is fatal.

Moderate Deficiency of Zinc

The manifestations of a moderate deficiency of zinc include growth retardation, male hypogonadism in adolescents, rough skin, poor appetite, mental lethargy, delayed wound healing, cell-mediated immune dysfunctions, and abnormal neurosensory changes. These manifestations have been reported in subjects with nutritional deficiency of zinc (Prasad et al., 1961, 1963; Prasad, 1993, 2013) and in subjects with acquired deficiency of zinc. It is now apparent that a nutritional deficiency of zinc in humans is prevalent throughout the world, particularly in areas where cereal proteins are primary in local diets. In Turkey, geophagia is also common and the majority of adolescents in the villages in Turkey with geophagia exhibit both iron and zinc deficiencies (Prasad, 1993, 2013; Cavdar et al., 1980). Cavdar et al. (1980) observed decreased plasma zinc levels in almost 30% of low socioeconomic status pregnant women in Turkey. Their diet consisted of mainly cereals. Maternal zinc deficiency was associated with severe congenital malformation of the central nervous system in the fetuses and maternal morbidity was increased.

Mild Deficiency of Zinc

Although the clinical, biochemical, and diagnostic aspects of severe and moderate levels of zinc deficiency in humans were well defined, the recognition of mild deficiency of zinc remains a difficult problem. We therefore developed an experimental model of zinc deficiency in humans to define mild deficiency of zinc. In a group of human volunteers, we induced a mild deficiency of zinc by dietary means. Adult male volunteers were kept in the clinical Research Center of the University of Michigan Medical School Hospital, Ann Arbor, Michigan. A semipurified diet which supplied approximately 3.0–5.0 mg of zinc daily was used to induce zinc deficiency (Prasad et al., 1978b, 1988; Beck et al., 1997a,b). The volunteers were given a hospital diet containing adequate animal protein daily for 4 weeks. This diet supplied approximated 12 mg of zinc daily consistent with the RDA. Following this, they received 3.0–5.0 mg of zinc daily while consuming a soy-protein based experimental diet. This regime was continued for 28 weeks. Following this, the volunteers received two cookies containing 27 mg of supplemental zinc. This supplementation was continued for 12 weeks. Throughout this study the level of all nutrients including protein, amino acids, vitamins, and minerals (both micro and macro elements) were kept constant, meeting RDA, except for zinc. By this technique, we were able to induce a specific mild deficiency of zinc in human volunteers. As a result of mild zinc deficiency, we observed decreased serum testosterone level, oligospermia, decreased natural killer (NK) cell lytic activity, decreased interleukin-2 (IL-2) activity of T helper cells, decreased serum thymulin activity, hyperammonemia, hypogeusia, decreased dark adaptation, and decreased lean body mass (Prasad et al., 1978b, 1988; Beck et al., 1997a,b). This study clearly established that even a mild deficiency of zinc in humans affects clinical, biochemical, and immunological functions adversely.

Zinc and Immune Cells

Zinc is a second messenger for immune cells, and intracellular zinc participates in signaling events (Hirano et al., 2008; Kitamura et al., 2006; Haase and Rink, 2007; Rosenkranz et al., 2011). Hirano et al. (Hirano et al., 2008; Kitamura et al., 2006; Haase and Rink, 2007; Rosenkranz et al., 2011) have shown that a decrease in intracellular free zinc is critical for lipopolysaccharide (LPS)-mediated CD^{4+} T-cell activation by dendritic cells (DCs). LPS binds to TLR4 on DCs and initiates Myd88 and TRIF-mediated signaling (domain containing adapter-inducing interferon-β) (Hirano et al., 2008; Kitamura et al., 2006; Haase and Rink, 2007; Rosenkranz et al., 2011). TRIF-mediated signaling increases ZNT-5 mRNA and decreases ZIP-6 mRNA, thus resulting in a decrease in the intracellular free zinc in DCs. Reduction in intracellular free zinc increases surface expression of major histocompatibility complex (MHC) Class II molecules, which is important for the activation of CD^{4+} T-cells (Hirano et al., 2008; Kitamura et al., 2006). Moreover, zinc affects the activity of monocytes and macrophages in several ways. Zinc is required in monocyte/macrophage development (Shankar and Prasad, 1998; Prasad et al., 2004, 2011; Bao et al., 2010, 2011) and it regulates various functions such as phagocytosis and proinflammatory cytokine production. LPS stimulation of zinc-sufficient monocytes results in downregulation of inflammatory cytokines such as tumor necrosis factor (TNF)-α, IL-1β, IL-6, and IL-8 (Shankar and Prasad, 1998; Prasad et al., 2004, 2011; Bao et al., 2010, 2011). Zinc inhibits the cell membrane phosphodiesterase, leading to elevated levels of the second messenger cGMP, which is followed by a subsequent suppression of the NF-κB-dependent mRNAs of TNF-α, IL-1β, and other inflammatory cytokines (Shankar and Prasad, 1998; Prasad et al., 2004, 2011; Bao et al., 2010, 2011). Additionally, zinc induces A-20 which inhibits NF-κB signaling via TNF receptor-associated pathways, resulting in downregulation of mRNAs encoding inflammatory cytokines (Prasad et al., 2004, 2011; Bao et al., 2010). Based on these findings, we propose that zinc is an important antiinflammatory agent.

Zinc deficiency also affects Th1 functions adversely in humans (Beck et al., 1997a,b; Prasad et al., 1988). Serum thymulin activity and generation of the Th1 cytokines, IL-2 and IFN-γ, were affected within 8–12 weeks of institution of zinc-restricted diet (3–5 mg daily) in humans, whereas plasma zinc decreased after 20–24 weeks of the institution of the experimental diet. This suggests that Th1 cells are very sensitive to zinc restriction. Th2 cytokines were not affected by zinc deficiency.

In Th0, a human malignant lymphoblastoid cell line, HUT-78 cells, we showed that in zinc-sufficient cells mRNA levels of IFN-γ, IL-12, Rβ2, and T-bet in phorbol myristate acetate (PMA)/phytohemagglutinin-p (PHA)-stimulated cells were increased in comparison to zinc-deficient cells (Bao et al., 2011). Although intracellular free zinc increased only slightly in PMA/PHA-stimulated cells, in Con A–stimulated cells in a zinc-sufficient medium, there was an increased sustained level of intracellular free zinc in comparison to the zinc-deficient cells (Bao et al., 2011). We concluded that stimulation of cells by Con A via T-cell receptor (TCR), there was a release of intracellular free zinc which functioned as a signal transduction molecule for generation of IFN-γ and T-bet and IL-12 Rβ2 mRNAs required for Th1 cell differentiation (Bao et al., 2011).

THERAPEUTIC IMPACT OF ZINC

Zinc and Infectious Diseases

Acute Diarrhea in Children

Supplementation with zinc has been shown to prevent and treat diarrhea among children under 5 years of age, decreasing both diarrhea and mortality (Sazawal et al., 1995; Fisher Walker et al., 2011). Zinc deficiency is also correlated with risk of respiratory tract infections, but the benefit of supplementation appears to be limited to more severe episodes and in populations with a high incidence of zinc deficiency (Fisher Walker et al., 2011). Diarrhea causes breakdown of the absorptive mucosa resulting in poor absorption of nutrients, including zinc. Prior studies linked diarrheal illness to the loss of endogenous zinc (Fisher Walker et al., 2011). Children with low plasma zinc were observed to be more susceptible to diarrhea, propagating a cycle of deficiency and infection. There is extensive evidence supporting the efficacy of zinc supplementation for the prevention of childhood diarrhea (Fisher Walker et al., 2011). In 2004, the WHO issued a global recommendation for the daily supplementation with 20 mg zinc in children ≥6 months and 10 mg of zinc in infants under 6 months for 10–14 days upon diarrheal onset.

Meta-analysis of routine supplementation for up to 3 months in seven studies providing one to two times the RDA for elemental zinc five to seven times per week found an 18% reduction in diarrheal incidence, a 25% decrease in diarrhea prevalence, and a 33% reduction in persistent diarrhea episodes among supplemented children compared to children who received placebo (Fisher Walker et al., 2011). A meta-analysis of three randomized controlled trials providing short course zinc supplementation with two to four times the daily RDA for 2 weeks following the onset of an episode of acute or persistent diarrhea was reported. The pooled analysis showed an 11% decrease in diarrhea incidence and a 34% decrease in diarrhea prevalence during a 3-month observation period.

Zinc for the Treatment of the Common Cold

The common cold is one of the most frequently occurring diseases in the world (Prasad et al., 2000, 2008). More than 20 viruses cause the common cold and these include rhinoviruses, corona viruses, adenoviruses, respiratory syncytial virus, and parainfluenza viruses. Annually, adults in the United States may suffer two to four times with the common cold and children may develop colds six to eight times in a year. The morbidity and subsequent financial loss resulting from absenteeism from work is substantial. Previously prescribed treatments have not provided a consistent relief of symptoms. We tested the efficacy of zinc acetate lozenges in the common cold in 50 volunteers who were recruited within 24 h of developing symptoms of the common cold and we carried out a randomized, double-blind, placebo-controlled trial (Prasad et al., 2000). Participants took one lozenge containing 12.8 mg zinc (as zinc acetate) or placebo every 2–3 h while awake as soon as they developed common cold symptoms. Subjective symptom scores for sore throat, nasal discharge, nasal congestion, sneezing, cough, scratchy throat, hoarseness, muscle ache, fever, and headache were recorded daily for 12 days. Plasma zinc and proinflammatory cytokines were assayed on day 1 and after participants were well. Twenty-five in the zinc group and 23 in the placebo group completed the study. Compared to the placebo group, the zinc group had shorter overall duration of cold symptoms (4.5 vs. 8.1 days; $p < .01$), cough (3.1 vs. 6.3 days; $p = .01$), and nasal discharge (4.1 vs. 5.8 days; $p = .02$), and decreased total severity scores for all symptoms ($p < .002$).

In another study, we recruited 50 ambulatory volunteers within 24 h of developing common cold symptoms for a randomized, double-blind, placebo-controlled trial of zinc (Prasad et al., 2008). Participants took one lozenge containing 13.3 mg of zinc (as zinc acetate) or placebo every 2–3 h while awake. The subjective scores of clinical symptoms were recorded daily. Plasma zinc, soluble interleukin (IL)-1 receptor antagonist (sIL-1ra), soluble tumor neurosis factor receptor 1, and soluble vascular endothelial cell adhesion molecule (sICAM-1) were assayed on days 1 and 5 (Prasad et al., 2008). Compared with the placebo group, the zinc group had a shorter mean overall duration of cold (4.0 vs. 7.1 days; $p = .001$), shorter duration of cough (2.1 vs. 5.0 days; $p < .001$), and nasal discharge (3.0 vs. 4.5 days; $p = .02$). Symptom severity scores were also significantly decreased in the zinc group ($p = .002$). The mean changes between zinc and placebo groups (before vs. after therapy) showed significant differences in sIL-1ra ($p = .033$) and sICAM-1 levels ($p = .04$). Both decreased in the zinc-treated group and the mean changes between zinc and placebo group (before vs. after therapy) showed a significant difference ($p < .001$).

Our results suggest that common cold viruses increase oxidative stress which activates macrophages and monocytes, and that zinc decreased activation of monocytes and macrophages by decreasing oxidative stress. We have previously shown that zinc functions as an antioxidant (Prasad et al., 2004; Bao et al., 2011).

Human rhinovirus type 24 "docks" with ICAM-1 on the surface of somatic cells (Prasad et al., 2000, 2008). Our results showed that zinc may act as an antiviral agent by reducing ICAM-1 levels. We have reported that zinc functions as a downregulator of NF-κB activity, which is involved in the gene expression of adhesion molecules such as ICAM-1 (Prasad et al., 2008).

We conclude that zinc acetate lozenges given within 24 h of the onset of common cold in proper dosages are very effective in decreasing the duration and severity of common cold. We propose that the beneficial effects seen in the zinc group were due to the antioxidant and antiinflammatory effects of zinc. We also suggest that a decrease in plasma ICAM-1 levels due to zinc therapy may have decreased the docking of the cold viruses on the surface of somatic cells.

A meta-analysis selected randomized, double-blind, placebo-controlled trials using zinc for at least 5 consecutive days to treat, or at least 5 months to prevent the common cold were included for analysis (Singh and Das, 2011). Thirteen therapeutic trials (966 participants) and two preventive trials (394 participants) were included for analysis. Studies reported that zinc significantly reduced the overall duration and severity of common cold symptoms if the therapy was started within 24 h of the onset of the cold.

It is critical that the solution chemistry of the zinc preparation and the dose must be proper. Zinc therapy must begin within 24 h of the onset of cold symptoms. The total daily dose of elemental zinc should be greater than 75 mg. The chemical formulation should be optimal so that zinc is ionized in the oral cavity at pH 7.4. Zinc acetate and zinc gluconate are good salts to use, however, if citric acid, glycine, tartarate, or other binders are present, zinc is prevented from ionization. Physicians and health practitioners must realize that one cannot treat common cold symptoms by swallowing zinc tablets, zinc syrup, or zinc lozenges. Zinc lozenges must be used orally and allowed to dissolve slowly in the mouth which will then allow ionic zinc to be released, absorbed, and transported to the virally infected nose.

Zinc Deficiency in Sickle Cell Disease

Our studies have documented the occurrence of zinc deficiency in adult sickle cell disease (SCD) patients (Prasad et al., 1999; Bao et al., 2008). Growth retardation, hypogonadism in males, hyperammonemia, abnormal dark adaptation, and cell-mediated immune dysfunction in SCD patients have been related to a deficiency of zinc. The biochemical evidences of zinc deficiency in SCD patients were: decreased levels of zinc in the plasma, erythrocytes, and hair; hyperzincuria; decreased activities of certain zinc-dependent enzymes, such as carbonic anhydrase in erythrocytes, alkaline phosphatase in the neutrophils, deoxythymidine kinase activity in newly synthesizing skin connective tissue and collagen; and hyperammonemia (Prasad et al., 1999; Bao et al., 2008). Inasmuch as zinc is known to be an inhibitor of ribonuclease (RNase), an increased activity of this enzyme in plasma was considered to be also an evidence of zinc deficiency. Zinc supplementation to SCD patients resulted in significant improvement in secondary sexual characteristics, normalization of plasma ammonia level, and correction of dark adaptation abnormality. Zinc supplementation also increased zinc levels in plasma, erythrocytes, and neutrophils. Expected response to zinc supplementation on enzyme activities was also observed. Increased longitudinal growth and body-weight in 14- to 18-year-old SCD patients were observed. Zinc supplementation also corrected impaired delayed type hypersensitivity and decreased NK cell lytic activity in SCD patients (Prasad et al., 1999; Bao et al., 2008).

A 3-month placebo-controlled zinc supplementation trial (25 mg zinc as zinc acetate three times a day) in 36 SCD patients showed that zinc-supplemented subjects had decreased incidences of infections, increased hemoglobin and hematocrit levels, increased plasma zinc and antioxidant power in comparison to the placebo group (Bao et al., 2008). Plasma nitrite and nitrate (NOx), lipid peroxidation products, DNA oxidation products, and soluble vascular cell adhesion molecule-1 (VCAM-1) decreased in the zinc-supplemented group in comparison to the placebo group. Zinc-supplemented subjects showed significant decreases in LPS-induced TNF-α, IL-1β mRNAs, and TNF-induced nuclear factor of κB-DNA binding in MNCs compared to the placebo group (Bao et al., 2008). Zinc supplementation also increased relative levels of IL-2 and IL-2Rα mRNAs in PHA-p stimulated MNCs (Bao et al., 2008). Moreover, a Cochrane Review (Swe et al., 2013) has concluded that zinc therapy is the only modality which is effective in decreasing incidence of infections and pain crises in SCD patients.

Zinc Therapy for Wilson's Disease

Wilson's disease is an inherited autosomal disorder of copper accumulation. The excretion of liver copper in the bile is decreased. This leads to failure of proper copper excretion in the stool and leads to accumulation of copper in the liver. Eventually, not only the liver but also the brain and other organs are damaged due to excess copper accumulation. Patients typically present with liver disease, neurological disease (movement disorder), or psychiatric disturbances in the second to fourth decades of life. In many cases, the diagnosis is either missed or delayed (Brewe and Yuzbasiyan-Gurkan, 1992; Brewer, 1995; Brewer et al., 1977). The gene for Wilson's disease has been now identified. The genetic mutation leads to defective production of a protein called ATP7B which is responsible for key step in biliary excretion of copper (Brewe and Yuzbasiyan-Gurkan, 1992; Brewer, 1995; Brewer et al., 1977). The disease is recessive, thus both copies of the ATP7B gene have to be mutated to cause a failure in biliary excretion of copper and produce the disease. A large number of mutations in this gene causing Wilson's disease have been identified.

Early diagnosis of Wilson's disease is important inasmuch as effective therapeutic measures may prevent accumulation of copper and serious damage to organs such as the liver and brain. Ninety percent of Wilson's disease patients have low levels of ceruloplasmin and ceruloplasmin-bound copper and nonceruloplasmin-bound copper is elevated in the plasma. Measurement of the 24 h urinary copper is a good diagnostic test; it is consistently elevated in these patients (Brewe and Yuzbasiyan-Gurkan, 1992; Brewer, 1995; Brewer et al., 1977). Urinary copper, however, may be elevated in patients with obstructive liver disease also who do not have Wilson's disease. A slit lamp examination for corneal copper deposits (Kayser–Fleischer rings) is a very useful noninvasive diagnostic test for Wilson's disease. This, however, is positive in only 50% of the cases. The initial treatment objective is to decrease copper burden. It is also desirable to prevent copper from shifting from one pool to the other while decoppering is being done. Initial copper control treatment may take 2–4 months (Brewe and Yuzbasiyan-Gurkan, 1992; Brewer, 1995; Brewer et al., 1977).

Several years ago, we were using 150 mg elemental zinc in six divided doses for the treatment of SCD patients (Prasad et al., 1978a). We observed that zinc was an effective antisickling drug. We observed, however, that at this level of zinc therapy, deficiency of copper was induced in our patients. This led Brewer et al. (Brewe and Yuzbasiyan-Gurkan, 1992; Brewer, 1995; Brewer et al., 1977) to develop zinc as an effective anticopper drug for Wilson's disease. Zinc competes with copper for similar binding sites and oral zinc efficiently decreases uptake of copper (Hall et al., 1979). Zinc may act by induction of intestinal cell metallothionein (MT). MT, once induced, has a high affinity for binding copper and prevents the serosal transfer of copper into the blood. The intestinal cells turn over rapidly and take the complexed copper into the stool for final excretion. Zinc not only blocks food copper but also the copper which is endogenously excreted via salivary, gastric, and other gastrointestinal juices. Thus, zinc is effective in producing a negative copper balance. Fifty milligram elemental zinc (as acetate) is given orally three times a day for management of Wilson's disease patients. Zinc is given in a fasting or postabsorptive state. The only side effect is that 10% of the subjects may have gastric discomfort. This is usually observed after the first morning dose and this can be avoided if zinc is administered between breakfast and lunch or after dinner before going to bed. Zinc is the drug of choice for maintenance therapy (Brewe and Yuzbasiyan-Gurkan, 1992; Brewer, 1995; Brewer et al., 1977). Zinc has no toxicity and is nonteratogenic, thus it can be given to subjects of all ages and even to pregnant women. Zinc has been approved by the FDA for the treatment of Wilson's disease patients.

Zinc and Age-Related Macular Degeneration

Age-related macular degeneration (AMD) affects nearly 25% of the subjects over 65 years of age and the late-stage disease accounts for nearly 50% of legal blindness in Europe and North America (Newsome et al., 1996). Newsome et al. (1996) demonstrated that concentrations of zinc are reduced in human eyes with signs of AMD and they suggested that zinc deficiency may have led to oxidative stress and retinal damage. The Age-Related Eye Disease Study Group (AREDS) supported by National Eye Institute, at the National Institutes of Health (NIH), conducted an 11-center double-masked clinical trial in patients with dry-type AMD (Age-Related Eye Disease Study Research group (AREDS Report No. 8), 2001). A total of 3640 participants were enrolled. Their ages ranged from 55 to 80 years and the average follow-up period was 6.3 years. Participants were randomly assigned to receive daily orally one of the following: (1) antioxidants (vitamin C 500 mg, Vitamin E 400 IU, and beta carotene 15 mg); (2) zinc, 80 mg as zinc oxide, and copper, 2 mg as copper oxide, to prevent copper deficiency induced by zinc; (3) antioxidants plus zinc; or (4) placebo. The group taking the antioxidant plus zinc reduced the risk of developing advanced AMD by about 25% and vision loss by about 19%. The group taking zinc alone reduced the risk of developing advanced AMD by about 21% and the vision loss by 11%. In the group taking the vitamins alone, the risk of developing advanced AMD was decreased by 17% and the vision loss was decreased by 10%. No significant side effects were noted in the group who received high levels of therapeutic zinc (Age-Related Eye Disease Study Research group (AREDS Report No. 8), 2001). Only the zinc-supplemented group showed increased longevity (AREDS Report No.13, 2004; Age-Related Eye Disease Study Research Group (AREDS Report No. 35), 2013). The risk of mortality was reduced by 27% in AREDS studies in subjects who received therapeutic zinc daily. No other micronutrient is known to have a similar effect on mortality. In a later publication, the AREDS group observed that a decrease in mortality was due to a decrease in cardiovascular events (AREDS Report No.13, 2004; Age-Related Eye Disease Study Research Group (AREDS Report No. 35), 2013).

Zinc Supplementation in the Elderly

The daily intake of zinc in elderly subjects in the Western world including the United States is only around 8–10 mg, whereas the RDA is 15 mg. Elderly subjects frequently do not eat the usual three meals a day and they may skip either breakfast or lunch. Many live alone and do not cook proper meals for themselves. Our study in the Detroit area has shown that 35% of the well-to-do ambulatory elderly subjects may have a deficiency of zinc. Results of the Third National Health and Nutrition Examination Survey (1988–1994) also reported that elderly persons >71 years were at the greatest risk of inadequate zinc intakes (Bao et al., 2010; Prasad et al., 1993, 2007). Oxidative stress and increased inflammatory cytokines have been recognized as important contributing factors for several chronic diseases attributed to aging, such as atherosclerosis and related cardiovascular disorders, mutagenesis and cancer, neurodegenerative disorders, type 2 diabetes, and Alzheimer's disease. Together, O^{2-}, H_2O_2, and OH radicals are known as reactive oxygen species (ROS) and excessive generation of ROS causes oxidative stress. Inflammatory cytokines such as TNF-α and IL-1β, generated by activated monocytes, are also known to generate greater levels of ROS. In the elderly, chronic inflammatory processes have been implicated as causing high cardiovascular mortality (Bao et al., 2010).

We have shown that zinc supplementation in subjects ages 20 to 50 decreased oxidative stress markers, such as malondehyde (MDA), 4-hydroxyalkenals, and 8-hydroxydeoxyguanine in the plasma, and also downregulated the ex vivo induction of TNF-α and IL-1β mRNA in MNCs by decreasing TNF-α induced NF-κβ activation (Bao et al., 2010; Prasad et al., 2007). We have also shown previously that in the promyelocytic leukemia cell line HL-60, which differentiates to a monocyte and macrophage phenotype in response to phorbol-12-myristate-13-acetate PHA, zinc upregulated the expression of A20, and the binding of A20 transactivating factor to DNA, which resulted in the inhibition of NF-κB activation (Bao et al., 2010; Prasad et al., 1993, 2007).

Inasmuch as zinc deficiency and susceptibility to infections due to cell-mediated immune dysfunctions have been observed in the elderly, we carried out a randomized, placebo-controlled trial of zinc supplementation in 50 healthy elderly subjects (55–87 years) of both sexes and all ethnic groups from St. Patrick's senior citizen center, Detroit, MI. One subject in the zinc group dropped out on the second day, thus we had complete data on 49 subjects (24 in zinc and 25 in placebo groups). Exclusion criteria were as follows: life expectancy of <8 months; progressive neoplastic disease; severe cardiac dysfunction; significant kidney disease; significant liver disease; and subjects who were not competent mentally. Zinc supplementation consisted of 45 mg elemental zinc (as gluconate) daily for 12 months. A comparison of the baseline data between the younger subjects (ages 18–54, $n = 31$) and the elderly subjects showed that the plasma zinc was lower and the percentage of cells producing IL-1β and TNF-α were significantly higher in the elderly subjects (Prasad et al., 2007). Intercellular adhesion molecules, vascular endothelial cell adhesion molecules, and E-selectin in the plasma also were significantly higher in the elderly. IL-10 generated by Th2 cells, which is known to negatively regulate IL-2 generation from TH1 cells, was significantly higher in the elderly. Moreover, oxidative stress markers were significantly higher in the elderly compared to the younger adults (Bao et al., 2010; Prasad et al., 2007). The mean incidence of infections per subject was lower ($p < .01$) in the zinc-supplemented group (1.4±0.95) versus the placebo group (20.29±0.46). Plasma zinc increased, and ex vivo generation of TNF-α and IL-10 decreased significantly in the zinc group in comparison to the placebo group (Prasad et al., 2007). Oxidative stress biomarkers in the plasma also decreased significantly in the zinc group in comparison to the placebo group (Bao et al., 2010). In MNCs isolated from zinc-deficient elderly subjects, zinc supplementation increased the ex vivo PHA-induced IL-2 mRNA expression, and plasma zinc concentration in comparison to the zinc-deficient subjects who received placebo (Prasad et al., 2007). Thus, our study showed that zinc supplementation (45 mg elemental zinc daily) to elderly subjects decreased the incidence of infection by nearly 66%. Following supplementation, oxidative stress markers, and the generation of inflammatory cytokines, which were increased prior to supplementation, decreased significantly. These are highly significant effects of zinc supplementation in the elderly and it may imply that zinc may prove to be an excellent agent for the prevention of some of the chronic diseases of aging.

BIOMARKERS OF ZINC DEFICIENCY IN EGYPT

In Egypt, we selected 17 dwarfs ages 16–19 years from villages around Cairo, Egypt who exhibited all the clinical features of the syndrome, as we described earlier (Prasad et al., 1963). Plasma zinc was measured by the dithizone technique. Extreme precautions were taken to avoid contamination. We assayed zinc in plasma, red blood cells, 24-h urine, and hair, and these were significantly decreased in comparison to the Egyptian controls of similar ages (Prasad et al., 1963). We utilized Zn^{65} to study zinc metabolism in these dwarfs. Plasma Zn^{65} disappearance curve was resolved into five phases; Phase I beginning with zero time and extending to 30 min; Phase II extending up to 60 min; Phase III extending up to 10 h; Phase IV up to 7 days; and Phase V extending beyond 7 days (Prasad et al., 1963; see Table 20.1). In the second and third phases, T ½ was shorter in the dwarfs as compared to the normal subjects and the plasma zinc turnover rate was greater in dwarfs in comparison to the normal subjects in the second phase. Twenty four-hour exchangeable pool was decreased in the dwarfs. The cumulative excretion of zinc in urine and stool at 13 days was also decreased in dwarfs, indicating body conservation of zinc in the zinc-deficient state. We concluded from these results that the dwarfs were zinc deficient. This was the first

TABLE 20.1 Zn^{65} Studies in Egypt

Time	Percent in Plasma			$T_{1/2}$	
	Normal	Dwarf	Comparison of Mean Values (p value)	Normal	Dwarf
30 min	1.24 ± 0.08	0.96 ± 0.14	<.01		
40 min	1.02 ± 0.08	0.72 ± 0.16	<.01	42 min	29 min
50 min	0.86 ± 0.08	0.58 ± 0.16	<.01		
60 min	0.76 ± 0.08	0.50 ± 0.14	<.01		
2 h	0.54 ± 0.08	0.40 ± 0.14	<.01		
4 h	0.42 ± 0.10	0.28 ± 0.12	<.01		
6 h	0.36 ± 0.12	0.22 ± 0.10	<.01	11.7 h	7.6 h
8 h	0.30 ± 0.08	0.20 ± 0.10	<.01		
10 h	0.30 ± 0.08	0.16 ± 0.10	<.01		
1 day	0.24 ± 0.04	0.12 ± 0.04	<.01		
3 days	0.16 ± 0.04	0.10 ± 0.04	<.01		
5 days	0.12 ± 0.02	0.08 ± 0.02	<.01	4.9 days	5.9 days
7 days	0.10 ± 0.02	0.06 ± 0.02	<.02		
8 days	0.08 ± 0.01	0.06 ± 0.02	=0.10		
10 days	0.08 ± 0.02	0.06 ± 0.02	=0.20		
12 days	0.08 ± 0.02	0.04 ± 0.02	=0.01		
14 days	0.06 ± 0.01	0.04 ± 0.02	=0.02		

Zn^{65} plasma concentration following intravenous administration. Values are expressed in percent of dose remaining in total plasma.
Reproduced from Prasad, A.S., Miale, A., Farid, Z., Schulert, A., Sandstead, H.H., 1963. Zinc metabolism in patients with the syndrome of iron deficiency anemia, hypogonadism and dwarfism. J. Lab. Clin. Med. 61, 537–549.

demonstration that zinc deficiency occurred in humans (Prasad et al., 1963; see Tables 20.1–20.3). We supplemented these dwarfs with 15 mg zinc as sulfate daily and we reported that the growth rate in these subjects was approximately 5–6 inches annually. They grew pubic hair and axillary hair within 3 months after zinc supplementation and their genitalia became adult-like within 6 months of zinc supplementation (Sandstead et al., 1967).

Measurement of Plasma Zinc by Atomic Absorption Spectrophotometery

From Egypt, I went to Wayne State University School of Medicine, Detroit, MI, as Chief of Hematology. Soon after my arrival, I got a call from Walter Slavin from Perkin–Elmer Corp., Norwalk, CT. He told me that now they have made available an atomic absorption spectrophotometer (AAS) which could make my life simpler and that I could measure zinc in

TABLE 20.2 Plasma Zinc Turnover Rate in Normal Subjects

No.	Plasma Zinc in µg Percent	Total Plasma Volume in mL	Weight in kg	$T_{1/2}$h for Second Part of Curve in minutes	Plasma Zinc Turnover Rate in mg/kg/day	Mean Plasma Zinc Turnover Rate in mg/kg/day
1	98	2740	56.8	48	0.896	
2	120	2168	57.7	45	0.994	
3	114	2339	55.4	48	0.965	
4	109	2207	44.0	48	1.127	
5	117	1935	50.0	40	1.138	1.00 ± 0.09
6	112	1885	54.5	40	0.952	
7	93	2519	58.1	40	1.032	
8	95	2139	48.6	40	1.035	
9	104	1784	51.8	37	0.866	

Reproduced from Prasad, A.S., Miale, A., Farid, Z., Schulert, A., Sandstead, H.H., 1963. Zinc metabolism in patients with the syndrome of iron deficiency anemia, hypogonadism and dwarfism. J. Lab. Clin. Med. 61, 537–549.

TABLE 20.3 Plasma Zinc Turnover Rate in Dwarfs

No.	Plasma Zinc in µg Percent	Total Plasma Volume in mL	Weight in kg	$T_{1/2}$ for Second Part of Curve in minutes	Plasma Zinc Turnover Rate in mg/kg/day	Mean Plasma Zinc Turnover Rate in mg/kg/day
1	75	1722	25.0	35.0	1.452	
2	66	2185	33.6	22.0	1.958	
3	72	1618	31.8	27.0	1.333	
4	71	2476	34.0	35.0	1.447	
5	63	1903	39.0	18.0	1.689	1.50 ± 0.29
6	63	1620	29.5	29.0	1.193	
7	51	1650	32.7	19.0	1.348	
8	55	2801	27.0	29.0	1.963	
9	66	2499	32.2	34.0	1.459	
10	65	2133	31.8	35.0	1.150	

Reproduced from Prasad, A.S., Miale, A., Farid, Z., Schulert, A., Sandstead, H.H., 1963. Zinc metabolism in patients with the syndrome of iron deficiency anemia, hypogonadism and dwarfism. J. Lab. Clin. Med. 61, 537–549.

the plasma without any difficulty. Not only that, he also offered to donate one instrument for my research. I was truly overwhelmed. I received the instrument but I could not assay zinc in the serum or plasma. Later, Walter told me that he sent the machine to me to develop methods for measurement of zinc in plasma and cells. We worked hard on the technique and finally succeeded. We published that method for measurement of zinc in plasma, red bold cells, and urine in 1965 (Prasad et al., 1965) and this technique is being used globally for measurement of plasma zinc even now. The problem was that plasma contained protein and salts which altered the flow of samples, and plasma zinc could not be measured and compared to the standard solutions of zinc which did not contain any interfering substances. Initially, we lyophilized the plasma, dissolved the lyophilized plasma in HCl and used trichloracetic acid to precipitate the proteins. This technique was carefully worked out and simplified and the method was then published (Prasad et al., 1965).

Now using flameless AAS, it is possible to use directly diluted plasma samples for zinc assay.

At present, plasma zinc is being widely used as a biomarker of zinc deficiency globally. However, AAS is an expensive instrument, needs careful maintenance, and is not available easily in developing countries. Furthermore, plasma zinc assay is not a specific biomarker for zinc deficiency in humans, inasmuch as the plasma zinc pool changes as a result of infections, exercise, and stress. Also, even slight hemolysis increases the plasma zinc level since red cells are rich in zinc.

Development of Biomarkers of Zinc Deficiency in Experimental Human Zinc Deficiency Model

In Detroit, we developed a human model that would allow us to study the effects of a mild zinc-deficient state in humans and also provide us with sensitive biomarkers of zinc deficiency. We recruited adult human volunteers for induction of dietary zinc deficiency. The details of selection of subjects and our protocol have been published earlier (Prasad et al., 1978b). The volunteers were kept on the metabolic ward in a Clinical Research Center at the University of Michigan Medical School and the study was supported by a grant from the NIH. They were monitored very closely by the physicians and clinical staff. A semipurified diet based on texturized soy protein was developed for this study. The diet provided adequate calcium, proteins, fats, and all essential nutrients according to the RDAs except for zinc (Prasad et al., 1978b). The experiment was designed to last for 56 weeks. Before the start of the experiment, subjects received normal hospital diets that provided 10 mg zinc/day for 4 weeks. Then for 8 weeks (stabilization phase), the semipurified diet was served. It was supplemented with 10 mg zinc as sulfate incorporated in cookies providing a total of 13.9 mg zinc/day. During the following 28 weeks, subjects entered the depletion phase, and the zinc supplementation was discontinued. Thus, during the depletion phase the dietary intake of zinc ranged from 3 to 5 mg/day. Following this in repletion phase for 20 weeks, subjects consumed a total of 30 mg zinc per day. In this model we studied several biomarkers of zinc deficiency. These included measurement of zinc in plasma, red blood cells, lymphocytes, granulocytes, and urine; assays for the following enzymes: deoxythymidine kinase activity in collagen connective tissue harvested following an implantation of sponge under the skin, 5'NT in lymphocytes (a marker of maturity of lymphocytes), and neutrophil alkaline phosphatase; serum active thymulin; generation of Th1 cytokines IL-2 and IFN-γ; and assay of mRNAs of Th1 cytokines in stimulated cells.

Zinc in Plasma and Blood Cells

We assayed plasma zinc by the AAS technique as published earlier (Prasad et al., 1965). The separation of platelets, lymphocytes, and granulocytes from whole blood required a careful procedure (Prasad et al., 1965). Platelets were removed first and plasma was removed from the platelet pool. Lymphocyte and granulocytes were separated by discontinuous Histopaque gradient (Wang et al., 1989). The lymphocyte pools are contaminated with platelets and therefore, careful steps were taken to remove the platelets from lymphocytes. Similarly, in the granulocyte pool, the problem was similar with respect to contamination with erythrocytes. We have published detailed methods and by this technique, we were able to measure zinc in platelets, lymphocytes, and granulocytes accurately (Wang et al., 1989). In the human experimental model of zinc deficiency, we observed that when the daily dietary zinc intake of our volunteers was around 3–5 mg, the plasma zinc decreased after 24 weeks. The decrease in zinc level of lymphocytes and granulocytes was observed after 20 weeks of zinc-deficient diet.

Changes in Zinc-Dependent Enzymes

We had reported earlier that deoxythymidine kinase is a zinc-dependent enzyme (Prasad and Oberleas, 1974). Zinc is required for the gene expression of this enzyme. Deoxythymidine kinase is required for DNA synthesis in S phase and is essential for cell division. To assay for this enzyme, we needed proliferating tissue. We, therefore, implanted a sponge under the skin in the volunteers, once at the end of the zinc-restricted period and again at the end of zinc repletion. The results showed that the assay of this enzyme is an excellent biomarker of zinc deficiency in humans (Prasad and Oberleas, 1974). The test is, however, not easy and impractical for routine assay (Prasad and Oberleas, 1974).

Ecto 5′ nucleotidase (5′ NT), a zinc-dependent enzyme, is an integral plasma membrane protein present in most mammalian cells. We assayed for 5′ NT activity in the lymphocytes of two groups of subjects (Meftah et al., 1991). The first group of subjects had a mild state of zinc deficiency, as assessed by zinc levels in lymphocytes, granulocytes, and platelets, but were healthy otherwise. We supplemented them with 50 mg zinc as acetate orally for 12 weeks. The second group of six subjects were normal human volunteers in whom a mild deficiency of zinc was induced by a dietary technique (4.2–5.6 mg zinc intake daily). For the assay of 5′ NT, intact lymphocytes were incubated with 8-C^{14}-labeled ionosine monophosphate as substrate. Product and substrate were separated by thin layer chromatography. In the first group of subjects with zinc deficiency, the decreased activity of 5′ NT was corrected and the cellular zinc levels normalized by zinc supplementation. In the second group of subjects, the baseline data were compared with those in early zinc depletion (4–8 weeks) and late depletion period (20 weeks). A decrease in the activity of 5′ NT was observed in the early depletion phase. Zinc levels in lymphocytes, granulocytes, and platelets decreased significantly only during the late zinc depletion phase. Plasma zinc level did not change even during the late zinc depletion phase. Our studies thus showed that 5′ NT activity is a sensitive and useful biomarker of human zinc deficiency. A decreased activity of 5′NT in zinc-deficient lymphocytes may be indicative of lymphocyte immaturity in human zinc deficiency (Meftah et al., 1991) (see Figs. 20.1–20.4).

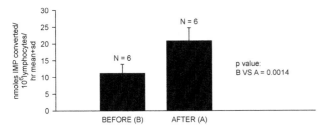

FIGURE 20.1   5′Nucleotidase in lymphocytes. 5′NT activity in lymphocytes (mean±SD) before and after zinc supplementation (experiment I). Values were as follows: Before (B). 10.93±2.74 η mol IMP converted per 10^5 lymphocytes per hour versus after (A). 20.39±4.77 η mol IMP converted per 10^5 lymphocytes per hour; $p=.0014$. *Reproduced from Meftah, S., Prasad, A.S., Lee, D.-Y., Brewer, G.J., 1991. Ecto 5′ nucleotidase (5′NT) as a sensitive indicator of human zinc deficiency. J. Lab. Clin. Med. 118, 309–316.*

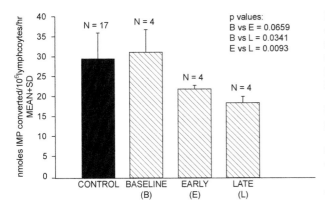

FIGURE 20.2   Changes in lymphocyte 5′NT activity during baseline, early zinc deficiency, and late zinc deficiency periods in experiment II. 5′NT activity (mean±SD) η moles IMP converted/10^6 lymphocytes/hour during baseline (B) versus early deficiency period (E) and late deficiency period (L) were as follows: B versus E 31.13 ± 5.56 η mol IMP converted per 10^6 lymphocytes per hour versus 21.95 ± 0.92 η mol IMP converted per 106 lymphocytes per hour versus 18.50±1.58 η mol IMP converted per 10^6 lymphocytes per hour, $p=.03$; E versus L, 21.95±0.92 η mol IMP converted per 10^6 lymphocytes per hour versus 18.50±1.58 η mol IMP converted per 10^6 lymphocytes per hour, $p=.009$. The values for 5′NT in normal control subjects are also shown (29.5 ± 6.53 η mol IMP converted per 10^6 lymphocytes per hour. *Reproduced from Meftah, S., Prasad, A.S., Lee, D.-Y., Brewer, G.J., 1991. Ecto 5′nucleotidase (5′NT) as a sensitive indicator of human zinc deficiency. J. Lab. Clin. Med. 118, 309–316.*

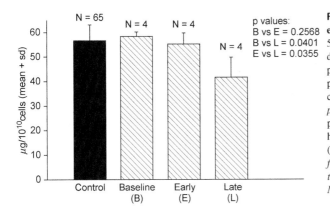

FIGURE 20.3   Changes in lymphocyte 5′NT activity during baseline, early zinc deficiency, and late zinc deficiency periods in experiment II. 5′NT activity (mean±SD η moles IMP converted/10^6 lymphocytes/hour) during baseline (B) versus early deficiency period (E) and late deficiency period (L) were as follows: B versus E 31.13±5.56 η mol IMP converted per 10^6 lymphocytes per 31.13±5.56 η mol IMP converted per 10^6 lymphocytes per hour versus 18.50±1.58 η mol IMP converted per 10^6 per hour, $p=.03$; E versus L, 21.95±0.92 η mol IMP converted per 10^6 lymphocytes per hour versus 18.50±1.58 η mol IMP converted per 10^6 lymphocytes per hour, $p=.009$. The values for 5′NT in normal control subjects are also shown (29.5±6.53 η mol IMP converted per 10^6 lymphocytes per hour). *Reproduced from Meftah, S., Prasad, A.S., Lee, D.-Y., Brewer, G.J., 1991. Ecto 5′nucleotidase (5′NT) as a sensitive indicator of human zinc deficiency. J. Lab. Clin. Med. 118, 309–316.*

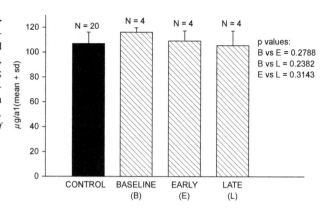

FIGURE 20.4 Changes in plasma zinc during early and late zinc deficiency periods in experiment II. Plasma zinc levels (mean ± SD) during baseline (B) versus early zinc deficiency period (E) and late zinc deficiency period (L) were as follows: B versus E, 116.20 ± 3.51 μg/dL versus 09.10 ± 8.30 μg/dL, $p = .27$; B versus L, 116.20 ± 3.51 μg/dL versus 105 ± 11.38 μg/dL, $p = .23$; and E versus L, 109.10 ± 8.30 μg/dL versus 105.38 μg/dL, $p = .31$. The values for plasma zinc in normal control subjects (mean ± SD) are also shown (107.26 ± 8.92 μg/dL). *Reproduced from Meftah, S., Prasad, A.S., Lee, D.-Y., Brewer, G.J., 1991. Ecto 5′ nucleotidase (5′NT) as a sensitive indicator of human zinc deficiency. J. Lab. Clin. Med. 118, 309–316.*

Serum Thymulin Activity as a Biomarker of Human Zinc Deficiency

Thymulin is a well-characterized thymic hormone with the following amino acid sequence: Pyro-Glu-Ala-Lys-Ser-Gln-Gly-Gly-Ser-Asn. Thymulin requires the presence of zinc to express its biological activity. It is known that two forms of thymulin exist; the first one deprived of zinc is biologically inactive, and the second one, containing zinc, is biologically active (Prasad et al., 1988). The zinc/thymulin relationship was first studied in zinc-deficient mice. Active thymulin levels in sera of mice subjected to a long-term marginal zinc deficiency decreased as early as 2 months after beginning of the diet (Prasad et al., 1988). However, these levels were corrected after in vitro addition of $ZnCl_2$. Similar observations were made with sera obtained from children suffering from nephrotic syndrome who were zinc deficient. The low level of active thymulin in sera was corrected after in vitro addition of $ZnCl_2$. These results confirm the presence of the inactive hormone in the serum of zinc-deficient subjects and its potential activation following in vitro zinc addition. The specificity of these results was confirmed by the lack of activation on experiments performed with sera from the thymectomized mice or patients with Di-George's syndrome, in whom the hormone is nonexistent.

The serum level of biologically active thymulin was evaluated by a rosette assay described elsewhere, and it was shown to be strictly thymus specific (Prasad et al., 1988). The assay analyzes the conversion of relatively azathioprine (Az)-resistant spleen cells of adult thymectomized mice to Θ-positive rosette-forming cells that are more sensitive to Az. In the presence of thymulin-containing sera, rosette formation was inhibited by Az. The results were expressed as the log 2 of the reciprocal of the highest serum dilution conferring sensitivity to Az inhibition upon spleen cells from adult thymectomized mice. To confirm the specificity of the biological activity measured, all the determinations were repeated after preincubation of the sera under study with an antithymulin monoclonal antibody or a specific antithymulin immune-adsorbent.

We also assayed serum thymulin activity in three models of mildly zinc-deficient subjects before and after zinc supplementation: (1) human volunteers in whom a mild deficiency of zinc was induced by dietary means; (2) zinc-deficient adult SCD patients; and (3) a few medical students who were only mildly zinc deficient inasmuch as their plasma zinc levels were within the normal range and zinc deficiency was diagnosed by assay of cellular zinc in lymphocytes, granulocytes, and platelets. In all of these subjects, the serum active thymulin was decreased, and this was corrected by both in vivo and in vitro zinc supplementation, suggesting that serum thymulin activity assay was a sensitive biomarker of zinc deficiency (Prasad et al., 1988). We also observed that T4+/T8+ ratio was decreased and the generation of IL-2 was decreased, and both of these were corrected following zinc supplementation. Inasmuch as thymulin is known to induce intra- and extrathymic T-cell differentiation, our studies provided a possible mechanism for the role of zinc on T-cell functions (Prasad et al., 1988; see Figs. 20.5–20.8).

Development of Immunological Biomarkers of Human Zinc Deficiency

Major manifestations of human zinc deficiency include growth retardation, immune deficiency, and cognitive impairment (Prasad et al., 1961, 1963; Sandstead et al., 1967). My experience in the Middle East showed that most of the zinc-deficient dwarfs died prior to the age of 25 years and these deaths were due to a variety of infections. This suggested to me that immune functions were sensitive to zinc status.

In our studies in the experimental human model of zinc deficiency, we showed that thymulin, a thymic hormone important for development, proliferation, and differentiation of T-cells, was affected adversely even when the deficiency of zinc was very mild (Prasad et al., 1988). In our experimental human model of zinc deficiency, when the dietary zinc intake

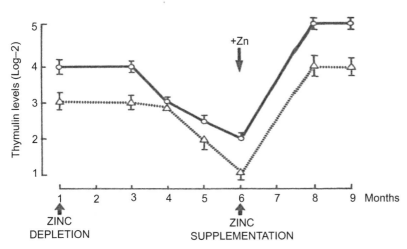

FIGURE 20.5   Thymulin activity. Levels of thymulin activity in sequential study of young human volunteers submitted to a zinc-restricted diet for six months followed by zinc supplementation are shown here. Results are expressed as log-2 reciprocal titers (mean ± SEM). Each determination was performed in triplicate. *Reproduced from Prasad, A.S., Meftah, S., Abdallah, J., Kaplan, J., Brewer, G.J., Bach, J.F., Dardenne, M., 1988. Serum thymulin in human zinc deficiency. J. Clin. Invest. 82, 1202–1210.*

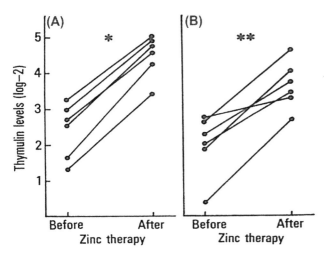

FIGURE 20.6   Thymulin activity. Effect of zinc therapy on the levels of thymulin activity in (A) SCA subjects and (B) non-SCA deficient patients are shown here. Results are expressed as log-2 reciprocal titers. Each point represents the mean of three determinations. *$p < .01$; **$p < .001$. *Reproduced from Prasad, A.S., Meftah, S., Abdallah, J., Kaplan, J., Brewer, G.J., Bach, J.F., Dardenne, M., 1988. Serum thymulin in human zinc deficiency. J. Clin. Invest. 82, 1202–1210.*

was restricted to 2–3.5 mg/day for 20–24 weeks, the plasma zinc declined (mean ± SD) from 109.8 ± 17 to 91.8 ± 10.5 μg/dL. Although the change in plasma zinc was small, a significant decrease in IL-2 and IFN-γ generated ex vivo by isolated MNCs was observed (Beck et al., 1997a,b; Prasad et al., 1988). Also, regeneration of new CD4+ T lymphocytes was decreased as a result of dietary zinc restriction. In another study, dietary zinc restriction was 4.2–5.6 mg/day. At this level, the plasma zinc decreased from (mean ± SD) 118.02 ± 5.3 to 108.0 ± 10.3 μg/dL at the end of 8 weeks in the zinc-restricted period. Moreover, the activity of lymphocyte 5′ NT was significantly decreased (Meftah et al., 1991). When the zinc intake in the volunteers averaged 3.0 mg/day, a decrease in serum thymulin activity was observed between 8 and 12 weeks, but no decrease was seen in this period in plasma zinc concentration (Prasad et al., 1988). Thus, it appears that significant changes in immunological functions due to mild zinc deficiency may be observed in the absence of significant decreases in the plasma zinc concentration, suggesting that the assay of zinc-dependent immunologic functions may be better indicators of human zinc deficiency.

A human Th0 malignant lymphoblastoid cell line, HUT-78, was used to study the effect of zinc on IL-2 production in PHA/PMA-activated T-cells (Prasad et al., 2002). The effect of zinc was at the transcriptional level and was specific for IL-2 (Prasad et al., 2002). A significant effect of zinc on the gene expression and generation of IL-2 and IL-2 receptors α and β was also demonstrated in that the expression of these genes was decreased in zinc-deficient cells. In another study, we reported that in zinc-deficient HUT-78 cells, phosphorylated IκB and IκK, ubiquitinated Iκβ, and binding of NF-κB to DNA were all significantly decreased in comparison to the zinc-sufficient cells (Prasad et al., 2001). Zinc increased the translocation of the NF-κB p50 subunit from cytosol to nucleus (Prasad et al., 2001). Also, the binding of recombinant NF-κB (p50)₂ to DNA in HUT-78 cells was zinc specific. Our studies also showed that the measurement of IL-2 mRNA in peripheral blood MNCs by RT-PCR was a very good indicator of zinc deficiency in humans (Prasad et al., 2006). In zinc-deficient cells, IL-2 mRNA was decreased in comparison to the zinc-sufficient cells, and if a physiological amount of

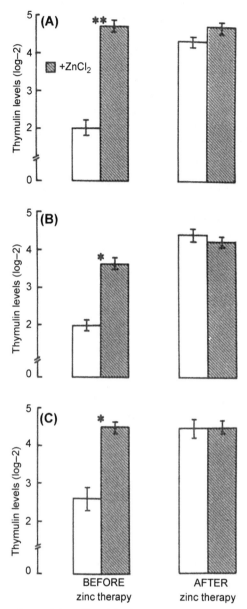

FIGURE 20.7 Thymulin activity. Restoration of normal thymulin activity in sera of zinc-deficient patients after in vitro addition of $ZnCl_2$ is shown. After chelation, $200\,\mu L$ of serum was incubated for 1 h at 37°C with 10 ng of $ZnCl_2$. Thymulin activity determination was performed before and after in vitro zinc addition on individual samples. *$p<.01$; **$p<.001$. (A) Healthy volunteers submitted to zinc restriction. (B) Non-SCA zinc-deficient subjects. (C) SCA zinc-deficient subjects. *Reproduced from Prasad, A.S., Meftah, S., Abdallah, J., Kaplan, J., Brewer, G.J., Bach, J.F., Dardenne, M., 1988. Serum thymulin in human zinc deficiency. J. Clin. Invest. 82, 1202–1210.*

zinc was added to the deficient cells ex vivo, IL-2 mRNA expression was normalized, thus providing a diagnosis for zinc deficiency. It was also demonstrated that the ex vivo addition of zinc effect was specific, inasmuch as no other essential trace element could correct IL-2 mRNA expression in zinc-deficient cases. Our studies also showed that the effect of zinc on the gene expression of IL-2 in the primary cells was because of its role on the activation of NF-κB (Prasad et al., 2006; see Figs. 20.9–20.11).

Endogenous Excretion of Zinc as a Biomarker of Zinc Deficiency

It appears that humans maintain zinc homeostasis by increasing efficiency of zinc absorption and decreasing endogenous excretion of zinc when they are subjected to short-term dietary zinc restriction; however, a mild deficiency of zinc in humans usually is an outcome of chronic exposure to low dietary zinc for many months and years. Therefore, it is important to determine whether or not the adapted zinc homeostasis during the short duration of dietary zinc deprivation is also

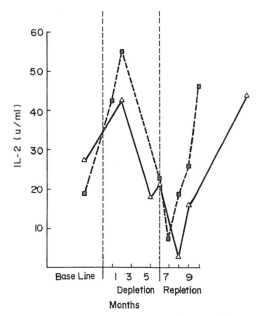

FIGURE 20.8   **Experimental human model interleukin-2 (IL-2) activity.** Changes in IL-2 activity as a result of zinc restriction and zinc repletion in the experimental human model subjects are shown here. Each data point represents the averages of two separate determinations. ■, subject 1; p=NS. Δ, subject 2; p=NS. *NS*, non-significant. *Reproduced from Prasad, A.S., Meftah, S., Abdallah, J., Kaplan, J., Brewer, G.J., Bach, J.F., Dardenne, M., 1988. Serum thymulin in human zinc deficiency. J. Clin. Invest. 82, 1202–1210.*

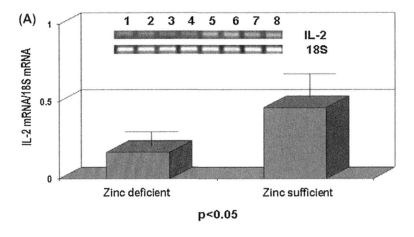

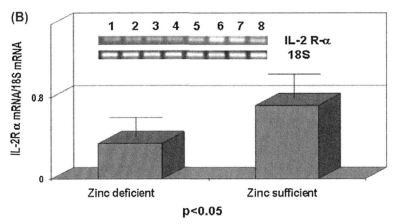

FIGURE 20.9   **IL-2 mRNA.** Relative levels of IL-2 and IL-2R mRNAs in zinc-deficient and zinc-sufficient elderly subjects. Human MNCs were isolated from zinc-deficient and zinc-sufficient elderly subjects and stimulated with PHA for 24 h. Total RNA was extracted, and samples were subjected to RT-PCR analysis. The data shown here are representative of 13 zinc-deficient elderly subjects and 18 zinc-sufficient elderly subjects. Lanes 1–4 (zinc deficient) and 5–8 (zinc sufficient) represent the samples. The results indicate that zinc-sufficient subjects had higher levels of IL-2 (A) and IL-2R (B) mRNAs in PHA-stimulated MNCs. *Reproduced from Prasad, A.S., Bao, B., Beck, F.W.J., Sarkar, F.H., 2006. Correction of IL-2 gene expression by in vitro zinc addition to MNC from zinc deficient human subjects: a specific test for zinc deficiency in humans. Transl. Res. 148, 325–333.*

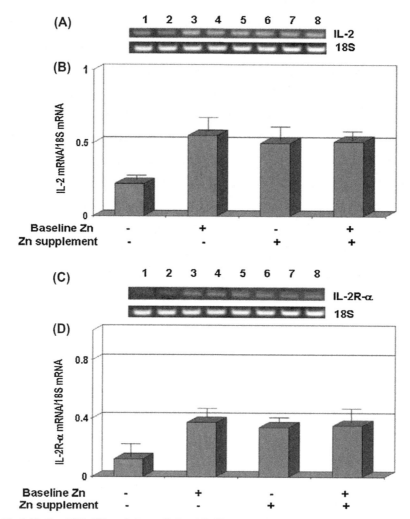

FIGURE 20.10 Interleukin-2 (IL-2) mRNA. Effect of zinc on IL-2 and IL-2R mRNAs in elderly subjects after 6 months of supplementation. Human MNCs were isolated from zinc-deficient and zinc-sufficient elderly subjects receiving either zinc supplementation or placebo for 6 months and stimulated with PHA for 24 h. Total RNA was extracted, and samples were subjected to RT-PCR analysis. Lanes one to two and three to four represent zinc-deficient and zinc-sufficient subjects receiving 6 months of placebo supplementation, respectively. Lanes five to six and seven to eight represent zinc-deficient and zinc-sufficient subjects receiving 6 months of zinc supplementation, respectively. The results indicate that, after 6 months, zinc supplementation increased IL-2 (A, B) and IL-2R (C, D) mRNAs in zinc-deficient elderly subjects, compared with placebo ($p = .05$). *Reproduced from Prasad, A.S., Bao, B., Beck, F.W.J., Sarkar, F.H., 2006. Correction of IL-2 gene expression by in vitro zinc addition to MNC from zinc deficient human subjects: a specific test for zinc deficiency in humans. Transl. Res. 148, 325–333.*

maintained during a prolonged period of dietary zinc restriction. We assessed the efficiency of zinc absorption as well as endogenous zinc excretion during a 6-month period of dietary zinc restriction (63.1 μm mol/day) in human volunteers by using a stable zinc isotope (Zn^{70}) (Lee et al., 1993). Our studies showed that the efficiency of zinc absorption was not sustained and decreased in the volunteers when the zinc-restricted diet was continued for six months. On the other hand, prolonged dietary zinc restriction did not impair the functional role of endogenous zinc excretion in zinc homeostasis. We observed a significant reduction of endogenous zinc excretion by restricting dietary zinc, and this continued at the end of the zinc-restricted period. The endogenous zinc excretion was 65.2 μm mol/day during the baseline period, and it gradually decreased to a mean level of 27.1 μm mol/day at the end of the 6th month of the dietary zinc restriction. When subjects received zinc supplementation, the endogenous zinc excretion increased to 60.1 μm mol/day. Our studies thus show that measurement of endogenous zinc excretion may also be a sensitive biomarker of human zinc deficiency (Lee et al., 1993) (see Figs. 20.12–20.14).

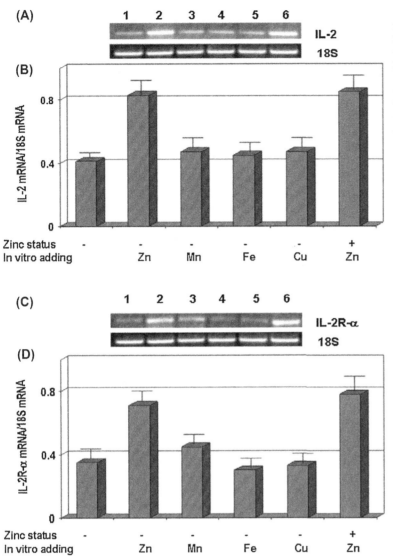

(A)

1 2 3 4 5 6

IL-2

18S

(B)

Zinc status
In vitro adding

(C)

1 2 3 4 5 6

IL-2R-α

18S

(D)

Zinc status
In vitro adding

FIGURE 20.11 **Interleukin-2 (IL-2) mRNA.** Effect of in vitro addition of Zn, Mn, Fe, and Cu on IL-2 and IL-2R mRNAs in elderly subjects. Human MNCs were isolated from zinc-deficient and zinc-sufficient elderly subjects. Either 15-μM Zn (as chloride), Mn (as chloride), Fe (as sulfate), or Cu (as sulfate) was added to isolated MNCs for 24 h, followed by PHA stimulation for another 24 h. Total RNA was extracted, and samples were subjected to RT-PCR analysis. The data shown here are representative of three subjects from each group. Lane one represents MNCs from zinc-deficient subjects. Lanes two to five represent MNCs from zinc-deficient subjects with either Zn, Mn, Fe, or Cu added to the media. Lane six is the MNCs from zinc-sufficient subjects with Zn addition in the media. The results indicate that zinc addition, but not Mn, Fe, and Cu, increased the IL-2 (A, B) and IL-2R (C, D) mRNAs from the MNCs of zinc-deficient subjects ($p = .05$). *Reproduced from Prasad, A.S., Bao, B., Beck, F.W.J., Sarkar, F.H., 2006. Correction of IL-2 gene expression by in vitro zinc addition to MNC from zinc deficient human subjects: a specific test for zinc deficiency in humans. Transl. Res. 148, 325–333.*

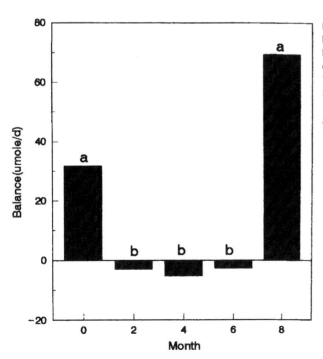

FIGURE 20.12 **Zn^{70} studies.** Effect of zinc depletion and repletion on zinc balance. Each data point represents average values of eight subjects during baseline and 2 months of zinc depletion, seven subject during 4 months and 6 months of zinc depletion, and three subjects during the zinc-repletion period. Values not shared by the same letter are significantly different at $p < .05$ by Scheffe contrast. *Reproduced from Lee, D.-Y., Prasad, A.S., Hydrick-Adair, C., Brewer, G.J., Johnson, P.E., 1993. Homeostasis of zinc in marginal human zinc deficiency: role of absorption and endogenous excretion of zinc. J. Lab. Clin. Med. 122, 549–556.*

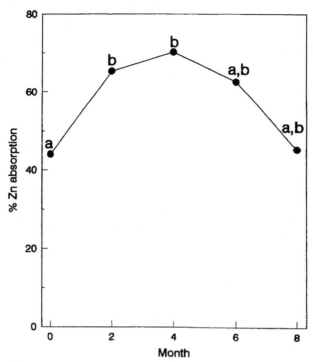

FIGURE 20.13 Zn[70] **studies.** Effect of zinc depletion and repletion on zinc absorption when using stable zinc isotope ([70]Zn). Each data point represents average values of eight subjects during baseline and 2 months of zinc depletion, seven subjects during 4 months and 6 months of zinc depletion, and three subjects during the zinc-repletion period. Values not shared by the same letter are significantly different at $p < .05$ by Scheffe contrast. *Reproduced from Lee, D.-Y., Prasad, A.S., Hydrick-Adair, C., Brewer, G.J., Johnson, P.E., 1993. Homeostasis of zinc in marginal human zinc deficiency: role of absorption and endogenous excretion of zinc. J. Lab. Clin. Med. 122, 549–556.*

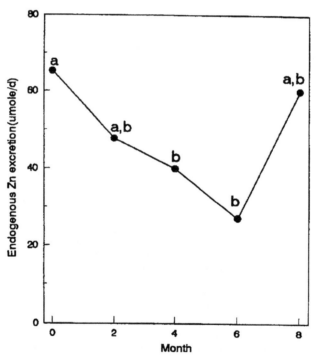

FIGURE 20.14 Zn[70] **studies.** Effect of zinc depletion and repletion on endogenous zinc excretion (see methods for calculation). Each data point represents average values of eight subjects during baseline and 2 months of zinc depletion, seven subjects during 4 months and 6 months of zinc depletion, and three subjects during the zinc-repletion period. Values not shared by the same letter are significantly different at $p < .05$ by Scheffe contrast. *Reproduced from Lee, D.-Y., Prasad, A.S., Hydrick-Adair, C., Brewer, G.J., Johnson, P.E., 1993. Homeostasis of zinc in marginal human zinc deficiency: role of absorption and endogenous excretion of zinc. J. Lab. Clin. Med. 122, 549–556.*

In summary, our studies in the experimental model of human zinc deficiency, suggest that measurement of serum active thymulin, lymphocyte 5′ NT enzyme activity, assays of Th1 cytokines and expression of their mRNAs, and endogenous excretion of zinc are very sensitive indicators of acute zinc deficiency status. Measurement of zinc levels in lymphocytes, granulocytes, and platelets are useful but these are less sensitive than the immunological assays mentioned in this chapter.

REFERENCES

Age-Related Eye Disease Study Research Group (AREDS Report No 35), 2013. Long term effects of vitamins C, E, beta-carotene and zinc in age related macular degeneration. Ophthalmology 120, 1604–1611.

Age-Related Eye Disease Study Research group (AREDS Report No 8), 2001. A randomized, placebo controlled, clinical trial of high-dose supplemented with vitamins C and E, beta-carotene, for age-related macular degeneration and vision loss. Archiv. Ophthalmol. 119, 1417–1436.

AREDS Report No. 13, 2004. Association of mortality with ocular disorders and an intervention of high dose antioxidants and zinc in the age-related eye disease study. Archiv. Ophthalmol. 122, 716–726.

Bao, B., Prasad, A.S., Beck, F.W.J., Snell, D., Sunega, A., Sarkar, F.H., Doshi, N., Fitzgerald, J.T., Swerdlow, P., 2008. Zinc supplementation decreased oxidative stress, incidence of infection and generation of inflammatory cytokines in sickle cell disease patients. Transl. Res. 152, 67–80.

Bao, B., Prasad, A.S., Beck, F.W.J., Fitzgerald, J.T., Snell, D., Bao, G.W., Singh, T., Cardozo, L.J., 2010. Zinc decreases C-Reactive protein, lipid peroxidation, and implication of zinc as an atheroprotective agent. Am. J. Clin. Nutr. 91, 1634–1641.

Bao, B., Prasad, A.S., Beck, W.J., Bao, G.W., Singh, T., Ali, S., Sarkar, F.H., 2011. Intracellular free zinc up-regulates IFN-γ and T-bet essential for Th1 differentiation in Con-A stimulated HUT-78 cells. Biochem. Biophys. Res. Commun. 407, 703–707.

Barnes, P.M., Moynahan, E.J., 1973. Zinc deficiency in acrodermatitis enteropathica. Proc. R. Soc. Med. 66, 327–329.

Beck, F.W.J., Kaplan, J., Fine, N., Handschu, W., Prasad, A.S., 1997a. Decreased expression of CD73 (ecto-5′-nucleotidase) in the CD8+ subset is associated with zinc deficiency in human patients. J. Lab. Clin. Med. 130, 147–156.

Beck, F.W.J., Prasad, A.S., Kaplan, J., Fitzgerald, J.T., Brewer, G.J., 1997b. Changes in cytokine production and T cell subpopulations in experimentally induced zinc deficient humans. Am. J. Physiol. Endocrinol. Metab. 272, 1002–1007.

Brewer, G.J., Yuzbasiyan-Gurkan, V., 1992. Wilson disease. Medicine 71, 139–164.

Brewer, G.J., Schoomaker, E.B., Leichtman, D.A., Kruckleberg, W.C., Brewer, L.F., Myers, N., 1977. The uses of pharmacologic doses of zinc in the treatment of sickle cell anemia. In: Brewer, G.J., Prasad, A.S. (Eds.), Zinc Metabolism: Current Aspects in Health and Disease. Allan R. Liss, Inc., New York, NY, pp. 241–258.

Brewer, G.J., 1995. Practical recommendations and new therapies for Wilson's disease. Drugs 2, 240–249.

Cavdar, A.O., Babacan, E., Arcasoy, A., Ertein, U., 1980. Effect of nutrition on serum zinc concentration during pregnancy in Turkish women. Am. J. Clin. Nutr. 33, 542–544.

Fisher, Walker, C.L., Lamberti, L., Roth, D., Black, R.E., 2011. In: Rink, L. (Ed.), Zinc in Human Health. IOS Press, Amsterdam, pp. 234–253.

Haase, H., Rink, L., 2007. Signal transduction in monocytes: the roll of zinc ions. Biometals 20, 579–585.

Hall, A.C., Young, B.W., Bremner, I., 1979. Intestinal metallothionein and the mutual antagonism between copper and zinc in the rat. J. Inorg. Biochem. 11, 57–66.

Hirano, T., Murakami, M., Fukada, T., Nishida, K., Yamasaki, S., Suzuki, T., 2008. Roles of zinc and zinc signaling in immunity: zinc as an intracellular signaling molecule. Adv. Immunol. 97, 149–176.

Kay, R.G., Tasman-Jones, C., 1975. Zinc deficiency and intravenous feeding. Lancet 2, 605–606.

Kitamura, H., Morikawa, H., Kamon, H., Iguchi, M., Hojyo, S., Fukada, T., Yamashita, S., Kaisho, T., Akiron, S., Murakami, M., Hirano, T., 2006. Toll-like receptor-mediated regulation of zinc homeostasis influences dentritic cell function. Nat. Immunol. 7, 971–977.

Klingberg, W.G., Prasad, A.S., Oberleas, D., 1976. Zinc deficiency following penicillamine therapy. In: Prasad, A.S. (Ed.), Trace Elements in Human Health and Disease, vol. 1. Academic Press, New York, pp. 51–65.

Lee, D.-Y., Prasad, A.S., Hydrick-Adair, C., Brewer, G.J., Johnson, P.E., 1993. Homeostasis of zinc in marginal human zinc deficiency: role of absorption and endogenous excretion of zinc. J. Lab. Clin. Med. 122, 549–556.

Meftah, S., Prasad, A.S., Lee, D.-Y., Brewer, G.J., 1991. Ecto 5′ nucleotidase (5′NT) as a sensitive indicator of human zinc deficiency. J. Lab. Clin. Med. 118, 309–316.

Newsome, D.A., Miceli, M.V., Tats, D.J., Alcock, N.W., Oliver, P.D., 1996. Zinc content of human retinal pigment epithelium decreases with age and macular degeneration but superoxide dismutase activity increases. J. Trace Elem. Exp. Med. 8, 193–199.

Okada, A., Takagi, Y., Itakura, T., Satani, M., Manabe, H., 1976. Skin lesions during intravenous hyperalimentation: zinc deficiency. Surgery 80, 629–635.

Prasad, A.S., Oberleas, D., 1974. Thymidine kinase activity and incorporation of thymidine into DNA in zinc-deficient tissue. J. Lab. Clin. Med. 83, 634–639.

Prasad, A.S., Halsted, J.A., Nadimi, M., 1961. Syndrome of iron deficiency anemia, hepatosplenomegaly, hypogonadism, dwarfism, and geophagia. Am. J. Med. 31, 532–546.

Prasad, A.S., Miale, A., Farid, Z., Schulert, A., Sandstead, H.H., 1963. Zinc metabolism in patients with the syndrome of iron deficiency anemia, hypogonadism and dwarfism. J. Lab. Clin. Med. 61, 537–549.

Prasad, A.S., Oberleas, D., Halsted, J.A., 1965. Determination of zinc in biological fluids by atomic absorption spectrophotometry in normal and cirrhotic subjects. J. Lab. Clin. Med. 66, 508–516.

Prasad, A.S., Brewer, G.J., Schoomaker, E.B., Rabbani, P., 1978a. Hypocupremia induced by zinc therapy in adults. J. Am. Med. Assoc. 240, 2166–2168.

Prasad, A.S., Rabbani, P., Abbasi, A., Bowersox, E., Spivey-Fox, M.R., 1978b. Experimental zinc deficiency in humans. Ann. Intern. Med. 89, 483–490.

Prasad, A.S., Meftah, S., Abdallah, J., Kaplan, J., Brewer, G.J., Bach, J.F., Dardenne, M., 1988. Serum thymulin in human zinc deficiency. J. Clin. Invest. 82, 1202–1210.

Prasad, A.S., Fitzgerald, J.T., Hess, J.W., Kaplan, J., Pelen, F., Dardenne, M., 1993. Zinc deficiency in the elderly patients. Nutrition 9, 218–224.

Prasad, A.S., Beck, F.W.J., Kaplan, J., Chandrasekar, P.H., Ortega, J., Fitzgerald, J.T., Swerdlow, P., 1999. Effect of zinc supplementation on incidence of infections and hospital admissions in sickle cell disease (SCD). Am. J. Hematol. 61, 194–202.

Prasad, A.S., Fitzgerald, J.T., Bao, B., Beck, W.J., Chandrasekar, P.H., 2000. Duration of symptoms and plasma cytokine levels in patients with the common cold treated with zinc acetate. Ann. Intern. Med. 133, 245–252.

Prasad, A.S., Bao, B., Beck, F.W.J., Sarkar, F.H., 2001. Zinc activates NF-kB in HUT-78 cells. J. Lab. Clin. Med. 138, 250–255.

Prasad, A.S., Bao, B., Beck, F.W.J., Sarkar, F.H., 2002. Zinc enhances the expression of IL-2 and IL-2 receptors in HUT-78 cells via NF-κB activation. J. Lab. Clin. Med. 140, 272–289.

Prasad, A.S., Bao, B., Beck, F.W.J., Kucuk, O., Sarkar, F.H., 2004. Antioxidant effect of zinc in humans. Free Radic. Biol. Med. 37, 1182–1190.

Prasad, A.S., Bao, B., Beck, F.W.J., Sarkar, F.H., 2006. Correction of IL-2 gene expression by in vitro zinc addition to MNC from zinc deficient human subjects: a specific test for zinc deficiency in humans. Transl. Res. 148, 325–333.

Prasad, A.S., Beck, F.W.J., Bao, B., Fitzgerald, J.T., Snell, D.C., Steinberg, J.D., Cardozo, L.J., 2007. Zinc supplementation decreases incidence of infections in the elderly: effect of zinc on generation of cytokines and oxidative stress. Am. J. Clin. Med. 85, 837–844.

Prasad, A.S., Beck, F.W.J., Bao, B., Snell, D., Fitzgerald, T., 2008. Duration and severity of symptoms and levels of plasma interleukin-1 receptor antagonist, soluble tumor necrosis factor receptor, and adhesion molecule in patients with common cold treated with zinc acetate. J. Infect. Dis. 197, 795–802.

Prasad, A.S., Bao, B., Beck, F.W.J., Sarkar, F.H., 2011. Zinc-suppressed inflammatory cytokines by induction of A20-mediated inhibition of nuclear factor-κB. Nutrition 27, 816–823.

Prasad, A.S., 1993. Biochemistry of Zinc. Plenum Press, New York.

Prasad, A.S., 2013. Discovery of human zinc deficiency: its impact on human health and disease. Adv. Nutr. 4, 176–190.

Raulin, J., 1869. Chemical studies on vegetation. Ann. Des. Sci. Nat. 11, 93–99 (in French).

Rosenkranz, E., Prasad, A.S., Rink, L., 2011. Immunobiology and hematology of zinc. In: Rink, L. (Ed.), Zinc and Human Health. IOS Press, Amsterdam, pp. 195–233.

Sandstead, H.H., Prasad, A.S., Schulert, A.R., Farid, Z., Miale Jr., A., Bassily, S., Darby, W.J., 1967. Human zinc deficiency, endocrine manifestations and response to treatment. Am. J. Clin. Nutr. 20, 422–442.

Sandstead, H.H., 2012. Zinc nutrition from discovery to global health impact. Adv. Nutr. 3, 718–719.

Sazawal, S., Black, R.E., Bhan, M.K., Bhandari, N., Sinha, A., Jalla, S., 1995. Zinc supplementation in young children with acute diarrhea in India. N. Engl. J. Med. 333, 839–844.

Shankar, A.H., Prasad, A.S., 1998. Zinc and immune function: the biological basis of altered resistance to infection. Am. J. Clin. Nutr. 68 (Suppl.), 447–463.

Singh, M., Das, R., 2011. Zinc for the common cold. Cochrane Database Syst. Rev. 2, 1–58 The Cochrane Collaboration. Published by John Wiley and Sons, Ltd. Issue.

Swe, K.M.M., Abas, A.B.L., Bhardwaj, A., Barua, A., Nair, N.S., 2013. Zinc supplementation for treating Thalassemia and sickle cell disease. Cochrane Rev. 1–36 The Cochrane Library Published by John Wiley and Sons, Ltd.

Todd, W.R., Elvehjem, C.A., Hart, E.B., 1933. Zinc in the nutrition of the rat. Am. J. Physiol. 107, 146–156.

Wang, H., Prasad, A.S., DuMouchelle, E.A., 1989. Zinc in platelets, lymphocytes and granulocytes by flameless atomic absorption spectrophotometry. J. Micronutr. Anal. 5, 181–190.

Wang, K., Zhou, B., Kuo, Y.M., Zemansky, J., Gitschier, J., 2002. A novel member of a zinc transporter family is defective in acrodermatitis enteropathica. Am. J. Hum. Genet. 71, 66–73.

Chapter 21

Zinc Signals and Immune Function

Hajo Haase

Berlin Institute of Technology, Berlin, Germany

SIGNAL TRANSDUCTION IN MAMMALIAN CELLS

In multicellular organisms, the division of functions between different tissues requires their coordination. This is of particular importance for the immune system, an organ that is distributed throughout the entire body. Messenger substances transport information on the systemic level (endocrine), or to cells in the vicinity (paracrine), or even signal back to the producing cell (autocrine). On the receiving end, most signals arrive at the cell surface where they are recognized by specific receptors. These receptors trigger biochemical events on the inside of the cell, controlling cellular behavior, such as gene expression or the activity of metabolic pathways (Fig. 21.1).

The major mechanism for intracellular signal transduction is phosphorylation, the transfer of γ-phosphate groups from adenosine triphosphate (ATP) to amino acid sidechains of serine, threonine, and tyrosine in proteins. This is controlled by two different types of enzymes: kinases, which mediate the transfer of phosphate onto the proteins; and phosphatases, which catalyze the opposite reaction. The activities of both kinases and phosphatases themselves are frequently regulated by phosphorylation, leading to complex cascades of subsequent phosphorylation reactions.

Signaling is not limited to phosphorylation. Second messengers are small molecules that transmit information by a change in their concentration. This can be cyclic nucleotides, such as 3′, 5′-cyclic adenosine monophosphate (cAMP) or its guanidine analog (cGMP). Alternatively, the free concentrations of metal ions can be second messengers. This has been intensely investigated for calcium ions (Ca^{2+}) (Chaigne-Delalande and Lenardo, 2014). Yet, zinc ions (Zn^{2+}) have also been shown to be involved in intracellular signal transduction, and many of these observations have been made in immune cells (Haase and Rink, 2014a).

As an endpoint for many signaling pathways, transcription factors transform signals into a regulation of mRNA expression. Many of these transcription factors contain secondary structural elements that are stabilized by Zn^{2+}, most importantly the "classic" zinc finger structures with a $CysX_{2-5}CysX_{12}HisX_{3-4}His$ signature motif (Maret, 2012). These motifs are critical elements of transcription factors as they are directly required for protein–DNA interaction.

THE IMMUNE SYSTEM

The immune system is an organ consisting of a myriad of cellular and soluble components that are distributed throughout the body. In case of an infection, several defense mechanisms against the particular kind of pathogen need to act in concert to successfully protect the host. This requires a particularly high level of coordination. The coordination between all these different cells is mediated by cytokines. These are defined as proteins secreted by various types of cells with the purpose of transmitting signals to or from immune cells by activating receptors on the plasma membrane.

In general, there are two major parts of the immune system: humoral and cellular immunity. Humoral immunity comprises several groups of soluble proteins. One is the complement system, a cascade of several proteins leading to the marking of pathogen surfaces, lysis of pathogens, and the release of inflammatory mediators. Another group is antibodies, which bind to their respective antigens on the pathogen's surface. This antigen–antibody complex is recognized by specific receptors on immune cell surfaces. Hereby, antibodies activate phagocytosis, the release of immune mediators, or killing of pathogens by a mechanism called antibody-dependent cellular cytotoxicity.

The cellular part of the immune system can be further distinguished by the receptors used for pathogen recognition. Some cells, designated as innate immune cells, have genetically encoded receptors (pattern recognition receptors [PRRs]) that recognize conserved pathogen structures, so-called pathogen-associated molecular patterns (PAMPs). In contrast, cells of the adaptive immune system have receptors with an amazing degree of variability, originating from somatic recombination of the encoding DNA-regions. This leads to the untargeted generation of millions of

Molecular, Genetic, and Nutritional Aspects of Major and Trace Minerals. http://dx.doi.org/10.1016/B978-0-12-802168-2.00021-X

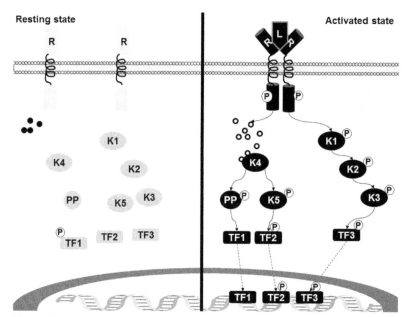

Resting state

Activated state

FIGURE 21.1 **Principles of intracellular signal transduction.** Recognition of ligands by their respective plasma membrane receptors leads to activation of intracellular signaling cascades, involving kinases (K), protein phosphatases (PP), transcription factors (TF), and second messengers (circles). Several additional PP, which negatively regulate kinase cascades, are not shown for reasons of clarity.

different receptor specificities. During an infection the few cells with receptors matching the invading pathogen are selectively activated and multiplied by proliferation. There are two major types of adaptive immune cells. B-cells produce antibodies, whereas T-cells have multiple functions. These include cytotoxic T-cells, which kill virus-infected or malignant cells. T-helper (T_H) cells support other immune cells, for example, by activating macrophages, facilitating the killing of intracellular pathogens, or by providing activation and direction for antibody production by B-cells. Finally, regulatory T-cells (T_{reg}) act to confine the immune reaction and prevent excessive immune reactions and autoimmunity.

THE PHYSIOLOGICAL ROLE OF ZINC IN THE IMMUNE SYSTEM

Virtually, all organs are negatively affected by zinc deficiency, underscoring the eminent importance of this trace element for human and animal health. The immune system is especially sensitive to zinc deficiency, indicating a paramount significance of Zn^{2+} for immune defense. Reduced immune function as a consequence of zinc deficiency manifests in thymic atrophy, impaired numbers and functions of lymphocytes, skewed balance between different subsets of T_H cells, and, consequently, a higher incidence of bacterial, viral, and fungal infections (Fraker and King, 2004; Haase and Rink, 2009a; Overbeck et al., 2008; Prasad, 2014a, 2007).

Zinc is of special importance in the elderly (Prasad, 2014b). From the third decade of life, serum Zn^{2+} levels decline with age (Hotz et al., 2003). Aging is frequently accompanied by immunological changes, so-called immunosenescence, including thymic atrophy, reduced responsiveness of lymphocytes toward activation, elevated inflammation, impaired cellular and humoral immune responses, and recurrent infections. Because of the similarities of the alterations in immune function during immunosenescence and zinc deficiency, it seems likely that reduced zinc levels contribute to age-related immune dysfunction (Haase and Rink, 2009b; Prasad, 2014b). This is backed up by multiple observations that symptoms of immunosenescence, such as elevated basal production of the cytokine interleukin (IL)-6, reduced cellular capability of responding to stimulation, and impaired effectiveness of vaccination, are ameliorated by zinc supplementation (Haase and Rink, 2009b).

Most zinc ions in cells are tightly bound to proteins. They serve important functions as catalytic or structural components of several thousand different proteins (Andreini et al., 2009; Brylinski and Skolnick, 2011). Certainly, a significant portion of the essentiality of zinc for the immune response is based on protein-bound zinc. However, many other functions of immune cells are controlled by zinc signals in which free or loosely bound Zn^{2+} serves as a second messenger. Several aspects of zinc signaling will be discussed in the next section and are summarized in Fig. 21.2.

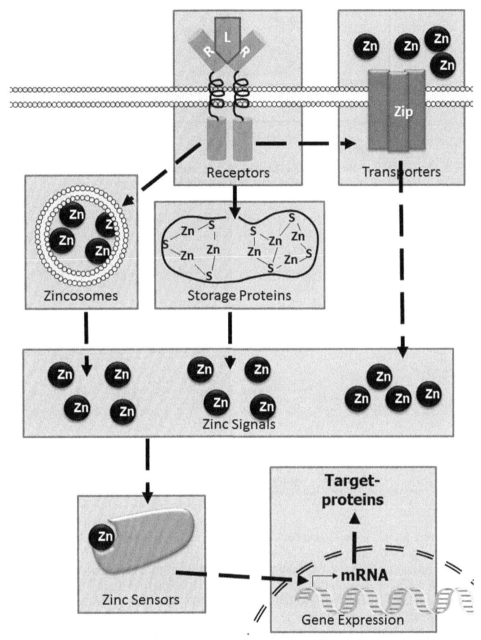

FIGURE 21.2   Overview of zinc signaling. Several molecular events contribute to zinc signaling. Most importantly, there are receptors that trigger zinc signals, various elements of zinc homeostasis (transport proteins, zincosomes, storage proteins), zinc signals themselves, sensors that translate zinc signals into biochemical events, and finally, functional targets that are regulated by zinc signals.

ZINC SIGNALS

Zinc signals are defined as changes in the intracellular concentration of free zinc ions. These changes can occur with different spatial and temporal properties. First, they occur on different timescales (Fig. 21.3).

Fast signals are observed within seconds after contact of the cell with the triggering event (Haase and Rink, 2014a). In contrast, the "zinc wave" has been introduced, which occurred considerably slower with a lag phase of several minutes (Yamasaki et al., 2012). The last kind of zinc signals develop in a matter of hours. Changes in the expression of proteins involved in zinc homeostasis, most importantly, zinc transporters and zinc binding proteins, lead to a change in the basal level of free zinc. Examples for all three kinds of zinc signals have been reported in the literature and are summarized in Table 21.1.

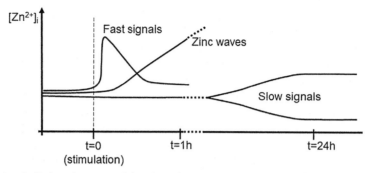

FIGURE 21.3   Types of zinc signals. To date, three types of zinc signals have been described. After stimulation of the receptor (t=0), they either occur immediately (fast signals), after a lag phase of several minutes (zinc wave), or after several hours (slow signals).

TABLE 21.1 Zinc Signals in Immune Cells

Receptor (Ligand)	Cell Type	Signal Type	References
TLR4 (LPS)	Monocytes	Fast signal	Haase et al. (2008)
CCR2/CCR4 (MCP-1)	THP-1 cell line	Fast signal	Kojima et al. (2007)
IL2-R (IL-2)	CTLL-2 cell line	Fast signal	Kaltenberg et al. (2010) and Plum et al. (2014)
FCε-R1 (antibodies[a])	Mast cells	Zinc wave	Yamasaki et al. (2007)
T-cell receptor (antibodies[a])	T-cells	Slow signal	Aydemir et al. (2009)
TLR4 (LPS)	Dendritic cells	Slow signal	Kitamura et al. (2006)
IL-1βR, TNF-αR, and IFN-γ-R (IL-1β, TNFα, IFN-γ)	Aortic endothelial cells	Slow signal (nuclear)	Spahl et al. (2003)
VDR (calcitriol)	HL-60 cell line	Slow signal	Dubben et al. (2010)

CCR, chemokine (C–C motif) receptor; *FCε-R*, immunoglobulin epsilon FC receptor; *IFN*, interferon; *IL*, interleukin; *LPS*, lipopolysaccharide; *MCP*, monocyte chemoattractant protein; *TLR*, Toll-like receptor; *TNF*, tumor necrosis factor; *VDR*, vitamin D receptor.
[a]*Receptors were activated by cross-linking with antibodies instead of their physiological ligands.*

Zinc signals may involve translocation of zinc between different cellular compartments. For example, release of lysosomal Zn^{2+} causes a rise in the cytosolic concentration in T-cells in response to stimulation of the IL-2 receptor (Kaltenberg et al., 2010). Moreover, zinc signals can also be limited to different parts of the cytosol, as demonstrated by the group of Goronzy for activation of the T-cell receptor (TCR; Yu et al., 2011). The signal, resulting from the influx of extracellular Zn^{2+}, is limited to the immediate vicinity of the cytosolic region of the TCR adjacent to the contact area between the T-cell and an antigen-presenting cell.

Detection of zinc signals is mainly achieved by fluorescent probes (Maret, 2015). Recently, the specificity of zinc probes has been a matter of debate; especially if they might, in fact, be detecting the well-known calcium signals. However, the majority of fluorescent probes designed for the detection of Zn^{2+} are sufficiently selective and do not detect Ca^{2+}. Yet, this specificity is not found with some of the most important probes for the detection of Ca^{2+} (Figueroa et al., 2014). In fact, it has already been shown that some calcium signals induced by thimerosal were actually based on the release of Zn^{2+} (Haase et al., 2009).

Zinc signals can result from translocation of Zn^{2+} from different places and are controlled by more than two dozen proteins. Cellular zinc homeostasis is maintained by a number of transporters from two families of solute-linked carrier proteins. SLC39A1-14, also known as Zrt, Irt-like protein (ZIP)1 to ZIP14, transports Zn^{2+} through lipid bilayers, such as the plasma membrane or membranes of cellular organelles, into the cytosol. The SLC30A1-10 carriers, also known as zinc transporter (ZnT)1 to ZnT10, move Zn^{2+} from the cytosol into intracellular compartments or to the extracellular space. A more detailed overview on zinc transport can be found in a number of excellent reviews (see, for example, Kambe et al., 2014; Lichten and Cousins, 2009). Dynamic regulation of the expression of zinc transporters has been shown in response to cytokines and hormones (Haase and Rink, 2014b). Moreover, altered expression of zinc transporters occurs during the maturation of dendritic cells (Kitamura et al., 2006) and the differentiation of myeloid precursor cells into monocytes (Dubben et al., 2010). Still, many signals occur so fast that they must involve posttranscriptional regulation. The molecular nature of these posttranscriptional regulatory events for zinc transporters in immune cell signaling is still unclear. Recent experiments with other cell types have shown that some of the transporters are regulated by phosphorylation. When ZIP7 is phosphorylated on serines 275 and 276 by casein kinase 2, this leads to a release of Zn^{2+} from the endoplasmic reticulum into the cytoplasm (Taylor et al., 2012). Tumor necrosis factor (TNF)α signaling in mammary epithelial cells leads to dephosphorylation of ZnT2, resulting in redistribution of the transporter to lysosomes, with subsequent accumulation of lysosomal Zn^{2+} (Hennigar and Kelleher, 2015). In addition to ZIPs and ZnTs, other cation transporters might also contribute to zinc signals. Recently, an involvement of L-type Ca^{2+} channels has been shown to be responsible for the zinc wave in mast cells (Yamasaki et al., 2012).

Zincosomes are membrane-enclosed structures rich in Zn^{2+}, which can be visualized with fluorescent probes (Beyersmann and Haase, 2001). An X-ray absorption near edge structure (XANES) analysis of isolated zincosomes from macrophages has shown that the average chemical environment of Zn^{2+} in these vesicles consists of 1 sulfur, 2.5 histidines, and 1 oxygen, suggesting that zincosomal Zn^{2+} is bound by an as-yet-unknown vesicular storage protein (Wellenreuther et al., 2009). In T-cells, zincosomes co-stain with lysosomes (Aydemir et al., 2009; Kaltenberg et al., 2010). However, in some other cell types zincosomes are clearly distinguishable from lysosomes and sometimes different fluorescent probes detect diverse types of zincosomes within the same cell (our unpublished observations).

Protein-bound Zn^{2+} is by far the largest pool of cellular Zn^{2+}. Part of it can be released, thereby contributing to zinc signaling. Metallothionein (MT) binds up to seven zinc ions with different affinities (Krezel and Maret, 2007). As shown in Fig. 21.4, MT, the apo-protein (thionein) and the oxidized form (thionin) represent just extreme forms of a protein that most likely exists as a partially metal-loaded, partially reduced, and potentially even polymerized protein (Haase and Maret, 2008). Changes in these equilibria will result in changes in the intracellular availability of free Zn^{2+}.

Regulation of MT expression was observed after treatment of monocytes with the PAMP lipopolysaccharide (LPS), a ligand for toll-like receptor (TLR)4, or with inflammatory cytokines (Leibbrandt and Koropatnick, 1994) and is induced by stimulating PRRs of primary human macrophages (Lahiri and Abraham, 2014). Cells isolated from MT knockout mice were shown to be functionally affected. MT$-$/$-$ cells showed reduced FCε-receptor-induced production of IL-4 (Ugajin et al., 2015) and macrophage colony stimulating factor (M-CSF) by lung fibroblasts (Kanekiyo et al., 2002a), and functional impairment and reduced production of TNFα by LPS-stimulated peritoneal macrophages (Itoh et al., 2005; Kanekiyo et al., 2002b).

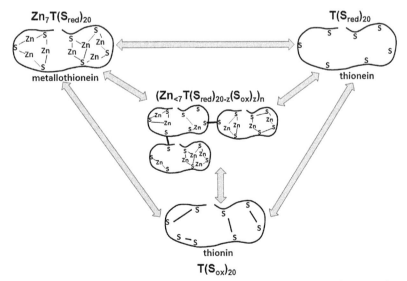

FIGURE 21.4 **Metallothionein.** Three chemically well-defined states of MT have been described. Complete saturation with seven Zn^{2+} is generally considered the classical form of metallothionein. The corresponding apo-protein is known as thionein. Oxidation of the thiol residues yields an oxidized version of the protein that is no longer capable of metal binding, referred to as thionin. In addition, intermolecular disulfides lead to polymerization. Within cells, MT probably occurs as a combination of intermediate forms that are partially metal saturated and oxidized.

SENSORS FOR ZINC SIGNALS

Multiple mechanisms have been shown to sense zinc signals. Many proteins involved in signal transduction contain inhibitory zinc sites that translate a change in free Zn^{2+} into altered enzymatic activity (Maret, 2013). First of all, several hydrolytic enzymes are targets for inhibition by Zn^{2+}. Protein tyrosine phosphatases (PTP) catalyze the cleavage of phosphate groups from tyrosine residues in proteins, during which the phosphate group is transferred onto a cysteine thiolate in the conserved $HC(X)_5R$ signature motif of the PTPs in a first step, followed by hydrolysis of the thioester (Tonks, 2006). Several members of the PTP family have been shown to be inhibited by physiological concentrations of free Zn^{2+}, especially PTP1B (Haase and Maret, 2003), mitogen-activated protein kinase phosphatases (MKPs) (Kaltenberg et al., 2010), and the lipid phosphatase "phosphatase and tensin homolog deleted on chromosome 10" (PTEN) (Plum et al., 2014). There is considerable variation between the inhibition constants (Table 21.2), showing that Zn^{2+} does not indiscriminately inhibit all cellular PTP activity. For example, phosphorylation reactions downstream of the IL-2 receptor show a differential pattern of reactivity toward zinc signals for the various tyrosine phosphorylation sites. Phosphorylations of the kinases JAK1 and JAK3, as well as the transcription factor STAT5, are unaffected by Zn^{2+}, whereas pathways involving the kinases ERK and PI3K are regulated via inhibition of MKPs and PTEN, respectively (Plum et al., 2014).

The inhibition of PTPs has been suggested to involve the active site, including the catalytically active cysteine thiolate (Haase and Maret, 2003). However, there seem to be differences with regard to the proposed mechanisms between PTPs. For PTEN, a conformational shift in the presence of Zn^{2+} has been shown, which is no longer observed after site-directed mutation of two cysteines, including the one in the catalytic site, to serine. This shift occurred even in the absence of substrate (Plum et al., 2014). In contrast, for PTP1B it has been reported that Zn^{2+} binds predominantly to the phosphointermediate form, inhibiting the second step of catalysis, the release of phosphate (Bellomo et al., 2014).

Cysteine-dependent aspartate-specific proteases (caspases) form a proteolytic cascade central to apoptosis. It is well documented that these enzymes are also highly sensitive to inhibition by Zn^{2+} (Maret et al., 1999). It has been suggested early on that catalytic activity of caspases might be regulated by binding of Zn^{2+} to the active site cysteine (Truong-Tran et al., 2001). However, there is still controversy about the exact mechanism of inhibition. Recently, it has been shown for caspase 9 that the ion binds primarily to the enzymes' active site, precisely to one histidine and two cysteine residues, potentially in combination with a nearby glutamate residue or water (Huber and Hardy, 2012). In contrast, Zn^{2+} is found distal from the active site in caspase 6, bound by lysine, glutamate, and histidine in a binding site that seems to be specific for this particular caspase (Velázquez-Delgado and Hardy, 2012). For caspase 3, yet another mechanism has been suggested in which the zinc binding site comprises the catalytic histidine and methionine, but not the catalytic cysteine (Daniel et al., 2014).

Other zinc-regulated enzymes include cyclic nucleotide phosphodiesterases (PDE), which cleave the second messengers cAMP and cGMP. In monocytes, PDE inhibition by Zn^{2+} with subsequent activation of protein kinase A is a negative regulator of LPS-induced cytokine secretion (von Bülow et al., 2005).

Another mechanism by which zinc ions can affect the activity of signaling proteins is the stabilization of structural elements, such as zinc fingers. Similar to all other cells, immune cells contain many different enzymes and zinc finger

TABLE 21.2 Inhibition Constants of PTPs

PTP	EC_{50} [nM]	References
LAR	20,000	Haase and Maret (2005)
T-cell PTP	200	Maret et al. (1999)
PTP 1C	98	Haase and Maret (2003)
PTP 1B	17	Haase and Maret (2003)
	16	Krezel and Maret (2008)
	5.6	Bellomo et al. (2014)
PTEN	0.59	Plum et al. (2014)
VE-PTP	0.021	Wilson et al. (2012)

LAR, leukocyte common antigen-related protein; *PTEN*, phosphatase and tensin homolog; *PTP*, protein tyrosine phosphatase; *VE-PTP*, vascular endothelial-PTP.

proteins with structural elements stabilized by Zn^{2+}. Consequently, at least a fraction of these may be targets for zinc signals. However, most zinc finger proteins have binding constants suggesting that their zinc fingers are permanently saturated with Zn^{2+} and therefore are not a target for zinc signals. For example, the concentrations and incubation times normally used for chelating cellular zinc signals with the high affinity chelator N,N,N',N'-tetrakis(2-pyridylmethyl)-1,2-ethylenediamine are not sufficient to remove Zn^{2+} from the zinc fingers of the transcription factor Sp1 (Rana et al., 2008). Moreover, an influence of Zn^{2+} on A20, a protein containing seven zinc fingers, has been shown. A20 is a deubiquitinase that cleaves ubiquitin from the signaling protein TRAF6, thereby acting as a negative regulator of TLR4 signaling (Wertz et al., 2004). Zn^{2+} enhances the expression of A20, thereby limiting the expression of TLR4 target genes such as the proinflammatory cytokines TNFα and IL-1β. Like for Sp1, there is no evidence that Zn^{2+} might control the activity of A20 by reversibly stabilizing its zinc finger structures. Even high concentrations of the zinc chelator 1,10-phenanthroline do not affect its enzymatic activity as a deubiquitinase (Prasad, 2014c, 2008). Hence, dynamic regulation of most zinc finger proteins by zinc signals seems unlikely. Nonetheless, the activity of the metal response element-binding transcription factor (MTF)-1 is regulated by Zn^{2+}, because two out of its six Cys2His2 zinc finger domains have a lower than usual affinity for Zn^{2+}. Consequently, MTF-1 DNA binding affinity, which is dependent on the saturation of all zinc fingers, is under the control of the availability of intracellular Zn^{2+} (Choi and Bird, 2014).

In addition to regulation via binding of Zn^{2+}, zinc fingers are targets for other mechanisms regulating transcription factor activity. This involves oxidation of the Zn^{2+}-binding cysteine thiols by oxidants such as nitric monoxide (NO), with subsequent release of Zn^{2+} (Kröncke, 2001). For example, NO inhibits the IL-1β-induced expression of IL-2 mRNA in T-cells by oxidizing at least two zinc finger-containing transcription factors, Sp1 and EGR-1 (Berendji et al., 1999).

Regulatory zinc binding sites do not necessarily have to be formed by amino acid residues from a single polypeptide chain. One example is the kinase Lck, which is essential during activation of the TCR. The TCR has no intrinsic kinase activity and therefore depends on phosphorylation by Lck to initiate signaling (Rossy et al., 2012). Lck contains two protein interface sites that are connected by Zn^{2+}. The first one bridges Zn^{2+} between the SH3 domains of two Lck molecules. This is thought to activate autophosphorylation on tyrosine 394 by inducing homodimerization of Lck (Romir et al., 2007). The second is a "zinc clasp" site that is formed by interaction of the N-terminal region of Lck, Zn^{2+}, and the cytoplasmic domains of the coreceptors CD4 and CD8. When CD4 or CD8 are recruited to the TCR signaling complex, they bring Lck into spatial proximity of its substrates. Here, it phosphorylates immunoreceptor tyrosine-based activation motifs (ITAM) of CD3 molecules, which are part of the TCR complex. The ITAMs serve as docking sites for scaffold proteins and the upstream kinase ZAP70, which is also phosphorylated by Lck (Huse et al., 1998; Kim et al., 2003; Lin et al., 1998). Lck is also one of the Src family of kinases involved in T-cell stimulation through the cell adhesion molecule CD44. The association between Lck and CD44 involves Zn^{2+}. This effect is specific for Lck, but it is not observed with the Src kinase Fyn, which also associates with CD44 (Lefebvre et al., 2010).

A regulation of Lck by zinc signals based on the reversible formation of protein interface sites is suggested by the observation that treatment of cells with extracellular Zn^{2+} stimulates Lck phosphorylation (Pernelle et al., 1991). It has been shown that during activation of the TCR by an antigen-presenting cell, an intracellular zinc signal is found right in the vicinity of the immunological synapse. The activation of Lck is affected by this zinc signal. However, the authors have found no indication for an impact on recruitment of Lck to CD4 or CD8 by immunoprecipitation of the TCR activation complex (Yu et al., 2011).

Zn^{2+} is secreted into the extracellular surroundings of mast cells upon their activation via cross-linking of the FCε-receptor (Nakashima-Kaneda et al., 2013). The physiological role for this release of zinc ions is still unclear, but could be a potential paracrine signaling function that participates in the immune response. However, no targets for extracellular zinc signals have been shown on immune cells, so far.

Zn^{2+} is well known as a regulator of ion channels in the central nervous system (Sensi et al., 2011) and ion channels are also important for immune cells. For example, extracellular ATP serves as a danger signal, and is recognized by P2 purinergic (P2X) receptors, which are widely expressed on immune cells (Di Virgilio and Vuerich, 2015). The P2X receptors are ligand-gated ion channels, and some of them, in particular $P2X_1$ and $P2X_7$, are inhibited in the presence of low micromolar concentrations of Zn^{2+} (Jiang, 2009). The G protein-coupled receptor (GPR39) is also known as "zinc sensing receptor," because Zn^{2+} is an effective activating ligand (Popovics and Stewart, 2011). While there are no reports on GPR39-mediated effects in immune cells, at least expression of the protein has been shown in hematopoietic cells of the bone marrow, nongerminal centers of secondary lymphoid tissue, and in alveolar macrophages (Uhlen et al., 2015).

An effect of Zn^{2+} on cell–cell adhesion has been shown in neuronal cells. For example, N-cadherin-mediated adhesion was modulated by Zn^{2+}, suggesting a role for extracellular zinc signals in synaptic plasticity (Heiliger et al., 2015). Zn^{2+} reduced the binding activity of N-cadherin. It would certainly be worthwhile to see if other forms of cadherins are also regulated by Zn^{2+}, especially E-cadherin, which is important for cell–cell contacts in epithelial tissue. If there was a comparable regulation by the free Zn^{2+} concentration, a release of Zn^{2+} by mast cells might influence the permeability of epithelia and thereby affect the influx of soluble factors as well as recruitment of additional immune cells by chemotaxis.

TARGETS OF ZINC SIGNALS

Genome-wide transcriptomic approaches to investigate the influence of Zn^{2+} on the mRNA expression in immune cells have shown that the expression of hundreds of genes is affected by the cells' zinc status, and that these genes differ with cell type (Cousins et al., 2003; Haase et al., 2007; Ryu et al., 2011). As mentioned before, T-cells are particularly sensitive to zinc deprivation (Fraker and King, 2004). This is at least partially based on a function of zinc signals in activation and cell cycle progression of T-cells (Kaltenberg et al., 2010; Plum et al., 2014; Yu et al., 2011). Similarly, development and activation of B-cells are controlled by the zinc transporter ZIP10 (Hojyo et al., 2014; Miyai et al., 2014).

Yet, the role of zinc signals is not limited to adaptive immune cells. Processes regulating the differentiation of myeloid cells into monocytes and maturation of dendritic cells are accompanied by changes in free Zn^{2+} and zinc transporter expression (Dubben et al., 2010; Kitamura et al., 2006). Maturation of dendritic cells can be triggered by stimulation of TLRs. TLRs are also involved in direct activation of innate immune cells. For monocytes, it has been shown that TLR-induced secretion of inflammatory cytokines depends on zinc signals (Haase et al., 2008). Within the cell, TLR-4 signaling depends on two major pathways. The myeloid differentiation primary response 88 (MyD88) pathway depends on zinc signals for its activation, whereas toll/IL-1 receptor domain–containing adaptor protein inducing IFN-β (TRIF)-mediated signaling is negatively regulated by Zn^{2+}. This indicates that zinc signals could be involved in balancing the two major TLR-dependent intracellular signaling pathways (Brieger et al., 2013). Finally, the formation of neutrophil extracellular traps, a matrix composed of DNA, chromatin, and granule proteins to capture extracellular bacteria, involves protein kinase C-mediated zinc signals (Hasan et al., 2012).

CONCLUSION

Zinc signaling is a complex event and only in recent years have we become aware of some of the molecular mechanisms that immune cells use for generation and recognition of zinc signals. Yet, some of these mechanisms, for example, the control of zinc transporter activity by posttranscriptional events, still remain largely unexplored. While some zinc-regulated signaling pathways are already known, there are probably several more for which a participation of zinc signals will be shown in the future. Taken together, all these pieces of information will finally allow a better understanding of zinc signaling in immune cells and will thus clarify the role of the essential trace element zinc in immunity and health.

REFERENCES

Andreini, C., Bertini, I., Rosato, A., 2009. Metalloproteomes: a bioinformatic approach. Acc. Chem. Res. 42, 1471–1479. http://dx.doi.org/10.1021/ar900015x.

Aydemir, T.B., Liuzzi, J.P., McClellan, S., Cousins, R.J., 2009. Zinc transporter ZIP8 (SLC39A8) and zinc influence IFN-gamma expression in activated human T cells. J. Leukoc. Biol. 86, 337–348. http://dx.doi.org/10.1189/jlb.1208759.

Bellomo, E., Massarotti, A., Hogstrand, C., Maret, W., 2014. Zinc ions modulate protein tyrosine phosphatase 1B activity. Metallomics 6, 1229–1239. http://dx.doi.org/10.1039/c4mt00086b.

Berendji, D., Kolb-Bachofen, V., Zipfel, P.F., Skerka, C., Carlberg, C., Kröncke, K.D., 1999. Zinc finger transcription factors as molecular targets for nitric oxide-mediated immunosuppression: inhibition of IL-2 gene expression in murine lymphocytes. Mol. Med. 5, 721–730.

Beyersmann, D., Haase, H., 2001. Functions of zinc in signaling, proliferation and differentiation of mammalian cells. BioMetals 14, 331–341. http://dx.doi.org/10.1023/A:1012905406548.

Brieger, A., Rink, L., Haase, H., 2013. Differential regulation of TLR-dependent MyD88 and TRIF signaling pathways by free zinc ions. J. Immunol. 191, 1808–1817. http://dx.doi.org/10.4049/jimmunol.1301261.

Brylinski, M., Skolnick, J., 2011. FINDSITE-metal: integrating evolutionary information and machine learning for structure-based metal-binding site prediction at the proteome level. Proteins 79, 735–751. http://dx.doi.org/10.1002/prot.22913.

Chaigne-Delalande, B., Lenardo, M.J., 2014. Divalent cation signaling in immune cells. Trends Immunol. 35, 332–344. http://dx.doi.org/10.1016/j.it.2014.05.001.

Choi, S., Bird, A.J., 2014. Zinc'ing sensibly: controlling zinc homeostasis at the transcriptional level. Metallomics 6, 1198–1215. http://dx.doi.org/10.1039/c4mt00064a.

Cousins, R.J., Blanchard, R.K., Popp, M.P., Liu, L., Cao, J., Moore, J.B., Green, C.L., 2003. A global view of the selectivity of zinc deprivation and excess on genes expressed in human THP-1 mononuclear cells. Proc. Natl. Acad. Sci. U.S.A. 100, 6952–6957. http://dx.doi.org/10.1073/pnas.0732111100.

Daniel, A.G., Peterson, E.J., Farrell, N.P., 2014. The bioinorganic chemistry of apoptosis: potential inhibitory zinc binding sites in caspase-3. Angew. Chem. Int. Ed. Engl. 53, 4098–4101. http://dx.doi.org/10.1002/anie.201311114.

Di Virgilio, F., Vuerich, M., 2015. Purinergic signaling in the immune system. Auton. Neurosci. http://dx.doi.org/10.1016/j.autneu.2015.04.011.

Dubben, S., Hönscheid, A., Winkler, K., Rink, L., Haase, H., 2010. Cellular zinc homeostasis is a regulator in monocyte differentiation of HL-60 cells by 1 alpha,25-dihydroxyvitamin D3. J. Leukoc. Biol. 87, 833–844. http://dx.doi.org/10.1189/jlb.0409241.

Figueroa, J.A.L., Vignesh, K.S., Deepe, G.S., Caruso, J., 2014. Selectivity and specificity of small molecule fluorescent dyes/probes used for the detection of Zn^{2+} and Ca^{2+} in cells. Metallomics 6, 301–315. http://dx.doi.org/10.1039/c3mt00283g.

Fraker, P.J., King, L.E., 2004. Reprogramming of the immune system during zinc deficiency. Annu. Rev. Nutr. 24, 277–298. http://dx.doi.org/10.1146/annurev.nutr.24.012003.132454.

Haase, H., Maret, W., 2003. Intracellular zinc fluctuations modulate protein tyrosine phosphatase activity in insulin/insulin-like growth factor-1 signaling. Exp. Cell Res. 291, 289–298.

Haase, H., Maret, W., 2005. Fluctuations of cellular, available zinc modulate insulin signaling via inhibition of protein tyrosine phosphatases. J. Trace Elem. Med. Biol. 19, 37–42.

Haase, H., Maret, W., 2008. Partial oxidation and oxidative polymerization of metallothionein. Electrophoresis 29, 4169–4176. http://dx.doi.org/10.1002/elps.200700922.

Haase, H., Rink, L., 2009a. Functional significance of zinc-related signaling pathways in immune cells. Annu. Rev. Nutr. 29, 133–152. http://dx.doi.org/10.1146/annurev-nutr-080508-141119.

Haase, H., Rink, L., 2009b. The immune system and the impact of zinc during aging. Immun. Ageing 6, 9. http://dx.doi.org/10.1186/1742-4933-6-9.

Haase, H., Rink, L., 2014a. Multiple impacts of zinc on immune function. Metallomics 6, 1175–1180. http://dx.doi.org/10.1039/c3mt00353a.

Haase, H., Rink, L., 2014b. Zinc signals and immune function. BioFactors 40, 27–40.

Haase, H., Mazzatti, D.J., White, A., Ibs, K.H., Engelhardt, G., Hebel, S., Powell, J.R., Rink, L., 2007. Differential gene expression after zinc supplementation and deprivation in human leukocyte subsets. Mol. Med. 13, 362–370. http://dx.doi.org/10.2119/2007–00049.

Haase, H., Ober-Blöbaum, J.L., Engelhardt, G., Hebel, S., Heit, A., Heine, H., Rink, L., 2008. Zinc signals are essential for lipopolysaccharide-induced signal transduction in monocytes. J. Immunol. 181, 6491–6502. http://dx.doi.org/10.4049/jimmunol.181.9.6491.

Haase, H., Hebel, S., Engelhardt, G., Rink, L., 2009. Zinc ions cause the thimerosal-induced signal of fluorescent calcium probes in lymphocytes. Cell Calcium 45, 185–191. http://dx.doi.org/10.1016/j.ceca.2008.09.003.

Hasan, R., Rink, L., Haase, H., 2012. Zinc signals in neutrophil granulocytes are required for the formation of neutrophil extracellular traps. Innate Immun. 19, 253–264. http://dx.doi.org/10.1177/1753425912458815.

Heiliger, E., Osmanagic, A., Haase, H., Golenhofen, N., Grabrucker, A.M., Weth, A., Baumgartner, W., 2015. N-Cadherin-mediated cell adhesion is regulated by extracellular Zn(2+). Metallomics 7, 355–362. http://dx.doi.org/10.1039/c4mt00300d.

Hennigar, S.R., Kelleher, S.L., 2015. TNFα post-translationally targets ZnT2 to accumulate zinc in lysosomes. J. Cell. Physiol. 230, 2345–2350. http://dx.doi.org/10.1002/jcp.24992.

Hojyo, S., Miyai, T., Fujishiro, H., Kawamura, M., Yasuda, T., Hijikata, A., Bin, B.-H., Irié, T., Tanaka, J., Atsumi, T., Murakami, M., Nakayama, M., Ohara, O., Himeno, S., Yoshida, H., Koseki, H., Ikawa, T., Mishima, K., Fukada, T., 2014. Zinc transporter SLC39A10/ZIP10 controls humoral immunity by modulating B-cell receptor signal strength. Proc. Natl. Acad. Sci. U.S.A. 111, 11786–11791. http://dx.doi.org/10.1073/pnas.1323557111.

Hotz, C., Lowe, N.M., Araya, M., Brown, K.H., 2003. Assessment of the trace element status of individuals and populations: the example of zinc and copper. J. Nutr. 133, 1563S–1568S.

Huber, K.L., Hardy, J.A., 2012. Mechanism of zinc-mediated inhibition of caspase-9. Protein Sci. 21, 1056–1065. http://dx.doi.org/10.1002/pro.2090.

Huse, M., Eck, M.J., Harrison, S.C., 1998. A Zn^{2+} ion links the cytoplasmic tail of CD4 and the N-terminal region of Lck. J. Biol. Chem. 273, 18729–18733.

Itoh, N., Shibayama, H., Kanekiyo, M., Namphung, D., Nakanishi, T., Matsuyama, A., Odani, T., Tanaka, K., 2005. Reduced bactericidal activity and nitric oxide production in metallothionein-deficient macrophages in response to lipopolysaccharide stimulation. Toxicology 216, 188–196. http://dx.doi.org/10.1016/j.tox.2005.08.005.

Jiang, L.-H., 2009. Inhibition of P2X(7) receptors by divalent cations: old action and new insight. Eur. Biophys. J. 38, 339–346. http://dx.doi.org/10.1007/s00249-008-0315-y.

Kaltenberg, J., Plum, L.M., Ober-Blöbaum, J.L., Hönscheid, A., Rink, L., Haase, H., 2010. Zinc signals promote IL-2-dependent proliferation of T cells. Eur. J. Immunol. 40, 1496–1503. http://dx.doi.org/10.1002/eji.200939574.

Kambe, T., Hashimoto, A., Fujimoto, S., 2014. Current understanding of ZIP and ZnT zinc transporters in human health and diseases. Cell. Mol. Life Sci. 71, 3281–3295. http://dx.doi.org/10.1007/s00018-014-1617-0.

Kanekiyo, M., Itoh, N., Kawasaki, A., Matsuda, K., Nakanishi, T., Tanaka, K., 2002a. Metallothionein is required for zinc-induced expression of the macrophage colony stimulating factor gene. J. Cell. Biochem. 86, 145–153. http://dx.doi.org/10.1002/jcb.10202.

Kanekiyo, M., Itoh, N., Kawasaki, A., Matsuyama, A., Matsuda, K., Nakanishi, T., Tanaka, K., 2002b. Metallothionein modulates lipopolysaccharide-stimulated tumour necrosis factor expression in mouse peritoneal macrophages. Biochem. J. 361, 363–369.

Kim, P.W., Sun, Z.-Y.J., Blacklow, S.C., Wagner, G., Eck, M.J., 2003. A zinc clasp structure tethers Lck to T cell coreceptors CD4 and CD8. Science 301, 1725–1728. http://dx.doi.org/10.1126/science.1085643.

Kitamura, H., Morikawa, H., Kamon, H., Iguchi, M., Hojyo, S., Fukada, T., Yamashita, S., Kaisho, T., Akira, S., Murakami, M., Hirano, T., 2006. Toll-like receptor-mediated regulation of zinc homeostasis influences dendritic cell function. Nat. Immunol. 7, 971–977. http://dx.doi.org/10.1038/ni1373.

Kojima, C., Kawakami, A., Takei, T., Nitta, K., Yoshida, M., 2007. Angiotensin-converting enzyme inhibitor attenuates monocyte adhesion to vascular endothelium through modulation of intracellular zinc. J. Pharmacol. Exp. Ther. 323, 855–860. http://dx.doi.org/10.1124/jpet.107.127944.

Krezel, A., Maret, W., 2007. Dual nanomolar and picomolar Zn(II) binding properties of metallothionein. J. Am. Chem. Soc. 129, 10911–10921. http://dx.doi.org/10.1021/ja071979s.

Krezel, A., Maret, W., 2008. Thionein/metallothionein control Zn(II) availability and the activity of enzymes. J. Biol. Inorg. Chem. 13, 401–409. http://dx.doi.org/10.1007/s00775-007-0330-y.

Kröncke, K.D., 2001. Zinc finger proteins as molecular targets for nitric oxide-mediated gene regulation. Antioxid. Redox Signal. 3, 565–575. http://dx.doi.org/10.1089/15230860152542934.

Lahiri, A., Abraham, C., 2014. Activation of pattern recognition receptors up-regulates metallothioneins, thereby increasing intracellular accumulation of zinc, autophagy, and bacterial clearance by macrophages. Gastroenterology 147, 835–846. http://dx.doi.org/10.1053/j.gastro.2014.06.024.

Lefebvre, D.C., Lai, J.C.Y., Maeshima, N., Ford, J.L., Wong, A.S.L., Cross, J.L., Johnson, P., 2010. CD44 interacts directly with Lck in a zinc-dependent manner. Mol. Immunol. 47, 1882–1889. http://dx.doi.org/10.1016/j.molimm.2010.03.018.

Leibbrandt, M.E., Koropatnick, J., 1994. Activation of human monocytes with lipopolysaccharide induces metallothionein expression and is diminished by zinc. Toxicol. Appl. Pharmacol. 124, 72–81. http://dx.doi.org/10.1006/taap.1994.1010.

Lichten, L.A., Cousins, R.J., 2009. Mammalian zinc transporters: nutritional and physiologic regulation. Annu. Rev. Nutr. 29, 153–176. http://dx.doi.org/10.1146/annurev-nutr-033009-083312.

Lin, R.S., Rodriguez, C., Veillette, A., Lodish, H.F., 1998. Zinc is essential for binding of p56(lck) to CD4 and CD8alpha. J. Biol. Chem. 273, 32878–32882.

Maret, W., Jacob, C., Vallee, B.L., Fischer, E.H., 1999. Inhibitory sites in enzymes: zinc removal and reactivation by thionein. Proc. Natl. Acad. Sci. U.S.A. 96, 1936–1940.

Maret, W., 2012. New perspectives of zinc coordination environments in proteins. J. Inorg. Biochem. 111, 110–116. http://dx.doi.org/10.1016/j.jinorgbio.2011.11.018.

Maret, W., 2013. Inhibitory zinc sites in enzymes. BioMetals 26, 197–204. http://dx.doi.org/10.1007/s10534-013-9613-7.

Maret, W., 2015. Analyzing free zinc(II) ion concentrations in cell biology with fluorescent chelating molecules. Metallomics 7, 202–211. http://dx.doi.org/10.1039/C4MT00230J.

Miyai, T., Hojyo, S., Ikawa, T., Kawamura, M., Irié, T., Ogura, H., Hijikata, A., Bin, B.-H., Yasuda, T., Kitamura, H., Nakayama, M., Ohara, O., Yoshida, H., Koseki, H., Mishima, K., Fukada, T., 2014. Zinc transporter SLC39A10/ZIP10 facilitates antiapoptotic signaling during early B-cell development. Proc. Natl. Acad. Sci. U.S.A. 111, 11780–11785. http://dx.doi.org/10.1073/pnas.1323549111.

Nakashima-Kaneda, K., Matsuda, A., Mizuguchi, H., Sasaki-Sakamoto, T., Saito, H., Ra, C., Okayama, Y., 2013. Regulation of IgE-dependent zinc release from human mast cells. Int. Arch. Allergy Immunol. 161 (Suppl.), 44–51. http://dx.doi.org/10.1159/000350359.

Overbeck, S., Rink, L., Haase, H., 2008. Modulating the immune response by oral zinc supplementation: a single approach for multiple diseases. Arch. Immunol. Ther. Exp. Warsz. 56, 15–30.

Pernelle, J.J., Creuzet, C., Loeb, J., Gacon, G., 1991. Phosphorylation of the lymphoid cell kinase p56lck is stimulated by micromolar concentrations of Zn^{2+}. FEBS Lett. 281, 278–282.

Plum, L.M., Brieger, A., Engelhardt, G., Hebel, S., Nessel, A., Arlt, M., Kaltenberg, J., Schwaneberg, U., Huber, M., Rink, L., Haase, H., 2014. PTEN-inhibition by zinc ions augments interleukin-2-mediated Akt phosphorylation. Metallomics 2, 1277–1287. http://dx.doi.org/10.1039/c3mt00197k.

Popovics, P., Stewart, A.J., 2011. GPR39: a Zn(2+)-activated G protein-coupled receptor that regulates pancreatic, gastrointestinal and neuronal functions. Cell. Mol. Life Sci. 68, 85–95. http://dx.doi.org/10.1007/s00018-010-0517-1.

Prasad, A.S., 2007. Zinc: mechanisms of host defense. J. Nutr. 137, 1345–1349.

Prasad, A.S., 2008. Zinc in human health: effect of zinc on immune cells. Mol. Med. 14, 353–357. http://dx.doi.org/10.2119/2008-00033.

Prasad, A.S., 2014a. Impact of the discovery of human zinc deficiency on health. J. Trace Elem. Med. Biol. 28, 357–363. http://dx.doi.org/10.1016/j.jtemb.2014.09.002.

Prasad, A.S., 2014b. Zinc: an antioxidant and anti-inflammatory agent: role of zinc in degenerative disorders of aging. J. Trace Elem. Med. Biol. 28, 364–371. http://dx.doi.org/10.1016/j.jtemb.2014.07.019.

Prasad, A.S., 2014c. Zinc is an antioxidant and anti-inflammatory agent: its role in human health. Front. Nutr. 1, 14. http://dx.doi.org/10.3389/fnut.2014.00014.

Rana, U., Kothinti, R., Meeusen, J., Tabatabai, N.M., Krezoski, S., Petering, D.H., 2008. Zinc binding ligands and cellular zinc trafficking: apo-metallothionein, glutathione, TPEN, proteomic zinc, and Zn-Sp1. J. Inorg. Biochem. 102, 489–499. http://dx.doi.org/10.1016/j.jinorgbio.2007.10.030.

Romir, J., Lilie, H., Egerer-Sieber, C., Bauer, F., Sticht, H., Muller, Y.A., 2007. Crystal structure analysis and solution studies of human Lck-SH3; zinc-induced homodimerization competes with the binding of proline-rich motifs. J. Mol. Biol. 365, 1417–1428. http://dx.doi.org/10.1016/j.jmb.2006.10.058.

Rossy, J., Williamson, D.J., Gaus, K., 2012. How does the kinase Lck phosphorylate the T cell receptor? Spatial organization as a regulatory mechanism. Front. Immunol. 3, 167. http://dx.doi.org/10.3389/fimmu.2012.00167.

Ryu, M.-S., Langkamp-Henken, B., Chang, S.-M., Shankar, M.N., Cousins, R.J., 2011. Genomic analysis, cytokine expression, and microRNA profiling reveal biomarkers of human dietary zinc depletion and homeostasis. Proc. Natl. Acad. Sci. U.S.A. 108, 20970–20975. http://dx.doi.org/10.1073/pnas.1117207108.

Sensi, S.L., Paoletti, P., Koh, J.-Y., Aizenman, E., Bush, A.I., Hershfinkel, M., 2011. The neurophysiology and pathology of brain zinc. J. Neurosci. 31, 16076–16085. http://dx.doi.org/10.1523/JNEUROSCI.3454-11.2011.

Spahl, D.U., Berendji-Grün, D., Suschek, C.V., Kolb-Bachofen, V., Kröncke, K.-D., 2003. Regulation of zinc homeostasis by inducible NO synthase-derived NO: nuclear metallothionein translocation and intranuclear Zn^{2+} release. Proc. Natl. Acad. Sci. U.S.A. 100, 13952–13957. http://dx.doi.org/10.1073/pnas.2335190100.

Taylor, K.M., Hiscox, S., Nicholson, R.I., Hogstrand, C., Kille, P., 2012. Protein kinase CK2 triggers cytosolic zinc signaling pathways by phosphorylation of zinc channel ZIP7. Sci. Signal. 5, ra11. http://dx.doi.org/10.1126/scisignal.2002585.

Tonks, N.K., 2006. Protein tyrosine phosphatases: from genes, to function, to disease. Nat. Rev. Mol. Cell Biol. 7, 833–846. http://dx.doi.org/10.1038/nrm2039.

Truong-Tran, A.Q., Carter, J., Ruffin, R.E., Zalewski, P.D., 2001. The role of zinc in caspase activation and apoptotic cell death. BioMetals 14, 315–330. http://dx.doi.org/10.1023/A:1012993017026.

Ugajin, T., Nishida, K., Yamasaki, S., Suzuki, J., Mita, M., Kubo, M., Yokozeki, H., Hirano, T., 2015. Zinc-binding metallothioneins are key modulators of IL-4 production by basophils. Mol. Immunol. 66, 180–188. http://dx.doi.org/10.1016/j.molimm.2015.03.002.

Uhlen, M., Fagerberg, L., Hallstrom, B.M., Lindskog, C., Oksvold, P., Mardinoglu, A., Sivertsson, A., Kampf, C., Sjostedt, E., Asplund, A., Olsson, I., Edlund, K., Lundberg, E., Navani, S., Szigyarto, C.A.-K., Odeberg, J., Djureinovic, D., Takanen, J.O., Hober, S., Alm, T., Edqvist, P.-H., Berling, H., Tegel, H., Mulder, J., Rockberg, J., Nilsson, P., Schwenk, J.M., Hamsten, M., von Feilitzen, K., Forsberg, M., Persson, L., Johansson, F., Zwahlen, M., von Heijne, G., Nielsen, J., Ponten, F., 2015. Proteomics. Tissue-based map of the human proteome. Science 347 (6220), 1260419. http://dx.doi.org/10.1126/science.1260419.

Velázquez-Delgado, E.M., Hardy, J.A., 2012. Zinc-mediated allosteric inhibition of caspase-6. J. Biol. Chem. 287, 36000–36011. http://dx.doi.org/10.1074/jbc.M112.397752.

Von Bülow, V., Rink, L., Haase, H., 2005. Zinc-mediated inhibition of cyclic nucleotide phosphodiesterase activity and expression suppresses TNF-alpha and IL-1 beta production in monocytes by elevation of guanosine 3′,5′-cyclic monophosphate. J. Immunol. 175 (7), 4697–4705. http://dx.doi.org/10.4049/jimmunol.175.7.4697.

Wellenreuther, G., Cianci, M., Tucoulou, R., Meyer-Klaucke, W., Haase, H., 2009. The ligand environment of zinc stored in vesicles. Biochem. Biophys. Res. Commun. 380, 198–203. http://dx.doi.org/10.1016/j.bbrc.2009.01.074.

Wertz, I.E., O'Rourke, K.M., Zhou, H., Eby, M., Aravind, L., Seshagiri, S., Wu, P., Wiesmann, C., Baker, R., Boone, D.L., Ma, A., Koonin, E.V., Dixit, V.M., 2004. De-ubiquitination and ubiquitin ligase domains of A20 downregulate NF-kappaB signalling. Nature 430, 694–699. http://dx.doi.org/10.1038/nature02794.

Wilson, M., Hogstrand, C., Maret, W., 2012. Picomolar concentrations of free zinc(II) ions regulate receptor protein-tyrosine phosphatase β activity. J. Biol. Chem. 287, 9322–9326. http://dx.doi.org/10.1074/jbc.C111.320796.

Yamasaki, S., Sakata-Sogawa, K., Hasegawa, A., Suzuki, T., Kabu, K., Sato, E., Kurosaki, T., Yamashita, S., Tokunaga, M., Nishida, K., Hirano, T., 2007. Zinc is a novel intracellular second messenger. J. Cell Biol. 177, 637–645. http://dx.doi.org/10.1083/jcb.200702081.

Yamasaki, S., Hasegawa, A., Hojyo, S., Ohashi, W., Fukada, T., Nishida, K., Hirano, T., 2012. A novel role of the L-type calcium channel α1D subunit as a gatekeeper for intracellular zinc signaling: zinc wave. PLoS One 7, e39654. http://dx.doi.org/10.1371/journal.pone.0039654.

Yu, M., Lee, W.-W., Tomar, D., Pryshchep, S., Czesnikiewicz-Guzik, M., Lamar, D.L., Li, G., Singh, K., Tian, L., Weyand, C.M., Goronzy, J.J., 2011. Regulation of T cell receptor signaling by activation-induced zinc influx. J. Exp. Med. 208, 775–785. http://dx.doi.org/10.1084/jem.20100031.

Chapter 22

Posttranslational Mechanisms of Zinc Signaling

Thirayost Nimmanon[1,2], Kathryn M. Taylor[1]

[1]Cardiff University, Cardiff, United Kingdom; [2]Phramongkutklao College of Medicine, Bangkok, Thailand

INTRODUCTION

Even though our body contains as little as 30 nmol or 2 g of zinc, thus classified as a trace element, we cannot live without it as it plays indispensable roles in various biological processes. Like calcium, zinc is now recognized as a second messenger that is released from cellular stores upon activation by an extracellular stimulus, producing the so-called "zinc wave," and subsequently triggering multiple cellular cascades (Yamasaki et al., 2007). These altered signaling pathways contribute to cellular changes such as proliferation and migration in breast cancer cells (Taylor et al., 2008, 2012). The zinc released from stores exerts its action, at least partly, through its strong reversible inhibitory effect on a family of enzymes called protein tyrosine phosphatases (PTP), resulting in widespread activation of cellular tyrosine kinases (Wilson et al., 2012). This remarkable early response to an extracellular stimulus, which becomes observable on a timescale of minutes, is referred to as the transcription-independent "fast" or "early" zinc signaling, in contrast to the transcription-dependent "late" zinc signaling in which cellular zinc homeostasis is established by changes in expression of zinc-transport proteins (Kambe et al., 2015; Yamasaki et al., 2007). These facts support the potential importance of zinc signaling in the control of cellular physiology on a scale comparable to that for calcium signaling. Nevertheless, in comparison to calcium, very little is known about the control of cellular zinc homeostasis.

ZINC-TRANSPORT PROTEINS

The charged divalent cation zinc is not capable of passively traversing cellular membranes and therefore the mobilization of zinc between cellular compartments requires dedicated transport proteins. It is perhaps surprising that there are 24 proteins designated as zinc-transport proteins when there is considered to be a small amount of zinc in the cell, with the total cellular zinc concentrations typically not exceeding a few hundred micromolar and the free zinc concentrations within the picomolar range (Maret, 2013). These zinc-transport proteins consist of 14 members of the Zrt- and Irt-like protein (ZIP) channel family (SLC39A) and 10 members of the zinc transporter (ZnT) transporter family (SLC30A). The ZIP channels increase cytoplasmic free zinc levels by importing zinc either from outside the cell or releasing zinc from cellular stores in, for example, the ER and Golgi, whereas the ZnT transporters counteract the ZIP channels by mobilizing zinc in the opposite direction (Kambe et al., 2015). Furthermore, it is noteworthy that besides these two families of zinc-transport proteins, zinc homeostasis also involves more than 12 members of the cysteine-rich metallothioneins, which help control free zinc levels within the appropriate range (Colvin et al., 2010).

Zrt- and Irt-Like Protein Channels

Human ZIP channels are phylogenetically grouped into four subfamilies: gufA (ZIP11), subfamily I (ZIP9), subfamily II (ZIP1–3), and the LIV-1 subfamily (ZIP4–8, ZIP10, and ZIP12–14) (Taylor and Nicholson, 2003; Fig. 22.1). All the ZIP channels are located on the plasma membrane, except ZIP7 which is in the ER and Golgi, and ZIP9, ZIP11, and ZIP13 which are in the Golgi (Fukada and Kambe, 2011), thereby capable of mobilizing zinc from the cellular stores rather than importing zinc from the extracellular space. According to computational analysis, ZIP channels are predicted to have eight transmembrane domains (TMs), with extracellular amino- and carboxyl-termini (Taylor and Nicholson, 2003). Between TM3 and TM4, there is a long cytoplasmic loop, which contains histidine-rich clusters predicted to be zinc-binding and important for zinc transport (Taylor and Nicholson, 2003).

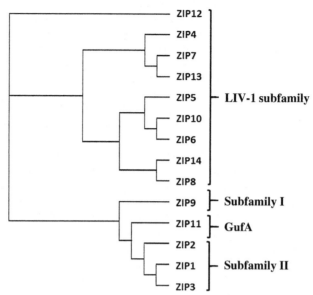

FIGURE 22.1 **Zrt- and Irt-like protein (ZIP) channels are classified into four subfamilies.** This unrooted phylogram shows classification of human ZIP channels into four subfamilies, consisting of gufA (ZIP11), subfamily I (ZIP9), subfamily II (ZIP1–3), and the LIV-1 subfamily (ZIP4–8, ZIP10, and ZIP12–14).

Specifically, for the LIV-1 subfamily of ZIP channels, the highly conserved TM5 contains the motif HEXPHEXGD (H=histidine, E=glutamate, X=any amino acid, P=proline, G=glycine, and D=aspartate), which is unique for this subfamily. Importantly, this motif matches the catalytic zinc-binding site of zincin and peptide deformylase (PDF; Taylor and Nicholson, 2003), the two members of the zinc metalloprotease superfamily, which also have a HEXXH motif as a signature zinc-binding sequence (Dardel et al., 1998; Hooper, 1994) that is essential for their function. Remarkably, ZIP8 and ZIP14, which are grouped in the same subbranch of the phylogenetic tree (Fig. 22.1), have the initial histidine residue of this motif (**H**EXPHEXGD) replaced by glutamine (both with sequence **E**EFPHELGD, where F=phenylalanine, L=leucine), suggesting that these two proteins may have preference for mobilizing other ions besides zinc (Taylor and Nicholson, 2003). Consistent with this prediction, both ZIP8 and ZIP14 have been shown to also transport cadmium and manganese (Fujishiro et al., 2012), as well as iron (Wang et al., 2012; Zhao et al., 2010). Furthermore, the presence of a conserved mixed charge region immediately upstream of TM4, a common feature of the LIV-1 subfamily (Taylor and Nicholson, 2003), suggests that this may be a reactive region with other proteins and therefore has a potential role in zinc transport.

Zinc Transporters

There are 10 members of ZnT transporters, which are grouped into subfamilies I to IV; however, ZnT9 has now been demonstrated to be a nuclear coactivator (Kambe, 2012). Although the mechanism of zinc transport is still obscure for the ZIP channels, the ZnT transporters are reported to mediate zinc transport by catalyzing H^+/Zn^{2+} exchange (Ohana et al., 2009). Most of the ZnT transporters are predicted to have six TMs, with the exception of ZnT5, which is thought to have 12 TM (Cousins et al., 2006). Both amino- and carboxyl-termini are on the cytoplasmic side, and the long intracytoplasmic loop between TM4 and TM5 contains a histidine-rich region (Palmiter and Huang, 2004), potentially important for zinc sensing and selectivity. This region is absent and replaced by a serine-rich region in ZnT6 (Seve et al., 2004), which may imply alternative functions for this transporter involving other metal ions. In contrast to other ZnT transporters, which are located intracellularly, ZnT1 is uniquely located on the plasma membrane and thereby crucially needed for cellular zinc export (Palmiter and Findley, 1995).

POSTTRANSLATIONAL MODIFICATION OF ZINC-TRANSPORT PROTEINS

Regulation at the transcriptional level in response to changes in body zinc status, hormones, or cytokines, which participates in the "late" zinc signaling, has been reported for various zinc-transport proteins, both ZnT transporters and ZIP channels (Cousins et al., 2006). Nevertheless, only a handful of data exists regarding the posttranslational control of zinc-transport proteins, which is described predominantly for the ZIP family of zinc channels.

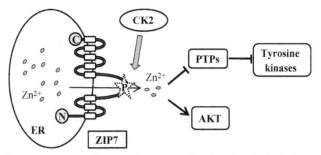

FIGURE 22.2   ZIP7-mediated zinc signaling is triggered by CK2 phosphorylation. ZIP7 is activated by CK2-mediated phosphorylation on S275 and S276, which are located in the cytoplasmic loop between TM3 and TM4, resulting in ZIP7-mediated zinc release from stores. The released zinc, in turn, indirectly causes widespread activation of cellular kinases through its inhibitory action on protein tyrosine phosphatases and also directly activates AKT.

Phosphorylation

In order for zinc-transport proteins to promptly participate in the "early" zinc signaling, they need to be regulated functionally or posttranslationally rather than transcriptionally. Our group has demonstrated a posttranslational modification that triggers ZIP7 activity in response to extracellular stimuli (Taylor et al., 2012). In this study, ZIP7 was shown to be activated by protein kinase CK2-mediated serine phosphorylation on S275 and S276, both of which match the CK2 consensus motif **S/T**-X-X-E (S = serine, T = threonine, E = glutamine, X = any amino acid) (Franchin et al., 2015), resulting in zinc release from ER stores (Taylor et al., 2012). This increase in cytosolic zinc bioavailability in turn causes a direct activation of AKT (Lee et al., 2009) and an indirect activation of ERK1/2 via zinc's inhibitory action on PTPs (Munkley et al., 2015; Wilson et al., 2012; Fig. 22.2), resulting in consequential promotion of cancer cell proliferation and migration due to increased activation of tyrosine kinases (Taylor et al., 2008). Given that CK2 is recognized as an oncogene upregulated in various primary tumors, these remarkable findings highlight that ZIP7-mediated zinc signaling triggered by phosphorylation is an important downstream cascade contributing to the CK2-associated cancer aggressive phenotypes and poor prognosis (Nimmanon and Taylor, 2015).

This discovery that ZIP7 function is posttranslationally regulated by phosphorylation allows us to consider whether other ZIP channels are controlled by a similar modification. Available online databases have reported multiple phosphorylation sites that have been confirmed in mammalian cells by mass spectrometry in some ZIP channels other than ZIP7 (Hornbeck et al., 2012; Fig. 22.3). Although extracellular kinases may be important (Yalak and Vogel, 2012), little is known about their identity and function to date. To discover new potential phosphorylation sites, the search was limited to the cytoplasmic loops (as for S275 and S276 of ZIP7), thereby assuming spatial approachability to cellular kinases (Fig. 22.3). Using this method, a number of potential serine/threonine and tyrosine phosphorylation sites can be predicted for ZIP3, ZIP6, ZIP8, and ZIP10 (Hogstrand et al., 2009; Fig. 22.3). All of the predicted sites are located in the large cytoplasmic loop between TM3 and TM4, except for one site in ZIP8 (T424) which is located in the small cytoplasmic loop between TM7 and TM8. Interestingly, among these potential phosphorylation sites, not only do residues S478 of ZIP6, S573 of ZIP10, and S129 of ZIP3 match the consensus sequence for CK2 binding, but CK2 is also the highest scoring kinase that is predicted to phosphorylate these sites according to the online phosphorylation site platforms PhosphoNet (Kinexus Bioinformatics Corporation) and PhosphoSitePlus (Hornbeck et al., 2012; Fig. 22.3).

It is noteworthy that potential phosphorylation sites are also detected in many ZnT transporters (Hogstrand et al., 2009). Interestingly, some potential sites in ZnT6 (Huang et al., 2002) and ZnT8 (Chimienti et al., 2004) are also predicted to be CK2-phosphorylated. However, whether phosphorylation has any functional relevance to the posttranslational regulation of this family of zinc transporters is yet to be experimentally determined. It is noteworthy that most consensus sequences for phosphorylation site predictions are quite simple, leading to the generation of false positives. Therefore, it is important that potential phosphorylation sites are properly examined experimentally.

Proteolytic Cleavage

Another mechanism that has been demonstrated to be required for localization and zinc-transporting function of zinc-transport proteins is proteolytic cleavage. During zinc deficiency, ZIP4 is proteolytically cleaved at the extracellular N-terminus immediately proximal to TM1 (Kambe and Andrews, 2009; Fig. 22.4A). The processed 37-kDa ZIP4 peptide, lacking the ectodomain and detected as the predominant immunoreactive form on a western blot, is then retained in the plasma membrane where it actively imports zinc, enhancing absorption from the intestinal lumen (Kambe and Andrews, 2009; Fig. 22.4A).

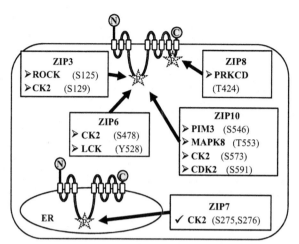

FIGURE 22.3 Phosphorylation sites are present in many zinc-transport proteins. Many phosphorylation sites have been reported for mammalian zinc-transport proteins (Hogstrand et al., 2009). Among these sites, only S275 and S276 of ZIP7 have been experimentally proven to be phosphorylated by CK2, triggering ZIP7 zinc-transport function.

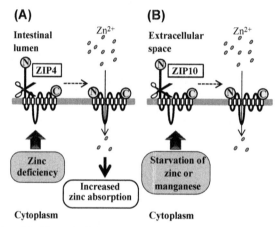

FIGURE 22.4 ZIP4 and ZIP10 are activated by N-terminal cleavage. ZIP4 (A) and ZIP10 (B) undergo N-terminal cleavage in response to starvation of zinc (ZIP4 and ZIP10) or manganese (only ZIP10), resulting in activation of their zinc-transport function.

Following the establishment of N-terminal cleavage as an activation process for ZIP4, similar processes, although in different contexts, have been implicated for two other members of the LIV-1 subfamily of ZIP channels. These are ZIP10 (Schmitt-Ulms et al., 2009; Fig. 22.4B) and ZIP6 (Hogstrand et al., 2013; Fig. 22.5), which are both grouped together in the same subbranch of the phylogenetic tree of the ZIP family of zinc-transport proteins (Fig. 22.1). Interestingly, both of these channels have been identified as candidate interactors with the cellular prion protein (PrP^C) according to interactome analyses (Watts et al., 2009). In response to deprivation of either zinc or manganese, ZIP10 is activated by shedding of its PrP-like N-terminal ectodomain (Schmitt-Ulms et al., 2009; Fig. 22.4B), thereby straightforwardly counteracting the improperly decreased availability of these metals. This phenomenon in ZIP10 is reminiscent of the prion protein, the gene of which descends from the LIV-1 subfamily of ZIP channels (Schmitt-Ulms et al., 2009) and the posttranslational proteolytic mechanism which also involves ectodomain shedding (Altmeppen et al., 2012).

Thereafter, our group observed a similar mechanism for ZIP6, which is involved in cell migration during both early embryonic development (Yamashita et al., 2004) and cancer progression (Hogstrand et al., 2013; Nimmanon and Taylor, 2014). ZIP6 is produced as a pro-protein in the endoplasmic reticulum (Hogstrand et al., 2013) where it remains until triggered by N-terminal cleavage to relocate to the plasma membrane (Fig. 22.5). It is the plasma membrane-located form that is then able to influx zinc into the cytoplasm from outside the cell (Fig. 22.5). This ZIP6-mediated zinc influx results in a consequential zinc-induced inactivating phosphorylation of GSK-3β, leading to nuclear retention of Snail (Zhou et al., 2004), down-regulation of E-cadherin, and cell detachment (Hogstrand et al., 2013; Nimmanon and Taylor, 2014; Fig. 22.5).

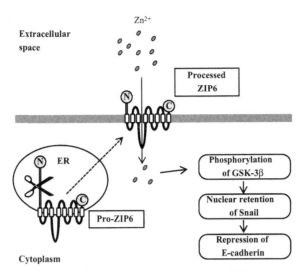

FIGURE 22.5 **ZIP6 plasma membrane relocation is triggered by N-terminal cleavage.** ZIP6 pro-protein is activated by N-terminal cleavage and relocation to the plasma membrane where it facilitates zinc uptake. The increased cytosolic zinc causes inhibitory phosphorylation of GSK-3β, resulting in nuclear retention of Snail, repression of E-cadherin, cell migration, and metastasis.

Collectively, these findings suggest that ectodomain proteolytic processing is an important mechanism for the functional regulation of ZIP channels as has been shown for ZIP4, ZIP10, and ZIP6. Whether this mechanism is important for other ZIP transporters will need to be experimentally confirmed.

Ubiquitination, Glycosylation, and SUMOylation

Ubiquitination (also known as ubiquitylation or ubiquitinylation) is a posttranslational modification in which the 8.5-kDa ubiquitously expressed globular protein ubiquitin is covalently conjugated to other proteins, typically leading to proteasomal degradation of the modified proteins (Weissman, 2001). Uniquely for plasma membrane proteins, this modification predestines the proteins to degradation in vacuoles or lysosomes instead of proteasomes (Weissman, 2001). This process has been implicated as a means whereby zinc-transport proteins are negatively controlled.

Zrt1, a primitive cellular zinc importer found in yeast, is the first zinc-transport protein that has been demonstrated to be negatively controlled by ubiquitination-mediated endocytosis and degradation in response to zinc and cadmium (Gitan et al., 2003). The ubiquitination of Zrt1 occurs at K195, which is located in the cytoplasmic loop between TM3 and TM4, the region where, like mammalian ZIP channels, the histidine-rich motif is present (Gitan et al., 2003). Even though the histidine-rich motif is necessary for Zrt1-mediated zinc transport, it has no known role in metal-responsive inactivation by ubiquitination (Gitan et al., 2003).

For mammalian ZIP channels, ZIP4 endocytosis and degradation has also been shown to be accelerated by ubiquitination in response to as little as low micromolar extracellular zinc concentrations, resulting in a decrease in zinc uptake (Mao et al., 2007). The ubiquitination is proposed to occur at residue K463, a highly conserved residue in human ZIP4, which is located in the cytoplasmic loop between TM3 and TM4, equivalent to the known ubiquitination site in Zrt1 (Mao et al., 2007). However, in contrast to Zrt1, this ubiquitination requires the histidine-rich cluster located in the same cytoplasmic loop as the ubiquitination site (Mao et al., 2007). Whether the histidine-rich region on this cytoplasmic loop is involved in regulating the interaction with ubiquitin in other ZIP channels is yet to be determined.

Ubiquitination has also been reported for ZIP13, and an excessive degradation of the pathogenic ZIP13 mutant G64D through the ubiquitin-proteasome pathway has been shown to explain the loss of function of the mutant in spondylocheirodysplastic form of Ehlers–Danlos syndrome, a genetic disorder of connective tissues, bones, and teeth (Bin et al., 2014). However, the ubiquitination site involved in the degradation process of ZIP13 has not yet been confirmed, and the precise molecular mechanism underlying degradation of this ZIP channel is yet to be elucidated. Additionally, our group has experimentally demonstrated the association of ZIP6 with ubiquitin, which may explain the constantly low expression of this ZIP channel (Taylor et al., 2003).

Remarkably, ubiquitination has been revealed to also be an important posttranslational mechanism regulating the function of the zinc exporter ZnT2, a member of the ZnT family of zinc-transport proteins, which plays an important role in the secretion of zinc into milk of the mammalian mammary gland (Seo et al., 2014). This study revealed that prolactin regulates

ZnT2 not only transcriptionally by enhancing its expression through the Jak2/STAT5 pathway but also posttranslationally by stimulating its ubiquitination (Seo et al., 2014). In contrast to ZIP channels, ZnT2 is shown to be ubiquitinated at the N-terminally located K4 and K6 residues (Seo et al., 2014). Interestingly, this modification initially results in a transient accumulation of ZnT2 in vesicles, thereby enhancing ZnT2-mediated zinc secretion, but eventually results in proteasome-dependent degradation of ZnT2 (Seo et al., 2014). These findings support the view that ubiquitination plays an important role in the posttranslational control of both ZIP and ZnT families of zinc-transport proteins, although there are variations between different proteins.

Glycosylation is a complex posttranslational modification that involves attachment of sugar molecules to a protein (Spiro, 2002). Many ZIP channels, such as ZIP4 (Kambe and Andrews, 2009), ZIP6 (Taylor et al., 2003), ZIP10 (Ehsani et al., 2012), and ZIP14 (Taylor et al., 2005), have been reported to be glycosylated. The role of this modification in the functional control of these proteins has, however, only been thoroughly investigated in ZIP14, which is capable of transporting iron and has been shown to be both glycosylated and ubiquitinated (Zhao et al., 2014). Interestingly, ZIP14 was also demonstrated to be degraded in response to iron deficiency through an alternative and novel proteasomal degradation pathway that is dependent upon glycosylation (Zhao et al., 2014). This removal mechanism is in contrast to the canonical endocytic pathway for the degradation of plasma membrane proteins, which primarily involves ubiquitination and lysosomal degradation of the proteins (Weissman, 2001).

Another posttranslational modification, SUMOylation, has been reported for ZIP7 (Rosas-Acosta et al., 2005). SUMOylation is a ubiquitination-like process in which the polypeptides called small ubiquitin-related modifiers (SUMOs) are attached to a protein (Wang and Dasso, 2009). However, the role of this modification in ZIP7 regulation has not been investigated.

These findings altogether suggest that a variety of posttranslational modifications of zinc-transport proteins, including ubiquitination, glycosylation, and SUMOylation, may serve as potential cellular protective mechanisms against alterations of cellular levels of zinc and other related metals as well as provide physiological control of zinc-transport protein function, such as ZnT2 function in lactation.

APPLICATIONS

The Use of the Phosphorylated Form of Zinc-Transport Proteins as Biomarkers

Many zinc-transport proteins have been linked to the pathogenesis of human diseases, including cancer (Nimmanon and Taylor, 2014). For example, an abnormal increase in ZIP7 contributes to breast cancer aggressiveness and resistance to antihormonal therapeutic agents such as tamoxifen (Taylor et al., 2007, 2008). Given that ZIP7 function is triggered by CK2-mediated phosphorylation, the phosphorylated form of ZIP7 should better correspond to ZIP7 activity than the total ZIP7 level, and thereby has higher potential to be used clinically as a biomarker of aggressive or endocrine-resistant breast cancer. Moreover, since ZIP7 is proved to be an important target of CK2 that mediates carcinogenic pathways (Nimmanon and Taylor, 2015; Taylor et al., 2012), the level of phosphorylated ZIP7 in cells should indicate the activation state of ZIP7 which in turn should predict the likelihood of successful treatment with a CK2 inhibitor. Based upon this hypothesis, a monoclonal antibody against ZIP7 when phosphorylated on residues S275 and S276 has recently been developed in our group and is currently under investigation in various cell types as well as clinical samples to determine its usefulness as a biomarker.

Additionally, the discovery that ZIP7 is phosphorylated may pave the way to further such discoveries relating to other zinc transporters. This can provide experimental tools such as antibodies that recognize the active and phosphorylated forms of these molecules, which should aid in biomarker detection in relevant disease states.

Modulating Modifications of Zinc-Transport Proteins as Therapeutic Strategies

It has been demonstrated that many zinc-transport proteins are regulated by different posttranslational mechanisms including phosphorylation, proteolytic cleavage, ubiquitination, and glycosylation. Manipulations of these processes, either by enhancing or inhibiting them, might therefore become potential therapeutic strategies for different disease states. For example, in the context of cancer, breast cancer metastasis may be prevented by impeding the N-terminal cleavage of ZIP6, which is required for plasma membrane relocation of the protein, which is obviously essential for its zinc-transporting function (Hogstrand et al., 2013), or, on the contrary, enhancing ubiquitination, which is the potential degradation mechanism for ZIP6 (Taylor et al., 2003). Furthermore, given that studies on ZIP4 have revealed many modifications such as ubiquitination (Mao et al., 2007) and ectodomain cleavage (Kambe and Andrews, 2009), which have been implicated in ZIP4 functional control, a method that successfully manipulates these processes may also benefit patients with acrodermatitis enteropathica, a rare genetic disorder associated with defective intestinal zinc uptake caused by ZIP4 mutation (Küry et al., 2002). Furthermore, this type of manipulation could also benefit pancreatic cancer patients as aberrant ZIP4 expression has been implicated in the carcinogenesis and progression of this disease (Li et al., 2007).

HOW OTHER MINERALS ARE AFFECTED

Many zinc-transport proteins, mainly but not limited to the ZIP family, have been reported to have other metals as substrates (Jeong and Eide, 2013). For example, ZIP7, which is known to be activated by phosphorylation, has been demonstrated to also transport manganese, as evidenced by the ability of ZIP7 (previously known as KE4) to compensate for the function of IAR1, a homolog of ZIP7 in Arabidopsis (Lasswell et al., 2000). Furthermore, many modifications have been reported to be a response of deprivation of metals other than zinc, such as the ZIP10 N-terminal cleavage which is triggered by manganese starvation (Ehsani et al., 2012) and ubiquitination-mediated inactivation of the yeast zinc transporter Zrt1, which is induced by cadmium. These findings imply that, regardless of the designation of zinc-transport proteins, some of these proteins may in fact also be involved in regulating homeostasis of other metals. The posttranslational mechanisms of these proteins may therefore be elucidated as important means for cellular controls of not only zinc but also these other metals. Additionally, given that zinc-transport proteins can be regulated posttranslationally, the transporters that are dedicated to mobilizing other metals may also be regulated in the same manner.

KEY FACTS

- Regulation of cellular zinc homeostasis involves two families of zinc-transport proteins: ZIP channels (SLC39A, zinc importers) and ZnT transporters (SLC30A, zinc exporters).
- ZIP channels mobilize zinc from either the extracellular space or the cellular stores to the cytoplasm, whereas ZnT transporters mobilize zinc in the opposite direction.
- Posttranslational mechanisms such as phosphorylation, proteolytic cleavage, ubiquitination, and glycosylation have been implicated for functional control of many zinc-transport proteins.
- The knowledge of posttranslational modifications of zinc-transport proteins may lead to invention of new biomarkers or therapeutic strategies for human diseases including cancer.
- Given that many zinc-transport proteins can mobilize other metals, posttranslational modifications of these proteins may therefore be involved in the regulation of these metals.

MINI DICTIONARY OF TERMS

Glycosylation Glycosylation is a complex posttranslational modification in which sugar molecules are attached to a protein. This modification involves many enzymatic steps and has many functions, thereby being considered as the most complicated posttranslational modification.

Phosphorylation Phosphorylation is a posttranslational mechanism in which a phosphate group is added to an organic molecule, such as a protein, an action which usually causes activation of the protein.

SUMOylation SUMOylation is a ubiquitination-like process in which the polypeptides called small ubiquitin-related modifiers (SUMOs) are attached to a protein.

Ubiquitination Ubiquitination, also known as ubiquitylation or ubiquitinylation, is a posttranslational mechanism in which the 8.5-kDa globular protein ubiquitin is covalently conjugated to a protein. Typically, the process leads the modified protein to proteasomal degradation.

ZIP ZIP, which stands for Zrt- and Irt-like proteins, is a family of zinc-transport proteins (SLC39A) that mobilize zinc from either the extracellular space or cellular stores to the cytoplasm.

ZnT ZnT is a family of zinc-transport proteins (SLC30A) that mobilize zinc from the cytoplasm to either the extracellular space or cellular stores.

REFERENCES

Altmeppen, H.C., Puig, B., Dohler, F., Thurm, D.K., Falker, C., Krasemann, S., Glatzel, M., 2012. Proteolytic processing of the prion protein in health and disease. Am. J. Neurodegener. Dis. 1, 15–31.

Bin, B.H., Hojyo, S., Hosaka, T., Bhin, J., Kano, H., Miyai, T., Ikeda, M., Kimura-Someya, T., Shirouzu, M., Cho, E.G., Fukue, K., Kambe, T., Ohashi, W., Kim, K.H., Seo, J., Choi, D.H., Nam, Y.J., Hwang, D., Fukunaka, A., Fujitani, Y., Yokoyama, S., Superti-Furga, A., Ikegawa, S., Lee, T.R., Fukada, T., 2014. Molecular pathogenesis of spondylocheirodysplastic Ehlers–Danlos syndrome caused by mutant ZIP13 proteins. EMBO Mol. Med. 6, 1028–1042.

Chimienti, F., Devergnas, S., Favier, A., Seve, M., 2004. Identification and cloning of a beta-cell-specific zinc transporter, ZnT-8, localized into insulin secretory granules. Diabetes 53, 2330–2337.

Colvin, R.A., Holmes, W.R., Fontaine, C.P., Maret, W., 2010. Cytosolic zinc buffering and muffling: their role in intracellular zinc homeostasis. Metallomics 2, 306–317.

Cousins, R.J., Liuzzi, J.P., Lichten, L.A., 2006. Mammalian zinc transport, trafficking, and signals. J. Biol. Chem. 281, 24085–24089.

Dardel, F., Ragusa, S., Lazennec, C., Blanquet, S., Meinnel, T., 1998. Solution structure of nickel-peptide deformylase. J. Mol. Biol. 280, 501–513.

Ehsani, S., Salehzadeh, A., Huo, H., Reginold, W., Pocanschi, C.L., Ren, H., Wang, H., So, K., Sato, C., Mehrabian, M., Strome, R., Trimble, W.S., Hazrati, L.N., Rogaeva, E., Westaway, D., Carlson, G.A., Schmitt-Ulms, G., 2012. LIV-1 ZIP ectodomain shedding in prion-infected mice resembles cellular response to transition metal starvation. J. Mol. Biol. 422, 556–574.

Franchin, C., Cesaro, L., Salvi, M., Millioni, R., Iori, E., Cifani, P., James, P., Arrigoni, G., Pinna, L., 2015. Quantitative analysis of a phosphoproteome readily altered by the protein kinase CK2 inhibitor quinalizarin in HEK-293T cells. Biochim. Biophys. Acta 1854, 609–623.

Fujishiro, H., Yano, Y., Takada, Y., Tanihara, M., Himeno, S., 2012. Roles of ZIP8, ZIP14, and DMT1 in transport of cadmium and manganese in mouse kidney proximal tubule cells. Metallomics 4, 700–708.

Fukada, T., Kambe, T., 2011. Molecular and genetic features of zinc transporters in physiology and pathogenesis. Metallomics 3, 662–674.

Gitan, R.S., Shababi, M., Kramer, M., Eide, D.J., 2003. A cytosolic domain of the yeast Zrt1 zinc transporter is required for its post-translational inactivation in response to zinc and cadmium. J. Biol. Chem. 278, 39558–39564.

Hogstrand, C., Kille, P., Nicholson, R.I., Taylor, K.M., 2009. Zinc transporters and cancer: a potential role for ZIP7 as a hub for tyrosine kinase activation. Trends Mol. Med. 15, 101–111.

Hogstrand, C., Kille, P., Ackland, M.L., Hiscox, S., Taylor, K.M., 2013. A mechanism for epithelial-mesenchymal transition and anoikis resistance in breast cancer triggered by zinc channel ZIP6 and STAT3 (signal transducer and activator of transcription). Biochem. J. 455, 229–237.

Hooper, N.M., 1994. Families of zinc metalloproteases. FEBS Lett. 354, 1–6.

Hornbeck, P.V., Kornhauser, J.M., Tkachev, S., Zhang, B., Skrzypek, E., Murray, B., Latham, V., Sullivan, M., 2012. PhosphoSitePlus: a comprehensive resource for investigating the structure and function of experimentally determined post-translational modifications in man and mouse. Nucleic Acids Res. 40, D261–D270.

Huang, L., Kirschke, C.P., Gitschier, J., 2002. Functional characterization of a novel mammalian zinc transporter, ZnT6. J. Biol. Chem. 277, 26389–26395.

Jeong, J., Eide, D.J., 2013. The SLC39 family of zinc transporters. Mol. Aspects Med. 34, 612–619.

Kambe, T., Andrews, G.K., 2009. Novel proteolytic processing of the ectodomain of the zinc transporter ZIP4 (SLC39A4) during zinc deficiency is inhibited by acrodermatitis enteropathica mutations. Mol. Cell. Biol. 29, 129–139.

Kambe, T., Tsuji, T., Hashimoto, A., Itsumura, N., 2015. The physiological, biochemical, and molecular roles of zinc transporters in zinc homeostasis and metabolism. Physiol. Rev. 95, 749–784.

Kambe, T., 2012. Molecular architecture and function of ZnT transporters. Curr. Top. Membr. 69, 199–220.

Küry, S., Dréno, B., Bézieau, S., Giraudet, S., Kharfi, M., Kamoun, R., Moisan, J.P., 2002. Identification of SLC39A4, a gene involved in acrodermatitis enteropathica. Nat. Genet. 31, 239–240.

Lasswell, J., Rogg, L.E., Nelson, D.C., Rongey, C., Bartel, B., 2000. Cloning and characterization of IAR1, a gene required for auxin conjugate sensitivity in Arabidopsis. Plant Cell 12, 2395–2408.

Lee, S., Chanoit, G., McIntosh, R., Zvara, D.A., Xu, Z., 2009. Molecular mechanism underlying Akt activation in zinc-induced cardioprotection. Am. J. Physiol. Heart Circ. Physiol. 297, H569–H575.

Li, M., Zhang, Y., Liu, Z., Bharadwaj, U., Wang, H., Wang, X., Zhang, S., Liuzzi, J.P., Chang, S.M., Cousins, R.J., Fisher, W.E., Brunicardi, F.C., Logsdon, C.D., Chen, C., Yao, Q., 2007. Aberrant expression of zinc transporter ZIP4 (SLC39A4) significantly contributes to human pancreatic cancer pathogenesis and progression. Proc. Natl. Acad. Sci. U.S.A. 104, 18636–18641.

Mao, X., Kim, B.E., Wang, F., Eide, D.J., Petris, M.J., 2007. A histidine-rich cluster mediates the ubiquitination and degradation of the human zinc transporter, hZIP4, and protects against zinc cytotoxicity. J. Biol. Chem. 282, 6992–7000.

Maret, W., 2013. Zinc biochemistry: from a single zinc enzyme to a key element of life. Adv. Nutr. 4, 82–91.

Munkley, J., Lafferty, N.P., Kalna, G., Robson, C.N., Leung, H.Y., Rajan, P., Elliott, D.J., 2015. Androgen-regulation of the protein tyrosine phosphatase PTPRR activates ERK1/2 signalling in prostate cancer cells. BMC Cancer 15, 9.

Nimmanon, T., Taylor, K., 2014. Zinc signalling and cancer. In: Fukada, T., Kambe, T. (Eds.), Zinc Signals in Cellular Functions and Disorders. Springer, Japan, Tokyo, pp. 285–313.

Nimmanon, T., Taylor, K., 2015. Cellular zinc signalling is triggered by CK2. In: Ahmed, K., Issinger, O.-G., Szyszka, R. (Eds.), Protein Kinase CK2 Cellular Function in Normal and Disease States. Springer International Publishing, Switzerland, pp. 141–157.

Ohana, E., Hoch, E., Keasar, C., Kambe, T., Yifrach, O., Hershfinkel, M., Sekler, I., 2009. Identification of the Zn^{2+} binding site and mode of operation of a mammalian Zn^{2+} transporter. J. Biol. Chem. 284, 17677–17686.

Palmiter, R.D., Findley, S.D., 1995. Cloning and functional characterization of a mammalian zinc transporter that confers resistance to zinc. EMBO J. 14, 639–649.

Palmiter, R.D., Huang, L., 2004. Efflux and compartmentalization of zinc by members of the SLC30 family of solute carriers. Pflugers Arch. 447, 744–751.

Rosas-Acosta, G., Russell, W.K., Deyrieux, A., Russell, D.H., Wilson, V.G., 2005. A universal strategy for proteomic studies of SUMO and other ubiquitin-like modifiers. Mol. Cell Proteomics 4, 56–72.

Schmitt-Ulms, G., Ehsani, S., Watts, J.C., Westaway, D., Wille, H., 2009. Evolutionary descent of prion genes from the ZIP family of metal ion transporters. PLoS One 4, e7208.

Seo, Y.A., Lee, S., Hennigar, S.R., Kelleher, S.L., 2014. Prolactin (PRL)-stimulated ubiquitination of ZnT2 mediates a transient increase in zinc secretion followed by ZnT2 degradation in mammary epithelial cells. J. Biol. Chem. 289, 23653–23661.

Seve, M., Chimienti, F., Devergnas, S., Favier, A., 2004. In silico identification and expression of SLC30 family genes: an expressed sequence tag data mining strategy for the characterization of zinc transporters' tissue expression. BMC Genomics 5, 32.

Spiro, R.G., 2002. Protein glycosylation: nature, distribution, enzymatic formation, and disease implications of glycopeptide bonds. Glycobiology 12, 43R–56R.

Taylor, K.M., Nicholson, R.I., 2003. The LZT proteins; the LIV-1 subfamily of zinc transporters. Biochim. Biophys. Acta 1611, 16–30.

Taylor, K.M., Morgan, H.E., Johnson, A., Hadley, L.J., Nicholson, R.I., 2003. Structure-function analysis of LIV-1, the breast cancer-associated protein that belongs to a new subfamily of zinc transporters. Biochem. J. 375, 51–59.

Taylor, K.M., Morgan, H.E., Johnson, A., Nicholson, R.I., 2005. Structure-function analysis of a novel member of the LIV-1 subfamily of zinc transporters, ZIP14. FEBS Lett. 579, 427–432.

Taylor, K.M., Morgan, H.E., Smart, K., Zahari, N.M., Pumford, S., Ellis, I.O., Robertson, J.F., Nicholson, R.I., 2007. The emerging role of the LIV-1 subfamily of zinc transporters in breast cancer. Mol. Med. 13, 396–406.

Taylor, K.M., Vichova, P., Jordan, N., Hiscox, S., Hendley, R., Nicholson, R.I., 2008. ZIP7-mediated intracellular zinc transport contributes to aberrant growth factor signaling in antihormone-resistant breast cancer cells. Endocrinology 149, 4912–4920.

Taylor, K.M., Hiscox, S., Nicholson, R.I., Hogstrand, C., Kille, P., 2012. Protein kinase CK2 triggers cytosolic zinc signaling pathways by phosphorylation of zinc channel ZIP7. Sci. Signal. 5, ra11.

Wang, Y., Dasso, M., 2009. SUMOylation and deSUMOylation at a glance. J. Cell Sci. 122, 4249–4252.

Wang, C.Y., Jenkitkasemwong, S., Duarte, S., Sparkman, B.K., Shawki, A., Mackenzie, B., Knutson, M.D., 2012. ZIP8 is an iron and zinc transporter whose cell-surface expression is up-regulated by cellular iron loading. J. Biol. Chem. 287, 34032–34043.

Watts, J.C., Huo, H., Bai, Y., Ehsani, S., Jeon, A.H., Won, A.H., Shi, T., Daude, N., Lau, A., Young, R., Xu, L., Carlson, G.A., Williams, D., Westaway, D., Schmitt-Ulms, G., 2009. Interactome analyses identify ties of PrP and its mammalian paralogs to oligomannosidic N-glycans and endoplasmic reticulum-derived chaperones. PLoS Pathog. 5, e1000608.

Weissman, A.M., 2001. Themes and variations on ubiquitylation. Nat. Rev. Mol. Cell Biol. 2, 169–178.

Wilson, M., Hogstrand, C., Maret, W., 2012. Picomolar concentrations of free zinc(II) ions regulate receptor protein-tyrosine phosphatase β activity. J. Biol. Chem. 287, 9322–9326.

Yalak, G., Vogel, V., 2012. Extracellular phosphorylation and phosphorylated proteins: not just curiosities but physiologically important. Sci. Signal 5, re7.

Yamasaki, S., Sakata-Sogawa, K., Hasegawa, A., Suzuki, T., Kabu, K., Sato, E., Kurosaki, T., Yamashita, S., Tokunaga, M., Nishida, K., Hirano, T., 2007. Zinc is a novel intracellular second messenger. J. Cell Biol. 177, 637–645.

Yamashita, S., Miyagi, C., Fukada, T., Kagara, N., Che, Y.S., Hirano, T., 2004. Zinc transporter LIVI controls epithelial-mesenchymal transition in zebrafish gastrula organizer. Nature 429, 298–302.

Zhao, N., Gao, J., Enns, C.A., Knutson, M.D., 2010. ZRT/IRT-like protein 14 (ZIP14) promotes the cellular assimilation of iron from transferrin. J. Biol. Chem. 285, 32141–32150.

Zhao, N., Zhang, A.S., Worthen, C., Knutson, M.D., Enns, C.A., 2014. An iron-regulated and glycosylation-dependent proteasomal degradation pathway for the plasma membrane metal transporter ZIP14. Proc. Natl. Acad. Sci. U.S.A. 111, 9175–9180.

Zhou, B.P., Deng, J., Xia, W., Xu, J., Li, Y.M., Gunduz, M., Hung, M.C., 2004. Dual regulation of Snail by GSK-3beta-mediated phosphorylation in control of epithelial-mesenchymal transition. Nat. Cell Biol. 6, 931–940.

Chapter 23

Zinc Transporters in Health and Disease

Taiho Kambe, Yukina Nishito, Kazuhisa Fukue
Kyoto University, Kyoto, Japan

INTRODUCTION

Zinc is the second-most abundant essential trace element found in the body after iron, and an adult human contains approximately 2–3 g of zinc (Kambe et al., 2015, 2014b). Zinc plays pivotal roles in various biological reactions by acting as a crucial structural, catalytic, and signaling component for proteins involved in cell proliferation, migration, survival, death, neuronal development, immunity, and signal transduction. Thus, zinc deficiency can lead to a broad range of consequences such as skin lesions, alopecia, diarrhea, immune dysfunction, taste disorders, loss of appetite, growth retardation, and neuropsychiatric change, some of which have life-threatening potential (Prasad, 1985).

Cellular and systemic zinc homeostasis is strictly maintained by regulated processes that monitor zinc transport. For example, dietary zinc enters intestinal epithelial cells from the apical side, is transferred to the basolateral side, and is released into the portal circulation. Pancreatic exocrine and biliary secretions, sloughing mucosal cells, and the integument excrete surplus zinc. During these processes, zinc transport proteins are essential for mobilizing zinc across all involved membranes.

In general, two zinc transporter families, including the Zrt- and Irt-like proteins (ZIP, solute carrier 39A [*SLC39A*]) and Zn transporters (ZnT, *SLC30A*), control zinc mobilization within cells (Fukada and Kambe, 2011; Huang and Tepaamorndech, 2013; Jeong and Eide, 2013; Lichten and Cousins, 2009). ZIP proteins transport zinc from the extracellular space or intracellular organelles to the cytosol while ZnT transporters efflux zinc in the opposite direction (Fig. 23.1).

Fourteen ZIP and nine ZnT transporter genes are encoded in the human genome. Most ZIP transporters are localized to the plasma membrane and most ZnT transporters are localized to intracellular compartmental membranes. This allows for controlled zinc homeostasis in the cytosol and in the lumens of involved organelles (Fig. 23.2).

Recent studies have revealed that dysfunction and deregulation of zinc transporter proteins can cause disease. Therefore, a better understanding of the molecular basis underlying zinc transport protein mechanisms, regulation of zinc transport protein expression, and zinc transport protein functions is crucial when discerning disease pathophysiology and treatment strategies. In this chapter, we provide an overview of the current understanding of human ZIP and ZnT transporters, with a focus on the evidence that links molecular function to health and disease.

ZIP AND ZNT TRANSPORTER BIOCHEMISTRY

ZIP transporters are predicted to contain eight transmembrane domains (TMDs) with extracellular amino- and carboxyl-termini (Fig. 23.1). This family is divided into four subfamilies (ZIP-I, ZIP-II, gufA, and LIV-1), and most characterized members belong to the LIV-1 subfamily, the largest subfamily of ZIP transporter proteins. ZIP transporters can form homodimers (Bin et al., 2011), and homodimerization is thought to be required for zinc transport. Two highly conserved histidine residues in TMDs IV and V are considered essential for zinc transport, although whether these two residues contribute to the formation of intramembranous zinc-binding sites has not been clarified (Fukada and Kambe, 2011; Kambe et al., 2015). Thus far, X-ray crystal structures of human ZIP family members or their homologs have not been determined and this limits a precise understanding of their molecular functions.

Zinc is likely the primary metal substrate for ZIP transporters, yet zinc mobilization by ZIP transporters is inhibited by various divalent cations, suggesting that ZIP transporters may be involved in the transport of other cations as well. For instance, it has been demonstrated that several ZIP transporters, including ZIP8 and ZIP14, are involved in the transport of iron, manganese, and cadmium (Fujishiro et al., 2012). Interestingly, the conserved histidine in TMD V, one of two conserved histidine residues described earlier, is replaced by glutamic acid in ZIP8 and ZIP14; this substitution may affect metal substrate specificity (Jenkitkasemwong et al., 2012).

Molecular, Genetic, and Nutritional Aspects of Major and Trace Minerals. http://dx.doi.org/10.1016/B978-0-12-802168-2.00023-3

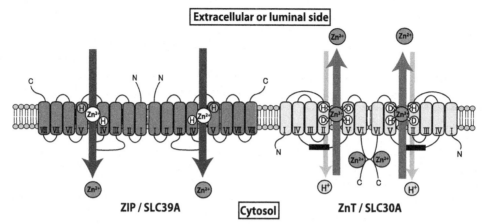

FIGURE 23.1 **Predicted ZIP and ZnT transporter protein structures.** Left, ZIP transporters move zinc from the extracellular space or the subcellular compartment lumen to the cytosol. Two histidine residues in the transmembrane domains (TMDs) may form part of an intramembranous zinc-binding site. The histidine (colored in aqua) in TMD V may be involved in determining metal specificity. Right, ZnTs move zinc from the cytosol to the extracellular space or the lumen of subcellular compartments. An intramembranous zinc-binding site consists of four hydrophilic residues (two histidine [H] and two aspartic acid [D] residues) in each protomer. ZnT transporters may operate as Zn^{2+}/H^+ exchangers.

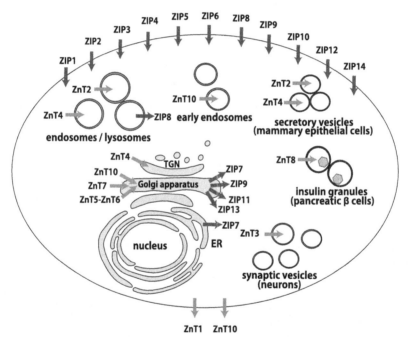

FIGURE 23.2 **Subcellular localization and direction of zinc transport for characterized ZIP and ZnT transporter proteins.** Known primary localization for ZnT (green (gray in print versions) arrows) and ZIP (red (dark gray in print versions) arrows) transporter proteins in human cells is depicted. However, this presents a static view of their localization and does not reflect dynamic changes of subcellular localization for these zinc transporters in response to various stimuli. *ER*, endoplasmic reticulum; *TGN*, Trans-Golgi network.

The molecular basis of zinc transport by ZIP transporters has not yet been elucidated. It has been suggested that ZIP transporters operate as Zn^{2+}/HCO_3^- symporters as zinc transport by some members is facilitated by HCO_3^- (He et al., 2006). However, it has been shown that at least one bacterial ZIP transporter homolog is a selective electrodiffusional ion channel protein, as demonstrated by nonsaturated zinc influx into proteoliposomes (Lin et al., 2010). Contrarily, zinc transport from the endoplasmic reticulum by ZIP7 is facilitated by phosphorylation, and this suggests that zinc may be transported via a phosphorylation-dependent channel mechanism (Taylor et al., 2012).

Most ZIP transporters control zinc uptake at the plasma membrane. Cell surface localization is likely upregulated under zinc-deficient conditions and downregulated via endocytosis in response to high zinc concentrations, followed by degradation. ZIP transporter endocytic and degradative pathways have been partially characterized. For ZIP1, a di-leucine (LL) sorting signal between the intracellular loop TMDs III and IV has been identified as critical for endocytosis and degradation (Huang and Kirschke, 2007). For ZIP4, a histidine-rich cluster in the intracellular loop between TMDs III and IV is involved in degradation, but not endocytosis, in response to high zinc concentrations (Mao et al., 2007). This indicates that separate regulatory processes are required for endocytosis and degradation of this particular transporter. The regulation of subcellular ZIP transporter localization may also prove to be important, but to our knowledge it has not yet been extensively investigated.

Structural data on ZnT transporters are derived from the Y-shaped dimeric forms of its bacterial homolog, YiiP. Elucidation of the YiiP crystal structure confirmed that ZnT transporters are six TMD membrane proteins (Fig. 23.1). Unlike ZIP transporters, however, both amino- and carboxyl-termini are located on the cytosolic side of the membrane (Lu and Fu, 2007). YiiP has four zinc-binding sites per protomer, one of which is formed by TMDs and is highly conserved among ZnT transporters. The intramembranous zinc-binding site is formed by three aspartic acid residues and one histidine residue found in YiiP TMDs II and V, yet two aspartic acid residues and two histidine residues are found in most vertebrate ZnT transporters. The zinc-binding site is indispensable for zinc transport across the biological membrane, as mutations introduced at any one of these four residues abolishes zinc transport activity (Kambe et al., 2015). These amino acid residues within the intramembranous zinc-binding site may also be important for metal substrate specificity, as substitution of histidine for aspartic acid in TMD II alters metal substrate specificity (Hoch et al., 2012). Finally, it has been suggested that the cytosolic histidine-rich loop between TMDs IV and V may be important for specifying substrate metals and sensing zinc concentrations, but these molecular functions have not yet been determined (Fukada and Kambe, 2011; Kambe et al., 2015).

In YiiP, it has been proposed that the cytosolic binuclear zinc-binding site, which is formed by a metallochaperone-like fold structure with an $\alpha\beta\beta\alpha$ fold (Lu and Fu, 2007), is a key site that induces a scissor-like movement of its dimers after binding zinc; thus, zinc transport may be partially regulated by an allosteric mechanism. However, this binuclear zinc-binding site may also function as a stabilizer for dimer formation in an alternative access mechanism in which inward- and outward-facing conformations are able to bind Zn^{2+} or H^+, and extracellular H^+ provides the driving force for exporting Zn^{2+} from the cytosol (Coudray et al., 2013; Gupta et al., 2014). Consistent with this transport mechanism, ZnT transporters operate as Zn^{2+}/H^+ exchangers (Kambe et al., 2012), for whom the stoichiometric ratio of Zn^{2+}/H^+ exchange is thought to be 1:1 (Chao and Fu, 2004).

Dimerization of ZnT transporters is a critical step for zinc transport. Most ZnT transporter proteins form homodimers although heterodimerization between ZnT5 and ZnT6 has also been reported (Fukunaka et al., 2009). Furthermore, studies suggest that other ZnT transporters can form heterodimers, and this phenomenon may affect functions (Golan et al., 2015). The physiological importance of ZnT heterodimerization needs to be clarified for a complete understanding of ZnT transporter functions under physiological and pathophysiological conditions.

ZIP AND ZNT TRANSPORTERS IN HEALTH AND DISEASE

ZIP and ZnT Transporter Involvement in Chronic Disease Pathogenesis

A growing body of evidence illustrates how ZIP and ZnT transporters play numerous and important physiological roles by mediating zinc homeostatic maintenance. For instance, zinc mobilization by ZIP and ZnT transporters expressed by immune cells contributes to immune cell functions by fine-tuning cellular signaling pathways via the dynamic regulation of cytosolic zinc concentration and distribution (Haase and Rink, 2009; Hirano et al., 2008). ZIP and ZnT transporters also contribute to nutritional immunity at the host–pathogen interface (Stafford et al., 2013). Another specific example is synaptic zinc regulation by ZnT3 that is thought to be necessary for normal activity of neuronal ion channels, transporters, and receptors in the synaptic cleft, thus modulating synaptic transmission and plasticity (Sensi et al., 2011; Takeda et al., 2013). Additionally, zinc mobilization by ZnT8 into insulin granules within pancreatic β cells is crucial for regulating zinc-dependent insulin clearance in hepatocytes (Tamaki et al., 2013). Furthermore, the involvement of ZIP LIV-1 subfamily members in the regulation of epithelial-mesenchymal transitioning is well-established (Fukada et al., 2011; Gumulec et al., 2011), and maternally-derived ZIP6 and ZIP10 drive mammalian oocyte-to-egg transition (Kong et al., 2014). Together, these simple examples clearly reveal the physiological importance of ZIP and ZnT transporters.

It has been widely reported that ectopic or deregulated expression of ZIP and ZnT transporters is involved in disease pathogenesis. ZIP and ZnT dysfunction or deregulation can negatively impact immunity (Haase and Rink, 2009; Hirano et al., 2008) while altered ZIP and ZnT transporter protein expression has been implicated in neurodegenerative diseases including Parkinson's disease, Alzheimer's disease, and amyotrophic lateral sclerosis (Szewczyk, 2013). Additionally, ectopic overexpression of LIV-1 subfamily transporters may contribute to cancer pathogenesis, progression, and invasive tumor behavior (Hogstrand et al., 2013) that is caused by unregulated activation of cellular signaling pathways via increased cytosolic zinc levels (Hogstrand et al., 2009). In contrast, decreased ZIP-I subfamily members are reportedly abnormally expressed in prostate cancer tissues (Gumulec et al., 2011). More evidence suggests that upregulated expression of *ZIP8* and a resulting increase in zinc influx is associated with osteoarthritis as this enhances the catabolic cascade by regulating expression of matrix-degrading enzymes in chondrocytes (Kim et al., 2014). As such, elucidating the molecular basis underlying altered ZIP and ZnT transporter expression may lead to the development of novel drugs that could treat all involved altered heath states, such as neurodegenerative diseases and certain cancer types, and could act as potential diagnostic and prognostic markers for these diseases.

Studies on ZIP and ZnT knockout mice contribute to much of our current understanding of their importance in health and disease (Table 23.1). For the most part, however, zinc transporter involvement in human diseases has not yet been clarified.

ZINC TRANSPORTERS AND INHERITED HUMAN DISEASES

A number of reports describe the link between mutated ZIP and ZnT transporter genes and inherited human diseases. Moreover, single nucleotide polymorphisms (SNPs) in both ZIP and ZnT genes have been associated with the onset of diabetes mellitus, hypertension, and schizophrenia. In this section, current information on characterized mutations and SNPs within ZIP and ZnT transporter genes is briefly summarized.

Acrodermatitis Enteropathica

Acrodermatitis enteropathica (AE) is a rare, inherited, and autosomal recessive disorder that results in defective intestinal zinc absorption (Andrews, 2008; Kambe et al., 2014b). AE prevalence is estimated at approximately 1 in 500,000, and alopecia, eczematous dermatitis, and diarrhea are recognized as a classic AE symptom triad that can disappear by administration of a daily oral zinc supplement (1–3 mg/kg/day elemental zinc). In 2002, the gene responsible for intestinal zinc absorption was mapped to chromosomal region 8q24.3, and mutations within the *SLC39A4/ZIP4* gene found within this region were identified in AE patients (Kury et al., 2002; Wang et al., 2002). Since then, more than 30 mutations have been mapped to the same locus (Andrews, 2008; Kambe et al., 2014a). AE-causing *ZIP4* mutations result in decreased expression, defects in protein trafficking to the plasma membrane, decreased zinc uptake (Wang et al., 2004), and defective proteolytic cleavage of the ZIP4 extracellular amino-terminal half (Kambe et al., 2014a). Elucidating how ZIP4 expression is regulated will further prove indispensable for unraveling AE pathogenesis.

Transient Neonatal Zinc Deficiency

Infants are vulnerable to zinc deficiency because large quantities of zinc are required for normal growth and development. Transient neonatal zinc deficiency (TNZD) is caused by low levels of zinc in breast milk, and thus occurs in breast-fed infants. TNZD is caused by mutations in the maternal *SLC30A2/ZnT2* gene that results in the production of low-zinc milk, and approximately 75% less than normal zinc levels have been reported (Chowanadisai et al., 2006; Itsumura et al., 2013; Lasry et al., 2012). While more than 30 mutations for *SLC39A4/ZIP4* have been implicated in AE, only five mutations have been identified in *SLC30A2/ZnT2* as TNZD-causing mutations (Chowanadisai et al., 2006; Itsumura et al., 2013; Lasry et al., 2012). Specifically, four missense mutations in *SLC30A2/ZnT2* result in decreased zinc transport activity, ZnT2 destabilization, and protein mislocalization. Importantly, TNZD infant patients develop erythematous and erosive dermatitis, persistent diarrhea, alopecia, and transient growth retardation. Unlike AE, however, TNZD symptoms only develop during breast-feeding. These symptoms are alleviated by zinc supplementation to the infant, but not to the mother. In mice, a similar zinc-deficiency phenotype caused by low-zinc milk has been characterized. The gene responsible for this lethal milk (lm) murine phenotype is *Slc30a4/Znt4* (Huang and Gitschier, 1997). Notably, there have been no reported mutations in the human *SLC30A4/ZnT4* gene in women who secrete low-zinc breast milk to date.

Ehlers–Danlos Syndrome: Spondylocheiro Dysplastic Form

Ehlers–Danlos syndrome (EDS) is a group of genetic disorders that affect connective tissue development and are classified into six different types. The spondylocheiro dysplastic form of EDS (SCD-EDS) is another disease variant caused by homozygous loss-of-function mutations in *SLC39A13/ZIP13* (Fukada et al., 2008; Giunta et al., 2008). Individuals affected with SCD-EDS have a short stature and abnormalities of skeletal and connective tissues, as well as other common EDS features including

TABLE 23.1 Tissue Specificity, Subcellular Localization, and Knockout/Mutant Mice Phenotypes of Zinc Transporter Proteins[a]

Protein (Gene Name)	Tissue Specificity	Subcellular Localization	Representative phenotypes
Zip1 (*Slc39a1*)	Ubiquitous	Plasma membrane	KO; abnormal embryonic development under a zinc-deficient diet
Zip2 (*Slc39a2*)	Liver, ovary, skin, and monocytes	Plasma membrane	KO; abnormal embryonic development under a zinc-deficient diet
Zip3 (*Slc39a3*)	Ubiquitous	Plasma membrane	KO; abnormal embryonic development under a zinc-deficient diet
Zip4 (*Slc39a4*)	Small intestine	Plasma membrane	KO; embryonic lethal, beta cell-specific KO; slightly improved glucose homeostasis in an oral glucose tolerance test (Hardy et al., 2015), intestine-specific KO; disruption of intestinal integrity
Zip5 (*Slc39a5*)	Small intestine, pancreas, and kidney	Plasma membrane	KO; zinc accumulation in the liver, intestine, and pancreas under a zinc-excess diet
Zip6 (*Slc39a6*)	Ubiquitous	Plasma membrane	–
Zip7 (*Slc39a7*)	Ubiquitous	ER and Golgi apparatus	–
Zip8 (*Slc3a8*)	Ubiquitous	Plasma membrane and lysosome	Hypomorph; neonatal lethal, chondrocyte-specific KO; suppression of osteoarthritis pathogenesis
Zip9 (*Slc39a9*)	Ubiquitous	Golgi apparatus and plasma membrane	–
Zip10 (*Slc39a10*)	Ubiquitous	Plasma membrane	B cell-specific KO; splenic atrophy and decreases of peripheral B cells
Zip11 (*Slc39a11*)	Gastrointestinal tract and testis	Golgi apparatus	–
Zip12 (*Slc39a12*)	Brain	Plasma membrane	–
Zip13 (*Slc39a13*)	Ubiquitous	Golgi apparatus and vesicles	KO; growth retardation and abnormal development of skeletal connective tissue
Zip14 (*Slc39a14*)	Ubiquitous	Plasma membrane	KO; growth retardation, osteodystrophy, and developmental disorder
Znt1 (*Slc30a1*)	Ubiquitous	Plasma membrane	KO; embryonic lethal
Znt2 (*Slc30a2*)	Ubiquitous	Endosomes/lysosomes and secretory vesicles	KO; mammary gland functional defect (Lee et al., 2015)
Znt3 (*Slc30a3*)	Brain	Synaptic vesicles	KO; hippocampal-dependent brain dysfunction
Znt4 (*Slc30a4*)	Ubiquitous	Endosomes/lysosomes, TGN, and secretory vesicles	Mutant; death before weaning
Znt5 (*Slc30a5*)	Ubiquitous	Golgi apparatus	KO; arrhythmia in males and growth retardation
Znt6 (*Slc30a6*)	Ubiquitous	Golgi apparatus	–
Znt7 (*Slc30a7*)	Ubiquitous	Golgi apparatus	KO; growth retardation, reduction of fat accumulation, and abnormal glucose tolerance
Znt8 (*Slc30a8*)	Pancreas	Insulin granules	KO; lacks dense core of zinc-insulin crystals, highly variable effects among mouse colonies
Znt10 (*Slc30a10*)	Ubiquitous	Golgi apparatus and recycling endosome, plasma membrane	–

KO, Knockout; *TGN*, Trans-Golgi network; -, No report available.
[a]*Select KO information taken from (Fukada and Kambe, 2011; Kambe, 2015).*

hyperelastic skin and hypermobile joints. SCD-EDS patients show normal collagen synthesis activity, yet hydroxyl collagen levels are decreased. SCD-EDS-causing mutations within *ZIP13* identified thus far result in accelerated ZIP13 destabilization and cause disturbed Smad translocation into the nucleus in bone morphogenic protein (BMP)/transforming growth factor β (TGF-β) signaling (Bin et al., 2014; Fukada et al., 2008). Therefore, inhibition of ZIP13 protein degradation could potentially improve SCD-EDS symptoms (Bin et al., 2014).

High Myopia

It was demonstrated that heterozygous *SLC39A5/ZIP5* mutations are associated with nonsyndromic high myopia (Guo et al., 2014), and these mutations may exert their effects in an autosomal-dominant or haploinsufficient manner. As high myopia with characteristic refractive error is one of the leading causes of blindness, these *SLC39A5/ZIP5* mutations merit further attention.

Parkinsonism

Recent findings have revealed that ZnT10 is involved in manganese transport (Leyva-Illades et al., 2014) and homozygous *SLC30A10/ZnT10* mutations cause hepatic cirrhosis, polycythemia, dystonia, and hypermanganesemia, all of which are characteristic of Parkinsonism (Quadri et al., 2012; Tuschl et al., 2012). Patients' symptoms are improved by a combination of chelation therapy and iron supplementation, which is a competitive inhibitor of intestinal manganese absorption. As described elsewhere (Kambe et al., 2015, 2014a), ZnT10 has several unique features when compared with other ZnT proteins that may enable this membrane protein to transport manganese. Elucidating the relationship between these features and manganese transport may facilitate our understanding of the molecular basis underlying ZnT transporter functions.

ZNT AND ZIP SINGLE NUCLEOTIDE POLYMORPHISMS

Recent genome-wide association studies suggest that SNPs within ZnT and ZIP transporter genes are associated with specific disease traits (Kambe et al., 2015, 2014b). Findings from these studies are summarized in Table 23.2.

SNPs in two specific zinc transporter genes have garnered much interest. SNP rs13107325 in *SLC39A8/ZIP8* causes a ZIP8 A391T substitution. The risk allele is reportedly associated with lower high-density lipoprotein cholesterol levels, high body mass index/obesity, high blood pressure, and increased schizophrenia risk (Kambe et al., 2015). SNP rs13266634 in *SLC30A8/ZnT8* is well characterized. It results in a ZnT8 R325W substitution and the major C allele encoding R residue appears to increase the risk of type 2 diabetes (Nicolson et al., 2009; Davidson et al., 2014). The C allele is also related to an unfavorable cardiometabolic lipid profile in HIV/hepatitis C virus-coinfected patients, as well as lower skeletal muscle strength and size (Sprouse et al., 2014). Additionally, this SNP is implicated in humoral autoreactivity to ZnT8 (Wenzlau et al., 2008). Finally, one study reported that single nucleotide variants resulting in truncated ZnT8 decreased the risk of type 2 diabetes for heterozygous individuals (Flannick et al., 2014). This finding illustrates a reverse relationship between the SNP at this locus (causing haploinsufficiency) and type 2 diabetes, which requires further investigation. Moreover, numerous SNPs of ZIP and ZnT transporter genes have been suggested as possible genetic links to schizophrenia and the autism spectrum (Kambe et al., 2015).

SNPs of ZIP and ZnT transporter genes that may be involved in human diseases are extensively reviewed elsewhere (Kambe et al., 2015). It is likely that planned genome-wide association studies will reveal other relationships between SNPs or copy number variants of ZIP and ZnT transporter genes and human diseases in the future.

CONCLUSIONS

Well-established evidence supports indispensable roles for zinc and zinc transporter proteins in human physiology. As such, ZIP and ZnT transporters have garnered much attention because they possess valuable therapeutic potential. The identification or design of specific small molecule agonists/antagonists that target individual ZIP and ZnT transporters may be of great interest and promise for treating neurodegenerative diseases, cancer, or diabetes. However, the molecular basis underlying relationships between zinc transporter protein function and healthy humans or disease states are not yet fully understood. In particular, much remains to be investigated regarding structure, zinc transport mechanisms, and regulation of activity for these proteins. Defining these issues will lead to an integrated understanding of zinc transporter functions in human health and some diseases that could achieve clinical applications.

TABLE 23.2 Mutations or Single Nucleotide Polymorphism (SNP) in Zinc Transporter Genes That Cause Inherited Diseases or Are Associated With Disease States

Gene	Inherited or Associated Disease	Mutation or SNP[a]	Chromosomal Location	MIM no.	Comments
SLC30A2/ZnT2	Transient neonatal zinc deficiency	Mutation	1p36.11	608,118	Erosive dermatitis around the mouth, genital region, and neck, as well as diarrhea and alopecia. Ameliorated with zinc supplementation during nursing. Mutations are maternal.
SLC30A8/ZnT8	NA	SNP	8q24.11	NA	Increased risk of type 2 diabetes and childhood onset type 1 diabetes (rs13266634, R325W). Decreased risk of type 2 diabetes in heterozygous individuals (rs200185429, R138X).
SLC30A10/ZnT10	Hypermanganesemia, hepatic cirrhosis, dystonia, and polycythemia	Mutation	1q41	613,280	Dysarthria, fine tremor, hypertonia, bradykinesia, and spastic paraparesis. Clinical symptoms are improved by metal chelation therapy.
SLC39A4/ZIP4	Acrodermatitis enteropathica	Mutation	8q24.3	201,100	Eczematous dermatitis on the perioral, perianal, and acral areas, as well as diarrhea and alopecia. Ameliorated with zinc supplementation.
SLC39A5/ZIP5	Nonsymptomatic high myopia	Mutation	12q13.3	615,946	Refractive error, tigroid, and focal atrophy of choroid.
SLC39A8/ZIP8	NA	SNP	4q24	NA	Hypertension, high body mass index/obesity, and decreased high-density lipoprotein cholesterol levels (rs13107325, A391T).
SLC39A13/ZIP13	Spondylocheiro dysplastic Ehlers-Danlos syndrome	Mutation	11p11.2	612,350	Connective tissue abnormalities, articular hypermobility, tapering fingers, and protruding eyes with bluish sclerae. Not ameliorated with zinc supplementation.

NA, Not available.
[a]Select SNP information taken from (Kambe et al., 2015).

ACKNOWLEDGMENTS

This work was supported by Grants-in-Aid for Challenging Exploratory Research and Scientific Research (B) from the Japan Society for the Promotion of Science (KAKENHI, Grant Nos. 26660086 and 15H04501).

REFERENCES

Andrews, G.K., 2008. Regulation and function of Zip4, the acrodermatitis enteropathica gene. Biochem. Soc. Trans. 36, 1242–1246.

Bin, B.H., Fukada, T., Hosaka, T., Yamasaki, S., Ohashi, W., Hojyo, S., Miyai, T., Nishida, K., Yokoyama, S., Hirano, T., 2011. Biochemical characterization of human ZIP13 protein: a homo-dimerized zinc transporter involved in the spondylocheiro dysplastic Ehlers-Danlos syndrome. J. Biol. Chem. 286, 40255–40265.

Bin, B.H., Hojyo, S., Hosaka, T., Bhin, J., Kano, H., Miyai, T., Ikeda, M., Kimura-Someya, T., Shirouzu, M., Cho, E.G., Fukue, K., Kambe, T., Ohashi, W., Kim, K.H., Seo, J., Choi, D.H., Nam, Y.J., Hwang, D., Fukunaka, A., Fujitani, Y., Yokoyama, S., Superti-Furga, A., Ikegawa, S., Lee, T.R., Fukada, T., 2014. Molecular pathogenesis of spondylocheirodysplastic Ehlers-Danlos syndrome caused by mutant ZIP13 proteins. EMBO Mol. Med. 6, 1028–1042.

Chao, Y., Fu, D., 2004. Kinetic study of the antiport mechanism of an Escherichia coli zinc transporter, ZitB. J. Biol. Chem. 279, 12043–12050.

Chowanadisai, W., Lonnerdal, B., Kelleher, S.L., 2006. Identification of a mutation in SLC30A2 (ZnT-2) in women with low milk zinc concentration that results in transient neonatal zinc deficiency. J. Biol. Chem. 281, 39699–39707.

Coudray, N., Valvo, S., Hu, M., Lasala, R., Kim, C., Vink, M., Zhou, M., Provasi, D., Filizola, M., Tao, J., Fang, J., Penczek, P.A., Ubarretxena-Belandia, I., Stokes, D.L., 2013. Inward-facing conformation of the zinc transporter YiiP revealed by cryoelectron microscopy. Proc. Natl. Acad. Sci. U.S.A. 110, 2140–2145.

Davidson, H.W., Wenzlau, J.M., O'Brien, R.M., 2014. Zinc transporter 8 (ZnT8) and beta cell function. Trends Endocrinol. Metab. 25, 415–424.

Flannick, J., Thorleifsson, G., Beer, N.L., Jacobs, S.B., Grarup, N., Burtt, N.P., Mahajan, A., Fuchsberger, C., Atzmon, G., Benediktsson, R., Blangero, J., Bowden, D.W., Brandslund, I., Brosnan, J., Burslem, F., Chambers, J., Cho, Y.S., Christensen, C., Douglas, D.A., Duggirala, R., Dymek, Z., Farjoun, Y., Fennell, T., Fontanillas, P., Forsen, T., Gabriel, S., Glaser, B., Gudbjartsson, D.F., Hanis, C., Hansen, T., Hreidarsson, A.B., Hveem, K., Ingelsson, E., Isomaa, B., Johansson, S., Jorgensen, T., Jorgensen, M.E., Kathiresan, S., Kong, A., Kooner, J., Kravic, J., Laakso, M., Lee, J.Y., Lind, L., Lindgren, C.M., Linneberg, A., Masson, G., Meitinger, T., Mohlke, K.L., Molven, A., Morris, A.P., Potluri, S., Rauramaa, R., Ribel-Madsen, R., Richard, A.M., Rolph, T., Salomaa, V., Segre, A.V., Skarstrand, H., Steinthorsdottir, V., Stringham, H.M., Sulem, P., Tai, E.S., Teo, Y.Y., Teslovich, T., Thorsteinsdottir, U., Trimmer, J.K., Tuomi, T., Tuomilehto, J., Vaziri-Sani, F., Voight, B.F., Wilson, J.G., Boehnke, M., McCarthy, M.I., Njolstad, P.R., Pedersen, O., Groop, L., Cox, D.R., Stefansson, K., Altshuler, D., 2014. Loss-of-function mutations in SLC30A8 protect against type 2 diabetes. Nat. Genet. 46, 357–363.

Fujishiro, H., Yano, Y., Takada, Y., Tanihara, M., Himeno, S., 2012. Roles of ZIP8, ZIP14, and DMT1 in transport of cadmium and manganese in mouse kidney proximal tubule cells. Metallomics 4, 700–708.

Fukada, T., Civic, N., Furuichi, T., Shimoda, S., Mishima, K., Higashiyama, H., Idaira, Y., Asada, Y., Kitamura, H., Yamasaki, S., Hojyo, S., Nakayama, M., Ohara, O., Koseki, H., Dos Santos, H.G., Bonafe, L., Ha-Vinh, R., Zankl, A., Unger, S., Kraenzlin, M.E., Beckmann, J.S., Saito, I., Rivolta, C., Ikegawa, S., Superti-Furga, A., Hirano, T., 2008. The zinc transporter SLC39A13/ZIP13 is required for connective tissue development; its involvement in BMP/TGF-beta signaling pathways. PLoS One 3, e3642.

Fukada, T., Kambe, T., 2011. Molecular and genetic features of zinc transporters in physiology and pathogenesis. Metallomics 3, 662–674.

Fukada, T., Yamasaki, S., Nishida, K., Murakami, M., Hirano, T., 2011. Zinc homeostasis and signaling in health and diseases : zinc signaling. J. Biol. Inorg. Chem. 16, 1123–1134.

Fukunaka, A., Suzuki, T., Kurokawa, Y., Yamazaki, T., Fujiwara, N., Ishihara, K., Migaki, H., Okumura, K., Masuda, S., Yamaguchi-Iwai, Y., Nagao, M., Kambe, T., 2009. Demonstration and characterization of the heterodimerization of ZnT5 and ZnT6 in the early secretory pathway. J. Biol. Chem. 284, 30798–30806.

Giunta, C., Elcioglu, N.H., Albrecht, B., Eich, G., Chambaz, C., Janecke, A.R., Yeowell, H., Weis, M., Eyre, D.R., Kraenzlin, M., Steinmann, B., 2008. Spondylocheiro dysplastic form of the Ehlers-Danlos syndrome–an autosomal-recessive entity caused by mutations in the zinc transporter gene SLC39A13. Am. J. Hum. Genet. 82, 1290–1305.

Golan, Y., Berman, B., Assaraf, Y.G., 2015. Heterodimerization, altered subcellular localization, and function of multiple zinc transporters in viable cells using bimolecular fluorescence complementation. J. Biol. Chem. 290, 9050–9063.

Gumulec, J., Masarik, M., Krizkova, S., Adam, V., Hubalek, J., Hrabeta, J., Eckschlager, T., Stiborova, M., Kizek, R., 2011. Insight to physiology and pathology of zinc(II) ions and their actions in breast and prostate carcinoma. Curr. Med. Chem. 18, 5041–5051.

Guo, H., Jin, X., Zhu, T., Wang, T., Tong, P., Tian, L., Peng, Y., Sun, L., Wan, A., Chen, J., Liu, Y., Li, Y., Tian, Q., Xia, L., Zhang, L., Pan, Y., Lu, L., Liu, Q., Shen, L., Xiong, W., Li, J., Tang, B., Feng, Y., Zhang, X., Zhang, Z., Pan, Q., Hu, Z., Xia, K., 2014. SLC39A5 mutations interfering with the BMP/TGF-beta pathway in non-syndromic high myopia. J. Med. Genet. 51, 518–525.

Gupta, S., Chai, J., Cheng, J., D'Mello, R., Chance, M.R., Fu, D., 2014. Visualizing the kinetic power stroke that drives proton-coupled zinc(ii) transport. Nature 512 (7512), 101–104.

Haase, H., Rink, L., 2009. Functional significance of zinc-related signaling pathways in immune cells. Annu. Rev. Nutr. 29, 133–152.

Hardy, A.B., Prentice, K.J., Froese, S., Liu, Y., Andrews, G.K., Wheeler, M.B., 2015. Zip4 mediated zinc influx stimulates insulin secretion in pancreatic Beta cells. PLoS One 10, e0119136.

He, L., Girijashanker, K., Dalton, T.P., Reed, J., Li, H., Soleimani, M., Nebert, D.W., 2006. ZIP8, member of the solute-carrier-39 (SLC39) metal-transporter family: characterization of transporter properties. Mol. Pharmacol 70, 171–180.

Hirano, T., Murakami, M., Fukada, T., Nishida, K., Yamasaki, S., Suzuki, T., 2008. Roles of zinc and zinc signaling in immunity: zinc as an intracellular signaling molecule. Adv. Immunol. 97, 149–176.

Hoch, E., Lin, W., Chai, J., Hershfinkel, M., Fu, D., Sekler, I., 2012. Histidine pairing at the metal transport site of mammalian ZnT transporters controls Zn^{2+} over Cd^{2+} selectivity. Proc. Natl. Acad. Sci. U.S.A. 109, 7202–7207.

Hogstrand, C., Kille, P., Ackland, M.L., Hiscox, S., Taylor, K.M., 2013. A mechanism for epithelial-mesenchymal transition and anoikis resistance in breast cancer triggered by zinc channel ZIP6 and STAT3 (signal transducer and activator of transcription 3). Biochem. J. 455, 229–237.

Hogstrand, C., Kille, P., Nicholson, R.I., Taylor, K.M., 2009. Zinc transporters and cancer: a potential role for ZIP7 as a hub for tyrosine kinase activation. Trends Mol. Med. 15, 101–111.

Huang, L., Gitschier, J., 1997. A novel gene involved in zinc transport is deficient in the lethal milk mouse. Nat. Genet. 17, 292–297.

Huang, L., Kirschke, C.P., 2007. A di-leucine sorting signal in ZIP1 (SLC39A1) mediates endocytosis of the protein. FEBS J 274, 3986–3997.

Huang, L., Tepaamorndech, S., 2013. The SLC30 family of zinc transporters – a review of current understanding of their biological and pathophysiological roles. Mol. Aspects Med. 34, 548–560.

Itsumura, N., Inamo, Y., Okazaki, F., Teranishi, F., Narita, H., Kambe, T., Kodama, H., 2013. Compound heterozygous mutations in SLC30A2/ZnT2 results in low milk zinc concentrations: a novel mechanism for zinc deficiency in a breast-fed infant. PLoS One 8, e64045.

Jenkitkasemwong, S., Wang, C.Y., Mackenzie, B., Knutson, M.D., 2012. Physiologic implications of metal-ion transport by ZIP14 and ZIP8. Biometals 25, 643–655.

Jeong, J., Eide, D.J., 2013. The SLC39 family of zinc transporters. Mol. Aspects Med. 34, 612–619.

Kambe, T., 2012. Molecular architecture and function of ZnT transporters. Curr. Top. Membr. 69, 199–220.

Kambe, T., Tsuji, T., Hashimoto, A., Itsumura, N., 2015. The physiological, biochemical and molecular roles of zinc transporters in zinc homeostasis and metabolism. Physiol. Rev. 95, 749–784.

Kambe, T., Hashimoto, A., Fujimoto, S., 2014a. Current understanding of ZIP and ZnT zinc transporters in human health and diseases. Cell Mol. Life Sci. 71, 3281–3295.

Kambe, T., Tsuji, T., Fukue, K., 2014b. Zinc transport proteins and zinc signaling. In: Fukada, T., Kambe, T. (Eds.), Zinc Signals in Cellular Functions and Disorders. Springer, Tokyo, pp. 27–53.

Kim, J.H., Jeon, J., Shin, M., Won, Y., Lee, M., Kwak, J.S., Lee, G., Rhee, J., Ryu, J.H., Chun, C.H., Chun, J.S., 2014. Regulation of the catabolic cascade in osteoarthritis by the zinc-ZIP8-MTF1 axis. Cell 156, 730–743.

Kong, B.Y., Duncan, F.E., Que, E.L., Kim, A.M., O'Halloran, T.V., Woodruff, T.K., 2014. Maternally-derived zinc transporters ZIP6 and ZIP10 drive the mammalian oocyte-to-egg transition. Mol. Hum. Reprod. 20, 1077–1089.

Kury, S., Dreno, B., Bezieau, S., Giraudet, S., Kharfi, M., Kamoun, R., Moisan, J.P., 2002. Identification of SLC39A4, a gene involved in acrodermatitis enteropathica. Nat. Genet. 31, 239–240.

Lasry, I., Seo, Y.A., Ityel, H., Shalva, N., Pode-Shakked, B., Glaser, F., Berman, B., Berezovsky, I., Goncearenco, A., Klar, A., Levy, J., Anikster, Y., Kelleher, S.L., Assaraf, Y.G., 2012. A dominant negative heterozygous G87R mutation in the zinc transporter, ZnT-2 (SLC30A2), results in transient neonatal zinc deficiency. J. Biol. Chem. 287, 29348–29361.

Lee, S., Hennigar, S.R., Alam, S., Nishida, K., Kelleher, S.L., 2015. Essential role for ZnT2-mediated zinc transport in mammary gland development and function during lactation. J. Biol. Chem. 290, 13064–13078.

Leyva-Illades, D., Chen, P., Zogzas, C.E., Hutchens, S., Mercado, J.M., Swaim, C.D., Morrisett, R.A., Bowman, A.B., Aschner, M., Mukhopadhyay, S., 2014. SLC30A10 is a cell surface-localized manganese efflux transporter, and parkinsonism-causing mutations block its intracellular trafficking and efflux activity. J. Neurosci. 34, 14079–14095.

Lichten, L.A., Cousins, R.J., 2009. Mammalian zinc transporters: nutritional and physiologic regulation. Annu. Rev. Nutr. 29, 153–176.

Lin, W., Chai, J., Love, J., Fu, D., 2010. Selective electrodiffusion of zinc ions in a Zrt-, Irt-like protein, ZIPB. J. Biol. Chem. 285, 39013–39020.

Lu, M., Fu, D., 2007. Structure of the zinc transporter YiiP. Science 317, 1746–1748.

Mao, X., Kim, B.E., Wang, F., Eide, D.J., Petris, M.J., 2007. A histidine-rich cluster mediates the ubiquitination and degradation of the human zinc transporter, hZIP4, and protects against zinc cytotoxicity. J. Biol. Chem. 282, 6992–7000.

Nicolson, T.J., Bellomo, E.A., Wijesekara, N., Loder, M.K., Baldwin, J.M., Gyulkhandanyan, A.V., Koshkin, V., Tarasov, A.I., Carzaniga, R., Kronenberger, K., Taneja, T.K., da Silva Xavier, G., Libert, S., Froguel, P., Scharfmann, R., Stetsyuk, V., Ravassard, P., Parker, H., Gribble, F.M., Reimann, F., Sladek, R., Hughes, S.J., Johnson, P.R., Masseboeuf, M., Burcelin, R., Baldwin, S.A., Liu, M., Lara-Lemus, R., Arvan, P., Schuit, F.C., Wheeler, M.B., Chimienti, F., Rutter, G.A., 2009. Insulin storage and glucose homeostasis in mice null for the granule zinc transporter ZnT8 and studies of the type 2 diabetes-associated variants. Diabetes 58, 2070–2083.

Prasad, A.S., 1985. Clinical and biochemical manifestations of zinc deficiency in human subjects. J. Am. Coll. Nutr. 4, 65–72.

Quadri, M., Federico, A., Zhao, T., Breedveld, G.J., Battisti, C., Delnooz, C., Severijnen, L.A., Di Toro Mammarella, L., Mignarri, A., Monti, L., Sanna, A., Lu, P., Punzo, F., Cossu, G., Willemsen, R., Rasi, F., Oostra, B.A., van de Warrenburg, B.P., Bonifati, V., 2012. Mutations in SLC30A10 cause parkinsonism and dystonia with hypermanganesemia, polycythemia, and chronic liver disease. Am. J. Hum. Genet. 90, 467–477.

Sensi, S.L., Paoletti, P., Koh, J.Y., Aizenman, E., Bush, A.I., Hershfinkel, M., 2011. The neurophysiology and pathology of brain zinc. J. Neurosci. 31, 16076–16085.

Sprouse, C., Gordish-Dressman, H., Orkunoglu-Suer, E.F., Lipof, J.S., Moeckel-Cole, S., Patel, R.R., Adham, K., Larkin, J.S., Hubal, M.J., Kearns, A.K., Clarkson, P.M., Thompson, P.D., Angelopoulos, T.J., Gordon, P.M., Moyna, N.M., Pescatello, L.S., Visich, P.S., Zoeller, R.F., Hoffman, E.P., Tosi, L.L., Devaney, J.M., 2014. SLC30A8 nonsynonymous variant is associated with recovery following exercise and skeletal muscle size and strength. Diabetes 63, 363–368.

Stafford, S.L., Bokil, N.J., Achard, M.E., Kapetanovic, R., Schembri, M.A., McEwan, A.G., Sweet, M.J., 2013. Metal ions in macrophage antimicrobial pathways: emerging roles for zinc and copper. Biosci. Rep. 33, e00049.

Szewczyk, B., 2013. Zinc homeostasis and neurodegenerative disorders. Front. Aging Neurosci. 5, 33.

Takeda, A., Nakamura, M., Fujii, H., Tamano, H., 2013. Synaptic Zn(2+) homeostasis and its significance. Metallomics 5, 417–423.

Tamaki, M., Fujitani, Y., Hara, A., Uchida, T., Tamura, Y., Takeno, K., Kawaguchi, M., Watanabe, T., Ogihara, T., Fukunaka, A., Shimizu, T., Mita, T., Kanazawa, A., Imaizumi, M.O., Abe, T., Kiyonari, H., Hojyo, S., Fukada, T., Kawauchi, T., Nagamatsu, S., Hirano, T., Kawamori, R., Watada, H., 2013. The diabetes-susceptible gene SLC30A8/ZnT8 regulates hepatic insulin clearance. J. Clin. Invest. 123, 4513–4524.

Taylor, K.M., Hiscox, S., Nicholson, R.I., Hogstrand, C., Kille, P., 2012. Protein kinase CK2 triggers cytosolic zinc signaling pathways by phosphorylation of zinc channel ZIP7. Sci. Signal. 5, ra11.

Tuschl, K., Clayton, P.T., Gospe Jr., S.M., Gulab, S., Ibrahim, S., Singhi, P., Aulakh, R., Ribeiro, R.T., Barsottini, O.G., Zaki, M.S., Del Rosario, M.L., Dyack, S., Price, V., Rideout, A., Gordon, K., Wevers, R.A., Chong, W.K., Mills, P.B., 2012. Syndrome of hepatic cirrhosis, dystonia, polycythemia, and hypermanganesemia caused by mutations in SLC30A10, a manganese transporter in man. Am. J. Hum. Genet. 90, 457–466.

Wang, F., Kim, B.E., Dufner-Beattie, J., Petris, M.J., Andrews, G., Eide, D.J., 2004. Acrodermatitis enteropathica mutations affect transport activity, localization and zinc-responsive trafficking of the mouse ZIP4 zinc transporter. Hum. Mol. Genet. 13, 563–571.

Wang, K., Zhou, B., Kuo, Y.M., Zemansky, J., Gitschier, J., 2002. A novel member of a zinc transporter family is defective in acrodermatitis enteropathica. Am. J. Hum. Genet. 71, 66–73.

Wenzlau, J.M., Liu, Y., Yu, L., Moua, O., Fowler, K.T., Rangasamy, S., Walters, J., Eisenbarth, G.S., Davidson, H.W., Hutton, J.C., 2008. A common nonsynonymous single nucleotide polymorphism in the SLC30A8 gene determines ZnT8 autoantibody specificity in type 1 diabetes. Diabetes 57, 2693–2697.

Chapter 24

Genetic Study of Zinc Transporters and Zinc Signaling

Toshiyuki Fukada[1,2,3]

[1]*Tokushima Bunri University, Tokushima, Japan;* [2]*Showa University, Tokyo, Japan;* [3]*RIKEN Center for Integrative Medical Sciences, Yokohama, Japan*

INTRODUCTION: ZINC AS AN ESSENTIAL TRACE ELEMENT

Zinc's (Zn) presence in living organisms was first reported in the 19th century when it was described in *Aspergillus niger* (Raulin, 1869). Another century passed, however, before Zn was found to be indispensable for human life (Prasad et al., 1961). Unlike iron and copper, Zn salts and compounds are normally colorless, which complicated early studies of this element. Recent advances in genetics and molecular biology, however, have permitted investigations of Zn's roles in mammalian biology (Fukada et al., 2011b; Maret, 2013; Fig. 24.1), and we now know that Zn is a structural constituent of many proteins and that it is often required for their biological activities (Prasad, 1995; Vallee and Falchuk, 1993).

Recently, a human genome bioinformatics study predicted that approximately 10% of the proteome possesses Zn-binding regions, further highlighting Zn's likely physiological importance (Andreini et al., 2006). Consistent with this idea, the intra- and extracellular Zn levels are tightly controlled by Zn transporters (Fig. 24.2) (Fukada and Kambe, 2011), which is probably essential for the function of Zn-binding proteins. In addition, free and labile Zn mediate many cellular responses: for example, Zn functions as a neurotransmitter at some synapses (Frederickson et al., 2005; Sensi et al., 2009), and as an extracellular signaling molecule to facilitate communication between pancreatic beta cells and hepatocytes for hepatic insulin clearance (Tamaki et al., 2013). These and additional findings suggest that when Zn is mobilized by Zn transporters, it acts as an inter- and intracellular signaling mediator, much like calcium; therefore, we refer to this aspect as "Zn signaling" (Fig. 24.3) (Fukada and Kambe, 2014).

Here, I review recent findings and updated information on the role of Zn transporters and Zn signaling from viewpoints of genetics, molecular biology, and medicine, mainly focusing on early development, nutritional deficiency, systemic growth, bone homeostasis, inflammation, and immunity.

ZINC TRANSPORTERS AND SIGNALING IN PHYSIOLOGY AND PATHOGENESIS

Zinc Transporters and Early Development

Zn homeostasis is controlled by the coordinated actions of Zn transporters (Fig. 24.2; Fukada and Kambe, 2011; Gaither and Eide, 2001; Kambe et al., 2004). Mammalian Zn transporters are divided into two families: the SLC39/ZIPs and SLC30/ZnTs. ZIP members increase cytoplasmic Zn by promoting its transport from extracellular fluid or intracellular organelles to the cytoplasm; ZnT members reduce cytoplasmic Zn by promoting its transport from the cytoplasm to the extracellular space or intracellular organelles (Fukada and Kambe, 2011). Sequence analyses of these transporters suggest that they transport Zn using ATP-independent processes, such as facilitated diffusion, secondary active transport, or symport (Fukada and Kambe, 2011).

Genetic studies in rodents, humans, and other organisms showing known roles of Zn transporters in physiological and developmental regulatory mechanisms are summarized in Table 24.1. Among them, some Zn transporters are crucial for early development. For example, embryos from single (*Zip1, Zip2,* or *Zip3*) and double (*Zip1/Zip2* or *Zip1/Zip3*) knockout (KO) mice exhibit abnormal development under Zn-limited conditions. In addition, the hypomorphic *Zip8* mutant mice, as well as the null *ZnT1* mutant mice, exhibit embryonic lethality (Andrews et al., 2004; Galvez-Peralta et al., 2012; Wang et al., 2011). Mouse ZIP4 is also essential for early embryonic development: it is proposed to transfer zinc from the visceral yolk sac into the embryo; later in development, it adopts its role in the uptake of dietary Zn as described later (Dufner-Beattie et al., 2007). These findings suggest that the regulation of Zn homeostasis by some Zn transporters is critical for normal embryonic development.

Molecular, Genetic, and Nutritional Aspects of Major and Trace Minerals. http://dx.doi.org/10.1016/B978-0-12-802168-2.00024-5

FIGURE 24.1 Chronology of seminal studies in zinc biology. The presence of zinc (Zn) in *Aspergillus niger*, the common bread mold, was first reported in 1869. Zn was recognized as being indispensable for humans in 1961 (Prasad et al., 1961). Biochemical studies revealed that Zn is a structural component of numerous enzymes, such as carbonic anhydrase. The Zn finger motif was discovered in 1985. Since the end of the 20th century, scientific advances have facilitated the characterization of Zn transporters and uncovered their physiological roles and disease associations.

Investigations of nutritional necessity	Investigations by biochemical approaches
1869: *Aspergillus niger*	
1934: Rat	
	1939: Zinc in carbonic anhydrase
1961: Human	
	1985: Discovery of zinc finger motif

1990~: **Investigation of zinc signaling by genetic and molecular approaches**

- Zinc transporters
- Human genome sequencing
- Bioinformatics
- Diseases caused by dysregulated zinc signaling

FIGURE 24.2 Topology and subcellular localization of zinc transporters. (A) Predicted topologies of the ZnT (left), ZIP (right), and transporters. The histidine and aspartic acid residues in transmembrane domains (TMD) 2 and 5 of ZnT transporters and the histidine residues in TMDs 4 and 5 of ZIP transporters are essential for Zn transport. ZnT transporters function as Zn^{2+}/H^+ exchangers, transporting Zn from the cytoplasm to the extracellular space or into intracellular compartments (Ohana et al., 2009). In contrast, ZIP transporters move zinc (Zn) from the extracellular space or intracellular compartments to the cytoplasm. The mechanisms for ZIP-mediated Zn transport are still unclear. (B) Localization and potential functions of the SLC39/ZIP (blue (light gray in print versions)) and SLC30/ZnT (red (gray in print versions)) Zn transporter families (Begum et al., 2002; Chimienti et al., 2004; Huang et al., 2005; Kambe et al., 2004; Kelleher and Lonnerdal, 2005; Taylor et al., 2005; Taylor and Nicholson, 2003; Wang et al., 2004). *Arrows* show the predicted direction of Zn mobilization. *ER*; endoplasmic reticulum.

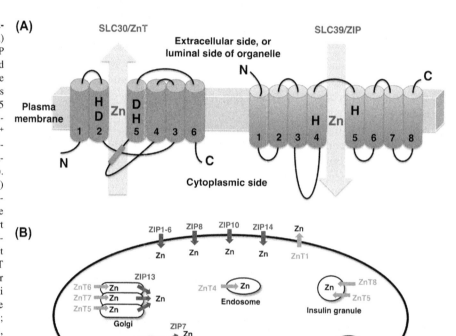

FIGURE 24.3 Zinc signaling by zinc transporters. Zinc (Zn) mobilized by Zn transporters and Zn-binding metallothioneins (MT) function as intra- and extracellular signaling molecules. Free or labile Zn interacts with and regulates the activity of target molecules, impacting cell signaling and downstream responses (see Fig. 24.4).

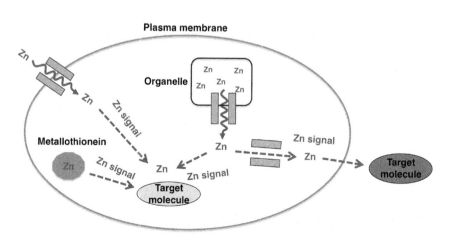

TABLE 24.1 Genetic Evidence for the Biological Significance of Zinc Transporters

Gene	Protein	Mutation Type	Phenotypes in Mouse, *Human, **Fruit Fly, and ***Nematode	References
Slc39a1	ZIP1	KO	Abnormal embryonic development	Dufner-Beattie et al. (2006)
Slc39a2	ZIP2	KO	Abnormal embryonic development	Peters et al. (2007)
Slc39a3	ZIP3	KO	Abnormal embryonic and T-cell development	Dufner-Beattie et al. (2005)
Slc39a4	ZIP4	KO *Mutation	Embryonic lethal Disruption of the intestinal stem cell niche *Acrodermatitis enteropathica	Dufner-Beattie et al. (2007), Kury et al. (2002), Wang et al. (2002)
Slc39a5	ZIP5	KO *Mutation	Impaired intestinal Zn excretion Abnormal pancreatic Zn accumulation *Nonsyndromic high myopia	Geiser et al. (2013), Guo et al. (2014)
Slc39a6	ZIP6	**Mutation (*Foi*)	Impaired gonad morphogenesis	Mathews et al. (2006), Van Doren et al. (2003)
Slc39a7	ZIP7	**Mutation (*Catsup*)	Semidominant lethals Impaired catecholamine metabolism Abnormal tracheal morphogenesis Impaired notch signaling by elevated endoplasmic reticulum (ER) stress	Groth et al. (2013), Hsouna et al. (2007), Stathakis et al. (1999)
Slc39a8	ZIP8	Hypomorphic mutation KO	Impaired multiple-organ organogenesis and hematopoiesis Abnormal innate immune function Osteoarthritis	Galvez-Peralta et al. (2012), Kim et al. (2014), Liu et al. (2013)
Slc39a10	ZIP10	KO	Impaired early B-cell development Abnormal B-cell receptor (BCR) signal transduction	Hojyo et al. (2015), Miyai et al. (2014)
Slc39a13	ZIP13	KO *Mutation	Abnormal bone, tooth, and connective tissue formation *Spondylocheiro dysplastic Ehlers–Danlos syndrome	Fukada et al. (2008), Giunta et al. (2008)
Slc39a14	ZIP14	KO	Impaired systemic growth Abnormal chondrocyte differentiation Abnormal iron homeostasis Changes in cadmium sensitivity	Hojyo et al. (2011), Jorge-Nebert et al. (2015), Nam et al. (2013)
Slc30a1	ZnT1	KO	Embryonic lethal ***Abnormal vulva formation ***Impaired Ras signaling	Andrews et al. (2004), Bruinsma et al. (2002), Jirakulaporn and Muslin (2004)
Slc30a2	ZnT2	*Mutation	*Low Zn in milk	Chowanadisai et al. (2006), Itsumura et al. (2013)
Slc30a3	ZnT3	KO	Prone to seizures Alzheimer's disease–like Impaired spatial memory and behavior dependent on the hippocampus and perirhinal cortex	Adlard et al. (2010), Cole et al. (1999), Martel et al. (2011)
Slc30a4	ZnT4	Mutation (Lethal milk)	Low Zn in milk	Huang and Gitschier (1997)
Slc30a5	ZnT5	KO	Growth retardation, osteopenia, hypodontia, and male-specific cardiac death Impaired mast cell functions	Inoue et al. (2002), Nishida et al. (2009)
Slc30a7	ZnT7	KO	Reduced body fat accumulation Susceptible to diet-induced glucose intolerance and insulin resistance	Huang et al. (2012), Huang et al. (2007)
Slc30a8	ZnT8	KO *SNP	Type II diabetes mellitus *Type I and II diabetes mellitus	Lemaire et al. (2009), Nicolson et al. (2009), Tamaki et al. (2013), Wenzlau et al. (2007)
Slc30a10	ZnT10	*Mutation	*Parkinsonism, dystonia, hypermanganesemia, polycythemia, chronic liver disease	Quadri et al. (2012), Tuschl et al. (2012)

Zinc Transporters and Zinc Nutritional Deficiency

Postneonatal Zn deficiency occurs when Zn levels are insufficient to support physiological requirements and can lead to abnormalities that include growth impedance, delayed sexual maturation, and increased susceptibility to infection. An estimated 2 billion people in developing countries suffer from various Zn deficiency-related diseases (Prasad, 1995), and Zn deficiency contributes to the death of approximately 800,000 children worldwide per year (Hambidge and Krebs, 2007). Experiments in mouse models have demonstrated multiple impacts of Zn deficiency (Prasad, 2008), including thymic atrophy and lymphopenia, compromised cell- and antibody-mediated responses, and increased infection rates and duration (Fernandes et al., 1979; Fraker and King, 2004; Keen and Gershwin, 1990; Prasad, 1991; Shankar and Prasad, 1998).

Functionally impaired Zn transporters can also cause Zn nutritional deficiency and physiological abnormalities (Table 24.1). For example, lethal milk mutant mice, which have a loss-of-function mutation in *ZnT4*, a mammary gland-expressed transporter, produce Zn-deficient milk (Huang and Gitschier, 1997). A similar abnormality is observed in patients carrying a loss-of-function mutation in *ZNT2* (Chowanadisai et al., 2006; Itsumura et al., 2013), confirming that Zn transporters in the mammary gland are required for Zn transfer into breast milk. In addition, intestinal Zn transporters, such as ZIP4 and ZIP5, promote the absorption of Zn from dietary sources. Once Zn is absorbed in the intestine, it is transported through the circulatory system for uptake by cells. ZIP4, encoded by the *Slc39a4* gene, is expressed on the apical membrane of polarized enterocytes in the intestine and is crucial for intestinal Zn transport (Dufner-Beattie et al., 2007). Disorders caused by ZIP4 deficiency include acrodermatitis enteropathica (AE), a recessive Zn deficiency disorder caused by a loss-of-function mutation in human *Slc39a4* (Kury et al., 2002; Wang et al., 2002). ZIP5, encoded by *Slc39a5*, is expressed in the basolateral membranes of intestinal enterocytes and pancreatic acinar cells. One study indicated that ZIP5 deficiency causes impaired intestinal Zn excretion and pancreatic toxicity in mice (Guo et al., 2014).

Thus, these reports suggest that the coordinated actions of the various Zn transporters are important for Zn uptake and excretion, and for Zn homeostasis throughout the body (Wang and Zhou, 2010). Moreover, their functions are indispensable for mammalian health.

Zinc Transporters in Systemic Growth, Bone Metabolism, and Cartilage Homeostasis

The concentration of Zn is relatively high in bone (Bergman and Soremark, 1968), and Zn has been reported to stimulate bone formation (Yamaguchi and Gao, 1998). However, the mechanism by which Zn regulates systemic growth and bone metabolism has been unclear. Analyses of ZIP8, ZIP13, and ZIP14 KO mice, together with human genetic studies, have confirmed that these transporters play important roles in growth and bone metabolism and uncovered the mechanism by which Zn contributes to normal bone maintenance and degeneration.

ZIP8/Slc39a8 in Cartilage Degeneration and Osteoarthritis

Osteoarthritis (OA) is a common musculoskeletal disorder, primarily characterized by cartilage degeneration. A report indicates that ZIP8, a plasma-membrane-localized Zn transporter, can promote OA in mice (Kim et al., 2014). ZIP8 expression levels and Zn influx are markedly increased in mouse articular chondrocytes following treatment with interleukin 1β, which promotes a cascade of inflammatory and catabolic events including the expression of cartilage degrading matrix metalloproteinases (MMP). In addition, Zn and ZIP8 expression levels are elevated in human and mouse OA-affected cartilage, consistent with their putative involvement in OA pathogenesis. ZIP8-mediated Zn influx induces the expression of a metal-regulatory transcription factor-1 (MTF-1), which regulates a number of target genes during cellular adaptation to heavy metal exposure (Laity and Andrews, 2007). MTF-1 in turn upregulates the expression of MMP3, MMP13, and ADAMTS5, which are Zn-binding, matrix-degrading enzymes that function as crucial effectors of OA cartilage destruction (Blom et al., 2007; Glasson et al., 2005; Little et al., 2009). The ectopic overexpression of ZIP8 in mouse cartilage tissue promotes OA-like pathogenesis, whereas *Zip8* cartilage-specific conditional KO mice are resistant to experimentally induced OA. These results clearly demonstrated that the "Zn-ZIP8-MTF1 signaling axis" in chondrocytes induces MMP upregulation, thereby leading to OA cartilage destruction (Fig. 24.4A) (Kim et al., 2014).

ZIP13/Slc39a13 in Bone and Connective Tissue Formation

ZIP13 is a homodimerized transporter, which transports Zn from the Golgi to the cytosol (Bin et al., 2011; Fukada and Kambe, 2011). *Zip13*-KO mice exhibit multiple phenotypes, including reduced growth, increased skin fragility, abnormal tooth and bone formation, and shortening of the long bones. *Zip13*-KO mice also exhibit irregularly organized proliferative

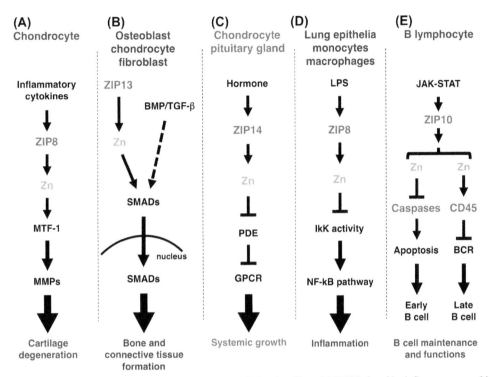

FIGURE 24.4 **Regulatory roles of selected zinc transporters in intracellular signaling.** (A) ZIP8 induced by inflammatory cytokines mediates zinc (Zn) signaling in chondrocytes, leading to MTF-1 activation and upregulated MMP expression, followed by cartilage degeneration and the development of OA. (B) ZIP13 is required for BMP/TGF-β-induced nuclear translocation of Smads, and is involved in bone and connective tissue development. (C) ZIP14 promotes GPCR signaling by inhibiting hormone-stimulated phosphodiesterase (PDE) in the pituitary gland, liver, and cartilage, and it is required for endocrine function and systemic growth. (D) Endotoxic stimulation induces ZIP8 expression and Zn signaling in lung epithelia, monocytes, and macrophages, leading to the negative regulation of IKKβ activity and NF-κB signaling. (E) ZIP10 inhibits caspase activity and regulates B-cell receptor (BCR) signaling in early and late B-cells, respectively; it is required for B-cell survival and functions.

chondrocytes in their growth plates and reduced osteoblast activity, indicating that ZIP13 controls the differentiation and function of chondrocytes and osteoblasts (Fukada et al., 2008, 2011a). The phenotypes of *Zip13*-KO mice are reminiscent of those associated with human Ehlers–Danlos syndrome (EDS), which represents a group of genetic disorders (Steinmann and Royce, 2002; Steinmann et al., 2002). Patients with spondylocheiro dysplastic EDS (SCD-EDS) possess homozygous mutations in *SLC39A13* (Fukada et al., 2008; Giunta et al., 2008), resulting in the expression of ZIP13 mutant proteins, which undergo proteasome-mediated degradation (Bin et al., 2014a,b). Further studies have indicated that the BMP/TGF-β signaling cascades, which coordinately regulate bone formation and growth (Akhurst, 2004; Cao and Chen, 2005; Cobourne and Sharpe, 2003), are significantly dysregulated in *Zip13*-KO cells (Fukada et al., 2008). These findings provide substantial evidence that intracellular ZIP13-regulated Zn signaling is critical for the control of mammalian bone development and growth (Fig. 24.4B).

ZIP14/Slc39a14 in Bone Homeostasis and Systemic Growth

ZIP14, encoded by *Slc39a14*, is a cell membrane-localized Zn transporter (Liuzzi et al., 2005) whose mRNA is highly expressed in bone tissues, particularly in proliferative zone of growth plate during endochondral ossification. *Zip14*-KO mice exhibit shortened long bones and dwarfism. The morphology of the *Zip14*-KO growth plate is a partial phenocopy of that observed in mice with a chondrocyte-specific deletion of parathyroid hormone 1 receptor (*Pth1r*; Chung et al., 1998; Kobayashi et al., 2002; Kronenberg, 2006), and parathyroid hormone-related peptide (PTHrP)-induced signaling is significantly reduced in the *Zip14*-KO chondrocytes. PTHrP binding to PTH1R, a G protein-coupled receptor (GPCR), leads to AC activation, and the production of high levels of cyclic (c)AMP, which in turn activate protein kinase A (PKA) by binding to its regulatory subunit (Ionescu et al., 2001), resulting in the phosphorylation of the cAMP response element-binding protein (CREB). *Zip14*-KO chondrocytes show substantially reduced basal cAMP levels and enhanced phosphodiesterase (PDE) activity. PDE is a critical regulator of cyclic nucleotide signaling and degrades cAMP. Zn treatment also reduces the cAMP-specific PDE activity and elevates cAMP levels in *Zip14*-KO chondrocytes, confirming that ZIP14 regulates the PTH1R-cAMP-CREB pathway by affecting Zn homeostasis. *Zip14*-KO mice show phenotypic similarity to growth hormone (GH)-mutant mice and human cases of GH deficiency (Procter et al., 1998). Indeed, the pituitary gland expresses *Zip14*, and the levels of pituitary cAMP and Zn are decreased in *Zip14*-KO mice. Growth hormone-releasing hormone (GHRH) acts on the GHRH receptor (GHRHR), a G protein-coupled receptor (GPCR) in the pituitary gland, to upregulate GH. Notably, the bolus administration of GHRH increases GH plasma levels in control mice, but has little effect in *Zip14*-KO mice. In addition, *Zip14*-KO mice exhibit reduced serum IGF-I and hepatic *Igf-I* expression levels, which are normally regulated by circulating GH. Collectively, these results demonstrate that ZIP14 functions as a positive GPCR regulator in both endochondral ossification and GH production, resulting in the concomitant regulation of vertebrate bone formation and systemic growth (Fig. 24.4C).

Taken together, these studies with genetics and molecular biological approaches on ZIP8, ZIP13, and ZIP14 uncovered the mechanisms by which abnormal Zn homeostasis causes impairments in bone regulation and systemic growth (Fukada et al., 2014).

Zinc Transporters in Inflammation and Immunity

Zn deficiency results in compromised cell and humoral responses to infection although the underlying mechanisms have been unclear. In the next two sections, I will review recent findings indicating that ZIP8 and ZIP10 are involved in regulating inflammation and immunity.

ZIP8/Slc39a8 in the Regulation of Inflammation

Although Zn inhibits NF-κB activation and suppresses inflammation, the underlying mechanism has been uncertain (Haase and Rink, 2009). However, as shown with chondroctyes' degeneration mentioned earlier (Fig. 24.4A), one study suggests that ZIP8 may function as a negative feedback regulator of NF-κB and innate immune activation in response to bacterial infection through the coordination of Zn metabolism (Liu et al., 2013). *Zip8* expression is upregulated NF-κB-dependently in vitro, in primary human lung epithelia, monocytes, and macrophages in response to TNFα or LPS, and in vivo, in response to sepsis. Intriguingly, ZIP8-mediated Zn transport is critical for attenuating NF-κB signaling, which involves the Zn-mediated inhibition of IKKβ kinase activity. Furthermore, *Slc39a8* hypomorphic mouse fetal fibroblasts, which exhibit a significant reduction in ZIP8 mRNA and protein expression, display a heightened response to proinflammatory cytokines, showing both elevated IKKβ kinase activity and increased expression of NF-κB-driven genes. The finding that ZIP8 is a negative feedback regulator of NF-κB and immune activation sheds new light on the mechanism by which Zn impacts innate immunity and suggests that ZIP8–Zn signaling may be a useful target in the treatment of inflammatory diseases (Fig. 24.4D) (Liu et al., 2013).

ZIP10/Slc39a10 in B-Cell Maintenance and Functions

ZIP10 is a cell membrane-localized Zn transporter that is expressed during both early and late B-cell stages, but its effects on B-cell development have been unclear. We found that the selective ablation of *Slc39a10* in early B-cells, under control of the *Mb-1* (Igα gene) promoter, severely reduces pro-B-cell survival in the bone marrow (Miyai et al., 2014). The inducible deletion of ZIP10 in pro-B-cells reduces intracellular Zn levels and increases constitutive caspase activity. Similar findings are observed in B-cells depleted of intracellular Zn by a chemical chelating agent. ZIP10 expression in early B-cells is regulated by cytokine-induced JAK-STAT activation, and the upregulated "ZIP10-Zn signaling axis" promotes early B-cell survival by inhibiting caspase activity (Miyai et al., 2014). Furthermore, ZIP10's expression correlates with STAT activation in human B-cell lymphoma, indicating that the "JAK-STAT-ZIP10-Zn signaling axis" influences B lymphocyte homeostasis, underscoring a definitive role for ZIP10 in the survival and maintenance of early B-cell populations (Fig. 24.4E, left; Miyai et al., 2014). Our finding, that ZIP10 ablation in early B-cells reduces intracellular Zn levels and substantially reduces total B-cell numbers, demonstrates an essential role for Zn homeostasis in B-cell development and provides a molecular basis for the immunodeficiency that results from Zn deficiency.

We also investigated the significance of ZIP10 in late-stage B-cells. We generated *Zip10*-conditional KO (*Zip10*-cKO) mice, in which the *Zip10* gene was ablated in antigen presenting cells (APCs), under the control of the invariant chain promoter. The APC-specific ablation of *Zip10* results in the cell-autonomous reduction of follicular B-cells. In addition, *Zip10*-cKO mice immunized with 4-hydroxy-3-nitrophenylacetyl (NP) conjugated to chicken gamma globulin exhibit a dramatic decrease in NP-specific IgM and IgG1 production that is B-cell intrinsic. Successful induction of high-affinity antibody responses require B-cell interactions with their cognate T helper cells, followed by germinal center (GC) reactions that involve class-switch recombination (Zotos and Tarlinton, 2012). GC formation is severely decreased in *Zip10*-cKO mice, confirming a critical role for the ZIP10-Zn signaling axis in regulating proper antibody responses. B-cell receptor (BCR)-mediated signaling is required for the full activation of B-cells during the effecter phase and for their interaction with cognate helper T-cells during GC formation (Kurosaki et al., 2010). Consistent with the impaired GC formation observed in vivo, the proliferation of *Zip10*-cKO B-cells is diminished after BCR cross-linking in vitro. Intriguingly, *Zip10*-cKO B-cells exhibit spleen tyrosine kinase (SYK) and Lck/Yes novel tyrosine kinase (LYN) hyperactivation after BCR cross-linking. CD45R is a key receptor-type protein tyrosine phosphatase (PTPase) that inhibits LYN activity (Shrivastava et al., 2004). We found that the overall CD45R PTPase activity in *Zip10*-cKO B-cells is downregulated, and the forced introduction of Zn suppresses BCR-induced LYN activation by upregulating the CD45R PTPase activity in *Zip10*-cKO B-cells, suggesting that the ZIP10-Zn signaling axis induces CD45R PTPase activity (Fig. 24.4E, right; Hojyo et al., 2014, 2015).

Collectively, these findings on the role of ZIP10 in early- and late-stage B-cells provide new insights into the relevance of Zn signaling in immuno-physiological events (Fig. 24.4E).

ZINC TRANSPORTERS AND THE "ZN SIGNALING AXIS"

We and others have demonstrated that the Zn mobilized by Zn transporters acts as a signaling molecule. This "Zn signal" is involved in a variety of signaling pathways (Fukada et al., 2011b; Maret, 2011), including growth factor- (Haase and Maret, 2003; Taylor et al., 2012), antigen receptor- (Hojyo et al., 2014; Nishida et al., 2009), NF-κB- (Liu et al., 2013), and JAK-STAT-mediated pathways (Hogstrand et al., 2013; Kitabayashi et al., 2010; Miyai et al., 2014; Yamashita et al., 2004). Consider the fact that not only the impaired Zn influx by ZIPs (such as *Zip8*, *Zip13*, and *Zip14*) but also the loss of Zn-efflux system by some ZnTs (like *ZnT5* and *ZnT7*) also cause growth retardation and abnormal bone homeostasis, which clearly suggests that each Zn transporter functions not merely to maintain zinc homeostasis, but to mobilize Zn for the controls of a variety of the regulatory pathways and cellular functions (Table 24.1; Fukada et al., 2008; Hojyo et al., 2011; Huang et al., 2007; Inoue et al., 2002; Miyai et al., 2014).

Importantly, each Zn transporter initiates a unique signal transduction cascade that regulates specific target molecules and cellular responses (Fig. 24.4; Fukada et al., 2011b). The term, "Zn signal axis," refers to the Zn regulatory system within each of these pathways (Fukada et al., 2013). Numbers of questions of the Zn signal axes still remain to be clarified, including the molecular basis of the Zn signal specificity and the downstream effects of the various Zn signals (Fig. 24.5).

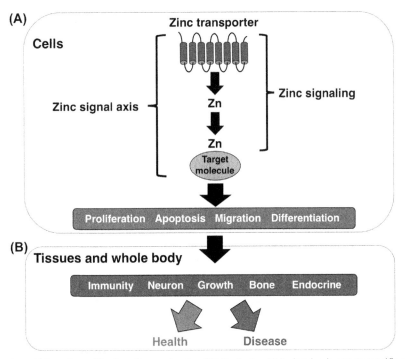

FIGURE 24.5 "Zinc signal axes" target specific signaling molecules. (A) Each zinc (Zn) signal axis targets specific molecules. Thus Zn signals impact a variety of cellular processes by selectively regulating distinct molecular events. (B) The disruption of nonredundant Zn signaling axes results in impaired cellular responses, which can prevent normal tissue development and/or function and have pathogenic consequences.

Further molecular and genetics-based studies on the function of Zn transporters will be required to fully understand Zn signaling.

FUTURE PERSPECTIVES

Many aspects of Zn transporter and Zn signaling function remain to be investigated, including the specific mechanisms for Zn transport across membranes, the molecular basis of Zn transport specificity, and the regulation of downstream molecular events. How Zn signaling and other metals affect each other is also an intriguing topic, since some Zn transporters mediate the transfer of additional metals, including iron (Jenkitkasemwong et al., 2015; Nam et al., 2013) and manganese or cadmium (Fujishiro et al., 2012; Jorge-Nebert et al., 2015; Nebert et al., 2012). Zn transporter–mediated Zn signaling may also have uncharacterized roles in intra- and intercellular signaling. Further molecular and genetic studies will be required to ascertain the precise mechanisms of Zn transport and to obtain a complete understanding of Zn transporters' regulatory roles in mammalian physiology.

ACKNOWLEDGMENTS

This work was supported by KAKENHI Grants-in-Aid from the Ministry of Education, Culture, Sports, Science, and Technology of Japan, and the SENSHIN Medical Research Foundation.

REFERENCES

Akhurst, R.J., 2004. TGF beta signaling in health and disease. Nat. Genet. 36, 790–792.

Andreini, C., Banci, L., Bertini, I., Rosato, A., 2006. Counting the zinc-proteins encoded in the human genome. J. Proteome Res. 5, 196–201.

Andrews, G.K., Wang, H., Dey, S.K., Palmiter, R.D., 2004. Mouse zinc transporter 1 gene provides an essential function during early embryonic development. Genesis 40, 74–81.

Adlard, P.A., Parncutt, J.M., Finkelstein, D.I., Bush, A.I., 2010. Cognitive loss in zinc transporter-3 knock-out mice: a phenocopy for the synaptic and memory deficits of Alzheimer's disease? J. Neurosci. 30, 1631–1636.

Begum, N.A., Kobayashi, M., Moriwaki, Y., Matsumoto, M., Toyoshima, K., Seya, T., 2002. Mycobacterium bovis BCG cell wall and lipopolysaccharide induce a novel gene, BIGM103, encoding a 7-TM protein: identification of a new protein family having Zn-transporter and Zn-metalloprotease signatures. Genomics 80, 630–645.

Bergman, B., Soremark, R., 1968. Autoradiographic studies on the distribution of zinc-65 in mice. J. Nutr. 94, 6–12.

Bin, B.H., Fukada, T., Hosaka, T., Yamasaki, S., Ohashi, W., Hojyo, S., Miyai, T., Nishida, K., Yokoyama, S., Hirano, T., 2011. Biochemical characterization of human ZIP13 protein: a homo-dimerized zinc transporter involved in the Spondylocheiro dysplastic Ehlers-Danlos syndrome. J. Biol. Chem. 286, 40255–40265.

Bin, B.H., Hojyo, S., Hosaka, T., Bhin, J., Kano, H., Miyai, T., Ikeda, M., Kimura-Someya, T., Shirouzu, M., Cho, E.G., et al., 2014a. Molecular pathogenesis of Spondylocheirodysplastic Ehlers-Danlos syndrome caused by mutant ZIP13 proteins. EMBO Mol. Med. 6, 1028–1042.

Bin, B.H., Hojyo, S., Lee, T.R., Fukada, T., 2014b. Spondylocheirodysplastic Ehlers-Danlos syndrome (SCD-EDS) and the mutant zinc transporter ZIP13. Rare Dis. 2, e974982.

Blom, A.B., van Lent, P.L., Libregts, S., Holthuysen, A.E., van der Kraan, P.M., van Rooijen, N., van den Berg, W.B., 2007. Crucial role of macrophages in matrix metalloproteinase-mediated cartilage destruction during experimental osteoarthritis: involvement of matrix metalloproteinase 3. Arthritis Rheum. 56, 147–157.

Bruinsma, J.J., Jirakulaporn, T., Muslin, A.J., Kornfeld, K., 2002. Zinc ions and cation diffusion facilitator proteins regulate Ras-mediated signaling. Dev. Cell 2, 567–578.

Cao, X., Chen, D., 2005. The BMP signaling and in vivo bone formation. Gene 357, 1–8.

Chimienti, F., Devergnas, S., Favier, A., Seve, M., 2004. Identification and cloning of a beta-cell-specific zinc transporter, ZnT-8, localized into insulin secretory granules. Diabetes 53, 2330–2337.

Chowanadisai, W., Lonnerdal, B., Kelleher, S.L., 2006. Identification of a mutation in SLC30A2 (ZnT-2) in women with low milk zinc concentration that results in transient neonatal zinc deficiency. J. Biol. Chem. 281, 39699–39707.

Chung, U.I., Lanske, B., Lee, K., Li, E., Kronenberg, H., 1998. The parathyroid hormone/parathyroid hormone-related peptide receptor coordinates endochondral bone development by directly controlling chondrocyte differentiation. Proc. Natl. Acad. Sci. USA 95, 13030–13035.

Cobourne, M.T., Sharpe, P.T., 2003. Tooth and jaw: molecular mechanisms of patterning in the first branchial arch. Arch. Oral Biol. 48, 1–14.

Cole, T.B., Wenzel, H.J., Kafer, K.E., Schwartzkroin, P.A., Palmiter, R.D., 1999. Elimination of zinc from synaptic vesicles in the intact mouse brain by disruption of the ZnT3 gene. Proc. Natl. Acad. Sci. USA 96, 1716–1721.

Dufner-Beattie, J., Weaver, B.P., Geiser, J., Bilgen, M., Larson, M., Xu, W., Andrews, G.K., 2007. The mouse acrodermatitis enteropathica gene Slc39a4 (Zip4) is essential for early development and heterozygosity causes hypersensitivity to zinc deficiency. Hum. Mol. Genet. 16, 1391–1399.

Dufner-Beattie, J., Huang, Z.L., Geiser, J., Xu, W., Andrews, G.K., 2005. Generation and characterization of mice lacking the zinc uptake transporter ZIP3. Mol. Cell Biol. 25, 5607–5615.

Dufner-Beattie, J., Huang, Z.L., Geiser, J., Xu, W., Andrews, G.K., 2006. Mouse ZIP1 and ZIP3 genes together are essential for adaptation to dietary zinc deficiency during pregnancy. Genesis 44, 239–251.

Fernandes, G., Nair, M., Onoe, K., Tanaka, T., Floyd, R., Good, R.A., 1979. Impairment of cell-mediated immunity functions by dietary zinc deficiency in mice. Proc. Natl. Acad. Sci. USA 76, 457–461.

Fraker, P.J., King, L.E., 2004. Reprogramming of the immune system during zinc deficiency. Annu. Rev. Nutr. 24, 277–298.

Frederickson, C.J., Koh, J.Y., Bush, A.I., 2005. The neurobiology of zinc in health and disease. Nat. Rev. Neurosci. 6, 449–462.

Fujishiro, H., Yano, Y., Takada, Y., Tanihara, M., Himeno, S., 2012. Roles of ZIP8, ZIP14, and DMT1 in transport of cadmium and manganese in mouse kidney proximal tubule cells. Metallomics 4, 700–708.

Fukada, T., Asada, Y., Mishima, K., Shimoda, S., Saito, I., 2011a. Slc39a13/Zip13: a crucial zinc transporter involved in tooth development and inherited disorders. J. Oral Biosci. 53, 1–12.

Fukada, T., Yamasaki, S., Nishida, K., Murakami, M., Hirano, T., 2011b. Zinc homeostasis and signaling in health and diseases: zinc signaling. J. Biol. Inorg. Chem. 16, 1123–1134.

Fukada, T., Civic, N., Furuichi, T., Shimoda, S., Mishima, K., Higashiyama, H., Idaira, Y., Asada, Y., Kitamura, H., Yamasaki, S., et al., 2008. The zinc transporter SLC39A13/ZIP13 is required for connective tissue development; its involvement in BMP/TGF-beta signaling pathways. PLoS One 3, e3642.

Fukada, T., Hojyo, S., Bin, B., 2014. Zinc signal in growth control and bone diseases. In: Fukada, T., Kambe, T. (Eds.), Zinc Signals in Cellular Functions and Disorders. Springer, Tokyo, pp. 249–267.

Fukada, T., Hojyo, S., Furuichi, T., 2013. Zinc signal: a new player in osteobiology. J. Bone Miner. Metab. 31, 129–135.

Fukada, T., Kambe, T., 2011. Molecular and genetic features of zinc transporters in physiology and pathogenesis. Metallomics 3, 662–674.

Fukada, T., Kambe, T., 2014. Zinc Signals in Cellular Functions and Disorders. Springer, Tokyo.

Gaither, L.A., Eide, D.J., 2001. Eukaryotic zinc transporters and their regulation. Biometals 14, 251–270.

Galvez-Peralta, M., He, L., Jorge-Nebert, L.F., Wang, B., Miller, M.L., Eppert, B.L., Afton, S., Nebert, D.W., 2012. ZIP8 zinc transporter: indispensable role for both multiple-organ organogenesis and hematopoiesis in utero. PLoS One 7, e36055.

Geiser, J., De Lisle, R.C., Andrews, G.K., 2013. The zinc transporter Zip5 (Slc39a5) regulates intestinal zinc excretion and protects the pancreas against zinc toxicity. PLoS One 8(11), e82149. http://dx.doi.org/10.1371/journal.pone.0082149.

Giunta, C., Elcioglu, N.H., Albrecht, B., Eich, G., Chambaz, C., Janecke, A.R., Yeowell, H., Weis, M., Eyre, D.R., Kraenzlin, M., et al., 2008. Spondylo-cheiro dysplastic form of the Ehlers-Danlos syndrome–an autosomal-recessive entity caused by mutations in the zinc transporter gene SLC39A13. Am. J. Hum. Genet. 82, 1290–1305.

Glasson, S.S., Askew, R., Sheppard, B., Carito, B., Blanchet, T., Ma, H.L., Flannery, C.R., Peluso, D., Kanki, K., Yang, Z., et al., 2005. Deletion of active ADAMTS5 prevents cartilage degradation in a murine model of osteoarthritis. Nature 434, 644–648.

Guo, H., Jin, X., Zhu, T., Wang, T., Tong, P., Tian, L., Peng, Y., Sun, L., Wan, A., Chen, J., Liu, Y., Li, Y., Tian, Q., Xia, L., Zhang, L., Pan, Y., Lu, L., Liu, Q., Shen, L., Xiong, W., Li, J., Tang, B., Feng, Y., Zhang, X., Zhang, Z., Pan, Q., Hu, Z., Xia, K., 2014. SLC39A5 mutations interfering with the BMP/TGF-beta pathway in non-syndromic high myopia. J. Med. Genet. 51, 518–525.

Groth, C., Sasamura, T., Khanna, M.R., Whitley, M., Fortini, M.E., 2013. Protein trafficking abnormalities in Drosophila tissues with impaired activity of the ZIP7 zinc transporter Catsup. Development 140, 3018–3027.

Haase, H., Maret, W., 2003. Intracellular zinc fluctuations modulate protein tyrosine phosphatase activity in insulin/insulin-like growth factor-1 signaling. Exp. Cell Res. 291, 289–298.

Haase, H., Rink, L., 2009. Functional significance of zinc-related signaling pathways in immune cells. Annu. Rev. Nutr. 29, 133–152.

Hambidge, K.M., Krebs, N.F., 2007. Zinc deficiency: a special challenge. J. Nutr. 137, 1101–1105.

Hogstrand, C., Kille, P., Ackland, M.L., Hiscox, S., Taylor, K.M., 2013. A mechanism for epithelial-mesenchymal transition and anoikis resistance in breast cancer triggered by zinc channel ZIP6 and STAT3 (signal transducer and activator of transcription 3). Biochem. J. 455, 229–237.

Hojyo, S., Fukada, T., Shimoda, S., Ohashi, W., Bin, B.H., Koseki, H., Hirano, T., 2011. The zinc transporter SLC39A14/ZIP14 controls G-protein coupled receptor-mediated signaling required for systemic growth. PLoS One 6, e18059.

Hojyo, S., Miyai, T., Fujishiro, H., Kawamura, M., Yasuda, T., Hijikata, A., Bin, B.H., Irie, T., Tanaka, J., Atsumi, T., et al., 2014. Zinc transporter SLC39A10/ZIP10 controls humoral immunity by modulating B-cell receptor signal strength. Proc. Natl. Acad. Sci. USA 111, 11786–11791.

Hojyo, S., Miyai, T., Fukada, T., 2015. B-cell receptor strength and zinc signaling: unraveling the role of zinc transporter ZIP10 in humoral immunity. Recept. Clin. Invest. 2, e387.

Huang, L., Gitschier, J., 1997. A novel gene involved in zinc transport is deficient in the lethal milk mouse. Nat. Genet. 17, 292–297.

Huang, L., Kirschke, C.P., Zhang, Y., Yu, Y.Y., 2005. The ZIP7 gene (Slc39a7) encodes a zinc transporter involved in zinc homeostasis of the Golgi apparatus. J. Biol. Chem. 280, 15456–15463.

Huang, L., Yu, Y.Y., Kirschke, C.P., Gertz, E.R., Lloyd, K.K., 2007. Znt7 (Slc30a7)-deficient mice display reduced body zinc status and body fat accumulation. J. Biol. Chem. 282, 37053–37063.

Hsouna, A., Lawal, H.O., Izevbaye, I., Hsu, T., O'Donnell, J.M., 2007. Drosophila dopamine synthesis pathway genes regulate tracheal morphogenesis. Dev. Biol. 308, 30–43.

Huang, L., Kirschke, C.P., Lay, Y.A., Levy, L.B., Lamirande, D.E., Zhang, P.H., 2012. Znt7-null mice are more susceptible to diet-induced glucose intolerance and insulin resistance. J. Biol. Chem. 287 (40), 33883–33896.

Inoue, K., Matsuda, K., Itoh, M., Kawaguchi, H., Tomoike, H., Aoyagi, T., Nagai, R., Hori, M., Nakamura, Y., Tanaka, T., 2002. Osteopenia and male-specific sudden cardiac death in mice lacking a zinc transporter gene, Znt5. Hum. Mol. Genet. 11, 1775–1784.

Ionescu, A.M., Schwarz, E.M., Vinson, C., Puzas, J.E., Rosier, R., Reynolds, P.R., O'Keefe, R.J., 2001. PTHrP modulates chondrocyte differentiation through AP-1 and CREB signaling. J. Biol. Chem. 276, 11639–11647.

Itsumura, N., Inamo, Y., Okazaki, F., Teranishi, F., Narita, H., Kambe, T., Kodama, H., 2013. Compound heterozygous mutations in SLC30A2/ZnT2 results in low milk zinc concentrations: a novel mechanism for zinc deficiency in a breast-fed infant. PLoS One 8, e64045.

Jenkitkasemwong, S., Wang, C.Y., Coffey, R., Zhang, W., Chan, A., Biel, T., Kim, J.S., Hojyo, S., Fukada, T., Knutson, M.D., 2015. SLC39A14 is required for the development of hepatocellular iron overload in murine models of hereditary hemochromatosis. Cell Metab. 22 (1), 138–150 (inpress).

Jorge-Nebert, L.F., Galvez-Peralta, M., Landero Figueroa, J., Somarathna, M., Hojyo, S., Fukada, T., Nebert, D.W., 2015. Comparing gene expression during cadmium uptake and distribution: untreated versus oral Cd-treated wild-type and ZIP14 knockout mice. Toxicol. Sci. 143, 26–35.

Jirakulaporn, T., Muslin, A.J., 2004. Cation diffusion facilitator proteins modulate Raf-1 activity. J. Biol. Chem. 279, 27807–27815.

Kambe, T., Yamaguchi-Iwai, Y., Sasaki, R., Nagao, M., 2004. Overview of mammalian zinc transporters. Cell Mol. Life Sci. 61, 49–68.

Keen, C.L., Gershwin, M.E., 1990. Zinc deficiency and immune function. Annu. Rev. Nutr. 10, 415–431.

Kelleher, S.L., Lonnerdal, B., 2005. Zip3 plays a major role in zinc uptake into mammary epithelial cells and is regulated by prolactin. Am. J. Physiol. Cell Physiol. 288, C1042–C1047.

Kim, J.H., Jeon, J., Shin, M., Won, Y., Lee, M., Kwak, J.S., Lee, G., Rhee, J., Ryu, J.H., Chun, C.H., et al., 2014. Regulation of the catabolic cascade in osteoarthritis by the zinc-ZIP8-MTF1 axis. Cell 156, 730–743.

Kitabayashi, C., Fukada, T., Kanamoto, M., Ohashi, W., Hojyo, S., Atsumi, T., Ueda, N., Azuma, I., Hirota, H., Murakami, M., et al., 2010. Zinc suppresses Th17 development via inhibition of STAT3 activation. Int. Immunol. 22, 375–386.

Kobayashi, T., Chung, U.I., Schipani, E., Starbuck, M., Karsenty, G., Katagiri, T., Goad, D.L., Lanske, B., Kronenberg, H.M., 2002. PTHrP and Indian hedgehog control differentiation of growth plate chondrocytes at multiple steps. Development 129, 2977–2986.

Kronenberg, H.M., 2006. PTHrP and skeletal development. Ann. NY Acad. Sci. 1068, 1–13.

Kurosaki, T., Shinohara, H., Baba, Y., 2010. B cell signaling and fate decision. Annu. Rev. Immunol. 28, 21–55.

Kury, S., Dreno, B., Bezieau, S., Giraudet, S., Kharfi, M., Kamoun, R., Moisan, J.P., 2002. Identification of SLC39A4, a gene involved in acrodermatitis enteropathica. Nat. Genet. 31, 239–240.

Laity, J.H., Andrews, G.K., 2007. Understanding the mechanisms of zinc-sensing by metal-response element binding transcription factor-1 (MTF-1). Arch. Biochem. Biophys. 463, 201–210.

Little, C.B., Barai, A., Burkhardt, D., Smith, S.M., Fosang, A.J., Werb, Z., Shah, M., Thompson, E.W., 2009. Matrix metalloproteinase 13-deficient mice are resistant to osteoarthritic cartilage erosion but not chondrocyte hypertrophy or osteophyte development. Arthritis Rheum. 60, 3723–3733.

Liu, M.J., Bao, S., Galvez-Peralta, M., Pyle, C.J., Rudawsky, A.C., Pavlovicz, R.E., Killilea, D.W., Li, C., Nebert, D.W., Wewers, M.D., et al., 2013. ZIP8 regulates host defense through zinc-mediated inhibition of NF-kappaB. Cell Rep. 3, 386–400.

Liuzzi, J.P., Lichten, L.A., Rivera, S., Blanchard, R.K., Aydemir, T.B., Knutson, M.D., Ganz, T., Cousins, R.J., 2005. Interleukin-6 regulates the zinc transporter Zip14 in liver and contributes to the hypozincemia of the acute-phase response. Proc. Natl. Acad. Sci. USA 102, 6843–6848.

Lemaire, K., Ravier, M.A., Schraenen, A., Creemers, J.W., Van de Plas, R., Granvik, M., Van Lommel, L., Waelkens, E., Chimienti, F., Rutter, G.A., Gilon, P., in't Veld, P.A., Schuit, F.C., 2009. Insulin crystallization depends on zinc transporter ZnT8 expression, but is not required for normal glucose homeostasis in mice. Proc. Natl. Acad. Sci. USA 106, 14872–14877.

Maret, W., 2011. Metals on the move: zinc ions in cellular regulation and in the coordination dynamics of zinc proteins. Biometals 24, 411–418.

Maret, W., 2013. Zinc biochemistry: from a single zinc enzyme to a key element of life. Adv. Nutr. 4, 82–91.

Miyai, T., Hojyo, S., Ikawa, T., Kawamura, M., Irie, T., Ogura, H., Hijikata, A., Bin, B.H., Yasuda, T., Kitamura, H., et al., 2014. Zinc transporter SLC39A10/ZIP10 facilitates antiapoptotic signaling during early B-cell development. Proc. Natl. Acad. Sci. USA 111, 11780–11785.

Martel, G., Hevi, C., Kane-Goldsmith, N., Shumyatsky, G.P., 2011. Zinc transporter ZnT3 is involved in memory dependent on the hippocampus and perirhinal cortex. Behav. Brain Res. 223, 233–238.

Mathews, W.R., Ong, D., Milutinovich, A.B., Van Doren, M., 2006. Zinc transport activity of fear of intimacy is essential for proper gonad morphogenesis and DE-cadherin expression. Development 133, 1143–1153.

Nam, H., Wang, C.Y., Zhang, L., Zhang, W., Hojyo, S., Fukada, T., Knutson, M.D., 2013. ZIP14 and DMT1 in the liver, pancreas, and heart are differentially regulated by iron deficiency and overload: implications for tissue iron uptake in iron-related disorders. Haematologica 98, 1049–1057.

Nebert, D.W., Galvez-Peralta, M., Hay, E.B., Li, H., Johansson, E., Yin, C., Wang, B., He, L., Soleimani, M., 2012. ZIP14 and ZIP8 zinc/bicarbonate symporters in Xenopus oocytes: characterization of metal uptake and inhibition. Metallomics 4, 1218–1225.

Nishida, K., Hasegawa, A., Nakae, S., Oboki, K., Saito, H., Yamasaki, S., Hirano, T., 2009. Zinc transporter Znt5/Slc30a5 is required for the mast cell-mediated delayed-type allergic reaction but not the immediate-type reaction. J. Exp. Med. 206, 1351–1364.

Nicolson, T.J., Bellomo, E.A., Wijesekara, N., Loder, M.K., Baldwin, J.M., Gyulkhandanyan, A.V., Koshkin, V., Tarasov, A.I., Carzaniga, R., Kronenberger, K., Taneja, T.K., da Silva Xavier, G., Libert, S., Froguel, P., Scharfmann, R., Stetsyuk, V., Ravassard, P., Parker, H., Gribble, F.M., Reimann, F., Sladek, R., Hughes, S.J., Johnson, P.R., Masseboeuf, M., Burcelin, R., Baldwin, S.A., Liu, M., Lara-Lemus, R., Arvan, P., Schuit, F.C., Wheeler, M.B., Chimienti, F., Rutter, G.A., 2009. Insulin storage and glucose homeostasis in mice null for the granule zinc transporter ZnT8 and studies of the type 2 diabetes-associated variants. Diabetes 58, 2070–2083.

Ohana, E., Hoch, E., Keasar, C., Kambe, T., Yifrach, O., Hershfinkel, M., Sekler, I., 2009. Identification of the Zn^{2+} binding site and mode of operation of a mammalian Zn^{2+} transporter. J. Biol. Chem. 284, 17677–17686.

Prasad, A.S., 1991. Discovery of human zinc deficiency and studies in an experimental human model. Am. J. Clin. Nutr. 53, 403–412.

Prasad, A.S., 1995. Zinc: an overview. Nutrition 11, 93–99.

Prasad, A.S., 2008. Zinc in human health: effect of zinc on immune cells. Mol. Med. 14, 353–357.

Prasad, A.S., Halsted, J.A., Nadimi, M., 1961. Syndrome of iron deficiency anemia, hepatosplenomegaly, hypogonadism, dwarfism and geophagia. Am. J. Med. 31, 532–546.

Procter, A.M., Phillips 3rd, J.A., Cooper, D.N., 1998. The molecular genetics of growth hormone deficiency. Hum. Genet. 103, 255–272.

Peters, J.L., Dufner-Beattie, J., Xu, W., Geiser, J., Lahner, B., Salt, D.E., Andrews, G.K., 2007. Targeting of the mouse Slc39a2 (Zip2) gene reveals highly cell-specific patterns of expression, and unique functions in zinc, iron, and calcium homeostasis. Genesis 45, 339–352.

Quadri, M., Federico, A., Zhao, T., Breedveld, G.J., Battisti, C., Delnooz, C., Severijnen, L.A., Di Toro Mammarella, L., Mignarri, A., Monti, L., Sanna, A., Lu, P., Punzo, F., Cossu, G., Willemsen, R., Rasi, F., Oostra, B.A., van de Warrenburg, B.P., Bonifati, V., 2012. Mutations in SLC30A10 cause parkinsonism and dystonia with hypermanganesemia, polycythemia, and chronic liver disease. Am. J. Hum. Genet. 90, 467–477.

Raulin, J., 1869. Etudes shimques sur la vegetation. Ann. des Sci. naturelles Botanique Biol. végétale 11, 93–345.

Sensi, S.L., Paoletti, P., Bush, A.I., Sekler, I., 2009. Zinc in the physiology and pathology of the CNS. Nat. Rev. Neurosci. 10, 780–791.

Shankar, A.H., Prasad, A.S., 1998. Zinc and immune function: the biological basis of altered resistance to infection. Am. J. Clin. Nutr. 68, 447S–463S.

Shrivastava, P., Katagiri, T., Ogimoto, M., Mizuno, K., Yakura, H., 2004. Dynamic regulation of Src-family kinases by CD45 in B cells. Blood 103, 1425–1432.

Steinmann, B., Royce, P.M. (Eds.), 2002. Connective Tissue and Its Heritable Disorders, second ed. Wiley-Liss, Inc., New York.

Steinmann, B., Royce, P.M., Superti-Furga, A., 2002. The Ehlers-Danlos syndrome. In: Steinmann, B., Royce, P.M. (Eds.), Connective Tissue and Its Heritable Disorders. Wiley-Liss, Inc., New York, pp. 431–524.

Stathakis, D.G., Burton, D.Y., McIvor, W.E., Krishnakumar, S., Wright, T.R., O'Donnell, J.M., 1999. The catecholamines up (Catsup) protein of *Drosophila melanogaster* functions as a negative regulator of tyrosine hydroxylase activity. Genetics 153, 361–382.

Tamaki, M., Fujitani, Y., Hara, A., Uchida, T., Tamura, Y., Takeno, K., Kawaguchi, M., Watanabe, T., Ogihara, T., Fukunaka, A., et al., 2013. The diabetes-susceptible gene SLC30A8/ZnT8 regulates hepatic insulin clearance. J. Clin. Invest. 123, 4513–4524.

Taylor, K.M., Hiscox, S., Nicholson, R.I., Hogstrand, C., Kille, P., 2012. Protein kinase CK2 triggers cytosolic zinc signaling pathways by phosphorylation of zinc channel ZIP7. Sci. Signal. 5, ra11.

Taylor, K.M., Morgan, H.E., Johnson, A., Nicholson, R.I., 2005. Structure-function analysis of a novel member of the LIV-1 subfamily of zinc transporters, ZIP14. FEBS Lett. 579, 427–432.

Taylor, K.M., Nicholson, R.I., 2003. The LZT proteins; the LIV-1 subfamily of zinc transporters. Biochim. Biophys. Acta 1611, 16–30.

Tuschl, K., Clayton, P.T., Gospe Jr., S.M., Gulab, S., Ibrahim, S., Singhi, P., Aulakh, R., Ribeiro, R.T., Barsottini, O.G., Zaki, M.S., Del Rosario, M.L., Dyack, S., Price, V., Rideout, A., Gordon, K., Wevers, R.A., Chong, W.K., Mills, P.B., 2012. Syndrome of hepatic cirrhosis, dystonia, polycythemia, and hypermanganesemia caused by mutations in SLC30A10, a manganese transporter in man. Am. J. Hum. Genet. 90, 457–466.

Vallee, B.L., Falchuk, K.H., 1993. The biochemical basis of zinc physiology. Physiol. Rev. 73, 79–118.

Van Doren, M., Mathews, W.R., Samuels, M., Moore, L.A., Broihier, H.T., Lehmann, R., 2003. Fear of intimacy encodes a novel transmembrane protein required for gonad morphogenesis in *Drosophila*. Development 130, 2355–2364.

Wang, B., He, L., Dong, H., Dalton, T.P., Nebert, D.W., 2011. Generation of a Slc39a8 hypomorph mouse: markedly decreased ZIP8 Zn(2)(+)/(HCO(3)(−))(2) transporter expression. Biochem. Biophys. Res. Commun. 410, 289–294.

Wang, F., Kim, B.E., Petris, M.J., Eide, D.J., 2004. The mammalian Zip5 protein is a zinc transporter that localizes to the basolateral surface of polarized cells. J. Biol. Chem. 279, 51433–51441.

Wang, K., Zhou, B., Kuo, Y.M., Zemansky, J., Gitschier, J., 2002. A novel member of a zinc transporter family is defective in acrodermatitis enteropathica. Am. J. Hum. Genet. 71, 66–73.

Wang, X., Zhou, B., 2010. Dietary zinc absorption: a play of Zips and ZnTs in the gut. IUBMB Life 62, 176–182.

Wenzlau, J.M., Juhl, K., Yu, L., Moua, O., Sarkar, S.A., Gottlieb, P., Rewers, M., Eisenbarth, G.S., Jensen, J., Davidson, H.W., Hutton, J.C., 2007. The cation efflux transporter ZnT8 (Slc30A8) is a major autoantigen in human type 1 diabetes. Proc. Natl. Acad. Sci. USA 104, 17040–17045.

Yamaguchi, M., Gao, Y.H., 1998. Potent effect of zinc acexamate on bone components in the femoral-metaphyseal tissues of elderly female rats. Gen. Pharmacol. 30, 423–427.

Yamashita, S., Miyagi, C., Fukada, T., Kagara, N., Che, Y.S., Hirano, T., 2004. Zinc transporter LIVI controls epithelial-mesenchymal transition in zebrafish gastrula organizer. Nature 429, 298–302.

Zotos, D., Tarlinton, D.M., 2012. Determining germinal centre B cell fate. Trends Immunol. 33, 281–288.

Part VI

Magnesium

Chapter 25

Magnesium: Basic Nutritional Aspects

Forrest Harold Nielsen

USDA, ARS, Grand Forks Human Nutrition Research Center, Grand Forks, ND, United States

INTRODUCTION

Before 1932, magnesium was considered an essential nutrient based on its abundant presence in the body of animals and humans and the finding that diets low in magnesium inhibited growth in rats and mice. In 1932, it was reported that severely magnesium-deficient rats had early violent deaths that were preceded by vasodilatation, hyperirritability of the nervous system, cardiac arrhythmia, spasticity, and tonic-clonic convulsions (Kruse et al., 1932). Although some early limited studies found that people with very low plasma magnesium exhibited twitching and convulsions, the incontrovertible signs of human magnesium deficiency were not described until the 1960s. In one study (Shils, 1969), subjects requiring feeding via a nasoesophageal tube were given a highly purified diet providing 6–10 mg magnesium/day until deficiency signs appeared (42–266 days). The subjects exhibited low plasma concentrations of calcium and potassium and neurological signs including Trousseau signs (carpopedal spasm) and electromyographic and electrocardiographic changes. The finding that magnesium deficiency could be induced in humans supported the conclusion that it was present in conditions that would inhibit its absorption (e.g., diarrhea, malabsorption syndromes), increase its excretion (e.g., alcoholism, diuretics), or cause endocrine disorders (e.g., diabetes mellitus, hypoparathyroidism). However, the occurrence of dietary magnesium deficiency severe enough to induce pathological signs has been considered rare (World Health Organization and Food and Agriculture Organization of the United Nations, 2004). The 2010 Dietary Guidelines for Americans (United States Department of Agriculture and Department of Human and Health Services, 2010) indicated that magnesium was not a major nutrient of concern for health and well-being beyond those exposed to medications and disorders that inhibit its absorption or induce its excretion. These opinions conflict with controlled metabolic unit experiments, described later in this chapter, and epidemiological and supplementation studies associating magnesium with chronic diseases. The associations, which are described in other chapters, support the opinion that dietary magnesium deficiency commonly occurs and it is a predisposing or risk factor for pathological conditions to which chronic inflammatory or oxidative stress contribute. The nutritional attributes of magnesium that have resulted in these conflicting views are reviewed. Because of a given limitation in references, reviews are often cited in this chapter. These reviews provide the original sources of the findings for the statements made.

TISSUE DISTRIBUTION

The human body contains about 760 mg of magnesium at birth, 5 g at age 4–5 months, and 25 g when adult (World Health Organization and Food and Agriculture Organization of the United Nations, 2002). Magnesium is the fourth-most abundant cation in the body and second-most abundant intracellular cation. There are at least three body pools of magnesium in humans (Volpe, 2012). One pool consisting mostly of extracellular magnesium has a turnover rate of less than 28 h. Blood contains 0.8–1.0% of magnesium in the body with about 0.3% in serum where about 65% exists in the free form. The second pool, consisting primarily of intracellular magnesium, has a turnover rate of about 11 days. About 40% of magnesium is found in soft tissues where it performs most of its essential functions. Skeletal magnesium is the site of the third pool, which has a slow turnover. About 60% of body magnesium is found in bone where two-third is within the hydration shell (bone mineral or poorly crystallized hydroxyapatite) and one-third is on the crystal surface (Rude and Shils, 2006). The surface magnesium is readily exchangeable with serum and serves as a reservoir that can be used to maintain magnesium for essential functions during deficient intakes. The magnesium in the hydration shell is not readily available and is responsible for the conclusion that the biologic half-life for the majority of magnesium in the body is about 1000 h (Elin, 2010).

Molecular, Genetic, and Nutritional Aspects of Major and Trace Minerals. http://dx.doi.org/10.1016/B978-0-12-802168-2.00025-7

BIOLOGICAL ROLES

It is beyond the scope of this review to describe in detail each of the extensive biochemical roles of magnesium, which include supporting enzyme–substrate interactions, complexing with highly charged anions such as polyphosphates and nucleic acids, and stabilizing the conformation of polymers (Volpe, 2012).

Magnesium is a cofactor for over 600 enzymatic reactions vital to metabolic pathways including DNA, RNA, protein, and adenosine 5'-triphosphate (ATP) synthesis; cellular energy production and storage; glycolysis; and cellular second messenger systems. The predominant enzymatic role of magnesium in these metabolic pathways involves ATP utilization with the enzyme interacting with MgATP to complete a reaction. Magnesium also binds directly to enzymes to produce conformational changes (allosteric activation) that results in a catalysis center for a reaction to occur.

The stabilization of ribonucleotides and deoxyribonucleotides results in physicochemical changes that allow for DNA duplication, transcription and maintenance, and transfer RNA function (Rude and Shils, 2006). The reaction of magnesium with phosphates and carboxylates stabilizes membranes, which affects their fluidity and permeability (Rude and Shils, 2006). As a result, magnesium influences cellular ion channels, transporters, and signaling.

Magnesium has been called nature's physiologic Ca^{2+} channel blocker. In magnesium deficiency, cellular Ca^{2+} increases through an influx from extracellular sources via slow Ca^{2+} transport channels, and the release from intracellular stores such as the sarcoplasmic reticulum. Besides the effects on Ca^{2+}, cellular K^+ is decreased in magnesium deficiency because of increased efflux of K^+ from cells via Mg^{2+}-sensitive K^+ channels (Rude and Shils, 2006). These channels normally allow K^+ to pass more readily inward than outward with magnesium apparently regulating outward movement. Thus, an increased amount of K^+ leaves the cell in magnesium deficiency (Rude and Shils, 2006). Through these effects on calcium and potassium movement, magnesium is a controlling factor in nerve transmission, skeletal and smooth muscle contraction, cardiac excitability, vasomotor tone, blood pressure, and bone turnover.

BODY HOMEOSTASIS

Absorption

The primary sites of absorption of magnesium are the jejunum and ileum but some may be absorbed at other sites including the colon (Rude and Shils, 2006). When dietary intake of magnesium is near requirements, 45–55% is absorbed. With intakes less than 50% of requirement, the percent absorbed may increase to 65–70%. As dietary intakes of magnesium increase over requirements, and with high doses of magnesium salts, the percent absorbed decreases. With intakes of about 40 mmol (972 mg), fractional absorption of only 11% has been found (Fine et al., 1991).

Magnesium is absorbed by both passive paracellular diffusion and an active transport mechanism. At high dietary intakes, passive diffusion accounts for about 90% of the intestinal magnesium absorption (Rude and Shils, 2006). The fractional amount of magnesium absorbed through the active transport system increases when dietary magnesium is deficient. The paracellular pathway between enterocytes is dependent upon a transepithelial electrical potential difference that is positive between the lumen and blood. In addition, there usually is a downward concentration gradient between the lumen and blood. Only free Mg^{2+} moves through the paracellular pathway. In the active transport system, a member of the melastatin subfamily of transient receptor potential, TRPM6, found in the colon has been determined to have a major role in the transport of magnesium across the apical side of intestinal epithelial cells (van der Wijst et al., 2009, 2014). Diet modulates the expression and activity of TRPM6. TRPM6 is a divalent cation channel that has a higher affinity for magnesium than calcium and its activity is tightly regulated by intracellular Mg^{2+} concentrations. TRPM6 has an α-kinase domain that apparently indirectly regulates channel activity by acting as an intracellular Mg^{2+} sensor (van der Wijst, 2009).

Upon reaching the basolateral side of the intestinal cell, magnesium is delivered to the blood via an Na^+-dependent extrusion mechanism called Na^+/Mg^{2+} exchanger. The extrusion mechanism apparently becomes active upon phosphorylation by cAMP, operates as an antiporter, and strictly requires extracellular Na^+ to be at a physiological concentration (Romani, 2013). If Na^+ is not optimal, Mg^{2+} apparently is extruded through another sodium-independent mechanism that involves both anions and cations for magnesium transport.

Because TRPM6 has an affinity for calcium that can replace magnesium, the active transport of magnesium across the gut may be affected by an increased calcium intake. This indicates that changes in dietary calcium would have the most discernable effect on magnesium absorption with deficient intakes when the active transport system has a greater role in magnesium absorption. An adequate intake of magnesium resulting in its predominantly passive absorption not affected by calcium explains why long-term balance studies with healthy people generally found that increasing the intake of calcium did not significantly affect magnesium absorption (Rude and Shils, 2006). A low magnesium intake resulting in TRPM6 having a role in absorption that could be inhibited by calcium is an explanation for the epidemiological and correlation findings showing that an increased ratio of calcium/magnesium in the diet is associated with an increased risk of pathological consequences to which magnesium deficiency contributes (Rosanoff et al., 2011).

Estrogen enhances transcriptional activity or mRNA stabilization of TRPM6, which stimulates TRPM6 activity. Thus, estrogen therapy may positively affect the amount of magnesium absorbed when its intake is low; whereas a low estrogen status may have a negative effect. Estrogen status affecting magnesium absorption (and excretion as described later) may explain the congruent findings concerning the effect of aging on absorption. The study that has been used (Martin, 1990) to suggest that aging results in a decline in the ability to absorb magnesium involved 43 women with an average age of 83 years (thus postmenopausal). These women had serum magnesium concentrations indicating magnesium deficiency. A deficient dietary intake most likely caused the deficiency. The deficiency may have been exacerbated by a decreased estrogen status after menopause hindering the use of the active transport of magnesium. Decreased TRPM6 activity caused by a decreased estrogen status is an explanation for this one study generating the opinion that the ability to absorb magnesium may be depressed in elderly people. The notion that age does not markedly affect magnesium absorption is supported by data from 27 tightly controlled metabolic ward studies that found age did not affect the point at which neutral balance between intake and excretion was achieved in men and women (Hunt and Johnson, 2006). Most of the time, the 243 subjects in these studies were consuming amounts of magnesium such that the passive paracellular route would be the primary mechanism for magnesium absorption instead of the active TRPM6 mechanism affected by estrogen.

Other dietary factors that may affect magnesium absorption are fiber and zinc. Diets high in fiber provided by fruits, vegetables, and grains have been reported to decrease magnesium absorption (Food and Nutrition Board, Institute of Medicine, 1997). The phytate in high fiber foods has been implicated as a factor that decreases magnesium absorption (Volpe, 2012). Binding to the phosphate groups in phytate has been suggested to decrease the bioavailability of magnesium. Two reports indicate that high dietary intakes of zinc decrease magnesium absorption and balance. In one study, zinc was increased from 12 to 142 mg/day (Spencer et al., 1994) and in the other, increased from 3 to 53 mg/day (Nielsen and Milne, 2004).

Low digestible carbohydrates, particularly fermentable polyols (sugar alcohols), enhance magnesium absorption. Different oligosaccharides including fructo-oligosaccharide, galacto-oligosaccharide, and lactitol-oligosaccharide have been found to significantly increase magnesium absorption (Coudray et al., 2003). Lactose and lactulose, which are poorly digestible and fermentable, also enhance magnesium absorption (Coudray et al., 2003; Seki et al., 2007). Organic acids produced by the catabolism of fermentable low digestible carbohydrates in the hindgut apparently are a major factor in enhancing magnesium absorption. A proposed mechanism for the enhancement of absorption is that acidification in the hindgut by the organic acids improves magnesium solubility (Coudray et al., 2003). Another proposal is that short-chain fatty acids produced by the fermentation activate an apical Mg^{2+}/H^+ antiport transport system by providing H^+ within the epithelial cell (Coudray et al., 2003).

Cellular Homeostasis and Transport

The magnesium concentration in cells is relatively constant but can vary some depending upon magnesium status; a range of 5–20 mM has been reported (Rude and Shils, 2006). Intracellular magnesium is tightly controlled by specific regulatory transport systems regulating its movement across the cell membrane and mechanisms involved in organelle compartmentalization and intracellular buffering. Cellular magnesium is mostly bound to proteins and negatively charged molecules and compartmentalized primarily in nuclei, mitochondria, and endo/sarcoplasmic reticulum (Touyz, 2008). In the cytosol, 90–95% of magnesium is bound to ligands such as ATP, ADP, citrate, proteins, and nucleic acids. Only 1–5% of intracellular magnesium exists as free Mg^{2+}. The major molecule in buffering magnesium is ATP.

The first magnesium influx transporter clearly identified as being involved in cellular magnesium homeostasis was the TRPM7 channel. Since then, several other magnesium transporters involved in cellular magnesium influx have been identified. These include mitochondrial RNA splicing 2 protein (Mrs2p), human solute carrier family 41 members 1 and 2 (SLC41A1, SLC41A2), ancient conserved domain protein 2 (ACDP2), and magnesium transporter 1 (MagT1) (Touyz, 2008). Of these, the one receiving the most attention is MagT1, which is the only known selective plasma membrane Mg^{2+} channel (Wolf and Trapani, 2011). Cellular magnesium homeostasis apparently is maintained through a complex balance of several transport systems that can compensate for each other when needed. For example, in TRPM7$^{-/-}$cells, MagT1 activity substantially increases (Wolf and Trapani, 2011).

Cellular magnesium efflux involves Na^+-dependent and Na^+-independent systems (Toyuz, 2008). Dependent efflux transport occurs via the Na^+/Mg^{2+}exchanger. The independent systems include Ca^{2+}/Mg^{2+}exchanger, Mn^{2+}/Mg^{2+} antiporter, and Cl^-/Mg^{2+} cotransporter (Touyz, 2008).

Excretion

The primary organ regulating magnesium homeostasis is the kidney where about 10% of total body magnesium is filtered daily through the glomeruli. Only about 5% of this filtered magnesium is excreted in the urine (Rude and Shils, 2006). About 15–20% of the filtered magnesium is reabsorbed in the proximal convoluted tubule, and 65–75% in the cortical thick ascending limb of Henle (Rude and Shils, 2006). About 90–95% of the magnesium reabsorbed at these sites occurs through a passive paracellular mechanism. Claudin 16 (originally called paracellin-1) and claudin 19 form the tight junction component in this mechanism that is controlled by the calcium sensing receptor and is influenced by the proper expression of the epidermal growth factor receptor (Romani, 2013).

The remaining 5–10% of filtered magnesium is reabsorbed in the distal convoluted tubule via an active transcellular mechanism. The epithelial Mg^{2+} channel TRPM6 found in the intestine also is the active transport channel in the kidney and is considered the major controller of magnesium balance in the body (van der Wijst et al., 2014). Several hormones, including estrogen and insulin, and epidermal growth factor (EGF) influence TRPM6 expression and channel function (van der Wijst et al., 2014). Thus, lack of insulin or insulin resistance, inhibition of EGF, and decreased estrogen, which would result in inhibited TRPM6 expression and function, can result in increased renal loss of magnesium leading to magnesium deficiency.

Sweat

Surface losses of magnesium can be a significant contributing factor for magnesium deficiency when dietary intake of magnesium is near minimal requirement levels and under conditions that increase sweating (e.g., heat, exercise, stress). The reported amount of magnesium loss in sweat varies markedly. This variation apparently is caused by the method used and the environmental conditions present for the collection of sweat. Reports up through 1997 stated that the concentration of magnesium in sweat was between 12 and 60 mg/L in a hot humid environment and 3.4 mg/L in a hot dry environment (Nielsen and Lukaski, 2006). A concentration of 7.3 mg/L was found in sweat of exercising males and this concentration decreased to 4.1 mg/L after 10 days of exercise-heat acclimation (Chinevere et al., 2008). Another study found a magnesium concentration of 1.4 mg/L in heat-acclimated subjects that declined only slightly to 1.16 mg/L after 5 × 60 min of treadmill exercise (1.56 m/s with 2% grade) with 20 min of rest between each exercise period (Montain et al., 2007). These studies suggest that surface losses of magnesium would be less than 5 mg daily for an average person not in an environment causing a high amount of sweating.

Tissues

Bone serves as an important reservoir for magnesium when dietary magnesium is inadequate and urinary magnesium is not decreased adequately to maintain extracellular magnesium. When dietary magnesium is inadequate, some extracellular magnesium may be obtained from muscle (Rude and Shils, 2006; Lukaski and Nielsen, 2002).

REQUIREMENTS

Since 1969, when magnesium deficiency signs were experimentally confirmed in humans (Shils, 1969), the setting of the requirement for magnesium has vacillated markedly. One early report suggested that 100 mg/day would meet the magnesium requirement of women (Marshall et al., 1976). In 1993, magnesium population reference intakes for the European Community were not set because it was decided that data were inadequate to do so. Instead, they gave an acceptable range of intakes for adults of 150–500 mg/day (Scientific Committee for Foods, Nutrient and Energy Intakes for the European Community, 1993). A review in 1995 found that the adult recommended dietary intakes for magnesium ranged from 220 to 400 mg/d throughout the world (Dreosti, 1995). In the United States and Canada, magnesium dietary reference intakes (DRIs) were set in 1997 (see Table 25.1). These were for adults aged between 19 and 30 years, and 31–70 years, respectively, recommended dietary allowances (RDAs) of 310 and 320 mg/d for women and 410 and 420 mg/d for men, and estimated average requirements (EARs) of 255 and 265 mg/day for women and 330 and 350 mg/day for men (Food and Nutrition Board, Institute of Medicine, 1997). The RDA is a daily intake determined to meet the requirement of 98% of the healthy population and the EAR is a daily intake determined to meet the requirement of 50% of the healthy population. In 2000, the French Society for Nutrition set an average requirement of 330 mg/day for women (Martin, 2000). In 2002, an expert committee (World Health Organization and Food and Agriculture Organization, 2004) concluded that evidence was lacking for nutritional magnesium deficiency occurring with the consumption of diets supplying a range of intakes sometimes considerably less than the US RDA. Thus, this committee subjectively set the recommended nutrient intakes for magnesium at 220 and 260 mg/day for adult women and men, respectively, aged 19–65 years.

The lack of a sensitive and reliable status indicator and, until recently, balance data from well-controlled studies has made it difficult to establish the dietary requirements for magnesium.

TABLE 25.1 The Magnesium Dietary Reference Intakes (DRIs, mg/day) Labeled as Estimated Average Requirement (EAR), Recommended Dietary Allowance (RDA), and Tolerable Upper Intake Level (UL) in the United States and Canada (Food and Nutrition Board, Institute of Medicine, 1997)

DRI	Children, Ages[a]			Males, Ages				Females, Ages				Pregnancy, Ages			Lactation, Ages		
	1–3	4–8	9–13	14–18	19–30	>30		14–18	19–30	>30		≤18	19–30	31–50	≤18	19–30	31–50
EAR	65	110	200	340	330	350		300	255	265		335	290	300	300	255	265
RDA	80	130	240	410	400	420		360	310	320		400	350	360	360	310	320
UL[b]	65	110	350	350	350	350		350	350	350		350	350	350	350	350	350

[a]EARs and UL were not set for infants aged 0–1 years. Adequate Intakes were set at 30 mg/day based on human milk content for infants aged 0–6 months, and 75 mg/day based on human milk and infant food content for infants aged 6–12 months.
[b]The UL is an amount that is supplemental to the intake by diet.

For example, the DRIs set for the United States and Canada in 1997 were based primarily on highly variable balance data from 34 men and women on self-selected diets that decreased in magnesium content during the balance periods (Lakshmanan et al., 1984). Based on the discussion in the homeostasis section, the decrease could have affected magnesium absorption and excretion. Some subjects with magnesium intakes less than 258 mg/day were in positive balance and some subjects with intakes greater than 299 mg/day were in negative balance. In addition, body weights of the subjects were not given, and as described later in this chapter, magnesium DRIs may vary with body weight.

Since 1997, improved balance data have been reported for the determination of magnesium requirements. These data include those from 27 tightly controlled metabolic ward studies that found neutral magnesium balance, without considering surface or phlebotomy losses, occurred at an intake of 165 mg/day with a 95% prediction interval of 113 and 237 mg/day (Hunt and Johnson, 2006). Using the upper 95% value of 237 mg/day and considering that 98% is the upper interval level used for setting RDAs, would result in an RDA of near 245 mg/day. Considering surface and phlebotomy losses in the balance studies would increase the RDA to 250 mg/day for a healthy 70 kg person.

The data from the metabolic ward studies that came from subjects with a mean weight of 76.3 ± 12.5 kg (range between 46 and 136 kg) also found a neutral magnesium balance per kg of body weight of 2.36 mg/day with a 95% prediction interval of 1.58 and 3.38 mg/day (Hunt and Johnson, 2006). These data indicate that individuals weighing more than 70 kg would have a higher magnesium requirement. For example, the RDAs for individuals weighing 70, 80, 90, or 100 kg would be about 250, 285, 320, and 355 mg/day; these values factor in a 98% upper interval level and surface and phlebotomy losses during the balance determinations. Likewise, using the lower prediction value would indicate EARs for 70, 80, 90, or 100 kg individuals of 170, 195, 220, and 245 mg/day. Basing magnesium requirements on body weight also would result in EARs less than 170 mg/day and RDAs less than 250 mg/day for individuals weighing less than 70 kg. For example, EARs for 40, 50, and 60 kg individuals would be about 100, 125, and 150 mg/day, and RDAs would be about 140, 175, and 210 mg/day. It should be noted that the balance data used to obtain these suggested EARs and RDAs were obtained from healthy volunteers under optimal dietary and environmental conditions in a metabolic ward. Thus, magnesium requirements may be increased by factors such as increased oxidative and inflammatory stress (Nielsen, 2010), foods that inhibit magnesium absorption, excessive alcohol intake resulting in hypermagnesuria, and medications that decrease magnesium absorption (e.g., proton pump inhibitors) or increase its excretion (e.g., diuretics, aminoglycosides, amphotericin B, cisplatin, and cyclosporine) (Rude and Shils, 2006).

Two other reports indicate that the RDAs and EARs based on the improved balance data are in the right range for healthy individuals. Balance and absorption data from German women and men indicated that normative magnesium requirements were satisfied by an intake of less than 200 mg/day for women and 250 mg/per day for men, or less than 3.0 mg/kg body weight/day (Anke et al., 2006). A study with female adolescents aged 11–15 years weighing about 55 kg gave findings indicating that a magnesium intake of 2.94 mg/kg body weight/day or 150 mg/day was inadequate, and balance data indicated 4.26 mg/kg body weight/day or 232 mg/day was adequate (Palacios et al., 2013).

Basing the dietary requirement for magnesium on body weight may be the reason the DRIs are usually higher for men than women; they are more likely to weigh more. Basing the requirement on body weight also implies increased DRIs for overweight and obese individuals and gives a basis for reports indicating that a low magnesium status occurs more often in obese than nonobese individuals (Nielsen, 2010; Shay et al., 2012; Guerrero-Romero and Rodriguez-Moran, 2013). Magnesium deficiency may be a contributing factor to the chronic low-grade inflammation often exhibited by obese individuals (Nielsen, 2010).

DIETARY DEFICIENCY OCCURRENCE

Based on the official DRIs established for the United States and Canada in 1997, magnesium deficiency in the range of 50–99% of the requirement commonly occurs throughout the world where unrefined foods of plant origin are not major components of the diet. Magnesium deficiency induced in animal models fed dietary magnesium in this range is considered moderate to marginal, or subclinical. This type of deficiency has also been designated as chronic latent magnesium deficiency (Elin, 2010). A United States survey performed in 2005–2006 found that 48% of all people over the age of 1 year consumed less than the EARs set for them in 1997 (Moshfegh et al., 2009). Magnesium intakes less than the EAR for females was 48% in ages 31–50, 55% in ages 51–70, and 70% in ages 71 and over; for males it was 45% in ages 31–50, 58% in ages 51–70, and 80% in ages 71 and over.

Even if the EAR is revised to 170 mg/day and the RDA to 250 mg/day for a healthy 70 kg person, as indicated by improved balance data, the 2005–2006 survey suggests that almost 50% of adult females and 25% of adult males would have intakes less than the revised RDA. Many individuals weigh more than 70 kg and thus it would not be surprising to find that about 25% of adults have usual intakes less than the EARs suggested by improved balance data. The likelihood of a large number of individuals consuming less than the revised RDA for magnesium resulting in chronic latent magnesium deficiency is supported by intakes of less than 250 mg/day being associated with chronic disease, which is described in other reviews (Nielsen and Lukaski, 2006; Del Gobbo et al., 2013; Nielsen, 2014) and chapters in this book. Two recent human studies also support a significant occurrence of chronic latent magnesium deficiency. Three 5-day food diaries indicated that 38% of 224 postmenopausal women had regular magnesium intakes less than 237 mg/day (Nielsen et al., 2011). These women had bone status measurements associating the low intake with decreased bone health. In the other study of 100 adults, 44 of 78 women older than 51 years had intakes less than 265 mg/day and 14 of 22 men had intakes less than 350 mg/day (the 1997 EARs) (Nielsen et al., 2010). Serum magnesium concentrations of less than 1.75 mg/dl (0.72 mmol/L) indicated that at least 37 of the 100 adults were magnesium deficient (Nielsen et al., 2010).

STATUS ASSESSMENT

Efforts to find an indicator of chronic latent magnesium deficiency have not yielded a cost-effective one that is dependable and well validated. Among the status indicators that have been evaluated are serum or plasma total or ionized magnesium, urinary magnesium excretion, sublingual cellular magnesium, erythrocyte magnesium, and the magnesium-load test.

At present, the magnesium-load test is thought to be the best method for determining total body magnesium status. This test determines the percentage of magnesium retained over a given period of time after the parenteral administration of a magnesium load (Elin, 2010; Rude and Shils, 2006). Retention of a greater percentage than that by individuals with adequate magnesium status (22–25%) indicates some body magnesium depletion (Rude and Shils, 2006). However, this test is invasive, time-consuming, and expensive; it also requires close supervision for about 24 h after magnesium infusion for the oversight of magnesium intake and urine collection.

Plasma or serum magnesium is an imperfect indicator of subclinical magnesium deficiency because exchangeable skeletal magnesium and urinary responses to changes in magnesium intake maintain extracellular magnesium at a rather constant level even while tissue magnesium is decreasing. Thus, normal serum magnesium concentrations are found in individuals with low magnesium in erythrocytes and tissues (Rude and Shils, 2006). Nonetheless, low serum total magnesium is the most common method for diagnosing severe magnesium deficiency. One review (Elin, 2010) stated that clinical magnesium deficiency was determined to be present in 90% of individuals with serum magnesium concentrations of 0.70 mmol/L (1.70 mg/dL) or less, 50% of individuals with concentrations 0f 0.75 mmol/L (1.82 mg/dL) or less, and 10% of individuals with concentrations of 0.80 mmol/L (1.94 mg/dL) or less. These findings indicate that in groups of people with serum magnesium concentrations of 0.75 mmol/L or less, a significant number would have chronic latent magnesium deficiency.

The ionized fraction (about 60%) of magnesium in serum has been suggested to be a better measure than total magnesium for status assessment because this is the physiologically active form that serves as a metabolic cofactor for many enzymatic reactions. However, this test has the same problem as total plasma or serum magnesium because the body also attempts to maintain plasma or serum ionized magnesium in a rather constant concentration.

A 24-h magnesium excretion in urine that is more than 10–15% of the amount ingested suggests adequate magnesium status (Rude and Shils, 2006). When deficient amounts of magnesium are ingested or absorbed, there is a rapid reduction within a few days in the urinary excretion of magnesium. Thus, a low urinary excretion of magnesium can occur while serum magnesium is normal and before total body reduction results in changes that become biochemically and clinically significant. Thus, urinary magnesium excretion is best used in connection with other measures indicating a subclinical magnesium deficiency such as dietary assessments indicating a routine intake less than the RDA or EAR or a serum magnesium less than 0.75 mmol/L.

Erythrocyte magnesium has been used to indicate a decrease in magnesium status in human magnesium depletion experiments (Lukaski and Nielsen, 2002). However, standard reference intervals have not been established that allow the use of this test for status assessment outside the research environment.

The measurement of magnesium in sublingual epithelial cells may be a method that can assess chronic latent magnesium deficiency (Silver, 2004). This expensive test that determines cellular magnesium by using energy-dispersive X-ray analysis requires skill in obtaining samples for analysis. Thus, it has been used mainly in research studies. It also needs to be validated by use of another test such as the magnesium-load test.

DEFICIENCY SIGNS

As described in other chapters in this book and in reviews (Nielsen, 2010, 2014; Volpe, 2012; Romani, 2013), low dietary magnesium intakes or low serum magnesium concentrations are associated with numerous pathological conditions including heart arrhythmias, hypertension, insulin resistance, and the metabolic syndrome; they also have been associated with chronic diseases such as cardiovascular disease, diabetes, osteoporosis, and some cancers. In spite of these associations, as indicated in the Introduction, pathological conditions attributed specifically to dietary magnesium deficiency alone are considered rare.

Probable reasons for the dichotomy about the incidence of dietary magnesium deficiency inducing pathological conditions are that evidence of magnesium deficiency is not consistently found in pathological conditions with which it has been associated, and not all individuals considered magnesium deficient consistently exhibit some type of pathology. This lack of consistency may result from the failure to consider experimental animal findings and human epidemiological and supplementation studies indicating that dietary magnesium deficiency may be a risk or disposing factor for pathological conditions through exacerbating chronic inflammatory stress that contribute to their development (Nielsen, 2010, 2014; Weglicki, 2012), and that this magnesium-induced inflammatory stress may be alleviated or prevented by other factors that have antiinflammatory actions. These factors include long-chain n-3 fatty acids and phytochemicals in foods such as blueberries, cherries, grape seed, and green tea (Nielsen, 2010). Such amelioration would prevent the appearance of an obvious pathological sign of magnesium deficiency. Another factor in the opinion that detrimental changes are not found consistently in individuals is that EARs and RDAs set in 1997, which improved balance data indicate are too high for many individuals, were used to determine deficiency. This resulted in some individuals being considered magnesium deficient when they actually were not. In addition, the use of the problematic indicator of status, serum, or plasma concentration of magnesium, which responds slowly to changes in deficient and adequate intakes (Elin, 2010; Lukaski and Nielsen, 2002; Nielsen et al., 2007), to determine magnesium status may have resulted in the conclusion that individuals were deficient but were not and vice versa.

A meta-analysis and systematic review of seven cross-sectional and five intervention studies (Dibaba et al., 2014) indicates the need for caution for using the 1997 DRIs, and to use the imperfect status indicator serum magnesium, to determine when an individual should exhibit a magnesium deficiency sign. Both deficient magnesium intakes (Dibaba et al., 2014) and serum magnesium concentrations (Guerrero-Romero and Rodriguez-Moran, 2002; Rodriguez-Moran and Guerrero-Romero, 2008) have been associated with elevated serum C-reactive protein (CRP), which is a widely used indicator of inflammatory stress. The systematic review and meta-analysis of studies that involved 32,198 individuals found that dietary magnesium was significantly and inversely associated with serum CRP concentration (Dibaba et al., 2014). The point at which individuals consistently had serum CRP concentrations greater than 3.0 mg/L, a threshold that has been used to indicate chronic inflammatory stress, was when intakes were less than 250 mg/day. Three of the five intervention studies evaluated showed that magnesium supplementation decreased serum CRP concentrations. These three studies involved participants with deficient serum magnesium concentrations (less than 0.75 mmol/L) that were increased by magnesium supplementation. The two studies that did not show a significant effect of magnesium supplementation on serum CRP involved participants without elevated concentrations of CRP, and magnesium supplementation did not elevate serum magnesium concentrations that were above 0.75 mmol/L at baseline. These two studies involved overweight individuals, who one might have expected to exhibit elevated serum CRP concentrations because numerous experimental, epidemiological, and clinical studies have shown that serum CRP concentrations increase with increasing body mass index (Nielsen, 2010). If the mean serum magnesium in the two studies had been less than 0.75 mmol/L, an elevated serum CRP affected by magnesium supplementation might have occurred based on the finding that the odds of chronic low-grade inflammation indicated by elevated serum tumor necrosis factor-α and CRP concentrations increased markedly in overweight and obese individuals when mean serum magnesium concentrations were below 0.75 mmol/L (Rodriguez-Moran and Guerrero-Romero, 2004). The CRP findings strongly support the concept that one consequence of chronic latent magnesium deficiency is that it predisposes chronic inflammatory stress and thus increases the risk for disease associated with that stress. This may be of special concern for obese individuals that findings indicate have increased magnesium requirements.

The mechanism through which magnesium deficiency becomes a risk factor for chronic disease could be its role in regulating cellular calcium. As stated earlier, one biological role of magnesium is the binding to gates of calcium channels to block calcium influx into the cell. Increased cellular calcium induced by magnesium deficiency results in signaling that releases inflammatory neuropeptides, cytokines, prostaglandins, and leukotrienes that cause chronic inflammatory and oxidative stress. One cellular site that has received much attention in establishing magnesium deficiency as a predisposing factor for inflammatory stress is the N-methyl-D-aspartate (NMDA) receptor (Weglicki, 2012). Magnesium deficiency lowers the threshold levels of excitatory amino acids (e.g., glutamate) needed to activate the NMDA receptor. Activation of this receptor allows the influx of calcium into and the release of substance P from cells in neuronal C fibers (Weglicki, 2012). Substance P induces the release of inflammatory factors. In rats, blockade of the NMDA receptor prevented substance P release, decreased proinflammatory prostaglandin E_2 and histamine in plasma, and inhibited cardiac inflammation indicators in the heart that were induced by magnesium deficiency (Weglicki, 2012).

In addition to contributing to the risk for some chronic diseases, controlled metabolic ward studies indicate that subclinical or chronic latent magnesium deficiency also can affect physical performance and heart function. Heart rate and oxygen consumption increased significantly during submaximal exercise when untrained postmenopausal women were fed 150 mg compared to 320 mg magnesium/day (Lukaski and Nielsen, 2002). These metabolic ward findings are consistent with those indicating that chronic latent magnesium deficiency can adversely affect athletic performance (Nielsen and Lukaski, 2006). In other metabolic ward studies, postmenopausal women fed subclinical magnesium-deficient diets exhibited heart arrhythmias including atrial fibrillation and flutter (Nielsen et al., 2007) and supraventricular ectopy (Klevay and Milne, 2002), and alterations in electroencephalograms indicating increased central nervous system excitability (Penland, 1995).

DIETARY SOURCES

Two reports indicated that foods of plant origin provide about 50% of the magnesium intake for adults on mixed diets (Hunt and Meacham, 2001; Anke et al., 2006). Rich sources of magnesium include cereals/grains, nuts, pulses, and some vegetables (See Table 25.2). Some of the magnesium provided by food of plant origin is consumed as beverages such as coffee, beer, and wine. Meats apparently supply about 14–16% of adult magnesium intakes, and dairy products (including milk) supply about 34% (Hunt and Meacham, 2001).

TOXICITY

Because the kidneys are very effective in the excretion of excessive magnesium, its toxicity rarely occurs. Severe toxicity resulting in high blood magnesium almost exclusively occurs in individuals with kidney dysfunction or failure and given magnesium salts or drugs such as laxatives or antacids. Concurrent with high blood magnesium are low blood calcium and high blood potassium (Romani, 2013; Rude and Shils, 2006; Volpe, 2012). Symptoms of toxicity include lethargy, confusion, nausea, diarrhea, impaired breathing, low blood pressure, muscle weakness, and heart arrhythmias (Romani, 2013). The heart rhythm changes are caused by abnormal electrical conduction at the nervous, muscular, and cardiac level.

Magnesium toxicity caused by excessive intake through food has not been reported. Intake of excessive amounts of supplemental magnesium available in forms such as aspartate, chelate, citrate, chloride, gluconate, lactate, and oxide can cause adverse effects. The major effects of excessive magnesium intake through supplement intake are diarrhea, abdominal cramping, and nausea. A tolerable upper intake level (UL), defined as the highest level of daily nutrient intake that is likely to pose no risks of adverse health effects to almost all individuals in the general population, was set at 350 mg of supplementary magnesium for adolescents and adults older than 8 years (Food and Nutrition Board, Institute of Medicine, 1997).

TABLE 25.2 Magnesium in Select Food

Food Group	Food	Description	Mg/100g	Food Group	Food	Description	Mg/100g
Grains/Cereals	Bran Flakes	Ready-to-eat	229	Nuts	Almonds	Dry-roasted	279
	Bread	Whole wheat	75		Cashews	Dry-roasted	260
	Toasted Oats	Ready-to-eat	114		Peanuts	Dry-roasted	178
	Oatmeal	Cooked	27		Pecans	Dry-roasted	132
	Shredded Wheat	Ready-to-eat	133		Walnuts	English	158
Pulses	Beans, Navy	Cooked	53	Vegetables	Corn, sweet	Cooked	26
	Beans, Pinto	Cooked	50		Peas, green	Cooked	39
	Cowpeas	Cooked	53		Potato with skin	Baked	28
	Lentils	Cooked	36		Potato, sweet	Baked	27
	Soybeans	Green, cooked	60		Spinach	Raw	79
Dairy	Milk	2%	11	Others	Avocado	Raw	27
	Cheese	Cheddar	27		Banana	Raw	29
	Yogurt	Low-fat, plain	17		Liver, beef	Pan-fried	22

USDA, ARS, National Nutrient Database for Standard Reference Release 27; http://ndb.nal.usda.gov/ndb/foods.

REFERENCES

Anke, M., Glei, M., Vormann, J., Müller, R., Hoppe, C., Schäfer, U., 2006. Magnesium in the nutrition of man. In: Porr, P.J., Nechifor, M., Durlach, J. (Eds.), Advances in Magnesium Research: New Data. John Libbey Eurotext, Montrouge, pp. 175–186.

Chinevere, T.D., Kenefick, R.W., Cheuvront, S.N., Lukaski, H.C., Sawka, M.N., 2008. Effect of heat acclimation on sweat minerals. Med. Sci. Sports Exerc. 40 (5), 886–891.

Coudray, C., Demigné, C., Rayssiguier, Y., 2003. Effects of dietary fibers on magnesium absorption in animals and humans. J. Nutr. 133, 1–4.

Del Gobbo, L.C., Imamura, F., Wu, J.H.Y., de Oliveira Otto, M.C., Chiuve, S.E., Mozaffarian, D., 2013. Circulating and dietary magnesium and risk of cardiovascular disease: a systematic review and meta-analysis of prospective studies. Am. J. Nutr. 98, 160–173.

Dreosti, I.E., 1995. Magnesium status and health. Nutr. Rev. 53, S23–S27.

Dibaba, D.T., Xun, P., He, K., 2014. Dietary magnesium intake is inversely associated with serum C-reactive protein levels: meta-analysis and systematic review. Eur. J. Clin. Nutr. 68, 510–516.

Elin, R.J., 2010. Assessment of magnesium status for diagnosis and therapy. Magnes. Res. 23, 1–5.

Fine, K.D., Santa Ana, C.A., Porter, J.L., Fordtran, J.S., 1991. Intestinal absorption of magnesium from food and supplements. J. Clin. Invest. 88, 396–402.

Food and Nutrition Board, Institute of Medicine, 1997. Dietary Reference Intakes for Calcium, Phosphorous, Magnesium, Vitamin D, and Fluoride. National Academies Press, Washington, DC.

Guerrero-Romero, F., Rodriguez-Moran, M., 2002. Relationship between serum magnesium levels and C-reactive protein concentrations, in non-diabetic, non-hypertensive obese subjects. Int. J. Obes. 26, 469–474.

Guerrero-Romero, F., Rodriguez-Moran, M., 2013. Serum magnesium in the metabolically-obese normal-weight and healthy-obese subjects. Eur. J. Intern. Med. 24, 639–643.

Hunt, C.D., Johnson, L.K., 2006. Magnesium requirements: new estimations for men and women by cross-sectional statistical analysis of metabolic magnesium balance data. Am. J. Clin. Nutr. 84, 843–852.

Hunt, C.D., Meacham, S.L., 2001. Aluminum, boron, calcium, copper, iron, magnesium, manganese, molybdenum, phosphorus, potassium, sodium, and zinc: concentrations in common western foods and estimated daily intakes by infants; toddlers; and male and female adolescents, adults, adults, and seniors in the United States. J. Am. Diet. Assoc. 101, 1058–1060.

Klevay, L.M., Milne, D.B., 2002. Low dietary magnesium increases supraventricular ectopy. Am. J. Clin. Nutr. 75, 550–554.

Kruse, H.D., Orent, E.R., McCollum, E.V., 1932. Studies on magnesium deficiency in animals. I. Symptomatology resulting from magnesium deprivation. J. Biol. Chem. 96, 510–539.

Lakshmanan, F.L., Rao, R.B., Kim, W.W., Kelsay, J.L., 1984. Magnesium intakes, balances, and blood levels of adults consuming self-selected diets. Am. J. Clin. Nutr. 40, 1380–1389.

Lukaski, H.C., Nielsen, F.H., 2002. Dietary magnesium depletion affects metabolic responses during submaximal exercise in postmenopausal women. J. Nutr. 132, 930–935.

Marshall, D.H., Nordin, B.E.C., Speed, R., 1976. Calcium, phosphorus and magnesium requirement. Proc. Nutr. Soc. 35, 163–173.

Martin, A. (coordinator), 2000. Apports nutritionnels conseillés pour la population Française. Technical Document. Lavoisier, Paris.

Martin, B.J., 1990. The magnesium load test: experience in elderly subjects. Aging (Milano) 2, 291–296.

Montain, S.J., Cheuvront, S.N., Lukaski, H.C., 2007. Sweat mineral-element responses during 7-h of exercise-heat stress. Int. J. Sport Nutr. Exerc. Metab. 17, 574–582.

Moshfegh, A., Goldman, J., Ahuja, J., Rhodes, D., LaComb, R., 2009. What We Eat in America. NHANES 2005–2006: Usual Nutrient Intakes from Food and Water Compared to 1997 Dietary Reference Intakes for Vitamin D, Calcium, Phosphorus, and Magnesium. United States Department of Agriculture, Agricultural Research Service. Available at: http://www.ars.usda.gov/ba/bhnrc/frsg.

Nielsen, F.H., 2010. Magnesium, inflammation, and obesity in chronic disease. Nutr. Rev. 68, 333–340.

Nielsen, F.H., 2014. Effects of magnesium depletion on inflammation in chronic disease. Curr. Opin. Clin. Nutr. Metab. Care. 17, 525–530.

Nielsen, F.H., Johnson, L.K., Zeng, H., 2010. Magnesium supplementation improves indicators of low magnesium status and inflammatory stress in adults older than 51 years with poor quality sleep. Magnes. Res. 23, 158–168.

Nielsen, F.H., Lukaski, H.C., 2006. Update on the relationship between magnesium and exercise. Magnes. Res. 19, 180–189.

Nielsen, F.H., Lukaski, H.C., Johnson, L.K., Roughead, Z.K., 2011. Reported zinc, but not copper, intakes influence whole-body bone density, mineral content and T score responses to zinc and copper supplementation in healthy postmenopausal women. Br. J. Nutr. 106, 1872–1879.

Nielsen, F.H., Milne, D.B., 2004. A moderately high intake compared to a low intake of zinc depresses magnesium balance and alters indices of bone turnover in postmenopausal women. Eur. J. Clin. Nutr. 58, 703–710.

Nielsen, F.H., Milne, D.B., Klevay, L.M., Gallagher, S., Johnson, L., 2007. Dietary magnesium deficiency induces heart rhythm changes, impairs glucose tolerance, and decreases serum cholesterol in postmenopausal women. J. Am. Coll. Nutr. 26, 121–132.

Palacios, C., Wigertz, K., Braun, M., Martin, B.R., McCabe, G.P., McCabe, L., Pratt, J.H., Peacock, M., Weaver, C.M., 2013. Magnesium retention from metabolic-balance studies in female adolescents: impact of race, dietary salt, and calcium. Am. J. Clin. Nutr. 97, 1014–1019.

Penland, J.G., 1995. Quantitative analysis of EEG effects following experimental marginal magnesium and boron deprivation. Magnes. Res. 8, 341–358.

Rodríguez-Morán, M., Guerrero-Romero, F., 2004. Elevated concentrations of TNF-alpha are related to low serum magnesium levels in obese subjects. Magnes. Res. 17, 189–196.

Rodriguez-Moran, M., Guerrero-Romero, F., 2008. Serum magnesium and C-reactive protein levels. Arch. Dis. Child. 93, 676–680.

Romani, A.M.P., 2013. Magnesium in health and disease. In: Sigel, A., Sigel, H., Sigel, R.K.O. (Eds.), Interelationships Between Essential Metal Ions and Human Diseases, Metal Ions in Life Sciences, vol. 13. Springer, Dordrecht, pp. 49–79.

Rosanoff, A., Weaver, C.M., Rude, R.K., 2011. Suboptimal magnesium status in the United States: are the health consequences underestimated? Nutr. Rev. 70, 153–164.

Rude, R.K., Shils, M.E., 2006. Magnesium. In: Shils, M.E., Shike, M., Ross, A.C., Caballero, B., Cousins, R.I. (Eds.), Modern Nutrition in Health and Disease, tenth ed. Lippincott Williams & Wilkins, Philiadelphia, pp. 223–247.

Scientific Committee for Foods, Nutrient and Energy Intakes for the European Community, 1993. Report of the Scientific Committee for Food, Thirty-first Series. European Commission, Brussels.

Seki, N., Hamano, H., Iiyama, Y., Asano, Y., Kokubo, S., Yamauchi, K., Tamura, Y., Uenishi, K., Kudou, H., 2007. Effect of lactulose on calcium and magnesium absorption: a study using stable isotopes in adult men. J. Nutr. Sci. Vitaminol. 53, 5–12.

Shay, C.M., Van Horn, L., Stamler, J., Dyer, A.R., Brown, I.J., Chan, Q., Miura, K., Zhao, L., Okuda, N., Daviglus, M.L., Elliot, P., for the INTERMAP Research Group, 2012. Food and nutrient intakes and their associations with lower BMI in middle-aged US adults: the International Study of Macro-/Micronutrients and Blood Pressure (INTERMAP). Am. J. Clin. Nutr. 96, 483–491.

Shils, M.E., 1969. Experimental human magnesium depletion. Medicine 48, 61–85.

Silver, B.B., 2004. Development of cellular magnesium nano-analysis in treatment of clinical magnesium deficiency. J. Am. Coll. Nutr. 23, 732S–737S.

Spencer, H., Norris, C., Williams, M.S., 1994. Inhibitory effects of zinc on magnesium balance and magnesium absorption in man. J. Am. Coll. Nutr. 13, 479–484.

Touyz, R.M., 2008. Transient receptor potential melastatin 6 and 7 channels, magnesium transport, and vascular biology: implications in hypertension. Am. J. Physiol. Heart Circ. Physiol. 294, H1103–H1118.

United States Department of Agriculture and Department of Human and Health Services, 2010. Dietary Guidelines of Americans. United States Government Printing Office, Washington, DC.

Volpe, S.L., 2012. Magnesium. In: Erdman Jr., J.W., Macdonald, I.A., Zeisel, S.H. (Eds.), Present Knowledge in Nutrition, tenth ed. Wiley-Blackwell, Oxford, pp. 459–474.

van der Wijst, J., Hoenderop, J.G.J., Bindels, R.J.M., 2009. Epithelial Mg^{2+} channel TRPM6: insight into the molecular regulation. Magnes. Res. 22, 127–132.

van der Wijst, J., Bindels, R.J.M., Hoenderop, J.G.J., 2014. Mg^{2+} homeostasis: the balancing act of TRPM6. Curr. Opin. Nephrol. Hypertens. 23, 361–369.

Weglicki, W.B., 2012. Hypomagnesemia and inflammation: clinical and basic aspects. Annu. Rev. Nutr. 32, 55–71.

Wolf, F.I., Trapani, V., 2011. MagT1: a highly specific magnesium channel with important roles beyond cellular magnesium homeostasis. Magnes. Res. 24, S86–S91.

World Health Organization and Food and Agriculture Organization of the United Nations, 2004. Vitamins and Mineral Requirements in Human Nutrition, second ed. World Health Organization, Geneva.

Chapter 26

Magnesium and the Immune Response

Carsten Schmitz[1,2,]*, Anne-Laure Perraud[2,1,]*

[1]*University of Colorado Denver, Denver, CO, United States;* [2]*National Jewish Health, Denver, CO, United States*

INTRODUCTION (BACKGROUND)

With a total cellular concentration between 14 and 20 mM, Mg^{2+} is the most abundant divalent cation in living cells. As opposed to calcium (Ca^{2+}), the consensus in the literature for several decades was that Mg^{2+} does not mediate cellular responses to signals by changes of its concentration within the cytosol. One main reason for this assumption is that free Mg^{2+} is kept in the cytosol and in extracellular fluids at a similar low millimolar level, whereas the cytosolic Ca^{2+} concentration is several orders of magnitude lower than extracellularly or in storing organelles, establishing a large concentration gradient across biological membranes for Ca^{2+}, which can be efficiently used for rapid information flow. Because of the initially high concentration of Mg^{2+}, a total decrease or increase in the cytosol equivalent to that occurring in relation to Ca^{2+} will result in negligible changes in free Mg^{2+}. Importantly, as for Ca^{2+}, only a small fraction of the total Mg^{2+} content, namely 0.5–0.8 mM, is actually free in solution, with most Mg^{2+} being bound intracellularly by biomacromolecules (ATP alone binds typically ~5 mM Mg^{2+}; Brandao et al., 2013; Mandel and Goodman, 1999; Perraud et al., 2004; Quamme, 2010; Wolf et al., 2003 and references therein). Within the cell Mg^{2+} appears to be evenly distributed among mitochondria, the nucleus, and endo(sarco)-plasmatic reticulum, with a concentration ranging from 14 to 18 mM in any of these organelles. It has been shown that some of the mitochondrial Mg^{2+} can be rapidly mobilized by an increase in cytosolic cyclic AMP (cAMP), indicating that these organelles can play the role of Mg^{2+} sinks (Romani et al., 1993; Romani and Scarpa, 2000).

Within a mammalian cell, not only do hundreds of enzymes require Mg^{2+} as an essential cofactor but also Mg^{2+} is crucial for the maintenance of the active conformation of macromolecules (i.e., DNA, RNA, ATP), for the regulation of lipid- and phosphoinositide-derived second messengers, for charge compensation, and for the regulation of various transporters and ion channels. Furthermore, Mg^{2+} is an important modulator of intracellular free Ca^{2+} concentration and intracellular pH, which are major determinants of cell contraction, secretion, motility, and proliferation. Studies focusing on the connection between cell proliferation and Mg^{2+} demonstrated that low Mg^{2+} inhibits cell cycle progression and might represent one crucial element in control of protein translation and cell proliferation (see Rubin, 2005 and references therein).

Clinical studies have documented the role of Mg^{2+} as an important contributor to the development or exacerbation of numerous pathologies, including asthma, diabetes mellitus, hyperlipidemia, atherosclerosis, and hypertension (reviewed in de Baaij et al., 2014). Mg^{2+} insufficiency is also associated with epilepsy, migraines, muscular dysfunction, and bone wasting, and up to 60% of critically ill patients show some degree of Mg^{2+} deficiency (Chernow, 1989; Escuela et al., 2005) as a consequence of losses from the kidneys often secondary to medication, predisposing these patients to serious, even life-threatening effects.

In the immune context, early observations revealed a direct correlation between Mg^{2+} deficiency and an increase in systemic inflammation based on elevated serum levels of tumor necrosis factor (TNF)-α and other proinflammatory cytokines and reduced concentrations of antiinflammatory cytokines (Long and Romani, 2015). Mg^{2+} deficiency can lead to different physiological stress reactions relevant to the immune response causing endothelial dysfunction and an inflammatory syndrome (as shown in rats fed a hypomagnesic diet) accompanied by leukocyte and macrophage activation as well as augmentation of proinflammatory cytokines, acute-phase proteins, and free radicals (Rayssiguier et al., 2002). Conversely, magnesium sulfate supplementation can mediate antiinflammatory effects via activation of phosphoinositide 3-kinase (PI3K) and inhibition of L-type ion channels (Su et al., 2013). Furthermore, Mg^{2+} deficiency affects mast cell proliferation

---

*Both the authors equally contributed

Molecular, Genetic, and Nutritional Aspects of Major and Trace Minerals. http://dx.doi.org/10.1016/B978-0-12-802168-2.00026-9

and function (histamine storing and secretion) and might be involved in mast cell–dependent hepatic fibrosis and steatosis (Kraeuter and Schwartz, 1980; Takemoto et al., 2014; Veilleux, 1975).

More recently, genetic mutations leading to functionally defective Mg^{2+} transporters have been linked to severe pathologies in humans. Mutations in the Mg^{2+} transporter MagT1 were shown to elicit specific immune defects in affected individuals (XMEN—*X*-linked immunodeficiency with *M*agnesium defect, chronic *E*pstein–Barr virus infections and *N*eoplasia), suggesting a defined signaling function for Mg^{2+} in this context (Chaigne-Delalande et al., 2013; Li et al., 2011). Patients with deficiencies in transient receptor channel potential melstatin member-6 (TRPM6) develop hypomagnesemia with secondary hypocalcemia (HSH), an autosomal-recessive disorder leading to muscle spasms, tetany, and seizures. The transient receptor channel potential melstatin member-7 (*TRPM7*) T1482I genetic variant exhibits an increased sensitivity to inhibition by intracellular Mg^{2+}, which is expected to result in the channel closing too soon to allow for sufficient Mg^{2+}-entry into the cell. Epidemiological and genetic studies have shown a connection between the TRPM7 T1482I genetic polymorphism and neurodegenerative disorders (gALS, gPD), as well as with colorectal cancer, when combined with an environment deficient in Ca^{2+} and Mg^{2+} (Dai et al., 2007; Hermosura et al., 2005). The role of TRPM7 as an important determinant in several types of cancer such as leukemia, retinoblastoma, and cancer of the pancreas, breast, ovaries, head, neck, and other organs is documented in a growing body of literature (Wolf and Trapani, 2008). As illustrated by the secondary hypocalcemia occurring in TRPM6-deficient HSH humans, when considering the impact of altered Mg^{2+} homeostasis, it is important to keep in mind the interdependence between Mg^{2+} and Ca^{2+} homeostasis and how changes in the Ca^{2+}/Mg^{2+} ratio might affect clinical outcomes and experimental results. For example, beyond its role in Mg^{2+} homeostasis, TRPM7 has been shown to influence Ca^{2+} signaling events, such as the spatial distribution of high-Ca^{2+} microdomains mediating directed fibroblast migration (Wei et al., 2009). Moreover, we found that in response to changing availability of environmental Mg^{2+}, TRPM7 can modulate phospholipase C (PLC)-γ2-mediated Ca^{2+} signaling via Ser/Thr phosphorylation, therefore providing novel molecular mechanisms underlying Ca^{2+}/Mg^{2+} interdependence (Deason-Towne et al., 2012).

Given the ubiquitous and abundant presence of Mg^{2+} in biology, and the lack of steep concentration gradients for free Mg^{2+} across biological membranes, it is often been perceived as a maintenance ion. As such, little attention was given to its role as a regulator of cellular physiology and signaling, and to molecular mechanisms of Mg^{2+} sensing and Mg^{2+} homeostasis regulation. In the next section, we will give brief insights into basic features of the Mg^{2+} ion itself and how these might shape the properties of molecules transporting it.

GENERAL STRUCTURAL FEATURES OF MG^{2+} PERMEABLE ION CHANNELS AND TRANSPORTERS

Ion transport across cellular membranes is thought to require the tailored binding of the ion at the ion selectivity filter of the pore region of the transporter, dehydration of the ion, and release of the ion on the other side of the membrane. The divalent cation Mg^{2+} is unique among biologically relevant ions because it possesses the smallest ionic radius and the largest hydration shell, which represents a challenge for its transport across biological membranes. Mg^{2+} tightly binds to its first hydration shell, consisting of six octahedrally coordinated water molecules, and holds on firmly to a second shell with 12–14 water molecules (Markham et al., 2002). Because a considerable amount of energy is needed to strip off the first water shell from Mg^{2+}, one possibility discussed in literature is that partially hydrated Mg^{2+} is transported through membranes (Dalmas et al., 2012).

In the past two decades the molecular discovery of multiple Mg^{2+} transporters in various life forms, from bacteria to plants and mammals, led to the conclusion that different structural properties are responsible to selectively transport Mg^{2+} into and out of the cells, or between cellular compartments. Recent research findings provided initial insights into the structure of the prokaryotic Mg^{2+} transporter CorA and MgtE, which are homologues of the eukaryotic Mg^{2+} transporters SLC41 and Mrs2/Alr1 (Moomaw and Maguire, 2008).

Although the N-terminal regulatory domains of MgtE are absent in the SLC41 transporters, human SLC41A2 can partially compensate and rescue cell growth of chicken DT40 B lymphocytes deficient in the Mg^{2+} channel TRPM7 (Sahni et al., 2007).

Recently published work about the structure of the two prokaryotic Mg^{2+} transporters MgtE and CorA revealed two different models of pore organization allowing for the formation of a filter highly selective toward Mg^{2+}. Whereas MgtE utilizes conserved acidic side chains to build its Mg^{2+}-specific selectivity filter, CorA uses conserved asparagines to recognize and accommodate hydrated Mg^{2+} cations within the CorA pore domain. Both proteins also have common structural motifs and function as Mg^{2+} sensors through distinct regulatory domains with multiple Mg^{2+} binding sites influencing changes in protein conformation and gating of the Mg^{2+} transporters, indicating that the sensing of Mg^{2+} concentrations defines the closing and opening of the Mg^{2+} transporter (Hattori et al., 2007; Takeda et al., 2005). The next section will summarize the potential capacity of Mg^{2+} in nutrient-dependent signal transduction.

MG^{2+} AND ITS POTENTIAL ROLES IN IMMUNORECEPTOR SIGNALING

Metabolic Enhancer and/or Second Messenger? The Potential Influence of Mg^{2+} on the Immune Response

Recent research studies indicate that Mg^{2+} could be a crucial and underestimated factor in nutritional sensing signaling cascades. Considering the high occurrence of environmental and nutritional deficiency in Ca^{2+} and Mg^{2+}, the emerging regulatory and signaling functions performed by Ca^{2+}or Mg^{2+} are very likely to be of public health relevance. Gaining new insights into the influence of Mg^{2+} on nutrient-sensitive signaling networks has the potential to allow development of new strategies to treat or prevent diseases ranging from immunodeficiencies to cancer.

Particularly interesting are the latest observations underscoring the importance of the immunoinflammatory processes in the pathology of acute Mg^{2+} deficiency in experimental animals in which it leads to inflammation, exacerbated immune stress responses, and a decrease of specific immune responses (Bussiere et al., 2002b; Malpuech-Brugere et al., 2000; Petrault et al., 2002; Zimowska et al., 2002). Mg^{2+} deficiency also results in a significant increase in the production of free radical species and subsequent tissue injury (Bussiere et al., 2002a).

Mg^{2+}-dependent stress also stimulates the sympathetic nervous system and the hypothalamic-pituitary-adrenal axis. Both systems influence hormone regulation, fat accumulation (insulin, aldosterone), and neuropeptide release, which can directly induce an immune response and lead to an inflammatory phenotype (Rayssiguier et al., 2002).

Furthermore, Mg^{2+} deficiency affects innate immunity in patients with Crohn's disease (CD), exhibiting a strong correlation with several proinflammatory cytokines such as TNF-α, interleukin (IL)-1, and IL-6. The acute-phase protein, C-reactive protein, one of the best established inflammatory indicators, also fluctuates up upon reduced Mg^{2+} levels in CD patients (Naser et al., 2014). Therefore Mg^{2+} is discussed as being anti-proinflammatory cytokine by controlling nuclear factor-κB (NF-κB) activity under physiological Mg^{2+} conditions, and leading to increased NF-κB activation and cytokine production when in suboptimal concentrations (Sugimoto et al., 2012).

Studies undertaken in B lymphocytes aiming at investigating the role of Mg^{2+} in modulating adaptive immunity demonstrated an increase in free intracellular Mg^{2+} concentration followed by an increase in intracellular Ca^{2+} concentration in response to B cell receptor (BCR) activation or ionophore treatment (Rijkers and Griffioen, 1993; Rijkers et al., 1993). Of special relevance in this context are studies showing the Mg^{2+} regulation of a phosphatidylinositol-specific PLC, and the Mg^{2+} regulated activation of 1,4,5-trisphosphate (InsP$_3$) phosphatase in B lymphocytes, suggesting a direct role for Mg^{2+} as a regulator of cellular levels of the second messenger diacylglycerol (Chien and Cambier, 1990). Importantly, there is another and potentially even more striking molecular connection between Mg^{2+} homeostasis regulation and PLC enzymes since it has been demonstrated that several PLC isoenzymes (β and γ1; γ2) associate with the kinase domain of the Mg^{2+} permeable channel kinase TRPM7 (Runnels et al., 2002). TRPM7 kinase was subsequently shown to phosphorylate PLCγ2, a pivotal element of the B cell signaling pathway after BCR ligation (Deason-Towne et al., 2012). This contributes to the Mg^{2+}-dependent modulation of the Ca^{2+} response elicited by the stimulation of the BCR, and provides the first defined molecular pathway underlying the Mg^{2+} sensitivity of immunoreceptor signaling responses (Deason-Towne et al., 2012).

In the past few years, new findings about the hereditary immunodeficiency disease XMEN revealed that it is caused by mutations in the gene encoding the MagT1 Mg^{2+} transporter, suggesting that Mg^{2+} could function as a second messenger in cellular signaling. Patients carrying MagT1 mutations exhibit impaired PLC signaling with reduced Ca^{2+}/Mg^{2+} responses after T cell receptor (TCR) stimulation and abrogated expression of the natural killer activating receptor NKG2D in natural killer and CD8$^+$ T cells (Chaigne-Delalande et al., 2013; Li et al., 2011). A more detailed overview of these exciting insights into MagT1 function will be provided in a different chapter of this book written by Dr. Chaigne-Delalande.

In summary, expanding our knowledge about the modulation of signaling in response to Mg^{2+} availability could allow for the development of unexplored strategies for therapeutic intervention in autoimmune diseases, immunodeficiencies, and lymphoma. In the following section the representation and potential roles of individual Mg^{2+} transporters and channels are discussed in more detail (see also Fig. 26.1 and Table 26.1).

SLC41A1 AND SLC41A2—RELATIVES OF THE BACTERIAL MGTE FAMILY

The human relatives of the bacterial MgtE family SLC41A1 and SLC41A2 (SLC: solute carrier) are members of a large group of membrane transporters classified into close to 50 subgroups with more than 300 transporters. SLC41A1 was

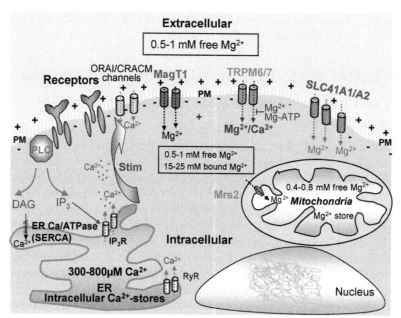

FIGURE 26.1 Mg^{2+} transporters in immune cells and their potential role in immunoreceptor signaling. A simplified scheme of an immunocyte is presented, showing the extra- and intracellular compartments divided by a plasma membrane (PM). Receptors, channels, and transporters are symbolized at the PM, the endoplasmic reticulum (ER), and the inner mitochondria, and the ion flow by *dashed lines*. Upon receptor stimulation, a cascade of phosphorylation events leads to phospholipase C (PLC) enzyme activation, the hydrolysis of phosphatidylinositol 4,5-bisphosphate to inositol triphosphate (IP$_3$) and diacylglycerol (DAG). Binding of IP$_3$ to the IP$_3$ receptors (IP$_3$R) at the ER causes Ca^{2+} store depletion with subsequent stromal interaction molecule-1 (Stim1)-mediated Ca^{2+} entry via ORAI/calcium release-activated calcium channel protein (CRACM) channels from the extracellular space. Mutations of Mg^{2+} transporter-1 (MagT1) linked to immunodeficient humans abolishes Mg^{2+} influx in T cells after T cell receptor stimulation followed by a delayed PLC activation and impaired Ca^{2+} response. Transient receptor potential melastatin cation channel-7 (TRPM7) currents are inhibited by intracellular Mg^{2+} or MgATP. TRPM7-deficient DT40 B-cells show reduced cell proliferation and Mg^{2+} uptake. Overexpression of MagT1 or the membrane protein solute carrier family 41 member 2 (SLC41A2) partially rescue cell growth and Mg^{2+} uptake in DT40 B lymphocytes lacking TRPM7.

identified by performing a modified version of the Signal Sequence Trap method designed to isolate transmembrane (TM) and secreted proteins and was recognized to bear homology to the prokaryotic Mg^{2+} transporter MgtE (Wabakken et al., 2003). SLC41A2 was discovered a few years later through a SLC41A1 homology search. SLC41A1 appears to be expressed ubiquitously. Its gene expression is strongest in heart and testis, and it is also present in all tissues of the hematopoietic compartments, albeit at lower levels, including in resting or activated lymphocytes. The SLC41A1 cDNA sequence was originally isolated from the leukemia cell line Tom-1, and by Northern blot a single SLC41A1 mRNA species was detected in the lymphoid cell lines Tom-1, BV173, Reh, and Jurkat (Wabakken et al., 2003). SLC41A2 also seems to be ubiquitously expressed and appears to be present in all immune cell lines from various lineages tested so far [Jurkat (human T cell leukemia), Raji, Ramos (humanBurkitt's lymphoma cell lines), Nalm-6 (human B cell precursor leukemia), OCI-Ly3 and OCI-Ly10 (B cell like DLBCL cell lines); Sahni et al., 2007].

The SLC41A1 and SLC41A2 proteins share 70% identity with 10–12 hypothetical TM domains as predicted by different computer protein structure algorithms. Membrane topology predictions support an 11-TM span model for SLC41A1 with a cytosolic N-terminal domain and an extracellular C-terminal region, which was supported by surface versus intracellular detection of SLC41A1 tagged with two different epitopes in avian DT40 B cells (Mandt et al., 2011). However, a different study using a split-ubiquitin assay in yeast has proposed a 10-TM model with both termini of SLC41A1 oriented intracellularly (Sponder et al., 2014). Given the difference in experimental approaches, it is conceivable that the cellular context (vertebrate vs. yeast) might explain this discrepancy. Regarding SLC41A2, the C- and N-termini show an inverse orientation as demonstrated in a different analysis (Goytain and Quamme, 2005; Mandt et al., 2011; Sahni et al., 2007; Wabakken et al., 2003).

The functional properties of both SLC41 proteins differ and depend on the chosen cellular context and the experimental conditions. Using a complementation approach in Salmonella cells deficient in the three Mg^{2+} transporters CorA, MgtA, and MgtB, overexpression of human SLC41A1 was found to rescue this bacterial mutant strain from the diminished growth phenotype they exhibit under low Mg^{2+} levels (Kolisek et al., 2008). Voltage-dependent Mg^{2+} currents were measured upon SLC41A1 overexpression in oocytes and found to be highly selective toward divalent cations (Goytain and Quamme, 2005),

TABLE 26.1 Mg^{2+} Transporters Expressed in the Immune System

Name	Structure–Function	Tissue Distribution	Remarks	References
MagT1	Four TM domains, N- and C-terminal tails extracellular or cytosolic? Cleavage of N-terminal signaling peptide; Mg^{2+} selective ion channel; K_M: 0.2 mM; Mg^{2+} transport is rheogenic, pH, and voltage dependent.	Widely expressed, spleen and thymus, B and T lymphocytes.	Mutations of MagT1 are linked to X-linked human T cell immunodeficiency and abrogate Mg^{2+} influx after TCR stimulation; potential role for regulating Mg^{2+} as second messenger. Magt1 overexpression partially rescues DT40 B lymphocytes lacking TRPM7.	Li et al. (2011), Chaigne-Delalande et al. (2013), Goytain and Quamme (2005), Deason-Towne et al. (2011)
Mrs2	Two TM domains, permeable to Mg^{2+} and Ni^{2+}; localized inner mitochondrial membrane.	Widely expressed, B lymphocytes.	Regulates Mg^{2+} uptake in mitochondria, which is a large Mg^{2+} store, but only 0.4–0.8 mM is free Mg.	Deason-Towne et al. (2012)
SLC41A1	Ten (11) TM domains with N-terminus extra- and C-terminus intracellular or vice versa? Carries preferably Mg^{2+}, K_M: 0.67 mM, entry of other divalents (Fe^{2+}, Ba^{2+}, Cn^{2+}, Zn^{2+}).	Ubiquitous.	Expressed in bone marrow, spleen, thymus, spleen, lymph nodes, and the lymphoid cell lines (Tom1, BV173, and Jurkat).	Wabakken et al. (2003), Mandt et al. (2011), Goytain and Quamme (2005)
SLC41A2	Eleven TM domains with N-terminus intra- and C-terminus extracellular(?); carries preferably Mg^{2+}, K_M: 0.34 mM, other divalents (Ba^{2+}, Ni^{2+}, Co^{2+}, Fe^{2+}, or Mn^{2+}).	Might be ubiquitously expressed in hematopoietic cells.	Expressed in lymphoid cell lines; Jurkat, Raji, Ramos, Nalm-6, OCI-Ly3, and OCI-Ly10. Partially compensates TRPM7 DT40 knockout phenotype.	Wabakken et al. (2003), Goytain and Quamme (2005), Kolisek et al. (2008)
TRPM6 (CHAK2)	Six TM domains; forms tetramers to build a functional pore; cytosolic N- and C-termini; intrinsic C-terminal Ser/Thr α-kinase; associates with TRPM7 (required for functional surface expression in some studies); highly Mg^{2+}- and Ca^{2+}-permeable channel.	Restricted expression (kidney, intestine), monocytes, peripheral blood lymphocytes (B but not T cells).	Mutations of TRPM6 are linked to human hypomagnesemia (autosomal-recessive disease), leading to seizures and death unless treated with a Mg^{2+}-rich diet. Many TRPM6 $^{-/-}$ mice are lethal at embryonic day 12.5; those that survive show neural tube defects. TRPM6 overexpression does not rescue DT40 B lymphocytes lacking TRPM7.	van der Wijst et al. (2009), Schlingmann et al. (2002), Walder et al. (2002), Schmitz et al. (2005), Brandao et al. (2014)
TRPM7 (LTRPC7, TRP-PLIK, CHAK1)	Six TM domains; forms tetramers to build a functional pore; cytosolic N- and C-termini; intrinsic C-terminal Ser/Thr α-kinase; phosphorylates annexin-1, myosin 2α, elongation factor 2-kinase, phospholipase, histones; highly Mg^{2+}- and Ca^{2+}-permeable channel, entry of other divalents (Ni^{2+}, Ba^{2+}, Co^{2+}, Mn^{2+}, Sr^{2+}, Cd^{2+}).	Ubiquitous.	TRPM7 $^{-/-}$ mice die at embryonic day 6.5, TRPM7 $^{+/-}$ mice hypomagnesic; $Trpm7^{-/fl}$ lck-Cre mice develop a T-cell block at double-negative stage; TRPM7 $^{-/-}$ DT40 B cells are lethal, unless medium supplemented with Mg^{2+} (10 mM), kinase not required for channel gating.	Perraud et al. (2004), Deason-Towne et al. (2012), Fleig and Chubanov (2014), Ryazanova et al. (2010), Runnels et al. (2001), Nadler et al. (2001), Ryazanov et al. (2002), Yamaguchi et al. (2001), Schmitz et al. (2003), Matsushita et al. (2005), Hofmann et al. (2010), Monteilh-Zoller et al. (2003), Prakriya and Lewis (2002), Kozak and Cahalan (2003), Jiang et al. (2003), Chokshi et al. (2012), Clark et al. (2008), Perraud et al. (2011), Krapivinsky et al. (2014), Clark et al. (2006), Dorovkov and Ryazanov (2004), Jia et al. (2011), Runnels et al. (2002), Langeslag et al. (2007), Brandao et al. (2014), Takezawa et al. (2004), Jin et al. (2008), Schilling et al. (2014), Wang et al. (2014)

but no Mg^{2+}-specific currents were detectable in human embryonic kidney 293 (HEK-293) cells overexpressing it (Kolisek et al., 2008). However, when these SLC41A1-overexpressing HEK-293 cells were cultured in Mg^{2+}-free media, it led to reduced intracellular Mg^{2+} levels, suggesting that SLC41A1 might be involved in Mg^{2+} efflux (40). A subsequent study was designed to explore the hypothesis that SLC41A1 could function as Na^+/Mg^{2+} exchanger and found convincing evidence that this might be the case, thus identifying SLC41A1 as the predominant Mg^{2+} efflux system (Kolisek et al., 2008).

Similar to the biophysical characterization of SLC41A1 in oocytes, the analysis of SLC41A2 demonstrated voltage-dependent Mg^{2+} uptake, confirming the hypothesis that SLC41A2 could function as a Mg^{2+} transporter (Goytain and Quamme, 2005). SLC41A2 differs from SLC41A1 in terms of its permeability profile toward divalent cations because it is able to transport Ba^{2+}, Ni^{2+}, Fe^{2+}, Co^{2+}, and Mn^{2+} but not Ca^{2+}, Cu^{2+}, or Zn^{2+} (Goytain and Quamme, 2005). High intracellular levels of Ca^{2+} were demonstrated to even inhibit transporter activity, a feature unique to SLC41A2 and not seen with SLC41A1 (Goytain and Quamme, 2005).

SLC41A1 and SLC41A2 could play an important role in regulating immune responses because they are widely expressed in various immune cell types, including B and T lymphocytes. Upon overexpression, SLC41A1 and SLC41A2 also share the ability to ameliorate the Mg^{2+} dependence of cell growth and to increase intracellular Mg^{2+} levels in DT40 B cells deficient in the cellular master regulator of Mg^{2+} homeostasis, TRPM7 (Mandt et al., 2011; Sahni et al., 2007). These observations suggest that both transporters can influence Mg^{2+} homeostasis in immune cells, and therefore immune functions, although future studies will need to address this question, particularly in vivo.

THE CHANNEL KINASES TRPM6 AND TRPM7—MG²⁺ SENSORS AND MASTER REGULATORS

TRPM7 and its closest homologue TRPM6 are unusual proteins because they represent the only known examples of ion channels linked to kinase domains (Fig. 26.2). Both ion channels are Mg^{2+} and Ca^{2+} permeable and belong to the TRPM subgroup of the TRP (transient receptor potential) ion channel family, which includes a total of eight members (TRPM1–8). TRPM stands for TRP *m*elastatin (now TRPM1), named after the first member of the family, which was originally discovered as a melanoma tumor suppressor (Duncan et al., 1998). TRPM7 and TRPM6 are essential regulators of Mg^{2+} homeostasis (Brandao et al., 2013; Fleig and Chubanov, 2014; Paravicini et al., 2012; van der Wijst et al., 2009) and can form functional heteromeric channels at the plasma membrane, regulating, for example, Mg^{2+} in the colon, thus affecting Mg^{2+}/Ca^{2+} homeostasis of the whole organism (Dai et al., 2007; Ryazanova et al., 2010). The C-terminal kinase domain of TRPM6/7 is an active Ser/Thr kinase belonging to a small family of atypical enzymes related to eukaryotic elongation factor-2 kinase (eEF2-k), a kinase involved in modulating rates of protein translational elongation in response to nutritional mTOR signaling. The functional implication of the fusion between the kinase and ion channel portions of TRPM6/7 has been a central aspect of the research effort aiming at clarifying the role of these two molecules in biology.

The Discovery and Primary Biophysical Characterization of TRPM7

Three different research laboratories cloned TRPM7 around the same time: (1) David Clapham's group performed a yeast two-hybrid screen using the C2 domain of PLCβ as a bait, and "fished" the kinase domain of TRPM7; (2) Alexey Ryazanov's group discovered TRPM7 kinase through a search for novel alpha kinase family members (eEF2-k family); and (3) Andrew Scharenberg and colleagues applied a bioinformatics approach to identify new ion entry pathways regulating ion homeostasis in immune cells with a special focus on Ca^{2+} signaling (Nadler et al., 2001; Runnels et al., 2001; Ryazanov, 2002).

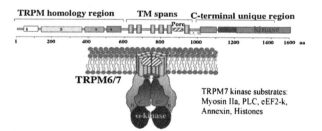

FIGURE 26.2 **Schematic representation of the chimeric channel kinases transient receptor channel potential melastatin members 6 (TRPM6) and 7 (TRPM7).** *Top:* Linear domain organization common to both channels. *TM*, transmembrane; *CCR*, coiled-coil region (putative association domain). *Bottom:* TRPM6/7 represented embedded in a cellular membrane as "half a pore". A complete functional pore is formed by tetramerization. *PLC*, phospholipase C; *eEF2-k*, eukaryotic elongation factor-2 kinase.

Similarly to other TRP channels, TRPM7 and its sister channel TRPM6 are thought to harbor six TM domains, resulting in both termini being intracellular (Fig. 26.2). TRP channels need to tetramerize to form a functional channel pore, which is thus lined by a total of 24 TM spans, a common structural feature of other ion channel families such as K_V or Ca_V channels. Large (>700 amino acid) cytosolic N-terminal domains are present in all eight members of the TRPM family, but they remain of unknown structure and function. It is assumed that gating is rather regulated by C-terminal regions although some of the TRPM proteins possess only very short sequences after their sixth TM span (e.g., TRPM4 and TRPM5). TRPM6 and TRPM7 both contain in their N-terminal portion a typical coiled-coil domain (demonstrated to mediate tetramerization of the TRPM subunits), followed by a 50-kD domain of uncharacterized function that is fused to the aforementioned Ser/Thr kinase (reviewed in Perraud et al., 2004).

The kinase domain of TRPM7 is homologous to eEF2-k, and although it does not exhibit protein sequence homology to other kinase families, it is structurally similar to protein kinase A (Yamaguchi et al., 2001). After some past controversy about the potential role of the kinase domain in TRPM7 channel gating, there is now agreement that the TRPM7 kinase domain is not required for channel gating (Matsushita et al., 2005; Schmitz et al., 2003), but it does have a modulatory effect on channel function (Schmitz et al., 2003). It is interesting to note that the TRPM6/7 kinase domain appears to be a more recent evolutionary addition because TRPM6/7 homologues in *Drosophila* and *Caenorhabditis elegans* do not include a kinase region, despite the channels playing a role in Ca^{2+}/Mg^{2+} homoeostasis (Hofmann et al., 2010).

The biophysical characterization of TRPM7 indicates the channel to be constitutively open under standard experimental conditions and under physiological ionic conditions to be selective toward most biologically relevant divalent cations, including Ca^{2+} and Mg^{2+}, but also Co^{2+}, Mn^{2+}, Ni^{2+}, or Zi^{2+} (Monteilh-Zoller et al., 2003; Nadler et al., 2001). Intracellular Mg^{2+} and MgATP (Nadler et al., 2001; Prakriya and Lewis, 2002) inhibit TRPM7 currents although the significance of Mg nucleotide gating is controversial (Kozak and Cahalan, 2003). Native TRPM7-like currents showing similar biophysical properties to heterologously expressed TRPM7 were measured in numerous cell types, including immune cells. These currents were named magnesium inhibited cation (MIC) or magnesium nucleotide regulated metal ion currents (MagNuM) (Jiang et al., 2003; Nadler et al., 2001; Prakriya and Lewis, 2002) based on their characteristic Mg^{2+} sensitivity and ionic selectivity profile.

A detailed analysis of native TRPM7 whole-cell currents in human Jurkat T cells conducted by Ashot Kozak's group showed a complex picture regarding the Mg^{2+} sensitivity of the channel. Inactivation of TRPM7 currents by Mg^{2+} was found to be biphasic (Chokshi et al., 2012), indicating the presence of two types of TRPM7 Mg^{2+} inhibitor sites: one high-affinity (IC_{50} ~10 µM Mg^{2+}) and one low-affinity site (IC_{50} ~165 µM Mg^{2+}). The same group recently conducted TRPM7 single-channel measurements in Jurkat cells, confirming the two different Mg^{2+} affinity sites, in which the involvement of the low-affinity site appears to be weaker and the modulation of Mg^{2+} sensitivity could also be regulated by TRPM7-associated molecules. Given the complexity of their findings, the authors discuss the possibility that "Mg^{2+} sensitivity is a dynamic property of this channel [that] varies depending on mitogenic activation of T cells." Future studies will need to investigate how the developmental and activation stage of immune cells might affect TRPM7's properties and therefore cellular ion homeostasis and signaling.

Phosphorylation studies of TRPM7 demonstrated (auto)phosphorylation of a Ser/Thr-rich domain directly upstream of the kinase domain (Clark et al., 2008). TRPM7's Mg^{2+} sensitivity is modulated through a functional coupling between channel gating and kinase activity (Schmitz et al., 2003). However, a structure extrinsic to the kinase domain mediates TRPM7's Mg^{2+}-sensitive gating (Schmitz et al., 2003). Moreover, the kinase domain of TRPM7 not only autophosphorylates channel residues, but also it has diverse intracellular substrates such as eEF2-k (Perraud et al., 2011); PLC (Deason-Towne et al., 2012); histones (Krapivinsky et al., 2014); the assembly domain of myosin IIA, IIB, and IIC (Clark et al., 2006); and annexin I (Dorovkov and Ryazanov, 2004; see Fig. 26.2). Therefore TRPM7 kinase appears to play an important role in chromatin modeling via histone phosphorylation (Krapivinsky et al., 2014), in cytoskeleton dynamics via myosin phosphorylation (Clark et al., 2006), and in regulating the rate of protein translation via eEF2-k phosphorylation under reduced availability of the nutrient Mg^{2+}, suggesting a direct link between Mg^{2+} levels, protein translation, and cell proliferation.

Because of the original identification of TRPM7 kinase via a yeast two-hybrid screen using the C2 domain of PLCβ as a bait (Runnels et al., 2001), several laboratories have analyzed the influence of phosphoinositide metabolism on TRPM6 and TRPM7 channel gating, leading to partly contradictory interpretations. Two studies claim that phosphatidylinositol 4,5-biphosphate (PIP_2) leads to inactivation of TRPM6 (Jia et al., 2011) or to TRPM7 inhibition through PLC-dependent PIP_2 hydrolysis (Runnels et al., 2002), which stands in contrast to investigations by Takezawa et al. (2004) suggesting that TRPM7 currents were weakened after carbachol stimulation via the G_s-cAMP signaling pathway, whereby PLC activation was not involved. Kees Jalink's group more recently discovered that modulation of TRPM7 currents by PLC is Mg^{2+} dependent, with PLC-mediated TRPM7 activation taking place under physiological Mg^{2+} conditions, whereas PLC had an inhibitory effect on TRPM7 under reduced Mg^{2+} levels (Langeslag et al., 2007). This further demonstrates that the regulation of

TRPM7 and its contribution to cellular functions need to be carefully studied in the context of the ionic environment and should in particular be put in relation to Mg^{2+} availability.

The Channel Kinases TRPM6/7 as Mg^{2+} Homeostasis Regulators and Components of the Nutritional Signaling Network

Our group chose a genetic approach to study TRPM7 and TRPM6 function using chicken DT40 B lymphocytes engineered to carry genomically disrupted TRPM7 sequences (Buerstedde and Takeda, 1991; Nadler et al., 2001). The DT40 cell line represents a convenient cellular system to generate targeted gene knockout because it exhibits a high rate of homologous recombination (Buerstedde et al., 1991). It has been extensively used to study signaling events after BCR activation (Kurosaki et al., 2000). TRPM7$^{-/-}$ DT40 B lymphocytes experience growth arrest within 24 h, which is unlike other known ion channel deficiencies in these cells. TRPM7$^{-/-}$ DT40 cells become Mg^{2+} deficient and die; however, their viability and proliferation are rescued by supplementing the growth media with millimolar Mg^{2+} amounts (but not with any other ion tested, including Mn^{2+}, Ca^{2+}, Zn^{2+}, and Ni^{2+}; Ryazanova et al., 2010; Schmitz et al., 2003). This phenotype is reminiscent of TRPM6-deficient humans who can live a normal life if provided with a Mg^{2+}-rich diet (Schlingmann et al., 2002; Walder et al., 2002), but can suffer from severe and even lethal seizures if not treated.

The expression patterns of TRPM6 and TRPM7 are not identical. TRPM6 is mainly expressed in kidney, colon, and intestine, three organs playing a key role in Mg^{2+} uptake and secretion, whereas TRPM7 is ubiquitously expressed. Cross-phosphorylation and coexpression studies of TRPM6 and TRPM7 indicate that TRPM6 phosphorylates TRPM7 but not vice versa (Brandao et al., 2014; Schmitz et al., 2005). TRPM6's kinase activity appears to modulate TRPM7 subcellular localization and to affect TRPM7-dependent cell growth under hypomagnesic conditions. This indicates that both channel kinases are in their own right important components of ion homeostasis regulation and pivotal signaling modules adjusting each other's biological activity as well as intracellular signaling events. These findings confirm that although the channel kinases TRPM6 and TRPM7 are both environmental Mg^{2+} sensors and regulators of Mg^{2+} homeostasis, they are not functionally redundant. This is further underscored by the phenotype of mice lacking either TRPM6 or TRPM7, or of TRPM6-deficient patients who are hypomagnesic despite having intact *Trpm7* alleles.

A main effort of TRPM7-related research, also in our group, is the elucidation of potential signaling functions of TRPM7 kinase in the context of TRPM7's ion transport and Mg^{2+} sensor activity. An attractive hypothesis is that the unique covalent bond between its kinase and channel portions ideally position TRPM6 and TRPM7 to operate as ionic monitoring systems that directly translate this information into the signaling networks of the cell via their kinases. As such, TRPM6/7 can be seen as nutritional sensing and signaling systems. In the context of Mg^{2+} homeostasis regulation, we were interested in investigating whether cellular effects caused by a hypomagnesic environment are actively influenced by TRPM7 and its kinase, and not just passively through the suboptimal levels of the biologically essential Mg^{2+} ion. To address this question we characterized biochemically potential substrates of TRPM7 kinase and tested the biological relevance of this substrate–kinase relationship by utilizing genetically manipulated cell lines, which were grown in defined media containing varying concentrations of Mg^{2+}. Using this approach, results from our laboratory showed that PLCγ2 can be phosphorylated by TRPM7 on a conserved serine residue, leading to modulation of BCR signaling strength in accordance with the availability of Mg^{2+}. As described in the previous section, multiple PLC isoforms have been shown to associate with TRPM7 kinase (Deason-Towne et al., 2012; Runnels et al., 2002), but subsequent studies have concentrated on the effect of phosphoinositide metabolism on TRPM7 function and did not investigate the reverse possibility that PLCs could be substrates of TRPM7 kinase (Deason-Towne et al., 2012).

PLC isoforms are multidomain signaling proteins that convert the membrane-bound phosphatidylinositol 4,5-bisphosphate to the second messengers diacylglycerol (DAG) and the soluble Ca^{2+} store mobilizing agent InsP$_3$. PLCγ isozymes are activated through tyrosine phosphorylation via tyrosine-kinase–linked receptors such as the TCR or BCR (Deason-Towne et al., 2012). B lymphocytes rely on PLCγ2 as their main BCR-mediated effector pathway, as highlighted by the phenotype of PLCγ2-deficient mouse models (Wang et al., 2000). Therefore PLC isozymes are at the core of immunoreceptor signaling and have a major influence on immune responses (Koss et al., 2014), but also on a multitude of other biological processes including G-protein–coupled receptor signaling. The modulation of PLC activity through Ser/Thr phosphorylation is very poorly characterized. Thus the finding that TRPM7 is involved in the Mg^{2+}-dependent modulation of PLCγ2 via Ser/Thr phosphorylation could have broader implications. To our knowledge this also represents the first molecular signaling mechanism shown to mediate the well-documented Mg^{2+} sensitivity of immune responses.

In addition, TRPM7 might affect receptor signaling through its impact on cytoskeleton regulation. It was demonstrated by several groups that TRPM7 is implicated in actomyosin contractility and cell adhesion (Clark et al., 2006), which might be at least partly achieved through phosphorylation of myosin IIA heavy chain by TRPM7 kinase (Clark et al., 2006).

B cell spreading and the formation of signaling BCR microclusters are critical to effective B cell activation (Treanor and Batista, 2010). These processes are dependent on the actin cytoskeleton, and there is a clear link between PLCγ2 activation and actin reorganization.

Dr. Treanor demonstrated using single-molecule tracking techniques that the actin cytoskeleton controls the diffusion dynamics of the BCR, and that simply altering the actin cytoskeleton is sufficient to induce signaling in B cells in the absence of ligand (Treanor et al., 2010). In addition to regulating BCR dynamics, the actin cytoskeleton plays an important role in the organization and signaling competence of ligand-induced receptor clustering (BCR microclusters; Treanor et al., 2010); however, the molecular pathways regulating actin reorganization during B cell activation remain undefined. Consistent with these reports, preliminary results we obtained suggest that the actin cytoskeleton is altered in TRPM7-deficient DT40 B cells upon BCR-induced cell spreading.

The nutritional signaling network is tightly intertwined with immune regulation. For example, a recent report has demonstrated that Ca^{2+} signaling promoting effector functions of tumor-specific T lymphocytes can be enhanced by increasing the levels of the glycolytic metabolite phosphoenolpyruvate (PEP). The survival of melanoma-bearing mice could be ameliorated by T cells overexpressing an enzyme producing PEP (Ho et al., 2015). Numerous studies in the past decade have highlighted the role of the master regulator of nutritional signaling mTOR in immune cell functions (Maciolek et al., 2014; Saleiro and Platanias, 2015). We found that eEF2-k, a substrate of mTOR, is also phosphorylated by TRPM7. eEF2-k is the only known kinase and regulator of eukaryotic elongation factor-2 (eEF2), which represents the main point of control of the highly energy-consuming protein translation elongation step. We have shown in B cells that the inhibitory serine phosphorylation of eEF2 is Mg^{2+} sensitive in a TRPM7-dependent manner, providing the first evidence of molecular signaling events underlying the well-characterized decrease in protein translation rate and cell growth in cells deprived of Mg^{2+} (Perraud et al., 2011). This also demonstrates that cells are proactively adjusting their physiology to a decline in Mg^{2+} availability, and not just passively as a result of insufficient Mg^{2+} levels to allow for optimal cellular functions because of defective ribosomes or low levels of MgATP. Thus, it is probable that TRPM7 impacts the extent and quality of immune functions by ensuring that immune cells respond in accordance with the availability of essential nutrient ions such as Mg^{2+}. Although there is an ongoing discussion about whether cytosolic Mg^{2+} functions as a true second messenger, over the past two decades a growing number of studies in various cellular systems have revealed that Mg^{2+} homeostasis is a very dynamic process with multiple and complex forms of regulation.

The Role of Mg^{2+} in Regulating the Immune Response via TRPM7 and TRPM6

Both channel kinases are expressed in cells of the hematopoietic compartment. As in other organs, TRPM7 appears to be ubiquitously represented in the immune compartment (Chokshi et al., 2012; Kerschbaum et al., 2003; Nadler et al., 2001), whereas the presence of TRPM6 is much more restricted. TRPM6 transcripts were discovered in human monocytes where it was upregulated during states of oxidative stress (Wuensch et al., 2010). In human peripheral blood lymphocytes, TRPM6 transcripts were detected in B but not in T cells (Wenning et al., 2011). On the basis of the expression profile of both channels and their ability to build TRPM6/7 heteromers, it is likely that both channels function in concert in regulating cellular Mg^{2+} homeostasis and signaling events in immune cells.

As mentioned before, TRPM7 DT40 knockout cells are hypomagnesic, and they become quiescent and eventually die, unless their media is supplemented with 5–10mM Mg^{2+}. Complementation studies with human TRPM7, wild-type, or a kinase-dead (K1648R) mutant have demonstrated that kinase-dead TRPM7 can restore cell growth and normal intracellular cellular Mg^{2+} levels under physiological Mg^{2+} cell culture conditions (1 mM) as efficiently as wild-type hTRPM7 (Schmitz et al., 2003). This is further supported by a recent report demonstrating that mice expressing a kinase-dead version of TRPM7 do not show developmental impairment and have normal levels of serum Mg^{2+}. Given our own findings in DT40 cells, it will be interesting to evaluate the effect of this kinase-dead mutant in vivo when the animals are fed a hypomagnesic diet (Kaitsuka et al., 2014). A different study reported that TRPM7$^{-/-}$ DT40 B cells exhibit downregulation of molecules involved in the phosphoinositide-3 kinase/Akt/mTOR signaling network, resulting in cell cycle arrest when cells were not supplemented with additional Mg^{2+}. On the basis of these results, Sahni et al. concluded that cells require Mg^{2+} for initiation of cell cycle reentry, which is dependent on TRPM7, but that can be provided through Mg^{2+} supplementation when TRPM7 is missing. This study provided the first insights into the role of Mg^{2+} and TRPM7 in cell division and B lymphocyte activation mediated via TRPM7 (Sahni and Scharenberg, 2008).

It is interesting to note that the controlled overexpression of transfected MagT1 in TRPM7-deficient DT40 B lymphocytes partially rescues the growth and the maintenance of intracellular levels of Mg^{2+}. Thus MagT1 and TRPM7 appear to play an important role in the regulation of intracellular Mg^{2+} levels and cell function (Deason-Towne et al., 2011).

David Clapham's group generated a mouse model with conditional, Cre-inducible Trpm7 deletion and produced a mouse with T-cell–restricted TRPM7 deficiency. These animals exhibit profoundly impaired T cell development, characterized by a block of thymocytes at the double-negative stage and a depletion of thymic medullary cells. It is surprising to note that no alteration in Ca^{2+} or Mg^{2+} homoeostasis (or in any other ions) was documented that could explain this strong phenotype (Jin et al., 2008). Ubiquitous deletion of TRPM7 in these mice resulted in lethality at embryonic day 7.5, a phenotype comparable to a constitutive $TRPM7^{-/-}$ mouse previously described (Ryazanova et al., 2010).

A few studies have begun to investigate the role of TRPM7 in innate immune functions. After the exposure of bone-marrow–derived macrophages (BMDMs) to IL-4 to induce polarization into antiinflammatory type-e macrophages, the current density of TRPM7-like currents was found to be strongly increased although mRNA levels did not change. Compounds known to block TRPM7 activity (NS8593 and FTY720) completely inhibited proliferation of these macrophages, which was not caused by an increase in apoptotic or necrotic cell death. The ability of BMDMs to differentiate into M2-type macrophages and of these cells to proliferate and to produce cytokines was also strongly impaired by pharmacological inhibitors of TRPM7 (Schilling et al., 2014). It will be interesting to further analyze macrophage function in vivo in genetic models with cell-type–restricted deletion of TRPM7 and of its kinase activity. In human neutrophils, TRPM7 was described to participate in the Ca^{2+}-dependent chemotaxis and in the increase in adhesion and invasiveness that is induced by CD147 signaling, a pathway also relevant in the context of rheumatoid arthritis (Wang et al., 2014).

The involvement of TRPM7 in modulating innate immunity is only starting to emerge. Because hypomagnesia is known to elicit a hyperinflammatory environment, future studies will need to determine how TRPM7 and its kinase might contribute to promote inflammation, which has relevance to a multitude of pathologies.

CONCLUSIONS

Recent advances in the field of Mg^{2+} research have led to the identification of unique molecules and signaling pathways (Fig. 26.1). The finding and primary characterization of these new proteins (Fig. 26.1 and Table 26.1) allowed us to gain insights into unexpected molecular mechanisms underlying the regulation of cellular and organism-wide Mg^{2+} homoeostasis. In the context of immune system functions, the acquired knowledge about the transport of Mg^{2+} and the role of this essential ion in signaling in immune cells holds the promise to help us understand its role in the development of many diseases such as immunodeficiencies and cancer.

ACKNOWLEDGMENT

This publication was supported by National Institutes of Health grants 5R21AI088421 (National Institute of Allergy and Infectious Diseases) and 5R01GM090123 (National Institute of General Medical Sciences, Office of Dietary Supplements) to A.L.P. and C.S.

REFERENCES

Brandao, K., Deason-Towne, F., Perraud, A.L., Schmitz, C., 2013. The role of Mg^{2+} in immune cells. Immunol Res 55, 261–269.

Brandao, K., Deason-Towne, F., Zhao, X., Perraud, A.L., Schmitz, C., 2014. TRPM6 kinase activity regulates TRPM7 trafficking and inhibits cellular growth under hypomagnesic conditions. Cell Mol Life Sci 71 (24), 4853–4867.

Buerstedde, J.M., Takeda, S., 1991. Increased ratio of targeted to random integration after transfection of chicken B cell lines. Cell 67, 179–188.

Bussiere, F.I., Gueux, E., Rock, E., Girardeau, J.P., Tridon, A., Mazur, A., Rayssiguier, Y., 2002a. Increased phagocytosis and production of reactive oxygen species by neutrophils during magnesium deficiency in rats and inhibition by high magnesium concentration. Br. J. Nutr. 87, 107–113.

Bussiere, F.I., Gueux, E., Rock, E., Mazur, A., Rayssiguier, Y., 2002b. Protective effect of calcium deficiency on the inflammatory response in magnesium-deficient rats. Eur. J. Nutr. 41, 197–202.

Chaigne-Delalande, B., Li, F.Y., O'Connor, G.M., Lukacs, M.J., Jiang, P., Zheng, L., Shatzer, A., Biancalana, M., Pittaluga, S., Matthews, H.F., Jancel, T.J., Bleesing, J.J., Marsh, R.A., Kuijpers, T.W., Nichols, K.E., Lucas, C.L., Nagpal, S., Mehmet, H., Su, H.C., Cohen, J.I., Uzel, G., Lenardo, M.J., 2013. Mg^{2+} regulates cytotoxic functions of NK and CD8 T cells in chronic EBV infection through NKG2D. Science 341 (6142), 186–191.

Chernow, B., 1989. Hypomagnesemia in intensive care. Correction of units. Chest 95, 1362.

Chien, M.M., Cambier, J.C., 1990. Divalent cation regulation of phosphoinositide metabolism. Naturally occurring B lymphoblasts contain a Mg(2+)-regulated phosphatidylinositol-specific phospholipase C. J. Biol. Chem. 265, 9201–9207.

Chokshi, R., Matsushita, M., Kozak, J.A., 2012. Detailed examination of Mg^{2+} and pH sensitivity of human TRPM7 channels. Am. J. Physiol. Cell Physiol.

Clark, K., Langeslag, M., van Leeuwen, B., Ran, L., Ryazanov, A.G., Figdor, C.G., Moolenaar, W.H., Jalink, K., van Leeuwen, F.N., 2006. TRPM7, a novel regulator of actomyosin contractility and cell adhesion. EMBO J. 25 (2), 290–301.

Clark, K., Middelbeek, J., Morrice, N.A., Figdor, C.G., Lasonder, E., van Leeuwen, F.N., 2008. Massive autophosphorylation of the Ser/Thr-rich domain controls protein kinase activity of TRPM6 and TRPM7. PLoS One 3 (3), e1876.

Dai, Q., Shrubsole, M.J., Ness, R.M., Schlundt, D., Cai, Q., Smalley, W.E., Li, M., Shyr, Y., Zheng, W., 2007. The relation of magnesium and calcium intakes and a genetic polymorphism in the magnesium transporter to colorectal neoplasia risk. Am. J. Clin. Nutr. 86, 743–751.

Dalmas, O., Sandtner, W., Medovoy, D., Frezza, L., Bezanilla, F., Perozo, E., 2012. A repulsion mechanism explains magnesium permeation and selectivity in CorA. Proc. Natl. Acad. Sci. U. S. A. 111, 3002–3007.

de Baaij, J.H., Hoenderop, J.G., Bindels, R.J., 2014. Magnesium in man: implications for health and disease. Physiol. Rev. 95, 1–46.

Deason-Towne, F., Perraud, A.L., Schmitz, C., 2011. The Mg(2+) transporter MagT1 partially rescues cell growth and Mg(2+) uptake in cells lacking the channel-kinase TRPM7. FEBS Lett. 585 (14), 2275–2278.

Deason-Towne, F., Perraud, A.L., Schmitz, C., 2012. Identification of Ser/Thr phosphorylation sites in the C2-domain of phospholipase C gamma2 (PLC-gamma2) using TRPM7-kinase. Cell Signal 24 (11), 2070–2075.

Dorovkov, M.V., Ryazanov, A.G., 2004. Phosphorylation of annexin I by TRPM7 channel-kinase. J. Biol. Chem. 279 (49), 50643–50646.

Duncan, L.M., Deeds, J., Hunter, J., Shao, J., Holmgren, L.M., Woolf, E.A., Tepper, R.I., Shyjan, A.W., 1998. Down-regulation of the novel gene melastatin correlates with potential for melanoma metastasis. Cancer Res. 58, 1515–1520.

Escuela, M.P., Guerra, M., Anon, J.M., Martinez-Vizcaino, V., Zapatero, M.D., Garcia-Jalon, A., Celaya, S., 2005. Total and ionized serum magnesium in critically ill patients. Intensive Care Med. 31, 151–156.

Fleig, A., Chubanov, V., 2014. Trpm7. Handb. Exp. Pharmacol. 222, 521–546.

Goytain, A., Quamme, G.A., 2005. Functional characterization of the mouse [corrected] solute carrier, SLC41A2. Biochem. Biophys. Res. Commun. 330 (3), 701–705.

Hattori, M., Tanaka, Y., Fukai, S., Ishitani, R., Nureki, O., 2007. Crystal structure of the MgtE Mg^{2+} transporter. Nature 448, 1072–1075.

Hermosura, M.C., Nayakanti, H., Dorovkov, M.V., Calderon, F.R., Ryazanov, A.G., Haymer, D.S., Garruto, R.M., 2005. A TRPM7 variant shows altered sensitivity to magnesium that may contribute to the pathogenesis of two Guamanian neurodegenerative disorders. Proc. Natl. Acad. Sci. U. S. A. 102, 11510–11515.

Ho, P.C., Bihuniak, J.D., Macintyre, A.N., Staron, M., Liu, X., Amezquita, R., Tsui, Y.C., Cui, G., Micevic, G., Perales, J.C., Kleinstein, S.H., Abel, E.D., Insogna, K.L., Feske, S., Locasale, J.W., Bosenberg, M.W., Rathmell, J.C., Kaech, S.M., 2015. Phosphoenolpyruvate is a metabolic checkpoint of anti-tumor T cell responses. Cell 162, 1217–1228.

Hofmann, T., Chubanov, V., Chen, X., Dietz, A.S., Gudermann, T., Montell, C., 2010. Drosophila TRPM channel is essential for the control of extracellular magnesium levels. PLoS One 5, e10519.

Jia, X., Baonan, S., Jianyang, D., Wenzhong, Y., Hsiang-Chin, C., Overton, J.D., Runnels, L.W., Yue, L., 2011. Phosphatidylinositol 4,5-bisphosphate (PIP2) controls magnesium gatekeeper TRPM6 activity. Sci. Rep. 1, 146.

Jiang, X., Newell, E.W., Schlichter, L.C., 2003. Regulation of a TRPM7-like current in rat brain microglia. J. Biol. Chem. 278 (44), 42867–42876.

Jin, J., Desai, B.N., Navarro, B., Donovan, A., Andrews, N.C., Clapham, D.E., 2008. Deletion of Trpm7 disrupts embryonic development and thymopoiesis without altering Mg^{2+} homeostasis. Science 322 (5902), 756–760.

Kaitsuka, T., Katagiri, C., Beesetty, P., Nakamura, K., Hourani, S., Tomizawa, K., Kozak, J.A., Matsushita, M., 2014. Inactivation of TRPM7 kinase activity does not impair its channel function in mice. Sci. Rep. 4, 5718.

Kerschbaum, H.H., Kozak, J.A., Cahalan, M.D., 2003. Polyvalent cations as permeant probes of MIC and TRPM7 pores. Biophys. J. 84, 2293–2305.

Kolisek, M., Launay, P., Beck, A., Sponder, G., Serafini, N., Brenkus, M., Froschauer, E.M., Martens, H., Fleig, A., Schweigel, M., 2008. SLC41A1 is a novel mammalian Mg^{2+} carrier. J. Biol. Chem. 283 (23), 16235–16247.

Koss, H., Bunney, T.D., Behjati, S., Katan, M., 2014. Dysfunction of phospholipase Cgamma in immune disorders and cancer. Trends Biochem. Sci. 39, 603–611.

Kozak, J.A., Cahalan, M.D., 2003. MIC channels are inhibited by internal divalent cations but not ATP. Biophys. J. 84 (2 pt 1), 922–927.

Kraeuter, S.L., Schwartz, R., 1980. Blood and mast cell histamine levels in magnesium-deficient rats. J. Nutr. 110, 851–858.

Krapivinsky, G., Krapivinsky, L., Manasian, Y., Clapham, D.E., 2014. The TRPM7 chanzyme is cleaved to release a chromatin-modifying kinase. Cell 157 (5), 1061–1072.

Kurosaki, T., Maeda, A., Ishiai, M., Hashimoto, A., Inabe, K., Takata, M., 2000. Regulation of the phospholipase C-gamma2 pathway in B cells. Immunol. Rev. 176, 19–29.

Langeslag, M., Clark, K., Moolenaar, W.H., van Leeuwen, F.N., Jalink, K., 2007. Activation of TRPM7 channels by phospholipase C-coupled receptor agonists. J. Biol. Chem. 282 (1), 232–239.

Li, F.Y., Chaigne-Delalande, B., Kanellopoulou, C., Davis, J.C., Matthews, H.F., Douek, D.C., Cohen, J.I., Uzel, G., Su, H.C., Lenardo, M.J., 2011. Second messenger role for Mg^{2+} revealed by human T-cell immunodeficiency. Nature 475 (7357), 471–476.

Long, S., Romani, A.M., 2015. Role of cellular magnesium in human diseases. Austin J. Nutr. Food Sci. 2.

Maciolek, J.A., Pasternak, J.A., Wilson, H.L., 2014. Metabolism of activated T lymphocytes. Curr. Opin. Immunol. 27, 60–74.

Malpuech-Brugere, C., Nowacki, W., Daveau, M., Gueux, E., Linard, C., Rock, E., Lebreton, J., Mazur, A., Rayssiguier, Y., 2000. Inflammatory response following acute magnesium deficiency in the rat. Biochim. Biophys. Acta 1501, 91–98.

Mandel, G., Goodman, R.H., 1999. Cell signalling. DREAM on without calcium. Nature 398, 29–30.

Mandt, T., Song, Y., Scharenberg, A.M., Sahni, J., 2011. SLC41A1 Mg(2+) transport is regulated via Mg(2+)-dependent endosomal recycling through its N-terminal cytoplasmic domain. Biochem. J. 439 (1), 129–139.

Markham, G.D., Glusker, J.P., Bock, C.W., et al., 2002. The arrangement of first- and second-sphere water molecules in divalent magnesium complexes: results from molecular orbital and density functional theory and from structural crystallography. J. Phys. Chem. B 106, 5118–5134.

Matsushita, M., Kozak, J.A., Shimizu, Y., McLachlin, D.T., Yamaguchi, H., Wei, F.Y., Tomizawa, K., Matsui, H., Chait, B.T., Cahalan, M.D., Nairn, A.C., 2005. Channel function is dissociated from the intrinsic kinase activity and autophosphorylation of TRPM7/ChaK1. J. Biol. Chem. 280 (21), 20793–20803.

Monteilh-Zoller, M.K., Hermosura, M.C., Nadler, M.J., Scharenberg, A.M., Penner, R., Fleig, A., 2003. TRPM7 provides an ion channel mechanism for cellular entry of trace metal ions. J. Gen. Physiol. 121 (1), 49–60.

Moomaw, A.S., Maguire, M.E., 2008. The unique nature of Mg^{2+} channels. Physiol. (Bethesda) 23, 275–285.

Nadler, M.J., Hermosura, M.C., Inabe, K., Perraud, A.L., Zhu, Q., Stokes, A.J., Kurosaki, T., Kinet, J.P., Penner, R., Scharenberg, A.M., Fleig, A., 2001. LTRPC7 is a Mg·ATP-regulated divalent cation channel required for cell viability. Nature 411 (6837), 590–595.

Naser, S.A., Abdelsalam, A., Thanigachalam, S., Naser, A.S., Alcedo, K., 2014. Domino effect of hypomagnesemia on the innate immunity of Crohn's disease patients. World J. Diabetes 5, 527–535.

Paravicini, T.M., Chubanov, V., Gudermann, T., 2012. TRPM7: a unique channel involved in magnesium homeostasis. Int. J. Biochem. Cell Biol. 44, 1381–1384.

Perraud, A.L., Knowles, H.M., Schmitz, C., 2004. Novel aspects of signaling and ion-homeostasis regulation in immunocytes. The TRPM ion channels and their potential role in modulating the immune response. Mol. Immunol. 41 (6-7), 657–673.

Perraud, A.L., Zhao, X., Ryazanov, A.G., Schmitz, C., 2011. The channel-kinase TRPM7 regulates phosphorylation of the translational factor eEF2 via eEF2-k. Cell Signal 23 (3), 586–593.

Petrault, I., Zimowska, W., Mathieu, J., Bayle, D., Rock, E., Favier, A., Rayssiguier, Y., Mazur, A., 2002. Changes in gene expression in rat thymocytes identified by cDNA array support the occurrence of oxidative stress in early magnesium deficiency. Biochim. Biophys. Acta 1586, 92–98.

Prakriya, M., Lewis, R.S., 2002. Separation and characterization of currents through store-operated CRAC channels and Mg^{2+}-inhibited cation (MIC) channels. J. Gen. Physiol. 119 (5), 487–507.

Quamme, G.A., 2010. Molecular identification of ancient and modern mammalian magnesium transporters. Am. J. Physiol. Cell Physiol. 298, C407–C429.

Rayssiguier, Y., Libako, P., Nowacki, W., Rock, E., 2002. Magnesium deficiency and metabolic syndrome: stress and inflammation may reflect calcium activation. Magnes. Res. 23, 73–80.

Rijkers, G.T., Griffioen, A.W., 1993. Changes in free cytoplasmic magnesium following activation of human lymphocytes. Biochem. J. 289, 373–377.

Rijkers, G.T., Henriquez, N., Griffioen, A.W., 1993. Intracellular magnesium movements and lymphocyte activation. Magnes. Res. 6, 205–213.

Romani, A., Marfella, C., Scarpa, A., 1993. Cell magnesium transport and homeostasis: role of intracellular compartments. Min. Electrolyte Metab. 19, 282–289.

Romani, A.M., Scarpa, A., 2000. Regulation of cellular magnesium. Front. Biosci. 5, D720–D734.

Rubin, H., 2005. Magnesium: the missing element in molecular views of cell proliferation control. Bioessays 27, 311–320.

Runnels, L.W., Yue, L., Clapham, D.E., 2001. TRP-PLIK, a bifunctional protein with kinase and ion channel activities. Science 291 (5506), 1043–1047.

Runnels, L.W., Yue, L., Clapham, D.E., 2002. The TRPM7 channel is inactivated by PIP(2) hydrolysis. Nat. Cell Biol. 4 (5), 329–336.

Ryazanov, A.G., 2002. Elongation factor-2 kinase and its newly discovered relatives. FEBS Lett. 514 (1), 26–29.

Ryazanova, L.V., Rondon, L.J., Zierler, S., Hu, Z., Galli, J., Yamaguchi, T.P., Mazur, A., Fleig, A., Ryazanov, A.G., 2010. TRPM7 is essential for Mg(2+) homeostasis in mammals. Nat. Commun. 1, 109.

Sahni, J., Nelson, B., Scharenberg, A.M., 2007. SLC41A2 encodes a plasma-membrane Mg^{2+} transporter. Biochem. J. 401, 505–513.

Sahni, J., Scharenberg, A.M., 2008. TRPM7 ion channels are required for sustained phosphoinositide 3-kinase signaling in lymphocytes. Cell Metab. 8, 84–93.

Saleiro, D., Platanias, L.C., 2015. Intersection of mTOR and STAT signaling in immunity. Trends Immunol. 36, 21–29.

Schilling, T., Miralles, F., Eder, C., 2014. TRPM7 regulates proliferation and polarisation of macrophages. J. Cell Sci. 127 (21), 4561–4566.

Schlingmann, K.P., Weber, S., Peters, M., Niemann Nejsum, L., Vitzthum, H., Klingel, K., Kratz, M., Haddad, E., Ristoff, E., Dinour, D., Syrrou, M., Nielsen, S., Sassen, M., Waldegger, S., Seyberth, H.W., Konrad, M., 2002. Hypomagnesemia with secondary hypocalcemia is caused by mutations in TRPM6, a new member of the TRPM gene family. Nat. Genet. 31 (2), 166–170.

Schmitz, C., Dorovkov, M.V., Zhao, X., Davenport, B.J., Ryazanov, A.G., Perraud, A.L., 2005. The channel kinases TRPM6 and TRPM7 are functionally nonredundant. J. Biol. Chem. 280 (45), 37763–37771.

Schmitz, C., Perraud, A.L., Johnson, C.O., Inabe, K., Smith, M.K., Penner, R., Kurosaki, T., Fleig, A., Scharenberg, A.M., 2003. Regulation of vertebrate cellular Mg^{2+} homeostasis by TRPM7. Cell 114 (2), 191–200.

Sponder, G., Rutschmann, K., Kolisek, M., 2014. "Inside-in" or "inside-out"? the membrane topology of SLC41A1. Magnes. Res. 26 (4), 176–181.

Su, N.Y., Peng, T.C., Tsai, P.S., Huang, C.J., 2013. Phosphoinositide 3-kinase/Akt pathway is involved in mediating the anti-inflammation effects of magnesium sulfate. J. Surg. Res. 185, 726–732.

Sugimoto, J., Romani, A.M., Valentin-Torres, A.M., Luciano, A.A., Ramirez Kitchen, C.M., Funderburg, N., Mesiano, S., Bernstein, H.B., 2012. Magnesium decreases inflammatory cytokine production: a novel innate immunomodulatory mechanism. J. Immunol. 188, 6338–6346.

Takeda, H., Hattori, M., Nishizawa, T., Yamashita, K., Shah, S.T., Caffrey, M., Maturana, A.D., Ishitani, R., Nureki, O., 2005. Structural basis for ion selectivity revealed by high-resolution crystal structure of Mg^{2+} channel MgtE. Nat. Commun. 5, 5374.

Takemoto, S., Yamamoto, A., Tomonaga, S., Funaba, M., Matsui, T., 2014. Magnesium deficiency induces the emergence of mast cells in the liver of rats. J. Nutr. Sci. Vitaminol. (Tokyo) 59, 560–563.

Takezawa, R., Schmitz, C., Demeuse, P., Scharenberg, A.M., Penner, R., Fleig, A., 2004. Receptor-mediated regulation of the TRPM7 channel through its endogenous protein kinase domain. Proc. Natl. Acad. Sci. U. S. A. 101 (16), 6009–6014.

Treanor, B., Batista, F.D., 2010. Organisation and dynamics of antigen receptors: implications for lymphocyte signalling. Curr. Opin. Immunol. 22, 299–307.

Treanor, B., Depoil, D., Gonzalez-Granja, A., Barral, P., Weber, M., Dushek, O., Bruckbauer, A., Batista, F.D., 2010. The membrane skeleton controls diffusion dynamics and signaling through the B cell receptor. Immunity 32, 187–199.

van der Wijst, J., Hoenderop, J.G., Bindels, R.J., 2009. Epithelial Mg^{2+} channel TRPM6: insight into the molecular regulation. Magnes. Res. 22 (3), 127–132.

Veilleux, R., 1975. Mast cell increase in the duodenum and kidney of magnesium-deficient rats. Lab. Invest. 33, 80–87.

Wabakken, T., Rian, E., Kveine, M., Aasheim, H.C., 2003. The human solute carrier SLC41A1 belongs to a novel eukaryotic subfamily with homology to prokaryotic MgtE Mg^{2+} transporters. Biochem. Biophys. Res. Commun. 306 (3), 718–724.

Walder, R.Y., Landau, D., Meyer, P., Shalev, H., Tsolia, M., Borochowitz, Z., Boettger, M.B., Beck, G.E., Englehardt, R.K., Carmi, R., Sheffield, V.C., 2002. Mutation of TRPM6 causes familial hypomagnesemia with secondary hypocalcemia. Nat. Genet. 31 (2), 171–174.

Wang, C.H., Rong, M.Y., Wang, L., Ren, Z., Chen, L.N., Jia, J.F., Li, X.Y., Wu, Z.B., Chen, Z.N., Zhu, P., 2014. CD147 up-regulates calcium-induced chemotaxis, adhesion ability and invasiveness of human neutrophils via a TRPM-7-mediated mechanism. Rheumatol. Oxf. 53 (12), 2288–2296.

Wang, D., Feng, J., Wen, R., Marine, J.C., Sangster, M.Y., Parganas, E., Hoffmeyer, A., Jackson, C.W., Cleveland, J.L., Murray, P.J., Ihle, J.N., 2000. Phospholipase Cgamma2 is essential in the functions of B cell and several Fc receptors. Immunity 13, 25–35.

Wei, C., Wang, X., Chen, M., Ouyang, K., Song, L.S., Cheng, H., 2009. Calcium flickers steer cell migration. Nature 457, 901–905.

Wenning, A.S., Neblung, K., Strauss, B., Wolfs, M.J., Sappok, A., Hoth, M., Schwarz, E.C., 2011. TRP expression pattern and the functional importance of TRPC3 in primary human T-cells. Biochim. Biophys. Acta 1813, 412–423.

Wolf, F.I., Torsello, A., Fasanella, S., Cittadini, A., 2003. Cell physiology of magnesium. Mol. Asp. Med. 24, 11–26.

Wolf, F.I., Trapani, V., 2008. Magnesium and its transporters in cancer: a novel paradigm in tumour development. Clin. Sci. (Lond) 123, 417–427.

Wuensch, T., Thilo, F., Krueger, K., Scholze, A., Ristow, M., Tepel, M., 2010. High glucose-induced oxidative stress increases transient receptor potential channel expression in human monocytes. Diabetes 59, 844–849.

Yamaguchi, H., Matsushita, M., Nairn, A.C., Kuriyan, J., 2001. Crystal structure of the atypical protein kinase domain of a TRP channel with phosphotransferase activity. Mol. Cell 7 (5), 1047–1057.

Zimowska, W., Girardeau, J.P., Kuryszko, J., Bayle, D., Rayssiguier, Y., Mazur, A., 2002. Morphological and immune response alterations in the intestinal mucosa of the mouse after short periods on a low-magnesium diet. Br. J. Nutr. 88, 515–522.

Chapter 27

Magnesium Intake and Chronic Disease in Humans

Ka He, Cari Lewis Tsinovoi

Indiana University, Bloomington, IN, United States

INTRODUCTION

Magnesium is the fourth most abundant mineral and it plays a role in over 300 metabolic reactions (Elin, 1994), related to reproduction; synthesis of DNA, RNA, and protein; production and storage of cellular energy; and the stabilization of mitochondrial membranes (Newhouse and Finstad, 2000). Additionally, magnesium is pivotal in the maintenance of normal heart rhythms, muscular contraction, muscle and nerve function, blood pressure control, neuromuscular conduction, glucose and insulin metabolism, and bone integrity (Paolisso and Barbagallo, 1997; Newhouse and Finstad, 2000). Because magnesium is not synthesized in the human body, it must be obtained via diet or supplementation. Even with the various dietary sources of magnesium, including green leafy vegetables, legumes, nuts, and whole grains, a survey of a representative sample of the US population indicates that the average magnesium intake is below the daily reference intake (Ford and Mokdad, 2003) and only half of Americans are achieving the recommended daily amount (Rosanoff et al., 2012). This may be because magnesium tends to be lost substantially during the cooking process or during food preparation. For these reasons, magnesium deficiency is not uncommon and has been implicated in a variety of chronic diseases and health conditions, including diabetes (Dong et al., 2011), metabolic syndrome (Dibaba et al., 2014a,b), hypertension (Huitrón-Bravo et al., 2015), cardiovascular disease (Larsson et al., 2012), and various cancers (Castiglioni and Maier, 2011). Furthermore, new evidence suggests that magnesium supplementation may even be an alternative therapy against diabetes (Guerrero-Romero et al., 2015), certain aspects of metabolic syndrome (Rodríguez-Moran and Guerrero-Romero, 2014), hypertension (Kass et al., 2012), cardiac arrhythmias (Gu et al., 2012), ischemic stroke (Saver et al., 2004), and atherosclerosis (Mortazavi et al., 2013). Finally, a review of the literature indicates that synergistic and competitive interactions with calcium and potassium should be taken into consideration when addressing potential associations between magnesium and chronic disease (Hardwick et al., 1991; Sellmeyer et al., 2002).

MAGNESIUM AND DIABETES MELLITUS

The important role of diet in relation to the risk of diabetes has been recognized. Magnesium, as an essential cofactor for numerous enzymes involved in glucose metabolism (Belin and He, 2007), has been hypothesized to provide beneficial effects on diabetes development, though pathophysiological mechanisms that are not fully understood. Experimental and metabolic studies suggest that intracellular magnesium deficiency may result in disorders of tyrosine kinase activity during insulin signaling and glucose-induced insulin secretion, leading to impaired insulin sensitivity in muscle cells and adipocytes (Suarez et al., 1995; Kandeel et al., 1996). In animal studies, magnesium supplementation has been shown to protect against fructose-induced insulin resistance (Balon et al., 1994) and reduce the development of diabetes (Balon et al., 1995). Also, evidence from human studies suggests that magnesium supplementation can help with glucose control in patients with type 2 diabetes (Song et al., 2006) and improve insulin sensitivity in nondiabetic individuals (Guerrero-Romero et al., 2004). In addition, studies indicate that magnesium intake is significantly and inversely associated with inflammation (Dibaba et al., 2014a,b), which is a risk factor of diabetes.

A number of epidemiological studies have examined magnesium intake in relation to risk of diabetes. A meta-analysis summarized data from 13 prospective cohort studies involving 536,318 individuals with 24,516 diabetic patients found an overall inverse association between magnesium intake and risk of diabetes (Dong et al., 2011). Compared to those in the lowest magnesium intake group, the risk of diabetes for individuals in the highest intake group was reduced by 22% [pooled relative risk (RR): 0.78, 95%; confidence interval (CI): 0.73–0.84]. For every 100mg/day increment in magnesium intake, the risk of diabetes was

Molecular, Genetic, and Nutritional Aspects of Major and Trace Minerals. http://dx.doi.org/10.1016/B978-0-12-802168-2.00027-0

lowered by 14% (RR: 0.86, 95%; CI: 0.82–0.89). This inverse association was not appreciably modified by geographic region, follow-up length, sex, or family history of diabetes. Findings from this meta-analysis are supported by a consortium study combining data from 15 studies including 52,684 participants of European descent, which found higher magnesium intake was associated with lower fasting glucose and insulin (Hruby et al., 2013). After adjustment for age, sex, total calorie intake, body mass index, and behavioral risk factors, higher magnesium intake was associated with lower fasting glucose ($\beta = -0.009$ mmol/L, 95%; CI: -0.013, -0.005, $P < .0001$) and insulin ($\beta = -0.020$ ln-pmol/L, 95%; CI: -0.024, -0.017, $P < .0001$).

Randomized clinical trials have been conducted to examine the effects of magnesium supplements on glucose metabolism, glycemic status, insulin sensitivity, and inflammation in various populations. The results are generally in concordance with findings from observational studies. For example, a double-blind placebo-controlled randomized trial in 116 men and women with prediabetes and hypomagnesemia reported that magnesium supplementation reduced plasma glucose levels and improved the glycemic status (Guerrero-Romero et al., 2015). In a similar clinical trial, magnesium supplementation was found to reduce the level of C-reactive protein (Simental-Mendia et al., 2014). Although the long-term effects of magnesium supplementation on glycemic control remain unclear, evidence from randomized double-blind controlled trials indicate that magnesium supplementation for 4–16 weeks may be effective in reducing plasma-fasting glucose levels in type 2 diabetic patients (Song et al., 2006).

In sum, a body of evidence from experimental studies suggests an essential role of magnesium in glucose metabolism and insulin sensitivity, which creates a plausible biological foundation for the hypothesis that magnesium intake may reduce the risk of diabetes. Accumulated evidence from epidemiological studies and randomized clinical trials provides strong evidence in support of the beneficial effects of magnesium intake on diabetes development among nondiabetic individuals or glycemic control in diabetic patients.

MAGNESIUM AND METABOLIC SYNDROME

The metabolic syndrome is a cluster of metabolic risk factors: abdominal obesity, hyperglycemia and/or hyperinsulinemia, dyslipidemia [high triglyceride and low high-density lipoprotein (HDL) cholesterol levels], and elevated blood pressure (Expert Panel on Detection, Evaluation, and Treatment of High Blood Cholesterol in Adults, 2001). This syndrome has become a worldwide health disorder that leads to increased risk of cardiovascular disease and diabetes mellitus (Eckel et al., 2005).

A few possible mechanisms have been proposed to explain the beneficial effects of magnesium intake on metabolic disorders, including improvement of glucose and insulin homeostasis (Paolisso and Ravussin, 1995) and lipid metabolism (Altura et al., 1990), as well as its antihypertensive (Touyz, 2003), antiinflammatory (Dibaba et al., 2014a,b), and antioxidant function (Barbagallo et al., 2003). For example, studies suggest that intracellular magnesium balance is important in maintaining peripheral glucose-mediated insulin action. Magnesium may preserve pancreatic β-cell function through its effects on cellular calcium homeostasis and/or oxidative stress (Barbagallo et al., 2003). Also, magnesium may affect lipid metabolism independent of its effects on insulin homeostasis. As a cofactor of many rate-limiting enzymes critical for lipid metabolism, magnesium has been shown to decrease low-density lipoprotein cholesterol and triglyceride levels and increase HDL cholesterol levels by inhibiting the activity of lecithin cholesterol acyl transferase (Itoh et al., 1997) and 3-hydroxy-3-methylglutaryl coenzyme A reductase, and stimulating lipoprotein lipase activity (Rayssiguier et al., 1991). In addition, it has been suggested that magnesium has multiple functions critical for its antihypertensive effects, including: inhibiting intracellular calcium mobilization as a calcium antagonist (Chakraborti et al., 2002); attenuating the adverse effect of sodium by stimulating activity of the sodium–potassium (Na–K) ATPase and increasing urinary excretion of sodium (Touyz, 2003); decreasing release of catecholamines (Chakraborti et al., 2002); and improving myocardial contractility (Chakraborti et al., 2002), systemic inflammation (Dibaba et al., 2014a,b), and endothelium-dependent vasodilation (Pearson et al., 1998). Moreover, studies indicate that magnesium may be involved in forming soaps with fatty acids in the intestine and may thus reduce the digestible energy content of the diet (Drenick, 1961). Thus, it may benefit weight maintenance.

Although many studies have examined associations between magnesium status and individual components of metabolic syndrome, no study directly related magnesium to the risk of metabolic syndrome until 2002 (Guerrero-Romero and Rodriguez-Moran, 2002). Since then, a number of observational studies on magnesium and metabolic syndrome have been published. In 2014, a meta-analysis quantitatively and systematically assessed the association between dietary magnesium intake and risk of metabolic syndrome (Dibaba et al., 2014a,b). After combining six cross-sectional studies including a total of 24,473 individuals and 6311 cases of metabolic syndrome, a weighted inverse association was observed between dietary magnesium intake and the risk of metabolic syndrome [pooled odds ratio (OR): 0.69, 95%; CI: 0.59–0.81] comparing the highest with the lowest group. For every 100 mg/day increment in magnesium intake, the overall risk of having metabolic syndrome was lowered by 17% (OR: 0.83, 95%; CI: 0.77–0.89). These findings were supported by two meta-analyses published more recently (Ju et al., 2014; La et al., 2015).

Data on magnesium supplementation and metabolic syndrome are sparse. A randomized double-blind placebo-controlled trial among 47 individuals with hypomagnesemia found that magnesium supplementation improves the metabolic profile including blood pressure, fasting glucose, and triglyceride levels (Rodriguez-Moran and Guerrero-Romero, 2014). Moreover, a number of clinical trials have been conducted on magnesium supplementation and individual components of metabolic syndrome including blood pressure (Kass et al., 2012), fasting glucose (Song et al., 2006), and lipid profile (Guerrero-Romero et al., 2015). The results are in concordance with findings from the epidemiological studies on magnesium intake and metabolic syndrome.

While data from longitudinal cohort studies and randomized placebo-controlled trials are limited, the existing literature suggests that dietary magnesium intake is inversely associated with the prevalence of metabolic syndrome. Further studies are needed for definitive conclusion.

MAGNESIUM AND HYPERTENSION

Hypertension is a significant risk factor for cardiovascular disease (Israili et al., 2007), affecting nearly 90% of adults by the age of 65 (Go et al., 2014). Hypertension is also associated with an elevated risk of incidence and mortality of coronary heart failure, coronary heart disease, cerebrovascular events, and end-stage renal disease. Because high blood pressure is often one of the first signs of the aforementioned diseases, it presents an early opportunity to correct lifestyle and, by extension, slow or halt further comorbid complications.

Magnesium works to lower blood pressure by being transported into the smooth muscle cells of the vascular system via transient receptor potential melastatin 6 and 7 cation transporter channels (Touyz, 2008). This action induces vasodilation, causes a rise in prostaglandin E (a platelet inhibitor and vasodilator), reduces intracellular sodium and calcium, cooperatively binds potassium, enhances endothelial dysfunction in hypertensive patients, and reduces blood pressure (McCarty, 1996; Barbagallo et al., 2010). Additionally, magnesium is a cofactor of delta-6-desaturase, an enzyme involved in the conversion of linoleic acid to gamma-linolenic acid. Gamma-linolenic acid eventually elongates to form dihomo-gamma-linoleic acid, a prostaglandin E precursor. Low-serum magnesium status leads to a deficiency in prostaglandin E, resulting in vasoconstriction and an elevated blood pressure (Das, 2010).

Evidence from observational studies indicates an overall inverse association of diets high in magnesium with blood pressure although these results are inconsistent (Sasaki et al., 2000; Bain et al., 2015). For example, a cohort study of an adult Mexican population revealed an inverse trend of diastolic blood pressure with increasing tertiles of dietary magnesium intake (Huitrón-Bravo et al., 2015). The Prevention of Renal and Vascular End-Stage Disease (PREVEND) population-based cohort study with a median follow-up of 7.6 years found a significant inverse relationship between urinary magnesium and risk of hypertension; every reduction of one unit in ln-transformed urinary magnesium resulted in a 21% lower risk of hypertension (Joosten et al., 2013).

Clinical trial data also show a promising therapeutic effect of magnesium on hypertension. In a double-blind placebo-controlled trial of 91 women showing mild to moderate levels of hypertension, taking 485 mg of magnesium aspartate-HCl daily for 6 months resulted in a significant decrease in both systolic and diastolic blood pressure (Witteman et al., 1994). A study of 60 patients by Kawano et al. (1998) demonstrated 20 mmol/day of magnesium oxide over 8 weeks significantly reduced office, home, and ambulatory blood pressure; individuals exhibiting the highest blood pressure at baseline showed the greatest reductions. Furthermore, 48 hypertensive patients given 600 mg/day of magnesium in addition to lifestyle changes showed significant decreases in systolic and diastolic blood pressure versus patients who changed lifestyle only (Hatzistavri et al., 2009). A comparative study of 40 hypertensive versus 40 nonhypertensive controls found that 600 mg/day of magnesium over an 8-week period resulted in a significant decrease in diastolic blood pressure in the hypertensive group (Annapurna et al., 2015).

The overall relationship between magnesium and hypertension seems to be a beneficial one. Combinations of magnesium and potassium may be more efficacious in reducing blood pressure than either alone. Magnesium also enhances the efficacy of antihypertensive drugs when taken concurrently. High intakes of magnesium as part of a healthy diet may help reduce or prevent hypertension and other comorbidities like cardiovascular disease, coronary heart disease, and stroke.

MAGNESIUM AND CARDIOVASCULAR DISEASE

Inadequate dietary magnesium intake and low-serum magnesium levels may contribute to the pathophysiology of cardiovascular diseases, such as myocardial infarction, cardiac arrhythmias, congestive heart failure, ischemic stroke, coronary heart disease, and atherosclerosis. Magnesium is known in cardiac medicine as an inducer of vascular smooth muscle–cell relaxation via its action as a physiological calcium blocker; an increase in extracellular magnesium results in a reduction of intracellular calcium (Mazur et al., 2007). Magnesium also has the ability to increase HDL and lower triglycerides by way of increased lipoprotein lipase activity, as well as by its action on HMGCoA reductase inhibition, an enzyme important for cholesterol synthesis (Rosanoff and Seelig, 2004; Bo and Pisu, 2008). Additionally, magnesium is thought to play a role in inflammation and has exhibited relationships with C-reactive protein (Dibaba et al., 2014a,b), interleukin-6 (IL-6) (Chacko et al., 2010), and tumor necrosis factor-alpha (TNF-α) (Chacko et al., 2010). Together, these mechanisms provide a basis for the notion that magnesium deficiency significantly contributes to cardiovascular disease.

Arrhythmias

The prevention and/or treatment of cardiac arrhythmias are the most widely practiced use of magnesium in cardiovascular medicine. In one study from the Framingham Offspring Study of 3530 participants, low-serum magnesium was associated with atrial fibrillation development in individuals with no history of cardiovascular disease (Khan et al., 2013). Atrial fibrillations tend to occur in 15–50% of patients following cardiac surgery and are associated with stroke, congestive heart failure, and prolonged hospitalizations (Hravnak et al., 2002). In a meta-analysis of randomized controlled trials, only 1 of 14 trials showed a potential benefit of magnesium chloride or magnesium sulfate in the significant reduction of postcardiac surgery atrial fibrillation. It should be noted, however, that the other 13 trials had serious methodological and study attrition limitations (Bradley et al., 2005). Inversely, a variety of other meta-analyses have shown intravenous magnesium to be an efficacious method of preventing postcardiac surgery and rapid onset atrial fibrillation (Henyan et al., 2005; Onalan et al., 2007; Gu et al., 2012). Overall, intravenous magnesium may be an effective therapy for atrial fibrillation and it has the ability to act synergistically with antiarrhythmic medications.

Congestive Heart Failure

Hypomagnesemia is common among patients with congestive heart failure. Patients with heart failure tend to have low-serum magnesium levels (Arinzon et al., 2010), and low-serum magnesium is associated with mortality among heart failure patients (Adamopoulos et al., 2009). In a cohort of 14,709 participants from the Atherosclerosis Risk in Communities (ARIC) Study followed for a median period of 20.6 years, serum magnesium was inversely associated with risk of heart failure (Lutsey et al., 2014). In a clinical trial of 22 heart failure patients, 800 mg daily of oral magnesium over a period of 3 months improved endothelial function and small arterial compliance (Fuentes et al., 2006). This is particularly important since a loss of magnesium in arteries with disrupted endothelium can result in vasoconstriction; this response is reversed when magnesium is replenished (Ku and Ann, 1987).

Ischemic Stroke

Magnesium may be effective in protecting against ischemic stroke. In a meta-analysis of eight prospective studies involving 241,378 participants, Larsson et al. (2012) demonstrated an inverse association between dietary magnesium and ischemic stroke risk. In a 2015 study, Bain et al. (2015) found a relationship between low dietary magnesium intake and stroke risk in men. In the ARIC study of 14,221 participants, low-serum magnesium showed a significant relation to ischemic stroke risk (Ohira et al., 2009). The use of magnesium in the treatment of stroke has yet to be confirmed, however. In a double-blind, placebo-controlled trial of 2589 stroke patients, intravenous magnesium therapy within 12 h of stroke onset was unable to significantly reduce the risk of mortality or disability over a period of 90 days (Muir et al., 2004).

Coronary Heart Disease

Low magnesium is hypothesized to play a role in coronary heart disease. A study by the Honolulu Heart Program based on 7172 men revealed a significant inverse relationship between dietary magnesium and risk of coronary heart disease after 15 years of follow-up (Abbott et al., 2003). The cohort of 39,633 men from the Health Professionals Follow-Up Study demonstrated similar results (Al-Delaimy et al., 2004). In the ARIC study of 13,992 men and women, low-serum magnesium was associated with a significantly greater risk of coronary heart disease (Liao et al., 1998). Dietary magnesium has also exhibited an inverse relationship with mortality from coronary heart disease in Japanese women (Zhang et al., 2012).

Atherosclerosis

Magnesium is considered a vascular calcification inhibitor (Luft, 2010), suggesting that it may be effective in the prevention and treatment of atherosclerosis. In a cross-sectional analysis from 2695 participants of the Framingham Heart Study, a 50 mg/day increment increase in dietary magnesium was associated with 12% lower abdominal aortic calcification and 22% lower coronary artery calcification. Clinical studies also reveal magnesium's potential efficacy as an atherosclerosis therapy. In the ARIC study, both low dietary and serum magnesium were associated with atherosclerosis development and an increase in carotid intima-media thickness (Ma et al., 1995).

Overall, the evidence seems to point to a significant role of magnesium in the prevention and treatment of cardiovascular disease. In the cases of ischemic stroke, congestive heart failure, and coronary heart disease, low dietary and/or serum magnesium is associated with increased risk. Additionally, intravenous magnesium therapy has shown to be an effective treatment for arrhythmias, ischemic stroke, congestive heart failure, and atherosclerosis, often working synergistically with and improving the performance of other standard pharmaceutical therapies.

MAGNESIUM AND CANCER

While age-adjusted cancer mortality rates have been steadily decreasing in the United States, the number of global cancer deaths is predicted to increase by 37% over the next 10 years (Ferlay et al., 2015). Epidemiological research suggests a significant percentage of cancer mortality cases can be attributed to diet (Lundstrom, 2012). While there is a general agreement about the benefits of consuming fruits, vegetables, and whole grains on cancer risk, the western diet is still relatively deficient in the many micronutrients derived from these food sources, including magnesium (Ford and Mokdad, 2003). Overall, the relationship between magnesium and the development of tumors is complex; however, a general consensus points to a prominent role of magnesium deficiency in cancer.

Magnesium is a participant in all major metabolic processes; therefore, it is not surprising that it plays a direct role in cell growth and survival. For example, high magnesium activates the Ser/Thr phosphatase PPM1D. PPM1D performs many roles, but of note is its ability to dephosphorylate and inactivate the p53 tumor suppressor gene, as well as complement many other oncogenes like Ras, HER-2/neu, and Myc (Bernards, 2004). Overall, high intracellular magnesium can create genome instability, insensitivity to antiproliferation signaling, activation of telomerase and unbounded replicative potential, and a reprogramming of cell metabolism (Hanahan and Weinberg, 2011). It is important to note that even though low extracellular magnesium causes neoplastic cells to become refractory toward proliferative inhibition initially, the avidity of the neoplastic cell to take up magnesium leads to an accumulation of high magnesium concentrations within the cell (Wolf et al., 2009). This is why neoplastic cells tend to have high concentrations of magnesium, despite the host having a magnesium deficiency overall. Finally, a link between inflammation and magnesium must also be considered. Increased inflammatory status is associated with a low magnesium status (Nielsen, 2010); the role of inflammation in cancer initiation and progression is well established (Colotta et al., 2009).

Colorectal Cancer

A variety of studies have shown an inverse association between low magnesium intake and colon cancer risk (Weyer et al., 2001; Larsson et al., 2005; Folsom and Hong, 2006; Van den Brandt et al., 2007; Wark et al., 2012). For example, in the Japan Public Health Center-Based Prospective Study, a large prospective study of 40,830 Japanese men and 46,287 women, high magnesium was significantly associated with colon cancer risk (Ma et al., 2010). The relationship between low magnesium and colon cancer has been attributed to the increased formation of carcinogenic N-nitroso compounds (Chiu et al., 2004). A meta-analysis of seven prospective cohort studies found a significant inverse nonlinear association between dietary magnesium and colorectal cancer, particularly when magnesium intake was between 200 and 270 mg/day (Qu et al., 2013).

Magnesium therapy has been employed in colorectal cancer treatment to reduce oxaliplatin-related neurotoxicity with varying results (Grothey et al., 2011; Knijn et al., 2011; Loprinzi et al., 2014). In two double-blind, placebo-controlled clinical trials, intravenous calcium and magnesium significantly reduced the incidence of chronic cumulative sensory neurotoxicity with no disruption of treatment effectiveness (Grothey et al., 2011; Knijn et al., 2011). Conversely, a phase III randomized, double-blind, placebo-controlled study found no neuropathy differences between the placebo and calcium/magnesium groups (Loprinzi et al., 2014). More trials needed before magnesium's role in colorectal cancer treatment can be definitively established.

Lung Cancer

The association between lung cancer and magnesium has shown to be inconsistent. A case-control study found a correlation between low dietary magnesium and increased risk of lung cancer, with a stronger association in those who were current smokers, elderly, drinkers, or in late-stage disease (Mahabir et al., 2008). A similar trend has been demonstrated between low magnesium levels in drinking water and lung cancer risk in women (Cheng et al., 2012). A prospective study of 288,257 men and 194,618 women, however, revealed no significant association between total dietary magnesium and increased lung cancer risk overall; magnesium intake did exhibit a direct association with risk in current smokers and men (Mahabir et al., 2010). This is in opposition to the inverse association seen previously.

Other Cancers

Aside from colorectal and lung cancers, studies between dietary magnesium and other types of cancers are scarce. In drinking water, magnesium seems to exhibit a protective effect against liver and esophageal cancers (Yang et al., 2002; Tukiendorf and Rybak, 2004), as well as prevent mortalities from prostate, ovarian, and breast cancers (Yang et al., 2000a,b; Chiu et al., 2004).

While the evidence for magnesium's role in carcinogenesis is still largely fragmented, most of the available evidence suggests magnesium may be an effective chemopreventive agent. Increasing magnesium intake, particularly in individuals with low magnesium status initially, may be a low-cost, preventive measure for preventing cancer. However, the effectiveness of treating cancer patients with intravenous magnesium therapy has yet to be clearly established. More clinical studies are needed before our understanding of magnesium's role in treatment can be advanced.

MAGNESIUM AND INTERACTIONS WITH OTHER MINERALS

When we discuss the role of magnesium in human metabolic function, a discussion of calcium cannot be neglected since they are equally important in regulation of cellular multiplication in normal and transformed cells (McKeehan and Ham, 1978). Calcium and magnesium have similar chemical properties and share the same homeostatic regulating system (Brown and MacLeod, 2001) involving a calcium-sensing receptor, as well as intestinal absorption and renal excretion (Hardwick et al., 1991). Experimental studies indicate that calcium and magnesium may directly or indirectly compete for intestinal absorption (Hardwick et al., 1991). In addition, studies found that potassium may help decrease calcium excretion, particularly in elderly women (Sellmeyer et al., 2002), and interact with magnesium in the pathogenesis of cardiovascular disease (Sheehan and Seelig, 1984). Moreover, magnesium, potassium, and calcium share common food sources such as dairy products, nuts, and seeds. Thus, while we study magnesium, calcium and potassium should be considered as either potential confounders or effect modifiers.

It has been suggested that the magnesium–calcium ratio may be an independent predictor of health end-points (Dai et al., 2013). The ratio of magnesium to calcium intake varies across populations. Whether the variation of magnesium and calcium ratio partially explains the different prevalence of chronic disease in diverse populations and what is the optimal ratio of magnesium to calcium are largely unknown. Further studies are warranted.

CONCLUSION

This review has indicated an overall role of magnesium deficiency in diabetes mellitus, metabolic syndrome, hypertension, various types of cardiovascular disease, and some cancers. Evidence also suggests an improvement across symptoms of diabetes, metabolic syndrome, hypertension, cardiac arrhythmias, ischemic stroke, and atherosclerosis when magnesium is administered as a supplemental therapy. While the research is not entirely consistent, the literature largely supports a beneficial influence of magnesium on human health and chronic disease. More large prospective studies and clinical trials addressing the dose-dependent effect of magnesium are necessary.

SUMMARY POINTS

- Magnesium is an essential mineral, which is very important for metabolic functions in the human body.
- Major dietary sources of magnesium include whole grains, legumes, nuts, and green leafy vegetables. Magnesium is easily lost through the cooking process.
- Only approximately half of Americans are achieving the recommended daily amount of magnesium.
- There are biologically plausible mechanisms supporting the beneficial effects of magnesium in human health.
- A large body of literature from experimental and epidemiological studies suggests that magnesium intake is inversely associated with the risk of chronic diseases, including diabetes, metabolic syndrome, cardiovascular disease, and cancer.

REFERENCES

Abbott, R.D., Ando, F., Masaki, K.H., Tung, K.-H., Rodriguez, B.L., Petrovitch, H., Yano, K., Curb, J.D., 2003. Dietary magnesium intake and the future risk of coronary heart disease (The Honolulu Heart Program). Am. J. Cardiol. 92 (6), 665–669.

Adamopoulos, C., Pitt, B., Sui, X., Love, T.E., Zannad, F., Ahmed, A., 2009. Low serum magnesium and cardiovascular mortality in chronic heart failure: a propensity-matched study. Int. J. Cardiol. 136 (3), 270–277.

Al-Delaimy, W.K., Rimm, E.B., Willett, W.C., Stampfer, M.J., Hu, F.B., 2004. Magnesium intake and risk of coronary heart disease among men. J. Am. Coll. Nutr. 23 (1), 63–70.

Altura, B.T., Brust, M., Bloom, S., Barbour, R.L., Stempak, J.G., Altura, B.M., 1990. Magnesium dietary intake modulates blood lipid levels and atherogenesis. Proc. Natl. Acad. Sci. U. S. A. 87 (5), 1840–1844.

Annapurna, D., Sujatha, P., Sudha, J., 2015. A study of oral magnesium supplementation in patients with essential hypertension. Hypertension 4 (10), 1591–1597.

Arinzon, Z., Peisakh, A., Schrire, S., Berner, Y.N., 2010. Prevalence of hypomagnesemia (HM) in a geriatric long-term care (LTC) setting. Arch. Gerontol. Geriatr. 51 (1), 36–40.

Bain, L.K., Myint, P.K., Jennings, A., Lentjes, M.A., Luben, R.N., Khaw, K.-T., Wareham, N.J., Welch, A.A., 2015. The relationship between dietary magnesium intake, stroke and its major risk factors, blood pressure and cholesterol, in the EPIC-Norfolk cohort. Int. J. Cardiol. 196, 108–114.

Balon, T.W., Gu, J.L., Tokuyama, Y., Jasman, A.P., Nadler, J.L., 1995. Magnesium supplementation reduces development of diabetes in a rat model of spontaneous NIDDM. Am. J. Physiol. 269 (4 Pt 1), E745–E752.

Balon, T.W., Jasman, A., Scott, S., Meehan, W.P., Rude, R.K., Nadler, J.L., 1994. Dietary magnesium prevents fructose-induced insulin insensitivity in rats. Hypertension 23 (6 Pt 2), 1036–1039.

Barbagallo, M., Dominguez, L.J., Galioto, A., Ferlisi, A., Cani, C., Malfa, L., Pineo, A., Busardo, A., Paolisso, G., 2003. Role of magnesium in insulin action, diabetes and cardio-metabolic syndrome X. Mol. Asp. Med. 24 (1–3), 39–52.

Barbagallo, M., Dominguez, L.J., Galioto, A., Pineo, A., Belvedere, M., 2010. Oral magnesium supplementation improves vascular function in elderly diabetic patients. Magnes. Res. 23 (3), 131–137.

Belin, R.J., He, K., 2007. Magnesium physiology and pathogenic mechanisms that contribute to the development of the metabolic syndrome. Magnes. Res. 20 (2), 107–129.

Bernards, R., 2004. Wip-ing out cancer. Nat. Genet. 36 (4), 319–320.

Bo, S., Pisu, E., 2008. Role of dietary magnesium in cardiovascular disease prevention, insulin sensitivity and diabetes. Curr. Opin. Lipidol. 19 (1), 50–56.

Bradley, D., Creswell, L.L., Hogue, C.W., Epstein, A.E., Prystowsky, E.N., Daoud, E.G., 2005. Pharmacologic prophylaxis: American College of Chest Physicians guidelines for the prevention and management of postoperative atrial fibrillation after cardiac surgery. Chest J. 128 (Suppl. 2), 39S–47S.

Brown, E.M., MacLeod, R.J., 2001. Extracellular calcium sensing and extracellular calcium signaling. Physiol. Rev. 81 (1), 239–297.

Castiglioni, S., Maier, J.A., 2011. Magnesium and cancer: a dangerous liason. Magnes. Res. 24 (3), 92–100.

Chacko, S.A., Song, Y., Nathan, L., Tinker, L., De Boer, I.H., Tylavsky, F., Wallace, R., Liu, S., 2010. Relations of dietary magnesium intake to biomarkers of inflammation and endothelial dysfunction in an ethnically diverse cohort of postmenopausal women. Diabetes Care 33 (2), 304–310.

Chakraborti, S., Chakraborti, T., Mandal, M., Mandal, A., Das, S., Ghosh, S., 2002. Protective role of magnesium in cardiovascular diseases: a review. Mol. Cell. Biochem. 238 (1–2), 163–179.

Cheng, M.-H., Chiu, H.-F., Tsai, S.-S., Chen, C.-C., Yang, C.-Y., 2012. Calcium and magnesium in drinking-water and risk of death from lung cancer in women. Magnes. Res. 25 (3), 112–119.

Chiu, H.F., Chang, C.C., Yang, C.Y., 2004. Magnesium and calcium in drinking water and risk of death from ovarian cancer. Magnes. Res. 17 (1), 28–34.

Colotta, F., Allavena, P., Sica, A., Garlanda, C., Mantovani, A., 2009. Cancer-related inflammation, the seventh hallmark of cancer: links to genetic instability. Carcinogenesis 30 (7), 1073–1081.

Dai, Q., Shu, X.-O., Deng, X., Xiang, Y.-B., Li, H., Yang, G., Shrubsole, M.J., Ji, B., Cai, H., Chow, W.-H., 2013. Modifying effect of calcium/magnesium intake ratio and mortality: a population-based cohort study. BMJ Open 3 (2), 1–14.

Das, U.N., 2010. Δ6 desaturase as the target of the beneficial actions of magnesium. Med. Sci. Monit. 16 (8), LE11–LE12.

Dibaba, D., Xun, P., Fly, A., Yokota, K., He, K., 2014a. Dietary magnesium intake and risk of metabolic syndrome: a meta-analysis. Diabet. Med. 31 (11), 1301–1309.

Dibaba, D.T., Xun, P., He, K., 2014b. Dietary magnesium intake is inversely associated with serum C-reactive protein levels: meta-analysis and systematic review. Eur. J. Clin. Nutr. 68 (4), 510–516.

Dong, J.-Y., Xun, P., He, K., Qin, L.-Q., 2011. Magnesium intake and risk of type 2 diabetes meta-analysis of prospective cohort studies. Diabetes Care 34 (9), 2116–2122.

Drenick, E.J., 1961. The influence of ingestion of calcium and other soap-forming substances on fecal fat. Gastroenterology 41, 242–244.

Eckel, R.H., Grundy, S.M., Zimmet, P.Z., 2005. The metabolic syndrome. Lancet 365 (9468), 1415–1428.

Elin, R., 1994. Magnesium: the fifth but forgotten electrolyte. Am. J. Clin. Pathol. 102 (5), 616–622.

Expert Panel on Detection, Evaluation, and Treatment of High Blood Cholesterol in Adults, 2001. Executive summary of the Third Report of the National Cholesterol Education Program (NCEP) Expert Panel on detection, evaluation, and treatment of high blood cholesterol in adults (adult treatment Panel III). JAMA 285 (19), 2486–2497.

Ferlay, J., Soerjomataram, I., Dikshit, R., Eser, S., Mathers, C., Rebelo, M., Parkin, D.M., Forman, D., Bray, F., 2015. Cancer incidence and mortality worldwide: sources, methods and major patterns in GLOBOCAN 2012. Int. J. Cancer 136 (5), E359–E386.

Folsom, A.R., Hong, C.-P., 2006. Magnesium intake and reduced risk of colon cancer in a prospective study of women. Am. J. Epidemiol. 163 (3), 232–235.

Ford, E.S., Mokdad, A.H., 2003. Dietary magnesium intake in a national sample of US adults. J. Nutr. 133 (9), 2879–2882.

Fuentes, J.C., Salmon, A.A., Silver, M.A., 2006. Acute and chronic oral magnesium supplementation: effects on endothelial function, exercise capacity, and quality of life in patients with symptomatic heart failure. Congest. Heart Fail. 12 (1), 9–13.

Go, A.S., Mozaffarian, D., Roger, V.L., Benjamin, E.J., Berry, J.D., Blaha, M.J., Dai, S., Ford, E.S., Fox, C.S., Franco, S., 2014. Heart disease and stroke statistics–2014 update: a report from the American Heart Association. Circulation 129 (3), e28.

Grothey, A., Nikcevich, D.A., Sloan, J.A., Kugler, J.W., Silberstein, P.T., Dentchev, T., Wender, D.B., Novotny, P.J., Chitaley, U., Alberts, S.R., 2011. Intravenous calcium and magnesium for oxaliplatin-induced sensory neurotoxicity in adjuvant colon cancer: NCCTG N04C7. J. Clin. Oncol. 29 (4), 421–427.

Gu, W.-J., Wu, Z.-J., Wang, P.-F., Aung, L., Yin, R.-X., 2012. Intravenous magnesium prevents atrial fibrillation after coronary artery bypass grafting: a meta-analysis of 7 double-blind, placebo-controlled, randomized clinical trials. Trials 13 (1), 1–8.

Guerrero-Romero, F., Rodriguez-Moran, M., 2002. Low serum magnesium levels and metabolic syndrome. Acta Diabetol. 39 (4), 209–213.

Guerrero-Romero, F., Simental-Mendía, L., Hernández-Ronquillo, G., Rodriguez-Morán, M., 2015. Oral magnesium supplementation improves glycaemic status in subjects with prediabetes and hypomagnesaemia: a double-blind placebo-controlled randomized trial. Diabetes Metab. 41 (3), 202–207.

Guerrero-Romero, F., Tamez-Perez, H.E., Gonzalez-Gonzalez, G., Salinas-Martinez, A.M., Montes-Villarreal, J., Trevino-Ortiz, J.H., Rodriguez-Moran, M., 2004. Oral magnesium supplementation improves insulin sensitivity in non-diabetic subjects with insulin resistance. A double-blind placebo-controlled randomized trial. Diabetes Metab. 30 (3), 253–258.

Hanahan, D., Weinberg, R.A., 2011. Hallmarks of cancer: the next generation. Cell 144 (5), 646–674.

Hardwick, L.L., Jones, M.R., Brautbar, N., Lee, D.B., 1991. Magnesium absorption: mechanisms and the influence of vitamin D, calcium and phosphate. J. Nutr. 121 (1), 13–23.

Hatzistavri, L.S., Sarafidis, P.A., Georgianos, P.I., Tziolas, I.M., Aroditis, C.P., Zebekakis, P.E., Pikilidou, M.I., Lasaridis, A.N., 2009. Oral magnesium supplementation reduces ambulatory blood pressure in patients with mild hypertension. Am. J. Hypertens. 22 (10), 1070–1075.

Henyan, N.N., Gillespie, E.L., White, C.M., Kluger, J., Coleman, C.I., 2005. Impact of intravenous magnesium on post-cardiothoracic surgery atrial fibrillation and length of hospital stay: a meta-analysis. Ann. Thorac. Surg. 80 (6), 2402–2406.

Hravnak, M., Hoffman, L.A., Saul, M.I., Zullo, T.G., Whitman, G.R., Griffith, B.P., 2002. Predictors and impact of atrial fibrillation after isolated coronary artery bypass grafting. Crit. Care Med. 30 (2), 330.

Hruby, A., Ngwa, J.S., Renstrom, F., Wojczynski, M.K., Ganna, A., Hallmans, G., Houston, D.K., Jacques, P.F., Kanoni, S., Lehtimaki, T., Lemaitre, R.N., Manichaikul, A., North, K.E., Ntalla, I., Sonestedt, E., Tanaka, T., van Rooij, F.J., Bandinelli, S., Djousse, L., Grigoriou, E., Johansson, I., Lohman, K.K., Pankow, J.S., Raitakari, O.T., Riserus, U., Yannakoulia, M., Zillikens, M.C., Hassanali, N., Liu, Y., Mozaffarian, D., Papoutsakis, C., Syvanen, A.C., Uitterlinden, A.G., Viikari, J., Groves, C.J., Hofman, A., Lind, L., McCarthy, M.I., Mikkila, V., Mukamal, K., Franco, O.H., Borecki, I.B., Cupples, L.A., Dedoussis, G.V., Ferrucci, L., Hu, F.B., Ingelsson, E., Kahonen, M., Kao, W.H., Kritchevsky, S.B., Orho-Melander, M., Prokopenko, I., Rotter, J.I., Siscovick, D.S., Witteman, J.C., Franks, P.W., Meigs, J.B., McKeown, N.M., Nettleton, J.A., 2013. Higher magnesium intake is associated with lower fasting glucose and insulin, with no evidence of interaction with select genetic loci, in a meta-analysis of 15 CHARGE Consortium Studies. J. Nutr. 143 (3), 345–353.

Huitrón-Bravo, G.G., Denova-Gutiérrez, E., de Jesús Garduño-García, J., Talavera, J.O., Herreros, B., Salmerón, J., 2015. Dietary magnesium intake and risk of hypertension in a Mexican adult population: a cohort study. BMC Nutr. 1 (1), 6.

Israili, Z.H., Hernández-Hernández, R., Valasco, M., 2007. The future of antihypertensive treatment. Am. J. Ther. 14 (2), 121–134.

Itoh, K., Kawasaka, T., Nakamura, M., 1997. The effects of high oral magnesium supplementation on blood pressure, serum lipids and related variables in apparently healthy Japanese subjects. Br. J. Nutr. 78 (5), 737–750.

Joosten, M.M., Gansevoort, R.T., Mukamal, K.J., Kootstra-Ros, J.E., Feskens, E.J., Geleijnse, J.M., Navis, G., Bakker, S.J., 2013. Urinary magnesium excretion and risk of hypertension the prevention of renal and vascular end-stage disease study. Hypertension 61 (6), 1161–1167.

Ju, S.Y., Choi, W.S., Ock, S.M., Kim, C.M., Kim, D.H., 2014. Dietary magnesium intake and metabolic syndrome in the adult population: dose-response meta-analysis and meta-regression. Nutrients 6 (12), 6005–6019.

Kandeel, F.R., Balon, E., Scott, S., Nadler, J.L., 1996. Magnesium deficiency and glucose metabolism in rat adipocytes. Metabolism 45 (7), 838–843.

Kass, L., Weekes, J., Carpenter, L., 2012. Effect of magnesium supplementation on blood pressure: a meta-analysis. Eur. J. Clin. Nutr. 66 (4), 411–418.

Kawano, Y., Matsuoka, H., Takishita, S., Omae, T., 1998. Effects of magnesium supplementation in hypertensive patients assessment by office, home, and ambulatory blood pressures. Hypertension 32 (2), 260–265.

Khan, A.M., Lubitz, S.A., Sullivan, L.M., Sun, J.X., Levy, D., Vasan, R.S., Magnani, J.W., Ellinor, P.T., Benjamin, E.J., Wang, T.J., 2013. Low serum magnesium and the development of atrial fibrillation in the community: the Framingham Heart Study. Circulation 127, 33–38.

Knijn, N., Tol, J., Koopman, M., Werter, M., Imholz, A., Valster, F., Mol, L., Vincent, A., Teerenstra, S., Punt, C., 2011. The effect of prophylactic calcium and magnesium infusions on the incidence of neurotoxicity and clinical outcome of oxaliplatin-based systemic treatment in advanced colorectal cancer patients. Eur. J. Cancer 47 (3), 369–374.

Ku, D., Ann, H., 1987. Magnesium deficiency produces endothelium-dependent vasorelaxation in canine coronary arteries. J. Pharmacol. Exp. Ther. 241 (3), 961–966.

La, S.A., Lee, J.Y., Kim, D.H., Song, E.L., Park, J.H., Ju, S.Y., 2015. Low magnesium levels in adults with metabolic syndrome: a meta-analysis. Biol. Trace Elem. Res.

Larsson, S.C., Bergkvist, L., Wolk, A., 2005. Magnesium intake in relation to risk of colorectal cancer in women. JAMA 293 (1), 86–89.

Larsson, S.C., Orsini, N., Wolk, A., 2012. Dietary magnesium intake and risk of stroke: a meta-analysis of prospective studies. Am. J. Clin. Nutr. 95 (2), 362–366.

Liao, F., Folsom, A.R., Brancati, F.L., 1998. Is low magnesium concentration a risk factor for coronary heart disease? the Atherosclerosis Risk in Communities (ARIC) Study. Am. Heart J. 136 (3), 480–490.

Loprinzi, C.L., Qin, R., Dakhil, S.R., Fehrenbacher, L., Flynn, K.A., Atherton, P., Seisler, D., Qamar, R., Lewis, G.C., Grothey, A., 2014. Phase III randomized, placebo-controlled, double-blind study of intravenous calcium and magnesium to prevent oxaliplatin-induced sensory neurotoxicity (N08CB/Alliance). J. Clin. Oncol. 32 (10), 997–1005.

Luft, F.C., 2010. Vascular calcification and magnesium. J. Mol. Med. 88 (5), 437–439.

Lundstrom, K., 2012. Cancer and diet. J. Pharm. Nutr. Sci. 2 (1).

Lutsey, P.L., Alonso, A., Michos, E.D., Loehr, L.R., Astor, B.C., Coresh, J., Folsom, A.R., 2014. Serum magnesium, phosphorus, and calcium are associated with risk of incident heart failure: the Atherosclerosis Risk in Communities (ARIC) Study. Am. J. Clin. Nutr. 100 (3), 756–764.

Ma, E., Sasazuki, S., Inoue, M., Iwasaki, M., Sawada, N., Takachi, R., Tsugane, S., 2010. High dietary intake of magnesium may decrease risk of colorectal cancer in Japanese men. J. Nutr. 140 (4), 779–785.

Ma, J., Folsom, A.R., Melnick, S.L., Eckfeldt, J.H., Sharrett, A.R., Nabulsi, A.A., Hutchinson, R.G., Metcalf, P.A., 1995. Associations of serum and dietary magnesium with cardiovascular disease, hypertension, diabetes, insulin, and carotid arterial wall thickness: the ARIC study. J. Clin. Epidemiol. 48 (7), 927–940.

Mahabir, S., Forman, M.R., Dong, Y.Q., Park, Y., Hollenbeck, A., Schatzkin, A., 2010. Mineral intake and lung cancer risk in the NIH-American association of Retired Persons diet and health study. Cancer Epidemiol. Biomarkers Prev. 19 (8), 1976–1983.

Mahabir, S., Wei, Q., Barrera, S.L., Dong, Y.Q., Etzel, C.J., Spitz, M.R., Forman, M.R., 2008. Dietary magnesium and DNA repair capacity as risk factors for lung cancer. Carcinogenesis 29 (5), 949–956.

Mazur, A., Maier, J.A., Rock, E., Gueux, E., Nowacki, W., Rayssiguier, Y., 2007. Magnesium and the inflammatory response: potential physiopathological implications. Arch. Biochem. Biophys. 458 (1), 48–56.

McCarty, M., 1996. Complementary vascular-protective actions of magnesium and taurine: a rationale for magnesium taurate. Med. Hypotheses 46 (2), 89–100.

McKeehan, W.L., Ham, R.G., 1978. Calcium and magnesium ions and the regulation of multiplication in normal and transformed cells. Nature 275 (5682), 756–758.

Mortazavi, M., Moeinzadeh, F., Saadatnia, M., Shahidi, S., McGee, J.C., Minagar, A., 2013. Effect of magnesium supplementation on carotid intima-media thickness and flow-mediated dilatation among hemodialysis patients: a double-blind, randomized, placebo-controlled trial. Eur. Neurol. 69 (5), 309–316.

Muir, K., Lees, K., Ford, I., Davis, S., 2004. Magnesium for acute stroke (Intravenous Magnesium Efficacy in Stroke trial): randomised controlled trial. Lancet 363 (9407), 439–445.

Newhouse, I.J., Finstad, E.W., 2000. The effects of magnesium supplementation on exercise performance. Clin. J. Sport Med. 10 (3), 195–200.

Nielsen, F.H., 2010. Magnesium, inflammation, and obesity in chronic disease. Nutr. Rev. 68 (6), 333–340.

Ohira, T., Peacock, J.M., Iso, H., Chambless, L.E., Rosamond, W.D., Folsom, A.R., 2009. Serum and dietary magnesium and risk of ischemic stroke the atherosclerosis risk in communities study. Am. J. Epidemiol. 169 (12), 1437–1444.

Onalan, O., Crystal, E., Daoulah, A., Lau, C., Crystal, A., Lashevsky, I., 2007. Meta-analysis of magnesium therapy for the acute management of rapid atrial fibrillation. Am. J. Cardiol. 99 (12), 1726–1732.

Paolisso, G., Barbagallo, M., 1997. Hypertension, diabetes mellitus, and insulin resistance: the role of intracellular magnesium. Am. J. Hypertens. 10 (3), 346–355.

Paolisso, G., Ravussin, E., 1995. Intracellular magnesium and insulin resistance: results in Pima Indians and Caucasians. J. Clin. Endocrinol. Metab. 80 (4), 1382–1385.

Pearson, P.J., Evora, P.R., Seccombe, J.F., Schaff, H.V., 1998. Hypomagnesemia inhibits nitric oxide release from coronary endothelium: protective role of magnesium infusion after cardiac operations. Ann. Thorac. Surg. 65 (4), 967–972.

Qu, X., Jin, F., Hao, Y., Zhu, Z., Li, H., Tang, T., Dai, K., 2013. Nonlinear association between magnesium intake and the risk of colorectal cancer. Eur. J. Gastroenterol. Hepatol. 25 (3), 309–318.

Rayssiguier, Y., Noe, L., Etienne, J., Gueux, E., Cardot, P., Mazur, A., 1991. Effect of magnesium deficiency on post-heparin lipase activity and tissue lipoprotein lipase in the rat. Lipids 26 (3), 182–186.

Rodríguez-Moran, M., Guerrero-Romero, F., 2014. Oral magnesium supplementation improves the metabolic profile of metabolically obese, normal-weight individuals: a randomized double-blind placebo-controlled trial. Arch. Med. Res. 45 (5), 388–393.

Rosanoff, A., Seelig, M.S., 2004. Comparison of mechanism and functional effects of magnesium and statin pharmaceuticals. J. Am. Coll. Nutr. 23 (5), 501S–505S.

Rosanoff, A., Weaver, C.M., Rude, R.K., 2012. Suboptimal magnesium status in the United States: are the health consequences underestimated? Nutr. Rev. 70 (3), 153–164.

Sasaki, S., Oshima, T., Matsuura, H., Ozono, R., Higashi, Y., Sasaki, N., Matsumoto, T., Nakano, Y., Ueda, A., Yoshimizu, A., 2000. Abnormal magnesium status in patients with cardiovascular diseases. Clin. Sci. 98 (2), 175–181.

Saver, J.L., Kidwell, C., Eckstein, M., Starkman, S., 2004. Prehospital neuroprotective therapy for acute stroke results of the Field Administration of Stroke Therapy–Magnesium (FAST–MAG) pilot trial. Stroke 35 (5), e106–e108.

Sellmeyer, D.E., Schloetter, M., Sebastian, A., 2002. Potassium citrate prevents increased urine calcium excretion and bone resorption induced by a high sodium chloride diet. J. Clin. Endocrinol. Metab. 87 (5), 2008–2012.

Sheehan, J.P., Seelig, M.S., 1984. Interactions of magnesium and potassium in the pathogenesis of cardiovascular disease. Magnesium 3 (4–6), 301–314.

Simental-Mendia, L.E., Rodriguez-Moran, M., Guerrero-Romero, F., 2014. Oral magnesium supplementation decreases C-reactive protein levels in subjects with prediabetes and hypomagnesemia: a clinical randomized double-blind placebo-controlled trial. Arch. Med. Res. 45 (4), 325–330.

Song, Y., He, K., Levitan, E.B., Manson, J.E., Liu, S., 2006. Effects of oral magnesium supplementation on glycaemic control in Type 2 diabetes: a meta-analysis of randomized double-blind controlled trials. Diabet. Med. 23 (10), 1050–1056.

Suarez, A., Pulido, N., Casla, A., Casanova, B., Arrieta, F.J., Rovira, A., 1995. Impaired tyrosine-kinase activity of muscle insulin receptors from hypomagnesaemic rats. Diabetologia 38 (11), 1262–1270.

Touyz, R.M., 2003. Role of magnesium in the pathogenesis of hypertension. Mol. Asp. Med. 24 (1–3), 107–136.

Touyz, R.M., 2008. Transient receptor potential melastatin 6 and 7 channels, magnesium transport, and vascular biology: implications in hypertension. Am. J. Physiol. Heart Circ. Physiol. 294 (3), H1103–H1118.

Tukiendorf, A., Rybak, Z., 2004. New data on ecological analysis of possible relationship between magnesium in drinking water and liver cancer. Magnes. Res. 17 (1), 46–52.

Van den Brandt, P., Smits, K., Goldbohm, R., Weijenberg, M., 2007. Magnesium intake and colorectal cancer risk in the Netherlands Cohort Study. Br. J. Cancer 96 (3), 510–513.

Wark, P.A., Lau, R., Norat, T., Kampman, E., 2012. Magnesium intake and colorectal tumor risk: a case-control study and meta-analysis. Am. J. Clin. Nutr. 96 (3), 622–631.

Weyer, P.J., Cerhan, J.R., Kross, B.C., Hallberg, G.R., Kantamneni, J., Breuer, G., Jones, M.P., Zheng, W., Lynch, C.F., 2001. Municipal drinking water nitrate level and cancer risk in older women: the Iowa Women's Health Study. Epidemiology 12 (3), 327–338.

Witteman, J., Grobbee, D.E., Derkx, F., Bouillon, R., de Bruijn, A.M., Hofman, A., 1994. Reduction of blood pressure with oral magnesium supplementation in women with mild to moderate hypertension. Am. J. Clin. Nutr. 60 (1), 129–135.

Wolf, F.I., Cittadini, A.R., Maier, J.A., 2009. Magnesium and tumors: ally or foe? Cancer Treat. Rev. 35 (4), 378–382.

Yang, C.-Y., Chiu, H.-F., Cheng, B.-H., Hsu, T.-Y., Cheng, M.-F., Wu, T.-N., 2000a. Calcium and magnesium in drinking water and the risk of death from breast cancer. J. Toxicol. Environ. Health Part A 60 (4), 231–241.

Yang, C.-Y., Chiu, H.-F., Tsai, S.-S., Cheng, M.-F., Lin, M.-C., Sung, F.-C., 2000b. Calcium and magnesium in drinking water and risk of death from prostate cancer. J. Toxicol. Environ. Health Part A 60 (1), 17–26.

Yang, C., Chiu, H., Tsai, S., Wu, T., Chang, C., 2002. Calcium and magnesium in drinking water and the risk of death from esophageal cancer. Magnes. Res. 15, 215–222.

Zhang, W., Iso, H., Ohira, T., Date, C., Tamakoshi, A., Group, J.S., 2012. Associations of dietary magnesium intake with mortality from cardiovascular disease: the JACC study. Atherosclerosis 221 (2), 587–595.

Chapter 28

Magnesium and Embryonic Development

Yuko Komiya, Loren Warren Runnels

Rutgers-Robert Wood Johnson Medical School, Piscataway, NJ, United States

INTRODUCTION

From the early embryo to adulthood, magnesium (Mg^{2+}), the second most abundant cellular cation, is essential for all stages of life. In humans, insufficient Mg^{2+} intake during pregnancy has been linked to an increased risk of preterm birth and low birth weight (Conradt et al., 1984; Doyle et al., 1989; Makrides and Crowther, 2001). These results may not be surprising given the wide array of functions Mg^{2+} plays in diverse cellular processes, ranging from DNA transcription and translation to cell metabolism and cell signaling (de Baaij et al., 2015; Romani, 2011). It was not until the transporters and channels that control cellular Mg^{2+} levels were identified and their effects on development analyzed that it was appreciated that Mg^{2+} may be playing more of a specific regulatory role during embryonic development than originally imagined. In this chapter, we review what has been learned about the function of Mg^{2+} and the transporters and channels that mediate this vital cation's cellular transport during embryogenesis.

CHEMICAL PROPERTIES OF MAGNESIUM AND ITS FUNCTION AND REGULATION IN CELLS

Before proceeding with a summary of Mg^{2+} functions in development, we will briefly describe the chemical properties of Mg^2 and its myriad important biological roles in cells. It has long been recognized that Mg^2 is an essential ion for human health (de Baaij et al., 2015; Romani, 2011). More importantly, Mg^{2+} is indispensable to life itself. The significance of Mg^{2+} and its relationship to the origin of life has been traced from the composition of the earth's crust (rich in iron-magnesium silicate) and the primeval ocean rich in Mg^{2+}, to the formation of chlorophyll with Mg^{2+} at the center of the molecule, and finally to its incorporation into the animal cell containing adenosine triphosphate (ATP) with its dependence on Mg^{2+} (Fawcett et al., 1999). Within the periodic table of elements, Mg has an atomic number of 12 and is classified as an alkaline earth element (group 2). Mg occurs in nature in three stable isotopes, ^{24}Mg, ^{25}Mg, and ^{26}Mg. ^{24}Mg is the most common isotope (78.99%) and has a relative atomic mass of 24.305 Da (de Baaij et al., 2015). In contrast to Ca^{2+}, Mg^{2+} is a smaller ion with a higher charge density that attracts water molecules more strongly, with a radius roughly 400 times larger than its dehydrated radius (de Baaij et al., 2015; Saris et al., 2000). Mg^{2+} binds weakly to proteins and enzymes ($K_a \leq 10^5 M^{-1}$) and can function as a Lewis acid in enzymatic reactions toward a bound substrate, allowing a large number of hydrolysis and condensation reactions to proceed under physiological conditions that would otherwise require extremes of pH (Cowan, 2002).

Mg^{2+} is a cofactor in hundreds of enzymatic reactions and it is especially important for those enzymes that use nucleotides as cofactors or substrates (de Baaij et al., 2015). Consequently, Mg^{2+} is involved in a wide variety of biological functions, including regulation of the cell cycle, ion channel function, ATPase activity, protein and nucleic acid synthesis, metabolism reactions, maintenance of body temperature, and neurotransmission in the nervous system (Romani, 2007). Indeed, it has been estimated that at least 600 enzymatic reactions are directly or indirectly regulated by Mg^{2+} (de Baaij et al., 2015). Mg^{2+} is necessary for the proper structure and activity of DNA and RNA polymerases, topoisomerases, helicases, and exonucleases, making Mg^{2+} an essential component of DNA replication, RNA transcription, amino acid synthesis, and protein formation. In addition, ATPases, guanylyl cyclases, adenylyl cyclases, and lipid and protein kinases, all depend on Mg-ATP for proper function. Given its broad cellular functions, a majority of mammalian tissues or cells hoard Mg^{2+}, with total Mg^{2+} concentration ranging between 14 and 20 mM (Romani, 2007). Greater than 90% of the cell magnesium is bound or sequestered to membranes, ATP, ribosomes, and polynucleotides (Murphy, 2000; Romani, 2007). However, unlike Ca^{2+}, whose cytosolic concentration can vary 10- to 100-fold, the free concentration of Mg^{2+} is comparably steady, hovering around 0.5–0.7 mM (Romani, 2007).

Molecular, Genetic, and Nutritional Aspects of Major and Trace Minerals. http://dx.doi.org/10.1016/B978-0-12-802168-2.00028-2

Mg^{2+} influx into the cell occurs mainly by diffusion from the slightly higher free concentration of Mg^{2+} (~1.2 mM) in the extracellular space. The Nernst potential for Mg^{2+} is approximately +10–20 mV and influx of Mg^{2+} is thus potentiated by the membrane potential, which is negative on the cytosolic side. To maintain the lower intracellular concentration of Mg^{2+} relative to its greater abundance in the extracellular space, extrusion of Mg^{2+} occurs against a steep electrochemical gradient, which in many cell types is achieved by antiport against Na^+ (Romani, 2011). The fact that the free concentration of cytosolic Mg^{2+} is relatively stable compared to the large span of concentrations that Ca^{2+} can assume, plus the fact that Mg^{2+} affects so many different cellular activities, has led some to question whether Mg^{2+} can function like Ca^{2+} as a second messenger and specifically target individual signaling pathways (Grubbs and Maguire, 1987). Nevertheless, rapid fluxes of Mg^{2+} into and out of cells have been detected in numerous cell types (Matsuura et al., 1993; Romani and Scarpa, 1990a,b). Disruption of one of these processes was shown to be responsible for a novel X-linked human immunodeficiency characterized by CD4 lymphopenia, severe chronic viral infections, and defective T-lymphocyte activation (Li et al., 2011). Mutations in the Mg^{2+} transporter MAGT1 abolished Mg^{2+} influx required for activation of phospholipase C-γ1 (PLC-γ1) and Ca^{2+} influx in T-cells. How Mg^{2+} is affecting PLC-γ1 activation, or for that matter signal transduction in general, is poorly understood. Therefore, the impact of Mg^{2+} may be underappreciated.

Cells work aggressively to maintain a relatively narrow range of intracellular free Mg^{2+} concentration, regardless of the extracellular Mg^{2+} concentration. This responsibility largely falls to Mg^{2+} transporters and ion channels. Magnesium transporters were originally identified in prokaryotes and protozoans (Quamme, 2010). In 1976, CorA was identified as a Mg^{2+} transporter in bacteria. CorA has two transmembrane domains and functions as a Mg^{2+} transporter when oligomerized (Park et al., 1976). CorA can permeate Co^{2+}, Ni^{2+} as well as Mg^{2+}, and CorA is the primary transporter responsible for Mg^{2+} intake in most prokaryotes. In an effort to search for new CorA family members and homologs, many other Mg^{2+} transporters such as MgtE, Alr1, and Mrs2 have been identified in bacteria, fungi, and yeast systems (Bui et al., 1999; Graschopf et al., 2001). These discoveries led to the identification of mammalian Mg^{2+} transporters, which were identified in part based on homologies to their ancestral counterparts. While some of the identified transporters exhibit relatively high specificity for Mg^{2+} transport, other molecules shown to play a role in Mg^{2+} homeostasis were surprisingly found to be relatively nonselective (Romani, 2007). Two members of the "melastatin" subgroup of transient receptor potential (TRP) ion channel family, TRPM7 and TRPM6, have been shown to play a vital role in both cellular and whole animal Mg^{2+} homeostasis (Ryazanova et al., 2010; Schlingmann et al., 2002; Schmitz et al., 2003; Walder et al., 2002). Later in this chapter, we will discuss the impact of Mg^{2+} transporters and the Mg^{2+}-permeant ion channels on embryonic development, but we first begin with a summary of the effect that Mg^{2+} itself has on embryogenesis.

MAGNESIUM'S IMPACT ON EMBRYOGENESIS

While Mg^{2+} is linked to a growing number of physiological and pathological roles in adult animals and humans, the cation's contribution to embryonic development is less understood. Unlike Ca^{2+}, whose influence over embryonic development is firmly established (Markova and Lenne, 2012; Webb and Miller, 2003), only a handful of studies have implicated Mg^{2+} in this process. Studies of the effects of gestational Mg^{2+} deficiency in rats were among the first reports to demonstrate that Mg^{2+} is required for embryonic development (Hurley et al., 1976; Wang et al., 1971). Pregnant females fed an Mg^{2+}-deficient diet between days 6 and 14 of gestation showed a high incidence of resorptions and gross malformation in the full-term fetuses (Hurley et al., 1976). Pups born from Mg^{2+}-deficient pregnant females also displayed growth retardation, an increased incidence of abnormal fat metabolism, insulin resistance, and diabetes (Venu et al., 2008). More recent epidemiological studies have associated Mg^{2+} deficiency with a negative impact on human fetal growth and development. A decrease in basal $[Mg^{2+}]_i$ from cord blood platelets is associated with small for gestational age (SGA) babies (Venu et al., 2008). Oral Mg^{2+} supplementation given before the 25th week of gestation is also associated with a lower frequency of preterm births, a lower frequency of low birth weight, and fewer SGA infants compared with placebo (Makrides and Crowther, 2001).

The recommended dietary allowance in the United States for magnesium is 410–420 mg per day for a male adult and 320–360 mg per day for a female; the magnesium requirement is increased during pregnancy and lactation (320–400 mg/day), depending on age (IOM, 1997). A maternal Mg^{2+} intake below 378 mg/day has been associated with a two- to three-fold higher risk for the neural tube defect, spina bifida, in offspring, suggesting a pivotal role for Mg^{2+} not just in embryonic growth but in morphogenesis as well (Groenen et al., 2004).

Perhaps the most extensive and detailed analysis of the impact of Mg^{2+} on embryonic development was conducted using *Xenopus laevis* embryos, presumably because the embryos can be developed ex vivo and the concentration of ions bathing the embryo can be easily manipulated. *X. laevis* embryos constitute a classic animal model to investigate early developmental processes. Many of the signaling molecules that regulate early embryogenesis were originally identified in *Xenopus*.

It is well known that key factors of early development, such as BMP, Wnt, and FGF signaling molecules, are functionally conserved between *Xenopus* and mammals. Since *Xenopus* eggs can be externally fertilized, it is easy to observe each step of embryogenesis. In 1977 Miller and Landesman performed the first comprehensive investigation of the significance of Mg^{2+} during *Xenopus* embryogenesis (Miller and Landesman, 1977). Typically, *Xenopus* embryos are cultured in a simple buffer such as 0.1×marc's modified ringers (MMR) solution that contains 0.1 mM $MgSO_4$. 0.1×MMR contains (in mM) 0.5 (4-(2-hydroxyethyl)-1-piperazineethanesulfonic acid) HEPES (pH 7.8), 10 NaCl, 0.2 KCl, 0.2 $CaCl_2$, and 0.1 $MgSO_4$. In addition to magnesium, this buffer also contains sodium, potassium, and calcium ions, which are also essential for embryonic development, since *Xenopus* embryos cannot survive in distilled water alone. In their experiments, the authors determined the impact of Mg^{2+} on embryogenesis by varying the concentration of Mg^{2+} in the culture buffer from 100 nM to 10 mM. When the Mg^{2+} concentration was lower than 10 µM, embryos exhibited reduced melanophores at stage 35–36. By stage 40, the embryos showed shorter tail expansion, abnormal coiling of the gut, slow head enlargement, restricted heart growth, and edema. In addition, abnormal somite development was observed, which may be related to paralysis of the Mg^{2+}-deficient embryos. In control embryos, the total Mg^{2+} content began to increase after hatching. By comparison, embryos cultured in an Mg^{2+}-deficient buffer maintained a very low level of total Mg^{2+} throughout development. These results suggested that proper Mg^{2+} uptake is a critical factor during early developmental processes in *Xenopus*.

We conducted experiments similar to those of Miller and Landesman and examined the effect of Mg^{2+} depletion from the culture medium on developmental processes of *X. laevis* (Komiya and Runnels, 2015). We observed the morphology of *Xenopus* embryos cultured in regular MMR solution of Holtfreter's buffer with or without 0.1 mM $MgSO_4$ (Fig. 28.1). Consistent with previous findings, early developmental processes such as cleavage, early gastrulation, and neurulation were not significantly affected by Mg^{2+} depletion in our experiments. At the tadpole stage, embryos cultured in Mg^{2+}-deficient media exhibited lighter pigmentation in eyes and body than control embryos reared in normal buffer (Komiya and Runnels, 2015). Moreover, all embryos cultured in Mg^{2+}-deficient media did not elongate their axis properly: the length of the anterior–posterior axis was significantly shorter than for control embryos. In addition, embryos cultured in Mg^{2+}-deficient media had spinal cords that were not straight, with some embryos exhibiting a mildly kinked axis. Many of the embryos cultured in Mg^{2+}-deficient media had a slightly smaller head size (Komiya and Runnels, 2015). Although we did not perform histological exams, the gut appeared abnormal in some of the embryos, consistent with the report from Miller and Landesman.

As stated earlier, we did not observe severe defects during early gastrulation and the reason for this is not clear. From egg to the stage 9 embryo, just before gastrulation begins, the intracellular concentration of Mg^{2+} in the embryo does not vary much from 14 mM (Slack et al., 1973). The rate at which the embryos lose Mg^{2+} in Mg^{2+}-deficient media is, however, not known, and could be too slow to produce a severe phenotype at earlier stages in development. More research on the regulation of Mg^{2+} homeostasis in the developing embryo is required to address this question. Nevertheless, culturing the embryos in Mg^{2+} deficient media did result in a shortened axis, which could be due to an inhibition of convergent extension cell movements. To directly determine whether Mg^{2+} influences gastrulation cell movements, we investigated

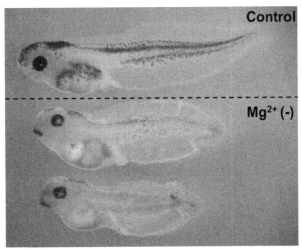

FIGURE 28.1 Magnesium depletion causes developmental abnormalities in *Xenopus laevis* embryos. The top embryo is an example of a wild type (control) embryo reared in normal MMR buffer. Embryos reared in Mg^{2+}-deficient MMR buffer showed pigmentation defects and a shortened anterior–posterior axis.

the effect of Mg^{2+} depletion on convergent extension using a Keller explant assay (Komiya and Runnels, 2015). After fertilization, embryos were grown in regular MMR or Mg^{2+}-free MMR until the early gastrula stage, after which the dorsal marginal zone was dissected. The isolated explants were then cultured in buffer containing Mg^{2+} or Mg^{2+}-free buffer. For the control condition (with Mg^{2+}), dorsal tissue explants underwent convergent extension movements and became well elongated. The explants that were precultured with Mg^{2+}-free MMR also elongated, but the length was shorter than the controls, indicating that elongation was impaired (Komiya and Runnels, 2015). These results suggest that the short axis phenotype in embryos cultured in Mg^{2+}-deficient media could be explained by mild inhibition of convergent extension cell movements. Thus, removal of Mg^{2+} by itself produces a wide range of developmental defects, including, as one of our studies showed, impairment of axial extension during gastrulation due to dysregulation of convergent extension cell movements.

FUNCTION OF TRPM7 AND TRPM6 CHANNELS DURING EMBRYOGENESIS

Regulation of Mg^{2+} homeostasis in mammals occurs in the intestine, kidney, and bone where the divalent cation is initially absorbed, excreted, and stored in its hydroxyapatite form, respectively (de Baaij et al., 2015). The kidney is believed to be the primary site for regulation of the whole-body Mg^{2+} homeostasis in humans where the balance of Mg^{2+} reabsorption and excretion is controlled (de Baaij et al., 2015). TRPM7 and TRPM6 are unique bifunctional proteins with ion channel and kinase domains (for a more detailed review, see Runnels, 2011). The channels are permeable to a wide range of divalent cations, including Mg^{2+}, Ca^{2+}, and Zn^{2+}(Li et al., 2006; Monteilh-Zoller et al., 2003; Voets et al., 2004). Mutations in TRPM6 have been clearly linked to familial hypomagnesemia with secondary hypocalcemia (HSH) (Schlingmann et al., 2002; Walder et al., 2002). HSH is characterized by very low Mg^{2+} and low Ca^{2+} serum levels. Shortly after birth, affected individuals exhibit neurologic symptoms of hypomagnesemic hypocalcemia, including seizures and muscle spasms, but otherwise develop normally if treated with magnesium supplements. Thus far, there has been no report of a mutation in the *TRPM7* gene for an HSH patient. While a majority of the mutations in individuals affected with HSH are either nonsense or frameshift mutations in *TRPM6* that are easily compatible with a loss-of-function phenotype, one missense mutation entails the exchange of a highly conserved serine for a leucine at amino acid position 141 (S141L), which disrupts the ability of TRPM6 to form heterooligomers with TRPM7 (Schlingmann et al., 2002). In two sets of studies, TRPM6 was reported to be dependent upon TRPM7 for cell surface expression (Chubanov et al., 2004; Schmitz et al., 2005). Full-length variants of TRPM6 failed to form functional channel complexes when heterologously expressed in HEK-293 cells and *Xenopus* oocytes (Chubanov et al., 2004). This data suggests that TRPM7 may also participate in whole-body Mg^{2+} homeostasis, either by itself or by forming heterooligomeric channels with TRPM6. Jin and colleagues generated conditional TRPM7 knockout mice using a tamoxifen-inducible system and multiple tissue-specific Cre recombinase lines (Jin et al., 2012). Tamoxifen-dependent deletion of TRPM7 at E7.5 to E8.5 caused embryonic lethality within 48–72 h, while the depletion at E14.5 did not cause embryonic lethality, with mutant mice developing normally. In the mouse embryo, expression of TRPM7 is dramatically increased from embryonic day 10.5 (E10.5) to E11.5 and global expression was observed through E14.5 (Jin et al., 2008). During mouse embryogenesis, the primitive streak is formed around E6.5, after which cells start moving dynamically to establish the anterior–posterior axis. From E8.5 to E15 the neural plate is formed and the neural tube closes. Taken together, results from TRPM7 mutant mice demonstrate a temporal requirement of TRPM7 during early gastrulation in mice. Surprisingly, knockout of TRPM6 in mice is also embryonically lethal (Walder et al., 2002; Woudenberg-Vrenken et al., 2011). In experiments reported by Walder et al. (2002), homozygous TRPM6 knockout embryos died by E12.5 and exhibited neural tube closure defects. Ten percent of TRPM6-null homozygotes had spina bifida occulta, 0% had spina bifida aperta, and 30% had exencephaly (Walder et al., 2009). The expression pattern of TRPM6 during embryogenesis demonstrated a significant increase at E10, suggesting that TRPM6 is also temporally required for early development. Heterozygous TRPM6 knockout mice exhibited mild hypomagnesemia (Walder et al., 2002; Woudenberg-Vrenken et al., 2011). Interestingly, in one of the studies, dams fed with a high-Mg^{2+} diet slightly suppressed the embryonic lethality caused by knockout of TRPM6, suggesting that embryonic lethality may in part be due to Mg^{2+} deficiency. However, in the second study by Woudenberg-Vrenken et al. (2011), a high-Mg^{2+} diet failed to rescue the lethality caused by homozygous deletion of TRPM6.

Studies from zebrafish also support a role for Mg^{2+} during early embryonic development. Zhou and Clapham (2009) have demonstrated that knockdown of the MagT1 and TUSC3 Mg^{2+} transporters in zebrafish embryos causes early developmental arrest, with embryos exhibiting an apparent defect in brain and eye development. Supplementation of Mg^{2+} in the growth media partially rescued the embryonic arrest caused by depletion of MagT1, demonstrating the importance of Mg^{2+} transporters and Mg^{2+} during embryogenesis. In another study, mutations in the gene encoding cyclin M2 (CNNM2) were demonstrated to be causative for mental retardation and seizures in patients with hypomagnesemia (Arjona et al., 2014). In patients with a recessive mode of inheritance, the intellectual disability is more severe and is accompanied by motor defects and brain malformations. In zebrafish, CNNM2 has two orthologues, CNNM2a and CNNM2b. Consistent with the human pathology, knockdown of CNNM2a in zebrafish resulted in reduced total magnesium content and in morphological phenotypes characterized by enlarged pericardial cavities and notochord defects. At 25 h postfertilization (hpf), abnormal development of the midbrain-hindbrain boundary (MHB) was observed. CNNM2b morphants also had lower total magnesium and were characterized by enlarged pericardial cavities, kidney cysts, and accumulation of cerebrospinal fluid in the cerebrum, and at 25 hpf, they also exhibited defects in development of the MHB. These studies highlight a particularly important role for CNNM2, and potentially for magnesium, in heart, kidney, and especially brain development.

Later in development, TRPM7 has also been demonstrated in mice to have critical functions during nephrogenesis, the development of neural-crest-derived pigment cells, and myocardial proliferation during early cardiogenesis (Jin et al., 2012; Sah et al., 2013). Studies in zebrafish have also highlighted roles for TRPM7 in pigmentation, but have also uncovered additional developmental roles in the development of other organs, including the pancreas (McNeill et al., 2007; Yee et al., 2011). Whether TRPM7's ability to permeate Mg^{2+} is involved in the channel's function later in development remains unknown. Nevertheless, investigation of TRPM7's early developmental roles in *X. laevis* have suggested that these channels' ability to permeate Mg^{2+} may be a key factor in their control over early embryogenesis.

TRPM7 REGULATES GASTRULATION DURING VERTEBRATE EMBRYOGENESIS

Following the blastula stage of development, coordinated movements of gastrulation begin, which transform a simple hollow ball of cells into a multilayered structure with a central gut tube and bilateral symmetry, containing the three germ layers from which the tissues of the adult vertebrate body are generated. During gastrulation, cells in the dorsal marginal zone undergo convergent extension movements in which polarized cell movements establish the physical body axis of the developing embryo (Keller, 2002). During gastrulation the noncanonical Wnt pathway (planar cell polarity pathway) regulates convergent extension movements through activation of the small GTPases Rho and Rac via the cytoplasmic phosphoprotein Disheveled (Dvl; Wallingford and Habas, 2005). Depletion of TRPM7 from developing *X. laevis* embryos using antisense morpholino technology resulted in defective gastrulation phenotypes (Liu et al., 2011). In *X. laevis*, TRPM7-depleted embryos' axial extension was impaired, resulting in a severe dorsal-flexure and failure of the blastopore and neural folds to close. In addition, anterior structures, including the head, eyes, and cement glands, were reduced or absent. One of the advantages of the *X. laevis* system is the ability to conduct gain-of-function analysis. Strikingly, the phenotypes caused by depletion of TRPM7 were prevented by Mg^{2+} supplementation, by expression of TRPM6, as well as by expression of the Mg^{2+} transporter SLC41A2. In contrast, elevating levels of Ca^{2+} in the media failed to rescue the phenotype caused by depletion of TRPM7. Surprisingly, no function for TRPM7's kinase domain was revealed in these studies.

As mentioned earlier, the noncanonical Wnt pathway regulates convergent extension movements through Dvl (Komiya and Habas, 2008). Activation of Dvl stimulates the coordinated activation of Rho and Rac to control cytoskeletal remodeling required for convergent extension cell movements (Habas et al., 2003). Expression of a constitutively active form of Dvl suppressed the gastrulation phenotype caused by depletion of TRPM7, indicating that TRPM7 is likely functioning upstream of Dvl to regulate convergent extension cell movements. Dvl controls activation of the Rho pathway through the formin protein Daam1 (Habas et al., 2001). A constitutively active form of Daam1 failed to rescue the gastrulation defects caused by depletion of TRPM7 in *Xenopus* embryos, indicating that TRPM7 does not influence convergent extension cell movements through Rho (Liu et al., 2011). Rather, Dvl's control of Rac, which regulates convergent extension movements via the c-Jun N-terminal kinase (JNK) pathway, was impaired in XTRPM7 depleted embryos.

Investigation of TRPM7's role in cell migration using fibroblasts gave an additional insight into the results obtained in *X. laevis*. Depletion of TRPM7 from fibroblasts interfered with cells' ability to polarize and migrate directionally (Su et al., 2011). Similar to what was observed in *X. laevis*, TRPM7's channel activity was essential for control of polarized cell movements via Rac. In addition, activation of the small GTPase Cdc42, which governs cell polarity, was impaired in TRPM7-depleted fibroblasts. In both systems, expression of the Mg^{2+} transporter SLC41A2 was effective in reversing the phenotype produced by reduced TRPM7 expression, indicating that TRPM7-mediated control of cellular magnesium homeostasis is critical to directional cell migration at both the cellular and organismal level. Studies in mice, however, have revealed that TRPM7 disruption can also result in the demise of embryonic stem cells and induced pluripotent stem cells, suggesting that TRPM7 could also be regulating early development through its ability to promote cell survival (Jin et al., 2012; Ryazanova et al., 2010). These results mirror those originally obtained by Schmitz et al. (2003) who found that deletion of TRPM7 from DT40 cells causes cell cycle arrest. Importantly, the loss of cell viability and cell cycle arrest in embryonic stem cells and DT40 cells in which TRPM7 is ablated can be suppressed by supplementation of the growth medium with Mg^{2+} (Ryazanova et al., 2010; Schmitz et al., 2003). Therefore, some of the effects of TRPM7 on early development may also be, in part, due to loss of populations of stem cells that depend on TRPM7 for survival. Furthermore, studies have also suggested that TRPM7's ability to permeate Ca^{2+} can contribute to the channel's control of cell adhesion and cell motility (Clark et al., 2006; Su et al., 2006; Visser et al., 2013; Wei et al., 2009). Thus, it still remains possible that both Mg^{2+} and Ca^{2+} are contributing to TRPM7's actions in early development through both of these cations' effects on cell adhesion (Su et al., 2010). For example, changes in intracellular Mg^{2+} could also potentially be affecting intracellular Ca^{2+} homeostasis and thus indirectly influencing cell migration (Volpe and Vezu, 1993). Another possibility is that Mg^{2+} may be affecting the concentration of reactive oxygen species (ROS) to augment signal transduction during cell migration and gastrulation cell movements during embryogenesis. In support of the latter hypothesis, depletion of TRPM7 in fibroblasts has been shown to lower the concentration of cellular ROS, which can be reversed by Mg^{2+} supplementation of the cellular growth media and by expression of the Mg^{2+} transporter SLC41A2 (Chen et al., 2012).

ROS are now well regarded as important regulators of cell migration and have been shown to act through several effectors (Hurd et al., 2012; Tochhawng et al., 2013). Redox signaling has been shown to modulate Wnt-β-catenin signaling through ROS-dependent binding of nucleoredoxin to Dvl (Funato et al., 2006). ROS could potentially be affecting Wnt signaling by targeting downstream effectors of Dvl such as JNK (Hoogeboom and Burgering, 2009). Interestingly, a connection between TRPM7, ROS, and JNK was already made. Overexpression of TRPM7 in HEK-293 cells activates JNK (Su et al., 2010), whereas depletion of the channel, which lowers ROS levels in an Mg^{2+}-dependent manner, suppresses JNK activity (Chen et al., 2012). This suggests a model by which Mg^{2+} could be affecting gastrulation and neural fold closure by augmenting ROS levels (Fig. 28.2). Finally, although studies in *Xenopus* and mice have failed to uncover a role for TRPM7's kinase domain, the kinase may still yet be shown to have a more subtle role during embryogenesis (Kaitsuka et al., 2014; Liu et al., 2011; Ryazanova et al., 2014). Nevertheless, even though a connection between Mg^{2+} and Wnt signaling has been made, much more research is needed to understand how the divalent cation is influencing signal transduction processes during development.

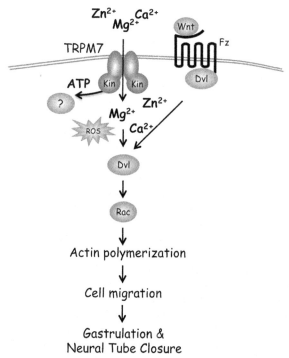

FIGURE 28.2 Model of TRPM7 in noncanonical Wnt signaling. Binding of Wnt ligands to Frizzled (Fz) receptors leads to the activation of Disheveled (Dvl). Dvl through Daam1 mediates activation of Rho, which stimulates Rho kinase (ROCK) (not shown). Dvl also controls activation of Rac to initiate cytoskeletal changes required for polarized cell movements during gastrulation and neural fold closure. TRPM7 is a nonselective divalent channel permeable to Mg^{2+}, Ca^{2+}, and Zn^{2+}; it has a kinase domain with no known function during embryogenesis. Shown is a model of how Mg^{2+} influx through TRPM7 could be controlling the activity of Rac. Mg^{2+} may be influencing Wnt signaling by regulating the levels of reactive oxygen species (ROS) to modulate Dvl to control Rac. Mg^{2+} may also be indirectly affecting the Wnt pathway by augmenting Ca^{2+} signaling.

CONCLUSIONS

Magnesium is an abundant and essential divalent cation in the body. Not surprisingly, disruption of Mg^{2+} homeostasis has the potential to contribute to birth defects as well as to many other human diseases. Knockout mouse studies have uncovered indispensable roles for TRPM7 and TRPM6 during early development. Experiments in zebrafish and *Xenopus* embryos have also yielded insight into the role that Mg^{2+} may be playing not only in gastrulation but also in body pigmentation, brain development, and body movement as well. Importantly, these observations may be relevant to the etiology of many human diseases, including neural tube closure defects and neurological disorders such as epilepsy and Parkinson's disease. The mechanisms by which Mg^{2+} exerts its effects during development and disease are still poorly understood. Mg^{2+} functions as a cofactor in many enzymatic processes and more work is required to understand how Mg^{2+} is able to influence signal transduction processes during development. However, studies in *X. laevis* have been the first to make a connection between Wnt signaling and alterations in magnesium homeostasis. Gain-of-function and loss-of-function studies, such as those that have been conducted in *Xenopus* and zebrafish, have the potential to uncover the mechanisms by which Mg^{2+} exerts its in vivo effects. Researches into the function of Mg^{2+} will give much needed insight into the approaches that may prevent or limit the devastating effects of the diseases caused or made worse by dysregulation of this vital divalent cation.

REFERENCES

Arjona, F.J., de Baaij, J.H., Schlingmann, K.P., Lameris, A.L., van Wijk, E., Flik, G., Regele, S., Korenke, G.C., Neophytou, B., Rust, S., Reintjes, N., Konrad, M., Bindels, R.J., Hoenderop, J.G., 2014. CNNM2 mutations cause impaired brain development and seizures in patients with hypomagnesemia. PLoS Genet. 10, e1004267.

Bui, D.M., Gregan, J., Jarosch, E., Ragnini, A., Schweyen, R.J., 1999. The bacterial magnesium transporter CorA can functionally substitute for its putative homologue Mrs2p in the yeast inner mitochondrial membrane. J. Biol. Chem. 274, 20438–20443.

Chen, H.C., Su, L.T., Gonzalez-Pagan, O., Overton, J.D., Runnels, L.W., 2012. A key role for Mg^{2+} in TRPM7's control of ROS levels during cell stress. Biochem. J.

Chubanov, V., Waldegger, S., Mederos y Schnitzler, M., Vitzthum, H., Sassen, M.C., Seyberth, H.W., Konrad, M., Gudermann, T., 2004. Disruption of TRPM6/TRPM7 complex formation by a mutation in the TRPM6 gene causes hypomagnesemia with secondary hypocalcemia. Proc. Natl. Acad. Sci. U.S.A. 101, 2894–2899.

Clark, K., Langeslag, M., van Leeuwen, B., Ran, L., Ryazanov, A.G., Figdor, C.G., Moolenaar, W.H., Jalink, K., van Leeuwen, F.N., 2006. TRPM7, a novel regulator of actomyosin contractility and cell adhesion. EMBO J. 25, 290–301.

Conradt, A., Weidinger, H., Algayer, H., 1984. Reduced frequency of gestoses in beta-mimetic treated risk pregnancies with added magnesium therapy. Geburtshilfe Frauenheilkd. 44, 118–123.

Cowan, J.A., 2002. Structural and catalytic chemistry of magnesium-dependent enzymes. Biometals 15, 225–235.

de Baaij, J.H., Hoenderop, J.G., Bindels, R.J., 2015. Magnesium in man: implications for health and disease. Physiol. Rev. 95, 1–46.

Doyle, W., Crawford, M.A., Wynn, A.H., Wynn, S.W., 1989. Maternal magnesium intake and pregnancy outcome. Magnes. Res. 2, 205–210.

Fawcett, W.J., Haxby, E.J., Male, D.A., 1999. Magnesium: physiology and pharmacology. Br. J. Anaesth. 83, 302–320.

Funato, Y., Michiue, T., Asashima, M., Miki, H., 2006. The thioredoxin-related redox-regulating protein nucleoredoxin inhibits Wnt-beta-catenin signalling through dishevelled. Nat. Cell Biol. 8, 501–508.

Graschopf, A., Stadler, J.A., Hoellerer, M.K., Eder, S., Sieghardt, M., Kohlwein, S.D., Schweyen, R.J., 2001. The yeast plasma membrane protein Alr1 controls Mg^{2+} homeostasis and is subject to Mg^{2+}-dependent control of its synthesis and degradation. J. Biol. Chem. 276, 16216–16222.

Groenen, P.M., van Rooij, I.A., Peer, P.G., Ocke, M.C., Zielhuis, G.A., Steegers-Theunissen, R.P., 2004. Low maternal dietary intakes of iron, magnesium, and niacin are associated with spina bifida in the offspring. J. Nutr. 134, 1516–1522.

Grubbs, R.D., Maguire, M.E., 1987. Magnesium as a regulatory cation: criteria and evaluation. Magnesium 6, 113–127.

Habas, R., Dawid, I.B., He, X., 2003. Coactivation of Rac and Rho by Wnt/Frizzled signaling is required for vertebrate gastrulation. Genes Dev. 17, 295–309.

Habas, R., Kato, Y., He, X., 2001. Wnt/Frizzled activation of Rho regulates vertebrate gastrulation and requires a novel Formin homology protein Daam1. Cell 107, 843–854.

Hoogeboom, D., Burgering, B.M., 2009. Should I stay or should I go: beta-catenin decides under stress. Biochim. Biophys. Acta 1796, 63–74.

Hurd, T.R., DeGennaro, M., Lehmann, R., 2012. Redox regulation of cell migration and adhesion. Trends Cell Biol. 22, 107–115.

Hurley, L.S., Cosens, G., Theriault, L.L., 1976. Teratogenic effects of magnesium deficiency in rats. J. Nutr. 106, 1254–1260.

IOM, 1997. Dietary Reference Intakes for Calcium, Phosphorus, Magnesium, Vitamin D, and Fluoride. Washington, DC.

Jin, J., Desai, B.N., Navarro, B., Donovan, A., Andrews, N.C., Clapham, D.E., 2008. Deletion of *Trpm7* disrupts embryonic development and thymopoiesis without altering Mg^{2+} homeostasis. Science 322, 756–760.

Jin, J., Wu, L.J., Jun, J., Cheng, X., Xu, H., Andrews, N.C., Clapham, D.E., 2012. The channel kinase, TRPM7, is required for early embryonic development. Proc. Natl. Acad. Sci. U.S.A. 109, E225–E233.

Kaitsuka, T., Katagiri, C., Beesetty, P., Nakamura, K., Hourani, S., Tomizawa, K., Kozak, J.A., Matsushita, M., 2014. Inactivation of TRPM7 kinase activity does not impair its channel function in mice. Sci. Rep. 4, 5718.

Keller, R., 2002. Shaping the vertebrate body plan by polarized embryonic cell movements. Science 298, 1950–1954.

Komiya, Y., Habas, R., 2008. Wnt signal transduction pathways. Organogenesis 4, 68–75.

Komiya, Y., Runnels, L.W., 2015. TRPM channels and magnesium in early embryonic development. Int. J. Dev. Biol.

Li, F.Y., Chaigne-Delalande, B., Kanellopoulou, C., Davis, J.C., Matthews, H.F., Douek, D.C., Cohen, J.I., Uzel, G., Su, H.C., Lenardo, M.J., 2011. Second messenger role for Mg^{2+} revealed by human T-cell immunodeficiency. Nature 475, 471–476.

Li, M., Jiang, J., Yue, L., 2006. Functional characterization of homo- and heteromeric channel kinases TRPM6 and TRPM7. J. Gen. Physiol. 127, 525–537.

Liu, W., Su, L.T., Khadka, D.K., Mezzacappa, C., Komiya, Y., Sato, A., Habas, R., Runnels, L.W., 2011. TRPM7 regulates gastrulation during vertebrate embryogenesis. Dev. Biol. 350, 348–357.

Makrides, M., Crowther, C.A., 2001. Magnesium supplementation in pregnancy. Cochrane Database Syst. Rev. CD000937.

Markova, O., Lenne, P.F., 2012. Calcium signaling in developing embryos: focus on the regulation of cell shape changes and collective movements. Semin. Cell Dev. Biol. 23, 298–307.

Matsuura, T., Kanayama, Y., Inoue, T., Takeda, T., Morishima, I., 1993. cAMP-induced changes of intracellular free Mg^{2+} levels in human erythrocytes. Biochim. Biophys. Acta 1220, 31–36.

McNeill, M.S., Paulsen, J., Bonde, G., Burnight, E., Hsu, M.Y., Cornell, R.A., 2007. Cell death of melanophores in zebrafish trpm7 mutant embryos depends on melanin synthesis. J. Invest. Dermatol. 127, 2020–2030.

Miller, J.C., Landesman, R., 1977. Magnesium deficiency in embryos of *Xenopus laevis*. J. Embryol. Exp. Morphol. 39, 97–113.

Monteilh-Zoller, M.K., Hermosura, M.C., Nadler, M.J., Scharenberg, A.M., Penner, R., Fleig, A., 2003. TRPM7 provides an ion channel mechanism for cellular entry of trace metal ions. J. Gen. Physiol. 121, 49–60.

Murphy, E., 2000. Mysteries of magnesium homeostasis. Circ. Res. 86, 245–248.

Park, M.H., Wong, B.B., Lusk, J.E., 1976. Mutants in three genes affecting transport of magnesium in *Escherichia coli*: genetics and physiology. J. Bacteriol. 126, 1096–1103.

Quamme, G.A., 2010. Molecular identification of ancient and modern mammalian magnesium transporters. Am. J. Physiol. Cell Physiol. 298, C407–C429.

Romani, A., 2007. Regulation of magnesium homeostasis and transport in mammalian cells. Arch. Biochem. Biophys. 458, 90–102.

Romani, A., Scarpa, A., 1990a. Hormonal control of Mg^{2+} transport in the heart. Nature 346, 841–844.

Romani, A., Scarpa, A., 1990b. Norepinephrine evokes a marked Mg^{2+} efflux from liver cells. FEBS Lett. 269, 37–40.

Romani, A.M., 2011. Cellular magnesium homeostasis. Arch. Biochem. Biophys. 512, 1–23.

Runnels, L.W., 2011. TRPM6 and TRPM7: a Mul-TRP-PLIK-cation of channel functions. Curr. Pharm. Biotechnol. 12, 42–53.

Ryazanova, L.V., Hu, Z., Suzuki, S., Chubanov, V., Fleig, A., Ryazanov, A.G., 2014. Elucidating the role of the TRPM7 alpha-kinase: TRPM7 kinase inactivation leads to magnesium deprivation resistance phenotype in mice. Sci. Rep. 4, 7599.

Ryazanova, L.V., Rondon, L.J., Zierler, S., Hu, Z., Galli, J., Yamaguchi, T.P., Mazur, A., Fleig, A., Ryazanov, A.G., 2010. TRPM7 is essential for Mg^{2+} homeostasis in mammals. Nat. Commun. 1, 109.

Sah, R., Mesirca, P., Mason, X., Gibson, W., Bates-Withers, C., Van den Boogert, M., Chaudhuri, D., Pu, W.T., Mangoni, M.E., Clapham, D.E., 2013. Timing of myocardial trpm7 deletion during cardiogenesis variably disrupts adult ventricular function, conduction, and repolarization. Circulation 128, 101–114.

Saris, N.E., Mervaala, E., Karppanen, H., Khawaja, J.A., Lewenstam, A., 2000. Magnesium. An update on physiological, clinical and analytical aspects. Clin. Chim. Acta 294, 1–26.

Schlingmann, K.P., Weber, S., Peters, M., Niemann Nejsum, L., Vitzthum, H., Klingel, K., Kratz, M., Haddad, E., Ristoff, E., Dinour, D., Syrrou, M., Nielsen, S., Sassen, M., Waldegger, S., Seyberth, H.W., Konrad, M., 2002. Hypomagnesemia with secondary hypocalcemia is caused by mutations in TRPM6, a new member of the TRPM gene family. Nat. Genet. 31, 166–170.

Schmitz, C., Dorovkov, M.V., Zhao, X., Davenport, B.J., Ryazanov, A.G., Perraud, A.L., 2005. The channel kinases TRPM6 and TRPM7 are functionally nonredundant. J. Biol. Chem. 280, 37763–37771.

Schmitz, C., Perraud, A.L., Johnson, C.O., Inabe, K., Smith, M.K., Penner, R., Kurosaki, T., Fleig, A., Scharenberg, A.M., 2003. Regulation of vertebrate cellular Mg^{2+} homeostasis by TRPM7. Cell 114, 191–200.

Slack, C., Warner, A.E., Warren, R.L., 1973. The distribution of sodium and potassium in amphibian embryos during early development. J. Physiol. 232, 297–312.

Su, L.T., Agapito, M.A., Li, M., Simonson, W.T., Huttenlocher, A., Habas, R., Yue, L., Runnels, L.W., 2006. TRPM7 regulates cell adhesion by controlling the calcium-dependent protease calpain. J. Biol. Chem. 281, 11260–11270.

Su, L.T., Chen, H.C., Gonzalez-Pagan, O., Overton, J.D., Xie, J., Yue, L., Runnels, L.W., 2010. TRPM7 activates m-calpain by stress-dependent stimulation of p38 MAPK and c-Jun N-terminal kinase. J. Mol. Biol. 396, 858–869.

Su, L.T., Liu, W., Chen, H.C., Gonzalez-Pagan, O., Habas, R., Runnels, L.W., 2011. TRPM7 regulates polarized cell movements. Biochem. J. 434, 513–521.

Tochhawng, L., Deng, S., Pervaiz, S., Yap, C.T., 2013. Redox regulation of cancer cell migration and invasion. Mitochondrion 13, 246–253.

Venu, L., Padmavathi, I.J., Kishore, Y.D., Bhanu, N.V., Rao, K.R., Sainath, P.B., Ganeshan, M., Raghunath, M., 2008. Long-term effects of maternal magnesium restriction on adiposity and insulin resistance in rat pups. Obesity (Silver Spring) 16, 1270–1276.

Visser, D., Langeslag, M., Kedziora, K.M., Klarenbeek, J., Kamermans, A., Horgen, F.D., Fleig, A., van Leeuwen, F.N., Jalink, K., 2013. TRPM7 triggers Ca^{2+} sparks and invadosome formation in neuroblastoma cells. Cell Calcium 54, 404–415.

Voets, T., Nilius, B., Hoefs, S., van der Kemp, A.W., Droogmans, G., Bindels, R.J., Hoenderop, J.G., 2004. TRPM6 forms the Mg^{2+} influx channel involved in intestinal and renal Mg^{2+} absorption. J. Biol. Chem. 279, 19–25.

Volpe, P., Vezu, L., 1993. Intracellular magnesium and inositol 1,4,5-trisphosphate receptor: molecular mechanisms of interaction, physiology and pharmacology. Magnes. Res. 6, 267–274.

Walder, R.Y., Landau, D., Meyer, P., Shalev, H., Tsolia, M., Borochowitz, Z., Boettger, M.B., Beck, G.E., Englehardt, R.K., Carmi, R., Sheffield, V.C., 2002. Mutation of TRPM6 causes familial hypomagnesemia with secondary hypocalcemia. Nat. Genet. 31, 171–174.

Walder, R.Y., Yang, B., Stokes, J.B., Kirby, P.A., Cao, X., Shi, P., Searby, C.C., Husted, R.F., Sheffield, V.C., 2009. Mice defective in Trpm6 show embryonic mortality and neural tube defects. Hum. Mol. Genet. 18, 4367–4375.

Wallingford, J.B., Habas, R., 2005. The developmental biology of dishevelled: an enigmatic protein governing cell fate and cell polarity. Development 132, 4421–4436.

Wang, F.L., Wang, R., Khairallah, E.A., Schwartz, R., 1971. Magnesium depletion during gestation and lactation in rats. J. Nutr. 101, 1201–1209.

Webb, S.E., Miller, A.L., 2003. Calcium signalling during embryonic development. Nat. Rev. Mol. Cell Biol. 4, 539–551.

Wei, C., Wang, X., Chen, M., Ouyang, K., Song, L.S., Cheng, H., 2009. Calcium flickers steer cell migration. Nature 457, 901–905.

Woudenberg-Vrenken, T.E., Sukinta, A., van der Kemp, A.W., Bindels, R.J., Hoenderop, J.G., 2011. Transient receptor potential melastatin 6 knockout mice are lethal whereas heterozygous deletion results in mild hypomagnesemia. Nephron Physiol. 117, p11–p19.

Yee, N.S., Zhou, W., Liang, I.C., 2011. Transient receptor potential ion channel Trpm7 regulates exocrine pancreatic epithelial proliferation by Mg^{2+} sensitive Socs3a signaling in development and cancer. Dis. Model. Mech. 4, 240–254.

Zhou, H., Clapham, D.E., 2009. Mammalian MagT1 and TUSC3 are required for cellular magnesium uptake and vertebrate embryonic development. Proc. Natl. Acad. Sci. U.S.A. 106, 15750–15755.

Chapter 29

Magnesium, Vascular Function, and Hypertension

Francisco J. Rios[1], Augusto C. Montezano[1], Tayze T. Antunes[2], Rhian M. Touyz[1]

[1]*University of Glasgow, Glasgow, Scotland;* [2]*University of Ottawa, Ottawa, ON, Canada*

INTRODUCTION

Magnesium (Mg^{2+}) is an essential ion that plays an important role in the physiological regulation of multiple organs, including the heart, vessels, brain, nerves, and skeletal muscle (de Baaij et al., 2015; Touyz, 2004). It is primarily an intracellular cation, and its levels are tightly regulated by various transporters. At the subcellular level, magnesium influences contractile proteins, modulates ion channels, acts as an essential cofactor in the activation of ATP, controls metabolic regulation of energy-dependent cytoplasmic and mitochondrial pathways, regulates glycolysis, influences DNA synthesis and transcription, induces protein synthesis, and promotes cell growth (Bowman et al., 2012; Touyz and Yao, 2003). As the second most abundant intracellular cation after potassium, magnesium influences more than 600 enzymes, among the most important of these being enzymes that hydrolyze and transfer phosphate groups, including those associated with reactions involving ATP (Pontes et al., 2015). All ATP-dependent enzymatic reactions have an absolute need for magnesium, which is essential for human life and health.

Because of the key functions of magnesium, tight regulation of plasma/serum concentrations (0.7–1.1 mmol/L) is critical. Body magnesium homeostasis is regulated by the kidney, gastrointestinal (GI) tract, and bone (de Baaij et al., 2015; Romani, 2011). Hypomagnesemia occurs when renal excretion of magnesium is increased and/or when GI absorption is reduced, and it is associated with a range of pathologies, including cardiovascular disease, neurological disorders, metabolic disorders, and immune disease (Hoorn and Zietse, 2013; Brandao et al., 2013). Symptoms associated with hypomagnesemia are nonspecific and include muscle spasm, fatigue, and depression. Major causes of hypomagnesemia include reduced dietary intake, drugs (e.g., diuretics, proton-pump inhibitors, immunosuppressants, antibiotics, anticancer drugs), increased loss (vomiting, diarrhea), and genetic hypomagnesemia (mutations of Mg^{2+} transporters; Konrad and Schlingmann, 2014; Janett et al., 2015). Hypomagnesemia may also occur as a consequence of underlying renal, GI, and bone disease (Pham et al., 2014). Hypermagnesemia is less common and is usually a direct consequence of administration of magnesium itself or magnesium-containing agents, such as Epsom salts, cathartics, or enemas. Hypermagnesemia may also occur in renal failure. Despite the clinical significance of disturbances of magnesium homeostasis, plasma magnesium levels, unlike potassium, sodium, and calcium, are not routinely assessed clinically; consequently, many patients may have undiagnosed hypomagnesemia.

In the cardiovascular system, small reductions in extracellular Mg^{2+} levels ($[Mg^{2+}]_e$) and/or intracellular free Mg^{2+} concentration ($[Mg^{2+}]_i$) can significantly affect cardiac and vascular function. In particular, hypomagnesemia influences excitation–contraction coupling, myocardial metabolism, vascular contractility, vasodilation, and endothelial function, processes also associated with increased intracellular free Ca^{2+} concentrations ($[Ca^{2+}]_i$; Altura and Altura, 1995; Altura et al., 1991). Magnesium has been considered as a natural calcium antagonist because it competes with calcium for binding sites on proteins, enzymes, and calcium transporters (Alhosaini and Leehey, 2015). Magnesium is also an important vasodilator; hence it regulates peripheral vascular resistance (Van Laecke et al., 2012). Thus magnesium plays an important physiological role in heart function and blood pressure regulation.

Over the past 2 decades, increasing clinical and experimental evidence indicates an association between low plasma/serum magnesium levels and cardiovascular diseases. Large epidemiological studies, meta-analyses, and prospective observational studies have demonstrated inverse correlations between plasma/serum magnesium levels and cardiovascular events in the general population (Touyz et al., 1992; Jiang et al., 2016; Lacson Jr. et al., 2015; Choi and Bae, 2015; Adebamowo et al., 2015). Perturbations in magnesium handling and homeostasis are now being considered as important pathophysiological processes associated with heart disease, hypertension, diabetes, and preeclampsia. The present

Molecular, Genetic, and Nutritional Aspects of Major and Trace Minerals. http://dx.doi.org/10.1016/B978-0-12-802168-2.00029-4

chapter discusses the role of magnesium in the cardiovascular system in health and disease, focusing specifically on the vascular system and hypertension.

MAGNESIUM METABOLISM

The US Food and Nutrition Board recommends a daily magnesium intake of 420 mg for men and 320 mg for women (Institute of Medicine). However, the reported dietary adult magnesium intake in the United States and Europe is less than this, in large part because of reduced consumption of fruits and vegetables (magnesium rich) and increased consumption of refined grains and processed foods (magnesium poor; Fine et al., 1991; Quann et al., 2015; An et al., 2015; Lichton, 1989; Lameris et al., 2015). Accordingly, a significant proportion of the population may have subclinical hypomagnesemia, and, as such, the importance of magnesium deficiency may clinically be underestimated.

Body magnesium homeostasis depends on the interactions among the GI system, responsible for magnesium absorption from dietary intake; the bone, which stores magnesium; and the kidneys, which regulate urinary excretion (Fine et al., 1991; Quann et al., 2015; An et al., 2015; Lichton, 1989; Lameris et al., 2015). Intestinal magnesium absorption is normally balanced by renal magnesium excretion. In patients consuming low dietary magnesium or those in negative magnesium balance, relative magnesium absorption and exchange from bone are increased as a compensation to restore normal magnesium balance and to maintain plasma/serum magnesium concentrations in the normal range (Fine et al., 1991).

MAGNESIUM ABSORPTION IN THE GASTROINTESTINAL TRACT

Approximately 40% of dietary magnesium is absorbed by the GI tract, primarily in the distal small intestine and in the colon (Blaine et al., 2015). Two types of magnesium transport systems have been identified in the intestine: (1) paracellular transport through paracellin-1 (claudin-16), which involves absorption through the tight junctions between epithelial cells and is passive (de Baaij et al., 2015; Hou et al., 2009; Kladnitsky et al., 2015) and (2) transcellular transport, which is an active process transporting magnesium from the intestinal lumen through the epithelial cells, which involves novel magnesium channels, namely transient receptor potential melastatin cation channels-6 and -7 (TRPM6 and TRPM7, respectively; Komiya and Runnels, 2015; Runnels et al., 2001). The exact physiologic factors that regulate GI magnesium absorption are unclear, but the dietary magnesium content, vitamin D metabolites, parathyroid hormone, calcitonin, calcium-sensing receptor, insulin, and vasoactive intestinal peptide may be important (Kladnitsky et al., 2015; Romani and Maguire, 2002; Bijvelds et al., 1998; Hardwick et al., 1991).

MAGNESIUM EXCRETION BY THE KIDNEY

Renal magnesium excretion, which reflects dietary intake and the amount of magnesium absorbed from the GI tract, involves filtration by the glomerulus and reabsorption in the nephron [especially the ascending limb; (An et al., 2015; Lichton, 1989; Lameris et al., 2015)]. The nephron recovers 95–99% of the daily magnesium filtered by the glomeruli; the remaining 1–5% of magnesium is excreted into the urine. Approximately 20% of magnesium is absorbed in the proximal tubule, 70% in the thick ascending limb, and 10% in the distal convoluted tubule (DCT). In the proximal tubule, magnesium reabsorption is primarily a passive, paracellular process. In the thick ascending limb of Henles' loop, magnesium reabsorption follows the paracellular pathway and depends on tight junction permeability, which is influenced by claudin proteins, particularly claudins 16 and 19. The DCT determines the final urinary magnesium concentration because this is the final site for magnesium reabsorption. Hormones regulating renal magnesium excretion include vasopressin, aldosterone, and thyroid hormone, which increase magnesuria, as well as parathyroid hormone, insulin, and vitamin D, which reduce magnesuria (Quamme, 1989, 1997; Felsenfeld et al., 2015).

COMPARTMENTALIZATION OF MAGNESIUM IN THE BODY

Magnesium is distributed in three major compartments of the body: approximately 65% in the mineral phase of bone, approximately 34% in muscle, and approximately 1% in plasma and interstitial fluid (de Baaij et al., 2015). Intracellular magnesium stores, particularly in the endoplasmic reticulum and mitochondria, are important in maintaining plasma levels, and when plasma levels are reduced, there is increased flux from intracellular stores. Unlike plasma calcium, where 40% is protein bound, only approximately 20% of plasma magnesium is protein bound. Accordingly, changes in plasma protein concentration have less effect on plasma magnesium than on plasma calcium levels.

REGULATION OF CELLULAR MAGNESIUM HOMEOSTASIS

For magnesium to regulate cellular processes, magnesium itself must be tightly controlled within the cell. Because it is a critical ion/cofactor in many biochemical reactions, even small changes in $[Mg^{2+}]_i$ can have significant effects on signaling pathways, which in vascular and cardiac cells influence contraction and relaxation.

Although magnesium is the most abundant cytosolic divalent cation, relatively little is known about intracellular magnesium homeostasis, and mechanisms controlling $[Mg^{2+}]_i$ are still poorly understood (Quamme, 2010; Goytain and Quamme, 2005a, 2008; Kraft and Harteneck, 2005). It is only since the early 2000s, when specific magnesium transporters were identified, that the understanding of cellular magnesium handling has become clearer. Magnesium enters cells along a concentration gradient and through magnesium channels (Quamme, 2010; Goytain and Quamme, 2005a, 2008; Kraft and Harteneck, 2005), and it is extruded from cells via magnesium exchangers, such as the Na^+-Mg^{2+} exchanger and the Mg^{2+}-ATPase.

Within cells magnesium is compartmentalized into mitochondria, endoplasmic/sarcoplasmic reticulum, and nuclei, and it is highly regulated by intracellular buffering processes as well as by hormones and vasoactive agents, such as glutathione, estrogen, endothelin-1 (ET-1), angiotensin II (Ang II), bradykinin, aldosterone, vasopressin, and norepinephrine (Herencia et al., 2015; Yogi et al., 2013).

CELLULAR MAGNESIUM INFLUX

Recent studies have now identified and characterized specific transmembrane magnesium transporters that are collectively responsible for magnesium influx (Fig. 29.1).

One of the first mammalian magnesium transporters to be identified was mitochondrial RNA splicing-2, which is responsible for mitochondrial magnesium uptake (Bui et al., 1999; Kolisek et al., 2003). Other proteins shown to regulate magnesium homeostasis include magnesium transporter subtype 1 (MagT1), the solute carrier (SLC) family 41 subtypes 1 and 2 (SLC41A1, SLAC41A2, respectively), ancient conserved domain protein-2, Huntington-interacting proteins (HIP14 and HIP14L), and non-imprinted in Prader–Willi/Angelman syndrome subtype 2 protein (Sahni et al., 2007; Goytain and Quamme, 2005b; Goytain et al., 2007, 2008). However, the physiological significance and mechanisms regulating these proteins still await clarification.

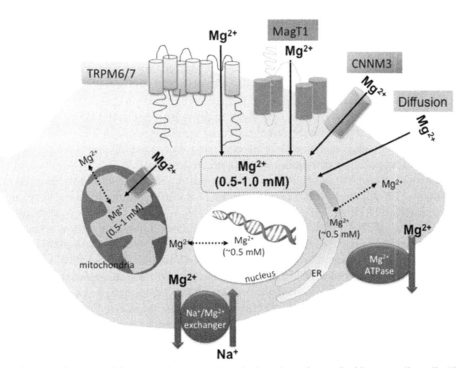

FIGURE 29.1 Diagram demonstrating some of the magnesium transporters that have been characterized in mammalian cells. These play an important role in cellular magnesium homeostasis. Major channels that transport magnesium into cells across the cell membrane include transient receptor potential melastatin cation channel-6 (TRPM6) and -7 (TRPM7), magnesium transporter subtype 1 (MagT1), and cyclin and CBS domain divalent metal cation transport mediator 3 (CNNM3). Within the cell, magnesium is compartmentalized in mitochondria, the endoplasmic reticulum (ER), and nucleus. Magnesium is transported out of cells through putative exchangers, including the Na^+/Mg^{2+} exchanger. The intracellular magnesium concentration is tightly regulated and maintained at 0.5–1.0 mmol/L.

Paracellin-1 (claudin-16), a member of the claudin family of tight-junction proteins (Simon et al., 1999), has been identified as a paracellular (between cells) Mg^{2+} transporter, particularly important in Mg^{2+} absorption in the GI tract (Simon et al., 1999). Claudins 16 and 19 form the pore-permitting paracellular magnesium reabsorption pathway in the thick ascending limb of the kidney (Yu, 2015; Naeem et al., 2011).

TRPM6 (and TRPM7) were identified as magnesium channels through genetic screening of patients with primary hypomagnesaemia and secondary hypocalcaemia (Schlingmann et al., 2002; Walder et al., 2002). TRPM6 and TRPM7 are ion channels covalently linked to a kinase domain; accordingly, they have been termed *chanzymes*. TRPM7 is a ubiquitously expressed divalent cation channel that is responsible for most of the magnesium entering cells (Nadler et al., 2001; Yogi et al., 2011a). TRPM6 is expressed primarily in the kidney and GI tract (Nadler et al., 2001). Whereas TRPM6 and TRPM7 are mainly magnesium channels, they can also transport other ions, such as calcium. The only known selective plasma membrane magnesium transporter is currently MagT1 (Wolf and Trapani, 2011). Of the numerous magnesium channels identified, only TRPM7, and to a lesser extent TRPM6 and MagT1, have been identified in the cardiovascular system; accordingly, only these transporters will be highlighted here.

TRPM6 AND TRPM7 CATION CHANNELS

TRPM6/7 proteins are members of the transient receptor potential superfamily (Liu and Montell, 2015) and have been implicated in cardiovascular pathologies (Watanabe et al., 2009; Montezano et al., 2010). TRPM6 and TRPM7 share approximately 50% amino acid sequence homology, are constitutively active, and contain protein kinase domains in their –COOH termini with sequence similarity to alpha kinases (Clark et al., 2008).

TRPM6 mRNA shows a limited expression pattern, with the highest levels in the intestine and the DCT of the kidney (Chubanov et al., 2005, 2007; Voets et al., 2004). The exact factors that regulate TRPM6 remain unclear, but 17 β-estradiol and $[Mg^{2+}]_i$ are important (Groenestege et al., 2006; Ferrè et al., 2011; Runnels, 2011). Epidermal growth factor (EGF) has been identified as an autocrine/paracrine magnesiotropic hormone (Groenestege et al., 2007; Thebault et al., 2009). EGF binds to its receptor (EGFR) at the basolateral membrane of the DCT and stimulates trafficking from the cytosol to the cell membrane, resulting in increased TRPM6 expression at the plasma membrane. The role of EGF/EGFR signaling in magnesium homeostasis is further supported by the clinical observation that patients treated with monoclonal antibodies directed against the EGFR (e.g., cetuximab) develop hypomagnesemia (Schrag et al., 2005; Tejpar et al., 2007; Costa et al., 2011).

TRPM7 is ubiquitously expressed and is found in cardiac and vascular tissue (Cuffe et al., 2015; Guo et al., 2014). Similar to TRPM6, TRPM7 is a tetrameric channel, with each subunit comprising six transmembrane regions with a pore region between the fifth and sixth transmembrane domains (Runnels, 2011). The –COOH cytosolic domain possesses a kinase domain that regulates autophosphorylation of the channel through, as yet, unknown processes. Although it is clear that the two domains are physically linked, it has recently been demonstrated that the kinase can be cleaved from TRPM7 by caspase-8 (Krapivinsky et al., 2014). However, the functional significance of this cleaved domain awaits clarification. TRPM7 preferentially transports magnesium and to a lesser extent calcium and other divalent cations. Targeted disruption of the TRPM7 gene in cell lines causes cell death, underpinning a nonredundant role of this channel in cell survival and health. TRPM7 is constitutively active, and similar to TRPM6, it is negatively regulated by changes in cytosolic magnesium, $[Mg^{2+}]_i$, or Mg^{2+}-ATP (Jin et al., 2008; Deason-Towne et al., 2011). Multiple stimulatory and inhibitory factors regulate TRPM7 activity (Fig. 29.2).

The unique characteristic of TRPM6/7 to act as a cation channel and at the same time as a kinase suggests that these proteins are involved in regulating transmembrane magnesium transport and intracellular signaling (Demeuse et al., 2006). Whether activation of the kinase is essential for channel activity is still unclear (Yamaguchi et al., 2001; Penner and Fleig, 2007; Matsushita et al., 2005; Yogi et al., 2011b) although it has recently been suggested that the kinase domain acts as a sensor of magnesium status and provides coordination of cellular and systematic responses to magnesium deprivation (Cuffe et al., 2015). This was demonstrated in transgenic mice, which have an inactive TRPM7 kinase but a functional channel, exhibit resistance to dietary magnesium deprivation, and have prolonged survival (Ryazanova et al., 2014).

Numerous TRPM7 kinase substrates, all of which are important in regulating vascular cell function, have been identified. These include annexin-1, myosin IIA heavy chain, and calpain (Dorovkov and Ryazanov, 2004; Clark et al., 2006; Su et al., 2006). Annexin-1 is an endogenous mediator of the antiinflammatory actions of glucocorticoids and is implicated in cell growth and apoptosis. Myosin II heavy chain is involved in cell migration, growth, apoptosis, contraction, and cytoskeletal organization. TRPM7 also activates calpain and initiates the disassembly or turnover of peripheral adhesion complexes, which play an important role in vascular cell adhesion, contraction, and migration, processes important for vascular remodeling in cardiovascular disease.

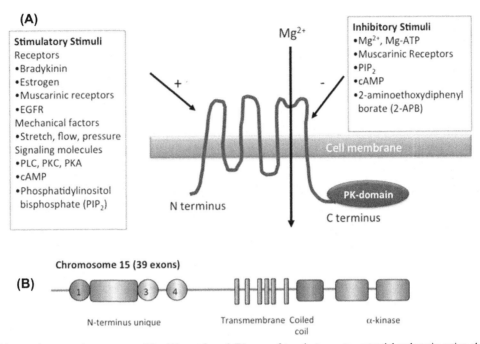

FIGURE 29.2   Diagram demonstrating structure of the (A) protein and (B) gene of transient receptor potential melastatin cation channel-7 (TRPM7). The TRPM7 gene is located on chromosome 15 and has 39 exons. TRPM7 contains a channel domain, linked to a protein kinase domain at the C-terminus. Multiple stimulatory (+) and inhibitory (−) factors regulate activity of TRPM7. *EGFR*, epidermal growth factor receptor; *PLC*, phospholipase C; *PKC*, protein kinase C; *PKA*, protein kinase A; *cAMP*, cyclic AMP.

MAGNESIUM TRANSPORTER SUBTYPE 1

MagT1 is a ubiquitously expressed magnesium channel that is essential for magnesium influx, cell growth, and development (Wolf and Trapani, 2011; Zhou and Clapham, 2009). Knockdown of MagT1 is embryonically lethal, indicating the critical role of this transporter in biological processes and development (Zhou and Clapham, 2009). In conditions in which TRPM7 is downregulated or deficient, cell viability and magnesium homeostasis can be rescued by increased expression of MagT1, indicating the similar functional roles of these transporters. Hence, when cells are exposed to suboptimal cellular magnesium levels, they respond by transcriptional upregulation of MagT1 to increase magnesium influx, possibly as a protective mechanism to prevent intracellular magnesium deficiency (Wolf and Trapani, 2011).

MagT1 has also been identified as an important regulator of magnesium influx in T lymphocytes. MagT1 induces a rapid magnesium influx in stimulated T cells and as such has been considered a major intracellular signaling mechanism in immune cells (Li et al., 2011). Patients with MagT1 mutations exhibit immunodeficiency characterized by CD4 lymphopenia, viral infections, and defective T lymphocyte activation due to abnormal magnesium signaling (Li et al., 2011; Wu and Veillette, 2011). Accordingly, MagT1 has been suggested to be a putative target for novel therapies in patients with defective T-lymphocyte function.

Despite its widespread distribution, there is a paucity of information about the expression of MagT1 in the cardiovascular system. A few studies demonstrated that MagT1 is present in the heart (Zhou and Clapham, 2009; Tashiro et al., 2013), but nothing is known about its expression in the vascular system, and the exact biological function of cardiac MagT1 is unclear.

MAGNESIUM EFFLUX FROM CELLS

Magnesium efflux occurs against an electrochemical gradient; therefore an energy-coupled mechanism for its extrusion must be present. Magnesium efflux appears to be regulated by at least two pathways: Na⁺-Mg²⁺ exchange driven by the sodium gradient and sodium-independent passive magnesium transport via Mg²⁺-permeable channels (Touyz and Schiffrin, 1999; Cefaratti and Romani, 2007). Sodium-dependent magnesium transport occurs mainly via the Na⁺/Mg²⁺ exchanger and has been demonstrated in many cell types, including vascular smooth muscle cells and cardiomyocytes (Touyz and Schiffrin, 1999; Cefaratti and Romani, 2007). On the other hand, sodium-independent transport, demonstrated mainly in erythrocytes and hepatic cells, involves calcium (Ca²⁺/Mg²⁺ exchanger), manganese (Mn²⁺/Mg²⁺ antiporter), and chloride

(Cl^-/Mg^{2+} cotransporter) dependent mechanisms (Fagan and Romani, 2000; Almulla et al., 2006). Although these transport processes were demonstrated more than 35 years ago at the functional level, the molecular identity of the proteins involved remains unknown.

MAGNESIUM AND VASCULAR FUNCTION

Magnesium directly influences vascular function as evidenced by clinical and experimental studies in which magnesium infusion reduced peripheral vascular resistance and blood pressure by promoting vasodilation (Dhungel et al., 2012; Murata et al., 2011). Decreasing the extracellular magnesium concentration has opposite effects—namely, promoting vasoconstriction (Laurant et al., 1997).

Mechanisms by which magnesium induces vasorelaxation are not fully understood, but they may relate to antagonistic effects on calcium because an increase in extracellular magnesium inhibits calcium-sensitive contraction of isolated vessels (Yoshimura et al., 1997; Gilbert D'Angelo et al., 1992). In vascular smooth muscle cells, magnesium acts extracellularly by inhibiting transmembrane calcium transport and calcium entry and decreasing contractile actions of vasoactive agents; it acts intracellularly as a calcium antagonist, thereby modulating the vasoconstrictor actions of increased $[Ca^{2+}]_i$ (Nakajima et al., 1997; McHugh and Beech, 1996).

Magnesium also influences vascular tone by modulating endothelial-derived nitric oxide (NO) production, a potent vasodilator, and by influencing vascular effects of vasoconstrictor and vasodilator agents (Basralı et al., 2015a). Magnesium stimulates activation of endothelial nitric oxide synthase (eNOS), which generates NO, leading to endothelium-dependent vasorelaxation. An acute reduction of extracellular magnesium leads to a transient vasodilation followed by sustained contraction. When there is endothelial damage, low magnesium induces a sustained contraction without the transient vasorelaxation phase (Ku and Ann, 1991; Gold et al., 1990). These findings suggest that magnesium could have a dual effect in the regulation of vascular reactivity, depending on the integrity of the endothelium. An intact endothelium protects against the detrimental effects of acute hypomagnesemia; however, in the presence of an injured endothelium, as is the case in many cardiovascular diseases, the compensatory vasodilatory effect is absent and low magnesium has a direct constrictor effect on vascular smooth muscle.

Magnesium may also influence production of certain vasoactive agents such as ET-1, a potent vasoconstrictor, and prostacyclin, an endothelial-derived vasodilator (Weglicki et al., 1992). In magnesium-deficient rats, plasma ET-1 levels are elevated, whereas in magnesium-supplemented rats plasma ET-1 levels are reduced (Weglicki et al., 1992). Increased magnesium attenuates ET-1- and Ang II-induced contraction whereas reduced magnesium levels augment contractile responses (Laurant and Berthelot, 1996; Kemp et al., 1993).

Another possible mechanism by which magnesium influences vascular function and contractility is via its antioxidant actions (Weglicki et al., 1996; Fiorani et al., 2015). Increasing evidence shows that the vasculature is a rich source of reactive oxygen species (ROS), which directly influences vascular smooth muscle cell function (Touyz, 2000). Magnesium has antioxidant properties that could attenuate the damaging actions of ROS in vascular cells. Moreover, magnesium may influence nicotinamide adenine dinucleotide phosphate (NADPH) oxidase-derived generation of ROS as well as influence antioxidant systems in vascular cells, such as Nrf-2, a master transcription factor that regulates expression of antioxidant genes (Kolisek et al., 2015; Hur et al., 2010). These effects may be particularly important in hypertension, diabetes, and atherosclerosis, in which $[Mg^{2+}]_i$ is reduced, NADPH oxidase activity is increased, oxidative stress is amplified, and Nrf-2 and antioxidant systems are downregulated (Kolisek et al., 2015). Low cellular magnesium levels also promote vascular inflammation and growth by increasing vascular smooth muscle cell activation of ceramides, which regulate cell differentiation, proliferation, and apoptosis, and through increased activation of nuclear factor kappa-light-chain-enhancer of activated B cells, which regulates expression of proinflammatory genes, including cytokines, chemokines, and adhesion molecules (Altura et al., 2012).

MAGNESIUM AND HYPERTENSION

Extensive experimental evidence demonstrates that magnesium deficiency is associated with increased blood pressure and that experimental models of hypertension display hypomagnesemia (Yogi et al., 2011b; Basralı et al., 2015b; Rondón et al., 2014; Chaumais et al., 2012). We showed that in spontaneously hypertensive rats, magnesium supplementation attenuated, but did not prevent, development of hypertension (Touyz and Milne, 1999). We also demonstrated that in mice with genetic magnesium deficiency, blood pressure is elevated (Yogi et al., 2011b). However, some studies failed to show a relationship between magnesium and hypertension in experimental models (Rayssiguier et al., 1992), possibly because of differences in compounds and concentrations of magnesium supplemented or because of disparities in the experimental models studied.

Clinical studies have also suggested important associations between magnesium and blood pressure. Many epidemiological and clinical observational studies demonstrated an inverse correlation between plasma/serum magnesium levels and blood pressure whereas others showed only a weak or modest relationship. This association has also been demonstrated in children where children with prehypertension or hypertension had lower serum magnesium levels than normotensive healthy counterparts (Guerrero-Romero et al., 2016). However, some clinical studies failed to demonstrate any changes in circulating magnesium in hypertension (Jee et al., 2002). In human investigations, studies as early as in the 1930s–1940s showed that intravenous $MgSO_4$ infusion induces a significant vasodilatory effect and blood pressure lowering (Winkler et al., 1940). In severe hypertension, $MgSO_4$ reduced blood pressure from approximately 250 mm Hg to approximately 160 mm Hg, indicating a profound hypotensive action (Winkler et al., 1940; Kelly et al., 1961).

Studies on dietary magnesium and risk of hypertension have been inconclusive, with some trials showing blood pressure lowering by oral magnesium supplementation whereas others failed to demonstrate any effect (Rosanoff and Plesset, 2013; Kass et al., 2012). However, most clinical dietary studies lacked accurate measures of actual dietary magnesium uptake. In the Prevention of Renal and Vascular End-Stage Disease (PREVEND) study, a large prospective study of 5511 normotensive participants, plasma and 24-h urinary magnesium levels were measured at baseline and subjects were followed for more than 9 years and tracked for development of incident hypertension (Joosten et al., 2013). Results demonstrated an inverse dose–response association between urinary magnesium excretion, an estimate of dietary absorption, and risk of hypertension. Higher urinary magnesium excretion (increased dietary absorption) was associated with a lower risk of hypertension, confirming other observations of inverse associations between magnesium intake and blood pressure (Joosten et al., 2013). However, the PREVEND study did not show any relationship between plasma magnesium and risk of hypertension, which may relate in part to the fact that circulating magnesium is not a good index of dietary magnesium intake. Moreover, serum/plasma magnesium levels do not necessarily reflect intracellular levels, which is the biologically significant fraction.

Regarding magnesium supplementation studies, a systematic review by the Cochrane Hypertension Group reported that magnesium supplementation causes a small reduction of diastolic blood pressure (Dickinson et al., 2006) whereas another meta-analysis showed a dose-dependent blood pressure lowering effect of magnesium supplementation (Yamamoto et al., 1995). A recent meta-analysis of 34 randomized trials involving 2028 participants showed that oral Mg supplementation (368 mg/day during 3 months) was associated with an increase in serum Mg levels and a significant reduction in both systolic and diastolic blood pressure (about 2.00 and 1.78 mm Hg, respectively) (Zhang et al., 2016). Several supplementation studies were negative and failed to show any antihypertensive effect of magnesium. Reasons for the conflicting data may relate to the heterogeneity of cohorts studied, challenges in accurately measuring magnesium status, and different preparations of magnesium supplements used.

The exact mechanisms by which magnesium influences the pathophysiology of hypertension are unclear, but effects on vascular function through its calcium antagonistic actions, stimulation of NO production, and antiinflammatory and antioxidant actions, together with improved endothelial function, might be important. In addition to promoting vasorelaxation, magnesium may affect vascular structure. We demonstrated that 6 months of magnesium supplementation in diuretic-treated hypertensive women was associated with decreased blood pressure, reduced aortic stiffness, and decreased carotid intima–media ratio, indicating improved vascular structure with associated blood pressure lowering by chronic magnesium supplementation (Cunha et al., 2013).

Taken together, the data are not consistent regarding the hypotensive actions of magnesium and additional well-controlled clinical studies are urgently needed. However, although not all hypertensive patients may benefit from magnesium supplementation/treatment, subgroups of patients may be magnesium sensitive, including in particular individuals who have proven magnesium deficiency, the elderly, African Americans, patients with resistant or severe hypertension, patients with/at risk for preeclampsia, and patients treated with magnesium-losing diuretics (Berhan and Berhan, 2015; Dørup et al., 1993).

TRPM7 AND CARDIOVASCULAR DISEASE

Potential causes for magnesium deficiency in hypertension may relate to reduced dietary intake, increased magnesium excretion, and possibly altered function of magnesium transporters. Studies in cell culture and experimental models suggest that TRPM7 dysregulation may play a role in endothelial dysfunction, arterial calcification, fibrosis, vascular remodeling, and hypertension (Paravicini et al., 2009; He et al., 2005). We and others demonstrated that endothelial cells and vascular smooth muscle cells from rodents and humans possess TRPM7 cation channels and that TRPM7 is critically involved in regulating magnesium influx, viability, proliferation, and contraction/dilation in vascular cells (Touyz et al., 2006; Sontia et al., 2008; Zholos et al., 2011; Yogi et al., 2009). TRPM7 is important in endothelial cell adhesion and vascular development (Zeng et al., 2015). Shear stress stimulates cytosol-to-membrane translocation of TRPM7 and increases TRPM7-like currents in vascular cells (Oancea et al., 2006). These observations suggest that TRPM7 may act as a mechanotransducer, which could be important in pathological responses to vascular injury, particularly in the context of cardiovascular disease. In addition, interaction between TRPM7 and macrophages promotes vascular adventitial remodeling in experimental aortic constriction (Li et al., 2014).

Ang II, bradykinin, and aldosterone, important vasoactive hormones, acutely regulate vascular TRPM7 by inducing phosphorylation and chronically regulate it by increasing expression at the mRNA and protein levels (Touyz et al., 2006; Sontia et al., 2008; Zholos et al., 2011; Yogi et al., 2009; Zeng et al., 2015; Oancea et al., 2006). Downregulation of vascular TRPM7 by small interfering RNA reduced basal $[Mg^{2+}]_i$ and Ang II-stimulated $[Mg^{2+}]_i$ transients and it attenuated growth in cultured vascular smooth muscle cells. These findings confirmed that TRPM7 is a key regulator of vascular cell magnesium homeostasis and that it plays a major role in vascular cell function (Zholos et al., 2011; Yogi et al., 2009). We showed that aldosterone, important in blood pressure regulation, regulates TRPM7-regulated magnesium influx and that this process is independent of the kinase domain because cells lacking the kinase domain exhibited similar magnesium fluxes as in cells with functionally intact TRPM7 kinase (Yogi et al., 2011, 2013). We also demonstrated that the kinase and channel domains of TRPM7 differentially regulate calpain, annexin-1, and NADPH oxidase activity, indicating the complexities of TRPM7-mediated cell signaling by aldosterone and other vasoactive agents (Yogi et al., 2013).

To examine the pathophysiological significance of TRPM7 in the development of hypertension, we studied blood pressure and vascular responses to Ang II, a potent prohypertensive peptide, in TRPM7 kinase-deficient mice (Antunes et al., 2016). Our studies demonstrated that TRPM7 kinase deficiency promotes development of hypertension and aggravates endothelial dysfunction and vascular inflammation, processes associated with significant hypomagnesemia (Antunes et al., 2016). These data identified TRPM7 as a new mechanism important in Ang II-mediated hypertension. Further studies are needed to fully elucidate the role of TRPM6/7 and other magnesium transporters in the pathogenesis of hypertension.

CONCLUSIONS

Magnesium is essential for cell viability and health. It plays an important role in regulating vascular function, through its effects on NO production, by influencing ATP-dependent enzymes and by modulating intracellular levels of calcium, sodium, and potassium, which are important players in the regulation of endothelial function and vascular smooth muscle contraction and relaxation. For magnesium to effectively exert its cellular effects, magnesium itself must be regulated. Recent studies have identified several channels responsible for transmembrane magnesium transport, such as TRPM6, TRPM7, and MagT1. There is still a paucity of information on the molecular characteristics of these transporters, especially in the cardiovascular system. Perturbations in magnesium homeostasis, possibly due to altered TRPM7 function, have been associated with altered vascular reactivity and arterial remodeling, processes important in blood pressure elevation. Growing epidemiological, clinical, and experimental evidence suggest a role for disturbed magnesium metabolism in hypertension, with numerous studies showing an inverse relationship between plasma/serum magnesium and blood pressure. However, data are inconsistent, with some studies failing to demonstrate any relationship. Additional studies are needed to further elucidate the exact role of magnesium in clinical hypertension, and the potential therapeutic potential of this important cation still awaits confirmation.

REFERENCES

Adebamowo, S.N., Spiegelman, D., Flint, A.J., Willett, W.C., Rexrode, K.M., 2015. Intakes of magnesium, potassium, and calcium and the risk of stroke among men. Int. J. Stroke 10, 1093–1100.

Altura, B.M., Altura, B., 1995. Magnesium in cardiovascular biology: an important link between cardiovascular risk factors and atherogenesis. Cell Mol. Bio Res. 41, 347–359.

Altura, B.M., Zhang, A., Altura, B.T., 1991. Magnesium, hypertensive vascular diseases, atherogenesis, subcellular compartmentation of Ca^{2+} and Mg^{2+} and vascular contractility. Min. Elect. Metab. 9, 323–336.

Alhosaini, M., Leehey, D.J., 2015. Magnesium and dialysis: the neglected cation. Am. J. Kidney Dis. 66, 523–531.

Almulla, H.A., Bush, P.G., Steele, M.G., Ellis, D., Flatman, P.W., 2006. Loading rat heart myocytes with Mg^{2+} using low-$[Na^+]$ solutions. J. Physiol. 575, 443–454.

Altura, B.M., Shah, N.C., Shah, G., Zhang, A., Li, W., Zheng, T., Perez-Albela, J.L., Altura, B.T., 2012. Short-term magnesium deficiency upregulates ceramide synthase in cardiovascular tissues and cells: cross-talk among cytokines, Mg^{2+}, NF-κB, and de novo ceramide. Am. J. Physiol. Heart Circ. Physiol. 302, H319–H332.

An, R., Chiu, C.Y., Andrade, F., 2015. Nutrient intake and use of dietary supplements among US adults with disabilities. Disabil. Health J. 8, 240–249.

Antunes, T.T., Callera, G.E., He, Y., Yogi, A., Ryazanov, A.G., Ryazanova, L.V., Zhai, A., Stewart, D.J., Shrier, A., Touyz, R.M., 2016. Transient receptor potential melastatin 7 cation channel (TRPM7) kinase: a new player in Ang II-induced hypertension. Hypertension 67(4), 763–773.

de Baaij, J.H., Hoenderop, J.G., Bindels, R.J., 2015. Magnesium in man: implications for health and disease. Physiol. Rev. 95, 1–46.

Basralı, F., Nasırcılar Ülker, S., Koçer, G., Ülker Karadamar, P., Özyurt, D., Cengiz, M., Kemal Şentürk, Ü., 2015a. Effect of magnesium on vascular reactivity in NOS inhibition-induced hypertension. Magnes. Res. 28, 64–74.

Basralı, F., Koçer, G., Ülker Karadamar, P., Nasırcılar Ülker, S., Satı, L., Özen, N., Özyurt, D., Şentürk, Ü.K., 2015b. Effect of magnesium supplementation on blood pressure and vascular reactivity in nitric oxide synthase inhibition-induced hypertension model. Clin. Exp. Hypertens. 37, 633–642.

Berhan, Y., Berhan, A., 2015. Should magnesium sulfate be administered to women with mild pre-eclampsia? A systematic review of published reports on eclampsia. J. Obstet. Gynaecol. Res. 41, 831–842.

Bijvelds, M.J., Flik, G., Kolar, Z.I., 1998. Cellular magnesium transport in the vertebrate intestine. Mag. Res. 11, 315–322.

Blaine, J., Chonchol, M., Levi, M., 2015. Renal control of calcium, phosphate, and magnesium homeostasis. Clin. J. Am. Soc. Nephrol. 10, 1257–1272.

Bowman, J.C., Lenz, T.K., Hud, N.V., Williams, L.D., 2012. Cations in charge: magnesium ions in RNA folding and catalysis. Curr. Opin. Struct. Biol. 22, 262–272.

Brandao, K., Deason-Towne, F., Perraud, A.L., Schmitz, C., 2013. The role of Mg²⁺ in immune cells. Immunol. Res. 55, 261–269.

Bui, D.M., Gregan, J., Jarosch, E., Ragnini, A., Schweyen, R.J., 1999. The bacterial magnesium transporter CorA can functionally substitute for its putative homologue Mrs2p in the yeast inner mitochondrial membrane. J. Biol. Chem. 274, 20438–20443.

Cefaratti, C., Romani, A.M., 2007. Functional characterization of two distinct Mg(2+) extrusion mechanisms in cardiac sarcolemmal vesicles. Mol. Cell Biochem. 303, 63–72.

Choi, M.K., Bae, Y.J., 2015. Association of magnesium intake with high blood pressure in Korean adults: Korea National health and Nutrition Examination Survey 2007–2009. PLoS One 10 (6), e0130405.

Clark, K., Middelbeek, J., Dorovkov, M.V., Figdor, C.G., Ryazanov, A.G., Lasonder, E., van Leeuwen, F.N., 2008. The alpha-kinases TRPM6 and TRPM7, but not eEF-2 kinase, phosphorylate the assembly domain of myosin IIA, IIB and IIC. FEBS Lett. 582, 2993–2997.

Chaumais, M.C., Lecerf, F., Fattal, S., Savale, L., Günther, S., Huertas, A., Montani, D., Perros, F., Humbert, M., German-Fattal, M., 2012. A study of magnesium deficiency in human and experimental pulmonary hypertension. Magnes. Res. 25, 21–27.

Chubanov, V., Gudermann, T., Schlingmann, K.P., 2005. Essential role for TRPM6 in epithelial magnesium transport and body magnesium homeostasis. Pflugers Arch. 451, 228–234.

Chubanov, V., Schlingmann, K.P., Waring, J., Heinzinger, J., Kaske, S., Waldegger, S., Schnitzler, M.M., Gudermann, T., 2007. Hypomagnesemia with secondary hypocalcemia due to a missense mutation in the putative pore-forming region of TRPM6. J. Biol. Chem. 282, 7656–7667.

Clark, K., Langeslag, M., van Leeuwen, B., Ran, L., Ryazanov, A.G., Figdor, C.G., Moolenaar, W.H., Jalink, K., van Leeuwen, F.N., 2006. TRPM7, a novel regulator of actomyosin contractility and cell adhesion. EMBO J. 25, 290–301.

Costa, A., Tejpar, S., Prenen, H., Van Cutsem, E., 2011. Hypomagnesaemia and targeted anti-epidermal growth factor receptor (EGFR) agents. Target Oncol. 6, 227–233.

Cuffe, J.S., Steane, S., Moritz, K.M., Paravicini, T.M., 2015. Differential mRNA expression and glucocorticoid-mediated regulation of TRPM6 and TRPM7 in the heart and kidney throughout murine pregnancy and development. PLoS One 10, e0117978.

Cunha, A.R., Medeiros, F., Umbelino, B., Oigman, W., Touyz, R.M., Neves, M.F., 2013. Altered vascular structure and wave reflection in hypertensive women with low magnesium levels. J. Am. Soc. Hypertens. 7, 344–352.

Deason-Towne, F., Perraud, A.L., Schmitz, C., 2011. The Mg²⁺ transporter MagT1 partially rescues cell growth and Mg²⁺ uptake in cells lacking the channel-kinase TRPM7. FEBS Lett. 585, 2275–2278.

Demeuse, P., Penner, R., Fleig, A., 2006. TRPM7 channel is regulated by magnesium nucleotides via its kinase domain. J. Gen. Physiol. 127, 421–434.

Dorovkov, M.V., Ryazanov, A.G., 2004. Phosphorylation of annexin I by TRPM7 channel-kinase. J. Biol. Chem. 279, 50643–50646.

Dhungel, K.U., Kim, T.W., Sharma, N., Bhattarai, J.P., Park, S.A., Han, S.K., Kim, C.J., 2012. Magnesium increases iberiotoxin-sensitive large conductance calcium activated potassium currents on the basilar artery smooth muscle cells in rabbits. Neurol. Res. 34, 11–16.

Dickinson, H.O., Nicolson, D.J., Campbell, F., Cook, J.V., Beyer, F.R., Ford, G.A., Mason, J., 2006. Magnesium supplementation for the management of essential hypertension in adults. Cochrane Database Syst. Rev. 3 CD004640.

Dørup, I., Skjaaa, K., Thybo, N.K., 1993. Oral magnesium supplementation restores the concentrations of magnesium, potassium and sodium-potassium pumps in skeletal muscle of patients receiving diuretic treatment. J. Intern Med. 233, 117–123.

Fine, K.D., Santa Ana, C.A., Porter, J.L., 1991. Intestinal absorption of magnesium from food supplements. J. Clin. Invest. 88, 396–400.

Felsenfeld, A.J., Levine, B.S., Rodriguez, M., 2015. Pathophysiology of calcium, phosphorus, and magnesium dysregulation in chronic kidney disease. Semin. Dial. 28, 564–577.

Ferrè, S., Hoenderop, J.G., Bindels, R.J., 2011. Insight into renal Mg²⁺ transporters. Curr. Opin. Nephrol. Hypertens. 20, 169–176.

Fagan, T.E., Romani, A., 2000. Activation of Na(+)- and Ca(2+)-dependent Mg(2+) extrusion by alpha(1)- and beta-adrenergic agonists in rat liver cells. Am. J. Physiol. Gastrointest. Liver Physiol. 279, G943–G950.

Fiorani, M., Azzolini, C., Cerioni, L., Scotti, M., Guidarelli, A., Ciacci, C., Cantoni, O., 2015. The mitochondrial transporter of ascorbic acid functions with high affinity in the presence of low millimolar concentrations of sodium and in the absence of calcium and magnesium. Biochim. Biophys. Acta 1848, 1393–1401.

Goytain, A., Quamme, G.A., 2008. Identification and characterization of a novel family of magnesium transporters, MMgT1 and MMgT2. Am. J. Physiol. Cell Physiol. 294, C495–C502.

Goytain, A., Quamme, G.A., 2005a. Identification and characterization of a novel mammalian Mg²⁺ transporter with channel-like properties. BMC Genomics 6, 48.

Goytain, A., Quamme, G.A., 2005b. Functional characterization of ACDP2 (ancient conserved domain protein), a divalent metal transporter. Physiol. Genomics 22, 382–389.

Goytain, A., Hines, R.M., Quamme, G.A., 2008. Huntingtin-interacting proteins, HIP14 and HIP14L, mediate dual functions: palmitoyl acyltransferase and Mg²⁺ transport. J. Biol. Chem. 283, 33365–33374.

Goytain, A., Hines, R.M., El-Husseini, A., Quamme, G.A., 2007. NIPA1 (SPG6), the basis for autosomal dominant form of hereditary spastic paraplegia encodes a functional Mg2 transporter. J. Biol. Chem. 282, 8060–8068.

Groenestege, W.M., Hoenderop, J.G., van den Heuvel, L., Knoers, N., Bindels, R.J., 2006. The epithelial Mg^{2+} channel transient receptor potential melastatin 6 is regulated by dietary Mg^{2+} content and estrogens. J. Am. Soc. Nephrol. 17, 1035–1043.

Groenestege, W.M., Thebault, S., van der Wijst, J., van den Berg, D., Janssen, R., Tejpar, S., van den Heuvel, L.P., van Cutsem, E., Hoenderop, J.G., Knoers, N.V., Bindels, R.J., 2007. Impaired basolateral sorting of pro-EGF causes isolated recessive renal hypomagnesemia. J. Clin. Invest. 117, 2260–2267.

Guo, J.L., Yu, Y., Jia, Y.Y., Ma, Y.Z., Zhang, B.Y., Liu, P.Q., Chen, S.R., Jiang, J.M., 2014. Transient receptor potential melastatin 7 (TRPM7) contributes to H2O2-induced cardiac fibrosis via mediating Ca(2+) influx and extracellular signal-regulated kinase 1/2 (ERK1/2) activation in cardiac fibroblasts. J. Pharmacol. Sci. 125, 184–192.

Gilbert D'Angelo, E.K., Singer, H.A., Rembold, C.M., 1992. Magnesium relaxes arterial smooth muscle by decreasing intracellular Ca^{2+} without changing intracellular Mg^{2+}. J. Clin. Invest. 89, 1988–1994.

Gold, M.E., Buga, G.M., Wood, K.S., Byrns, R.E., Chadhuri, G., Ignarro, L.J., 1990. Antagonistic modulatory roles of magnesium and calcium on release of endothelium-derived relaxing factor and smooth muscle tone. Circ. Res. 66, 355–366.

Guerrero-Romero, F., Rodríguez-Morán, M., Hernández-Ronquillo, G., Gómez-Díaz, R., Pizano-Zarate, M.L., Wacher, N.H., Mondragón-González, R., Simental-Mendia, L.E., Network of Childhood Obesity of the Mexican Social Security Institute, 2016. Low serum magnesium levels and its association with high blood pressure in children. J. Pediatr. 168, 93–98.

Hoorn, E.J., Zietse, R., 2013. Disorders of calcium and magnesium balance: a physiology-based approach. Pediatr. Nephrol. 28, 1195–1206.

Hou, J., Renigunta, A., Gomes, A.S., Hou, M., Paul, D.L., Waldegger, S., Goodenough, D.A., 2009. Claudin-16 and claudin-19 interaction is required for their assembly into tight junctions and for renal reabsorption of magnesium. Proc. Natl. Acad. Sci. U.S.A. 106, 15350–15355.

Hardwick, L.L., Jones, M.R., Brauter, N., 1991. Magnesium absorption: mechanics and the influence of vitamin D, calcium and phosphate. J. Nutr. 121, 13–20.

Herencia, C., Rodríguez-Ortiz, M.E., Muñoz-Castañeda, J.R., Martinez-Moreno, J.M., Canalejo, R., Montes de Oca, A., Díaz-Tocados, J.M., Peralbo-Santaella, E., Marín, C., Canalejo, A., Rodriguez, M., Almaden, Y., 2015. Angiotensin II prevents calcification in vascular smooth muscle cells by enhancing magnesium influx. Eur. J. Clin. Invest. 45, 1129–1144.

Hur, K.Y., Kim, S.H., Choi, M.A., Williams, D.R., Lee, Y.H., Kang, S.W., Yadav, U.C., Srivastava, S.K., Jung, M., Cho, J.W., Kim, S.G., Kang, E.S., Lee, E.J., Lee, H.C., 2010. Protective effects of magnesium lithospermate B against diabetic atherosclerosis via Nrf2-ARE-NQO1 transcriptional pathway. Atherosclerosis 211, 69–76.

He, Y., Yao, G., Savoia, C., Touyz, R.M., 2005. Transient receptor potential melastatin 7 ion channels regulate magnesium homeostasis in vascular smooth muscle cells: role of angiotensin II. Circ. Res. 96, 207–215.

Janett, S., Camozzi, P., Peeters, G.G., Lava, S.A., Simonetti, G.D., Goeggel Simonetti, B., Bianchetti, M.G., Milani, G.P., 2015. Hypomagnesemia induced by long-term treatment with proton-pump inhibitors. Gastroenterol. Res. Pract. 2015, 951768.

Jiang, L., He, P., Chen, J., Liu, Y., Liu, D., Qin, G., Tan, N., 2016. Magnesium levels in drinking water and coronary heart disease mortality risk: a meta-analysis. Nutrients 2 (1), 8.

Jin, J., Desai, B.N., Navarro, B., Donovan, A., Andrews, N.C., Clapham, D.E., 2008. Deletion of Trpm7 disrupts embryonic development and thymopoiesis without altering Mg^{2+} homeostasis. Science 322, 756–760.

Jee, S.H., Miller 3rd, E.R., Guallar, E., Singh, V.K., Appel, L.J., Klag, M.J., 2002. The effect of magnesium supplementation on blood pressure: a meta-analysis of randomized clinical trials. Am. J. Hypertens. 15, 691–696.

Joosten, M.M., Gansevoort, R.T., Mukamal, K.J., Kootstra-Ros, J.E., Feskens, E.J., Geleijnse, J.M., Navis, G., Bakker, S.J., PREVEND Study Group, 2013. Urinary magnesium excretion and risk of hypertension: the prevention of renal and vascular end-stage disease study. Hypertension 61, 1161–1167.

Konrad, M., Schlingmann, K.P., 2014. Inherited disorders of renal hypomagnesaemia. Nephrol. Dial. Transpl. 29, iv63–71.

Kladnitsky, O., Rozenfeld, J., Azulay-Debby, H., Efrati, E., Zelikovic, I., 2015. The claudin-16 channel gene is transcriptionally inhibited by 1,25-dihydroxyvitamin D. Exp. Physiol. 100, 79–94.

Komiya, Y., Runnels, L.W., 2015. TRPM channels and magnesium in early embryonic development. Int. J. Dev. Biol. 59, 281–288.

Kraft, R., Harteneck, C., 2005. The mammalian melastatin-related transient receptor potential cation channels: an overview. Pflugers Arch. 451, 204–211.

Kolisek, M., Zsurka, G., Samaj, J., Weghuber, J., Schweyen, R.J., Schweigel, M., 2003. Mrs2p is an essential component of the major electrophoretic Mg^{2+} influx system in mitochondria. EMBO J. 22, 1235–1244.

Krapivinsky, G., Krapivinsky, L., Manasian, Y., Clapham, D.E., 2014. The TRPM7 chanzyme is cleaved to release a chromatin-modifying kinase. Cell 157 (5), 1061–1072.

Ku, D.D., Ann, H., 1991. Differential effect of magnesium on basal and agonist-induced EDRF relaxation in canine coronary arteries. J. Cardiovasc. Pharmacol. 17, 999–1006.

Kemp, P.A., Gardiner, S.M., Bennett, T., Rubin, P.C., 1993. Magnesium sulphate reverses the carotid vasoconstriction caused by endothelin-I, angiotensin II and neuropeptide-Y, but not that caused by NG-nitro-L-arginine methyl ester, in conscious rats. Clin. Sci. (Lond) 85, 175–181.

Kolisek, M., Montezano, A.C., Sponder, G., Anagnostopoulou, A., Vormann, J., Touyz, R.M., Aschenbach, J.R., 2015. PARK7/DJ-1 dysregulation by oxidative stress leads to magnesium deficiency: implications in degenerative and chronic diseases. Clin. Sci. (Lond) 129, 1143–1150.

Kelly, H.G., Turton, M.R., Hatcher, J.D., 1961. Renal and cardiovascular effects induced by intravenous infusion of magnesium chelate. Can. Med. Assoc. J. 84, 1124–1128.

Kass, L., Weekes, J., Carpenter, L., 2012. Effect of magnesium supplementation on blood pressure: a meta-analysis. Eur. J. Clin. Nutr. 66, 411–418.

Lacson Jr., E., Wang, W., Ma, L., Passlick-Deetjen, J., 2015. Serum magnesium and mortality in hemodialysis patients in the United States: a cohort study. Am. J. Kidney Dis. 66, 1056–1066.

Lichton, I.J., 1989. Dietary intake levels and requirements of Mg and Ca for different segments of the U.S. population. Magnesium 8, 117–123.

Lameris, A.L., Nevalainen, P.I., Reijnen, D., Simons, E., Eygensteyn, J., Monnens, L., Bindels, R.J., Hoenderop, J.G., 2015. Segmental transport of Ca^{2+} and Mg^{2+} along the gastrointestinal tract. Am. J. Physiol. Gastrointest. Liver Physiol. 308, G206–G216.

Liu, C., Montell, C., 2015. Forcing open TRP channels: mechanical gating as a unifying activation mechanism. Biochem. Biophys. Res. Commun. 460, 22–25.

Li, F.Y., Chaigne-Delalande, B., Kanellopoulou, C., Davis, J.C., Matthews, H.F., Douek, D.C., Cohen, J.I., Uzel, G., Su, H.C., Lenardo, M.J., 2011. Second messenger role for Mg^{2+} revealed by human T-cell immunodeficiency. Nature 475, 471–476.

Laurant, P., Touyz, R.M., Schiffrin, E.L., 1997. Effect of magnesium on vascular tone and reactivity in pressurized mesenteric arteries from SHR. Can. J. Physiol. Pharmacol. 5, 293–300.

Laurant, P., Berthelot, A., 1996. Endothelin-1-induced contraction in isolated aortae from normotensive and DOCA-salt hypertensive rats: effect of magnesium. Br. J. Pharmacol. 119, 1367–1374.

Li, Y., Jiang, H., Ruan, C., Zhong, J., Gao, P., Zhu, D., Niu, W., Guo, S., 2014. The interaction of transient receptor potential melastatin 7 with macrophages promotes vascular adventitial remodeling in transverse aortic constriction rats. Hypertens. Res. 37, 35–42.

Montezano, A.C., Zimmerman, D., Yusuf, H., Burger, D., Chignalia, A.Z., Wadhera, V., van Leeuwen, F.N., Touyz, R.M., 2010. Vascular smooth muscle cell differentiation to an osteogenic phenotype involves TRPM7 modulation by magnesium. Hypertension 56, 453–462.

Matsushita, M., Kozak, J.A., Shimizu, Y., McLachlin, D.T., Yamaguchi, H., Wei, F.Y., Tomizawa, K., Matsui, H., Chait, B.T., Cahalan, M.D., Nairn, A.C., 2005. Channel function is dissociated from the intrinsic kinase activity and autophosphorylation of TRPM7/ChaK1. J. Biol. Chem. 280, 20793–20803.

Murata, T., Horiuchi, T., Goto, T., Li, Y., Hongo, K., 2011. Vasomotor response induced by change of extracellular potassium and magnesium in cerebral penetrating arterioles. Neurosci. Res. 70, 30–34.

McHugh, D., Beech, D.J., 1996. Modulation of Ca^{2+} channel activity by ATP metabolism and internal Mg^{2+} in guinea-pig basilar artery smooth muscle cells. J. Physiol. 492, 359–376.

Naeem, M., Hussain, S., Akhtar, N., 2011. Mutation in the tight-junction gene claudin 19 (CLDN19) and familial hypomagnesemia, hypercalciuria, nephrocalcinosis (FHHNC) and severe ocular disease. Am. J. Nephrol. 34, 241–248.

Nadler, M.J., Hermosura, M.C., Inabe, K., Perraud, A.L., Zhu, Q., Stokes, A.J., Kutosaki, T., Kinet, J.P., Penner, R., Scharenberg, A.M., Fleig, A., 2001. LTRPC7 is a Mg ATP-regulated divalent cation channel required for cell viability. Nature 411, 590–595.

Nakajima, T., Iwasawa, K., Hazama, H., Asano, M., Okuda, Y., Omata, M., 1997. Extracellular Mg^{2+} inhibits receptor-mediated Ca^{2+}-permeable non-selective cation currents in aortic smooth muscle cells. Eur. J. Pharmacol. 320, 81–86.

Oancea, E., Wolfe, J.T., Clapham, D.E., 2006. Functional TRPM7 channels accumulate at the plasma membrane in response to fluid flow. Circ. Res. 98, 245–253.

Pontes, M.H., Sevostyanova, A., Groisman, E.A., 2015. When too much ATP is bad for protein synthesis. J. Mol. Biol. 427, 2586–2594.

Pham, P.C., Pham, P.A., Pham, S.V., Pham, P.T., Pham, P.M., Pham, P.T., 2014. Hypomagnesemia: a clinical perspective. Int. J. Nephrol. Renov. Dis. 7, 219–230.

Penner, R., Fleig, A., 2007. The Mg^{2+} and Mg(2+)-nucleotide-regulated channel-kinase TRPM7. Handb. Exp. Pharmacol. 179, 313–328.

Paravicini, T.M., Yogi, A., Mazur, A., Touyz, R.M., 2009. Dysregulation of vascular TRPM7 and annexin-1 is associated with endothelial dysfunction in inherited hypomagnesemia. Hypertension 53, 423–429.

Quann, E.E., Fulgoni 3rd, V.L., Auestad, N., 2015. Consuming the daily recommended amounts of dairy products would reduce the prevalence of inadequate micronutrient intakes in the United States: diet modeling study based on NHANES 2007–2010. Nutr. J. 14, 90.

Quamme, G.A., 1997. Renal magnesium handling: new insights in understanding old problems. Kidney Int. 52, 1180–1195.

Quamme, G.A., 1989. Control of magnesium transport in the thick ascending limb. Am. J. Physiol. 256, F197–F202.

Quamme, G.A., 2010. Molecular identification of ancient and modern mammalian magnesium transporters. Am. J. Physiol. Cell Physiol. 298, C407–C429.

Romani, A.M., 2011. Cellular magnesium homeostasis. Arch. Biochem. Biophys. 512, 1–23.

Runnels, L.W., Yue, L., Clapham, D.E., 2001. TRP-PLIK, a bifunctional protein with kinase and ion channel activities. Science 291, 1043–1047.

Romani, A.M., Maguire, M.E., 2002. Hormonal regulation of Mg^{2+} transport and homeostasis in eukaryotic cells. Biometals 15, 271–283.

Runnels, L.W., 2011. TRPM6 and TRPM7: a Mul-TRP-PLIK-cation of channel functions. Curr. Pharm. Biotechnol. 12, 42–53.

Ryazanova, L.V., Hu, Z., Suzuki, S., Chubanov, V., Fleig, A., Ryazanov, A.G., 2014. Elucidating the role of the TRPM7 alpha-kinase: TRPM7 kinase inactivation leads to magnesium deprivation resistance phenotype in mice. Sci. Rep. 4, 7599.

Rondón, L.J., Marcano, E., Rodríguez, F., del Castillo, J.R., 2014. Blood pressure, magnesium and other mineral balance in two rat models of salt-sensitive, induced hypertension: effects of a non-peptide angiotensin II receptor type 1 antagonist. Magnes. Res. 27, 113–130.

Rayssiguier, Y., Mbega, J.D., Durlach, V., Gueux, E., Durlach, J., Giry, J., Dalle, M., Mazur, A., Laurant, P., Berthelot, A., 1992. Magnesium and blood pressure. I. Animal studies. Magnes. Res. 5, 139–146.

Rosanoff, A., Plesset, M.R., 2013. Oral magnesium supplements decrease high blood pressure (SBP>155 mm Hg) in hypertensive subjects on antihypertensive medications: a targeted meta-analysis. Magnes. Res. 26, 93–99.

Sahni, J., Nelson, B., Scharenberg, A.M., 2007. SLC41A2 encodes a plasma-membrane Mg^{2+} transporter. Biochem. J. 401, 505–513.

Simon, D.B., Lu, Y., Choate, K.A., Velazquez, H., Al-Sabban, E., Praga, M., Casari, G., Bettinelli, A., Colussi, G., Rodriguez-Soriano, J., McCredie, D., Milford, D., Sanjad, S., Lifton, R.P., 1999. Paracellin-1, a renal tight junction protein required for paracellular Mg^{2+} resorption. Science 285, 103–106.

Schlingmann, K.P., Weber, S., Peters, M., Niemann Nejsum, L., Vitzthum, H., Klingel, K., Kratz, M., Haddad, E., Ristoff, E., Dinour, D., Syrrou, M., Nielsen, S., Sassen, M., Waldegger, S., Seyberth, H.W., Konrad, M., 2002. Hypomagnesemia with secondary hypocalcemia is caused by mutations in TRPM6, a new member of the TRPM gene family. Nat. Genet. 31, 166–170.

Schrag, D., Chung, K.Y., Flombaum, C., Saltz, L., 2005. Cetuximab therapy and symptomatic hypomagnesemia. J. Natl. Cancer Inst. 97, 1221–1224.

Su, L.T., Agapito, M.A., Li, M., Simonson, W.T., Huttenlocher, A., Habas, R., Yue, L., Runnels, L.W., 2006. TRPM7 regulates cell adhesion by controlling the calcium-dependent protease calpain. J. Biol. Chem. 281, 11260–11270.

Sontia, B., Montezano, A.C., Paravicini, T., Tabet, F., Touyz, R.M., 2008. Downregulation of renal TRPM7 and increased inflammation and fibrosis in aldosterone-infused mice: effects of magnesium. Hypertension 51, 915–921.

Touyz, R.M., 2004. Magnesium in clinical medicine. Front. Biosci. 9, 1278–1293.

Touyz, R.M., Yao, G., 2003. Modulation of vascular smooth muscle cell growth by magnesium-role of mitogen-activated protein kinases. J. Cell Physiol. 197, 326–335.

Touyz, R.M., Milne, F.J., Reinach, S.G., 1992. Intracellular Mg^{2+}, Ca^{2+}, Na^+ and K^+ in platelets and erythrocytes of essential hypertensive patients: relation to blood pressure. Clin. Exp. Hypertens. A14, 1189–1209.

Thebault, S., Alexander, R.T., Tiel Groenestege, W.M., Hoenderop, J.G., Bindels, R.J., 2009. EGF increases TRPM6 activity and surface expression. J. Am. Soc. Nephrol. 20, 78–85.

Tejpar, S., Piessevaux, H., Claes, K., Piront, P., Hoenderop, J.G., Verslype, C., Van Cutsem, E., 2007. Magnesium wasting associated with epidermal growth- factor receptor-targeting antibodies in colorectal cancer: a prospective study. Lancet Oncol. 8, 387–394.

Tashiro, M., Inoue, H., Konishi, M., 2013. Magnesium homeostasis in cardiac myocytes of Mg-deficient rats. PLoS One 8, e73171.

Touyz, R.M., Schiffrin, E.L., 1999. Activation of the Na^+/H^+ exchanger modulates angiotensin II-stimulated Na^+-dependent Mg^{2+} transport in vascular smooth muscle cells from spontaneously hypertensive rats. Hypertension 34, 442–449.

Touyz, R.M., 2000. Oxidative stress in vascular damage in hypertension. Curr. Hypert Rep. 2, 98–105.

Touyz, R.M., Milne, F.J., 1999. Magnesium supplementation attenuates, but does not prevent, development of hypertension in spontaneously hypertensive rats. Am. J. Hypertens. 12, 757–765.

Touyz, R.M., He, Y., Montezano, A.C., Yao, G., Chubanov, V., Gudermann, T., Callera, G.E., 2006. Differential regulation of transient receptor potential melastatin 6 and 7 cation channels by Ang II in vascular smooth muscle cells from spontaneously hypertensive rats. Am. J. Physiol. Regul. Integr. Comp. Physiol. 290, R73–R78.

Van Laecke, S., Van Biesen, W., Vanholder, R., 2012. Hypomagnesaemia, the kidney and the vessels. Nephrol. Dial. Transpl. 27, 4003–4010.

Voets, T., Nilius, B., Hoefs, S., van der Kemp, A.W., Droogmans, G., Bindels, R.J., Hoenderop, J.G., 2004. TRPM6 forms the Mg^{2+} influx channel involved in intestinal and renal Mg^{2+} absorption. J. Biol. Chem. 279, 19–25.

Walder, R.Y., Landau, D., Meyer, P., Shalev, H., Tsolia, M., Borochowitz, Z., Boettger, M.B., Beck, G.E., Englehardt, R.K., Carmi, R., Sheffield, V.C., 2002. Mutation of TRPM6 causes familial hypomagnesemia with secondary hypocalcemia. Nat. Genet. 31, 171–174.

Watanabe, H., Murakami, M., Ohba, T., Ono, K., Ito, H., 2009. The pathological role of transient receptor potential channels in heart disease. Circ. J. 73, 419–427.

Weglicki, W.B., Phillips, T.M., Freedman, A.M., Cassidy, M.M., Dickens, B.F., 1992. Magnesium-deficiency elevates circulating levels of inflammatory cytokines and endothelin. Mol. Cell Biochem. 110, 169–173.

Weglicki, W.B., Mak, I.T., Kramer, J.H., Dickens, B.F., Cassidy, M.M., Stafford, R.E., Phillips, T.M., 1996. Role of free radicals and substance P in magnesium deficiency. Cardiovasc Res. 31, 677–682.

Winkler, A.W., Hoff, H.E., Smith, P.K., 1940. Cardiovascular effects of potassium, calcium, magnesium, and barium: an experimental study of toxicity and rationale of use in therapeutics. Yale J. Biol. Med. 13, 123–132.

Wolf, F.I., Trapani, V., 2011. Mag T1: a highly specific magnesium channel with important roles beyond cellular magnesium homeostasis. Magnes. Res. 24, S86–S91.

Wu, N., Veillette, A., 2011. Immunology: magnesium in a signalling role. Nature 475, 462–463.

Yogi, A., Callera, G.E., O'Connor, S., Antunes, T.T., Valinsky, W., Miquel, P., Montezano, A.C., Perraud, A.L., Schmitz, C., Shrier, A., Touyz, R.M., 2013. Aldosterone signaling through transient receptor potential melastatin 7 cation channel (TRPM7) and its α-kinase domain. Cell Signal 25, 2163–2175.

Yamaguchi, H., Matsushita, M., Nairn, A.C., Kuriyan, J., 2001. Crystal structure of the atypical protein kinase domain of a TRP channel with phosphotransferase activity. Mol. Cell 7, 1047–1057.

Yamamoto, M.E., Applegate, W.B., Klag, M.J., Borhani, N.O., Cohen, J.D., Kirchner, K.A., Lakatos, E., Sacks, F.M., Taylor, J.O., Hennekens, C.H., 1995. Lack of blood pressure effect with calcium and magnesium supplementation in adults with high-normal blood pressure. Results from Phase I of the Trials of Hypertension Prevention (TOHP). Trials of Hypertension Prevention (TOHP) Collaborative Research Group. Ann. Epidemiol. 5, 96–107.

Yogi, A., Callera, G.E., Tostes, R., Touyz, R.M., 2009. Bradykinin regulates calpain and proinflammatory signaling through TRPM7-sensitive pathways in vascular smooth muscle cells. Am. J. Physiol. Regul. Integr. Comp. Physiol. 296, R201–R207.

Yogi, A., Callera, G.E., Antunes, T.T., Tostes, R.C., Touyz, R.M., 2011a. Transient receptor potential melastatin 7 (TRPM7) cation channels, magnesium and the vascular system in hypertension. Circ. J. 75, 237–245.

Yogi, A., Callera, G.E., O'Connor, S.E., He, Y., Correa, J.W., Tostes, R.C., Mazur, A., Touyz, R.M., 2011b. Dysregulation of renal transient receptor potential melastatin 6/7 but not paracellin-1 in aldosterone-induced hypertension and kidney damage in a model of hereditary hypomagnesemia. J. Hypertens. 29, 1400–1410.

Yoshimura, M., Oshima, T., Matsuura, H., Ishida, T., Kambe, M., Kajiyama, G., 1997. Extracellular Mg^{2+} inhibits capacitance Ca^{2+} entry in vascular smooth muscle cells. Circulation 95, 2567–2572.

Yu, A.S., 2015. Claudins and the kidney. J. Am. Soc. Nephrol. 26, 11–19.

Zeng, Z., Inoue, K., Sun, H., Leng, T., Feng, X., Zhu, L., Xiong, Z.G., 2015. TRPM7 regulates vascular ndothelial cell adhesion and tube formation. Am. J. Physiol. Cell Physiol. 308, C308–C318.

Zhang, X., Li, Y., Del Gobbo, L.C., Rosanoff, A., Wang, J., Zhang, W., Song, Y., 2016. Effects of magnesium supplementation on blood pressure a meta-analysis of randomized double-blind placebo-controlled trials. Hypertension 68, 324–333.

Zholos, A., Johnson, C., Burdyga, T., Melanaphy, D., 2011. TRPM channels in the vasculature. Adv. Exp. Med. Biol. 704, 707–729.

Zhou, H., Clapham, D.E., 2009. Mammalian MagT1 and TUSC3 are required for cellular magnesium uptake and vertebrate embryonic development. Proc. Natl. Acad. Sci. U. S. A. 106, 15750–15756.

Part VII

Manganese

Chapter 30

Nutritional, Genetic, and Molecular Aspects of Manganese Intoxication

Tanara Vieira Peres, Michael Aschner

Albert Einstein College of Medicine, New York, NY, United States

ESSENTIALITY AND TOXICITY OF MANGANESE

Metals are classified as essential when they participate in biological processes, acting as enzyme cofactors or functional groups of proteins, and are required in small quantities. Manganese (Mn) is classified as an essential metal for its participation in various physiological processes. Mn acts in gluconeogenesis as an activator of pyruvate carboxylase and in the Krebs cycle as a cofactor for isocitrate dehydrogenase. In the antioxidant defense system, Mn acts as a cofactor for superoxide dismutase (SOD). Moreover, Mn acts on the central nervous system (CNS) as a cofactor for glutamine synthetase, which is preferentially localized in astrocytes (Santamaria, 2008).

The chemical element Mn is a heavy metal that occurs naturally, being the 5th most abundant metal in the Earth's crust and the 12th most abundant element as a whole. Mn is usually found in the environment, forming oxides, carbonates, and silicates. The human population is readily exposed to Mn because of its prevalence in the air, soil, and waterways secondary to natural erosion and from industrial sources. However, the main route of human exposure to Mn is from food intake. Whole grains, vegetables, rice, and nuts contain the highest levels of Mn, which can also be found in chocolate, tea, green leafy vegetables, and fruits. The range of food sources containing Mn allows humans to easily obtain adequate levels of Mn (2.3 mg/day for men and 1.8 mg/day for adult women; ATSDR, 2012).

The typical adult human ingests less than 5 mg Mn per kilogram. Of the ingested Mn, approximately 1–5% is absorbed through the gastrointestinal tract. Radiolabeled [54]Mn uptake studies show that for a meal containing 1 mg Mn, adult males absorb $1.35 \pm 0.51\%$ whereas adult females absorb $3.55 \pm 2.11\%$ (Davis et al., 1993; Finley et al., 1994). There is rapid turnover of ingested Mn, and homeostatic Mn levels are maintained through proper biliary excretion. The majority of the excreted Mn is conjugated with bile in the liver and secreted into the intestines for elimination in feces (Davis et al., 1993; Malecki et al., 1996).

Mn neurotoxicity was first described in 1837 by John Couper, who observed a parkinsonian syndrome characterized by slowness of movement (bradykinesia) and gait (postural instability) in occupationally exposed individuals. Therefore it has been observed that despite its essentiality, Mn could be neurotoxic at high levels (Aschner et al., 2009). The mechanisms involved in Mn neurotoxicity have not been fully elucidated; however, dopamine (DA) metabolism disorders (Guilarte, 2013; Guilarte et al., 2008), mitochondrial dysfunction (Gunter et al., 2010; Malecki, 2001), and induction of reactive oxygen species (ROS; Milatovic et al., 2009) and nitrogen species (RNS; Moreno et al., 2011) that lead to cell death have all been considered important aspects of Mn's neurotoxicity.

Chronic exposure to high doses by different exposure routes can lead to accumulation of Mn in the CNS. Excess Mn preferentially accumulates in the basal ganglia structures, especially in the striatum (caudate nucleus, putamen, and nucleus accumbens), globus pallidus, and substantia nigra (Guilarte et al., 2006a,b). The substantia nigra pars compacta was recently identified as a site of Mn accumulation in rats exposed intraperitoneally (Robison et al., 2015). The neurodegenerative process induced by accumulation of Mn is called manganism. Manganism is a syndrome similar to Parkinson's disease (PD), characterized by psychiatric and cognitive deficits and motor impairment (Bowler et al., 2006a,b). Symptoms caused by the accumulation of Mn include dystonia, bradykinesia, and rigidity due to damage to dopaminergic neurons and gliosis (Cersosimo and Koller, 2006; Gunter et al., 2006). The period of occupational exposure that can lead to manganism is 6 months to 2 years. The motor and neuropsychiatric symptoms may remain even 14 years after the end of exposure to Mn (Bouchard et al., 2007).

Occupational exposure to Mn occurs in the mining, welding, battery industries, and from the use of fungicides containing the metal, such as maneb and mancozeb (Aschner et al., 2009; Cersosimo and Koller, 2006; Josephs et al., 2005;

Molecular, Genetic, and Nutritional Aspects of Major and Trace Minerals. http://dx.doi.org/10.1016/B978-0-12-802168-2.00030-0

Sriram et al., 2015). Concern has also mounted because of the increased release of Mn into the atmosphere secondary to the use of the gasoline additive methylcyclopentadienyl manganese tricarbonyl (Gulson et al., 2006). The availability of the metal in the environment, water, or food containing high levels of Mn represents a source of contamination for the general population (Oulhote et al., 2014). Mn poisoning increased in users of the injectable drug methcathinone because of the use of potassium permanganate in the compound synthesis (Stepens et al., 2008). In particular, patients with hepatic impairment and those receiving total parenteral nutrition (TPN), especially newborns, are also susceptible to metal accumulation (Aschner and Aschner, 2005; Aschner et al., 2009; Boggio Bertinet et al., 2000; Nagatomo et al., 1999). Furthermore, Mn is present in levels considered excessive in children's TPN formula (Aschner and Aschner, 2005).

Mn is a trace element present in parenteral nutrition. When a solution for parenteral nutrition is prepared, it needs to be supplemented with an essential trace element solution. However, infants and children are particularly vulnerable to inappropriate supplementation of Mn, which in some cases may lead to hypermanganesemia depending on the duration of the treatment. This occurs because parenteral administration bypasses the regulatory mechanisms of the gastrointestinal tract. The bioavailability of Mn in parenteral fluid is 100%, compared with only 3–5% for enteral dietary Mn. The Mn burden derived from parenteral nutrition can be 100 times greater than human milk (Aschner and Aschner, 2005; Boggio Bertinet et al., 2000; Dobson et al., 2004; Santos et al., 2014).

The immature CNS is more susceptible to Mn neurotoxicity than the adult. Evidence suggests that exposure to Mn during development may affect neurological function in adulthood (Beaudin et al., 2013; Kern and Smith, 2011; Kern et al., 2010; Moreno et al., 2009). Metal absorption in the intestine is generally high during the first week of life, followed by a steady decline with age. These factors contribute to the significantly increased risk of neurotoxicity for newborns when exposed to excess Mn (Molina et al., 2011). It is unclear at this point whether other factors, environmental or genetic, are responsible for the alterations caused in the CNS in the presence of Mn or if the metal itself can produce these effects (ATSDR, 2012). Overall, genetic factors are responsible for 30–40% of cases of neurodevelopmental disorders. Thus nongenetic factors such as environmental exposures are likely involved in the cause of these disorders through interaction with genetically inherited predispositions (Grandjean and Landrigan, 2006, 2014).

MECHANISMS OF MN TRANSPORT TO THE CENTRAL NERVOUS SYSTEM

Because Mn homeostasis is essential for living organisms, levels of the metal are tightly controlled via several transport mechanisms (Dobson et al., 2004; Takeda, 2003; Tuschl et al., 2013; Yokel, 2009; Yokel et al., 2003). Among the mechanisms described for Mn transport are (1) diffusion (Yokel et al., 2003), (2) calcium channels Crossgrove and Yokel, 2005; Roth et al., 2002b), (3) divalent metal-ion transporter-1 (DMT-1; also known as macrophage natural resistance-associated protein-2 and solute carrier family 11, member 2 (SLC11A2); Kim et al., 2013; Roth et al., 2002b; Thompson et al., 2007), (4) carrier (Crossgrove et al., 2003), and (5) transport dependent on transferrin (Tf; Aschner and Aschner, 1990; Gunter et al., 2013; Roth et al., 2002b). Proposed Mn exporters include ferroportin (also known as FPN or solute carrier family 40, member 1 (SLC40A1); Madejczyk and Ballatori, 2012) and zinc transporter 10 (ZNT10) [also known as SLC30A10 (solute carrier family 30, member 10); Tuschl et al., 2012; Fig. 30.1]. Mn importers and exporters were recently reviewed by Chen et al. (2015a).

When ingested, Mn can be transported from the gastrointestinal tract through the intestinal mucosa into the blood stream via DMT-1 and Tf, similarly to the transport of iron (Fe). Only 3–5% of ingested Mn is absorbed. Mn excess is effectively excreted by the liver into the bile through the portal circulation (Roth, 2006). Using a cell line that mimics the intestinal absorptive epithelium (Caco-2 cells), it was possible to study the kinetics of Mn absorption and export. It was found that Mn is absorbed across the apical surface of intestinal cells and is exported via a combination of carrier-mediated processes and diffusion through tight junctions and desmosomes, depending on Mn concentration. DMT-1 expression was reduced in Caco-2 cells exposed to Mn, possibly as a mechanism to prevent the accumulation of toxic levels of Mn in the cells. At the same time, the level of FPN1, a Mn-exporting protein, was increased in Caco-2 cells exposed to Mn, providing evidence that the expression of FPN is responsive to Mn levels (Li et al., 2013).

The homeostasis of Fe and that of Mn are interdependent because of their chemical similarities and their shared transport mechanisms (Gunshin et al., 1997; Illing et al., 2012). This occurs in humans and rodent models (Claus Henn et al., 2011). Fe deficiency leads to increased expression of DMT-1 and Tf. Consequently, in the case of anemia, for example, the absence of Fe to compete for the same carriers could result in the accumulation of Mn (Erikson et al., 2002a, 2004; Garcia et al., 2007; Roth and Garrick, 2003). There is also in vitro evidence that the presence of Fe influences the absorption and biological actions of Mn in PC12 cells (Roth et al., 2002a). Animal models have proven to be efficient for the study of Fe and Mn DMT-1–dependent homeostasis: the Belgrade rat with Fe deficiency (Burdo et al., 2001; Chua and Morgan, 1997; Fleming et al., 1998) and the microcytic anemia mice (Canonne-Hergaux et al., 2000), both of which are characterized by loss of DMT-1 function due to a missense mutation that causes a substitution of glycine for arginine (G185R; Fleming et al., 1998). This mutation impairs the transport of Fe bound to Tf

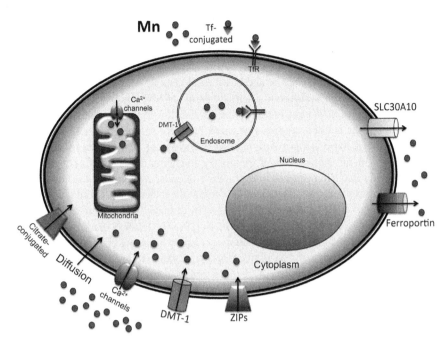

FIGURE 30.1   Proteins that participate in manganese (Mn) transport across the plasma membrane. When conjugated with transferrin (Tf) and recognized by the transferrin receptor (TfR), Mn is endocytosed and can access the cytoplasm through divalent metal transporter-1 (DMT-1). Mn can also be conjugated to citrate in the plasma, which facilitates its crossing the plasma membrane. Another possible mechanism for Mn to cross the plasma membrane is diffusion. Mn can cross calcium (Ca^{2+}) channels, including the ones present in mitochondria, given that both ions are divalent. DMT-1 is the main transporter for Mn and other divalent ions. ZIPs are zinc transporters that also show affinity for Mn. Among the mechanisms for Mn export is SLC30A10 (solute carrier family 30, member 10), a newly characterized Mn exporter and ferroportin, which has also been described to export Mn to the extracellular space.

from endocytic vesicles into the cytoplasm (Garrick et al., 1999). In the Belgrade rat model, it was demonstrated that Fe and Mn levels are concomitantly reduced. Thus this model has helped to elucidate the role of DMT-1 in the transport and homeostasis of Fe and Mn (Chua and Morgan, 1997). Furthermore, the experiments performed with the Belgrade rat indicate that DMT-1 also participates in Mn transport through the olfactory epithelium (Thompson et al., 2007), reviewed by Fitsanakis et al. (2010).

Because of the relationship between Fe and Mn homeostasis, it was hypothesized that Fe supplementation would protect from Mn accumulation in the brain. This hypothesis was tested in rats given Fe-deficient or Fe-supplemented diets combined with low-dose intravenous Mn injections. Data from magnetic resonance imaging and graphite furnace atomic absorption spectroscopy indicated that even with Fe supplementation, Mn accumulated in different regions of the rat brain. This effect can be explained by the action of another important element in the mechanism of Fe/Mn transport—FPN. This cytoplasmic protein is responsible for the export of Fe to the extracellular space (plasma) and can be degraded in the event of Fe overload. Thus, with Fe supplementation, FPN can be degraded, resulting in less Fe export to the extracellular space. With lower levels of extracellular Fe, there is less competition for Tf, resulting in increased Mn accumulation via Tf/transferrin receptor (TfR; Fitsanakis et al., 2011). FPN has also shown affinity for Mn attenuating Mn toxicity in vitro (Madejczyk and Ballatori, 2012). In vivo, mice acutely exposed to Mn via subcutaneous injection show increased protein level of FPN in the cortex and cerebellum (Yin et al., 2010). Contrary to Madejczyk and Ballatori (2012), Mitchell et al. (2014) did not find Mn to be a physiological substrate of FPN. Although both studies used the same cell model, Mn exposure and incubation times varied. The latter authors state that further testing is needed to clarify FPN's role in Mn export.

The transporter SLC30A10 has been identified for its role in the transport of Mn in humans. This protein is widely expressed in the basal ganglia and is localized to the cell surface where it mediates Mn efflux. Mutations in the gene encoding this carrier cause a hypermanganesemia syndrome with high Mn accumulation in the liver and brain without previous exposure to high levels of Mn (Leyva-Illades et al., 2014; Quadri et al., 2012; Tuschl et al., 2012). This suggests that SLC30A10 plays a role in Mn efflux. The solute carrier family is conserved among species, and some family members are expressed in the CNS of rats (Sreedharan et al., 2011).

The exchange of Mn from the plasma into the brain occurs through the blood–brain barrier (BBB) or the blood–cerebrospinal fluid barrier (Li et al., 2006). The Mn transport mechanism across the BBB involves various carriers, such as TfR-mediated endocytosis (Aschner and Aschner, 1990), Tf-independent mechanisms (Malecki et al., 1999), and Ca channels (Crossgrove and Yokel, 2005). When Mn is conjugated to citrate, it may be transported by members of the family of organic anion transporting polypeptides or the ABC-type transporters (ATP-binding cassette; Crossgrove et al., 2003).

MOLECULAR MECHANISMS OF MN NEUROTOXICITY

In vitro studies using cell culture models have contributed largely to the elucidation of Mn mechanisms of action. The use of rodent models is also an important tool for the study of the mechanisms of Mn toxicity. Studies using these models specifically focused on metal transport, homeostasis, and behavioral effects. Animal models facilitate the use of different routes of exposure to Mn as well as the use of different chemical forms of Mn, which may mimic environmental or occupational exposure. Several animal models have contributed to the elucidation of the mechanisms involved in Mn neurotoxicity. Much of what is known about the cellular response to stress induced by toxic concentrations of metals was obtained in mammalian models. Furthermore, there is a growing interest in the use of new animal models for these studies. In this regard, because of its homology with the mammalian biochemistry, the nematode worm *Caenorhabditis elegans* has proven to be useful in this regard. Several effects in response to exposure to metals, especially those involving gene expression and intracellular signaling pathways, have been reported using the nematode as a model (Caito et al., 2012).

A prominent feature of Mn intoxication is oxidative stress. Mn induces the formation of hydrogen peroxide (H_2O_2) in vitro (Posser et al., 2009) and hydroxyl radical ($HO\cdot$) production in vivo (Bałasz et al., 2015), in addition to interacting with extracellular DA, facilitating its oxidation and generation of reactive quinones (Benedetto et al., 2010). Several studies have reported changes in antioxidant enzymes and oxidative stress parameters induced by Mn (Deng et al., 2015; Donaldson et al., 1982; Maddirala et al., 2015; Milatovic et al., 2009). The main antioxidant strategies studied include the enzymes catalase, glutathione reductase, glutathione peroxidase, SOD, and the tripeptide F067-glutamyl cysteine or glutathione (GSH; Limón-Pacheco and Gonsebatt, 2009).

The production of antioxidant enzymes is largely under control of the transcription factor nuclear factor erythroid-2 related factor 2 (Nrf-2). Nrf-2 is located in the cytoplasm in complex with an inhibitory protein Keap-1 (Kelch like ECH associated protein 1), which in addition to retaining it in the cytoplasm marks Nrf-2 for proteosomal degradation. Stressful stimuli cause conformational changes in Keap-1 and induce Nrf-2 phosphorylation. This leads to the release of Nrf2 and allows its migration to the nucleus where it binds to the regulatory activation sequence, called antioxidant response element, located in the promoter region of genes encoding antioxidant enzymes (Kensler et al., 2007). Exposure to Mn induced alterations in Keap-1 and Nrf-2 levels in different studies (Deng et al., 2015; Li et al., 2011). It has been reported that different signaling pathways can activate the translocation of Nrf2. Mitogen-activated protein kinase, phosphoinositol-3-kinase/Akt (also known as protein kinase B or PKB), and protein kinase C pathways facilitate nuclear translocation of Nrf2. On the other hand, glycogen synthase kinase-3β negatively regulates Nrf2 signaling through phosphorylation of specific residues (Salazar et al., 2006; Sun et al., 2009).

Another transcription factor known for its role in cellular responses, including glucose metabolism, cell cycle progression, detoxification of ROS, DNA repair, and apoptosis, belongs to the family of Forkhead-box class O (FOXO) proteins. Mammals have four isoforms: FOXO1, FOXO3, FOXO4, and FOXO6. Three of these isoforms— FOXO1, FOXO3, and FOXO4—are regulated by Akt-dependent phosphorylation of three specific sites in response to stimulation by insulin and growth factors. Insulin/insulin-like growth factor-1 and growth factors inhibit FOXO by phosphorylation at specific sites through Akt. In response to stress, c-Jun N-terminal kinase (JNK) phosphorylates FOXO, directly activating its nuclear localization (Wang et al., 2014b). Mn has been shown to alter Akt phosphorylation in rats, as well as DAF-16 (the sole ortholog of the FOXO family of transcription factors in the nematode *C. elegans*) nuclear localization in *C. elegans* (Cordova et al., 2012; Gubert et al., 2016).

Oxidative stress occurs when the production of ROS or RNS exceeds the ability of cells to neutralize their effects. As a consequence, damage to cell membranes, DNA, and proteins; changes in intracellular signaling pathways; and eventually cell death occur. Brain tissue is particularly vulnerable to oxidative stress because of its high oxygen consumption, relatively low antioxidant capacity, and high levels of polyunsaturated fatty acids present in cell membranes (Halliwell, 2006). Thus strategies that have been tested in vitro and in vivo, involving antioxidant agents, are promising in attenuating Mn effects in the CNS, including Trolox, an analog of vitamin E (Cordova et al., 2012, 2013; Milatovic et al., 2011), organoselenium and organotellurium compounds (Avila et al., 2012), Brazilian fruit açaí (*Euterpe oleracea*) methanolic extract (da Silva Santos et al., 2014), ebselen and *para*-aminosalicylic acid (Santos et al., 2012a; Wang et al., 2014a), GSH and *N*-acetylcysteine (Stephenson et al., 2013), and *N*-acetylcysteineamide (Maddirala et al., 2015). It remains to be elucidated whether the beneficial effects of these compounds may also be related to their chelating properties.

Organelles such as the endoplasmic reticulum (ER) and mitochondria may be distressed by excess intracellular Mn. Activation of the ER stress-mediated apoptotic pathway in the striatum of Mn-exposed rats has been reported, with consequent apoptosis of neurocytes (Wang et al., 2015). ROS formation may also impair mitochondrial function. Mn may induce mitochondrial dysfunction with membrane permeability transition in astrocytes and neuroblastoma SH-SY5Y cells. This effect is due to calcium accumulation in the mitochondria, which leads to inner membrane collapse and energy failure (Maddirala et al., 2015; Yin et al., 2008; Zhang et al., 2004, 2008).

Mn-induced alteration of behavioral patterns (i.e., the motor incoordination or emotional and cognitive dysfunction observed in patients and/or animal models) may be associated with neurotransmitter metabolism disruption. Impaired neurotransmitter signaling may occur via different mechanisms, such as inhibition of neurotransmitter release, alterations in neurotransmitter clearance from the synaptic cleft, or modulation of receptor levels or action.

The main neurotransmitter system studied in Mn neurotoxicity is the dopaminergic system. In contrast to PD, in manganism, DA production may not be compromised. However, evidence from nonhuman primate studies shows that DA release from presynaptic terminals may be impaired. This is likely due to inhibition of proteins that function in neurotransmitter release by unknown mechanisms. For this reason, treatment with levodopa (a DA precursor widely used to treat PD) does not appear efficient in manganism patients (Guilarte et al., 2008). In a rat model, oral exposure to Mn during development (postnatal day 1–21) induced reduction of DA transporter levels on postnatal day 90. Moreover, the efflux of [^{3}H]DA in striatum and nucleus accumbens was impaired. This indicates that Mn exposure during development persistently impairs nigro-striatal DA signaling (McDougall et al., 2008).

F067-Aminobutyric acid (GABA) is the main inhibitory neurotransmitter in the brain, modulating excitatory signals within the basal ganglia. Thus GABA has a critical role in the coordination of motor function. The specific mechanisms by which Mn influences GABA activity are unclear, but Mn appears to target GABA uptake, leading to a greater than threefold increase in extracellular GABA. Astrocytes play a major role in GABA clearance from the synaptic cleft after neurotransmitter release. Astrocytes import GABA mainly through the GABA transporter (GAT3). Mn exposure inhibits astrocytic GABA uptake by a mechanism that involves Mn accumulation in the plasma membrane fraction of astrocytes, altering membrane dynamics by increment in fatty acid content and inhibition of GAT3 (Fordahl and Erikson, 2014).

Glutamate (Glu) is an excitatory neurotransmitter. Several transporters participate in Glu clearance to avoid excitotoxic injury (Danbolt, 2001). Numerous studies have established that Mn is capable of disrupting Glu transporting systems, leading to a reduction in Glu uptake either by regulating Glu transporter function (Lee et al., 2009) or reducing transporter expression (Erikson et al., 2002b), leading to an elevation in extracellular Glu levels. Moreover, Mn-induced ROS generation may inhibit astrocytic Glu transporters and contribute to excitotoxicity (da Silva Santos et al., 2014). A comprehensive review on Mn actions on the glutamine/Glu-GABA cycle was published by Sidoryk-Wegrzynowicz and Aschner (2013). Furthermore, evidence supports the notion that Mn at neurotoxic levels also affects the cholinergic system and inhibits acetylcholinesterase activity (Finkelstein et al., 2007; Santos et al., 2012b).

GENETIC ASPECTS OF MN INTOXICATION

Several of PD are idiopathic (of unknown cause). However, we cannot disregard an environmental component in the development of PD. Mn can be considered a risk factor for the development of PD given that there is a strong correlation between occupational Mn exposure and an increased risk of PD (Gorell et al., 1999). Environmental Mn exposure has also been associated with a higher prevalence of Parkinsonian disturbances (Lucchini et al., 2007). PD and Mn-induced neurotoxicity have common mechanistic features, including mitochondrial impairment and oxidative stress. The main differences between these disorders relate to the area of the brain that degenerates: in PD it is mainly the dopaminergic neurons of the substantia nigra pars compacta whereas in manganism the GABAergic cells of the striatum and globus pallidus are predominantly affected. PD patients respond to levodopa treatment whereas manganism is more effectively treated with chelating agents such as ethylenediametetraacetic acid or *para*-aminosalicylic acid (Chen et al., 2014; Erikson et al., 2004; Guilarte et al., 2006a; Roth, 2014).

Given the similarities between manganism and PD, understanding the molecular aspects of Mn transport and toxicity may help to understand the mechanisms underlying the etiology of PD. Recent discoveries show that certain PD-related genes [*parkin (PARK2), DJ-1 (PARK7), PINK1 (PTEN-induced putative kinase-1, PARK6), ATP13A2 (PARK9)*, and *SLC30A10* as well as *LRRK2 (leucine-rich repeat kinase-2, PARK8)* and *VPS35 (vacuolar protein sorting-associated protein-35, PARK17)*] are involved in Mn transport and/or toxicity. A review on the properties of genes involved in Mn toxicity and its association with PD onset has recently been published (Roth, 2014).

Modeling the complex gene–environment interaction in the simple model organism *C. elegans* represents a promising path to understanding the etiology and pathogenesis of PD and parkinsonism. Mutations in α-*synuclein, parkin*, and *dj-1*, all of which are associated with early-onset PD, have been shown to alter Mn transport. Bornhorst et al. (2014) recently demonstrated in *C. elegans* that *pdr-1* (worm homolog of *parkin*) and DJ-1-related (djr) gene *djr-1.1* (worm homolog of *DJ-1*) mutants show enhanced Mn accumulation and increased oxidative stress, and these effects may be reduced by expression of wild-type (WT) human α-synuclein. It is interesting to note that WT α-synuclein also protected against dopaminergic neurodegeneration induced by Mn exposure in *pdr-1* (the homolog of mammalian *parkin/PARK2*) mutant worms (Bornhorst et al., 2014). In the same model organism, the authors demonstrated that FPN might facilitate Mn export from the cells.

Worms overexpressing FPN in the background of *pdr-1* deletion displayed attenuation of several Mn-induced toxicity endpoints, such as survival, dopaminergic integrity, and accumulation of prooxidant metals (Chakraborty et al., 2015).

DJ-1 is thought to confer antioxidant protection to dopaminergic neurons by a not yet described mechanism. Mutations in *dj-1* cause autosomal-recessive, early-onset PD (Alvarez-Castelao et al., 2012; Taira et al., 2004). In *C. elegans* the deletion of *djr-1.2* decreased survival and lifespan after Mn exposure. Lifespan reduction resulting from *djr-1.2* deletion could be restored to normal by overexpression of either DJR-1.2 or DAF-16 (the worm homologue of FOXO), suggesting that DAF-16 mediates DJR-1.2 effects on lifespan upon Mn exposure. Moreover, dopaminergic signaling was disrupted by *djr1.2* deletion and improved by DAF-16 overexpression. Thus these authors established a clear interaction among three established risk factors for PD/parkinsonism: aging, Mn exposure, and loss-of-function alleles of DJ-1 (Chen et al., 2015b).

Huntington's disease (HD) is an autosomal-dominant, inherited neurodegenerative illness characterized by motor decline including chorea (a dance-like involuntary movement) and deterioration of coordination and motor skills. HD is caused by an abnormal expansion of a CAG repeat located in exon 1 of the gene encoding for the Huntingtin (HTT) protein. The striatum, an area that accumulates Mn, is also the main brain structure affected in HD (Walker, 2007). Several Mn-dependent enzymes, including arginase, glutamine synthetase, pyruvate decarboxylase, and Mn SOD, are altered in the postmortem brains of HD patients and pharmacological models of the disease. Moreover, intracellular Mn levels are significantly reduced in a striatal cell line (STHdh) expressing mutant HTT (Williams et al., 2010). This indicates that in HD the disease-causing allele of HTT may lead to alteration of Mn homeostasis, which in turn may compromise enzyme function and contribute to neurodegeneration.

CONCLUDING REMARKS

Mn neurotoxicity has been recognized for more than a century, and Mn neurotoxic actions toward the CNS have been extensively studied over the past decades. Progress has been achieved in the elucidation of mechanisms of Mn transport and homeostasis. The participation of proteins such as DMT-1, Tf, and FPN in these processes have been demonstrated. Novel Mn transporters have also been identified, such as SLC30A10. Several mechanisms of Mn action at the molecular level have been described, including alterations in neurotransmitter levels and the role of extracellular DA in the generation of oxidative stress induced by Mn. Nonetheless, there is still much to be learned about Mn transport and homeostasis/dyshomeostasis. Accordingly, Mn participation in the environmental component of neurodegenerative disease etiology, such as PD and HD, is poorly understood to date. Thus the field of toxicology is continuously being enriched by the study of Mn toxicity.

ACKNOWLEDGMENT

This manuscript was supported by the National Institute of Health grants R01 ES10563, R01 ES10563S1, R01 ES07331, and R01 ES16931.

REFERENCES

Agency for Toxic Substances and Disease Registry (ATSDR), 2012. Toxicological Profile for Manganese. U.S. Department of Health and Human Services, Public Health Service, Atlanta, GA.

Alvarez-Castelao, B., Muñoz, C., Sánchez, I., Goethals, M., Vandekerckhove, J., Castaño, J.G., 2012. Reduced protein stability of human DJ-1/PARK7 L166P, linked to autosomal recessive Parkinson disease, is due to direct endoproteolytic cleavage by the proteasome. Biochim. Biophys. Acta (BBA) – Mol. Cell Res. 1823, 524–533.

Aschner, J.L., Aschner, M., 2005. Nutritional aspects of manganese homeostasis. Mol. Aspects Med. 26, 353–362.

Aschner, M., Aschner, J.L., 1990. Manganese transport across the blood–brain barrier: relationship to iron homeostasis. Brain Res. Bull. 24, 857–860.

Aschner, M., Erikson, K., Hernández, E., Tjalkens, R., 2009. Manganese and its role in Parkinson's disease: from transport to neuropathology. NeuroMol. Med. 11, 252–266.

Avila, D.S., Benedetto, A., Au, C., Manarin, F., Erikson, K., Soares, F.A., Rocha, J.B.T., Aschner, M., 2012. Organotellurium and organoselenium compounds attenuate Mn-induced toxicity in *Caenorhabditis elegans* by preventing oxidative stress. Free Radic. Biol. Med. 52, 1903–1910.

Bałasz, M., Szkilnik, R., Brus, R., Malinowska-Borowska, J., Kasperczyk, S., Nowak, D., Kostrzewa, R., Nowak, P., 2015. Perinatal manganese exposure and hydroxyl radical formation in rat brain. Neurotox. Res. 27, 1–14.

Beaudin, S.A., Nisam, S., Smith, D.R., 2013. Early life versus lifelong oral manganese exposure differently impairs skilled forelimb performance in adult rats. Neurotoxicol. Teratol. 38, 36–45.

Benedetto, A., Au, C., Avila, D.S., Milatovic, D., Aschner, M., 2010. Extracellular dopamine potentiates Mn-induced oxidative stress, lifespan reduction, and dopaminergic neurodegeneration in a BLI-3-dependent manner in *Caenorhabditis elegans*. PLoS Genet. 6 (8), e1001084. http://dx.doi.org/10.1371/journal.pgen.1001084.

Boggio Bertinet, D., Tinivella, M., Alessandro Balzola, F., de Francesco, A., Davini, O., Rizzo, L., Massarenti, P., Antonietta Leonardi, M., Balzola, F., 2000. Brain manganese deposition and blood levels in patients undergoing home parenteral nutrition. J. Parenter. Enter. Nutr. 24, 223–227.

Bornhorst, J., Chakraborty, S., Meyer, S., Lohren, H., Brinkhaus, S.G., Knight, A.L., Caldwell, K.A., Caldwell, G.A., Karst, U., Schwerdtle, T., Bowman, A., Aschner, M., 2014. The effects of pdr1, djr1.1 and pink1 loss in manganese-induced toxicity and the role of α-synuclein in *C. elegans*. Metallomics 6, 476–490.

Bouchard, M., Mergler, D., Baldwin, M., Panisset, M., Bowler, R., Roels, H.A., 2007. Neurobehavioral functioning after cessation of manganese exposure: a follow-up after 14 years. Am. J. Industrial Med. 50, 831–840.

Bowler, R.M., Gysens, S., Diamond, E., Nakagawa, S., Drezgic, M., Roels, H.A., 2006a. Manganese exposure: neuropsychological and neurological symptoms and effects in welders. NeuroToxicology 27, 315–326.

Bowler, R.M., Koller, W., Schulz, P.E., 2006b. Parkinsonism due to manganism in a welder: neurological and neuropsychological sequelae. NeuroToxicology 27, 327–332.

Burdo, J.R., Menzies, S.L., Simpson, I.A., Garrick, L.M., Garrick, M.D., Dolan, K.G., Haile, D.J., Beard, J.L., Connor, J.R., 2001. Distribution of divalent metal transporter 1 and metal transport protein 1 in the normal and Belgrade rat. J. Neurosci. Res. 66, 1198–1207.

Caito, S., Fretham, S., Martinez-Finley, E., Chakraborty, S., Avila, D., Chen, P., Aschner, M., 2012. Genome wide analyses of metal responsive genes in *Caenorhabditis elegans*. Front. Genet. 3, 52. http://dx.doi.org/10.3389/fgene.2012.00052.

Canonne-Hergaux, F., Fleming, M.D., Levy, J.E., Gauthier, S., Ralph, T., Picard, V., Andrews, N.C., Gros, P., 2000. The Nramp2/DMT1 iron transporter is induced in the duodenum of microcytic anemia mk mice but is not properly targeted to the intestinal brush border. Blood 96 (12), 3964–3970.

Cersosimo, M.G., Koller, W.C., 2006. The diagnosis of manganese-induced parkinsonism. NeuroToxicology 27, 340–346.

Chakraborty, S., Chen, P., Bornhorst, J., Schwerdtle, T., Schumacher, F., Kleuser, B., Bowman, A.B., Aschner, M., 2015. Loss of pdr-1/parkin influences Mn homeostasis through altered ferroportin expression in *C. elegans*. Metallomics 7, 847–856.

Chen, P., Chakraborty, S., Mukhopadhyay, S., Lee, E., Paoliello, M.M.B., Bowman, A.B., Aschner, M., 2015a. Manganese homeostasis in the nervous system. J. Neurochem. 134, 601–610.http://dx.doi.org/10.1111/jnc.13170.

Chen, P., DeWitt, M.R., Bornhorst, J., Soares, F.A., Mukhopadhyay, S., Bowman, A.B., Aschner, M., 2015b. Age- and manganese-dependent modulation of dopaminergic phenotypes in a *C. elegans* DJ-1 genetic model of Parkinson's disease. Metallomics 7, 289–298.

Chen, P., Parmalee, N., Aschner, M., August 4,2014. Genetic factors and manganese-induced neurotoxicity. Front. Genet. 5, 265. http://dx.doi.org/10.3389/fgene.2014.00265. [eCollection 2014].

Chua, A.C.G., Morgan, E.H., 1997. Manganese metabolism is impaired in the Belgrade laboratory rat. J. Comp. Physiol. B 167, 361–369.

Claus Henn, B., Kim, J., Wessling-Resnick, M., Tellez-Rojo, M., Jayawardene, I., Ettinger, A., Hernandez-Avila, M., Schwartz, J., Christiani, D., Hu, H., Wright, R., 2011. Associations of iron metabolism genes with blood manganese levels: a population-based study with validation data from animal models. Environ. Health 10, 97.

Cordova, F., Aguiar Jr., A., Peres, T., Lopes, M., Gonçalves, F., Pedro, D., Lopes, S., Pilati, C., Prediger, R.S., Farina, M., Erikson, K., Aschner, M., Leal, R., 2013. Manganese-exposed developing rats display motor deficits and striatal oxidative stress that are reversed by Trolox. Arch. Toxicol. 1–14.

Cordova, F.M., Aguiar Jr., A.S., Peres, T.V., Lopes, M.W., Gonçalves, F.M., Remor, A.P., Lopes, S.C., Pilati, C., Latini, A.S., Prediger, R.D.S., Erikson, K.M., Aschner, M., Leal, R.B., 2012. In vivo manganese exposure modulates Erk, Akt and Darpp-32 in the striatum of developing rats, and impairs their motor function. PLoS One 7, e33057.

Crossgrove, J.S., Allen, D.D., Bukaveckas, B.L., Rhineheimer, S.S., Yokel, R.A., 2003. Manganese distribution across the blood–brain barrier: I. Evidence for carrier-mediated influx of manganese citrate as well as manganese and manganese transferrin. NeuroToxicology 24, 3–13.

Crossgrove, J.S., Yokel, R.A., 2005. Manganese distribution across the blood–brain barrier: IV. Evidence for brain influx through store-operated calcium channels. NeuroToxicology 26, 297–307.

da Silva Santos, V., Bisen-Hersh, E., Yu, Y., Cabral, I.S., Nardini, V., Culbreth, M., Teixeira da Rocha, J.B., Barbosa, F., Aschner Jr., M., 2014. Anthocyanin-rich acai (*Euterpe oleracea* Mart.) extract attenuates manganese-induced oxidative stress in rat primary astrocyte cultures. J. Toxicol. Environ. Health A 77, 390–404.

Danbolt, N.C., 2001. Glutamate uptake. Prog. Neurobiol. 65, 1–105.

Davis, C.D., Zech, L., Greger, J.L., 1993. Manganese metabolism in rats: an improved methodology for assessing gut endogenous losses. Proc. Soc. Exp. Biol. Med. Soc. Exp. Biol. Med. 202, 103–108.

Deng, Y., Jiao, C., Mi, C., Xu, B., Li, Y., Wang, F., Liu, W., Xu, Z., 2015. Melatonin inhibits manganese-induced motor dysfunction and neuronal loss in mice: involvement of oxidative stress and dopaminergic neurodegeneration. Mol. Neurobiol. 51, 68–88.

Dobson, A.W., Erikson, K.M., Aschner, M., 2004. Manganese neurotoxicity. Ann. N. Y. Acad. Sci. 1012, 115–128.

Donaldson, J., McGregor, D., LaBella, F., 1982. Manganese neurotoxicity: a model for free radical mediated neurodegeneration? Can. J. Physiol. Pharmacol. 60, 1398–1405.

Erikson, K., Shihabi, Z., Aschner, J., Aschner, M., 2002a. Manganese accumulates in iron-deficient rat brain regions in a heterogeneous fashion and is associated with neurochemical alterations. Biol. Trace Elem. Res. 87, 143–156.

Erikson, K.M., Suber, R.L., Aschner, M., 2002b. Glutamate/Aspartate transporter (GLAST), taurine transporter and metallothionein mRNA levels are differentially altered in astrocytes exposed to manganese chloride, manganese phosphate or manganese sulfate. NeuroToxicology 23, 281–288.

Erikson, K.M., Syversen, T., Steinnes, E., Aschner, M., 2004. Globus pallidus: a target brain region for divalent metal accumulation associated with dietary iron deficiency. J. Nutr. Biochem. 15, 335–341.

Finkelstein, Y., Milatovic, D., Aschner, M., 2007. Modulation of cholinergic systems by manganese. NeuroToxicology 28, 1003–1014.

Finley, J.W., Johnson, P.E., Johnson, L.K., 1994. Sex affects manganese absorption and retention by humans from a diet adequate in manganese. Am. J. Clin. Nutr. 60, 949–955.

Fitsanakis, V., Zhang, N., Garcia, S., Aschner, M., 2010. Manganese (Mn) and iron (Fe): interdependency of transport and regulation. Neurotox. Res. 18, 124–131.

Fitsanakis, V.A., Zhang, N., Avison, M.J., Erikson, K.M., Gore, J.C., Aschner, M., 2011. Changes in dietary iron exacerbate regional brain manganese accumulation as determined by magnetic resonance imaging. Toxicol. Sci. 120, 146–153.

Fleming, M.D., Romano, M.A., Su, M.A., Garrick, L.M., Garrick, M.D., Andrews, N.C., 1998. Nramp2 is mutated in the anemic Belgrade (b) rat: evidence of a role for Nramp2 in endosomal iron transport. Proc. Natl. Acad. Sci. 95, 1148–1153.

Fordahl, S.C., Erikson, K.M., 2014. Manganese accumulation in membrane fractions of primary astrocytes is associated with decreased γ-aminobutyric acid (GABA) uptake, and is exacerbated by oleic acid and palmitate. Environ. Toxicol. Pharmacol. 37, 1148–1156.

Garcia, S.J., Gellein, K., Syversen, T., Aschner, M., 2007. Iron deficient and manganese supplemented diets alter metals and transporters in the developing rat brain. Toxicol. Sci. 95, 205–214.

Garrick, L.M., Dolan, K.G., Romano, M.A., Garrick, M.D., 1999. Non-transferrin-bound iron uptake in Belgrade and normal rat erythroid cells. J. Cell. Physiol. 178, 349–358.

Gorell, J.M., Johnson, C.C., Rybicki, B.A., Peterson, E.L., Kortsha, G.X., Brown, G.G., Richardson, R.J., 1999. Occupational exposure to manganese, copper, lead, iron, mercury and zinc and the risk of Parkinson's disease. NeuroToxicology 20, 239–247.

Grandjean, P., Landrigan, P.J., 2006. Developmental neurotoxicity of industrial chemicals. Lancet 368, 2167–2178.

Grandjean, P., Landrigan, P.J., 2014. Neurobehavioural effects of developmental toxicity. Lancet Neurol 13, 330–338.

Gubert, P., Puntel, B., Lehmen, T., Bornhorst, J., Avila, D.S., Aschner, M., Soares, F.A.A., 2016. Reversible reprotoxic effects of manganese through DAF-16 transcription factor activation and vitellogenin downregulation in *Caenorhabditis elegans*. Life Sci. 151, 218–223.

Guilarte, T.R., 2013. Manganese neurotoxicity: new perspectives from behavioral, neuroimaging, and neuropathological studies in humans and non-human primates. Front. Aging Neurosci. 5, 23.

Guilarte, T.R., Burton, N.C., McGlothan, J.L., Verina, T., Zhou, Y., Alexander, M., Pham, L., Griswold, M., Wong, D.F., Syversen, T., Schneider, J.S., 2008. Impairment of nigrostriatal dopamine neurotransmission by manganese is mediated by pre-synaptic mechanism(s): implications to manganese-induced parkinsonism. J. Neurochem. 107, 1236–1247.

Guilarte, T.R., Chen, M.-K., McGlothan, J.L., Verina, T., Wong, D.F., Zhou, Y., Alexander, M., Rohde, C.A., Syversen, T., Decamp, E., Koser, A.J., Fritz, S., Gonczi, H., Anderson, D.W., Schneider, J.S., 2006a. Nigrostriatal dopamine system dysfunction and subtle motor deficits in manganese-exposed non-human primates. Exp. Neurol. 202, 381–390.

Guilarte, T.R., McGlothan, J.L., Degaonkar, M., Chen, M.-K., Barker, P.B., Syversen, T., Schneider, J.S., 2006b. Evidence for cortical dysfunction and widespread manganese accumulation in the nonhuman primate brain following chronic manganese exposure: a 1H-MRS and MRI study. Toxicol. Sci. 94, 351–358.

Gulson, B., Mizon, K., Taylor, A., Korsch, M., Stauber, J., Davis, J.M., Louie, H., Wu, M., Swan, H., 2006. Changes in manganese and lead in the environment and young children associated with the introduction of methylcyclopentadienyl manganese tricarbonyl in gasoline–preliminary results. Environ. Res. 100, 100–114.

Gunshin, H., Mackenzie, B., Berger, U.V., Gunshin, Y., Romero, M.F., Boron, W.F., Nussberger, S., Gollan, J.L., Hediger, M.A., 1997. Cloning and characterization of a mammalian proton-coupled metal-ion transporter. Nature 388, 482–488.

Gunter, T.E., Gavin, C.E., Aschner, M., Gunter, K.K., 2006. Speciation of manganese in cells and mitochondria: a search for the proximal cause of manganese neurotoxicity. NeuroToxicology 27, 765–776.

Gunter, T.E., Gerstner, B., Gunter, K.K., Malecki, J., Gelein, R., Valentine, W.M., Aschner, M., Yule, D.I., 2013. Manganese transport via the transferrin mechanism. NeuroToxicology 34, 118–127.

Gunter, T.E., Gerstner, B., Lester, T., Wojtovich, A.P., Malecki, J., Swarts, S.G., Brookes, P.S., Gavin, C.E., Gunter, K.K., 2010. An analysis of the effects of Mn^{2+} on oxidative phosphorylation in liver, brain, and heart mitochondria using state 3 oxidation rate assays. Toxicol. Appl. Pharmacol. 249, 65–75.

Halliwell, B., 2006. Oxidative stress and neurodegeneration: where are we now? J. Neurochem. 97, 1634–1658.

Illing, A.C., Shawki, A., Cunningham, C.L., Mackenzie, B., 2012. Substrate profile and metal-ion selectivity of human divalent metal-ion transporter-1. J. Biol. Chem. 287, 30485–30496.

Josephs, K.A., Ahlskog, J.E., Klos, K.J., Kumar, N., Fealey, R.D., Trenerry, M.R., Cowl, C.T., 2005. Neurologic manifestations in welders with pallidal MRI T1 hyperintensity. Neurology 64, 2033–2039.

Kensler, T.W., Wakabayashi, N., Biswal, S., 2007. Cell survival responses to environmental stresses via the Keap1-Nrf2-ARE pathway. Annu. Rev. Pharmacol. Toxicol. 47, 89–116.

Kern, C.H., Smith, D.R., 2011. Preweaning Mn exposure leads to prolonged astrocyte activation and lasting effects on the dopaminergic system in adult male rats. Synapse 65, 532–544.

Kern, C.H., Stanwood, G.D., Smith, D.R., 2010. Preweaning manganese exposure causes hyperactivity, disinhibition, and spatial learning and memory deficits associated with altered dopamine receptor and transporter levels. Synapse 64, 363–378.

Kim, J., Buckett, P.D., Wessling-Resnick, M., May 21, 2013. Absorption of manganese and iron in a mouse model of hemochromatosis. PLoS One 8(5), e64944. http://dx.doi.org/10.1371/journal.pone.0064944.

Lee, E.-S.Y., Sidoryk, M., Jiang, H., Yin, Z., Aschner, M., 2009. Estrogen and tamoxifen reverse manganese-induced glutamate transporter impairment in astrocytes. J. Neurochem. 110, 530–544.

Leyva-Illades, D., Chen, P., Zogzas, C.E., Hutchens, S., Mercado, J.M., Swaim, C.D., Morrisett, R.A., Bowman, A.B., Aschner, M., Mukhopadhyay, S., 2014. SLC30A10 is a cell surface-localized manganese efflux transporter, and parkinsonism-causing mutations block its intracellular trafficking and efflux activity. J. Neurosci. 34, 14079–14095.

Li, G.J., Choi, B.-S., Wang, X., Liu, J., Waalkes, M.P., Zheng, W., 2006. Molecular mechanism of distorted iron regulation in the blood–CSF barrier and regional blood–brain barrier following in vivo subchronic manganese exposure. NeuroToxicology 27, 737–744.

Li, H., Wu, S., Shi, N., Lian, S., Lin, W., 2011. Nrf2/HO-1 pathway activation by manganese is associated with reactive oxygen species and ubiquitin–proteasome pathway, not MAPKs signaling. J. Appl. Toxicol. 31, 690–697.

Li, X., Xie, J., Lu, L., Zhang, L., Zhang, L., Zou, Y., Wang, Q., Luo, X., Li, S., 2013. Kinetics of manganese transport and gene expressions of manganese transport carriers in Caco-2 cell monolayers. BioMetals 26, 941–953.

Limón-Pacheco, J., Gonsebatt, M.E., 2009. The role of antioxidants and antioxidant-related enzymes in protective responses to environmentally induced oxidative stress. Mutat. Res./Genet. Toxicol. Environ. Mutagen 674, 137–147.

Lucchini, R.G., Albini, E., Benedetti, L., Borghesi, S., Coccaglio, R., Malara, E.C., Parrinello, G., Garattini, S., Resola, S., Alessio, L., 2007. High prevalence of parkinsonian disorders associated to manganese exposure in the vicinities of ferroalloy industries. Am. J. Industrial Med. 50, 788–800.

Maddirala, Y., Tobwala, S., Ercal, N., 2015. N-acetylcysteineamide protects against manganese-induced toxicity in SHSY5Y cell line. Brain Res. 1608, 157–166.

Madejczyk, M.S., Ballatori, N., 2012. The iron transporter ferroportin can also function as a manganese exporter. Biochim. Biophys. Acta (BBA) – Biomembr. 1818, 651–657.

Malecki, E.A., 2001. Manganese toxicity is associated with mitochondrial dysfunction and DNA fragmentation in rat primary striatal neurons. Brain Res. Bull. 55, 225–228.

Malecki, E.A., Cook, B.M., Devenyi, A.G., Beard, J.L., Connor, J.R., 1999. Transferrin is required for normal distribution of 59Fe and 54Mn in mouse brain. J. Neurol. Sci. 170, 112–118.

Malecki, E.A., Radzanowski, G.M., Radzanowski, T.J., Gallaher, D.D., Greger, J.L., 1996. Biliary manganese excretion in conscious rats is affected by acute and chronic manganese intake but not by dietary fat. J. Nutr. 126, 489–498.

McDougall, S.A., Reichel, C.M., Farley, C.M., Flesher, M.M., Der-Ghazarian, T., Cortez, A.M., Wacan, J.J., Martinez, C.E., Varela, F.A., Butt, A.E., Crawford, C.A., 2008. Postnatal manganese exposure alters dopamine transporter function in adult rats: potential impact on nonassociative and associative processes. Neuroscience 154, 848–860.

Milatovic, D., Gupta, R.C., Yu, Y., Zaja-Milatovic, S., Aschner, M., 2011. Protective effects of antioxidants and anti-inflammatory agents against manganese-induced oxidative damage and neuronal injury. Toxicol. Appl. Pharmacol. 256, 219–226.

Milatovic, D., Zaja-Milatovic, S., Gupta, R.C., Yu, Y., Aschner, M., 2009. Oxidative damage and neurodegeneration in manganese-induced neurotoxicity. Toxicol. Appl. Pharmacol. 240, 219–225.

Molina, R.M., Phattanarudee, S., Kim, J., Thompson, K., Wessling-Resnick, M., Maher, T.J., Brain, J.D., 2011. Ingestion of Mn and Pb by rats during and after pregnancy alters iron metabolism and behavior in offspring. NeuroToxicology 32, 413–422.

Moreno, J.A., Streifel, K.M., Sullivan, K.A., Hanneman, W.H., Tjalkens, R.B., 2011. Manganese-induced NF-κB activation and nitrosative stress is decreased by estrogen in juvenile mice. Toxicol. Sci. 122, 121–133.

Moreno, J.A., Yeomans, E.C., Streifel, K.M., Brattin, B.L., Taylor, R.J., Tjalkens, R.B., 2009. Age-dependent susceptibility to manganese-induced neurological dysfunction. Toxicol. Sci. 112, 394–404.

Nagatomo, S., Umehara, F., Hanada, K., Nobuhara, Y., Takenaga, S., Arimura, K., Osame, M., 1999. Manganese intoxication during total parenteral nutrition: report of two cases and review of the literature. J. Neurol. Sci. 162, 102–105.

Oulhote, Y., Mergler, D., Barbeau, B., Bellinger, D.C., Bouffard, T., Brodeur, M.E., Saint-Amour, D., Legrand, M., Sauve, S., Bouchard, M.F., 2014. Neurobehavioral function in school-age children exposed to manganese in drinking water. Environ. Health Perspect. 122(12), 1343–1350. http://dx.doi.org/ 10.1289/ehp.1307918.

Posser, T., Franco, J.L., Bobrovskaya, L., Leal, R.B., Dickson, P.W., Dunkley, P.R., 2009. Manganese induces sustained Ser40 phosphorylation and activation of tyrosine hydroxylase in PC12 cells. J. Neurochem. 110, 848–856.

Quadri, M., Federico, A., Zhao, T., Breedveld, G.J., Battisti, C., Delnooz, C., Severijnen, L.-A., Di Toro Mammarella, L., Mignarri, A., Monti, L., Sanna, A., Lu, P., Punzo, F., Cossu, G., Willemsen, R., Rasi, F., Oostra, B.A., van de Warrenburg, B.P., Bonifati, V., 2012. Mutations in SLC30A10 cause parkinsonism and dystonia with hypermanganesemia, polycythemia, and chronic liver disease. Am. J. Hum. Genet. 90, 467–477.

Robison, G., Sullivan, B., Cannon, J.R., Pushkar, Y., May 2015. Identification of dopaminergic neurons of the substantia nigra pars compacta as a target of manganese accumulation. Metallomics 7(5), 748–755. http://dx.doi.org/10.1039/c5mt00023h.

Roth, J.A., 2006. Homeostatic and toxic mechanisms regulating manganese uptake, retention, and elimination. Biol. Res. 39, 45–57.

Roth, J.A., 2014. Correlation between the biochemical pathways altered by mutated parkinson-related genes and chronic exposure to manganese. NeuroToxicology 44, 314–325.

Roth, J.A., Feng, L., Dolan, K.G., Lis, A., Garrick, M.D., 2002a. Effect of the iron chelator desferrioxamine on manganese-induced toxicity of rat pheochromocytoma (PC12) cells. J. Neurosci. Res. 68, 76–83.

Roth, J.A., Horbinski, C., Higgins, D., Lein, P., Garrick, M.D., 2002b. Mechanisms of manganese-induced rat pheochromocytoma (PC12) cell death and cell differentiation. NeuroToxicology 23, 147–157.

Roth, J.A., Garrick, M.D., 2003. Iron interactions and other biological reactions mediating the physiological and toxic actions of manganese. Biochem. Pharmacol. 66, 1–13.

Salazar, M., Rojo, A.I., Velasco, D., de Sagarra, R.M., Cuadrado, A., 2006. Glycogen synthase kinase-3β inhibits the xenobiotic and antioxidant cell response by direct phosphorylation and nuclear exclusion of the transcription factor Nrf2. J. Biol. Chem. 281, 14841–14851.

Santamaria, A.B., 2008. Manganese exposure, essentiality & toxicity. Indian J. Med. Res. 128, 484–500.

Santos, A.P., Lucas, R.L., Andrade, V., Mateus, M.L., Milatovic, D., Aschner, M., Batoreu, M.C., 2012a. Protective effects of ebselen (Ebs) and para-aminosalicylic acid (PAS) against manganese (Mn)-induced neurotoxicity. Toxicol. Appl. Pharmacol. 258, 394–402.

Santos, D., Milatovic, D., Andrade, V., Batoreu, M.C., Aschner, M., Marreilha dos Santos, A.P., 2012b. The inhibitory effect of manganese on acetylcholinesterase activity enhances oxidative stress and neuroinflammation in the rat brain. Toxicology 292, 90–98.

Santos, D., Batoreu, C., Mateus, L., Marreilha Dos Santos, A.P., Aschner, M., 2014. Manganese in human parenteral nutrition: considerations for toxicity and biomonitoring. NeuroToxicology 43, 36–45.

Sidoryk-Wegrzynowicz, M., Aschner, M., 2013. Manganese toxicity in the central nervous system: the glutamine/glutamate-γ-aminobutyric acid cycle. J. Intern. Med. 273, 466–477.

Sreedharan, S., Stephansson, O., Schiöth, H.B., Fredriksson, R., 2011. Long evolutionary conservation and considerable tissue specificity of several atypical solute carrier transporters. Gene 478, 11–18.

Sriram, K., Lin, G.X., Jefferson, A.M., Stone, S., Afshari, A., Keane, M.J., McKinney, W., Jackson, M., Chen, B.T., Schwegler-Berry, D., Cumpston, A., Cumpston, J.L., Roberts, J.R., Frazer, D.G., Antonini, J.M., 2015. Modifying welding process parameters can reduce the neurotoxic potential of manganese-containing welding fumes. Toxicology 328, 168–178.

Stepens, A., Logina, I., Liguts, V., Aldiņš, P., Ekšteina, I., Platkājis, A., Mārtiņsone, I., Tērauds, E., Rozentāle, B., Donaghy, M., 2008. A parkinsonian syndrome in methcathinone users and the role of manganese. N. Engl. J. Med. 358, 1009–1017.

Stephenson, A.P., Schneider, J.A., Nelson, B.C., Atha, D.H., Jain, A., Soliman, K.F., Aschner, M., Mazzio, E., Renee Reams, R., 2013. Manganese-induced oxidative DNA damage in neuronal SH-SY5Y cells: attenuation of thymine base lesions by glutathione and N-acetylcysteine. Toxicol. Lett. 218, 299–307.

Sun, Z., Huang, Z., Zhang, D.D., 2009. Phosphorylation of Nrf2 at multiple sites by MAP kinases has a limited contribution in modulating the Nrf2-dependent antioxidant response. PLoS One 4, e6588.

Taira, T., Saito, Y., Niki, T., Iguchi-Ariga, S.M.M., Takahashi, K., Ariga, H., 2004. DJ-1 has a role in antioxidative stress to prevent cell death. EMBO Rep. 5 (2), 213–218.

Takeda, A., 2003. Manganese action in brain function. Brain Res. Rev. 41, 79–87.

Thompson, K., Molina, R.M., Donaghey, T., Schwob, J.E., Brain, J.D., Wessling-Resnick, M., 2007. Olfactory uptake of manganese requires DMT1 and is enhanced by anemia. FASEB J. 21, 223–230.

Tuschl, K., Clayton, P.T., Gospe Jr., Sidney, M., Gulab, S., Ibrahim, S., Singhi, P., Aulakh, R., Ribeiro, R.T., Barsottini, O.G., Zaki, M.S., Rosario, D., Maria, L., Dyack, S., Price, V., Rideout, A., Gordon, K., Wevers, R.A., Chong, W.K., Mills, P.B., 2012. Syndrome of hepatic cirrhosis, dystonia, polycythemia, and hypermanganesemia caused by mutations in SLC30A10, a manganese transporter in man. Am. J. Hum. Genet. 90, 457–466.

Tuschl, K., Mills, P.B., Clayton, P.T., 2013. Chapter twelve – manganese and the brain. In: Kailash, P.B., Susanne, A.S. (Eds.), International Review of Neurobiology. Academic Press, pp. 277–312.

Walker, F.O., 2007. Huntington's disease. Lancet 369, 218–228.

Wang, F., Wang, C., Jiang, Y., Deng, X., Lu, J., Ou, S., 2014a. Protective role of sodium para-amino salicylic acid against manganese-induced hippocampal neurons damage. Environ. Toxicol. Pharmacol. 37, 1071–1078.

Wang, T., Li, X., Yang, D., Zhang, H., Zhao, P., Fu, J., Yao, B., Zhou, Z., 2015. ER stress and ER stress-mediated apoptosis are involved in manganese-induced neurotoxicity in the rat striatum in vivo. NeuroToxicology 48, 109–119.

Wang, Y., Zhou, Y., Graves, D.T., 2014b. FOXO transcription factors: their clinical significance and regulation. BioMed Res. Int. 2014, 925350.

Williams, B.B., Li, D., Wegrzynowicz, M., Vadodaria, B.K., Anderson, J.G., Kwakye, G.F., Aschner, M., Erikson, K.M., Bowman, A.B., 2010. Disease-toxicant screen reveals a neuroprotective interaction between Huntington's disease and manganese exposure. J. Neurochem. 112, 227–237.

Yin, Z., Aschner, J.L., dos Santos, A.P., Aschner, M., 2008. Mitochondrial-dependent manganese neurotoxicity in rat primary astrocyte cultures. Brain Res. 1203, 1–11.

Yin, Z., Jiang, H., Lee, E.-S.Y., Ni, M., Erikson, K.M., Milatovic, D., Bowman, A.B., Aschner, M., 2010. Ferroportin is a manganese-responsive protein that decreases manganese cytotoxicity and accumulation. J. Neurochem. 112, 1190–1198.

Yokel, R., 2009. Manganese flux across the blood–brain barrier. NeuroMol. Med. 11, 297–310.

Yokel, R.A., Crossgrove, J.S., Bukaveckas, B.L., 2003. Manganese distribution across the blood–brain barrier: ii. manganese efflux from the brain does not appear to be carrier mediated. NeuroToxicology 24, 15–22.

Zhang, F., Xu, Z., Gao, J., Xu, B., Deng, Y., 2008. In vitro effect of manganese chloride exposure on energy metabolism and oxidative damage of mitochondria isolated from rat brakin. Environ. Toxicol. Pharmacol. 26, 232–236.

Zhang, S., Fu, J., Zhou, Z., 2004. In vitro effect of manganese chloride exposure on reactive oxygen species generation and respiratory chain complexes activities of mitochondria isolated from rat brain. Toxicol. Vitro 18, 71–77.

Chapter 31

Manganese and Nutritional Immunity

Lillian J. Juttukonda, Eric P. Skaar

Vanderbilt University Medical Center, Nashville, TN, United States

INTRODUCTION

Transition metals are required for life. Bacteria require transition metals to fulfill specific physiological requirements. Metals stabilize the structure of proteins, nucleic acids, and small molecules; donate and accept electrons in oxidation/reduction chemistry; and form electrostatic interactions during enzymatic catalysis (Hood and Skaar, 2012). Because metals are essential for bacterial survival and replication, bacterial pathogens must acquire metals during infection to cause disease (Hood and Skaar, 2012). Vertebrate hosts, including humans, are a reservoir of essential nutrients, which are required for eukaryotic cell function. To prevent bacterial acquisition of metals during invasive infections, vertebrates have evolved mechanisms to manipulate metal availability as a component of the immune response (Hood and Skaar, 2012). Hosts deplete essential metals at infection sites to "starve" pathogens while directing an influx of highly reactive (and therefore highly toxic) metals to the same sites (Hood and Skaar, 2012). Host redirection of essential metals as an antibacterial strategy is known as nutritional immunity (Hood and Skaar, 2012). Nutritional immunity was originally used to describe host restriction of iron. However, it is now known that other metals, including manganese, are also sequestered during infection.

Manganese is an essential nutrient for organisms in all domains of life. Manganese is a first-row transition metal most commonly in a 2+ cationic state. Manganese has a notable role in oxidative stress resistance. Similar to iron, manganese is capable of redox cycling; however, manganese is much less susceptible than iron to participate in toxic Fenton chemistry when exposed to hydrogen peroxide (Aguirre and Culotta, 2012; Lisher and Giedroc, 2013). Therefore manganese can serve as an antioxidant instead of promoting the toxicity of reactive oxygen species (Aguirre and Culotta, 2012).

Manganese is a cofactor for many enzymes. In humans, manganese is a required cofactor for superoxide dismutase, pyruvate carboxylase, and arginase (Bowman et al., 2011). Manganese less specifically activates hydrolases, phosphatases, transferases, dehydrogenases, kinases, peptidases, decarboxylases, and sugar transferases (Aschner and Aschner, 2005). As a result, manganese is important for bone formation, fat and carbohydrate metabolism, and blood sugar regulation (Bowman et al., 2011). In bacteria, manganese can serve as a cofactor for specific superoxide dismutases, ribonucleotide reductases, adhesion proteins, and antibiotic resistance enzymes, among others (Lin et al., 2011; Papp-Wallace and Maguire, 2006). Therefore manganese confers activity to several important physiological processes in the host and microbial invaders.

Several host mechanisms deplete manganese at the site of infection to starve pathogens of this nutrient, contributing to nutritional immunity specific to manganese. Successful pathogens have developed mechanisms that allow them to compete for manganese and therefore subvert nutritional immunity. In this chapter we describe host mechanisms of manganese nutritional immunity as well as the mechanisms utilized by bacterial pathogens to combat manganese restriction. We follow with a brief summary of bacterial processes that utilize manganese. Finally, we propose areas of future interest to the field of manganese nutritional immunity.

MANGANESE DISTRIBUTION

Manganese primarily enters the body through the gastrointestinal tract after ingestion of the metal in the diet (Fig. 31.1; Aschner and Aschner, 2005). Absorption occurs predominantly in the proximal small intestine (i.e., the duodenum). An average of 2–4 mg of manganese is consumed by humans on a daily basis, but only 1–3.5% is ultimately absorbed (Bowman et al., 2011). Manganese is absorbed by enterocytes via divalent metal-ion transporter-1 (DMT-1; Aschner and Aschner, 2005). After absorption, manganese is transported to the liver, where it enters the bloodstream for serum transport to other tissues (Aschner and Aschner, 2005). Most manganese in plasma is in complex with β1-microglobulin and albumin (Bowman et al., 2011). A small percentage of plasma manganese is oxidized to Mn^{3+} and bound to transferrin (Aschner and Aschner, 2005). Cells throughout the body take up manganese, utilizing several different transport mechanisms. These

Molecular, Genetic, and Nutritional Aspects of Major and Trace Minerals. http://dx.doi.org/10.1016/B978-0-12-802168-2.00031-2

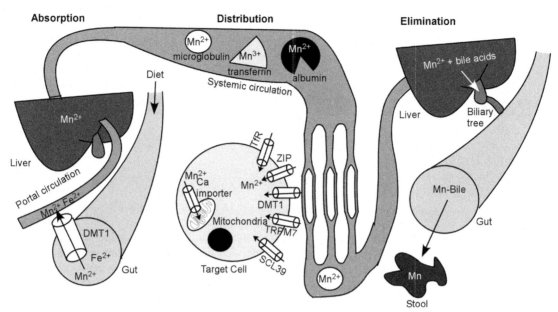

FIGURE 31.1 **Manganese homeostasis in vertebrates.** Manganese is absorbed in the gut by divalent metal-ion transporter-1 (DMT1) and transported to the liver. From the liver, manganese is distributed to the body in the bloodstream, where it is carried bound to β1-microglobulin, albumin, and transferrin. Manganese is brought into cells by various transporters throughout the body and sequestered in the mitochondria. Manganese is excreted from the body in bile; Mn interacts with bile salts, which limits its reabsorption; thus it is excreted in feces. *SLC39*, solute carrier-39; *TRPM7*, transient receptor potential melastatin-7; *ZIP*, zinc-interacting protein.

include DMT-1, transferrin receptor-mediated endocytosis, zinc-interacting proteins (ZIPs) ZIP8 and ZIP14, the transient receptor potential melastatin-7, and the solute carrier-39 family of zinc transporters (Bowman et al., 2011). All of these transporters have numerous metal substrates, with manganese being intermediate in affinity. Within cells, manganese is concentrated in the mitochondria by the calcium uniporter (Bowman et al., 2011). Finally, manganese is excreted in bile and removed from the body in the stool (Bowman et al., 2011). It is notable that no mechanisms have been identified for manganese-specific transport or storage throughout the body; instead, it appears that manganese usurps iron, zinc, calcium, and magnesium transport mechanisms. Whether any of these transport mechanisms have a role in nutritional immunity pertaining to manganese remains unclear. Given the importance of manganese in the host–pathogen interaction, as discussed throughout this chapter, future work should investigate whether any of these transport mechanisms are altered by cytokine profiles and serve to redistribute manganese as an antibacterial strategy.

It is unclear whether global mechanisms exist to redistribute manganese during infection. Iron and zinc are redistributed during infection, and this is thought to serve a functional role in nutritional immunity (King, 2011; Nairz et al., 2014). In a healthy adult, manganese levels vary between tissues, with relatively low levels in muscle, heart, and brain (Schroeder et al., 1966). Higher levels of manganese are found in the pancreas and the liver; both organs participate in bile production and manganese excretion (Schroeder et al., 1966). The distribution of manganese in the mouse appears to be similar (Kehl-Fie et al., 2013). Intriguingly, infection with *Staphylococcus aureus* leads to dramatic redistribution of manganese in the mouse kidney (Kehl-Fie et al., 2013). This finding suggests that mechanisms exist to restructure tissue and body manganese during an acute inflammatory response. However, the specific mechanisms remain obscure and should be a focus of future research.

CALPROTECTIN: CHELATING MANGANESE AND ZINC AT THE SITE OF INFECTION

Upon entering the host, most bacterial pathogens elicit an inflammatory response that attempts to kill or restrict growth of the invading pathogen and quell the infection. The cells and products of the innate immune system are vital for this inflammatory response. Neutrophils are an essential component of the innate immune response to extracellular bacteria (Mantovani et al., 2011). After rapid recruitment to the site of bacterial invasion, neutrophils kill bacteria via phagocytosis and through the release of antibacterial cellular contents (Mantovani et al., 2011). These contents include proteases, reactive oxygen species, histones, and metal-sequestering proteins (Mantovani et al., 2011). One such metal-binding protein is calprotectin (Fig. 31.2), an extracellular chelator of manganese and zinc that is highly antimicrobial (Zackular et al., 2015). Because of its abundance at the site of infection and its high-affinity manganese-binding properties, calprotectin is an important component of manganese nutritional immunity.

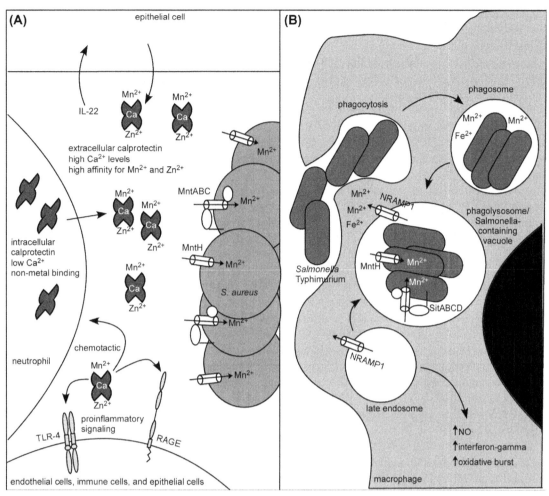

FIGURE 31.2 Calprotectin, natural resistance-associated macrophage protein-1 (NRAMP1), and nutritional immunity. (A) Calprotectin is constitutively produced by neutrophils or it is induced in epithelial cells by inflammatory stimuli. After being released from the cell, calprotectin binds calcium (Ca) to become activated. Calprotectin then binds manganese (Mn) and zinc (Zn), sequestering these metals during infection. Calprotectin also promotes the proinflammatory response by binding to various receptors. *Staphylococcus aureus* imports manganese through MntH and MntABC. This allows *S. aureus* to compete with calprotectin for manganese in the liver and the kidney. (B) NRAMP1 is expressed in late endosomes and lysosomes in macrophages. After a pathogen is phagocytosed, NRAMP1 is recruited to the phagosome where it serves to deplete the phagosome of manganese and other metals. This causes phagocytosed bacteria to be manganese starved and more susceptible to other macrophage killing strategies. NRAMP1 also promotes antibacterial functions of macrophages. *Salmonella enterica* serovar Typhimurium imports manganese through MntH and SitABCD. This permits *S.* Typhimurium to compete with NRAMP1 for manganese in the *Salmonella*-containing vacuole within macrophages and disseminate throughout the body. *IL-22*, Interleukin-22; *RAGE*, receptor for advanced glycation end products; *TLR-4*, toll-like receptor-4.

Calprotectin is a heterodimer of S100A8 and S100A9. The S100 family of proteins is found in vertebrates and characterized by two helix-loop-helix (EF-hand) calcium-binding domains (Zackular et al., 2015). Although most S100 proteins form homodimers, S100A8 and S100A9 instead form the heterodimer calprotectin (also known as MRP8/14 and calgranulin A/B; Zackular et al., 2015). Calprotectin is expressed in multiple cell types, including monocytes, endothelial cells, and keratinocytes (Zackular et al., 2015). Neutrophils constitutively produce calprotectin, where it makes up approximately 40% of the protein content in the cytoplasm (Zackular et al., 2015). Because neutrophils produce high quantities of calprotectin, they are typically considered the most important cellular source of calprotectin during infection. In support of this, experimental infection of mice with the extracellular bacterium *S. aureus* leads to high concentrations of calprotectin in staphylococcal lesions in a neutrophil-dependent manner (Corbin et al., 2008). *S100A8* and *S100A9* gene expression is constitutive in neutrophils but induced by proinflammatory stimuli in various epithelial cell lines. Inducers include the bacterial surface components flagella and lipopolysaccharide (LPS) as well as the cytokines interleukin (IL)-6, IL-1α, and IL-22 (Behnsen et al., 2014; Goyette and Geczy, 2011). It is an intriguing possibility that calprotectin may be produced by epithelial cells or endothelial cells to chelate manganese and zinc very early in infection before neutrophils have accumulated at an infectious site, but this has not been investigated. In summary, calprotectin is produced in high amounts by inflammatory cells as part of the innate immune response.

Calprotectin has two transition-metal binding sites, site 1 (S1) and site 2 (S2). S2 is an S100 protein canonical metal binding site that coordinates Zn ($K_d = 8.2$ nM) by three histidines and an aspartate (Damo et al., 2013). S1 is a noncanonical binding site that consists of six histidines: two from S100A8 and four from S100A9 (Damo et al., 2013). S1 is capable of binding zinc ($K_d = 3.4$ nM) or manganese ($K_d = 5.8$ nM; Damo et al., 2013). Manganese bound to S1 at this position has been confirmed by X-ray crystallography (Damo et al., 2013). Thus calprotectin is capable of simultaneously binding two Zn atoms or one Zn and one Mn atom.

The metal-chelating properties of calprotectin are regulated by calcium. As previously mentioned, all S100 proteins bind calcium with EF-hand domains (Zackular et al., 2015). Calcium activation of calprotectin is necessary for high-affinity manganese binding (Hood et al., 2012). This posttranslational regulation of calprotectin metal binding ensures that calprotectin only serves as a metal chelator in the extracellular environment because cytoplasmic calcium concentrations are low (Juttukonda and Skaar, 2015).

Calprotectin is an antimicrobial protein that inhibits growth of diverse microbial species through metal chelation. This has been demonstrated by experiments in liquid and solid culture where microbes are grown in the presence of purified recombinant human calprotectin. These experiments have shown that calprotectin inhibits growth of the human bacterial pathogens *S. aureus*, *Acinetobacter baumannii*, *Helicobacter pylori*, *Salmonella enterica*, *Staphylococcus epidermidis*, *Staphylococcus lugdunensis*, *Enterococcus faecalis*, *Pseudomonas aeruginosa*, and *Shigella flexneri*, as well as the fungal pathogens *Candida albicans*, *Aspergillus fumigatus*, and *Cryptococcus neoformans* (Amich et al., 2014; Damo et al., 2013; Juttukonda and Skaar, 2015; Liu et al., 2012; Mambula et al., 2000; Zackular et al., 2015). To determine whether the antibacterial properties of calprotectin require the manganese- or zinc-binding sites, point mutations were introduced in S1 and/or S2 and the resulting metal-binding–deficient forms of calprotectin were tested for antimicrobial activity (Damo et al., 2013). Metal-binding deficient forms of calprotectin retain their structural integrity (Damo et al., 2013). Calprotectin that cannot bind metal loses all antimicrobial capabilities in liquid culture growth, proving that growth inhibition by calprotectin is due to metal chelation (Damo et al., 2013). Strikingly, calprotectin engineered to lack the manganese/zinc-binding SI is highly attenuated for antimicrobial activity (Damo et al., 2013). In contrast, calprotectin engineered to lack the zinc-binding S2 still inhibits growth, underscoring the physiological importance of manganese chelation as an antibacterial strategy (Damo et al., 2013). Indeed, calprotectin inhibits manganese-dependent superoxide dismutase activity in *S. aureus* (Juttukonda and Skaar, 2015).

Mice that are calprotectin deficient (*S100a9* constitutive knockout) are more susceptible to infection with a range of bacterial pathogens. Calprotectin-deficient mice have higher bacterial burdens in the liver after intravenous *S. aureus* infection (Corbin et al., 2008), higher bacterial burdens in the lung and liver after *A. baumannii* intranasal infection (Hood et al., 2012), decreased survival and increased bacterial dissemination in a *Klebsiella pneumonia* model of pneumonia (Juttukonda and Skaar, 2015), and higher fungal burdens and decreased survival in a subcutaneous *C. albicans* model (Urban et al., 2009). These findings imply that calprotectin-mediated metal sequestration may play an important part in extracellular pathogen defense in vivo. In contrast to this, calprotectin deficiency is protective in some mouse infection models, including *S. enterica* serovar Typhimurium infection (Liu et al., 2012), *Streptococcus pneumoniae* intranasal inoculation (Juttukonda and Skaar, 2015), and *H. pylori* gastric infection (Gaddy et al., 2014). For *Salmonella*, the presence of calprotectin allows *Salmonella* to outcompete gut microbes because *Salmonella* possesses a high-affinity zinc importer (Liu et al., 2012). For *Helicobacter*, calprotectin activates a virulence program of gene expression (Gaddy et al., 2014). Calprotectin metal sequestration contributes to nutritional immunity, but some pathogens have evolved adaptive responses to metal limitation that increase virulence.

Calprotectin also contributes to alterations in tissue metal distribution that are dramatically evident after infection. Infection with *S. aureus* leads to dramatic alterations in the distribution of manganese and other metals in the liver and kidney (Corbin et al., 2008; Kehl-Fie et al., 2013). *S. aureus* infection causes visible lesions that contain bacterial microcolonies surrounded by immune cells. In the kidneys and livers of wild-type mice, *S. aureus* lesions are devoid of manganese and zinc relative to the surrounding tissue (Corbin et al., 2008; Kehl-Fie et al., 2013). On the other hand, *S. aureus* lesions in calprotectin-deficient mice contain levels of manganese consistent with surrounding healthy tissue (Corbin et al., 2008; Kehl-Fie et al., 2013). Therefore calprotectin is an important factor in producing a manganese-depleted *S. aureus* lesion.

Calprotectin has multiple other functions in addition to metal sequestration. Calprotectin serves as a proinflammatory molecule, activating the transcription factor nuclear factor-κB by acting as an agonist for the receptor for advanced glycation end products and toll-like receptor-4 (Zackular et al., 2015). Calprotectin is also predicted to serve as an oxidant sink because the S100A8 protomer of calprotectin isolated from asthmatic sputa exhibits oxidative modifications on cysteine residues, methionine residues, and a tryptophan residue (Gomes et al., 2013). These proposed functions for calprotectin may or may not intersect with its function as a metal-chelating protein.

In conclusion, the S100A8/A9 heterodimer calprotectin is an extracellular chelator of manganese. An important function of calprotectin is to inhibit bacterial growth during infection via manganese and zinc limitation, and calcium binding regulates its metal binding such that calprotectin is a chelator only in the extracellular environment. In vitro evidence suggests that manganese binding may be more important than zinc binding for the antibacterial functions of calprotectin, although experiments in wild-type and calprotectin-deficient mice have demonstrated that zinc binding is also important in vivo. Taken together, it is clear that calprotectin is a vital component of nutritional immunity with respect to manganese.

NATURAL RESISTANCE-ASSOCIATED MACROPHAGE PROTEIN-1/*SLC11A1*: STARVING BACTERIA IN MACROPHAGES

The previous section detailed a mechanism for starving extracellular bacteria of manganese. A distinct mechanism also exists for depleting the phagolysosome of manganese. Phagolysosomal depletion is performed by natural resistance-associated macrophage protein-1 (NRAMP1), encoded by the gene *SLC11A1*, which is a member of the solute carrier family of transporters. NRAMP1 is a phagocyte-specific protein that is expressed in late endosomes, lysosomes, and phagolysosomes, where it transports metals to the cytoplasm. NRAMP1 function is important for the defense against several pathogens in mice and humans, including *Mycobacterium* species.

NRAMP1 expression patterns suggest that its functional role is limited to classically activated, M1-type macrophages and neutrophils; both cell types have vital roles in antibacterial defense (Italiani and Boraschi, 2014; Mantovani et al., 2011). Classically activated macrophage subsets are important for antibacterial defense and are categorized by production of nitric oxide (Italiani and Boraschi, 2014). In neutrophils NRAMP1 is limited to gelatinase-positive tertiary granules, but its function in this cell type is not understood (Canonne-Hergaux et al., 2002). However, NRAMP1 function in macrophages has been extensively characterized. NRAMP1 contains a conserved N-terminal lysosomal targeting motif and is expressed in late endosomes and lysosomes of resting macrophages (Lam-Yuk-Tseung et al., 2006). The cellular targeting of NRAMP1 to the lysosome was confirmed by colocalization with the lysosomal marker Lamp1 (Cellier et al., 2007). After phagocytosis, NRAMP1 is recruited to the membrane of phagosomes as the phagolysosome is formed (Cellier et al., 2007). This expression pattern poises NRAMP1 to antagonize bacteria contained within the vacuole. NRAMP1 expression is induced by macrophage classical activators, including interferon (IFN)-γ, IL-1, IL-6, IL-10, tumor necrosis factor (TNF)-α, and exposure to bacteria or bacterial products such as LPS (Cellier et al., 2007). The expression patterns of NRAMP1 place NRAMP1 in a suitable position to aid in macrophage killing of phagocytosed bacteria.

NRAMP1 transports divalent cations across the phagosomal membrane (Lam-Yuk-Tseung et al., 2006). Much of what is known about NRAMP1 metal transport has been inferred by comparison with the other SLC11 transporter in humans, DMT-1 (also called SLC11A2 and NRAMP2), which is ubiquitously expressed (Gruenheid et al., 1999). NRAMP1 and DMT-1 are highly similar, with 12 predicted transmembrane domains and 78% amino acid identity across the hydrophobic core (Gruenheid et al., 1999). DMT-1 is responsible for iron uptake in the duodenum and iron import at the plasma membrane in some, but not all, peripheral tissues, as well as for iron transport in transferrin-recycling endosomes (Nevo and Nelson, 2006). In vitro studies have demonstrated DMT-1 transport of iron and manganese (as well as other divalent cations) in a pH-dependent fashion, symporting the metals with protons (Illing et al., 2012; Nevo and Nelson, 2006). On the basis of the discovery that DMT-1 transports metals, it was hypothesized that NRAMP1 also functions as a metal transporter. Indeed, NRAMP1 pH-dependent metal transport has been confirmed by studies utilizing isolated oocytes and peritoneal macrophages (Goswami et al., 2001; Jabado et al., 2000). Similar to DMT-1, NRAMP1 also transports other divalent cations (Forbes and Gros, 2003; Goswami et al., 2001; Nevo and Nelson, 2006). However, unlike DMT-1, NRAMP1 transports manganese preferentially over iron (Forbes and Gros, 2003). In summary, NRAMP1 is a divalent cation transporter with a wide range of metal substrates, but a preference for manganese.

Although the transport properties of NRAMP1 are largely agreed upon, the direction of NRAMP1 transport has been more difficult to establish. Some arguments have been made that NRAMP1 imports metals into the phagosome to intoxicate bacteria via the Fenton and Haber–Weiss reactions (Nairz et al., 2014). However, most evidence suggests that NRAMP1 depletes the phagolysosome of divalent cations (Nairz et al., 2014). Expression of NRAMP1 decreases phagosome manganese levels as measured by a fluorescent probe that is quenched by manganese (Forbes and Gros, 2003; Jabado et al., 2000). When NRAMP1 is engineered to be expressed at the plasma membrane, it inserts in the membrane in the same orientation and transports metals in the same direction as DMT-1 (Forbes and Gros, 2003). Macrophages expressing NRAMP1 starve *S.* Typhimurium for iron, whereas macrophages expressing nonfunctional NRAMP1 do not iron starve the bacterium (Nairz et al., 2009). Finally, functional NRAMP1 expression in peritoneal macrophages enhances killing of *S.* Typhimurium defective for manganese transport; the same *S.* Typhimurium strain is not killed by macrophages with a nonfunctional NRAMP1 (Zaharik and Finlay, 2004). Taken together, these results support a model by which the antibacterial mechanism of NRAMP1 is metal depletion, not metal intoxication. NRAMP1 has a function during nutritional immunity: to starve bacteria that have been phagocytosed, increasing bacterial susceptibility to macrophage-killing mechanisms.

The nutritional immunity imposed by NRAMP1 increases resistance to numerous pathogens in mice. NRAMP1 was discovered as the locus that determines susceptibility of different inbred mouse strains to specific infections. Many common mouse lines, including C57BL/6 and BALB/C, encode an *SLC11A1* allele (the gene encoding NRAMP1) with an amino acid substitution, G169D (Vidal et al., 1996). This mutation occurs in the fourth transmembrane domain, and the addition of the negatively charged aspartate results in a nonfunctional protein that is rapidly degraded (Vidal et al., 1996). Mouse strains expressing the nonfunctional Asp169 allele are susceptible to several intracellular pathogens, meaning that the pathogens grow to high numbers in the liver and spleen early in infection (Govoni et al., 1996). For *S.* Typhimurium infection, susceptible mice are unable to clear the bacterium and succumb to disease (Govoni et al., 1996). To demonstrate that susceptibility is specifically due to the *SLC11A1* allele, congenic mouse lines that express the functional Gly169 *SLC11A1* allele were engineered and found to be protected from *S.* Typhimurium and *Mycobacterium bovis* (Govoni et al., 1996). In addition to these pathogens, NRAMP1 increases resistance to *Mycobacterium intracellulare*, *Mycobacterium lepraemurium*, and *Campylobacter jejuni* as well as the parasite *Leishmania donovani* (Blackwell et al., 2000; Champion et al., 2008; Govoni et al., 1996). Therefore NRAMP1 is a vital component of the immune response to certain bacteria.

There is additional evidence demonstrating that the role of NRAMP1 in defense against intracellular pathogens is due to nutritional immunity. Mice expressing nonfunctional NRAMP1 or the functional NRAMP1 allele were infected with *S.* Typhimurium strains with different inactivating mutations in metal transport systems (Boyer et al., 2002). *S.* Typhimurium has two divalent transporters, MntH and SitABCD, both of which exhibit higher affinity for Mn than Fe in vitro (Boyer et al., 2002). Mice with the nonfunctional Asp169 NRAMP1 are equally susceptible to wild-type bacteria and a strain lacking *mntH*, and deletion of *sitABCD* modestly decreases virulence (Boyer et al., 2002). However, when mice express functional Gly169 NRAMP1, single and double knockouts of *mntH* and *sitABCD* exhibit decreased virulence relative to wild-type bacteria (Zaharik and Finlay, 2004). This establishes that host NRAMP1 is required to stress the bacterium by metal depletion during intracellular infection.

NRAMP1 is encoded by the gene *SLC11A1*. Intriguingly, *SLC11A1* promoter alleles are linked with human disease susceptibility. Sequencing analysis identified four promoter alleles for *SLC11A1*, two of which are extremely rare (Blackwell et al., 2000). Allele 3 is highly active and is present in approximately three-quarters of the population (Blackwell et al., 2000). Allele 2 is a very low-expressing promoter and is present in approximately one-quarter of the population (Blackwell et al., 2000). Intriguingly, the high-expressing allele 3 is associated with autoimmunity and poor infectious disease outcomes when the disease is driven by immune-mediated pathology (i.e., meningococcal meningitis; Blackwell et al., 2003). In contrast, the low-expressing allele 2 is associated with diseases that require activated macrophages for clearance (i.e., *M. tuberculosis*; Blackwell et al., 2003). *SLC11A1* haplotypes are also associated with human immunodeficiency virus, leprosy, and protozoal infections (Blackwell et al., 2003). The genetic associations between *SLC11A1* and infectious disease susceptibility in humans support an important role for NRAMP1 in human disease.

The role for NRAMP1 in immunity is not limited to metal deprivation. NRAMP1 influences production of several important macrophage molecules, including nitric oxide, IFN-γ, IL-6, IL-1β, major histocompatibility complex-II, and TNF-α (Blackwell et al., 2000). NRAMP1 also alters L-arginine flux, oxidative burst, and processing of antigen (Blackwell et al., 2000). As a result, expression of the functional Gly169 NRAMP1 allele biases the immune response toward a Th1-type response, whereas expression of the nonfunctional Asp169 allele biases the immune response toward a Th2-type response (Blackwell et al., 2000). The mechanistic basis for the differences in macrophage function is not understood, but it is an intriguing idea to presume that metal flux from the phagosome is an intracellular stimulus that alters gene expression or transcript stability.

In summary, NRAMP1 is important for nutritional immunity in the phagosome. NRAMP1 localizes to lysosomes and phagolysosomes within macrophages and neutrophils where phagocytosed bacteria are subjected to intense stressors, including reactive oxygen species and proteases. NRAMP1 depletes the phagolysosome of essential metals, especially manganese. Manganese starvation renders bacteria susceptible to the onslaught of reactive oxygen species because many antioxidant mechanisms require manganese for function. Thus NRAMP1 is an essential component of macrophage defense during bacterial infection.

BACTERIAL MANGANESE IMPORTERS: ACQUIRING MANGANESE DURING INFECTION

As detailed in the previous two sections, vertebrate hosts have multiple strategies that limit the concentration of available manganese for an invading bacterium. Bacteria possess high-affinity manganese importers that allow them to combat host nutritional immunity. Bacterial manganese importers are essentially ubiquitous because they allow cells to survive in low-manganese environments or conditions that require high amounts of manganese, such as oxidative stress (Juttukonda and Skaar, 2015). Because of its ionic nature, manganese cannot pass through membranes via diffusion; instead, a pore must be formed in the membrane (Ferguson and Deisenhofer, 2004). To import manganese, Gram-positive bacteria must transport this cation across a single lipid bilayer whereas Gram-negative bacteria must transport the metal across two lipid bilayers. Transport mechanisms across the Gram-negative inner membrane and Gram-positive cytoplasmic membrane are homologous and are generally well understood. Manganese transport mechanisms across the Gram-negative outer membrane are poorly defined. Manganese requirements vary drastically between species. For instance, *Borrelia burgdorferi*, the causative agent of Lyme disease, has tremendously high requirements for manganese because it uses manganese in the place of iron (Posey and Gherardini, 2000). In contrast, *Escherichia coli* requires very low levels of manganese and only expresses a single manganese importer under conditions of oxidative stress (Anjem et al., 2009). To provide examples as to how manganese transporters compete with host NRAMP1 and calprotectin during the battle for nutrient metal, two pathogens will be discussed in depth: *S. aureus* and *S.* Typhimurium.

S. aureus is a Gram-positive pathogen that is a leading cause of skin and soft-tissue infection, endocarditis (heart valve infection), osteomyelitis (bone infection), and bacteremia in the United States and throughout the world (Tong et al., 2015). *S. aureus* asymptomatically colonizes the anterior nares (nose) of many adults, but it is able to infect nearly any organ in the body if the physical or immunological barriers to infection are breached (Tong et al., 2015). Numerous virulence factors and nutrient acquisition systems enable *S. aureus* to cause invasive infections (Tong et al., 2015). *S. aureus* expresses two manganese transporters, an NRAMP family importer and an ABC family importer (Horsburgh et al., 2002).

The NRAMP family of manganese transporters in bacteria is homologous to the SLC11 family of transporters previously discussed in eukaryotes. This family of transporters possesses 10–12 transmembrane domains and utilizes proton gradients to provide the energy needed to transport manganese against a concentration gradient (Ehrnstorfer et al., 2014). Many NRAMP family transporters in bacteria are somewhat promiscuous, transporting multiple divalent cations with different affinities (Ehrnstorfer et al., 2014). In *S. aureus*, the NRAMP family transporter is named MntH (Horsburgh et al., 2002).

The ABC transport family in bacteria is a broad family of transporters with many different substrates. In *S. aureus*, the manganese ABC transporter is named MntABC; the genes *mntABC* are transcribed as an operon but translated into three separate proteins with three separate functions (Kehl-Fie et al., 2013). MntB forms the transmembrane channel that permits passage of manganese across the cytoplasmic membrane (Kehl-Fie et al., 2013). MntA is a cytoplasmic protein that hydrolyzes ATP to generate the energy required for transport against a concentration gradient (Kehl-Fie et al., 2013). Finally, MntC is a substrate binding protein that is on the extracellular side of the cytoplasmic membrane (Kehl-Fie et al., 2013). MntC binds manganese and transfers manganese to MntB for transport (Kehl-Fie et al., 2013). MntH and MntABC are important for manganese import under conditions of host nutritional immunity.

S. aureus relies on MntH and MntABC for subversion of calprotectin-mediated nutritional immunity. In liquid growth, inactivation of *mntH*, *mntC*, or both *mntH* and *mntC* causes *S. aureus* to be more susceptible to calprotectin (Kehl-Fie et al., 2013). In a mouse model of systemic *S. aureus* infection, inactivation of *mntH* and *mntC* decreases the ability of *S. aureus* to colonize the kidney and the liver (Kehl-Fie et al., 2013). It is interesting to note that this is only true for mice that express calprotectin because calprotectin-deficient animals have similar numbers of wild-type bacteria and the *mntH/mntC* mutant strain recovered (Kehl-Fie et al., 2013). This suggests that calprotectin is required to manganese starve *S. aureus*, and the MntH and MntABC transporters combat manganese starvation.

S. Typhimurium is a Gram-negative bacteria that causes food poisoning in humans and a mouse disease that mimics typhoid fever in humans (House et al., 2001). Typhoid fever is a life-threatening illness caused by systemic spread of *Salmonella* from the gut, causing high fever and abdominal pain (House et al., 2001). To cause typhoid fever, *Salmonella* subverts the antibacterial functions of macrophages, surviving within the macrophage to be carried from the gut to the liver, spleen, bone marrow, and lymph nodes (House et al., 2001). Therefore surviving within the macrophage is critical to *Salmonella* pathogenesis. The importance of macrophage NRAMP1, which depletes the phagosome of manganese to starve bacteria, was discussed previously in this chapter. In fact, the presence of a functional NRAMP1 is required for mice to survive *S.* Typhimurium infection (Govoni et al., 1996). *S.* Typhimurium expresses two manganese importers, MntH and SitABCD (Boyer et al., 2002). MntH is an NRAMP family transporter and SitABCD is an ABC family manganese transporter (Boyer et al., 2002). Both of these proteins exhibit higher affinity for Mn than Fe in vitro (Boyer et al., 2002). Confirming their importance in the virulence of *Salmonella*, inactivation of *mntH*, *sitABCD*, or both decreases virulence of the bacterium in *SLC11A1* congenic mice (Zaharik et al., 2004). However, MntH is not important in mice that lack functional NRAMP1 (Boyer et al., 2002). This establishes that *Salmonella* importers are required to combat nutritional immunity of the macrophage imposed by NRAMP1.

Manganese acquisition systems may be nearly universally required for complete bacterial fitness in the host. In addition to these three examples, manganese acquisition systems are important for the pathogenesis of several other bacterial pathogens, including *Brucella abortus* infection in mice, *Yersinia pseudotuberculosis* in a *Galleria mellonella* larvae infection, *B. burgdorferi* infection of ticks, *Streptococcus mutans* in a rat model of endocarditis, *Yersinia pestis* in murine bubonic plague, *Streptococcus suis* in a murine peritoneal infection, *Streptococcus sanguinis* in a rabbit model of endocarditis, and avian pathogenic *E. coli* in a chick air sac infection model (Anderson et al., 2009; Juttukonda and Skaar, 2015). The fact that this requirement for manganese is conserved points to the importance of manganese restriction in the host and the attractiveness of manganese acquisition systems as novel drug targets for multidrug-resistant pathogens.

MANGANESE-REQUIRING PROCESSES IN BACTERIAL PATHOGENS

Bacteria import manganese because it is required for specific bacterial processes. As of now a limited number of enzymes have been identified that require manganese for activity. Most prominently, manganese is useful as a cofactor in defense against oxidative stress in superoxide dismutase, catalase, Dps, and as a replacement for iron in mononuclear enzymes when hydrogen peroxide is high (Juttukonda and Skaar, 2015). Manganese metabolic enzymes include phosphoglyceromutase, enolase, pyruvate kinase, phosphoenolpyruvate carboxylase, and phosphoenolpyruvate carboxykinase (Papp-Wallace and Maguire, 2006). Manganese is a cofactor for specific ribonucleotide reductases (Juttukonda and Skaar, 2015). Certain proteins important for cell signaling utilize manganese, including phosphodiesterases and ppGpp hydrolases (Papp-Wallace and Maguire, 2006). *Clostridium difficile* encodes a fibronectin-binding protein, Fbp68, which requires manganese for the conformational change that stabilizes binding (Lin et al., 2011). The *S. aureus* fosfomycin resistance protein FosB is activated by manganese (Roberts et al., 2013). Overall, manganese is involved in diverse cellular processes within bacteria.

INTERSECTION OF MANGANESE NUTRITIONAL IMMUNITY WITH OTHER MINERALS

Nutritional immunity of manganese intersects with the homeostasis of many other minerals. First, this occurs in mechanisms of manganese homeostasis. Absorption and distribution of manganese exploits transport mechanisms for iron, calcium, and magnesium. This may mean that intake of high amounts of manganese could cause manganese to outcompete iron, causing functional iron deficiency. Second, interactions with other metals occur as the proteins involved in nutritional immunity of manganese also bind other metals. For instance, the manganese chelating protein calprotectin binds zinc at the same site as manganese and serves as a zinc chelator; binding affinities for either metal are much higher when calcium is at high levels. Therefore high concentrations of zinc or low concentrations of calcium could alter the manganese binding potential of calprotectin. In another example, NRAMP1 depletes the phagosome of manganese, but it also transports iron, cadmium, and cobalt. It is likely that starving phagocytosed bacteria of multiple metals is a better antibacterial strategy than removing one metal alone. Third, manganese import proteins in bacteria may be promiscuous, transporting multiple metals in addition to manganese. The potential consequences of this remain unclear. In summary, the manganese handling mechanisms described in this chapter certainly do not exist in a vacuum devoid of other metals; rather, the relative levels of each metal and their interactions with each other are an important component of nutritional immunity.

AREAS FOR FUTURE STUDY

Nutritional immunity of manganese is an important aspect of the host–pathogen dynamic, but many questions remain regarding manganese restriction during infection. In addition to calprotectin and NRAMP1, what other host mechanisms exist to limit manganese during infection? What are the mechanisms permitting global tissue redistribution in manganese levels during infection? Do alterations in manganese status caused by variations in dietary manganese intake alter the risk of bacterial infectious disease? How does the microbiota shape manganese homeostasis? Does restriction of manganese have a functional consequence in viral infection? Better understanding vertebrate manganese homeostasis will shape our understanding of these processes during infection. On the bacterial side, several important questions remain to be answered. What are the subcellular destinations of manganese in the bacterial cell? How is manganese trafficked within the cell? What speciation does excess or stored manganese take within the cell? Improved identification of bacterial manganese-binding proteins and small molecules will aid in answering these questions.

One important long-term goal for the field of nutritional immunity is to use this knowledge to guide therapeutic interventions. Before any such therapies may be derived based on NRAMP1 and calprotectin, several unanswered mechanistic questions should be answered. NRAMP1 has a well-understood role in macrophages, but it is also expressed in neutrophils. What is the function of NRAMP1 in nutritional immunity of neutrophils? Likewise, although calprotectin is produced in high levels in neutrophils, it is produced by many other cell types upon inflammatory stimulation. Is calprotectin produced in high enough quantities in these cells to sequester manganese? In addition, calprotectin and NRAMP1 have roles in proinflammatory cytokine production. Is this tied to the metal-handling properties of these proteins? Can these two functions be distinguished in vivo? Finally, calprotectin and NRAMP1 both have significant associations with autoimmune and allergic disease. Do the metal binding properties of these proteins promote the generation of autoreactive cells or inhibit immune tolerance mechanisms? It is intriguing to hypothesize that manganese levels and/or flux may be key signals for orchestrating inflammatory responses.

CONCLUSIONS

Manganese is an essential nutrient for bacteria because it serves as a cofactor for numerous cellular processes. When a pathogen enters a vertebrate host, it faces a hostile environment with minimal bioavailable manganese. The host establishes this environment by sequestering manganese at the focus of the infection. In the extracellular environment, calprotectin, a protein produced by neutrophils, tightly binds manganese. Inside of the phagosomes of macrophages, NRAMP1 transports manganese out into the cytosol, starving bacteria within the vacuole and contributing to intracellular killing. To combat these strategies, pathogenic bacteria express high-affinity manganese importers that are important for the virulence of numerous bacteria. The host–pathogen interface is a battle for nutrient manganese, and this battle shapes the pathogenesis of infection.

SUMMARY POINTS

- Hosts restrict essential metals, including manganese, as an antibacterial strategy.
- Calprotectin is a manganese- and zinc-binding protein that starves extracellular bacteria for metals. It is made by neutrophils, epithelial cells, and other cells.
- NRAMP1 is a metal transporter that pumps manganese and other metals out of the phagosome. It starves bacteria that have been engulfed by macrophages so that the bacteria cannot survive inside of the cell.
- Bacteria are able to acquire manganese during infection because they make manganese transport proteins. These transporters are required for some bacteria, including *S. aureus* and *Salmonella*, to cause disease.

KEY FACTS

- Bacteria require manganese to be fully virulent during infections.
- Hosts restrict manganese to limit bacterial virulence.
- Calprotectin sequesters manganese away from extracellular bacteria.
- NRAMP1 manganese starves bacteria that live in macrophage phagosomes.
- Bacteria fight back by importing manganese through high-affinity importers.

MINI DICTIONARY OF TERMS

ABC-family transporters ATP-binding cassette family transporters are a large and diverse family of transporters in bacteria. Manganese-transporting ABC family transporters are important for the fitness of multiple pathogens.

Calprotectin A heterodimer of S100A8 and S100A9 that is produced by neutrophils and other cell types. Calprotectin has two transitional metal-binding sites: one that binds zinc and one that binds zinc or manganese. Calprotectin is an antimicrobial protein that extracellularly chelates manganese and zinc. Also known as calgranulin A/B and MRP8/14.

DMT-1 Divalent metal-ion transporter-1 is a solute-linked carrier family member of transporter in vertebrates. DMT-1 is essential for iron homeostasis because it absorbs iron from the lumen of the gut, imports iron into individual cells throughout the body, and participates in transferrin–iron recycling. DMT-1 also transports manganese as well as other cations. DMT-1 is a highly similar homolog to NRAMP1. Also known as NRAMP2 and SLC11A2.

NRAMP1 Natural resistance-associated macrophage protein-1 is a membrane transporter in vertebrates. NRAMP1 is only expressed in late endosome, lysosomes, and phagosomes. NRAMP1 transports metals, especially manganese, out of the phagosome to starve intracellular bacteria. Also known as SLC11A1.

NRAMP-family transporters Bacterial homologs of the SLC11 family of transporters. NRAMP family transporters include manganese transporters, iron transporters, and magnesium transporters. NRAMP family importers are important for the fitness of multiple pathogens.

Nutritional immunity The concept that hosts restrict essential metals during infection. Host mechanisms include metal transporters, metal-binding proteins, and global redistributions of metals during infection. Nutritional immunity has been predominantly studied for the metals iron, zinc, and manganese. Copper has also been studied as a metal for which toxicity has been directed against bacteria by the host.

ACKNOWLEDGMENT

The authors thank the members of the Skaar laboratory for critical review of the manuscript. Work in the Skaar laboratory is supported by grant numbers AI101171, AI107233, and AI069233 from the National Institutes of Health; grant number INFB-024-13F from Veterans Affairs (VA); and the Defense Advanced Research Projects Agency (DARPA). Lillian J. Juttukonda has received support from grant number 15PRE25060007 from the American Heart Association and Public Health Service award T32 GM07347 from the National Institute of General Medical Studies for the Vanderbilt Medical-Scientist Training Program.

REFERENCES

Aguirre, J.D., Culotta, V.C., 2012. Battles with iron: manganese in oxidative stress protection. J. Biol. Chem. 287, 13541–13548.

Amich, J., Vicentefranqueira, R., Mellado, E., Ruiz-Carmuega, A., Leal, F., Calera, J.A., 2014. The ZrfC alkaline zinc transporter is required for *Aspergillus fumigatus* virulence and its growth in the presence of the Zn/Mn-chelating protein calprotectin. Cell. Microbiol. 16, 548–564.

Anderson, E.S., Paulley, J.T., Gaines, J.M., Valderas, M.W., Martin, D.W., Menscher, E., Brown, T.D., Burns, C.S., Roop 2nd, R.M., 2009. The manganese transporter MntH is a critical virulence determinant for *Brucella abortus* 2308 in experimentally infected mice. Infect. Immun. 77, 3466–3474.

Anjem, A., Varghese, S., Imlay, J.A., 2009. Manganese import is a key element of the OxyR response to hydrogen peroxide in *Escherichia coli*. Mol. Microbiol. 72, 844–858.

Aschner, J.L., Aschner, M., 2005. Nutritional aspects of manganese homeostasis. Mol. Aspects Med. 26, 353–362.

Behnsen, J., Jellbauer, S., Wong, C.P., Edwards, R.A., George, M.D., Ouyang, W., Raffatellu, M., 2014. The cytokine IL-22 promotes pathogen colonization by suppressing related commensal bacteria. Immunity 40, 262–273.

Blackwell, J.M., Searle, S., Goswami, T., Miller, E.N., 2000. Understanding the multiple functions of Nramp1. Microbes Infect. 2, 317–321.

Blackwell, J.M., Searle, S., Mohamed, H., White, J.K., 2003. Divalent cation transport and susceptibility to infectious and autoimmune disease: continuation of the Ity/Lsh/Bcg/Nramp1/Slc11a1 gene story. Immunol. Lett. 85, 197–203.

Bowman, A.B., Kwakye, G.F., Herrero Hernandez, E., Aschner, M., 2011. Role of manganese in neurodegenerative diseases. J. Trace Elem. Med. Biol. 25, 191–203.

Boyer, E., Bergevin, I., Malo, D., Gros, P., Cellier, M.F., 2002. Acquisition of Mn(II) in addition to Fe(II) is required for full virulence of *Salmonella enterica* serovar Typhimurium. Infect. Immun. 70, 6032–6042.

Canonne-Hergaux, F., Calafat, J., Richer, E., Cellier, M., Grinstein, S., Borregaard, N., Gros, P., 2002. Expression and subcellular localization of NRAMP1 in human neutrophil granules. Blood 100, 268–275.

Cellier, M.F., Courville, P., Campion, C., 2007. Nramp1 phagocyte intracellular metal withdrawal defense. Microbes Infect. 9, 1662–1670.

Champion, O.L., Valdez, Y., Thorson, L., Guttman, J.A., Menendez, A., Gaynor, E.C., Finlay, B.B., 2008. A murine intraperitoneal infection model reveals that host resistance to *Campylobacter jejuni* is Nramp1 dependent. Microbes Infect. 10, 922–927.

Corbin, B.D., Seeley, E.H., Raab, A., Feldmann, J., Miller, M.R., Torres, V.J., Anderson, K.L., Dattilo, B.M., Dunman, P.M., Gerads, R., Caprioli, R.M., Nacken, W., Chazin, W.J., Skaar, E.P., 2008. Metal chelation and inhibition of bacterial growth in tissue abscesses. Science 319, 962–965.

Damo, S.M., Kehl-Fie, T.E., Sugitani, N., Holt, M.E., Rathi, S., Murphy, W.J., Zhang, Y., Betz, C., Hench, L., Fritz, G., Skaar, E.P., Chazin, W.J., 2013. Molecular basis for manganese sequestration by calprotectin and roles in the innate immune response to invading bacterial pathogens. Proc. Natl. Acad. Sci. U.S.A. 110, 3841–3846.

Ehrnstorfer, I.A., Geertsma, E.R., Pardon, E., Steyaert, J., Dutzler, R., 2014. Crystal structure of a SLC11 (NRAMP) transporter reveals the basis for transition-metal ion transport. Nat. Struct. Mol. Biol. 21, 990–996.

Ferguson, A.D., Deisenhofer, J., 2004. Metal import through microbial membranes. Cell 116, 15–24.

Forbes, J.R., Gros, P., 2003. Iron, manganese, and cobalt transport by Nramp1 (Slc11a1) and Nramp2 (Slc11a2) expressed at the plasma membrane. Blood 102, 1884–1892.

Gaddy, J.A., Radin, J.N., Loh, J.T., Piazuelo, M.B., Kehl-Fie, T.E., Delgado, A.G., Ilca, F.T., Peek, R.M, Cover, T.L., Chazin, W.J., Skaar, E.P., Scott Algood, H.M., 2014. The host protein calprotectin modulates the *Helicobacter pylori cag* type IV secretion system via zinc sequestration. PLoS Pathog 10, e1004450.

Gomes, L.H., Raftery, M.J., Yan, W.X., Goyette, J.D., Thomas, P.S., Geczy, C.L., 2013. S100A8 and S100A9-oxidant scavengers in inflammation. Free Radic. Biol. Med. 58, 170–186.

Goswami, T., Bhattacharjee, A., Babal, P., Searle, S., Moore, E., Li, M., Blackwell, J.M., 2001. Natural-resistance-associated macrophage protein 1 is an H+/bivalent cation antiporter. Biochem. J. 354, 511–519.

Govoni, G., Vidal, S., Gauthier, S., Skamene, E., Malo, D., Gros, P., 1996. The Bcg/Ity/Lsh locus: genetic transfer of resistance to infections in C57BL/6J mice transgenic for the Nramp1 Gly169 allele. Infect. Immun. 64, 2923–2929.

Goyette, J., Geczy, C.L., 2011. Inflammation-associated S100 proteins: new mechanisms that regulate function. Amino Acids 41, 821–842.

Gruenheid, S., Canonne-Hergaux, F., Gauthier, S., Hackam, D.J., Grinstein, S., Gros, P., 1999. The iron transport protein NRAMP2 is an integral membrane glycoprotein that colocalizes with transferrin in recycling endosomes. J. Exp. Med. 189, 831–841.

Hood, M.I., Mortensen, B.L., Moore, J.L., Zhang, Y., Kehl-Fie, T.E., Sugitani, N., Chazin, W.J., Caprioli, R.M., Skaar, E.P., 2012. Identification of an *Acinetobacter baumannii* zinc acquisition system that facilitates resistance to calprotectin-mediated zinc sequestration. PLoS Pathog. 8, e1003068.

Hood, M.I., Skaar, E.P., 2012. Nutritional immunity: transition metals at the pathogen-host interface. Nat. Rev. Microbiol. 10, 525–537.

Horsburgh, M.J., Wharton, S.J., Cox, A.G., Ingham, E., Peacock, S., Foster, S.J., 2002. MntR modulates expression of the PerR regulon and superoxide resistance in *Staphylococcus aureus* through control of manganese uptake. Mol. Microbiol. 44, 1269–1286.

House, D., Bishop, A., Parry, C., Dougan, G., Wain, J., 2001. Typhoid fever: pathogenesis and disease. Curr. Opin. Infect. Dis. 14, 573–578.

Illing, A.C., Shawki, A., Cunningham, C.L., Mackenzie, B., 2012. Substrate profile and metal-ion selectivity of human divalent metal-ion transporter-1. J. Biol. Chem. 287, 30485–30496.

Italiani, P., Boraschi, D., 2014. From monocytes to M1/M2 macrophages: phenotypical vs. functional differentiation. Front. Immunol. 5, 514.

Jabado, N., Jankowski, A., Dougaparsad, S., Picard, V., Grinstein, S., Gros, P., 2000. Natural resistance to intracellular infections: natural resistance-associated macrophage protein 1 (Nramp1) functions as a pH-dependent manganese transporter at the phagosomal membrane. J. Exp. Med. 192, 1237–1248.

Juttukonda, L.J., Skaar, E.P., 2015. Manganese homeostasis and utilization in pathogenic bacteria. Mol. Microbiol.

Kehl-Fie, T.E., Zhang, Y., Moore, J.L., Farrand, A.J., Hood, M.I., Rathi, S., Chazin, W.J., Caprioli, R.M., Skaar, E.P., 2013. MntABC and MntH contribute to systemic *Staphylococcus aureus* infection by competing with calprotectin for nutrient manganese. Infect. Immun. 81, 3395–3405.

King, J.C., 2011. Zinc: an essential but elusive nutrient. Am. J. Clin. Nutr. 94, 679S–684S.

Lam-Yuk-Tseung, S., Picard, V., Gros, P., 2006. Identification of a tyrosine-based motif (YGSI) in the amino terminus of Nramp1 (Slc11a1) that is important for lysosomal targeting. J. Biol. Chem. 281, 31677–31688.

Lin, Y.P., Kuo, C.J., Koleci, X., McDonough, S.P., Chang, Y.F., 2011. Manganese binds to *Clostridium difficile* Fbp68 and is essential for fibronectin binding. J. Biol. Chem. 286, 3957–3969.

Lisher, J.P., Giedroc, D.P., 2013. Manganese acquisition and homeostasis at the host-pathogen interface. Front. Cell. Infect. Microbiol. 3, 91.

Liu, J.Z., Jellbauer, S., Poe, A.J., Ton, V., Pesciaroli, M., Kehl-Fie, T.E., Restrepo, N.A., Hosking, M.P., Edwards, R.A., Battistoni, A., Pasquali, P., Lane, T.E., Chazin, W.J., Vogl, T., Roth, J., Skaar, E.P., Raffatellu, M., 2012. Zinc sequestration by the neutrophil protein calprotectin enhances *Salmonella* growth in the inflamed gut. Cell Host Microbe 11, 227–239.

Mambula, S.S., Simons, E.R., Hastey, R., Selsted, M.E., Levitz, S.M., 2000. Human neutrophil-mediated nonoxidative antifungal activity against *Cryptococcus neoformans*. Infect. Immun. 68, 6257–6264.

Mantovani, A., Cassatella, M.A., Costantini, C., Jaillon, S., 2011. Neutrophils in the activation and regulation of innate and adaptive immunity. Nat. Rev. Immunol. 11, 519–531.

Nairz, M., Fritsche, G., Crouch, M.L., Barton, H.C., Fang, F.C., Weiss, G., 2009. Slc11a1 limits intracellular growth of *Salmonella enterica* sv. Typhimurium by promoting macrophage immune effector functions and impairing bacterial iron acquisition. Cell. Microbiol. 11, 1365–1381.

Nairz, M., Haschka, D., Demetz, E., Weiss, G., 2014. Iron at the interface of immunity and infection. Front. Pharmacol. 5, 152.

Nevo, Y., Nelson, N., 2006. The NRAMP family of metal-ion transporters. Biochim. Biophys. Acta 1763, 609–620.

Papp-Wallace, K.M., Maguire, M.E., 2006. Manganese transport and the role of manganese in virulence. Annu. Rev. Microbiol. 60, 187–209.

Posey, J.E., Gherardini, F.C., 2000. Lack of a role for iron in the Lyme disease pathogen. Science 288, 1651–1653.

Roberts, A.A., Sharma, S.V., Strankman, A.W., Duran, S.R., Rawat, M., Hamilton, C.J., 2013. Mechanistic studies of FosB: a divalent-metal-dependent bacillithiol-S-transferase that mediates fosfomycin resistance in *Staphylococcus aureus*. Biochem. J. 451, 69–79.

Schroeder, H.A., Balassa, J.J., Tipton, I.H., 1966. Essential trace metals in man: manganese. A study in homeostasis. J. Chronic Dis. 19, 545–571.

Tong, S.Y., Davis, J.S., Eichenberger, E., Holland, T.L., Fowler Jr., V.G., 2015. *Staphylococcus aureus* infections: epidemiology, pathophysiology, clinical manifestations, and management. Clin. Microbiol. Rev. 28, 603–661.

Urban, C.F., Ermert, D., Schmid, M., Abu-Abed, U., Goosmann, C., Nacken, W., Brinkmann, V., Jungblut, P.R., Zychlinsky, A., 2009. Neutrophil extracellular traps contain calprotectin, a cytosolic protein complex involved in host defense against *Candida albicans*. PLoS Pathog. 5, e1000639.

Vidal, S.M., Pinner, E., Lepage, P., Gauthier, S., Gros, P., 1996. Natural resistance to intracellular infections: Nramp1 encodes a membrane phosphoglycoprotein absent in macrophages from susceptible (Nramp1 D169) mouse strains. J. Immunol. 157, 3559–3568.

Zackular, J.P., Chazin, W.J., Skaar, E.P., 2015. Nutritional immunity: S100 proteins at the host-pathogen interface. J. Biol. Chem.

Zaharik, M.L., Cullen, V.L., Fung, A.M., Libby, S.J., Kujat Choy, S.L., Coburn, B., Kehres, D.G., Maguire, M.E., Fang, F.C., Finlay, B.B., 2004. The *Salmonella enterica* serovar Typhimurium divalent cation transport systems MntH and SitABCD are essential for virulence in an Nramp1G169 murine typhoid model. Infect. Immun. 72, 5522–5525.

Zaharik, M.L., Finlay, B.B., 2004. Mn^{2+} and bacterial pathogenesis. Front. Biosci. 9, 1035–1042.

Chapter 32

Manganese and Mitochondrial Function

Thomas E. Gunter

University of Rochester, Rochester, NY, United States

INTRODUCTION

Manganese (Mn) is a transition metal adjacent to iron (Fe) in the periodic table. It is a biologically essential element, and similar to Fe it can transition between the 2^+ and 3^+ oxidation states and serve biological oxidation/reduction functions. In the 2^+ state it has a half-filled 3d shell; therefore it is spherically symmetric, similar to Ca^{2+} and Mg^{2+} (Armstrong, 2008; Gunter and Pfeiffer, 1990). Mn^{2+} also has an ionic radius intermediate between those of Ca^{2+} and Mg^{2+}; therefore it can easily bind to Ca^{2+} or Mg^{2+} binding sites. Because Mn readily binds to Ca^{2+}, Mg^{2+}, and Fe^{2+} or Fe^{3+} binding sites, it has the potential to bind to and affect myriad biological processes, greatly complicating the understanding of its effects. Because of its partially filled 3d shell, it is paramagnetic and can be studied using electron paramagnetic resonance. The electron paramagnetism also causes Mn to have strong relaxation effects on nearby nuclear spins such as protons, making Mn amenable to be studied using nuclear magnetic resonance (NMR) as well as techniques used to study other ions, such as atomic absorption and counting of radioactive tracer isotopes.

Although small amounts of Mn are biologically essential, large amounts can lead to neurodegeneration. This Mn-induced neurodegenerative syndrome shows similarities to Parkinson's disease (PD) and has been reported to be a risk factor in idiopathic PD (Benedetto et al., 2009; Roth, 2014). The symptoms of Mn-induced neurodegeneration correlate with Mn accumulation in the globus pallidus, striatum, caudate-putamen, and subthalamic nuclei (Aschner et al., 1999; Eriksson et al., 1992; Newland et al., 1989). At the organ and cellular level the accumulation of Mn in these brain regions also correlates with cell death in the internal segment of the globus pallidus, a decrease of dopamine in the striatum, and gliosis and swelling in the posterior limb of the internal capsule (Aschner et al., 1999). The globus pallidus and striatum have been reported to be the brain areas most sensitive to Mn-induced damage (Aschner et al., 1999, 2007).

The most frequently reported types of initial biochemical damage produced by Mn accumulation in the brain have been (1) mitochondrial effects, including impaired energy metabolism and ATP production, production of reactive oxygen species (ROS), and apoptotic effects including activation of proteases, such as caspases 3, 7, 8, 9, and 12 discussed in section The Mitochondrial Role in Apoptosis (Brouillet et al., 1993; Galvani et al., 1995; Gavin et al., 1992; Malecki, 2001; Roth et al., 2000, 2002; Zwingmann et al., 2003; Zhang et al., 2004; Marreilha Dos Santos et al., 2010; Alaimo et al., 2013; Gonzalez et al., 2008) and (2) pathological changes in the levels of neurotransmitters such as dopamine, γ-aminobutyric acid, and glutamate (Aschner et al., 1999, 2007; Benedetto et al., 2009; Martinez-Finley et al., 2013). Inside of the cell, Mn ions are found primarily in the nucleus and inside of energized mitochondria. Accumulation in the nucleus is because Mn^{2+}, similar to Mg^{2+}, can electrostatically associate with nucleic acids. Mn^{2+} readily enters mitochondria via the Ca^{2+} uniporter, moves down the internally negative mitochondrial membrane potential to the interior, and remains there as long as the membrane potential is maintained. If the mitochondrial membrane potential falls, then Mn^{2+} can leave the mitochondria via reverse uniport (Gunter and Pfeiffer, 1990). We will focus here on the mitochondrial effects of Mn.

MANGANESE SPECIATION: DOES MN^{2+} OR MN^{3+} CAUSE THE MOST DAMAGE?

To cause damage within a living system, Mn ions must be free to bind to and affect important sites or they must be so reactive and present in high enough concentrations to cause wholesale damage to the system's components. Although stable millimolar concentrations of Mn^{2+} are easy to produce, free Mn^{3+} is not stable in aqueous solution near neutral pH and readily disproportionates into Mn^{2+} and MnO_2 in the presence of oxygen (Armstrong, 2008; Gunter et al., 2006). Mn^{3+}, but not Mn^{2+}, is a strong oxidizing agent (Armstrong, 2008; Gunter et al., 2006). Unquestionably on a molar basis, free Mn^{3+} could cause much more damage to biological tissue than Mn^{2+}; however, only very small amounts of any Mn^{3+} complex have been reported in biological tissue, cells, or mitochondria, and those reports found the Mn^{3+} in complexes

Molecular, Genetic, and Nutritional Aspects of Major and Trace Minerals. http://dx.doi.org/10.1016/B978-0-12-802168-2.00032-4

stable in aqueous solution not as free Mn^{3+} (Gunter et al., 2006). XANES spectroscopy (X-ray absorption near edge structure) has many advantages in experiments designed to determine speciation of Mn ions inside of biological tissues and cells. XANES spectra in the optimum energy range for Mn detection can easily differentiate between the spectra of Mn^{2+} and Mn^{3+} complexes because of an energy shift between their spectra. Experimental XANES measurements show a composite spectrum that is the sum of the spectra of each of the types of Mn ions in the sample each weighted by its concentration. This technique can be sensitive enough to show endogenous Mn superoxide dismutase within neuron-like cells, astrocytes, and mitochondria (Gunter et al., 2006). Most mechanisms that transport Mn ions into cells appear to transport Mn^{2+} (Gunter et al., 2013). Whereas Mn^{3+} can be transported into brain cells via the transferrin mechanism, because of reduction by STEAP (six transmembrane epithelial antigen of the prostate) proteins inside of endosomes, it ends up as Mn^{2+} in the cell cytosol or inside mitochondria (Gunter et al., 2013). Although the mitochondrial Ca^{2+} uniporter is known to transport Mn^{2+} into mitochondria, no mechanism has been shown to transport Mn^{3+} into mitochondria (Gunter and Pfeiffer, 1990; Gunter et al., 2010). It was hypothesized during the 1980s that a superoxide radical such as that produced by mitochondria through metabolic reactions could oxidize Mn^{2+} to Mn^{3+} (Archibald and Tyree, 1987); however, extensive experimentation using XANES spectroscopy has found no evidence supporting this hypothesis (Gunter et al., 2006). Today we know that Mn^{2+} inhibits ATP production, increases ROS production, and induces many characteristics of apoptosis via the mitochondrial pathway, as will be described in section The Mitochondrial Role in Apoptosis. Although it is impossible to show that no free Mn^{3+} exists in biological tissue, experimental results show that the levels of the reactive free Mn^{3+} in biological tissue and cells are very small and below those detectable by XANES spectroscopy (Gunter et al., 2006). Therefore the damage detected is probably caused by Mn^{2+}.

MITOCHONDRIA

Mitochondria, the organelles responsible for oxidative phosphorylation, are found in almost every animal or plant cell as well as in many single-celled species. It is well known that mitochondria contain their own DNA and ribosomes and produce some of their own peptides; however, the peptides that they produce represent only a very small fraction of the total number of different types of peptides and protein complexes found inside of mitochondria. The majority of the peptides and proteins found in mitochondria are coded by nuclear DNA in the cells in which they are found; therefore they are subject to differences in the nuclear DNA read in these different types of cells. For example, in placental mammals mitochondrial DNA codes only 13 of the approximately 3000 different types of peptides and proteins that have been reported to be inside such mitochondria (Nicholls, 1982). This has important implications for toxicologists. The most commonly isolated types of mitochondria, liver and heart mitochondria, are not always good models for mitochondria from the target tissue for Mn toxicology, the brain (Fig. 32.1). Different types of mitochondria are known to behave differently. For example, brain cells and mitochondria primarily metabolize sugars whereas heart cells and mitochondria primarily metabolize fats. These mitochondrial differences are caused by the presence of different complexes in some mitochondria and certainly in differences in the amounts and activities of some important metabolic complexes. Even different segments of the brain may have mitochondria that function somewhat differently. Therefore Mn may interfere with different processes in different types of mitochondria. Furthermore, even brain mitochondrial preparations are mixtures of the mitochondria found in the different types of cells found in the functioning brain, such as neurons and astrocytes. Because there is evidence that Mn-induced damage to astrocytes may be an important cause of Mn neurotoxicity, the subsequent mitochondrial differences could become important (Aschner et al., 1992; Gonzalez et al., 2008; Hazell, 2002).

Mitochondria perform many functions in addition to their well-known function of producing more than 90% of the cell's ATP by oxidative phosphorylation. Among these other functions are cytosolic Ca^{2+} buffering by sequestration of Ca^{2+} from the cytosol (Gunter and Pfeiffer, 1990; Gunter and Sheu, 2009) and partial control of apoptosis (Czabotar et al., 2014; Elmore, 2007), both of which could couple Mn-induced mitochondrial effects to processes that could damage or kill cells. Ca^{2+} sequestration by mitochondria not only activates the ATP production necessary for the life of the cell, but it also stimulates production of ROS and potentiates the mitochondrion to a permeability transition that could lead to cell death either via apoptosis or necrosis (Gunter and Sheu, 2009; Martinez-Finley et al., 2013).

Mitochondria produce by far the majority of the ROS found in the typical cell. Most mitochondrial ROS are produced as the membrane impermeable superoxide radical ($O_2^{-\cdot}$) at complexes I and III of the mitochondrial electron transport chain (Goncalves et al., 2014; Lambert and Brand, 2009). This superoxide radical can be converted into the membrane-permeable H_2O_2 by superoxide dismutase. In the presence of Fe or Mn, the H_2O_2 can be converted into the very reactive hydroxyl radical (OH$^{\cdot}$; Goldstein et al., 1993). ROS production has been reported to be increased by mitochondrial Ca^{2+} uptake and by induction of the Ca^{2+}-induced mitochondrial permeability transition (MPT; Gunter and Sheu, 2009). Both an increase in intramitochondrial Ca^{2+} and ROS production increase the probability of induction of the MPT (Gunter and Sheu, 2009).

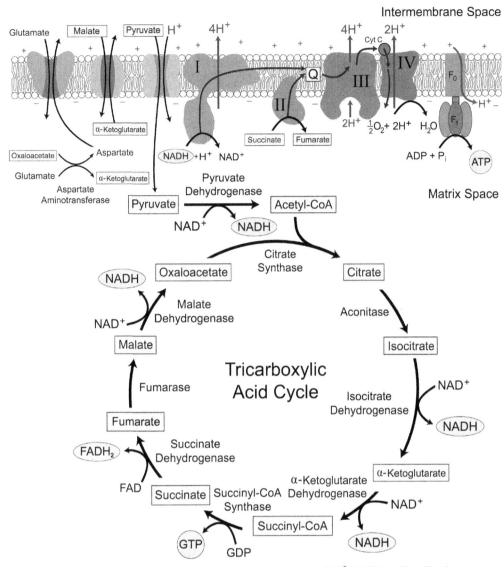

FIGURE 32.1 Portions of the mitochondrial metabolic pathways showing reported sites of Mn^{2+} inhibition. (Top) The electron transport chain and the F_1F_0 ATP synthase, transporters, and aspartate aminotransferase. (Bottom) The tricarboxylic acid (TCA) cycle. Mn^{2+} has been found to inhibit the following steps: the F_1F_0 ATP synthase in heart and liver mitochondria (Gunter et al., 2010); complex II or succinate dehydrogenase in brain mitochondria (Gunter et al., 2010), cultures of neurons and astrocytes (Zwingmann et al., 2003), and in striatal neurons (Malecki, 2001); α-ketoglutarate dehydrogenase as a purified enzyme (Gunter et al., 2010); aconitase in rat brain mitochondria (Zheng et al., 1998); complex I in PC12 cells (Galvani et al., 1995); and to possibly inhibit the glutamate/aspartate exchanger or aspartate aminotransferase in brain mitochondria (Gunter et al., 2010). *CoA*, coenzyme A; *FAD*, flavin adenine dinucleotide; *NAD*, nicotinamide adenine dinucleotide. *Printed with permission from Toxicol. Appl. Pharmacol 249, 65–75, 2010.*

THE MITOCHONDRIAL ROLE IN APOPTOSIS

The mitochondrion's role and function in control of apoptosis is perhaps second in importance only to oxidative phosphorylation. Apoptosis is a topic that is very relevant to the mitochondrial role in Mn neurotoxicity, and we must briefly review some of its properties; however, it is far too complex to be covered adequately here. Fortunately, there are many good reviews available (Czabotar et al., 2014; Elmore, 2007; Skulachev, 2006). It used to be thought that there were two completely distinct forms of cell death: necrosis and apoptosis. Necrosis is caused by stresses induced by stress agents such as mechanical damage, high temperature, chemical damage, and toxins, and it represents a form of cell death characterized by loss of cell components, swelling, and inflammation of nearby tissue. Apoptosis, or programmed cell death, is a conservative process, characterized by contraction of cell volume rather than swelling, chromatin condensation, nuclear fragmentation, blebbing of the plasma membrane, and the formation of apoptotic vesicles containing cellular material that is taken up and used by surrounding cells, especially macrophages (Czabotar et al., 2014; Elmore, 2007). Now many people believe that cell death that is partially necrotic and partially apoptotic can occur (Skulachev, 2006). Many papers have discussed two general pathways of apoptosis: the extrinsic or death receptor pathway and the intrinsic or mitochondrial pathway. Both pathways lead to the activation of proteases called caspases (cysteine-containing proteases that cleave at aspartate residues) that induce the characteristics of apoptosis (Czabotar et al., 2014; Elmore, 2007). Now, some papers describe a third pathway as well, involving stresses at the endoplasmic reticulum (Czabotar et al., 2014; Elmore, 2007; Oubrahim et al., 2002; Yoon et al., 2011). The death receptor pathway is thought to act through caspase (casp)-8 to activate what are called "executioner" or "effector" caspases such as casp 3 or casp 7. The mitochondrial pathway is thought to act through casp 9 to activate the executioner caspases whereas the endoplasmic reticulum pathway is thought to act through casp 12 to activate the executioner caspases. The executioner caspases take the cell apart so as to cause the characteristics of apoptosis. Many complex feedback and feed-forward regulatory steps operate within and between these pathways, providing multiple layers of control over this very complex process (Czabotar et al., 2014). Our main focus here will be on the mitochondrial pathway, but we will also consider how this may overlap with the endoplasmic reticulum pathway.

Many of the control steps involve members of what is called the Bcl-2 protein family, which can be divided into three separate groups. All of these proteins contain what are called BH domains. There are "initiator" proapoptotic BH3-only proteins (such as Bim, Bad, tBid), "guardian" antiapoptotic multidomain prosurvival proteins (such as Bcl-2, Bcl-XL, Bcl-W), and "effector" multidomain proapoptotic proteins (such as Bax, Bak; Czabotar et al., 2014). Although this may sometimes be an oversimplification, generally a decision for or against apoptosis is made on the basis of whether pro- or antiapoptotic Bcl-2 family proteins dominate (Czabotar et al., 2014). The mitochondrial outer membrane and intermembrane space contain proapoptotic Bcl-2 family members and other proapoptotic factors such as cytochrome c (cyt c), which can be released by permeabilization of the outer membrane as occurs following induction of the MPT. The MPT is induced by the opening of a pore through the inner membrane referred to as the permeability transition pore (PTP; Bernardi, 1999; Halestrap, 2010). In addition, the proapoptotic Bcl-2 family members can be released by apoptotic regulation independently of the opening of the PTP (Czabotar et al., 2014). Many reports in the early literature emphasize that apoptosis requires ATP whereas necrosis does not; however, apoptosis induced by a fall in ATP levels in the cell has been reported. The differences between apoptosis and necrosis now do not seem as distinct as was originally thought, and cases describing admixtures of apoptotic and necrotic characteristics have been reported (Roth et al., 2002; Skulachev, 2006). Often the form of cell death observed depends on the intensity of the stress causing it.

MANGANESE INHIBITION OF OXIDATIVE PHOSPHORYLATION

One of the most frequently reported Mn-induced mitochondrial effects is a decrease in energy metabolism (Brouillet et al., 1993; Galvani et al., 1995; Gavin et al., 1992; Malecki, 2001; Roth et al., 2000, 2002; Zwingmann et al., 2003), and this is sometimes explicitly identified as a decrease in ATP production by oxidative phosphorylation (Gavin et al., 1992; Gunter et al., 2010; Roth et al., 2000; Zwingmann et al., 2003). Because Mn^{2+} binds so readily to Ca^{2+}, Mg^{2+}, and Fe sites, many steps in the overall process of oxidative phosphorylation or, for that matter, any process could be affected by Mn^{2+}, and several steps in the metabolic pathways have been found to be inhibited by Mn^{2+}. This is especially true in the metabolic pathways leading to ATP production because intramitochondrial $[Ca^{2+}]$ is the leading controller of the rate of ATP production (Balaban, 2002); therefore substitution of Mn^{2+} for Ca^{2+} could have very significant effects on ATP production. Gavin and coworkers (Gavin et al., 1992) using liver and brain mitochondria found that treatment with Mn^{2+} strongly inhibited ADP-stimulated O_2 use rates in both types of mitochondria and concluded that Mn^{2+} directly inhibited oxidative phosphorylation. Roth and coworkers used the luciferin/luciferase assay to measure ATP production and showed that Mn^{2+} inhibited oxidative phosphorylation in PC12 cells (Roth et al., 2000). Using NMR spectroscopy, Zwingmann and coworkers found that a 5-day exposure of neurons to $100\,\mu M$ Mn^{2+} led to a 50% decrease in the ATP/ADP ratio (Zwingmann et al., 2003). Gunter and coworkers, continuing the approach of Gavin et al., showed a Mn^{2+}-induced decrease in oxidation rate and ATP production in liver, heart, and brain mitochondria (Gunter et al., 2010). Brouillet et al. injected $2\,\mu mol$ of Mn^{2+} in the left striatum and used the right striatum as a control. Animals were sacrificed 1 week after Mn^{2+} injection, and the striata were surgically removed and quick frozen to permit metabolic analysis. Results showed that the ATP levels in the Mn^{2+}-injected rats had fallen to 51% of control (Brouillet et al., 1993). Galvani et al. found that Mn^{2+} treatment inhibited mitochondrial complex I in PC12 cells (Galvani et al., 1995). Using primary cultures of striatal neurons, Malecki found that 48-h treatment with micromolar concentrations of Mn^{2+} led to decreases of mitochondrial membrane potential and complex II activity (Malecki, 2001).

Several sites within the metabolic pathways have been identified as being inhibited by Mn^{2+} (Galvani et al., 1995; Gunter et al., 2010; Malecki, 2001; Zheng et al., 1998; Zwingmann et al., 2003). However, just because Mn^{2+} inhibits a component of the metabolic pathways does not mean that inhibition at this site inhibits ATP production by oxidative phosphorylation itself. To inhibit ATP production, the inhibitory step must be rate limiting for the entire process of oxidative phosphorylation. Furthermore, as previously discussed, because the amounts and activities of components of the metabolic pathways vary between different types of mitochondria, what is rate limiting in one type of mitochondria may not be rate limiting in another type. A new type of experiment was developed to determine which of these steps was actually rate limiting for the overall process of oxidative phosphorylation. To elucidate the mechanisms by which Mn^{2+} inhibits mitochondrial energy production, we need a convenient way to measure the rate of ATP synthesis. Fortunately, the rate of O_2 use by mitochondria is known to correlate closely with the rate of ATP production and it is easy to measure (Chance and Williams, 1955). Gunter and coworkers used this approach to measure the inhibitory effects of Mn^{2+} on isolated liver, heart, and brain mitochondria. Conditions were carefully chosen so that the metabolic pathways, including the ATP synthase, but not substrate concentration, O_2, etc., were rate limiting. Measurements were made using three sets of substrates under coupled and uncoupled conditions (uncoupling takes the ATP synthase out of the process). It was important to use different sets of substrates because these different substrates utilize different portions of the metabolic pathways in such a way as to allow us to identify important aspects of the process.

Results showed the ATP synthase was the rate-limiting step of oxidative phosphorylation caused by Mn^{2+} inhibition in liver and heart mitochondria, but not in brain mitochondria. In brain mitochondria, rate limitation was at complex II, which contains succinate dehydrogenase, or fumarase and the glutamate/aspartate exchanger. Mn^{2+} inhibition of ATP production was approximately 3 times stronger in brain mitochondria than in liver or heart mitochondria. There were metabolic sites that had been shown to be inhibited by Mn^{2+} that did not show up as rate limiting to the overall process (Gunter et al., 2010). Although these inhibitions could be important to some other process, they were not rate limiting for oxidative phosphorylation under conditions of rapid ATP production. Interestingly, Zwingmann and coworkers using the completely independent technique, NMR, also found inhibition of ATP production in cell suspensions containing neurons and astrocytes to be at complex II (Zwingmann et al., 2003). NMR is able to measure the amounts of succinate in the samples; therefore it could show an increase in [succinate] with increasing [Mn^{2+}], indicating an inhibition at complex II under relatively physiological conditions. The fact that these two very different experimental approaches and technologies identify the same enzymatic step as being rate limiting strengthens the case for complex II being one of the rate-limiting steps in ATP production induced by Mn^{2+}. In addition, another laboratory used two-dimensional polyacrylamide gel electrophoresis (PAGE) and matrix-assisted laser desorption/ionization time-of-flight mass spectroscopy of rat brain samples with and without exposure to Mn treatment at the intact animal level to identify proteins that were significantly affected by the Mn treatment. They found that the PAGE spots for both the F_1F_0 ATP synthase and succinate dehydrogenase (complex II) were decreased by Mn treatment (Zhang et al., 2005). The agreement of these different types of experimental approaches is reassuring and supports the validity of each of the approaches.

MANGANESE-INDUCED INCREASES IN MITOCHONDRIAL PRODUCTION OF REACTIVE OXYGEN SPECIES

Published results from many laboratories over the past 2 decades have shown that treatment of various cells and tissues with Mn^{2+} often induces a significant increase in ROS production, signs of the mitochondrial MPT, and many characteristics of apoptosis. It must be kept in mind that these responses are not independent of each other and do not necessarily indicate different modes of cell death. For example, increased ROS production is well known to increase the probability of MPT induction (Bernardi, 1999; Halestrap, 2010) and MPT induction usually leads to release of proapoptotic factors such as cyt c from the mitochondrial intermembrane space followed by apoptosis via the intrinsic pathway (Czabotar et al., 2014). In addition, apoptosis is mediated by a complex network of reactions that, as was found with Mn inhibition of oxidative phosphorylation, undoubtedly vary from one cell type and set of experimental conditions to another. These considerations suggest that although we should consider all of these results, it is most important to consider data obtained in cells or tissues most like the target tissues involved with Mn neurotoxicity under conditions similar to typical exposure conditions in vivo. Although it is likely that astrocytes in the globus pallidus and striatum may be the cells most affected by Mn effects, that is not certain. Furthermore, it is very difficult while working at the cellular level to duplicate the long-term, low concentration exposure of Mn that usually induces Mn toxicity in the workplace.

Evidence for increased oxidative stress or ROS production has been shown in a wide range of mitochondria, cells, and tissues including liver and brain mitochondria (Zhang et al., 2004), Hela cells (Oubrahim et al., 2001), neural stem cells (NSCs; Tamm et al., 2008), *Caenorhabditis elegans* (Settivari et al., 2009), rat brain endothelial cells (RBE4; Marreilha Dos Santos et al., 2010), to human neuroblastoma cells (Yoon et al., 2011). Using the dichloro-dihydro-fluorescein diacetate (DCFH-DA) technique, Zhang and coworkers showed that Mn^{2+} increased ROS production that was sensitive to treatment with antioxidants (Zhang et al., 2004). In work with Hela cells, Mn^{2+} treatment not only increased ROS as measured using DCFH-DA but caused an increase in mitochondrial membrane potential as measured using tetra methyl rhodamine ethyl ester (TMRE; Oubrahim et al., 2001). The authors speculated that the increase might be due to an increase in cardiolipin; however, it might also be caused by inhibition of the F_1F_0 ATP synthase by Mn^{2+} in this cell type (Gunter et al., 2010) that could cause the membrane potential to increase because proton flow through the synthase could significantly decrease. Work with NSCs not only showed an increase in oxidative stress induced by Mn^{2+}, but it also showed characteristics of the mitochondrial pathway of apoptosis including Bax activation and oligomerization, cyt c release, and a significant increase in casp 3 activation that was insensitive to blockage of casp 8 activation (Tamm et al., 2008). Work with *C. elegans* showed that Mn^{2+} induced an increase in ROS measured using DCFH-DA and a decrease in mitochondrial membrane potential measured using TMRE (Settivari et al., 2009). Work with RBE4 cells not only showed that Mn^{2+} induced a decrease in cell viability using the MTT (3-(4,5-dimethylthiazol-2-yl)-2,5-diphenyltetrazolium bromide) assay but also that it increased ROS as indicated by the F2-isoprostane assay and decreased mitochondrial membrane potential using TMRE (Marreilha Dos Santos et al., 2010). Work with SK-N-MC human neuroblastoma cells showed that Mn^{2+} induced an increase in ROS as measured using DCFH-DA and that 3-, 4-, or 5-amino salicylic acids could inhibit the effects of Mn^{2+} (Yoon et al., 2011). Furthermore, a review of work on rat brains after exposure to inhaled Mn was performed that showed evidence for increased oxidative stress in selected brain tissues (Taylor et al., 2006). The tissues selected included cerebellum, olfactory bulb, striatum, hippocampus, and hypothalamus. In addition, work with cultured astrocytes has shown clear evidence for induction of the mitochondrial MPT after exposure to Mn^{2+} (Rama Ran and Norenberg, 2004).

EVIDENCE FOR INDUCTION OF APOPTOSIS BY MANGANESE

Many characteristics of apoptosis have been reported in a wide range of cell types, including striatal neurons and astrocytes. After treatment with Mn^{2+}, human B cells from the immune system showed characteristics of apoptosis in a time- and dose-dependent manner. Activation of casp 1 followed by activation of casp 3 was observed and could be suppressed by the general casp inhibitor zVAD and partially inhibited by the casp 3 inhibitor DEV-cmk. Cleavage of poly ADP ribose polymerase (PARP), another standard assay for apoptosis, was also observed (Schrantz et al., 1999). When PC12 cells were treated with $300 \mu M$ Mn^{2+}, Roth and coworkers (Roth et al., 2000) found that viability fell to approximately 30% of control and ATP production fell to approximately 45% of control. Although casp-3–like activity was induced, neither the general casp inhibitor Z-VAD fmk nor the casp 3 inhibitor DEVD-amc decreased cell death. In another study using PC12 cells, it was shown that cyclopentenone prostaglandins could inhibit DNA fragmentation and that casp 9 activation was induced by Mn^{2+} (Shibata et al., 2009). In an unusual result using Hela cells, Oubrahim and coworkers showed that although Mn^{2+} did induce increased ROS production and apoptosis as shown by casp 3 activation, there was no evidence for release of cyt c or activation of casp 9 (Oubrahim et al., 2001). It was concluded that although Mn^{2+} treatment did induce apoptosis, it was not via the mitochondrial pathway. Tamm and coworkers (Tamm et al., 2008) found that after treatment with Mn^{2+}, NCSs showed evidence for induction of the mitochondrial pathway of apoptosis and insensitivity to blockage of casp 8 activation. Using primary cultures of striatal neurons that had been treated with Mn^{2+}, Malecki (2001) found evidence for DNA fragmentation indicative of apoptosis, inhibition at mitochondrial complex II, and a decrease in mitochondrial membrane potential. Using cells from the Gli36 cell line, which have similarities with astrocytes, Alaimo and coworkers (2013) found many signs of apoptosis via the mitochondrial pathway. They found significant increases in the Bax/Bcl-2 ratio, activation of casp 3 and casp 7, PARP-1 cleavage, cell and mitochondrial fragmentation, and disruption of the mitochondrial network. Using primary cultures of cortical neurons Gonzalez and coworkers (Gonzalez et al., 2008) found clear evidence for apoptosis. They observed condensed and fragmented nuclei, decreased mitochondrial membrane potential, cyt c release, activation of casp 3 and casp 7, PARP-1 cleavage, and an increase in Bax and in the Bax to Bcl-2 ratio. In most of the cases cited here, the data suggest apoptosis via the mitochondrial or intrinsic pathway; however, activation of casp 8 (El Mchichi et al., 2007) and casp 12 (Oubrahim et al., 2002; Yoon et al., 2011) have also been reported. Furthermore, although many characteristics of apoptosis were found in PC12 cells after exposure to Mn^{2+}, inhibitors of apoptosis did not delay cell death, suggesting that death may not have depended on apoptotic mechanisms in this case (Roth et al., 2000). It is important that in the work reported here on cells most similar to neurons (Malecki, 2001) and particularly astrocytes in vivo (Alaimo et al., 2013; Gonzalez et al., 2008), the apoptotic characteristics observed were most similar to those of the mitochondrial or intrinsic pathway.

CONCLUSIONS

Because of similarity in binding to Ca^{2+}, Mg^{2+}, Fe^{2+}, and Fe^{3+}, Mn ions can bind to and affect many critical cellular processes. It is readily sequestered by energized mitochondria, and there is a strong case that in the mitochondria it can inhibit the production of ATP by oxidative phosphorylation, induce ROS production or oxidative stress, and induce apoptosis. The evidence suggests that the damage is caused by Mn^{2+} and not Mn^{3+}. Differences in mitochondria from different types of tissue cause the rate-limiting step in inhibition of oxidative phosphorylation to vary with the tissue; however, inhibition of the F_1F_0 ATP synthase is rate limiting in some mitochondria, such as liver and heart, whereas other loci, particularly complex II, is rate limiting in brain mitochondria. This Mn inhibition of ATP production is approximately 3 times as strong in brain mitochondria as that in liver and heart mitochondria. Mn clearly enhances the mitochondrial production of ROS and oxidative stress, and this undoubtedly contributes to induction of the MPT and apoptosis. In cells most similar to those of the target tissues of Mn toxicity, the induced effects seem to follow the mitochondrial or intrinsic apoptotic pathway. Although there has been considerable progress in understanding the biological effect of Mn, there is still much to be done. Whether the observed decrease in ATP production is a cause of increased ROS production or induction of apoptosis needs to be clarified. How the decrease in ATP production, increase in ROS production, and induction of apoptosis relate to changes in neurotransmitters in the target tissues should be determined. The goal must be to understand in detail how the mitochondrial effects, described herein, and the other observed cellular effects of Mn on biological tissues fit into the explanation of the observed signs and symptoms of Mn neurotoxicity, for only then will we be able to identify the most effective loci at which intervention can reverse or otherwise improve clinical outcomes.

ACKNOWLEDGMENT

The author thanks Dr. Karlene Gunter for help with references and for help in formatting the paper. The author also thanks Ms. Kerstin Navik for reading and copy editing the paper.

REFERENCES

Alaimo, A., Gorojod, R.M., Miglietta, E.A., Villarreal, A., Ramos, A.J., Kotler, M.L., 2013. Manganese induces mitochondrial dynamics impairment and apoptotic cell death: a study in human gli36 cells. Neurosci. Lett. 554, 76–81.

Archibald, F.S., Tyree, C., 1987. Manganese poisoning and the attack of trivalent manganese upon catecholamines. Arch. Biochem. Biophys. 256, 638–650.

Armstrong, F.A., 2008. Why did nature choose manganese to make oxygen? Phil. Trans. R. Soc. B 363, 1263–1270.

Aschner, M., Gannon, M., Kimelberg, H.K., 1992. Manganese uptake and efflux in cultured rat astrocytes. J. Neurochem. 58, 730–735.

Aschner, M., Vrana, K.E., Zheng, W., 1999. Manganese uptake and distribution in the central nervous system (CNS). Neurotoxicology 20, 173–180.

Aschner, M., Guilarte, T.R., Schneider, J.S., Zheng, W., 2007. Manganese: recent advances in understanding its transport and neurotoxicity. Toxicol. Appl. Pharmacol. 221, 131–147.

Balaban, R.S., 2002. Cardiac energy metabolism homeostasis: role of cytosolic calcium. J. Mol. Cell. Cardiol. 34, 1259–1271.

Benedetto, A., Au, C., Aschner, M., 2009. Manganese-induced dopaminergic neurodegeneration: insight into mechanisms and genetics shared with Parkinson's disease. Chem. Rev. 109, 4862–4884.

Bernardi, P., 1999. Mitochondrial transport of cations: channels, exchangers, and permeability transition. Physiol. Rev. 79, 1127–1155.

Brouillet, E.P., Shinobu, L., Mcgarvey, U., Hochberg, F., Beal, M.F., 1993. Manganese injection into the rat striatum produces excitotoxic lesions by impairing energy metabolism. Exp. Neurol. 120, 89–94.

Chance, B., Williams, G.R., 1955. Respiratory enzymes in oxidative phosphorylation. I. Kinetics of oxygen utilization. J. Biol. Chem. 217, 383–393.

Czabotar, P.E., Lessene, G., Strasser, A., Adams, J.M., 2014. Control of apoptosis by the Bcl-2 protein family: implications for physiology and therapy. Nat. Rev. Mol. Cell Biol. 15, 49–63.

El Mchichi, B., Hadji, A., Vazquez, A., Leca, G., 2007. P38 MAPk and MSk1 mediate caspase-8 activation in manganese-induced mitochondrial-dependent cell death. Cell Death Differ. 14, 1826–1836.

Elmore, S., 2007. Apoptosis: a review of programmed cell death. Toxicol. Pathol. 35, 495–516.

Eriksson, H., Tedroff, J., Thomas, K.A., Aquilonius, S.M., Hartvig, P., Fasth, K.J., Bjurling, P., Langstrom, B., Hedstrom, K.G., Heilbronn, E., 1992. Manganese induced brain lesions in Macaca fascicularis as revealed by positron emission tomography and magnetic resonance imaging. Arch. Toxicol. 66, 403–407.

Galvani, P., Fumagalli, P., Santagostino, A., 1995. Vulnerability of mitochondrial complex i in PC12 cells exposed to manganese. Eur. J. Pharmacol. 293, 377–383.

Gavin, C.E., Gunter, K.K., Gunter, T.E., 1992. Mn^{2+} sequestration by mitochondria and inhibition of oxidative phosphorylation. Toxicol. Appl. Pharmacol. 115, 1–5.

Goldstein, S., Meyerstein, D., Czapski, G., 1993. The Fenton reagents. Free Radic. Biol. Med. 15, 435–445.

Goncalves, R.L.S., Quinlan, C.L., Perevoshchikova, I.V., Hey-Morgensen, M., Brand, M.D., 2014. Sites of superoxide and hydrogen peroxide production by muscle mitochondria assessed ex vivo under conditions mimicking rest and exercise. J. Biol. Chem. 290, 209–227.

Gonzalez, L.E., Juknat, A.A., Venosa, A.J., Verrengia, N., Kotler, M.L., 2008. Manganese activates the mitochondrial apoptotic pathway in rat astrocytes by modulating the expression of proteins of the bcl-2 family. Neurochem. Int. 53, 408–415.

Gunter, T.E., Pfeiffer, D.R., 1990. Mechanisms by which mitochondria transport calcium. Am. J. Physiol. 258, C755–C786.

Gunter, T.E., Gavin, C.E., Aschner, M., Gunter, K.K., 2006. Speciation of manganese in cells and mitochondria: a search for the proximal cause of manganese neurotoxicity. Neurotoxicology 27, 765–776.

Gunter, T.E., Sheu, S.-S., 2009. Characteristics and possible functions of mitochondrial Ca^{2+} transport mechanisms. Biochim. Biophys. Acta 1787, 1291–1308.

Gunter, T.E., Gerstner, B., Lester, T., Wojtovich, A.P., Malecki, J., Swarts, S.G., Brookes, P.S., Gavin, C.E., Gunter, K.K., 2010. An analysis of the effects of Mn^{2+} on oxidative phosphorylation in liver, brain, and heart mitochondria using state 3 oxidation rate assays. Toxicol. Appl. Pharmacol. 249, 65–75.

Gunter, T.E., Gerstner, B., Gunter, K.K., Malecki, J., Gelein, R., Valentine, W.M., Aschner, A., Yule, D.I., 2013. Manganese transport via the transferrin mechanism. Neurotoxicology 34, 118–127.

Halestrap, A.P., 2010. A pore way to die: the role of mitochondria in reperfusion injury and cardioprotection. Biochem. Soc. Trans. 38, 841–860.

Hazell, A.S., 2002. Astrocytes and manganese neurotoxicity. Neurochem. Int. 41, 271–277.

Lambert, A.J., Brand, M.D., 2009. Reactive oxygen species production by mitochondria. Methods Mol. Biol. 554, 165–181.

Malecki, E.A., 2001. Manganese toxicity is associated with mitochondrial dysfunction and DNA fragmentation in rat primary striatal neurons. Brain Res. Bull. 55, 225–228.

Marreilha Dos Santos, A.P., Milatovic, D., Au, C., Yin, Z., Batoreu, M.C.C., Aschner, M., 2010. Rat brain endothelial cells are a target of manganese toxicity. Brain Res. 1326, 152–161.

Martinez-Finley, E.J., Gavin, C.E., Aschner, M., Gunter, T.E., 2013. Manganese neurotoxicology and the role of reactive oxygen species. Free Radic. Biol. Med. 62, 65–75.

Newland, M.C., Ceckler, T.L., Kordower, J.H., Weiss, B., 1989. Visualizing manganese in the primate basal ganglia with magnetic resonance imaging. Exp. Neurol. 106, 251–258.

Nicholls, D.G., 1982. Bioenergetics, an Introduction to the Chemiosmotic Theory. Academic Press, London, New York.

Oubrahim, H., Stadtman, E.R., Chock, P.B., 2001. Mitochondria play no roles in Mn(II)-induced apoptosis in hela cells. Proc. Natl. Acad. Sci. U.S.A. 98, 9505–9510.

Oubrahim, H., Chock, P.B., Stadman, E.R., 2002. Manganese (II) induces apoptotic cell death in NIH 3T3 cells via a caspase-12-dependent pathway. J. Biol. Chem. 277, 20135–20138.

Rama Ran, K.V., Norenberg, M.D., 2004. Manganese induces the mitochondrial permeability transition in cultured astrocytes. J. Biol. Chem. 279, 32333–32338.

Roth, J.A., Feng, L., Walowitz, J., Browne, R.W., 2000. Manganese-induced rat pheochromocytoma (PC12) cell death is independent of caspase activation. J. Neurosci. Res. 61, 162–171.

Roth, J.A., Horbinski, C., Higgins, D., Lein, P., Garrick, M.D., 2002. Mechanisms of manganese-induced rat pheochromocytoma (PC12) cell death and cell differentiation. Neurotoxicology 23, 147–157.

Roth, J.A., 2014. Correlation between the biochemical pathways altered by mutated parkinson-related genes and chronic exposure to manganese. Neurotoxicology 44, 314–325.

Schrantz, N., Blanchard, D.A., Kimelberg, H.K., 1999. Manganese induces apoptosis of human b cells: caspase-dependent cell death blocked by Bcl-2. Cell Death Differ. 6, 445–453.

Settivari, J., Levora, J., Nass, R., 2009. The divalent metal transporter homologues SMF-1/2 mediate dopamine neuron sensitivity in caenorhabditis elegans. J. Biol. Chem. 284, 35758–35768.

Shibata, M., Maeda, K., Furuta, M., Suzuki, K., Oh-Hashi, K., Kiuchi, K., Hirata, Y., 2009. Neuroprotective effects of (arylthio)cyclopentenone derivatives on manganese-induced apoptosis in PC12 cells. Brain Res. 1294, 218–225.

Skulachev, V.P., 2006. Bioenergetic aspects of apoptosis, necrosis and mitoptosis. Apoptosis 11, 473–485.

Tamm, C., Sabri, F., Ceccatelli, S., 2008. Mitochondrial-mediated apoptosis in neural stem cells exposed to manganese. Toxicol. Sci. 101, 310–320.

Taylor, M.D., Erikson, K.M., Dobson, A.W., Fitsanakis, V.A., Dorman, D.C., Aschner, M., 2006. Effects of inhaled manganese on biomarkers of oxidative stress in the rat brain. Neurotoxicology 27, 788–797.

Yoon, G.H., Lee, D.S., Kim, D.S., Kim, K.W., Kim, H.R., Chae, H.J., 2011. The effects of 3, 4, or 5 amino salicylic acids on manganese-induced neuronal death: Er stress and mitochondrial complexes. Toxicol. In Vitro 25, 1259–1268.

Zhang, S., Fu, J., Zhou, Z., 2004. In vitro effect of manganese chloride exposure on reactive oxygen species generation and respiratory chain complexes activities of mitochondria isolated from rat brain. Toxicol. In Vitro 18, 71–77.

Zhang, S., Fu, J., Zhou, Z., 2005. Changes in the brain mitochondrial proteome of male Sprague-Dawley rats treated with manganese chloride. Toxicol. Appl. Pharm 202, 13–17.

Zheng, W., Ren, S., Graziano, J.H., 1998. Manganese inhibits mitochondrial aconitase: a mechanism of manganese neurotoxicity. Brain Res. 799, 334–342.

Zwingmann, C., Leibfritz, D., Hazell, A.S., 2003. Energy metabolism in astrocytes and neurons treated with manganese: relation among cell-specific energy failure, glucose metabolism, and intercellular trafficking using multinuclear nmr-spectroscopic analysis. J. Cereb. Blood Flow. Metab. 23, 756–771.

Part VIII

Molybdenum

Chapter 33

Molybdenum Cofactor in Humans: Health, Disease, and Treatment

Abdel A. Belaidi[1], Guenter Schwarz[2]

[1]The University of Melbourne, Parkville, VIC, Australia; [2]University of Cologne, Cologne, Germany

INTRODUCTION

Molybdenum in Biological Systems

Molybdenum (Mo) is the only trace metal of the second row of the periodic table that exhibits biological activity when it is ligated to a cofactor. In nature it is highly abundant in seawater, with an average concentration of 100 nM, and it is mostly found in the form of the oxyanion molybdate (MoO_4^{2-}), which is the only known source of Mo that can be taken up by organisms (Belaidi and Schwarz, 2013a; Lindsay, 1979). In prokaryotes, molybdate is transported across the cell membrane through proteins of the ATP-binding cassette transporter family whereas in eukaryotes two transporters within the family of sulfate transporters and major facilitator superfamily were identified and named molybdate transporter type-1 (MOT1) and type-2 (MOT2; Tejada-Jimenez et al., 2011, 2007).

More than 50 molybdenum enzymes have been described in nature, catalyzing key redox reactions in the global carbon, sulfur, and nitrogen cycles (Schwarz et al., 2009). The overall reaction is characterized by the transfer of an oxygen atom to or from a substrate in a two-electron transfer reaction (Hille, 2002). With the exception of nitrogenase, in all molybdenum enzymes molybdenum is chelated and activated by the molybdenum cofactor (Moco), consisting of Mo covalently bound via the dithiolate moiety of a conserved pterin backbone commonly referred to as molybdopterin (MPT; Rajagopalan and Johnson, 1992) or metal-binding pterin (Schwarz et al., 2000).

Molybdenum Cofactor Biosynthesis

In all organisms studied so far, Moco is synthesized by a conserved biosynthetic pathway that can be divided into four steps (Mendel and Schwarz, 2011) according to the biosynthetic intermediates cyclic pyranopterin monophosphate (cPMP), MPT, and adenylated MPT (MPT-AMP; Fig. 33.1). Moco biosynthesis starts with the conversion of GTP into cPMP in a complex rearrangement reaction catalyzed in humans by two proteins: MOCS1A and MOCS1AB (Arenas et al., 2009; Hänzelmann et al., 2002). [1]H-Nuclear magnetic resonance (NMR) by ^{13}C-NMR studies unraveled the structure of cPMP as a fully reduced tetrahydropyranopterin with a terminal cyclic phosphate and a C1 geminal diol (Santamaria-Araujo et al., 2004, 2011). In the second step of Moco biosynthesis, two sulfur atoms are transferred to cPMP to form the MPT dithiolate by the enzyme MPT-synthase, a heterotetrameric complex of two small (MOCS2B) and two large (MOCS2A) subunits in humans (Gutzke et al., 2001; Wuebbens and Rajagopalan, 2003). The third and fourth steps of Moco biosynthesis consist of two succeeding reactions, resulting in the adenylation of MPT and subsequent molybdenum insertion, respectively (Belaidi and Schwarz, 2013a; Llamas et al., 2004). The multidomain proteins Cnx1 (Cofactor for nitrate reductase and xanthine dehydrogenase 1) in plants and gephyrin in humans catalyze both reactions, whereas in bacteria two separate proteins are required (Belaidi and Schwarz, 2013a; Llamas et al., 2006; Nichols and Rajagopalan, 2002; Fig. 33.1). In addition, the human protein gephyrin has an essential function in the central nervous system where it clusters inhibitory glycine and γ-aminobutyric acid (GABA) type A receptors at postsynaptic synapses (Fritschy et al., 2008; Stallmeyer et al., 1999). Upon completion of Moco biosynthesis, the cofactor is immediately incorporated into the *apo*-enzyme because Moco is labile and oxygen sensitive.

Molybdenum Enzymes

Most Mo-containing enzymes are found in bacteria whereas in humans only four are known so far, which are divided into two families according to the origin of the third sulfur ligand of molybdenum in Moco, which is either a cysteine in the

Molecular, Genetic, and Nutritional Aspects of Major and Trace Minerals. http://dx.doi.org/10.1016/B978-0-12-802168-2.00033-6

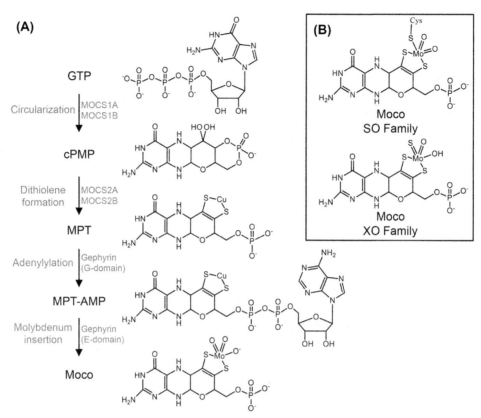

FIGURE 33.1 Molybdenum cofactor biosynthesis. (A) The synthesis of Moco is divided in four steps consisting of GTP circularization into cyclic pyranopterin monophosphate (cPMP), dithiolene formation, adenylation, and molybdenum insertion. The structures of the cofactor intermediates are shown in (A), and the human proteins and the corresponding catalyzed steps are highlighted in gray. (B) Chemical structure of molybdenum cofactor (Moco) in the two Mo enzyme families showing a cysteine ligand in the sulfite oxidase (SO) family and a sulfide ligand in the xanthine oxidase (XO) family. *MPT*, molybdoterin; *MPT-AMP*, adenylated MPT.

sulfite oxidase family or a sulfido ligand in the xanthine oxidase (XO) family (Hille et al., 2011; Fig. 33.1). The sulfite oxidase family includes sulfite oxidase (SO), which catalyzes the oxidation of sulfite to sulfate (Mudd et al., 1967), and mitochondrial amidoxime reducing component (mARC), which is believed to have detoxifying functions in the metabolism of prodrugs (Gruenewald et al., 2008). It is interesting to note that there is no significant homology of the Moco binding domains of SO and mARC, but because in the algae orthologue of mARC a cysteine residue has been found to be crucial for activity (Chamizo-Ampudia et al., 2011), mARC could be classified to the SO family. The XO family is formed by two homologous members: XO/xanthine dehydrogenase (XDH), which is involved in purine catabolism (Minoshima et al., 1995) and aldehyde oxidase (AO), which oxidizes various aldehydes (Garattini et al., 2009). The isolated deficiency of each enzyme or the combined deficiency of all enzymes in Moco deficiency will be discussed in the following sections.

DEFICIENCIES IN MOLYBDENUM ENZYMES

Xanthine Dehydrogenase and Oxidase

XDH and XO (EC 1.17.1.4) play important roles in the terminal step of purine catabolism, and they catalyze the last two oxidative reactions converting hypoxanthine to xanthine and xanthine to uric acid, which is finally excreted in urine (Hille et al., 2011). Eukaryotic XDH and XO originate from the same gene, and both forms can be termed xanthine oxidoreductase (XOR). XDH is converted to XO through the reversible oxidation of two conserved cysteine residues, whereas proteolytic cleavage of XDH converts the enzyme irreversibly to the XO form (Amaya et al., 1990; Nishino, 1997).

XOR deficiency results in the accumulation of xanthine in urine, leading to a disease termed xanthinuria, which exists in two forms: xanthinuria type 1 and type 2 (Dent and Philpot, 1954). Xanthinuria type 1 is caused by the loss of activity of XOR, resulting in an accumulation of xanthine (and to a lesser extent hypoxanthine; Ichida et al., 1997). In contrast, xanthinuria type 2 is caused by the simultaneous loss of activity of XOR and AO, which is caused by mutations in the

MCSU gene, which encodes for a protein necessary for the sulfuration of Moco in enzymes of the XO family (Ichida et al., 2001). In both types of xanthinuria a very low level of plasma uric acid and high levels of xanthine in plasma are hallmarks of the disease. Patients of both groups have similar clinical presentation, mostly because of increased xanthine deposition (Simmonds et al., 1995). However, the mechanism involved in the disease is less clear because some patients may develop symptoms, which lead to acute renal failure, whereas others remain asymptomatic.

Aldehyde Oxidase

AO (EC 1.2.3.1) is structurally very similar to XO and displays an active site that can be superimposed on that of XO. AO catalyzes the oxidation of aldehydes into carboxylic acid, and similar to XO, during the AO reaction, superoxide and hydrogen peroxide are also produced (Garattini et al., 2009). However, AO exhibits broader substrate specificity than XO, mediating oxidation of heterocycles, purines, pteridines. AO also plays an important role in the metabolism of numerous drugs. Because of this and its high expression in the liver, AO is a very attractive target for drug development and pharmacokinetics (Garattini and Terao, 2013). In addition to its relevance for drug metabolism in the liver, there is lack of knowledge about the physiological significance of this enzyme, and neither endogenous substrates nor human deficiency cases have been described in the literature (Cerqueira et al., 2015; Marelja et al., 2014).

Mitochondrial Amidoxime-Reducing Component

mARC is the molybdenum enzyme that has been discovered most recently in humans, and two different isoforms have been described: mARC1 and mARC2 (Havemeyer et al., 2006). Similar to AO, mARC proteins are involved in drug metabolism because they were found to metabolize several N-hydroxylated compounds commonly used as prodrugs (Plitzko et al., 2013). mARC proteins are monomeric proteins, which require the formation of complexes with cytochrome b_5 and nicotinamide adenine dinucleotide hydride/cytochrome b_5 reductase for catalytic activity (Havemeyer et al., 2011; Klein et al., 2012). However, no physiological substrates and no mARC deficiency cases have been identified in humans so far.

Sulfite Oxidase

Cysteine Catabolism to Sulfite

Cysteine plays a crucial role in maintaining proper cellular functions, and because of its thiol side chain, it participates in many enzymatic reactions such as the synthesis of iron–sulfur clusters, which are essential cofactors for various metalloproteins. The thiol group of cysteine is also essential for protein structure and folding because of its susceptibility to oxidation and formation of disulfide bonds in proteins (Fig. 33.2). Furthermore, and perhaps most importantly, cysteine is the limiting amino acid in the synthesis of glutathione (GSH), the major antioxidant reservoir in animals (Lyons et al., 2000; Wu et al., 2004). Although considered a nonessential amino acid, cysteine can become essential under certain conditions such as limited dietary intake of methionine. In plasma, cysteine is present in its oxidized form cystine, which is formed by the condensation of two cysteine residues through disulfide bridge formation. In hepatocytes, where most cysteine catabolism takes place, abundant levels of GSH allow reduction of cystine to cysteine, which is then imported across the plasma membrane and catabolized.

In mammals, the main route of cysteine catabolism follows an oxidative pathway, which is initiated by the irreversible oxidation of cysteine to cysteine sulfinic acid (CSA), a reaction catalyzed by the cytosolic enzyme cysteine dioxygenase (CDO; EC 1.13.11.20; Arjune et al., 2015; Stipanuk and Ueki, 2011; Fig. 33.2). The second step involves either a decarboxylation reaction catalyzed by CSA decarboxylase (EC 4.1.1.29) yielding taurine as a final product or a CSA deamination reaction catalyzed by an aspartate aminotransferase (2.6.1.1), yielding sulfite as a terminal product (Stipanuk and Ueki, 2011; Fig. 33.2). Taurine is the most abundant amino acid derivative in the body, and it is involved in several biological processes (Huxtable, 1989; Jacobsen and Smith, 1968), whereas sulfite is toxic and must be further oxidized to sulfate by the mitochondrial enzyme SO (see next section).

Cysteine is also catabolized via a nonoxidative pathway resulting in the formation of hydrogen sulfide (H_2S) through the action of three different enzymes: cystathionine γ-lyase (CSE; EC 4.4.1.1), cystathionine β-synthase (EC 4.2.1.22), and 3-mercaptopyruvate sulfurtransferase (EC 2.8.1.2; Kabil and Banerjee, 2010; Fig. 33.2). H_2S has been attributed several essential functions in the periphery and in the brain, where significant levels were detected and a functional role as a neural messenger was proposed (Baranano et al., 2001). CSE appears to be the rate-limiting enzyme in H_2S production in the brain and has been recently implicated in the neurodegenerative process associated with Huntington's disease (Paul et al., 2014).

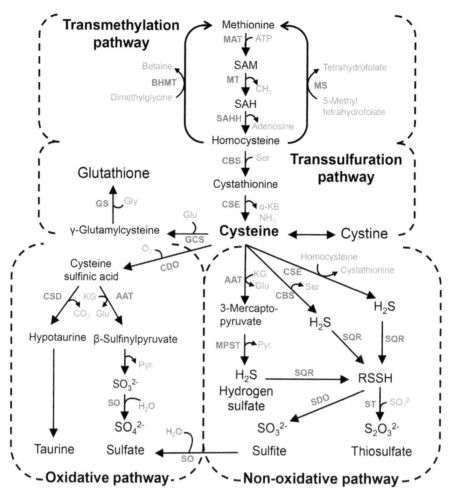

FIGURE 33.2 Cysteine catabolism in humans. Cysteine can be absorbed from the diet or synthesized from serine, with the sulfur atom being derived from methionine through several intermediates of the transmethylation pathway. Cysteine is the precursor of glutathione synthesis in two enzymatic reactions catalyzed successively by the enzymes, γ-glutamylcysteine synthetase (GCS) and glutathione synthetase (GS). The catabolism of cysteine follows two major routes: an oxidative pathway leading to the formation of sulfate and taurine as end sulfur-containing products or a nonoxidative pathway leading to hydrogen sulfide formation, which can be further converted to thiosulfate. The enzyme abbreviations used are *AAT*, aspartate aminotransferase; *BHMT*, betaine-homocysteine methyl transferase; *CBS*, cystathionine β-synthase; *CDO*, cysteine dioxygenase; *CSD*, cysteine sulfinate decarboxylase; *CSE*, cystathionine γ-lyase (cystathionase); *MAT*, methionine-S-adenosyl transferase; *MPST*, 3-mercaptopyruvate sulfurtransferase; *MS*, methionine synthase; *MT*, methyl transferase; *SAHH*; S-adenosylhomocysteine hydrolase; *SDO*, sulfur dioxygenase; *SO*, sulfite oxidase; *SQR*, quinone oxidoreductase; *ST*, sulfur transferase.

Finally, H_2S can be further metabolized through sequential reactions involving several mitochondrial enzymes yielding the final product thiosulfate (Hildebrandt and Grieshaber, 2008). In mammals, cysteine is mainly catabolized through the oxidative pathway because an increase in dietary intake of cysteine is usually accompanied by an increase in CDO protein levels in the liver. Moreover, only under conditions of CDO deficiency is an increase in the products of the nonoxidative catabolism of cysteine observed (Stipanuk et al., 2009; Ueki et al., 2011).

Sulfite Oxidase Function

Sulfite oxidation represents the final step of oxidative cysteine catabolism, which is catalyzed by SO (EC 1.8.3.1). Vertebrate SO is a dimeric protein with each monomer harboring an N-terminal cytochrome b_5-heme domain, a central catalytic Moco domain, and a C-terminal dimerization domain (Kisker et al., 1997; Klein and Schwarz, 2012). Vertebrate SO is localized in the mitochondrial intermembrane space, where it catalyzes the oxidation of sulfite to sulfate in a catalytic cycle involving electron transfer from sulfite to the Moco domain, followed by two individual electron transfer steps from heme to the terminal electron acceptor cytochrome C. SO from *Gallus gallus* was the first crystallized eukaryotic Mo enzyme and depicted the presence of a large distance of 30 Å between the heme and Moco domains, which would not support the high electron transfer rate observed between the two domains by electrochemical methods (Johnson-Winters et al., 2010; Kisker et al., 1997). Therefore it was hypothesized that Moco and the heme domain undergo conformational changes that bring both redox centers in close proximity, allowing efficient electron transfer to take place, a hypothesis that was later confirmed using biochemical and site-directed mutagenesis methods (Feng et al., 2002; Johnson-Winters et al., 2010). In plants, a homologous protein to vertebrate SO was discovered in *Arabidopsis thaliana*. However, plant SO lacks the heme domain and catalyzes sulfite oxidation using molecular oxygen as an electron acceptor, a reaction that takes place in peroxisomes and leads to the formation of hydrogen peroxide as a final product (Eilers et al., 2001; Hansch et al., 2006).

Sulfite Oxidase Deficiency

Isolated sulfite oxidase deficiency (SOD) is a rare inborn error of metabolism caused by the loss of SO activity due to mutations in the *SUOX* gene leading to sulfite accumulation and cell toxicity. Sulfite accumulation is also observed in Moco deficiency, which is caused by mutations in any of the genes encoding proteins in the Moco biosynthesis pathway (Johnson and Duran, 2001; Schwarz, 2005). Several mutations within the *SUOX* gene have been described and biochemically characterized in the literature (Tan et al., 2005). SOD is less frequent than Moco deficiency, with approximately 30 cases known so far. However, SO and Moco deficiencies are clinically very similar, which qualifies SO as the most important Mo enzyme in humans (Tan et al., 2005). Indeed, patients of both groups are characterized by a severe neurodegenerative phenotype starting shortly after birth and manifested by intractable seizures, brain and head dysmorphism due to neurodegeneration, and developmental delay: in the absence of treatment, death in infancy has been the usual outcome (see next section).

MOLYBDENUM COFACTOR DEFICIENCY

Genetic Background of Molybdenum Cofactor Deficiency

Molybdenum cofactor deficiency (MoCD) is a rare inherited metabolic disorder caused by defects in the biosynthesis of Moco leading to the simultaneous loss of activity of all Mo-dependent enzymes (Johnson and Duran, 2001). Affected patients exhibit similar symptoms to SO deficiency, including severe neurological abnormalities, such as microcephaly and seizures leading to death in infancy (Johnson and Duran, 2001). Three types of MoCD are classified according to the underlying genetic defect (Fig. 33.3). Type A deficiency affects two-thirds of all patients and is caused by mutations in the *MOCS1* gene, impairing the synthesis of the first Moco intermediate cPMP (Reiss and Johnson, 2003). Type B patients accumulate cPMP due to defects in the *MOCS2* gene (Reiss et al., 1999), and type C deficiency is caused by mutations in the *GPHN* gene (Fig. 33.3). Numerous mutations or deletions causing MoCD have been described in the *MOCS1* and *MOCS2* genes (Reiss, 2000; Reiss and Hahnewald, 2011; Reiss and Johnson, 2003) whereas only two cases of type C deficiency have been reported due to mutations in the *GPHN* gene (Reiss et al., 2001, 2011). Because of its additional function in neuronal synaptic transmission, mutations within the *GPHN* gene are also associated with impaired synaptic inhibition, leading to neuronal hyperexcitability and seizures, including different types of epilepsy. The incidence of MoCD is expected to be low; however, many missed and nondiagnosed cases are suspected and therefore a solid basis of data for determining prevalence is limited (Mendel and Schwarz, 2011).

Biochemical and Clinical Presentation of Patients With Molybdenum Cofactor Deficiency

The first case of a MoCD patient was reported in 1978 by Duran et al., describing a neonatal patient with initial feeding difficulties, therapy-resistant seizures, lens dislocation of the eyes, dysmorphic facial features, and severe neurodegenerative abnormalities (Duran et al., 1978). Since then more than 100 cases have been described in the literature, all sharing severe neuronal cell death as a major disease feature, including progressive cerebral palsy, microencephaly, and disordered autonomic function (Reiss and Hahnewald, 2011). In general, maternal clearance of toxic metabolites such as sulfite is believed to suppress prenatal brain damage in the fetus; therefore the first clinical symptoms are observed in most cases shortly after birth and are first depicted by feeding difficulties followed by intractable repetitive seizures and exaggerated startle reactions (Schwarz et al., 2009). In the absence of effective treatments, patients develop major severe neurological damage and usually die within the first years of life.

Biochemically, the loss of SO activity in MoCD (and SOD) results in the accumulation of toxic sulfite in plasma and urine, which is usually detected in fresh urine using a sulfite dipstick. However, this method is not reliable because urinary sulfite is very unstable and false-negative and false-positive results may occur depending on storage conditions or the presence of drugs with free reactive sulfhydryl groups such as *N*-acetylcysteine, mercaptamine, and dimercaprol (Kutter and Humbel, 1969; Wadman et al., 1983). Sulfite accumulation is accompanied by the formation of secondary metabolites

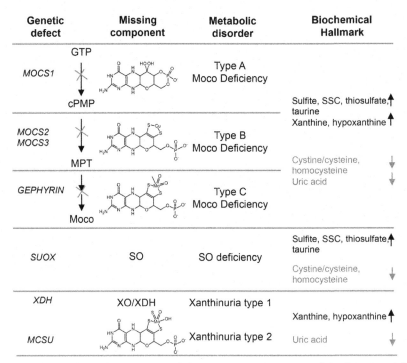

Genetic defect	Missing component	Metabolic disorder	Biochemical Hallmark

FIGURE 33.3 Classification and characteristics of molybdenum cofactor (Moco) and Mo enzyme deficiencies. Moco deficiency is classified in three types A, B, and C caused by mutations in the *MOCS1*, *MOCS2/3*, and *GPHN* genes, respectively. Deficiencies of Mo enzymes are separated in sulfite oxidase (SO) deficiency and xanthinuria caused by mutations of the *SUOX* and *XDH* genes, respectively. In addition, xanthinuria can also be caused by mutations in the *MCSU* gene, which encodes a protein necessary for cofactor sulfuration in the xanthine oxidase (XO) family. The missing cofactor intermediate or protein and the biochemical hallmark of the corresponding disorder are summarized. *SSC*, S-sulfocysteine.

such as taurine, thiosulfate, and S-sulfocysteine (SSC; Belaidi and Schwarz, 2013b; Fig. 33.3). The latter is very abundant in MoCD patients, and its excretion in urine is detectable shortly after birth. In the past we reported the development of a fast, sensitive, and reproducible high-performance liquid chromatography method for SSC quantification in urine, which has been validated and used for diagnosis and treatment monitoring of MoCD patients (Belaidi et al., 2012). Thus elevated urinary sulfite, thiosulfate, and SSC as well as reduced homocysteine levels are common biochemical indicators of MoCD and SOD (Johnson and Duran, 2001; Sass et al., 2004; Fig. 33.3). SOD can be distinguished from MoCD either genetically through gene sequencing methods or biochemically if the purine levels (xanthine, uric acid) or urothione, which is a specific metabolic degradation product of Moco (Bamforth et al., 1990), are in a normal range (Fig. 33.3).

Molecular Basis of Neurodegeneration in Molybdenum Cofactor Deficiency

MoCD is clinically indistinguishable from SOD, implicating impaired SO activity, which is required for detoxifying endogenously generated sulfite into sulfate, as a major underlying cause for neurodegeneration in MoCD (and SOD). However, little is known about the mechanisms of how sulfite accumulation causes brain cell loss in MoCD. Sulfite is a highly reactive molecule that breaks disulfide bridges in proteins and reduces cystine, thereby affecting various protein and cellular functions (Zhang et al., 2004). Excess sulfite reacts with cysteine, generating SSC and depleting cystine (the major transport form of cysteine) from plasma, which may have a dual detrimental consequence, especially in neuronal tissues (Fig. 33.4). First, SSC is structurally very similar to glutamate, and early experiments in rats suggested that it may bind to neuronal glutamate receptors of the *N*-methyl-D-aspartate type (Olney et al., 1975), thus causing excess excitation of the brain, termed excitotoxicity. SSC, which is present at very low levels in healthy individuals (Belaidi et al., 2012; Pitt et al., 2002), is one of the most abundant metabolites in MoCD patients, which may explain the observed seizures, convulsions, contractions, and twitching associated with MoCD (Kurlemann et al., 1996). Second, cysteine is the rate-limiting substrate for the major neuronal antioxidant GSH, and cystine depletion from plasma is expected to further exaggerate the vulnerability of the brain toward oxidative stress (Fig. 33.4). However, information on the impact of SSC on neuronal tissue as well as on GSH levels in MoCD patients is still missing, and further studies will be required to further dissect the contribution of different metabolites toward the mechanisms involved in neuronal cell death in MoCD.

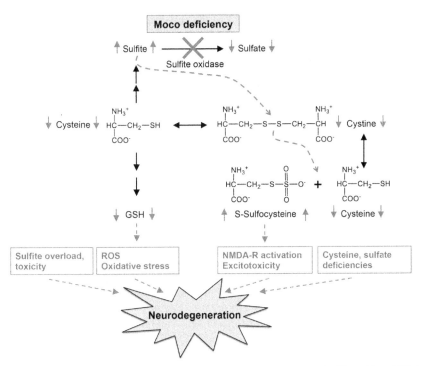

FIGURE 33.4 Proposed biochemical and pathological pathway in molybdenum cofactor (Moco) and sulfite oxidase (SO) deficiencies. Moco and SO deficiencies both result in the accumulation of sulfite, which results in sulfite toxicity and sulfate deficiency. In addition, accumulated sulfite reacts with cystine (the oxidation product of cysteine) leading to two major events: First, the formation of S-sulfocysteine, which can cause excitotoxicity because of its structural similarity to glutamate and consequently overactivation of NMDA receptors. Second, cysteine deficiency may lead to glutathione depletion and consequently to an increase in oxidative stress. Because of the high metabolic requirement of the brain, glutathione depletion and S-sulfocysteine–dependent excitotoxicity are believed to be responsible for neuronal cell death in Moco and SO deficiencies. *GSH*, glutathione; *NMDA*, N-methyl-D-aspartate; *ROS*, reactive oxygen species.

TREATMENT OF MOLYBDENUM COFACTOR DEFICIENCY

Dietary Restriction

To date, there is no approved treatment for MoCD or SOD, and patients usually die in early childhood, except for a very few mild cases with late onset of symptoms (Hughes et al., 1998; Johnson et al., 2001). Early attempts to reduce the severity of disease symptoms have mainly focused on restricting the dietary intake of sulfur-containing amino acids in an attempt to reduce sulfite formation and toxicity in patients. However, inconsistent outcomes have been reported in the literature. In a single-patient study, Boles et al. reported a short-term, rapid decrease in urinary sulfite after methionine restriction and cysteine supplementation in MoCD, whereas cysteine supplementation therapy in an SOD patient resulted in an increase in urinary sulfite with no improvement, and treatment was discontinued after 4 weeks (Boles et al., 1993; Sass et al., 2010). In 2000, Touati et al. reported a favorable development in two patients with a mild form of SOD after dietary restriction of both sulfur amino acids methionine and cysteine, which was effective in reducing sulfite, thiosulfate, and SSC with no apparent neurological deterioration (Touati et al., 2000). Similar results were also reported in an SOD patient with a mild phenotype (Del Rizzo et al., 2013). Thus reducing dietary intake of methionine and cysteine may be considered as a treatment option for MoCD and SOD patients. However, it should be noted that sulfur amino acids are required for many biological processes, including the synthesis of sulfate, taurine, and GSH, as well as Fe–S clusters, just to name a few, and future therapies may consider supplementation with those compounds to reduce possible side effects resulting from their deficiencies.

Cyclic Pyranopterin Monophosphate Substitution Therapy

In 2002, Lee et al. reported the generation of a first mouse model of human MoCD type A by disrupting the *mocs1* gene in mice and thus impairing the synthesis of the first Moco intermediate cPMP. Homozygous animals displayed a severe phenotype, which reflected all biochemical characteristics of human MoCD and died within the first 11 days of life with an average life span of 7.5 days (Lee et al., 2002). On the basis of the universal structure of cPMP, a bacterial fermentation procedure was developed for cPMP production in *Escherichia coli* (Santamaria-Araujo et al., 2004), and a first successful treatment of *mocs1* knockout mice was reported in 2004 with a complete restoration of activity of all Mo enzymes. cPMP-treated mice reached adulthood, were fertile, and were not distinguishable from their wild-type littermates (Schwarz et al., 2004).

In 2010 the first treatment of a human MoCD type A patient with purified cPMP was reported (Veldman et al., 2010). Treatment of the patient started on day 36 of life, and within days after cPMP treatment was started the patient showed a remarkable normalization of MoCD biomarkers including sulfite, SSC, xanthine, and uric acid, which returned to almost normal levels. Clinically, the patient became more alert a few days after the treatment started; convulsions and twitchings disappeared within the first 2 weeks and epileptic discharges were markedly reduced (Veldman et al., 2010). On the basis of the first successful treatment, a treatment plan was developed for cPMP substitution therapy, and the favorable biochemical and clinical improvement of other patients has been recently reported in a prospective study of 11 MoCD type A patients (Schwahn et al., 2015). The study represents the first effective therapy for patients with MoCD type A, leading to an improved neurodevelopmental outcome of all treated patients. In addition, the study also reported the requirement to avoid any delay in the treatment of patients to maximize treatment benefit and avoid irreversible damage.

Enzyme Substitution Therapy

To date, cPMP is the only stable Moco intermediate that can be isolated and used in a substitution therapy for MoCD type A. Thus similar therapies for MoCD type B and C patients are not feasible (Schwarz and Belaidi, 2013). Knowing that loss of SO activity is the major cause of the neurological features of Moco and SO deficiencies, an enzyme replacement therapy with human SO appears to be a promising therapeutic approach for the treatment of both deficiencies, including MoCD type B and C. However, such a therapy would require translocation of vertebrate SO into mitochondria, which is indispensable for catalytic activity. Knowing that the mitochondrial import machinery requires protein unfolding, which in the case of Mo enzymes would implicate the loss of Moco, because free Moco is very unstable, the development of such a therapy is not possible. In a recent study, functional characterization of mammalian SO suggested that truncated SO variants may be used for sulfite oxidation within the blood using molecular oxygen as an electron acceptor, resulting in an extracellular formation of H_2O_2, which requires peroxidation by catalase (Belaidi et al., 2015). Future animal studies are required to investigate the potential of such an enzyme replacement therapy toward sulfite toxicity disorders.

ASSOCIATION OF MOLYBDENUM WITH OTHER METALS AND DISORDERS

In 2004, Kuper et al. reported the finding of an unexpected copper ion bound to the dithiolate group of two Moco intermediates: MPT and MPT-AMP (Kuper et al., 2004). Further in vitro studies showed an inhibition of Moco synthesis in the presence of copper, suggesting that Moco biosynthesis might be affected under conditions when cellular copper concentrations are increased and providing a possible link between Mo and copper metabolism. In nature, it is known that molybdenum may act as an antagonist to copper. In Australian farmland molybdate overload due to excessive fertilization of the soil caused pathologic symptoms of molybdenosis in animals, which in particular in ruminants triggered secondary copper deficiency due to the copper chelation capacity of molybdate (Mason, 1986). Indeed, tetrathiomolybdate is a known potent copper chelator, which is used to treat Wilson's disease and several other disorders that are linked to copper homeostasis (Brewer, 2009).

Mutations in gephyrin are associated with multiple neuropsychiatric disorders, with several mutations not affecting its catalytic function in Moco biosynthesis but leading to an impairment in inhibitory synaptic transmission, and are linked to various disorders such as hyperekplexia, autism, schizophrenia, and temporal lobe epilepsy (Dejanovic et al., 2014; Forstera et al., 2010; Harvey et al., 2008). Ethylmalonic encephalopathy, similar to MoCD, is an autosomal-recessive inborn error of metabolism caused by defects in the *ETHE1* gene, which is a mitochondrial sulfur dioxygenase involved in hydrogen sulfide metabolism. Loss of function of the *ETHE1* gene product leads to an accumulation of hydrogen sulfide and inhibition of cytochrome c oxidase and short-chain fatty acid oxidation (Tiranti et al., 2009). Affected patients develop symptoms shortly after birth, including typical neurological features such as delayed development, encephalopathy, and seizures as well as microangiopathy, hypotonia, and chronic diarrhea (Tiranti et al., 2009). A common biochemical hallmark of MoCD and ethylmalonic encephalopathy relies on the excessive secretion of thiosulfate in urine, whereas sulfite excretion is only observed in MoCD. Therefore investigation of Moco biosynthesis may be beneficial for the development of future biomarkers and diagnosis tools for neurodegenerative disorders related to other metals such as copper or to neuropsychiatric disorders related to the function of gephyrin or other associated proteins such as glycine or GABA receptors.

CONCLUSION AND FUTURE PERSPECTIVES

Mo forms the active center in all Mo enzymes, catalyzing key redox reactions in the global cycles of carbon, sulfur, and nitrogen metabolism. Moco deficiency represents an inborn error of metabolism resulting in the loss of activity of all Mo enzymes, and it is characterized by a severe and progressive neurodegeneration leading to death in infancy. Loss of SO has been identified as the disease-causing event in MoCD, pointing to sulfite, SSC, and oxidative stress as major events implicated in the pathogenesis of the disease. However, future studies are needed to understand how neuronal cell death is initiated and which pathways are activated to identify key players in metabolism that may help to develop future diagnostic tools and treatment plans. The recent reports of the favorable outcome of several Moco-deficient type A patients treated with cPMP substitution therapy suggest that such a therapy may be approved and made available for all MoCD type A patients in the near future, which represents a milestone in the treatment of MoCD and may trigger the development of future enzyme replacement therapies for the other subtypes of the disease.

LIST OF ABBREVIATIONS

AO Aldehyde oxidase
CDO Cysteine dioxygenase
cPMP Cyclic pyranopterin monophosphate
GSH Glutathione
mARC Mitochondrial amidoxime reducing component
MoCD Molybdenum cofactor deficiency
Moco Molybdenum cofactor
MPT Metal-binding pterin (or molybdopterin)
MRI Magnetic resonance imaging
SO Sulfite oxidase
SOD Sulfite oxidase deficiency
SSC S-sulfocysteine
XDH Xanthine dehydrogenase
XO Xanthine oxidase

ACKNOWLEDGMENT

Research funding by the German Research Foundation, the Federal Ministry of Education and Research, the Fonds der Chemischen Industrie, and the Center for Molecular Medicine Cologne are gratefully acknowledged.

REFERENCES

Amaya, Y., Yamazaki, K., Sato, M., Noda, K., Nishino, T., 1990. Proteolytic conversion of xanthine dehydrogenase from the NAD-dependent type to the O_2-dependent type. Amino acid sequence of rat liver xanthine dehydrogenase and identification of the cleavage sites of the enzyme protein during irreversible conversion by trypsin. J. Biol. Chem. 265, 14170–14175.

Arenas, M., Fairbanks, L.D., Vijayakumar, K., Carr, L., Escuredo, E., Marinaki, A.M., 2009. An unusual genetic variant in the *MOCS1* gene leads to complete missplicing of an alternatively spliced exon in a patient with molybdenum cofactor deficiency. J. Inherit. Metab. Dis. 32, 560–569.

Arjune, S., Schwarz, G., Belaidi, A.A., 2015. Involvement of the Cys-Tyr cofactor on iron binding in the active site of human cysteine dioxygenase. Amino Acids 47, 55–63.

Bamforth, F.J., Johnson, J.L., Davidson, A.G., Wong, L.T., Lockitch, G., Applegarth, D.A., 1990. Biochemical investigation of a child with molybdenum cofactor deficiency. Clin. Biochem. 23, 537–542.

Baranano, D.E., Ferris, C.D., Snyder, S.H., 2001. Atypical neural messengers. Trends Neurosci. 24, 99–106.

Belaidi, A.A., Arjune, S., Santamaria-Araujo, J.A., Sass, J.O., Schwarz, G., 2012. Molybdenum cofactor deficiency: a new HPLC method for fast quantification of S-sulfocysteine in urine and serum. JIMD Rep. 5, 35–43.

Belaidi, A.A., Roper, J., Arjune, S., Krizowski, S., Trifunovic, A., Schwarz, G., 2015. Oxygen reactivity of mammalian sulfite oxidase provides a concept for the treatment of sulfite oxidase deficiency. Biochem. J. 469, 211–221.

Belaidi, A.A., Schwarz, G., 2013a. Metal insertion into the molybdenum cofactor: product-substrate channelling demonstrates the functional origin of domain fusion in gephyrin. Biochem. J. 450, 149–157.

Belaidi, A.A., Schwarz, G., 2013b. Molybdenum cofactor deficiency: metabolic link between taurine and S-sulfocysteine. Adv. Exp. Med. Biol. 776, 13–19.

Boles, R.G., Ment, L.R., Meyn, M.S., Horwich, A.L., Kratz, L.E., Rinaldo, P., 1993. Short-term response to dietary therapy in molybdenum cofactor deficiency. Ann. Neurol. 34, 742–744.

Brewer, G.J., 2009. The use of copper-lowering therapy with tetrathiomolybdate in medicine. Expert Opin. Investig. Drugs 18, 89–97.

Cerqueira, N.M., Coelho, C., Bras, N.F., Fernandes, P.A., Garattini, E., Terao, M., Romao, M.J., Ramos, M.J., 2015. Insights into the structural determinants of substrate specificity and activity in mouse aldehyde oxidases. J. Biol. Inorg. Chem. 20, 209–217.

Chamizo-Ampudia, A., Galvan, A., Fernandez, E., Llamas, A., 2011. The *Chlamydomonas reinhardtii* molybdenum cofactor enzyme crARC has a Zn dependent activity and protein partners similar to those of its human homologue. Eukaryot. Cell 10, 1270–1282.

Dejanovic, B., Lal, D., Catarino, C.B., Arjune, S., Belaidi, A.A., Trucks, H., Vollmar, C., Surges, R., Kunz, W.S., Motameny, S., Altmuller, J., Kohler, A., Neubauer, B.A., Epicure, C., Nurnberg, P., Noachtar, S., Schwarz, G., Sander, T., 2014. Exonic microdeletions of the gephyrin gene impair GABAergic synaptic inhibition in patients with idiopathic generalized epilepsy. Neurobiol. Dis. 67, 88–96.

Del Rizzo, M., Burlina, A.P., Sass, J.O., Beermann, F., Zanco, C., Cazzorla, C., Bordugo, A., Giordano, L., Manara, R., Burlina, A.B., 2013. Metabolic stroke in a late-onset form of isolated sulfite oxidase deficiency. Mol. Genet. Metab. 108, 263–266.

Dent, C.E., Philpot, G.R., 1954. Xanthinuria – an inborn error (or deviation) of metabolism. Lancet 1, 182–185.

Duran, M., Beemer, F.A., van de Heiden, C., Korteland, J., de Bree, P.K., Brink, M., Wadman, S.K., Lombeck, I., 1978. Combined deficiency of xanthine oxidase and sulphite oxidase: a defect of molybdenum metabolism or transport? J. Inherit. Metab. Dis. 1, 175–178.

Eilers, T., Schwarz, G., Brinkmann, H., Witt, C., Richter, T., Nieder, J., Koch, B., Hille, R., Hansch, R., Mendel, R.R., 2001. Identification and biochemical characterization of *Arabidopsis thaliana* sulfite oxidase. A new player in plant sulfur metabolism. J. Biol. Chem. 276, 46989–46994.

Feng, C., Kedia, R.V., Hazzard, J.T., Hurley, J.K., Tollin, G., Enemark, J.H., 2002. Effect of solution viscosity on intramolecular electron transfer in sulfite oxidase. Biochemistry 41, 5816–5821.

Forstera, B., Belaidi, A.A., Juttner, R., Bernert, C., Tsokos, M., Lehmann, T.N., Horn, P., Dehnicke, C., Schwarz, G., Meier, J.C., 2010. Irregular RNA splicing curtails postsynaptic gephyrin in the cornu ammonis of patients with epilepsy. Brain 133, 3778–3794.

Fritschy, J.M., Harvey, R.J., Schwarz, G., 2008. Gephyrin: where do we stand, where do we go? Trends Neurosci. 31, 257–264.

Garattini, E., Fratelli, M., Terao, M., 2009. The mammalian aldehyde oxidase gene family. Hum. Genomics 4, 119–130.

Garattini, E., Terao, M., 2013. Aldehyde oxidase and its importance in novel drug discovery: present and future challenges. Expert Opin. Drug Discov. 8, 641–654.

Gruenewald, S., Wahl, B., Bittner, F., Hungeling, H., Kanzow, S., Kotthaus, J., Schwering, U., Mendel, R.R., Clement, B., 2008. The fourth molybdenum containing enzyme mARC: cloning and involvement in the activation of N-hydroxylated prodrugs. J. Med. Chem. 51, 8173–8177.

Gutzke, G., Fischer, B., Mendel, R.R., Schwarz, G., 2001. Thiocarboxylation of molybdopterin synthase provides evidence for the mechanism of dithiolene formation in metal-binding pterins. J. Biol. Chem. 276, 36268–36274.

Hansch, R., Lang, C., Riebeseel, E., Lindigkeit, R., Gessler, A., Rennenberg, H., Mendel, R.R., 2006. Plant sulfite oxidase as novel producer of H_2O_2: combination of enzyme catalysis with a subsequent non-enzymatic reaction step. J. Biol. Chem. 281, 6884–6888.

Hänzelmann, P., Schwarz, G., Mendel, R.R., 2002. Functionality of alternative splice forms of the first enzymes involved in human molybdenum cofactor biosynthesis. J. Biol. Chem. 277, 18303–18312.

Harvey, R.J., Topf, M., Harvey, K., Rees, M.I., 2008. The genetics of hyperekplexia: more than startle!. Trends Genet. 24, 439–447.

Havemeyer, A., Bittner, F., Wollers, S., Mendel, R., Kunze, T., Clement, B., 2006. Identification of the missing component in the mitochondrial benzamidoxime prodrug-converting system as a novel molybdenum enzyme. J. Biol. Chem. 281, 34796–34802.

Havemeyer, A., Lang, J., Clement, B., 2011. The fourth mammalian molybdenum enzyme mARC: current state of research. Drug Metab. Rev. 43, 524–539.

Hildebrandt, T.M., Grieshaber, M.K., 2008. Three enzymatic activities catalyze the oxidation of sulfide to thiosulfate in mammalian and invertebrate mitochondria. FEBS J. 275, 3352–3361.

Hille, R., 2002. Molybdenum and tungsten in biology. Trends Biochem. Sci. 27, 360–367.

Hille, R., Nishino, T., Bittner, F., 2011. Molybdenum enzymes in higher organisms. Coord. Chem. Rev. 255, 1179–1205.

Hughes, E.F., Fairbanks, L., Simmonds, H.A., Robinson, R.O., 1998. Molybdenum cofactor deficiency-phenotypic variability in a family with a late-onset variant. Dev. Med. Child Neurol. 40, 57–61.

Huxtable, R.J., 1989. Taurine in the central nervous system and the mammalian actions of taurine. Prog. Neurobiol. 32, 471–533.

Ichida, K., Amaya, Y., Kamatani, N., Nishino, T., Hosoya, T., Sakai, O., 1997. Identification of two mutations in human xanthine dehydrogenase gene responsible for classical type I xanthinuria. J. Clin. Invest. 99, 2391–2397.

Ichida, K., Matsumura, T., Sakuma, R., Hosoya, T., Nishino, T., 2001. Mutation of human molybdenum cofactor sulfurase gene is responsible for classical xanthinuria type II. Biochem. Biophys. Res. Commun. 282, 1194–1200.

Jacobsen, J.G., Smith, L.H., 1968. Biochemistry and physiology of taurine and taurine derivatives. Physiol. Rev. 48, 424–511.

Johnson, J.L., Coyne, K.E., Rajagopalan, K.V., Van Hove, J.L., Mackay, M., Pitt, J., Boneh, A., 2001. Molybdopterin synthase mutations in a mild case of molybdenum cofactor deficiency. Am. J. Med. Genet. 104, 169–173.

Johnson, J.L., Duran, M., 2001. Molybdenum cofactor deficiency and isolated sulfite oxidase deficiency. In: Scriver, C., Beaudet, A., Sly, W., Valle, D. (Eds.), The Metabolic and Molecular Bases of Inherited Disease, Eighth ed. McGraw-Hill, New York, pp. 3163–3177.

Johnson-Winters, K., Tollin, G., Enemark, J.H., 2010. Elucidating the catalytic mechanism of sulfite oxidizing enzymes using structural, spectroscopic, and kinetic analyses. Biochemistry 49, 7242–7254.

Kabil, O., Banerjee, R., 2010. Redox biochemistry of hydrogen sulfide. J. Biol. Chem. 285, 21903–21907.

Kisker, C., Schindelin, H., Rees, D.C., 1997. Molybdenum-cofactor-containing enzymes: structure and mechanism. Annu. Rev. Biochem. 66, 233–267.

Klein, J.M., Busch, J.D., Potting, C., Baker, M.J., Langer, T., Schwarz, G., 2012. The mitochondrial amidoxime-reducing component (mARC1) is a novel signal-anchored protein of the outer mitochondrial membrane. J. Biol. Chem. 287, 42795–42803.

Klein, J.M., Schwarz, G., 2012. Cofactor-dependent maturation of mammalian sulfite oxidase links two mitochondrial import pathways. J. Cell Sci. 125, 4876–4885.

Kuper, J., Llamas, A., Hecht, H.J., Mendel, R.R., Schwarz, G., 2004. Structure of the molybdopterin-bound Cnx1G domain links molybdenum and copper metabolism. Nature 430, 803–806.

Kurlemann, G., Debus, O., Schuierer, G., 1996. Dextromethorphan in molybdenum cofactor deficiency. Eur. J. Pediatr. 155, 422–423.

Kutter, D., Humbel, R., 1969. Screening for sulfite oxidase deficiency. Clin. Chim. Acta 24, 211–214.

Lee, H.-J., Adham, I.M., Schwarz, G., Kneussel, M., Sass, J.-O., Engel, W., Reiss, J., 2002. Molybdenum cofactor-deficient mice resemble the phenotype of human patients. Hum. Mol. Gen. 11, 3309–3317.

Lindsay, W.L., 1979. Chemical Equilibria in Soils. John Wiley & Sons, New York.

Llamas, A., Mendel, R.R., Schwarz, G., 2004. Synthesis of adenylated molybdopterin: an essential step for molybdenum insertion. J. Biol. Chem. 279, 55241–55246.

Llamas, A., Otte, T., Multhaup, G., Mendel, R.R., Schwarz, G., 2006. The mechanism of nucleotide-assisted molybdenum insertion into molybdopterin. A novel route toward metal cofactor assembly. J. Biol. Chem. 281, 18343–18350.

Lyons, J., Rauh-Pfeiffer, A., Yu, Y.M., Lu, X.M., Zurakowski, D., Tompkins, R.G., Ajami, A.M., Young, V.R., Castillo, L., 2000. Blood glutathione synthesis rates in healthy adults receiving a sulfur amino acid-free diet. Proc. Natl. Acad. Sci. U.S.A. 97, 5071–5076.

Marelja, Z., Dambowsky, M., Bolis, M., Georgiou, M.L., Garattini, E., Missirlis, F., Leimkuhler, S., 2014. The four aldehyde oxidases of *Drosophila melanogaster* have different gene expression patterns and enzyme substrate specificities. J. Exp. Biol. 217, 2201–2211.

Mason, J., 1986. Thiomolybdates: mediators of molybdenum toxicity and enzyme inhibitors. Toxicology 42, 99–109.

Mendel, R.R., Schwarz, G., 2011. Molybdenum cofactor biosynthesis in plants and humans. Coord. Chem. Rev. 255, 1145–1158.

Minoshima, S., Wang, Y., Ichida, K., Nishino, T., Shimizu, N., 1995. Mapping of the gene for human xanthine dehydrogenase (oxidase) (XDH) to band p23 of chromosome 2. Cytogenet. Cell Genet. 68, 52–53.

Mudd, S.H., Irreverre, F., Laster, L., 1967. Sulfite oxidase deficiency in man: demonstration of the enzymatic defect. Science 156, 1599–1602.

Nichols, J., Rajagopalan, K.V., 2002. *Escherichia coli* MoeA and MogA. Function in metal incorporation step of molybdenum cofactor biosynthesis. J. Biol. Chem. 277, 24995–25000.

Nishino, T., 1997. The conversion from the dehydrogenase type to the oxidase type of rat liver xanthine dehydrogenase by modification of cysteine residues with fluorodinitrobenzene. J. Biol. Chem. 272, 29859–29864.

Olney, J.W., Misra, C.H., de Gubareff, T., 1975. Cysteine-S-sulfate: brain damaging metabolite in sulfite oxidase deficiency. J. Neuropathol. Exp. Neurol. 34, 167–177.

Paul, B.D., Sbodio, J.I., Xu, R., Vandiver, M.S., Cha, J.Y., Snowman, A.M., Snyder, S.H., 2014. Cystathionine γ-lyase deficiency mediates neurodegeneration in Huntington's disease. Nature 509, 96–100.

Pitt, J.J., Eggington, M., Kahler, S.G., 2002. Comprehensive screening of urine samples for inborn errors of metabolism by electrospray tandem mass spectrometry. Clin. Chem. 48, 1970–1980.

Plitzko, B., Ott, G., Reichmann, D., Henderson, C.J., Wolf, C.R., Mendel, R., Bittner, F., Clement, B., Havemeyer, A., 2013. The involvement of mitochondrial amidoxime reducing components 1 and 2 and mitochondrial cytochrome b_5 in N-reductive metabolism in human cells. J. Biol. Chem.

Rajagopalan, K.V., Johnson, J.L., 1992. The pterin molybdenum cofactors. J. Biol. Chem. 267, 10199–10202.

Reiss, J., 2000. Genetics of molybdenum cofactor deficiency. Hum. Genet. 106, 157–163.

Reiss, J., Dorche, C., Stallmeyer, B., Mendel, R.R., Cohen, N., Zabot, M.T., 1999. Human molybdopterin synthase gene: genomic structure and mutations in molybdenum cofactor deficiency type B. Am. J. Hum. Genet. 64, 706–711.

Reiss, J., Gross-Hardt, S., Christensen, E., Schmidt, P., Mendel, R.R., Schwarz, G., 2001. A mutation in the gene for the neurotransmitter receptor–clustering protein gephyrin causes a novel form of molybdenum cofactor deficiency. Am. J. Hum. Genet. 68, 208–213.

Reiss, J., Hahnewald, R., 2011. Molybdenum cofactor deficiency: mutations in GPHN, MOCS1, and MOCS2. Hum. Mutat. 32, 10–18.

Reiss, J., Johnson, J.L., 2003. Mutations in the molybdenum cofactor biosynthetic genes MOCS1, MOCS2, and GEPH. Hum. Mutat. 21, 569–576.

Reiss, J., Lenz, U., Aquaviva-Bourdain, C., Joriot-Chekaf, S., Mention-Mulliez, K., Holder-Espinasse, M., 2011. A GPHN point mutation leading to molybdenum cofactor deficiency. Clin. Genet. 80, 598–599.

Santamaria-Araujo, J.A., Fischer, B., Otte, T., Nimtz, M., Mendel, R.R., Wray, V., Schwarz, G., 2004. The tetrahydropyranopterin structure of the sulfur-free and metal-free molybdenum cofactor precursor. J. Biol. Chem. 279, 15994–15999.

Santamaria-Araujo, J.A., Wray, V., Schwarz, G., 2011. Structure and stability of the molybdenum cofactor intermediate cyclic pyranopterin monophosphate. J. Biol. Inorg. Chem.

Sass, J.O., Gunduz, A., Araujo Rodrigues Funayama, C., Korkmaz, B., Dantas Pinto, K.G., Tuysuz, B., Yanasse Dos Santos, L., Taskiran, E., de Fatima Turcato, M., Lam, C.W., Reiss, J., Walter, M., Yalcinkaya, C., Camelo Junior, J.S., 2010. Functional deficiencies of sulfite oxidase: differential diagnoses in neonates presenting with intractable seizures and cystic encephalomalacia. Brain Dev. 32, 544–549.

Sass, J.O., Nakanishi, T., Sato, T., Shimizu, A., 2004. New approaches towards laboratory diagnosis of isolated sulphite oxidase deficiency. Ann. Clin. Biochem. 41, 157–159.

Schwahn, B.C., Van Spronsen, F.J., Belaidi, A.A., Bowhay, S., Christodoulou, J., Derks, T.G., Hennermann, J.B., Jameson, E., Konig, K., McGregor, T.L., Font-Montgomery, E., Santamaria-Araujo, J.A., Santra, S., Vaidya, M., Vierzig, A., Wassmer, E., Weis, I., Wong, F.Y., Veldman, A., Schwarz, G., 2015. Efficacy and safety of cyclic pyranopterin monophosphate substitution in severe molybdenum cofactor deficiency type A: a prospective cohort study. Lancet.

Schwarz, G., 2005. Molybdenum cofactor biosynthesis and deficiency. Cell Mol. Life Sci. 62, 2792–2810.

Schwarz, G., Belaidi, A.A., 2013. Molybdenum in human health and disease. Met. Ions Life Sci. 13, 415–450.

Schwarz, G., Mendel, R.R., Ribbe, M.W., 2009. Molybdenum cofactors, enzymes and pathways. Nature 460, 839–847.

Schwarz, G., Santamaria-Araujo, J.A., Wolf, S., Lee, H.J., Adham, I.M., Grone, H.J., Schwegler, H., Sass, J.O., Otte, T., Hanzelmann, P., Mendel, R.R., Engel, W., Reiss, J., 2004. Rescue of lethal molybdenum cofactor deficiency by a biosynthetic precursor from *Escherichia coli*. Hum. Mol. Genet. 13, 1249–1255.

Schwarz, G., Schulze, J., Bittner, F., Eilers, T., Kuper, J., Bollmann, G., Nerlich, A., Brinkmann, H., Mendel, R.R., 2000. The molybdenum cofactor biosynthetic protein Cnx1 complements molybdate-repairable mutants, transfers molybdenum to the metal binding pterin, and is associated with the cytoskeleton. Plant Cell 12, 2455–2472.

Simmonds, H.A., Reiter, S., Nishino, T., 1995. Hereditary xanthinuria. In: Scriver, C.R., Beaudet, A.L., Sly, W.S., Valle, D. (Eds.), The Metabolic and Molecular Bases of Inherited Disease, seventh ed. McGraw-Hill, New York, pp. 1781–1797.

Stallmeyer, B., Schwarz, G., Schulze, J., Nerlich, A., Reiss, J., Kirsch, J., Mendel, R.R., 1999. The neurotransmitter receptor-anchoring protein gephyrin reconstitutes molybdenum cofactor biosynthesis in bacteria, plants, and mammalian cells. Proc. Natl. Acad. Sci. U.S.A. 96, 1333–1338.

Stipanuk, M.H., Ueki, I., 2011. Dealing with methionine/homocysteine sulfur: cysteine metabolism to taurine and inorganic sulfur. J. Inherit. Metab. Dis. 34, 17–32.

Stipanuk, M.H., Ueki, I., Dominy Jr., J.E., Simmons, C.R., Hirschberger, L.L., 2009. Cysteine dioxygenase: a robust system for regulation of cellular cysteine levels. Amino Acids 37, 55–63.

Tan, W.H., Eichler, F.S., Hoda, S., Lee, M.S., Baris, H., Hanley, C.A., Grant, P.E., Krishnamoorthy, K.S., Shih, V.E., 2005. Isolated sulfite oxidase deficiency: a case report with a novel mutation and review of the literature. Pediatrics 116, 757–766.

Tejada-Jimenez, M., Galvan, A., Fernandez, E., 2011. Algae and humans share a molybdate transporter. Proc. Natl. Acad. Sci. U.S.A. 108, 6420–6425.

Tejada-Jimenez, M., Llamas, A., Sanz-Luque, E., Galvan, A., Fernandez, E., 2007. A high-affinity molybdate transporter in eukaryotes. Proc. Natl. Acad. Sci. U.S.A. 104, 20126–20130.

Tiranti, V., Viscomi, C., Hildebrandt, T., Di Meo, I., Mineri, R., Tiveron, C., Levitt, M.D., Prelle, A., Fagiolari, G., Rimoldi, M., Zeviani, M., 2009. Loss of ETHE1, a mitochondrial dioxygenase, causes fatal sulfide toxicity in ethylmalonic encephalopathy. Nat. Med. 15, 200–205.

Touati, G., Rusthoven, E., Depondt, E., Dorche, C., Duran, M., Heron, B., Rabier, D., Russo, M., Saudubray, J.M., 2000. Dietary therapy in two patients with a mild form of sulphite oxidase deficiency. Evidence for clinical and biological improvement. J. Inherit. Metab. Dis. 23, 45–53.

Ueki, I., Roman, H.B., Valli, A., Fieselmann, K., Lam, J., Peters, R., Hirschberger, L.L., Stipanuk, M.H., 2011. Knockout of the cysteine dioxygenase gene results in severe impairment in taurine synthesis and increased catabolism of cysteine to hydrogen sulfide. Am. J. Physiol. Endocrinol. Metab.

Veldman, A., Santamaria-Araujo, J.A., Sollazzo, S., Pitt, J., Gianello, R., Yaplito-Lee, J., Wong, F., Ramsden, C.A., Reiss, J., Cook, I., Fairweather, J., Schwarz, G., 2010. Successful treatment of molybdenum cofactor deficiency type A with cPMP. Pediatrics 125, e1249–1254.

Wadman, S.K., Cats, B.P., Debree, P.K., 1983. Sulfite oxidase deficiency and the detection of urinary sulfite. Eur. J. Pediatr. 141, 62–63.

Wu, G., Fang, Y.Z., Yang, S., Lupton, J.R., Turner, N.D., 2004. Glutathione metabolism and its implications for health. J. Nutr. 134, 489–492.

Wuebbens, M.M., Rajagopalan, K.V., 2003. Mechanistic and mutational studies of *Escherichia coli* molybdopterin synthase clarify the final step of molybdopterin biosynthesis. J. Biol. Chem. 278, 14523–14532.

Zhang, X., Vincent, A.S., Halliwell, B., Wong, K.P., 2004. A mechanism of sulfite neurotoxicity: direct inhibition of glutamate dehydrogenase. J. Biol. Chem. 279, 43035–43045.

Part IX

Phosphorus

Chapter 34

Phosphorus: Basic Nutritional Aspects

Pawel R. Kiela, Vijayababu M. Radhakrishnan, Fayez K. Ghishan
University of Arizona, Tucson, AZ, United States

INTRODUCTION

Elemental phosphorus is a white or yellow waxy substance that is highly reactive and burns on contact with air. This is the reason why it is most commonly found in nature in a pentavalent form in combination with oxygen as phosphate $(PO_4{}^{3-})$. Phosphorus is an essential constituent of all living cells, and its content (although not its chemical form) is uniform across most plant and animal tissues. Its contribution to the biology of all living organisms cannot be overestimated because its functions range from energy storage, cellular signaling networks, and forming the building blocks of nucleic acids and lipid membranes to the formation of skeletal structure and teeth. Therefore it is not surprising that systemic phosphate homeostasis is under complex control by at least three families of phosphate transport proteins under tight regulation by an endocrine network.

Primary dietary phosphorus deficiency is very rare because phosphorus is readily available in the food supply. However, genetic, metabolic, and medication-related disturbances in intestinal and renal phosphate (re)absorption can contribute to very serious disorders. Moreover, excess levels of phosphate in the blood have been associated with diseases that require more careful monitoring of phosphate intake or outright phosphate sequestration. Nutritional management of at-risk patient populations that could benefit from controlled phosphate intake is hindered by the unaccounted for amounts of inorganic phosphate (P_i) used as food additives and the lack of US Food and Drug Administration (FDA)-mandated reporting of phosphorus content in processed foods. In this chapter we overview the physiological roles of phosphorus, means of intestinal and renal phosphate (re)absorption and homeostatic regulation, food sources and dietary recommendations, health outcomes associated with increased phosphorus intake and hyperphosphatemia, and disorders associated with inadequate systemic phosphate levels.

Physiological Roles of Phosphorus

Phosphorus makes up approximately 0.5% of the newborn infant body and 0.65–1.1% of the adult body. Of the total body stores of phosphate, 85% are deposited in bones and teeth, where it associates with calcium in the form of hydroxyapatite; 14% of phosphate resides within cells; and phosphate in serum and extracellular fluids accounts for only 1% of total body phosphate. In soft tissues and body fluids, because of its high reactivity, phosphorus is rarely found as a free element and usually exists as the phosphate $(PO_4{}^{3-})$ ion and its esters. In this form it is a component of lipids, proteins, carbohydrates, and nucleic acids. In addition to its structural functions, phosphorous is also a mediator of energy transfer and is involved in a wide variety of metabolic reactions in cells. Energy derived from and fueling various metabolic processes is stored primarily in the high-energy phosphate bonds of creatine phosphate and ATP. Phosphorus is needed for the growth, maintenance, and repair of all tissues and cells and for the production of the genetic building blocks—DNA and RNA. Because of the critical role of phosphorous as P_i and phosphate esters in cell physiology, humans have developed extensive mechanisms for extracting phosphate from the diet and for the conservation of phosphate by the kidneys. As a consequence, plasma phosphate concentrations are maintained within a relatively narrow range and are modulated by fairly complex homeostatic mechanisms. Negative P_i balance may result from improperly regulated intestinal absorption, systemic utilization, and renal excretion (Fig. 34.1). Although dietary P_i deficiency is a relatively infrequent clinical problem, dysregulation of P_i homeostasis can result in a spectrum of debilitating diseases manifested acutely (e.g., myopathy, cardiac dysfunction, abnormal neutrophil function, platelet dysfunction, or erythrocyte membrane fragility) as well as chronically (e.g., by impaired bone mineralization, rickets, or osteomalacia).

P_i Absorption and Homeostatic Regulation

Three major mechanisms are responsible for the maintenance of systemic phosphate homeostasis: intestinal uptake, retention or release from the bone, and renal reabsorption. Most (~70%) dietary phosphate is absorbed through the intestine. Although it has been thought that systemic balance is primarily regulated through changes in renal fractional excretion,

Molecular, Genetic, and Nutritional Aspects of Major and Trace Minerals. http://dx.doi.org/10.1016/B978-0-12-802168-2.00034-8

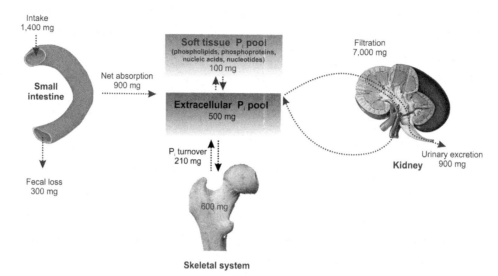

FIGURE 34.1 Systemic phosphate balance in an adult human. *P_i*, inorganic phosphate. *Adapted from Kiela, P.R., Ghishan, F.K., 2009. Recent advances in the renal-skeletal-gut axis that controls phosphate homeostasis. Lab. Invest. 89, 7–14.*

more recent evidence suggests a prominent role for regulation of intestinal phosphate uptake as well as for the coupling of phosphate absorption with the release of hormones that signal the kidney and bone (Kiela and Ghishan, 2009). There is remarkable diversity in the routes and regulation of intestinal and renal phosphate absorption.

Intestinal P_i Absorption

Intestinal phosphate absorption in mammals occurs primarily in the small intestine, with relative absorptive efficiency generally decreasing in more distal parts of the small bowel: duodenum > jejunum > ileum (Kayne et al., 1993; Walton and Gray, 1979). The bulk of phosphate absorption occurs in the jejunum (Borowitz and Ghishan, 1989) because of its longer length and increased transit time of the digesta through this gut segment. In experimental rodents the contribution of proximal and distal small intestinal segments to P_i absorption varies by species. In mice, ileal P_i absorption is a significant contributor (Radanovic et al., 2005; Stauber et al., 2005), whereas in rats, maximal absorption occurs in the duodenum with very little absorption occurring in the ileum, which is similar to the pattern reported in humans. However, in both rodent species only the jejunum shows an increase in P_i absorption in response to 1,25-dihydroxy-vitamin D_3 (1,25(OH)$_2$D$_3$; Marks et al., 2006). Phosphate absorption in the colon is considered physiologically irrelevant under most settings and can be observed only under conditions of extremely high luminal phosphate concentrations, such as through the use of phosphate enemas, which can result in luminal phosphate concentrations 1500 times greater than in the blood.

Intestinal P_i absorption involves two processes: a transcellular route, which requires sodium-dependent components, and a paracellular route, which takes place by a sodium-independent mechanism. McHardy and Parsons (1956) assessed serum phosphate levels after gavaging rats with P_i and found that the rates of phosphate absorption directly correlated with luminal phosphate concentrations without apparent saturation, thereby leading to the discovery of the passive diffusion mechanism. However, later studies using in vivo and in vitro approaches revealed that the rate of phosphate absorption was also dependent on the sodium concentration, providing the first line of evidence for a sodium-dependent pathway (Eto et al., 2006; Kayne et al., 1993; Walton and Gray, 1979). Perfusion studies with the human jejunum demonstrated that the active transport of phosphate is maximal at low phosphate intake levels and conversely that nonsaturable, passive diffusion occurs with high phosphate concentrations (Walton and Gray, 1979). We will discuss these two principal transport components in more detail in the following.

At a low luminal phosphate concentration, simple diffusion across the intestinal mucosa through the paracellular space is very limited because of the negative charge of the proteins forming the semiselective pores (occludins and claudins). The transmembrane potential difference between the interior of the cell and the intestinal lumen also creates a powerful barrier to the transport of negatively charged $H_2PO_4^-$ or $H_2PO_4^{2-}$ ions across the brush-border membrane. $H_2PO_4^{2-}$ ions, which can accumulate to high concentrations at acidic pH in the proximal small intestine, are more readily transported against this electronegative gradient. $H_2PO_4^{2-}$ ions, which predominate at the alkaline pH of the jejunum and ileum, require an active, energy-dependent process to ensure adequate absorption (Danisi et al., 1984). The potential difference across the basolateral membrane of the intestinal epithelial cells does not inhibit the transport of negatively charged, ionic phosphate molecules out of the cell into the interstitial space. Because the P_i concentrations in the cell and plasma are approximately 2.0 and 1.0 mM, respectively, passive efflux of phosphate across the basolateral membrane down an electrochemical gradient is less restrictive than paracellular transport or diffusion across the brush-border membrane. Although the presence of basolateral carrier-mediated phosphate transport has also been described (Ghishan et al., 1987; Kikuchi and Ghishan, 1987), it remains relatively uncharacterized. In general, the rate of passive transfer depends on ingested phosphate load and tightness of the junctions. Despite the growing understanding of how cellular regulatory mechanisms can influence passive transport, specificity for phosphate in association with specific tight junction proteins has not been reported so far. Paracellular absorption is likely favored after ingestion of a phosphate-containing meal (Williams and DeLuca, 2007), consistent with other in vivo findings that demonstrated that total phosphate absorption is Na^+ independent under normal dietary conditions. Precise characterization of the mechanism(s) that control transepithelial movement of P_i in the intestine is further complicated by the fact that P_i becomes rapidly integrated into various metabolic processes once it enters enterocytes. As such, the rate and control of phosphate absorption is not only a function of the brush-border and basolateral membrane transport processes, but it is also dependent on the rate of utilization and esterification of phosphate within the intracellular space. Three different Na^+-dependent phosphate transporter families have been identified: type I, II and type III transporters. They have been more exhaustively reviewed elsewhere (Biber et al., 2013; Forster et al., 2013; Wagner et al., 2014).

Type I transporters, which include sodium–phosphate cotransporter (NPT)-1 [solute-carrier family (SLC)-17A1/NaP_i-1], NPT3 (SLC17A2), NPT4 (SLC17A3), and NPT5 (SLC17A4), belong to SLC17, which also consists of a lysosomal acidic sugar transporter (sialin; SLC17A5), vesicular glutamate transporters (SLC17A7, SLC17A6, and SLC17A8, respectively), and a vesicular nucleotide transporter (SLC17A9). Although the first known member, NaP_i-1 was cloned as a Pi transporter (Murer et al., 2000), the SLC17 family is now recognized as multifunctional organic anion transporters, or as an anion:cation symporter family (Reimer, 2013). As an example, expression of NPT5 has been identified in the intestine, but it was postulated to facilitate intestinal ureate excretion (Togawa et al., 2012).

Type II Na^+-dependent phosphate transporters belong to the SLC34 family, comprising electrogenic and electroneutral isoforms expressed in epithelial cells. Thus far, NaP_i-IIa (Npt2a; SLC34A1), NaP_i-IIb (SLC34A2), and NaP_i-IIc (SLC34A3) isoforms have been identified in this class. In addition to the small intestine and kidney, these transporters are also expressed in other organs such as lungs, colon, salivary and mammary glands, liver, and testes (Wagner et al., 2014). All three have a preference for divalent $\left(HPO_4^{2-}\right)$ phosphate and are inhibited by phosphonoformic acid (PFA); NaP_i-IIa and NaP_i-IIb are electrogenic and transport phosphate with a stoichiometry of 3:1 Na^+: $\left(HPO_4^{2-}\right)$ whereas NaP_i-IIc is electroneutral with a 2:1 Na^+: $\left(HPO_4^{2-}\right)$ stoichiometry. Null knockout of NaP_i-IIb results in embryonic lethality, suggesting that this isoform is crucial for P_i absorption from the maternal circulation across the placenta (Sabbagh et al., 2009). In parallel studies, Sabbagh et al. (2009) showed that tamoxifen-inducible epithelial-specific deletion of NaP_i-IIb led to increased fecal phosphate excretion and hypophosphaturia despite a compensatory increase in renal NaP_i-IIa expression. The same report estimated that NaP_i-IIb contributes more than 90% of total Na^+-dependent active phosphate absorption in the mouse ileum. NaP_i-IIb is also expressed in lung alveolar type II cells, which are responsible for surfactant production. In humans, polymorphisms in the SLC34A2 gene have been associated with pulmonary alveolar microlithiasis (calcium phosphate deposition in the alveolar air spaces), and less commonly in testicular microlithiasis (Corut et al., 2006).

Type III transporters include the PiT1 (*SLC20A1*) and PiT2 (*SLC20A2*) genes from the ubiquitously expressed SLC20 family. They were originally identified as retroviral receptors (called Glvr-1 and Ram-1) and later shown to function as sodium–phosphate cotransporters (Miller et al., 1994). PiT1 and PiT2 are expressed broadly across various cell types and are generally considered to be involved in supplying cells with inorganic phosphorus to meet the needs of individual cell functions. The type III family of transporters have distinct characteristics: they prefer monovalent phosphate ($H_2PO_4^-$), they are all electrogenic and are not inhibited by PFA, and they have a reported stoichiometry of 2:1 Na^+: $H_2PO_4^-$ (Virkki et al., 2007). PiT1 and PiT2 are expressed in the small intestine (Bai et al., 2000; Giral et al., 2009). Although their contribution to intestinal phosphate absorption is largely uncharacterized, it is estimated to be at a modest 5–10% level because the remainder of the Na^+-dependent transport could not be attributed to NaP_i-IIb. However, systemic relevance of PiT1 is high as demonstrated by embryonic lethality of the *Slc20a1* knockout mice, attributed to abnormal liver development (Beck et al., 2010). In addition to liver development, PiT1 is further involved in vascular and bone physiology as well as regulation of parathyroid cell function. PiT2 has a more prominent role in renal P_i reabsorption, and polymorphisms within the *SLC20A2* gene have been associated with a brain disorder characterized by basal ganglia calcification, a finding confirmed in *Slc20a2*-deficient mice (Jensen et al., 2013). Wang et al. (2012) postulated that a functional loss of *SLC20A2* in the brain may result in regional accumulation of P_i in the extracellular matrix, causing calcium–phosphate deposition.

Renal Inorganic Phosphate Reabsorption

The filtered load of phosphate is approximately 4–8 g/day (130–194 mmol/day). For example, if the glomerular filtration rate (GFR) is 180 L/day (125 mL/min) and the phosphate concentration is 4 mg/dL (1.3 mmol/L), then the filtered load will be 7.2 g/day. Only 5–20% of the filtered phosphate is normally excreted. Approximately 80% of total phosphate reabsorption occurs in the proximal convoluted tubule (segments S1–S2), 15–20% in the proximal straight tubule (segment S3), and the remaining 5–10% in the distal convoluted tubule. Renal reabsorption of phosphate occurs against an electrochemical gradient and is dependent on active Na^+-dependent transporters. In S1–S3 segments of the proximal tubules, NaP_i-IIa expressed at the apical membrane is largely responsible for the majority (~80%) of phosphate reabsorption, a function supported by the coexpressed NaP_i-IIc and PiT2 transporters in the S1 and S2 segments. The relative importance of these transporters varies among species. Mechanisms of renal epithelial phosphate transport and its regulation have been the subject of many excellent in-depth reviews (Biber et al., 2013; Murer and Biber, 2010).

Sodium Phosphate Cotransporters Also Play Critical Roles in Bone, Choroid Plexus, and Vascular Physiology and Pathophysiology

The control of systemic inorganic P_i levels is known to be vital for bone formation, not only because it is an integral component of hydroxyapatite crystals but also because it regulates the rates of bone matrix formation and resorption. Carefully regulated phosphate transport via PiT1 is required for bone mineralization independently of its effects on cellular proliferation and differentiation. Reduced P_i transport in osteoblasts stimulates expression of stanniocalcin-1, which via an autocrine or paracrine effect stimulates the expression of PiT1 and promotes mineralization (Yoshiko et al., 2007).

A major role for the brain choroid plexus (CP), also sometimes referred to as the "kidney of the brain," is to regulate the exchange of solutes between the blood plasma and the cerebrospinal fluid (CSF) using selective transporters. The P_i concentration in human ventricular CSF ($[P_i]_{CSF}$) is maintained at approximately 0.4 mM compared with plasma at approximately 1.8 mM. A recent finding suggested that the CP actively transports phosphate from the CSF to blood utilizing the PiT2 transporter and perhaps to a lesser extent PiT1 because both transporters are expressed in the CP (Guerreiro et al., 2014).

Vascular calcification, or the deposition of calcium phosphate in the blood vessels, myocardium, and cardiac valves, is an extremely common and deleterious condition. P_i plays a major role as a signaling molecule with the ability to initiate phenotypic change and mineralization in vascular smooth muscle cells (VSMCs). For example, VSMCs do not spontaneously mineralize when they are grown under standard conditions, but when grown with elevated extracellular phosphate levels VSMCs undergo calcification, a phenotypic transition characterized by loss of smooth muscle-specific gene expression and upregulation of genes commonly associated with bone differentiation including osteocalcin, osteopontin, and Runx2. This effect is dependent on the expression and function of the PiT1 transporter (Giachelli, 2004). Hyperphosphatemia, a disorder discussed in more detail later in this chapter, is highly correlated with calcification of coronary and peripheral arteries as well as cardiac valves. Several large epidemiological studies have found that elevated serum phosphate is a major nontraditional risk factor for cardiovascular mortality in end-stage renal disease (ESRD) patients (Block et al., 1998; Raggi et al., 2002).

Solute-Carrier Family-37 Family of Intracellular Sugar-Phosphate/Phosphate Exchangers

Another group of phosphate transporters deserving a brief discussion is the SLC37 family members. These are four endoplasmic reticulum (ER), membrane-bound, sugar-phosphate/phosphate exchangers (SPX): SLC37A1 (SPX1), SLC37A2 (SPX2), SLC37A3 (SPX3), and SLC37A4 [SPX4, glucose 6-phosphate (G6P) transporter]. Their primary functions have been identified as P_i-linked G6P antiporters catalyzing $G6P:P_i$ and $P_i:P_i$ exchanges. All members perform these functions except for SLC37A3, the function of which is not yet clear. Chou and Mansfield (2014) recently reviewed the molecular physiology of these transport proteins in greater detail. Clinically, SPX4 (G6PT) may be the most recognized member of this family. Expression of two alternative splice variants varies in tissue specificity, but they serve the same basic function to translocate G6P from the cytoplasm into the lumen of the ER, where it is hydrolyzed by glucose-6-phosphatases into glucose and P_i. To date, 91 separate mutations have been identified in the *SLC37A4* gene of patients with type Ib glycogen storage disease, an autosomal-recessive disorder with an overall incidence of approximately 1 in 100,000 patients.

Phosphate Homeostasis

Maintenance of phosphate homeostasis involves three key elements: bone phosphate handling and intestinal and renal (re) absorption. The major identifiable regulatory substances are dietary phosphate itself, parathyroid hormone (PTH), vitamin D, and fibroblast growth factor-23 (FGF23) along with its required cofactor, the glucuronidase Klotho. Disorders affecting these organs and/or regulatory factors are frequently accompanied by significant hyper- or hypophosphatemia. We will only briefly review these mechanisms here to facilitate understanding of related disorders and dietary phosphate intake, and the reader is referred to more comprehensive review articles and references therein (Kiela and Ghishan, 2009; Kuro-o, 2012; Lederer, 2014).

Diet

Extremes in dietary phosphate content affect serum phosphate levels in otherwise healthy individuals. Portale et al. (1987) studied the effects of low- (500 mg/day), normal- (1500 mg/day), and high- (3000 mg/day) phosphate diets on serum phosphate concentration and diurnal variations. Phosphorus restriction resulted in a 40% reduction whereas phosphorus supplementation for 10 days led to a 14% increase in the 24-h mean serum phosphorus levels. Fasting usually does not cause a decrease in serum phosphorus concentration and even patients with frank malnutrition may have normal serum phosphorus levels. However, severely malnourished individuals, such as patients with eating disorders, may exhibit profound hypophosphatemia with enteral or parenteral feeding, the most consistent component of the so-called refeeding syndrome (Skipper, 2012). A low-phosphate diet stimulates transepithelial transport of P_i in the small intestine and in the renal proximal tubules, a response likely resulting from an increased abundance of the apical NaP_i-IIb (intestine) and NaP_i-IIa, NaP_i-IIc, and PiT-2 (kidney). Low-P_i diet stimulates the activity of the renal 1α-hydrolase (CYP27B1) and leads to an increase in the level of 1,25-$(OH)_2D_3$. However, the intestinal and renal adaptive increase in phosphate transport cannot be fully explained by the 1,25$(OH)_2D_3$–vitamin D receptor (VDR) axis because it is equally efficient in VDR knockout mice (Capuano et al., 2005; Segawa et al., 2004).

Parathyroid Hormone

PTH is an 84-amino–acid peptide hormone produced by chief cells in the parathyroid gland. Two types of PTH receptors differ in tissue expression: PTH1R is expressed predominantly in bone and kidney whereas PTH2R is expressed mainly in the central nervous system, pancreas, testes, and placenta. Although the primary role of PTH is to regulate calcium homeostasis, PTH secretion is also modulated by serum phosphate, FGF23, and 1,25$(OH)_2D_3$. PTH secretion is indirectly stimulated by elevated serum phosphate, which complexes with serum calcium, forming calcium phosphate. Ca-sensitive receptors on chief cells do not sense this form of calcium and "interpret" this as a reduced Ca^{2+} stimulation, thereby triggering an increase in PTH secretion. Increased circulating PTH has opposing effects on phosphate balance. Because of PTH-induced "coupled" release of phosphorus and calcium during bone resorption, as well as because of increased synthesis of 1,25$(OH)_2D_3$ in the kidney and the resulting increased intestinal phosphate absorption, PTH would be expected to increase the circulating phosphorus pool. However, these effects are counterweighed by the phosphaturic effect of PTH, which decreases the expression of NaP_i-IIa at the protein and mRNA levels in the renal proximal tubule and increases phosphate excretion. PTH also upregulates expression of FGF23 in bone, and as a humoral factor, FGF23 inhibits renal phosphate reabsorption (see Section Fibroblast Growth Factor-23 and Klotho below).

Vitamin D_3

The role of vitamin D_3 as a hormone-regulating systemic phosphate homeostasis is complex, highly integrated with other key players, and not fully understood. Circulating 1,25$(OH)_2D_3$ is a result of its renal production (hydroxylation), which is regulated by several factors including PTH, FGF23, phosphate, calcium, acid–base balance, and the hormone itself. Phosphate transport activity and NaP_i-IIb expression were reduced in the proximal, but not distal small intestine of VDR null knockout mice (Kaneko et al., 2011). However, the effects of vitamin D_3 are clearly proabsorptive (Ghishan, 1992), and it was shown to increase NaP_i-IIb expression in a VDR-independent manner (Brown et al., 2012). These effects are much more prominent in young animals than in adults. In general, high 1,25$(OH)_2D_3$ levels are associated with hyperphosphatemia whereas low 1,25$(OH)_2D_3$ levels are associated with hypophosphatemia, albeit only at extremes of vitamin D_3 concentrations. It is unclear to what extent vitamin D_3 affects serum phosphate directly versus indirectly through modulating PTH and FGF23 levels. High vitamin D_3 levels suppress PTH secretion directly and through elevation of serum calcium, thus leading to increased reabsorption and decreased renal phosphate excretion. On the other hand, low vitamin D_3 levels would lead to higher PTH and FGF23 levels and phophaturia and a lower serum phosphate concentration.

Fibroblast Growth Factor-23 and Klotho

FGF23 is a member of the FGF family, initially described as a regulator of phosphate homeostasis in studies related to congenital forms of hypophosphatemic rickets and oncogenic osteomalacia (Consortium, 2000). FGF23 is produced predominantly in the bone by osteocytes and osteoblasts and is regulated by serum phosphate levels and dietary phosphate intake. FGF23 represents an important element of a very complex regulatory network with implications for inherited disorders of bone/kidney mineral homeostasis, chronic kidney disease (CKD), CKD–mineral and bone disorder, obesity, diabetes, and osteoporosis (Kiela and Ghishan, 2009; Rowe, 2015). In the kidney, FGF23 signaling is dependent on the presence of the glucuronidase Klotho, which serves as an obligatory coreceptor to downregulate the expression of NaP$_i$-IIa and NaP$_i$-IIc. Klotho protein exists in membrane-bound as well as in a circulating soluble form. In addition to its other systemic effects (Kuro-o, 2012), this single transmembrane protein associates with FGF receptors in the kidney, enabling FGF23 binding and inhibition of phosphate transport. Klotho plays a similar role in the parathyroid gland, where it mediates the inhibitory effects of FGF23 on PTH mRNA expression. The absence of Klotho leads to a syndrome similar to FGF23 deficiency (i.e., hyperphosphatemia with an absence of phosphaturia). Klotho deficiency leads to premature aging symptoms in mice, including osteoporosis and soft tissue calcification. These symptoms have been attributed to uncontrolled overproduction of 1,25(OH)$_2$D$_3$ and could be largely eliminated by crossing Klotho$^{-/-}$ mice with 1α-hydroxylase (CYP27B1)-deficient mice (Ohnishi et al., 2009). FGF23 also decreases intestinal absorption of phosphate at least in part by inhibition of renal 1α-hydroxylase (CYP27B1) expression and decreased synthesis of 1,25(OH)$_2$D$_3$. Collectively, all of the effects of FGF23 promote a decrease in serum phosphate levels. Although the mechanisms are not clear, PTH appears to play a permissive role in the phosphaturic effect of FGF23.

Other Hormones

Through an uncharacterized mechanism, dietary phosphate can stimulate renal dopamine synthesis, which in turn decreases phosphate reabsorption in the proximal tubule (Bansal et al., 2012; Sizova et al., 2013). Altered dopamine signaling in the renal proximal tubule has been implicated in certain forms of hypertension and in aging. Estrogen also decreases renal phosphate reabsorption. Angiotensin II and insulin-like growth factor stimulate phosphate absorption in the renal proximal tubule (Caverzasio et al., 1990; Xu et al., 2004). The postnatal developmental pattern of expression and activity of renal phosphate transporters has been postulated to be controlled by the thyroid and triiodothyronine (Alcalde et al., 1999). Moreover, insulin was shown to stimulate renal phosphate reabsorption (Hammerman et al., 1984), and in diabetes phosphaturia may contribute to some of the pathophysiologic complications.

Phosphorus Intake: Sources and Recommendations

Dietary Sources and Forms of Phosphorus

In rapidly growing infants, maternal milk supplies adequate amounts of P$_i$. One day after birth the concentration of phosphate in milk is 0.26 ± 0.16 mM (mean ± SEM) whereas 4 days later it increases 6.6-fold to 1.69 ± 0.11 mM (Kent et al., 1992). This translates to relatively low intakes of phosphorus in breast-fed infants. However, it has been postulated that this may actually confer an advantage by which low residual phosphorus in the intestine, especially the colon, lowers fecal pH, which in turn may reduce proliferation of potentially pathogenic microorganisms, thus providing an immunoprotective effect. The efficiency of absorption is highest from human milk (85–90%; Williams et al., 1970), lower from cows milk (72%; Williams et al., 1970; Ziegler and Fomon, 1983), and lowest from soy formulas (59%), which contain the less-bioavailable form, phytic acid (Ziegler and Fomon, 1983). However, cow-milk– and soy-based infant formulas contain substantially greater amounts of total phosphorus than human milk, which offsets the lower bioavailability compared with human milk. Later in life dietary phosphorus is obtained from organic phosphorus compounds found naturally in foods and from P$_i$ in food additives. In general, protein-rich foods, such as meat, poultry, fish, eggs, dairy products, nuts, and legumes, are good sources of phosphorus. Approximately 70–80% of dietary phosphorus is derived from milk products, grains, and meats, with dairy products constituting a major source of dietary phosphorus. Most phosphorus found naturally in foods is present as organic compounds. These compounds must be enzymatically hydrolyzed to be absorbed, thus making them less bioavailable, with approximately 40–60% absorption rates. Plant-derived phosphorus, especially in beans, seeds, and nuts, is found mainly in the form of phytate, which is even less bioavailable than phosphorus derived from animal-based foods. The human digestive system cannot hydrolyze phytic acid (Fig. 34.2), but phytases of plant and microbial origin (colonic bacteria) are responsible for phosphate release. Some foods contain considerable phytase activity (wheat, wheat bran, rye, barley) whereas others have little or no phytase activity (corn, oats, sorghum, and oilseeds). Optimal temperatures of plant phytase activity (45–60°C or 113–140°F) represent a physiological limitation. Moreover, plant phytases may be partially or totally inactivated by overheating, such as during cooking or baking.

Several microbial organisms have been shown to express phytases. Examples include *Escherichia coli*, *Bacillus subtilis*, *Klebsiella terringa*, *Pseudomonas* spp., or the yeast *Saccharomyces cerevisiae*. Although plant phytases typically hydrolyze the phosphate at the six position of the phytin molecule, most microbial or fungal phytases typically hydrolyze phosphate at the three position. After releasing the first phosphate group, the five remaining phosphate groups can be sequentially released from phytin by phytase and nonspecific acid phosphatases, which are present in large quantities in the digestive tract. Complete hydrolysis yields one molecule of myo-inositol and six soluble phosphate groups.

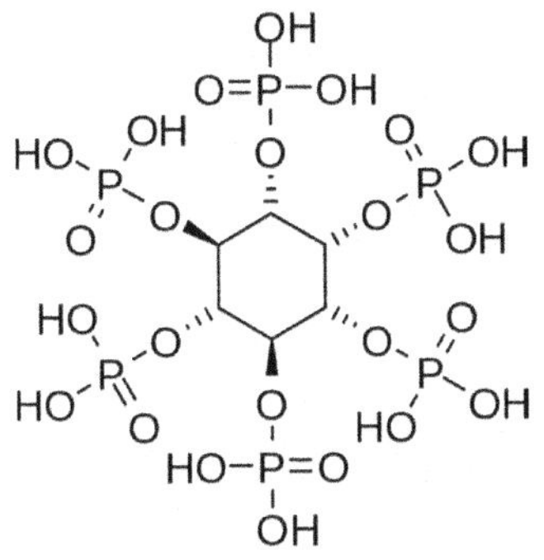

FIGURE 34.2 Chemical structure of phytic acid.

US diets are replete with food additives that contribute a considerable (~30%) amount of inorganic forms of phosphorus in the diet. Compared with organic forms of phosphorus, inorganic forms are easily absorbable, with nearly 100% bioavailability. Although total phosphorus contents in soft drinks, especially colas, is not very high (55 mg/12-oz can) as compared, for example, with milk (230 mg/8-oz), its molecular form (phosphoric acid) may be responsible for the reported associations between cola consumption and decreased bone density or renal diseases. Although clinical observations are still lacking, consumption of phosphoric acid has been implicated as leading to reduced calcium absorption and reduced levels of 25-hydroxyvitamin D_3 and $1,25(OH)_2D_3$ levels in young but not adult rats (Amato et al., 1998). The phosphorus content of the US food supply continues to increase as food manufacturers strive to improve taste, speed of preparation, shelf life, and convenience of products through the addition of phosphate ingredients. In an effort to help consumers determine the nutritional composition of foods, the FDA developed the daily value for the Nutrition Facts panel on processed foods. However, manufacturers are not required to include phosphorus content on these labels. In cases when it is included, it is expressed as a percentage of the daily value, which was somewhat arbitrarily established at 1000 mg/day, regardless of age. For individuals who need to limit phosphorus intake, this can be a significant source of confusion because the Nutrition Facts panel guideline is 300 mg higher than the adult recommended daily allowance (RDA).

Recommendations

The most recent comprehensive review and evaluation of nutritional requirements for dietary phosphorus was published in 1997 by the Institute of Medicine (IOM; United States) Standing Committee on the Scientific Evaluation of Dietary Reference Intakes (IOM, 1997). More recently, current dietary phosphorus intake and potential implications for public health were discussed during a dedicated meeting organized by the New York Academy of Sciences and summarized in an online eBriefing.[1] Although the precise requirement for phosphorus is difficult to establish, a dietary intake of approximately 700 mg/day is considered to be required to maintain a positive phosphate balance in adults (Table 34.1). Intestinal absorption of phosphorus increases by approximately 10% during pregnancy (Heaney and Skillman, 1971), and it is thought to be sufficient to provide the necessary increase in phosphorus needed for fetal growth. Therefore there is no further need to increase the RDA during pregnancy above the level recommended during the nonpregnant state. RDAs for dietary phosphorous are listed in Table 34.1. Most people do not need to take phosphorus supplements.

TABLE 34.1 Dietary Reference Intake of Phosphorus for Individuals According to the Report of the Institute of Medicine (United States) Standing Committee on the Scientific Evaluation of Dietary Reference Intakes (IOM, 1997)

	Infants (Months)		Children (Years)		Adults (Years)		Pregnancy (Years)		Lactation (Years)	
Age	0–6	7–12	1–3	4–8	9–18	≥19	14–18	19–50	14–18	19–50
Phosphorus (mg/day)	100*	275*	**460**	**500**	**1250**	**700**	**1250**	**700**	**1250**	**700**

Asterisks (*) indicate adequate intake whereas numbers in bold show recommended dietary allowances.

1. http://www.nyas.org/Publications/Ebriefings/Detail.aspx?cid=c491ca87-2379-4f82-a217-78af7876d954.

Phosphorus Intake: Risks of Overdose and Health Outcomes Associated With Increased Phosphorus Intake and Hyperphosphatemia

The phosphorus intake in the United States, as well as the intake of other essential nutrients, is monitored in the National Health and Nutrition Education Surveys (NHANES) conducted by the Centers for Disease Control and Prevention (CDC). The latest dataset has been analyzed and recently published by Lee and Cho (2015) and showed that among 13- to 19-year-old participants, daily phosphorus intake was 1382 ± 14 mg/day (LSM, least square mean $\pm$ SEM), close to the RDA values recommended by the IOM report for this age group (IOM, 1997). LSM for the age group of 20–99 years was reported as 1602 ± 7 in males and 1145 ± 4 in females, approximately double the RDA for this population, but significantly short of reaching the tolerable upper intake level (UL) set at 4000 mg/day. However, the calculated daily phosphorus intake is likely significantly underestimated. This notion is supported by the discrepancy between the evidence of increased phosphorus content of the food supply over the years but relatively flat intake values reported by the NHANES surveys (Calvo et al., 2014) and suggests the need for updated analysis of additional categories of processed foods.

Although phosphorus supplements are not widely used in the United States because of the potential side effects and interactions with prescription and nonprescription medications, one should take dietary supplements only under the supervision of a knowledgeable health-care provider. Too much phosphate can be toxic. It can cause diarrhea and calcification (hardening) of organs and soft tissue, and it can interfere with the body's ability to use iron, calcium, magnesium, and zinc. Athletes and others taking supplements that contain phosphate should only do so occasionally and with the guidance and direction of a health-care provider. Nutritionists recommend a balance of calcium and phosphorus in the diet. However, the typical Western diet contains roughly 2–4 times more phosphorus than calcium. Meat and poultry contain 10–20 times as much phosphorus as calcium, and carbonated beverages such as colas have as much as 500 mg of phosphorus in one serving. When there is more phosphorus than calcium in the body, the body will use calcium stored in bones. This can cause osteoporosis (brittle bones) and lead to gum and tooth problems. A balance of dietary calcium and phosphorus can lower the risk of osteoporosis.

Hyperphosphatemia is defined by serum phosphate concentration greater than 4.5 mg/dL (>1.4 mm/L). When measuring the serum phosphate level, it is important to remember that higher phosphate levels could be spurious (pseudohyperphosphatemia) and may be secondary to hyperlipidemia, hyperbilirubinemia, paraproteinemia, or when the blood is sampled from a line containing heparin. The clinical manifestations of true hyperphosphatemia are not very specific. Indeed, many of these patients are asymptomatic. However, some patients may occasionally experience and report fatigue, muscle cramps, bone and joint pain, and perioral numbness. These symptoms could be secondary to hypocalcemia accompanying the hyperphosphatemia. Laboratory investigations of patients who have hyperphosphatemia should include determination of serum calcium, phosphate levels, magnesium levels, PTH, blood urea, creatinine levels, and vitamin D_3 and its metabolites. The causes of true hyperphosphatemia could be secondary to increased intake, decreased excretion, and shifts of intracellular phosphate to the extracellular space.

Increased Intake

The phosphorus content of the American diet has been increasing, mostly because of the consumption of foods processed with phosphate additives (Calvo and Uribarri, 2013). Although it is unambiguously established that excessive phosphorus intake is detrimental in selected at-risk populations (e.g., in patients with CKD), it remains unclear whether phosphorus is a passive marker for adverse events or a true toxin for the general population. Erring on the side of caution, it is generally considered that a dietary phosphorus intake beyond nutritional needs disrupts phosphate homeostasis even in healthy individuals and that it may have significant potential public health consequences in terms of bone, cardiovascular, and kidney disease. Other clinically relevant scenarios that may lead to increased phosphate load include excessive oral or rectal use of phosphate-based laxatives, vitamin D intoxication, milk alkali syndrome (almost always caused by taking too many calcium supplements), or excessive parenteral administration of phosphate.

Decreased Renal Excretion

As we discuss in more detail in Section Phosphorus in Chronic Kidney Disease, hyperphosphatemia is associated with significant pathophysiology in CKD and renal failure. Electrolyte disorders, such as hypomagnesemia and hypokalemia, can also be associated with chronic hyperphosphatemia. Other clinical scenarios that may be accompanied by hyperphosphatemia include hypoparathyroidism and other hormonal disturbances, such as Cushing syndrome and hypothyroidism, or tumoral calcinosis. Medication-induced hyperphosphatemias are discussed in more detail in Section Phosphorus–Drug Interactions later in this chapter.

Shifts From Intracellular Phosphate to the Extracellular Space

Massive cellular shift of phosphate out of the cells is a relatively rare cause of hyperphosphatemia. However, conditions such as tumor lysis syndrome (TLS), rhabdomyolysis, acute hemolysis, or some forms of metabolic acidosis are known to cause elevated serum phosphate levels and may lead to acute kidney injury. TLS is an oncologic emergency that is caused by massive tumor cell lysis with the release of large amounts of potassium, phosphate, and nucleic acids into the systemic circulation. An excess of nucleic acids is catabolized to uric acid, which leads to hyperuricemia, precipitation of uric acid in the renal tubules and renal vasoconstriction, decreased renal flow, oxidation, and inflammation, all resulting in acute kidney injury. High concentrations of both uric acid and phosphate potentiate the risk of acute kidney injury through calcium phosphate deposition in the renal tubules, which further aids uric acid precipitation. Rhabdomyolysis is a syndrome caused by injury to skeletal muscle, and similar to TLS, it involves leakage of large quantities of potentially toxic intracellular contents into plasma. Direct nephrotoxicity (ischemia and tubular injury), intrarenal vasoconstriction, and acute kidney injury are also typical consequences of this disorder. A similar extracellular shift of phosphate may occur as a result of severe hemolytic anemia, a disorder related to intravascular or extravascular hemolysis (abnormal breakdown of red blood cells). Hyperphosphatemia can be observed in patients with acute metabolic acidosis, especially with lactic acidosis and diabetic ketoacidosis. In addition to promoting a shift of phosphate out of cells, metabolic acidosis can diminish glycolysis and therefore cellular phosphate utilization, thus leading to an increase in serum phosphate concentration. It is interesting to note that not all forms of acidosis equally affect serum phosphate levels. In lactic acidosis, tissue hypoxia and cell death are likely to further contribute to a decreased consumption and increased release of phosphate from cells. In the case of diabetic ketoacidosis, decreased cellular phosphate uptake due to insulin deficiency may contribute to hyperphosphatemia, although hyperglycemia of diabetes leads to urinary phosphate losses due to osmotic diuresis. This state may turn into phosphate deficiency (hypophosphatemia) by the administration of insulin.

Management of Hyperphosphatemia

Determining the cause of hyperphosphatemia is the cornerstone of its management. Hyperphosphatemia secondary to excess intake requires limiting intake of phosphate and/or the use of phosphate binders. The following agents that bind phosphate have been used clinically:

1. *Aluminum-containing phosphate binders*: These compounds are no longer in use because of aluminum toxicity.
2. *Calcium-containing phosphate binders*: These compounds include calcium carbonate, calcium citrate, and calcium acetate. Although effective as phosphate absorption inhibitors, they may result in hypercalcemia.
3. *Other phosphate binders*: These compounds include sucroferric oxyhydroxide (velphoro), sevelamer (renagel), lanthanum carbonate, and ferrous citrate. Sucroferric oxyhydroxide is an iron-based absorbent of phosphate. Side effects associated with its use have been noted, such as diarrhea, discoloration of the stools, and nausea. Sevelamer is a polymeric amine that binds phosphate. Side effects include constipation and black stools due to gastrointestinal bleeding. Lanthanum carbonate is a phosphate binder and prevents phosphate absorption. Side effects include gastrointestinal upsets, myalgia, and muscular cramping. Ferrous citrate also interferes with phosphate absorption, and the side effects are related to increased iron absorption.

Renal failure is the most common cause of hyperphosphatemia that requires the expertise of a nephrologist. TLS typically responds to saline diuresis to enhance urinary phosphate excretion. Enhanced renal excretion in the setting of normal renal function requires volume repletion coupled with loop diuretics such as furosemide, which inhibits the reabsorption of sodium and chloride in the loop of Henle as well as in the proximal and distal tubules of the kidney.

Phosphorus in Chronic Kidney Disease

Animal studies showed that high dietary phosphorus can initiate and/or worsen progression of kidney dysfunction whereas dietary phosphate restriction prevents or reverses it. Although calcification is certainly an important mode of tissue injury, other mechanisms such as podocyte or endothelial dysfunction likely contribute to renal phosphate toxicity. Although observational in nature, many well-designed studies with patients with CKD consistently demonstrated that elevated serum phosphate is a risk factor for progression to ESRD (Nadkarni and Uribarri, 2014). It is interesting to note that elevated FGF23 has also been positively associated with the risk of ESRD. Clinical trials in which reducing dietary phosphorus intake was used as the primary intervention in CKD are still lacking. The closest to it was the modification of diet in renal disease study, a large multicenter clinical trial in which modification of intake of dietary protein and phosphorus was the primary intervention. Although the study did not achieve its endpoint of reducing the

GFR loss in CKD patients, a significant reduction in the risk of ESRD or death was observed during extended follow-up in patients assigned to the reduced dietary protein group (Klahr et al., 1994). A meta-analysis of five studies of nondiabetic renal disease reported similar findings (Klahr et al., 1994). Although these studies causally implicate phosphorus, they consistently demonstrate a beneficial effect of protein restriction and, by extension, of phosphorus restriction on CKD-associated mortality. However, limitation of phosphate intake in the CKD patients, as well as other at-risk groups, which could benefit from lower P_i intake, is not easy in practice. The extra phosphorus added during food processing remains unaccounted for, and listing of phosphorus content on the product's Nutrition Facts panel would provide an important tool needed to control dietary phosphorus intake. Considering the size of the CKD population, estimated at approximately 16 million, it is hoped that the FDA will eventually enact similar rules for phosphorus as for other essential nutrients and minerals to define the actual content of this mineral in food in an effort to limit dietary phosphorus intake in the United States.

Roles of Phosphorus in Cardiovascular Disease and Vascular Calcification

It has been well recognized that in CKD patients, especially in the ESRD stage, hyperphosphatemia causes soft tissue calcification, including vascular calcifications. In 1998 Block and colleagues showed an independent association of serum phosphorus concentrations of 5.5 mg/dL or greater with all-cause and cardiovascular mortality in 6000 US hemodialysis patients (Block et al., 1998). More recent evidence also demonstrated the epidemiologic link between elevated serum phosphorus and adverse outcomes in patients with mild to moderate kidney dysfunction or even in patients with apparent normal renal function. As an example, the Cholesterol and Recurrent Events study showed a graded independent relation between elevated phosphorus and risk of death in people with prior myocardial infarction (Tonelli et al., 2005). Although, as in CKD, the primary mechanism behind the association between hyperphosphatemia and adverse cardiovascular outcomes appears to be the phosphorus-induced vascular calcification, other mechanisms likely contribute. Some of the examples of other possible mechanisms are acute endothelial dysfunction, decreased flow-mediated arterial dilation, and increased reactive oxygen species production, which leads to inhibition of endothelial nitric oxide (NO) synthase and reduced NO production. These acute events suggest that in susceptible individuals, even transient postprandial elevation of serum phosphorus may be a contributor to the pathogenesis of CVD as significant as the average phosphorus serum concentration. Phosphorus might also exert direct actions on the myocardium, inducing fibrosis and left ventricular hypertrophy in CKD and ESRD. Increased FGF23 concentrations have also been associated with higher risk of heart failure, stroke, and death among individuals with normal renal function, although it remains unclear whether the reported effects of elevated phosphorus on CVD risks can be explained by elevated serum FGF23 concentrations.

Disorders Associated with Inadequate Systemic P_i Levels (Hypophosphatemia)

Hypophosphatemia is defined as a serum phosphate level of less than 2.5 mg/dL (0.8 mmol/L). Three major causes of hypophosphatemia include (1) inadequate intake/decreased intestinal absorption of phosphate, (2) excessive losses with increased urinary phosphate excretion, and (3) redistribution of phosphate from the extracellular fluid into the cells. Inadequate intake/decreased absorption may be secondary to poor dietary intake, alcoholism, eating disorders, malabsorption syndromes (e.g., in Crohn's disease or steatorrhea), or vitamin D deficiency. Excessive urinary phosphate losses may be the result of genetic disorders or due to acquired phosphate wasting. Genetic disorders leading to hypophosphatemia include X-linked hypophosphatemic rickets (XLH), autosomal-dominant hypophosphatemic rickets (ADHR), type 1 and type 2 vitamin D-dependent rickets (VDDR), hereditary hypophosphatemic rickets with hypercalciuria (HHRH), hypophosphatemic nephrolithiasis/osteoporosis-1, and McCune–Albright syndrome (MAS). We will briefly discuss these genetic disorders later in this section. Acquired phosphate wasting encompasses vitamin D deficiency, primary hyperparathyroidism, drugs such as loop diuretics, heavy metal intoxication after kidney transplant, and oncogenic osteomalacia. Lastly, hypophosphatemia may develop as a consequence of intracellular shifts with a redistribution of phosphate from the extracellular fluid to the cells. Such a scenario has been demonstrated for increased insulin secretion, particularly during refeeding, acute respiratory alkalosis, hungry bone syndrome, salicylate poisoning, or sepsis. Acute respiratory alkalosis results in the rise of intracellular pH, which stimulates phosphofructokinase activity, which in turn stimulates glycolysis. In addition, extreme hyperventilation in normal subjects can lower serum phosphate to below 1 mg/dL. This is probably the most common cause of marked hypophosphatemia in hospitalized patients. Hungry bone syndrome can occur in patients with preexisting osteopenia who undergo parathyroidectomy. It can result in marked deposition of calcium and phosphate in the bone, which may be associated with marked hypocalcemia and hypophosphatemia.

Nutritional Rickets

The term rickets comes from the old English *wrick*, meaning "to twist," indicating bending of the bones. A dietary deficiency of calcium or phosphorus can rarely cause rickets, but the most common cause of rickets in children is vitamin D deficiency. Vitamin D deficiency leads to decreased absorption of calcium and phosphate from the intestine and kidney with decreased calcification of bone, primarily at the metaphyseal growing ends of bones, but also throughout the osteoid of the skeleton. Vitamin D deficiency also results in hypercalcemia, which stimulates PTH secretion resulting in phosphaturia and eventually in hypophosphatemia, which is the hallmark of rickets. Serum alkaline phosphatase will be elevated from leakage of bone secondary to overactive osteoblasts cells. Anticonvulsant drugs accelerate metabolism of vitamin D and may lead to malabsorption syndrome. In some cases liver and kidney diseases may also lead to nutritional rickets. The manifestations of rickets include muscular hypotonia, frontal bossing, and delay in the closure of the anterior fontanelle. Bone abnormalities are common, such as bowing of the legs and rachitic rosary of the costochondrial junctions. Kyphosis can be seen as a manifestation of rickets, and X rays of the bone show fraying and cupping of the metaphysis. Treatment of rickets includes vitamin D, either as a single dose of 600,000 IU or a daily dose of 5000 IU for 3 months.

Genetic Disorders Leading to Hypophosphatemia

There is a considerable genetic heterogeneity of hypophosphatemic rickets, with mutations affecting several key players in systemic phosphate homeostasis. The frequently severe consequences of these mutations indicate that despite the complexity of the regulatory network, there is limited redundancy, which makes the system prone to dysregulation with pathogenic sequelae.

X-Linked Hypophosphatemic Rickets

Patients with low serum phosphate levels and rickets that are resistant to vitamin D are commonly diagnosed with XLH, a disorder that accounts for 80% of all familial hypophosphatemias. The disorder is caused by mutations in the phosphate-regulating gene homologous to endopeptidases on the X-chromosome (PHEX). PHEX regulates FGF23 expression directly and indirectly. Loss of PHEX activity leads to elevated expression of FGF23 by osteoblasts/osteocytes and to increased levels of circulating FGF23. Patients with XLH usually present with growth retardation associated with bowing of long bones. Dentations may be absent or delayed. Laboratory tests show low phosphate levels, normal or low calcium serum levels, and high alkaline phosphatase. $1,25(OH)_2D_3$ levels are inappropriately low, whereas PTH is normal or slightly elevated. In these patients, renal tubular reabsorption of phosphate is typically 60% (>90% is normal). The primary treatment is calcitriol ($1,25(OH)_2D_3$). Amiloride or thiazide diuretics may be used to enhance calcium reabsorption and reduce risk of nephrocalcinosis. Dietary phosphate supplementation is achieved with a mix of dibasic sodium phosphate, monobasic potassium phosphate, and monobasic sodium phosphate (e.g., K-Phos Neutral).

Autosomal-Dominant Hypophosphatemic Rickets

Patients display similar clinical manifestations as in XLH, with laboratory findings of hypophosphatemia with inappropriately low $1,25(OH)_2D_3$. However, this disorder is caused by mutations in *FGF23* that result in resistance to FGF23 protein degradation and high circulating levels of FGF23, which lead to hyperphosphaturia and hypophosphatemia.

Hereditary Hypophosphatemic Rickets with Hypercalciuria

HHRH is a rare disorder resulting from inactivating mutations in *SLC34A3* coding for NaP$_i$-IIc, a renal sodium phosphate cotransporter. Patients have hypophosphatemia, phosphaturia, hypercalciuria, bone pain, rickets, osteomalacia, muscle weakness, and high $1,25(OH)_2D_3$. It is interesting to note that mice lacking *Slc34a3* have similar biochemical features of HHRH but do not develop rickets or osteomalacia.

Autosomal-Recessive Hypophosphatemic Rickets Types 1 and 2

ARHR-1 and ARHR-2 at their roots have elevated FGF23 levels, along with hypophosphatemia and phosphaturia. In ARHR-1, the mutations occur in the dentin matrix acidic phosphoprotein-1 (DMP1) gene, coding for a member of the small integrin-binding ligand N-linked glycoprotein family, which is highly expressed in osteocytes. The hypophosphatemic phenotype observed in *Dmp1$^{-/-}$* mice was rescued by the disruption of FGF23 expression, suggesting that an abnormally high expression of FGF23 was also a contributing factor in the phenotypes of ARHR patients. However, the functional link between *DMP1* and *FGF23* remains unclear. ARHR-2 is caused by a homozygous mutation in the *ENPP1* gene, which encodes nucleotide pyrophosphatase phosphodiesterase-1 and in some cases results in symptoms resembling idiopathic infantile arterial calcification. *Enpp1*$^{-/-}$ mice also develop hypophosphatemia and hypocalcemia, and they exhibit abnormalities including reduced bone mass and strength, articular cartilage and arterial calcification, and elevated serum and bone expression of FGF23. Similar to *DMP1*, the mechanistic link between *ENPP1* and *FGF23* is not yet known.

Vitamin D-Dependent Rickets

The condition, also known as pseudovitamin D deficiency, is split into two major types. The first type, VDDR1 or vitamin D hydroxylation-deficient rickets, is caused by mutations in the CYP27B1 gene coding for renal 1α-hydroxyase, which is responsible for the synthesis of 1,25(OH)$_2$D$_3$ (type 1A) or in the CYP2R1 gene encoding vitamin D 25-hydroxylase (type 1B). Type 2 VDDR is secondary to organ resistance to 1,25(OH)$_2$D$_3$, resulting from mutations in the VDR (type VDDR2a). VDDR2b, with normal VDR, is an unusual form of rickets with normal ligand-receptor binding of 1,25(OH)$_2$D$_3$ but no nuclear localization of the receptor complex. This has been attributed to an overexpression of two proteins from the family of heterogeneous nuclear ribonucleoproteins, hnRNP C1 and C2, which block binding of VDR/RXR heterodimers to DNA response elements located on vitamin-D–regulated genes (Chen et al., 2003).

McCune–Albright Syndrome

This complex disorder results from early embryonic postzygotic somatic activating mutations in the *GNAS1* gene. Hypophosphatemic rickets is a known complication of this disorder, when polyostotic fibrous dysplasia is observed. Serum levels of *FGF23* are elevated in these patients with renal phosphate wasting. The production of *FGF23* by fibrodysplastic tissue plays a role in renal wasting and is the source of *FGF23*.

Phosphorus–Drug Interactions

In addition to dietary habits and metabolic/genetic disorders, several other factors, including popular medications, can interfere with phosphate absorption and systemic homeostasis and may require a judicial phosphorus repletion/supplementation. Alcohol abuse may lead to hypophosphatemia. This may be related to hypomagnesemia, metabolic acidosis or alkalosis, or a proximal tubular defect in phosphate transport. In addition, alcohol withdrawal syndrome, associated with respiratory alkalosis and diarrhea, may lead to phosphate deficiency. Phosphate-binding antacids containing aluminum, calcium, or magnesium (such as Mylanta, Amphojel, Maalox, Riopan, and Alternagel) can sequester phosphate in the intestine and limit its absorption. Long-term use of these antacids can cause hypophosphatemia. Prolonged therapy with certain anticonvulsants (e.g., phenobarbital, carbamazepine, or Tegretol) may lower phosphorus absorption by induction of cytochrome CYP450 and increased vitamin D$_3$ catabolism as well as by reduced calcium absorption. Bile acid sequestrants, such as cholestyramine or Colestipol used as cholesterol-lowering drugs, may also decrease intestinal phosphate absorption. Oral phosphate supplements are recommended at least 1 h before or 4 h after these drugs. Corticosteroids, such as prednisone or methylprednisolone, may increase urinary phosphorus losses, presumably through decreased renal expression of NaP$_i$-IIa. High doses of insulin may lower blood levels of phosphorus in people with diabetic ketoacidosis. Using phosphorus supplements along with potassium supplements or potassium-sparing diuretics (e.g., spironolactone) may result in hyperkalemia, which may lead to life-threatening cardiac arrhythmia. Angiotensin-converting enzyme inhibitors, used to treat high blood pressure, may lower serum phosphate levels.

These are examples of a larger array of drugs that may cause hypophosphatemia and associated symptoms. Other drugs that have been reported to cause similar side effects include cyclosporine (immunosuppressant), cardiac glycosides (digoxin or lanoxin), heparins (blood-thinning drugs), nonsteroidal antiinflammatory drugs (e.g., ibuprofen), and theophylline overdose. In addition, salt substitutes containing high levels of potassium may lower phosphorus levels if used long term. Medication-induced hypophosphatemia has been reviewed in greater detail by Liamis et al. (2010).

HOW OTHER MINERALS ARE AFFECTED OR BEHAVE

Many case reports documented a decrease in serum phosphate in individuals ingesting large quantities of calcium supplements, presumably through intestinal phosphate binding. Therefore the Ca:P ratio in the diet has been considered to be a critical determinant of bioavailability of either nutrient, particularly in infant nutrition. This concept may be especially important under conditions of rapid growth. The estimates of optimal Ca:P intake ratios have frequently been based on the calcium and phosphorus needs of bone mineralization. However, there are complicating factors to consider. The molar ratio of Ca:P in synthetic hydroxyapatite is 1.67:1 whereas it is closer to 1.5:1 in the actual bone mineral and even lower in the amorphous calcium phosphate (1.3:1), which is the first mineral deposited at the mineralizing site. One has to consider the accretion of phosphorus in soft tissue during growth, which accounts for approximately 1 mmol P$_i$ for every 5 mmol added to bone. Considering negligible soft tissue accretion of calcium compared with the skeleton, it has been estimated that the absorbed Ca:P molar ratio sufficient to support the sum of skeletal and soft tissue growth would be approximately 1.3:1. Another complicating factor is presented by discrepancies in absorption efficiencies for dietary calcium and phosphorus. In infants the respective net absorption efficiency of calcium and phosphorus is approximately 60% and 80%, respectively; therefore the corrected optimal Ca:P ratio would be approximately 2:1. However, this value is higher than the Ca:P molar ratio of human milk (1.5:1), which one would presume to be optimal for the infant's nutritional needs. The shortcomings of these calculations and predictions of optimal Ca:P ratio is further magnified by nonparallel age-related changes in calcium and phosphate absorption, and relative changes in the utilization of food for energy and growth into adulthood. Therefore current nutritional concepts do not heavily rely on the dietary Ca:P molar ratio. This is supported by findings from nutritional balance studies in adult humans, in which Ca:P molar ratios ranging from 0.08:1 to 2.40:1 (a 30-fold range) had no effect on calcium balance or absorption.

The dietary source/form of phosphorus may be related to the utilization of other nutrients. For example, phytic acid has a strong binding affinity for calcium, magnesium, iron, copper, and zinc. This results in precipitation, making the minerals less available for intestinal absorption, which caused phytic acid to be classified as an antinutrient. Phosphorus is also needed to help balance and use other vitamins and minerals, including vitamin D, iodine, magnesium, and zinc.

SUMMARY POINTS

- Dietary P_i deficiency is a relatively rare clinical problem, but dysregulation of P_i homeostasis can result in a spectrum of debilitating diseases acutely or chronically manifested.
- The RDA, 700 mg/day of phosphorus for healthy adults, is meant to maintain serum phosphorus concentrations within the physiologic range of 2.5–4.5 mg/dL. The tolerable UL for phosphorus intake is 4000 mg/day for healthy individuals.
- Protein-rich foods, such as meat, poultry, fish, eggs, and dairy products, are good sources of phosphorus. Approximately 70–80% of dietary phosphorus is derived from milk products and meats, with dairy products constituting a major source of dietary phosphorus. Plant-derived phosphorus, especially in beans, seeds, and nuts, is found mainly in the form of phytate, which is less bioavailable than phosphorus derived from animal-based foods.
- Estimates of dietary phosphorus intakes in the US population is generally inaccurate because food additives contain considerable amounts of P_i and are not always included on the nutrition fact label, which is used to compute nutrient intakes.
- Hyperphosphatemia has been associated with CKD and renal failure, tumoral calcinosis, TLS, rhabdomyolysis, metabolic acidosis, and acute hemolysis.
- Three major causes of hypophosphatemia include the following: (1) inadequate intake/decreased intestinal absorption of phosphate, (2) excessive losses with increased urinary phosphate excretion, and (3) redistribution of phosphate of the extracellular fluid into the cells.
- The calcium:phosphate ratio of the diet has been traditionally considered to be a critical determinant of bioavailablity of either nutrient. Although this may be the case during rapid growth, the value of calculating the dietary Ca:P ratio in adults is questionable.
- Tighter control of dietary phosphorus intake in at-risk patient populations remains difficult because the extra phosphorus added for food processing remains unaccounted for. Listing of phosphorus content on the product's Nutrition Facts panel, as required for other essential nutrients and minerals, would provide an important tool needed to control dietary phosphorus intake.

KEY FACTS

- Phosphorus is an essential component for biological processes, including growth, bone mineralization, maintenance and repair of all tissues and cells, energy production, cell signaling, regulation of acid–base homeostasis, and genetic building blocks—DNA and RNA.
- Three major mechanisms are responsible for the maintenance of systemic phosphate homeostasis: intestinal uptake, retention or release from the bone, and renal reabsorption.
- The major regulatory factors involved in systemic phosphate homeostasis are dietary phosphate itself, PTH, vitamin D, and FGF23 along with its required cofactor, Klotho, all of which regulate tissue-specific membrane phosphate transport proteins.
- Excessive phosphorus intake is detrimental and may have significant potential public health consequences in terms of bone, cardiovascular, and kidney disease. High serum phosphorus concentrations have been associated with increased mortality rates of cardiovascular disease patients with or without kidney disease.

MINI DICTIONARY OF TERMS

Hyperphosphatemia A metabolic condition in which there is an abnormally high accumulation of phosphorus in blood (>4.5 mg/dL).

Hypophosphatemia A clinical state defined by a serum phosphate level of less than 2.5 mg/dL.

Phosphaturia A condition in which there is too much phosphorus in urine, which causes the urine to appear cloudy or murky in color. This condition is typically assessed through 24-h urine collection. It is defined as exceeding the typical range of 170–1200 mg/day in females and 360–1600 mg/day in males.

Rickets Bone disorder in which bones soften and become prone to fractures and deformity.

REFERENCES

Alcalde, A.I., Sarasa, M., Raldua, D., Aramayona, J., Morales, R., Biber, J., Murer, H., Levi, M., Sorribas, V., 1999. Role of thyroid hormone in regulation of renal phosphate transport in young and aged rats. Endocrinology 140, 1544–1551.

Amato, D., Maravilla, A., Montoya, C., Gaja, O., Revilla, C., Guerra, R., Paniagua, R., 1998. Acute effects of soft drink intake on calcium and phosphate metabolism in immature and adult rats. Rev. Invest. Clin. 50, 185–189.

Bai, L., Collins, J.F., Ghishan, F.K., 2000. Cloning and characterization of a type III Na-dependent phosphate cotransporter from mouse intestine. Am. J. Physiol. Cell Physiol. 279, C1135–C1143.

Bansal, N., Hsu, C.Y., Whooley, M., Berg, A.H., Ix, J.H., 2012. Relationship of urine dopamine with phosphorus homeostasis in humans: the heart and soul study. Am. J. Nephrol. 35, 483–490.

Beck, L., Leroy, C., Beck-Cormier, S., Forand, A., Salaun, C., Paris, N., Bernier, A., Urena-Torres, P., Prie, D., Ollero, M., et al., 2010. The phosphate transporter PiT1 (Slc20a1) revealed as a new essential gene for mouse liver development. PLoS One 5, e9148.

Biber, J., Hernando, N., Forster, I., 2013. Phosphate transporters and their function. Annu. Rev. Physiol. 75, 535–550.

Block, G.A., Hulbert-Shearon, T.E., Levin, N.W., Port, F.K., 1998. Association of serum phosphorus and calcium × phosphate product with mortality risk in chronic hemodialysis patients: a national study. Am. J. Kidney Dis. 31, 607–617.

Borowitz, S.M., Ghishan, F.K., 1989. Phosphate transport in human jejunal brush-border membrane vesicles. Gastroenterology 96, 4–10.

Brown, A.J., Zhang, F., Ritter, C.S., 2012. The vitamin D analog ED-71 is a potent regulator of intestinal phosphate absorption and NaPi-IIb. Endocrinology 153, 5150–5156.

Calvo, M.S., Moshfegh, A.J., Tucker, K.L., 2014. Assessing the health impact of phosphorus in the food supply: issues and considerations. Adv. Nutr. 5, 104–113.

Calvo, M.S., Uribarri, J., 2013. Public health impact of dietary phosphorus excess on bone and cardiovascular health in the general population. Am. J. Clin. Nutr. 98, 6–15.

Capuano, P., Radanovic, T., Wagner, C.A., Bacic, D., Kato, S., Uchiyama, Y., St-Arnoud, R., Murer, H., Biber, J., 2005. Intestinal and renal adaptation to a low-Pi diet of type II NaPi cotransporters in vitamin D receptor- and 1alphaOHase-deficient mice. Am. J. Physiol. Cell Physiol. 288, C429–C434.

Caverzasio, J., Montessuit, C., Bonjour, J.P., 1990. Stimulatory effect of insulin-like growth factor-1 on renal Pi transport and plasma 1,25-dihydroxyvitamin D3. Endocrinology 127, 453–459.

Chen, H., Hewison, M., Hu, B., Adams, J.S., 2003. Heterogeneous nuclear ribonucleoprotein (hnRNP) binding to hormone response elements: a cause of vitamin D resistance. Proc. Natl. Acad. Sci. U.S.A. 100, 6109–6114.

Chou, J.Y., Mansfield, B.C., 2014. The SLC37 family of sugar-phosphate/phosphate exchangers. Curr. Top. Membr. 73, 357–382.

Consortium, A., 2000. Autosomal dominant hypophosphataemic rickets is associated with mutations in FGF23. Nat. Genet. 26, 345–348.

Corut, A., Senyigit, A., Ugur, S.A., Altin, S., Ozcelik, U., Calisir, H., Yildirim, Z., Gocmen, A., Tolun, A., 2006. Mutations in SLC34A2 cause pulmonary alveolar microlithiasis and are possibly associated with testicular microlithiasis. Am. J. Hum. Genet. 79, 650–656.

Danisi, G., Murer, H., Straub, R.W., 1984. Effect of pH on phosphate transport into intestinal brush-border membrane vesicles. Am. J. Physiol. 246, G180–G186.

Eto, N., Tomita, M., Hayashi, M., 2006. NaPi-mediated transcellular permeation is the dominant route in intestinal inorganic phosphate absorption in rats. Drug Metab. Pharmacokinet. 21, 217–221.

Forster, I.C., Hernando, N., Biber, J., Murer, H., 2013. Phosphate transporters of the SLC20 and SLC34 families. Mol. Aspects Med. 34, 386–395.

Ghishan, F.K., 1992. Phosphate transport by plasma membranes of enterocytes during development: role of 1,25-dihydroxycholecalciferol. Am. J. Clin. Nutr. 55, 873–877.

Ghishan, F.K., Kikuchi, K., Arab, N., 1987. Phosphate transport by rat intestinal basolateral-membrane vesicles. Biochem. J. 243, 641–646.

Giachelli, C.M., 2004. Vascular calcification mechanisms. J. Am. Soc. Nephrol. 15, 2959–2964.

Giral, H., Caldas, Y., Sutherland, E., Wilson, P., Breusegem, S., Barry, N., Blaine, J., Jiang, T., Wang, X.X., Levi, M., 2009. Regulation of rat intestinal Na-dependent phosphate transporters by dietary phosphate. Am. J. Physiol. Renal Physiol. 297, F1466–F1475.

Guerreiro, P.M., Bataille, A.M., Parker, S.L., Renfro, J.L., 2014. Active removal of inorganic phosphate from cerebrospinal fluid by the choroid plexus. Am. J. Physiol. Renal Physiol. 306, F1275–F1284.

Hammerman, M.R., Rogers, S., Hansen, V.A., Gavin 3rd, J.R., 1984. Insulin stimulates Pi transport in brush border vesicles from proximal tubular segments. Am. J. Physiol. 247, E616–E624.

Heaney, R.P., Skillman, T.G., 1971. Calcium metabolism in normal human pregnancy. J. Clin. Endocrinol. Metab. 33, 661–670.

IOM, 1997. Dietary Reference Intakes for Calcium, Phosphorus, Magnesium, Vitamin D, and Fluoride. Washington, DC.

Jensen, N., Schroder, H.D., Hejbol, E.K., Fuchtbauer, E.M., de Oliveira, J.R., Pedersen, L., 2013. Loss of function of Slc20a2 associated with familial idiopathic basal ganglia calcification in humans causes brain calcifications in mice. J. Mol. Neurosci. 51, 994–999.

Kaneko, I., Segawa, H., Furutani, J., Kuwahara, S., Aranami, F., Hanabusa, E., Tominaga, R., Giral, H., Caldas, Y., Levi, M., et al., 2011. Hypophosphatemia in vitamin D receptor null mice: effect of rescue diet on the developmental changes in renal Na$^+$-dependent phosphate cotransporters. Pflügers Arch. 461, 77–90.

Kayne, L.H., D'Argenio, D.Z., Meyer, J.H., Hu, M.S., Jamgotchian, N., Lee, D.B., 1993. Analysis of segmental phosphate absorption in intact rats. A compartmental analysis approach. J. Clin. Invest. 91, 915–922.

Kent, J.C., Arthur, P.G., Retallack, R.W., Hartmann, P.E., 1992. Calcium, phosphate and citrate in human milk at initiation of lactation. J. Dairy Res. 59, 161–167.

Kiela, P.R., Ghishan, F.K., 2009. Recent advances in the renal-skeletal-gut axis that controls phosphate homeostasis. Lab. Invest. 89, 7–14.

Kikuchi, K., Ghishan, F.K., 1987. Phosphate transport by basolateral plasma membranes of human small intestine. Gastroenterology 93, 106–113.

Klahr, S., Levey, A.S., Beck, G.J., Caggiula, A.W., Hunsicker, L., Kusek, J.W., Striker, G., 1994. The effects of dietary protein restriction and blood-pressure control on the progression of chronic renal disease. Modification of Diet in Renal Disease Study Group. N. Engl. J. Med. 330, 877–884.

Kuro-o, M., 2012. Klotho in health and disease. Curr. Opin. Nephrol. Hypertens. 21, 362–368.

Lederer, E., 2014. Regulation of serum phosphate. J. Physiol. 592, 3985–3995.

Lee, A.W., Cho, S.S., 2015. Association between phosphorus intake and bone health in the NHANES population. Nutr. J. 14, 28.

Liamis, G., Milionis, H.J., Elisaf, M., 2010. Medication-induced hypophosphatemia: a review. QJM 103, 449–459.

Marks, J., Srai, S.K., Biber, J., Murer, H., Unwin, R.J., Debnam, E.S., 2006. Intestinal phosphate absorption and the effect of vitamin D: a comparison of rats with mice. Exp. Physiol. 91, 531–537.

McHardy, G.J.R., Parsons, D.S., 1956. The absorption of inorganic phosphate from the small intestine of the rat. Exp. Physiol. 41, 398–409.

Miller, D.G., Edwards, R.H., Miller, A.D., 1994. Cloning of the cellular receptor for amphotropic murine retroviruses reveals homology to that for gibbon ape leukemia virus. Proc. Natl. Acad. Sci. U.S.A. 91, 78–82.

Murer, H., Biber, J., 2010. Phosphate transport in the kidney. J. Nephrol. 23 (Suppl. 16), S145–S151.

Murer, H., Hernando, N., Forster, I., Biber, J., 2000. Proximal tubular phosphate reabsorption: molecular mechanisms. Physiol. Rev. 80, 1373–1409.

Nadkarni, G.N., Uribarri, J., 2014. Phosphorus and the kidney: what is known and what is needed. Adv. Nutr. 5, 98–103.

Ohnishi, M., Nakatani, T., Lanske, B., Razzaque, M.S., 2009. Reversal of mineral ion homeostasis and soft-tissue calcification of klotho knockout mice by deletion of vitamin D 1alpha-hydroxylase. Kidney Int. 75, 1166–1172.

Portale, A.A., Halloran, B.P., Morris Jr., R.C., 1987. Dietary intake of phosphorus modulates the circadian rhythm in serum concentration of phosphorus. Implications for the renal production of 1,25-dihydroxyvitamin D. J. Clin. Invest. 80, 1147–1154.

Radanovic, T., Wagner, C.A., Murer, H., Biber, J., 2005. Regulation of intestinal phosphate transport. I. Segmental expression and adaptation to low-P(i) diet of the type IIb Na$^+$-P(i) cotransporter in mouse small intestine. Am. J. Physiol. Gastrointest. Liver Physiol. 288, G496–G500.

Raggi, P., Boulay, A., Chasan-Taber, S., Amin, N., Dillon, M., Burke, S.K., Chertow, G.M., 2002. Cardiac calcification in adult hemodialysis patients. A link between end-stage renal disease and cardiovascular disease? J. Am. Coll. Cardiol. 39, 695–701.

Reimer, R.J., 2013. SLC17: a functionally diverse family of organic anion transporters. Mol. Aspects Med. 34, 350–359.

Rowe, P.S., 2015. A unified model for bone-renal mineral and energy metabolism. Curr. Opin. Pharmacol. 22, 64–71.

Sabbagh, Y., O'Brien, S.P., Song, W., Boulanger, J.H., Stockmann, A., Arbeeny, C., Schiavi, S.C., 2009. Intestinal npt2b plays a major role in phosphate absorption and homeostasis. J. Am. Soc. Nephrol. 20, 2348–2358.

Segawa, H., Kaneko, I., Yamanaka, S., Ito, M., Kuwahata, M., Inoue, Y., Kato, S., Miyamoto, K., 2004. Intestinal Na-P(i) cotransporter adaptation to dietary P(i) content in vitamin D receptor null mice. Am. J. Physiol. Renal Physiol. 287, F39–F47.

Sizova, D., Velazquez, H., Sampaio-Maia, B., Quelhas-Santos, J., Pestana, M., Desir, G.V., 2013. Renalase regulates renal dopamine and phosphate metabolism. Am. J. Physiol. Renal Physiol. 305, F839–F844.

Skipper, A., 2012. Refeeding syndrome or refeeding hypophosphatemia: a systematic review of cases. Nutr. Clin. Pract. 27, 34–40.

Stauber, A., Radanovic, T., Stange, G., Murer, H., Wagner, C.A., Biber, J., 2005. Regulation of intestinal phosphate transport. II. Metabolic acidosis stimulates Na$^+$-dependent phosphate absorption and expression of the Na$^+$-P(i) cotransporter NaPi-IIb in small intestine. Am. J. Physiol. Gastrointest. Liver Physiol. 288, G501–G506.

Togawa, N., Miyaji, T., Izawa, S., Omote, H., Moriyama, Y., 2012. A Na$^+$-phosphate cotransporter homologue (SLC17A4 protein) is an intestinal organic anion exporter. Am. J. Physiol. Cell Physiol. 302, C1652–C1660.

Tonelli, M., Sacks, F., Pfeffer, M., Gao, Z., Curhan, G., Cholesterol and Recurrent Events Trial Investigators, 2005. Relation between serum phosphate level and cardiovascular event rate in people with coronary disease. Circulation 112, 2627–2633.

Virkki, L.V., Biber, J., Murer, H., Forster, I.C., 2007. Phosphate transporters: a tale of two solute carrier families. Am. J. Physiol. Renal Physiol. 293, F643–F654.

Wagner, C.A., Hernando, N., Forster, I.C., Biber, J., 2014. The SLC34 family of sodium-dependent phosphate transporters. Pflügers Arch. 466, 139–153.

Walton, J., Gray, T.K., 1979. Absorption of inorganic phosphate in the human small intestine. Clin. Sci. (London) 56, 407–412.

Wang, C., Li, Y., Shi, L., Ren, J., Patti, M., Wang, T., de Oliveira, J.R., Sobrido, M.J., Quintans, B., Baquero, M., et al., 2012. Mutations in *SLC20A2* link familial idiopathic basal ganglia calcification with phosphate homeostasis. Nat. Genet. 44, 254–256.

Williams, K.B., DeLuca, H.F., 2007. Characterization of intestinal phosphate absorption using a novel in vivo method. Am. J. Physiol. Endocrinol. Metab. 292, E1917–E1921.

Williams, M.L., Rose, C.S., Morrow 3rd, G., Sloan, S.E., Barness, L.A., 1970. Calcium and fat absorption in neonatal period. Am. J. Clin. Nutr. 23, 1322–1330.

Xu, L., Dixit, M.P., Chen, R., Dixit, N.M., Collins, J.F., Ghishan, F.K., 2004. Effects of angiotensin II on NaPi-IIa co-transporter expression and activity in rat renal cortex. Biochim. Biophys. Acta 1667, 114–121.

Yoshiko, Y., Candeliere, G.A., Maeda, N., Aubin, J.E., 2007. Osteoblast autonomous P$_i$ regulation via Pit1 plays a role in bone mineralization. Mol. Cell. Biol. 27, 4465–4474.

Ziegler, E.E., Fomon, S.J., 1983. Lactose enhances mineral absorption in infancy. J. Pediatr. Gastroenterol. Nutr. 2, 288–294.

Chapter 35

Molecular Mechanisms of Adverse Health Effects Associated With Excess Phosphorus Intake

Jaime Uribarri[1], Mona S. Calvo[2]

[1]The Icahn School of Medicine at Mount Sinai, New York, NY, United States; [2]U.S. Food and Drug Administration, Laurel, MD, United States

INTRODUCTION

Health-care professionals involved in the care of chronic kidney disease (CKD) patients have long been aware of the health risks associated with the body's retention of dietary phosphorus as renal disease progresses. Therefore it is not unexpected that most of the knowledge in the area of phosphorus metabolism has come from the field of nephrology. In the recent past, new endocrine pathways essential in phosphorus homeostasis, such as fibroblast growth factor-23 (FGF23), have been identified that have contributed to our understanding that, although phosphorus is an essential nutrient, its excess could be linked to tissue damage by various mechanisms. This becomes particularly important in view of our modern eating habits that have increased cumulative use of phosphorus ingredients in food processing. Therefore the potential toxicity of an essential mineral such as phosphorus when taken in excess needs to be considered in detail, not just in relation to the increasing population with CKD but also in the general population. In this review we will briefly summarize these different points of view regarding phosphate balance.

MOLECULAR MECHANISMS REGULATING PHOSPHORUS HOMEOSTASIS

Organ-Level Phosphorus Balance

Most (85%) of the phosphorous in humans (~700g in a 70-kg man) is stored in bone and teeth, 14% in soft tissue, and the remaining 1% in the extracellular space. Therefore circulating levels of phosphate represent only a minute portion of the overall body pool of this mineral. Serum phosphate at any particular time represents the balance between gastrointestinal absorption, intra- or extracellular shift, and urinary excretion of the mineral. In an adult in the steady state, without any net shift between intra- and extracellular compartments, daily urinary excretion should equal net gastrointestinal absorption of phosphorus (Fig. 35.1; Uribarri, 2007). Any excess dietary phosphorous will be handled by increased renal excretion in the presence of normal kidney function; however, this becomes a problem in patients with CKD. An average daily dietary intake of phosphorus is approximately 1400 mg. However, there is suggestion that phosphorus intake is on the rise as a result of the growing consumption of highly processed foods, especially restaurant, fast, and convenience foods, and that phosphorus intake levels are systematically underestimated in national surveys (Calvo and Uribarri, 2013a).

Cellular Mechanisms of Phosphorus Regulation

Absorption of dietary phosphorus occurs in the small intestine via passive paracellular diffusion and actively across cell walls via sodium phosphate (NaPi) cotransporter type IIb (NaPi-IIb; Fig. 35.2). Approximately 60% of dietary phosphorus is usually absorbed and the remainder comes out in the stool. However, this percentage is very variable depending on the kind of food, the degree of additional phosphate added during food processing, and the relative amount of calcium in the diet. Of clinical importance, these phosphorus transporters are inhibited by nicotinamide, and clinical trials with oral nicotinamide administration to dialysis patients showed a significant reduction of serum phosphate (Rennick et al., 2013). Some phosphorus transport may also occur through members of the solute carrier-20 family or type NaPi-III cotransporters (PiT-1, PiT-2; Fig. 35.2; John et al., 2011).

Molecular, Genetic, and Nutritional Aspects of Major and Trace Minerals. http://dx.doi.org/10.1016/B978-0-12-802168-2.00035-X

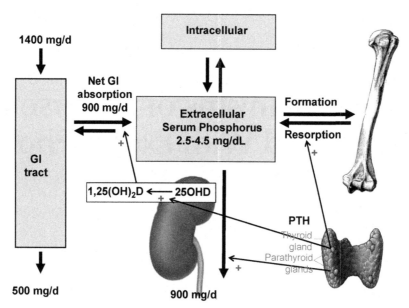

FIGURE 35.1 Phosphorus Homeostasis in the Normal subject. This figure describes the relationships between Net GI (gastrointestinal) absorption, movements into and out of the bones and intracellular space, and renal handling of this mineral. The importance of PTH (parathyroid hormone) on regulating bone shift of phosphorus as well as on activating vitamin D at the level of the kidneys is also illustrated.

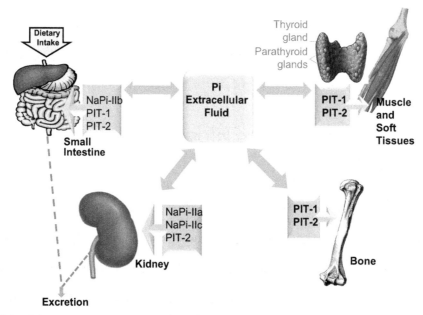

FIGURE 35.2 Multiplicity of Phosphorus Transporters throughout the body. As described within the text, this figure illustrates the different Pi (phosphorus) transporters in the different tissues, including NaPi (sodium phosphate co-transporters IIa, IIb, and IIc) and PiT-1 and PiT-2 (sodium phosphate III transporters).

Phosphorous excretion is predominantly via the kidneys. Phosphorus is filtered in the glomerulus, with 75% reabsorbed in the proximal tubule, 10% in the distal tubule, and 15% lost in urine. Two main transporters have been identified as responsible for phosphate reabsorption: NaPi-IIa and NaPi-IIc. NaPi-IIa is responsible for 70% of reabsorption, with 30% handled by NaPi-IIc; a small amount is reabsorbed by the PiT-2 phosphate transporter (Fig. 35.2). All of these transporters are located on the apical brush border membrane of renal proximal tubule cells. Phosphorus distribution between the extracellular compartment and other body tissues, including bone, may be mediated by the PiT-1 and PiT-2 cotransporters (Fig. 35.2).

Molecular Levels of Phosphorus Regulation

Phosphorus balance is regulated by the interplay of several hormonal factors, some of which are well studied. Increased net gastrointestinal absorption of phosphorus will produce a mild increase in postprandial serum phosphate, which in turn will initiate several hormonal changes, including increased release of parathyroid hormone (PTH) and FGF23 produced by osteocytes and secreted from bone. The effect of phosphate on PTH is both directly through the parathyroid gland and indirectly by decreasing serum calcium, which in turn is a strong stimulus for PTH release. Both of these hormones will increase urinary excretion of phosphorus, with FGF23 acting through Klotho, a transmembrane protein that functions as an obligatory coreceptor for FGF23, helping to promote negative phosphorus balance (John et al., 2011). When the extracellular portion of the Klotho protein is cleaved, a soluble form circulates, which can modulate NaPi-IIa cotransporters as well as calcium and potassium channels in some tissues. Klotho action in the kidneys will help to eliminate the excess phosphorus absorbed from the intestine as long as kidney function is still significant. Of interest, although both hormones, PTH and FGF23, have the same effect on urinary phosphorus excretion, their effects on the renal 25-hydroxyvitamin D (25-(OH)D) activating enzyme, 1-α-hydroxylase, are opposite. PTH stimulates 25-(OH) D activation whereas FGF23 suppresses formation of 1,25-dihydroxyvitamin D (1,25-$(OH)_2$D). In turn, the activated hormone will increase gastrointestinal absorption of phosphorus but suppress release of PTH.

Disruption of Phosphorus Regulation With Increased Intake

As described earlier, an increase in dietary phosphorus intake will have a significant effect, disrupting all of the regulatory mechanisms, mainly leading to an increase in PTH, FGF23, and Klotho, with a variable effect on 1,25-$(OH)_2$D.

In CKD, this phosphorus dysregulation may occur even in the presence of a normal phosphorus intake. In the presence of decreased glomerular filtration rate (GFR), the hallmark of CKD, urinary excretion of phosphorus will initially tend to decrease, but increased levels of PTH and FGF23 will compensate (Uribarri, 2007). As a result, although a tendency for phosphorus retention occurs early on in CKD, actual fasting hyperphosphatemia does not develop until advanced disease, when GFR is markedly decreased. Thus early increases in serum phosphorus can often go undetected, with most physicians drawing blood samples in the morning fasted state.

EXCESS PHOSPHORUS INTAKE ASSOCIATED WITH DISRUPTION OF REGULATORY MECHANISMS RELATED TO PHOSPHORUS HOMEOSTASIS AND DISEASE RISK

Hyperphosphatemia and Disease Risk

Phosphorus is an essential mineral needed for cell structure, signaling pathways, and metabolic functions; however, when dietary intakes exceed adult requirements, the delicate balance in the tightly regulated serum phosphate pool may be disrupted (Calvo and Uribarri, 2013a,b; Pizzorno, 2014). Acute and sustained elevation in serum phosphate may disrupt the endocrine regulation of this serum phosphate pool in healthy individuals by triggering the secretion of hormones for which the actions can damage tissue and contribute to disease risk (Calvo and Uribarri, 2013a). Much of what is known about the disease risks of excess phosphorus intake stems from studies of patients with kidney disease who became hyperphosphatemic over time because of their progressively impaired renal function and inability to excrete phosphorus. It is well established that the high phosphorus content of the modern, highly processed Western diet, containing a growing number of readily absorbed inorganic phosphate food additives, can be toxic to people with impaired ability to excrete excess dietary phosphate (Hong et al., 2015; Calvo et al., 2014; Winger et al., 2012). Greater understanding of the mechanisms involved in hyperphosphatemia of CKD and its link to specific diseases and increased mortality has led to ever-increasing evidence that excess phosphorus intake may also contribute to cardiovascular, bone, cancer, and kidney disease risk in healthy individuals without impaired kidney function. Examples of the emerging evidence from studies exploring the health impact of high phosphorus intake on these disease states are discussed in the next sections. For information on the current level of phosphorus intake in Western cultures and the recommended intake guidelines, see Calvo and Lamberg-Allardt (2015).

Chronic Kidney Disease

Dietary phosphorus intake is particularly important in patients with CKD and/or end-stage renal disease (ESRD) in which it may be a risk factor for cardiovascular disease (CVD), mortality, bone disease, and even progression of renal disease itself.

In 1998, Block et al., using the United States Renal Data System data, showed that hemodialysis patients with serum phosphate greater than 6.5 mg/dL had a higher relative risk of death compared with those with serum phosphate levels between 2.7 and 6.5 mg/dL (Block et al., 1998). The same association between serum phosphate and mortality was later confirmed by the same authors in another group of 40,538 hemodialysis patients (Block et al., 2004).

Chue and associates demonstrated that elevated serum phosphate levels are associated with changes in left ventricular mass as defined by cardiac magnetic resonance, in the CKD population (Chue et al., 2012). This finding has also been documented in hemodialysis patients. This association is important because left ventricular hypertrophy (LVH) is a known marker of CVD.

In contrast to these findings, a review of a CKD population of 10,672 patients over a 2-year follow-up period did not find an increased risk of all-cause mortality when adjusted for confounders (Mehrotra et al., 2013). Of note, these were patients with early-stage CKD, with serum creatinine averaging 1.3 mg/dL. However, in a review of data from 6370 CKD patients with slightly higher serum creatinine (mean value 1.7 mg/dL), Kestenbaum et al. (2005) found that patients with serum phosphate levels greater than 3.5 mg/dL had an increased risk of death. Moreover, elevated phosphate levels have been associated with progression of kidney disease itself. This was first demonstrated in animals that developed impaired kidney function on a high-phosphorus diet. More recently, several epidemiological studies have supported this association. In a group of 985 Veterans Administration patients, the likelihood of either reaching ESRD or doubling the serum creatinine was higher in those with higher serum phosphate (Schwarz et al., 2006). Zoccali et al. (2011) evaluated the association of serum phosphorus and decline of renal function in the prospective Ramipril Efficacy in Nephropathy trial. They found that independent of treatment, patients with serum phosphate levels in the highest two quartiles progressed significantly faster either to ESRD or to a composite endpoint of doubling of serum creatinine or ESRD compared with patients with serum phosphate levels below the median (Zocalli et al., 2011). Others have shown a good correlation between serum phosphate at baseline and the rate of decline of GFR in a group of 225 CKD patients in the United Kingdom (Chue et al., 2011). Further evidence linking high serum phosphorus and declining renal function is presented in an analysis of 2269 non-CKD patients from the Framingham Heart Study and 13,372 patients from the Third National Health and Nutrition Examination Survey (NHANES III), which showed a significant association between baseline serum phosphate levels and the risk of incident CKD or ESRD (O'Seaghdha et al., 2011).

A recent meta-analysis reviewing studies that focused on health changes over time and serum phosphorus in 12 cohort studies (total of 25,546 participants) reported that 1442 (8.8%) developed kidney failure and 3089 (13.6%) died. Overall, every 1-mg/dL increase in serum phosphate level was independently associated with increased risk of developing kidney failure [hazard ratio, 1.36; 95% confidence interval (CI), 1.20–1.55] and mortality (hazard ratio, 1.20; 95% CI, 1.05–1.37). In this analysis, kidney failure was defined as doubling of serum creatinine level, 50% decline in estimated GFR, or development of ESRD (Da et al., 2015).

Excess Phosphorus Intake and Bone Health

Adequate phosphorus and calcium intake are essential for normal bone mineralization, but an imbalance between calcium and phosphorus intake that can be brought about by diets high in phosphorus and low in calcium can lead to bone loss (Takeda et al., 2014). Dietary surveys show a gradual increase in total phosphorus intake over time in Western cultures, and more importantly, national surveys reveal that approximately 25% of the US population has a calcium-to-phosphorus intake ratio of less than 0.5 in contrast to intake guidelines that recommend ratios of 1–1.2 (Calvo et al., 2014). Many ascribe the increasing dietary phosphorus content to the growing availability of and preference for highly processed foods. Several different phosphate additives are used in processing a wide variety of foods (Calvo and Park, 1996; Winger et al., 2012). When diets are chronically low in calcium and high in phosphorus, more phosphorus than calcium is absorbed because of the high efficiency of phosphorus absorption, especially from inorganic phosphates that characterize the majority of approved phosphorus-containing additives (Takeda et al., 2014; Calvo et al., 2014; Gutierrez et al., 2015). Lower serum calcium and high serum phosphate will increase PTH secretion, which will in turn stimulate bone resorption, thus liberating both minerals from bone into the circulation. Mineral balance is restored in the serum pool through PTH action in the kidneys, which increases urinary phosphate excretion. This homeostatic mechanism has been shown to occur acutely with phosphate loading in young adults (Karkkainen and Lamberg-Allardt, 1996), in dietary intervention studies examining foods containing phosphate additives (Calvo et al., 1990; Gutierrez et al., 2015), and in large cross-sectional population studies (Kemi et al., 2009; Pinheiro et al., 2009). Bone disease can also occur in CKD and is associated with disordered phosphorus homeostasis related to hyperphosphatemia and the impact of sustained FGF23 inhibition of PTH action on bone and other organs (Uribarri, 2007; Gutierrez et al., 2015).

Cardiovascular Disease Risk With High Phosphorus Intake

Our understanding of the CVD risks associated with high-phosphorus intake and hyperphosphatemia comes from studying the events occurring with the changes in phosphate retention during the progression of CKD (Gonzalez-Parra et al., 2012; Menon and Ix, 2013; Kendrick et al., 2011). Hyperphosphatemia in CKD patients has been shown to be significantly associated with CVD and increased mortality and is characterized by changes in compensatory hormones, specifically FGF23, Klotho, and PTH, which have been shown to induce cardiac tissue damage (Gutierrez, 2013a,b). Dietary phosphorus loading in healthy men has been shown to increase serum FGF23 concentrations (Antoniucci et al., 2006) and to induce left ventricular hypertrophy (Faul et al., 2011), a cardiovascular risk factor that has been linked with higher dietary phosphorus intakes in a general population cross-sectional study (Yamamoto et al., 2013). These hormones have been proposed as reliable biomarkers in CKD for testing the efficacy of dietary phosphorus restriction in preventing or delaying the onset of cardiovascular risk factors associated with the high phosphorus content of the Western diet (Gutierrez, 2013a,b; Menon and Ix, 2013).

More recent studies have shown high serum phosphate concentrations to be significantly associated with several risk factors for CVD or mortality in healthy adults without renal impairment (Dhingra et al., 2007; Foley et al., 2009). Cardiac disease risk factors such as carotid-intima-medial thickness and left ventricular hypertrophy were first associated with high serum phosphate in CKD patients (Ishimura et al., 2005; Faul et al., 2011); however, growing evidence supports a link between high phosphorus intake and endothelial dysfunction (Shuto et al., 2009), left ventricular hypertrophy (Yamamoto et al., 2013) and carotid-intima-media thickness (Itkonen et al., 2013) in the general population without renal impairment. Important epidemiologic evidence recently demonstrated significant associations of well-known risk factors for cardiovascular events and mortality with serum phosphate concentrations in the general population. Exploring a large British cohort from the Quality Improvement in Chronic Kidney Disease survey follow-up over 2.5 years, McGovern et al. (2013) reported a significant association in cardiovascular events in people with and without renal impairment. In a prospective study in adults from the NHANES III (1988–1994), Chang et al. (2014) reported a strong association of high phosphorus intake with increased mortality in renal-impaired and functional adults. Seeking fuller understanding of the mechanisms involved in the association between high dietary phosphorus and CVD in CKD patients, Gluba-Brzozka et al. (2016) is studying additional risk factors, including measurement of cardiac contractile dysfunction, mitral and aortic valve calcification, and changes in hypertension.

Foremost among the concerns of nephrologists is a way to identify the source of dietary phosphorus, which may occur naturally in food (organic) or be added during processing (inorganic), which is more rapidly and efficiently absorbed. Because of the difficulty in determining the phosphate additive content of foods, few studies have directly examined the contribution to total phosphorus intake and the health impact of phosphate food additives on the progression of CVD or renal disease (Itkonen et al., 2013; Moore et al., 2015). Establishing the safe cumulative level of these processing ingredients should be a research priority.

Excess Phosphorus Intake and Cancer Risk

In contrast to cardiovascular and bone diseases, understanding and identifying the high dietary intake of phosphorus as a risk factor for cancer did not arise from studies of hyperphosphatemia in CKD. Inorganic phosphates used in food processing were identified as a potential cancer-modifying factor largely through in vitro cell-based studies and those utilizing murine cancer models (Camalier et al., 2013; Anderson, 2013). The changes in paracrine and endocrine cell signaling factors (e.g., osteopontin, FGF23, Klotho) with diets high in inorganic phosphate have been linked to aspects of cancer development and progression. Elevated phosphorus concentrations in cell cultures have been shown to directly activate specific metabolic pathways including ERK1/2-protein kinase C and AKT (protein kinase B)-ERK signaling that promote cell transformation and tumorigenesis in skin and lungs (Camalier et al., 2013; Camalier et al., 2010; Jin et al., 2009; Jin et al., 2007). Alternatively, high dietary inorganic phosphate can indirectly affect lung cancer by altering paracrine or endocrine factors such as osteopontin. Inorganic phosphate consumption has been shown to regulate osteopontin, the action of which induces cancer cell-mediated angiogenesis, an action that supports tumor growth (Beck and Knecht, 2003; Lin et al., 2015). The ability to suppress lung tumorigenesis induced by high-phosphate diets in NaPi-IIb knockout mice further implicates inorganic phosphorus intake as a potential modifiable risk factor for lung cancer prevention (Hong et al., 2013).

Epidemiologists are now beginning to study the association between inorganic phosphate and cancer risks. Overall higher cancer risks with increasing serum phosphate levels have been reported for Swedish men, but a negative association was reported for Swedish women. However, the risks of breast, endometrial, and other endocrine-related cancers were lower in both men and women with higher serum phosphate concentrations (Wulaningsih et al., 2013). In a recent 24-year follow-up study of the association between phosphate intake and advanced-stage and high-grade prostate cancer, Wilson et al. (2015) reported an independent associated increased risk of advanced-stage and high-grade disease 0–8 years after exposure. These early findings from follow-up studies need to be interpreted cautiously because they are limited by inaccurate measurement of phosphorus intake attributed to the incomplete data describing phosphorus content of foods in the nutrient composition databases (Calvo et al., 2014; Wilson et al., 2015).

SUMMARY AND CONCLUSIONS

Phosphorus balance is tightly regulated by the interplay of several hormonal factors, which has recently been more clearly defined. Excess phosphorus intake disrupts this balance, leading to compensatory mechanisms that by themselves may contribute to health hazards, including CVD, bone disease, progression of kidney disease, and even cancer. The negative effect of excess dietary phosphorus on health has been well identified in the field of CKD patients, and maintenance of adequate phosphorus balance is a mainstay in the clinical care of these patients. The recently recognized excessive levels of phosphorus in the food supply together with large cohort epidemiological data linking phosphorus balance and health hazards suggest that this should also be a concern for the general population with healthy kidney function.

DISCLAIMER

The issues and opinions presented in this chapter are strictly those of the author and do not necessarily represent the views and opinions of the US Food and Drug Administration (FDA). Mention of trade names, product labels, or food manufacturers does not constitute endorsement or recommended use by the FDA.

REFERENCES

Anderson, J.J.B., 2013. Potential health concerns of dietary phosphorus: cancer, obesity, and hypertension. Ann. NY Acad. Sci. 1301, 1–8.

Antoniucci, D.M., Yamashita, T., Portale, A.A., 2006. Dietary phosphorus regulates serum fibroblast growth factor-23 concentrations in healthy men. J. Clin. Endocrinol. Metab. 91, 3144–3149.

Beck Jr., G.R., Knecht, N., 2003. Osteopontin regulation by inorganic phosphate is ERK1/2-, protein kinase C-, and proteasome-dependent. J. Biol. Chem. 278 (43), 41921–41929.

Block, G.A., Hulbert-Shearon, T.E., Levin, N.W., Port, F.K., 1998. Association of serum phosphorus and calcium x phosphate product with mortality risk in chronic hemodialysis patients: a national study. Am. J. Kidney Dis. 31 (4), 607–617.

Block, G.A., Klassen, P.S., Lazarus, J.M., Ofsthun, N., Lowrie, E.G., Chertow, G.M., 2004. Mineral metabolism, mortality, and morbidity in maintenance hemodialysis. J. Am. Soc. Nephrol. 15 (8), 2208–2218.

Calvo, M.S., Kumar, R., Heath III, H., 1990. Persistently elevated parathyroid hormone secretion and action in young women after four weeks of ingesting high phosphorus, low calcium diets. J. Clin. Endocrinol. Metab. 70, 1334–1340.

Calvo, M.S., Park, Y.K., 1996. Changing phosphorus content of the U.S. diet: potential for adverse effects on bone. J. Nutr. 126, 1168s–1180s.

Calvo, M.S., Uribarri, J., 2013a. Contributions to total phosphorus intake: all sources considered. Semin. Dial. 26, 54–61.

Calvo, M.S., Uribarri, J., 2013b. Public impact of dietary phosphorus excess on bone and cardiovascular health in the general population. Am. J. Clin. Nutr. 98 (1), 6–15.

Calvo, M.S., Moshfegh, A.J., Tucker, K.L., 2014. Assessing the health impact of phosphorus in the food supply: issues and considerations. Adv. Nutr. 5 (1), 104–113.

Calvo, M.S., Lamberg-Allardt, C.J., 2015. Phosphorus. Adv. Nutr. 6 (6), 860–862.

Camalier, C.E., Young, M.R., Bobe, G., Perella, C.M., Colburn, N.H., Beck Jr., G.R., 2010. Elevated phosphate activates N-ras and promotes cell transformation and skin tumorigenesis. Cancer Prev. Res. 3 (3), 359–370.

Camalier, C.E., Yi, M., Yu, L.R., Hood, B.L., Conrads, K.A., Lee, Y.J., Lin, Y., Garneys, L.M., Bouloux, G.F., Young, M.R., Veenstra, T.D., Stephens, R.M., Colburn, N.H., Conrads, T.P., Beck Jr., G.R., 2013. An integrated understanding of the physiological response to elevated extracellular phosphate. J. Cell Physiol. 228 (7), 1536–1550.

Chang, A.R., Lazo, M., Appel, L.J., Gutierrez, O.M., Grams, M.E., 2014. High dietary intake is associated with all-cause mortality: results from NHANES III. Am. J. Clin. Nutr. 99 (2), 320–327.

Chue, C.D., Edwards, N.C., Davis, L.J., Steeds, R.P., Townend, J.N., Ferro, C.J., 2011. Serum phosphate but not pulse wave velocity predicts decline in renal function in patients with early chronic kidney disease. Nephrol. Dial. Transplant. 26 (8), 2576–2582.

Chue, C.D., Edwards, N.C., Moody, W.E., Steeds, R.P., Townend, J.N., Ferro, C.J., 2012. Serum phosphate is associated with left ventricular mass in patients with chronic kidney disease: a cardiac magnetic resonance study. Heart 98 (3), 219–224.

Da, J., Xie, X., Wolf, M., Disthabanchong, S., Wang, J., Zha, Y., Lv, J., Zhang, L., Wang, H., 2015. Serum phosphorus and progression of CKD and mortality: a meta-analysis of cohort studies. Am. J. Clin. Dis. 66, 258–265.

Dhingra, R., Sullivan, L.M., Fox, C.S., Wang, T.J., D'Agostino Sr., R.B., Gaziano, J.M., Vason, R.S., 2007. Relations of serum phosphorus and calcium levels to the incidence of cardiovascular disease in the community. Arch. Intern. Med. 167, 879–885.

Faul, C., Amaral, A.P., Oskouei, B., Hu, M.C., Sloan, A., Isakova, T., Gutierrez, O.M., Aguillion-Prada, R., Lincoln, J., Hare, J.M., et al., 2011. FGF-23 induces left ventricular hypertrophy. J. Clin. Invest. 121, 4393–4408.

Foley, R.N., Collins, A.J., Herzog, C.A., Ishani, A., Kalra, P.A., 2009. Serum phosphorus levels associated with coronary atherosclerosis in young adults. J. Am. Soc. Nephrol. 20, 397–404.

Gluba-Brzozka, A., Michalska-Kasiczak, M., Franczyk, B., Nocun, M., Toth, P., Banach, M., Rysz, J., 2016. Markers of increased atherosclerotic risk in patients with chronic kidney disease: a preliminary study. Lipids Health Dis. 15 (22). http://dx.doi.org/10.1186/s12944-016-0191-x.

Gonzalez-Parra, E., Tunon, J., Egido, J., Ortiz, A., 2012. Phosphate a stealthier killer than previously thought. Cardiovasc. Pathol. 21 (5), 372–381.

Gutierrez, O.M., 2013a. Sodium and phosphorus–based food additives: persistent but surmountable hurdles in the management of nutrition in chronic kidney disease. Adv. Chronic Kidney Dis. 20, 150–156.

Gutierrez, O.M., 2013b. The connection between dietary phosphorus, cardiovascular disease, and mortality: where we stand and what we need to know. Adv. Nutr. 4, 723–729.

Gutierrez, O.M., Luzuriaga-McPherson, A., Lin, Y., Gilbert, L.C., Ha, S.-W., Beck Jr., G.R., 2015. Impact of phosphorus-based food additives on bone and mineral metabolism. J. Clin. Endocrinol. Metab. 100, 4264–4271.

Hong, S.H., Minai-Tehrani, A., Chang, S.H., Jiang, H.L., Lee, S., Lee, A.Y., Seo, H.W., Chae, C., Beck Jr., G.R., Cho, M.H., 2013. Knockdown of the sodium-dependent phosphate co-transporter 2b (NPT2b) suppresses lung tumorigenesis. PLoS One 8 (10), e77121.

Hong, S.-H., Park, S.-Y., Kim, S., Cho, M.-H., 2015. Biological effects of inorganic phosphate: potential signals of toxicity. J. Toxicol. Sci. 40 (1), 55–69.

Itkonen, S.T., Karp, H., Kemi, V.E., Kokkonen, E.M., Saarnio, E.M., Pekkinen, M.H., Karkkainen, M.U.M., Laitinen, E.K., Turanlahti, M.I., Lamberg-Allardt, C.J.E., 2013. Associations among total and food additive phosphorus intake and carotid intima-media thickness—across-sectional study in a middle-aged population in Southern Finland. Nutr. J. 12, 94.

Ishimura, E., Taniwaki, H., Tabata, T., Tsujimoto, Y., Jono, S., Wmoto, M., Shoji, T., Inaba, M., Inoue, T., Nishizawa, Y., 2005. Cross-sectional association of serum phosphate with carotid intima-medial thickness in hemodialysis patients. Am. J. Kidney Dis. 45, 859–865.

Jin, H., Chang, S.H., Xu, C.X., Shin, J.Y., Chung, Y.S., Park, S.J., Lee, Y.S., An, G.H., Lee, K.H., Cho, M.H., 2007. High dietary inorganic phosphate affects lung through altering protein translation, cell cycle and angiogenesis in developing mice. Toxicol. Sci. 100 (1), 215–223.

Jin, H., Xu, C.X., Lim, H.T., Park, S.J., Shin, J.Y., Chung, Y.S., Park, S.C., Chang, S.H., Youn, H.J., Lee, K.H., Lee, Y.S., Ha, Y.C., Chae, C.H., Beck, G.R., Cho, M.H., 2009. High dietary inorganic phosphate increases lung tumorigenesis and alters AKT signaling. Am. J. Respir. Crit. Care Med. 179 (1), 59–68.

John, G.B., Cheng, C.Y., Kuro-o, M., 2011. Role of Klotho in aging, phosphate metabolism, and CKD. Am. J. Kidney Dis. 58, 127–134.

Karkkainen, M., Lamberg-Allardt, C., 1996. An acute intake of phosphate increases parathyroid hormone secretion and inhibits bone formation on young women. J. Bone Miner. Res. 11 (12), 1905–1912.

Kemi, V.E., Rita, H.R., Karkkainen, M.U.M., Viljakainen, H.T., Laaksonen, M.M., Outila, T.A., Lamberg-Allardt, C.J.E., 2009. Habitual high phosphorus intakes and foods with phosphate additives negatively affect serum parathyroid hormone concentrations: a cross-sectional study on healthy premenopausal women. Public Health Nutr. 12, 1885–1892.

Kendrick, J., Kestenbaum, B., Chonchol, M., 2011. Phosphate and cardiovascular disease. Adv. Chronic Kidney Dis. 18 (2), 113–119.

Kestenbaum, B., Sampson, J.N., Rudser, K.D., Patterson, D.J., Seliger, S.I., Young, B., Sherrad, D.J., Andress, D.L., 2005. Serum phosphate levels and mortality risk among people with chronic kidney disease. J. Am. Soc. Nephrol. 16, 520–528.

Lin, Y., McKinnon, K.E., Ha, S.W., Beck Jr., G.R., 2015. Inorganic phosphate induces cancer cell mediated angiogenesis dependent on forkhead box protein C2(FOXC2) regulated osteopontin expression. Mol. Carcinog. 54 (9), 926–934.

McGovern, A.P., de Lusignan, S., Vlymen, J.V., Liyanage, H., Tomson, C.R., Gallagher, H., Rafiq, M., Jones, S., 2013. Serum phosphate as a risk factor for cardiovascular events in people with and without chronic kidney disease: a large community based cohort study. PLoS One 8 (9), e74996.

Mehrotra, R., Peralta, C.A., Chen, S.C., Li, S., Sachs, M., Shah, A., et al., 2013. No independent association of serum phosphorus with risk for death or progression to end-stage renal disease in a large screen for chronic kidney disease. Kidney Int. 84 (5), 989–997.

Menon, M.C., Ix, J.H., 2013. Dietary phosphorus, serum phosphorus and cardiovascular disease. Ann. NY Acad. Sci. 1301, 21–26.

Moore, L.W., Note, J.V., Gaber, A.O., Suki, W., 2015. Association of dietary phosphate and serum phosphorus concentration by levels of kidney function. Am. J. Clin. Nutr. 102, 444–453.

O'Seaghdha, C.M., Hwang, S.J., Muntner, P., Melamed, M.L., Fox, C.S., 2011. Serum phosphorus predicts incident chronic kidney disease and end-stage renal disease. Nephrol. Dial. Transplant. 26, 2885–2890.

Pinheiro, M.M., Schuch, N.J., Genaro, P.S., Ciconelli, R.M., Ferraz, M.B., Martini, L.A., 2009. Nutrient intakes related to osteoporotic fractures in men and women – the Brazilian Osteoporosis Study (BRAZOS). Nutr. J. 8, 6–8.

Pizzorno, L., 2014. Canaries in the phosphate-toxicity coal mines. Integr. Med. 13 (6), 24–32.

Rennick, A., Kalakeche, R., Seel, L., Shepler, B., 2013. Nicotinic acid and nicotinamide: a review of their use for hyperphosphatemia in dialysis patients. Pharmacotherapy 33, 683–690.

Schwarz, S., Trivedi, B.K., Kalantar-Zadeh, K., Kovesdy, C.P., 2006. Association of disorders in mineral metabolism with progression of chronic kidney disease. Clin. J. Am. Soc. Nephrol. 1 (4), 825–831.

Shuto, E., Taketani, Y., Tanaka, R., Harada, N., Isshiki, M., Saro, M., Nashiki, K., Amo, K., Yamamoto, H., Higashi, Y., Nakaya, Y., Takeda, E., 2009. Dietary phosphorus acutely impairs endothelial function. J. Am. Soc. Nephol. 20 (7), 1504–1512.

Takeda, E., Yamamoto, H., Yamanaka-Okumura, H., Taketani, Y., 2014. Increasing dietary phosphorus intake from food additives: potential for negative impact on bone health. Adv. Nutr. 5, 92–97.

Uribarri, J., 2007. Phosphorus homeostasis in normal and chronic kidney disease patients with special emphasis on dietary phosphorus intake. Semin. Dial. 20 (4), 295–301.

Wilson, K.M., Shui, I.M., Lorelei, A.M., Giovannucci, E., 2015. Calcium and phosphorus intake and prostate cancer risk: a 24- year follow-up study. Am. J. Clin. Nutr. 101, 173–183.

Winger, R.J., Uribarri, J., Lloyd, L., 2012. Phosphorus-containing food additives: an insidious danger for people with chronic kidney disease. Trends Food Sci. Technol. 24, 92–102.

Wulaningsih, W., Michaelsson, K., Garmo, H., Hammar, N., Jungner, I., Walldius, G., Holmberg, L., Van Hemelrijck, M., 2013. Inorganic phosphate and the risk of cancers in the Swedish Amoris study. BMC Cancer. 13, 257. http://www.biomedcentral.com1471-2407/13/257.

Yamamoto, K.T., Robinson-Cohen, C., de Oliveira, M.C., Kostina, A., Mettleton, J.A., Ix, J.H., Ha, N., Eng, J., Lima, J.A.C., Siscovick, D., Weiss, N.S., Kestenbaum, B., 2013. Dietary phosphorus is associated with a significant increase in left ventricular mass. Kidney Int. 83 (4), 707–714.

Zoccali, C., Ruggenenti, P., Perna, A., Leonardis, D., Tripepi, R., Tripepi, G., Mallamaci, F., Remuzzi, G., 2011. Phosphate may promote CKD progression and attenuate reno-protective effect of ACE inhibition. J. Am. Soc. Nephrol. 22, 1923–1930.

Chapter 36

Transcriptional Regulation of Sodium-Phosphate Cotransporter Gene Expression

Ichiro Kaneko[1], Hironori Yamamoto[2], Kayo Ikuta[1], Sawako Tatsumi[1], Hiroko Segawa[1], Ken-ichi Miyamoto[1]

[1]*Tokushima University, Tokushima, Japan;* [2]*Jin-ai University, Fukui, Japan*

INTRODUCTION

Inorganic phosphate (Pi) is an essential nutrient for skeletal development, intracellular signal transduction, and various cellular functions as a component of nucleic acids, and is involved in energy-production mechanisms (Biber et al., 2009; Murer et al., 2000; Tenenhouse, 2007). Most Pi in the body is stored in the bone as hydroxyapatite (up to 85%). The remainder is stored intracellularly as a phospholipid component of cell membranes, nucleic acids, ATP, or ADP (~15%). The small fraction of Pi exists mainly as circulating phosphate in the serum (<1%). Pi in mammals generally consists of HPO_4^{2-} and HPO_4^- (4:1 ratio at pH 7.40; Iheagwara et al., 2013). Both of these Pi forms are physiologically active and are typically monitored by routine blood tests in patients with various disorders, such as inherited rickets or chronic kidney disease. In addition, the Pi contained in most food preservatives is highly absorbable and can lead to deleterious consequences. A high dietary phosphorus burden may cause or aggravate diseases of the endocrine system, bones, and kidneys, and may be a risk factor for cardiovascular disease and cancer (Chang et al., 2014; Kalantar-Zadeh, 2013; Watari et al., 2015). Therefore, extracellular and blood Pi levels are maintained within an optimal range through the interplay of multiple mechanisms, such as renal tubular reabsorption, intestinal absorption, and exchange with intracellular or bone storage. The crucial regulation of blood Pi levels is mediated by sodium-dependent Pi transport in the brush-border membrane (BBM) of epithelial cells of the intestine and renal proximal tubules (Fig. 36.1). The sodium-dependent Pi cotransporters are classified into three categories: type I (SLC17A1/NaPi-I), type II (SLC34A1/NaPi-IIa, SLC34A2/NaPi-IIb, and SLC34A3/NaPi-IIc), and type III (SLC20A1/PiT1, and SLC20A2/PiT2) (Biber et al., 2013; Forster et al., 2013; Miyamoto et al., 2011). Here, we briefly characterize type I–III NaPi cotransporters (Fig. 36.1 and Table 36.1). In addition, we discuss the transcriptional regulation of type II NaPi cotransporters in the kidney and small intestine.

REGULATORY FACTORS OF TYPE II NAPI COTRANSPORTERS

Physiologic and pathophysiologic regulation of renal and intestinal epithelial transport of Pi occurs through alterations of the levels of type II sodium-dependent Pi cotransporters (NaPi-II; SLC34 family). Parathyroid hormone (PTH) rapidly inhibits renal NaPi cotransport through endocytic removal of the transporter from the BBM of epithelial cells of the renal proximal tubules through the PTH receptor pathway (cAMP-dependent protein kinase, protein kinase A, and protein kinase C). This rapid regulation (1–4 h) is not affected by the transcriptional levels of NaPi-II. In contrast, more long-term regulation (12–24 h or more) generally involves transcriptional regulation of NaPi-II transporters (Biber et al., 2000; Segawa et al., 2007; Traebert et al., 2000). Transcriptional regulation of NaPi-II is mediated by FGF23, 1, 25 $(OH)_2$ vitamin D_3, insulin, growth hormones (e.g., EGF), sex hormones (e.g., estrogen, androgen), calcitonin, glucagon, stanniocalcin, atrial natriuretic peptide, glucocorticoids, dopamine, retinoic acid, and liver X receptor analogs. Nonhormonal factors, including fasting, plasma calcium, pH, and volume expansion, also contribute to transcriptional regulation (Table 36.2; Berndt and Kumar, 2007; Murer et al., 2000). In the next section, we describe the influence of these hormonal and nonhormonal factors on the NaPi-II gene promoter and gene expression.

TRANSCRIPTIONAL REGULATION OF SLC34A1/NAPI-IIA GENE

Due to its physiologic importance, the transcriptional mechanisms of the renal NaPi-IIa gene are well investigated. The promoter includes a typical TATA box and various cis-acting elements, and a cAMP-responsive element, as well

Molecular, Genetic, and Nutritional Aspects of Major and Trace Minerals. http://dx.doi.org/10.1016/B978-0-12-802168-2.00036-1

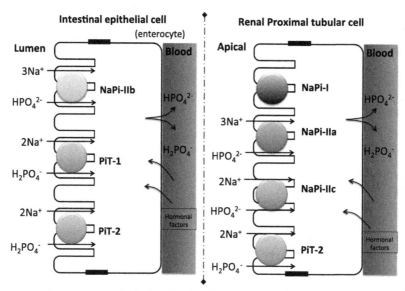

FIGURE 36.1 Sodium-dependent Pi cotransporters. In the intestine, NaPi-IIb predominantly regulates Pi absorption on the luminal brush-border membrane (BBM). PiT1 and PiT2 also transport Pi across the BBM. In the renal proximal tubules, NaPi-IIa, NaPi-IIc, and PiT2 mediate Pi reabsorption across the BBM of renal epithelial cells. The contribution of these transporters might differ according to region or among species.

TABLE 36.1 Features of NaPi Transporters

NaPi Transporters	Substrate	Ratio of Na$^+$-Pi	Km (Pi) pH Dependent	Tissue Distribution (Mainly)
SLC17A1/NaPi-I/OATv1	Na$^+$, HPO$_4^{2-}$, organic anion, Cl$^-$	Not determined	3–6 mM	Kidney, liver, brain
SLC34A1/NaPi-IIa	Na$^+$, HPO$_4^{2-}$, arsenate	3:1	0.1–0.2 mM pH 5.5 < pH 7.5	Kidney
SLC34A2/NaPi-IIb	Na$^+$, HPO$_4^{2-}$, arsenate	3:1	0.05 mM pH 5.5 > pH 7.5	Lung, intestine, colon, liver, testis
SLC34A3/NaPi-IIc	Na$^+$, HPO$_4^{2-}$, arsenate	2:1 electroneutral	0.1–0.2 mM pH 5.5 < pH 7.5	Kidney
SLC20A1/PiT-1 SLC20A2/PiT-2	Na$^+$, HPO$_4^-$, Li$^+$, arsenate	2:1	0.025 mM decreased at high pH	Ubiquitous

as AP-1, AP-2, and SP-1 binding sites upstream of the transcription initiation site (Fig. 36.2) (Shachaf et al., 2000; Taketani et al., 1997).

Chronic dietary Pi restriction leads to an increased NaPi cotransportation rate, along with increased NaPi-IIa protein stability (Markovich et al., 1995) and mRNA synthesis (Takahashi et al., 1998). The phosphate responsive element (PRE) of the NaPi-IIa gene was investigated by electrophoretic mobility-shift assay (EMSA) or luciferase assays. EMSA demonstrated that a signal for the binding element is markedly increased in nuclear protein isolated from the kidney cortex of the mice fed a low-Pi diet (Kido et al., 1999). This binding protein recognizes the consensus sequence 5'-CACGTG-3' (-1012 to -1017 in relation to the TSS) (Fig. 36.2). It is predicted that transcription factor μE3 (TFE3) is the predominant factor in renal extracts that directly binds to the PRE. These findings suggested that TFE3 mediates transcriptional regulation of the NaPi-IIa gene by dietary Pi. This 5'-CACGTG-3' motif is thus thought to be a transcriptional activation region that senses dietary phosphate. TFE3 could work the Pi sensing (Fig. 36.2).

The biologically active form of vitamin D, 1, 25 (OH)$_2$ vitamin D$_3$, regulates NaPi-IIa gene expression in the kidney. In vitamin D–deficient rats, NaPi-IIa protein and mRNA expression are significantly downregulated in the kidney cortex compared with normal rats. Administration of 1, 25 (OH)$_2$ vitamin D$_3$ to vitamin D–deficient rats upregulates the initial rate of Pi uptake as well as the amounts of NaPi-IIa mRNA and protein in the BBM of cortical epithelial cells. Transcriptional

TABLE 36.2 Transcriptional Regulators of Type II NaPi Cotransporters in Mammals

	Promotion	Repression
NaPi-IIa (kidney)	1,25D	High-Pi diet
	Low-Pi diet	FGF23
	T3	EGF
	Retinoic acid	Stanniocalcin
		Estrogen
		LXR agonist
		Potassium-deficiency
NaPi-IIb (intestine)	1,25D	High-Pi diet
	Estrogen	EGF
	Low-Pi diet	Glucocorticoid
		LXR agonist
NaPi-IIc (kidney)	1,25D (human only)	High-Pi diet
	Low-Pi diet	FGF23
	Retinoic acid	LXR agonist
		Potassium-deficiency

activity of a luciferase reporter plasmid containing the promoter region of the human NaPi-IIa gene is markedly increased by $1, 25\,(OH)_2$ vitamin D_3 exposure in COS-7 cells expressing the human vitamin D response (VDR). Deletion and mutation analysis led to the identification of the vitamin D response element (VDRE) as the sequence 5′-GGGGCAgcaAGGGCA-3′ (direct repeat 3 [DR3]) at −1977/−1963 from the transcription initiation site. EMSA revealed that a heterodimer of the VDR and retinoid X receptor (RXR) directly and specifically binds to the DR3 element (Fig. 36.2; Taketani et al., 1998).

The thyroid hormone 3,3′,5-tri-iodothyronine (T_3) upregulates renal NaPi-IIa protein and mRNA expression in rats (Alcalde et al., 1999). Also, hypothyroid mice exhibit low serum Pi and a significant decrease in renal NaPi-IIa protein in the BBM of renal epithelial cells. NaPi-IIa-KO mice have impaired serum Pi responsiveness to T_3 compared with wild-type mice. Functional promoter analysis with a luciferase assay revealed that the transcriptional activity of a reporter gene containing the NaPi-IIa promoter and intron 1 is dependent upon thyroid hormone receptors and is specifically activated by T_3 exposure in OK cells. T_3 enhances NaPi-IIa transcriptional activity through thyroid hormone receptor binding sites on the promoter in a renal cell-specific manner. Renal Pi homeostasis and its activation by T_3 are transcriptionally regulated by thyroid hormone receptor-RXR through a T_3 response region, including multiple DR4 motifs, located in intron 1 of the NaPi-IIa gene (Ishiguro et al., 2010).

Estrogen replacement therapy in postmenopausal women prevents bone resorption and osteoporosis. This therapy, however, also leads to hypophosphatemia in postmenopausal women (Uemura et al., 2000). In ovariectomized rates, estrogen causes hypophosphatemia with hyperphosphaturia (Faroqui et al., 2008). Estrogen directly downregulates NaPi-IIa mRNA and protein levels in the proximal tubules in ovariectomized rat kidney as well as human bone osteosarcoma epithelial (U2OS) cells. This is independent of circulating factors, such as PTH or FGF23. Renal NaPi-IIc and PiT2 are not altered by estrogen (Burris et al., 2015). Interestingly, the effect of estrogen on NaPi-IIa in the rat kidney proximal tubules requires the activation of both estrogen receptor ERα and ERβ. Further studies are needed to determine how ERα and ERβ form functional heterodimers in the proximal tubules or activate synergistic pathways for NaPi-IIa expression. Further, ER binding sites in the NaPi-IIa gene/promoter must be identified.

EGF, a growth hormone, is also an important physiologic regulator of Pi homeostasis. EGF broadly affects cell division, DNA synthesis, cellular differentiation and proliferation, and nutrient absorption (Carpenter and Cohen, 1990). In hyperphosphatemia induced by intestinal ischemia/injury, EGF production and utilization as well as serum Pi levels are increased (Alcalde et al., 1999). EGF reduces NaPi transport, and decreases NaPi-IIa mRNA in OK cells by the phosphorylation of phospholipase C. These effects are blocked by treatment with actinomycin D (transcription

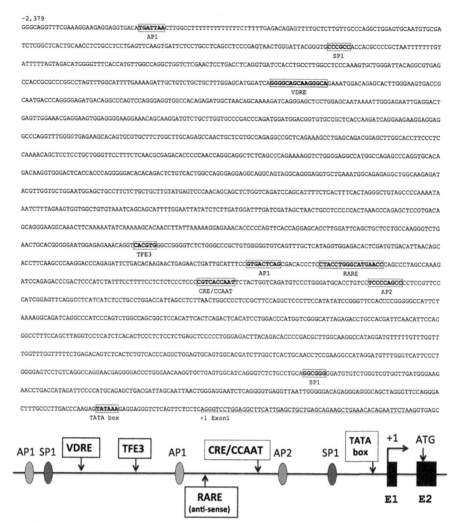

FIGURE 36.2 Locations of transcriptional regulome on human NaPi-IIa gene upstream. Various well-established cis-acting elements or major transcriptional elements are indicated by bold type and boxes (Kido et al., 1999; Masuda et al., 2010; Taketani et al., 1997; Taketani et al., 1998). *Modified from Taketani, Y., Miyamoto, K., Tanaka, K., Katai, K., Chikamori, M., Tatsumi, S., Segawa, H., Yamamoto, H., Morita, K., Takeda, E., 1997. Gene structure and functional analysis of the human Na⁺/phosphate co-transporter. Biochem. J. 324 (Pt 3), 927–934.*

inhibitor). In vivo, EGF reduces renal NaPi transport in rats as well as in OK cells. Inhibition of renal NaPi transport by EGF is associated with decreased NaPi-IIa protein in rat renal epithelial cells. In contrast, EGF does not affect NaPi-IIa mRNA expression in vivo. The physiologic effects on NaPi cotransport would be posttranscriptional (Arar et al., 1999).

Recently, our group showed that an all transretinoic acid (ATRA) and an ATRA receptor (RAR)-specific agonist, TTNPB {4-[*E*-2-(5,6,7,8-tetrahydro-5,5,8,8-tetra-methyl-2-naphtalenyl)-1-propenyl] benzoic acid}, induces both NaPi-IIa mRNA expression and promoter activity in OK cells (Masuda et al., 2010). We also identified retinoic acid responsive elements (RARE) in the human NaPi-IIa promoter, located at bps −855/−839, 5′-GGTTCAtgcccAGGTAG-3′ (DR5) (Fig. 36.2). Since vitamin A–deficient diet reduces the renal expression of NaPi-IIa genes in rats (Masuda et al., 2010), the transcriptional control of NaPi-IIa genes through RARE by vitamin A is physiologically important.

Caldas et al. (2011) revealed that administration of liver X receptor agonists (DMHCA or TO901317) directly reduces renal NaPi-IIa, NaPi-IIc, and intestinal NaPi-IIb mRNA and protein expression with reduced NaPi cotransport in mice. Further studies are needed to identify the liver X receptor responsive element around these genes.

As described earlier, many factors control NaPi-IIa transcriptional expression. The network of nuclear receptor superfamily through enhancer/silencer regions seems to be key for the regulation of phosphate homeostasis through NaPi-IIa expression.

TRANSCRIPTIONAL REGULATION OF SLC34A2/NAPI-IIB GENE

Intestinal absorption of Pi has been characterized in some mammalian and avian species (Murer et al., 2004). Transepithelial Na-dependent Pi cotransport across the BBM of intestinal enterocytes is primarily mediated by NaPi-IIb (Hattenhauer et al., 1999; Radanovic et al., 2005; Sabbagh et al., 2009). Intestinal Pi absorption declines with age in several mammalian species. NaPi-IIb mRNA and protein levels also decrease with age (Arima et al., 2002), and this observation is consistent with functional studies of NaPi cotransport (Ghishan and Arab, 1988; Taufiq et al., 1997). Thus, it seems likely that NaPi-IIb expression contributes to the changes in intestinal Pi absorption observed during development. Methylprednisolone (a synthetic glucocorticoid) treatment reduces intestinal NaPi uptake approximately threefold in 14-day-old mice and this reduction correlates with reduced NaPi-IIb mRNA and protein expression (Arima et al., 2002). Thus, glucocorticoids (GCs) may contribute to the changes in age-dependent NaPi-IIb expression.

A typical TATA box sequence is not found on the human NaPi-IIb gene promoter, but there is a GC-rich region present within the proximal promoter region located 181 bp upstream from the transcription initiation site. A cluster of transcription factor-binding motifs, including motifs for the nuclear factor 1 protein family, was detected within 100 bp of the proximal promoter region (Xu et al., 2001). Nuclear factor 1 interacts with basal transcription factors, such as human transcriptional initiation factor and yeast TATA box-binding protein, to activate transcription. Thus, nuclear factor 1 might contribute to the basal transcriptional activity of the NaPi-IIb promoter without TATA box-binding protein. Other basal transcription factors would be important for NaPi-IIb basal expression.

The effect of 1, 25 $(OH)_2$ vitamin D_3 was investigated in rat intestinal epithelial (RIE) cells. NaPi-IIb gene activity increases ~25% with 1, 25 $(OH)_2$ vitamin D_3 treatment and is blocked with actinomycin D treatment. In vivo studies in suckling rats and in vitro studies in RIE cells show that 1, 25 $(OH)_2$ vitamin D_3 administration increases NaPi-IIb mRNA expression approximately twofold (Xu et al., 2002). Subsequent studies with human NaPi-IIb promoter constructs show that 1, 25 $(OH)_2$ vitamin D_3 increases NaPi-IIb promoter (−1103 bp) activity up to 1.6-fold in RIE cells. Together, these data indicate that the effect of 1, 25 $(OH)_2$ vitamin D_3 on intestinal NaPi-IIb gene expression could be directly mediated by VDR binding to a VDRE in the NaPi-IIb gene promoter. The VDRE is predicted to exist within 1103 bp upstream of the transcriptional initiation site (Xu et al., 2002). VDREs have been identified in many genes, including NaPi-IIa/IIc, osteocalcin, osteopontin, calbindin D-9k, and PTH (Haussler et al., 2013). No classical VDRE sequences (DR3) were identified in the human NaPi-IIb gene promoter region upstream to −1103 bp, even though the NaPi-IIb promoter is responsive to 1, 25 $(OH)_2$ vitamin D_3, suggesting that a novel VDRE may exist in this gene promoter, or alternatively, the 1, 25 $(OH)_2$ vitamin D_3 response could be mediated by an unidentified transacting factor (Saini et al., 2013). On the other hand, our group demonstrated that the enhanced intestinal NaPi absorption caused by 1, 25 $(OH)_2$ vitamin D_3 in adult rodents is mediated by an increase in NaPi-IIb protein, but changes NaPi-IIb mRNA expression were not seen (Katai et al., 1999). Also, NaPi-IIb mRNA levels do not change in mature VDR-KO mice compared with wild-type mice (Segawa et al., 2004). These findings suggest the effect of 1, 25 $(OH)_2$ vitamin D_3 on NaPi-IIb gene expression may be age-specific. Moreover, it is currently unknown if there is nongenomic regulation or indirect effects that control NaPi-IIb protein expression by 1, 25 $(OH)_2$ vitamin D_3.

Dietary Pi also induces transcriptional changes in the NaPi-IIb gene. A low-Pi diet increases NaPi-IIb mRNA levels, whereas a high-Pi diet decreases NaPi-IIb mRNA in the rat and mouse intestine (Giral et al., 2009; Hattenhauer et al., 1999; Katai et al., 1999; Radanovic et al., 2005). NaPi-IIb expression controlled by dietary Pi was previously thought to be induced through a VDR-transcriptional mechanism after an increase in renal 1, 25 $(OH)_2$ vitamin D_3 production. In VDR-KO mice as well as wild-type mice administered a low-Pi diet, however, NaPi-IIb mRNA and protein levels are increased (Capuano et al., 2005; Segawa et al., 2004). In addition, a low-Pi diet should not reduce blood FGF23 levels because blood FGF23 levels are quite low (undetectable) in VDR-KO mice (Kaneko et al., 2011). These findings indicate that dietary Pi controls NaPi-IIb transcriptional expression independent of the 1, 25 $(OH)_2$ vitamin D_3/VDR system and FGF23.

Estrogen treatment also stimulates intestinal NaPi cotransport. Transfection with several NaPi-IIb gene promoter constructs in intestinal cells revealed significant reporter activity. Only the longest promoter construct (−2783 bp) is weakly responsive to estrogen treatment (36% increase compared with vehicle) (Xu et al., 2003b). Only one ER half-binding site has been identified, however, which is located in the −2613/−2609 bp region. This finding implies that a novel estrogen responsive element exists around the gene, or, alternatively, that the estrogen response could be mediated by a transacting factor that acts independently (secondary effects) of the ER in the same cell. Estrogen administration increases intestinal Na-dependent Pi absorption in rats, at least partially through increasing NaPi-IIb mRNA and protein abundance. Estrogen directly increases NaPi-IIb mRNA expression in Caco-2 (human colon) cells. As actinomycin D treatment blocks estrogen-induced NaPi-IIb mRNA expression in Caco-2 cells, estrogen is likely involved in the transcriptional mechanism of NaPi-IIb expression in the intestine (Xu et al., 2003b).

EGF is also an important physiologic regulator of intestinal Pi absorption through modulation of NaPi cotransporter activity. EGF treatment reduces intestinal NaPi-IIb mRNA abundance by 40–50% in rat jejunal mucosa and human intestinal cells

(Caco-2). The promoter activity reduced by EGF treatment is blocked by actinomycin D treatment in Caco-2 cells (Xu et al., 2001). The EGF responsive region is confined to −751 to −727 bp based on EMSA and a reporter gene analysis in Caco-2 cells. According to transcription factor–binding motif searches, the sequence (5′-AACTGG-3′) identified in the NaPi-IIb promoter is recognized by the c-myb transcription factor. Additionally, EMSA analysis with an anti-c-myb antibody indicated that c-myb could directly/specifically bind to this EGF responsive sequence in the NaPi-IIb promoter (Xu et al., 2003a). As described earlier, EGF reduces renal Pi reabsorption by modulating NaPi-IIa protein levels. Taken together, EGF acts to negatively regulate Pi homeostasis by inhibiting both intestinal and renal Pi (re)absorption. It would be independent action of phosphate regulatory factors, such as dietary phosphate, PTH, FGF23 or vitamin D. We need further study to reveal the regional or systemic role.

The pregnane X receptor (PXR) controls phosphate homeostasis. PXR-KO mice exhibit an inherited phenotype of hypophosphatemia with an associated bone disorder. The lack of PXR results in severe suppression of NaPi-IIb mRNA and protein expression, thus leading to the development of hypophosphatemia in PXR-KO mice (Konno et al., 2010). Subsequent studies showed that PXR and its agonists, rifampicin or pregnenolone 16α-carbonitrile, activate the human and mouse NaPi-IIb (human; −2.9 kb, mouse; −1.4 kb) promoters and enhance mRNA levels for 24 h in Caco-2 cells. Hence, PXR seems to directly activate transcription of the NaPi-IIb gene via an ER-6 motif located at −1202/−1185 bp promoter region. On the other hand, sulforaphane, a PXR antagonist, suppresses NaPi-IIb gene expression in human colon epithelial cells (Chen and Jiang, 2014). Such natural nutritional factors that act to reduce NaPi-IIb expression could be utilized therapeutically to inhibit hyperphosphatemia in end-stage renal disease.

TRANSCRIPTIONAL REGULATION OF THE SLC34A3/NAPI-IIC GENE

In rats and mice, NaPi-IIc expression (mRNA and protein) is highest just after weaning at four weeks of age (growth stage), then expression decreases gradually as the rats and mice age (Kaneko et al., 2011; Segawa et al., 2002). The mechanisms underlying the developmental changes are unclear, but growth factors might be involved in this phenomenon. A chronic low-Pi diet in rats leads to increased levels of NaPi-IIc transporter protein and mRNA, whereas a chronic high-Pi diet suppresses NaPi-IIc expression. In contrast to NaPi-IIa protein levels, acute dietary-induced changes in the NaPi-IIc transporter are not observed (Segawa et al., 2005).

ATRA addition increases the rate of Na-dependent Pi transport in OK cells. NaPi-IIc mRNA and protein levels are clearly decreased in rats fed a vitamin A–deficient diet for seven weeks, similar to NaPi-IIa expression in the kidney. The transcriptional activity of a luciferase reporter coupled to the promoter region of the human NaPi-IIc (−900 bp) gene was markedly increased by exposure to ATRA or the retinoic acid analog TTNPB in OK cells expressing RAR and RXR, in a dose-dependent manner. Further, two functional RAREs were identified in the NaPi-IIc promoter region; a proximal RARE located at −338/−322 (DR-5; 5′-GTGTGAcccaaGAGACA-3′), and a distal RARE located at −558/−575 (DR-6; 5′-GGTTCAgatccgGGTTCA-3′) (Masuda et al., 2010).

Interestingly, the half-site sequence (5′-GGTTCA-3′: −563 to −558) of distal RARE overlaps that of VDRE (DR-3; 5′-AGGT-CAgagGGTTCA-3′) in the human NaPi-IIc gene. The overlapping half-site sequences (5′-GGTTCA-3′) of both are functionally active motifs for transcriptional regulation of the human NaPi-IIc gene by ATRA and 1, 25 (OH)$_2$ vitamin D$_3$ in both independent and additive regulation. A subsequent study revealed that the VDRE (−563/−549) in the human NaPi-IIc promoter is not conserved between human and mouse. In animal experiments, renal NaPi-IIc mRNA in conventional VDR-KO mice is not altered at any age compared with WT mice (Kaneko et al., 2011). Also, renal NaPi-IIc mRNA is not altered in wild-type rat or mice administered 1, 25 (OH)$_2$ vitamin D$_3$. Human NaPi-IIc must be properly controlled by 1, 25 (OH)$_2$ vitamin D$_3$ because humans must be able to adapt to wide variations in the diet or environment. The impact of 1, 25 (OH)$_2$ vitamin D$_3$ on renal NaPi-IIc is not fully understood. NaPi-IIc-KO mice exhibit no phosphate imbalance or rickets/osteomalacia, but have high serum calcium and circulating 1, 25 (OH)$_2$ vitamin D$_3$ levels, and high urinary calcium excretion during growth periods, suggesting that in mice, NaPi-IIc is more relevant to calcium homeostasis, with phosphate reabsorption predominantly controlled by NaPi-IIa (Segawa et al., 2009a,b). On the other hand, mice with kidney-specific NaPi-IIc-KO do not have normal calcium homeostasis (Myakala et al., 2014). This suggests that the calcium-related phenotypes observed in conventional KO mice are not caused by the inactivation of renal NaPi-IIc. Nevertheless, NaPi-IIc is thought to have a main role in renal phosphate reabsorption in humans. Further elucidation of the details of NaPi-IIc gene regulation are needed.

CONCLUSION AND PERSPECTIVE

This chapter highlights the transcriptional regulation of type II NaPi cotransporters. Dysfunction of renal phosphate excretion increases serum Pi levels and induces a premature-aging phenotype, like in the klotho-KO mouse (Kuro-o et al., 1997; Kurosu et al., 2005) and FGF23-KO mouse (Shimada et al., 2004). Klotho contributes renal phosphate excretion since it mediates FGF23 signal as the co-receptor with FGF receptor at kidney. The aging-like phenotypes in klotho-KO and FGF23-KO mice are primarily caused by toxicity due to abnormally high phosphate, calcium, and/or 1, 25 $(OH)_2$ vitamin D_3 levels. A low phosphate diet can rescue these aging-like phenotypes despite its activity to further increase already high serum calcium and vitamin D levels, suggesting that excess Pi in the body causes the aging-like phenotypes in both klotho-KO and FGF23-KO mice (Ohnishi et al., 2009; Ohnishi and Razzaque, 2010; Streicher et al., 2012). In some mammals, longevity and serum Pi levels are negatively correlated ($R^2 = 0.8942$) (Kuro-o, 2010). Life span in klotho-transgenic mice is ~20–30% longer than that in wild-type mice (Kurosu et al., 2005). It may be that our intake of phosphate is excessive due to the consumption of processed food with additives. Elucidating the mechanisms of Pi homeostasis is very important in maintaining health. And, to boost klotho expression may lead increased longevity (Chang et al., 2014).

The development of new technologies, such as a Next Generation Sequencers, has enabled molecular transcriptional studies. Chromatin Immunoprecipitation (ChIP)-Sequence analyses will reveal more genome-wide information for transcriptional gene control using Next Generation Sequencers. Several binding or interacting proteins are also present on genomic DNA and these proteins or their modifications determine the cell-specific and gene-specific actions of ligands. Pike et al. used ChIP to reveal various patterns of the VDRE region at 5′ upstream and 3′ downstream gene regions. They also demonstrated that CYP24A1, which is well known as a gene encoded enzyme for degradation of 1, 25 $(OH)_2$ vit D_3 and also most effective target gene for 1, 25 $(OH)_2$ vit D_3, is controlled by several hundreds of distant enhancers (Meyer et al., 2010). The physical distance between a gene and an enhancer/silencer based on the three-dimensional genome conformation and not simply the DNA base pair distance, may be important. Transcription analysis should be investigated more widely and systematically, and gradually shifted from in vitro to in vivo analysis.

Recent progress in chromatin modification or epigenomic gene control studies has demonstrated that chromatin reorganization (three-dimensional conformation) is an essential process for dynamic transcriptional control. For example, various posttranslational modifications of DNA-binding proteins, such as phosphorylation, methylation, acetylation, ubiquitylation, deamination, ADP ribosylation, SUMOylation, and proline isomerization, are important for the conformation of the chromatin structure (Kato et al., 2011). Additionally, insulator proteins (CTCF-binding proteins, cohesion, etc.) are involved in DNA looping and chromatin conformation on genomic DNA (Amouyal, 2010; Herold et al., 2012; Merkenschlager and Odom, 2013). The properties of regulators and their interactions in these systems are mostly unknown. Further studies are needed to determine the precise roles of these coregulators in chromatin reorganization, and how these factors may influence transcription of genes encoding NaPi cotransporters.

REFERENCES

Alcalde, A.I., Sarasa, M., Raldua, D., Aramayona, J., Morales, R., Biber, J., Murer, H., Levi, M., Sorribas, V., 1999. Role of thyroid hormone in regulation of renal phosphate transport in young and aged rats. Endocrinology 140, 1544–1551.

Amouyal, M., 2010. Gene insulation. Part II: natural strategies in vertebrates. Biochem. Cell Biol. 88, 885–898.

Arar, M., Zajicek, H.K., Elshihabi, I., Levi, M., 1999. Epidermal growth factor inhibits Na-Pi cotransport in weaned and suckling rats. Am. J. Physiol. 276, F72–F78.

Arima, K., Hines, E.R., Kiela, P.R., Drees, J.B., Collins, J.F., Ghishan, F.K., 2002. Glucocorticoid regulation and glycosylation of mouse intestinal type IIb Na-P(i) cotransporter during ontogeny. Am. J. Physiol. Gastrointest. Liver Physiol. 283, G426–G434.

Berndt, T., Kumar, R., 2007. Phosphatonins and the regulation of phosphate homeostasis. Annu. Rev. Physiol. 69, 341–359.

Biber, J., Hernando, N., Forster, I., 2013. Phosphate transporters and their function. Annu. Rev. Physiol. 75, 535–550.

Biber, J., Hernando, N., Forster, I., Murer, H., 2009. Regulation of phosphate transport in proximal tubules. Pflugers Arch. 458, 39–52.

Biber, J., Hernando, N., Traebert, M., Volkl, H., Murer, H., 2000. Parathyroid hormone-mediated regulation of renal phosphate reabsorption. Nephrol. Dial. Transpl. 15 (Suppl. 6), 29–30.

Burris, D., Webster, R., Sheriff, S., Faroqui, R., Levi, M., Hawse, J.R., Amlal, H., 2015. Estrogen directly and specifically downregulates NaPi-IIa through the activation of both estrogen receptor isoforms (ERalpha and ERbeta) in rat kidney proximal tubule. Am. J. Physiol. Ren. Physiol. 308, F522–F534.

Caldas, Y.A., Giral, H., Cortazar, M.A., Sutherland, E., Okamura, K., Blaine, J., Sorribas, V., Koepsell, H., Levi, M., 2011. Liver X receptor-activating ligands modulate renal and intestinal sodium-phosphate transporters. Kidney Int. 80, 535–544.

Capuano, P., Radanovic, T., Wagner, C.A., Bacic, D., Kato, S., Uchiyama, Y., St-Arnoud, R., Murer, H., Biber, J., 2005. Intestinal and renal adaptation to a low-Pi diet of type II NaPi cotransporters in vitamin D receptor- and 1alphaOHase-deficient mice. Am. J. Physiol. Cell Physiol. 288, C429–C434.

Carpenter, G., Cohen, S., 1990. Epidermal growth factor. J. Biol. Chem. 265, 7709–7712.

Chang, A.R., Lazo, M., Appel, L.J., Gutierrez, O.M., Grams, M.E., 2014. High dietary phosphorus intake is associated with all-cause mortality: results from NHANES III. Am. J. Clin. Nutr. 99, 320–327.

Chen, L., Jiang, H., 2014. Potential therapeutic target for hyperphosphatemia: sulforaphane reduces the expression of the SLC34a2 gene through pregnane X receptor. J. Am. Soc. Nephrol. 25 SA-PO048.

Faroqui, S., Levi, M., Soleimani, M., Amlal, H., 2008. Estrogen downregulates the proximal tubule type IIa sodium phosphate cotransporter causing phosphate wasting and hypophosphatemia. Kidney Int. 73, 1141–1150.

Forster, I.C., Hernando, N., Biber, J., Murer, H., 2013. Phosphate transporters of the SLC20 and SLC34 families. Mol. Asp. Med. 34, 386–395.

Ghishan, F.K., Arab, N., 1988. Phosphate transport by intestinal endoplasmic reticulum during maturation. Pediatr. Res. 23, 612–615.

Giral, H., Caldas, Y., Sutherland, E., Wilson, P., Breusegem, S., Barry, N., Blaine, J., Jiang, T., Wang, X.X., Levi, M., 2009. Regulation of rat intestinal Na-dependent phosphate transporters by dietary phosphate. Am. J. Physiol. Ren. Physiol. 297, F1466–F1475.

Hattenhauer, O., Traebert, M., Murer, H., Biber, J., 1999. Regulation of small intestinal Na-P(i) type IIb cotransporter by dietary phosphate intake. Am. J. Physiol. 277, G756–G762.

Haussler, M.R., Whitfield, G.K., Kaneko, I., Haussler, C.A., Hsieh, D., Hsieh, J.C., Jurutka, P.W., 2013. Molecular mechanisms of vitamin D action. Calcif. Tissue Int. 92, 77–98.

Herold, M., Bartkuhn, M., Renkawitz, R., 2012. CTCF: insights into insulator function during development. Development 139, 1045–1057.

Iheagwara, O.S., Ing, T.S., Kjellstrand, C.M., Lew, S.Q., 2013. Phosphorus, phosphorous, and phosphate. Hemodial. Int. 17, 479–482.

Ishiguro, M., Yamamoto, H., Masuda, M., Kozai, M., Takei, Y., Tanaka, S., Sato, T., Segawa, H., Taketani, Y., Arai, H., Miyamoto, K., Takeda, E., 2010. Thyroid hormones regulate phosphate homoeostasis through transcriptional control of the renal type IIa sodium-dependent phosphate co-transporter (NaPi-IIa) gene. Biochem. J. 427, 161–169.

Kalantar-Zadeh, K., 2013. Patient education for phosphorus management in chronic kidney disease. Patient Prefer. Adherence 7, 379–390.

Kaneko, I., Segawa, H., Furutani, J., Kuwahara, S., Aranami, F., Hanabusa, E., Tominaga, R., Giral, H., Caldas, Y., Levi, M., Kato, S., Miyamoto, K., 2011. Hypophosphatemia in vitamin D receptor null mice: effect of rescue diet on the developmental changes in renal Na+ -dependent phosphate cotransporters. Pflugers Arch. 461, 77–90.

Katai, K., Miyamoto, K., Kishida, S., Segawa, H., Nii, T., Tanaka, H., Tani, Y., Arai, H., Tatsumi, S., Morita, K., Taketani, Y., Takeda, E., 1999. Regulation of intestinal Na+-dependent phosphate co-transporters by a low-phosphate diet and 1,25-dihydroxyvitamin D3. Biochem. J. 343 (Pt 3), 705–712.

Kato, S., Yokoyama, A., Fujiki, R., 2011. Nuclear receptor coregulators merge transcriptional coregulation with epigenetic regulation. Trends Biochem. Sci. 36, 272–281.

Kido, S., Miyamoto, K., Mizobuchi, H., Taketani, Y., Ohkido, I., Ogawa, N., Kaneko, Y., Harashima, S., Takeda, E., 1999. Identification of regulatory sequences and binding proteins in the type II sodium/phosphate cotransporter NPT2 gene responsive to dietary phosphate. J. Biol. Chem. 274, 28256–28263.

Konno, Y., Moore, R., Kamiya, N., Negishi, M., 2010. Nuclear xenobiotic receptor PXR-null mouse exhibits hypophosphatemia and represses the Na/Pi-cotransporter SLC34A2. Pharmacogenet. Genomics 20, 9–17.

Kuro-o, M., 2010. Klotho. Pflugers Arch. 459, 333–343.

Kuro-o, M., Matsumura, Y., Aizawa, H., Kawaguchi, H., Suga, T., Utsugi, T., Ohyama, Y., Kurabayashi, M., Kaname, T., Kume, E., Iwasaki, H., Iida, A., Shiraki-Iida, T., Nishikawa, S., Nagai, R., Nabeshima, Y.I., 1997. Mutation of the mouse klotho gene leads to a syndrome resembling ageing. Nature 390, 45–51.

Kurosu, H., Yamamoto, M., Clark, J.D., Pastor, J.V., Nandi, A., Gurnani, P., McGuinness, O.P., Chikuda, H., Yamaguchi, M., Kawaguchi, H., Shimomura, I., Takayama, Y., Herz, J., Kahn, C.R., Rosenblatt, K.P., Kuro-o, M., 2005. Suppression of aging in mice by the hormone Klotho. Science 309, 1829–1833.

Markovich, D., Verri, T., Sorribas, V., Forgo, J., Biber, J., Murer, H., 1995. Regulation of opossum kidney (OK) cell Na/Pi cotransport by Pi deprivation involves mRNA stability. Pflugers Arch. 430, 459–463.

Masuda, M., Yamamoto, H., Kozai, M., Tanaka, S., Ishiguro, M., Takei, Y., Nakahashi, O., Ikeda, S., Uebanso, T., Taketani, Y., Segawa, H., Miyamoto, K., Takeda, E., 2010. Regulation of renal sodium-dependent phosphate co-transporter genes (NaPi-IIa and NaPi-IIc) by all-trans-retinoic acid and its receptors. Biochem. J. 429, 583–592.

Merkenschlager, M., Odom, D.T., 2013. CTCF and cohesin: linking gene regulatory elements with their targets. Cell 152, 1285–1297.

Meyer, M.B., Goetsch, P.D., Pike, J.W., 2010. A downstream intergenic cluster of regulatory enhancers contributes to the induction of CYP24A1 expression by 1alpha,25-dihydroxyvitamin D3. J. Biol. Chem. 285, 15599–15610.

Miyamoto, K., Haito-Sugino, S., Kuwahara, S., Ohi, A., Nomura, K., Ito, M., Kuwahata, M., Kido, S., Tatsumi, S., Kaneko, I., Segawa, H., 2011. Sodium-dependent phosphate cotransporters: lessons from gene knockout and mutation studies. J. Pharm. Sci. 100, 3719–3730.

Murer, H., Forster, I., Biber, J., 2004. The sodium phosphate cotransporter family SLC34. Pflugers Arch. 447, 763–767.

Murer, H., Hernando, N., Forster, I., Biber, J., 2000. Proximal tubular phosphate reabsorption: molecular mechanisms. Physiol. Rev. 80, 1373–1409.

Myakala, K., Motta, S., Murer, H., Wagner, C.A., Koesters, R., Biber, J., Hernando, N., 2014. Renal-specific and inducible depletion of NaPi-IIc/Slc34a3, the cotransporter mutated in HHRH, does not affect phosphate or calcium homeostasis in mice. Am. J. Physiol. Ren. Physiol. 306, F833–F843.

Ohnishi, M., Nakatani, T., Lanske, B., Razzaque, M.S., 2009. Reversal of mineral ion homeostasis and soft-tissue calcification of klotho knockout mice by deletion of vitamin D 1alpha-hydroxylase. Kidney Int. 75, 1166–1172.

Ohnishi, M., Razzaque, M.S., 2010. Dietary and genetic evidence for phosphate toxicity accelerating mammalian aging. FASEB J. 24, 3562–3571.

Radanovic, T., Wagner, C.A., Murer, H., Biber, J., 2005. Regulation of intestinal phosphate transport. I. Segmental expression and adaptation to low-P(i) diet of the type IIb Na(+)-P(i) cotransporter in mouse small intestine. Am. J. Physiol. Gastrointest. Liver Physiol. 288, G496–G500.

Sabbagh, Y., O'Brien, S.P., Song, W., Boulanger, J.H., Stockmann, A., Arbeeny, C., Schiavi, S.C., 2009. Intestinal NaPi-IIb plays a major role in phosphate absorption and homeostasis. J. Am. Soc. Nephrol. 20, 2348–2358.

Saini, R.K., Kaneko, I., Jurutka, P.W., Forster, R., Hsieh, A., Hsieh, J.C., Haussler, M.R., Whitfield, G.K., 2013. 1,25-dihydroxyvitamin D(3) regulation of fibroblast growth factor-23 expression in bone cells: evidence for primary and secondary mechanisms modulated by leptin and interleukin-6. Calcif. Tissue Int. 92, 339–353.

Segawa, H., Kaneko, I., Takahashi, A., Kuwahata, M., Ito, M., Ohkido, I., Tatsumi, S., Miyamoto, K., 2002. Growth-related renal type II Na/Pi cotransporter. J. Biol. Chem. 277, 19665–19672.

Segawa, H., Kaneko, I., Yamanaka, S., Ito, M., Kuwahata, M., Inoue, Y., Kato, S., Miyamoto, K., 2004. Intestinal Na-P(i) cotransporter adaptation to dietary P(i) content in vitamin D receptor null mice. Am. J. Physiol. Ren. Physiol. 287, F39–F47.

Segawa, H., Onitsuka, A., Furutani, J., Kaneko, I., Aranami, F., Matsumoto, N., Tomoe, Y., Kuwahata, M., Ito, M., Matsumoto, M., Li, M., Amizuka, N., Miyamoto, K., 2009a. NaPi-IIa and NaPi-IIc in mice play distinct and synergistic roles in inorganic phosphate metabolism and skeletal development. Am. J. Physiol. Ren. Physiol. 297, F671–F678.

Segawa, H., Onitsuka, A., Kuwahata, M., Hanabusa, E., Furutani, J., Kaneko, I., Tomoe, Y., Aranami, F., Matsumoto, N., Ito, M., Matsumoto, M., Li, M., Amizuka, N., Miyamoto, K., 2009b. Type IIc sodium-dependent phosphate transporter regulates calcium metabolism. J. Am. Soc. Nephrol. 20, 104–113.

Segawa, H., Yamanaka, S., Ito, M., Kuwahata, M., Shono, M., Yamamoto, T., Miyamoto, K., 2005. Internalization of renal type IIc Na-Pi cotransporter in response to a high-phosphate diet. Am. J. Physiol. Ren. Physiol. 288, F587–F596.

Segawa, H., Yamanaka, S., Onitsuka, A., Tomoe, Y., Kuwahata, M., Ito, M., Taketani, Y., Miyamoto, K., 2007. Parathyroid hormone-dependent endocytosis of renal type IIc Na-Pi cotransporter. Am. J. Physiol. Ren. Physiol. 292, F395–F403.

Shachaf, C., Skorecki, K.L., Tzukerman, M., 2000. Role of AP2 consensus sites in regulation of rat Npt2 (sodium-phosphate cotransporter) promoter. Am. J. Physiol. Ren. Physiol. 278, F406–F416.

Shimada, T., Kakitani, M., Yamazaki, Y., Hasegawa, H., Takeuchi, Y., Fujita, T., Fukumoto, S., Tomizuka, K., Yamashita, T., 2004. Targeted ablation of Fgf23 demonstrates an essential physiological role of FGF23 in phosphate and vitamin D metabolism. J. Clin. Invest. 113, 561–568.

Streicher, C., Zeitz, U., Andrukhova, O., Rupprecht, A., Pohl, E., Larsson, T.E., Windisch, W., Lanske, B., Erben, R.G., 2012. Long-term Fgf23 deficiency does not influence aging, glucose homeostasis, or fat metabolism in mice with a nonfunctioning vitamin D receptor. Endocrinology 153, 1795–1805.

Takahashi, F., Morita, K., Katai, K., Segawa, H., Fujioka, A., Kouda, T., Tatsumi, S., Nii, T., Taketani, Y., Haga, H., Hisano, S., Fukui, Y., Miyamoto, K.I., Takeda, E., 1998. Effects of dietary Pi on the renal Na$^+$-dependent Pi transporter NaPi-2 in thyroparathyroidectomized rats. Biochem. J. 333 (Pt 1), 175–181.

Taketani, Y., Miyamoto, K., Tanaka, K., Katai, K., Chikamori, M., Tatsumi, S., Segawa, H., Yamamoto, H., Morita, K., Takeda, E., 1997. Gene structure and functional analysis of the human Na$^+$/phosphate co-transporter. Biochem. J. 324 (Pt 3), 927–934.

Taketani, Y., Segawa, H., Chikamori, M., Morita, K., Tanaka, K., Kido, S., Yamamoto, H., Iemori, Y., Tatsumi, S., Tsugawa, N., Okano, T., Kobayashi, T., Miyamoto, K., Takeda, E., 1998. Regulation of type II renal Na$^+$-dependent inorganic phosphate transporters by 1,25-dihydroxyvitamin D3. Identification of a vitamin D-responsive element in the human NAPi-3 gene. J. Biol. Chem. 273, 14575–14581.

Taufiq, S., Collins, J.F., Ghishan, F.K., 1997. Posttranscriptional mechanisms regulate ontogenic changes in rat renal sodium-phosphate transporter. Am. J. Physiol. 272, R134–R141.

Tenenhouse, H.S., 2007. Phosphate transport: molecular basis, regulation and pathophysiology. J. Steroid Biochem. Mol. Biol. 103, 572–577.

Traebert, M., Volkl, H., Biber, J., Murer, H., Kaissling, B., 2000. Luminal and contraluminal action of 1-34 and 3-34 PTH peptides on renal type IIa Na-P(i) cotransporter. Am. J. Physiol. Ren. Physiol. 278, F792–F798.

Uemura, H., Irahara, M., Yoneda, N., Yasui, T., Genjida, K., Miyamoto, K.I., Aono, T., Takeda, E., 2000. Close correlation between estrogen treatment and renal phosphate reabsorption capacity. J. Clin. Endocrinol. Metab. 85, 1215–1219.

Watari, E., Taketani, Y., Kitamura, T., Tanaka, T., Ohminami, H., Abuduli, M., Harada, N., Yamanaka-Okumura, H., Yamamoto, H., Takeda, E., 2015. Fluctuating plasma phosphorus level by changes in dietary phosphorus intake induces endothelial dysfunction. J. Clin. Biochem. Nutr. 56, 35–42.

Xu, H., Bai, L., Collins, J.F., Ghishan, F.K., 2002. Age-dependent regulation of rat intestinal type IIb sodium-phosphate cotransporter by 1,25-(OH)(2) vitamin D(3). Am. J. Physiol. Cell Physiol. 282, C487–C493.

Xu, H., Collins, J.F., Bai, L., Kiela, P.R., Ghishan, F.K., 2001. Regulation of the human sodium-phosphate cotransporter NaP(i)-IIb gene promoter by epidermal growth factor. Am. J. Physiol. Cell Physiol. 280, C628–C636.

Xu, H., Inouye, M., Hines, E.R., Collins, J.F., Ghishan, F.K., 2003a. Transcriptional regulation of the human NaPi-IIb cotransporter by EGF in Caco-2 cells involves c-myb. Am. J. Physiol. Cell Physiol. 284, C1262–C1271.

Xu, H., Uno, J.K., Inouye, M., Xu, L., Drees, J.B., Collins, J.F., Ghishan, F.K., 2003b. Regulation of intestinal NaPi-IIb cotransporter gene expression by estrogen. Am. J. Physiol. Gastrointest. Liver Physiol. 285, G1317–G1324.

Part X

Selenium

Chapter 37

Selenium: Basic Nutritional Aspects

Wen-Hsing Cheng[1], Xin Gen Lei[2]

[1]*Mississippi State University, Mississippi, MS, United States;* [2]*Cornell University, Ithaca, NY, United States*

Selenium is an essential micronutrient, and its nutritional essentiality is exerted mainly through functions of multiple seleno-proteins and/(or) small selenium metabolites in the body. This metalloid element is incorporated into plants in either inorganic (e.g., selenide, selenite, and selenate) or organic forms (e.g., selenomethionine, selenocysteine, and methylseleninic acid). Accumulation of selenium in plants is usually dependent on soil selenium contents that, for example, range from 0.02 to 1500 mg/kg in the United States (Lakin, 1961). Thus, grazing animals and living humans may suffer from selenium deficiency or toxicity, depending on their geographic locations and food sources. Meanwhile, supranutritional or pharmacological levels of selenium intake may improve or damage health, effects varying with specific physiological or pathological conditions.

Selenium is present in three types of proteins (Fig. 37.1). The first group, selenoproteins, contain selenocysteine, the so-called 21st amino acid (detailed later in the chapter). The second group, selenium-binding proteins, include selenium-binding protein-1 and 14-kDa fatty acid binding protein. These proteins have no selenocysteine residues, but are able to bind selenium specifically (Bansal et al., 1989a,b). The third group comprises proteins typically rich in cysteine and methionine whose sulfur can be replaced by selenium in a nonspecific manner.

NUTRITIONAL ESSENTIALITY AND METABOLISM OF SELENIUM

After selenium was initially discovered during the study of sulfuric acid production (Berzelius and de Lettre, 1818), it remained as a toxic element to living organisms for about 140 years. Although two earlier studies demonstrated beneficial roles of selenium in improving chick growth (Poley et al., 1941) and dehydrogenase activity of *Escherichia coli* (Pinsent, 1954), Schwarz and Foltz (1957) were given the credit for discovering the nutritional essentiality of selenium by producing liver necrosis in selenium-deficient rats. Subsequently, symptoms of dietary selenium/vitamin E deficiencies were observed in other species. These deficiencies caused exudative diathesis (Patterson et al., 1957; Schwarz et al., 1957) and pancreatic atrophy (Whitacre et al., 1987) in chicks; liver necrosis (Michel et al., 1969) and mulberry heart disease in pigs (Van Vleet et al., 1970); and muscular dystrophy, also known as "white muscle disease" or "stiff lamb disease," a myopathy found in lambs (Hogue, 1958; Muth et al., 1958). In 1972, the Food and Drug Administration approved sodium selenite and selenate as feed additives to meet selenium requirements in all these species. These additives were equivalent to 0.1 mg selenium/kg of feed for cattle, sheep, chickens, ducks, and swine, and 0.2 mg/kg for turkeys, as administered through injections, salt-mineral supplements, or selenium-fortified feeds (Ullrey et al., 1983). Recently, selenium-enriched yeast has been approved as a feed additive, and 0.3 mg Se/kg of diet has been allowed for several species.

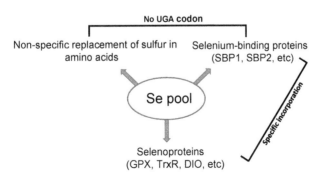

FIGURE 37.1 **Categories of Selenium-Containing Proteins.** Selenium can replace sulfur on cysteine and methionine residues of any proteins nonspecifically. Based on poorly understood mechanisms, selenium can specifically bind a group of proteins known as selenium-binding proteins. Selenium can be specifically incorporated into selenoproteins via a cotranslational mechanism through decoding UGA.

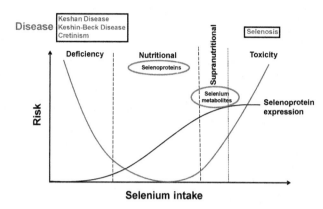

FIGURE 37.2 The "U-shape" response of disease risks to various levels of dietary selenium intake.

A selenium deficiency–related human disease Keshan cardiomyopathy, was formally reported in 1979 by the Keshan Disease Research Group in China, followed by reports of Kashin–Beck disease (Moreno-Reyes et al., 1998; Yang et al., 1988) and cretinism (Vanderpas et al., 1990). In the USA, the recommended dietary allowances for selenium are 55 µg/day for both males and females at 14 years of age or older, increased to 60 and 70 µg/day during pregnancy and lactation, respectively. Selenium can exert distinctive health impacts at deficient, nutritional, supranutritional, and toxic levels of intake (Fig. 37.2). Selenium status in many species can be assessed by plasma or serum selenium concentration, glutathione peroxidase (GPX) activity, and selenoprotein P (Sepp1) concentrations (Combs et al., 2011).

Selenomethionine represents the major dietary source of selenium in foods for humans and in foodstuffs for free-living and domestic animals. Absorption of selenium is believed to take place in the small intestine via simple or carrier-mediated diffusion and/(or) active transporters. Nonetheless, the exact mechanisms remain unclear. However, selenomethionine may be absorbed through methionine transporters, as selenomethionine competes with methionine for intracellular transport in Caco-2 colorectal cells (Thiry et al., 2013). The chemical forms of selenium affect its bioavailability through not only absorption but also subsequent metabolism and retention (Van Dael et al., 2002; Wastney et al., 2011). Ingesting selenomethionine elevates plasma selenium concentrations to greater extents than inorganic selenium in subjects over a wide range body of selenium status (Burk et al., 2006; Xia et al., 2005). The nonspecific incorporation of selenium into sulfur-containing amino acids is largely attributed to the similarity in chemical features of the two elements. While the majority of selenium at nutritional intakes is retained in the body, a significant amount can be excreted through feces and urine. At levels above the nutritional requirement, urinary loss becomes the main excretory route for excess selenium (Pedrosa et al., 2012). Chemical forms of fecal and urinary selenium differ; the latter metabolites enter the urine after being metabolized to methylated products (Zeng and Combs, 2008). The major forms of urinary selenium vary with species: the trimethylselenonium ion ($[CH_3]_3Se^+$) in rats and the selenosugar (1β-methylseleno-N-acetyl-D-galactosamine) in humans being prominent examples (Kobayashi et al., 2002). Despite the complexity of selenium metabolism and multiple forms of selenium ingested, modeling has been employed in an attempt to effectively describe the absorption, distribution, and retention of total selenium (Patterson et al., 1989; Wastney et al., 2011). This is typically derived from isotopic analyses of blood, urinary, and fecal samples collected from participants given oral organic and inorganic selenium stable isotope tracers, such as [74]Se as selenomethionine and [76]Se as sodium selenite.

INCORPORATION OF SELENIUM INTO SELENOPROTEINS

Although selenium was suggested as a "toxicant" presented in the protein fraction of plants (Franke, 1934), it was found in 1972 to be the prosthetic group of glutathione peroxidase 1 (GPX1). This milestone discovery was based on the migration of injected ^{75}Se to a tissue protein fraction containing GPX activity (Flohe et al., 1973; Rotruck et al., 1973). Later, two additional selenium-containing proteins, formate dehydrogenase and "protein A" in the glycine reductase system, were identified in *Clostridium thermoaceticum* (Andreesen and Ljungdahl, 1973; Turner and Stadtman, 1973). After selenocysteine was found in protein A of *C. Thermoaceticum* (Cone et al., 1976), the incorporation of selenium into this novel amino acid was shown to be cotranslational and directed by a unique genetic codon UGA, a stop codon in most genes (Chambers et al., 1986). Selenocysteine biosynthesis and its insertion into selenoproteins in prokaryotes is mediated by products of four genes: *selA, selB, selC,* and *selD* that encode selenocysteine synthase, a specialized translation factor, selenocysteine tRNA, and selenophosphate synthetase (Böck, 2001). Mammalian biosynthesis of selenocysteine follows a similar pattern to that of bacteria (Labunskyy et al., 2014), but the exact mechanism is not fully understood. Two essential advances toward understanding of that mechanism are the identification of a selenocysteine tRNA that decodes UGA and the selenocysteine insertion sequence (SECIS) in the 3′-untranslated region of all selenoprotein mRNAs (Berry et al., 1991a; Lee et al., 1989).

PHYSIOLOGICAL FUNCTIONS OF SELENOPROTEINS

There are up to 24 or 25 selenoproteins identified in mammalian species by a search for selenocysteine codon appearance in whole genomes (Kryukov et al., 2003). Selenocysteine is usually located in active sites of selenoenzymes where it participates in redox reactions. Structurally and functionally, selenoproteins are categorized into three main enzyme families, the thioredoxin-like, endoplasmic reticulum proteins (Sep15 and selenoprotein M), and others with functions related to selenium transport and storage. Because selenocysteine tRNA, the gene product of *Trsp*, is necessary for the expression of all selenoproteins, *Trsp*$^{-/-}$ mice are embryonically lethal (Bosl et al., 1997). Genetic overexpression or knockout of individual selenoprotein genes in mice has been applied for study of their physiological functions.

Glutathione Peroxidases

GPX was discovered as an erythrocyte enzyme that catalyzes the reduction of H_2O_2 by glutathione in 1957 (Mills, 1957), but the presence of selenium in the enzyme and its importance for the peroxidase activity were not confirmed until 1972 (Flohe et al., 1973; Rotruck et al., 1973). The GPX1 enzyme isolated from bovine erythrocyte exhibits "ping-pong" kinetics, without saturation with glutathione and thus infinite for extrapolated V_{max} and no true K_m (Flohé et al., 1972; Günzler et al., 1972). While GPX1–3 and GPX6 are homotetramers, GPX4 is a monomer. Mammalian GPX5 is not a selenoprotein. All GPXs use glutathione as a reducing agent, and the peroxide substrate preference for GPX1-3 and GPX4 is H_2O_2 and organic hydroperoxides, respectively (Brigelius-Flohe and Maiorino, 2013). Only GPX4 can catalyze the reduction of phospholipid hydroperoxides. Gene expression and protein production of the GPX family exhibits a tissue- and organelle-specificity (Brigelius-Flohe and Maiorino, 2013). In essence, GPX1 resides mainly in cytosol of virtually all types of cells in all tissues, with the highest abundance found in liver, kidney, erythrocyte, and placenta. Also known as gastrointestinal GPX, GPX2 occurs almost exclusively in intestines but also exists in liver and mammary gland (Chu et al., 1993). Extracellular GPX3 is known as plasma GPX and it is highly expressed in and secreted from kidney and epididymis (Burk et al., 2011). Phospholipid hydroperoxide GPX (GPX4) is expressed highly in testis, and has three isoforms in cytosol, mitochondria, and the nucleus (Brigelius-Flohe and Maiorino, 2013; Maiorino et al., 2003; Pushpa-Rekha et al., 1995). Although human GPX6 is expressed at high levels in embryos and the olfactory epithelium, its cellular localization is unknown (Dear et al., 1991; Kryukov et al., 2003).

GPX1 accounts for 60% of total liver selenium and protects against acute oxidative stress through redox protection of lipids and proteins in mice (Cheng et al., 1998, 1999; Fu et al., 1999). While *Gpx1*$^{-/-}$ mice develop normally, they have been widely used to study roles of GPX1 in cardiovascular disease, neurodegeneration, and autoimmune disorders (Lei et al., 2007). Global overexpression of GPX1 in mice induces a type 2 diabetes–like phenotype (McClung et al., 2004). Insulin resistance produced by consumption of a high-fat diet is attenuated in *Gpx1*$^{-/-}$ mice (Loh et al., 2009). This paradoxical role of GPX1 is attributed to the physiological needs for maintaining appropriate intracellular H_2O_2 levels to support insulin signaling. In contrast, β-cell specific overexpression of GPX1 in leptin receptor defective *db/db* obese mice rescues the intrinsic β-cell dysfunction and prevents diabetes at 20 weeks of age (Guo et al., 2013; Harmon et al., 2009). Apparently, GPX1 exerts different impacts on the physiological and pathophysiological responses to glucose homeostasis under distinctive circumstances.

Reminiscent of *Gpx1*$^{-/-}$ mice, *Gpx2*$^{-/-}$ mice are apparently normal, but *Gpx1*$^{-/-}$*Gpx2*$^{-/-}$ mice develop intestinal tumors with associated lipid peroxidation (Esworthy et al., 2001). GPX1 expression is upregulated in the colon and ileum of *Gpx2*$^{-/-}$ mice (Florian et al., 2010), suggesting a functional interaction between the two selenoperoxidases in protecting against oxidative intestinal insults. A similar role of Gpx3 in protecting against reactive oxygen species-related pathophysiology exists, as the cerebral infarction symptom in *Gpx3*$^{-/-}$ mice can be alleviated by antioxidant treatment (Jin et al., 2011). As a direct consequence of *Gpx4*$^{-/-}$ mice displaying embryonic lethality (Yant et al., 2003), various mouse lines of inducible or conditional knockout of Gpx4 have been generated to study its physiological roles in male fertility (Imai et al., 2009; Schneider et al., 2009), maturation of photoreceptor cells (Ueta et al., 2012b), and prevention of neurodegeneration (Seiler et al., 2008).

Thioredoxin Reductases

Thioredoxin (Trx) reductases (TrxR) belong to a family of pyridine nucleotide-disulfide oxidoreductase that recycle oxidized Trx. There are three TrxRs in mammals: cytosolic TrxR1 (Tamura and Stadtman, 1996), mitochondrial TrxR2 (Miranda-Vizuete et al., 1999), and TrxR3, also known as thioredoxin glutathione reductase, expressed mainly in spermatids (Su et al., 2005). $Txnrd1^{-/-}$ and $Txnrd2^{-/-}$ mice, similar to $Gpx4^{-/-}$ mice, display embryonic lethality (Bondareva et al., 2007; Conrad et al., 2004; Jakupoglu et al., 2005). Consequently, many mouse conditional knockouts of $Txnrd1$ and $Txnrd2$ have been generated for functional studies of the two selenoproteins in the pathogenesis of cancer and neurodegeneration (Carlson et al., 2012; Jakupoglu et al., 2005; Soerensen et al., 2008).

Iodothyronine Deiodinases

Iodothyronine deiodinases (DIOs), a family of selenoenzymes (Arthur et al., 1990; Berry et al., 1991b), catalyze the transformations between thyroxine (T_4), 3,5,3′-triiodothyronine (T_3) and reverse triiodothyronine (rT_3). The less active T_4 requires DIO1 or DIO2 to remove one iodine to produce the active T_3, followed by secretion from the thyroid gland or other organs containing deiodinases into the circulation. Although $Dio1^{-/-}$ or $Dio2^{-/-}$ mice do not display abnormal serum T_3 levels (Galton et al., 2009), $Dio2^{-/-}$ mice retain higher T_4 levels in the serum (St Germain et al., 2009). This suggests enzymatic redundancy between DIO1 and DIO2 for T_3 production, as well as elevated T_4 expression under DIO1 or DIO2 deficiency. In addition to the catalysis of thyroid hormone maturation, recent advances indicate additional physiological functions of DIOs. These include: (1) hearing loss and impaired thermogenesis and cognition in $Dio2^{-/-}$ mice (St Germain et al., 2009); (2) susceptibility to ventilator-induced lung injury in $Dio2^{-/-}$ mice (Barca-Mayo et al., 2011); (3) defective cerebellar development and restrictive cardiomyopathy in $Dio3^{-/-}$ mice (Peeters et al., 2013; Ueta et al., 2012a); and (4) glucose intolerance, insulin resistance, and/or impaired glucose-stimulated insulin secretion in $Dio2^{-/-}$ and $Dio3^{-/-}$ mice (Castillo et al., 2011; Marsili et al., 2011; Medina et al., 2011). Furthermore, all three DIO enzymes are implicated in carcinogenesis as thyroid hormones are involved in the signaling events of differentiation, proliferation, and apoptosis (Casula and Bianco, 2012; Piekielko-Witkowska and Nauman, 2011).

Selenoproteins Responsible for Protein Quality Control

Structural analyses establish 15-kDa selenoprotein (Sep15) and selenoprotein M as members of the Trx1-like protein family that possess thiol-disulfide oxidoreductase activity and control the quality of protein folding in the endoplasmic reticulum (Ferguson et al., 2006). Furthermore, Sep15 is abundant in mouse liver, kidney, testis, and prostate, and it regulates protein folding via cooperation with the chaperon protein, UDP-glucose: glycoprotein glucosyltransferase (Davis et al., 2012; Korotkov et al., 2001). Despite Sep15 protection against protein misfolding, a study of $Sep15^{-/-}$ mice has suggested an unexpected contribution of Sep15 to the promotion and metastasis of colon cancer (Irons et al., 2010). Similarly, $Selm^{-/-}$ mice are obese but display normal motor and cognitive functions, even though selenoprotein M is highly expressed in the brain (Pitts et al., 2013) and can sequester zinc in Alzheimer's brain (Du et al., 2013). Methionine sulfoxide reductase B (MsrB1), also known as selenoprotein R or X, is a Trx-dependent protein in the cytosol and nucleus that reduces oxidized methionine-R-sulfoxide on proteins. $MsrB1^{-/-}$ mice have increased malondialdehyde, protein carbonyls, and methionine sulfoxide, and oxidized glutathione in the liver and kidney, but not in the heart, testis, and brain (Fomenko et al., 2009). Furthermore, there is a need for conserving reduced methionine in actin by MsrB1 for proper assembly to support innate immunity through macrophages (Lee et al., 2013).

Other Selenoproteins

Sepp1 is an extracellular selenoprotein containing 10 selenocysteine residues. Mainly being synthesized in the liver, it plays important roles in body selenium homeostasis (Burk and Hill, 1994; Himeno et al., 1996). In particular, Sepp1 is thought to transport selenium to the kidney through a receptor protein called megalin and to testis and brain through apolipoprotein E receptor-2 (apoER2) (Hill et al., 2003; Olson et al., 2007; Schomburg et al., 2003). In addition, Sepp1 is implicated in spermatogenesis and the prevention of brain degeneration (Burk and Hill, 2009). It is noteworthy that Sepp1 can bind zinc and prevent amyloid-β peptide aggregation in the brain (Du et al., 2013). Sepp1 offers an example of UGA being a wobble codon for either decoding selenocysteine or translation termination, as the three shortened Sepp1 isoforms can arise from alternatively using the second, third, and seventh UGAs as stop signals (Burk and Hill, 2005). Moreover, selenoprotein W (SelW) is expressed mainly in the muscle, heart, spleen, and brain of mammals and is suggested to play important roles in immune responses and Trx-dependent pathways (Whanger, 2009).

Selenoprotein H (SelH) exhibits Trx-like and DNA-binding domains and GPX activity (Novoselov et al., 2007), and it suppresses replicative and stress-induced senescence of MRC-5 normal lung fibroblasts through inhibition of DNA damage and oxidative stress (Wu et al., 2014). The mRNA expression of SelH is highest in mouse embryo, brain, thymus, testis, and uterus (Novoselov et al., 2007). Selenoprotein K (Fredericks et al., 2014) and selenoprotein N (Arbogast and Ferreiro, 2010; Castets et al., 2012) are implicated in calcium homeostasis. There are comprehensive reviews of the physiological functions, molecular mechanisms, and health implications associated with expression of all selenoproteins (Labunskyy et al., 2014; Lei et al., 2016).

REGULATION OF SELENOPROTEIN EXPRESSION

Expression of the 24 or 25 mammalian selenoproteins in various species and organs responds to dietary selenium supply fluctuations differentially. Understanding of this area has advanced rapidly in the last decade, largely as a result of the application of quantitative polymerase chain reactions to quantify mRNA expression levels. While this replaced the more time-consuming and labor-intensive Northern blot analyses, global analysis of protein expression levels is still incomplete largely due to the limited availability of high-quality antibodies. It is becoming clear that the expression of selenoproteins is not always in parallel with changes in selenium status, which differs by species, tissues, and pathophysiological conditions such as cancer (Huang et al., 2015; Liu et al., 2012; Sun et al., 2013; Sunde et al., 2009). Consistent with this, a simplified view of selenium regulation of selenoprotein expression, in which the expression increases with selenium supply and reaches a plateau when nutritional needs are met, does not always hold true, and available evidence collectively points to at least five differing scenarios. First, as originally proposed, selenoprotein mRNA expression is decreased by limited selenium supply. Most selenoproteins with this regulation are highly abundant and the precise control depends on tissue and species (see examples in mice [Sunde et al., 2009], pigs [Liu et al., 2012; Zhou et al., 2009], and chicks [Huang et al., 2015]). Second, dietary selenium deficiency upregulates mRNA levels of selenoprotein-encoding genes. Gene expression in this group, such as selenophosphate synthetase-2 in rats and pigs and six others in pigs (Barnes et al., 2009; Liu et al., 2012), is presumed to be prioritized under limited selenium supply. Selenophosphate synthetase-2 catalyzes the formation of selenophosphate, a donor of selenium for selenocysteine synthesis, and thus facilitates the expression of all selenoproteins. Third, selenium exceeding adequate levels can further upregulate selenoprotein mRNA expression. This mode of regulation again differs by selenoprotein and cell type (Sun et al., 2013). Fourth, excess dietary selenium downregulates mRNA levels of selenoprotein-encoding genes. Tissues with this pattern of control are mainly endocrine, such as testes, thyroid, and pituitary with altered *Sep15* expression (Liu et al., 2012). Selenoproteins in this group may play important homeostatic roles in metabolism. Fifth, selenoprotein expression is unaffected by dietary selenium deficiency or excess. This mode can exist for all selenoproteins but primarily occurs in tissues that are relatively resistant to dietary selenium changes, such as hypothalamus and protein-rich muscle tissues (Liu et al., 2012; Zhou et al., 2009). Furthermore, forms of selenium can also affect selenoprotein mRNA expression in a manner depending on type of cancer cells (Sun et al., 2013).

Similarly, the mRNA and protein expression of selenoproteins do not always go in parallel, as previously mentioned. First, protein expression is further upregulated despite a plateau of mRNA at adequate dietary levels. Examples of this include Sepp1 in pig thyroid and testes, SelH in liver, SelS in thyroid (Liu et al., 2012), and Sep15 in human DU145 prostate cancer cells (Sun et al., 2013). Second, mRNA expression responds to selenium changes when there is no change in protein level. An example of this is SelS in pig heart (Liu et al., 2012). Third, mRNA and protein changes can go in opposite directions. While the expression of porcine SelS mRNA is decreased by dietary selenium excess (3.0 mg/kg versus basal diet) in the kidney, protein expression is increased (Liu et al., 2012). Although detailed mechanisms for such unexpected expression patterns remain unclear, additional molecular events must regulate selenoprotein expression. These mechanisms differ by selenoprotein, tissue, species, and disease conditions. Furthermore, metals may upregulate selenoprotein expression through binding to metal response elements in the promoters of GPX3, SelW, SelH, and TrxR2 (Bierl et al., 2004; Stoytcheva et al., 2010; Wimmer et al., 2005).

PHARMACOLOGICAL EFFECTS OF SELENIUM

Above nutritional requirements, only minimal amounts of selenium may be additionally incorporated into selenoproteins, leaving selenium compounds and/(or) their metabolites to possible pharmacological reactions. August von Wassermann showed a chemotherapeutic effect of selenite in the suppression of Ehrlich sarcoma in mice (von Wassermann et al., 1911). The Nutritional Prevention of Cancer (NPC) trial used selenium-enriched yeast that contained 65–85% selenomethionine, with the rest being >20 unidentified selenium compounds. Although the NPC trial revealed supranutritional selenium as an effective chemoprevention agent against prostate, colorectal, and lung cancers (Clark et al., 1996), the follow-up Selenium and Vitamin E Cancer Prevention Trial (SELECT, 100% selenomethionine) did not replicate or extend the same outcome of selenium supplementation (Lippman et al., 2009). Nonetheless, a plausible notion is that the consequence of selenium supplementation in cancer prevention and treatment varies with the type and stage of tumorigenesis, body selenium status, gender, and age of the subjects, in addition to the form or source of selenium supplements (Steinbrenner et al., 2013). Two mouse models of prostate tumorigenesis collectively have shown that methylseleninic acid, but not selenomethionine, effectively suppresses carcinogenesis (Li et al., 2008; Wang et al., 2009). As a chemoprevention reagent, it is more logical for selenium to stifle tumorigenesis at a stage before full-blown cancer rather than at later stages. Major barriers to tumorigenesis at early stages include prevention of DNA damage and senescence responses, and both of which require ataxia telangiectasia mutated (ATM) kinase (Bartkova et al., 2005, 2006; Gorgoulis et al., 2005). In this regard, methylseleninic acid and methylselenocysteine compounds induce oxidative stress and ATM-dependent senescence response in noncancerous but not in cancerous cells (Wu et al., 2010).

Although an insulin-mimetic function of selenium is seen in cultured cells and islets, the micronutrient has a dual role in glucose homeostasis and lipid metabolism in animals and humans. Compromised body selenium status is associated with onset and development of diabetes (Furnsinn et al., 1995; Mueller and Pallauf, 2006; Becker et al., 1996; McNeill et al., 1991). Selenium has antidiabetic effects at high doses in obese Zucker rats (Furnsinn et al., 1995), diabetic *db/db* mice (Mueller and Pallauf, 2006), and streptozotocin-treated rats (Becker et al., 1996; McNeill et al., 1991). However, overdosing mice, rats, and pigs with selenium (0.4–3 mg/kg of diet) caused diabetogenic responses, consistent with the type 2 diabetes–like phenotype induced by overexpression of GPX1 in mice and adverse effects of Sepp1 on glucose homeostasis and insulin resistance (Zhou et al., 2013). Likewise, diabetogenic risks of supplemental selenium have been seen in a number of major human studies. Elevated selenium intake or status may alter expression and/or function of genes and proteins involved in insulin synthesis, secretion, signaling, glycolysis, gluconeogenesis, lipogenesis, lipolysis, and protein synthesis (Zhou et al., 2015). Apparently, the underlying mechanism for a complex or dual role of selenium in tumorigenesis and energy metabolism deserves further investigation.

Pharmacologic influences of selenium have also been proposed and/or tested in cardiovascular diseases such as stroke and atherosclerosis, infectious conditions such as AIDS and sepsis, neurodegenerations such as Alzheimer's disease and amyotrophic lateral sclerosis, arthritis, and autoimmune thyroiditis (Burbano et al., 2002; Dharmasena, 2014; Huang et al., 2013a; Sanmartin et al., 2011). However, these epidemiological and clinical data show inconclusive or contradictory effects of selenium. Thus, meta-analysis may offer a platform to authenticate such claimed effects with better statistical power. Of particular importance is autoimmune thyroiditis, because human thyroid contains the highest concentration of selenium in the body (Dickson and Tomlinson, 1967) and a meta-analysis of nine clinical studies enlisting patients diagnosed with autoimmune thyroiditis has revealed an efficacy of supranutritional selenium in counteracting the disease (Fan et al., 2014). Additional preclinical and clinical studies are needed to clarify the conflicting results of the selenium supplementation effects and to help understand the underlying mechanisms for the effects.

SELENIUM TOXICITY AND INTERACTIONS WITH VITAMIN E AND MINERALS

During the first 125 years after its discovery, selenium had been considered a toxic element or a drug with the potential to shrink tumors. Chronic selenium toxicity, known as "alkali disease," or "change hoof disease," in farm animals was described in 1934 by Kurt Franke who produced selenium toxicity in rats fed seleniferous plants grown on the lands rich in selenium in the Northern Plains and Western states (Franke, 1934). Other seleniferous areas include Enshi in China, parts of Ireland, Northern and South America, and India. In particular, the minority autonomous prefecture of Enshi in Hubei, China, has the only selenium mines in the world and produces local foods exceptionally high in the micronutrient, offering extraordinary opportunities to study selenium toxicity in humans. With an estimated intake of selenium of up to 4990 µg/day in Enshi, there were 477 cases of selenosis reported over a period of 60 years (1923–88). Based on these data, nutritionists have determined the maximal tolerable level of selenium intake as 400 µg/day in humans (Huang et al., 2013b). Signs of chronic selenosis in humans include garlic-like odor of the breath, hair and nail damage or loss, fatigue, and gastrointestinal and neurological disorders.

Although the classical selenium-deficiency diseases in farm animals are often associated with combined vitamin E deficiency, these two nutrients exhibit somewhat different efficacy in alleviating various symptoms of those diseases. Whereas liver necrosis in rats and exudative diathesis and pancreatic atrophy in chicks can be prevented by supplemental selenium or vitamin E, the latter is thought to be more effective than the former in preventing white muscle disease in sheep (Noguchi et al., 1973; Schwarz et al., 1957; Schwarz and Foltz, 1957; Whanger et al., 1977; Whitacre et al., 1987). In contrast, up to 100-fold of required intakes of vitamin E failed to replace the role of the selenium-dependent GPX1 in protecting against paraquat-induced lethality (Cheng et al., 1999). However, ebselen, a GPX-mimic, prevented the

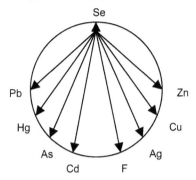

FIGURE 37.3 **Antagonistic interactions of selenium with other elements.** Selenium (Se) toxicity may be affected or attenuated by other elements including Pb (lead), Hg (mercury), As (Arsenic), Cd (cadmium), F (fluorine), Ag (silver), Cu (copper), and Zn (zinc).

signs of nutritional selenium deficiency in vitamin E-deficient chicks with exudative diathesis (Mercurio and Combs, 1986). Activity of GPX4 can be mimicked by adding tocopherol to peroxidizing microsomes (Barsacchi et al., 1984). The mRNA expression of two selenoproteins (*Selk* and *Selm*) in muscle was increased, whereas four (*Sepx1*, *Sepw1*, *Sep15*, and *Gpx7*) in muscle and four (*Gpx1*, *Seli*, *Txnrd1*, and *Txnrd2*) in liver were decreased by dietary vitamin E in chicks (Huang et al., 2011, 2015).

Predatory fish and some other seafood species such as shrimp and shellfish often accumulate high amounts of heavy metals such as mercury, arsenic, lead, and cadmium, which in principle can be detoxified by selenium through a potential chelating effect. For instance, selenium negatively associates with mercury and lead toxicity in fish-eating populations in the Amazon (Lemire et al., 2011). Furthermore, selenium can alleviate mercury toxicity based on clinical studies conducted in Wanshan, China (Li et al., 2012). Other examples of metals whose toxicity can be alleviated by selenium include mercury in quails (El-Begearmi et al., 1977) and silver in rats (Wagner et al., 1975). Reciprocally, selenium toxicity can be attenuated by other minerals, such as copper or silver in chicks (Jensen, 1975a) and fluoride in rats (Yu et al., 1992). Consequently, high levels of dietary copper or zinc can induce exudative diathesis and muscular dystrophy in chicks fed a selenium-adequate diet (Jensen, 1975b). In contrast to antagonistic effects, minerals synergistic with selenium exist. In rats, hepatic GPX activity is decreased by either selenium or copper deficiency (Arthur et al., 1987; Jenkinson et al., 1982). Although the mechanisms remain largely unknown, physiological and pathophysiological evidence clearly supports both antagonistic (Fig. 37.3) and synergistic interactions of selenium with other elements.

NEW METHODS AND MODELS AND FUTURE PERSPECTIVES

High-resolution synchrotron X-ray fluorescence microscopy has facilitated the visualization of selenium distribution in tissues (Malinouski et al., 2012). RNA interference (RNAi)-based models have been used for mechanistic studies of selenoprotein function. Genome-wide RNAi screening has been combined with ICP-MS detection of minerals for studying the regulation of selenium metabolism by selenoproteins and the selenocysteine incorporation machinery (Malinouski et al., 2014). Genetic knockout mice continue to serve as an important tool for physiological evaluation of selenium function. As mice with targeted deletions of *Trsp*, *Gpx4*, *Txnrd1*, and *Txnrd2* are embryonically lethal, conditional knockouts have been generated to assess tissue-specific functions of *Gpx1*, *TrxR1*, and *TrxR2* (Lei et al., 2016). Furthermore, microbiome analyses have been applied to determine effects of dietary selenium status on composition and colonization of intestinal microflora (Kasaikina et al., 2011). These models, together with the more recently developed tools such as CRISPR/Cas9 system for genome engineering (Shalem et al., 2015), will continue to help advance our understanding of selenium and selenoproteins at molecular, cellular, and physiological levels. Another underexplored area is the profiling of the spectrum of selenium compounds in selenium-enriched products with chemoprevention potential. In addition, selenium metabolites involved in chronic diseases such as diabetes and insulin resistance remain to be identified and may well give clues as to potential treatments for these disorders.

ACKNOWLEDGMENTS

The authors are most grateful to Professor Emeritus John Arthur for reviewing and editing this chapter. The research in the author's (XL) Laboratory was supported in part by NIH DK 53018 and a CNSF Major International (Regional) Joint Research Program of China (Grant No. 31320103920).

REFERENCES

Andreesen, J.R., Ljungdahl, L.G., 1973. Formate dehydrogenase of *Clostridium thermoaceticum*: incorporation of selenium-75, and the effects of selenite, molybdate, and tungstate on the enzyme. J. Bacteriol. 116, 867–873.

Arbogast, S., Ferreiro, A., 2010. Selenoproteins and protection against oxidative stress: selenoprotein N as a novel player at the crossroads of redox signaling and calcium homeostasis. Antioxid. Redox Signal. 12, 893–904.

Arthur, J.R., Morrice, P.C., Nicol, F., Beddows, S.E., Boyd, R., Hayes, J.D., Beckett, G.J., 1987. The effects of selenium and copper deficiencies on glutathione S-transferase and glutathione peroxidase in rat liver. Biochem. J. 248, 539–544.

Arthur, J.R., Nicol, F., Beckett, G.J., 1990. Hepatic iodothyronine 5'-deiodinase. The role of selenium. Biochem. J. 272, 537–540.

Bansal, M.P., Cook, R.G., Danielson, K.G., Medina, D., 1989a. A 14-kilodalton selenium-binding protein in mouse liver is fatty acid-binding protein. J. Biol. Chem. 264, 13780–13784.

Bansal, M.P., Oborn, C.J., Danielson, K.G., Medina, D., 1989b. Evidence for two selenium-binding proteins distinct from glutathione peroxidase in mouse liver. Carcinogenesis 10, 541–546.

Barca-Mayo, O., Liao, X.H., DiCosmo, C., Dumitrescu, A., Moreno-Vinasco, L., Wade, M.S., Sammani, S., Mirzapoiazova, T., Garcia, J.G., Refetoff, S., Weiss, R.E., 2011. Role of type 2 deiodinase in response to acute lung injury (ALI) in mice. Proc. Natl. Acad. Sci. U.S.A. 108, E1321–E1329.

Barnes, K.M., Evenson, J.K., Raines, A.M., Sunde, R.A., 2009. Transcript analysis of the selenoproteome indicates that dietary selenium requirements of rats based on selenium-regulated selenoprotein mRNA levels are uniformly less than those based on glutathione peroxidase activity. J. Nutr. 139, 199–206.

Barsacchi, R., Pelosi, G., Camici, P., Bonaldo, L., Maiorino, M., Ursini, F., 1984. Glutathione depletion increases chemiluminescence emission and lipid peroxidation in the heart. Biochim. Biophys. Acta – Bioenerg. 804, 356–360.

Bartkova, J., Horejsi, Z., Koed, K., Kramer, A., Tort, F., Zieger, K., Guldberg, P., Sehested, M., Nesland, J.M., Lukas, C., Orntoft, T., Lukas, J., Bartek, J., 2005. DNA damage response as a candidate anti-cancer barrier in early human tumorigenesis. Nature 434, 864–870.

Bartkova, J., Rezaei, N., Liontos, M., Karakaidos, P., Kletsas, D., Issaeva, N., Vassiliou, L.V., Kolettas, E., Niforou, K., Zoumpourlis, V.C., Takaoka, M., Nakagawa, H., Tort, F., Fugger, K., Johansson, F., Sehested, M., Andersen, C.L., Dyrskjot, L., Orntoft, T., Lukas, J., Kittas, C., Helleday, T., Halazonetis, T.D., Bartek, J., Gorgoulis, V.G., 2006. Oncogene-induced senescence is part of the tumorigenesis barrier imposed by DNA damage checkpoints. Nature 444, 633–637.

Becker, D.J., Reul, B., Ozcelikay, A.T., Buchet, J.P., Henquin, J.C., Brichard, S.M., 1996. Oral selenate improves glucose homeostasis and partly reverses abnormal expression of liver glycolytic and gluconeogenic enzymes in diabetic rats. Diabetologia 39, 3–11.

Berry, M.J., Banu, L., Chen, Y.Y., Mandel, S.J., Kieffer, J.D., Harney, J.W., Larsen, P.R., 1991a. Recognition of UGA as a selenocysteine codon in type I deiodinase requires sequences in the 3' untranslated region. Nature 353, 273–276.

Berry, M.J., Banu, L., Larsen, P.R., 1991b. Type I iodothyronine deiodinase is a selenocysteine-containing enzyme. Nature 349, 438–440.

Berzelius, J.J., de Lettre, M., 1818. Berzelius a M Berthollet sur deux metaux nouveaux. Ann. Chim. Phys. 199–202.

Bierl, C., Voetsch, B., Jin, R.C., Handy, D.E., Loscalzo, J., 2004. Determinants of human plasma glutathione peroxidase (GPx-3) expression. J. Biol. Chem. 279, 26839–26845.

Böck, A., 2001. Selenium metabolism in bacteria. In: Hatfield, D.L. (Ed.), Selenium: Its Molecular Biology and Role in Human Health. Kluwer Academic Publishers, Dordrecht, The Netherlands, pp. 7–22.

Bondareva, A.A., Capecchi, M.R., Iverson, S.V., Li, Y., Lopez, N.I., Lucas, O., Merrill, G.F., Prigge, J.R., Siders, A.M., Wakamiya, M., Wallin, S.L., Schmidt, E.E., 2007. Effects of thioredoxin reductase-1 deletion on embryogenesis and transcriptome. Free Radic. Biol. Med. 43, 911–923.

Bosl, M.R., Takaku, K., Oshima, M., Nishimura, S., Taketo, M.M., 1997. Early embryonic lethality caused by targeted disruption of the mouse selenocysteine tRNA gene (Trsp). Proc. Natl. Acad. Sci. U.S.A. 94, 5531–5534.

Brigelius-Flohe, R., Maiorino, M., 2013. Glutathione peroxidases. Biochim. Biophys. Acta 1830, 3289–3303.

Burbano, X., Miguez-Burbano, M.J., McCollister, K., Zhang, G., Rodriguez, A., Ruiz, P., Lecusay, R., Shor-Posner, G., 2002. Impact of a selenium chemoprevention clinical trial on hospital admissions of HIV-infected participants. HIV Clin. Trials 3, 483–491.

Burk, R.F., Hill, K.E., 1994. Selenoprotein P. A selenium-rich extracellular glycoprotein. J. Nutr. 124, 1891–1897.

Burk, R.F., Hill, K.E., 2005. Selenoprotein P: an extracellular protein with unique physical characteristics and a role in selenium homeostasis. Annu. Rev. Nutr. 25, 215–235.

Burk, R.F., Hill, K.E., 2009. Selenoprotein P-expression, functions, and roles in mammals. Biochim. Biophys. Acta 1790, 1441–1447.

Burk, R.F., Norsworthy, B.K., Hill, K.E., Motley, A.K., Byrne, D.W., 2006. Effects of chemical form of selenium on plasma biomarkers in a high-dose human supplementation trial. Cancer Epidemiol. Biomarkers Prev. 15, 804–810.

Burk, R.F., Olson, G.E., Winfrey, V.P., Hill, K.E., Yin, D., 2011. Glutathione peroxidase-3 produced by the kidney binds to a population of basement membranes in the gastrointestinal tract and in other tissues. Am. J. Physiol. Gastrointest. Liver Physiol. 301, G32–G38.

Carlson, B.A., Yoo, M.H., Tobe, R., Mueller, C., Naranjo-Suarez, S., Hoffmann, V.J., Gladyshev, V.N., Hatfield, D.L., 2012. Thioredoxin reductase 1 protects against chemically induced hepatocarcinogenesis via control of cellular redox homeostasis. Carcinogenesis 33, 1806–1813.

Castets, P., Lescure, A., Guicheney, P., Allamand, V., 2012. Selenoprotein N in skeletal muscle: from diseases to function. J. Mol. Med. Berl. 90, 1095–1107.

Castillo, M., Hall, J.A., Correa-Medina, M., Ueta, C., Kang, H.W., Cohen, D.E., Bianco, A.C., 2011. Disruption of thyroid hormone activation in type 2 deiodinase knockout mice causes obesity with glucose intolerance and liver steatosis only at thermoneutrality. Diabetes 60, 1082–1089.

Casula, S., Bianco, A.C., 2012. Thyroid hormone deiodinases and cancer. Front. Endocrinol. (Lausanne) 3, 74.

Chambers, I., Frampton, J., Goldfarb, P., Affara, N., McBain, W., Harrison, P.R., 1986. The structure of the mouse glutathione peroxidase gene: the selenocysteine in the active site is encoded by the 'termination' codon, TGA. EMBO J. 5, 1221–1227.

Cheng, W.H., Combs Jr., G.F., Lei, X.G., 1998. Knockout of cellular glutathione peroxidase affects selenium-dependent parameters similarly in mice fed adequate and excessive dietary selenium. Biofactors 7, 311–321.

Cheng, W.H., Valentine, B.A., Lei, X.G., 1999. High levels of dietary vitamin E do not replace cellular glutathione peroxidase in protecting mice from acute oxidative stress. J. Nutr. 129, 1951–1957.

Chu, F.F., Doroshow, J.H., Esworthy, R.S., 1993. Expression, characterization, and tissue distribution of a new cellular selenium-dependent glutathione peroxidase, GSHPx-GI. J. Biol. Chem. 268, 2571–2576.

Clark, L.C., Combs Jr., G.F., Turnbull, B.W., Slate, E.H., Chalker, D.K., Chow, J., Davis, L.S., Glover, R.A., Graham, G.F., Gross, E.G., Krongrad, A., Lesher Jr., J.L., Park, H.K., Sanders Jr., B.B., Smith, C.L., Taylor, J.R., 1996. Effects of selenium supplementation for cancer prevention in patients with carcinoma of the skin. A randomized controlled trial. Nutritional Prevention of Cancer Study Group. JAMA 276, 1957–1963.

Combs Jr., G.F., Watts, J.C., Jackson, M.I., Johnson, L.K., Zeng, H., Scheett, A.J., Uthus, E.O., Schomburg, L., Hoeg, A., Hoefig, C.S., Davis, C.D., Milner, J.A., 2011. Determinants of selenium status in healthy adults. Nutr. J. 10, 75.

Cone, J.E., Del Rio, R.M., Davis, J.N., Stadtman, T.C., 1976. Chemical characterization of the selenoprotein component of clostridial glycine reductase: identification of selenocysteine as the organoselenium moiety. Proc. Natl. Acad. Sci. U.S.A. 73, 2659–2663.

Conrad, M., Jakupoglu, C., Moreno, S.G., Lippl, S., Banjac, A., Schneider, M., Beck, H., Hatzopoulos, A.K., Just, U., Sinowatz, F., Schmahl, W., Chien, K.R., Wurst, W., Bornkamm, G.W., Brielmeier, M., 2004. Essential role for mitochondrial thioredoxin reductase in hematopoiesis, heart development, and heart function. Mol. Cell. Biol. 24, 9414–9423.

Davis, C.D., Tsuji, P.A., Milner, J.A., 2012. Selenoproteins and cancer prevention. Annu. Rev. Nutr. 32, 73–95.

Dear, T.N., Campbell, K., Rabbitts, T.H., 1991. Molecular cloning of putative odorant-binding and odorant-metabolizing proteins. Biochemistry 30, 10376–10382.

Dharmasena, A., 2014. Selenium supplementation in thyroid associated ophthalmopathy: an update. Int. J. Ophthalmol. 7, 365–375.

Dickson, R.C., Tomlinson, R.H., 1967. Selenium in blood and human tissues. Clin. Chim. Acta 16, 311–321.

Du, X., Li, H., Wang, Z., Qiu, S., Liu, Q., Ni, J., 2013. Selenoprotein P and selenoprotein M block Zn2+ -mediated Abeta42 aggregation and toxicity. Metallomics 5, 861–870.

El-Begearmi, M.M., Sunde, M.L., Ganther, H.E., 1977. A mutual protective effect of mercury and selenium in Japanese quail. Poult. Sci. 56, 313–322.

Esworthy, R.S., Aranda, R., Martin, M.G., Doroshow, J.H., Binder, S.W., Chu, F.F., 2001. Mice with combined disruption of Gpx1 and Gpx2 genes have colitis. Am. J. Physiol. Gastrointest. Liver Physiol. 281, G848–G855.

Fan, Y., Xu, S., Zhang, H., Cao, W., Wang, K., Chen, G., Di, H., Cao, M., Liu, C., 2014. Selenium supplementation for autoimmune thyroiditis: a systematic review and meta-analysis. Int. J. Endocrinol. 2014, 904573.

Ferguson, A.D., Labunskyy, V.M., Fomenko, D.E., Arac, D., Chelliah, Y., Amezcua, C.A., Rizo, J., Gladyshev, V.N., Deisenhofer, J., 2006. NMR structures of the selenoproteins Sep15 and SelM reveal redox activity of a new thioredoxin-like family. J. Biol. Chem. 281, 3536–3543.

Flohe, L., Gunzler, W.A., Schock, H.H., 1973. Glutathione peroxidase: a selenoenzyme. FEBS Lett. 32, 132–134.

Flohé, L., Loschen, G., Günzler, W.A., Eichele, E., 1972. Glutathione peroxidase, V. The kinetic mechanism. Hoppe-Seyler´ s Z. für Physiol. Chem. 353, 987–1000.

Florian, S., Krehl, S., Loewinger, M., Kipp, A., Banning, A., Esworthy, S., Chu, F.F., Brigelius-Flohe, R., 2010. Loss of GPx2 increases apoptosis, mitosis, and GPx1 expression in the intestine of mice. Free Radic. Biol. Med. 49, 1694–1702.

Fomenko, D.E., Novoselov, S.V., Natarajan, S.K., Lee, B.C., Koc, A., Carlson, B.A., Lee, T.H., Kim, H.Y., Hatfield, D.L., Gladyshev, V.N., 2009. MsrB1 (methionine-R-sulfoxide reductase 1) knock-out mice: roles of MsrB1 in redox regulation and identification of a novel selenoprotein form. J. Biol. Chem. 284, 5986–5993.

Franke, K.W., 1934. A new toxicant occurring naturally in certain samples of plant foodstuffs. II. The occurrence of the toxicant in the protein fraction. J. Nutr. 8, 609–613.

Fredericks, G.J., Hoffmann, F.W., Rose, A.H., Osterheld, H.J., Hess, F.M., Mercier, F., Hoffmann, P.R., 2014. Stable expression and function of the inositol 1,4,5-triphosphate receptor requires palmitoylation by a DHHC6/selenoprotein K complex. Proc. Natl. Acad. Sci. U.S.A. 111, 16478–16483.

Fu, Y., Cheng, W.H., Porres, J.M., Ross, D.A., Lei, X.G., 1999. Knockout of cellular glutathione peroxidase gene renders mice susceptible to diquat-induced oxidative stress. Free Radic. Biol. Med. 27, 605–611.

Furnsinn, C., Leitner, G., Roden, M., Osterode, W., Waldhausl, W., 1995. Improved glucose tolerance by acute vanadate but not by selenate exposure in genetically obese rats (fa/fa). Int. J. Obes. Relat. Metab. Disord. 19, 458–463.

Galton, V.A., Schneider, M.J., Clark, A.S., St Germain, D.L., 2009. Life without thyroxine to 3,5,3′-triiodothyronine conversion: studies in mice devoid of the 5′-deiodinases. Endocrinology 150, 2957–2963.

Gorgoulis, V.G., Vassiliou, L.V., Karakaidos, P., Zacharatos, P., Kotsinas, A., Liloglou, T., Venere, M., Ditullio Jr., R.A., Kastrinakis, N.G., Levy, B., Kletsas, D., Yoneta, A., Herlyn, M., Kittas, C., Halazonetis, T.D., 2005. Activation of the DNA damage checkpoint and genomic instability in human precancerous lesions. Nature 434, 907–913.

Günzler, W.A., Vergin, H., Müller, I., Flohé, L., 1972. Glutathione peroxidase, VI. Die Reaktion der Glutathion Peroxydase mit verschiedenen Hydroperoxyden. Hoppe-Seyler´ s Z. für Physiol. Chem. 353, 1001–1004.

Guo, S., Dai, C., Guo, M., Taylor, B., Harmon, J.S., Sander, M., Robertson, R.P., Powers, A.C., Stein, R., 2013. Inactivation of specific beta cell transcription factors in type 2 diabetes. J. Clin. Invest. 123, 3305–3316.

Harmon, J.S., Bogdani, M., Parazzoli, S.D., Mak, S.S., Oseid, E.A., Berghmans, M., Leboeuf, R.C., Robertson, R.P., 2009. beta-Cell-specific overexpression of glutathione peroxidase preserves intranuclear MafA and reverses diabetes in db/db mice. Endocrinology 150, 4855–4862.

Hill, K.E., Zhou, J., McMahan, W.J., Motley, A.K., Atkins, J.F., Gesteland, R.F., Burk, R.F., 2003. Deletion of selenoprotein P alters distribution of selenium in the mouse. J. Biol. Chem. 278, 13640–13646.

Himeno, S., Chittum, H.S., Burk, R.F., 1996. Isoforms of selenoprotein P in rat plasma. Evidence for a full-length form and another form that terminates at the second UGA in the open reading frame. J. Biol. Chem. 271, 15769–15775.

Hogue, D.E., 1958. Vitamin E, selenium and other factors related to nutritional muscular dystrophy in lambs. Cornell Nutrition Conference for Feed Manufacturers Proceedings. 32–39.

Huang, J.Q., Li, D.L., Zhao, H., Sun, L.H., Xia, X.J., Wang, K.N., Luo, X., Lei, X.G., 2011. The selenium deficiency disease exudative diathesis in chicks is associated with downregulation of seven common selenoprotein genes in liver and muscle. J. Nutr. 141, 1605–1610.

Huang, J.Q., Ren, F.Z., Jiang, Y.Y., Xiao, C., Lei, X.G., 2015. Selenoproteins protect against avian nutritional muscular dystrophy by metabolizing peroxides and regulating redox/apoptotic signaling. Free Radic. Biol. Med. 83, 129–138.

Huang, T.S., Shyu, Y.C., Chen, H.Y., Lin, L.M., Lo, C.Y., Yuan, S.S., Chen, P.J., 2013a. Effect of parenteral selenium supplementation in critically ill patients: a systematic review and meta-analysis. PLoS One 8, e54431.

Huang, Y., Wang, Q., Gao, J., Lin, Z., Banuelos, G.S., Yuan, L., Yin, X., 2013b. Daily dietary selenium intake in a high selenium area of Enshi, China. Nutrients 5, 700–710.

Imai, H., Hakkaku, N., Iwamoto, R., Suzuki, J., Suzuki, T., Tajima, Y., Konishi, K., Minami, S., Ichinose, S., Ishizaka, K., Shioda, S., Arata, S., Nishimura, M., Naito, S., Nakagawa, Y., 2009. Depletion of selenoprotein GPx4 in spermatocytes causes male infertility in mice. J. Biol. Chem. 284, 32522–32532.

Irons, R., Tsuji, P.A., Carlson, B.A., Ouyang, P., Yoo, M.H., Xu, X.M., Hatfield, D.L., Gladyshev, V.N., Davis, C.D., 2010. Deficiency in the 15-kDa selenoprotein inhibits tumorigenicity and metastasis of colon cancer cells. Cancer Prev. Res. (Phila) 3, 630–639.

Jakupoglu, C., Przemeck, G.K., Schneider, M., Moreno, S.G., Mayr, N., Hatzopoulos, A.K., de Angelis, M.H., Wurst, W., Bornkamm, G.W., Brielmeier, M., Conrad, M., 2005. Cytoplasmic thioredoxin reductase is essential for embryogenesis but dispensable for cardiac development. Mol. Cell. Biol. 25, 1980–1988.

Jenkinson, S.G., Lawrence, R.A., Burk, R.F., Williams, D.M., 1982. Effects of copper deficiency on the activity of the selenoenzyme glutathione peroxidase and on excretion and tissue retention of 75SeO3(2-). J. Nutr. 112, 197–204.

Jensen, L.S., 1975a. Modification of a selenium toxicity in chicks by dietary silver and copper. J. Nutr. 105, 769–775.

Jensen, L.S., 1975b. Precipitation of a selenium deficiency by high dietary levels of copper and zinc. Proc. Soc. Exp. Biol. Med. 149, 113–116.

Jin, R.C., Mahoney, C.E., Coleman Anderson, L., Ottaviano, F., Croce, K., Leopold, J.A., Zhang, Y.Y., Tang, S.S., Handy, D.E., Loscalzo, J., 2011. Glutathione peroxidase-3 deficiency promotes platelet-dependent thrombosis in vivo. Circulation 123, 1963–1973.

Kasaikina, M.V., Kravtsova, M.A., Lee, B.C., Seravalli, J., Peterson, D.A., Walter, J., Legge, R., Benson, A.K., Hatfield, D.L., Gladyshev, V.N., 2011. Dietary selenium affects host selenoproteome expression by influencing the gut microbiota. FASEB J. 25, 2492–2499.

Kobayashi, Y., Ogra, Y., Ishiwata, K., Takayama, H., Aimi, N., Suzuki, K.T., 2002. Selenosugars are key and urinary metabolites for selenium excretion within the required to low-toxic range. Proc. Natl. Acad. Sci. U.S.A. 99, 15932–15936.

Korotkov, K.V., Kumaraswamy, E., Zhou, Y., Hatfield, D.L., Gladyshev, V.N., 2001. Association between the 15-kDa selenoprotein and UDP-glucose:glycoprotein glucosyltransferase in the endoplasmic reticulum of mammalian cells. J. Biol. Chem. 276, 15330–15336.

Kryukov, G.V., Castellano, S., Novoselov, S.V., Lobanov, A.V., Zehtab, O., Guigo, R., Gladyshev, V.N., 2003. Characterization of mammalian selenoproteomes. Science 300, 1439–1443.

Labunskyy, V.M., Hatfield, D.L., Gladyshev, V.N., 2014. Selenoproteins: molecular pathways and physiological roles. Physiol. Rev. 94, 739–777.

Lakin, H.W., 1961. Vertical and lateral distribution of selenium in sedimentary rocks of Western United States. No. 200. In: Selenium in Agriculture Agriculture Handbook, pp. 12–24.

Lee, B.C., Peterfi, Z., Hoffmann, F.W., Moore, R.E., Kaya, A., Avanesov, A., Tarrago, L., Zhou, Y., Weerapana, E., Fomenko, D.E., Hoffmann, P.R., Gladyshev, V.N., 2013. MsrB1 and MICALs regulate actin assembly and macrophage function via reversible stereoselective methionine oxidation. Mol. Cell 51, 397–404.

Lee, B.J., Worland, P.J., Davis, J.N., Stadtman, T.C., Hatfield, D.L., 1989. Identification of a selenocysteyl-tRNA(Ser) in mammalian cells that recognizes the nonsense codon, UGA. J. Biol. Chem. 264, 9724–9727.

Lei, X.G., Cheng, W.H., McClung, J.P., 2007. Metabolic regulation and function of glutathione peroxidase-1. Annu. Rev. Nutr. 27, 41–61.

Lei, X.G., Zhu, J.H., Cheng, W.H., Bao, Y., Ho, Y.S., Reddi, A.R., Holmgren, A., Arner, E.S., 2016. Paradoxical roles of antioxidant enzymes: basic mechanisms and health implications. Physiol. Rev. 96, 307–364.

Lemire, M., Fillion, M., Frenette, B., Passos, C.J., Guimaraes, J.R., Barbosa Jr., F., Mergler, D., 2011. Selenium from dietary sources and motor functions in the Brazilian Amazon. Neurotoxicology 32, 944–953.

Li, G.X., Lee, H.J., Wang, Z., Hu, H., Liao, J.D., Watts, J.C., Combs Jr., G.F., Lu, J., 2008. Superior in vivo inhibitory efficacy of methylseleninic acid against human prostate cancer over selenomethionine or selenite. Carcinogenesis 29, 1005–1012.

Li, Y.F., Dong, Z., Chen, C., Li, B., Gao, Y., Qu, L., Wang, T., Fu, X., Zhao, Y., Chai, Z., 2012. Organic selenium supplementation increases mercury excretion and decreases oxidative damage in long-term mercury-exposed residents from Wanshan, China. Environ. Sci. Technol. 46, 11313–11318.

Lippman, S.M., Klein, E.A., Goodman, P.J., Lucia, M.S., Thompson, I.M., Ford, L.G., Parnes, H.L., Minasian, L.M., Gaziano, J.M., Hartline, J.A., Parsons, J.K., Bearden 3rd, J.D., Crawford, E.D., Goodman, G.E., Claudio, J., Winquist, E., Cook, E.D., Karp, D.D., Walther, P., Lieber, M.M., Kristal, A.R., Darke, A.K., Arnold, K.B., Ganz, P.A., Santella, R.M., Albanes, D., Taylor, P.R., Probstfield, J.L., Jagpal, T.J., Crowley, J.J., Meyskens Jr., F.L., Baker, L.H., Coltman Jr., C.A., 2009. Effect of selenium and vitamin E on risk of prostate cancer and other cancers: the Selenium and Vitamin E Cancer Prevention Trial (SELECT). JAMA 301, 39–51.

Liu, Y., Zhao, H., Zhang, Q., Tang, J., Li, K., Xia, X.J., Wang, K.N., Li, K., Lei, X.G., 2012. Prolonged dietary selenium deficiency or excess does not globally affect selenoprotein gene expression and/or protein production in various tissues of pigs. J. Nutr. 142, 1410–1416.

Loh, K., Deng, H., Fukushima, A., Cai, X., Boivin, B., Galic, S., Bruce, C., Shields, B.J., Skiba, B., Ooms, L.M., Stepto, N., Wu, B., Mitchell, C.A., Tonks, N.K., Watt, M.J., Febbraio, M.A., Crack, P.J., Andrikopoulos, S., Tiganis, T., 2009. Reactive oxygen species enhance insulin sensitivity. Cell Metab. 10, 260–272.

Maiorino, M., Scapin, M., Ursini, F., Biasolo, M., Bosello, V., Flohe, L., 2003. Distinct promoters determine alternative transcription of gpx-4 into phospholipid-hydroperoxide glutathione peroxidase variants. J. Biol. Chem. 278, 34286–34290.

Malinouski, M., Hasan, N.M., Zhang, Y., Seravalli, J., Lin, J., Avanesov, A., Lutsenko, S., Gladyshev, V.N., 2014. Genome-wide RNAi ionomics screen reveals new genes and regulation of human trace element metabolism. Nat. Commun. 5, 3301.

Malinouski, M., Kehr, S., Finney, L., Vogt, S., Carlson, B.A., Seravalli, J., Jin, R., Handy, D.E., Park, T.J., Loscalzo, J., Hatfield, D.L., Gladyshev, V.N., 2012. High-resolution imaging of selenium in kidneys: a localized selenium pool associated with glutathione peroxidase 3. Antioxid. Redox Signal. 16, 185–192.

Marsili, A., Aguayo-Mazzucato, C., Chen, T., Kumar, A., Chung, M., Lunsford, E.P., Harney, J.W., Van-Tran, T., Gianetti, E., Ramadan, W., Chou, C., Bonner-Weir, S., Larsen, P.R., Silva, J.E., Zavacki, A.M., 2011. Mice with a targeted deletion of the type 2 deiodinase are insulin resistant and susceptible to diet induced obesity. PLoS One 6, e20832.

McClung, J.P., Roneker, C.A., Mu, W., Lisk, D.J., Langlais, P., Liu, F., Lei, X.G., 2004. Development of insulin resistance and obesity in mice overexpressing cellular glutathione peroxidase. Proc. Natl. Acad. Sci. U.S.A. 101, 8852–8857.

McNeill, J.H., Delgatty, H.L., Battell, M.L., 1991. Insulinlike effects of sodium selenate in streptozocin-induced diabetic rats. Diabetes 40, 1675–1678.

Medina, M.C., Molina, J., Gadea, Y., Fachado, A., Murillo, M., Simovic, G., Pileggi, A., Hernandez, A., Edlund, H., Bianco, A.C., 2011. The thyroid hormone-inactivating type III deiodinase is expressed in mouse and human beta-cells and its targeted inactivation impairs insulin secretion. Endocrinology 152, 3717–3727.

Mercurio, S.D., Combs Jr., G.F., 1986. Synthetic seleno-organic compound with glutathione peroxidase-like activity in the chick. Biochem. Pharmacol. 35, 4505–4509.

Michel, R.L., Whitehair, C.K., Keahey, K.K., 1969. Dietary hepatic necrosis associated with selenium-vitamin E deficiency in swine. J. Am. Vet. Med. Assoc. 155, 50–59.

Mills, G.C., 1957. Hemoglobin catabolism. I. Glutathione peroxidase, an erythrocyte enzyme which protects hemoglobin from oxidative breakdown. J. Biol. Chem. 229, 189–197.

Miranda-Vizuete, A., Damdimopoulos, A.E., Pedrajas, J.R., Gustafsson, J.A., Spyrou, G., 1999. Human mitochondrial thioredoxin reductase cDNA cloning, expression and genomic organization. Eur. J. Biochem. 261, 405–412.

Moreno-Reyes, R., Suetens, C., Mathieu, F., Begaux, F., Zhu, D., Rivera, M.T., Boelaert, M., Neve, J., Perlmutter, N., Vanderpas, J., 1998. Kashin-Beck osteoarthropathy in rural Tibet in relation to selenium and iodine status. N. Engl. J. Med. 339, 1112–1120.

Mueller, A.S., Pallauf, J., 2006. Compendium of the antidiabetic effects of supranutritional selenate doses. In vivo and in vitro investigations with type II diabetic db/db mice. J. Nutr. Biochem. 17, 548–560.

Muth, O.H., Oldfield, J.E., Remmert, L.F., Schubert, J.R., 1958. Effects of selenium and vitamin E on white muscle disease. Science 128, 1090.

Noguchi, T., Cantor, A.H., Scott, M.L., 1973. Mode of action of selenium and vitamin E in prevention of exudative diathesis in chicks. J. Nutr. 103, 1502–1511.

Novoselov, S.V., Kryukov, G.V., Xu, X.M., Carlson, B.A., Hatfield, D.L., Gladyshev, V.N., 2007. Selenoprotein H is a nucleolar thioredoxin-like protein with a unique expression pattern. J. Biol. Chem. 282, 11960–11968.

Olson, G.E., Winfrey, V.P., Nagdas, S.K., Hill, K.E., Burk, R.F., 2007. Apolipoprotein E receptor-2 (ApoER2) mediates selenium uptake from selenoprotein P by the mouse testis. J. Biol. Chem. 282, 12290–12297.

Patterson, B.H., Levander, O.A., Helzlsouer, K., McAdam, P.A., Lewis, S.A., Taylor, P.R., Veillon, C., Zech, L.A., 1989. Human selenite metabolism: a kinetic model. Am. J. Physiol. 257, R556–R567.

Patterson, E.L., Milstrey, R., Stokstad, E.L., 1957. Effect of selenium in preventing exudative diathesis in chicks. Proc. Soc. Exp. Biol. Med. 95, 617–620.

Pedrosa, L.F., Motley, A.K., Stevenson, T.D., Hill, K.E., Burk, R.F., 2012. Fecal selenium excretion is regulated by dietary selenium intake. Biol. Trace Elem. Res. 149, 377–381.

Peeters, R.P., Hernandez, A., Ng, L., Ma, M., Sharlin, D.S., Pandey, M., Simonds, W.F., St Germain, D.L., Forrest, D., 2013. Cerebellar abnormalities in mice lacking type 3 deiodinase and partial reversal of phenotype by deletion of thyroid hormone receptor alpha1. Endocrinology 154, 550–561.

Piekielko-Witkowska, A., Nauman, A., 2011. Iodothyronine deiodinases and cancer. J. Endocrinol. Invest. 34, 716–728.

Pinsent, J., 1954. The need for selenite and molybdate in the formation of formic dehydrogenase by members of the coli-aerogenes group of bacteria. Biochem. J. 57, 10–16.

Pitts, M.W., Reeves, M.A., Hashimoto, A.C., Ogawa, A., Kremer, P., Seale, L.A., Berry, M.J., 2013. Deletion of selenoprotein M leads to obesity without cognitive deficits. J. Biol. Chem. 288, 26121–26134.

Poley, W.E., Moxon, A.L., Taylor, J.B., Wilson, W.O., 1941. The effect of selenized grains on the rate of growth of chicks. Poult. Sci. 20, 171–179.

Pushpa-Rekha, T.R., Burdsall, A.L., Oleksa, L.M., Chisolm, G.M., Driscoll, D.M., 1995. Rat phospholipid-hydroperoxide glutathione peroxidase. cDNA cloning and identification of multiple transcription and translation start sites. J. Biol. Chem. 270, 26993–26999.

Rotruck, J.T., Pope, A.L., Ganther, H.E., Swanson, A.B., Hafeman, D.G., Hoekstra, W.G., 1973. Selenium: biochemical role as a component of glutathione peroxidase. Science 179, 588–590.

Sanmartin, C., Plano, D., Font, M., Palop, J.A., 2011. Selenium and clinical trials: new therapeutic evidence for multiple diseases. Curr. Med. Chem. 18, 4635–4650.

Schneider, M., Forster, H., Boersma, A., Seiler, A., Wehnes, H., Sinowatz, F., Neumuller, C., Deutsch, M.J., Walch, A., Hrabe de Angelis, M., Wurst, W., Ursini, F., Roveri, A., Maleszewski, M., Maiorino, M., Conrad, M., 2009. Mitochondrial glutathione peroxidase 4 disruption causes male infertility. FASEB J. 23, 3233–3242.

Schomburg, L., Schweizer, U., Holtmann, B., Flohe, L., Sendtner, M., Kohrle, J., 2003. Gene disruption discloses role of selenoprotein P in selenium delivery to target tissues. Biochem. J. 370, 397–402.

Schwarz, K., Bieri, J.G., Briggs, G.M., Scott, M.L., 1957. Prevention of exudative diathesis in chicks by factor 3 and selenium. Proc. Soc. Exp. Biol. Med. 95, 621–625.

Schwarz, K., Foltz, C.M., 1957. Selenium as an integral part of factor 3 against dietary necrotic liver degeneration. J. Am. Chem. Soc. 79, 3292–3293.

Seiler, A., Schneider, M., Forster, H., Roth, S., Wirth, E.K., Culmsee, C., Plesnila, N., Kremmer, E., Radmark, O., Wurst, W., Bornkamm, G.W., Schweizer, U., Conrad, M., 2008. Glutathione peroxidase 4 senses and translates oxidative stress into 12/15-lipoxygenase dependent- and AIF-mediated cell death. Cell Metab. 8, 237–248.

Shalem, O., Sanjana, N.E., Zhang, F., 2015. High-throughput functional genomics using CRISPR-Cas9. Nat. Rev. Genet. 16, 299–311.

Soerensen, J., Jakupoglu, C., Beck, H., Forster, H., Schmidt, J., Schmahl, W., Schweizer, U., Conrad, M., Brielmeier, M., 2008. The role of thioredoxin reductases in brain development. PLoS One 3, e1813.

St Germain, D.L., Galton, V.A., Hernandez, A., 2009. Minireview: defining the roles of the iodothyronine deiodinases: current concepts and challenges. Endocrinology 150, 1097–1107.

Steinbrenner, H., Speckmann, B., Sies, H., 2013. Toward understanding success and failures in the use of selenium for cancer prevention. Antioxid. Redox Signal. 19, 181–191.

Stoytcheva, Z.R., Vladimirov, V., Douet, V., Stoychev, I., Berry, M.J., 2010. Metal transcription factor-1 regulation via MREs in the transcribed regions of selenoprotein H and other metal-responsive genes. Biochim. Biophys. Acta (BBA) – Bioenerg. 1800, 416–424.

Su, D., Novoselov, S.V., Sun, Q.A., Moustafa, M.E., Zhou, Y., Oko, R., Hatfield, D.L., Gladyshev, V.N., 2005. Mammalian selenoprotein thioredoxin-glutathione reductase. Roles in disulfide bond formation and sperm maturation. J. Biol. Chem. 280, 26491–26498.

Sun, L.H., Li, J.G., Zhao, H., Shi, J., Huang, J.Q., Wang, K.N., Xia, X.J., Li, L., Lei, X.G., 2013. Porcine serum can be biofortified with selenium to inhibit proliferation of three types of human cancer cells. J. Nutr. 143, 1115–1122.

Sunde, R.A., Raines, A.M., Barnes, K.M., Evenson, J.K., 2009. Selenium status highly regulates selenoprotein mRNA levels for only a subset of the selenoproteins in the selenoproteome. Biosci. Rep. 29, 329–338.

Tamura, T., Stadtman, T.C., 1996. A new selenoprotein from human lung adenocarcinoma cells: purification, properties, and thioredoxin reductase activity. Proc. Natl. Acad. Sci. U.S.A. 93, 1006–1011.

Thiry, C., Ruttens, A., Pussemier, L., Schneider, Y.J., 2013. An in vitro investigation of species-dependent intestinal transport of selenium and the impact of this process on selenium bioavailability. Br. J. Nutr. 109, 2126–2134.

Turner, D.C., Stadtman, T.C., 1973. Purification of protein components of the clostridial glycine reductase system and characterization of protein A as a selenoprotein. Arch. Biochem. Biophys. 154, 366–381.

Ueta, C.B., Oskouei, B.N., Olivares, E.L., Pinto, J.R., Correa, M.M., Simovic, G., Simonides, W.S., Hare, J.M., Bianco, A.C., 2012a. Absence of myocardial thyroid hormone inactivating deiodinase results in restrictive cardiomyopathy in mice. Mol. Endocrinol. 26, 809–818.

Ueta, T., Inoue, T., Furukawa, T., Tamaki, Y., Nakagawa, Y., Imai, H., Yanagi, Y., 2012b. Glutathione peroxidase 4 is required for maturation of photoreceptor cells. J. Biol. Chem. 287, 7675–7682.

Ullrey, D., Combs, G., Conrad, H., Hoekstra, W., Jenkins, K., Levander, O., Whanger, P., 1983. Selenium in Nutrition, revised ed. NAS-NRC, Washington, DC.

Van Dael, P., Davidsson, L., Ziegler, E.E., Fay, L.B., Barclay, D., 2002. Comparison of selenite and selenate apparent absorption and retention in infants using stable isotope methodology. Pediatr. Res. 51, 71–75.

Van Vleet, J.F., Carlton, W., Olander, H.J., 1970. Hepatosis dietetica and mulberry heart disease associated with selenium deficiency in Indiana swine. J. Am. Vet. Med. Assoc. 157, 1208–1219.

Vanderpas, J.B., Contempre, B., Duale, N.L., Goossens, W., Bebe, N., Thorpe, R., Ntambue, K., Dumont, J., Thilly, C.H., Diplock, A.T., 1990. Iodine and selenium deficiency associated with cretinism in northern Zaire. Am. J. Clin. Nutr. 52, 1087–1093.

von Wassermann, A.V., Keysser, F., Wassermann, M., 1911. Beiträge zum Problem: Geschwülste von der Blubahn aus therapeutisch zu beeinflussen. Dtsch. Med. Wschr 37, 223–232.

Wagner, P.A., Hoekstra, W.G., Ganther, H.E., 1975. Alleviation of silver toxicity by selenite in the rat in relation to tissue glutathione peroxidase. Proc. Soc. Exp. Biol. Med. 148, 1106–1110.

Wang, L., Bonorden, M.J., Li, G.X., Lee, H.J., Hu, H., Zhang, Y., Liao, J.D., Cleary, M.P., Lu, J., 2009. Methyl-selenium compounds inhibit prostate carcinogenesis in the transgenic adenocarcinoma of mouse prostate model with survival benefit. Cancer Prev. Res. Phila. Pa. 2, 484–495.

Wastney, M.E., Combs Jr., G.F., Canfield, W.K., Taylor, P.R., Patterson, K.Y., Hill, A.D., Moler, J.E., Patterson, B.H., 2011. A human model of selenium that integrates metabolism from selenite and selenomethionine. J. Nutr. 141, 708–717.

Whanger, P.D., 2009. Selenoprotein expression and function-selenoprotein W. Biochim. Biophys. Acta 1790, 1448–1452.

Whanger, P.D., Weswig, P.H., Schmitz, J.A., Oldfield, J.E., 1977. Effects of selenium and vitamin E deficiencies on reproduction, growth, blood components, and tissue lesions in sheep fed purified diets. J. Nutr. 107, 1288–1297.

Whitacre, M.E., Combs Jr., G.F., Combs, S.B., Parker, R.S., 1987. Influence of dietary vitamin E on nutritional pancreatic atrophy in selenium-deficient chicks. J. Nutr. 117, 460–467.

Wimmer, U., Wang, Y., Georgiev, O., Schaffner, W., 2005. Two major branches of anti-cadmium defense in the mouse: MTF-1/metallothioneins and glutathione. Nucleic Acids Res. 33, 5715–5727.

Wu, M., Kang, M.M., Schoene, N.W., Cheng, W.H., 2010. Selenium compounds activate early barriers of tumorigenesis. J. Biol. Chem. 285, 12055–12062.

Wu, R.T., Cao, L., Chen, B.P., Cheng, W.H., 2014. Selenoprotein H suppresses cellular senescence through genome maintenance and redox regulation. J. Biol. Chem. 289, 34378–34388.

Xia, Y., Hill, K.E., Byrne, D.W., Xu, J., Burk, R.F., 2005. Effectiveness of selenium supplements in a low-selenium area of China. Am. J. Clin. Nutr. 81, 829–834.

Yang, F.Y., Lin, Z.H., Li, S.G., Guo, B.Q., Yin, Y.S., 1988. Keshan disease—an endemic mitochondrial cardiomyopathy in China. J. Trace Elem. Electrolytes Health Dis. 2, 157–163.

Yant, L.J., Ran, Q., Rao, L., Van Remmen, H., Shibatani, T., Belter, J.G., Motta, L., Richardson, A., Prolla, T.A., 2003. The selenoprotein GPX4 is essential for mouse development and protects from radiation and oxidative damage insults. Free Radic. Biol. Med. 34, 496–502.

Yu, Q., Cerklewski, F.L., Whanger, P.D., Hedstrom, O., Ridlington, J.W., 1992. Effect of dietary fluoride on selenite toxicity in the rat. Biol. Trace Elem. Res. 34, 265–278.

Zeng, H., Combs Jr., G.F., 2008. Selenium as an anticancer nutrient: roles in cell proliferation and tumor cell invasion. J. Nutr. Biochem. 19, 1–7.

Zhou, J., Huang, K., Lei, X.G., 2013. Selenium and diabetes—evidence from animal studies. Free Radic. Biol. Med. 65, 1548–1556.

Zhou, J.C., Steinbrenner, H., Rayman, M.P., Lei, X.G., 2015. Multifaceted and intriguing effects of selenium and selenoproteins on glucose metabolism and diabetes. In: Brigelius-Flohe, R., Sies, H. (Eds.), Diversity of Selenium Functions in Health and Disease. CRC Press, pp. 217–246.

Zhou, J.C., Zhao, H., Li, J.G., Xia, X.J., Wang, K.N., Zhang, Y.J., Liu, Y., Zhao, Y., Lei, X.G., 2009. Selenoprotein gene expression in thyroid and pituitary of young pigs is not affected by dietary selenium deficiency or excess. J. Nutr. 139, 1061–1066.

Chapter 38

Selenium and Cancer

Dolph Lee Hatfield[1], Bradley Allen Carlson[1], Petra Akiko Tsuji[2], Ryuta Tobe[1], Vadim N. Gladyshev[3]

[1]National Institutes of Health, Bethesda, MD, United States; [2]Towson University, Towson, MD, United States; [3]Harvard Medical School, Boston, MA, United States

INTRODUCTION

Selenium has a long and fascinating history manifesting multifarious roles in health and disease in mammals. Selenium was initially identified as a toxin and detrimental to health (Franke, 1934), but it was later found to be an essential dietary component for mammals (Schwarz and Foltz, 1958). The first recorded report of selenium as a therapeutic agent in cancer appeared 100 years ago (Walker and Klein, 1915). Another seminal study suggested that this element was a carcinogen, when it was found to promote liver cell adenoma and carcinoma in rats (Nelson, 1943), whereas an additional early study provided evidence that selenium had a role in cancer prevention. Rats treated with 3-methyl-4-dimethylaminobenzene were protected from colon cancer when fed high levels of selenium (Clayton and Baumann, 1949). This intriguing element, indeed, had diversified effects on health serving as a toxin, a carcinogen, and a cancer chemopreventive agent.

Between 1957 and early 2000 many studies appeared attributing health benefits to selenium. These included serving as a cancer chemopreventive agent; preventing heart disease and numerous other cardiovascular diseases and muscle disorders; and having roles in mammalian development, slowing the aging process, inhibiting viral expression, and boosting immune function (Hatfield, 2001). Of all of the health benefits attributed to selenium, its role in preventing cancer, and more recently in promoting cancer, have by far received the most attention, which are subjects in this chapter.

SELENIUM AND EPIDEMIOLOGIC STUDIES

Epidemiology is considered by many researchers to be one of the cornerstones of public health. Epidemiologic studies have provided an important tool in assessing the significance of selenium in cancer (see reviews in Steinbrenner et al., 2013; Vinceti et al., 2013; Wallenberg et al., 2014, and references therein). There are several epidemiologic studies that we would like to briefly mention. In an early study the intake of selenium was estimated in the diets in 27 countries and reported significant inverse correlations for numerous cancers, such as lung, breast, ovary, prostate, the large intestine, rectum, and leukemia, and weak inverse associations with cancers of the pancreas, skin, and bladder (Schrauzer et al., 1977).

In a human clinical trial involving 1312 patients with a prior history of skin cancer, the subjects were administered a yeast extract rich in selenomethionine (SeMet; 200 μg of selenium) or a placebo for a mean period (SD) of 4.5 (2.8) years and the patients were followed for another 6.4 (2.0) years (Clark et al., 1996). The incidence of skin cancer was unaffected, but significant reductions in mortality, lung, prostate, and colorectal cancers were recorded. Although subsequent analyses of the data from this trial supported the findings of a reduction in prostate cancer, but not in lung and colorectal cancers (Duffield-Lillico et al., 2002), it marked the first human trial involving selenium that received enormous attention in the scientific community and the general population.

An interesting study that examined whether low selenium levels were associated with increased lung cancer incidence was performed by measuring serum selenium levels in 3333 males over a 16-year period (Suadicani et al., 2012). Although low levels of serum selenium were not found to be associated with increased lung cancer risk, a significantly greater risk from lung cancer mortality was observed among heavy smokers with higher selenium levels.

Another epidemiological study has recently cautioned against elevated uses of selenium in the diet. Men ($n = 4459$) that had been diagnosed with nonmetastatic prostate cancer were followed for a 22-year period, and it was found that supplementing their diets with 140 μg/day or more of selenium appeared to enhance mortality (Kenfield et al., 2015).

Although epidemiologic studies can be instrumental in pinpointing selenium, or a specific molecular form of selenium (e.g., SeMet), as a dietary component having a major impact in preventing and/or promoting disease, a major drawback in such studies is that they provide little insight into how selenium acts at the molecular level. For example, it is not known whether small-molecular–weight selenocompounds (designated hereafter as selenocompounds), selenoproteins, or both may be responsible for the dual personality of selenium involving the beneficial and detrimental effects of this essential nutrient on health. We will next turn our attention to these two groups of selenium species to further examine their roles in cancer prevention and/or promotion.

SELENOCOMPOUNDS OR SELENOPROTEINS?

A major debate occurred in the selenium field in the late 1900s and early 2000s with regard to which selenium species were responsible for the health benefits of selenium, including its role as an important cancer chemopreventive agent (Combs and Gray, 1998; Ganther, 1999). Selenocompounds, such as triphenylselenonium chloride (TPSC), selenoamino acids, and methylated Se-metabolites [e.g., Se-methylselenocysteine (MeSecys) and SeMet], have been shown to have roles in providing health benefits in animal cancer models (Abdulah et al., 2005; El-Bayoumy, 2001). Little was known about the functions of selenoproteins before early 2000, which prompted numerous laboratories to elucidate their roles in health and development as discussed in the next section.

Selenium and Selenocompounds

Adequate and supranutritional selenium levels have been shown to be beneficial with respect to cancer prevention. Selenite and selenate are major inorganic forms of selenium, whereas SeMet, selenocysteine (Sec), and MeSecys are common organic sources in the diet (Rayman, 2012). Other forms of selenocompounds present in food include selenocysteine (CysSeSeCys), selenoneine, and γ-glutamyl-Se-methylselenocysteine (Rayman, 2012). Selenocompounds found in dietary supplements include selenite and selenate, SeMet, MeSecys, and selenium-enriched yeast, which has been found to contain primarily SeMet (El-Bayoumy, 2001). Selenocompounds such as selenobetaine, methyl selenocyanate, and MeSecys, which generate methylated selenium metabolites, have been shown to be more protective against cancer than selenite or SeMet (see (Weekley and Harris, 2013) and references therein). Se-allylselenocysteine, Se-propylselenocysteine, benzyl selenocyanate, and phenylenebis(methylene) selenocyanate have been reported as having even greater cancer chemopreventive effects (Brigelius-Flohé, 2008). In addition, supplementation in rats with the compound TPSC was shown to suppress mammary tumorigenesis, while having little effect on tissue selenium or selenoprotein levels (Ip et al., 2000).

Human clinical studies have primarily used selenium in the form of sodium selenite, SeMet, or Se-enriched yeast (reviewed in Brigelius-Flohé, 2008; Rayman, 2012). At nutritional levels, these compounds have been considered to have antioxidant functions, including glutathione peroxidase (GPx)-like radical scavenging activity and binding metals to prevent oxidative DNA damage; they are also further metabolized for use in the expression of selenoproteins with antioxidant properties (Weekley and Harris, 2013). Thus they function in protecting cells from oxidative damage, which is known to contribute to carcinogenesis. At higher or supranutritional levels, selenium may function as a prooxidant (Fernandes and Gandin, 2014), and this anticancer, prooxidant role of selenocompounds is most likely due to generation of reactive oxygen species, oxidation of thiols, and increases in DNA damage leading to apoptosis (Fernandes and Gandin, 2014; Weekley and Harris, 2013). Many cancer cells are known to be more sensitive to selenium than normal cells (Fernandes and Gandin, 2014; Wallenberg et al., 2014). Selenocompounds, such as methylseleninic acid, MeSecys, and CysSeSeCys, have also been shown to sensitize cancer cells to chemotherapeutics (Weekley and Harris, 2013), whereas others, such as selenite, are more toxic to drug-resistant cancer cells (Fernandes and Gandin, 2014). These results are most likely due to cancer cells having a greater sensitivity to oxidative stress. The roles of these selenocompounds have also been reviewed elsewhere (Fernandes and Gandin, 2014; Wallenberg et al., 2014; Weekley and Harris, 2013).

Selenoproteins

Two important approaches used in studying the molecular biology of selenoproteins that provided avenues to identification and elucidating their roles in health, development, and disease were a bioinformatics approach that identified selenoprotein genes in genomes (Lobanov et al., 2009) and mouse models involving alterations in expression of the two subclasses of selenoproteins, housekeeping selenoproteins and stress-related selenoproteins (Carlson et al., 2015), and individual selenoproteins (Conrad and Schweizer, 2010).

Bioinformatics and Identification of Selenoproteins

To understand the roles of selenoproteins in cancer, and more generally their roles in biology, it was important to determine the identity of selenoprotein genes in the genomes of organisms, including the human genome. For most trace elements this is a challenge because some trace-element–containing proteins may be of low abundance, expressed only under certain conditions or present only in certain cell types. However, because Sec insertion is cotranslational, information on selenoprotein genes is faithfully represented in genomes in the form of in-frame UGA codons and *Sec insertion sequence* (SECIS) elements in the 3′-untranslated regions (UTRs). Indeed, it was hypothesized early on that these features can be used for selenoprotein gene identification.

The most efficient way to identify selenoprotein genes is to search genomes for SECIS elements (Kryukov et al., 1999). Application of tools, such as SECISearch, allowed identification of all or almost all selenoprotein genes encoded in various genomes (Kryukov et al., 2003). However, a challenge has been to develop tools specific for a particular domain of life because SECIS elements are different in bacteria, archaea, and eukaryotes (Kryukov et al., 2003; Kryukov and Gladyshev, 2004; Zhang and Gladyshev, 2005). An independent approach to selenoprotein identification has been to search genomes for open reading frames (ORFs) containing in-frame UGA codons that correspond to cysteine codons in homologs in the same or other organisms. This method took advantage of the observation that almost all selenoproteins have homologs in which Cys replaces Sec. This approach can be used to identify selenoproteins and determine the location of the Sec in these proteins, which in turn is indicative of the redox function of Cys-containing orthologs and indicates the location of redox-active Cys residues that align with Sec (Fomenko et al., 2007).

Interestingly, the two methods (SECISearch based and the Sec/Cys homology based) yielded very similar sets of selenoprotein genes in organisms. Thus both methods work well in identifying selenoprotein genes. More recently, much more sophisticated methods have been developed, which improved SECIS analyses and integrated them with prediction of ORFs. These tools—SECISearch 3, Seblastian (Mariotti et al., 2013) and Seprofiles (Mariotti and Guigo, 2010)—as well as databases relevant to selenoprotein research—SelenoDB (Romagne et al., 2014), dbTEU (Zhang and Gladyshev, 2010), and Recode 2 (Bekaert et al., 2010)—provide the necessary tools for the analyses of selenoproteomes, which represent full sets of selenoproteins in an organism. A summary of the selenoprotein genes identified are shown in Fig. 38.1.

Selenoproteomes allow linking the genomic information with the specific biology of selenium. For example, the human selenoproteome is encoded by 25 known genes, whereas the mouse selenoproteome by 24 genes (Kryukov et al., 2003). Thus, the essential nature of selenium in these organisms must be due to some of these genes and the function of their products. Moreover, expression and localization of these proteins can be linked with various biological processes that depend on selenium. Other mammalian selenoproteomes feature similar numbers of genes, suggesting general conservation of selenium biology in these organisms. Interestingly, the selenoproteomes of fish are typically larger (e.g., zebrafish has 38 genes) whereas those of invertebrates are generally smaller (e.g., there are three selenoprotein genes in *Drosophila melanogaster*, and only one such gene, thioredoxin reductase (TrxR), in *Caenorhabditis elegans*).

The origin and evolution of each human selenoprotein have been traced within vertebrates (Mariotti et al., 2013). Interestingly, some mammals (e.g., platypus) have additional selenoprotein genes, such as Selenoprotein U, Selenoprotein Pb, and a second copy of selenophosphate synthetase 2, whereas some selenoproteins were converted during evolution to Cys-containing proteins (e.g., GPx6) in some rodents and some primates.

A particularly interesting case is presented by selenoprotein P (Sepp1), a protein with many Sec residues (7–15 Sec in mammals, 0–27 Sec in animals). The first Sec in this protein is strictly conserved and located in the N-terminal thioredoxin (Trx)-like domain, whereas the rest of the Sec residues are characterized by variable conservation and reside in the C-terminal Sec-rich sequence. As such, this protein offers an interesting evolutionary model for future studies.

Selenoproteomes may be adjusted during evolution. For example, it was found that terrestrial organisms generally have smaller selenoproteomes and aquatic organisms larger selenoproteomes (Lobanov et al., 2007). Some terrestrial animals lost all selenoproteins. This is especially common among insects, but all selenoproteins were also lost in higher plants, fungi, and many unicellular eukaryotes. It is also clear that mammals have reduced the utilization of Sec during evolution (Lobanov et al., 2008). Rapid accumulation of genome sequences provides new opportunities to examine the evolution of selenium utilization in mammals and other organisms.

A list of the selenoproteins in mammals, their functions, or plausible functions, and the position of the UGA Sec codon within the coding region are shown in Fig. 38.2. Three of the selenoproteins shown in Fig. 38.1, TrxR1, Sep15, and GPx2, have dual personalities in health because they have roles in preventing and promoting cancer. The role of GPx2 in this regard has been reviewed elsewhere (Kipp, 2015) and will not be further discussed herein. However, before examining the roles of TrxR1 and Sep15 in cancer prevention and promotion, the mouse model approach that involved manipulating the expression of Sec tRNA (designated Sec tRNA[Ser]Sec) will be discussed.

Mouse Models

Mouse models to elucidate the roles of stress-related and housekeeping selenoproteins in health, development, and disease were generated by taking advantage of the fact that selenoprotein synthesis is dependent on Sec tRNA[Ser]Sec

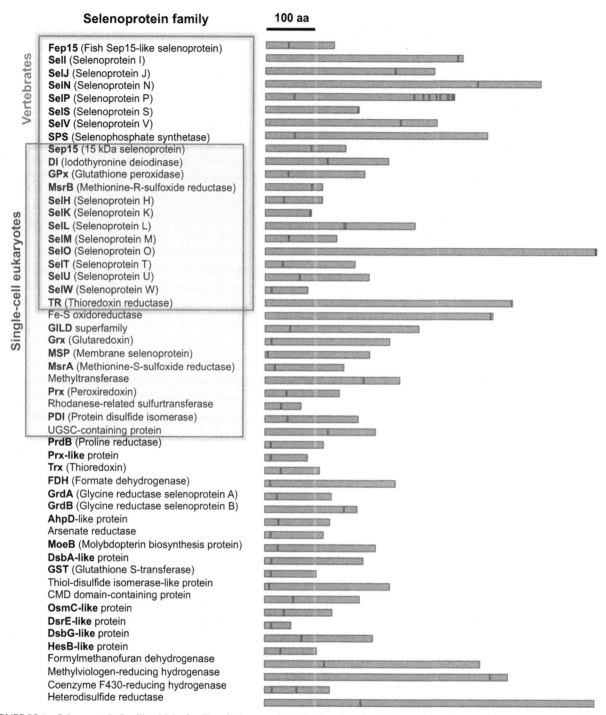

FIGURE 38.1 **Selenoprotein families.** Major families of selenoproteins in the three domains of life are shown. Proteins found in vertebrates and single-cell eukaryotes are indicated by *boxes*. The panel on the right shows relative length of selenoproteins with the position of selenocysteine (Sec) indicated by *red marks*. Some rare selenoproteins are not shown in the figure.

expression (see Carlson et al., 2015; Hatfield et al., 2006 and references therein). Two Sec tRNA[Ser]Sec isoforms occur in higher eukaryotes—5-methylcarboxymethyluridine (mcmU) and 5-methylcarboxymethyluridine-2′-*O*-methylribose (mcmUm)—that differ by a single methyl group on the 2′-*O*-ribosyl moiety at position 34, designated Um34 (Carlson et al., 2015; Hatfield et al., 2006). Um34 in mcmUm influences secondary and tertiary structure and decoding properties of Sec-tRNA[Ser]Sec and is involved in the downregulation of Um34 expression and concomitantly in the downregulation of stress-related selenoprotein synthesis (Carlson et al., 2015; Hatfield et al., 2006). mcmU is involved in the synthesis of

Selenoprotein name	Abbreviations (synonyms)	Function	Sec location/Protein size
15 kDa selenoprotein	Sep15	Quality control of protein folding in ER (putative)	93/162
Iodothyronine deiodinase 1	DI1 (D1, Dio1)	Thyroid hormone-activating deiodinase	126/249
Iodothyronine deiodinase 2	DI2 (D2, Dio2)	Tissue-specific thyroid hormone-activating deiodinase	133,266/273
Iodothyronine deiodinase 3	DI3 (D3, Dio3)	Tissue-specific thyroid hormone-deactivating deiodinase	144/278
Glutathione peroxidase 1	GPx1	Cytosolic glutathione peroxidase	47/201
Glutathione peroxidase 2	GPx2	Gastrointestinal glutathione peroxidase	40/190
Glutathione peroxidase 3	GPx3	Plasma glutathione peroxidase	73/226
Glutathione peroxidase 4	GPx4	Phospholipid hydroperoxide glutathione peroxidase	73/197
Glutathione peroxidase 6	GPx6	Olfactory glutathione peroxidase	73/221
Methionine sulfoxide reductase B1	MsrB1 (SelR, SelX)	Reduction of oxidized methionine residues	95/116
Selenophosphate synthetase 2	SPS2	Synthesis of selenophosphate	60/448
Selenoprotein H	SelH	Unknown	38/116
Selenoprotein I	SelI	Unknown	387/397
Selenoprotein K	SelK	ER-associated degradation (putative)	92/94
Selenoprotein M	SelM	Unknown	48/145
Selenoprotein N	SelN (SEPN1, SepN)	Muscle development (putative)	428/556
Selenoprotein O	SelO	Unknown	667/669
Selenoprotein P	SelP (Sepp1)	Selenium transport	59,300,318,330,345,352,369,376,378/381
Selenoprotein S	SelS (SEPS1, VIMP)	ER-associated degradation (putative)	188/189
Selenoprotein U	SelT	Unknown	36/182
Selenoprotein V	SelV	Unknown (testis-specific)	273/346
Selenoprotein W	SelW (SEPW1)	Unknown	13/87
Thioredoxin reductase 1	TR1 (TrxR1, TxnRd1)	Reduction of cytosolic thioredoxin	498/499
Thioredoxin/glutathione reductase	TGR (TR2, TrxR3, TxnRd3)	Testis-specific thioredoxin reductase	655/656
Thioredoxin reductase 3	TR3 (TrxR2, TxnRd2)	Reduction of mitochondrial thioredoxin and glutaredoxin	522/523

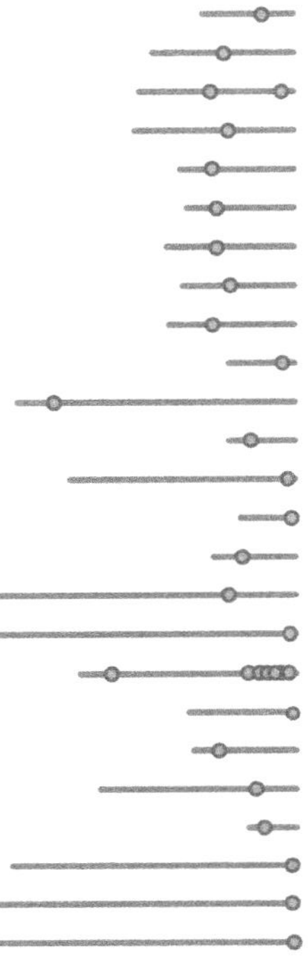

FIGURE 38.2   The human selenoproteome. The human genome codes for 25 selenoprotein genes as shown in the figure. Selenocysteine (Sec) location is designated at the amino acid position within the protein and protein size is the total number of amino acids. A panel on the right shows the relative length of selenoproteins and the location of Sec (*red circle*). Selenoprotein P (Sepp1) has 10 Sec residues.

TABLE 38.1 Mouse Models that Demonstrate Roles of Selenoproteins in Cancer

Model Description	Major Findings[a,b]
TrsptG37 Colon targeted by azoxymethane exposure.	Enhanced azoxymethane-induced aberrant crypt formation (a preneoplastic lesion for colon cancer). First study suggesting that stress-related selenoproteins reduce colon cancer incidence (Irons et al., 2006).
Trsp[tG37]/Tag	[C3(1)/Tag] is a prostate cancer driver gene. Accelerated development of lesions associated with prostate cancer progression. First study suggesting stress-related selenoproteins have a role in preventing prostate cancer (Diwadkar-Navsariwala et al., 2006).
Trsp[tG37]	Higher incidence of micronuclei formation in erythrocytes after exposure to X rays. Data indicate a role of stress-related selenoproteins in protecting DNA from damage (Baliga et al., 2008).[c]
Trsp[tG37] Varying dietary selenium levels. Liver targeted by diethylnitrosamine exposure.	Mice maintained on selenium-deficient, -adequate, and -supplemented diets. Increased incidence of liver tumors in *Trsp*[tG37] mice on selenium-adequate diets, in which selenium-deficient and supplemented levels protected against tumor formation. Selenium-deficient *Trsp*[tG37] mice manifested neurological phenotype (Kasaikina et al., 2013).
Trsp[tG37] *Trsp*[tG37]/TGFα Varying dietary selenium levels or TPSC.	Mice maintained on selenium-deficient, -adequate, and -supplemented diets or diets with TPSC, a nonmetabolized selenium compound. Widespread pyogranuloma formation, severe neurological phenotype associated with early morbidity and mortality in *Trsp*[tG37], and bitransgenic mice on selenium-deficient or TPSC diets. Liver tumors significantly enhanced in *TGFα* (liver cancer driver gene) mice irrespective of selenium or selenoprotein status (Moustafa et al., 2013).
MMTV-Cre; Trsp[fl/fl] Mammary-specific knockout of *Trsp*	Mice lacking *Trsp* in mammary epithelium mice treated with 7,12-dimethylbenz[a]anthracene had significantly more tumors suggesting that selenoproteins protect against carcinogen-induced mammary cancer (Hudson et al., 2012).
PB-Cre4; Trsp[fl/fl] Prostate-specific knockout of *Trsp*	Mice develop prostatic intraepithelial neoplasia (PIN)-like lesions and microinvasive carcinoma by 24 weeks that was associated with loss of basement membrane and increased cell cycle and apoptotic activity (Luchman et al., 2014).

[a]The studies shown in this column describe the major findings observed in selenoprotein altered mice, and the findings described are relative to the corresponding control mice in the study.
[b]The text in the table was taken virtually verbatim from that shown in Tables 1 and 2 in Carlson et al., 2015.
[c]This study did not involve cancer per se, but such changes as shown are known to lead to cancer.

housekeeping selenoproteins. Examples of stress-related selenoproteins are GPx1 and SelW. Selenoproteins in this subclass are nonessential to the animal's survival, whereas examples of housekeeping selenoproteins are TrxR1 and TrxR2 and are essential.

The mouse models consisted of three major groups: (1) transgenes encoding either the wild-type Sec tRNA gene, designated *Trsp*[t], or a mutant form of *Trsp* at position 37, A37→G37 or T34→A34, designated *Trsp*[tG37] and *Trsp*[tA34], respectively; (2) conditional knockout of *Trsp*, designated *Trsp*[cΔ], which targeted specific organs and tissues; and (3) total knockout of *Trsp*, designated *Trsp*Δ, which was complemented by *Trsp*[t] or by *Trsp*[tG37] or *Trsp*[tA34] (Carlson et al., 2015; Hatfield et al., 2006). Synthesis of the modified base, N^6-isopentyladenosine, at position 37 on mcmU was prevented by mutating A to G, which also precluded Um34 synthesis inhibiting stress-related selenoprotein expression. These mouse models provided a major tool in elucidating the roles of selenoproteins in many disorders and other metabolic functions (Carlson et al., 2015), including cancer prevention (Table 38.1).

One disadvantage of the mouse models involving *Trsp* is that they do not distinguish which selenoproteins within either subclass are responsible for specific cellular functions. Therefore, knockout of specific selenoprotein genes has also provided a highly significant tool in elucidating the roles of individual selenoproteins (Conrad and Schweizer, 2010). As noted earlier, there are three selenoproteins—GPx2, TrxR1, and Sep15—that have been examined extensively regarding their roles in preventing and promoting cancer. GPx2 has recently been reviewed (Kipp, 2015), and TrxR1 and Sep15 are discussed in the next sections.

TrxR1

TrxR is found in all examined organisms and all cell types. It exists as either a Sec- or Cys- (in place of Sec) containing enzyme and occurs in organisms within all three domains of life—Eubacteria, Archaea, and Eukaryota (Labunskyy et al., 2014). TrxR1 is found in the cytoplasm and the nucleus in eukaryotes, and its principal function in normal cells is to control the redox state and, therefore, the function of Trx (see Arnér, 2009; Labunskyy et al., 2014 and references therein). This selenoenzyme is described as one of the most important antioxidants and redox regulators in mammalian cells; it also has cellular functions other than reducing Trx (see Arnér, 2009; Arnér and Holmgren, 2000; Labunskyy et al., 2014 and references therein). By maintaining Trx in the reduced state, TrxR1 also has roles in cell proliferation, DNA repair and replication, angiogenesis, transcription, and resistance to oxidative stress, and it serves as a redox regulator in cell signaling (Arnér, 2009; Arnér and Holmgren, 2000; Carlson, 2015). In addition, TrxR1 has a role in activating the P53 and other tumor cancer suppressors and is specifically targeted by carcinogenic electrophilic compounds such as cyclopentenone prostaglandins and lipid aldehydes and ketones with electrophilic β-carbons (Moos et al., 2003). Thus TrxR1 is a key protein in redox homeostasis.

Many cancer cells suffer from oxidative stress, in part because of accumulation of deleterious mutations in cancer genomes. These cells become even more dependent on the antioxidant properties of TrxR1 to drive the malignancy. Indeed, this selenoenzyme is frequently overexpressed in cancer cells, a property that has been linked to tumor aggressiveness and apoptosis resistance (see Arnér, 2009; Arnér and Holmgren, 2000; Carlson, 2015; Hedstrom et al., 2009; Moos et al., 2003 and references therein). Because TrxR1 is such an important protein for cancer cells, it has been targeted by many investigators as a tool in cancer therapy (Arnér, 2009; Arnér and Holmgren, 2000; Carlson, 2015). For example, its knockdown in a mouse lung cancer cell line in vitro has shown that the cancer was dependent on TrxR1 (Yoo et al., 2006), whereas its knockout in mouse liver showed that TrxR1 protects hepatocytes exposed to the liver carcinogen, diethylnitrosamine (DEN), from cancer development (Carlson et al., 2012). Loss of TrxR1 in mouse liver resulted in the elevation of Nrf2 activity that in turn induced the expression of GPx2 and several other enzymes in the glutathione (GSH) system under the influence of this transcription factor (Carlson et al., 2012; Suvorova et al., 2009).

The synergistic effect of the two major antioxidant systems, Trx and GSH, in driving malignancy in mammalian cells was first recognized in 2010 (Mandal et al., 2010), in which the downregulation of the Trx system alone had no effect on cancer cells because of the enhancement of the GSH system; however, inhibition of both antioxidant systems resulted in tumor death. Subsequently, it was reported that the GSH and Trx systems synergize in cancer initiation and progression, in which the role of GSH in driving malignancy was found to become dispensable after cancer initiation. Interestingly, however, the inhibition of both systems was required to cause cancer death in vivo and in vitro. The two latter studies provide alternative approaches for targeting therapeutic intervention (Harris et al., 2015).

Sep15

Similar to TrxR1, Sep15 belongs to the group of thiol-oxidoreductase-like selenoproteins with a Trx-like fold in which Sec is the functional residue. Sep15 was initially purified from human T cells and subsequently sequenced, molecularly characterized, and found to be located on chromosome locus 1p31 in humans (Gladyshev et al., 1998). This locus is commonly deleted or mutated in human cancer (Nasr et al., 2003). It may be involved in the quality control of posttranslational protein folding or secretion of specific glycoproteins because of its formation of a 1:1 complex with the UDP-glucose:glycoprotein glucosyltransferase, which is responsible for sensing the status of structural maturation of N-glycosylated proteins in the endoplasmic reticulum (Gladyshev et al., 1998; Nasr et al., 2003).

Low expression of Sep15 mRNA has been observed in malignant tissues of lung, breast, prostate, and liver; in leukocytes of male patients with bladder cancer compared with controls; and in 14 of 23 malignant mesothelioma–derived cell lines and 5 malignant mesothelioma tumor specimens (see Davis et al., 2012 and references therein). The correlation between the expression of Sep15 and colon cancer is less clear, and mechanistic studies utilizing targeted gene downregulation through RNA interference in mammalian cells in culture and development of systemic Sep15 knockout mice suggest that SEP15 mRNA expression is downregulated in many, but not all tumors. Analysis of the 60 human tumor cell lines in the National Cancer Institute's Developmental Therapeutics Program database (http://dtp.nci.nih.gov/mtweb/index.jsp) demonstrated increased Sep15 expression in human colon cancer cell lines compared with other cancer cell lines as well as compared with other selenoproteins. Unlike observations in lung cancer cells, decreased expression of Sep15 was associated with protection against colon cancer, which was mediated, at least in part, by cell cycle arrest in mouse colon CT26 cells and two human colon cancer cell lines, HCT116 and HT29 (see Davis et al., 2012 and references therein). In addition, functional analyses in these mouse colon cancer cell lines using microarrays suggested a possible link between Sep15 expression and interferon-γ-regulated proteins.

Systemic knockout of Sep15 in mice has also been shown to influence colon cancer susceptibility in vivo (Tsuji et al., 2012). This protective effect appears to likely be mediated through altered inflammatory pathways, including upregulation of interferon-γ and interferon-γ-regulated guanylate binding proteins (GBPs) in Sep15 knockout mice. In colorectal cancer patients, upregulation of GBP-1 has been associated with a highly significant, enhanced survival rate (Naschberger et al., 2008).

Interestingly, colon cancer cells that lack Sep15 and TrxR1 do not demonstrate additive or synergistic effects in terms of reversal of the cancer phenotype. A double knockdown of Sep15 and TrxR1 in mouse colon cancer CT26 cells reversed the effects seen in single knockdowns (Tsuji et al., 2015), including a lack of cell cycle arrest that had been observed in cells that lacked either Sep15 or TrxR1. Furthermore, the expression of interferon-γ-regulated Gbp-1, the top upregulated gene in short hairpin Sep15 (shSep15) knockdown cells, was not affected in colon cancer cells lacking Sep15 and TrxR1. The molecular mechanisms behind this unexpected interplay between Sep15 and TrxR1 remain to be elucidated.

POLYMORPHISMS IN SELENOPROTEINS AND RELEVANCE TO CANCER

Epidemiological studies have suggested a relationship between single-nucleotide polymorphisms (SNPs) in selenoprotein genes and cancer risk. Because Sec incorporation requires specific RNA structures, SNPs occurring within the 3′-UTR may alter selenoprotein expression. Variations in genes encoding GPx1-4, selenoproteins H (SelH), N (SelN), R (SelR), S (SelS), W (SelW), P (Sepp1), Sep15, and TrxR1-3 appear to have functional consequences.

The GPx1 variant (rs1050450), which results in a substitution of proline with leucine, significantly links to GPx activity (Karunasinghe et al., 2012) and associates with risk of breast, bladder, or lung cancer (Ravn-Haren et al., 2006). An SNP (T/C) in GPx4 at nucleotide 718 linked with a significantly higher proportion of individuals with C/C genotype and adenocarcinomas (Bermano et al., 2007). Other studies revealed a rare variant in the 5′-UTR of Sepp1 and three SNPs located in Sepp1's 3′-UTR to be significantly associated with increased colorectal adenoma risk (Méplan and Hesketh, 2012). In contrast, the variant Sepp1 (44,321 bp 3′ of short tandem repeat C>T) was associated with reduced adenoma risk. For TrxR1, an 80% reduction for advanced colorectal adenoma risk was observed for carriers of the variant allele IVS1-181C>G. Two SNPs in TrxR3 (rs11718498 and rs9637365) and four SNPs in SelN (rs11247735, rs2072749, rs4659382, rs718391) were associated with rectal cancer. Furthermore, various SNPs in TrxR1-3, and one SNP in SelS, Sep15, and SelW, were found to interact with nonsteroidal antiinflammatory drugs or with estrogen to modify either colon or rectal cancer risk. A promoter polymorphism in SelS was associated with gastric cancer and found to modulate disease risk, with the T allele of rs34713741 increasing colorectal cancer risks in Czech and Korean populations (see Méplan and Hesketh, 2012 and references herein).

Two linked variants in the 3′-UTR of Sep15 cause a C-T or G-A substitution at positions 811 and 1125, respectively. They are in linkage disequilibrium, with one haplotype being relatively rare, occurring in only 7% of Caucasians, albeit in 31% of African Americans (Hu et al., 2001). In reporter gene experiments, these polymorphisms decreased the efficiency of the SECIS element at higher selenium concentrations. The influence of Sep15 genotype in a Polish population and lung cancer as a function of selenium status indicated that those with GA and GG genotypes at position 1125 showed increased risk, but AA genotype benefitted from higher serum selenium concentrations (Jablonska et al., 2008). Furthermore, the Sep15 rs5845 minor T allele was found to be associated with a higher risk for benign prostate disease but a lower risk of developing malignant disease (reviewed in Davis et al., 2012). In a Korean population, the minor alleles for either rs5845 (GG-GA) or rs5859 (CC-CT) were associated with a borderline increased risk of rectal cancer in men, and an association between a different Sep15 polymorphism (rs9433110) and survival after diagnosis with colorectal cancer was observed in two US populations (reviewed in Davis et al., 2012).

The effect of selenoprotein polymorphisms on cancer incidence and mortality appears to depend on the ethnic population studied, variations in selenium status, and cancer type investigated (Davis et al., 2012). Recent advances into the functional consequences of many of these polymorphisms will benefit from the understanding of the mechanism of several of these selenoproteins and provide tools in cancer treatment and prevention.

CONCLUDING REMARKS

The dietary reference intake for selenium in the United States is currently set for men and women with recommended dietary allowances of 55 μg/day (Institute of Medicine, 2000), which is often exceeded through consumption of selenium-rich foods such as fruits, vegetables, fish and red meats, and/or dietary supplements. Because there are many unknowns about the role of selenium in disease, and particularly in preventing and promoting and/or sustaining cancer, supplementation of the diet with selenium is not recommended without careful consideration. Maintaining a good healthy diet will provide sufficient selenium to an individual.

Because oxidative stress is common in cells stressed by malignancy, and selenoproteins help maintain redox homeostasis, it is not surprising that many selenoproteins are present in higher levels in many cancer cells. Humans, who prefer to maintain an enriched diet with selenium, have been reported to do so by ingesting one to two Brazil nuts each day (Thomson et al., 2008). They are seleniferous and contain upward of 60–80 μg of selenium per kernel (ATSDR, 2003; Thomson et al., 2008).

Selenocompounds and selenoproteins have been implicated as the components providing the wide range of health benefits of selenium in mammals. However, the emphasis has shifted dramatically over the past 10 or so years in favor of selenoproteins being the selenium-containing components. This was spurred largely by identification of the mammalian selenoproteome (Kryukov et al., 2003) and characterization of the roles of housekeeping and stress-related selenoproteins (Carlson et al., 2015; Hatfield et al., 2006), as well as individual selenoproteins (Conrad and Schweizer, 2010), in health and disease. The interrelationships between different selenoproteins (Tsuji et al., 2015) and between selenoproteins and other components of cellular metabolism are in the very early stages of being understood (Labunskyy et al., 2014). Other factors that must be taken into account include the effects of selenium deficiency and excess on malignancy. For example, selenium deficiency has been implicated in reducing peritoneal plasmacytoma incidence in mice (Felix et al., 2004) and hepatocarcinomas in transforming growth factor (TGF)-α/c-Myc transgenic mice (Novoselov et al., 2005). The incidence of hepatocarcinogenesis in tumors induced by TGFα/c-Myc was also reduced in the presence of high levels of selenium in the diet. The contrasting effects of deficient and excess selenium levels on hepatocarcinogenesis were also observed in mice exposed to DEN (Kasaikina et al., 2013). In addition, studies have appeared on the role of another selenoprotein, GPx4, in cancer (Harris et al., 2015; Yang et al., 2014). Thus, many investigations have elucidated the role of selenium, and particularly selenoproteins, in cancer, but there is much to be done to further clarify their protective and promoting roles.

ACKNOWLEDGMENT

This work was supported by the Intramural Research Program of the National Institutes of Health (NIH), National Cancer Institute, Center for Cancer Research to D.L.H.; NIH grants CA080946, GM061603, and GM065204 to V.N.G.; and Towson University's Jess and Mildred Fisher College of Science and Mathematics Endowed Chair to P.A.T. The authors thank Dr. Alexey Lobanov for his generous help in preparing the figures.

REFERENCES

Abdulah, R., Miyazaki, K., Nakazawa, M., Koyama, H., 2005. Chemical forms of selenium for cancer prevention. J. Trace Elem. Med. Biol. 19, 141–150.

Arnér, E.S., 2009. Focus on mammalian thioredoxin reductases—important selenoproteins with versatile functions. Biochim. Biophys. Acta 1790, 495–526.

Arnér, E.S., Holmgren, A., 2000. Physiological functions of thioredoxin and thioredoxin reductase. Eur. J. Biochem. 267, 6102–6109.

ATSDR, 2003. Toxicological Profile for Selenium. Center for Disease Control and Prevention, Atlanta, GA.

Baliga, M.S., Diwadkar-Navsariwala, V., Koh, T., Fayad, R., Fantuzzi, G., Diamond, A.M., 2008. Selenoprotein deficiency enhances radiation-induced micronuclei formation. Mol. Nutr. Food Res. 52, 1300–1304.

Bekaert, M., Firth, A.E., Zhang, Y., Gladyshev, V.N., Atkins, J.F., Baranov, P.V., 2010. Recode-2: new design, new search tools, and many more genes. Nucleic Acids Res. 38, D69–D74.

Bermano, G., Pagmantidis, V., Holloway, N., Kadri, S., Mowat, N.A., Shiel, R.S., Arthur, J.R., Mathers, J.C., Daly, A.K., Broom, J., Hesketh, J.E., 2007. Evidence that a polymorphism within the 3'UTR of glutathione peroxidase 4 is functional and is associated with susceptibility to colorectal cancer. Genes Nutr. 2, 225–232.

Brigelius-Flohé, R., 2008. Selenium compounds and selenoproteins in cancer. Chem. Biodivers. 5, 389–395.

Carlson, B.A., 2015. Thioredoxin reductase 1. In: Flohé-Brigelius, R., Sies, H. (Eds.), Diversity of Selenium Functions in Health and Disease. Taylor & Francis, pp. 173–188.

Carlson, B.A., Tobe, R., Tsuji, P.A., Yoo, M.H., Feigenbaum, L., Tessarollo, L., Lee, B.J., Schweizer, U., Gladyshev, V.N., Hatfield, D.L., 2015. Seleno-cysteine tRNA[Ser]Sec: the central component of selenoprotein biosynthesis. In: Flohé-Brigelius, R., Sies, H. (Eds.), Diversity of Selenium Functions in Health and Disease. Taylor & Francis, pp. 55–79.

Carlson, B.A., Yoo, M.H., Tobe, R., Mueller, C., Naranjo-Suarez, S., Hoffmann, V.J., Gladyshev, V.N., Hatfield, D.L., 2012. Thioredoxin reductase 1 protects against chemically induced hepatocarcinogenesis via control of cellular redox homeostasis. Carcinogenesis 33, 1806–1813.

Clark, L.C., Combs Jr., G.F., Turnbull, B.W., Slate, E.H., Chalker, D.K., Chow, J., Davis, L.S., Glover, R.A., Graham, G.F., Gross, E.G., Krongrad, A., Lesher Jr., J.L., Park, H.K., Sanders Jr., B.B., Smith, C.L., Taylor, J.R., 1996. Effects of selenium supplementation for cancer prevention in patients with carcinoma of the skin. A randomized controlled trial. Nutritional Prevention of cancer study group. JAMA 276, 1957–1963.

Clayton, C.C., Baumann, C.A., 1949. Diet and azo dye tumors; effect of diet during a period when the dye is not fed. Cancer Res. 9, 575–582.

Combs Jr., G.F., Gray, W.P., 1998. Chemopreventive agents: selenium. Pharmacol. Ther. 79, 179–192.

Conrad, M., Schweizer, U., 2010. Unveiling the molecular mechanisms behind selenium-related diseases through knockout mouse studies. Antioxid. Redox Signal 12, 851–865.

Davis, C.D., Tsuji, P.A., Milner, J.A., 2012. Selenoproteins and cancer prevention. Annu. Rev. Nutr. 32, 73–95.

Diwadkar-Navsariwala, V., Prins, G.S., Swanson, S.M., Birch, L.A., Ray, V.H., Hedayat, S., Lantvit, D.L., Diamond, A.M., 2006. Selenoprotein deficiency accelerates prostate carcinogenesis in a transgenic model. Proc. Natl. Acad. Sci. U. S. A. 103, 8179–8184.

Duffield-Lillico, A.J., Reid, M.E., Turnbull, B.W., Combs Jr., G.F., Slate, E.H., Fischbach, L.A., Marshall, J.R., Clark, L.C., 2002. Baseline characteristics and the effect of selenium supplementation on cancer incidence in a randomized clinical trial: a summary report on the Nutritional Prevention of Cancer Trial. Cancer Epidemiol. Biomarkers Prev. 11, 630–639.

El-Bayoumy, K., 2001. The protective role of selenium on genetic damage and on cancer. Mutat. Res. 475, 123–139.

Felix, K., Gerstmeier, S., Kyriakopoulos, A., Howard, O.M., Dong, H.F., Eckhaus, M., Behne, D., Bornkamm, G.W., Janz, S., 2004. Selenium deficiency abrogates inflammation-dependent plasma cell tumors in mice. Cancer Res. 64, 2910–2917.

Fernandes, A.P., Gandin, V., 2014. Selenium compounds as therapeutic agents in cancer. Biochim. Biophys. Acta (in press).

Fomenko, D.E., Xing, W., Adair, B.M., Thomas, D.J., Gladyshev, V.N., 2007. High-throughput identification of catalytic redox-active cysteine residues. Science 315, 387–389.

Franke, K.W., 1934. A new toxicant occurring naturally in certain samples of plant foodstuffs I. Results obtained in preliminary feeding trials. J. Nutr. 8, 597–608.

Ganther, H.E., 1999. Selenium metabolism, selenoproteins and mechanisms of cancer prevention: complexities with thioredoxin reductase. Carcinogenesis 20, 1657–1666.

Gladyshev, V.N., Jeang, K.T., Wootton, J.C., Hatfield, D.L., 1998. A new human selenium-containing protein. Purification, characterization, and cDNA sequence. J. Biol. Chem. 273, 8910–8915.

Harris, I.S., Treloar, A.E., Inoue, S., Sasaki, M., Gorrini, C., Lee, K.C., Yung, K.Y., Brenner, D., Knobbe-Thomsen, C.B., Cox, M.A., Elia, A., Berger, T., Cescon, D.W., Adeoye, A., Brustle, A., Molyneux, S.D., Mason, J.M., Li, W.Y., Yamamoto, K., Wakeham, A., Berman, H.K., Khokha, R., Done, S.J., Kavanagh, T.J., Lam, C.W., Mak, T.W., 2015. Glutathione and thioredoxin antioxidant pathways synergize to drive cancer initiation and progression. Cancer Cell 27, 211–222.

Hatfield, D.L., 2001. Selenium: Its Molecular Biology and Role in Human Health. Kluwer Academic Publishers, Norwell, MA.

Hatfield, D.L., Carlson, B.A., Xu, X.M., Mix, H., Gladyshev, V.N., 2006. Selenocysteine incorporation machinery and the role of selenoproteins in development and health. Prog. Nucleic Acid. Res. Mol. Biol. 81, 97–142.

Hedstrom, E., Eriksson, S., Zawacka-Pankau, J., Arner, E.S., Selivanova, G., 2009. p53-dependent inhibition of TrxR1 contributes to the tumor-specific induction of apoptosis by RITA. Cell Cycle 8, 3584–3591.

Hu, Y.J., Korotkov, K.V., Mehta, R., Hatfield, D.L., Rotimi, C.N., Luke, A., Prewitt, T.E., Cooper, R.S., Stock, W., Vokes, E.E., Dolan, M.E., Gladyshev, V.N., Diamond, A.M., 2001. Distribution and functional consequences of nucleotide polymorphisms in the 3′-untranslated region of the human Sep15 gene. Cancer Res. 61, 2307–2310.

Hudson, T.S., Carlson, B.A., Hoeneroff, M.J., Young, H.A., Sordillo, L., Muller, W.J., Hatfield, D.L., Green, J.E., 2012. Selenoproteins reduce susceptibility to DMBA-induced mammary carcinogenesis. Carcinogenesis 33, 1225–1230.

Institute of Medicine, 2000. Dietary References Intakes for Vitamin C, Vitamin E, Selenium, Carotenoids. Institute of Medicine. National Academy Press, Washington, DC.

Ip, C., Lisk, D.J., Ganther, H.E., 2000. Chemoprevention with triphenylselenonium chloride in selenium-deficient rats. Anticancer Res. 20, 4179–4182.

Irons, R., Carlson, B.A., Hatfield, D.L., Davis, C.D., 2006. Both selenoproteins and low molecular weight selenocompounds reduce colon cancer risk in mice with genetically impaired selenoprotein expression. J. Nutr. 136, 1311–1317.

Jablonska, E., Gromadzinska, J., Sobala, W., Reszka, E., Wasowicz, W., 2008. Lung cancer risk associated with selenium status is modified in smoking individuals by Sep15 polymorphism. Eur. J. Nutr. 47, 47–54.

Karunasinghe, N., Han, D.Y., Zhu, S., Yu, J., Lange, K., Duan, H., Medhora, R., Singh, N., Kan, J., Alzaher, W., Chen, B., Ko, S., Triggs, C.M., Ferguson, L.R., 2012. Serum selenium and single-nucleotide polymorphisms in genes for selenoproteins: relationship to markers of oxidative stress in men from Auckland, New Zealand. Genes Nutr. 7, 179–190.

Kasaikina, M.V., Turanov, A.A., Avanesov, A., Schweizer, U., Seeher, S., Bronson, R.T., Novoselov, S.N., Carlson, B.A., Hatfield, D.L., Gladyshev, V.N., 2013. Contrasting roles of dietary selenium and selenoproteins in chemically induced hepatocarcinogenesis. Carcinogenesis 34, 1089–1095.

Kenfield, S.A., Van Blarigan, E.L., DuPre, N., Stampfer, M.J., Giovannucci, L.E., Chan, J.M., 2015. Selenium supplementation and prostate cancer mortality. J. Natl. Cancer Inst. 107, 360.

Kipp, A.P., Muller, M.F., 2015. Glutathione peroxidase 2. In: Brigelius-Flohé, R., Sies, H. (Eds.), Diversity of Selenium Functions in Health and Disease. Taylor & Francis, pp. 189–202.

Kryukov, G.V., Castellano, S., Novoselov, S.V., Lobanov, A.V., Zehtab, O., Guigo, R., Gladyshev, V.N., 2003. Characterization of mammalian selenoproteomes. Science 300, 1439–1443.

Kryukov, G.V., Gladyshev, V.N., 2004. The prokaryotic selenoproteome. EMBO Rep. 5, 538–543.

Kryukov, G.V., Kryukov, V.M., Gladyshev, V.N., 1999. New mammalian selenocysteine-containing proteins identified with an algorithm that searches for selenocysteine insertion sequence elements. J. Biol. Chem. 274, 33888–33897.

Labunskyy, V.M., Hatfield, D.L., Gladyshev, V.N., 2014. Selenoproteins: molecular pathways and physiological roles. Physiol. Rev. 94, 739–777.

Lobanov, A.V., Fomenko, D.E., Zhang, Y., Sengupta, A., Hatfield, D.L., Gladyshev, V.N., 2007. Evolutionary dynamics of eukaryotic selenoproteomes: large selenoproteomes may associate with aquatic life and small with terrestrial life. Genome Biol. 8, R198.

Lobanov, A.V., Hatfield, D.L., Gladyshev, V.N., 2008. Reduced reliance on the trace element selenium during evolution of mammals. Genome Biol. 9, R62.

Lobanov, A.V., Hatfield, D.L., Gladyshev, V.N., 2009. Eukaryotic selenoproteins and selenoproteomes. Biochim. Biophys. Acta 1790, 1424–1428.

Luchman, H.A., Villemaire, M.L., Bismar, T.A., Carlson, B.A., Jirik, F.R., 2014. Prostate epithelium-specific deletion of the selenocysteine tRNA gene Trsp leads to early onset intraepithelial neoplasia. Am. J. Pathol. 184, 871–877.

Mandal, P.K., Schneider, M., Kolle, P., Kuhlencordt, P., Forster, H., Beck, H., Bornkamm, G.W., Conrad, M., 2010. Loss of thioredoxin reductase 1 renders tumors highly susceptible to pharmacologic glutathione deprivation. Cancer Res. 70, 9505–9514.

Mariotti, M., Guigo, R., 2010. Selenoprofiles: profile-based scanning of eukaryotic genome sequences for selenoprotein genes. Bioinformatics 26, 2656–2663.

Mariotti, M., Lobanov, A.V., Guigo, R., Gladyshev, V.N., 2013. SECISearch3 and Seblastian: new tools for prediction of SECIS elements and selenoproteins. Nucleic Acids Res. 41, e149.

Méplan, C., Hesketh, J., 2012. The influence of selenium and selenoprotein gene variants on colorectal cancer risk. Mutagenesis 27, 177–186.

Moos, P.J., Edes, K., Cassidy, P., Massuda, E., Fitzpatrick, F.A., 2003. Electrophilic prostaglandins and lipid aldehydes repress redox-sensitive transcription factors p53 and hypoxia-inducible factor by impairing the selenoprotein thioredoxin reductase. J. Biol. Chem. 278, 745–750.

Moustafa, M.E., Carlson, B.A., Anver, M.R., Bobe, G., Zhong, N., Ward, J.M., Perella, C.M., Hoffmann, V.J., Rogers, K., Combs Jr., G.F., Schweizer, U., Merlino, G., Gladyshev, V.N., Hatfield, D.L., 2013. Selenium and selenoprotein deficiencies induce widespread pyogranuloma formation in mice, while high levels of dietary selenium decrease liver tumor size driven by TGFalpha. PLoS One 8, e57389.

Naschberger, E., Croner, R.S., Merkel, S., Dimmler, A., Tripal, P., Amann, K.U., Kremmer, E., Brueckl, W.M., Papadopoulos, T., Hohenadl, C., Hohenberger, W., Sturzl, M., 2008. Angiostatic immune reaction in colorectal carcinoma: impact on survival and perspectives for antiangiogenic therapy. Int. J. Cancer 123, 2120–2129.

Nasr, M.A., Hu, Y.J., Diamond, A.M., 2003. Allelic loss at the SEP15 locus in breast cancer. Cancer Ther. 1, 293–298.

Nelson, A.A., Fitzhugh, O.G., Calvery, H.O., 1943. Liver tumors following cirrhosis caused by selenium in rats. Cancer Res. 3, 230–236.

Novoselov, S.V., Calvisi, D.F., Labunskyy, V.M., Factor, V.M., Carlson, B.A., Fomenko, D.E., Moustafa, M.E., Hatfield, D.L., Gladyshev, V.N., 2005. Selenoprotein deficiency and high levels of selenium compounds can effectively inhibit hepatocarcinogenesis in transgenic mice. Oncogene 24, 8003–8011.

Ravn-Haren, G., Olsen, A., Tjonneland, A., Dragsted, L.O., Nexo, B.A., Wallin, H., Overvad, K., Raaschou-Nielsen, O., Vogel, U., 2006. Associations between GPX1 Pro198Leu polymorphism, erythrocyte GPX activity, alcohol consumption and breast cancer risk in a prospective cohort study. Carcinogenesis 27, 820–825.

Rayman, M.P., 2012. Selenium and human health. Lancet 379, 1256–1268.

Romagne, F., Santesmasses, D., White, L., Sarangi, G.K., Mariotti, M., Hubler, R., Weihmann, A., Parra, G., Gladyshev, V.N., Guigo, R., Castellano, S., 2014. SelenoDB 2.0: annotation of selenoprotein genes in animals and their genetic diversity in humans. Nucleic Acids Res. 42, D437–D443.

Schrauzer, G.N., White, D.A., Schneider, C.J., 1977. Cancer mortality correlation studies–III: statistical associations with dietary selenium intakes. Bioinorg. Chem. 7, 23–31.

Schwarz, K., Foltz, C.M., 1958. Factor 3 activity of selenium compounds. J. Biol. Chem. 233, 245–251.

Steinbrenner, H., Speckmann, B., Sies, H., 2013. Toward understanding success and failures in the use of selenium for cancer prevention. Antioxid. Redox Signal 19, 181–191.

Suadicani, P., Hein, H.O., Gyntelberg, F., 2012. Serum selenium level and risk of lung cancer mortality: a 16-year follow-up of the Copenhagen Male Study. Eur. Respir. J. 39, 1443–1448.

Suvorova, E.S., Lucas, O., Weisend, C.M., Rollins, M.F., Merrill, G.F., Capecchi, M.R., Schmidt, E.E., 2009. Cytoprotective Nrf2 pathway is induced in chronically txnrd 1-deficient hepatocytes. PLoS One 4, e6158.

Thomson, C.D., Chisholm, A., McLachlan, S.K., Campbell, J.M., 2008. Brazil nuts: an effective way to improve selenium status. Am. J. Clin. Nutr. 87, 379–384.

Tsuji, P.A., Carlson, B.A., Naranjo-Suarez, S., Yoo, M.H., Xu, X.M., Fomenko, D.E., Gladyshev, V.N., Hatfield, D.L., Davis, C.D., 2012. Knockout of the 15 kDa selenoprotein protects against chemically induced aberrant crypt formation in mice. PLoS One 7, e50574.

Tsuji, P.A., Carlson, B.A., Yoo, M., Naranjo-Suarez, S., Xu, X., He, Y., Asaki, E., Seifried, H.E., Reinhold, W.C., Davis, C.D., Gladyshev, V.N., Hatfield, D.L., 2015. The 15kDa selenoprotein and thioredoxin reductase 1 promote colon cancer by different pathways. PLoS One 10, e0124487.

Vinceti, M., Crespi, C.M., Malagoli, C., Del Giovane, C., Krogh, V., 2013. Friend or foe? The current epidemiologic evidence on selenium and human cancer risk. J. Environ. Sci. Health C Environ. Carcinog. Ecotoxicol. Rev. 31, 305–341.

Walker, C.H., Klein, F., 1915. Selenium—its therapeutic value, especially in cancer. Am. Med. J. 1915, 628–629.

Wallenberg, M., Misra, S., Bjornstedt, M., 2014. Selenium cytotoxicity in cancer. Basic Clin. Pharmacol. Toxicol. 114, 377–386.

Weekley, C.M., Harris, H.H., 2013. Which form is that? The importance of selenium speciation and metabolism in the prevention and treatment of disease. Chem. Soc. Rev. 42, 8870–8894.

Yang, W.S., SriRamaratnam, R., Welsch, M.E., Shimada, K., Skouta, R., Viswanathan, V.S., Cheah, J.H., Clemons, P.A., Shamji, A.F., Clish, C.B., Brown, L.M., Girotti, A.W., Cornish, V.W., Schreiber, S.L., Stockwell, B.R., 2014. Regulation of ferroptotic cancer cell death by GPX4. Cell 156, 317–331.

Yoo, M.H., Xu, X.M., Carlson, B.A., Gladyshev, V.N., Hatfield, D.L., 2006. Thioredoxin reductase 1 deficiency reverses tumor phenotype and tumorigenicity of lung carcinoma cells. J. Biol. Chem. 281, 13005–13008.

Zhang, Y., Gladyshev, V.N., 2005. An algorithm for identification of bacterial selenocysteine insertion sequence elements and selenoprotein genes. Bioinformatics 21, 2580–2589.

Zhang, Y., Gladyshev, V.N., 2010. dbTEU: a protein database of trace element utilization. Bioinformatics 26, 700–702.

Chapter 39

Could Selenium Be a Double-Edged Sword?

Nishi Karunasinghe, Lynnette Robyn Ferguson
University of Auckland, Auckland, New Zealand

INTRODUCTION

Selenium (Se) is an essential micronutrient in animals, plants, and microorganisms, but is harmful in excess. Of all the elements in nature, Se has the narrowest range between deficiency and toxicity for humans (<40 to >400 μg/day; World Health Organization, 1996). Its deficiencies are implicated in human and animal health, and several pathophysiological conditions. Human and animal Se requirements are generally supplied through the food chain originating from plant sources. Inorganic Se is available in nature in four known oxidative states: selenate, selenite, elemental Se, and selenide (Mangiapane et al., 2014). Metalloid Se in soil generally varies between 0.01 and 2.0 mg/kg, with a mean of 0.4 mg/kg. Plant Se levels vary considerably based on absorbable levels from soil (Mehdi et al., 2013). For example, wheat grain produced in United States may contain 0.37–0.457 mg/kg (Hahn et al., 1981; Wolnik et al., 1983) while that from United Kingdom is between 0.025 and 0.033 mg/kg (Adams et al., 2002). Plants convert absorbed Se into seleno amino acids and they can get further metabolized to other Se moieties including elemental Se (Pilon et al., 2003) and dimethylselenide (Carvalho et al., 2001). In wheat grains, more than 50% of seleno amino acids are selenomethionine (SeMe) (Wolf and Goldschmidt, 2007). Various other Se moieties, including selenomethionine Se-oxide and Se-methyl-selenocysteine, are also recoded from root and xylem extracts of wheat plants (Li et al., 2008). Seleno amino acids are incorporated nonspecifically into proteins by the replacement of the amino acid with its' seleno amino acid counterpart (e.g., methionine by SeMe), or specifically in a genetically coded pathway that inserts selenocysteine (Sec) into synthesized polypeptide chains to form selenoproteins (Mangiapane et al., 2014; Varlamova et al., 2013; Burk et al., 2001). It is believed that selenoproteins have been independently lost in higher plants, fungi, and some animal species (Novoselov et al., 2002).

Sec is present in three evolutionary lines of descent: eukarya, archaea, and eubacteria (Labunskyy et al., 2014). Sec incorporation into polypeptide chains is very different from that of the other 20 amino acids. This incorporation involves an in-frame stop codon UGA on the mRNA that will be recognized by the specific tRNA (Shetty and Copeland, 2015). Sec is the only known amino acid in eukaryotes that has a unique tRNA of its own, the tRNA[Ser]Sec (Lee et al., 1989). Each selenoprotein molecule carries at least one inserted Sec (Labunskyy et al., 2014). Kitajima and Chiba (2013) discuss the possibility of Sec incorporation in polypeptide chains as a mechanism of trapping toxic Sec in proteins, and preventing effects of free selenol.

There are 25 selenoproteins known to date with additional isoforms made through RNA splicing (Mariotti et al., 2012). This shows the importance of Se as a micronutrient. To overcome Se nutrition deficiencies in certain parts of the world, attempts are being made to produce Se-enriched plant or yeast types. However, Se enrichment in plants is a formidable task as the majority of plants, except for rare Se-hyper-accumulating species, show Se toxicities at higher levels (Van Huysen et al., 2003).

SOURCES OF ANIMAL-BASED SE FOR HUMAN CONSUMPTION

Fish is considered as having high levels of bioavailable Se (Fox et al., 2004) and the level of fish consumption has shown correlations with plasma Se, selenoprotein P (Sepp1), and glutathione peroxidase (GPX) levels (Hagmar et al., 1998). Cunnane and Crawford (2014) and Cunnane (2006) discuss the possibility of access to marine and fresh water food to early hominin lineage as responsible for human brain expansion and evolution. Although fish provides a Se source for humans, it has been shown that Se, mercury, and arsenic levels in umbilical cord blood, mothers blood, and in breast milk correlate with fish consumption (Miklavcic et al., 2013). The recorded Se levels have varied between 0.17 mg/kg dry matter (dm) in farmed Atlantic salmon to 8.23 mg/kg dm in blue mussel (Bryszewska and Mage, 2015).

Molecular, Genetic, and Nutritional Aspects of Major and Trace Minerals. http://dx.doi.org/10.1016/B978-0-12-802168-2.00039-7

The kidney and liver of vertebrates are rich sources of Se due to the fact Se is metabolized in these organs (Levander, 1987). Red meat is also a good source of Se (Ferguson, 2010). A study in Greek adults has shown that consumption of red meat has a significant correlation with serum Se levels, while other Se-containing foods, including fish, dairy, cereals, and vegetables, did not show as significant a correlation (Letsiou et al., 2010). Se availability from meats produced through farming will vary based on geographical location and feed types used. It is reported that 30–35 μg of Se per day can be provided by eggs produced by hens fed Se-enriched diets containing 0.3–0.5 mg/kg Se. This process of producing Se-enriched eggs has been adopted by many countries the world over (Fisinin et al., 2009).

The biological activities of Se depend on the speciation of Se metabolites between times of Se consumption until all or part of it is removed from the body. Removal of Se from the body takes place through urine (as moieties including trimethylselenonium [TMSe+] and seleno sugars), or eliminated through respiration or removed through feces as elemental Se (Weekley et al., 2013, 2014; Tsuji et al., 2009; Davis et al., 2013; Suzuki et al., 2013). The intermediates between ingestion and removal include selenoproteins, proteins with seleno amino acids in place of sulfur-containing amino acids as well as other seleno metabolites including dimethylselenide, selenomethionine Se-oxide, and Se-methyl-selenocysteine.

SELENOPROTEINS AND THEIR FUNCTIONS

There are a variety of selenoproteins carrying out diverse biological functions through various molecular pathways. Sepp1 is the main selenoprotein produced in the liver of mammals and distributed throughout the body for various functions. Burk et al. has studied the structure and function of Sepp1 in greater detail (Burk and Hill, 1994, 2005, 2009). Well-known roles of selenoproteins include antioxidant function, immune regulation, hormone regulation, and reproduction. Selenoprotein functions have been reviewed previously (Labunskyy et al., 2014; Ferguson et al., 2012; Raymond et al., 2014; Hatfield et al., 2014; Gromer et al., 2005; Steinbrenner et al., 2015; Zhang et al., 2014; Pillai et al., 2014). Five selenoproteins, GPX4, thioredoxin reductases 1 & 2 (Txnrd1, Txnrd2), iodothyronine deiodinase 3 (Dio3), and Sepp1, are considered "essential" based on the triage theory, where preference is given for short-term survival and/or reproduction (McCann and Ames, 2011). McCann and Ames have grouped GPX1, GPX 2, GPX 3, Dio1, Dio2, methionine sulfoxide reductase B1, and selenoprotein N (SelN) as "non-essential selenoproteins," the activities and concentrations of which are preferentially lost in a modest Se deficiency. They also associate age-related diseases such as cancer, immune dysfunction, and heart disease as partly due to modest Se deficiencies that deprive the activities and concentrations of such "non-essential selenoproteins." Both in vitro and in vivo experiments have shown that Se depletion causes variable reduction of either protein or mRNA levels or both in various tissues types and various seleno molecules to varying degrees (Bermano et al., 1995; Villette et al., 1998). Varying roles of Dio2 and Dio3 with regards to conversion of the prohormone thyroxine T_4 to the active thyroid hormone T_3, and inactivation of both T_4 and T_3 are previously reported (Bianco et al., 2002). Meanwhile, Dios are also known to take part in tumorigenesis and cancer growth (Casula and Bianco, 2012).

MECHANISMS OF SELENIUM/SELENOPROTEIN ACTION

Selenoproteins and/or various selenium compounds exert their action through various mechanisms and a few are described here.

Epigenetic Modulation

Histone acetylation effects of selenite on primary and immortalized macrophages have been recorded (Narayan et al., 2015). In this study, acetylation of histone H4 at K12 and K16 in the COX-2 and TNFα promoters, and of the p65 subunit of the redox sensitive transcription factor, nuclear factor kappa-light-chain-enhancer of activated B cells (NF-κB), was decreased significantly by selenite but only weakly by SeMe (Narayan et al., 2015). These authors further recorded a similar effect of selenite on histone H4 at K12 and K16 on the HIV-1 promoter in HIV-1-infected human monocytes, implying a down regulation of proviral expression. It has been shown that esophageal squamous cell carcinoma cells treated with methylseleninic acid decrease the activity level of histone deacetylases (HDAC), with a significant increase of H3K9 acetylation (Hu et al., 2014).

Apoptosis

Okuno et al. have shown that both methylseleninic acid (MSA) (2.5×10^{-6} mol/L) and sodium selenite (7.5×10^{-6} mol/L) cause apoptosis in human lung cancer A549 cells. Under depletion of intracellular reduced glutathione (GSH), sodium selenite has shown increased apoptosis levels while apoptosis by MSA was only slightly suppressed (Okuno et al., 2014). These studies could potentially be showing apoptosis being induced by sodium selenite under stress situations where GSH is depleted. Our own studies have shown that Se supplementation as selenized yeast enhances caspase-cleaved keratin 18 levels in men (Karunasinghe et al., 2016). While this increase is beneficial in eliminating damaged DNA, it could also have implications in pancreatic beta cells that require reactive oxygen species (ROS) stimulation for insulin production (Zhou et al., 2013). Using data from molecular and morphological alterations of Se treatment, it has been shown that selenite produces necrosis/apoptosis-like cell death while selenodiglutathione produces apoptotic cell death in HeLa cells at IC_{50} of $5 \mu M$. In contrast, Sec at IC_{50} of $100 \mu M$ has produced apoptosis-like cell death as well as paraptosis-like cell death involving the production of large vacuoles, endoplasmic reticulum stress, and unfolded protein response (Wallenberg et al., 2014). These experiments have also shown that selenite produces mitochondrial DNA condensation faster than selenodiglutathione, while the latter produces a faster nuclear DNA condensation.

Cell Cycle Arrest

Variation of effects of Se towards cell cycle arrest in two metastatic adenocarcinoma breast cancer cell lines has been shown (Burke et al., 1978; Park et al., 2015). Okuno et al. (2014) reported that in human lung cancer A549 cells, cell cycle arrest at G1 was suppressed by sodium selenite but was improved by MSA in the presence of reduced intracellular GSH levels. It is possible that under stressed circumstances where GSH is depleted, G1 arrest is favored in the presence of MSA.

Induction of Reactive Oxygen Species

It is reported that both selenite and MSA induce ROS as a measure of growth inhibition and apoptosis (Liu et al., 2008, 2012). Wallenberg et al. (2014) have shown that selenite induces significant superoxide production in HeLa cells while Sec does not. It has also been shown that both selenite and selenodiglutathione act as substrates for glutaredoxins, producing ROS (Wallenberg et al., 2010).

Glutathionylation of Proteins

Wallenberg et al. have shown that the thiol groups of cell surface proteins can undergo glutathionylation by selenodiglutathione and selenite. The authors suggest that this can lead to the activation of extrinsic pathways leading to cell death (Wallenberg et al., 2010, 2014).

Post-transcriptional Modifications

It has been shown that sodium selenite and various seleno-carbamates increase the activity of Dio1, Dio2, and Dio3, without any significant change to the corresponding mRNA levels (Stoedter et al., 2015). These data indicate post-transcriptional modifications are taking place due to these Se moieties.

NUTRITIONAL REQUIREMENTS OF SELENIUM

Apart from the five prioritized micronutrients (folic acid, vitamin B_{12}, iodine, iron, and zinc) recognized by the European Micronutrient Recommendations Aligned Network, Se has also been recognized as essential in adequate or supranutritional levels among certain subpopulations (Hurst et al., 2013). These subpopulations include elderly >50 years, patients with HIV infections, men at risk of prostate cancer, and for maintaining immune function and fertility in men. The well-known Nutritional Prevention of Cancer (NPC) study showed the benefit of 200 μg/day Se as selenized yeast in significantly reducing cancer incidence in a cohort in the United States (Clark et al., 1996, 1998). The subsequent Selenium and Vitamin E Cancer Trial (SELECT) that supplemented 200 μg/day Se as SeMe in a larger cohort from the United States could not reproduce the benefits of Se supplementation (Lippman et al., 2005, 2009). Both these trials also raised concerns with regards to possible increased risk of type 2 diabetes, alopecia, and dermatitis associated with Se supplementation (Vinceti et al., 2014). These authors also indicate that, although observational studies record Se supplements for benefits of cancer prevention and cancer-related mortality, randomized control studies do not show such benefits. The general serum Se level beneficial for health has been assessed previously as 120–150 ng/mL (Karunasinghe et al., 2013a; Rayman, 2012). Our studies have also indicated that Se requirements vary considerably with demographic and lifestyle status, health issues, genotype, and inherent DNA stability (Karunasinghe et al., 2012, 2013a; Ferguson et al., 2012).

Demographic Factors

A healthy, male cohort from Auckland, New Zealand, showed no significant change in serum Se level with increasing age or body mass index (BMI). However, both GPX and Txnrd enzyme activities significantly increased both with increasing age and BMI in this cohort (Ferguson et al., 2012; Karunasinghe et al., 2012). Among men and women within a 75–84 years age range in a Tasmanian population, the highest Se levels were recorded among young men in the age range 25–34 years, and in women, the highest levels were recorded between 55 and 74 years (Beckett and Ball, 2011). In a group of British people >65 years, plasma Se concentrations have shown an inverse correlation with age (Bates et al., 2002). A small cohort (n = 83) of men and women between the ages of 45 and 85 years from Brazil, having euthyroid, dyslipidemic, and hypertension health conditions and taking lipid lowering medication, have been assessed for plasma Se levels. In this group the third tertile plasma Se level (>94 µg/L) was associated with those with a mean BMI of around 30.7 kg/m^2 compared to second tertile plasma Se level between 80 and 94 µg/L for those with a BMI of 27.7 kg/m^2. Those at the first tertile of plasma Se showed a mean BMI of 29.0 kg/m^2 implying a "U shaped pattern" of BMI distribution between Se tertiles (Carvalho et al., 2015).

A 200 µg/day Se supplementation to our Auckland, New Zealand, cohort as selenized yeast for six months produced a significant increase in serum Se levels with increasing age. With this supplementation, the correlation of GPX activity with age and BMI seen at the baseline was lost. Supplementation still supported a significant increase in Txnrd activity with age, but not with BMI (Ferguson et al., 2012). Those data imply that both an increase in age and BMI require more of the stress modulating seleno-antioxidant enzymes.

Lifestyle Factors

Seasonal variations in plasma Se levels have been reported in a British study (Bates et al., 2002). Lower plasma Se levels have also been recorded from people in lower socioeconomic strata and those with tobacco-smoking habits (Karunasinghe et al., 2012; Bates et al., 2002). A survey carried out in non-affluent women of the southern United States, including both Caucasian and African-Americans, has shown that 60% of these women had selenium intakes below recommended levels (Lewis et al., 2003). Similarly, 60% of female adolescent ballet dancers living in New Zealand also had a below adequate level of dietary Se intakes. These authors have also recorded similar inadequacies of calcium, folate, and magnesium in this group (Beck et al., 2014). Using data from over 8000 participants in the National Health and Nutrition Examination Survey (NHANES), it has been reported that those inclined to take dietary-mineral supplements also have a higher dietary intake of other minerals compared to non-supplement users. These higher dietary intakes were particularly high for magnesium, copper, potassium, and selenium among males. For females too, these levels were higher, except for the levels of Se (Bailey et al., 2011). According to the NHANES study assessments with over 16,000 subjects, less than 6% of subjects have recorded Se dietary intakes below the estimated average requirement (EAR) (Fulgoni et al., 2011). A study carried out in New Zealand with pregnant women indicated that 89% have taken dietary Se below the EAR of 55 µg/day (Watson and McDonald, 2009). Among adult Arctic indigenous populations in Canada, Se intake was either adequate or exceeded the upper tolerable intake level along with iron, zinc, and vitamins A and D during the 1993 to 1999 period (Kuhnlein et al., 2008). Even among Arctic indigenous children of 10–12 years, Se nutrition along with protein, carbohydrate, iron, copper, zinc, manganese, riboflavin, and vitamins B_6 and C were adequate, in a survey conducted in the 2000–01 period. A Canadian study has shown that those with food addiction have a higher intake of Se alongside other nutrients compared to non-foodaddicted individuals (Pedram and Sun, 2015).

Our studies have shown that current tobacco smokers have lower levels of serum Se compared to those who never smoked (Karunasinghe et al., 2012). Those with a current/past smoking habit had an almost significant lower level of GPX activity compared to never smokers. However, upon Se supplementation, these men showed a significant increase in their GPX activity (Ferguson et al., 2012). This implies that current or past smokers could benefit from Se supplementation. Dietary supplement users in this Auckland cohort also had a significantly higher level of serum Se with a mean serum level at 116 ng/mL, if the supplements contained Se (Karunasinghe et al., 2012). Interestingly, one group of men with a mean serum Se level of 110 ng/mL had a significant increase in serum Se upon supplementation, while the other with a mean of 138 ng/mL had a significant decrease. Upon Se supplementation, the former group had no significant increase in mean GPX activity while showing a significant increase in Txnrd activity; while the latter group had a significant decrease in GPX activity with no change in mean Txnrd activity (Karunasinghe et al., 2013a). Our unpublished data indicates that the latter group had a significantly higher BMI compared to the former. It is possible that the latter obese group utilizes supplemented Se to overcome obesity-related stress at the expense of seleno enzymes. We have also reported that Se supplementation significantly reduces inherent DNA damage in leukocytes in those in the third tertile of baseline DNA damage, while those in the first tertile get an increased damage by supplementation (Karunasinghe et al., 2013a). These data imply that supplementation is beneficial for those with higher levels of inherent DNA damage, while detrimental for those with lower levels of standing DNA damage.

Selenium Requirements in Disease

Selenium and Cardiovascular Risk

The association of cardiovascular disorders with decreased plasma Se levels is discussed by Joseph and Loscalzo (2013) and references therein. Flores-Mateo et al. have conducted a meta-analysis into the association of Se deficiency and coronary disease. They reported that in predominantly Se-deficient populations, a 50% increase in blood or toenail Se decreased the coronary disease risk by 24% (Flores-Mateo et al., 2006). Meanwhile the US Physicians study shows no association between low Se and myocardial infarction (Salvini et al., 1995). The NHANES 1988–94 analysis has shown that serum Se level has a "U shaped" relationship with cardiovascular mortality, with a nadir at 120 ng/mL (Bleys et al., 2008). This beneficial nadir is similar to beneficial levels for optimum health assessed by other studies including ours (Karunasinghe et al., 2012, 2013; Rayman, 2012). This pattern is similar to the relationship of plasma Se level with DNA stability reported for male dogs by Waters et al. (2005). Selenium levels are also negatively correlated with the diameter of abdominal aortic aneurysms with a greatest correlation seen with height-adjusted diameters (Witkowska et al., 2006). Among a dyslipidemic patient cohort of both males and females from Brazil, taking lipid lowering medication, a mean plasma Se level of 88.7 µg/L was recorded (Carvalho et al., 2015), which is considered low for optimal health benefits (Karunasinghe et al., 2012). In our studies, men with hypertension and/or hyperlipidemia, or those taking related medication, showed a significantly higher level of GPX activity in red blood cells (RBC) compared to men with no such health issue, implying a health/medication-related stress. However, after receiving a 200 µg/day Se as selenized yeast for six months, levels showed no significant difference compared to those with no health issues. Upon receiving these supplements, a significant lowering of peroxide-induced DNA damage in white blood cells was recorded in this group with cardiovascular issues compared to control men (Ferguson et al., 2012). Moreover, an endemic cardiomyopathy (Keshan disease) from certain areas in China has been long associated with Se deficiencies, and Se supplementation has proven effective in modulation of this condition (Chen, 2012). However, indirect causes associated with Se deficiency are also reported to play a role in this association. These include the conversion of an amyocarditic strain of coxsackievirus B3 to a pathogenic strain during Se deficiency (Ren et al., 2004; Beck, 1997; Beck et al., 1994) and deficiency of GPX activity, family history, and genetic predisposition to this disease (Lei et al., 2011). Although the Lei et al. (2011) study has shown blood Se has a significant association with Keshan disease by univariate analysis, multivariate analysis did not reveal a significant association with this disease.

Selenium and Diabetes

A meta-analysis has shown no significant benefit of Se supplementation for the prevention of type 2 diabetes mellitus (T2DM) in Caucasian populations (Mao et al., 2014). Women with gestational diabetes mellitus are reported to have a lower serum Se level compared to normal pregnant women especially in the third trimester (Askari et al., 2015). Among T2DM patients from Iran with macroalbuminuria, plasma Se and GPX levels were significantly lower than that of healthy controls and patients with and without microalbuminuria (Sedighi et al., 2014). In a group of female patients with type 1 diabetes mellitus from Sardinia, Italy, Se levels were positively associated with triglyceride levels (Peruzzu et al., 2015).

Selenium and Cancer

Our studies have indicated a significantly lower level of serum Se levels associated with prostate cancer compared to age matched controls from Auckland, New Zealand (Karunasinghe et al., 2013b). However, whether it is the cause or the effect of the disease is not yet known. Indeed, the effect of Se in modulating aggressive prostate cancer is controversial. Plasma Se was not associated with aggressive prostate cancer or its recurrence in a study conducted at the University of California, San Francisco (Gerstenberger et al., 2015). However, a prospective study from the Netherlands indicated that aggressive prostate cancer risk had an inverse relationship with toenail Se levels (Geybels et al., 2014). Dietary intakes of antioxidant minerals Se, zinc, and copper were found to be significantly lower among a group of nonmelanoma skin cancer patients compared to controls (De Almendra Freitas et al., 2015). Moreover, a study from Iran has reported that higher levels of esophageal cancer is associated with areas with high soil Se and where rice produced has significantly higher Se levels (Rahimzadeh-Barzoki et al., 2014; Semnani et al., 2010). However, these studies have not ruled out genetic and lifestyle variability in these regions, although people in high esophageal cancer regions had a higher mean serum Se level of 155 µg/L compared to 82–123 µg/L in other regions. The benefit of Se as selenized yeast for relative risk reduction of prostate, lung, and colon cancer was shown by Clark et al. (1996). Although the subsequent SELECT study proved no benefit of selenomethionine on prostate cancer risk reduction, the difference in the forms of Se used may be critical (Lippman et al., 2009). According to Nolfo et al. (2013), the effect of Se on colon cancer risk reduction is controversial and inconclusive. A meta-analysis has shown an inverse relationship between serum or toenail Se level and breast cancer risk (Babaknejad et al., 2014). Lener et al. (2015) considers Se as a marker for risk surveillance for cancer.

Selenium Requirements for Critical-Care Patients

It is stressed that patients with major burns cared for at critical-care units require maintenance of sufficient blood Se levels (to maintain sufficient antioxidant GPX activity) to avoid infectious complications (Gagnon et al., 2014). Sakr et al. have reported that critically ill surgical patients have significantly lower plasma Se levels as compared to controls. They have also reported that the minimum Se level carries a strong predictive value (0.87) for ICU mortality (Sakr et al., 2007). Patients with severe sepsis and septic shock and those with ischemia-reperfusion from aortic cross-clamping reportedly have lower levels (0.20–0.72 μmol/L) of plasma Se (Forceville et al., 1998). Similarly, ICU patients with systemic inflammatory response syndrome (SIRS) also have lower plasma Se levels than those without SIRS (Forceville et al., 1998). Both GPX3 activity and plasma Se levels were lower than that of controls among ICU patients with SIRS and multiple organ dysfunction syndrome (Manzanares et al., 2009). However, pharmaconutrition with high-dose intravenous Se in the management of severe sepsis and septic shock patients is still controversial (Dellinger et al., 2013; Manzanares et al., 2013; Hardy et al., 2012).

Selenium and Mental Health

Recent studies by Conner et al. have evaluated the serum Se requirement for optimal mental health in young adults in the age range of 17–25 from New Zealand. Their study showed that both higher and lower levels of serum Se elevated depression/impaired mood symptoms compared to those in a midrange of 82 ng/mL (Conner et al., 2015). In this young adult cohort, it is possible that they require a lower level of Se as compared to older individuals for optimal health. Our studies have previously shown that those recording anxiety, depression, bipolar disorder, or schizophrenia, or those taking related medication have a non-significant lower level of serum Se, RBC GPX, and Txnrd activities compared to men with no health concern. When they were supplemented with 200 μg/day of Se as selenized yeast, 24% dropped out of the study without giving a reason for pulling out. We assume that at least one of the reasons for this could be additional Se having negative impacts on their health. We also noted that those with mental illness had a significantly higher peroxide-induced DNA damage, both before and after Se supplementation, compared to that of healthy controls, associated with an increase in the *GPX1* rs1050450 *C* allele (Ferguson et al., 2012). It is possible that their physiology is such that these men react faster to oxidative stress resulting in increased apoptotic DNA damage.

Selenium Requirements With Other Disease Conditions

The dietary Se intake was shown to be similar among HIV-infected children and controls, although plasma Se levels were shown to be significantly lower among HIV-infected children (Pugliese et al., 2014). Dietary intakes of Se were shown to be lower among children taking a gluten-free diet as compared to controls taking a normal diet (Ohlund et al., 2010). A group of patients with thalassemia from hematology outpatient clinics from the United States and Canada were assessed for their dietary intakes as they are generally lower in their nutrient profiles. Although they were lower in their intakes of vitamin A, D, E, K, folate, calcium, and magnesium, they had adequate intakes of riboflavin, vitamin B_{12}, and selenium (Fung et al., 2012). A study in China records that hantavirus infection leading to hemorrhagic fever with renal syndrome (HFRS) is significantly prevalent in areas where crop Se levels are lower (≤0.02 ppm) compared to areas producing higher crop Se levels of >0.06 ppm. They have further recorded that both plasma GPX3 levels and GPX activity levels were higher in men with HFRS compared to healthy men. In women from the HFRS group, although the Gpx3 levels were higher, the activity level was similar to that of control women.

Selenium Requirement Variation With Genotypes

Our studies with a healthy male cohort from Auckland, New Zealand have shown that the maximal beneficial level of serum Se for DNA integrity varies with genotype. Beyond this beneficial serum Se level, DNA damage starts to increase. Those carrying the *GPX1* rs1050450 *CC* genotype benefit by increasing serum Se up to a maximum of 116 ng/mL. Similarly, those with the *GPx4* rs713041 *TT* genotype benefit up to a serum Se level of 149 ng/mL, and thereafter, their DNA damage starts to increase (Karunasinghe et al., 2012). This study has also shown significant variability of RBC GPX, and Txnrd activities with seleno genotypes. All men carrying homozygous major alleles or heterozygous alleles of these genotypes showed a significant (p = .0003–.049) correlation between serum Se levels and GPX activity. In a prospective study of men from the Physicians' Health Study, Li et al. have reported the protective feature of higher baseline plasma Se level against both prostate cancer and aggressive prostate cancer risk in interaction with genotype. They reported that only those carrying the *CC* genotype of the manganese superoxide dismutase (SOD2) gene rs4880 polymorphism show benefits (Li et al., 2005). Evaluation of endogenous lymphocyte DNA damage assessed in a young, healthy, non-smoking male cohort from Newcastle upon Tyne, United Kingdom, has shown similar associations with the same genetic variability (Caple et al., 2010). The interaction of seleno genotypes and Se status influencing the risk of various cancers is also reported (Meplan and Hesketh, 2014). Apart from that, various seleno genotypes are implicated in various disease conditions including cancer (Gerstenberger et al., 2015; Bermano et al., 2007; Meplan et al., 2013; Meulenbelt et al., 2011; Peters et al., 2008; Sutherland et al., 2010; Van Blarigan et al., 2014; Pellatt et al., 2013).

TYPES OF SELENIUM SUPPLEMENTS

Among the well-known Se supplements are sodium selenite, sodium selanate, selenized yeast, and SeMe. Although SeMe is mostly used in Se supplements, multivitamin/mineral preparations, infant formulas, protein mixtures, weight-loss products, and animal feed use either sodium selenite or sodium selenate in their preparations (Schrauzer, 2001). The level of the dominant ingredient SeMe in commercial yeast preparations also varies (Fagan et al., 2015). There is also variability of the requirement of each type of these supplements for optimal health benefits. Studies assessing the optimization of Sepp1 in a Chinese cohort, have shown that this is achievable with 50 µg/day selenized yeast on top of a habitual intake of approximately 55 µg/day. However, in a Se-depleted group, this was achieved by 49 µg/day SeMe (Hurst et al., 2010; Xia et al., 2010). The major Se constituent of selenized yeast is SeMe, although other constituents such as selenomethylselenocysteine (SeMCys) and γ-glu-SeMCys are also recorded (Amoako et al., 2009; Block et al., 2004; Chen et al., 2011; Ip et al., 2000; Rao et al., 2010). Compared to normal yeast types, selenized yeast is reported to have higher levels of oxidized and protein-bound thiols including protein-bound, reduced glutathione (Rao et al., 2010, 2012). These could be exerting toxicities and redox imbalance upon selenized yeast supplementation (Mapelli et al., 2011). However, according to our studies, of those who received 200 µg/day Se as selenized yeast (as SEL-PLEX supplements from Alltech, Dunboyne, Ireland) for six months, less adverse events such alopecia, dermatitis, nail discoloration and pitting, and fatigue were reported compared to those reported in the SELECT study (Karunasinghe et al., 2013a).

DUAL ROLE OF CANCER PREVENTION AND PROMOTION

According to Hatfield et al. (2009), the role of Se as an antioxidant is oversimplified as some selenoproteins carry dual personality. This group has highlighted the role of Txnrd and selenoprotein 15 (Sel15) having both cancer-protective and cancer-promoting properties. Txnrd is regarded as having multiple cancer-preventive functions (Ferguson et al., 2012). However, it is also known that Txnrd is a target for therapy in controlling various cancers (Casini and Messori, 2011; Jackson-Rosario and Self, 2010; Pennington et al., 2007) due to the pro-survival support tendered by Txnrd to the tumor environment (Pennington et al., 2007). When bound to electrophilic factors such as those provided by prostaglandins, Txnrd is known to affect p53 function (Cassidy et al., 2006). A Txnrd knocked-down lung cancer cell line has shown cancer modulating properties such as reverting cancer morphology back to normal, reduced anchorage-independent growth, and reduced expression of two cancer-related genes (Yoo et al., 2006). A Sel15 knockdown colorectal cancer cell line has also shown reversion to normal characteristics (Irons et al., 2010). A comparison of normal and Sel15 knockdown mice has shown that aberrant crypt formation induced by azoxymethane is significantly reduced in the latter (Tsuji et al., 2012). Moreover, Tsuji et al. (2015) have shown that knockdown of both Txnrd1 and Sel15 in CT26 mouse colon carcinoma cells affects both anchorage-dependent or anchorage-independent growth and experimental measures of metastases.

INTERACTION OF SELENIUM WITH OTHER MICRO NUTRIENTS

Copper and Se dietary intakes among HIV-infected children and controls have been reported to be similar; yet plasma levels of Se are lower and plasma levels of copper are higher among HIV-infected than the control subjects (Pugliese et al., 2014). An analysis of qualitative food frequency questionnaire data in a Saudi population indicates an inverse association between vitamin D and Se intakes (Al-Daghri et al., 2015). Wadhwa et al. and references therein discuss the effect of carcinogenic processes affecting trace element balance in individuals. They have also shown that the levels of essential elements Se and Zinc in hair samples of women with breast, ovarian, cervix, and oral cancers are significantly lower than that of controls. They have further shown that levels of toxic elements Arsenic, Nickel, and Cadmium were significantly higher in these samples as compared to controls (Wadhwa et al., 2015). Weekley et al. (2011) have shown that accumulation of Se in A549 human lung cancer cells incubated with selenite resulted in a twofold increase in copper levels, but without resulting in Se–copper bonding.

CONCLUSION

Se is an essential micronutrient to all living organisms, although requirements vary a great deal, as discussed in this chapter. The best source of Se nutrition is through dietary means; however, when this cannot provide Se in adequate levels due to various circumstances, supplementation comes to the rescue. Moreover, unless a complete understanding of the type of requirement as well as the source of supplement is known, Se supplementation can be detrimental, and thus Se is a double-edged sword.

REFERENCES

Adams, L.A., et al., 2002. Evidence of low selenium concentrations in Uk bread-making wheat grain. J. Sci. Food Agric. 82, 1160–1165.

Al-Daghri, N.M., et al., 2015. Dietary products consumption in relation to serum 25-hydroxyvitamin D and selenium level in Saudi children and adults. Int. J. Clin. Exp. Med. 8 (1), 1305–1314.

Amoako, P.O., Uden, P.C., Tyson, J.F., 2009. Speciation of selenium dietary supplements; formation of S-(methylseleno)cysteine and other selenium compounds. Anal. Chim. Acta 652 (1–2), 315–323.

Askari, G., et al., 2015. The association between serum selenium and gestational diabetes mellitus: a systematic review and meta-analysis. J. Trace Elem. Med. Biol. 29, 195–201.

Babaknejad, N., et al., 2014. The relationship between selenium levels and breast cancer: a systematic review and meta-analysis. Biol. Trace Elem. Res. 159 (1–3), 1–7.

Bailey, R.L., et al., 2011. Dietary supplement use is associated with higher intakes of minerals from food sources. Am. J. Clin. Nutr. 94 (5), 1376–1381.

Bates, C.J., et al., 2002. Selenium status and its correlates in a British national diet and nutrition survey: people aged 65 years and over. J. Trace Elem. Med. Biol. 16 (1), 1–8.

Beck, M.A., et al., 1994. Increased virulence of a human enterovirus (coxsackievirus B3) in selenium-deficient mice. J. Infect. Dis. 170 (2), 351–357.

Beck, K.L., et al., 2014. Dietary intake, anthropometric characteristics, and iron and vitamin D status of female adolescent ballet dancers living in New Zealand. Int. J. Sport Nutr. Exercise Metab. 25 (4), 335–343.

Beck, M.A., 1997. Increased virulence of coxsackievirus B3 in mice due to vitamin E or selenium deficiency. J. Nutr. 127 (5 Suppl.), 966S–970S.

Beckett, J.M., Ball, M.J., 2011. Marginal selenium status in northern Tasmania. Br. J. Nutr. 106 (5), 718–724.

Bermano, G., et al., 1995. Tissue-specific regulation of selenoenzyme gene expression during selenium deficiency in rats. Biochem. J. 311, 425–430.

Bermano, G., et al., 2007. Evidence that a polymorphism within the 3′UTR of glutathione peroxidase 4 is functional and is associated with susceptibility to colorectal cancer. Genes Nutr. 2, 227–232.

Bianco, A.C., et al., 2002. Biochemistry, cellular and molecular biology, and physiological roles of the iodothyronine selenodeiodinases. Endocr. Rev. 23 (1), 38–89.

Bleys, J., Navas-Acien, A., Guallar, E., 2008. Serum selenium levels and all-cause, cancer, and cardiovascular mortality among US adults. Arch. Intern. Med. 168 (4), 404–410.

Block, E., et al., 2004. Identification and synthesis of a novel selenium-sulfur amino acid found in selenized yeast: rapid indirect detection NMR methods for characterizing low-level organoselenium compounds in complex matrices. J. Agric. Food Chem. 52 (12), 3761–3771.

Bryszewska, M.A., Mage, A., 2015. Determination of selenium and its compounds in marine organisms. J. Trace Elem. Med. Biol. 29, 91–98.

Burk, R.F., Hill, K.E., 2009. Selenoprotein P – expression, functions, and roles in mammals. Biochim. Biophys. Acta 1790 (11), 1441–1447.

Burk, R.F., Hill, K.E., 2005. Selenoprotein P: an extracellular protein with unique physical characteristics and a role in selenium homeostasis. Annu. Rev. Nutr. 25, 215–235.

Burk, R.F., Hill, K.E., Motley, A.K., 2001. Plasma selenium in specific and nonspecific forms. Biofactors 14 (1–4), 107–114.

Burk, R.F., Hill, K.E., 1994. Selenoprotein P. A selenium-rich extracellular glycoprotein. J. Nutr. 124, 1891–1897.

Burke, R.E., Harris, S.C., McGuire, W.L., 1978. Lactate dehydrogenase in estrogen-responsive human breast cancer cells. Cancer Res. 38 (9), 2773–2776.

Caple, F., et al., 2010. Inter-individual variation in DNA damage and base excision repair in young, healthy nonsmokers: effects of dietary supplementation and genotype. Br. J. Nutr. 103 (11), 1585–1593.

Carvalho, R.F., et al., 2015. The association of selenium status with thyroid hormones and anthropometric values in dyslipidemic patients. Nutr. Hosp. 31 (4), 1832–1838.

Carvalho, K.M., McGettigan, M.J., Martin, D.F., 2001. GC/MS analysis of volatile organic selenium species produced during phytoremediation. J. Environ. Sci. Health A Tox Hazard Subst. Environ. Eng. 36 (7), 1403–1409.

Casini, A., Messori, L., 2011. Molecular mechanisms and proposed targets for selected anticancer gold compounds. Curr. Top. Med. Chem. 11 (21), 2647–2660.

Cassidy, P.B., et al., 2006. Thioredoxin reductase is required for the inactivation of tumor suppressor p53 and for apoptosis induced by endogenous electrophiles. Carcinogenesis 27 (12), 2538–2549.

Casula, S., Bianco, A.C., 2012. Thyroid hormone deiodinases and cancer. Front. Endocrinol. (Lausanne) 3, 74.

Chen, B., et al., 2011. Ionic liquids improved reversed-phase HPLC on-line coupled with ICP-MS for selenium speciation. Talanta 83 (3), 724–731.

Chen, J., 2012. An original discovery: selenium deficiency and Keshan disease (an endemic heart disease). Asia Pac. J. Clin. Nutr. 21 (3), 320–326.

Clark, L.C., et al., 1996. Effects of selenium supplementation for cancer prevention in patients with carcinoma of the skin. A randomized controlled trial. Nutritional Prevention of Cancer Study Group. JAMA 276 (24), 1957–1963.

Clark, L.C., et al., 1998. Decreased incidence of prostate cancer with selenium supplementation: results of a double-blind cancer prevention trial. Br. J. Urol. 81 (5), 730–734.

Conner, T.S., Richardson, A.C., Miller, J.C., 2015. Optimal serum selenium concentrations are associated with lower depressive symptoms and negative mood among young adults. J. Nutr. 145 (1), 59–65.

Cunnane, S.C., Crawford, M.A., 2014. Energetic and nutritional constraints on infant brain development: implications for brain expansion during human evolution. J. Hum. Evol. 77, 88–98.

Cunnane, S.C., 2006. Survival of the fattest: the key to human brain evolution. Med. Sci. (Paris) 22 (6–7), 659–663.

Davis, T.Z., et al., 2013. Evaluation of the respiratory elimination kinetics of selenate and Se-methylselenocysteine after oral administration in lambs. Res. Vet. Sci. 95 (3), 1163–1168.

De Almendra Freitas, B.E., et al., 2015. Antioxidant capacity total in nonmelanoma skin cancer and its relationship with food consumption of antioxidant nutrients. Nutr. Hosp. 31 (4), 1682–1688.

Dellinger, R.P., et al., 2013. Surviving Sepsis Campaign: international guidelines for management of severe sepsis and septic shock, 2012. Intensive Care Med. 39 (2), 165–228.

Fagan, S., et al., 2015. Biochemical comparison of commercial selenium yeast preparations. Biol. Trace Elem. Res. 166 (2), 245–259.

Ferguson, L.R., et al., 2012. Selenium and its' role in the maintenance of genomic stability. Mutat. Res. 733 (1–2), 100–110.

Ferguson, L.R., Karunasinghe, N., Zhu, S., Han, D.Y., Triggs, Christopher, M., Wang, A.H., Masters, J.G., 2012. Understanding heterogeneity in supplementation effects of selenium in men: a study of stratification variables and human genetics in a prospective sample from New Zealand. Curr. Pharmacogenomics Pers. Med. 10 (3), 204–216.

Ferguson, L.R., 2010. Meat and cancer. Meat Sci. 84 (2), 308–313.

Fisinin, V.I., Papazyan, T.T., Surai, P.F., 2009. Producing selenium-enriched eggs and meat to improve the selenium status of the general population. Crit. Rev. Biotechnol. 29 (1), 18–28.

Flores-Mateo, G., et al., 2006. Selenium and coronary heart disease: a meta-analysis. Am. J. Clin. Nutr. 84 (4), 762–773.

Forceville, X., et al., 1998. Selenium, systemic immune response syndrome, sepsis, and outcome in critically ill patients. Crit. Care Med. 26 (9), 1536–1544.

Fox, T.E., et al., 2004. Bioavailability of selenium from fish, yeast and selenate: a comparative study in humans using stable isotopes. Eur. J. Clin. Nutr. 58 (2), 343–349.

Fulgoni 3rd, V.L., et al., 2011. Foods, fortificants, and supplements: where do Americans get their nutrients? J. Nutr. 141 (10), 1847–1854.

Fung, E.B., et al., 2012. Inadequate dietary intake in patients with thalassemia. J. Acad. Nutr. Diet. 112 (7), 980–990.

Gagnon, G., et al., 2014. Trace element monitoring in the ICU: quality and economic impact of a change in sampling practice. Clin. Nutr. 34 (3), 422–427.

Gerstenberger, J.P., et al., 2015. Selenoprotein and antioxidant genes and the risk of high-grade prostate cancer and prostate cancer recurrence. Prostate 75 (1), 60–69.

Geybels, M.S., et al., 2014. Selenoprotein gene variants, toenail selenium levels, and risk for advanced prostate cancer. J. Natl. Cancer Inst. 106 (3), dju003.

Gromer, S., et al., 2005. Human selenoproteins at a glance. Cell Mol. Life Sci. 62 (21), 2414–2437.

Hagmar, L., et al., 1998. Plasma levels of selenium, selenoprotein P and glutathione peroxidase and their correlations to fish intake and serum levels of thyrotropin and thyroid hormones: a study on Latvian fish consumers. Eur. J. Clin. Nutr. 52 (11), 796–800.

Hahn, M.H., et al., 1981. Determination of trace amounts of selenium in corn, lettuce, potatoes, soybeans, and wheat by hydride generation/condensation and flame atomic absorption spectrometry. J. Agric. Food Chem. 29, 792–796.

Hardy, G., Hardy, I., Manzanares, W., 2012. Selenium supplementation in the critically ill. Nutr. Clin. Pract. 27 (1), 21–33.

Hatfield, D.L., et al., 2009. Selenoproteins that function in cancer prevention and promotion. Biochim. Biophys. Acta 1790 (11), 1541–1545.

Hatfield, D.L., et al., 2014. Selenium and selenocysteine: roles in cancer, health, and development. Trends Biochem. Sci. 39 (3), 112–120.

Hu, C., et al., 2014. Upregulation of KLF4 by methylseleninic acid in human esophageal squamous cell carcinoma cells: modification of histone H3 acetylation through HAT/HDAC interplay. Mol. Carcinog. 54 (10), 1051–1059.

Hurst, R., et al., 2010. Establishing optimal selenium status: results of a randomized, double-blind, placebo-controlled trial. Am. J. Clin. Nutr. 91 (4), 923–931.

Hurst, R., et al., 2013. EURRECA-Estimating selenium requirements for deriving dietary reference values. Crit. Rev. Food Sci. Nutr. 53 (10), 1077–1096.

Ip, C., et al., 2000. Chemical speciation influences comparative activity of selenium-enriched garlic and yeast in mammary cancer prevention. J. Agric. Food Chem. 48 (6), 2062–2070.

Irons, R., et al., 2010. Deficiency in the 15-kDa selenoprotein inhibits tumorigenicity and metastasis of colon cancer cells. Cancer Prev. Res. (Phila) 3 (5), 630–639.

Jackson-Rosario, S.E., Self, W.T., 2010. Targeting selenium metabolism and selenoproteins: novel avenues for drug discovery. Metallomics 2 (2), 112–116.

Joseph, J., Loscalzo, J., 2013. Selenistasis: epistatic effects of selenium on cardiovascular phenotype. Nutrients 5 (2), 340–358.

Karunasinghe, N., et al., 2012. Serum selenium and single-nucleotide polymorphisms in genes for selenoproteins: relationship to markers of oxidative stress in men from Auckland, New Zealand. Genes Nutr. 7 (2), 179–190.

Karunasinghe, N., et al., 2013a. Effects of supplementation with selenium, as selenized yeast, in a healthy male population from New Zealand. Nutr. Cancer 65 (3), 355–366.

Karunasinghe, N., et al., 2013b. Prostate disease risk factors among a New Zealand cohort. J. Nutr. Nutr. 5 (6), 339–351.

Karunasinghe, N., et al., April 27, 2016. Benefits of selenium supplementation on leukocyte DNA integrity interact with dietary micronutrients: a short communication. Nutrients 8 (5), pii: E249. http://dx.doi.org/10.3390/nu8050249.

Kitajima, T., Chiba, Y., 2013. Selenomethionine metabolism and its toxicity in yeast. Biomol. Concepts 4 (6), 611–616.

Kuhnlein, H.V., et al., 2008. Unique patterns of dietary adequacy in three cultures of Canadian Arctic indigenous peoples. Public Health Nutr. 11 (4), 349–360.

Labunskyy, V.M., Hatfield, D.L., Gladyshev, V.N., 2014. Selenoproteins: molecular pathways and physiological roles. Physiol. Rev. 94 (3), 739–777.

Lee, B.J., et al., 1989. Identification of a selenocysteyl-tRNA(Ser) in mammalian cells that recognizes the nonsense codon, UGA. J. Biol. Chem. 264 (17), 9724–9727.

Lei, C., et al., 2011. Is selenium deficiency really the cause of Keshan disease? Environ. Geochem. Health 33 (2), 183–188.

Lener, M., et al., 2015. Selenium as a marker of cancer risk and of selection for control examinations in surveillance. Contemp. Oncol. (Pozn.) 19 (1A), A60–A61.

Letsiou, S., et al., 2010. Dietary habits of Greek adults and serum total selenium concentration: the ATTICA study. Eur. J. Nutr. 49 (8), 465–472.

Levander, O.A., 1987. Assessing the bioavailability of selenium in foods. In: Combs Jr., G.F., Spallholz, J.E., Oldfield, J.E. (Eds.), Selenium in Biology and Medicine. AVI Books, New York, pp. 403–412.

Lewis, S.M., et al., 2003. Assessment of antioxidant nutrient intake of a population of southern US African-American and Caucasian women of various ages when compared to dietary reference intakes. J. Nutr. Health Aging 7 (2), 121–128.

Li, H., et al., 2005. Manganese superoxide dismutase polymorphism, prediagnostic antioxidant status, and risk of clinical significant prostate cancer. Cancer Res. 6, 2498–2504.

Li, H.F., McGrath, S.P., Zhao, F.J., 2008. Selenium uptake, translocation and speciation in wheat supplied with selenate or selenite. New Phytol. 178 (1), 92–102.

Lippman, S.M., et al., 2005. Designing the selenium and vitamin E cancer prevention trial (SELECT). J. Natl. Cancer Inst. 97 (2), 94–102.

Lippman, S.M., et al., 2009. Effect of selenium and vitamin E on risk of prostate cancer and other cancers: the Selenium and Vitamin E Cancer Prevention Trial (SELECT). JAMA 301 (1), 39–51.

Liu, C., et al., 2008. Selenium compounds induce ROS in human high-metastatic large cell lung cancer cell line L9981. Zhongguo Fei Ai Za Zhi 11 (3), 354–358.

Liu, C., et al., 2012. Intracellular glutathione content influences the sensitivity of lung cancer cell lines to methylseleninic acid. Mol. Carcinog. 51 (4), 303–314.

Mangiapane, E., Pessione, A., Pessione, E., 2014. Selenium and selenoproteins: an overview on different biological systems. Curr. Protein Pept. Sci. 15 (6), 598–607.

Manzanares, W., et al., 2009. Serum selenium and glutathione peroxidase-3 activity: biomarkers of systemic inflammation in the critically ill? Intensive Care Med. 35 (5), 882–889.

Manzanares, W., Langlois, P.L., Hardy, G., 2013. Selenium pharmaconutrition in sepsis: to give or not to give? Is this still the question? Nutrition 29 (11–12), 1429–1430.

Mao, S., Zhang, A., Huang, S., 2014. Selenium supplementation and the risk of type 2 diabetes mellitus: a meta-analysis of randomized controlled trials. Endocrine 47 (3), 758–763.

Mapelli, V., et al., 2011. Metabolic and bioprocess engineering for production of selenized yeast with increased content of seleno-methylselenocysteine. Metab. Eng. 13 (3), 282–293.

Mariotti, M., et al., 2012. Composition and evolution of the vertebrate and mammalian selenoproteomes. PLoS One 7 (3), e33066.

McCann, J.C., Ames, B.N., 2011. Adaptive dysfunction of selenoproteins from the perspective of the triage theory: why modest selenium deficiency may increase risk of diseases of aging. FASEB J. 25 (6), 1793–1814.

Mehdi, Y., et al., 2013. Selenium in the environment, metabolism and involvement in body functions. Molecules 18 (3), 3292–3311.

Meplan, C., et al., 2013. Association between polymorphisms in glutathione peroxidase and selenoprotein P genes, glutathione peroxidase activity, HRT use and breast cancer risk. PLoS One 8 (9), e73316.

Meplan, C., Hesketh, J., 2014. Selenium and cancer: a story that should not be forgotten-insights from genomics. Cancer Treat. Res. 159, 145–166.

Meulenbelt, I., et al., 2011. Meta-analyses of genes modulating intracellular T3 bio-availability reveal a possible role for the DIO3 gene in osteoarthritis susceptibility. Ann. Rheum. Dis. 70 (1), 164–167.

Miklavcic, A., et al., 2013. Mercury, arsenic and selenium exposure levels in relation to fish consumption in the Mediterranean area. Environ. Res. 120, 7–17.

Narayan, V., et al., 2015. Epigenetic regulation of inflammatory gene expression in macrophages by selenium. J. Nutr. Biochem. 26 (2), 138–145.

Nolfo, F., et al., 2013. Pharmacological and dietary prevention for colorectal cancer. BMC Surg. 13 (Suppl. 2), S16.

Novoselov, S.V., et al., 2002. Selenoproteins and selenocysteine insertion system in the model plant cell system, Chlamydomonas reinhardtii. EMBO J. 21 (14), 3681–3693.

Ohlund, K., et al., 2010. Dietary shortcomings in children on a gluten-free diet. J. Hum. Nutr. Diet. 23 (3), 294–300.

Okuno, T., et al., 2014. Glutathione-dependent cell cycle g1 arrest and apoptosis induction in human lung cancer a549 cells caused by methylseleninic acid: comparison with sodium selenite. Biol. Pharm. Bull. 37 (11), 1831–1837.

Park, S.O., et al., 2015. Effects of combination therapy of docetaxel with selenium on the human breast cancer cell lines MDA-MB-231 and MCF-7. Ann. Surg. Treat. Res. 88 (2), 55–62.

Pedram, P., Sun, G., 2015. Hormonal and dietary characteristics in obese human subjects with and without food addiction. Nutrients 7 (1), 223–238.

Pellatt, A.J., et al., 2013. SEPP1 influences breast cancer risk among women with greater native american ancestry: the breast cancer health disparities study. PLoS One 8 (11), e80554.

Pennington, J.D., et al., 2007. Thioredoxin and thioredoxin reductase as redox-sensitive molecular targets for cancer therapy. Curr. Pharm. Des. 13 (33), 3368–3377.

Peruzzu, A., et al., 2015. Association of trace elements with lipid profiles and glycaemic control in patients with type 1 diabetes mellitus in northern Sardinia, Italy: an observational study. Chemosphere 132, 101–107.

Peters, U., et al., 2008. Variation in the selenoenzyme genes and risk of advanced distal colorectal adenoma. Cancer Epidemiol. Biomarkers Prev. 17 (5), 1144–1154.

Pillai, R., Uyehara-Lock, J.H., Bellinger, F.P., 2014. Selenium and selenoprotein function in brain disorders. IUBMB Life 66 (4), 229–239.

Pilon, M., et al., 2003. Enhanced selenium tolerance and accumulation in transgenic Arabidopsis expressing a mouse selenocysteine lyase. Plant Physiol. 131 (3), 1250–1257.

Pugliese, C., et al., 2014. Assessment of antioxidants status and superoxide dismutase activity in HIV-infected children. Braz J. Infect. Dis. 18 (5), 481–486.

Rahimzadeh-Barzoki, H., et al., 2014. Selenium levels in rice samples from high and low risk areas for esophageal cancer. Saudi Med. J. 35 (6), 617–620.

Rao, Y., et al., 2010. Mapping of selenium metabolic pathway in yeast by liquid chromatography-Orbitrap mass spectrometry. Anal. Chem. 82 (19), 8121–8130.

Rao, Y., McCooeye, M., Mester, Z., 2012. Mapping of sulfur metabolic pathway by LC Orbitrap mass spectrometry. Anal. Chim. Acta 721, 129–136.

Rayman, M.P., 2012. Selenium and human health. Lancet 379 (9822), 1256–1268.

Raymond, L.J., Deth, R.C., Ralston, N.V., 2014. Potential role of selenoenzymes and antioxidant metabolism in relation to autism etiology and pathology. Autism Res. Treat. 2014, 164938.

Ren, L.Q., et al., 2004. Coxsackievirus B3 infection and its mutation in Keshan disease. World J. Gastroenterol. 10 (22), 3299–3302.

Sakr, Y., et al., 2007. Time course and relationship between plasma selenium concentrations, systemic inflammatory response, sepsis, and multiorgan failure. Br. J. Anaesth. 98 (6), 775–784.

Salvini, S., et al., 1995. Plasma levels of the antioxidant selenium and risk of myocardial infarction among U.S. physicians. Am. J. Cardiol. 76 (17), 1218–1221.

Schrauzer, G.N., 2001. Nutritional selenium supplements: product types, quality, and safety. J. Am. Coll. Nutr. 20 (1), 1–4.

Sedighi, O., et al., 2014. Association between plasma selenium and glutathione peroxidase levels and severity of diabetic nephropathy in patients with type two diabetes mellitus. Nephrourol. Mon. 6 (5), e21355.

Semnani, S., et al., 2010. Soils selenium level and esophageal cancer: an ecological study in a high risk area for esophageal cancer. J. Trace Elem. Med. Biol. 24 (3), 174–177.

Shetty, S.P., Copeland, P.R., 2015. Selenocysteine incorporation: a trump card in the game of mRNA decay. Biochimie 114, 97–101.

Steinbrenner, H., et al., 2015. Dietary selenium in adjuvant therapy of viral and bacterial infections. Adv. Nutr. 6 (1), 73–82.

Stoedter, M., et al., 2015. Strong induction of iodothyronine deiodinases by chemotherapeutic selenocompounds. Metallomics 7 (2), 347–354.

Sutherland, A., et al., 2010. Polymorphisms in the selenoprotein S and 15-kDa selenoprotein genes are associated with altered susceptibility to colorectal cancer. Genes Nutr. 5 (3), 215–223 Epub 2010 May 13.

Suzuki, Y., et al., 2013. Selenium metabolism and excretion in mice after injection of (82)Se-enriched selenomethionine. Metallomics 5 (5), 445–452.

Tsuji, Y., et al., 2009. Selenium metabolism in rats with long-term ingestion of Se-methylselenocysteine using enriched stable isotopes. J. Toxicol. Sci. 34 (2), 191–200.

Tsuji, P.A., et al., 2012. Knockout of the 15kDa selenoprotein protects against chemically-induced aberrant crypt formation in mice. PLoS One 7 (12), e50574.

Tsuji, P.A., et al., 2015. The 15kDa selenoprotein and thioredoxin reductase 1 promote colon cancer by different pathways. PLoS One 10 (4), e0124487.

Van Blarigan, E.L., et al., 2014. Plasma antioxidants, genetic variation in SOD2, CAT, GPX1, GPX4, and prostate cancer survival. Cancer Epidemiol. Biomarkers Prev. 23 (6), 1037–1046.

Van Huysen, T., et al., 2003. Overexpression of cystathionine-gamma-synthase enhances selenium volatilization in Brassica juncea. Planta 218 (1), 71–78.

Varlamova, E.G., et al., 2013. Biosynthesis and mechanism of selenocysteine incorporation into synthesized proteins. Mol. Biol. Mosk. 47 (4), 558–567.

Villette, S., et al., 1998. Thyroid stimulating hormone and selenium supply interact to regulate selenoenzyme gene expression in thyroid cells (FRTL-5) in culture. FEBS Lett. 438 (1–2), 81–84.

Vinceti, M., et al., 2014. Selenium for preventing cancer. Cochrane Database Syst. Rev. 3, CD005195.

Wadhwa, S.K., et al., 2015. Interaction between carcinogenic and anti-carcinogenic trace elements in the scalp hair samples of different types of Pakistani female cancer patients. Clin. Chim. Acta 439, 178–184.

Wallenberg, M., et al., 2010. Selenium compounds are substrates for glutaredoxins: a novel pathway for selenium metabolism and a potential mechanism for selenium-mediated cytotoxicity. Biochem. J. 429 (1), 85–93.

Wallenberg, M., et al., 2014. Selenium induces a multi-targeted cell death process in addition to ROS formation. J. Cell Mol. Med. 18 (4), 671–684.

Waters, D.J., et al., 2005. Prostate cancer risk and DNA damage: translational significance of selenium supplementation in a canine model. Carcinogenesis 26 (7), 1256–1262.

Watson, P.E., McDonald, B.W., 2009. Major influences on nutrient intake in pregnant New Zealand women. Matern. Child. Health J. 13 (5), 695–706.

Weekley, C.M., et al., 2011. Metabolism of selenite in human lung cancer cells: X-ray absorption and fluorescence studies. J. Am. Chem. Soc. 133 (45), 18272–18279.

Weekley, C.M., et al., 2013. Selenium metabolism in cancer cells: the combined application of XAS and XFM techniques to the problem of selenium speciation in biological systems. Nutrients 5 (5), 1734–1756.

Weekley, C.M., et al., 2014. XAS studies of Se speciation in selenite-fed rats. Metallomics 6 (12), 2193–2203.

Witkowska, A.M., Borawska, M.H., Gacko, M., 2006. Relationship among TNF-alpha, sICAM-1, and selenium in presurgical patients with abdominal aortic aneurysms. Biol. Trace Elem. Res. 114 (1–3), 31–40.

Wolf, W.R., Goldschmidt, R.J., 2007. Updated estimates of the selenomethionine content of NIST wheat reference materials by GC-IDMS. Anal. Bioanal. Chem. 387 (7), 2449–2452.

Wolnik, K.A., et al., 1983. Elements in major raw agricultural crops in the United States. 2. Other elements in lettuce, peanuts, potatoes, soybeans, sweet corn, and wheat. J. Agric. Food Chem. 31, 1244–1249.

World Health Organization, 1996. Trace Elements in Human Nutrition and Health. World Health Organization, Geneva, p. 361.

Xia, Y., et al., 2010. Optimization of selenoprotein P and other plasma selenium biomarkers for the assessment of the selenium nutritional requirement: a placebo-controlled, double-blind study of selenomethionine supplementation in selenium-deficient Chinese subjects. Am. J. Clin. Nutr. 92 (3), 525–531 Epub 2010 Jun 23.

Yoo, M.H., et al., 2006. Thioredoxin reductase 1 deficiency reverses tumor phenotype and tumorigenicity of lung carcinoma cells. J. Biol. Chem. 281 (19), 13005–13008.

Zhang, Z., Zhang, J., Xiao, J., 2014. Selenoproteins and selenium status in bone physiology and pathology. Biochim. Biophys. Acta 1840 (11), 3246–3256.

Zhou, J., Huang, K., Lei, X.G., 2013. Selenium and diabetes–evidence from animal studies. Free Radic. Biol. Med. 65, 1548–1556.

Part XI

Electrolytes

Chapter 40

Sodium: Basic Nutritional Aspects

Daniel Laubitz, Fayez K. Ghishan, Pawel R. Kiela

University of Arizona, Tucson, AZ, United States

INTRODUCTION

Sodium is an essential nutrient and the principal extracellular cation that plays critical roles in many life-sustaining processes. It regulates and maintains blood and cell volume, maintains membrane potential, and plays key roles in neural excitation, nerve conductance, and muscle contraction. It is absorbed through various membrane transport proteins along the entire length of the gastrointestinal tract, and under homeostatic conditions systemic Na^+ excess is readily compensated for by increased renal excretion, a process tightly controlled by complex neuroendocrine mechanisms. The Institute of Medicine (IOM) of the National Academies recommended daily sodium intake to not exceed 2300 mg per day (~1 teaspoon of salt). Although this amount is considerably higher than the minimal safe amount needed by the body for proper function (500 mg/day), it is still greatly exceeded in the United States and in other populations, where it contributes to chronic morbidity and overall mortality.

In this chapter we provide a brief review of the physiological importance of sodium ions, mechanisms of transepithelial absorption in the gut and reabsorption in the kidneys, dietary sources and recommendations for intake levels, and consequences of hyponatremia and high sodium intake in individuals and in populations.

Physiological Roles of Na^+

Sodium is a key chemical element that controls blood pressure (BP) and regulates the function of nerves and muscles, which is why sodium concentration in the body fluid is precisely controlled and all significant departures from homeostasis may lead to pathogenic consequences. In this section we will briefly review the physiological functions of Na^+ to put them into a context of clinical consequences of inadequate or excessive sodium intake. More detailed analysis of the role of sodium in health and disease can be found in the Michel Burnier monograph (Burnier, 2008). Sodium, together with potassium, is a principal cation that functions as a regulator of extracellular fluid volume and cellular volume. Extracellular fluid is the major sodium reservoir (142 mmol/L versus 10 mmol/L in the cytosol in a resting state). The exact opposite is true for potassium, in which the extracellular fluid contains only a small quantity (4 mmol/L) in comparison to the intracellular fluid (140 mmol/L). The large amounts of extracellular sodium are accompanied by large quantities of chloride (103 mmol/L), which is present only at 4 mmol/L in the cytosol. These ions play a principal role as osmotic determinants in regulating cell volume and systemic electrolyte balance.

Osmoregulation

Life has evolved in the aqueous milieu, and water is the most abundant constituent of all cells and multicellular organisms. It has been proposed that salinity of the ancient oceans decreased over time to the present value (Knauth, 1998) and that the earliest cells had to use mechanisms to regulate water and ion flow to protect them from high NaCl concentration (Kultz, 2001). This statement is supported by the fact that all life forms have very conserved osmoregulatory strategies, including cell volume regulation, regulation of water and ion permeability, regulation of active ion transport, and intracellular organic osmolyte systems.

The composition of the solutes in the extracellular and intracellular fluid is not identical; however, the combined concentrations of dissolved particles in both fluids are equal because of the cell membrane's permeability to water. This concentration, proportional to the number of particles and independent of the size or their nature, per 1 kg of solvent is called osmolality and is expressed as milliosmols per kilogram. In a clinical laboratory osmolality is measured using an osmometer. The typical range for serum osmolality is 282–295 mOsm/kg and, on average, 500–800 mOsm/kg for urine.

Molecular, Genetic, and Nutritional Aspects of Major and Trace Minerals. http://dx.doi.org/10.1016/B978-0-12-802168-2.00040-3

Another measurement that refers to the concentration of solutes is called osmolarity, and it is a measure of the osmoles per liter of solution. Because the volume of fluid changes with pressure or temperature, osmolarity is difficult to determine and osmolality is easier to measure; therefore it is commonly used and preferred. Expected osmolarity is a common bedside calculation used in practice in clinical settings. It is most commonly calculated based on the measured concentrations of NaCl, glucose, and blood urea nitrogen (BUN). Because NaCl dissociates completely in water to form Na^+ and Cl^-, each mole of NaCl becomes two osmoles. Glucose and BUN concentrations are typically expressed as milligrams per deciliter, with the required conversion to millimoles per liter, the formula for expected osmolarity appears as follows:

$$\text{Plasma osmilarity} \left(\frac{\text{mOsm}}{\text{L}}\right) = 2\,[Na^+]\left(\frac{\text{mmole}}{\text{L}}\right) + \frac{[\text{Glucose}]\left(\frac{\text{mg}}{\text{dL}}\right)}{18} + \frac{[\text{BUN}]\left(\frac{\text{mg}}{\text{dL}}\right)}{2.8}$$

The normal range of plasma osmolarity is 270–300 mOsm/L. Clinically, calculation of the osmolality gap (difference between measured osmolality and calculated osmolarity) is an important indicator of the amount of unmeasured solutes in the blood (alcohols, sugars, lipids, and proteins) and of potential toxicity or other physiological disturbances.

Any discrepancy in osmolality between the intra- and extracellular compartments causes rapid water flux until the two compartments achieve equilibrium (equal osmolality). Extreme variations in osmolality lead to cell shrinking or swelling and result in damage to cell structure and function. Systemically, regulation of osmolality must be integrated with blood volume regulation and is achieved by balancing the intake and excretion of sodium with that of water.

When extracellular osmolality increases (hypertonicity) because of loss of water and/or accumulation of solutes, the osmotic equilibrium is attained by changes in water flux, thus reducing cell volume. This situation occurs when solute entry into cells is limited (mannitol, sodium, glucose). However, when solutes can pass the membrane barrier (ethanol, urea) the osmotic equilibrium is achieved simply by diffusion of the solutes across the membrane and no changes in cell volume occur.

Cells subjected to a hypertonic challenge respond to volume changes by activation of membrane electrolyte transporters to prevent excessive volume perturbations. Ions including Na^+, K^+, and Cl^- are transported into the cell by many transporters such as the Na^+–K^+–Cl^- cotransporter and the Na^+/H^+ and Cl^-/HCO_3^- exchangers. Accumulation of these ions in the cell, accompanied by decreased cell volume and crowding of the intracellular molecules, leads to deleterious effects, including protein misfolding and denaturation, DNA damage, disruption of mitochondrial structure and function, and finally in apoptotic cell death (Cheung and Ko, 2013). All mammalian cells have adaptive mechanisms in which accumulated intracellular ions are replaced by uncharged small organic molecules including sorbitol, betaine, myo-inositol, taurine, and glycerophosphocholine (Garcia-Perez and Burg, 1991; Nakanishi and Takamitsu, 1996). These organic molecules can accumulate to high levels without negative consequences for the cell. Expression of osmoprotective genes encoding enzymes and transporters responsible for the control of organic osmolyte flux is regulated by enhancers known as osmotic-response elements (OREs) or tonicity-responsive enhancers (TonEs) located in the regulatory regions of these genes (Ferraris et al., 1999). NFAT5 [known also as ORE-binding protein (OREBP) or TonE-binding protein (TonEBP)], a member of the nuclear factor of activated T cells family, is an example of a transcription factor that binds to the ORE/TonE elements and regulates expression of osmoprotective genes upon hyperosmotic challenge. NFAT5 activity is regulated at multiple levels in a tonicity-dependent (Cheung and Ko, 2013) and tonicity-independent manner (Halterman et al., 2012).

Neuronal Excitability

As mentioned earlier, the cytosolic ionic composition differs greatly from that of the surrounding fluid. The diffusion potential, which is driven mostly by K^+ diffusion (100 times greater than Na^+ diffusion) and by ion-selective transport against their concentration gradients, creates a difference in voltage across the plasma membrane. The inside is slightly more negative relative to the outside, and this gives a net resting membrane potential of approximately −70 mV (Lodish, 2000). Thus the cell membrane acts similar to a capacitor that can store positive charge on one side and negative charge on the other. These ionic gradients and generated electric potential drive many biological processes. Some cells such as nerve and muscle cells are "excitable," meaning that they are able to generate electrochemical impulses at their membrane and to transmit this signal along the membranes.

Any event able to cause disruption of the resting membrane potential leads to development of an action potential. The action potential, an electrical signal, is generated near the neuron cell body portion of the axon and propagated along the neuron. Action potentials are very rapid changes in the membrane potential from −70 mV toward zero. Among the ion channels, voltage-gated sodium channels (VGSCs) have a clear role in action potential generation. When a polarized neuron in its resting stage is stimulated by a neurotransmitter in a chemical synapse or via current flow through gap junctions in an electrical synapse, it becomes partially depolarized. If this depolarization reaches the neuron's threshold (−60 to −50 mV), then VGSCs change their conformation by flipping to the open position, allowing Na⁺ to pour inward. In this depolarization stage, the membrane permeability for sodium ions increases 500- to 5000-fold. In many neurons, the membrane potential overshoots beyond zero and becomes positive (up to +35 mV). The depolarization stage is followed by a repolarization stage, when VGSCs are closed and potassium channels open, allowing rapid diffusion of K⁺ to the exterior, which restores the resting membrane potential.

As an example, VGSCs, as well as background (or leak) channels, contribute to the regulation of the excitability of pain receptors (nociceptors). Dysregulated expression of these channels can cause neuronal hyperexcitability and may result in hypersensitivity and allodynia (pain due to a stimulus that does not usually provoke pain). Recently, painful peripheral neuropathy has been linked to different types of VGSC mutations (Hoeijmakers et al., 2015).

Muscle Contraction

The same VGSCs present in neurons are responsible for the generation and conduction of action potentials in cardiac myocytes and skeletal muscles. These Na⁺ channels along with K⁺ channels regulate polarization of the sarcolemma, the cell membrane of the muscle fiber. A signal from motor nerves releases acetylcholine, a neurotransmitter that docks with its receptor in the muscle cell membrane. Acetylcholine acts on a local area of the membrane, causing localized cell membrane depolarization via the action of acetylcholine-gated sodium channels. When the potential difference reaches the threshold, the VGSCs open for extracellular sodium and an action potential is initiated in the muscle fiber. The action potential travels along the muscle fiber and causes the sarcoplasmic reticulum to release large quantities of stored calcium ions. Ca^{2+} is required to initiate attractive forces between actin and myosin filaments, causing them to slide together and initiating the contractile process.

Abnormal activity of VGSCs can result in disorders that manifest as cardiac arrhythmias, disturbed skeletal muscle contraction or relaxation, epilepsy, migraines, or neuropathic pain (Hoeijmakers et al., 2015). Although sodium deficiency does not generally result from inadequate dietary intake, even with very low-salt diet, a systemic loss of sodium (e.g., because of the use of natriuretic drugs) can result in an imbalance that may contribute to muscle cramps. The risk of developing muscle cramps as a result of hyponatremia (sodium concentration <136 mmol/L) is higher in individuals exposed to excessive and prolonged exercise and in those who work in high temperatures as a result of excessive sweating. Hyponatremia has also been associated with an inappropriate secretion of antidiuretic hormone because of the use of certain medications (e.g., furosemide, nonsteroidal antiinflammatory drugs, carbamazepine, oxytocin), in some renal diseases, and with prolonged vomiting or diarrhea. A lower sodium level has been also associated the disrupted circadian rhythm of urinary Na⁺ excretion and abnormal muscle contraction in idiopathic restless leg syndrome (Tribl et al., 2005).

Epithelial Na⁺ Absorption

Polarized intestinal and renal epithelial cells are equipped with differentially expressed and specifically regulated transport proteins at the apical and basolateral domains. Although apically expressed transporters (symporters, exchangers, and channels) may uniquely define a specific segment of the gut or the nephron, basolateral Na⁺ extrusion is universally and predominantly supported by the Na⁺/K⁺ ATPase (sodium pump). Discovered by Skou in 1957 (Skou, 1989), the Na⁺/K⁺ ATPase is a member of the P-type ATPase family, which is capable of transporting sodium and potassium ions across the cell membrane against their concentration gradients utilizing energy released by ATP hydrolysis. The Na⁺/K⁺ ATPase consists of two noncovalently linked α and β subunits. Humans express four α isoforms, which are expressed in a tissue-specific manner. Whereas the α1 isoform is expressed ubiquitously, the α2 and α3 isoforms are mainly expressed in skeletal muscle, neuronal tissue, and cardiac myocytes, and the expression of the α4 isoform is limited to the testes, where it is involved in the regulation of sperm motility. An auxiliary, regulatory γ subunit (a member of the FXYD protein family) is associated with the Na⁺/K⁺ ATPase in a tissue-specific manner and regulates the function of the enzyme. It is important to note that the sodium pump is not only involved in transepithelial electrolyte (re)absorption, but it also serves as a signaling molecule, a function best described in the renal epithelia, which alters cellular function by modulating the activities of the Src kinase family (Xie et al., 2013).

Apical Na⁺ Transport in the Small Intestine

The surface area of the human small intestinal mucosa averages 30 square meters. In healthy adult humans, intestinal fluid load is a combination of 1.5–2 L from ingested food and drinks and 8–10 L of fluid (containing 2000–3000 mg of sodium) secreted by the gastrointestinal tract every day. The small intestinal epithelium faces the challenge to reduce this volume to the typical ileocecal daily flow of 2 L. The net fluid movement across the gastrointestinal epithelium is primarily the result of active transport of Na^+, Cl, and HCO_3^-. Disturbances of these coordinated transport events, either a result of genetic mutations, pathogenic infections, or dysregulation (e.g., during inflammation), often result in diarrhea, especially when colonic capacity for compensatory fluid absorption is overwhelmed. Three mechanisms contribute to apical Na^+ transport in the small intestine: (1) nutrient-coupled Na^+ absorption mediated by several families of Na^+-dependent nutrient transporters, (2) electroneutral NaCl absorption, and (3) electrogenic Na^+ absorption by epithelial Na^+ channels (ENaC).

Nutrient-coupled Na^+ absorption was discovered in the 1960s when transepithelial sugar and amino acid transport was measured using the short-circuit current technique in Ussing chambers. These studies found that absorption is dependent on extracellular (luminal) Na^+, a finding critical for the later development of oral rehydration solution to control mortality associated with cholera and diarrheal diseases (Hirschhorn et al., 1968). This was hailed as "potentially the most important medical advance" in the 20th century (Carpenter, 1990). The molecular basis for oral rehydration therapy became apparent with the cloning and functional characterization of the Na^+/glucose cotransporter SGLT1 by Ernest M. Wright (Hediger et al., 1987; Loo et al., 1996). The basolateral Na^+/K^+ ATPase creates an electrical and chemical "downhill" Na^+ gradient across the epithelium. SGLT1 expressed on the apical membrane of enterocytes uses this gradient to transport Na^+ and glucose at a 2:1 ratio against a glucose concentration gradient. In each transport cycle, each sugar molecule is cotransported with Na^+ across the cell, which is accompanied by 260 water molecules (Loo et al., 1996). This mechanism was calculated to account for 5 L of water absorption per day in the human intestine.

The same early studies with the Ussing chambers and the short-circuit current technique also identified a component of basal NaCl transport that was not associated with transepithelial currents. This electroneutral NaCl absorption was later attributed to members of the solute carrier family (SLC)-9 family of Na^+/H^+ exchangers (NHEs; Donowitz et al., 2013; Kiela and Ghishan, 2012). Three NHE isoforms have been identified to be expressed on the apical membrane of enterocytes: NHE2, NHE3, and NHE8 (Bookstein et al., 1994; Tse et al., 1993; Xu et al., 2005). Of the three, NHE3 contributes most significantly to small intestinal Na^+ and water absorption, as exemplified by the intestinal electrolyte and fluid absorptive defects in NHE3-deficient mice (Schultheis et al., 1998b). Because congenital sodium diarrhea (CSD), an autosomal-recessive disorder, has been long associated with defective Na^+/H^+ exchange mechanism, apical NHEs have been thought to be the main culprits. Although Müller et al. showed that CSD was not related to mutations in the NHE isoforms known at that time (Muller et al., 2000), missense mutations in the SCL9A3 gene coding for NHE3 have been very recently described in a small cohort of CSD patients (Yin, 2015). Reduced expression and mistargeting of the NHE3 protein in enterocytes is also believed to be partially responsible for diarrhea in patients with microvillous inclusion disease, a rare genetic disorder associated with mutations in Myo5B and Syntaxin 3 (Ameen and Salas, 2000; Knowles et al., 2014; Michail et al., 1998). However, there are scenarios when NHE3 inhibition may be clinically beneficial. A recent study with a novel, poorly bioavailable, orally administered NHE3 inhibitor, tenapanor, showed that reduction of NHE3-mediated intestinal Na^+ absorption was beneficial in a rat model of chronic kidney disease (CKD), where it reduced extracellular fluid volume, left ventricular hypertrophy, albuminuria, and BP (Spencer et al., 2014).

Although nutrient-coupled Na^+ transport is often an electrogenic process, the true rheogenic or "current generating" Na^+ absorption is most commonly attributed to the activity of mineralocorticoid-regulated apical ENaCs. This mechanism of intestinal Na^+ absorption is essentially limited to the colon (see Apical Na^+ Transport in the Colon), although it may contribute to a small degree to ileal Na^+ transport, particularly as an adaptive response after proctocolectomy (surgical removal of colon and rectum; Koyama et al., 1999).

Apical Na^+ Transport in the Colon

In addition to bacterial fermentation of dietary materials that are not absorbed in the small intestine, the colon plays a critical role in extracting salt and water from the feces before they are eliminated from the body as solid waste. The human colon has a nominal mucosal surface area of approximately $2000\,cm^2$, although the total absorptive area is even greater because colonic crypt cells are capable of absorption as well as secretion. The colon displays a segmental heterogeneity in its capacity for sodium and water absorption because of the predominant transport mechanisms involved. Electrogenic Na^+ absorption is present throughout the human colon, although it provides the predominant route of Na^+ absorption in the distal segment. It is mediated by Na^+ channels, located predominantly in the apical membrane of surface colonocytes, which are composed of three subunits (designated α-, β-, and γ-hENaC). One of the key functional characteristics of ENaC is its exquisite sensitivity to the inhibitory effects of a pyrazine diuretic, amiloride (Benos et al., 1995), and stimulation by a mineralocorticoid hormone, aldosterone (Rossier, 2014). Electrogenic but amiloride-insensitive Na^+ absorption has been also described in the proximal colon (Sandle, 1989). Although the physiological mechanism for this transport is not clearly defined, it may be related to the function of Na^+/HCO_3 cotransporters (NBCs), particularly encoded by NBCe1B/C (Barmeyer et al., 2013).

A substantial fraction of net colonic Na^+ absorption, particularly in the proximal colon, is mediated by electroneutral NaCl transport. It is now generally accepted that this process is a reflection of the activities of Na^+/H^+ and Cl^-/HCO_3^- exchangers [the latter mechanism mediated by putative anion transporter 1 (*SLC26A6*) and downregulated in adenoma DRA (*SLC26A3*)] operating in parallel in the apical membrane. The same three major NHEs are expressed on the apical surface of colonocytes as in the small intestine.

Diarrhea is one of the common symptoms in patients with inflammatory bowel diseases (IBDs), and it occurs in approximately 50% of acute flare-ups of Crohn's disease and in nearly all patients with ulcerative colitis (Seidler et al., 2006). It can be at least partially attributed to impaired epithelial Na^+/H^+ exchange activity, which has been demonstrated in patients and in animal models of IBD, more recently pinpointed to either decreased expression and/or activity of NHE3 (Ghishan and Kiela, 2014). Intriguingly, loss of NHE3 activity in mice also exacerbates colitis (Kiela et al., 2009; Larmonier et al., 2011; Laubitz et al., 2008) and influences intestinal microbial ecology (Engevik et al., 2013; Larmonier et al., 2013). These findings imply a profound role for colonic epithelial Na^+ transport not only in water homeostasis but also in regulating mucosal immune and microbial responses during intestinal inflammation.

Renal Na+ Reabsorption

The kidneys, in concert with neural and endocrine input, regulate volume and osmolality of the extracellular fluid by altering the amount of sodium and water excreted. This is accomplished primarily through alterations in sodium and water reabsorption, the mechanisms of which differ within each nephron segment. Of active Na^+ reabsorption, 65% occurs in the renal proximal tubules, 0% in the descending limb of the loop of Henle, 25% in the ascending limb of the loop of Henle (thick and thin), 5% in the distal convoluted tubules, and 4–5% in the collecting duct system. In the proximal tubule, basolateral Na^+/K^+ ATPase provides the downhill Na^+ gradient utilized by the apically expressed NHEs, especially NHE3 and to a lesser extent NHE2 and NHE8. Na^+-nutrient symporters, such as Na^+-amino acid cotransporter B0AT1, Na^+-glucose cotransporters SGLT1 and SGLT-2, and Na^+-phosphate cotransporters (Npt2a, Npt2c, and Pit-2), also contribute to the net Na^+ reabsorption in the proximal convoluted tubules. Although the descending loop of Henle is permeable to water, it shows no significant Na^+ absorption. In the ascending limb, which is less permeable to water, active reabsorption of Na^+ is performed primarily via the $Na^+/K^+/2Cl^-$ cotransporters NKCC1 and NKCC2. NKCC2 reabsorbs approximately 20–30% of the NaCl filtered by the glomerulus in the thick ascending limb of Henle. NKCC2 is the target of the thiazide furosemide and other "loop diuretics" used in the treatment of hypertension. Loss-of-function mutations in the SLC12A1 gene encoding NKCC2 result in Bartter's syndrome, an autosomal-recessive disease characterized by plasma volume reduction, polyuria, hyponatremia, hypotension, hypochloremia, hypokalemia, magnesuria, metabolic alkalosis, and hypercalciuria (Simon et al., 1996a). Conversely, enhanced activity of NKCC2 has been linked to hypertension and hypertensive disorders (Capasso et al., 2005; Trepiccione et al., 2012). In the distal convoluted tubules, secondary active Na^+ reabsorption is performed primarily via another thiazide-sensitive Na^+–Cl^- cotransporter (NCC, coded by SLC12A3 gene) as well as via an electrogenic ENaC-mediated transport. Similar to ENaC, NCC is a target of aldosterone, the main hormone controlling renal Na^+ transport, and inactivating mutations in NCC cause Gitelman syndrome, an inherited recessive disease characterized by low BP (Simon et al., 1996b). Heterozygous inactivating NCC mutations confer a low BP or protection against arterial hypertension (Ji et al., 2008). On the other hand, excessive NCC activity is responsible for the symptoms of familial hyperkalemic hypertension, also known as Gordon's syndrome or pseudohypoaldosteronism type II, a rare inherited disease characterized by hypertension that is highly sensitive to thiazide compounds (Hadchouel et al., 2006).

Dietary Na+ Intake: Sources and Recommendations

Sodium is an essential nutrient, and its homeostasis is tightly controlled. Without substantial sweating, total obligatory sodium losses are very small, up to 180 mg/day (Dahl, 1958). Under steady-state conditions of sodium and fluid balance, urinary losses of sodium are approximately equal to its intake (90–95% of total intake). Fecal excretion of Na^+ is minimal, accounting for a maximum of 5% of total intake under the highest (8 g/day) intake levels (Allsopp et al., 1998; Dahl, 1958). However, chronic excessive consumption of salt can easily overwhelm the available homeostatic mechanism and contribute to multiple forms of pathogenesis.

Monitoring population-wide sodium intake is critical as an assessment and monitoring tool to guide public health initiatives aimed at sodium reduction. There are several methods for assessment of dietary sodium intake in a population, a topic recently reviewed by McLean (2014) and Mente et al. (2015). Losses of Na+ through the skin and sweat vary widely and depend on intake levels, temperature, physical activity, and time allowed for acclimatization to temperature changes.

Sodium is primarily consumed as salt (NaCl). As a food ingredient, salt is used in curing meat, baking, masking off-flavors, retaining moisture, and enhancing flavor. Salt added at the table and in cooking is thought to provide only a small proportion of the total sodium that Americans consume. Most sodium comes from salt added during food processing. When reading a Nutrition Facts panel on a food product, foods that are low in sodium (<140 mg or 5% of the daily value) can be considered "low-salt."

The National Heart Lung and Blood Association reported that 500 mg is a safe daily minimum intake of sodium in young adults and is sufficient to maintain all organismal functions dependent on Na+. In an average temperate climate, a normal adult may be able to thrive with as little as 115 mg of sodium each day. Because most foods contain at least a small amount of natural sodium, it is difficult to drop below the minimum recommendation even on an extremely low-sodium diet. However, the US Food and Drug Administration estimates that Americans eat on average approximately 3300 mg of sodium a day (~8.4 g of NaCl) and that 75% of dietary sodium comes from eating processed, frozen, or restaurant food. This is far in excess of the adequate intake (AI) amount set by the IOM of the National Academies in the most recent *Dietary Reference Intakes for Water, Potassium, Sodium, Chloride, and Sulfate* report (IOM, 2005). The AI for sodium is set for young adults at 1.5 g (65 mmol)/day (3.8 g of NaCl). This AI does not apply to individuals who lose large volumes of sodium in sweat, such as competitive athletes and workers exposed to extreme heat stress (e.g., foundry workers and fire fighters). The AI for sodium for older adults and the elderly is even lower and is set at 1.3 g (55 mmol)/day for men and women 50–70 years of age and at 1.2 g (50 mmol)/day for those 71 years of age and older. For adolescents and adults of all ages (14 years and older), the IOM set the tolerable upper intake level (UL) at 2300 mg per day. The UL is the highest daily nutrient intake level that is likely to pose no risk of adverse health effects (in this case, primarily increased BP) to almost all individuals in the general population. These limits and positions have been endorsed in the latest eighth issue (2015) of the *Dietary Guidelines for Americans*[1], a research-based guide published jointly by the US Department of Health and Human Services and the US Department of Agriculture (USDA) every 5 years. This report concluded that 63–91% of females and 81–97% of males consumed more than the UL for sodium in particular age groups and designated sodium as a "nutrient of public health concern" due to the critical link of sodium intake to health and its wide overconsumption across the entire US population. The World Health Organization report on *Sodium Intake for Adults and Children*[2] published in 2012 contains a strong recommendation to limit sodium consumption in adults even further, to less than 2000 mg sodium/day (5 g salt/day). This optimal individual level of dietary sodium was reaffirmed in two systematic reviews (Aburto et al., 2013; He et al., 2013).

Additional restrictions apply to specific populations. Some individuals tend to be more salt sensitive than others, including people with hypertension, African Americans, and middle-aged and older adults. In these cases, sodium intake should not exceed 1500 mg per day. According to the USDA, Black nonHispanics have a relatively low intake of potassium and a high prevalence of elevated BP and salt sensitivity (see Blood Pressure); this population subgroup may especially benefit from an increased dietary intake of potassium and reduced intake of sodium. Although salt substitutes containing potassium chloride may be useful for some individuals, they may be harmful to people with certain medical conditions such as kidney disease, diabetes, heart disease, Addison's disease, or stomach ulcers. These individuals should consult a health-care provider before using salt substitutes.

Hyponatremia

Insufficient sodium intake in the diet alone is very rarely the cause of hyponatremia, a disorder defined as serum sodium concentrations less than 135 mmol/L. This condition is more commonly caused by solute dilution resulting from excessive consumption of water. Water intoxication, also known as hypotonic (dilutional) hyponatremia, develops when the intake of water exceeds the kidney's ability to eliminate it. Causes of dilutional hyponatremia include psychiatric disorders, forced water intake as a form of child abuse (Joo and Kim, 2013), and iatrogenic infusion of excessive hypotonic fluid. Dilutional hyponatremia can also occur as a complication in the patient's postoperative state, drug toxicity, or renal and hormonal disorders (Adrogue and Madias, 2000, 2014; Sterns, 2015). A common clinical problem, hyponatremia is relatively frequent in hospitalized patients (Anderson, 1986). Although the severity of morbidity related to hyponatremia varies widely, serious complications can arise from the disorder itself as well as from errors in management.

---

1. http://health.gov/dietaryguidelines/2015/guidelines/.
2. http://apps.who.int/iris/bitstream/10665/77985/1/9789241504836_eng.pdf?ua=1&ua=1.

The nonhypotonic hyponatremias are represented by hypertonic (or translocational) hyponatremia, isotonic hyponatremia, and pseudohyponatremia. Translocational hyponatremia is a direct result of a water flux from cells to the extracellular fluid that is driven by the accumulation of osmolytes confined in the extracellular compartment (e.g., in hyperglycemia or after administration of hypertonic mannitol) and results in cell dehydration. Isotonic hyponatremia is a state related to the retention of large volumes of isotonic fluids in the extracellular space that do not contain sodium (e.g., mannitol) and which does not generate transcellular shifts of water. Pseudohyponatremia is essentially a misleading form of isoosmolar and isotonic hyponatremia identified when Na^+ concentration is measured by flame photometry in patients with severe hypertriglyceridemia or paraproteinemia. This laboratory artifact has been eliminated by the use of ion-specific electrodes (Adrogue and Madias, 2000).

Exercise-associated hyponatremia (EAH) has been described after sustained physical exertion during marathons, triathlons, and other endurance athletic events. Symptoms may vary from isolated serum sodium levels less than 135 mmol/L with no phenotopic consequences to confusion, seizures, and altered mental status consistent with EAH encephalopathy. The pathogenesis of EAH is complex and multifactorial. Overhydration and dilutional hyponatremia, as well as sodium losses in sweat, are certainly two factors in EAH pathogenesis. Medication, especially nonsteroidal antiinflammatory drugs, has also been associated with decreased renal filtration, which impairs the urine-diluting capacity of the kidney in endurance athletes by potentiating the effects of arginine vasopressin (AVP). In fact, inappropriate secretion of AVP has been postulated to be the main causative factor in the pathogenesis of EAH. This seemed counterintuitive because AVP is normally stimulated by volume contraction whereas most athletes with EAH finish events with an increase in body weight and possibly an expanded plasma volume. One unifying hypothesis to explain this posits that initial AVP secretion may be stimulated by volume contraction whereas excessively consumed water accumulates in the gut lumen because of impaired gastrointestinal blood flow and water absorption during exercise. At the end of the exercise, hypotonic fluid absorption in the gastrointestinal tract may increase rapidly, and coupled with impaired free water excretion would lead to a rapid decrease in serum (especially arterial) sodium levels. More detailed descriptions of the problem and the pathogenesis of EAH have been published elsewhere (Rosner, 2009).

Health Outcomes Associated With Increased Na^+ Intake

Two relatively recent and comprehensive reviews that analyzed the evidence for associations between dietary sodium intake and health outcomes published in the peer-reviewed literature are the IOM report on the *Dietary Reference Intakes for Water, Potassium, Sodium, Chloride, and Sulfate* (IOM, 2005) and the 2015 report of the Dietary Guidelines Advisory Committee.[3] The Committee on the Consequences of Sodium Reduction in Populations of the IOM published another comprehensive summary of the findings and conclusions about the associations between sodium intake and risk of cardiovascular disease (CVD)-related events and mortality (Strom BL, 2013)[4]. Here, we summarize those and some additional recent findings, and the reader is referred to these three excellent and comprehensive reports for more details.

Hypertension is a multifactorial disease linking genetics; environmental factors; and neural, mechanical, and hormonal perturbations, as recently summarized by Padmanabhan et al. (2015). It is important to note that BP is a biomarker of CVD and needs to be considered as a modifiable risk factor for coronary heart disease, stroke, chronic renal diseases, metabolic syndrome, and diabetes.

Blood Pressure

Sodium is a key chemical element that controls BP. An individual's BP response to salt is heterogeneous and possibly related to inherited susceptibility. Those who respond to reduced salt intake with lower BP are called "salt-sensitive" individuals and those who experience little or no change are considered "salt-resistant" individuals. However, it is difficult to assess salt sensitivity because there is no universal definition or commonly accepted methods to determine it. There is considerable heterogeneity within the published studies and among the studied population groups with regard to a relationship between sodium intake and BP. Although skeptical reports have been published as well (Stolarz-Skrzypek and Staessen, 2015), collective analysis by the IOM and USDA (IOM, 2005; USDA, 2010) supported the evidence for the link between excessive dietary sodium intake and elevated BP, especially in at risk subgroups, such as individuals with hypertension or prehypertension. Reducing salt intake can lead to small reductions in systolic and diastolic BP, albeit a recent metaanalysis showed weak evidence of benefit for cardiovascular mortality in normotensive and hypertensive patients (Adler et al., 2014). On the other

---

3. http://health.gov/dietaryguidelines/2015/guidelines/.
4. http://www.ncbi.nlm.nih.gov/books/NBK201517/.

hand, in the Chinese adult population, with mean national consumption at more than 12 g/day of NaCl, salt restriction more significantly lowered mean BP, with the strongest effect among hypertensive participants (Wang et al., 2015). Comorbidities and life quality changes have not been evaluated in this metaanalysis.

The mechanism of salt-induced hypertension is not fully recognized. Hypertension can be induced in response to high dietary salt intake or/and altered renal sodium excretion. Genetic factors, particularly single-nucleotide polymorphisms (SNPs) in genes participating in the renin–angiotensin–aldosterone system (RAAS), an important regulator of renal sodium retention, are associated with salt sensitivity. Beeks et al., summarized genetic polymorphisms that are related to sodium transport, salt sensitivity, and changes in BP (Beeks et al., 2004). These polymorphisms include α-adducin, G-protein β3, angiotensin II type 1 receptor, angiotensinogen, aldosterone synthase (CYP11B2), 11β-hydroxysteroid dehydrogenase type 2, and γENaC. A strong association among salt sensitivity, BP, and SNPs in the sodium-bicarbonate cotransporter gene SLC4A5 has also been reported (Carey et al., 2012). Similar genetic association with BP has also been identified for adrenomedullin (ADM) and ADM2/intermedin, which are genes associated with the activity of glucocorticoid receptor and impaired sodium excretion. Dopamine produced in the kidney also regulates ion transport, and polymorphisms related to salt sensitivity have been described in genes encoding dopamine receptors (e.g., dopamine receptor D1) and proteins involved in recycling of dopamine receptors such as G-protein–coupled receptor kinases (e.g., G-protein–coupled receptor kinase-4; Sanada et al., 2011).

Cardiovascular Disease, Stroke, and Mortality

As mentioned earlier, BP is used as a surrogate marker for CVD, stroke, and mortality risk, especially in populations already at risk of disease. Projections based solely on CVD suggest that decreasing salt intake by 3 g/day (to 1200 mg sodium/day) could reduce health-care costs by $10–24 billion per year (Bibbins-Domingo et al., 2010). The Committee on the Consequences of the Sodium Reduction in Populations (Strom et al., 2013) reviewed the evidence on the relationship between sodium intake and direct health outcomes, with an emphasis on intervention studies. The committee specifically focused on studies done on the general population; in populations older than 51 years of age; and in populations with CKD, CVD, prehypertension, diabetes, and congestive heart failure (CHF). In the general population the committee found an association between excessive sodium intakes and increased risk of CVD, particularly for stroke. With sodium intakes below 2300 mg per day, the results were less consistent, with results ranging from lower, similar, or higher risk of CVD, stroke, or mortality, including all-cause mortality. In specific populations with preexisting conditions, the findings were very heterogeneous, likely because of varying methods of sodium assessment, with the hazard ratio values ranging between 0.11 and 3.54. However, studies in prehypertensive patients with multiple 24-h urine collections (considered the best available method to measure sodium intake) found a significant 25% reduction in CVD incidence, as well as a nonsignificant 20% reduction in total mortality when average levels of sodium intake were decreased from approximately 3.6 to 2.3 g per day (Strom et al., 2013). The committee also considered three randomized clinical trials with CHF patients from one center in Italy, which consistently demonstrated higher adverse events (hospital readmission and mortality) associated with sodium restriction from 2870 to 1840 mg/day. Although no specific weaknesses in the study designs were identified, the committee pointed to the low rates of β-blocker use and the use of high-dose furosemide diuretic combined with significant fluid restriction, which is not consistent with current US management of patients with CHF. In general, the committee's findings reflected insufficient evidence in the general population and population subgroups and strongly indicated the need for further research to better define relationships between sodium intake and risk of CVD, stroke, and mortality, particularly at the lower range of sodium intake within the US population.

Chronic Kidney Disease

An estimated 6 million American adults have CKD, and millions of others are at increased risk, with diabetes and high BP responsible for up to two-thirds of the cases. Studies with CKD populations that utilized 24-h urine collection for sodium intake analysis indicate intakes comparable or higher than in the general population (Humalda and Navis, 2014). The 2012 Kidney Disease Improving Global Outcomes guideline recommends reduction of daily sodium intake to less than 2000 mg/day in CDK patients (KDIGO, 2012). Moderate restriction of dietary sodium can substantially improve the protective effects of therapeutic targeting of the RAAS system in CKD, including decreased urinary protein losses. However, some studies showed a J-curve[5] between sodium intake and renal and cardiovascular outcome, which has

---

5. In this case J-curve refers to a graph in which the x-axis measures dietary Na^+ intake whereas the y-axis measures the chance that a patient will develop cardiovascular or renal disease. The J-shape of the curve indicates that up to a point, dietary Na^+ restriction can be beneficial, but past some point, it may be associated with higher disease incidence.

raised concerns about the safety of rigorous sodium restriction. In addition, animal studies demonstrated that a RAAS blockade combined with rigorous sodium restriction might be detrimental to kidney function, possibly leading to tubulointerstitial damage. Although these findings are of significant concern, in CKD patients with habitually high sodium intake a moderate restriction of salt consumption remains a critical part of health and lifestyle management in this population (Humalda and Navis, 2014).

Metabolic Syndrome and Diabetes

Metabolic syndrome is a disorder characterized by the cooccurrence of three of five of the following medical conditions: abdominal (central) obesity, hypertension, elevated fasting plasma glucose, high serum triglycerides, and low high-density lipoprotein levels. Estimated to affect 34% of the US population, the mechanisms leading to metabolic syndrome are not fully understood, but several complementary hypotheses have been proposed, including insulin resistance, adipose tissue dysregulation, inadequate aldosterone suppression, increased cortisol production, chronic subclinical inflammation, and changes in the gut microbial ecology. Several studies that attempted to identify the link between sodium intake and metabolic syndrome reported a high prevalence of sodium sensitivity in the studied populations. Some reported not only a higher Na^+ consumption and urinary excretion but also a statistical association with higher incidence of hypertension and cardiovascular events in metabolic syndrome. It remains unclear whether males or females with metabolic syndrome are more susceptible to the detrimental effects of dietary Na^+, with contrary reports published.

Gastric Cancer

Stomach cancer is the fourth most common cancer, but its geographical variations in the prevalence rates indicate that dietary differences play a role, including a range of food groups to which salt and/or nitrates have been added. Many studies showed that food rich in salt, nitrite, and N-nitroso compounds are associated with an increased risk of gastric cancer (GC). There is relatively strong epidemiological evidence for the contribution of high salt intake as an environmental component of the overall risk for GC, including the INTERSALT study, which included 24 countries and showed a significant direct association between 24-h urinary sodium excretion (as a marker of dietary salt intake) and mortality from GC (Joossens et al., 1996). Epidemiological studies in Japan, where the GC incidence rate is the highest in the world, have shown that high salt intake was positively associated with a risk for GC.

High Na^+ intake may also facilitate gastric colonization with *Helicobacter pylori* (HP), and act synergistically to promote the development of GC (Tsugane, 2005). The international EUROGAST study found a positive relationship between HP infection rates and urinary sodium excretion, which was evident in almost all age and gender groups (Beevers et al., 2004). Single-dose oral exposure to a saturated salt solution causes reversible histopathological damage to the gastric mucosa. These changes are characterized by marked edema and loss of the outer epithelial cell layer, which is soon covered by a thick mucoid layer followed by increased cell proliferation (Cohen and Roe, 1997). Although the exact biological mechanisms of salt on carcinogenesis is not known, damage to epithelial barrier (e.g., due to high osmolarity) may lead to inflammatory responses, which, in conjunction with chronic HP infection, may increase epithelial cell proliferation and increase probability of accumulation of somatic mutations. High salt may also increase expression of the HP gene *cagA*. *CagA* is associated with the intensity of gastric inflammation, mucosal atrophy, and loss of runt-related transcription factor-3, a tumor suppressor gene (Loh et al., 2007). The relationship between dietary Na^+ and GC is covered in greater detail elsewhere (D'Elia et al., 2014).

Other Health Outcomes

Aberrations in sodium transport/metabolism can lead to many disorders, including a major impediment to childhood development. Loss-of-function mutations in endosomal Na^+/H^+ exchangers (eNHEs) have been linked to various neurological conditions, including autism, attention deficit hyperactivity disorder, intellectual disability, and epilepsy. Two of the eNHE isoforms, NHE6 and NHE9, are highly expressed in brain, including hippocampus and cortex. Therefore NHE-mediated transport, luminal cation content, and pH in the neurons should be taken into account when considering synaptic function, neuronal plasticity, and differentiation. New-generation drugs restoring normal function of the sodium and associated proton transport should minimize or restore neuronal functions (Kondapalli et al., 2014). For instance, studies with organ-specific knockout mouse models showed that specific sodium/calcium exchangers (NCX) variants contribute to heart contractility, arrhythmia, ischemic damage, smooth muscle vasoconstriction, BP regulation, and brain activities (hippocampal long-term potentiation in learning, cerebral reperfusion damage, stroke, preconditioning, etc.; Khananshvili, 2014). In addition, NHE-deficient mice, depending on the isoform targeted, develop ataxia, growth retardation, and seizures (NHE1; Bell et al., 1999);

gastritis (NHE2; Boivin et al., 2000; Schultheis et al., 1998a); defects in intestinal sodium and water absorption, low BP, systemic acidosis, colitis, and gut microbial dysbiosis (NHE3; Larmonier et al., 2013; Laubitz et al., 2008; Schultheis et al., 1998b); or infertility, gastric ulceration, or dry eye syndrome (NHE8; Xu et al., 2013, 2015a,b).

HOW OTHER MINERALS ARE AFFECTED OR BEHAVE

Membrane sodium transport is frequently coupled with transport of other ions, a process accomplished by antiporter systems that utilize the electrochemical gradient of Na^+ to catalyze extrusion of Ca^{2+}, H^+, or K^+. These transporters are members of the SLC gene families: SLC5, sodium glucose transporters; SLC6, sodium/chloride-dependent small amino acid or amino acid-like substrate transporters; SLC8, NCX; SLC9, NHEs; SLC10, sodium/bile acid transporters; SLC13, sodium-sulfate/carboxylate transporters; SLC20, sodium-dependent phosphate transporters; SLC23, sodium-dependent ascorbic acid transporters; SLC24, sodium/calcium-potassium exchangers; SLC28, sodium-coupled nucleoside transporters; SLC34, type II sodium/phosphate cotransporters; and SLC38, sodium-coupled neutral amino acid transporters. Comprehensive descriptions of these transporters are beyond the scope of this chapter. However, this highlights the importance of sodium ions in other membrane transport processes in many other aspects of cellular and systemic physiology.

Potassium and calcium are related to renal Na^+ handling. Administration of potassium bicarbonate and potassium chloride in patients leads to increased urinary sodium excretion. This may be an important component of the antihypertensive effect of potassium, particularly in patients with hypertension. The mechanism is likely similar to that shown for the effects of hypovolemia and hyperkalemia, where the kidneys maintain homeostasis by Na^+ retention and inhibition of the NCC in the distal convoluted tubules (van der Lubbe et al., 2013). In a recent study, potassium supplementation, on top of a relatively low-sodium diet, reduced BP in hypertensive individuals (Gijsbers et al., 2015). Moreover, high Na^+ intake results in increased urinary calcium losses, which may contribute to negative calcium balance and increased bone turnover, which may be especially detrimental in postmenopausal women. On the other hand, calcium supplementation has limited effects on renal Na^+ excretion, and although some animal studies suggested positive effects on BP, clinical studies have not been conclusive. Selected patient groups may benefit from Ca^{2+} supplementation, such as in pregnancy-induced hypertension or in hypertensive patients with inadequate Ca^{2+} intake.

SUMMARY POINTS

- Sodium is an essential nutrient, and the principal extracellular ion that plays critical roles in regulating blood and cell volume; maintaining membrane potential; and in neural excitation, nerve conductance, and muscle contraction.
- Systemic sodium homeostasis is regulated at the level of intestinal and renal membrane transporters, which are frequent targets of pharmacological interventions. Genetic polymorphisms in those transporters and related regulatory factors increase risk for developing disorders associated with inappropriate sodium handling.
- Multiple negative health outcomes are strongly associated with high dietary sodium intake.
- Current recommendations of a generalized and indiscriminate reduction of salt intake in the general population level are not well supported by scientific evidence, although dietary sodium restriction is of value in specific at risk patients.
- The current state of knowledge strongly indicates further need for additional well-controlled population-based studies to understand and resolve the existing controversies regarding sodium intake and development of chronic disease in humans.

KEY FACTS

- Although 500 mg is considered to be a safe daily minimum intake of sodium, sufficient to maintain all organismal Na^+-dependent functions, the vast majority of the US population exceeds the UL set at 2300 mg per day.
- With segmental differences, electroneutral Na^+/H^+ exchange and electrogenic Na^+ channels are responsible for the bulk of intestinal and renal sodium (re)absorption.
- Very low and high sodium intake have been associated with pathological states, although the evidence is the strongest for the positive association between high Na^+ intake and hypertension, CVDs, stroke, CKDs, metabolic syndrome, diabetes, and GC.
- Reduction of dietary sodium intake is more beneficial in patients at risk for or diagnosed with these conditions than in the general population.

MINI DICTIONARY OF TERMS

CHF Congestive heart failure.

CKD Chronic kidney disease.

CVD Cardiovascular disease.

ENaCs Epithelial Na^+ channels.

Osmolality Concentration of solutes, proportional to the number of particles and independent of the size or their nature, expressed per 1 kg of solvent (expressed as a mOsmol/kg).

Osmolarity Similar to osmolality but expressed as the numbers of osmoles per liter of solution (mOsmol/L).

RAAS Renin–angiotensin–aldosterone system.

VGSCs Voltage-gated sodium channels.

REFERENCES

Aburto, N.J., Ziolkovska, A., Hooper, L., Elliott, P., Cappuccio, F.P., Meerpohl, J.J., 2013. Effect of lower sodium intake on health: systematic review and meta-analyses. BMJ 346, f1326.

Adler, A.J., Taylor, F., Martin, N., Gottlieb, S., Taylor, R.S., Ebrahim, S., 2014. Reduced dietary salt for the prevention of cardiovascular disease. Cochrane Database Syst. Rev. 12, CD009217.

Adrogue, H.J., Madias, N.E., 2000. Hyponatremia. N. Engl. J. Med. 342, 1581–1589.

Adrogue, H.J., Madias, N.E., 2014. Diagnosis and treatment of hyponatremia. Am. J. Kidney Dis. 64, 681–684.

Allsopp, A.J., Sutherland, R., Wood, P., Wootton, S.A., 1998. The effect of sodium balance on sweat sodium secretion and plasma aldosterone concentration. Eur. J. Appl. Physiol. Occup. Physiol. 78, 516–521.

Ameen, N.A., Salas, P.J., 2000. Microvillus inclusion disease: a genetic defect affecting apical membrane protein traffic in intestinal epithelium. Traffic 1, 76–83.

Anderson, R.J., 1986. Hospital-associated hyponatremia. Kidney Int. 29, 1237–1247.

Barmeyer, C., Ye, J.H., Soroka, C., Geibel, P., Hingsammer, L.M., Weitgasser, L., Atway, D., Geibel, J.P., Binder, H.J., Rajendran, V.M., 2013. Identification of functionally distinct $Na–HCO_3$ co-transporters in colon. PLoS One 8, e62864.

Beeks, E., Kessels, A.G., Kroon, A.A., van der Klauw, M.M., de Leeuw, P.W., 2004. Genetic predisposition to salt-sensitivity: a systematic review. J. Hypertens. 22, 1243–1249.

Beevers, D.G., Lip, G.Y., Blann, A.D., 2004. Salt intake and *Helicobacter pylori* infection. J. Hypertens. 22, 1475–1477.

Bell, S.M., Schreiner, C.M., Schultheis, P.J., Miller, M.L., Evans, R.L., Vorhees, C.V., Shull, G.E., Scott, W.J., 1999. Targeted disruption of the murine *Nhe1* locus induces ataxia, growth retardation, and seizures. Am. J. Physiol. 276, C788–C795.

Benos, D.J., Awayda, M.S., Ismailov, I.I., Johnson, J.P., 1995. Structure and function of amiloride-sensitive Na^+ channels. J. Membr. Biol. 143, 1–18.

Bibbins-Domingo, K., Chertow, G.M., Coxson, P.G., Moran, A., Lightwood, J.M., Pletcher, M.J., Goldman, L., 2010. Projected effect of dietary salt reductions on future cardiovascular disease. N. Engl. J. Med. 362, 590–599.

Boivin, G.P., Schultheis, P.J., Shull, G.E., Stemmermann, G.N., 2000. Variant form of diffuse corporal gastritis in NHE2 knockout mice. Comp. Med. 50, 511–515.

Bookstein, C., DePaoli, A.M., Xie, Y., Niu, P., Musch, M.W., Rao, M.C., Chang, E.B., 1994. Na+/H+ exchangers, NHE-1 and NHE-3, of rat intestine. Expression and localization. J. Clin. Invest. 93, 106–113.

Burnier, M., 2008. Sodium in Health and Disease. Informa Healthcare, New York, NY, USA.

Capasso, G., Rizzo, M., Evangelista, C., Ferrari, P., Geelen, G., Lang, F., Bianchi, G., 2005. Altered expression of renal apical plasma membrane Na^+ transporters in the early phase of genetic hypertension. Am. J. Physiol. Ren. Physiol. 288, F1173–F1182.

Carey, R.M., Schoeffel, C.D., Gildea, J.J., Jones, J.E., McGrath, H.E., Gordon, L.N., Park, M.J., Sobota, R.S., Underwood, P.C., Williams, J., et al., 2012. Salt sensitivity of blood pressure is associated with polymorphisms in the sodium-bicarbonate cotransporter. Hypertension 60, 1359–1366.

Carpenter, C.C., 1990. The erratic evolution of cholera therapy: from folklore to science. Clin. Ther. 12 (Suppl. A), 22–27 discussion 28.

Cheung, C.Y., Ko, B.C., 2013. NFAT5 in cellular adaptation to hypertonic stress—regulations and functional significance. J. Mol. Signal. 8, 5.

Cohen, A.J., Roe, F.J., 1997. Evaluation of the aetiological role of dietary salt exposure in gastric and other cancers in humans. Food Chem. Toxicol. 35, 271–293.

D'Elia, L., Galletti, F., Strazzullo, P., 2014. Dietary salt intake and risk of gastric cancer. Cancer Treat. Res. 159, 83–95.

Dahl, L.K., 1958. Salt intake and salt need. N. Engl. J. Med. 258, 1152–1157 contd.

Donowitz, M., Ming Tse, C., Fuster, D., 2013. SLC9/NHE gene family, a plasma membrane and organellar family of Na+/H+ exchangers. Mol. Asp. Med. 34, 236–251.

Engevik, M.A., Aihara, E., Montrose, M.H., Shull, G.E., Hassett, D.J., Worrell, R.T., 2013. Loss of NHE3 alters gut microbiota composition and influences *Bacteroides thetaiotaomicron* growth. Am. J. Physiol. Gastrointest. Liver Physiol. 305, G697–G711.

Ferraris, J.D., Williams, C.K., Ohtaka, A., Garcia-Perez, A., 1999. Functional consensus for mammalian osmotic response elements. Am. J. Physiol. 276, C667–C673.

Garcia-Perez, A., Burg, M.B., 1991. Renal medullary organic osmolytes. Physiol. Rev. 71, 1081–1115.

Ghishan, F.K., Kiela, P.R., 2014. Epithelial transport in inflammatory bowel diseases. Inflamm. Bowel Dis. 20, 1099–1109.

Gijsbers, L., Dower, J.I., Mensink, M., Siebelink, E., Bakker, S.J., Geleijnse, J.M., 2015. Effects of sodium and potassium supplementation on blood pressure and arterial stiffness: a fully controlled dietary intervention study. J. Hum. Hypertens. 10, 592–598.

Hadchouel, J., Delaloy, C., Faure, S., Achard, J.M., Jeunemaitre, X., 2006. Familial hyperkalemic hypertension. J. Am. Soc. Nephrol. 17, 208–217.

Halterman, J.A., Kwon, H.M., Wamhoff, B.R., 2012. Tonicity-independent regulation of the osmosensitive transcription factor TonEBP (NFAT5). Am. J. Physiol. Cell Physiol. 302, C1–C8.

He, F.J., Li, J., Macgregor, G.A., 2013. Effect of longer term modest salt reduction on blood pressure: Cochrane systematic review and meta-analysis of randomised trials. BMJ 346, f1325.

Hediger, M.A., Coady, M.J., Ikeda, T.S., Wright, E.M., 1987. Expression cloning and cDNA sequencing of the Na$^+$/glucose co-transporter. Nature 330, 379–381.

Hirschhorn, N., Kinzie, J.L., Sachar, D.B., Northrup, R.S., Taylor, J.O., Ahmad, S.Z., Phillips, R.A., 1968. Decrease in net stool output in cholera during intestinal perfusion with glucose-containing solutions. N. Engl. J. Med. 279, 176–181.

Hoeijmakers, J.G., Faber, C.G., Merkies, I.S., Waxman, S.G., 2015. Painful peripheral neuropathy and sodium channel mutations. Neurosci. Lett. 596, 51–59.

Humalda, J.K., Navis, G., 2014. Dietary sodium restriction: a neglected therapeutic opportunity in chronic kidney disease. Curr. Opin. Nephrol. Hypertens. 23, 533–540.

IOM, 2005. Dietary Reference Intakes for Water, Potassium, Sodium, Chloride, and Sulfate. The National Academies Press.

Ji, W., Foo, J.N., O'Roak, B.J., Zhao, H., Larson, M.G., Simon, D.B., Newton-Cheh, C., State, M.W., Levy, D., Lifton, R.P., 2008. Rare independent mutations in renal salt handling genes contribute to blood pressure variation. Nat. Genet. 40, 592–599.

Joo, M.A., Kim, E.Y., 2013. Hyponatremia caused by excessive intake of water as a form of child abuse. Ann. Pediatr. Endocrinol. Metab. 18, 95–98.

Joossens, J.V., Hill, M.J., Elliott, P., Stamler, R., Lesaffre, E., Dyer, A., Nichols, R., Kesteloot, H., 1996. Dietary salt, nitrate and stomach cancer mortality in 24 countries. European Cancer Prevention (ECP) and the INTERSALT Cooperative Research Group. Int. J. Epidemiol. 25, 494–504.

KDIGO, C.W.G., 2012. KDIGO 2012 clinical practice guideline for the evaluation and management of chronic kidney disease. Kidney Int. Suppl. 3, 1–150.

Khananshvili, D., 2014. Sodium–calcium exchangers (NCX): molecular hallmarks underlying the tissue-specific and systemic functions. Pflugers Arch. 466, 43–60.

Kiela, P.R., Ghishan, F.K., 2012. Na$^+$/H$^+$ exchange in mammalian digestive tract. In: Johnson, F.K.G.L., Kaunitz, J., Merchant, J., Said, H., Wood, J.D. (Eds.), Physiology of the Gastrointestinal Tract. Academic Press, Waltham, MA, pp. 1781–1818.

Kiela, P.R., Laubitz, D., Larmonier, C.B., Midura-Kiela, M.T., Lipko, M.A., Janikashvili, N., Bai, A., Thurston, R., Ghishan, F.K., 2009. Changes in mucosal homeostasis predispose NHE3 knockout mice to increased susceptibility to DSS-induced epithelial injury. Gastroenterology 137, 965–975 975 e961–910.

Knauth, L.P., 1998. Salinity history of the Earth's early ocean [letter]. Nature 395, 554–555.

Knowles, B.C., Roland, J.T., Krishnan, M., Tyska, M.J., Lapierre, L.A., Dickman, P.S., Goldenring, J.R., Shub, M.D., 2014. Myosin Vb uncoupling from RAB8A and RAB11A elicits microvillus inclusion disease. J. Clin. Invest. 124, 2947–2962.

Kondapalli, K.C., Prasad, H., Rao, R., 2014. An inside job: how endosomal Na$^+$/H$^+$ exchangers link to autism and neurological disease. Front. Cell Neurosci. 8, 172.

Koyama, K., Sasaki, I., Naito, H., Funayama, Y., Fukushima, K., Unno, M., Matsuno, S., Hayashi, H., Suzuki, Y., 1999. Induction of epithelial Na$^+$ channel in rat ileum after proctocolectomy. Am. J. Physiol. 276, G975–G984.

Kultz, D., 2001. Cellular osmoregulation: beyond ion transport and cell volume. Zoology 104, 198–208.

Larmonier, C.B., Laubitz, D., Hill, F.M., Shehab, K.W., Lipinski, L., Midura-Kiela, M.T., McFadden, R.M., Ramalingam, R., Hassan, K.A., Golebiewski, M., et al., 2013. Reduced colonic microbial diversity is associated with colitis in NHE3-deficient mice. Am. J. Physiol. Gastrointest. Liver Physiol. 305, G667–G677.

Larmonier, C.B., Laubitz, D., Thurston, R.D., Bucknam, A.L., Hill, F.M., Midura-Kiela, M., Ramalingam, R., Kiela, P.R., Ghishan, F.K., 2011. NHE3 modulates the severity of colitis in IL-10-deficient mice. Am. J. Physiol. Gastrointest. Liver Physiol. 300, G998–G1009.

Laubitz, D., Larmonier, C.B., Bai, A., Midura-Kiela, M.T., Lipko, M.A., Thurston, R.D., Kiela, P.R., Ghishan, F.K., 2008. Colonic gene expression profile in NHE3-deficient mice: evidence for spontaneous distal colitis. Am. J. Physiol. Gastrointest. Liver Physiol. 295, G63–G77.

Lodish, H., Berk, A., Zipursky, S.L., Matsudaira, P., Baltimore, D., Darnell, J., 2000. Intracellular ion environment and membrane electric potential. In: Freeman, W.H. (Ed.), Molecular Cell Biology. W. H. Freeman and Company.

Loh, J.T., Torres, V.J., Cover, T.L., 2007. Regulation of *Helicobacter pylori cagA* expression in response to salt. Cancer Res. 67, 4709–4715.

Loo, D.D., Zeuthen, T., Chandy, G., Wright, E.M., 1996. Cotransport of water by the Na$^+$/glucose cotransporter. Proc. Natl. Acad. Sci. U. S. A. 93, 13367–13370.

McLean, R.M., 2014. Measuring population sodium intake: a review of methods. Nutrients 6, 4651–4662.

Mente, A., O'Donnell, M.J., Yusuf, S., 2015. Measuring sodium intake in populations: simple is best? Am. J. Hypertens. 11, 1303–1305.

Michail, S., Collins, J.F., Xu, H., Kaufman, S., Vanderhoof, J., Ghishan, F.K., 1998. Abnormal expression of brush-border membrane transporters in the duodenal mucosa of two patients with microvillus inclusion disease. J. Pediatr. Gastroenterol. Nutr. 27, 536–542.

Muller, T., Wijmenga, C., Phillips, A.D., Janecke, A., Houwen, R.H., Fischer, H., Ellemunter, H., Fruhwirth, M., Offner, F., Hofer, S., et al., 2000. Congenital sodium diarrhea is an autosomal recessive disorder of sodium/proton exchange but unrelated to known candidate genes. Gastroenterology 119, 1506–1513.

Nakanishi, T., Takamitsu, Y., 1996. Renal concentrating defect and organic osmolytes. Paradoxical changes of renal medullary taurine contents in potassium-depleted rats. Adv. Exp. Med. Biol. 403, 193–201.

Padmanabhan, S., Caulfield, M., Dominiczak, A.F., 2015. Genetic and molecular aspects of hypertension. Circ. Res. 116, 937–959.

Rosner, M.H., 2009. Exercise-associated hyponatremia. Semin. Nephrol. 29, 271–281.

Rossier, B.C., 2014. Epithelial sodium channel (ENaC) and the control of blood pressure. Curr. Opin. Pharmacol. 15, 33–46.

Sanada, H., Jones, J.E., Jose, P.A., 2011. Genetics of salt-sensitive hypertension. Curr. Hypertens. Rep. 13, 55–66.

Sandle, G.I., 1989. Segmental heterogeneity of basal and aldosterone-induced electrogenic Na transport in human colon. Pflugers Arch. 414, 706–712.

Schultheis, P.J., Clarke, L.L., Meneton, P., Harline, M., Boivin, G.P., Stemmermann, G., Duffy, J.J., Doetschman, T., Miller, M.L., Shull, G.E., 1998a. Targeted disruption of the murine Na$^+$/H$^+$ exchanger isoform 2 gene causes reduced viability of gastric parietal cells and loss of net acid secretion. J. Clin. Invest. 101, 1243–1253.

Schultheis, P.J., Clarke, L.L., Meneton, P., Miller, M.L., Soleimani, M., Gawenis, L.R., Riddle, T.M., Duffy, J.J., Doetschman, T., Wang, T., et al., 1998b. Renal and intestinal absorptive defects in mice lacking the NHE3 Na$^+$/H$^+$ exchanger. Nat. Genet. 19, 282–285.

Seidler, U., Lenzen, H., Cinar, A., Tessema, T., Bleich, A., Riederer, B., 2006. Molecular mechanisms of disturbed electrolyte transport in intestinal inflammation. Ann. N. Y. Acad. Sci. 1072, 262–275.

Simon, D.B., Karet, F.E., Hamdan, J.M., DiPietro, A., Sanjad, S.A., Lifton, R.P., 1996a. Bartter's syndrome, hypokalaemic alkalosis with hypercalciuria, is caused by mutations in the Na–K–2Cl cotransporter *NKCC2*. Nat. Genet. 13, 183–188.

Simon, D.B., Nelson-Williams, C., Bia, M.J., Ellison, D., Karet, F.E., Molina, A.M., Vaara, I., Iwata, F., Cushner, H.M., Koolen, M., et al., 1996b. Gitelman's variant of Bartter's syndrome, inherited hypokalaemic alkalosis, is caused by mutations in the thiazide-sensitive Na–Cl cotransporter. Nat. Genet. 12, 24–30.

Skou, J.C., 1989. The influence of some cations on an adenosine triphosphatase from peripheral nerves. Biochim. Biophys. Acta.

Spencer, A.G., Labonte, E.D., Rosenbaum, D.P., Plato, C.F., Carreras, C.W., Leadbetter, M.R., Kozuka, K., Kohler, J., Koo-McCoy, S., He, L., et al., 2014. Intestinal inhibition of the Na$^+$/H$^+$ exchanger 3 prevents cardiorenal damage in rats and inhibits Na$^+$ uptake in humans. Sci. Transl. Med. 6, 227ra236.

Sterns, R.H., 2015. Disorders of plasma sodium–causes, consequences, and correction. N. Engl. J. Med. 372, 55–65.

Stolarz-Skrzypek, K., Staessen, J.A., 2015. Reducing salt intake for prevention of cardiovascular disease—times are changing. Adv. Chronic Kidney Dis. 22, 108–115.

Strom, B.L., Yaktine, A.L., Oria, M., 2013. In: Strom, B.L., Yaktine, A.L., Oria, M. (Eds.), Sodium Intake in Populations: Assessment of Evidence (Washington, DC).

Trepiccione, F., Zacchia, M., Capasso, G., 2012. The role of the kidney in salt-sensitive hypertension. Clin. Exp. Nephrol. 16, 68–72.

Tribl, G.G., Waldhauser, F., Druml, W., Sycha, T., Endler, G., Zeitlhofer, J., Auff, E., 2005. Loss of normal circadian profile of urine excretion in idiopathic restless legs syndrome. Sleep. Med. 6, 391–398.

Tse, C.M., Levine, S.A., Yun, C.H., Montrose, M.H., Little, P.J., Pouyssegur, J., Donowitz, M., 1993. Cloning and expression of a rabbit cDNA encoding a serum-activated ethylisopropylamiloride-resistant epithelial Na$^+$/H$^+$ exchanger isoform (NHE-2). J. Biol. Chem. 268, 11917–11924.

Tsugane, S., 2005. Salt, salted food intake, and risk of gastric cancer: epidemiologic evidence. Cancer Sci. 96, 1–6.

USDA, 2010. Report of the Dietary Guidelines Advisory Committee on the Dietary Guidelines for Americans, 2010, to the Secretary of Agriculture and the Secretary of Health and Human Services. USDA/ARS, Washington, DC.

van der Lubbe, N., Moes, A.D., Rosenbaek, L.L., Schoep, S., Meima, M.E., Danser, A.H., Fenton, R.A., Zietse, R., Hoorn, E.J., 2013. K$^+$-induced natriuresis is preserved during Na$^+$ depletion and accompanied by inhibition of the Na$^+$-Cl- cotransporter. Am. J. Physiol. Ren. Physiol. 305, F1177–F1188.

Wang, M., Moran, A.E., Liu, J., Qi, Y., Xie, W., Tzong, K., Zhao, D., 2015. A meta-analysis of effect of dietary salt restriction on blood pressure in Chinese adults. Glob. Heart S2211–S8160.

Xie, J.X., Li, X., Xie, Z., 2013. Regulation of renal function and structure by the signaling Na/K-ATPase. IUBMB Life 65, 991–998.

Xu, H., Chen, H., Li, J., Zhao, Y., Ghishan, F.K., 2015a. Disruption of NHE8 expression impairs Leydig cell function in the testes. Am. J. Physiol. Cell Physiol. 308, C330–C338.

Xu, H., Chen, R., Ghishan, F.K., 2005. Subcloning, localization, and expression of the rat intestinal sodium–hydrogen exchanger isoform 8. Am. J. Physiol. Gastrointest. Liver Physiol. 289, G36–G41.

Xu, H., Li, J., Chen, H., Wang, C., Ghishan, F.K., 2013. NHE8 plays important roles in gastric mucosal protection. Am. J. Physiol. Gastrointest. Liver Physiol. 304, G257–G261.

Xu, H., Zhao, Y., Li, J., Wang, M., Lian, F., Gao, M., Ghishan, F.K., 2015b. Loss of NHE8 expression impairs ocular surface function in mice. Am. J. Physiol. Cell Physiol. 308, C79–C87.

Yin, J.S.,R., Yang, J., Heinz-Erian, P., Müller, T., Zoller, H., Tse, M., Janecke, A.R., Donowitz, M., 2015. 110 Missense mutations of SLC9A3 in patients with congenital sodium diarrhea are associated with reduced Na$^+$/H$^+$ exchanger 3 (NHE3) activity: identification of the cause of the phenotype. Gastroenterology 148, S-29.

Chapter 41

Potassium Channel Mutations and Human Disease: Focus on Adrenal Hypertension

Michael Stowasser

The University of Queensland, School of Medicine, Brisbane, QLD, Australia

INTRODUCTION

With more than 80 known genes encoding their subunits, potassium channels form the largest family of ion channels and are involved in the regulation of a diverse array of physiological processes (Coetzee et al., 1999). This diversity of structure and function is further enhanced by alternative splicing, regulatory subunits, and heteromerization. Therefore it is not surprising that mutations in potassium channel-encoding genes have been associated with a wide variety of phenotypic manifestations, affecting multiple organ systems (including cardiovascular, neural, sensory, renal, musculoskeletal, and endocrine). Although most of these appear to be inherited (germline) mutations and rare, recently described somatic mutations in one potassium channel (*KCNJ5*) have been found to be relatively common among patients with hypertension due to adrenal aldosterone-producing adenoma (APA), with mutations shown to be present in approximately 40% of these tumors (Azizan et al., 2012b; Boulkroun et al., 2012; Choi et al., 2011; Scholl et al., 2012b). Given that hypertension affects approximately 30% of adults in Western societies, and APA possibly accounts for up to 3% of hypertension, somatic *KCNJ5* mutations may be much more common than any of the inherited potassium channel mutations so far described, and they may affect a sizable number of individuals (possibly up to 3–4 per 1000).

Because it is beyond the scope of this chapter to describe all conditions associated with potassium channel mutations in detail, a brief overview will be provided of the major classes of potassium channels known to exist and the disease states associated with mutations of genes encoding some of their members. As an illustration of the potential impact of these mutations on human health, and of the advances in knowledge of cell physiology and pathophysiology that have resulted from their discovery and subsequent analysis, the rest of the chapter will focus on one of the more important of these, *KCNJ5*.

BROAD CLASSIFICATION OF POTASSIUM CHANNELS

On the basis of their structure, activity, and mode of activation, potassium channels can be broadly divided into four classes: (1) inwardly rectifying potassium channels (K_{ir}), (2) voltage-gated potassium channels (K_V), (3) tandem pore domain potassium channels (K_{2P}), and (4) calcium-activated potassium channels (K_{Ca}; Giudicessi and Ackerman, 2012; Lason et al., 2013; Shimizu et al., 1984).

Inwardly Rectifying Potassium Channels

K_{ir} channels promote inward current (positive charge into the cell), which is the opposite of leak channels. In this way they correct hyperpolarization by allowing the flow of positively charged K^+ ions into the cell, pushing the membrane potential back to the resting potential. During an action potential (AP) the channels close, prolonging the AP (Hibino et al., 2010). The phenomenon of inward rectification of K_{ir} channels is the result of high-affinity block by endogenous polyamines, namely spermine, as well as magnesium ions, which plug the channel pore at positive potentials, resulting in a decrease in outward currents (Hibino et al., 2010).

K_{ir} channels are composed of four subunits and can be homo- or heterotetrameric. They have only two transmembrane segments, a single pore with a selectivity filter, and they lack the voltage sensor (i.e., they are not voltage gated). All K_{ir} channels require phosphatidylinositol 4,5-bisphosphate (PIP_2) for activation. In this regard K_{ir} channels are PIP_2 ligand-gated ion channels (Hibino et al., 2010).

Molecular, Genetic, and Nutritional Aspects of Major and Trace Minerals. http://dx.doi.org/10.1016/B978-0-12-802168-2.00041-5

Voltage-Gated Potassium Channels

K_V channels are voltage sensitive. They play a vital role in AP initiation and propagation, shaping of APs, and regulation of AP firing patterns. Delayed rectifier and A-type (rapidly inactivating) potassium channels, which account for the majority of the K_V class, are typically responsible for repolarizing a cell after an AP. Unlike K_{ir} channels, these more typical potassium channels preferentially carry outward (rather than inward) potassium currents at depolarized membrane potentials; hence they may be thought of as outwardly rectifying. They have six transmembrane segments and one pore. K_V channels have α and β subunits. α subunits form the conductance pore and are grouped into 12 families based on sequence homology of the hydrophobic transmembrane cores. These typically form tetramers, which are usually homo- (but can be hetero-) tetrameric. β subunits are auxiliary proteins that associate with α subunits and modulate the activity of the channel (Norris et al., 2010).

Tandem Pore Domain Potassium Channels

Otherwise known as leak potassium channels, K_{2P} channels are constitutively open or possess high basal activation and thereby set the negative membrane potential of cells (meaning that there is a higher negative charge inside compared with outside of the cell) by leaking potassium out of the cell. The α subunits have four transmembrane segments and, unlike other potassium channels, two pore loops and form dimers rather than tetramers in the cell membrane (Noel et al., 2011).

Calcium-Activated Potassium Channels

K_{Ca} channels open in response to the presence of intracellular calcium ions or other signaling molecules. They are subclassified according to whether they demonstrate small (SK), intermediate (IK), or large (BK) conductance. The SK and IK subclasses are voltage insensitive whereas the BK channels are voltage activated in addition to being activated by Ca^{2+} (Coetzee et al., 1999; Hoshi et al., 2013). Some of these channels are also responsive to intracellular Na^+ and Cl^-. As in K_V channels, the K_{Ca} channel α subunits have six transmembrane segments with the exception of $K_{Ca}1$, which has seven. The α subunits make homo- and heterotetrameric complexes (Coetzee et al., 1999; Hoshi et al., 2013).

KNOWN MUTATED POTASSIUM CHANNEL GENES AND ASSOCIATED PHENOTYPES

Inwardly Rectifying Potassium Channels (Table 41.1)

KCNJ1 encodes the renal outer medullary potassium (ROMK) channel (Kir1.1), which recycles K^+ from the cell of the thick ascending limb (TAL) of the loop of Henle within the nephron back into the tubule lumen, which (1) sets a positive transmembrane potential, important for paracellular reabsorption of cations, and (2) maintains an adequate supply of K^+ in the TAL to preserve the turnover of the sodium chloride cotransporter NKCC2 (Hibino et al., 2010). Without ROMK, the TAL cannot absorb salt. Loss-of-function mutations of *KCNJ1* cause one form of Bartter syndrome (Simon et al., 1996), which is characterized by renal loss of potassium (causing hypokalemia) and sodium (leading to hypotension). Ji et al. (2008) found rare variants of ROMK (and other Bartter's-mutated channels), known or predicted (and later confirmed in the majority) to induce loss of function, to confer resistance to hypertension among a general population from the Framingham Heart Study.

Encoded by *KCNJ2*, Kir2.1 is a component of the inward rectifier current, which provides repolarizing (outward) current to the most terminal phase of cardiac repolarization and is the primary conductance controlling the diastolic membrane potential (Hibino et al., 2010). Loss-of-function mutations cause prolongation of the cardiac AP manifested as a long QT interval on the surface electrocardiogram (ECG) and lead to the LQTS7 form of congenital long QT syndrome (LQTS, Romano–Ward syndrome), which predisposes to "torsade de pointes" arrhythmia, the signature form of ventricular tachycardia in LQTS, which can degenerate into fatal ventricular fibrillation (Giudicessi and Ackerman, 2012; Shimizu and Horie, 2011). Loss-of-function mutations of *KCNJ2* also cause Andersen–Tawil syndrome (lengthening of the ventricular AP and QT interval, cardiac arrhythmias, periodic paralysis, short stature, and dysmorphic features that include cleft palate, low-set ears, dental abnormalities, and limb abnormalities including syndactyly, brachydactyly, and clinodactyly; Plaster et al., 2001). Gain-of-function mutations are associated with the SQTS3 form of short QT syndrome (SQTS), which predisposes to atrial fibrillation, ventricular fibrillation, and sudden cardiac death (Giudicessi and Ackerman, 2012; Shimizu and Horie, 2011).

KCNJ5 encodes G-protein-activated inwardly rectifying K+ channel 4 (GIRK4 or Kir 3.2), which is one of the Kir3 family of inwardly directing potassium channels that are G-protein–coupled receptor-regulated. GIRK4 is sensitive to

TABLE 41.1 Disease States Associated With Germline or Somatic Mutations in Genes Encoding Inwardly Rectifying Potassium Channels

Gene	Protein	Effect of Mutation	Disease
KCNJ1	Kir1.1 (ROMK)	Loss of function	Bartter syndrome
KCNJ2	Kir2.1	Loss of function	LQTS7; Andersen–Tawil syndrome
		Gain of function	SQTS3
KCNJ5	Kir3.4 (GIRK4)	Loss of function	LQTS13
		Gain of function	Familial hyperaldosteronism type III; APA[a]
KCNJ6	Kir3.2 (GIRK2)	Uncertain	Keppen–Lubinsky syndrome
KCNJ8	Kir6.1	Loss of function	SIDS
		Gain of function	BrS8; ERS
KCNJ10	Kir4.1	Loss of function	SeSAME syndrome; EAST syndrome
		Gain of function	Autism, seizures, intellectual disability
KCNJ11	Kir6.2	Loss of function	CHI
		Gain of function	Neonatal DM; DEND syndrome
SUR1	ABCC8	Loss of function	CHI
		Gain of function	Neonatal DM; DEND syndrome
SUR2	ABCC9	Loss of function	AF; DCM
		Gain of function	Cantu syndrome
KCNJ13	Kir7.1	Loss of function	SVD; LCA
KCNJ15	Kir4.2	Gain of function	Type II DM
KCNJ18	Kir2.6	Loss of function	Thyrotoxic HPP

ABCC, ATP-binding cassette subfamily C; AF, atrial fibrillation; APA, aldosterone-producing adenoma; BrS, Brugada syndrome; CHI, congenital hyperinsulinemia; DCM, dilated cardiomyopathy; DEND, developmental delay, epilepsy, and neonatal DM; DM, diabetes mellitus; EAST, epilepsy, ataxia, sensineural deafness, and tubulopathy; ERS, early repolarization syndrome; GIRK4, G-protein-activated inwardly rectifying K+ channel 4; HPP, hypokalemic periodic paralysis; Kir, inwardly rectifying potassium channel; LCA, Leber congenital amaurosis; LQTS, long-QT syndrome; ROMK, renal outer medullary potassium channel; SeSAME, seizures, sensineural deafness, ataxia, mental retardation and electrolyte imbalance; SIDS, sudden infant death syndrome; SQTS, short QT syndrome; SVD, snowflake vitreoretinal degeneration.
[a]Somatic mutations only (all others refer to germline mutations).

acetylcholine (K_{Ach}; Hibino et al., 2010). As discussed in detail later in this chapter, inherited gain-of-function mutations cause familial hyperaldosteronism (FH)-III and somatic mutations are found in APAs, benign tumors within the adrenal cortex that secrete aldosterone (salt-retaining hormone) autonomously and excessively and thereby cause hypertension (Choi et al., 2011). Loss-of-function mutations cause the LQTS13 form of congenital LQTS (Romano–Ward syndrome; Giudicessi and Ackerman, 2012; Shimizu and Horie, 2011).

GIRK2 (Kir 3.2), encoded by *KCNJ6*, is also G-protein coupled and is sensitive to increases in intracellular ATP, which lead to channel closure (Hibino et al., 2010). Mutations in the selectivity filter cause Keppen–Lubinsky syndrome, a very rare disease in which clinical features are highly distinct and include severe developmental delay, intellectual disability, hypertonia, hyperreflexia, growth parameters above the 50th–75th percentile at birth and below the 5th percentile by the age of 6–9 months, microcephaly, large prominent eyes, a narrow nasal bridge, a tented upper lip, a high palate, an open mouth, tightly adherent skin, and an aged appearance, in keeping with a generalized lipodystrophy disorder (Masotti et al., 2015).

KCNJ8 encodes Kir6.1, another ATP-sensitive channel in which ATP closes and ADP opens the channel (Hibino et al., 2010). Gain-of-function mutations have been found in patients with ventricular arrhythmias due to early repolarization syndrome and Brugada syndrome (BrS; diagnosed by coved-typed ST elevation in the right precordial leads of the ECG, V1-3, and an episode of ventricular fibrillation in the absence of structural heart disease) type 8 (BrS8; Giudicessi and Ackerman, 2012; Shimizu and Horie, 2011). Loss-of-function mutations may be a cause of sudden infant death syndrome (Tester et al., 2011).

Loss-of-function mutations of *KCNJ10* (Kir4.1) cause SeSAME syndrome (seizures, sensorineural deafness, ataxia, mental retardation, and electrolyte imbalance) or EAST syndrome (epilepsy, ataxia, sensorineural deafness, and tubulopathy; Scholl et al., 2012a). Gain-of-function mutations may cause autism with seizures and intellectual disability (Sicca et al., 2011).

Encoded by *KCNJ11*, Kir6.2 is another ATP-sensitive channel. Expressed in the pancreatic β islet cells, high glucose leads to high ATP, which closes the channel, resulting in an increase in cytosolic Ca^{2+}, which stimulates insulin secretion. Low glucose leads to increased ADP, which opens the channel, resulting in decrease in cytosolic Ca^{2+} and inhibition of insulin secretion (Hibino et al., 2010). Gain-of-function mutations cause neonatal diabetes, either permanent or transient, or as part of the DEND syndrome (developmental delay, epilepsy, and neonatal diabetes). Loss-of-function mutations cause neonatal hypoglycemia due to congenital hyperinsulinism (CHI). A mild gain-of-function mutation is associated with type II diabetes (Ashcroft and Rorsman, 2013).

ATP-binding cassette (ABC) transporter subfamily C, member 8, encoded by *SUR1*, is the predominant regulatory sulfonylurea receptor (SUR) subunit of Kir6.2, but it is also expressed elsewhere (Hibino et al., 2010). Gain-of-function mutations cause neonatal diabetes (subtypes as previously listed for Kir6.2) and loss-of-function mutations cause neonatal hypoglycemia due to CHI. Similar to *KCNJ11*, a mild gain-of-function mutation of *SUR1* is associated with type II diabetes (Ashcroft and Rorsman, 2013).

Similar to *SUR1*, *SUR2* encodes a regulatory SUR subunit (ABCC9) that forms a heterooctamer with Kir6.2 (mainly in cardiac muscle) and 6.1 (mainly smooth muscle) to comprise an ATP-sensitive channel. Two splice variants exist: SUR2A (mostly cardiac) and SUR2B (mostly smooth muscle; Hibino et al., 2010). A patient with paroxysmal atrial fibrillation had a loss-of-function mutation (Olson et al., 2007), as did two patients with heart failure due to idiopathic dilated cardiomyopathy (Bienengraeber et al., 2004). Gain-of-function mutations cause Cantu syndrome, characterized by hypertrichosis, multiple dysmorphic features, and cardiomegaly (Harakalova et al., 2012).

Loss-of-function mutations of *KCNJ13* (Kir7.1, which is found in epithelial cells of the choroid plexus and retina) cause visual impairment due to snowflake vitreoretinal degeneration (Hejtmancik et al., 2008) and Leber congenital amaurosis (Sergouniotis et al., 2011). Gain-of-function mutations of *KCNJ15* (Kir4.2), expressed in β cells of the pancreas, are associated with type II diabetes (Okamoto et al., 2010). Loss-of-function mutations of *KCNJ18* (which encodes the skeletal muscle-specific channel Kir2.6) lead to susceptibility to thyrotoxic hypokalemic periodic paralysis (Cheng et al., 2011).

Voltage-Gated Potassium Channels (Table 41.2)

The gene product of *KCNQ1* (K_v7.1) forms the α-subunit of the channel that mediates the repolarizing slow delayed rectifier (I_{Ks}) current of the cardiac AP and for which the β-subunit (minK) is encoded by *KCNE1* (Maljevic et al., 2010). Heterozygous loss-of-function mutations cause the LQTS1 form of LQTS (Romano–Ward syndrome). Homozygous loss-of-function mutations cause congenital LQTS and deafness (Jervell and Lange–Nielsen syndrome). Gain-of-function mutations are associated with short-QT syndrome (SQTS)-2, atrial fibrillation, and sinus bradycardia (Maljevic et al., 2010). Polymorphisms of *KCNQ1* are also associated with type II diabetes through uncertain mechanisms (Wang et al., 2014). Heterozygous loss-of-function mutations of *KCNE1* (minimal potassium channel subunit or minK, the β-subunit for I_{Ks}) cause the LQTS5 form of Romano–Ward syndrome and homozygous loss-of-function mutations cause Jervell and Lange–Nielsen syndrome (Maljevic et al., 2010).

KCNH2 encodes "Human-Ether-a-go-go-Related" (hERG1) protein (K_v11.1), which mediates the repolarizing I_{Kr} (rapid delayed rectifier) current (Sanguinetti, 2010). Loss-of-function mutations cause the LQTS2 form of LQTS and have been implicated in epilepsy. A gain-of-function mutation has been described in SQTS1 and another in familial atrial fibrillation (Sanguinetti, 2010).

KCNE2 encodes MinK-related peptide (MiRP)-1, an accessory β-subunit that interacts with *KCNQ1* and *KCNH2*. Loss-of-function mutations cause the LQTS6 form of Romano–Ward syndrome. Gain-of-function mutations are associated with atrial fibrillation.

Encoded by *KCND3*, K_v4.3 is an A-type (rapidly inactivating) potassium ion channel that mediates the transient outward K^+ current, which is the main contributing current to phase I repolarization of the cardiac AP. MiRP2 (encoded by *KCNE3*) is an accessory β-subunit that interacts with K_v4.3. Gain-of-function mutations of both genes are associated with BrS (BrS11 and BrS6, respectively) and atrial fibrillation (Delpon et al., 2008; Giudicessi et al., 2011; Lundby et al., 2008; Olesen et al., 2013).

K_v1.5, encoded by *KCNA5*, is a delayed rectifier that mediates the ultra-rapid repolarizing current that is specific for the atrium. Loss-of-function mutations cause familial atrial fibrillation (Giudicessi and Ackerman, 2012).

TABLE 41.2 Disease States Associated With Germline Mutations in Genes Encoding Voltage-Gated Potassium Channels

Gene	Protein	Effect of Mutation	Disease
KCNQ1	$K_v7.1$	Loss of function	LQTS1; Jervell and Lange–Nielsen syndrome
		Gain of function	SQTS2; AF; sinus bradycardia
KCNE1	minK	Loss of function	LQTS5; Jervell and Lange–Nielsen syndrome
KCNH2	$K_v11.1$ (hERG)	Loss of function	LQTS2; epilepsy
		Gain of function	SQTS1; AF
KCNE2	MiRP1	Loss of function	LQTS6
		Gain of function	AF
KCND3	$K_v4.3$	Gain of function	BrS 11; AF
KCNE3	MiRP2	Gain of function	BrS 6; AF
KCNA5	$K_v1.5$	Loss of function	AF
AKAP-9	A-kinase anchor protein 9	Loss of function	LQTS11
KCNQ2	$K_v7.2$	Loss of function	Benign neonatal convulsions; neonatal epileptic encephalopathy; neuromyotonia
KCNQ3	$K_v7.3$	Loss of function	Benign neonatal convulsions; neonatal epileptic encephalopathy
KCNQ4	$K_v7.4$	Loss of function	Congenital deafness
KCTD7	K^+ channel tetramerization domain containing 7	Loss of function	Progressive myoclonic epilepsy, ataxia, and dementia
KCNA1	$K_v1.1$	Loss of function	Episodic ataxia type 1; seizures
KCNC3	$K_v3.3$	Loss of function	Spino-cerebellar ataxia type 13

AF, atrial fibrillation; *BrS*, Brugada syndrome; *hERG1*, Human-Ether-a-go-go-Related; *KV*, voltage-gated potassium channel; *LQTS*, long-QT syndrome; *MiRP*, minimal potassium channel subunit or *minK*-related peptide; *SQTS*, short QT syndrome.

AKAP-9 encodes A-kinase anchor protein-9, which interacts with *KCNQ1*. Loss-of-function mutations cause the LQTS11 form of LQTS (Giudicessi and Ackerman, 2012).

Heteromeric complexes of $K_v7.2$ and $K_v7.3$, encoded by *KCNQ2* and *KCNQ3*, respectively, regulate (inhibit) neuronal excitability and repeated neuronal discharges via generation of the muscarinic-regulated outward potassium current (M-current), which is important in the modulation of the resting membrane potential (Lason et al., 2013). Loss-of-function mutations typically cause benign familial neonatal convulsions, a syndrome characterized by the occurrence of seizures in the first few days of infancy and remitting within a few weeks, but also severe neonatal epileptic encephalopathy (Maljevic et al., 2010). Peripheral nerve hyperexcitability (neuromyotonia) has also been described for *KCNQ2* mutations (Maljevic et al., 2010).

KCNQ4 ($K_v7.4$) is expressed in the outer hair cells of the inner ear. Loss-of-function mutations cause congenital deafness (Kubisch et al., 1999).

The *KCTD7* gene product hyperpolarizes the cell membrane and, in experimental studies, reduces the excitability of neurons transfected with *KCTD7* through mechanisms that are not well understood (Azizieh et al., 2011). Loss-of-function mutations cause progressive myoclonic epilepsy, associated with progressive ataxia and dementia (Blumkin et al., 2012).

KCNA1 encodes $K_v1.1$, a delayed rectifier that plays a role in repolarization of membranes. Loss-of-function mutations cause episodic ataxia type 1, presenting as brief episodes of cerebellar dysfunction with persistent neuromyotonia, and may also be a risk factor for seizures (Tomlinson et al., 2009).

KCNC3 encodes an A-type (rapidly inactivating) potassium channel ($K_v3.3$) that normally opens and closes very quickly and is important in "fast-bursting neurons." Loss-of-function mutations cause an autosomal dominantly inherited form of spino-cerebellar ataxia, spino-cerebellar ataxia-13 (Figueroa et al., 2010).

TABLE 41.3 Disease States Associated With Germline Mutations in Genes Encoding Tandem Pore Domain and Calcium-Activated Potassium Channels

Gene	Protein	Effect of Mutation	Disease
K_{2P} Channels			
KCNK3	TASK-1	Loss of function	Pulmonary arterial hypertension
KCNK18	TRESK	Loss of function	Migraine
K_{Ca} Channels			
KCNMA1	BK_{Ca}	Loss of function	TLE, tonic-clonic seizures; alcohol withdrawal seizures
		Gain of function	Primary absence epilepsy

K_{Ca}, calcium-activated potassium channel; K_{2P}, tandem pore domain potassium channel; TASK, TWIK-related acid sensitive; TLE, temporal lobe epilepsy; TRESK, TWIK-related spinal cord K^+

Tandem Pore Domain Potassium Channels (Table 41.3)

TASK (tandem of P domains in a weak inwardly rectifying (TWIK)-related acid sensitive)-1 channels (encoded by *KCNK3*) contribute substantially to the resting membrane potential in human pulmonary artery smooth muscle cells, modulating vascular tone and diameter. Heterozygous loss-of-function mutations of *KCNK3* are associated with familial pulmonary arterial hypertension (Ma et al., 2013).

TRESK (TWIK-related spinal cord K^+) channels (encoded by *KCNK18*) are thought to play a major role in the functioning of primary sensory neurons of dorsal root and trigeminal ganglia and to be involved in the transmission of pain signals to the brain. Heterozygous loss-of-function mutations of *KCNK18* are a rare cause of migraine (Silberstein and Dodick, 2013).

K_{Ca} Channels (Table 41.3)

BK_{Ca} channels (encoded by *KCNMA1*) open in response to membrane depolarization and an increase in intracellular Ca^{2+}. They control neuronal hyperexcitability because the large outward current hyperpolarizes the neuronal membrane (N'Gouemo, 2011). Loss-of-function mutations cause temporal lobe epilepsy, tonic-clonic seizures, and alcohol withdrawal seizures. Gain-of-function mutations can cause primary absence epilepsy (N'Gouemo, 2011).

PRIMARY ALDOSTERONISM: A COMMON CAUSE OF HYPERTENSION ASSOCIATED WITH *KCNJ5* MUTATIONS

Interest in potassium channel mutations has been reinvigorated by the discovery of their association with primary aldosteronism (PA), a specifically treatable and potentially curable form of human hypertension. In PA, aldosterone production is excessive for the body's prevailing sodium and volume status and relatively autonomous of its main chronic regulator, renin-angiotensin II. Over time, the resulting excessive retention of sodium at the distal tubule leads to the development of hypertension. In exchange for the retained sodium, potassium and hydrogen ions are excreted, and, if this is prolonged and severe enough, hypokalemia and metabolic alkalosis may occur (Conn, 1966). Case detection is of considerable potential benefit to affected individuals in that unilateral adrenalectomy results in cure or improvement in hypertension and correction of hypokalemia (when present) in patients with unilateral PA (usually due to APA) and an improvement in quality of life (which is often marked; Celen et al., 1996; Rutherford et al., 1998; Sukor et al., 2010). Agents that antagonize aldosterone action have beneficial effects on control of hypertension and hypokalemia in medically treated patients with PA (mostly those with bilateral forms such as bilateral adrenal hyperplasia; Lim et al., 1999a, 2001; Stowasser et al., 2003). Although once thought to account for less than 1% of hypertension, evidence first published in 1992 by the Greenslopes Hospital HT Unit in Brisbane, Australia suggested that the prevalence is much higher (5–10%) and most patients are normokalemic and hence masquerade as if they have "essential" hypertension (Gordon et al., 1992), findings that were subsequently supported by additional reports from other centers (Lim et al., 1999b; Mulatero et al., 2004; Rossi et al., 2006a).

The clinical relevance of PA is further enhanced by the fact that, in addition to causing hypertension, aldosterone excess in PA induces injury (inflammation, remodeling, and fibrosis) in cardiovascular and renal tissues in ways that are at least partly independent of its effect on blood pressure (BP). Patients with PA have been reported to demonstrate increased left ventricular dimensions (Pimenta et al., 2011a; Rossi et al., 2003) and myocardial backscatter (Kozakova et al., 2003; as a marker of fibrosis) on echocardiography; increased carotid intima-media thickness (Bernini et al., 2008; Holaj et al., 2007), femoral pulse wave velocity (Bernini et al., 2008), and reduced endothelial function (Tsuchiya et al., 2009); greater degrees of proteinuria and albuminuria (Pimenta et al., 2011b; Rossi et al., 2006b); and, most importantly, increased rates of cardiovascular events including arrhythmias, myocardial infarctions, strokes, and mortality (Catena et al., 2008; Milliez et al., 2005; Reincke et al., 2012) when compared with matched essential hypertensives. PA also appears to bring about adverse psychological effects and is associated with reduced quality of life (Ahmed et al., 2011; Sonino et al., 2011; Sukor et al., 2010). All of these adverse manifestations of aldosterone excess can be ameliorated by specific surgical or medical treatment (Ahmed et al., 2011; Catena et al., 2008; Sukor et al., 2010), making it an important diagnosis not to miss.

Rationale for Potassium Channels as Candidate Genes for PA

In normal individuals, the main chronic regulators of aldosterone synthesis are angiotensin II and plasma potassium. Angiotensin II signaling via its G-protein–coupled receptor, and increased plasma potassium, lead to cell membrane depolarization and opening of voltage-gated membrane Ca^{2+} channels with activation of intracellular calcium signaling pathways, promoting aldosterone synthesis (Lotshaw, 2001; Spat and Hunyady, 2004). Adrenal zona glomerulosa (ZG) cells maintain a hyperpolarized resting membrane potential of −70mV, largely set by potassium channel activity (Czirjak and Enyedi, 2002a,b; Spat and Hunyady, 2004). Therefore genetic mutations in genes encoding channels predisposing to cell membrane depolarization are highly plausible candidates for the pathogenesis of PA.

Early Studies Implicating Potassium Channel Mutations in PA

In a study involving human adrenal cortical cells reported by Nogueira et al. (2010), *KCNK3* (encoding TASK-1) appeared to be the most highly expressed among genes of the tandem two-pore ("leak") potassium channel family in human adrenal cortical cells. Knockdown (with small interfering RNA) of *KCNK3* in vitro led to increased intracellular calcium, induced expression of the steroidogenic acute regulatory protein and aldosterone synthase (*CYP11B2*), and stimulated pregnenolone and aldosterone production (Nogueira et al., 2010). However, normal adrenal cortex and APA did not appear to differ in terms of *KCNK3* expression, and there have been no reports of either germline or somatic (within APAs) mutations in *KCNK3* or *KCNJ9* (TASK-3) in patients with PA.

KCNH2 (hERG1 gene) encodes hERG1 protein, the pore-forming region of a four-subunit potassium channel responsible for the rapid component of the delayed rectifying current (I_{Kr}), which facilitates, through efflux of K^+, phase III repolarization of cardiac myocytes (Curran et al., 1995; Itoh et al., 1998). *KCNH2* is expressed in normal adrenal tissue and, in one Italian study (Sarzani et al., 2006), was expressed in all 17 APAs analyzed. *KCNH2* somatic mutations were not detected. However, analysis of peripheral blood DNA revealed a *KCNH2* variant (897T) to be significantly more common among patients with APA (41%) than in patients with moderate to severe essential hypertension (20%, $P=.007$) or in a normotensive control population (12%, $P<.0001$). The 897T/T genotype was present in 24% of the APA patients compared with only 7% ($P = .040$) and 3% ($P = .001$) in the essential hypertension group and control population, respectively. The authors concluded that the common functional *KCNH2* variant 897T may predispose to the development of aldosteronoma (Sarzani et al., 2006). However, these findings have yet to be confirmed in larger numbers of patients with PA.

KCNJ5 and PA

Initial Report

In 2011, Choi et al. reported somatic mutations (G151R and L168R) in *KCNJ5*, encoding a K_{ir} channel (GIRK4 or Kir3.4), in 8 of 22 APAs removed from Swedish patients with apparently nonfamilial PA. All but 1 (88%) of those with *KCNJ5* mutations were female, compared with 9 (64%) of the 14 without. The mean plasma aldosterone/renin ratio was slightly but significantly ($P<.05$) higher in those with mutations, suggesting somewhat more severe PA.

In the same report, a third, germline mutation (T158A) in *KCNJ5* was identified in an American family with a familial form of PA first reported by this group in 2008 (Geller et al., 2008). The affected family members demonstrated very florid PA with severe childhood-onset hypertension, hypokalemia, and markedly elevated aldosterone and suppressed renin levels. Remarkably, 18-hydroxy- and 18-oxo-cortisol levels were elevated as in a previously described, glucocorticoid-suppressible familial form of PA (FH-I), caused by inheritance of an adrenocorticotrophin-regulated "hybrid" gene mutations composed of regulatory sequences from *CYP11B1* (11β-hydroxylase) and coding sequences from *CYP11B2* (aldosterone synthase; Lifton et al., 1992). However, the levels of these so-called "hybrid" steroids were elevated to a much greater degree than in FH-I, the PA was not glucocorticoid suppressible (dexamethasone actually increased aldosterone and BP), and the hybrid gene mutation was not present. Resected adrenals showed marked, diffuse hyperplasia of zona fasciculata (ZF), the combined adrenal weight in one patient reaching 81 g (normal <12 g; Geller et al., 2008). In normal adrenals, expression of KCNJ5 appeared to be restricted to ZG cells (Choi et al., 2011).

All three amino acid substitutions reported by Choi et al. (2011) were in close proximity to the channel's selectivity filter. Expressed in human embryonic kidney (HEK) 293T cells all were associated with a loss in channel selectivity for K^+ with increased permeability to Na^+ compared with the wild-type channel and less negative reversal potentials. These properties would be expected, as the authors proposed, to predispose to chronic adrenocortical cell membrane depolarization and consequently to influx of Ca^{2+}, in turn leading to upregulation of enzymes involved in aldosterone synthesis and possibly cell proliferation (Choi et al., 2011).

Other Reports of Somatic and Germline KCNJ5 Mutations in Familial Primary Aldosteronism

Mulatero et al. (2012) subsequently reported one each of the same 3 somatic KCNJ5 mutations in 3 of 11 APAs removed from patients with FH-II (defined as familial PA with FH-I excluded by hybrid gene testing). They also described a novel germline mutation (G151E), which again demonstrated loss of K^+/Na^+ selectivity when expressed in HEK cells, in genomic DNA from two affected members in 1 of 21 families studied. Hence genomic KCNJ5 mutations appeared to be an uncommon cause of familial PA. Both had early-onset hypertension and moderately severe biochemical PA, but, unlike the family studied by Choi et al., the adrenals appeared normal on computed tomography (CT) scans and hybrid steroid levels were only slightly higher than those for apparently nonfamilial PA patients, indicating a milder phenotype associated with this mutation (Mulatero et al., 2012). This was supported by a report by Scholl et al. (2012b) in which two more families with germline G151E mutations demonstrated early-onset but mild PA that was usually easily controlled with spironolactone and not associated with detectable abnormal adrenal morphology on CT scanning, whereas two other pedigrees with germline G151R mutations showed a much more severe phenotype, characterized by very early-onset and severe PA that was poorly responsive to spironolactone. As a result, most of these subjects required bilateral adrenalectomy before the age of 5 years, resulting in control of hypertension and hypokalemia. Surprisingly, the G151E mutated channels demonstrated markedly higher (i.e., more abnormal) sodium conductance than G151R mutated channels, and HEK 293T cells transfected with G151E mutated KCNJ5 demonstrated markedly reduced survival compared with cells with wild-type KCNJ5 whereas those with G151R mutated KCNJ5 showed only a modest reduction in survival. Survival was greatly enhanced by incubating in a low-sodium medium, suggesting that the greater conductance of sodium was causal in conferring the greater cell lethality associated with the G151E mutation compared with G151R. The authors speculated that this increased lethality may explain the lack of adrenal hyperplasia occurring clinically in patients bearing germline G151E (vs. G151R) mutations (Scholl et al., 2012b).

Similar to G151R, another novel germline KCNJ5 mutation (I157S) detected in a single family with PA was associated with a very severe phenotype, with early-onset resistant hypertension and bilateral massive adrenal hyperplasia (Charmandari et al., 2012).

We have not identified germline KCNJ5 mutations among any of our patients with FH-II (>30 families so far studied). However, in a study of 251 Caucasian subjects with PA (almost all apparently nonfamilial), we identified 3 heterozygous missense mutations (R52H, E246K, and G247R) and found that 12 (5% of the cohort) were carriers for the rare nonsynonymous single-nucleotide polymorphism rs7102584, causing E282Q substitution (Murthy et al., 2014). Unlike previously described mutations associated with PA, these three mutations and rare polymorphism were all remote from the potassium channel selectivity filter. In vitro studies showed that R52H, E246K, and E282Q (but not G247) substitutions to be functional, affecting the inward rectification, enhancing the ability of the GIRK4 channels to conduct Na^+ currents, and increasing angiotensin-II–induced aldosterone release from transfected H295R cells (Murthy et al., 2014). Hence germline variation in the KCNJ5 gene may have a role to play in the common sporadic form as well as the much rarer syndromic forms of PA.

Other Reports of Somatic KCNJ5 Mutations in Apparently Nonfamilial Primary Aldosteronism

Somatic KCNJ5 mutations (G151R or L168R) were found in 129 (34%) of 380 APAs removed from patients studied by a European consortium of investigators (Boulkroun et al., 2012). They were significantly more prevalent in females (49%) than males (19%) and were associated with a younger mean age and higher preoperative aldosterone levels. Another large study, which included centers from Europe, the United States, and Australia, found somatic G151R and L168R mutations in 155 (44%) of 348 APAs but in none of 130 nonaldosterone-secreting lesions (Akerstrom et al., 2012). A further two APAs showed a novel E145Q mutation located near the GIRK4 selectivity filter. This study again demonstrated a greater prevalence of mutations among APAs from females (63%) than males (24%). Males, but not females, with somatic mutations were significantly younger and had larger tumors than those without.

Among a cohort of 47 adrenals collected from patients with APA from centers in the United States, Italy, and Japan, somatic *KCNJ5* mutations were found in 18 (38%), all of them G151R or L168R (Monticone et al., 2012). Again, females were overrepresented among patients with mutated APAs (71% vs. males 29%, $P < .05$). GIRK4 immunostaining was highest in the ZG in normal adrenals but also seen in the outer part of the ZF and was heterogeneous in APA tissue. *KCNJ5* gene expression was fourfold higher in APA compared with normal adrenals and higher than the nontumorous cortex in adrenals containing an APA, but there was no difference in expression between APAs with or without *KCNJ5* mutations. *KCNJ5*-mutated APAs had lower serum K^+ levels than those with nonmutated tumors. Despite similar plasma aldosterone levels, mutated APAs showed threefold higher *CYP11B2* expression. Human adrenocortical carcinoma (HAC15) cells transfected with and overexpressing *KCNJ5* bearing G151R- or L168R-producing mutations showed greater expression of *CYP11B2* and of two key regulators of *CYP11B2* transcription (*NR4A2* and *NR4A3*), and greater production of aldosterone, than cells transfected with wild-type *KCNJ5* (Monticone et al., 2012). Greater production of aldosterone was also reported by Oki and coworkers for HAC15 cells transfected with the *KCNJ5* bearing the T158A-producing mutation (Oki et al., 2012).

We reported (in collaboration with Kevin O'Shaughnessy and Morris Brown in Cambridge, United Kingdom) somatic *KCNJ5* mutations in 30 (41%) of 73 APAs removed from Australian and UK patients (Azizan et al., 2012b), a frequency similar to that of other reported series (Boulkroun et al., 2012; Choi et al., 2011; Mulatero et al., 2012). In addition to the G151R and L168R ($n = 10$) mutations described by Choi et al., a novel deletion (delI157) was detected in a single UK APA. When expressed in Xenopus oocytes, it again demonstrated loss of K^+/Na^+ selectivity. Similar to several other series, we found mutated APAs to be more common among females (60% vs. males 26%; $P < .01$) and to be associated with significantly younger mean age (46 ± 11 SD vs. 51 ± 9 years; $P < .05$). Mutation-bearing APAs were larger on average than nonmutation bearing APAs, but there was considerable overlap (Azizan et al., 2012b).

Gordon et al. (Gordon et al., 1987; Tunny et al., 1991) previously described two distinct forms of APA: one (angiotensin-II–unresponsive APA) composed predominantly of ZF-like cells and associated with elevated hybrid steroid levels and lack of responsiveness of aldosterone to upright posture or angiotensin II infusion and one (angiotensin-II–responsive APA) with predominantly non-ZF (ZG or hybrid ZF/ZG)-like morphology in which hybrid steroid levels were usually normal and aldosterone responsive to upright posture or angiotensin II. Although Choi's group reported that *KCNJ5* expression in normal adrenals was restricted to the ZG (Choi et al., 2011), the massive hyperplasia in patients with the genomic *KCNJ5* mutation involved ZF (Geller et al., 2008). The markedly elevated hybrid steroid levels in those patients suggests that, as in FH-I, these cells were expressing both aldosterone synthase (*CYP11B2*, which is normally restricted to ZG) and 17α-hydroxylase (*CYP17A1*, normally restricted to ZF) activity; thus they had molecular characteristics of ZG and ZF. It is interesting to note that among the 26 Australian patients with APA from our collaborative (with Cambridge University, United Kingdom) study who had undergone preoperative posture studies, all 10 with somatic *KCNJ5* mutations (vs. only 6 of 16 without; $P < .05$) showed lack of posture responsiveness of plasma aldosterone to upright posture in keeping with the angiotensin-II–unresponsive form of APA (Azizan et al., 2012b). Brown et al. subsequently reported that among the 46 APAs removed from the UK patients, those harboring *KCNJ5* mutations had predominantly ZF-like cells and increased expression of *CYP17A1* compared with wild-type tumors (Azizan et al., 2012a). Therefore it seems likely that somatic *KCNJ5* mutations are associated with the development of APAs with the angiotensin-II–unresponsive phenotype (see Fig. 41.1 for an example). In keeping with this, we have previously noted that patients angiotensin-II–unresponsive APAs are on average younger, more often

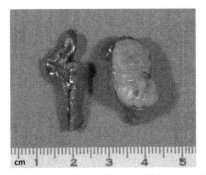

FIGURE 41.1 Aldosterone-producing adenoma (APA) removed from a 33-year-old female with hypertension and hypokalemia due to florid biochemical primary aldosteronism. Preoperatively, plasma aldosterone failed to increase in response to upright posture (i.e., the APA was angiotensin-II unresponsive). The tumor (shown on the right, with adjacent normal adrenal on the left for comparison) was relatively large at 1.9 × 1.0 cm (average maximal diameter of APAs in our center is ~1.0 cm) and composed predominantly of cells resembling the zona fasciculata. Genetic testing of DNA extracted from the APA revealed a somatic mutation (G151R) of *KCNJ5*. The mutation was not detected in peripheral blood DNA or adjacent nontumorous adrenal cortex. The patient was postoperatively cured of hypertension and hypokalemia.

female, have more florid biochemical PA, demonstrate higher hypertension cure rates postadrenalectomy, and on average have larger tumors than those with angiotensin-II–responsive APAs (Stowasser et al., 1994), features that have been reported by several groups for patients with somatic *KCNJ5* mutations.

In a Japanese study, G151R (*n* = 8) and L168R (*n* = 3) *KCNJ5* mutations were seen in a higher proportion of APA (15 of 23 = 65%) than in other reported series to date (Taguchi et al., 2012). Reasons for this high prevalence are not clear, but this observation may help to explain the relatively high rate of APA among Japanese patients with hypertension (8% vs. approximately 4% in Western countries) and among those with PA (>80% compared with ~0% in our own experience; Mulatero et al., 2004; Omura et al., 2004, 2006). Again, patients with mutations showed a female preponderance, were of younger mean age, and had biochemical evidence of more severe PA with higher plasma aldosterone and lower serum K$^+$ levels than those without, but there was no difference between the two groups in terms of tumor size (Taguchi et al., 2012).

Somatic mutations within APAs have since also been described in other ion channels and pumps—namely *ATP1A1* (encoding the α-subunit of Na$^+$/K$^+$ ATPase), *ATP2B3* (encodes a Ca^{2+} ATPase calcium channel), and *CACNA1D* (encodes a voltage-gated calcium channel)—but with much lower frequency (Azizan et al., 2013; Beuschlein et al., 2013; Fernandes-Rosa et al., 2014; Scholl et al., 2013; Williams et al., 2014). In vitro studies suggest that these mutations, similar to those in *KCNJ5*, also lead to ZG cell membrane depolarization, resulting in increased aldosterone production (Azizan et al., 2013; Beuschlein et al., 2013; Scholl et al., 2013; Williams et al., 2014). However, unlike patients with *KCNJ5*-mutated APAs, those with APAs carrying the newer mutations have tended to be more commonly males (Azizan et al., 2013; Beuschlein et al., 2013), with smaller APAs (Azizan et al., 2013; Fernandes-Rosa et al., 2014) containing predominantly ZG-like cells (Azizan et al., 2013).

CONCLUSIONS

Mutations involving almost 30 genes encoding potassium channel subunits have so far have been reported to result in human disease states involving the cardiovascular, central and peripheral nervous, sensory, musculoskeletal, renal, and endocrine systems. Mutations in *KCNJ5* are associated with PA, the most common secondary endocrine form of hypertension. Germline *KCNJ5* mutations (T158A, G151E, G151R, and I157S) cause bilateral, familial PA, which appears less severe for G151E than the other reported mutations. The prevalence rate for somatic *KCNJ5* mutations in APAs is approximately 40% among Caucasians, but it may be higher in Japanese subjects. Patients with somatic *KCNJ5* mutations tend to be younger; more often female; and demonstrate more severe PA, lack of responsiveness of plasma aldosterone to upright posture, and ZF histology. Of the five so far described, G151R and L168R are by far the most common. By reducing K$^+$/Na$^+$ channel selectivity and resting cell membrane potential, *KCNJ5* mutations lead to chronic ZG cell depolarization, resulting in increased calcium influx, expression of genes involved in aldosterone synthesis, and aldosterone production by adrenocortical cells. These findings have greatly enhanced understanding of potassium channel and general cell physiology and pathophysiology and provide a foundation for the development of new treatment modalities.

REFERENCES

Ahmed, A.H., Gordon, R.D., Sukor, N., Pimenta, E., Stowasser, M., 2011. Quality of life in patients with bilateral primary aldosteronism before and during treatment with spironolactone and/or amiloride, including a comparison with our previously published results in those with unilateral disease treated surgically. J. Clin. Endocrinol. Metab. 96 (9), 2904–2911.

Akerstrom, T., Crona, J., Delgado, V.A., Starker, L.F., Cupisti, K., Willenberg, H.S., Knoefel, W.T., Saeger, W., Feller, A., Ip, J., Soon, P., Anlauf, M., Alesina, P.F., Schmid, K.W., Decaussin, M., Levillain, P., Wangberg, B., Peix, J.L., Robinson, B., Zedenius, J., Backdahl, M., Caramuta, S., Iwen, K.A., Botling, J., Stalberg, P., Kraimps, J.L., Dralle, H., Hellman, P., Sidhu, S., Westin, G., Lehnert, H., Walz, M.K., Akerstrom, G., Carling, T., Choi, M., Lifton, R.P., Bjorklund, P., 2012. Comprehensive re-sequencing of adrenal aldosterone producing lesions reveal three somatic mutations near the KCNJ5 potassium channel selectivity filter. PLoS One 7 (7), e41926.

Ashcroft, F.M., Rorsman, P., 2013. K(ATP) channels and islet hormone secretion: new insights and controversies. Nat. Rev. Endocrinol. 9 (11), 660–669.

Azizan, E.A., Lam, B.Y., Newhouse, S.J., Zhou, J., Kuc, R.E., Clarke, J., Happerfield, L., Marker, A., Hoffman, G.J., Brown, M.J., 2012a. Microarray, qPCR, and *KCNJ5* sequencing of aldosterone-producing adenomas reveal differences in genotype and phenotype between zona glomerulosa- and zona fasciculata-like tumors. J. Clin. Endocrinol. Metab. 97 (5), E819–E829.

Azizan, E.A., Murthy, M., Stowasser, M., Gordon, R., Kowalski, B., Xu, S., Brown, M.J., O'Shaughnessy, K.M., 2012b. Somatic mutations affecting the selectivity filter of KCNJ5 are frequent in 2 large unselected collections of adrenal aldosteronomas. Hypertension 59 (3), 587–591.

Azizan, E.A., Poulsen, H., Tuluc, P., Zhou, J., Clausen, M.V., Lieb, A., Maniero, C., Garg, S., Bochukova, E.G., Zhao, W., Shaikh, L.H., Brighton, C.A., Teo, A.E., Davenport, A.P., Dekkers, T., Tops, B., Kusters, B., Ceral, J., Yeo, G.S., Neogi, S.G., McFarlane, I., Rosenfeld, N., Marass, F., Hadfield, J., Margas, W., Chaggar, K., Solar, M., Deinum, J., Dolphin, A.C., Farooqi, I.S., Striessnig, J., Nissen, P., Brown, M.J., 2013. Somatic mutations in *ATP1A1* and *CACNA1D* underlie a common subtype of adrenal hypertension. Nat. Genet. 45 (9), 1055–1060.

Azizieh, R., Orduz, D., Van Bogaert, P., Bouschet, T., Rodriguez, W., Schiffmann, S.N., Pirson, I., Abramowicz, M.J., 2011. Progressive myoclonic epilepsy-associated gene KCTD7 is a regulator of potassium conductance in neurons. Mol. Neurobiol. 44 (1), 111–121.

Bernini, G., Galetta, F., Franzoni, F., Bardini, M., Taurino, C., Bernardini, M., Ghiadoni, L., Bernini, M., Santoro, G., Salvetti, A., 2008. Arterial stiffness, intima-media thickness and carotid artery fibrosis in patients with primary aldosteronism. J. Hypertens. 26 (12), 2399–2405.

Beuschlein, F., Boulkroun, S., Osswald, A., Wieland, T., Nielsen, H.N., Lichtenauer, U.D., Penton, D., Schack, V.R., Amar, L., Fischer, E., Walther, A., Tauber, P., Schwarzmayr, T., Diener, S., Graf, E., Allolio, B., Samson-Couterie, B., Benecke, A., Quinkler, M., Fallo, F., Plouin, P.F., Mantero, F., Meitinger, T., Mulatero, P., Jeunemaitre, X., Warth, R., Vilsen, B., Zennaro, M.C., Strom, T.M., Reincke, M., 2013. Somatic mutations in *ATP1A1* and *ATP2B3* lead to aldosterone-producing adenomas and secondary hypertension. Nat. Genet. 45 (4), 444e1–444e2 440–444.

Bienengraeber, M., Olson, T.M., Selivanov, V.A., Kathmann, E.C., O'Cochlain, F., Gao, F., Karger, A.B., Ballew, J.D., Hodgson, D.M., Zingman, L.V., Pang, Y.P., Alekseev, A.E., Terzic, A., 2004. *ABCC9* mutations identified in human dilated cardiomyopathy disrupt catalytic K(ATP) channel gating. Nat. Genet. 36 (4), 382–387.

Blumkin, L., Kivity, S., Lev, D., Cohen, S., Shomrat, R., Lerman-Sagie, T., Leshinsky-Silver, E., 2012. A compound heterozygous missense mutation and a large deletion in the KCTD7 gene presenting as an opsoclonus-myoclonus ataxia-like syndrome. J. Neurol. 259 (12), 2590–2598.

Boulkroun, S., Beuschlein, F., Rossi, G.P., Golib-Dzib, J.F., Fischer, E., Amar, L., Mulatero, P., Samson-Couterie, B., Hahner, S., Quinkler, M., Fallo, F., Letizia, C., Allolio, B., Ceolotto, G., Cicala, M.V., Lang, K., Lefebvre, H., Lenzini, L., Maniero, C., Monticone, S., Perrocheau, M., Pilon, C., Plouin, P.F., Rayes, N., Seccia, T.M., Veglio, F., Williams, T.A., Zinnamosca, L., Mantero, F., Benecke, A., Jeunemaitre, X., Reincke, M., Zennaro, M.C., 2012. Prevalence, clinical, and molecular correlates of KCNJ5 mutations in primary aldosteronism. Hypertension 59 (3), 592–598.

Catena, C., Colussi, G., Nadalini, E., Chiuch, A., Baroselli, S., Lapenna, R., Sechi, L.A., 2008. Cardiovascular outcomes in patients with primary aldosteronism after treatment. Arch. Intern Med. 168 (1), 80–85.

Celen, O., O'Brien, M.J., Melby, J.C., Beazley, R.M., 1996. Factors influencing outcome of surgery for primary aldosteronism. Arch. Surg. 131 (6), 646–650.

Charmandari, E., Sertedaki, A., Kino, T., Merakou, C., Hoffman, D.A., Hatch, M.M., Hurt, D.E., Lin, L., Xekouki, P., Stratakis, C.A., Chrousos, G.P., 2012. A novel point mutation in the KCNJ5 gene causing primary hyperaldosteronism and early-onset autosomal dominant hypertension. J. Clin. Endocrinol. Metab. 97 (8), E1532–E1539.

Cheng, C.J., Lin, S.H., Lo, Y.F., Yang, S.S., Hsu, Y.J., Cannon, S.C., Huang, C.L., 2011. Identification and functional characterization of Kir2.6 mutations associated with non-familial hypokalemic periodic paralysis. J. Biol. Chem. 286 (31), 27425–27435.

Choi, M., Scholl, U.I., Yue, P., Bjorklund, P., Zhao, B., Nelson-Williams, C., Ji, W., Cho, Y., Patel, A., Men, C.J., Lolis, E., Wisgerhof, M.V., Geller, D.S., Mane, S., Hellman, P., Westin, G., Akerstrom, G., Wang, W., Carling, T., Lifton, R.P., 2011. K+ channel mutations in adrenal aldosterone-producing adenomas and hereditary hypertension. Science 331 (6018), 768–772.

Coetzee, W.A., Amarillo, Y., Chiu, J., Chow, A., Lau, D., McCormack, T., Moreno, H., Nadal, M.S., Ozaita, A., Pountney, D., Saganich, M., Vega-Saenz de Miera, E., Rudy, B., 1999. Molecular diversity of K+ channels. Ann. N. Y. Acad. Sci. 868, 233–285.

Conn, J.W., 1966. The evolution of primary aldosteronism: 1954–1967. Harvey Lect. 62, 257–291.

Curran, M.E., Splawski, I., Timothy, K.W., Vincent, G.M., Green, E.D., Keating, M.T., 1995. A molecular basis for cardiac arrhythmia: *HERG* mutations cause long QT syndrome. Cell 80 (5), 795–803.

Czirjak, G., Enyedi, P., 2002a. Formation of functional heterodimers between the TASK-1 and TASK-3 two-pore domain potassium channel subunits. J. Biol. Chem. 277 (7), 5426–5432.

Czirjak, G., Enyedi, P., 2002b. TASK-3 dominates the background potassium conductance in rat adrenal glomerulosa cells. Mol. Endocrinol. 16 (3), 621–629.

Delpon, E., Cordeiro, J.M., Nunez, L., Thomsen, P.E., Guerchicoff, A., Pollevick, G.D., Wu, Y., Kanters, J.K., Larsen, C.T., Hofman-Bang, J., Burashnikov, E., Christiansen, M., Antzelevitch, C., 2008. Functional effects of *KCNE3* mutation and its role in the development of Brugada syndrome. Circ. Arrhythm. Electrophysiol. 1 (3), 209–218.

Fernandes-Rosa, F.L., Williams, T.A., Riester, A., Steichen, O., Beuschlein, F., Boulkroun, S., Strom, T.M., Monticone, S., Amar, L., Meatchi, T., Mantero, F., Cicala, M.V., Quinkler, M., Fallo, F., Allolio, B., Bernini, G., Maccario, M., Giacchetti, G., Jeunemaitre, X., Mulatero, P., Reincke, M., Zennaro, M.C., 2014. Genetic spectrum and clinical correlates of somatic mutations in aldosterone-producing adenoma. Hypertension 64 (2), 354–361.

Figueroa, K.P., Minassian, N.A., Stevanin, G., Waters, M., Garibyan, V., Forlani, S., Strzelczyk, A., Burk, K., Brice, A., Durr, A., Papazian, D.M., Pulst, S.M., 2010. KCNC3: phenotype, mutations, channel biophysics-a study of 260 familial ataxia patients. Hum. Mutat. 31 (2), 191–196.

Geller, D.S., Zhang, J., Wisgerhof, M.V., Shackleton, C., Kashgarian, M., Lifton, R.P., 2008. A novel form of human mendelian hypertension featuring nonglucocorticoid-remediable aldosteronism. J. Clin. Endocrinol. Metab. 93 (8), 3117–3123.

Giudicessi, J.R., Ackerman, M.J., 2012. Potassium-channel mutations and cardiac arrhythmias—diagnosis and therapy. Nat. Rev. Cardiol. 9 (6), 319–332.

Giudicessi, J.R., Ye, D., Tester, D.J., Crotti, L., Mugione, A., Nesterenko, V.V., Albertson, R.M., Antzelevitch, C., Schwartz, P.J., Ackerman, M.J., 2011. Transient outward current (I_{to}) gain-of-function mutations in the *KCND3*-encoded Kv4.3 potassium channel and Brugada syndrome. Heart Rhythm 8 (7), 1024–1032.

Gordon, R.D., Hamlet, S.M., Tunny, T.J., Klemm, S.A., 1987. Aldosterone-producing adenomas responsive to angiotensin pose problems in diagnosis. Clin. Exp. Pharmacol. Physiol. 14 (3), 175–179.

Gordon, R.D., Klemm, S.A., Tunny, T.J., Stowasser, M., 1992. Primary aldosteronism: hypertension with a genetic basis. Lancet 340 (8812), 159–161.

Harakalova, M., van Harssel, J.J., Terhal, P.A., van Lieshout, S., Duran, K., Renkens, I., Amor, D.J., Wilson, L.C., Kirk, E.P., Turner, C.L., Shears, D., Garcia-Minaur, S., Lees, M.M., Ross, A., Venselaar, H., Vriend, G., Takanari, H., Rook, M.B., van der Heyden, M.A., Asselbergs, F.W., Breur, H.M., Swinkels, M.E., Scurr, I.J., Smithson, S.F., Knoers, N.V., van der Smagt, J.J., Nijman, I.J., Kloosterman, W.P., van Haelst, M.M., van Haaften, G., Cuppen, E., 2012. Dominant missense mutations in *ABCC9* cause Cantu syndrome. Nat. Genet. 44 (7), 793–796.

Hejtmancik, J.F., Jiao, X., Li, A., Sergeev, Y.V., Ding, X., Sharma, A.K., Chan, C.C., Medina, I., Edwards, A.O., 2008. Mutations in *KCNJ13* cause autosomal-dominant snowflake vitreoretinal degeneration. Am. J. Hum. Genet. 82 (1), 174–180.

Hibino, H., Inanobe, A., Furutani, K., Murakami, S., Findlay, I., Kurachi, Y., 2010. Inwardly rectifying potassium channels: their structure, function, and physiological roles. Physiol. Rev. 90 (1), 291–366.

Holaj, R., Zelinka, T., Wichterle, D., Petrak, O., Strauch, B., Widimsky Jr., J., 2007. Increased intima-media thickness of the common carotid artery in primary aldosteronism in comparison with essential hypertension. J. Hypertens. 25 (7), 1451–1457.

Hoshi, T., Pantazis, A., Olcese, R., 2013. Transduction of voltage and Ca^{2+} signals by Slo1 BK channels. Physiology 28 (3), 172–189.

Itoh, T., Tanaka, T., Nagai, R., Kamiya, T., Sawayama, T., Nakayama, T., Tomoike, H., Sakurada, H., Yazaki, Y., Nakamura, Y., 1998. Genomic organization and mutational analysis of HERG, a gene responsible for familial long QT syndrome. Hum. Genet. 102 (4), 435–439.

Ji, W., Foo, J.N., O'Roak, B.J., Zhao, H., Larson, M.G., Simon, D.B., Newton-Cheh, C., State, M.W., Levy, D., Lifton, R.P., 2008. Rare independent mutations in renal salt handling genes contribute to blood pressure variation. Nat. Genet. 40 (5), 592–599.

Kozakova, M., Buralli, S., Palombo, C., Bernini, G., Moretti, A., Favilla, S., Taddei, S., Salvetti, A., 2003. Myocardial ultrasonic backscatter in hypertension: relation to aldosterone and endothelin. Hypertension 41 (2), 230–236.

Kubisch, C., Schroeder, B.C., Friedrich, T., Lutjohann, B., El-Amraoui, A., Marlin, S., Petit, C., Jentsch, T.J., 1999. KCNQ4, a novel potassium channel expressed in sensory outer hair cells, is mutated in dominant deafness. Cell 96 (3), 437–446.

Lason, W., Chlebicka, M., Rejdak, K., 2013. Research advances in basic mechanisms of seizures and antiepileptic drug action. Pharmacol. Rep. 65 (4), 787–801.

Lifton, R.P., Dluhy, R.G., Powers, M., Rich, G.M., Cook, S., Ulick, S., Lalouel, J.M., 1992. A chimaeric 11 beta-hydroxylase/aldosterone synthase gene causes glucocorticoid-remediable aldosteronism and human hypertension. Nature 355 (6357), 262–265.

Lim, P.O., Jung, R.T., MacDonald, T.M., 1999a. Raised aldosterone to renin ratio predicts antihypertensive efficacy of spironolactone: a prospective cohort follow-up study. Br. J. Clin. Pharmacol. 48 (5), 756–760.

Lim, P.O., Rodgers, P., Cardale, K., Watson, A.D., MacDonald, T.M., 1999b. Potentially high prevalence of primary aldosteronism in a primary-care population. Lancet 353 (9146), 40.

Lim, P.O., Young, W.F., MacDonald, T.M., 2001. A review of the medical treatment of primary aldosteronism. J. Hypertens. 19 (3), 353–361.

Lotshaw, D.P., 2001. Role of membrane depolarization and T-type Ca^{2+} channels in angiotensin II and K^+ stimulated aldosterone secretion. Mol. Cell. Endocrinol. 175 (1–2), 157–171.

Lundby, A., Ravn, L.S., Svendsen, J.H., Hauns, S., Olesen, S.P., Schmitt, N., 2008. KCNE3 mutation V17M identified in a patient with lone atrial fibrillation. Cell. Physiol. Biochem. 21 (1–3), 47–54.

Ma, L., Roman-Campos, D., Austin, E.D., Eyries, M., Sampson, K.S., Soubrier, F., Germain, M., Tregouet, D.A., Borczuk, A., Rosenzweig, E.B., Girerd, B., Montani, D., Humbert, M., Loyd, J.E., Kass, R.S., Chung, W.K., 2013. A novel channelopathy in pulmonary arterial hypertension. N. Engl. J. Med. 369 (4), 351–361.

Maljevic, S., Wuttke, T.V., Seebohm, G., Lerche, H., 2010. KV7 channelopathies. Pflugers Arch. 460 (2), 277–288.

Masotti, A., Uva, P., Davis-Keppen, L., Basel-Vanagaite, L., Cohen, L., Pisaneschi, E., Celluzzi, A., Bencivenga, P., Fang, M., Tian, M., Xu, X., Cappa, M., Dallapiccola, B., 2015. Keppen-Lubinsky syndrome is caused by mutations in the inwardly rectifying K^+ channel encoded by *KCNJ6*. Am. J. Hum. Genet. 96 (2), 295–300.

Milliez, P., Girerd, X., Plouin, P.F., Blacher, J., Safar, M.E., Mourad, J.J., 2005. Evidence for an increased rate of cardiovascular events in patients with primary aldosteronism. J. Am. Coll. Cardiol. 45 (8), 1243–1248.

Monticone, S., Hattangady, N.G., Nishimoto, K., Mantero, F., Rubin, B., Cicala, M.V., Pezzani, R., Auchus, R.J., Ghayee, H.K., Shibata, H., Kurihara, I., Williams, T.A., Giri, J.G., Bollag, R.J., Edwards, M.A., Isales, C.M., Rainey, W.E., 2012. Effect of KCNJ5 mutations on gene expression in aldosterone-producing adenomas and adrenocortical cells. J. Clin. Endocrinol. Metab. 97 (8), E1567–E1572.

Mulatero, P., Stowasser, M., Loh, K.C., Fardella, C.E., Gordon, R.D., Mosso, L., Gomez-Sanchez, C.E., Veglio, F., Young Jr., W.F., 2004. Increased diagnosis of primary aldosteronism, including surgically correctable forms, in centers from five continents. J. Clin. Endocrinol. Metab. 89 (3), 1045–1050.

Mulatero, P., Tauber, P., Zennaro, M.C., Monticone, S., Lang, K., Beuschlein, F., Fischer, E., Tizzani, D., Pallauf, A., Viola, A., Amar, L., Williams, T.A., Strom, T.M., Graf, E., Bandulik, S., Penton, D., Plouin, P.F., Warth, R., Allolio, B., Jeunemaitre, X., Veglio, F., Reincke, M., 2012. *KCNJ5* mutations in European families with nonglucocorticoid remediable familial hyperaldosteronism. Hypertension 59 (2), 235–240.

Murthy, M., Xu, S., Massimo, G., Wolley, M., Gordon, R.D., Stowasser, M., O'Shaughnessy, K.M., 2014. Role for germline mutations and a rare coding single nucleotide polymorphism within the KCNJ5 potassium channel in a large cohort of sporadic cases of primary aldosteronism. Hypertension 63 (4), 783–789.

N'Gouemo, P., 2011. Targeting BK (big potassium) channels in epilepsy. Expert Opin. Ther. Targets 15 (11), 1283–1295.

Noel, J., Sandoz, G., Lesage, F., 2011. Molecular regulations governing TREK and TRAAK channel functions. Channels 5 (5), 402–409.

Nogueira, E.F., Gerry, D., Mantero, F., Mariniello, B., Rainey, W.E., 2010. The role of TASK1 in aldosterone production and its expression in normal adrenal and aldosterone-producing adenomas. Clin. Endocrinol. (Oxf) 73 (1), 22–29.

Norris, A.J., Foeger, N.C., Nerbonne, J.M., 2010. Neuronal voltage-gated K^+ (Kv) channels function in macromolecular complexes. Neurosci. Lett. 486 (2), 73–77.

Okamoto, K., Iwasaki, N., Nishimura, C., Doi, K., Noiri, E., Nakamura, S., Takizawa, M., Ogata, M., Fujimaki, R., Grarup, N., Pisinger, C., Borch-Johnsen, K., Lauritzen, T., Sandbaek, A., Hansen, T., Yasuda, K., Osawa, H., Nanjo, K., Kadowaki, T., Kasuga, M., Pedersen, O., Fujita, T., Kamatani, N., Iwamoto, Y., Tokunaga, K., 2010. Identification of *KCNJ15* as a susceptibility gene in Asian patients with type 2 diabetes mellitus. Am. J. Hum. Genet. 86 (1), 54–64.

Oki, K., Plonczynski, M.W., Luis, L.M., Gomez-Sanchez, E.P., Gomez-Sanchez, C.E., 2012. Potassium channel mutant KCNJ5 T158A expression in HAC-15 cells increases aldosterone synthesis. Endocrinology 153 (4), 1774–1782.

Olesen, M.S., Refsgaard, L., Holst, A.G., Larsen, A.P., Grubb, S., Haunso, S., Svendsen, J.H., Olesen, S.P., Schmitt, N., Calloe, K., 2013. A novel KCND3 gain-of-function mutation associated with early-onset of persistent lone atrial fibrillation. Cardiovasc. Res. 98 (3), 488–495.

Olson, T.M., Alekseev, A.E., Moreau, C., Liu, X.K., Zingman, L.V., Miki, T., Seino, S., Asirvatham, S.J., Jahangir, A., Terzic, A., 2007. KATP channel mutation confers risk for vein of Marshall adrenergic atrial fibrillation. Nat. Clin. Pract. Cardiovasc. Med. 4 (2), 110–116.

Omura, M., Saito, J., Yamaguchi, K., Kakuta, Y., Nishikawa, T., 2004. Prospective study on the prevalence of secondary hypertension among hypertensive patients visiting a general outpatient clinic in Japan. Hypertens. Res. 27 (3), 193–202.

Omura, M., Sasano, H., Saito, J., Yamaguchi, K., Kakuta, Y., Nishikawa, T., 2006. Clinical characteristics of aldosterone-producing microadenoma, macroadenoma, and idiopathic hyperaldosteronism in 93 patients with primary aldosteronism. Hypertens. Res. 29 (11), 883–889.

Pimenta, E., Gordon, R.D., Ahmed, A.H., Cowley, D., Leano, R., Marwick, T.H., Stowasser, M., 2011a. Cardiac dimensions are largely determined by dietary salt in patients with primary aldosteronism: results of a case-control study. J. Clin. Endocrinol. Metab. 96 (9), 2813–2820.

Pimenta, E., Gordon, R.D., Ahmed, A.H., Cowley, D., Robson, D., Kogovsek, C., Stowasser, M., 2011b. Unilateral adrenalectomy improves urinary protein excretion but does not abolish its relationship to sodium excretion in patients with aldosterone-producing adenoma. J. Hum. Hypertens. 25 (10), 592–599.

Plaster, N.M., Tawil, R., Tristani-Firouzi, M., Canun, S., Bendahhou, S., Tsunoda, A., Donaldson, M.R., Iannaccone, S.T., Brunt, E., Barohn, R., Clark, J., Deymeer, F., George Jr., A.L., Fish, F.A., Hahn, A., Nitu, A., Ozdemir, C., Serdaroglu, P., Subramony, S.H., Wolfe, G., Fu, Y.H., Ptacek, L.J., 2001. Mutations in Kir2.1 cause the developmental and episodic electrical phenotypes of Andersen's syndrome. Cell 105 (4), 511–519.

Reincke, M., Fischer, E., Gerum, S., Merkle, K., Schulz, S., Pallauf, A., Quinkler, M., Hanslik, G., Lang, K., Hahner, S., Allolio, B., Meisinger, C., Holle, R., Beuschlein, F., Bidlingmaier, M., Endres, S., 2012. Observational study mortality in treated primary aldosteronism: the German Conn's registry. Hypertension 60 (3), 618–624.

Rossi, G.P., Bernini, G., Caliumi, C., Desideri, G., Fabris, B., Ferri, C., Ganzaroli, C., Giacchetti, G., Letizia, C., Maccario, M., Mallamaci, F., Mannelli, M., Mattarello, M.J., Moretti, A., Palumbo, G., Parenti, G., Porteri, E., Semplicini, A., Rizzoni, D., Rossi, E., Boscaro, M., Pessina, A.C., Mantero, F., 2006a. A prospective study of the prevalence of primary aldosteronism in 1,125 hypertensive patients. J. Am. Coll. Cardiol. 48 (11), 2293–2300.

Rossi, G.P., Bernini, G., Desideri, G., Fabris, B., Ferri, C., Giacchetti, G., Letizia, C., Maccario, M., Mannelli, M., Matterello, M.J., Montemurro, D., Palumbo, G., Rizzoni, D., Rossi, E., Pessina, A.C., Mantero, F., 2006b. Renal damage in primary aldosteronism: results of the PAPY Study. Hypertension 48 (2), 232–238.

Rossi, G.P., Cesari, M., Pessina, A.C., 2003. Left ventricular changes in primary aldosteronism. Am. J. Hypertens. 16 (1), 96–98.

Rutherford, J.C., Taylor, W.L., Stowasser, M., Gordon, R.D., 1998. Success of surgery for primary aldosteronism judged by residual autonomous aldosterone production. World J. Surg. 22 (12), 1243–1245.

Sanguinetti, M.C., 2010. HERG1 channelopathies. Pflugers Arch. 460 (2), 265–276.

Sarzani, R., Pietrucci, F., Corinaldesi, C., Francioni, M., Letizia, C., D'Erasmo, E., Dessi-Fulgheri, P., Rappelli, A., 2006. The functional HERG variant 897T is associated with Conn's adenoma. J. Hypertens. 24 (3), 479–487.

Scholl, U.I., Dave, H.B., Lu, M., Farhi, A., Nelson-Williams, C., Listman, J.A., Lifton, R.P., 2012a. SeSAME/EAST syndrome–phenotypic variability and delayed activity of the distal convoluted tubule. Pediatr. Nephrol. 27 (11), 2081–2090.

Scholl, U.I., Goh, G., Stolting, G., de, O.R.C., Choi, M., Overton, J.D., Fonseca, A.L., Korah, R., Starker, L.F., Kunstman, J.W., Prasad, M.L., Hartung, E.A., Mauras, N., Benson, M.R., Brady, T., Shapiro, J.R., Loring, E., Nelson-Williams, C., Libutti, S.K., Mane, S., Hellman, P., Westin, G., Akerstrom, G., Bjorklund, P., Carling, T., Fahlke, C., Hidalgo, P., Lifton, R.P., 2013. Somatic and germline *CACNA1D* calcium channel mutations in aldosterone-producing adenomas and primary aldosteronism. Nat. Genet. 45 (9), 1050–1054.

Scholl, U.I., Nelson-Williams, C., Yue, P., Grekin, R., Wyatt, R.J., Dillon, M.J., Couch, R., Hammer, L.K., Harley, F.L., Farhi, A., Wang, W.H., Lifton, R.P., 2012b. Hypertension with or without adrenal hyperplasia due to different inherited mutations in the potassium channel *KCNJ5*. Proc. Natl. Acad. Sci. U. S. A. 109 (7), 2533–2538.

Sergouniotis, P.I., Davidson, A.E., Mackay, D.S., Li, Z., Yang, X., Plagnol, V., Moore, A.T., Webster, A.R., 2011. Recessive mutations in *KCNJ13*, encoding an inwardly rectifying potassium channel subunit, cause leber congenital amaurosis. Am. J. Hum. Genet. 89 (1), 183–190.

Shimizu, W., Horie, M., 2011. Phenotypic manifestations of mutations in genes encoding subunits of cardiac potassium channels. Circ. Res. 109 (1), 97–109.

Shimizu, Y., Hanew, K., Sato, S., Sasaki, A., Murakami, O., Yoshinaga, K., 1984. Evidence for endogenous dopaminergic control of mineralocorticoids secretion in normal subjects and in patients with hyperaldosteronism. Tohoku J. Exp. Med. 144 (1), 33–42.

Sicca, F., Imbrici, P., D'Adamo, M.C., Moro, F., Bonatti, F., Brovedani, P., Grottesi, A., Guerrini, R., Masi, G., Santorelli, F.M., Pessia, M., 2011. Autism with seizures and intellectual disability: possible causative role of gain-of-function of the inwardly-rectifying K^+ channel Kir4.1. Neurobiol. Dis. 43 (1), 239–247.

Silberstein, S.D., Dodick, D.W., 2013. Migraine Genet. Part II. Headache 53 (8), 1218–1229.

Simon, D.B., Karet, F.E., Rodriguez-Soriano, J., Hamdan, J.H., DiPietro, A., Trachtman, H., Sanjad, S.A., Lifton, R.P., 1996. Genetic heterogeneity of Bartter's syndrome revealed by mutations in the K^+ channel, ROMK. Nat. Genet. 14 (2), 152–156.

Sonino, N., Tomba, E., Genesia, M.L., Bertello, C., Mulatero, P., Veglio, F., Fava, G.A., Fallo, F., 2011. Psychological assessment of primary aldosteronism: a controlled study. J. Clin. Endocrinol. Metab. 96 (6), E878–E883.

Spat, A., Hunyady, L., 2004. Control of aldosterone secretion: a model for convergence in cellular signaling pathways. Physiol. Rev. 84 (2), 489–539.

Stowasser, M., Gordon, R.D., Gunasekera, T.G., Cowley, D.C., Ward, G., Archibald, C., Smithers, B.M., 2003. High rate of detection of primary aldosteronism, including surgically treatable forms, after 'non-selective' screening of hypertensive patients. J. Hypertens. 21 (11), 2149–2157.

Stowasser, M., Klemm, S.A., Tunny, T.J., Storie, W.J., Rutherford, J.C., Gordon, R.D., 1994. Response to unilateral adrenalectomy for aldosterone-producing adenoma: effect of potassium levels and angiotensin responsiveness. Clin. Exp. Pharmacol. Physiol. 21 (4), 319–322.

Sukor, N., Kogovsek, C., Gordon, R.D., Robson, D., Stowasser, M., 2010. Improved quality of life, blood pressure, and biochemical status following laparoscopic adrenalectomy for unilateral primary aldosteronism. J. Clin. Endocrinol. Metab. 95 (3), 1360–1364.

Taguchi, R., Yamada, M., Nakajima, Y., Satoh, T., Hashimoto, K., Shibusawa, N., Ozawa, A., Okada, S., Rokutanda, N., Takata, D., Koibuchi, Y., Horiguchi, J., Oyama, T., Takeyoshi, I., Mori, M., 2012. Expression and mutations of KCNJ5 mRNA in Japanese patients with aldosterone-producing adenomas. J. Clin. Endocrinol. Metab. 97 (4), 1311–1319.

Tester, D.J., Tan, B.H., Medeiros-Domingo, A., Song, C., Makielski, J.C., Ackerman, M.J., 2011. Loss-of-function mutations in the *KCNJ8*-encoded Kir6.1 K(ATP) channel and sudden infant death syndrome. Circ. Cardiovasc. Genet. 4 (5), 510–515.

Tomlinson, S.E., Hanna, M.G., Kullmann, D.M., Tan, S.V., Burke, D., 2009. Clinical neurophysiology of the episodic ataxias: insights into ion channel dysfunction in vivo. Clin. Neurophysiol. 120 (10), 1768–1776.

Tsuchiya, K., Yoshimoto, T., Hirata, Y., 2009. Endothelial dysfunction is related to aldosterone excess and raised blood pressure. Endocr. J. 56 (4), 553–559.

Tunny, T.J., Gordon, R.D., Klemm, S.A., Cohn, D., 1991. Histological and biochemical distinctiveness of atypical aldosterone-producing adenomas responsive to upright posture and angiotensin. Clin. Endocrinol. (Oxf) 34 (5), 363–369.

Wang, J., Zhang, J., Shen, J., Hu, D., Yan, G., Liu, X., Xu, X., Pei, L., Li, Y., Sun, C., 2014. Association of KCNQ1 and KLF14 polymorphisms and risk of type 2 diabetes mellitus: a global meta-analysis. Hum. Immunol. 75 (4), 342–347.

Williams, T.A., Monticone, S., Schack, V.R., Stindl, J., Burrello, J., Buffolo, F., Annaratone, L., Castellano, I., Beuschlein, F., Reincke, M., Lucatello, B., Ronconi, V., Fallo, F., Bernini, G., Maccario, M., Giacchetti, G., Veglio, F., Warth, R., Vilsen, B., Mulatero, P., 2014. Somatic *ATP1A1*, *ATP2B3*, and *KCNJ5* mutations in aldosterone-producing adenomas. Hypertension 63 (1), 188–195.

Part XII

Nonessentials

Chapter 42

Chromium: Basic Nutritional and Toxicological Aspects

John Bertram Vincent

The University of Alabama, Tuscaloosa, AL, United States

INTRODUCTION

Currently, in the United States and Canada chromium is formally considered an essential element. In 2002 the National Academies of Sciences (United States) established an adequate intake (AI) value of 25 µg/d for adult women and 35 µg/d for adult males, a reduction from a suggested intake of 50–200 mg daily (Institute of Medicine, 2001). By definition, a recommended daily allowance (RDA) indicates that more than 98% of Americans have an AI. Because the AI is established when not enough scientific data exist to establish an RDA, the AI is set more conservatively than the RDA; thus more than 98% cannot be considered Cr deficient. As will be discussed, the next analysis of the status of chromium should formally remove any suggested intake. In the European Union, the European Food Safety Authority has already determined that chromium is not an essential element for animals or humans (European Food Safety Authority, 2014; European Food Safety Authority, 2009). To understand the nutritional biochemistry of chromium and why the positions on the essentiality of chromium currently conflict, one must look at the misdirections of the past (Vincent, 2013).

CHROMIUM IS NOT AN ESSENTIAL ELEMENT

More than 5 decades ago, chromium was suggested to be an essential trace element in the mammalian diet (Vincent, 2013). Mertz and Schwarz (1955) fed rats a torula yeast-based diet, which compromised the health of the rats. The rats developed necrotic liver degeneration and apparently impaired glucose tolerance in response to an intravenous glucose load. Selenium (Se) was discovered to reverse the former (which ultimately led to the demonstration that Se was an essential trace element) but not the glucose intolerance; thus, a new dietary requirement was proposed and coined glucose tolerance factor (GTF). This requirement was believed to be absent from the torula yeast-based diet and responsible for the glucose intolerance (Schwarz and Mertz, 1957). To identify this missing dietary component, the diet was supplemented with various chemicals and some foods. Inorganic compounds containing more than 40 different elements (200–500 mg element/kg body mass) could not restore glucose tolerance whereas several inorganic Cr(III) complexes (200 mg Cr/kg body mass) restored glucose tolerance (Schwarz and Mertz, 1959). Of the foods, brewer's yeast and acid-hydrolyzed porcine kidney powder were identified as natural sources of the missing dietary component and were found to contain appreciable quantities of Cr (Mertz and Schwarz, 1955). When given by stomach tube (500–1000 mg/kg body mass), brewer's yeast, porcine kidney powder, and concentrates made from them restored proper glucose metabolism in rats on the torula yeast-based diet (Schwarz and Mertz, 1959). Thus, the active ingredient of GTF was proposed to be trivalent chromium, making Cr a potential new essential trace element (Schwarz and Mertz, 1959).

From the advantage of hindsight, this is unfortunately not what these studies established. The chromium content of a regular laboratory rat diet and the torula yeast-based diet were not determined; thus, the rats were not shown to actually receive a diet lacking chromium, only adding chromium to the diet was shown lead to potential effects on apparent glucose intolerance. Subsequently, the chromium content of torula yeast has been determined, but the content ranges significantly in value (Shepherd et al., 1992; Anderson et al., 1978), possibly from variations in growth conditions. Hence the content of the original diet cannot be determined. In addition, the rats were housed in wire mesh cages, possibly with stainless steel components (the metal composition of the wire was not reported), possibly allowing the rats to obtain chromium by gnawing on these components. Thus, no deficiency was established. Another possible explanation of the results is possible, one that was not originally considered—the supplemental chromium was having a pharmacological or therapeutic effect.

To examine this possibility, the chromium doses utilized need to be put into perspective. An American consuming a nutritionist-designed diet (Anderson and Kozlovsky, 1985) or a self-selected diet (Anderson et al., 1993) consumes approximately 30 μg of Cr each day. This value is the average AI set by the Food and Nutrition Board of the Institute of Medicine of the National Academy of Sciences (United States; Institute of Medicine, 2001); the AI being more conservative than the RDA suggests that more than 98% of the population receiving this quantity of an item display no health problems from deficiency. Using the average body mass of a human, 65 kg, gives an adequate chromium intake of less than 0.5 μg Cr/kg per day. Rats on the torula yeast-based diet supplemented with chromium(III) compounds received 400 times this quantity, a supranutritional dose (assuming that the biochemistry of Cr is similar in rodents and primates). Statistical concerns have also been raised about these studies (Woolliscroft and Barbosa, 1977); however, other studies in the 1960s apparently observed altered carbohydrate metabolism in rats fed torula yeast-based diets (for a review, cf Vincent, 2001).

Subsequently, establishing the essentiality of chromium using nutritionally compromised diets supplemented with chromium has been attempted (Striffler et al., 1995, 1998, 1999); however, these studies suffer from the same basic flaw as the initial studies. Rats were fed a high-sugar or high-fat diet (supposedly a "low-Cr" diet with ~30 μg Cr/kg diet) for 24 weeks, which led to the rat's lipid and carbohydrate metabolism being compromised. Addition of chromium(III) (5 ppm) to the stressed rat's drinking water resulted in plasma insulin levels tending to be higher in intravenous glucose tolerance tests after the 24-week period (Striffler et al., 1995). However, based on the typical food intake of a rat, the diet provided 5 μg Cr/kg body mass per day, 10 times what the typical American human intakes. Therefore, the low-Cr diet was not deficient, unless rats require more than 10 times the chromium dose that humans do (alternative adjustments for differences between rats and humans using surface area or metabolic rate rather than the body mass would provide similar results because these adjustments would change the value less than 10-fold). The lowering of plasma insulin levels by supplemental chromium cannot be considered a nutritionally relevant effect.

Similar complications arise for studies on patients on total parenteral nutrition (TPN); these studies have often been cited as evidence for chromium being an essential trace. The TPN solutions initially provided 2–6 μg Cr/d in the cases in which the concentration was reported (Vincent, 2013). Because TPN is an intravenous diet, all of the Cr in the TPN is introduced into the bloodstream, whereas only approximately 0.5% of Cr in a regular human diet is absorbed into the bloodstream (Vincent, 2013). Consequently, the average 30 μg of Cr in a typical daily American diet presents only approximately 0.15 μg Cr to the bloodstream. Therefore, the TPN solutions provided roughly an order of magnitude more Cr than the typical American diet, which is chromium sufficient. Subjects were treated with TPN supplemented with an additional 125–250 μg Cr/d (Freund et al., 1979; Jeejeebhoy et al., 1977; Verhage et al., 1996), a 1000-fold increase compared with Cr provided to the bloodstream from the typical diet to alleviate their conditions. This undoubtedly represents pharmacological doses. This provides no evidence for Cr being an essential element, but it suggests that large doses of Cr may have pharmacological effects in humans with altered glucose and carbohydrate metabolism, not just in rodents.

Finally, a study has unambiguously demonstrated that chromium has a pharmacological rather than a nutritional effect (Bona et al., 2011). The study used carefully controlled metal-free conditions and a series of purified diets of varying chromium content. Male lean Zucker rats housed in the metal-free cages for 6 months were fed the AIN-93G diet with no added chromium in the mineral mix component (16 μg Cr/kg diet), the standard AIN-93G diet (containing added 1000 μg Cr/kg), the standard AIN-93G diet supplemented with 200 μg Cr/kg, or the standard AIN-93G diet supplemented with 1000 μg Cr/kg. The chromium content of the diet had no effect on the body mass, food intake, or glucose levels in glucose tolerance or insulin tolerance tests. However, a statistically significant lowering of insulin areas under the curve after a glucose challenge was observed with increasing chromium content in the diet. The diet with as little chromium as reasonably possible had no effect on body composition, glucose metabolism, or insulin sensitivity compared with a "Cr-sufficient" diet; however, pharmacological quantities of chromium had a concentration-dependent effect on lowering insulin levels in glucose tolerance tests, indicating a pharmacological effect on increasing insulin sensitivity in the healthy rats (Bona et al., 2011). This represents a complete paradigm shift in the nutritional biochemistry of chromium, in which for more than 4 decades chromium had been considered to have only a nutritional effect, not a pharmacological effect. In summary, nutritional studies have failed to provide evidence that Cr is an essential element. Without conclusive positive evidence, chromium cannot be considered an essential trace element.

Despite the lack of evidence for chromium essentiality, textbooks in recent decades have occasionally contained a figure of the proposed structure of GTF, the "biologically active form of Cr." This three-dimensional structure displays two trans N-bound nicotinic acid ligands with amino acids occupying the remaining four sites of an octahedron around a Cr^{3+} ion (Mertz et al., 1974). The basis for this structure derives from the apparent isolation of brewer's yeast GTF reported by Mertz and coworkers in 1977 (Toepfer et al., 1977). However, examination of the procedures used in the isolation and characterization of the material reveals that a naturally occurring form of chromium was neither isolated nor characterized. The isolation involved rupturing the yeast, binding its contents on activated charcoal, removing the adhered material using a 50:50

mixture of concentrated ammonia and diethyl ether, and refluxing the product of the extraction in 5 M HCl for 18 h (Toepfer et al., 1977). Any proteins, complex carbohydrates, or nucleic acids would have been hydrolyzed (i.e., destroyed; however, it should be noted that the authors were primarily looking for what they believed would be a heat-stable, acid-stable vitamin or analogue). Nicotinic acid was reportedly sublimed from the material and identified (although no data or experimental details were presented for the mass spectral, sublimation, or extraction studies; Toepfer et al., 1977). Amino acid analyses revealed the presence of glycine, glutamic acid, and cysteine as well as other amino acids, although the relative amounts were not reported. On the basis of these results, GTF was claimed to be a complex of Cr, nicotinate, glycine, cysteine, and glutamate. In subsequent paper chromatography experiments described in the same paper, the material produced several Cr-containing spots; however, only one was active in bioassays (Toepfer et al., 1977). Consequently, the isolated species used for the composition studies was not pure. The bioassays measured the ability of a material to activate the metabolism of glucose by adipose tissue from rats on the torula yeast diet in the presence of insulin. Chromium in the active band actually represented only 6% of the total chromium (Toepfer et al., 1977). No experiments were performed to otherwise characterize the bioactive Cr-containing component; thus even if Cr were essential, so that GTF actually existed and was not an artifact generated during the harsh isolation conditions, this study would offer no information on its composition. Several laboratories in trying to reproduce this work have separated Cr^{3+} from components in brewer's yeast responsible for in vitro stimulation of glucose metabolism in adipocytes (for a review, cf Vincent, 2013). Therefore the component of yeast that is active in the bioassays does not even appear to contain Cr. In addition, potentially how such chromium-containing artifacts could be formed under the harsh isolation conditions has been shown in the other suggested source of GTF—acid-hydrolyzed porcine kidney powder (Sumrall and Vincent, 1997).

Despite the numerous publications refuting the existence of GTF, these studies on GTF unfortunately continue to be cited as evidence for the essential role of Cr. The use of the term GTF should be discontinued.

CHROMIUM(III) TOXICITY AND THE FATE OF CHROMIUM SUPPLEMENTS IN VIVO

Since the early 1990s, the best-selling and most studied Cr(III) compound as a nutritional supplement is Cr(III) picolinate $[Cr(pic)_3]$ (Fig. 42.1). The toxicity of $[Cr(pic)_3]$ has been an area of significant research and debate; the question recently appears to have been answered, although understanding the chemistry of the complex under physiological conditions was required (Vincent, 2013). $[Cr(pic)_3]$ has a solubility of only 600 μM at near-neutral pH. The solubility decreases as pH is lowered until the complex hydrolyzes to release picolinic acid (Research Triangle Institute, 2002). The picolinate ligands alter the redox properties of the Cr center such that it was more susceptible to undergoing redox chemistry in the body than hexaaquachromium(III), $[Cr(H_2O)_6]^{3+}$ (Parajon-Costa et al., 2003). This results from the binding of the imine nitrogens of the picolinate ligands; the Cr center of $[Cr(pic)_3]$ is bound to three imine ligands. Cr(III)-imine complexes such as $[Cr(2,2'-bipyridine)_3]^{3+}$, in which the Cr center is bound to six imine ligands, are known to be mutagenic and toxic because the Cr center in this environment can enter into redox chemistry with biomolecules. $[Cr(pic)_3]$'s unique ability to enter into redox chemistry compared with other forms of Cr used or proposed as nutrition supplements is believed to be responsible for its greater ability to cleave DNA in vitro and its greater toxicity in cell culture studies than those of the other supplements.

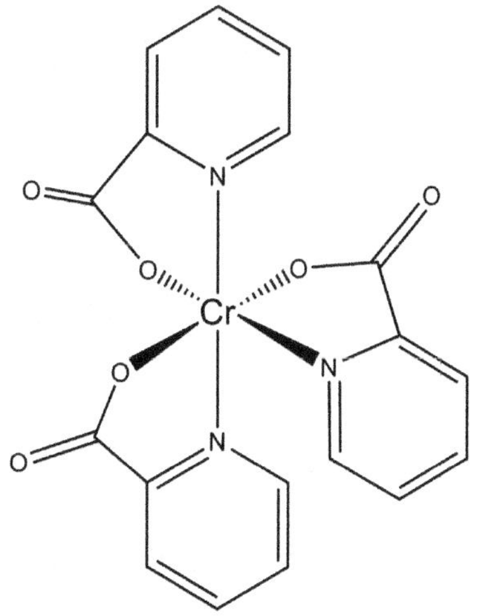

FIGURE 42.1 Structure of chromium picolinate.

In Ames assays, [Cr(pic)$_3$] has shown no deleterious effects (Whitaker et al., 2005). However, in contrast to these bacterial cell assays, mammalian cell culture studies and studies in which the complex was given intravenously found that [Cr(pic)$_3$] was toxic and mutagenic, unlike other commercial forms of Cr(III) (Stearns, 2007; Stallings and Vincent, 2006). Not all cell studies have identified toxic effects. Studies funded by a company that sells the supplement observed no effects in cell culture studies (Slesinski et al., 2005; Gudi et al., 2005); however, subsequent research has shown that dimethylsulfoxide, used as a solvent for the supplement and that can serve as a radical trap, quenching deleterious effects (Coryell and Stearns, 2006). [Cr(pic)$_3$] has been demonstrated to be a potent mutagen in fruit flies, whereas other complexes of Cr(III) tested were not (Hepburn et al., 2003; Stallings et al., 2006). [Cr(pic)$_3$] apparently is not toxic or a mutagen or carcinogen when given orally to mammals. A large study conducted by the National Toxicology Program division of the National Institutes of Health on the effects of [Cr(pic)$_3$] at doses up to 5% of the diet (by mass) on male and female rats and mice for up to 2 years found no harmful effects. The only potential exception was that data for one type of carcinogenicity (preputial gland adenoma) in male rats were ambiguous. No changes in body mass in either sex of rats or mice were observed (Stout et al., 2009), strongly refuting previous claims that chromium nutritional supplements could be used for body mass reduction. Overall, chromium supplements appear to be safe for consumption; however, no reason exists to take the supplements and incur any potential risk until a clear need or benefit from taking them is demonstrated.

Most data on the fate of the chromium in the body from chromium supplementation come from studies on [Cr(pic)$_3$] and CrCl$_3$. Over the past 3 decades, numerous claims have been made that [Cr(pic)$_3$] is absorbed better than the inorganic forms of Cr (e.g., CrCl$_3$) used to model dietary Cr. However, [Cr(pic)$_3$], Cr nicotinate (the second most popular form of Cr sold as a nutritional supplement), and CrCl$_3$ have been found to be absorbed to a similar degree in rats in well-designed studies; these studies examined the fate of ^{51}Cr from labeled compounds in tissues and body fluids (Olin et al., 1994; Anderson et al., 1996). Contradictory results have been obtained when absorption is estimated using a more limited dataset, such as just examining urine.

[Cr(pic)$_3$] is stable in H$_2$O (Chakov et al., 1999) such that it should not break down appreciably when dissolved or slurred in buffered H$_2$O. Using X-ray absorption studies, Lay and coworkers (Nguyen et al., 2008) have found that the complex was stable in simulated gastric juice and simulated intestinal fluid for 24 h (Nguyen et al., 2008). Only approximately one-fourth and one-half of the compound degraded in artificial gastric juice with added components of a semisynthetic meal over 1 and 3 h, respectively (Nguyen et al., 2008). In high-performance liquid chromatography (HPLC) studies using ^{14}C-labeled picolinate, [Cr(pic)$_3$] has been shown to be stable for 2 h when added to excised stomach contents of rats and for 4 h when added to excised small intestine contents (Research Triangle Institute, 2002).

In rats, only 1% of absorbed Cr from [Cr(pic)$_3$] (again followed by HPLC using ^{14}C-labeled picolinate) was found in the bloodstream as the intact complex, suggesting that little of the intact complex is absorbed (Research Triangle Institute, 2002). Double-labeling experiments with ^{51}Cr- and ^{3}H-labeled [Cr(pic)$_3$] have shown that gavaged [Cr(pic)$_3$] breaks down rapidly in the gastrointestinal tract, probably primarily in the stomach, because the fate of the labels diverged rapidly in the gastrointestinal tract and body of rats (Gullick and Vincent, unpublished results; Fig. 42.2). Thus, when ingested, the complex probably hydrolyzes near the stomach lining, releasing Cr and picolinic acid, which are subsequently absorbed independently, on the basis of the greatly different extent of absorbance of Cr and picolinic acid, even at the earliest time points in Fig. 42 2. Thus [Cr(pic)$_3$] would appear to be a source of chromic ions for absorption, not unique from using CrCl$_3$ or Cr nicotinate, although small differences in the rate at which ligand loss occurs may exist. Once in the bloodstream, [Cr(pic)$_3$] appears to degrade rapidly, so that what might be absorbed intact would half a short lifetime. The complex appears to be less stable in rat blood serum than in buffer, with approximately 50%, 70%, and 85% degradation after 1, 6, and 24 h, respectively (Nguyen et al., 2008). In tissues the [Cr(pic)$_3$] appears to extremely rapidly break down (Hepburn and Vincent, 2003). Hepatocyte microsomes have been shown to degrade [Cr(pic)$_3$] in vitro (Kareus et al., 2001); however, the major product derived from the picolinate ligand was different from that observed when [Cr(pic)$_3$] was administered by gavage to rats (Coryell and Stearns, 2006).

These data on the fate of [Cr(pic)$_3$] can be used to reconcile the results of the studies investigating the toxicity of [Cr(pic)$_3$] (Vincent, 2013). When [Cr(pic)$_3$] is intact in the presence of appropriate redox active species, oxidative damage is expected. Thus, [Cr(pic)$_3$] dissolved in H$_2$O or buffer in a test tube is susceptible to generating this type of chemistry. The compound does not cause damage to bacterial cells in the Ames assays because it probably cannot diffuse into the cells. Mammalian cells with less robust outer membranes allow the diffusion of [Cr(pic)$_3$] into the cells; thus, the mammalian cell culture studies almost uniformly observe oxidative or DNA damage. Likewise, if [Cr(pic)$_3$] is intravenously administered, the intact complex has the brief ability to diffuse into cells and generate damage. In contrast, when given orally to rodents and humans, [Cr(pic)$_3$] readily breaks down in the gastrointestinal tract so that only approximately 1% of the small quantity of absorbed Cr exists as intact [Cr(pic)$_3$]. Thus, when administered orally, the animal is fortuitously

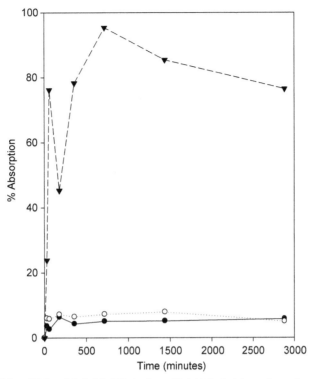

FIGURE 42.2 Percentage absorbed ^{51}Cr or ^{3}H from labeled Cr(III) picolinate [Cr(pic)$_3$] administrated to male rats by gavage. The dose was 3 µg Cr/kg body mass for the ^{3}H and low-dose ^{51}Cr experiments and 250 µg Cr/kg body mass for the high-dose ^{51}Cr experiments. *Dashed line*, [Cr([^{3}H(pic)$_3$]; *dotted line*, low-dose [^{51}Cr(pic)$_3$]; *solid line*, high-dose [^{51}Cr(pic)$_3$]. Extent of absorption was calculated by subtracting the radiolabel content in feces, urine, body fluids, and tissues from the radiolabel content of the oral dose. Error is approximately ±5% absorption.

not exposed to the form capable of entering into the redox chemistry leading to oxidative damage and DNA cleavage. In the very different digestive system of fruit flies, the complex, when absorbed intact, is a potent clastogen, mutagen, and developmental toxin.

PHARMACOLOGICALLY ACTIVE CHROMIUM

After absorption, chromium in the bloodstream binds to the iron-transport protein transferrin (for a review, cf Vincent, 2013). Transferrin then carries chromium to the tissues via endocytosis. The fate of chromium bound to transferrin has been monitored utilizing ^{51}Cr-labeled transferrin. Injection of ^{51}Cr-labeled transferrin into the bloodstream results in a rapid and insulin-sensitive movement of chromium into the tissues as Cr–transferrin; the majority of the labeled chromium is transported to the tissues within 30 min (Clodfelder and Vincent, 2005). Tissue levels of chromium are maximal 30 min after injection whereas decreases in tissue chromium with time are mirrored by increases in urine chromium. Insulin treatment concurrent with injection of ^{51}Cr-labeled transferrin significantly increases the rate of chromium movement, which is not surprising because insulin results in a movement of transferrin receptor in cells to the plasma membrane. The removal of ^{51}Cr from the blood is faster than the appearance of ^{51}Cr in the urine; the lag in time indicates that the Cr–transferrin in the blood and chromium in the urine are not in direct equilibrium and that intermediates in the transport of chromium must be involved (Clodfelder and Vincent, 2005).

When chromium is taken orally, only one other biomolecule with bound chromium is found to an appreciable extent; the biomolecule coelutes on size exclusion columns with the peptide low-molecular–weight chromium-binding substance (LMWCr, or chromodulin). LMWCr has been isolated from tissues of reptiles, birds, and mammals and from human urine. Thus the peptide carries chromium from the tissues to the bloodstream and then ultimately to the urine for elimination. Thus transferrin and LMWCr provide a mechanism for the removal of small amounts of absorbed chromium from the body (for a review, cf Vincent, 2013).

A mechanism for the pharmacological effects of chromium has recently been proposed in which chromium potentially acts as if it were a second messenger (Fig. 42.3; Vincent, 2000, 2015). An increase in insulin, such as would result after

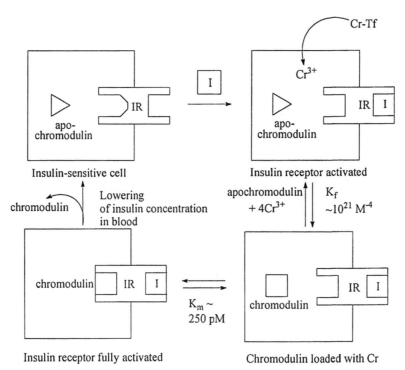

FIGURE 42.3 Proposed mechanism for the activation of insulin receptor kinase activity by low-molecular–weight chromium-binding substance (LMWCr; chromodulin) in response to insulin. The inactive form of the insulin receptor (IR) is converted into the active form by binding insulin (I). This triggers a movement of Cr (presumably in the form of Cr transferrin, Cr–Tf) from the blood into insulin-dependent cells, which in turn results in a binding of Cr to apoLMWCr (apochromodulin; *triangle*). Finally, the holoLMWCr (holochromodulin; *square*) binds to the insulin receptor, further activating the receptor kinase activity. ApoLMWCr is unable to bind to the insulin receptor and activate kinase activity. When the insulin concentration drops, holo-LMWCr is released from the cell to relieve its effects. *Reproduced from Vincent, J., 2000. The biochemistry of chromium. J. Nutr. 130, 715–718;* © *2000 ASN Journals.*

consuming a meal, results in an increased movement of chromium from the bloodstream to the tissues via transferrin endocytosis. When followed using ^{51}Cr-labeled [Cr(pic)$_3$], approximately 60% of chromium entering the tissues is transported to the skeletal muscle (Gullick and Vincent, unpublished results), an insulin-sensitive tissue largely responsible for glucose uptake and metabolism. This results in the generation of Cr-loaded LMWCr. In in vitro assays, Cr-loaded LMWCr has been found to tightly bind to the insulin receptor, stimulating its tyrosine kinase activity. Hence pharmacological doses of chromium could potentially result in abnormally high concentrations of LMWCr, which at these concentrations could enhance insulin signaling and increase insulin sensitivity.

This model is supported by studies in humans. For example, in euglycemic hyperinsulinemic clamp studies, Morris et al. have shown that increases in blood insulin concentrations after an oral glucose load result in significant decreases in plasma Cr levels; a subsequent infusion of insulin led to further Cr losses (Morris et al., 1993). Within 90 min after the increases of blood insulin concentrations, blood Cr levels started to recover. Patients also showed increased urinary Cr losses during the course of the experiments, with the amount of Cr lost roughly corresponding to the amount of chromium estimated to be removed from the intravascular space (Morris et al., 1993). Numerous other studies have demonstrated that Cr is released in urine within 90 min of a dietary stress such as high sugar intake (for a review, cf Vincent, 2013). As glucose tolerance as a result of repeated application of carbohydrate stress decreases, the successive mobilizations of Cr and resulting Cr losses have been shown to decrease (Kozlovsky et al., 1986). These homoeostasis and urinary output studies suggest that Cr is stored or maintained in the blood and can be mobilized in response to increases in blood insulin concentrations where it ultimately appears in the urine. In addition, as previously noted, numerous studies of rodent models of insulin insensitivity have observed increased insulin sensitivity from Cr administration.

Unfortunately, the mode of action of chromium for enhancing insulin sensitivity in insulin-sensitive tissues has not been clearly established at a molecular level. In fact, attempting to examine this in cell culture has generated conflicting results based on the source of Cr used and other variables (for a review, cf Vincent, 2013). Because cells are normally exposed to Cr^{3+} released from transferrin, the exposure of cells to chromium compounds dissolved in the buffer or media should not mimic their in vivo exposure. For example, it clearly does not when [Cr(pic)$_3$] is used. As previously described, [Cr(pic)$_3$]

is toxic and mutagenic to cultured mammalian cells (and in animals as complex as fruit flies), whereas in mammals the compound breaks down in the gastrointestinal tract so that the intact molecule is not presented to cells. Cell culture studies using Cr–transferrin as the chromium source are sorely needed. Thus establishing whether LMWCr or some other chromium-binding molecule(s) is responsible for the pharmacological action of chromium requires more study. The sequencing of the contiguous peptide component of LMWCr and the demonstration that it tightly binds chromium and is biologically active (Chen et al., 2011) may provide an avenue for further study. In addition, given that chromium administration has been shown in rodent models to have effects on insulin sensitivity and cholesterol levels could suggest multiple mechanisms of chromium action. In fact, numerous mechanisms have been proposed (Vincent, 2013; Hua et al., 2012), although none is well substantiated by in vivo studies.

Another issue that the field must address is that clinical studies using chromium supplements have failed to result in beneficial effects in diabetic subjects (Bailey, 2014), in stark contrast to studies with model rodents. However, clinical studies were performed under the assumption that chromium was an essential element and used "nutritionally" relevant doses of chromium, far below those used in the rodent studies when scaled based on body mass (if even different metabolic rates, surface areas, etc., are considered). The effects reported in the case studies of subjects on TPN are consistent with higher doses of chromium potentially having beneficial effects in humans. The Food Standard Agency (United Kingdom) has determined that doses up to 10 mg chromium daily should be safe for humans (Expert Group on Vitamins and Minerals, 2003). Although effects of chromium could be different between rodents and humans, well-designed clinical trials using doses approaching 10 mg chromium daily are needed to establish whether chromium has a future as a pharmacological agent to treat insulin resistance and related conditions in humans.

REFERENCES

Anderson, R.A., Kozlovsky, A.S., 1985. Chromium intake, absorption and excretion of subjects consuming self-selected diets. Am. J. Clin. Nutr. 41, 1177–1183.

Anderson, R.A., Brantner, J.H., Polansky, M.M., 1978. An improved assay for biologically active chromium. J. Agric. Food Chem. 26, 1219–1221.

Anderson, R.A., Bryden, N.A., Polansky, M.M., 1993. Dietary intake of calcium, chromium, copper, iron, magnesium, manganese, and zinc: duplicate plate values corrected using derived nutrient intake. J. Am. Diet. Assoc. 93, 462–464.

Anderson, R.A., Bryden, N.A., Polansky, M.M., Gautschi, K.J., 1996. Dietary chromium effects on tissue chromium concentrations and chromium absorption in rats. Trace Elem. Exp. Med. 9, 11–25.

Bailey, C.H., 2014. Improved meta-analytic methods show no effect of chromium supplements on fasting glucose. Biol. Trace Elem. Res. 157, 1–8.

Di Bona, K.R., Love, S., Rhodes, N.R., McAdory, D., Sinha, S.H., Kern, N., Kent, J., Strickland, J., Wilson, A., Beaird, J., Ramage, J., Rasco, J.F., Vincent, J.B., 2011. Chromium is not an essential trace element for mammals: effects of a "low-chromium" diet. J. Biol. Inorg. Chem. 16, 381–390.

Chakov, N.E., Collins, R.A., Vincent, J.B., 1999. A re-investigation of the electronic spectra of chromium(III) picolinate complexes and high yield synthesis and characterization Cr$_2$(μ-OH)$_2$(pic)$_4$·5H$_2$O (Hpic = picolinic acid). Polyhedron 18, 2891–2897.

Chen, Y., Watson, H.M., Gao, J., Sinha, S.H., Cassady, C.J., Vincent, J.B., 2011. Characterization of the organic component of low-molecular-weight chromium-binding substance and its binding of chromium. J. Nutr. 141, 1225–1232.

Clodfelder, B.J., Vincent, J.B., 2005. The time-dependent transport of chromium in adult rats from the bloodstream to the urine. J. Biol. Inorg. Chem. 10, 383–393.

Coryell, V.H., Stearns, D.M., 2006. Molecular analysis of HPRT mutations induced by chromium picolinate in CHO AA8 cells. Mutat. Res. 610, 114–123.

European Food Safety Authority, 2009. Scientific opinion of the panel on additives and products used in animal feed (FEEDAP) on a request from the European Commission on the safety and efficacy of chromium methione (Avail®Cr) as feed additive for all species. EFSA J. 1043, 1–69.

European Food Safety Authority, 2014. Scientific opinion on dietary reference values for chromium. EFSA J. 12, 3845.

Expert Group on Vitamins and Minerals, 2003. Safe Upper Levels for Vitamins and Minerals. Food Standards Agency, London.

Freund, H., Atamian, S., Fischer, J.E., 1979. Chromium deficiency during total parenteral nutrition. JAMA 241, 496–498.

Gudi, R., Slesinski, R.S., Clarke, J.J., San, R.H., 2005. Chromium picolinate does not produce chromosome damage in CHO cells. Mutat. Res. 587, 140–146.

Hepburn, D.D.D., Vincent, J.B., 2003. Tissue and subcellular distribution of chromium picolinate with time after entering the bloodstream. J. Inorg. Biochem. 94, 86–93.

Hepburn, D.D.D., Xiao, J., Bindom, S., Vincent, J.B., O'Donnell, J., 2003. Nutritional supplement chromium picolinate causes sterility and lethal mutations in Drosophila melanogaster. Proc. Natl. Acad. Sci. U. S. A. 100, 3766–3771.

Hua, Y., Clark, S., Ren, J., Sreejayan, N., 2012. Molecular mechanisms of chromium in alleviating insulin resistance. J. Nutr. Biochem. 23, 313–319.

Institute of Medicine, 2001. Dietary Reference Intakes for Vitamin A, Arsenic, Boron, Chromium, Copper, Iodine, Iron, Manganese, Molybdenum, Nickel, Silicon, Vanadium, and Zinc. A Report of the Panel of Micronutrients, the Subcommittee on Upper Levels of Nutrients and Interpretations and Uses of Dietary Reference Intakes, and the Standing Committee on the Scientific Evaluation of Dietary Reference Intakes. Food and Nutrition Board, Institute of Medicine, National Academies of Sciences, Washington, DC.

Jeejeebhoy, K.N., Chu, R.C., Marliss, E.B., Greenberg, G.R., Bruce-Robertson, A., 1977. Chromium deficiency, glucose intolerance, and neuropathy reversed by chromium supplementation, in a patient receiving long term total parenteral nutrition. Am. J. Clin. Nutr. 30, 531–538.

Kareus, S.A., Kelley, C., Walton, H.S., Sinclair, P.C., 2001. Release of Cr(III) from Cr(III) picolinate upon metabolic activation. J. Hazard. Mat. B84, 163–174.

Kozlovsky, A.S., Moser, P.B., Reisner, S., Anderson, R.A., 1986. Effects of diets high in simple sugars on urinary chromium losses. Metabolism 35, 515–518.

Mertz, W., Schwarz, K., 1955. Impaired intravenous glucose tolerance as an early sign of dietary necrotic liver degeneration. Arch. Biochem. Biophys. 58, 504–506.

Mertz, W., Toepfer, E.W., Roginski, E.E., Polansky, M.M., 1974. Present knowledge of the role of chromium. Fed. Proc. 33, 2275–2280.

Morris, B.W., MacNeil, S., Stanley, K., Gray, T.A., Fraser, R., 1993. The inter-relationship between insulin and chromium in hyperinsulinaemic euglycaemic clamps in healthy volunteers. J. Endocrinol. 139, 339–345.

Nguyen, A., Mulyani, I., Levina, A., Lay, P.A., 2008. Reactivity of chromium(III) nutritional supplements in biological media: an X-ray absorption spectroscopic study. Inorg. Chem. 47, 4299–4309.

Olin, K.L., Stearns, D.M., Armstrong, W.H., Keen, C.L., 1994. Comparative retention/absorption of chromium (^{51}Cr) from ^{51}Cr chloride, ^{51}Cr nicotinate and ^{51}Cr picolinate in a rat model. Trace Elem. Electrolytes 11, 182–186.

Parajon-Costa, B.S., Wagner, C.C., Baran, E.J., 2003. Voltammetric and spectroscopic study of chromium(III)/picolinate complexes. Z. Anorg. Allg. Chem. 629, 1085–1090.

Research Triangle Institute, 2002. Project Report, "[^{14}C] Chromium Picolinate Monohydrate: Disposition and Metabolism in Rats and Mice", Submitted to National Institutes of Environmental Health Sciences, 2002.

Schwarz, K., Mertz, M., 1957. A glucose tolerance factor and its differentiation from factor 3. Arch. Biochem. Biophys. 72, 515–518.

Schwarz, K., Mertz, W., 1959. Chromium(III) and the glucose tolerance factor. Arch. Biochem. Biophy. 85, 292–295.

Shepherd, P.R., Elwood, C., Buckley, P.D., Blackwell, L.F., 1992. Glucose tolerance factor potentiation of insulin action in adipocytes from rats raised on a Torula yeast diet cannot be attributed to a deficiency of chromium. Biol. Trace Elem. Res. 32, 109–113.

Slesinski, R.S., Clarke, J.J., San, R.H.C., Gudi, R., 2005. Lack of mutagenicity of chromium picolinate in the hypoxanthine phosphoribosyltransferase gene assay in Chinese hamster ovary cells. Mutat. Res. 585, 86–95.

Stallings, D., Vincent, J.B., 2006. Chromium: a case study on how not to perform nutraceutical research. Curr. Top. Nutraceutical Res. 4, 89–111.

Stallings, D.M., Hepburn, D.D., Hannah, M., Vincent, J.B., O'Donnell, J.D., 2006. Nutritional supplement chromium picolinate generates chromosomal aberrations and impedes progeny development in *Drosophila melanogaster*. Mutat. Res. 610, 101–113.

Stearns, D.M., 2007. Evaluation of Cr(III) genotoxicity with cell culture and in vitro assays. In: Vincent, J.B. (Ed.), The Nutritional Biochemistry of Chromium(III). Elsevier, Amsterdam, pp. 209–224.

Stout, M.D., Nyska, A., Collins, B.J., Witt, K.L., Kissling, G.E., Malarkey, D.E., Hooth, M.J., 2009. Chronic toxicity and carcinogenicity studies of chromium picolinate monohydrate administered in feed to F344/N rats and B6C3F1 mice for 2 years. Food Chem. Toxicol. 47, 729–733.

Striffler, J.S., Law, J.S., Polansky, M.M., Bhathena, S.J., Anderson, R.A., 1995. Chromium improves insulin response to glucose in rats. Metabolism 44, 1314–1320.

Striffler, J.S., Polansky, M.M., Anderson, R.A., 1998. Dietary chromium decreases insulin resistance in rats fed a high fat mineral imbalanced diet. Metabolism 47, 396–400.

Striffler, J.S., Polansky, M.M., Anderson, R.A., 1999. Overproduction of insulin in the chromium-deficient rat. Metabolism 48, 1063–1068.

Sumrall, K.H., Vincent, J.B., 1997. Is glucose tolerance factor an artifact produced by acid hydrolysis of low-molecular-weight chromium-binding substance? Polyhedron 16, 4171–4177.

Toepfer, E.W., Mertz, W., Polansky, M.M., Roginski, W.W., Wolf, W.R., 1977. Preparation of chromium-containing material of glucose tolerance factor activity from brewer's yeast extracts and by synthesis. J. Agric. Food Chem. 25, 162–166.

Verhage, A.H., Cheong, W.K., Jeejeebhoy, K.N., 1996. Neurologic symptoms due to possible chromium deficiency in long-term total parenteral nutrition that closely mimic metronidazole-induced syndromes. JPEN 20, 123–127.

Vincent, J., 2000. The biochemistry of chromium. J. Nutr. 130, 715–718.

Vincent, J.B., 2001. The bioinorganic chemistry of chromium. Polyhedron 20, 1–26.

Vincent, J.B., 2013. The Bioinorganic Chemistry of Chromium. John Wiley & Sons, Chichester.

Vincent, J.B., 2015. Is the pharmacological mode of action of chromium(III) as a second messenger? Biol. Trace Elem. Res. 166, 7–12.

Whitaker, P., San, R.H.C., Clark, J.J., Seifried, H.E., Dunkel, V.C., 2005. Mutagenicity of chromium picolinate and its components in *Salmonella typhimurium* and L5178Y mouse lymphoma cells. Food Chem. Toxicol. 43, 1619–1625.

Woolliscroft, J., Barbosa, J., 1977. Analysis of chromium induced carbohydrate intolerance in the rat. J. Nutr. 107, 1702–1706.

Chapter 43

Nonessential Trace Minerals: Basic Nutritional and Toxicological Aspects

Forrest Harold Nielsen

USDA, ARS, Grand Forks Human Nutrition Research Center, Grand Forks, ND, United States

INTRODUCTION

During the 1960s and 1970s, several definitions of essentiality for minerals appeared that often included several criteria. One criterion that was included in all definitions was that a dietary deficiency had to consistently and adversely change a biological function from optimal, and this change was preventable or reversible by physiological or nutritional amounts of the mineral. In the 1980s and 1990s, establishing essentiality on this basis was challenged when many trace minerals were suggested to be essential based on some small change in a physiological or biochemical variable in an experimental animal model supposedly fed a diet deficient in a specific mineral. Doubts were raised about the "deficient diets" causing suboptimal function. It was suggested that trace mineral supplementation of the diet might be having a pharmacological effect such as alleviating a pathological condition caused by a poor diet or environmental conditions, substituting for another essential nutrient provided in a deficient amount, or having an effect on intestinal organisms in a manner that was beneficial to the host. As a result, if the lack of a trace mineral cannot be shown to cause death or interrupt the life cycle, it is now not considered essential unless it has a defined biochemical function in higher animals and humans. Trace minerals that were once suggested to be essential based mostly on an apparent beneficial action in animals, but that cannot meet the current definition of essentiality, include aluminum, arsenic, boron, bromine, cadmium, chromium, fluoride, germanium, lead, lithium, nickel, rubidium, silicon, strontium, tin, and vanadium (Nielsen, 2012).

The US Food and Nutrition Board, Institute of Medicine (1994), has suggested that the reduction in the risk of chronic diseases that can disable and/or lead to premature death could be used in the formulation of dietary reference intakes (DRIs) when sufficient data for efficacy and safety exist. This suggestion apparently was used for setting the DRIs for the essential nutrient potassium (Food and Nutrition Board, 2005). Recommended dietary allowances (RDAs) were not established for potassium because of insufficient dose–response data to determine an estimated average requirement (EAR). The RDA is the average daily dietary nutrient intake sufficient to meet the nutrient requirement of nearly all (98%) healthy individuals in a particular life stage and gender group. The EAR is an intake estimated to meet the nutritional requirement of half of the healthy individuals in a particular life stage or gender group. When EARs and RDAs cannot be established for an essential nutrient, an adequate intake (AI) is determined based on the approximate or estimated intake by a group, or groups, of healthy individuals assumed to be adequate with respect to a particular nutrient (Food and Nutrition Board, Institute of Medicine, 2001). Instead of the usual method for determining the AI for potassium, it was set on the basis of amounts that were found to beneficially affect blood pressure, reduce the risk for kidney stones, and possibly reduce bone loss. On the basis of a dietary intake survey (Food and Nutrition Board, Institute of Medicine, 2005), the potassium intakes of only 10% of adult men and less than 1% of adult women meet the AI.

AIs also have been set for the nonessential trace minerals chromium and fluoride because in supranutritional amounts fluoride prevents the pathological demineralization of calcified tissue and chromium promotes insulin action. The nutritional and toxicological aspects of these trace minerals are described in other chapters in this book. Setting AIs for chromium and fluoride is the precedence for providing dietary guidance for nonessential trace minerals. However, of all of the other nonessential trace minerals listed, only boron, bromine, silicon, strontium, and vanadium have been shown to have beneficial effects in humans. The beneficial actions of three of these minerals have been used or have been proposed to be used in pharmacological applications. Bromine has been used in insomnia medications, strontium in the ranelate form for the treatment of postmenopausal osteoporosis, and vanadium has insulin-like action (Nielsen, 2012). Nutritional aspects of these trace minerals are too limited to be included in this chapter. Thus the nutritional and toxicological aspects of only boron and silicon will be presented here. However, it should be noted that some of the trace

Molecular, Genetic, and Nutritional Aspects of Major and Trace Minerals. http://dx.doi.org/10.1016/B978-0-12-802168-2.00043-9

minerals not discussed have been found to be essential components of enzymes in lower forms of life (Nielsen, 2012). These include arsenic, nickel, and vanadium. These minerals might be found essential or beneficial in nutritional or trace amounts for humans in the future.

Because of a given limitation in references, reviews are often cited in this chapter. These reviews provide the original sources for the statements made.

BORON

Historical Aspects

In the 1870s it was discovered that sodium borate and boric acid could be used to preserve foods (Nielsen and Meacham, 2011). For almost 50 years, borate addition was considered one of the best methods for preserving and extending the palatability of foods such as meat and dairy products. Boron had a vital role as a preservative in preventing food crises during World Wars I and II. Use of boron in this manner declined when it was reported in 1904 that boric acid in doses greater than 500 mg/day (77 mg boron) for 50 days resulted in disturbances in appetite, digestion, and health in human volunteers. It also was concluded that boric acid at 4000 mg/day (699 mg boron) was the limit beyond which harm to humans would occur. By the 1950s boron as a food preservative was essentially forbidden throughout the world.

After boron was determined essential for plants in the 1920s, several unsuccessful attempts were made in the next 20 years to show that boron was essential for higher animals. The lack of success resulted in generations of students in nutrition being taught that boron was unique in that it was essential for plants but not for animals or humans. In 1981 it was suggested that boron could alleviate arthritis and that boron deprivation exacerbated gross bone abnormalities in chicks fed marginal amounts of vitamin D. Since then an increasing number of reports have shown that nutritional or physiological amounts of boron have numerous beneficial effects in higher animals and humans.

Beneficial Actions

Boron has been shown to be essential for the completion of the life cycle (i.e., deficiency causes impaired growth, development, or maturation such that procreation is prevented) for organisms in all phylogenetic systems. Higher animals for which boron has been found essential are frogs and zebrafish (Nielsen and Meacham, 2011). Boron-deprived male frogs exhibited atrophied testes, depressed sperm counts, and sperm dysmorphology. Female frogs exhibited atrophied ovaries and impaired oocyte maturation. Boron deprivation induced high mortality in frog and zebrafish embryos. The biochemical basis for boron deficiency causing the inability to complete the life cycle of these higher animals has not been definitively established. Because boron does not have a defined biochemical function in higher animals, it generally is not classified as essential for humans. However, cell, epidemiological, supplementation, and deprivation studies have shown that boron probably should receive dietary guidance attention.

Arthritis and Inflammation

Since 1981 occasional reports have appeared suggesting that boron can ameliorate or prevent arthritis (Nielsen and Meacham, 2011; Nielsen, 2014a). These studies usually involved a few subjects, subjective indicators of amelioration, and/or lack of blinding of subjects. However, two recent studies support the suggestion that boron is beneficial to arthritic subjects through alleviating inflammatory effects. The effect of boron supplementation as calcium fructoborate supplementation on primary knee osteoarthritis in middle-aged patients was evaluated in a double-blind, placebo-controlled pilot study (Scorei et al., 2011). The study was completed by 60 of 72 subjects in which groups of 15 were supplemented with a placebo or boron at 3, 6, or 12 mg/day for 15 days. When all boron-supplemented subjects were grouped together, the inflammatory stress biomarkers serum C-reactive protein, plasma fibrinogen, and erythrocyte sedimentation rate were significantly improved compared with the placebo group. Furthermore, in a cross-sectional study that enrolled 107 rheumatoid arthritis patients and 214 controls matched in age and sex, serum boron concentrations were significantly lower in the arthritic patients and negatively correlated with rheumatoid factor titer (Al-Rawi et al., 2013).

The finding that boron reduces elevated mediators of inflammation in some in vitro and supplementation studies provides further support for the suggestion that boron may be alleviating inflammation in arthritic subjects. Calcium fructoborate treatment of lipopolysaccharide-stimulated murine macrophage RAW 264.7 cells upregulated tumor necrosis factor-α and decreased interleukin-1β and interleukin-6 release (Scorei et al., 2010). In a double-blind experiment involving 3 groups of 28 subjects with blood C-reactive protein concentrations greater than 3.0 mg/L (indicating chronic inflammatory stress), each subject received a supplement of calcium fructoborate containing 1.4 or 2.8 mg of boron per day or a placebo for

30 days (Roogoveanu et al., 2015). The 2.8-mg supplement significantly reduced blood C-reactive protein and interleukin-1β levels. Both boron supplements also significantly reduced interleukin-6 and monocyte chemoattractant protein-1.

Bone

Considerable evidence has appeared that indicates that boron in nutritional amounts beneficially affects trabecular and alveolar bone growth and maintenance in animal models (Nielsen, 2014a). Recent studies with bioactive glasses, which are used for bone tissue engineering and in situ bone tissue regeneration, provide supporting evidence that boron is beneficial for bone formation. Bone formation is enhanced when bioactive glasses are modified to contain boron (Nielsen, 2014a). Some of this enhancement might be the result of increased angiogenesis. The stimulation of human umbilical vein endothelial cell (HUVEC) proliferation and migration by boron-doped bioactive glass has been attributed to the boron in the ionic dissolution products from the glass. The stimulation was associated with phosphorylation of extracellular signal–related kinase (ERK) 1/2, focal adhesion kinase, and the p38 proteins (Haro Durand et al., 2014). In addition, the ionic dissolution products enhanced in vitro HUVEC tubule formation and secretion of interleukin-6 and basic fibroblast growth factor.

Three other cell culture experiments have indicated that boron is beneficial for bone growth and maintenance in humans. Boron supplementation at 1 or 10 ng/mL increased mineralized nodule formation and mineralized tissue-associated mRNA expression of type 1 collagen, osteopontin, bone sialoprotein, osteocalcin, and runt-related transcription factor 2 by cultured osteoblasts (MC3T3-E1; Hakki et al., 2010). In addition, boron supplementation increased bone morphogenetic protein-4, -6, and -7 levels. Boron at 1 and 10 ng/mL increased calcium deposition in cultured human bone marrow stromal cells (Ying et al., 2011). Boron at 10 and 100 ng/mL also increased mRNA expression of alkaline phosphatase, osteocalcin, collagen type 1, and bone morphogenetic protein-7. Boron also was found to increase alkaline phosphatase activity and enhance the expression of osteogenic markers collagen type 1 and osteocalcin in human tooth germ stem cells in vitro (Taşh et al., 2013).

In addition to the cell culture experiments, a supplementation study found that boron could be beneficial for bone maintenance. Six months of providing 226 mg of calcium fructoborate per day incorporated into margarine improved bone density in 66 of 100 patients with osteoporosis (Scorei and Rotaru, 2011). Because the supplement provided only 20 mg/day calcium, the improvement was attributed to the additional 5.65 mg/day boron.

Central Nervous System

Under well-controlled dietary conditions, boron supplementation (3 mg/day) to older men and women after consuming diets providing approximately 0.25 mg/boron per 2000 kcal for approximately 63 days altered electroencephalograms (EEGs) such that there was a shift toward less activity in the low frequencies and more activity in the high, dominant frequencies of the EEG spectrum (Nielsen and Meacham, 2011; Nielsen, 2014a). Increased low-frequency activity is typical of states of reduced behavioral activation and has been associated with reduced performance of psychomotor tasks. Decreased high-frequency activity has been associated with impaired memory performance. The changes in EEGs supported findings that the subjects supplemented with boron after deprivation exhibited improved psychomotor skills of motor speed and dexterity and cognitive processes of attention and short-term memory. These findings have not been confirmed by human experiments elsewhere. However, animal experiments comparing boron deprivation to boron supplementation in nutritional amounts produced results supporting those found in the human experiments (Nielsen and Meacham, 2011; Nielsen, 2014a).

Cancer

One of the most recent suggested beneficial effects of boron is a reduced risk for some types of cancer. This suggested benefit was initiated by an epidemiological study that found an inverse association between dietary boron intake levels and prostate cancer. Since then, several studies have shown that boron inhibits the growth of cultured prostate cells and human prostate adenocarcinoma tumors in nude mice (Nielsen and Meacham, 2011; Nielsen, 2014a). The growth of cultured breast cancer cells also has been shown to be inhibited by boron (Nielsen and Meacham, 2011). In addition to prostate cancer, boron has been inversely associated with cervical and lung cancer. A study of cervical smears from 472 women with a mean boron intake of 8.41 mg/day and 587 with a mean intake of 1.26 mg/day found 15 cases of cytopathological indication of cervical cancer in the boron-low women and none in the boron-high women (Korkmaz et al., 2007). In a study of 763 women with lung cancer and 838 matched healthy controls, boron intake was inversely associated with the incidence of lung cancer (Mahabir et al., 2008). The odds increased substantially if the women were not on hormone replacement therapy.

Other

Numerous animal studies indicate that boron can beneficially affect the function of various hormones, including vitamin D, estrogen, thyroid hormone, insulin, and progesterone (Nielsen and Meacham, 2011; Nielsen, 2014a). However, human findings showing such effects are sparse. In older men and women, boron supplementation (3 mg/day) after 63 days of boron deprivation (0.25 mg/day) increased serum 25-hydroxy-vitamin D concentrations (Nielsen, 1996). In postmenopausal women, the increases in serum 17β-estradiol and plasma copper induced by estrogen therapy were significantly higher when the women consumed 3.25 mg of boron per day instead of 0.25 mg/day (Nielsen, 1994). Boron supplementation (2.5 mg/kg diet) for 90 days decreased serum triiodothyronine in perimenopausal women after consuming a placebo for 90 days (Nielsen and Penland, 1999). An effect on insulin utilization might have been the basis for the observation that boron deprivation induced a modest but significantly increased fasting serum glucose concentration in older men and women fed a low-magnesium, marginal copper diet (Nielsen, 1994). Moreover, hair boron concentrations have been inversely associated with homeostasis assessment model-insulin resistance (Choi et al., 2014).

In addition to affecting markers of immune response and chronic inflammatory stress (described earlier), boron has been shown to modify reactive oxygen species metabolism. Boron supplementation (3.0 mg/day) significantly increased erythrocyte superoxide dismutase concentrations in boron-deprived (0.25 mg/day) men and women (Nielsen, 1996). Low doses of boron promoted antioxidant enzyme activities, including superoxide dismutase and catalase, in human blood cultures, and they were chemoprotective against aflatoxin B1 genetic damage in cultured human lymphocytes (Turkez et al., 2012). Reactive oxygen species and lipid peroxidation are considered to be a major cause for the pathological consequences of aflatoxin B1. Furthermore, calcium fructoborate decreased the intracellular production or amount of superoxide ions in cultured keratinocytes exposed to oxidative stress (Scorei and Rotaru, 2011). Obesity often is associated with oxidative and inflammatory stress. A low boron status might contribute to this stress because a reverse relationship between blood boron concentration and body mass index has been reported (Hasbahceci et al., 2013).

Plausible Mechanisms of Action

Biochemistry

The diverse reported beneficial effects of boron suggest that they are secondary to boron influencing a cell signaling system or the formation and/or activity of an entity that is involved in many biochemical processes. The biochemistry of boron gives some clues about the possible basis for its beneficial bioactivity.

Boron biochemistry is essentially that of boric acid, which is a Lewis acid that accepts an electron pair from a base (H_2O) to form tetracovalent compounds such as $B(OH)_4^-$. At the pH of blood (7.4), the reaction

$$B(OH)_3 + H_2O \leftrightarrow B(OH)_4^- + H^+$$

results in a dilute aqueous solution composed of $B(OH)_3$ (boric acid) and $B(OH)_4^-$ (borate). Because the pK_a of boric acid is 9.25, the abundance of these two species in blood should be 98.4% and 1.6%, respectively. Boric acid forms ester complexes with hydroxyl groups of organic compounds; this preferably occurs when the hydroxyl groups are adjacent and in *cis* orientation. This property results in boron as boric acid forming complexes with several biologically important compounds that are involved in numerous body functions as described next.

Adenosine Compound Bioactivity

Boron forms ester complexes with compounds that contain ribose (which has a *cis*-hydroxyl group), including adenosine (Nielsen and Meacham, 2011; Nielsen, 2014a). S-adenosylmethionine and diadenosine phosphates have higher affinities for boron than any other recognized boron ligands in animal tissues. Thus boron might exert some of its beneficial effects by modifying the utilization or metabolism of these compounds. Diadenosine phosphates are present in all animal cells and function as signal nucleotides involved with neuronal response. S-adenosylmethionine is one of the most frequently used enzyme substrates in the body. Approximately 95% of S-adenosylmethionine is used in methylation reactions, which influence the activity of DNA, RNA, proteins, phospholipids, hormones, and cell transmitters. The methylation reactions result in the formation of S-adenosylhomocysteine, which can be hydrolyzed into homocysteine. In rats, boron deprivation increased plasma homocysteine and decreased liver S-adenosylmethionine (Nielsen and Meacham, 2011; Nielsen, 2014a). In humans with blood C-reactive protein concentrations greater than 3.0 mg/L, boron supplemented at 2.8 mg/day as calcium fructoborate modestly but significantly decreased blood homocysteine (Roogoveanu et al., 2015). High circulating homocysteine and depleted S-adenosylmethionine have been implicated as causative for many pathological conditions (e.g., arthritis, osteoporosis, cancer, diabetes, and impaired brain function) that have been alleviated by nutritional intakes (~2–5 mg/day) of boron.

Cell Membrane Function Modification

Another adenosine compound to which boron strongly binds is oxidized nicotinamide adenine dinucleotide (NAD+). One role of extracellular NAD+ is binding to the plasma membrane receptor CD38, an adenosine diphosphate ribosyl cyclase that converts NAD+ to cyclic ADP ribose. Cyclic ADP ribose is released intracellularly and binds the ryanodine receptor, which induces the release of calcium ions from the endoplasmic reticulum. Cell culture studies show that boron binds to and is a reversible inhibitor of cyclic ADP ribose (Eckhert, 2006). Boron in concentrations that are found in blood decreases Ca^{2+} release from ryanodine receptor endoplasmic reticulum stores. A rat experiment also indicated that boron deprivation may increase, or boron supplementation may decrease, intracellular Ca^{2+} levels. Calcium released from stores after activation with thrombin was decreased in platelets from rats fed low-boron diets supplemented with boron (Nielsen, 1994). Increased cellular calcium is a signal that causes releases of inflammatory neuropeptides, cytokines, prostaglandins, and leukotrienes, which contribute to chronic inflammatory and oxidative stress. Amelioration of this stress through decreasing intracellular calcium is a plausible mechanism of the action for boron beneficially affecting bone formation, immune response, insulin release, and brain function.

Studies with plants have supported the suggestion that boron could affect the transport of Ca^{2+} across cell membranes. These studies indicate that boron may form diester borate complexes with cellular membrane phosphoinositides, glycoproteins, and glycolipids. Such diester borate polyol complexes, which occur in human and animal cellular membranes, have been suggested to act as calcium chelators and/or redox modifiers that affect membrane integrity and thus the transduction of regulatory or signaling ions in and out cells (Wimmer et al., 2009).

Microbiome

Nutrients consumed affect the numbers and types of microorganisms found in the human gastrointestinal tract. These microorganisms generate compounds from food components that might affect the susceptibility to some of the pathological conditions affected by or associated with nutritional intakes of boron. The bacterial quorum-sensing signal molecule autoinducer-2 is a furanosyl borate ester synthesized from S-adenosylmethionine (Chen et al., 2002). Quorum sensing is the cell-to-cell communication between bacteria accomplished through the exchange of extracellular signaling molecules (autoinducers). It was recently found that the amount of autoinducer-2 in the gut can alter the composition of the microbiome. Boosting the levels of this boron-containing autoinducer helped in reestablishing the beneficial bacterial populations in the guts of antibiotic-treated mice (Thompson et al., 2015).

Body Homeostasis and Tissue Distribution

Approximately 85% of ingested boron is absorbed and then efficiently excreted via the urine, where it is found mainly as boric acid (Nielsen, 2012). The Sodium-coupled borate cotransporter (NaBC1) described herein and/or the transport of boric acid through aquaporins might be involved in the transport of boron into the urine. As a result, urinary boron mirrors boron intake. During transport in the body, boron most likely is weakly attached to organic molecules containing *cis*-hydroxyl groups.

A mammalian boron transporter, NaBC1, has been described (Park et al., 2004) that has substantial homology to the boron transporters discovered in *Arabidopsis thaliana* and *Saccharomyces cerevisiae*. NaBC1 is ubiquitously expressed and a unique transporter because it behaves differently in the absence and presence of borate (Park et al., 2004). In the absence of borate, NaBC1 acts as a transporter permeable to Na^+ and H^+. In the presence of borate, NaBC1 acts as a selective, electrogenic Na^+-B$(OH)_4^-$ cotransporter (Park et al., 2004). NaBC1 is essential for boron homeostasis, growth, and proliferation of mammalian HEK293 cells (Park et al., 2004). A boron transporter might have been responsible for the observation that RAW 264.7 and HL60 cells accumulate boron against a concentration gradient (Ralston and Hunt, 2004).

It has been hypothesized that the entry of boron into cells in the forms of boric acid and fructoborate occurs through a different mechanism. Scorei and Scorei (2013) suggested that boric acid enters the cell via passive transport through aquaporins, whereas boron from fructoborate enters as the borate ion through the NaBC1 transporter. Scorei and Scorei (2013) also suggested that the entry of boric acid would increase cellular acidity whereas entry of borate and its conversion to boric acid would increase alkalinity, and that the different pH changes might have an impact on the movement of signaling ions across cell membranes.

Boron is distributed throughout soft tissues at concentrations mostly between 1.39 and 1.85 μmol/kg fresh tissue (Nielsen, 2012). On the basis of studies with postmenopausal women, normal fasting plasma boron concentrations range from 3.14 to 8.79 mmol/L (34–95 ng/mL).

Dietary Considerations

Beneficial Intakes

In human depletion-repletion experiments, participants responded to a 3-mg/day boron supplement after consuming a diet supplying only 0.2–0.4 mg boron/day for 63 days (Nielsen, 1994, 1996). Extrapolations from animal experiments indicate that to achieve optimal benefits of boron, intakes greater than 0.5 mg/day are needed and that boron supplementation is unlikely to elicit a response in individuals consuming at least 1 mg boron/day (World Health Organization, 1996). Thus if an AI level is ever established for the health benefits of boron, it is likely to be between 0.5 and 1.0 mg/day.

Safe Tolerable Upper Intake Level

In the United States and Canada, an RDA or AI has not been set for boron. However, tolerable upper intake levels (ULs) were set; these were 3 mg/day for children 1–3 years of age; 6 mg/day for children 4–8 years of age; 17 mg/day for adolescents 9–18 years of age; and 20 mg/day for adults (Food and Nutrition Board, Institute of Medicine, 2001). The World Health Organization (1996) first suggested that 13 mg/day would be a safe upper intake level but later increased this to 0.4 mg/kg body weight (World Health Organization, International Program on Chemical Safety, 1998) or approximately 28 mg/day for a 70-kg person. The European Food Safety Authority (2004) established a UL for boron based on body weight that results in approximately 10 mg/day. The lack of finding adverse effects because of high boron in drinking water supports the establishment of these relatively high ULs (Nielsen and Meacham, 2011). In Turkey, populations consuming high amounts of boron via drinking water and consequently food do not exhibit any apparent adverse effects. For example, in a population exposed to drinking water up to 29 mg boron/L and to boron mining and production, no adverse effects on health and fertility were found over three generations. In another study, no adverse effects were found in 66 men (mean age of 39 years) residing in a high-boron area for 36 years with a calculated mean boron excretion of 6.77 mg/L. The drinking water from where the men resided had boron concentrations that ranged from 2.05 to 29.00 mg/L, with a mean of 10.2 ± 4.1 mg/L.

Assessment of Dietary Intakes

In the United States a survey conducted between 1994 and 1996 indicated that boron intakes ranged from a low of 0.35 mg/day to a high of 3.25 mg/day for adults. The median intakes for various age groups of adults ranged from 0.87 to 1.13 mg/day (Food and Nutrition Board, Institute of Medicine, 2001). A study involving 43 postmenopausal women in eastern North Dakota found that average urinary excretion of boron (indicator of dietary intake) was less than 0.5 mg/day for 2 women and between 0.5 and 1.0 mg/day for 14 women (Nielsen and Penland, 1999). These findings suggest that a significant number of people could benefit from an increased intake of boron. Foods that provide rich amounts of boron include fruits, leafy vegetables, nuts, and legumes. Beverages based on fruits and grains, such as wine, beer, and cider, also are good sources of boron. Examples of high concentrations of boron found in foods and beverages (mg/kg fresh weight) are avocado (14.3), peanut butter (5.9), prune juice (5.6), chocolate powder (4.3), wine (3.6), grape juice (3.4), and pecans (2.6).

SILICON

Historical Aspects

Since 1972, reports about silicon having beneficial effects, especially on connective tissue and bone formation, in higher animals have appeared. Initial experiments performed in the 1970s used supranutritional amounts of silicon (100 and 500 mg/kg diet) to prevent abnormalities in animal models fed low-silicon diets of questionable nutritional quality based on growth data. Experiments performed since 2000 have indicated that only nutritional amounts of silicon are needed to prevent bone, hexosamine, and collagen abnormalities (similar to, but of less magnitude, than those reported in the 1970s) in animal models fed apparently nutritionally adequate diets low in silicon (Nielsen, 2014b). Except for a few cell culture and epidemiological studies, most reports indicating that nutritional amounts of silicon have beneficial effects in humans have appeared in the 21st century.

Beneficial Actions

Bone and Connective Tissue

Recent epidemiological studies have indicated that nutritional intakes of silicon are beneficial for bone health. Dietary silicon was positively associated with bone mineral density in four hip sites of men and premenopausal women in the Framingham

Offspring Cohort study of 1251 men and 1596 women (Jugdaohsingh, 2007). Large differences of up to 10% were found between the highest (>40 mg/day) and lowest (<14 mg/day) quintiles of silicon intake in this cohort. Also in the Framingham Offspring Cohort study, the beneficial effect of a moderate consumption of beer on hip and spine bone mineral density was associated with the silicon in the beer (Tucker et al., 2009). In the Aberdeen Prospective Osteoporosis Screening Study, silicon intake was positively associated with bone mineral density at the femur neck in late premenopausal women and postmenopausal women on hormone replacement therapy (Macdonald et al., 2012). In this study the lowest quartile of silicon intake was 16 mg/day and the highest quartile was 31.5 mg/day. Increasing quartiles of energy-adjusted silicon intakes were negatively associated with urinary pyridinoline and deoxypyridinoline crosslinks, markers of bone resorption, and positively associated with the serum N-terminal propeptide of type 1 collagen, a marker of bone formation.

A limited number of reports have indicated that silicon supplementation can beneficially affect bone and connective tissue in humans. Two studies performed before 2000 indicated that silicon increased bone mineral density in women with postmenopausal osteoporosis (Jugdaohsingh, 2007). A dose of 50 mg of silicon as monomethyl trisilanol administered intramuscularly twice a week for 4 months to 8 women significantly improved femoral bone density by a mean of 4.7%. In the other study, osteoporotic women consuming their normal diets were parenterally administered 16.5 mg silicon/week for 4 months ($n = 16$) or orally supplemented with 16.5 mg silicon/week for 3 months ($n = 16$) and compared with controls not given treatment. The two groups receiving supplemental silicon exhibited increased trabecular bone volume in iliac crest biopsies. Furthermore, a 28-mg/day silicon supplement for 12 weeks was found to increase spine bone density by 2.5% in 6 women with low bone mass (Jugdaohsingh, 2007). A double-blind, placebo-controlled study was performed in which 136 women with low bone mass supplemented daily with 1 g calcium and 20 µg vitamin D were also supplemented daily with 3, 6, or 12 mg silicon as choline-stabilized orthosilicic acid or a placebo for 12 months (Spector et al., 2008). The 6- and 12-mg supplements significantly increased the bone formation marker of type 1 collagen at 12 months. Another study supplementing silicon as choline-stabilized orthosilicic acid at a dose of 10 mg/day for 12 months improved photodamaged skin surface and mechanical properties and decreased hair and nail brittleness (Barel et al., 2007). In contrast to these positive reports, consuming 1 L of artesian water containing 86 mg/L or purified water daily for 12 weeks to 9 and 8 women with low bone mass, respectively, did not yield any significant differences in bone turnover markers procollagen type 1, N-terminal propeptide, bone-specific alkaline phosphatase, and osteocalcin within or between groups (Li et al., 2010).

In vitro studies also indicate silicon is beneficial for bone formation. Orthosilicic acid in physiological concentrations was found to stimulate collagen type 1 synthesis in human osteoblast-like cells and enhance osteoblastic differentiation (Jugdaohsingh, 2007). Silicon-substituted hydroxyapatites and Bioglass implants have been shown to bond better to bone than their nonsilicon-containing counterparts (Jugdaohsingh, 2007). Dissolution of the implants yielding silicon has been shown to upregulate genes encoding bone formation proteins, stimulate osteoblast proliferation and differentiation, and increase type 1 collagen and apatite formation (Jugdaohsingh, 2007).

Mental Health

To date, only association studies between silicon in drinking water and mental health have been reported. In 1996 it was reported that an association between cognitive impairment and aluminum in drinking water depended upon the silica concentration in the water (Gillette-Guyonnet et al., 2007). High levels of aluminum were associated with deleterious effects on cognitive function when the silica concentration was low, but not when the silica concentration was high. In 2000 it was reported that the 8-year follow-up of subjects in the Paquid cohort found that exposure to drinking water with silica concentrations 11.25 mg/L or greater was associated with a reduced risk for developing Alzheimer's disease (Gillette-Guyonnet et al., 2007). Higher silica in drinking water was also associated with a reduced risk of developing Alzheimer's disease in the 7-year follow-up of 1462 women (aged 75 years or greater) participating in the Epidemiology of Osteoporosis Study (Gillette-Guyonnet et al., 2007).

Immune and Inflammatory Response

Experiments with animal models indicate that silicon may modulate immune and inflammatory responses (Nielsen, 2014b). Human studies that suggest such a possibility have been limited to experiments with cell cultures. Silicon was suggested to have a regulatory role in the cell cycle of lymphocytes because monomethyl silanetriol at an optimal concentration of 10 mg/L silicon in the culture media stimulated peripheral lymphocyte proliferation and decreased lymphoblast proliferation (Henrotte et al., 1988). Sodium metasilicate at concentrations of 1, 5, 25, and 50 µM suppressed lipopolysaccharide-induced expression of mRNA encoding tumor necrosis factor-α, inducible nitric oxide synthase, and cyclooxygenase-2 in RAW 264.7 murine macrophage cells (Kim et al., 2013). These findings resulted in the suggestion that silicon has the ability to suppress the production of inflammatory cytokines and mediators.

Plausible Mechanisms of Action

Silicon easily forms stable complexes with polyols that have at least four hydroxyl groups (Nielsen, 2014b). Such polyols include hexosamine and ascorbate used to form glycosaminoglycans, mucopolysaccharides, and collagen, which are involved in connective tissue formation and stabilization and bone formation. Thus silicon may be having its beneficial effects through the formation and utilization of these polyols. Support for this plausible mechanism of action includes the finding of significant silicon concentrations in connective tissue (Carlisle, 1997). In bone, silicon is highly concentrated in immature osteoid but declines as calcium content increases in mature bone, which suggests that silicon is involved in the initiation of mineralization (Carlisle, 1997).

The finding that the beneficial effect of silicon on bone is more pronounced in premenopausal women and postmenopausal women on hormone therapy suggests that estrogen enhances the beneficial effect of silicon on bone health (Macdonald et al., 2012). The basis for this potential enhancement is unclear.

In plants, silicon apparently binds hydroxyl groups of proteins involved in signal transduction (Nielsen, 2014b). A similar action in higher animals and humans might be a plausible explanation for associating silicon with mental function, inflammatory and immune responses, and gene expression of factors involved in osteoblastogenesis and osteoclastogenesis.

Silicon may be beneficial through altering the absorption and utilization of other mineral elements involved in bone metabolism, immune or inflammatory response, or cognitive function. Supranutritional amounts of silicon have been found to facilitate the absorption, retention, and/or utilization of copper and magnesium in animal models (Nielsen, 2006, 2014b). The apparent beneficial effect of silicon on cognitive function has been postulated to occur because a reaction between silicon as silicic acid and aluminum compounds such aluminum hydroxide forms aluminosilicate (Gillette-Guyonnet et al., 2007). Formation of aluminosilicate prevents aluminum absorption and retention; thus it prevents chronic aluminum accumulation that may cause or enhance neurodegeneration in the brain (Gillette-Guyonnet et al., 2007). It also has been proposed that the formation of aluminosilicates may prevent aluminum from competing for iron-binding sites that result in decreased functions related to iron (Nielsen, 2006). These functions include the prolyl hydroxylase activity that has a role in collagen formation.

Body Homeostasis and Tissue Distribution

Most of the recent human studies related to silicon have involved its dietary intake, absorption, transport, retention, and excretion. These studies indicate that homeostatic controls are in place with regards to overall silicon metabolism (Jugdaohsingh, 2007; Nielsen, 2014b). Silicon is relatively well absorbed from foods and drinks. One study found that an average of 41% of the silicon in food was excreted in urine, which is an adequate indicator of absorption (Jugdaohsingh, 2007). The determination of silicon absorption from various foods, drinks, and supplements found that it was highest from monomethyl silanetriol and beer (64% of dose), followed by green beans (44%), orthosilicic acid solution (43%), choline-stabilized orthosilicic acid (17%), and bananas (4%); (Sripanyakorn et al., 2009). An in vitro method found that the availability of silicon from various foods and drinks ranged from 0.6% from beans to 100% from beer (Robberecht et al., 2009). These findings indicate that the absorption of silicon is strongly influenced by the solubility of the silicon compounds consumed or compounds formed after consumption.

Some dietary factors also may affect the absorption of silicon (Jugdaohsingh, 2007). A high-fiber diet was reported to decrease the gastrointestinal uptake and balance of silicon in humans. Silicon supplementation is more effective in producing beneficial effects when dietary calcium is low, which suggests that calcium is antagonistic to silicon absorption, possibly by forming insoluble calcium silicate. Magnesium also might reduce the bioavailability of silicon by forming insoluble silicates.

The absorption mechanism has not been clearly defined. It has been suggested that simple uncharged silicon species such as orthosilicic acid are readily and rapidly absorbed predominantly in the proximal small intestine through a paracellular or small-pore transcellular pathway (Jugdaohsingh, 2007). Charged polymeric silica species are less mobile and/or too large to easily cross the gastrointestinal tract. Thus silicon in polymeric and colloidal forms that are not readily broken down in the gastrointestinal tract will be excreted in the feces (Jugdaohsingh, 2007). High dietary calcium and magnesium, through the formation of insoluble silicates and high fiber through binding silicon to make it unavailable, may decrease the absorption of silicon (Jugdaohsingh, 2007).

Silicon is not protein bound in plasma, where it is believed to exist mainly as a neutral orthosilicic acid species that readily diffuses into erythrocytes and other tissues (Jugdaohsingh, 2007). Evidence that silicon entering the bloodstream is efficiently transferred to tissues and urine is that the silicon concentration in blood remains relatively constant over a range of dietary intakes. Most reported human serum concentrations are between 10 and 31 µg/dL (Nielsen, 2009). Connective tissues, including aorta, bone, skin, tendon, trachea, and fingernails, contain much of the silicon that is retained in the body (Carlisle, 1997).

Absorbed silicon is rapidly excreted in the urine (Jugdaohsingh, 2007), where it likely exists as orthosilicic acid and/ or magnesium orthosilicate (Nielsen, 2006). Silicon is readily filtered by the renal glomerulus and eliminated with little tubular reabsorption (Jugdaohsingh, 2007). The upper limits of urinary excretion apparently are set by the rate and extent of silicon absorption and not by the excretory ability of the kidney because peritoneal injection of silicon can elevate urinary excretion above the upper limit achieved by dietary intake (Nielsen, 2006, 2009). This indicates that silicon homeostasis is controlled by absorption and excretory mechanisms.

Dietary Considerations

Beneficial Intakes

The Food and Nutrition Board, Institute of Medicine (2001) judged that animal and human data were too limited to justify establishing DRIs for silicon. In the Framingham Offspring Cohort study, the highest bone mineral density found was in the highest silicon quintile with intakes ranging from 30.2 to 63.2 mg/day and 34.4–118.0 mg/day compared with the lowest quintile with intakes ranging from 7.1 to 16.7 and 7.6–18.8 mg/day for premenopausal women and men, respectively (Jugdaohsingh et al., 2004). Thus, on the basis of these findings, extrapolations from animal data (Nielsen, 2006), and inconclusive balance data from humans (Nielsen, 2006), a beneficial intake for silicon may be near 25 mg/day.

Safe Upper Tolerable Intake Level

Long-term use of silicate-containing drugs and chronic hemodialysis may cause silicon accumulation in the kidney that induces nephropathy (Jugdaohsingh, 2007). However, there are no reports of silicon toxicity through excessive dietary intake for people with normal renal function. Thus it is not possible to suggest a reasonable upper level of tolerable intake for silicon.

Assessment of Dietary Intake

The dietary intake of silicon is between 15 and 50 mg/day for most Western populations (Nielsen, 2006). For example, in the Framingham Offspring Cohort study, most silicon intakes were within a relatively narrow range of 23.6 ± 8.9 mg/day for women and 27.5 ± 10.7 mg/day for men (Jugdaohsingh et al., 2004). A duplicate meal study in Belgium (Robberecht et al., 2009) and a food frequency study in the United Kingdom (McNaughton et al., 2005) both found a mean daily silicon intake of 18.6 mg. Higher intakes (140–240 mg/day) have been reported for China and India, where plant-based foods are a major part of the diet (Jugdaohsingh, 2007). If an optimal intake of silicon is near 25 mg/day, then intake findings suggest that an increased consumption of silicon-rich foods would be beneficial for a significant number of individuals in Western countries.

A provisional food database indicated that the highest concentrations of silicon are found in cereal and cereal products, especially less refined cereals and oat-based products (Powell et al., 2005). The silicon in barley and hops is solubilized during the beer-making process, which makes this beverage a rich source of silicon. Fruit and vegetables were found to be variable sources of silicon (Powell et al., 2005). Substantial amounts were found in Kenyan beans, French beans, runner beans, spinach, dried fruit, bananas, and red lentils. Undetectable amounts of silicon were found in tomatoes, oranges, and onions. Drinking water can be a significant source of water. The silicon content in lowland Britain and European mineral waters were found to contain 2.8–16 mg/L; higher concentrations (30–40 mg/L) were found in Malaysian Spritzer and Fiji mineral waters (Jugdaohsingh, 2007).

It should be noted that silicon is added to processed food in various forms for anticaking, thickening, and stabilizing purposes (Jugdaohsingh, 2007). These forms include calcium silicate, sodium aluminosilicate, magnesium hydrogen metasilicate (talc), magnesium trisilicate, calcium aluminum silicate, bentonite, and kaolin, which are considered to be inert and not readily absorbed from the gastrointestinal tract (Jugdaohsingh, 2007).

CONCLUSION

Substantial evidence exists indicating that boron and silicon in nutritional amounts and by involving plausible mechanisms of action have beneficial effects in humans. An intake of boron of 1 mg/day or more has been found to be beneficial for bone growth and maintenance, central nervous system function, and hormone action. Silicon intakes greater than 15 mg/day have been associated with bone health. Consideration should be given for providing dietary guidance for these trace minerals. An appropriate dietary recommendation would be the consumption of foods that result in intakes of boron and silicon that have been found to be beneficial for health and well-being.

REFERENCES

Al-Rawi, Z.S., Gorial, F.I., Al-Shammary, W.A., Muhsin, F., Al-Naaimi, A.S., Kareem, S., 2013. Serum boron concentration in rheumatoid arthritis: correlation with disease activity, functional class, and rheumatoid factor. J. Exp. Integr. Med. 3, 9–15.

Barel, A., Calomme, M., Timchenko, A., Paepe, K.D., Demeester, N., Rogiers, V., Clarys, P., Vanden Berghe, D., 2007. Effect of oral intake of choline-stabilized orthosilicic acid on skin, nails, and hair in women with photodamaged skin. Arch. Dermatol. Res. 297, 147–153.

Carlisle, E.M., 1997. Silicon. In: O'Dell, B.L., Sunde, R.A. (Eds.), Handbook of Nutritionally Essential Minerals. Marcel Dekker, New York, pp. 603–618.

Chen, X., Schauder, S., Potier, N., van Dorsselaer, A., Pelczer, I., Bassier, B.L., Hughson, F.M., 2002. Structural identification of a bacterial quorum-sensing signal containing boron. Nature 415, 545–549.

Choi, W.-C., Kim, S.-H., Chung, J.-H., 2014. Relationships of hair mineral concentrations with insulin resistance in metabolic syndrome. Biol. Trace Elem. Res. 158, 323–329.

Eckhert, C.D., 2006. Other trace elements. In: Shils, M.E., Shike, M., Ross, A.C., Caballero, B. (Eds.), Modern Nutrition in Health and Disease, tenth ed. Lippincott Williams & Wilkins, Philiadelphia, pp. 338–350.

European Food Safety Authority, 2004. Opinion of the scientific panel on dietetic products, nutrition, and allergies on a request from the commission related to the tolerable upper intake level of boron (sodium borate and boric acid). Eur. Food Saf. Auth. J. 237, 1–8.

Food and Nutrition Board, Institute of Medicine, 1994. How Should the Recommended Dietary Allowances Be Revised? National Academies Press, Washington, DC.

Food and Nutrition Board, Institute of Medicine, 2001. Dietary Reference Intakes for Vitamin A, Vitamin K, Arsenic, Boron, Chromium, Copper, Iodine, Iron, Manganese, Molybdenum, Nickel, Silicon, Vanadium, and Zinc. National Academies Press, Washington, DC.

Food and Nutrition Board, Institute of Medicine, 2005. Dietary Reference Intakes for Water, Potassium, Sodium, Chloride, and Sulfate. National Academies Press, Washington, DC.

Gillette-Guyonnet, S., Andrieu, S., Vellas, B., 2007. The potential influence of silica present in drinking water on Alzheimer's disease and associated disorders. J. Nutr. Health Aging 11, 119–124.

Hakki, S.S., Bozkurt, B.S., Hakki, E.E., 2010. Boron regulates mineralized tissue-associated proteins in osteoblasts (MC3T3-E1). J. Trace Elem. Med. Biol. 24, 243–250.

Hasbahceci, M., Cipe, G., Kadioglu, H., Aysan, E., Muslumanoglu, M., 2013. Reverse relationship between blood boron level and body mass index in humans: does it matter for obesity? Biol. Trace Elem. Res. 153, 141–144.

Haro Durand, L.A.H., Góngora, A., López, J.M.P., Boccaccini, A.R., Zago, M.P., Baldi, A., Gorustovich, A., 2014. In vitro endothelial cell response to ionic dissolution products from boron-doped bioactive glass in the SiO_2-CaO-P_2O_5-Na_2O system. J. Mater. Chem. B 2, 7620–7630.

Henrotte, J.-G., Viza, D., Vich, J.M., Gueyne, J., 1988. Le role régulateur du silicium dans la division cellulaire. Comptes Rendes l'Académie Sci. Paris 306, 525–528 Série III.

Jugdaohsingh, R., 2007. Silicon and bone health. J. Nutr. Health Aging 11, 99–110.

Jugdaohsingh, R., Tucker, K.L., Qiao, N., Cuppoes, L.A., Kiel, D.P., Powell, J.J., 2004. Dietary silicon intake is positively associated with bone mineral density in men and premenopausal women of the Framingham Offspring Cohort. J. Bone Miner. Res. 19, 297–307.

Kim, E.-J., Bu, S.-Y., Sung, M.-K., Kang, M.-H., Choi, M.-K., 2013. Analysis of antioxidant and anti-inflammatory activity of silicon in murine marcrophages. Biol. Trace Elem. Res. 156, 329–337.

Korkmaz, M., Uzgören, E., Bakirdere, S., Aydin, F., Yavuz Ataman, O., 2007. Effects of dietary boron on cervical cytopathology and on micronucleus frequency in exfoliated buccal cells. Environ. Toxicol. 22, 17–25.

Li, Z., Karp, H., Zerlin, A., Lee, W.Y.A., Carpenter, C., Heber, D., 2010. Absorption of silicon from artesian aquifer water and its impact on bone health in postmenopausal women: a 12 week pilot study. Nutr. J. 9, 44.

Macdonald, H.M., Hardcastle, A.C., Jugdaohsingh, R., Fraser, W.D., Reid, D.M., Posell, J.J., 2012. Dietary silicon interacts with oestrogen to influence bone health: evidence from the Aberdeen prospective osteoporosis screening study. Bone 50, 681–687.

Mahabir, S., Spitz, M.R., Barrera, S.L., Dong, Q., Eastham, C., Forman, M.R., 2008. Dietary boron and hormone replacement therapy as risk factors for lung cancer in women. Am. J. Epidemiol. 167, 1070–1080.

McNaughton, S.A., Bolton-Smith, C., Mishra, G.D., Jugdaohsingh, R., Powell, J.J., 2005. Dietary silicon intake in post-menopausal women. Br. J. Nutr. 94, 813–817.

Nielsen, F.H., 1994. Biochemical and physiologic consequences of boron deprivation in humans. Environ. Health Perspect. 102 (Suppl. 7), 59–63.

Nielsen, F.H., 1996. Evidence for the nutritional essentiality of boron. J. Trace Elem. Exp. Med. 9, 215–229.

Nielsen, F.H., 2006. Boron, manganese, molybdenum, and other trace elements. In: Bowman, B.A., Russell, R.M. (Eds.), Present Knowledge in Nutrition, vol. 1. ninth ed. ILSI Press, Washington, DC, pp. 506–526.

Nielsen, F.H., 2009. Micronutrients in parenteral nutrition: boron, silicon, and fluoride. Gastroenterology 137, S55–S60.

Nielsen, F.H., 2012. Manganese, molybdenum, boron, chromium, and other trace elements. In: Erdman Jr., J.W., Macdonald, I.A., Zeisel, S.H. (Eds.), Present Knowledge in Nutrition, tenth ed. Wiley-Blackwell, Oxford, pp. 586–607.

Nielsen, F.H., 2014a. Update on human health effects of boron. J. Trace Elem. Med. Biol. 28, 383–387.

Nielsen, F.H., 2014b. Update on the possible nutritional importance of silicon. J. Trace Elem. Med. Biol. 28, 379–382.

Nielsen, F.H., Meacham, S.L., 2011. Growing evidence for human health benefits of boron. J. Evid. Based Complement. Altern. Med. 16, 169–180.

Nielsen, F.H., Penland, J.G., 1999. Boron supplementation of peri-menopausal women affects boron metabolism and indices associated with macromineral metabolism, hormonal status, and immune function. J. Trace Elem. Exp. Med. 12, 251–261.

Park, M., Li, Q., Shcheynikov, N., Zeng, W., Muallem, S., 2004. NaBC1 is a ubiquitous electrogenic Na^+-coupled borate transporter essential for cellular boron homeostasis and cell growth and proliferation. Mol. Cell 16, 331–341.

Powell, J.J., McNaughton, S.A., Jugdaohsingh, R., Anderson, S.H.C., Dear, J., Khot, F., Mowatt, L., Gleason, K.L., Sykes, M., Thompson, R.P.H., Bolton-Smith, C., Hodson, M.J., 2005. A provisional database for the silicon content of foods in the United Kingdom. Br. J. Nutr. 94, 804–812.

Ralston, N.V.C., Hunt, C.D., 2004. Transmembrane patitioning of boron and other elements in RAW 264.7 and HL60 cell cultures. Biol. Trace Elem. Res. 98, 181–191.

Robberecht, H., Van Cauwenbergh, R., Van Vlaslaer, V., Hermans, N., 2009. Dietary silicon intake in Belgium: sources, availability from foods, and human serum levels. Sci. Total Environ. 407, 4777–4782.

Roogoveanu, O.-C., Mogoşanu, G.D., Bejenaru, C., Bejenaru, L.E., Croitoru, O., Neamţu, J., Pietrzkowski, Z., Reyes-Izquierdo, T., Biţă, Scorei, I.D., Scorei, R.I., 2015. Effects of calcium fructoborate on levels of C-reactive protein, total cholesterol, low-density lipoprotein, triglycerides, IL-1β, IL-6, and MCP-1: a double-blind, placebo-controlled clinical study. Biol. Trace Elem. Res. 163, 124–131.

Scorei, R.I., Rotaru, P., 2011. Calcium fructoborate – potential anti-inflammatory agent. Biol. Trace Elem. Res. 143, 1223–1238.

Scorei, I.D., Scorei, R.I., 2013. Calcium fructoborate helps control inflammation associated with diminished bone health. Biol. Trace Elem. Res. 155, 315–321.

Scorei, R.I., Ciofrangeanu, C., Ion, R., Cimpean, A., Galateanu, B., Mitran, V., Iordachescu, D., 2010. In vitro effects of calcium fructoborate upon production of inflammatory mediators by LPS-stimulated RAW 264.7 macrophages. Biol. Trace Elem. Res. 135, 334–344.

Scorei, R., Mitrut, P., Petrisor, I., Scorei, I., 2011. A double-blind, placebo-controlled pilot study to evaluate the effect of calcium fructoborate on systemic inflammation and dyslipidemia markers for middle-aged people with primary osteoarthritis. Biol. Trace Elem. Res. 144, 253–263.

Spector, T.D., Calomme, M.R., Anderson, S.H., Clement, G., Bevan, L., Demeester, N., Swaminathan, R., Jugdaohsingh, R., Berghe, D.A.V., Powell, J.J., 2008. Choline-stabilized orthosilicic acid supplementation as an adjunct to calcium/vitamin D3 stimulates markers of bone formation in osteopenic females: a randomized, placebo-controlled study. BMC Musculoskelet. Disord. 9;85.

Sripanyakorn, S., Jugdaohsingh, R., Dissayabutr, W., Anderson, S.H.C., Thompson, R.P.H., Powell J.J., 2009. The comparative absorption of silicon from different foods and food supplements. Brit. J. Nutr 102, 825–834.

Taşh, P.N., Doğan, A., Demirci, S.M., Şahin, F., 2013. Boron enhances odontogenic and osteogenic differentiation of human tooth germ stem cells (hTGSCs) in vitro. Biol. Trace Elem. Res. 153, 419–427.

Thompson, J.A., Oliveira, R.A.m, Djukovic, A., Ubeda, C., Xavier, K.B., 2015. Manipulation of the quorum-sensing signal AI-2 affects the antibiotic-treated gut microbiota. Cell Rep. 10, 1861–1871.

Tucker, K.L., Jugdaohsingh, R., Powell, J.J., Qiao, N., Hannan, M.T., Sripanyakorn, S., Cupples, L.A., Kiel, D.P., 2009. Effects of beer, wine, and liquor intakes on bone mineral density in older men and women. Am. J. Clin. Nutr. 89, 1188–1196.

Turkez, H., Geyikoğlu, F., Dirican, E., Tatar, A., 2012. In vitro studies on chemoprotective effect of borax against aflatoxin B1-induced genetic damage in human lymphocytes. Cytotechnology 64, 607–612.

Wimmer, M.A., Lochnit, G., Bassil, E., Mühling, K.H., Goldbach, H.E., 2009. Membrane-associated, boron-interacting proteins isolated by boronate affinity chromatography. Plant Cell Physiol. 50, 1292–1304.

World Health Organization, 1996. Boron. In: Trace Elements in Human Nutrition and Health. World Health Orgainzation, Geneva, pp. 175–179.

World Health Organization, International Programme on Chemical Safety, 1998. Environmental Health Criteria 204 Boron. World Health Organization, Geneva.

Ying, X., Cheng, S., Wang, W., Lin, Z., Chen, Q., Zhang, W., Kou, D., Shen, Y., Cheng, X., Rompis, F.N., Peng, L., Lu, C.Z., 2011. Effect of boron on osteogenic differentiation of human bone marrow stromal cells. Biol. Trace Elem. Res. 144, 306–315.

Chapter 44

Fluoride: Intake and Metabolism, Therapeutic and Toxicological Consequences

Fatemeh Vida Zohoori[1], Ralph Marsland Duckworth[1,2]

[1]Teesside University, Middlesbrough, United Kingdom; [2]Newcastle University, Newcastle-upon-Tyne, United Kingdom

CHEMICAL ORIGIN AND PROPERTIES

Fluorine and fluoride are generic terms that are frequently used interchangeably in the literature. Fluorine, the first element of the halogen group (group 17) in the periodic table, has atomic number 9 and a relative atomic mass of 19. It is a pale, yellow-green gas at room temperature and is one of the few elements that can form diatomic molecules.

Being extremely electronegative, fluorine scarcely appears in its free elemental form in nature, appearing rather as fluoride compounds. The fluoride ion forms compounds with all elements apart from argon, helium, and neon (Banks et al., 1994), and because it has the same charge and nearly the same radius as the hydroxyl ion, they may replace each other in mineral structures (Hem, 1985). Fluorine is the 13th most copious element in the environment and represents approximately 0.06–0.07% by mass of the Earth's crust (Jaccaud et al., 2000). More than 50 fluoride-containing minerals have been identified, such as fluorspar, cryolite, fluorapatite, and mica (Murray, 1986). However, only three are commercially important: cryolite used for the production of aluminum; rock phosphates used for the production of fertilizers; and apatite, the mineral of bones and teeth (Murray, 1986). The availability of free fluoride ions in the soil is governed by various factors, including the acidity of the soil, the presence of other minerals or chemical compounds, and the natural solubility of the fluoride compound in question (World Health Organization, 1994).

SOURCES OF FLUORIDE EXPOSURE

Air

Airborne fluoride can be in particulate or gaseous forms, which may be moved by wind over large distances. The natural sources of fluoride in air are dusts of fluoride-containing soils, gases emitted in areas of volcanic activity, and volcanic dust. However, most airborne fluoride is generated through human activity. The most prevalent form of atmospheric fluoride is (highly toxic) hydrogen fluoride, which is rapidly absorbed from the lungs (Whitford, 1996). In general, the concentration of fluoride in air is very low ($0.05-1.90\,\mu g/m^3$) in nonindustrial areas (Murray, 1986). However, the amount of fluoride in air is increased with increasing urbanization and industrialization. The aluminum industry is a major source of fluoride contamination of the air, whilst steel production plants, superphosphate plants, and ceramic factories also emit F.

For an individual who lives in a nonindustrial area, the contribution of atmospheric fluoride to total fluoride exposure is small (0.01 mg/day; Hodge and Smith, 1977). However, high levels of fluoride exposure by the inhalation route have been reported in polluted areas such as some provinces of China, where the fluoride concentration in indoor air ranged from 16 to $46\,\mu g/m^3$ because of the burning of domestic high-fluoride coal for cooking.

Water

Because fluorides are universally present in the Earth's crust, almost all water contains fluoride at various concentrations. The level of fluoride in sea water is usually 0.8–1.4 mg/L, whereas the fluoride content of water obtained from lakes, rivers, or artesian wells is, for the most part, below 0.5 mg/L (World Health Organization, 1994).

Molecular, Genetic, and Nutritional Aspects of Major and Trace Minerals. http://dx.doi.org/10.1016/B978-0-12-802168-2.00044-0

Waters at the foot of volcanic mountains and in areas with geological deposits of marine origin usually contain high concentrations of fluoride. One such geological belt can be found stretching from Turkey through Iraq, Iran and Afghanistan to India, Northern Thailand and China (World Health Organization, 1994). Similar areas can be found in the southern parts of the United States, Southern Europe, and Japan. The most distinct area linked with volcanic activity trails the East African Rift system from the Jordan valley down through Sudan, Ethiopia, Uganda, Kenya, and the United Republic of Tanzania. The fluoride concentration of many of the Rift Valley lakes is extremely high, with up to 2800 mg/L in the Kenyan Lake Nakuru (Nair et al., 1984).

The concentration of fluoride in ground waters may be affected by several factors, such as availability and solubility of fluoride-containing minerals, porosity of the rocks or soils through which the water passes, residence time, temperature, pH, and the presence of other elements that may complex with fluoride. Ground waters from calcium-deficient aquifers may contain higher fluoride concentrations because the absence of calcium in solution allows higher stability of the fluoride ion (Edmunds and Smedley, 1996).

Foods and Beverages

Table 44.1 lists the fluoride content of various foods and beverages. The fluoride contents of unprocessed foods and beverages are usually low except for tea. The tea plant (*Camellia sinensis*) is the only plant known to take up fluoride in considerable amounts from soil and accumulate it in its leaves. Therefore tea, after water, could potentially be the second most important source of dietary fluoride intake, particularly in areas where tea drinking is common. The level of fluoride in tea is highly variable and depends on the plant's botanical variety, the area in which it is grown, and the procedure by which the tea infusion is prepared. The fluoride content of tea leaves ranges from 26 to 820 mg/kg whilst that of tea infusions ranges from 0.29 to 8.85 mg/L (Chan et al., 2013; Koblar et al., 2012).

In general, the levels of fluoride in milk are relatively low—0.02 mg/L in human breast milk (Koparal et al., 2000; Sener et al., 2007) and 0.02–0.05 mg/L in cow's milk (Koparal et al., 2000). However, the fluoride content of soy milks is often higher, up to 0.96 mg/L (Lal et al., 2014). The fluoride content of powdered infant milk formulas ranges from 0.01 to 3.71 mg/kg. Such formulas prepared with nonfluoride water have F concentrations ranging from 0.02 to 0.53 mg/L whereas infant milk formulas made with fluoridated water (typically containing 1 mg F/L) have a level of fluoride ranging from 0.49 to 1.53 mg/L (Nohno et al., 2011; Silva and Reynolds, 1996; Zohoori et al., 2012b). Fluoride levels of ready-to-feed infant milks are significantly lower, up to 0.030 mg/L (Maguire et al., 2012).

Fresh fruit juices and carbonated soft drinks are low in fluoride, with contents ranging from 0.009 to 0.931 mg/L (Nanda, 1972; Pang et al., 1992; Taves, 1983; U.S. Department of Agriculture, 2005; Zohoori and Maguire, 2015, 2016). However, fluoride levels in reconstituted fruit juices and other drinks are based on the fluoride contents of the water used for processing.

Raw vegetables and fruits are insignificant sources of fluoride, even when grown in fluoride-rich soils. However, the concentration of fluoride is higher in plants grown in acidic soil than in those grown in alkaline soil. In comparison with other vegetables, leafy vegetables such as lettuce and celery contain more fluoride, up to 0.13 mg/kg (U.S. Department of Agriculture, 2005; Zohoori and Maguire, 2015). The fluoride content of meat is low, although fish products generally contain more fluoride than other types of meat.

The fluoride content of processed foods may be influenced by many factors. For example, the fluoride content of boiled rice when cooked in fluoridated water is approximately 11% higher than that of unboiled rice. However, there is not a proportional increase in the fluoride level of prepared food with increasing fluoride concentration of the water used for food preparation. For example, in Hungary, Schamschula et al. (1988b) found that despite an increased water fluoride concentration of approximately 21 times for boiling rice, the fluoride content of cooked rice increased just 5 times on average. The material and composition of the cooking vessel may also influence the fluoride content of cooked food. Using aluminum pots, the fluoride content of food is diminished because of formation of aluminum fluoride, but in steel and Pyrex no changes have been reported (Full and Parkins, 1975). The method of preparing food may also alter the amount of fluoride in the food. Polished rice contains 50 times more fluoride than unpolished rice (Oelschlager, 1970). Fish and shellfish free from bones and skin contain less than 1 mg/kg fluoride but more than 10 mg/kg when these components are present (Ekstrand et al., 1988; Walters et al., 1983).

In Western countries the contribution of drinking water to the total dietary fluoride intake of children ranges from 4% in nonfluoridated areas to 54% in communities with optimally fluoridated water (Ophaug et al., 1985; Schamschula et al., 1988a; Zohoori et al., 2006); the contribution of tea is negligible (Zohoori et al., 2006). The reported percentage of the dietary fluoride intake contributed by "meat, fish and poultry" ranged from 1% in a group of British children (Zohoori et al., 2006) to 17% in US children (Ophaug et al., 1985).

TABLE 44.1 Fluoride Content (µg/100 g) of Various Foods and Beverages

	United Kingdom[a]	United States[b]
Beverages and Drinks		
Freshly squeezed juice	2.30–17.20	
Fruit juice, ready-to-drink	4.80–10.50	3–287
Carbonated, cola	0.09	0–121
Fruit juices, reconstituted with deionized water	5.00–6.00	
Tea, black, infused	193	159–533
Coffee	83	0–134
Fruit and Nuts		
Raw fruits	0.77–6.80	1–27
Dried fruits	19.00	–
Nuts	9.20	–
Fish		
Cod	358.40	–
Mackerel	87.70	–
Salmon	7.90	–
Sardine in can	1054.20	–
Herring in can	319.00	–
Crab, canned	–	210
Shrimp, canned	–	201
Tuna, canned in oil	–	31
Meat		
Beef	5.80	4–72
Chicken	2.50	4–25
Lamb	23.35	6–36
Pork	3.50	4–60
Turkey	3.70	21
Milk and Milk Products		
Milk, cow	0.80	2–5
Milk, soy	32.7	–
Butter	0.05	1–4
Processed cheese, smoked	58.20	–
Miscellaneous Foods		
Bread	3.80–56.10	28–67
Eggs	0.57–0.88	2–12
Honey	14.65	7
Sugar, brown	23.93	–

Continued

TABLE 44.1 Fluoride Content (µg/100 g) of Various Foods and Beverages—cont'd

	United Kingdom[a]	United States[b]
Sugar, white	1.20	–
Breakfast cereals	3.30–75.30	4–201
Vegetables		
Carrots	0.95	2–6
Celery	1.40	
Cucumber	1.00	1–2
Lettuce	5.45	0–13
Baby Food/Drink		
Baby cereals, ready-to-eat	3.00–73.10	1–31
Baby fruit juices, ready-to-drink	5.00–8.30	4–133
Infant milk formula (cow-based), powdered	1.04–25.20	–
Infant milk formula (soy-based), powdered	6.80	–
Infant milk formula, ready-to-drink	0.90–2.70	–
Milk-based dessert, ready-to-eat	3.80–24.50	4

[a]Zohoori and Maguire (2015).
[b]US Department of Agriculture (2005).

Dental Care Products

Dentifrices and other dental-care products can be a source of fluoride ingestion. The contribution of toothpaste ingestion to total daily fluoride intake could be up to 87% in children younger than 6 years of age (Zohoori et al., 2013a). This is mainly because young children are not in full control of their swallowing reflex (MCHB Expert Panel, 2007). Toothpastes usually contain 1000–1450 mg/kg of fluoride; therefore 1 g of toothpaste contains approximately 1.0–1.5 mg of fluoride. On average, children ingest 0.13–0.59 mg of fluoride per tooth-brushing session (Cochran et al., 2004; de Almeida et al., 2007; Maguire et al., 2007; Zohoori et al., 2012a, 2013b, 2014), which depends on the age of the child, the amount of dentifrice used, and rinsing habits. Use of fluoridated mouth rinses by young children is less common and not recommended for children younger than 6 years old. With young children it is recommended to dispense a small amount of paste on the toothbrush and be supervised by parents when brushing their teeth to avoid ingestion of large amounts of toothpaste.

Other Sources

Artificially fluoridated water, milk, and salt as well as nondietary fluoride supplements and treatments such as fluoride tablets and fluoride varnishes are other sources of systemic fluoride intake.

FLUORIDE METABOLISM

Several features of fluoride metabolism, including gastric absorption, distribution, and renal excretion, are pH dependent. As a weak acid with a pK_a of 3.4, half of the hydrofluoric acid is in the dissociated or ionic form (F^-) at pH 3.4 and the other half in the undissociated form (HF), which is especially relevant in acidic environments (e.g., the gastrointestinal tract). Since the coefficient of permeability of lipid bilayer membranes to uncharged HF is 10^6 times higher than that of negatively charged fluoride ions (Gutknecht and Walter, 1981), fluoride crosses cell membranes as HF.

Fluoride Absorption

Absorption of orally taken fluoride starts in the oral cavity. Fluoride from some dental products with acidic formulations, such as acidulated phosphate fluoride gels and SnF_2 solutions, may be absorbed systemically even when they are not swallowed (Whitford, 1996).

Approximately 80–90% of orally ingested fluoride is absorbed from the gastrointestinal tract by passive diffusion with a half-life of approximately 30 min (Whitford, 1996). Up to 40% of ingested fluoride may be absorbed in the stomach, and the extent of absorption is inversely related to the pH of the stomach contents. Unabsorbed fluoride in the stomach is then absorbed from the upper small intestine, which has a huge capacity for fluoride absorption (Nopakun et al., 1989). Although the permeation of fluoride through the gastric mucosa depends on the acidity of the stomach content, the absorption of fluoride from the small intestine is independent of pH and happens largely as the fluoride ion, which crosses the leaky epithelia through the paracellular tight junction channels between epithelial cells (Nopakun et al., 1989).

Because an individual inhales almost 20 m^3 of air per day, the lungs could be another route of fluoride absorption under certain extreme environmental conditions, such as in heavy dust-polluted industrial areas (Hodge and Smith, 1977).

Fluoride in Blood Plasma

After absorption, fluoride is distributed very quickly throughout the body (Buzalaf and Whitford, 2011; Maguire et al., 2005). An increase in plasma fluoride concentration is detected within 10 min after fluoride ingestion. Peak plasma fluoride concentration occurs within 20–60 min after ingestion, depending on the time, amount, and type of food ingested beforehand. Plasma F returns to predigestion levels within 3–11 h.

Human plasma contains two forms of fluoride: ionic (inorganic or free fluoride) and nonionic (organic or bound fluoride). Ionic fluoride is significant in dentistry, medicine, and public health. Nonionic fluoride consists of several lipid-soluble organic fluorocompounds, such as perfluorooctanoic acid, bound to plasma proteins (Singer and Ophaug, 1982; Whitford, 1996). The biological function (if any) of these compounds is not yet understood (Buzalaf and Whitford, 2011).

In contrast to most other biologically relevant ions, the concentration of plasma ionic fluoride is not homeostatically regulated, which means it increases or decreases depending on the amount of fluoride ingestion, deposition in or removal from the soft and hard tissues, and excretion (Buzalaf and Whitford, 2011; Whitford, 1996).

Fluoride Distribution

Soft Tissues

Fluoride is fairly rapidly distributed from plasma to all tissues and organs in the body, which contain less than 1% of body fluoride. The rate of blood flow to different tissues governs the rate of fluoride distribution (Ekstrand, 1996). Because fluoride diffuses across cell membranes in the form of HF, it moves from a relatively acidic to a more alkaline environment; therefore it accumulates in the more alkaline compartments in response to a pH gradient (Buzalaf and Whitford, 2011). Short-term laboratory animal experiments with radioactive fluoride have shown 10–50% lower fluoride concentrations in intracellular fluid than those in extracellular fluid and plasma, which is caused by the extracellular fluid being more alkaline than the cytosol of mammalian cells (Buzalaf and Whitford, 2011).

Specialized Body Fluids

Fluoride concentrations of specialized body fluids differ from those in plasma. The fluoride concentration of human milk is less than 50% of that of simultaneously collected plasma (Whitford, 1996). Therefore human breast milk is a negligible source of fluoride in breast-fed infants.

Compared with plasma fluoride concentration, ductal saliva fluoride concentrations are slightly lower whereas the fluoride concentration of gingival crevicular fluid is marginally higher (Buzalaf and Whitford, 2011). The fluoride concentration of whole saliva is variable and higher than that in plasma (Whitford, 1996) because of topically applied fluoride retained in the mouth.

Mineralized Tissues

After absorption from the gastrointestinal tract, fluoride is rapidly incorporated into calcified tissues, which contain 99% of body fluoride. Because of the regular uptake of fluoride during life, the fluoride content of bones tends to increase with age (Parkins et al., 1974; Richards et al., 1994; Weidmann and Weatherell, 1959). Concentrations of fluoride in bone vary; compact bone has a lower fluoride concentration than cancellous bone, and in long bones the periosteal and endosteal regions have higher fluoride concentrations (Weidmann and Weatherell, 1959).

The fluoride contents of bones and dentin are similar, whereas the enamel fluoride content is lower on average (Buzalaf and Whitford, 2011). Dentin fluoride concentrations are higher close to the pulp and decrease progressively toward the dentin–enamel junction. The fluoride concentration of tooth enamel generally reflects the level of fluoride exposure during its formation.

Fluoride Excretion

Fluoride is excreted in the urine and feces and through sweat. The kidneys are the principal route for fluoride removal from the body. In healthy adults, the renal clearance (i.e., removal from a volume of plasma per unit time) of fluoride is approximately 35 mL/min, which is higher than the clearance of other halides, which is usually less than 1–2 mL/min (Buzalaf and Whitford, 2011). However, there is a large variation between different age groups and among individuals depending on the degree of active bone growth as well as other factors such as differences in glomerular filtration rate, urinary pH, and flow rate (Ekstrand et al., 1982; Spak et al., 1985; Whitford et al., 1976). Under normal conditions, almost 45% of absorbed fluoride is excreted in urine in healthy children and 65% excreted by adults (Villa et al., 2010). The concentration of ionic fluoride in the glomerular filtrate is the same as that in plasma. After ionic fluoride enters the renal tubules, 10–90% of the ion is reabsorbed and returned to the systemic circulation and the remainder is excreted in urine.

Fecal fluoride mainly consists of the fraction of fluoride that was not absorbed from the gastrointestinal tract. Ekstrand et al. (1984, 1994) suggested that almost 10% of the ingested fluoride is excreted through feces. However, laboratory animal studies suggest a net fluoride flux from the extracellular fluids into the lumen of the gastrointestinal tract under certain conditions such as high plasma concentration and when consuming a diet containing more than 1% calcium (Whitford, 1996).

Fluoride concentrations in sweat are 1–3 μmol/L (19–57 ng/mL), similar to those in plasma (Whitford, 1996). Although elimination of fluoride through sweat is negligible compared with that through urine in temperate climates, it has been suggested in a tropical climate or during heavy and prolonged exercise that the loss of fluoride with sweat might be significant.

Factors Affecting Fluoride Metabolism

The fluoride balance in the human body is illustrated schematically in Fig. 44.1. Because of pH dependency, the gastric absorption, distribution, and renal excretion of fluoride could be affected by changes in the acid–base equilibrium in the body. The body maintains acid–base balance by blood and tissue buffering, excretion of CO_2 by the lungs, and renal excretion of H^+ and regeneration of HCO_3^- (Welch et al., 2008). The acid–base status can be altered by several systemic and metabolic factors, including various metabolism and respiratory disorders, certain drugs, the level of physical activity, altitude of residence, as well as the composition of the diet, which is normally the controlling factor of acid–base status and urinary pH (Whitford, 1996).

Diet influences acid–base balance through the supply of acid and alkaline precursors from foods (Remer, 2001). Protein-containing foods such as meat, cereals and dairy foods generate H^+ ions through hepatic oxidation of the S-containing

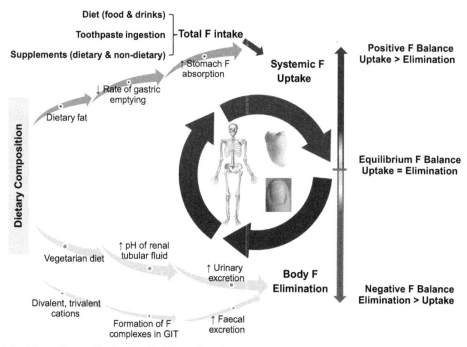

FIGURE 44.1 Dietary factors affecting fluoride balance in the human body (Maguire and Zohoori, 2013). *GIT*, gastrointestinal tract.

amino acids cysteine and methionine present (Frassetto et al., 1998; Remer and Manz, 1995). However, most fruit and vegetables supply base precursors as well as large amounts of Mg and K in the diet. The potential acid renal load of food has been estimated to range from an average maximum of 23.6 mEq/100 g for certain hard cheeses to an average minimum of −3 mEq/100 g for fruits and vegetables (Remer and Manz, 1995). Therefore vegetarianism might lead to an increase in urinary pH, increasing renal fluoride excretion and consequently decreasing body fluoride concentration.

The extent of fluoride absorption could also be influenced by the composition of the diet and intake with other foods. In the absence of a significant amount of divalent or trivalent cations, fluoride is rapidly absorbed and the extent of absorption is nearly 100%. However, ions such as calcium, magnesium and aluminum form insoluble complexes or precipitates with fluoride and consequently reduce its degree of absorption. High levels of dietary fats may enhance fluoride absorption, although the mechanism is unclear.

FUNCTION

Fluoride has an important nutritional and public health impact because of its role in the mineralization of bones and teeth, which contain more than 99% of total body fluoride content (Palmer et al., 2005).

Cariostatic Effect

The use of fluorides has had a dramatic effect in reducing dental decay over past decades (e.g., Switzerland, Fig. 44.2) and, as a public health measure, fluoride still has a valuable role in improving the oral health of communities. Untreated dental decay can lead to an inability to bite and chew and consequently to poor nutrition and poor health.

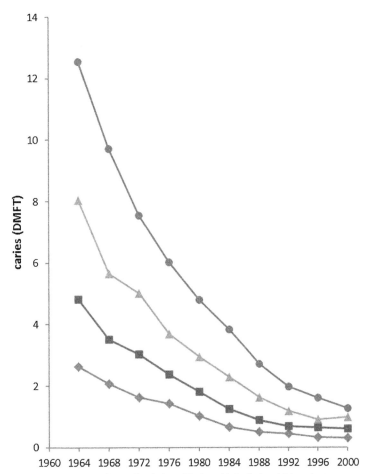

FIGURE 44.2 Decline in dental caries, mean DMFT (decayed, missing, filled teeth), in Swiss children (permanent residents of the Canton of Zurich) since 1964, largely as a result of the introduction of fluoridated toothpaste (Marthaler, 2004). Age: ◆, 8 years; ■, 10 years; ▲, 12 years; ●, 14 years, respectively.

The benefits of fluoride are achieved by systemic and topical effects. During tooth development, ingested (systemic) fluoride is incorporated into the mineralizing structure of the developing tooth, which helps in enhancing resistance to acid demineralization. After tooth eruption, fluoride retained in the saliva, after use of fluoridated toothpaste, mouthwash, etc., provides topical protection. Saliva, which contains water, protein, bicarbonate, fluoride, phosphate, calcium, and immuno-globulin, is important for maintaining the integrity of tooth enamel, dilution and buffering of acid produced by oral bacteria after meals, and oral clearance of food debris. It is now believed that topical (posteruptive) fluoride is the major factor by which fluoride provides optimal protection against dental caries (Singh and Spencer, 2004).

The topical effects of fluoride are independent of its systemic effect and benefit people of all age groups. Therefore the presence of low, constant levels of fluoride in the fluid phase at the tooth enamel surface is important in controlling tooth decay (Fejerskov et al., 1981; Featherstone, 1999). Fluoride beneficially influences the so-called demineralization–remineralization balance of the tooth in several ways (Shellis and Duckworth, 1994; ten Cate, 1999; Buzalaf et al., 2011). F helps decrease demineralization of tooth enamel and dentin hypersensitivity not only by reducing the acid solubility of enamel but, more importantly, by encouraging the uptake of tooth minerals (calcium and phosphate) and the precipitation of fluoridated hydroxyapatite within the enamel, and consequently by lessening the net rate of transport of minerals out of the enamel. Fluoride also affects the oral plaque bacteria that cause tooth decay. By fermenting carbohydrates, oral bacteria produce acid, which can dissolve teeth. Fluoride enters bacterial cells and is able to interfere with their acid production and, consequently, it decreases potential enamel destruction.

Effects on Bones

The actions of fluoride on bones appear to be through its physicochemical effects on bone crystals as well as biological effects on bone cells (Grynpas, 1990; Mousny et al., 2008). Fluoride has been recognized as one of only a few ions that can stimulate bone cell (osteoblast) proliferation and increase new mineral deposition in cancellous bone (Palmer et al., 2005). Moreover, fluoride can convert carbonated hydroxyapatite into carbonated fluorapatite, which is more stable and resistant to acid dissolution (Grynpas and Cheng, 1988). While fluoride exerts anabolic effects on bone and thus enhances bone mass (Kleerekoper and Balena, 1991), the normal structure and strength of the newly formed bone might be affected (Sogaard et al., 1994).

Fluoride as a therapeutic agent for the treatment of postmenopausal osteoporosis has been thoroughly investigated. However, the effect of fluoride exposure in drinking water on fracture risk and osteoporosis is still equivocal, showing an increased risk (Danielson et al., 1992; Jacobsen et al., 1990; Karagas et al., 1996) to no effect (Arnala et al., 1986; Cooper et al., 1990; Madans et al., 1983) to a decreased risk (Jacobsen et al., 1993; Li et al., 2001; Simonen and Laitinen, 1985).

EXCESSIVE INTAKE AND TOXICITY

Fluoride exhibits contrasting actions: it can be mitogenic at low concentrations but toxic at very high concentrations (Everett, 2011). Long-term exposure to relatively high fluoride concentrations can lead to dental and skeletal fluorosis (see, e.g., Fig. 44.3).

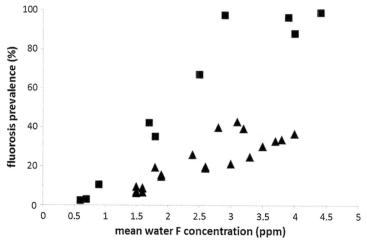

FIGURE 44.3 Effects of fluoride on the bones and teeth of people living in regions with high levels of fluoride in the local water supply: ■, dental fluorosis in 853 child residents of 10 US cities (Dean and Elvove, 1936); ▲, skeletal fluorosis in 2303 adult residents of 21 Indian villages (Choubisa, 2001).

Dental Fluorosis

Chronic exposure to greater-than-optimal systemic fluoride concentrations during critical periods of amelogenesis may result in the development of dental fluorosis (known also as mottling of tooth enamel). Dental fluorosis is characterized by opaque white spots of the teeth in mild forms to staining and pitting in severe forms due to increased subsurface porosity (Den Besten and Li, 2011). Dental fluorosis can only arise when teeth are forming. The first 3 years of life, especially between 6 and 24 months, are most important in dental fluorosis development in permanent incisors as well as the first permanent molars (Hong et al., 2006), whilst excessive fluoride intake from 3 to 6 years can also put the later developing permanent canines, premolars, and second molars at risk (Levy et al., 2003).

Fluoride is not the only cause of dental enamel defects. Enamel opacities similar to dental fluorosis are associated with other conditions, such as malnutrition, and with deficiencies of vitamins D and A or a low-protein–energy diet.

Skeletal Fluorosis

Skeletal fluorosis is a serious condition, resulting from chronic ingestion of large amounts of fluoride over many years during periods of bone modeling (growth) and/or remodeling. In skeletal fluorosis the bones are generally weaker than normal with stiffness and pain in the joints as the early symptoms. In severe cases, muscles are impaired and bones in the central skeleton are irregularly thickened due to periosteal sleeves of abnormally structured osseous tissue, osteophytosis, mineralization of tendons and muscle attachments, and bridging between the edges of the vertebral bodies (Vieira et al., 2005). Deformities, particularly of the weight-bearing bones, are characteristically evident in severe cases (Krishnamachari, 1986).

Acute Toxicity

Similar to many other minerals, fluoride can be toxic when ingested as a large amount in a single dose or in multiple doses within a few hours. Because the first organ to be affected by systemic acute exposure is the stomach, clinical systemic toxicity starts with gastric signs and symptoms, ranging from some degree of nausea to abdominal pain, hemorrhagic gastroenteritis, vomitus, and diarrhea (Whitford, 2011). The "probably toxic dose" of fluoride is 5 mg/kg body weight (11 mg/kg body weight of NaF). Death can occur with a fluoride ingestion of 16 mg/kg body weight (World Health Organisation, 1984).

Influence of Genetics on the Effect of Fluoride

The significant variation in the prevalence and severity of dental or skeletal fluorosis seen within and between populations with comparable fluoride exposures (Butler et al., 1985; Choubisa et al., 2001; Yoder et al., 1998) indicates a genetic involvement in the absorption and metabolism of fluoride. Mousny et al. (2006) found that genetic factors influenced the bone response to fluoride exposure in mice. Genetic studies in nematodes (*Caenorhabditis elegans*) have led to the identification of novel fluoride-resistance (*flr*) genes, *flr1*, *flr3* and *flr4* (Katsura, 1993; Katsura et al., 1994).

Differential physiological responses to ingested fluoride have also been observed in different mouse strains: the A/J mouse strain exhibited high susceptibility with a rapid onset and development of severe dental fluorosis whereas the 129P3/J mouse strain showed resistance with minimal dental fluorosis (Everett et al., 2002). Metabolic handling of fluoride between the two strains also differs; the susceptible A/J strain excretes a significantly higher amount of fluoride in urine (Carvalho et al., 2009) and presents with a significant decrease in ultimate load and stiffness of the femoral and vertebral bodies with increasing fluoride dose (Mousny et al., 2006). In humans, an increased risk of dental fluorosis has been reported in children carrying a single-nucleotide polymorphism (SNP) in the collagen type 1 α2 gene (SNP database number rs414408) compared with those not carrying the variation in an endemic fluorosis village in Henan Province, China (Huang et al., 2008).

ADEQUATE DAILY INTAKE AND TOLERABLE UPPER INTAKE LEVEL

Insufficient evidence is available to develop a recommended daily allowance for fluoride. Suggested adequate intakes (AIs) for fluoride are based on estimated intakes that have been shown to reduce the incidence of dental caries while minimizing unwanted health effects. The AI for infants during the first 6 months of life is recommended as 0.01 mg/day, in line with the amount of fluoride that a breast-fed infant receives from human milk (Institute of Medicine, 1997). For children older than 6 months and adults, the AI for fluoride from all sources (dietary and nondietary) is set at 0.05 mg/kg body weight per day (Institute of Medicine, 1997).

The fluoride concentration of human milk is very low and is relatively unchanged in response to variations in the fluoride intake of the mother. There are also no data on fluoride metabolism from human studies to suggest a modification to the AI for fluoride during pregnancy and lactation. Therefore the AI for pregnant women and lactating mothers is set at the same level as that for nonpregnant women.

The tolerable upper intake level (UL) for fluoride is established as 0.1 mg/kg body weight per day for infants and children up to 8 years old. This suggested UL was based on a high degree of certainty that a chronic systemic fluoride ingestion of less than that by children at risk of dental fluorosis was linked with a low prevalence (<10%) to the milder forms of dental fluorosis (Institute of Medicine, 1997). For children older than 8 years and adults, a UL of 10 mg F/day has been determined (Institute of Medicine, 1997) based on data from fluoride exposure studies reporting a small risk of development of preclinical or stage 1 skeletal fluorosis in individuals who had received 10 mg F/day for 10 or more years (Hodge and Smith, 1977; Institute of Medicine, 1997).

ACKNOWLEDGMENTS

The authors thank the *British Dental Journal* and Karger AG, Basel for permission to reproduce Figs. 44.1 and 44.2, respectively.

REFERENCES

Arnala, I., Alhava, E.M., Kivivuori, R., Kauranen, P., 1986. Hip fracture incidence not affected by fluoridation. Osteofluorosis studied in Finland. Acta Orthop. Scand. 57, 344–348.

Banks, R., Smart, B., Bruce, E., 1994. Organofluorine Chemistry: Principles and Commercial Applications. Plenum Press, New York.

Butler, W.J., Segreto, V., Collins, E., 1985. Prevalence of dental mottling in school-aged lifetime residents of 16 Texas communities. Am. J. Public Health 75, 1408–1412.

Buzalaf, M.A., Pessan, J.P., Honório, H.M., ten Cate, J.M., 2011. Mechanisms of action of fluoride for caries control. In: Buzalaf, M.A.R. (Ed.), Fluoride and the Oral Environment. Monographs in Oral Science, vol. 22. Karger, Basel, Switzerland, 97–114.

Buzalaf, M.A., Whitford, G.M., 2011. Fluoride metabolism. In: Buzalaf, M.A.R. (Ed.), Fluoride and the Oral Environment. Monographs in Oral Science, vol. 22. Karger, Basel, Switzerland, 20–36.

Carvalho, J.G., Leite, A.L., Yan, D., Everett, E.T., Whitford, G.M., Buzalaf, M.A.R., 2009. Influence of genetic background on fluoride metabolism in mice. J. Dent. Res. 88, 1054–1058.

Chan, L., Mehra, A., Saikat, S., Lynch, P., 2013. Human exposure assessment of fluoride from tea (*Camellia sinensis* L.): a UK based issue? Food Res. Int. 51, 564–570.

Choubisa, S.L., 2001. Endemic fluorosis in southern Rajasthan, India. Fluoride 34, 61–70.

Choubisa, S.L., Choubisa, L., Choubisa, D.K., 2001. Endemic fluorosis in Southern Rajasthan, India. Indian J. Environ. Health 43, 177–189.

Cochran, J.A., Ketley, C.E., Duckworth, R.M., van Loveren, C., Holbrook, W.P., Seppa, L., Sanches, L., Polychronopoulou, A., O'Mullane, D.M., 2004. Development of a standardized method for comparing fluoride ingested from toothpaste by 1.5–3.5-year-old children in seven European countries. Part 2: Ingestion results. Community Dent. Oral Epidemiol. 32 (Suppl. 1), 47–53.

Cooper, C., Wickham, C., Lacey, R.F., Barker, D.J., 1990. Water fluoride concentration and fracture of the proximal femur. J. Epidemiol. Community Health 44, 17–19.

Danielson, C., Lyon, J.L., Egger, M., Goodenough, G.K., 1992. Hip fractures and fluoridation in Utah's elderly population. J. Am. Med. Assoc. 268, 746–748.

de Almeida, B., da Silva Cardoso, V., Buzalaf, M., 2007. Fluoride ingestion from toothpaste and diet in 1- to 3-year-old Brazilian children. Community Dent. Oral Epidemiol. 35, 53–63.

Dean, H.T., Elvove, E., 1936. Some epidemiological aspects of chronic endemic dental fluorosis. Am. J. Public Health 26, 567–575.

Den Besten, P.K., Li, W., 2011. Chronic fluoride toxicity: dental fluorosis. In: Buzalaf, M.A.R. (Ed.), Fluoride Intake and the Oral Environment. Monographs in Oral Science, vol. 22. Karger, Basel, Switzerland, pp. 81–96.

Edmunds, W.M., Smedley, P.L., 1996. Groundwater geochemistry and health: an overview. In: Appleton, J.D., Fuge, R., McCall, G.J.H. (Eds.), Environmental Geochemistry and Health. Geological Society Special Publication No. 113, pp. 91–105.

Ekstrand, J., 1996. Fluoride metabolism. In: Fejerskov, O., Ekstrand, J., Burt, B.A. (Eds.), Fluoride in Dentistry, second ed. Munksgaard, Copenhagen, pp. 55–68.

Ekstrand, J., Fejerskov, O., Silverstone, L.M., 1988. Fluoride in Dentistry, first ed. Munksgaard, Copenhagen.

Ekstrand, J., Fomon, S.J., Ziegler, E.E., Nelson, S.E., 1994. Fluoride pharmacokinetics in infancy. Pediatr. Res. 35, 157–163.

Ekstrand, J., Hardell, L.I., Spak, C.J., 1984. Fluoride balance studies on infants in a 1-ppm-water-fluoride area. Caries Res. 18, 87–92.

Ekstrand, J., Spak, C.J., Ehrnebo, M., 1982. Renal clearance of fluoride in a steady state condition in man: influence of urinary flow and pH changes by diet. Acta Pharmacol. Toxicol. 50, 321–325.

Everett, E.T., 2011. Fluoride's effects on the formation of teeth and bones, and the influence of genetics. J. Dent. Res. 90, 552–560.

Everett, E.T., McHenry, M.A., Reynolds, N., Eggertsson, H., Sullivan, J., Kantmann, C., Martinez-Mier, E.A., Warrick, J.M., Stookey, G.K., 2002. Dental fluorosis: variability among different inbred mouse strains. J. Dent. Res. 81, 794–798.

Featherstone, J.D.B., 1999. Prevention and reversal of dental caries: role of low level fluoride. Community Dent. Oral Epidemiol. 27, 31–40.

Fejerskov, O., Thylstrup, A., Larsen, M.J., 1981. Rational use of fluorides in caries prevention. A concept based on possible cariostatic mechanisms. Acta Odontol. Scand. 39, 241–249.

Frassetto, L.A., Todd, K.M., Morris, R.C., Sebastian, A., 1998. Estimation of net endogenous noncarbonic acid production in humans from diet potassium and protein contents. Am. J. Clin. Nutr. 68, 576–583.

Full, C.A., Parkins, F.M., 1975. Effect of cooking vessel composition on fluoride. J. Dent. Res. 54, 192.

Grynpas, M.D., 1990. Fluoride effects on bone crystals. J. Bone Miner. Res. 5 (Suppl. 1), S169–S175.

Grynpas, M.D., Cheng, P.T., 1988. Fluoride reduces the rate of dissolution of bone. Bone Miner. 5, 1–9.

Gutknecht, J., Walter, A., 1981. Hydrofluoric and nitric acid transport through lipid bilayer membranes. Biochim. Biophys. Acta 644, 153–156.

Hem, J.D., 1985. Study and Interpretation of the Chemical Characteristics of Natural Water, Water Supply Paper 2254, third ed. US Geological Survey, Washington, DC.

Hodge, H.C., Smith, F.A., 1977. Occupational fluoride exposure. J. Occup. Med. 19, 12–39.

Hong, L., Levy, S.M., Broffitt, B., Warren, J.J., Kanellis, M.J., Wefel, J.S., Dawson, D.V., 2006. Timing of fluoride intake in relation to development of fluorosis on maxillary central incisors. Community Dent. Oral Epidemiol. 34, 299–309.

Huang, H., Ba, Y., Cui, L., Cheng, X., Zhu, J., Zhang, Y., Yan, P., Zhu, C., Kilfoy, B., Zhang, Y., 2008. COL1A2 gene polymorphisms (*Pvu* II and *Rsa* I), serum calciotropic hormone levels, and dental fluorosis. Community Dent. Oral Epidemiol. 36, 517–522.

Institute of Medicine, 1997. Dietary Reference Intakes for Calcium, Magnesium, Vitamin D, and Fluoride. National Academy Press, Washington, DC.

Jaccaud, M., Faron, R., Devilliers, D., Romano, R., 2000. Fluorine. In: Ullmann, F. (Ed.), Ullmann's Encyclopedia of Industrial Chemistry, vol. 15. Wiley-VCH, Weinheim, pp. 381–395.

Jacobsen, S.J., Goldberg, J., Miles, T.P., Brody, J.A., Stiers, W., Rimm, A.A., 1990. Regional variation in the incidence of hip fracture. US white women aged 65 years and older. J. Am. Med. Assoc. 264, 500–502.

Jacobsen, S.J., O'Fallon, W.M., Melton 3rd, L.J., 1993. Hip fracture incidence before and after the fluoridation of the public water supply, Rochester, Minnesota. Am. J. Public Health 83, 743–745.

Karagas, M.R., Baron, J.A., Barrett, J.A., Jacobsen, S.J., 1996. Patterns of fracture among the United States elderly: geographic and fluoride effects. Ann. Epidemiol. 6, 209–216.

Katsura, I., 1993. In search of new mutants in cell-signaling systems of the nematode *Caenorhabditis elegans*. Genetica 88, 137–146.

Katsura, I., Kondo, K., Amano, T., Ishihara, T., Kawakami, M., 1994. Isolation, characterization and epistasis of fluoride-resistant mutants of *Caenorhabditis elegans*. Genetics 136, 145–154.

Kleerekoper, M., Balena, R., 1991. Fluorides and osteoporosis. Annu. Rev. Nutr. 11, 309–324.

Koblar, A., Tavčar, G., Ponikvar-Svet, M., 2012. Fluoride in teas of different types and forms and the exposure of humans to fluoride with tea and diet. Food Chem. 130, 286–290.

Koparal, E., Ertugrul, F., Oztekin, K., 2000. Fluoride levels in breast milk and infant foods. J. Clin. Pediatr. Dent. 24, 299–302.

Krishnamachari, K.A., 1986. Skeletal fluorosis in humans: a review of recent progress in the understanding of the disease. Prog. Food Nutr. Sci. 10, 279–314.

Lal, H., Zohoori, F.V., Omid, N., Valentine, R., Maguire, A., 2014. The fluoride contents of commercially-available soya milks in the UK. Br. Dent. J. 217, E8.

Levy, S.M., Warren, J.J., Broffitt, B., 2003. Patterns of fluoride intake from 36 to 72 months of age. J. Public Health Dent. 63, 211–220.

Li, Y.M., Liang, C.K., Slemenda, C.W., Ji, R.D., Sun, S.Z., Cao, J.X., Emsley, C.L., Ma, F., Wu, Y.P., Ying, P., Zhang, Y., Gao, S.J., Zhang, W., Katz, B.P., Niu, S.R., Cao, S.R., Johnston, C.C., 2001. Effect of long-term exposure to fluoride in drinking water on risks of bone fractures. J. Bone Miner. Res. 16, 932–939.

Madans, J., Kleinman, J.C., Cornoni-Huntley, J., 1983. The relationship between hip fracture and water fluoridation: an analysis of national data. Am. J. Public Health 73, 296–298.

Maguire, A., Omid, N., Abuhaloob, L., Moynihan, P.J., Zohoori, F.V., 2012. Fluoride content of ready-to-feed (RTF) infant food and drinks in the UK. Community Dent. Oral Epidemiol. 40, 26–36.

Maguire, A., Zohoori, F.V., 2013. Fluoride balance in infants and young children in the U.K. and its clinical relevance for the dental team. Br. Dent. J. 214, 587–593.

Maguire, A., Zohouri, F.V., Hindmarch, P.N., Hatts, J., Moynihan, P.J., 2007. Fluoride intake and urinary excretion in 6- to 7-year-old children living in optimally, sub-optimally and non-fluoridated areas. Community Dent. Oral Epidemiol. 35, 479–488.

Maguire, A., Zohouri, F.V., Mathers, J.C., Steen, I.N., Hindmarch, P.N., Moynihan, P.J., 2005. Bioavailability of fluoride in drinking water: a human experimental study. J. Dent. Res. 84, 989–993.

Marthaler, T.M., 2004. Changes in dental caries 1953–2003. Caries Res. 38, 173–181.

MCHB Expert Panel, 2007. Topical Fluoride Recommendations for High-Risk Children Development of Decision Support Matrix. Recommendations From Maternal and Child Health Bureau (MCHB) Expert Panel. Altarum Institute, Washington, DC.

Mousny, M., Banse, X., Wise, L., Everett, E.T., Hancock, R., Vieth, R., Devogelaer, J.P., Grynpas, M.D., 2006. The genetic influence on bone susceptibility to fluoride. Bone 39, 1283–1289.

Mousny, M., Omelon, S., Wise, L., Everett, E.T., Dumitriu, M., Holmyard, D.P., Banse, X., Devogelaer, J.P., Grynpas, M.D., 2008. Fluoride effects on bone formation and mineralization are influenced by genetics. Bone 43, 1067–1074.

Murray, J.J., 1986. Appropriate Use of Fluorides for Human Health. World Health Organization, Geneva.

Nair, K.R., Manji, F., Gitonga, J.N., 1984. The occurrence and distribution of fluoride in groundwaters of Kenya. East Afr. Med. J. 61, 503–512.

Nanda, R.S., 1972. Fluoride content of North Indian foods. Indian J. Med. Res. 60, 1470–1482.

Nohno, K., Zohoori, F.V., Maguire, A., 2011. Fluoride intake of Japanese infants from infant milk formula. Caries Res. 45, 486–493.

Nopakun, J., Messer, H.H., Voller, V., 1989. Fluoride absorption from the gastrointestinal tract of rats. J. Nutr. 119, 1411–1417.

Oelschlager, W., 1970. Fluoride in food. Fluoride 3, 6–11.

Ophaug, R.H., Singer, L., Harland, B.F., 1985. Dietary fluoride intake of 6-month and 2-year-old children in four dietary regions of the United States. Am. J. Clin. Nutr. 42, 701–707.

Palmer, C., Wolfe, S.H., American Dietetic Association, 2005. Position of the American Dietetic Association: the impact of fluoride on health. J. Am. Diet. Assoc. 105, 1620–1628.

Pang, D., Phillips, C., Bawden, J., 1992. Fluoride intake from beverage consumption in a sample of North Carolina children. J. Dent. Res. 71, 1382–1388.

Parkins, F.M., Tinanoff, N., Moutinho, M., Anstey, M.B., Waziri, M.H., 1974. Relationships of human plasma fluoride and bone fluoride to age. Calcif. Tissue Res. 16, 335–338.

Remer, T., 2001. Influence of nutrition on acid-base balance – metabolic aspects. Eur. J. Nutr. 40, 214–220.

Remer, T., Manz, F., 1995. Potential renal acid load of foods and its influence on urine pH. J. Am. Diet. Assoc. 95, 791–797.

Richards, A., Mosekilde, L., Sogaard, C.H., 1994. Normal age-related changes in fluoride content of vertebral trabecular bone – relation to bone quality. Bone 15, 21–26.

Schamschula, R.G., Duppenthaler, J.L., Sugar, E., Un, P.S., Toth, K., Barmes, D.E., 1988a. Fluoride intake and utilization by Hungarian children: associations and interrelationships. Acta Physiol. Hung. 72, 253–261.

Schamschula, R.G., Un, P.S., Sugar, E., Duppenthaler, J.L., Toth, K., Barmes, D.E., 1988b. The fluoride content of selected foods in relation to the fluoride concentration of water. Acta Physiol. Hung. 72, 217–227.

Sener, Y., Tosun, G., Kahvecioglu, F., Gokalp, A., Koc, H., 2007. Fluoride levels of human plasma and breast milk. Eur. J. Dent. 1, 21–24.

Shellis, R.P., Duckworth, R.M., 1994. Studies on the cariostatic mechanisms of fluoride. Int. Dent. J. 44, 263–273.

Silva, M., Reynolds, E.C., 1996. Fluoride content of infant formulae in Australia. Aust. Dent. J. 41, 37–42.

Simonen, O., Laitinen, O., 1985. Does fluoridation of drinking-water prevent bone fragility and osteoporosis? Lancet 2, 432–434.

Singer, L., Ophaug, R., 1982. Ionic and nonionic fluoride in plasma (or serum). Crit. Rev. Clin. Lab. Sci. 18, 111–140.

Singh, K.A., Spencer, A.J., 2004. Relative effects of pre- and post-eruption water fluoride on caries experience by surface type of permanent first molars. Community Dent. Oral Epidemiol. 32, 435–446.

Sogaard, C.H., Mosekilde, L., Richards, A., Mosekilde, L., 1994. Marked decrease in trabecular bone quality after five years of sodium fluoride therapy – assessed by biomechanical testing of iliac crest bone biopsies in osteoporotic patients. Bone 15, 393–399.

Spak, C.J., Berg, U., Ekstrand, J., 1985. Renal clearance of fluoride in children and adolescents. Pediatrics 75, 575–579.

Taves, D.R., 1983. Dietary intake of fluoride ashed (total fluoride) v. unashed (inorganic fluoride) analysis of individual foods. Br. J. Nutr. 49, 295–301.

ten Cate, J.M., 1999. Current concepts on the theories of the mechanism of action of fluoride. Acta Odontol. Scand. 57, 325–329.

U.S. Department of Agriculture, 2005. USDA National Fluoride Database of Selected Beverages and Foods. Prepared by Nutrient Data Laboratory, Beltsville Human Nutrition Research Center, Agricultural Research Service and U.S. Department of Agriculture, Beltsville, Maryland.

Vieira, A.P.G.F., Mousny, M., Maia, R., Hancock, R., Everett, E.T., Grynpas, M.D., 2005. Assessment of teeth as biomarkers for skeletal fluoride exposure. Osteoporos. Int. 16, 1576–1582.

Villa, A., Anabalon, M., Zohouri, V., Maguire, A., Franco, A.M., Rugg-Gunn, A., 2010. Relationships between fluoride intake, urinary fluoride excretion and fluoride retention in children and adults: an analysis of available data. Caries Res. 44, 60–68.

Walters, C.B., Sherlock, J.C., Evans, W.H., Read, J.I., 1983. Dietary intake of fluoride in the United Kingdom and fluoride content of some foodstuffs. J. Sci. Food Agric. 34, 523–528.

Weidmann, S.M., Weatherell, J.A., 1959. The uptake and distribution of fluorine in bones. J. Pathol. Bacteriol. 78, 243–255.

Welch, A.A., Mulligan, A., Bingham, S.A., Khaw, K.T., 2008. Urine pH is an indicator of dietary acid-base load, fruit and vegetables and meat intakes: results from the European Prospective Investigation into Cancer and Nutrition (EPIC)-Norfolk population study. Br. J. Nutr. 99, 1335–1343.

Whitford, G.M., 1996. The Metabolism and Toxicity of Fluoride. Monographs in Oral Science, vol. 16. Karger, Basel.

Whitford, G.M., 2011. Acute toxicity of ingested fluoride. In: Buzalaf, M.A.R. (Ed.), Fluoride Intake and the Oral Environment. Monographs in Oral Science, vol. 22. Karger, Basel, Switzerland, pp. 66–80.

Whitford, G.M., Pashley, D.H., Stringer, G.I., 1976. Fluoride renal clearance: a pH-dependent event. Am. J. Physiol. 230, 527–532.

World Health Organisation, 1984. Fluorine and Fluorides, Environmental Health Criteria, 36. International Programme on Chemical Safety, Geneva.

World Health Organization, 1994. Fluorides and oral health. Report of a WHO expert committee on oral health status and fluoride use, WHO Tech. Rep. Ser. 836, 1–37.

Yoder, K.M., Mabelya, L., Robison, V.A., Dunipace, A.J., Brizendine, E.J., Stookey, G.K., 1998. Severe dental fluorosis in a Tanzanian population consuming water with negligible fluoride concentration. Community Dent. Oral Epidemiol. 26, 382–393.

Zohoori, F.V., Buzalaf, M.A., Cardoso, C.A., Olympio, K.P., Levy, F.M., Grizzo, L.T., Mangueira, D.F., Sampaio, F.C., Maguire, A., 2013a. Total fluoride intake and excretion in children up to 4 years of age living in fluoridated and non-fluoridated areas. Eur. J. Oral Sci. 121, 457–464.

Zohoori, F.V., Walls, R., Teasdale, L., Landes, D., Steen, I.N., Moynihan, P., Omid, N., Maguire, A., 2013b. Fractional urinary fluoride excretion of 6–7-year-old children attending schools in low-fluoride and naturally fluoridated areas in the UK. Br. J. Nutr. 109, 1903–1909.

Zohoori, F.V., Duckworth, R.M., Omid, N., O'Hare, W.T., Maguire, A., 2012a. Fluoridated toothpaste: usage and ingestion of fluoride by 4- to 6-yr-old children in England. Eur. J. Oral Sci. 120, 415–421.

Zohoori, F.V., Moynihan, P.J., Omid, N., Abuhaloob, L., Maguire, A., 2012b. Impact of water fluoride concentration on the fluoride content of infant foods and drinks requiring preparation with liquids before feeding. Community Dent. Oral Epidemiol. 40, 432–440.

Zohoori, F.V., Maguire, A., 2015. Database of the Fluoride (F) Content of Selected Drinks and Foods in the UK. Teesside University and Newcastle University. http://tees.openrepository.com/tees/handle/10149/581272. http://eprint.ncl.ac.uk/pub_details2.aspx?pub_id=216711.

Zohoori, F.V., Maguire, A., 2016. Development of a database of the fluoride content of selected drinks and foods in the UK. Caries Res 50, 331–336. http://dx.doi.org/10.1159/000445981.

Zohoori, F.V., Maguire, A., Moynihan, P.J., 2006. Sources of dietary fluoride intake in 6–7 year old English children receiving optimally, sub-optimally, and non-fluoridated water. J. Public Health Dent. 66, 227–234.

Zohoori, F.V., Whaley, G., Moynihan, P.J., Maguire, A., 2014. Fluoride intake of infants living in non-fluoridated and fluoridated areas. Br. Dent. J. 216, E3.

Index

Printed in the United States
By Bookmasters